CALCULUS I

WITH INTEGRATED PRECALCULUS

LAURA TAALMAN

James Madison University

W. H. Freeman and Company
New York

Senior Publisher: Ruth Baruth
Executive Editor: Terri Ward
Marketing Manager: Steve Thomas
Market Development Manager: Steven Rigolosi
Developmental Editors: Leslie Lahr, Katrina Wilhelm
Senior Media Editor: Laura Judge
Associate Editor: Jorge Amaral
Editorial Assistant: Liam Ferguson
Photo Editor: Ted Szczepanski
Cover Photo Researcher: Elyse Rieder
Cover Designer: Vicki Tomaselli
Text Designer: Marsha Cohen
Illustrations: Network Graphics
Illustration Coordinator: Bill Page
Production Coordinator: Susan Wein
Project Management and Composition: Aptara
Printing and Binding: RR Donnelley

Library of Congress Control Number: 2012947835

ISBN-13: 978-1-4292-4073-4
ISBN-10: 1-4292-4073-3

Printed in the United States of America

W. H. Freeman and Company
41 Madison Avenue
New York, NY 10010
Houndmills, Basingstoke RG21 6XS, England
www.whfreeman.com

To all the students who struggle, fall,
and get right back up again

—Laura Taalman

CONTENTS

*Starred titles indicate optional material

I DIFFERENTIAL CALCULUS

II EXPLORING FUNCTIONS WITH CALCULUS

III INTEGRAL CALCULUS

Laura Taalman is a professor of mathematics at James Madison University. She has taught hybrid calculus/precalculus courses to students who need additional algebra and precalculus help for nearly 20 years. She received her undergraduate degree from the University of Chicago and master's and Ph.D. degrees in mathematics from Duke University. Her research includes singular algebraic geometry, knot theory, and the mathematics of games and puzzles. She is a recipient of both the Alder Award and the Trevor Evans award from the Mathematical Association of America, and the author of five books on Sudoku and the mathematics of Sudoku. In her spare time, she enjoys being a geek.

Calculus books have become full of clutter, distracting margin notes, and unneeded features. This calculus book clears out that clutter so that students can focus on the important ideas of calculus. The goal was to create a clean, streamlined calculus book that is accessible and readable for students while still upholding the standards required in science, mathematics, and engineering programs, and that is flexible enough to accommodate different teaching and learning styles.

This particular book is an alternative edition of Taalman/Kohn *Calculus* that covers Calculus I and part of what is usually taught as Calculus II, together with precalculus and algebra material. Students who need to strengthen their precalculus and algebra backgrounds while taking calculus can start with this book for an extended Calculus I course and then proceed to a Calculus II course using any other calculus text (although this book is structured to work particularly well with a continuation of the course using Taalman/Kohn *Calculus*).

Calculus before Calculations

The primary way that this book differs from Taalman/Kohn *Calculus* and other calculus books is that it begins by using relatively simple functions such as polynomials and simple algebraic functions to develop all of the main ideas of differential calculus. After students have gained a firm grounding in the ideas of calculus, that grounding is used to explore more complicated algebraic functions as well as exponential, logarithmic, and trigonometric functions. Students who struggle with precalculus and algebra benefit from this approach by being able to understand calculus first and then build their calculational skills afterwards.

A Full Calculus Course

Everything in Chapters 0 through 6 of Taalman/Kohn *Calculus* is included in this textbook, although the material appears in a different order and the book is supplemented with additional algebra and precalculus material. This means that students who learn from *Calculus I with Integrated Precalculus* are fully prepared to join their peers in Calculus II, and it is easy to run parallel sections of a course using *Calculus I with Integrated Precalculus* alongside sections of an existing Calculus course, while being confident that students leave with the same background and understanding of the material.

Linear Flow with Clean Margins

One thing that is distinctive about this calculus book is that it follows a linear writing style. Figures and equations flow with the text as part of a clear, structured exposition instead of being scattered about in the margins. This approach greatly increases the clarity of the book and encourages focused reading.

Exposition before Calculation

Another distinctive feature of this book is that in each section the exposition and illustrative examples are separated from the longer, more complicated calculational examples. Including these longer examples separately from the exposition increases flexibility: Students who want to read and understand the development of the material can do so without being bogged down or distracted by large examples, while students who want to use the book as a reference for looking up examples that are similar to homework problems can also do that.

Examples to Learn From

Within the exposition of each section are short examples that quickly illustrate the concepts being developed. Following the exposition is a set of detailed, in-depth examples that explore both calculations and concepts. Many steps and illustrations are provided in each example in order to aid the student, including details about how to get started on a problem and choose an appropriate solution method. One of the elements of the book that I am most proud of is the "Checking the Answer" feature, which is included after selected examples to encourage students to learn how to check their own answers.

Building Mathematics

This book approaches mathematics as a discipline that is developed logically, theorem by theorem. Whenever possible, theorems are followed by proofs that are written to be understood by students. These proofs are included because they are part of the logical development of the material, but each proof is clearly labeled and indented to indicate that it can be covered or skipped, according to instructor preference. Each exercise set contains an optional subsection of proofs, many of which are accessible even to beginning students. In addition, the interconnections among topics are emphasized by "Thinking Back" and "Thinking Forward" exercises that appear in each section and "Capstone" problems at the end of each chapter.

Consistency and Reliability

Another improvement in this book is that it has a consistent and predictable structure. For example, instructors can rely on every section concluding with a "Test Your Understanding" feature which includes five questions that students can use to self-test and that instructors can choose to use as pre-class questions. The exercises are always consistently split into subsections of different types of problems: "Thinking Back," "Concepts," "Skills," "Applications," "Proofs," and "Thinking Forward." In addition, the "Concepts" subsection always begins with a summary exercise, eight true/false questions, and three example construction exercises. Instructors and students alike can rely on this consistent structure when assigning exercises and choosing a path of study.

Flexibility

I recognize that instructors use calculus books in many different ways and that the real direction of a calculus course comes from the instructor, not any book. The streamlined, consistent structure of this book makes it easy to use with a wide variety of courses and pedagogical styles. In particular, instructors will find it easy to include or omit sections, proofs, examples, and exercises consistently according to their preferences and course requirements. Students can focus on mathematical development or on examples and calculations as they need to throughout the course. Later, they can use the book as a reliable reference.

It will be immediately clear to anyone opening either Taalman/Kohn *Calculus* or *Calculus I with Integrated Precalculus* that these books are substantially different from the other calculus books on the market today while still following the standard topics taught in most modern science, mathematics, and engineering calculus courses. My hope is that faculty who use either book will find it flexible for different pedagogical approaches and that students will be able to read it on different levels as they learn to understand the beauty of calculus. In addition, I hope that instructors find *Calculus I with Integrated Precalculus* to be a simple, streamlined way to address the needs of students who require additional help with algebra and precalculus in their first college calculus course.

Comparison of *Calculus I with Integrated Precalculus* with Taalman/Kohn *Calculus*

Calculus I with Integrated Precalculus includes all of the material in Chapters 0–6 of Taalman/Kohn *Calculus,* but in a different order and with supplementary precalculus and algebra material.

▶ Chapters 0–3 of *Calculus I with Integrated Precalculus* cover the same development of differential calculus topics as Taalman/Kohn *Calculus* but the more complicated calculational examples are deferred to later chapters.

▶ Chapters 4–6 of *Calculus I with Integrated Precalculus* revisit differential calculus through the lens of studying progressively more challenging types of functions. Any exercises or examples from Taalman/Kohn *Calculus* that were left out of Chapters 0–3 of *Calculus I with Integrated Precalculus* are included in Chapters 4–6. The requisite background precalculus and algebra material is built from the ground up.

▶ Chapters 7–9 of *Calculus I with Integrated Precalculus* are identical to Chapters 4–6 of Taalman/Kohn *Calculus* and cover all topics from integral calculus.

Instructors using *Calculus I with Integrated Precalculus* for a two-semester course in which they wish to cover all of integration might find it convenient to cover Chapters 0–4 in the first semester and Chapters 5–9 in the second semester. For *Calculus I with Integrated Precalculus* courses that do not need to cover as much integration, I suggest covering Chapters 0–3 in the first semester and Chapters 4–7 in the second semester. It is also possible to use *Calculus I with Integrated Precalculus* in an extended-hours one-semester course.

Students who learn Calculus I from *Calculus I with Integrated Precalculus* can continue with Calculus II using Taalman/Kohn *Calculus* or any other calculus textbook.

For an examination copy of Taalman/Kohn *Calculus*, please contact your local W. H. Freeman & Company representative.

For Instructors

Instructor's Solutions Manual

ISBN: 1-4292-4188-8

Contains worked-out solutions to all exercises in the text.

Test Bank

Computerized (CD-ROM), ISBN: 1-4292-5509-9

Includes multiple-choice and short-answer test items.

Instructor's Resource Manual

ISBN: 1-4641-5024-9

Provides suggested class time, key points, lecture material, discussion topics, class activities, worksheets, and group projects corresponding to each section of the text.

Instructor's Resource CD-ROM

ISBN: 1-4292-4189-6

Search and export all resources by key term or chapter. Includes text images, Instructor's Solutions Manual, Instructor's Resource Manual, and Test Bank.

For Students

Student Solutions Manual

ISBN: 1-4292-4187-X

Contains worked-out solutions to all odd-numbered exercises in the text.

Software Manuals

Maple™ and Mathematica® software manuals are available within CalcPortal. Printed versions of these manuals are available through custom publishing. They serve as basic introductions to popular mathematical software options and guides for their use with *Calculus I with Integrated Precalculus*.

Book Companion Web Site at www.whfreeman.com/integcalculus

For students, this site serves as a FREE 24–7 electronic study guide, and it includes such features as self-quizzes and interactive applets.

Online Homework Options

WebAssign. *Premium* www.webassign.net/whfreeman

WebAssign Premium integrates the book's exercises into the world's most popular and trusted online homework system, making it easy to assign algorithmically generated homework and quizzes. Algorithmic exercises offer the instructor optional algorithmic solutions. WebAssign Premium also offers access to resources, including the new Dynamic Figures, CalcClips whiteboard videos, tutorials, and "Show My Work" feature. In addition,

WebAssign Premium is available with a fully customizable e-Book option that includes links to interactive applets and projects.

calcportal www.yourcalcportal.com

CalcPortal combines a fully customizable e-Book, exceptional student and instructor resources, and a comprehensive online homework assignment center. Included are algorithmically generated exercises, as well as Precalculus diagnostic quizzes, Dynamic Figures, interactive applets, CalcClips whiteboard videos, student solutions, online quizzes, Mathematica and Maple manuals, and homework management tools, all in one affordable, easy-to-use, and fully customizable learning space.

WeBWorK webwork.maa.org

W. H. Freeman offers approximately 2,500 algorithmically generated questions (with full solutions) through this free, open-source online homework system at the University of Rochester. Adopters also have access to a shared national library test bank with thousands of additional questions, including 1,500 problem sets matched to the book's table of contents.

Additional Media

SolutionMaster

This easy-to-use Web-based version of the Instructor's Solutions Manual allows instructors to generate a solution file for any set of homework exercises. Solutions can be downloaded in PDF format for convenient printing and posting.

Interactive e-Book at ebooks.bfwpub.com/integcalculus

The Interactive e-Book integrates a complete and customizable online version of the text with its media resources. Students can quickly search the text, and they can personalize the e-Book just as they would the print version, with highlighting, bookmarking, and note-taking features. Instructors can add, hide, and reorder content, integrate their own material, and highlight key text.

Course Management Systems

W. H. Freeman and Company provides courses for Blackboard, WebCT (Campus Edition and Vista), Angel, Desire2Learn, Moodle, and Sakai course management systems. These are completely integrated solutions that you can easily customize and adapt to meet your teaching goals and course objectives. Visit www.macmillanhighered.com/catalog/other/coursepack for more information.

i-clicker

This two-way radio frequency classroom response solution was developed by educators for educators. University of Illinois physicists Tim Stelzer, Gary Gladding, Mats Selen, and Benny Brown created the i-clicker system after using competing classroom responses and discovering that they were neither appropriate for the classroom nor friendly to the student. Each step of i-clicker's development has been informed by teaching and learning. i-clicker is superior to other systems from both a pedagogical and a technical standpoint. To learn more about packaging i-clicker with this textbook, contact your local sales representative or visit www.iclicker.com.

Each section opens with a **list of the three main section topics**. The list provides a focus and highlights key concepts.

3.3 THE SECOND DERIVATIVE AND CURVE SKETCHING

▶ Using first and second derivatives to define and detect concavity

▶ The behavior of the first and second derivatives at inflection points

▶ Using the second-derivative test to determine whether critical points are maxima, minima, or neither

---

Definitions are clearly boxed, numbered, and labeled for easy reference. To reinforce their importance and meaning, definitions are followed by brief, often illustrated, examples.

DEFINITION 3.9   **Formally Defining Concavity**

Suppose f and f' are both differentiable on an interval I.

(a) f is **concave up** on I if f' is increasing on I.

(b) f is **concave down** on I if f' is decreasing on I.

How does this formal definition of concavity correspond with our intuitive notion of concavity? Consider the functions graphed next. On each graph four slopes are illustrated and estimated. Notice that when f is concave up, its slopes increase from left to right, and when f is concave down, its slopes decrease from left to right.

Slopes increase when f is concave up   *Slopes decrease when f is concave down*

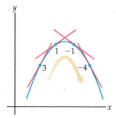

Theorems are developed intuitively before they are stated formally, and simple examples inform the discussion. **Proofs** follow most theorems, although they are optional, given instructor preference.

THEOREM 3.4   **Rolle's Theorem**

If f is continuous on $[a, b]$ and differentiable on (a, b), and if $f(a) = f(b) = 0$, then there exists at least one value $c \in (a, b)$ for which $f'(c) = 0$.

Actually, Rolle's Theorem also holds in the more general case where $f(a)$ and $f(b)$ are equal to each other (not necessarily both zero). For example, Rolle's Theorem is also true if $f(a) = f(b) = 5$, or if $f(a) = f(b) = -3$, and so on, because vertically shifting a function by adding a constant term does not change its derivative. However, the classic way to state Rolle's Theorem is with $f(a)$ and $f(b)$ both equal to zero.

> **Proof.** Rolle's Theorem is an immediate consequence of the Extreme Value Theorem from Section 1.4 and the fact that every extremum is a critical point. Suppose f is continuous on the closed interval $[a, b]$ and differentiable on the open interval (a, b), with $f(a) = f(b) = 0$. By the Extreme Value Theorem, we know that f attains both a maximum and a minimum value on $[a, b]$. If one of these extreme values occurs at a point $x = c$ in the interior (a, b) of the interval, then $x = c$ is a local extremum of f. By the previous theorem, this means that $x = c$ is a critical point of f. Since f is assumed to be differentiable at $x = c$, it follows that $f'(c) = 0$ and we are done.
>
> It remains to consider the special case where all of the maximum and minimum values of f on $[a, b]$ occur at the endpoints of the interval (i.e., at $x = a$ or at $x = b$). In this case, since $f(a) = f(b) = 0$, the maximum and minimum values of $f(x)$ must both equal zero. For all x in $[a, b]$ we would have $0 \le f(x) \le 0$, which means that f would have to be the constant function $f(x) = 0$ on $[a, b]$. Since the derivative of a constant function is always zero, in this special case we have $f'(x) = 0$ for *all* values of c in (a, b), and we are done. ∎

Cautions are appropriately placed at points in the exposition where students typically have questions about the nuances of mathematical thinking, processes, and notation.

 CAUTION | It is important to note that although we use the notation x^{-1} to denote the reciprocal $\frac{1}{x}$, the notation f^{-1} does *not* stand for the reciprocal $\frac{1}{f}$ of f. The notation f^{-1} used in Definition 0.10 is pronounced "f inverse." We are now using the same notation for two very different things, but it should be clear from the context which one we mean.

Every section includes short, illustrative examples as part of the discussion and development of the material. Once the groundwork has been laid, more complex **examples** and calculations are provided. Students find this approach easier to handle because the difficult calculations do not interfere with the development of why things work. Example **solutions** are explained in detail and include all the steps necessary for student comprehension.

EXAMPLE 4 Using critical points and Rolle's Theorem to find local extrema

The function $f(x) = x(x-1)(x-3)$ is a cubic polynomial with one local maximum and one local minimum. Use Rolle's Theorem to identify intervals on which these extrema exist. Then use derivatives to find the exact locations of these extrema.

SOLUTION

The roots of $f(x) = x(x-1)(x-3)$ are $x = 0$, $x = 1$, and $x = 3$. Since f is a polynomial, it is continuous and differentiable everywhere. Therefore Rolle's Theorem applies on the intervals $[0, 1]$ and $[1, 3]$, and it tells us that at least one critical point must exist inside each of these intervals.

The critical points of f are the possible locations of the local extrema that we seek. To find the critical points we must solve the equation $f'(x) = 0$. It is simpler to do some algebra before differentiating:

$$f'(x) = \frac{d}{dx}(x(x-1)(x-3)) = \frac{d}{dx}(x^3 - 4x^2 + 3x) = 3x^2 - 8x + 3.$$

By the quadratic formula, we have $f'(x) = 0$ at the points

$$x = \frac{-(-8) \pm \sqrt{8^2 - 4(3)(3)}}{2(3)} = \frac{8 \pm \sqrt{28}}{6} = \frac{4 \pm \sqrt{7}}{3}.$$

These x-values are approximately $x \approx 0.451$ and $x \approx 2.215$. If we look at the graph of f, then we can see that the smaller of these two x-values is the location of the local maximum and the larger is the location of the local minimum; see the figure that follows. ☐

Following many example solutions, **Checking the Answer** encourages students to learn to check their work, using technology such as a graphing calculator when appropriate.

 | The graph of $f(x) = x(x-1)(x-3)$ is shown next. Notice that the local extrema do seem to occur at the values we just found.

Extrema at $x \approx 0.451$ and $x \approx 2.215$

Each section closes with five **Test Your Understanding** questions that test students on the concepts and reading presented in the section. Because answers are not provided, instructors may choose to use these questions for discussion or assessment.

> **TEST YOUR UNDERSTANDING**
>
> ► Why could we not give a precise mathematical definition of concavity before this section?
>
> ► The domain points $x = c$ where $f''(c) = 0$ or where $f''(c)$ does not exist are the critical points of the function f'. Why?
>
> ► Why is it not clear to say a sentence such as "Because it is positive, it is concave up"? How could this information be conveyed more precisely?
>
> ► Why does it make sense that f' is increasing when f'' is positive?
>
> ► Suppose $x = c$ is a critical point with $f'(c) = 0$. Why does it make graphical sense that f has a local minimum at $x = c$ when f is concave up in a neighborhood around $x = c$?

Section Exercises are provided in a consistent format that offers the same types of exercises within each section. This approach allows instructors to tailor assignments to their course, goals, and student audience.

Thinking Back exercises ask students to review relevant concepts from previous sections and lessons.

Concepts exercises are consistently formatted to start with the following three problems:

- Problem 0 tests understanding.
- Problem 1 consists of eight true/false questions.
- Problem 2 asks the student to create examples based on their understanding of the reading.

Skills exercises offer ample practice, grouped into varying degrees of difficulty.

Applications exercises contain at least two in-depth real-world problems.

Proofs exercises can be completed by students in non-theoretical courses. Hints are often provided, and many exercises mimic work presented in the reading and examples. Often, these exercises are a continuation of a proof offered as a road map in the narrative.

Thinking Forward exercises plant seeds of concepts to come. In conjunction with the Thinking Back exercises, they tie together past and future topics, thereby providing a seamless flow of concepts.

Chapter Review, Self-Test, and Capstones, found at the end of each chapter, present the following categories:

Definitions exercises prompt students to recall definitions and give an illustrative example.

Theorems exercises ask students to complete fill-in-the-blank theorem statements.

Formulas, Notation, and/or Rules ask students to show a working understanding of important formulas, equations, notation, and rules.

Skill Certification exercises provide practice with basic computations from the chapter.

Capstone Problems pull together the essential ideas of the chapter in more challenging mathematical and application problems.

There are many people whose contributions to this project have made it immeasurably better. I am grateful to the many instructors from across the United States and Canada who have offered comments that assisted in the development of this book:

Jorge Balbas, *California State University, Northridge*

Rann Bar-On, *Duke University*

Michael Berg, *Loyola Marymount University*

Jay Bookman, *Duke University*

Jennifer Bowen, *The College of Wooster*

Priscilla Bremser, *Middlebury College*

Elaine Brooks Rinaldo, *Hudson Valley Community College*

Monica L. Brown, *St. Catherine University*

Katherine S. Byler Kelm, *California State University, Fresno*

Leonard Chastkovsky, *University of Georgia*

Marty Cook, *Furman University*

Barbara Cortzen, *DePaul University*

David Cox, *Amherst College*

David Cruz-Uribe, *Trinity College*

Charles Cunningham, *James Madison University*

David Erwin, *Trinity College*

Frances Gulick, *University of Maryland*

James Hartman, *The College of Wooster*

Slawomir Kwasik, *Tulane University*

Hong-Jian Lai, *West Virginia University*

Major Darryl Lanford, *U.S. Military Academy*

Carmen Latterell, *University of Minnesota, Duluth*

Carl Lutzer, *Rochester Institute of Technology*

Jeffery McLean, *University of St. Thomas*

Norma Noguera, *California State University, Long Beach*

Russell W. Park, *Gustavus Adolphus College*

Melissa Shepard Loe, *University of St. Thomas*

Melanie Stein, *Trinity College*

Elaine Terry, *St. Joseph's University*

Mary Vlastnik Armon, *Knox College*

Barbara Wahl, *Hanover College*

George Brock Williams, *Texas Tech University*

My students and colleagues at James Madison University have used preliminary versions of this text for the past two years and have helped to clarify the exposition and remove ambiguities. I would particularly like to thank my colleagues Elizabeth Theta Brown, David Carothers, Heena Chalishajar, Charles Cunningham, Rickie Domangue, Brett Enge, Rebecca Field, Jeannie Fitzgerald, Hasan Hamdan, Robert Hanson, John Johnson, Brant Jones, Muhammad Khan, Peter Kohn, Hun Kwon, James Liu, Caroline Lubert, Stephen Lucas, Carter Lyons, John Marafino, Carla Martin, Jason Martin, Edwin O'Shea, Ed Parker, Gary Peterson, Katie Quertermous, Jason Rosenhouse, Hamidreza Semiyari, Jim Sochacki, Eva Strawbridge, Roger Thelwell, Anthony Tongen, Leonard Van Wyk, Brian Walton, Debra Warne, and Cassandra Williams for class-testing the book and for their helpful feedback. During the class-testing at JMU, hundreds of students provided feedback, made suggestions that improved the book, and, of course, showed me how they learned from the book! Thank you to all of these students, especially Lane O'Brien and Melissa Moxie for their meticulous review of an earlier draft of the text.

Chris Brazfield, now at Carroll Community College, helped with the initial development of the text and the ideas behind it. Kevin Cooper of Washington State University contributed many interesting and challenging real-world applications. Laura Person provided valuable class-testing feedback that helped improve the text. I owe all of them great thanks for their expertise.

I also owe thanks to all of the people at W. H. Freeman who helped with the development of this text. Developmental editor Leslie Lahr has been with this project from the beginning. Even under pressure, Leslie always maintains a positive attitude and finds a way for us to move forward. Without her support, I would not have made it through the rocky patches. Executive editor Terri Ward and developmental editor Katrina Wilhelm helped keep me on track while I wrote, rewrote, revised, revised, and revised some more, and I thank them for their support and patience. Brian Baker, meticulous copy editor, made significant improvements to the text. Misplaced commas, dangling modifiers, and run-on sentences didn't stand a chance under his scrutiny. Ron Weickart and his team at Network Graphics took my graphs and sketches and turned them into the beautiful artwork in the text. Sherrill Redd and the compositing team at Aptara did a great job implementing all of the design elements from my crazy LaTeX files.

Finally, I would like to thank all of my students who showed determination and courage while tackling the monumental task of learning calculus, precalculus, and algebra all at the same time. I am so proud of all of you.

Learning something new can be both exciting and daunting. To gain a full understanding of the material in this text, you will have to read, you will have to think about the connections between the new topics and the topics that were previously presented, and you will have to work problems—many, many problems.

The structure of this text should help you understand the material. The material is laid out in a linear fashion that I think will facilitate your understanding. Each section is separated into two main parts: first, a presentation of new material and then second, a set of *Examples and Explorations*, where you will find problems that are carefully worked through. Working through these examples on your own, as you read the steps for guidance, will help prepare you for the exercises.

Reading a mathematics book isn't like reading a novel: You may have to read some parts more than once, and you may need to make notes or work things out on paper. Pay special attention to the "Checking Your Answer" features, so that you can learn how to check your own answers to many types of questions.

To succeed in calculus, you need to do homework exercises. The exercises in every section of this text are broken into six categories: "Thinking Back," "Concepts," "Skills," "Applications," "Proofs," and "Thinking Forward."

- As the title suggests, the *Thinking Back* problems are intended to tie the current material to material you've seen in previous sections or even previous courses.
- The *Concepts* problems are designed to help you understand the main ideas of the section, which are presented without a lot of calculation. Every group of *Concepts* exercises begins by asking you to summarize the section, continues with eight true–false questions, and then asks for three examples illustrating ideas from the section.
- The bulk of the exercises in each section consists of *Skills* problems that may require more calculation.
- The *Applications* exercises use the concepts from the section in "real-world" problems.
- The *Proofs* exercises ask you to prove some basic theory from the section.
- Finally, the *Thinking Forward* questions use current ideas to introduce topics that you will see in subsequent sections.

I hope this structure allows you to tie together the material as you work through the book. I have supplied the answers to the odd-numbered exercises, but don't restrict yourself to those problems. You can check answers to even-numbered questions by hand or by using a calculator or an online tool such as wolframalpha.com. After all, on a quiz or test you won't have the answers, so you'll have to know how to decide for yourself whether or not your answers are reasonable.

Some students may like to work through each section "backwards," starting by attempting the exercises, then checking back to the examples as needed when they get stuck, and, finally, using the exposition as a reference when they want to see the big picture. That is fine; although I recommend that you at least try reading through the sections in order to see how things work for you. Either way, I hope that the separation of examples from exposition and the division of homework problems into subsections will help make the process of learning this beautiful subject easier. I have written this text with you, the student, in mind. I hope you enjoy using it!

Functions and Precalculus

0.1   NUMBERS AND SETS

▶   Number systems, sets, set membership, and set notation

▶   The real number line, absolute value, distance, and the Cartesian plane

▶   Finding unions and intersections of sets on the real number line and in the Cartesian plane

Real Numbers

Mathematics is a language. In order to understand it, we have to learn how to read it and speak it with the correct vocabulary. Since calculus is at its heart the study of functions of real numbers, the universe we will spend most of our time exploring is the set of **real numbers** and the relationships between sets of real numbers. Therefore we must begin by setting out the mathematical language that describes numbers and sets of real numbers. Once we all speak the same language, we can start building the theory of functions and calculus.

The real numbers can be represented graphically on a **number line** like the one illustrated next. Every real number can be represented as a point on a number line. For example, the real number 1.3 is marked on the number line shown. The arrows at the ends of the number line indicate that the number line continues indefinitely in either direction.

We have different names for different types of real numbers. For example, you probably already know that the positive and negative whole numbers

$$\ldots, -3, -2, -1, 0, 1, 2, 3, \ldots$$

are called the **integers**. In this course we will be concerned not just with the integers, but with all the real numbers in between. Calculus is fundamentally the study of the entire continuum of real numbers, which we write as $\mathbb{R}$. Because we will work primarily with real numbers in this text, when we say "a number," we will always mean "a *real* number," unless we specify otherwise.

The real numbers split into two important categories of numbers:

DEFINITION 0.1   **Rational and Irrational Numbers**

(a) A **rational number** is a real number that can be written as a quotient of the form $\frac{p}{q}$ for some integers p and q, with $q \neq 0$.

(b) An **irrational number** is a real number that cannot be written in the form of a rational number.

In a previous course, you likely thought of rational numbers as terminating or non-repeating decimals, such as 0.25 or 0.33$\overline{3}$. This definition of rational numbers is equivalent to Definition 0.1, but in this course the quotient definition will prove to be far more useful. You should think of rational numbers as quotients of integers and irrational numbers, like π or $\sqrt{2}$, as numbers that cannot be written as quotients of integers.

We can also classify real numbers according to their **signs**, that is, according to whether they are positive or negative.

DEFINITION 0.2

Positive and Negative Numbers

A real number is said to be

(a) *positive* if greater than 0

(c) *nonnegative* if greater than or equal to 0

(b) *negative* if less than 0

(d) *nonpositive* if less than or equal to 0

Here are some examples of real numbers and their classifications into the preceding categories:

▶ the real number 3 is positive, nonnegative, rational, and an integer;

▶ the real number $-\frac{11}{7}$ is negative, nonpositive, and rational;

▶ the real number π is positive, nonnegative, and irrational;

▶ the real number 0 is nonnegative, nonpositive, rational, and an integer.

Sets and Intervals of Real Numbers

In this course we will often be interested in certain connected subsets of the real number line, such as the set of real numbers greater than 0 or the set of real numbers between 0 and 2. Such subsets are called *intervals*. Following are four examples of intervals, together with their descriptions in inequality notation on the left and their expressions in *interval notation* on the right:

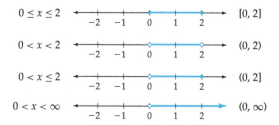

$$0 \le x \le 2 \qquad [0, 2]$$

$$0 < x < 2 \qquad (0, 2)$$

$$0 < x \le 2 \qquad (0, 2]$$

$$0 < x < \infty \qquad (0, \infty)$$

The first interval, $[0, 2]$, includes its endpoints 0 and 2, so we say that this is a *closed interval*. The second interval, $(0, 2)$, does not include either endpoint 0 or 2, so we say that this is an *open interval*. The third interval, $(0, 2]$, is neither closed nor open. (We could call this interval half closed, or half open, or—and this is actually true—some people call it a *clopen interval*.) The fourth interval, $(0, \infty)$, is said to be an *unbounded interval* because it contains numbers that are arbitrarily large. Note that we use an open bracket after the ∞ because "∞" is not a real number and therefore is not included in this interval.

Since open and closed intervals will be of particular interest to us in this course, we should state their definitions precisely:

DEFINITION 0.3

Closed and Open Intervals of Real Numbers

An *interval* of real numbers is a connected subset of the real number line.

(a) A *closed interval* $[a, b]$ is the set of real numbers x with $a \le x \le b$.

(b) An *open interval* (a, b) is the set of real numbers x with $a < x < b$.

There are other types of intervals besides those just discussed. For example, the entire real number line $\mathbb{R}$ is the interval $(-\infty, \infty)$. As another example, the set {2} containing only the number 2 is an interval, since it is a connected subset of the real numbers. Perhaps the strangest subset of real numbers is the set containing no numbers at all. This set is called the ***empty set*** or the ***null set***, and is denoted by the symbol ∅. For example, the set of real numbers that are both negative and positive is empty, since no real number is both strictly less than and strictly greater than zero.

Intervals are a special type of ***subset*** of the set of real numbers. There are many other types of subsets of real numbers—for example, the set containing the four real numbers 0, 3, 6, and 7. Sets such as this one that consist of a finite collection of real numbers can be described simply by listing the elements in curly brackets:

$$\{0, 3, 6, 7\}.$$

If a set S has an infinite number of elements, or a large number of elements, we can list them all only if they have a recognizable pattern; for example, the set of all positive even numbers can be written as

$$\{2, 4, 6, 8, 10, 12, \ldots\}.$$

Here is a more interesting set of real numbers to consider: The set of real numbers x for which x^2 is greater than 4. The real numbers that satisfy the property $x^2 > 4$ are the numbers x with $x > 2$ and also the numbers x with $x < -2$ (since, for example, $(-3)^2 = 9$ is greater than 4). This set is not an interval, because it is not a connected subset of the real number line. We need a compact way to write such sets in mathematical notation.

In general, we can describe sets compactly by using ***set notation***, which specifies two things: first, a *category* (or larger set) of objects to consider; second, a *test* that describes when an object in that category is in the set.

DEFINITION 0.4

Set Notation and Elements of a Set

Schematically, ***set notation*** for a set S is written in the form

$$S = \{x \in category \mid test\ determining\ whether\ x\ is\ in\ the\ set\,\}.$$

An object x is an ***element*** of a set S if it is contained in S, that is, if it passes the test written in the second half of the preceding notation. When this happens, we write $x \in S$.

For example, the set consisting of all real numbers x whose squares are greater than 4 can be written as

$$\{x \in \mathbb{R} \mid x^2 > 4\}.$$

We pronounce this notation as "the set of real numbers x such that x^2 is greater than 4." If we call the set in this example S, then, since $x = 5$ and $x = -3$ pass the test "$x^2 > 4$," we say that $5 \in S$ and $-3 \in S$. In contrast, since $x = 1$ does not pass the test "$x^2 > 4$," we say that $1 \notin S$.

Considering our earlier discussion, we could also write the preceding set as

$$\{x \in \mathbb{R} \mid x > 2 \text{ or } x < -2\}.$$

As we will see in a moment, with a little bit more notation we can express this set even more compactly as a combination of the intervals $(-\infty, -2)$ and $(2, \infty)$.

Unions and Intersections of Sets

Throughout our study of calculus we will often have to consider combinations of intervals of real numbers. The **union** of two sets A and B is the set of elements that are in *either A or B*, and the **intersection** of A and B is the set of elements that are in *both A and B*.

DEFINITION 0.5

Union and Intersection

Suppose A and B are sets.

(a) The **union** of A and B is the set $A \cup B = \{x \mid x \in A \text{ or } x \in B\}$.

(b) The **intersection** of A and B is the set $A \cap B = \{x \mid x \in A \text{ and } x \in B\}$.

It may help you distinguish $\cup$ and $\cap$ if you notice that the symbol $\cup$ looks like "u" for "union" and the symbol $\cap$ looks like lowercase "n" for "intersection." For example, the set of real numbers whose square is greater than 4 is a union of intervals:

$$\{x \in \mathbb{R} \mid x^2 > 4\} = \{x \in \mathbb{R} \mid x > 2 \text{ or } x < -2\} = (-\infty, -2) \cup (2, \infty).$$

Here is another simple example: Consider the intervals $[-3, 1]$ and $(-1, 3)$ shown next at the left. The intersection $[-3, 1] \cap (-1, 3)$ is the set of real numbers that are contained in both $[-3, 1]$ and $(-1, 3)$. This intersection is the interval $(-1, 1]$ and is shown in the middle figure, which is obtained by looking at where the two original intervals overlap.

$[-3, 1]$ *and* $(-1, 3)$ $[-3, 1] \cap (-1, 3)$ $[-3, 1] \cup (-1, 3)$

On the other hand, the union $[-3, 1] \cup (-1, 3)$ is the set of real numbers that are contained in either $[-3, 1]$ or $(-1, 3)$, or both. The right-hand figure shows this union, $[-3, 3)$, obtained by looking at the real numbers that fall under the shadow of either of the two original intervals in the left-hand figure.

⊗ CAUTION In daily language we often use **exclusive "or,"** in which "P or Q" is true if either P, or Q, but not both, is true. For example, if Phil says, "I'll meet you in the lobby or at the coffee shop," clearly he means that he will be in one or the other of these places, but not both. In mathematics, however, we always mean **inclusive "or,"** where "P or Q" is true if P is true, if Q is true, or if both P and Q are true. For example, "-2 is negative or an integer" is a true statement.

Another way to visualize intersections and unions is with **Venn diagrams**, by thinking of sets as shaded regions defined by circles. Suppose we represent two sets A and B by circles, as in the first two figures that follow Then the set $A \cap B$ is the intersection of the two circles, and the set $A \cup B$ is their union; see the third and fourth figures.

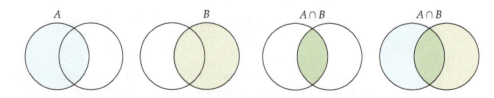

A B $A \cap B$ $A \cap B$

Absolute Value and Distance

Absolute value and distance give us ways to measure the relative positions of numbers on the real number line. Intuitively, the **absolute value** of a real number a is the magnitude, or size, of a and is denoted $|a|$. For example, $|3| = 3$ and $|-3| = 3$. The absolute value of a "makes the number a positive." More precisely, if a is positive or zero, then taking the absolute value should do nothing: In this case $|a| = a$. However, if a is negative, then taking the absolute value should change the sign: In this case $|a| = -a$; for example $|-3| = -(-3) = 3$.

DEFINITION 0.6   **Absolute Value**

The **absolute value** of a real number a is $|a| = \begin{cases} a, & \text{if } a \geq 0 \\ -a, & \text{if } a < 0. \end{cases}$

 CAUTION The quantity $-a$ is not necessarily negative; it depends on whether the number a is positive or negative. For example, if $a = -3$, then $-a = -(-3) = 3$ is positive. In Definition 0.6, $|a| = -a$ only if a is negative, in which case $-a$ is a *positive* number.

Absolute values behave well with respect to products and quotients; for example, the absolute value of a product is simply the product of the absolute values.

THEOREM 0.7   **Absolute Values Commute with Products and Quotients**

Given any real numbers a and b, $|ab| = |a||b|$ and $\left| \dfrac{a}{b} \right| = \dfrac{|a|}{|b|}$.

For example, $|(-2)(3)| = |-6| = 6$ is equal to $|-2||3| = (2)(3) = 6$. Note that absolute values do *not* behave well with sums; for example $|-2 + 3| = 1$ is not equal to $|-2| + |3| = 5$.

The absolute value $|a|$ of any real number a is the distance between a and 0 on the real number line. Geometrically, the distance between any two real numbers a and b is the length of the line segment from a to b on the number line. Algebraically, this distance is given by $|b - a|$, regardless of the signs of a and b. For example, the diagram that follows shows that the distance between $a = 2$ and $b = -1$ is 3 units. Algebraically, $|b - a| = |-1 - 2| = |-3| = 3$. We will take the algebraic formulation as the definition of distance, and in Exercise 14 we will show that this defnition is equivalent to our geometric interpretation.

DEFINITION 0.8   **Distance Between Two Real Numbers**

The **distance** between two real numbers a and b is $\text{dist}(a, b) = |b - a|$.

The Cartesian Plane

We now take our knowledge of the one-dimensional real number line and extend it to a two-dimensional plane where points are represented by pairs of real numbers. An **ordered pair** of real numbers x and y is denoted (x, y). The order matters; for example, the ordered pair $(2, 5)$ is different from the ordered pair $(5, 2)$. Notice that the notation $(2, 5)$ now has two possible meanings, depending on the context. Earlier we used this notation to represent the open interval from 2 to 5; now we are using it to represent an ordered pair of numbers.

The set of all ordered pairs forms the ***Cartesian plane***, named for the famous mathematician and philosopher René Descartes. This is a two-dimensional plane with two perpendicular axes, one horizontal and one vertical, that intersect at a point called the ***origin***. Every ordered pair (x, y) represents one point in the Cartesian plane, where the first coordinate, x, in the pair represents a horizontal signed distance to the left or right of the origin, and the second coordinate, y, represents a vertical signed distance up or down from the origin. When we denote coordinates with the letters x and y, we refer to the horizontal axis as the ***x-axis***, the vertical axis as the ***y-axis***, and the plane itself as the ***xy-plane***. Shown next are four examples of ordered pairs as points in the Cartesian plane. Notice that the Cartesian plane is naturally divided into four quadrants, depending on the signs of x and y.

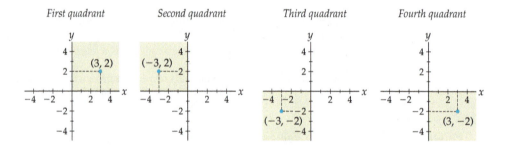

Geometrically, the distance between two ordered pairs of real numbers (x_1, y_1) and (x_2, y_2) is the length of the line segment connecting these two points in the plane. Algebraically, this distance is given by the formula that follows. In Section 0.7 we will use the Pythagorean theorem to show that these two interpretations of distance are equivalent.

DEFINITION 0.9

Distance Between Two Points in the Plane

The ***distance*** between two points $P = (x_1, y_1)$ and $Q = (x_2, y_2)$ in the plane is given by the ***distance formula***:

$$\text{dist}(P, Q) = \sqrt{(x_2 - x_1)^2 + (y_2 - y_1)^2}.$$

In the real number line, the midpoint between two points a and b is the midpoint of the line segment from a to b, which algebraically is simply the average of the numbers a and b. In the Cartesian plane, the midpoint between two points (x_1, y_1) and (x_2, y_2) is the point (m_1, m_2), that is, at the midpoint of the line segment connecting the two points. We can find this midpoint algebraically by taking the averages of the x- and y-coordinates of the two points.

DEFINITION 0.10

Midpoint Between Two Points in the Plane

The coordinates of the ***midpoint*** between two points $P = (x_1, y_1)$ and $Q = (x_2, y_2)$ in the plane are given by the ***midpoint formula***:

$$\text{midpoint}(P, Q) = \left(\frac{x_1 + x_2}{2}, \frac{y_1 + y_2}{2} \right).$$

Examples and Explorations

EXAMPLE 1   Identifying rational numbers

Which of the numbers that follow are rational numbers? Express each rational number in the form $\frac{p}{q}$ for some integers p and q.

(a) 0.001   **(b)** 5   **(c)** 3.258   **(d)** $-\frac{11}{7}$ **(e)** $0.\overline{333}$

SOLUTION

All of the preceding numbers are rational. The first number is one one-thousandth and is thus represented by the fraction $\frac{1}{1000}$. The number 5 can be written as $\frac{5}{1}$ and is thus a rational number. The third number is the sum of 3 and 0.258, or $\frac{3000}{1000} + \frac{258}{1000} = \frac{3258}{1000}$. The fourth number could be represented in the form $\frac{p}{q}$ either by $\frac{-11}{7}$ or by $\frac{11}{-7}$. Finally, the last number is represented by the fraction $\frac{1}{3}$. (Try dividing 3 into 1 to see why.)   □

EXAMPLE 2   Unions and intersections of intervals

Consider the intervals $A = [-3, -1)$, $B = [0, 2]$, and $C = [-2, 1)$. Express each of the following in interval notation:

(a) $A \cap B$ **(b)** $A \cup B$ **(c)** $(A \cup B) \cap C$

SOLUTION

The diagram that follows shows the given intervals above and below a real number line. To think about unions and intersections you should think about the "shadows" of these intervals on the real number line and where they overlap.

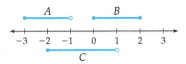

(a) The intervals A and B do not overlap at all, so their intersection is empty; in other words, $A \cap B = \emptyset$.

(b) The union of A and B is $[-3, -1) \cup [0, 2]$. This is a disconnected subset of the real number line, and there is no simpler way to express $A \cup B$ in interval notation.

(c) To find $(A \cup B) \cap C$ we must first consider the union of A and B, and then find the subset of those real numbers which are also contained in C. By looking at the number line and where these intervals overlap, we see that $(A \cup B) \cap C = [-2, 1) \cup [0, 1)$.   □

EXAMPLE 3   Expressing distance statements with absolute values

Suppose x is a real number with the property that the distance between x and 2 is less than 3. Express this statement in mathematical notation with an inequality and absolute value. Then use a number line to find the interval of values of x that satisfy the given property.

SOLUTION

Since the distance between two real numbers a and b is given by $|b - a|$, our real number x satisfies the inequality $|x - 2| < 3$ (or equivalently, $|2 - x| < 3$). The following real number line shows that this means that x must be between -1 and 5, or in other words, x must be in the interval $(-1, 5)$.

□

A punctured interval defined by an absolute value statement

Let A be the set of all real numbers x that satisfy the double inequality $0 < |x - 2| < 1$. Interpret A in terms of distances, sketch A on a real number line, and express A in interval notation.

SOLUTION

The statement $0 < |x - 2| < 1$ says that the distance between x and 2 is greater than 0 and less than 1. This means that x is a real number which is not equal to 2, but is within distance 1 of 2. The set of such values of x is an open interval with a hole in the middle:

Since this is not a connected subset of the real number line, A is not a single interval. We can, however, express A in interval notation as a union of intervals, as $(1, 2) \cup (2, 3)$. We call this the ***punctured open interval*** of radius 1 around 2.

□

Using the distance and midpoint formulas in the Cartesian plane

Plot the points $P = (-1, 3)$ and $Q = (4, -2)$ on a set of axes, and sketch the line segment connecting these two points. Use the distance formula to find the length of this line segment, and use the midpoint formula to find the midpoint of the line segment.

SOLUTION

The points $P = (-1, 3)$ and $Q = (4, -2)$, and the line segment between them, are shown as follows:

The distance between P and Q is

$$\text{dist}(P, Q) = \sqrt{(4 - (-1))^2 + (-2 - 3)^2} = \sqrt{5^2 + (-5)^2} = \sqrt{50} \approx 7.071 \text{ units.}$$

The midpoint of the line segment from P to Q has coordinates

$$\text{midpoint}(P, Q) = \left(\frac{-1+4}{2}, \frac{3+(-2)}{2}\right) = \left(\frac{3}{2}, \frac{1}{2}\right).$$

CHECKING THE ANSWER

We can check whether our answer seems correct by plotting the point $(3/2, 1/2)$ in the plane, as follows:

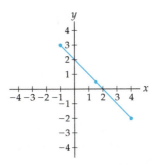

From the graph it seems right that the midpoint of the line segment is at the point $(3/2, 1/2)$. Moreover, it does not seem unreasonable that this line segment is just over 7 units long.

EXAMPLE 6   **Intersections of subsets of the Cartesian plane**

Sketch the points (x, y) in the Cartesian plane that simultaneously satisfy the three inequalities $x \leq 2$, $y > -1$, and $y \leq x$.

SOLUTION

Each of the three inequalities defines a subset of the plane, as shown in the three figures that follow. For example, the inequality $y \leq x$ defines the set of points (x, y) whose y-coordinates are less than or equal to their x-coordinates. The graphs of the given inequalities are shown in the figures that follow, where in each case we shade in the region of points (x, y) that satisfy the inequality.

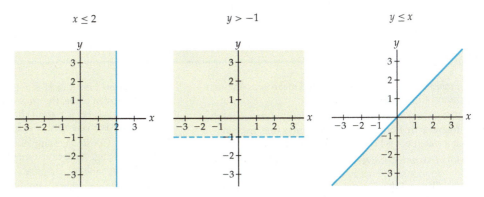

We seek the set of points (x, y) contained in all three regions. The figure that follows at the left shows the three boundary lines, and the shaded region shown at the right is the desired intersection. Notice that the $x = -1$ boundary is dashed, which means that the points on that line are not part of the intersection.

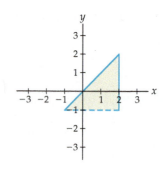

TEST YOUR UNDERSTANDING

▶ Without referring to decimal notation, what is the condition for a real number to be rational? Irrational?

▶ What is the algebraic definition of the absolute value of a real number? What does this definition mean in words? Why do we need to consider two cases?

▶ In what sense does the intersection symbol ∩ represent "and"? Is $A \cap B$ the set of all elements of A and all elements of B, or is it the set of all elements that are both in A and in B?

▶ How do you find the distance and midpoint between two points in the Cartesian plane?

▶ Define each of the following: open interval, absolute value, second quadrant, union of sets, intersection of sets.

EXERCISES 0.1

Thinking Back

▶ *Properties of real numbers:* You may remember from a previous math course that the operations of addition and multiplication of real numbers are associative and commutative. What does this mean? Are subtraction and division also associative and/or commutative operations?

▶ *Properties of integers:* Is the sum of two integers always an integer? What about products and quotients of integers?

▶ *Sums and products of positive and negative numbers:* If a and b are positive numbers, what can you say about their product and their sum? What if a and b are both negative? What if one is negative and one is positive?

Concepts

0. *Problem Zero:* Read the section and make your own summary of the material.

1. *True/False:* Determine whether each of the statements that follow is true or false. If a statement is true, explain why. If a statement is false, provide a counterexample.

   (a) *True or False:* Every real number is either rational or irrational, but not both.

   (b) *True or False:* Every nonpositive number is less than zero.

   (c) *True or False:* $1.5 \in \{x \in \mathbb{R} \mid x - 1 > 0\}$.

   (d) *True or False:* If $a \in \mathbb{R}$, then $-a$ is negative.

   (e) *True or False:* $|0| = 0$.

   (f) *True or False:* The distance between any point in the coordinate plane and itself is zero.

   (g) *True or False:* If (a, b) is in the second quadrant of the Cartesian plane, then $a > 0$ and $b < 0$.

   (h) *True or False:* If $x \in A \cap B$, then x is an element of A and x is an element of B, but not both.

2. *Examples:* Construct examples of the thing(s) described in the following. Try to find examples that are different than any in the reading.

   (a) A number a that is rational but not an integer, and a number b that is real but not rational.

   (b) An open interval, a closed interval, and an unbounded interval, in both interval notation and as pictures on the real number line.

   (c) Two intervals whose intersection is the empty set, and two intervals whose union is all of $\mathbb{R}$.

3. Is every integer a rational number? Is every rational number a real number? Is every irrational number a real number? Is every real number a complex number?

4. Is it possible to tell with a calculator that the number $\sqrt{2}$ is irrational? Why or why not?

5. Assuming that the pattern continues, is the number $0.61661166611166661111\ldots$ a rational number?

6. Suppose a is a real number whose first 24 decimal places are $a = 0.123123123123123123123123\ldots$. Explain why this information does not necessarily show that a is a rational number.

7. Is $\dfrac{1.7}{2.3}$ a rational number? If not, why not? If so, write it in the form $\dfrac{p}{q}$ for some integers p and q.

8. It is an interesting but unusual fact that $0.99\overline{9}$ and 1 represent the same number.

   (a) Use long division to argue that $0.33\overline{3} = \dfrac{1}{3}$.

   (b) Use part (a) to show that $0.99\overline{9} = 1$.

   (c) What other real numbers have two different decimal representations?

9. What do the following symbols represent?

   (a) $\in$ (b) $\cap$ (c) $\cup$ (d) $\emptyset$

10. List all of the elements of $\{x \in \mathbb{R} \mid x^2 = x\}$.

11. Out of context, the notation $(3, 5)$ could mean two different things. What are those two things?

12. There are 11 possible types of intervals of real numbers. If a and b are any real numbers with $a < b$, these intervals are $[a, b]$, $[a, b)$, $(a, b]$, (a, b), $[a, \infty)$, (a, ∞), $(-\infty, b]$, $(-\infty, b)$, $(-\infty, \infty)$, $\{a\}$, and $\emptyset$. Choose two real numbers a and b, and represent each of the 11 types of intervals in set notation, in words, and on a real number line.

13. Is the set $\{x \mid x^2 \geq 4\}$ an interval of real numbers? Why or why not? Include a sketch of this set on the real number line as part of your answer.

14. For each of the values of a and b that follow, sketch a and b on the real number line and visually find the distance between a and b. Then calculate $|b - a|$ and verify that it is equal to the geometric distance.

    (a) $a = 3$, $b = 5$ (b) $a = -3$, $b = 5$

    (c) $a = 3$, $b = -5$ (d) $a = -3$, $b = -5$

15. Verify that $|b - a| = |a - b|$ for each of the pairs of values of a and b given in Exercise 14.

16. Express the punctured interval of radius 3 around 1 in four ways:

    (a) As a picture on the real number line.

    (b) Using interval notation.

    (c) In words, using the concept of distance.

    (d) With a double inequality involving an absolute value.

17. Suppose $P = (u, v)$ and $Q = (x, y)$ are two points in the Cartesian plane. Describe the distance between P and Q in two ways: geometrically and algebraically.

18. Suppose $P = (u, v)$ and $Q = (x, y)$ are two points in the Cartesian plane. Describe the midpoint between P and Q in two ways: geometrically and algebraically.

19. If A and B are any two sets, is $A \cap B$ necessarily a subset of A? Is $A \cap B$ necessarily a subset of $A \cup B$? Is A a subset of $A \cap B$?

20. Use set notation to describe (a) the intersection $A \cap B$ and (b) the union $A \cup B$ of two sets A and B.

21. Is the union of two intervals always an interval? What about the intersection? Why or why not?

22. For each of the following pairs of sets A and B, describe $A \cap B$ and $A \cup B$:

    (a) $A = \{\text{red pants}\}$, $B = \{\text{leather pants}\}$.

    (b) $A = \{\text{cats}\}$, $B = \{\text{dogs}\}$.

    (c) $A = \{\text{posters}\}$, $B = \{\text{Elvis posters}\}$.

Skills

For Exercises 23–32, express each given rational number in the form $\dfrac{p}{q}$ for some integers p and q.

23. 0.17

24. 0.26

25. 0.00001

26. 5.31

27. -11.7

28. -62.14

29. 0

30. $-\dfrac{4}{13}$

31. $-0.\overline{66}$

32. $0.\overline{99}$

Assuming that a is a positive number and b is a negative number, write each of the expressions in Exercises 33–38 without absolute values, if possible. Your answers may involve a and b.

33. $|2b|$

34. $|8a|$

35. $|ab^3|$

36. $|b + 2|$

37. $|ab - 1|$

38. $|b^2 + 1|$

Describe each of the sets in Exercises 39–52, using (a) set notation, (b) interval notation (if possible), and (c) the real number line or the Cartesian plane.

39. The set of real numbers greater than -7 and less than 1.

40. The set of real numbers greater than 1.75 and less than or equal to 3.8.

41. The set of all real numbers whose product with zero is zero.

42. The set of real numbers that are both greater than 5 and less than 1.

43. The set of real numbers that are either greater than 5 or less than 1.

44. The set containing only -3, 4, 7, and 102.

45. The set of all odd negative integers.

46. The set of real numbers whose squares are less than 4.

47. The set of real numbers x with $\text{dist}(x, 3) = 1$.

48. The set of real numbers x with $\text{dist}(x, 3) > 1$.

49. The set of real numbers x that satisfy the double inequality $0 < |x - 3| < 5$.

50. The set of real numbers x that satisfy the double inequality $0 < |x + 2| < 3$.

51. The set of points (x, y) in the Cartesian plane that satisfy $y > x + 1$.

52. The set of points (x, y) in the Cartesian plane that are a distance of 1 from the origin.

Each subset of the real numbers shown in Exercises 53–56 can be interpreted in terms of distance. Express these subsets with absolute value inequalities.

53.

54.

55.

56.

Sketch each pair of points P and Q in Exercises 57–60 in the Cartesian plane. Calculate the length and midpoint of the line segment from P to Q, and use your picture to check whether your answers are reasonable.

57. $P = (-2, -1)$, $Q = (1, -3)$

58. $P = (-1, 0)$, $Q = (3, -2)$

59. $P = (-5, 2)$, $Q = (2, 5)$

60. $P = (-3, 4)$, $Q = (-3, -6)$

Each of the sets in Exercises 61–72 is some combination of the following six sets:

$$A = [-2, 3) \qquad D = \{x \mid x \geq -1\}$$
$$B = (-\infty, 0] \qquad E = \{x \mid 0 \leq x < 2\}$$
$$C = (-1, 3] \qquad F = \{x \mid x < -2\}$$

Write each set three ways: (a) in set notation; (b) in interval notation; and (c) on a real number line.

61. $C \cap D$

62. $A \cup B$

63. $C \cup D$

64. $E \cup F$

65. $B \cap E$

66. $C \cap F$

67. $C \cup E \cup F$

68. $A \cap B \cap C$

69. $(B \cap C) \cup F$

70. $(E \cup F) \cap B$

71. $(B \cup E) \cap (D \cup F)$

72. $(A \cap C) \cap (D \cap E)$

In Exercises 73–76, sketch the points (x, y) in the Cartesian plane that simultaneously satisfy all the given conditions.

73. $y \leq 3$, $y \geq x$, and $y \geq -x$

74. $y > 1$, $y < x$, and $x < 4$

75. $x \geq 0$ and $\text{dist}((x, y), (0, 0)) \leq 4$

76. $y > 0$, $y > x$, and $\text{dist}((x, y), (0, 0)) < 4$

Applications

77. Suppose a taxi driver's territory is along just one east-west street called Main Street. The taxi dispatch center is at 5th and Main. If the taxi travels from the dispatch center on Main Street to the corners of 9th Street, then 3rd Street, then 12th Street, then 8th Street, then 1st Street, and back to the dispatch center, how many total blocks did the taxi travel? Give not only your final answer, but also an expression of how to solve the problem with absolute values.

78. Starting from your apartment, you drive 6 blocks east to the grocery, then 5 blocks north to the coffee shop, then 17 blocks west to the laundromat, then 9 blocks south to the bakery. How far away are you from your apartment (as the crow flies)?

79. Anne, Bekah, and Calvin are going to lunch. Anne is willing to spend between \$9 and \$17 on her lunch, Bekah between \$3 and \$6, and Calvin between \$5 and \$12. Is there a lunch price that will work for all three people? What lunch prices are acceptable to at least two people?

80. A room at an art gallery measures 20 feet east-west and 15 feet north-south. An artist paints the floor red in a 9×9 square aligned with the southwest corner and blue in a 13×13 square aligned with the northeast corner. What is the area of the purple region where the red and blue squares intersect? (*Recall that the area of a rectangle is found by multiplying its length and width.*)

Proofs

81. Argue that all integers are rational numbers. (*Hint: Given an integer n, describe how n can be written in the form $\frac{p}{q}$ for some integers p and q.*)

82. Argue that the sum of two rational numbers is always a rational number. (*Hint: Given two rational numbers $\frac{p}{q}$ and $\frac{r}{s}$, what is their sum?*)

83. Argue that the product of two rational numbers is always a rational number. (*Hint: Given two rational numbers $\frac{p}{q}$ and $\frac{r}{s}$, what is their product?*)

84. The points (x, y) that lie on a circle of radius 1 centered at the origin satisfy the equation $x^2 + y^2 = 1$. Use the distance formula to show that any point (x, y) satisfying this equation is a distance of 1 from the origin.

Thinking Forward

▶ *Absolute value inequalities with "less than":* The absolute value inequality $|x-3| < 7$ describes the real numbers x that are within 7 units of the number 3. Equivalently, this subset of real numbers can be described as those that satisfy both $x - 3 < 7$ and $x - 3 > -7$. Illustrate this equivalence on a real number line.

▶ *Absolute value inequalities with "greater than":* The absolute value inequality $|x - 3| > 7$ describes the real numbers x that are more than 7 units from the number 3. Equivalently, this subset of real numbers can be described as those that satisfy either $x - 3 > 7$ or $x - 3 < -7$. Illustrate this equivalence on a real number line.

▶ *Areas under curves:* The set of points (x, y) that satisfy the equation $y = x^2$ form a curve in the plane, as shown in the figure that follows. Shade the subset of points (x, y) in the plane that satisfy $0 < y < x^2$ and $-2 < x < 3$. In later chapters we will learn how to find the areas of such regions.

Points (x, y) satisfying $y = x^2$

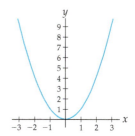

0.2   EQUATIONS

▶ Solving equations and systems of equations

▶ Simple factoring techniques and using products to solve equations

▶ Basic algebra skills involving fractions and dealing with quotients in equations

Equations and Solutions

An **equation** consists of two mathematical expressions connected by an equals sign. For example,

$$x^2 - 1 = 0, \qquad x^2 - x = y + 2, \qquad \text{and} \qquad 3u + 2v - 8x = 5$$

are equations. We will be concerned primarily with equations involving one or two **variables.** Usually we will use letters near the end of the alphabet, such as x, y, z, u, and v, to represent variables and will reserve letters near the beginning of the alphabet, such as a, b, and c, for **constants.** A variable represents a changing or unknown quantity, whereas a constant represents a fixed number.

An equation may be true for some values of the variables involved and false for others. For example, the equation $x^2 - 1 = 0$ is clearly not true for all values of x. In fact, this equation is true only when $x = 1$ or when $x = -1$. The set of values for which an equation is true is called the **solution set** for that equation (or simply the **solution** of that equation). Thus the solution of the equation $x^2 - 1 = 0$ is the set $\{1, -1\}$. A **system of equations** is simply a list of two or more equations. A solution of a system of equations is a solution of each of the equations simultaneously.

If an equation involves two variables, then its solution set consists of ordered pairs of numbers. For example, solutions of the two-variable equation $x^2 - 1 = y$ are ordered pairs (x, y) that make the equation true. In this case, $(x, y) = (1, 2)$ is not a solution (since $1^2 - 1 \neq 2$) and $(x, y) = (-2, 3)$ is a solution (since $(-2)^2 - 1 = 3$). Interestingly, if we draw a point in the Cartesian plane for each solution (x, y) to the equation $x^2 - 1 = y$, we get

Fractions and Quotients

Many equations we will be interested in will involve fractions. The rules for adding, multiplying, and dividing fractions are given in Theorem 0.14.

THEOREM 0.14   **Algebraic Rules for Fractions**

Let a, b, c, and d be any real numbers. Then (assuming that no denominator in any expression is zero)

$$\frac{a}{b} + \frac{c}{d} = \frac{ad + bc}{bd} \qquad \left(\frac{a}{b}\right)\left(\frac{c}{d}\right) = \frac{ac}{bd} \qquad \frac{(a/b)}{(c/d)} = \left(\frac{a}{b}\right)\left(\frac{d}{c}\right) = \frac{ad}{bc}$$

You can also add fractions by forming the lowest common denominator. As a simple example, note that it is easier to calculate $\frac{1}{6} + \frac{2}{3} = \frac{1}{6} + \frac{4}{6} = \frac{5}{6}$ than it is to calculate $\frac{1}{6} + \frac{2}{3} = \frac{(1)(3) + (6)(2)}{(6)(3)} = \frac{3 + 12}{18} = \frac{15}{18} = \frac{5}{6}$.

We will accept the rules of fractions in Theorem 0.14 as basic properties of the real numbers. As immediate consequences we obtain the following additional rules (see also Exercises 85 and 86):

THEOREM 0.15   **More Algebraic Rules for Fractions**

Let a, b, c, and d be any real numbers. Then (assuming that no denominator in any expression is zero)

$$\frac{a}{b} - \frac{c}{d} = \frac{ad - bc}{bd} \qquad c\left(\frac{a}{b}\right) = \frac{ac}{b} \qquad \frac{(a/b)}{c} = \frac{a}{bc} \qquad \frac{a}{(b/c)} = \frac{ac}{b}$$

If A is a real number, then $\frac{A}{A}$ is equal to 1 as long as A is not equal to zero. If $A = 0$, then the expression $\frac{A}{A}$ is undefined (since we cannot divide zero by zero). This tells us when we can cancel a **common factor** in the numerator and the denominator of a fraction. The key point is that if the factor you are canceling involves a variable, then the cancellation is not valid when that factor is equal to zero.

THEOREM 0.16   **Cancellation when Variables Are Involved**

Given any real expression A that depends on a variable x,

$$\frac{A}{A} = \begin{cases} 1, & \text{for values of } x \text{ that make } A \neq 0 \\ \text{undefined}, & \text{for values of } x \text{ that make } A = 0. \end{cases}$$

For example, you can cancel a common factor $x - c$ from the numerator and denominator of a fraction, and the cancelled expression will equal the original expression for all values of x except $x = c$. One illustration of this theorem is the cancellation

$$\frac{x^2 - 1}{x + 1} = \frac{(x - 1)(x + 1)}{x + 1} = \begin{cases} x - 1, & \text{if } x \neq -1 \\ \text{undefined}, & \text{if } x = -1. \end{cases}$$

Theorem 0.11 told us when a product AB can be equal to zero. The next theorem tells us when a quotient $\frac{A}{B}$ can be equal to zero. This theorem will help us solve equations that can be written as quotients.

THEOREM 0.17   **When Is a Quotient Equal to Zero?**
If A and B are any real expressions, then $\frac{A}{B} = 0$ if and only if $A = 0$ and $B \neq 0$.

For example, the equation $\frac{x^2 - 1}{x + 1} = 0$ is true if and only if $x^2 - 1 = 0$ (which means that $x = 1$ or $x = -1$) but $x + 1 \neq 0$ (which means that $x \neq -1$). Therefore there is only one solution of this equation, and it is $x = 1$.

⊗ **CAUTION** | Note in particular that $B = 0$ does *not* imply that $\frac{A}{B} = 0$. In fact, if $B = 0$, then $\frac{A}{B}$ is undefined. That is why we require not only that $A = 0$ but also that $B \neq 0$. Note also that Theorem 0.17 is valid only for equations that are set to zero.

Examples and Explorations

EXAMPLE 1   **Solving equations by factoring**

Solve the following equations by using Theorems 0.11–0.13:

(a) $2x^3 - 5x^2 - 3x = 0$ **(b)** $3x^2 = 7x - 1$ **(c)** $2x^5 - 32x = 0$

SOLUTION

(a) As we will see in Chapter 4, the number of real number solutions of a polynomial equation is at most the value of the **degree**, or highest power, of the polynomial. Therefore we can expect the equation $2x^3 - 5x^2 - 3x = 0$ to have at most three real number solutions. The right-hand side of this equation can be easily factored as:

$$2x^3 - 5x^2 - 3x = x(2x^2 - 5x - 3) = x(2x + 1)(x - 3).$$

By Theorem 0.11, the expression $2x^3 - 5x^2 - 3x$ is zero if and only if one of x, $2x + 1$, or $x - 3$ is zero. In other words, the solutions of the equation $2x^3 - 5x^2 - 3x = 0$ are $x = 0$, $x = -\frac{1}{2}$, and $x = 3$.

(b) We first need to write the equation $3x^2 = 7x - 1$ in the general form of a quadratic equation: $3x^2 - 7x + 1 = 0$. This equation cannot be easily factored with the reverse "FOIL" (first-outside-inside-last) method, so we'll apply the quadratic formula. In this example we have $a = 3$, $b = -7$, and $c = 1$, so the solutions of $3x^2 - 7x + 1 = 0$ are

$$x = \frac{-(-7) \pm \sqrt{(-7)^2 - 4(3)(1)}}{2(3)} = \frac{7 \pm \sqrt{49 - 12}}{6} = \frac{7 \pm \sqrt{37}}{6}.$$

Therefore, the solutions of $3x^2 - 7x + 1 = 0$ are $x = \frac{1}{6}(7 + \sqrt{37})$ and $x = \frac{1}{6}(7 - \sqrt{37})$. Clearly we could not have easily figured that out by doing the "FOIL" method backwards!

(c) This time the factoring will involve two applications of a formula from Theorem 0.13:

$$2x^5 - 32x = 0$$
$$2x(x^4 - 16) = 0$$
$$2x(x^2 - 4)(x^2 + 4) = 0 \qquad \leftarrow \text{formula for } a^2 - b^2 \text{ with } a = x^2 \text{ and } b = 4$$
$$2x(x - 2)(x + 2)(x^2 + 4) = 0 \qquad \leftarrow \text{formula for } a^2 - b^2 \text{ with } a = x \text{ and } b = 2$$

Thus $2x^5 - 32x = 0$ whenever $2x = 0$, $x - 2 = 0$, $x + 2 = 0$, or $x^2 + 4 = 0$. Note that $x^2 + 4 = 0$ has no real solutions, because there is no real number that satisfies $x^2 = -4$. Therefore the real-number solution set of the original equation $2x^5 - 32x = 0$ is $\{-2, 0, 2\}$. □

CHECKING THE ANSWER

To check the answers in Example 1, simply substitute each proposed solution into the original equation. Each solution should satisfy the equation. For example, to check that $x = -2$, $x = 0$, and $x = 2$ are solutions in part (c) of the example we note that

$$2(-2)^5 - 32(-2) = 0, \qquad \leftarrow \text{evaluate equation at } x = -2$$
$$2(0)^5 - 32(0) = 0, \qquad \leftarrow \text{evaluate equation at } x = 0$$
$$2(2)^5 - 32(2) = 0. \qquad \leftarrow \text{evaluate equation at } x = 2$$

Of course, this will not tell you whether you have *missed* any solutions, but it will tell you whether the solutions you found are correct.

EXAMPLE 2

A simple system of equations

Solve the system of equations $\begin{cases} x^2 - 1 = 0 \\ x^2 - x - 2 = 0. \end{cases}$

SOLUTION

The first equation can be written as $x^2 = 1$, so its solutions are $x = \pm\sqrt{1}$, that is, $x = 1$ and $x = -1$. The second equation can be factored as $(x - 2)(x + 1) = 0$, and thus its solutions are $x = 2$ and $x = -1$. The only number that is a solution of *both* equations is $x = -1$, so the solution set for the system of equations is $\{-1\}$. □

EXAMPLE 3

Solving a two-variable system of equations

Solve the systems of equations $\begin{cases} x^2 + y = 3 \\ 2x + y = 0. \end{cases}$

SOLUTION

In solving a system of equations, sometimes it is a good strategy to use one of the equations to solve for one variable in terms of the other variables and then use the information obtained to simplify the remaining equations. We'll start by using the second equation to solve for y in terms of x. Since $2x + y = 0$, it follows that $y = -2x$. Now we can substitute this into the first equation (note that the resulting equation has only one variable) and solve:

$$x^2 + y = 3 \qquad \leftarrow \text{first equation}$$
$$x^2 - 2x = 3 \qquad \leftarrow \text{substitute } y = -2x$$
$$x^2 - 2x - 3 = 0 \qquad \leftarrow \text{collect terms to left side}$$
$$(x - 3)(x + 1) = 0 \qquad \leftarrow \text{factor}$$

The last equation clearly has solutions $x = 3$ and $x = -1$, so *any* solution (x, y) of the original system has the property that $x = 3$ or $x = -1$. Keeping in mind the fact that $y = -2x$, we can use these values of x to solve for the corresponding values of y. If $x = 3$, then $y = -2(3) = -6$. If $x = -1$, then $y = -2(-1) = 2$. Thus the solutions of the system of equations are $(x, y) = (3, -6)$ and $(x, y) = (-1, 2)$. □

CHECKING THE ANSWER

To check our solutions, we need only substitute them into each equation in the system of equations to see whether they work. For example, $(x, y) = (3, -6)$ satisfies both of the equations in the system, because $3^2 + (-6) = 3$ and $2(3) + (-6) = 0$. In a similar fashion we can verify that $(x, y) = (-1, 2)$ is a solution to the system. This won't tell us if we have missed any solutions, but it will verify that the solutions we found are, in fact, correct.

We can also do a graphical check, by graphing the equations in the system and looking at where their solution sets intersect. If we solve the equations in our system for y we obtain $y = 3 - x^2$ and $y = -2x$ as shown in the following figure. Note that there are two intersection points, namely, at $(3, -6)$ and at $(-1, 2)$.

$$y = 3 - x^2 \text{ and } y = -2x$$

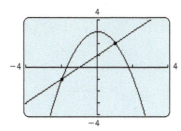

EXAMPLE 4   Solving a system by adding one equation to another

Solve the system of equations $\begin{cases} a + 2b - c = 0 \\ a + b = c - 1 \\ a - 2b + c = 2. \end{cases}$

SOLUTION

This system of equations has three equations and three variables, a, b, and c. The solutions of the system will be the triples (a, b, c) that are solutions of all three equations.

Since the first and last equations look somewhat similar, we'll try adding them together; in other words, we will add the left-hand side of the last equation to the left-hand side of the first equation and, similarly, add the right-hand side of the last equation to the right-hand side of the first equation. (Why is this valid?) If we do so, we get the new equation

$$(a + 2b - c) + (a - 2b + c) = 0 + 2 \qquad \leftarrow \text{add first and third equations}$$
$$2a = 2 \qquad \leftarrow \text{simplify}$$
$$a = 1.$$

This calculation immediately tells us that any solution (a, b, c) of our original system must have $a = 1$. Replacing a with 1 in the second equation gives us $1 + b = c - 1$, which implies that $b = c - 2$. Using $a = 1$ and $b = c - 2$, we can rewrite the first equation so that it involves only one variable:

$$a + 2b - c = 0 \qquad \leftarrow \text{first equation}$$
$$1 + 2(c - 2) - c = 0 \qquad \leftarrow \text{substitute } a = 1 \text{ and } b = c - 2$$
$$c - 3 = 0 \qquad \leftarrow \text{simplify}$$
$$c = 3.$$

Now, because $b = c - 2$, we know that $b = 3 - 2 = 1$. There is only one solution of the system of equations, namely, $(a, b, c) = (1, 1, 3)$. $\qquad \square$

EXAMPLE 5   Writing an expression as a simple fraction

Write the following expression in the form of a **_simple fraction_**, that is, in the form $\dfrac{A}{B}$ for some expressions A and B that do not themselves involve fractions and do not have any common factors:

$$\frac{\dfrac{3}{x + 2} - \dfrac{1}{x}}{x - 1}$$

SOLUTION

Using the rules for fractions, we obtain

$$\frac{\dfrac{3}{x+2}-\dfrac{1}{x}}{x-1}=\frac{\dfrac{3x-1(x+2)}{(x+2)(x)}}{x-1} \qquad \leftarrow \text{ combine numerator using common denominator}$$

$$=\frac{\dfrac{2x-2}{x(x+2)}}{x-1} \qquad \leftarrow \text{ simplify numerator of numerator}$$

$$=\frac{2x-2}{x(x+2)(x-1)} \qquad \leftarrow \dfrac{A}{B} \text{ divided by } C \text{ is } \dfrac{A}{BC}$$

$$=\frac{2(x-1)}{x(x+2)(x-1)} \qquad \leftarrow \text{ factor numerator}$$

$$=\frac{2}{x(x+2)}. \qquad \leftarrow \text{ cancel common factor of } x-1$$

The cancellation in the last equality is true only for $x \neq 1$. Can you list each of the algebraic rules used in the calculation?

There will often be more than one correct way to simplify an expression. In this example we could have instead used the technique of **_clearing denominators_**, in which we multiply the numerator and denominator of the original expression by something that cancels out all the denominators in the sub-expressions. Since the sub-expressions have denominators of $x+2$ and x in this example, we multiply numerator and denominator by $x(x+2)$:

$$\frac{\dfrac{3}{x+2}-\dfrac{1}{x}}{x-1}\cdot\frac{x(x+2)}{x(x+2)}=\frac{(3x)-(x+2)}{x(x+2)(x-1)} \qquad \leftarrow \text{ multiply by } \dfrac{x(x+2)}{x(x+2)}=1$$

$$=\frac{2x-2}{x(x+2)(x-1)} \qquad \leftarrow \text{ simplify numerator}$$

$$=\frac{2(x-1)}{x(x+2)(x-1)} \qquad \leftarrow \text{ factor}$$

$$=\frac{2}{x(x+2)}. \qquad \leftarrow \text{ cancel common factor of } x-1$$

Once again the cancellation in the last step is true only for $x \neq 1$, and we obtain the same answer as before.　　　　□

EXAMPLE 6   Solving equations involving quotients

Find the solution sets of the following equations:

(a) $\dfrac{x+1}{x-1}=4$ 　　　　　　　　**(b)** $\dfrac{x^3+x^2-6x}{x^2-3x+2}=0$

SOLUTION

(a) One way to solve this equation is to rewrite it so that one side of the equation is zero. We could do that by subtracting the 4 from both sides of the equation and simplifying, but it is easier to clear denominators first by multiplying both sides of the equation by $x-1$. As long as $x \neq -1$, this will produce an equivalent equation that can be solved easily:

$$x+1=4(x-1) \qquad \leftarrow \text{ multiply both sides by } x=1$$

$$x+1=4x-4 \qquad \leftarrow \text{ expand}$$

$$5=3x \qquad \leftarrow \text{ collect } x \text{ terms to right-hand side}$$

$$x=\frac{5}{3}. \qquad \leftarrow \text{ solve for } x$$

Thus the solution set of the equation $\frac{x+1}{x-1}=4$ is $\left\{\dfrac{5}{3}\right\}$.

(b) We will use Theorem 0.17 to solve this equation, but first we need to factor the numerator and the denominator:

$$\frac{x^3 + x^2 - 6x}{x^2 - 3x + 2} = \frac{x(x^2 + x - 6)}{x^2 - 3x + 2} = \frac{x(x-2)(x+3)}{(x-2)(x-1)}.$$

This expression is equal to zero only when the numerator is equal to zero (and the denominator is *not* zero). The numerator of the expression is zero when $x = 2$, but since the denominator is also zero when $x = 2$, the entire expression is undefined when $x = 2$. The solution of the original equation consists only of those values of x for which the numerator is zero while the denominator is nonzero, namely, $x = 0$ and $x = -3$. □

? TEST YOUR UNDERSTANDING

▶ What do we mean when we say that $x = 2$ and $x = -2$ are "solutions" to the equation $x^2 - 4 = 0$?

▶ How can factoring help us solve an equation? Why is it often easier to solve a factored equation than a non-factored equation?

▶ What is the quadratic formula for? What does it help you find?

▶ Explain why the expression $\frac{(x-2)(x+1)}{x-2}$ is not equivalent to the expression $x + 1$ for all values of x. For which value(s) of x do they differ, and why?

▶ Define each of the following: one-variable equation, identity, factoring, quadratic equation, discriminant.

EXERCISES 0.2

Thinking Back

Finding a common denominator: Compute each of the following quantities by finding a common denominator.

▶ $\dfrac{1}{8} + \dfrac{3}{4}$ ▶ $\dfrac{3}{5} - \dfrac{11}{2}$ ▶ $7 + \dfrac{3}{7}$

Common algebra mistakes: Each of the false statements that follow illustrates a common algebra mistake. For each false statement, find a number x for which the right-hand side is not equal to the left-hand side.

▶ *False:* $\dfrac{1}{x} + \dfrac{x}{2} = \dfrac{1+x}{x+2}$

▶ *False:* $\dfrac{2x+1}{2x-1} = \dfrac{x+1}{x-1}$

▶ *False:* $\dfrac{3x}{x^2-1} = \dfrac{3}{x-1}$

Factoring simple quadratics: Factor each of the quadratic expressions that follow. Consult your old algebra textbook or an online resource if you need help remembering how to do this.

▶ $x^2 + 2x - 15$ ▶ $x^2 - 9x + 18$

▶ $2x - 48 + x^2$ ▶ $3x^2 - 11x - 4$

▶ $6x^2 - x - 2$ ▶ $-3 - 5x + 12x^2$

Using a graphing utility: If a graphing utility is allowed in your course, then you should be familiar with using it. Practice by using a graphing utility to graph each of the following equations in the given graphing window.

▶ $y = 2 - x + x^4$, in window $-4 \le x \le 4$, $0 \le y \le 10$

▶ $y = x(x^2 - 1)$, in window $-3 \le x \le 3$, $-3 \le y \le 3$

▶ $y - 3x = 6 - x^2$, in window $-3 \le x \le 5$, $-3 \le y \le 10$

Concepts

0. *Problem Zero:* Read the section and make your own summary of the material.

1. *True/False:* Determine whether each of the statements that follow is true or false. If a statement is true, explain why. If a statement is false, provide a counterexample.

(a) *True or False:* If $AB = 0$, then $A = 0$ and $B = 0$.

(b) *True or False:* $\dfrac{A}{B} = 0$ if and only if $A = 0$.

(c) *True or False:* $\dfrac{A}{A} = 1$ for every real number A.

(d) *True or False:* Every quadratic equation has exactly two real solutions.

(e) *True or False:* An identity is an equation that is always true.

(f) *True or False:* For all x, $\dfrac{(x+1)(x+2)}{x+1} = x + 2$.

(g) *True or False:* If $b, d \neq 0$, then $3\left(\dfrac{a}{b} + \dfrac{c}{d}\right) = \dfrac{3ad + 3bc}{bd}$.

(h) *True or False:* If $a, b \neq 0$, then $\dfrac{12}{(a/b)} = \dfrac{12b}{a}$.

2. *Examples:* Construct examples of the thing(s) described in the following. Try to find examples that are different than any in the reading.

   (a) A two-variable equation.

   (b) An equation that is an identity.

   (c) A three-variable system of equations.

3. Suppose you solve an equation and want to check your answers. Describe two ways (one algebraic, one graphical) that you can check your work.

4. Suppose you solve a system of equations and want to check your answers. Describe two ways (one algebraic, one graphical) that you can check your work.

In Exercises 5 and 6, a graph describes the points (x, y) that are solutions of some equation involving x and y. Use the graph to estimate three points (x, y) that are solutions and three points that are not.

5.

6.
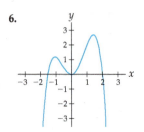

For each equation in Exercises 7 and 8, (a) determine whether $x = 1$ and $x = 2$ are solutions and explain why or why not, and then (b) use a graphing utility to estimate all of the solutions of the equation.

7. $x^3 - 4x^2 + x + 6 = 0$

8. $x^3 - 3x^2 - 6x + 8 = 0$

For each equation in Exercises 9 and 10, (a) determine whether $(x, y) = (1, 4)$ and $(x, y) = (2, 1)$ are solutions and explain why or why not, (b) use the equation to find three additional solutions, and then (c) use a graphing utility to graph the equation and verify your solutions.

9. $y = 2 - x + 3x^5$

10. $y - 1 = (x - 2)(x - 3)$

For each system of equations in Exercises 11 and 12, (a) determine whether $(1, -1)$ and $(-2, -1)$ are solutions and explain why or why not, and then (b) use a graphing utility to approximate all of the solutions of the system.

11. $\begin{cases} x - y = -1 \\ y + x^2 = 3 \end{cases}$

12. $\begin{cases} y + x^3 = 0 \\ y - x^2 + 2x = 0 \end{cases}$

13. The equation $9x^2 + 3x - 2 = 0$ can be rewritten as $3x(3x + 1) - 2 = 0$. Would the rewritten equation be considered factored? Why or why not?

14. The equation $9x^2 + 3x - 2 = 0$ can be rewritten as $3x(3x + 1) = 2$. Can Theorem 0.11 be applied to this rewritten equation? Why or why not?

15. If a quadratic equation has two distinct roots r_1 and r_2 given by the quadratic formula, what are the possible factored forms of the equation? What if there is only one root r_1 given by the quadratic formula?

16. Why does the discriminant of a quadratic equation determine the number of real solutions? Think about square roots and the cases when the discriminant is positive, negative, and zero.

17. What is the value of $\dfrac{x+2}{x^2-4}$ when $x = -2$? After factoring and cancelling, this expression becomes $\dfrac{1}{x-2}$. What is the value of the cancelled expression when $x = -2$? Is it the same as the value of the original expression when $x = -2$? Why or why not?

18. The numerator of $\dfrac{x+2}{x^2-4}$ is zero when $x = -2$. Does this mean that the entire expression is equal to zero when $x = -2$? Why or why not?

Skills

Factor each of the expressions in Exercises 19–32 as much as possible.

19. $6x^3 + 2x^2 - 20x$

20. $8x^2 - 50$

21. $4x^2 - 81$

22. $x^2 + 16$

23. $81x^4 - 1$

24. $16 - x^4$

25. $x^3 + 3x^2 + 3x + 1$

26. $x^4 + 81x$

27. $27 - 64x^3$

28. $27x^3 - 8$

29. $x^5 - 1$

30. $x^5 - 243$

31. $t^7 - x^7$

32. $x^8 - a^8$

Write each of the expressions in Exercises 33–38 in the form of a simple fraction. For which values of x, if any, does your simplified expression differ from the original expression?

33. $\dfrac{x^2 - 1}{\dfrac{1}{x+1} - \dfrac{1}{x}}$

34. $\dfrac{\dfrac{x+1}{x} - \dfrac{1}{x-1}}{x+1}$

35. $\dfrac{\left(\dfrac{x^2-4}{x+1}\right)}{\left(\dfrac{x+2}{x-1}\right)}$

36. $\dfrac{x^2}{x-2}\left(\dfrac{x+1}{x-2} + \dfrac{3}{x+1}\right)$

37. $3x\left(\dfrac{2-x}{x^2-1}\right) - \dfrac{x}{x-1}$

38. $\dfrac{\dfrac{1}{x^2-4} - \dfrac{3}{x}}{\dfrac{x-2}{x^2-x}}$

Find the solution set of each equation in Exercises 39–68. When possible, solve by factoring and using Theorem 0.11 and Theorem 0.17. Some equations may not factor easily and will require the quadratic formula.

39. $6 - 7x + x^2 = 0$

40. $x^2 + 2x - 8 = 0$

41. $4x(x^2 - 1) = 6x^2$

42. $3x^2 + 5x = -2$

43. $2x^2 - x - 3 = 0$

44. $2x^2 + 4x + 2 = 0$

45. $4x^2 - 12x + 9 = 0$

46. $x^2 + 2x + 3 = 0$

47. $10x = 32x^2 - 6x^3$

48. $6x^2 + 3 = -11x$

49. $5 - 2x - x^2 = 0$

50. $x - 2 - 3x^2 = 0$

51. $2x(x - 1) = 3$

52. $x^2 - 2x - 4 = 0$

53. $(x - 2)^2 = 2$

54. $4x^2 - 12x = -9$

55. $x^3 - 9x = 0$

56. $8 - x^3 = 0$

57. $16x^4 - 81 = 0$

58. $x^3 - 3x^2 + 3x = 1$

59. $4x^3 - 9x = 0$

60. $\dfrac{x^2 - 1}{x^2 - 4} = 0$

61. $\dfrac{x - 1}{x + 1} = 3$

62. $\dfrac{2x - 1}{x - 3} = \dfrac{x - 2}{1 + x}$

63. $\dfrac{x^3 + x^2 - 2x}{x^2 - 4x + 3} = 0$

64. $\dfrac{4x^3 - 1}{2x^2 + x - 1} = 0$

65. $\dfrac{2}{x^2 - x} = \dfrac{1}{1 - x}$

66. $\dfrac{(x - 2)^2}{3x - 2} = x - 5$

67. $\dfrac{1}{x} = \dfrac{1}{x + 1}$

68. $\dfrac{1}{3x^2 - 2x - 1} = 2x - 1$

Find the solution sets of each of the systems of equations in Exercises 69–78.

69. $\begin{cases} 2x + y = 4 \\ 3x + 2y = 0 \end{cases}$

70. $\begin{cases} x^2 + y = 0 \\ y - x = 2 \end{cases}$

71. $\begin{cases} 3a + b = -1 \\ a + 5b = 3 \end{cases}$

72. $\begin{cases} r^2 - s = 0 \\ r^2 + s = 1 \end{cases}$

73. $\begin{cases} 3t - u^2 = -3 \\ u - 1 = t \end{cases}$

74. $\begin{cases} 3x + y - z = 2 \\ x - y + 2z = 0 \end{cases}$

75. $\begin{cases} 2a + 3b - c = 4 \\ a - b - c = 3 \\ 2a + b - c = 2 \end{cases}$

76. $\begin{cases} r + s + t = 1 \\ r - s - t = 0 \\ r - s + t = 2 \end{cases}$

77. $\begin{cases} x + 2y = 1 \\ 2y = 1 - 3x \\ 3x - y = 2 \end{cases}$

78. $\begin{cases} x - 2y + z = 0 \\ x + y - 3z = 1 \\ 3x - 2y + z = 2 \end{cases}$

Applications

79. Len works for a company that produces cans that are 6 inches tall and vary in radius. The cost of producing a can with height 6 inches and radius r is $C = 10\pi r^2 + 24\pi r$ cents. If Len wants to spend 300 cents on each can, how large should he make the radius? (*Hint: Set up and solve an equation; you will need the quadratic formula.*)

80. Suppose x and y are positive real numbers whose sum is 25 and whose product is 136. Construct and solve a system of equations to find x and y.

81. Three real numbers have the property that when you add them in pairs, you get 19, 30, and 35. Construct and solve a system of equations to find these three real numbers.

82. A sheet of paper with height h and width w is to have 2-inch margins on all four sides. Suppose the perimeter of the paper is 38 inches and the area of the printable part of the paper (inside the margins) is 45 square inches. Construct and solve a system of equations to find h and w.

Proofs

83. The second part of Theorem 0.13 says that for any real numbers a and b,

$$a^3 - b^3 = (a - b)(a^2 + ab + b^2).$$

Prove this identity by multiplying out the right-hand side and then showing that it is equal to the left-hand side.

84. The third part of Theorem 0.13 says that for any real numbers a and b, and any positive integer n,

$$a^n - b^n = (a - b)(a^{n-1} + a^{n-2}b + \cdots + ab^{n-2} + b^{n-1}).$$

Prove this identity by multiplying out the right-hand side and then showing that it is equal to the left-hand side.

85. The second part of Theorem 0.15 says that for any real numbers a, b, and c with $b \neq 0$,

$$c\left(\frac{a}{b}\right) = \frac{ac}{b}.$$

Show that this statement is true by using the algebraic rules for fractions from Theorem 0.14. (*Hint: How can you write the real number c as a fraction?*)

86. The third part of Theorem 0.15 says that for any real numbers a, b, and c with b and c nonzero,

$$\frac{(a/b)}{c} = \frac{a}{bc}.$$

Show that this statement is true by using the algebraic rules for fractions from Theorem 0.14.

Thinking Forward

▶ *Polynomials, roots, and factors:* Quadratic equations are examples of **polynomials**, which are expressions built by adding up terms of the form ax^k where k is a nonnegative integer; for example, $-x^7 + 2x^3 - 5x + 4$ is a polynomial. Write down a polynomial that has roots $x = -2$, $x = -1$, $x = 3$, $x = 5$, and $x = 11$. (*Hint: Write the polynomial in factored form.*)

▶ *Difficulties with solving inequalities:* Consider the inequality $\dfrac{5}{x-2} < 3$. If this were an equation with an equals sign instead of a less-than sign, our first step in solving the equation would be to multiply both sides by $x - 2$. Unfortunately, in the next section we will see that this method will not work with an inequality. To illustrate, find a value of x for which $\dfrac{5}{x-2} < 3$ is true but $5 < 3(x - 2)$ is false.

Finding where an expression has roots or does not exist: In later chapters we will be interested in finding the values (called **roots**) that make an expression equal to zero and the values for which an expression does not exist. Find these values for each of the following expressions:

▶ $\dfrac{4x^2 - 1}{x^3 - x^2 - 2x}$

▶ $\dfrac{3x^2 - x - 2}{x^2 - x}$

▶ $\dfrac{1}{x^4 - 5x^2 + 4}$

▶ $\dfrac{x^3 + x^2}{x^2 - 1}$

0.3   INEQUALITIES

▶ Algebraic rule for solving inequalities

▶ Solving inequalities by examining signs of factors

▶ Turning absolute value inequalities into pairs of simpler inequalities

Basic Properties of Inequalities

An **inequality** consists of two mathematical expressions related by a greater-than ($>$), less-than ($<$), greater-than-or-equal-to ($\geq$), or less-than-or-equal-to ($\leq$) sign. The **solution set** of an inequality is the set of all values that make the inequality true. Whereas the solution set of an equation often consists of a finite number of points, the solution set of an inequality is often an *interval* of real numbers.

The next theorem describes what happens to inequalities when they are inverted or multiplied by a nonzero constant. This information will be useful for solving inequalities.

THEOREM 0.18   **Algebraic Rules for Inequalities**

Suppose a, b, and c are nonzero real numbers.

(a) If $a < b$ and $c > 0$, then $ac < bc$.   **(b)** If $a < b$ and $c < 0$, then $ac > bc$.

(c) If $0 < a < b$, then $\dfrac{1}{a} > \dfrac{1}{b}$.   **(d)** If $a < b$, then $a + c < b + c$.

In particular, notice that part (b) implies that multiplying an inequality by a negative number "flips" the inequality; for example, if $a < b$, then $-a > -b$. Part (d) illustrates that we may add any number to, or subtract any number from, both sides of an inequality without changing the direction of the inequality. Theorem 0.18 is written in terms of **strict inequalities** $>$ and $<$, but similar results apply for the inequalities $\geq$ and $\leq$.

CAUTION | Since multiplying both sides of an inequality by a negative number "flips" the inequality, you cannot multiply both sides of an inequality by a quantity whose sign you do not know. For example, you can't multiply both sides of the inequality $\frac{3}{x} > 2$ by x, because x may be positive or negative. Therefore, you don't know whether or not multiplying both sides of an inequality by x will change the direction of the inequality.

A key property of the real numbers is that a product of two real numbers is positive if the numbers have the same sign and negative if the numbers have opposite signs. This property suggests the following theorem, which can help us solve certain factored inequalities:

THEOREM 0.19

The Sign of a Product Depends on the Signs of Its Factors

If A and B are any real expressions, then

(a) $AB > 0$ if and only if $(A > 0$ and $B > 0)$ or $(A < 0$ and $B < 0)$.

(b) $AB < 0$ if and only if $(A > 0$ and $B < 0)$ or $(A < 0$ and $B > 0)$.

An easy corollary of this theorem is that $AB \geq 0$ if and only if either A and B have the same sign or one of A and B is zero. Theorem 0.19 also holds for quotients: $\frac{A}{B}$ is greater than zero if A and B have the same sign and is less than zero if A and B have opposite signs. We can also generalize this theorem to products with more than two factors:

THEOREM 0.20

The Sign of a Product Depends on the Number of Its Negative Factors

A product or quotient of nonzero real numbers is positive if and only if it has an even number of negative factors and is negative if and only if it has an odd number of negative factors.

For example, the product $(-2)(-1)(2)(-6)$ is negative because it has an odd number of negative factors. In contrast, $\frac{(-3)(-2)(-8)}{(7)(-4)}$ is positive because it has an even number of negative factors.

Solving Inequalities That Involve Products or Quotients

Theorem 0.19 reduces the problem of solving a factored inequality to the problem of solving a collection of related simpler inequalities and using intersections and unions as appropriate. For example, the inequality $x^2 - x < 0$ factors as $x(x-1) < 0$. The product of x and $x-1$ is negative exactly when x and $x - 1$ have opposite signs, which means that the solution of the original inequality is the solution of:

$$(x > 0 \text{ and } x - 1 < 0) \text{ or } (x < 0 \text{ and } x - 1 > 0).$$

In interval notation, and using intersections for "and" and unions for "or," this expression is equal to

$$((0, \infty) \cap (-\infty, 1)) \cup ((-\infty, 0) \cap (1, \infty)).$$

This expression in turn is equal to $(0, 1) \cup \emptyset = (0, 1)$, so the solution of the original inequality $x(x - 1) < 0$ is the interval $(0, 1)$.

Although the method of breaking up a factored inequality into a union and intersection of a lot of simple inequalities is not difficult, it can become tedious when there are more than two factors involved. The next theorem will allow us a quicker way to solve inequalities that feature **polynomials**. A polynomial is an expression that can be written as a sum of terms of the form ax^k, where a is a real number and k is a nonnegative integer. For example, $x^4 + x^2 - 2$ is a polynomial, that, in factored form, can be written as $(x^2 + 2)(x + 1)(x - 1)$. We will say that an expression **changes sign** at $x = c$ if, as we consider x varying from just less than c to just greater than c, the expression changes sign from positive to negative or from negative to positive. For example, the expression $x - 5$ changes sign (from negative to positive) at $x = 5$.

THEOREM 0.21

When Can a Polynomial or Quotient of Polynomials Change Sign?

A polynomial can change sign only at values where it is equal to zero. A quotient of polynomials can change sign only at values where its numerator or denominator is equal to zero.

This theorem is a special case of the Intermediate Value Theorem, which we will prove in Chapter 1.

As an illustration of the power of Theorem 0.21, consider the problem of solving the inequality $(x^2 + 2)(x + 1)(x - 1) < 0$. We could examine all the possible cases when an odd number of the three factors are negative, but this would be quite tedious. Instead, let's apply Theorem 0.21. Clearly the expression $(x^2 + 2)(x + 1)(x - 1)$ is zero only when $x = -1$ or $x = 1$. (Note that $x^2 + 2$ is never zero.) Therefore these are the only two x-values at which the expression can change sign. That means that the expression has the same sign on the entire interval $(-\infty, -1)$, and the same sign on all of $(-1, 1)$, and the same sign on all of $(1, \infty)$. To determine what the sign is in each case we need only check the sign of the expression at one point in each interval; for example, at $x = -2$, $x = 0$, and $x = 2$:

$$((-2)^2 + 2)(-2 + 1)(-2 - 1) = (\text{pos})(\text{neg})(\text{neg}) = \text{positive},$$
$$(0^2 + 2)(0 + 1)(0 - 1) = (\text{pos})(\text{pos})(\text{neg}) = \text{negative},$$
$$(2^2 + 2)(2 + 1)(2 - 1) = (\text{pos})(\text{pos})(\text{pos}) = \text{positive}.$$

We can mark this information on a number line with a **sign chart** by making tick marks at each x-value that made the expression zero and writing $+$ or $-$ above each interval we tested, as shown in the sign chart that follows. Note that we include tick marks *only* at the potential sign-changing points.

Finally, we can conclude that $(x^2 + 2)(x + 1)(x - 1)$ is less than zero exactly when $x \in (-1, 1)$, and we have solved the inequality.

⊗ CAUTION | The sign charts in this text will follow the convention that tick marks are only marked at the locations where the quantity being measured is zero or fails to exist. Do not make additional tick marks on your number lines if you follow this convention. Or, if you prefer to use additional tick marks as reference points, then you must label all of the relevant tick marks with "0" or "DNE" as needed.

Inequalities and Absolute Values

In Section 0.1 we saw that the distance between two real numbers x and c is by definition $|x - c|$. Therefore, the solutions of inequalities of the form $|x - c| < \delta$ and $|x - c| > \delta$ describe how close or far away x is from c. (The symbol δ is the Greek letter **delta** and is one of the traditional letters used to represent a distance.) On the real number line, we can graphically represent the solution sets to $|x - c| < \delta$ and $|x - c| > \delta$ as shown in the two figures that follow. Remember that the blue shaded areas are the locations on the real number line where a number x satisfies the inequality.

The preceding discussion suggests the following theorem, which will allow us to immediately solve certain simple absolute value inequalities:

THEOREM 0.22   **Expressing Distances with Inequalities and Intervals**

Suppose x and c are any real numbers and δ is any positive real number. Then

(a) The following statements are equivalent:

- ▶ the distance between x and c is less than δ;
- ▶ $|x - c| < \delta$;
- ▶ x satisfies both $x > c - \delta$ and $x < c + \delta$;
- ▶ $x \in (c - \delta, c + \delta)$.

(b) The following statements are equivalent:

- ▶ the distance between x and c is greater than δ;
- ▶ $|x - c| > \delta$;
- ▶ x satisfies either $x < c - \delta$ or $x > c + \delta$;
- ▶ $x \in (-\infty, c - \delta) \cup (c + \delta, \infty)$.

Theorem 0.22 also holds for "greater than or equal to" and "less than or equal to," with appropriate changes. With this theorem we can immediately solve simple absolute value inequalities involving $|x - c|$. For example, if x satisfies the inequality $|x - 2| \geq 5$, then the distance between x and 2 is greater than or equal to 5. This means that x is greater than or equal to $2 + 5 = 7$ and less than or equal to $2 - 5 = -3$. Therefore $x \in (-\infty, -3] \cup [7, \infty)$.

For more complicated inequalities, such as $|2x - 5| > 3$ and $|x^2 - 4| < 12$, the following theorem is helpful:

THEOREM 0.23   **Replacing an Absolute Value Inequality with Two Simpler Inequalities**

Suppose δ is any positive real number and A is any real expression. Then

(a) $|A| < \delta$ if and only if $A < \delta$ and $A > -\delta$ (that is, $-\delta < A < \delta$);

(b) $|A| > \delta$ if and only if $A > \delta$ or $A < -\delta$.

The words "and" and "or" in the theorem are essential. The "and" represents an intersection, whereas the "or" represents a union. As usual, this theorem also holds when $<$ and $>$ are replaced by $\leq$ and $\geq$. We will prove Theorem 0.23 in Section 0.7.

Theorem 0.23 allows us to trade an absolute value inequality for a pair of simple inequalities. For example, the absolute value inequality $|2x + 5| < 3$ becomes a pair of inequalities that do not involve absolute values, namely, $2x+5 < 3$ and $2x+5 > -3$. These inequalities are easy to solve, and the intersection of their solutions, $(-\infty, -1) \cap (-4, \infty) = (-4, -1)$, is the solution of the original inequality.

Examples and Explorations

 EXAMPLE 1   **Solving a simple inequality**

Find the solution set of the inequality $3 - 2x > 7$.

SOLUTION

This is a simple calculation that makes use of parts (b) and (d) of Theorem 0.18. Note the inequality "flip" in the last step:

$$3 - 2x > 7 \implies -2x > 7 - 3 \implies -2x > 4 \implies x < -2.$$

(The symbol $\implies$ used in the calculation above denotes that the inequality to the left of the symbol logically implies the inequality to the right of the symbol.) Therefore the solution set of the original inequality is $(-\infty, -2)$. □

✔ CHECKING THE ANSWER

You can use a graphing utility to check the solution we just found by graphing $y = 3 - 2x$ (the diagonal line) and $y = 7$ (the dashed line), as shown next. The values of x for which the line described by $3 - 2x$ has a y-coordinate greater than 7 are those to the left of $x = -2$.

EXAMPLE 2   **Solving an inequality by examining signs of factors**

Solve the inequality $\dfrac{3}{x+1} > 2$ by applying Theorem 0.19.

SOLUTION

Although it is tempting, we can't multiply both sides of the inequality by $x + 1$, because $x+1$ may be positive or may be negative (depending on the value of x). Instead, we will use some algebra to turn this inequality into one that involves a product or quotient of factors on one side and zero on the other:

$$\frac{3}{x+1} > 2 \qquad \leftarrow \text{original inequality}$$

$$\frac{3}{x+1} - 2 > 0 \qquad \leftarrow \text{collect terms so 0 is on right-hand side}$$

$$\frac{3}{x+1} - \frac{2(x+1)}{x+1} > 0 \qquad \leftarrow \text{find common denominator}$$

$$\frac{3 - 2(x+1)}{x+1} > 0 \qquad \leftarrow \text{combine fractions}$$

$$\frac{3 - 2x - 2}{x+1} > 0 \qquad \leftarrow \text{simplify numerator}$$

$$\frac{1 - 2x}{x+1} > 0 \qquad \leftarrow \text{simplify numerator some more}$$

The expression $\frac{1-2x}{x+1}$ is positive if $1 - 2x$ and $x + 1$ both have the same sign; in other words, the solution set of this inequality consists of all values x for which

$$(1 - 2x > 0 \text{ and } x + 1 > 0) \quad \text{or} \quad (1 - 2x < 0 \text{ and } x + 1 < 0).$$

Solving these four simple inequalities, we get

$$(x < 1/2 \text{ and } x > -1) \quad \text{or} \quad (x > 1/2 \text{ and } x < -1).$$

Therefore the solution of the original inequality is

$$((-\infty, 1/2) \cap (-1, \infty)) \cup ((1/2, \infty) \cap (-\infty, -1)),$$

which is equal to $(-1, 1/2) \cup \emptyset = (-1, 1/2)$. □

EXAMPLE 3

Solving an inequality by testing points between sign changes

Solve the inequality $\dfrac{3}{x+1} > 2$ again, this time by using the "point-testing" method allowed by Theorem 0.21.

SOLUTION

We begin the same way as we did in Example 2, by writing the inequality in factored form with zero on one side:

$$\frac{1 - 2x}{x+1} > 0.$$

By Theorem 0.21, the expression $\frac{1-2x}{x+1}$ can change sign only at $x = 1/2$ and $x = -1$. We now find the sign of the expression at representative points $x = -2$, $x = 0$, and $x = 1$ between each of these potential sign-changing points:

$$\frac{1 - 2(-2)}{-2 + 1} = \frac{\text{pos}}{\text{neg}} = \text{negative},$$

$$\frac{1 - 2(0)}{0 + 1} = \frac{\text{pos}}{\text{pos}} = \text{positive},$$

$$\frac{1 - 2(1)}{1 + 1} = \frac{\text{neg}}{\text{pos}} = \text{negative}.$$

We can record this information on a number line as follows:

This means that $\dfrac{1-2x}{x+1}$ is greater than zero only on the interval $(-1, 1/2)$.   □

EXAMPLE 4   Solving an inequality that involves three factors

Find the solution set of the inequality $\dfrac{1}{x-2} \leq \dfrac{1}{6-x}$.

SOLUTION

Be careful; we cannot cross multiply here, because we do not know whether $x-2$ and $6-x$ are positive or negative. Instead we rewrite:

$$\frac{1}{x-2} \leq \frac{1}{6-x} \qquad \leftarrow \text{original inequality}$$

$$\frac{1}{x-2} - \frac{1}{6-x} \leq 0 \qquad \leftarrow \text{collect terms so 0 is on right-hand side}$$

$$\frac{(6-x)-(x-2)}{(x-2)(6-x)} \leq 0 \qquad \leftarrow \text{combine fractions using common denominator}$$

$$\frac{8-2x}{(x-2)(6-x)} \leq 0 \qquad \leftarrow \text{simplify numerator}$$

$$\frac{2(4-x)}{(x-2)(6-x)} \leq 0. \qquad \leftarrow \text{simplify numerator some more}$$

There are three factors in the numerator and denominator of the last expression, namely, $4-x$, $x-2$, and $6-x$. We could solve this inequality by examining all the cases where an odd number of these factors is negative, but it will be faster to use the "point-testing" method of the previous example.

The only places where $\dfrac{2(4-x)}{(x-2)(6-x)}$ can change sign are at $x=4$, $x=2$, and $x=6$. Testing the sign of this expression at representative x-values between these potential sign-changing points (say, at $x=0$, $x=3$, and $x=5$, for example) results in the following sign chart:

From this information we can see exactly what happens in the intervals between the points $x=2$, $x=4$, and $x=6$. But we have to be careful about the value of the expression *at* those points. We are looking for the places where $\dfrac{2(4-x)}{(x-2)(6-x)}$ is greater than *or equal to* zero, so we must include the point $x=4$ in our solution. We also must be careful to exclude the points $x=2$ and $x=6$ from our solution, for at these values of x the expression is not defined at all. Putting this all together with our sign chart, we see that the solution is $(-\infty, 2) \cup [4, 6)$.   □

CHECKING THE ANSWER

To test whether the solution we found is reasonable, try evaluating the inequality at a few values in the solution set and at a few values outside the solution set. For example, the inequality should be true for $x = 0$ and $x = 5$ (these are in the solution set) but false for $x = 3$ and for $x = 8$. You can also check the answer by using a graphing utility to graph the equation $y = \dfrac{1}{x-2} - \dfrac{1}{6-x}$. The graph should have y-values that are less than or equal to zero for $x \in (-\infty, 2) \cup [4, 6)$. For all other values of x, the y-values should be positive. Verify that this is the case for the following graph (the graph of the equation has three components, and the dashed lines at $x = 2$ and $x = 6$ are not part of the graph):

EXAMPLE 5

Using cases to solve inequalities that involve absolute values

Solve the following inequalities:

(a) $|2x - 5| > 3$ (b) $|x^2 - 4| \leq 12$

SOLUTION

(a) By Theorem 0.23, $|2x - 5| > 3$ if and only if

$$
\begin{array}{ccc}
2x - 5 > 3 & & 2x - 5 < -3 \\
2x > 8 & \text{or} & 2x < 2 \\
x > 4 & & x < 1
\end{array}
$$

The solution set of the original inequality is the union of the two solutions just found: $(-\infty, 1) \cup (4, \infty)$.

(b) By Theorem 0.23, $|x^2 - 4| \leq 12$ if and only if

$$
\begin{array}{ccc}
x^2 - 4 \leq 12 & & x^2 - 4 \geq -12 \\
x^2 \leq 16 & \text{and} & x^2 \geq -8
\end{array}
$$

The solution set of the original inequality is the intersection of the solutions of these two inequalities. The first inequality, $x^2 \leq 16$, is true for values of x that are inclusively between -4 and 4. The second inequality, $x^2 \geq -8$, is *always* true. Therefore the solution of the original inequality is $[-4, 4] \cap (-\infty, \infty) = [-4, 4]$. □

TEST YOUR UNDERSTANDING

▶ Why you can't solve the inequality $\dfrac{2}{x+1} < 7$ by multiplying both sides of the inequality by $x + 1$?

▶ What are the possible cases to consider when solving an inequality of the form $AB > 0$ or $AB < 0$?

▶ What are the possible cases to consider when solving an inequality of the form $\dfrac{A}{B} \geq 0$ or $\dfrac{A}{B} \leq 0$?

▶ What are the possible cases to consider when solving an inequality of the form $|A| \leq \delta$ or $|A| \geq \delta$?

▶ Define each of the following: strict inequality, sign of a real number, factor, polynomial.

EXERCISES 0.3

Thinking Back

Algebra review of fractions: Write each of the following expressions as a simple fraction.

▶ $x - \dfrac{1}{x+5}$

▶ $\dfrac{1}{x-1} - \dfrac{1}{x+1}$

▶ $3 - \dfrac{1}{x^2-4}$

▶ $\dfrac{1}{x-2} - \dfrac{x-2}{x-1}$

Algebra review of factoring: Factor each expression to identify the values of x for which the expression is zero or does not exist.

▶ $\dfrac{4x^2-9}{x-1}$

▶ $\dfrac{1}{x^4-16}$

▶ $\dfrac{x^2+x-2}{x^2+2x-3}$

▶ $\dfrac{6x^2-11x+3}{x^3-7x^2+10x}$

Concepts

0. *Problem Zero:* Read the section and make your own summary of the material.

1. *True/False:* Determine whether each of the statements that follow is true or false. If a statement is true, explain why. If a statement is false, provide a counterexample.

   (a) *True or False:* If $a \le b$, then $-3a \le -3b$.
   (b) *True or False:* If $a < b$, then $-4a < b$.
   (c) *True or False:* If $x < -2$, then $\dfrac{1}{x} < -\dfrac{1}{2}$.
   (d) *True or False:* If a, b, and c are negative real numbers, then $-2abc$ is negative.
   (e) *True or False:* $|x - c| > \delta$ if and only if $x < c - \delta$ and $x > c + \delta$.
   (f) *True or False:* The set $\{x \mid x > 4 \text{ and } x < 6\}$ is equal to the set $(4, \infty) \cup (-\infty, 6)$.
   (g) *True or False:* For any real numbers a and b, $|a + b| \le |a| + |b|$.
   (h) *True or False:* For any real numbers a and b, $|a - b| \le |a| - |b|$.

2. *Examples:* Construct examples of the thing(s) described in the following. Try to find examples that are different than any in the reading.

   (a) A polynomial that changes sign at $x = -2$ and $x = 5$.
   (b) A real number x for which $|x - 7| > 2$ is true but $x - 7 > 2$ is not, and a real number x for which $x - 7 > 2$ is true but $|x - 7| > 2$ is not.
   (c) Real numbers a and b for which $|a + b|$ is strictly less than $|a| + |b|$.

3. Answer the following two questions without solving any inequalities or using a graphing utility: Is $x = 2$ a solution of the inequality $x^4 \le 11x - 1$? What about $x = 3$? Explain your answers.

4. Consider the inequality $(x - 2)^2(x - 1) > 0$. The left-hand side of this inequality has three factors. If you were to solve the inequality by examining the signs of the factors, would you need to look at the signs of *all* of the factors? Why or why not? *(Hint: Think about the sign of $(x - 2)^2$.)*

5. Explain the difference between the sets $\{x \mid x < 2$ or $x > 0\}$ and $\{x \mid x < 2$ and $x > 0\}$. Use a number line and the words "union" and "intersection" in your explanation.

6. Can you solve the inequality $x^2 > 4$ by taking the square root of both sides? Why or why not? What is the solution of this inequality?

In Exercises 7–12, A, B, C and D are real expressions and $\delta > 0$. For each inequality, write a list of simpler inequalities with "and" and "or" that together are equivalent to the original inequality.

7. $AB \ge 0$
8. $AB \le 0$
9. $ABC < 0$
10. $ABCD > 0$
11. $|A| \ge \delta$
12. $|A| \le \delta$

13. Suppose A, B, C, and D are real numbers and that $A \le B < C = D$. What can we say about how A is related to D and why? What if we knew instead that $A < B = C \ge D$?

Exercises 14–18 each involve something that is either (a) an absolute value inequality, (b) a combination of simple inequalities, (c) a set in interval notation, (d) a statement involving distance, or (e) a picture on the real number line. Whichever way the statement is given, express the statement in the remaining four ways listed.

14. $|x - 2| < 0.1$
15. $x \ge -1$ and $x \le 4$
16. $(-\infty, 4] \cup [3, \infty)$
17. The distance between x and -3 is greater than 2.
18.

Skills

Find the solution sets of the inequalities in Exercises 19–40 by testing signs between potential sign-changing points. Illustrate your results on a labeled number line. When you are done, check your answers with a graphing utility.

19. $2x + 1 \ge 4$
20. $1 - 5x > 11$
21. $0 \le \dfrac{1}{x} \le 3$
22. $(x + 1)(x - 2) > 0$
23. $2x^2 - 7x + 3 \le 0$
24. $x^3 - x \ge 0$
25. $x^3 \le 4x$
26. $6x^2 > x^3 + 9x$
27. $x^4 > 16$
28. $x^5 > 32$

29. $\dfrac{x-1}{x+2} \geq 0$

30. $\dfrac{x-3}{x+5} \geq 0$

31. $\dfrac{5}{x-2} \leq 1$

32. $\dfrac{2}{x+1} \geq 3$

33. $\dfrac{x^2-1}{x-3} < 0$

34. $\dfrac{x(x-1)}{x^2-4} \geq 0$

35. $\dfrac{3x^2-2x-1}{x^2+x+6} \leq 0$

36. $\dfrac{x^2+7x+10}{4x^2+6x+2}$

37. $x+2 \leq \dfrac{1}{x+2}$

38. $\dfrac{1}{1+x} < \dfrac{1}{x+x^2}$

39. $\dfrac{1}{x-1} < \dfrac{2}{x+2}$

40. $\dfrac{1}{x-1} \geq \dfrac{x-1}{x+5}$

Find the solution sets of the absolute value inequalities in Exercises 41–52.

41. $|x+2| > 4$

42. $|x-3| < 8$

43. $|x+8| \leq 1$

44. $|3x+1| \geq 5$

45. $|4-2x| > 6$

46. $|3x+1| \leq 4$

47. $|x^8 - 3x + 1| \geq -2$

48. $|x(x-1)| < 8$

49. $|x^2 - 1| \geq 3$

50. $|x^2 - 4| \geq 5$

51. $|x^3 + x^2 - 3x| \leq 3x$

52. $|x^2 + 4x - 1| > 4$

Solve each of the inequalities in Exercises 53–58 by factoring it and then reducing it to a system of simpler inequalities combined with unions and intersections.

53. $x^2 \geq 16$

54. $x^2 - 7 \leq 0$

55. $3x^2 - 5x - 2 < 0$

56. $6x^2 - 5x > -1$

57. $x^3 - x^2 - 6x \leq 0$

58. $x^3 - 4x^2 - 5x \geq 0$

Use a graphing utility to approximate the solution sets of the inequalities in Exercises 59–62.

59. $x^5 \leq 2x^3 + 5$

60. $x^3 - 4x^2 > 1$

61. $3x^4 - 1 \geq x^5$

62. $\dfrac{x-1}{x^2+5} < \dfrac{x}{x+1}$

Applications

63. The formula for converting degrees Fahrenheit (F) to degrees Celsius (C) is $C = \dfrac{5}{9}(F-32)$.

    (a) For water to be between its freezing and boiling points, its temperature must be between $32°$ and $212°$ Fahrenheit. What is the corresponding temperature range in degrees Celsius?

    (b) Ella says that tomorrow's temperature will be between 20 degrees Celsius and 25 degrees Celsius. What is the corresponding range of temperatures in degrees Fahrenheit?

64. The instructions that came with Byron's stereo claim that the optimal volume range v for his speakers satisfies the inequality $|v - 85| < 35$, where v is measured in decibels. What is the lowest optimal volume for Byron's speakers? The highest?

65. For his summer job, Frank fills large buckets with pre-sliced pickles. The label on each bucket indicates that it contains 20 pounds of pickles. His boss says it is okay to be over or under by as much as half a pound. Use an absolute value inequality to express the acceptable weight W of a full bucket of pickles.

66. The cylindrical cans that Len's company produces are 6 inches tall and vary in radius. The cost of producing a can with height 6 inches and radius r is $C = 10\pi r^2 + 24\pi r$ cents.

    (a) Len's boss wants him to construct the cans so that the cost is between 250 and 450 cents. Write this condition as an absolute value inequality.

    (b) Given the cost restrictions in part (a), what is the acceptable range of values for r?

Proofs

67. Prove that for any $c > 0$, the solution set of the inequality $x^2 < c^2$ is $(-c, c)$.

68. Prove that for any $c > 0$, the solution set of the inequality $x^2 > c^2$ is $(-\infty, -c) \cup (c, \infty)$.

Thinking Forward

Conditional inequalities: In Chapter 1 we will be interested in how the range of values of a one-variable expression depends on a range of values for its variable. Each of the problems that follow involves two related absolute value inequalities. State these absolute value inequalities and then use them to solve the problems.

▶ Show that if x is within a distance of 1 from 2, then $3x + 1$ is within a distance of 3 from 7.

▶ Show that if x is within a distance of 0.25 from 1, then $4x - 2$ is within distance 1 of 2.

▶ If we want to require that x^2 is within a distance of 3 from 0, how close must x be to 0?

▶ If we want to require that $1 - 2x$ is within a distance of 0.5 from -2, how close must x be to 3?

▶ Use the following graph of $y = 1 - x^2$ to illustrate that $1 - x^2$ is within a distance of 4 from 1 if and only if x is within a distance of 2 from 0:

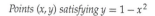

Points (x, y) satisfying $y = 1 - x^2$

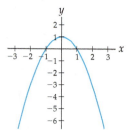

0.4   FUNCTIONS AND GRAPHS

▶ Definition of functions and their domain and range

▶ Graphs, horizontal and vertical line tests, and one-to-one-functions

▶ Graphical properties and features, asymptotes, and average rate of change

What Is a Function?

Functions and their properties will be at the core of everything we study in this text. In previous courses you likely encountered functions that were given in terms of formulas, such as

$$y(x) = x^2,$$

that relate two variables x and y. To set the stage for studying such functions, we must first be more precise about what functions are. Instead of thinking of functions merely as formulas, think of them as describing a certain kind of rule, relationship, or mapping from the elements of one set to the elements of another set.

DEFINITION 0.24   **Functions**

A *function* f from a set A to a set B is an assignment f that associates to each element x of the *domain* set A exactly one element $f(x)$ of the *codomain*, or *target*, set B.

We will use the notation

$$f : A \to B$$

to represent a function f together with its domain set A and target set B. This notation is pronounced "f from A to B." If x and y are variables that represent elements of the sets A and B, respectively, then we say that y is a function of x and write $y = f(x)$ or $y(x)$.

The variable x is called the *independent variable* and represents the "input" of the function. The function f sends each input x to one and only one "output," some value of the *dependent variable* y. Notice that y depends on x, according to the assignment defined by the function f.

For example, the assignment $f : \mathbb{R} \to \mathbb{R}$ that squares real numbers is a function, since each real number x is assigned to one and only one real-number square x^2. Here $\mathbb{R}$ denotes the set of all real numbers, and f assigns each real-number input to exactly one real-number output. Some real numbers (such as 3 and -3) get sent to the same square ($f(-3) = f(3) = 9$), but this does not violate the definition of function. You can think of a function as a machine that takes any given input value x and produces exactly one output value $f(x)$ (pronounced "f of x"), shown as follows:

So, what *isn't* a function? If a rule assigns a real-number input to more than one output, then that rule is not a function. For example, consider the formula $y = \pm\sqrt{x}$. This assignment does not define y as a function of x, because the input $x = 4$ corresponds to two different y-values, both $y = -2$ and $y = 2$. In the "function machine" type of illustration just shown, the number 4 would go into the machine and two numbers, -2 and 2, would come out at once as outputs. This situation is not allowed for functions.

Returning to the squaring function $y = x^2$, notice that some real numbers can never serve as outputs, because squares of real numbers can never be negative. The *range*, or set of possible outputs, of the squaring function is $[0, \infty)$. In this text we will usually work with functions whose domains and ranges are unspecified subsets of real numbers and whose rules are given by formulas such as $f(x) = x^2$.

DEFINITION 0.25

Domain and Range of a Function

If f is a function between unspecified subsets of $\mathbb{R}$, then we will take the **domain** of f to be the largest subset of $\mathbb{R}$ for which f is defined:

$$\text{Domain}(f) = \{x \in \mathbb{R} \mid f(x) \text{ is defined}\}.$$

The **range** of such a function is the set of all possible outputs that it can attain:

$$\text{Range}(f) = \{y \in \mathbb{R} \mid \text{ there is some } x \in \text{Domain}(f) \text{ for which } f(x) = y\}.$$

For example, the function $f(x) = \sqrt{x-1}$ is defined only when $x \geq 1$, and therefore $f(x) = \sqrt{x-1}$ has domain $[1, \infty)$. When we write the square root symbol without the "$\pm$" before it, we always mean the positive square root. This means that $f(x) = \sqrt{x-1}$ can attain only nonnegative y-values. In fact, every nonnegative value can be expressed in the form $\sqrt{x-1}$ for some value of x, and therefore the function $f(x) = \sqrt{x-1}$ has range $[0, \infty)$.

TECHNICAL POINT   The name of a function is usually a single letter, such as "f." The name of the output of a function f evaluated at an input x is "$f(x)$." In this situation f is a function, or relationship, and $f(x)$ is a number that represents the output of the function at the input value x. However, it is sometimes convenient to write $f(x)$ (the name of the output of the function) instead of f (the name of the function itself). This allows us to indicate the name we are using for the independent variable when we reference the function. We may also write things like "consider the function $f(x) = x^2 + 1$," by which we mean "consider the function f whose output at a real number x is $f(x) = x^2 + 1$."

Vertical and Horizontal Line Tests

A function whose domain and range are sets of real numbers can be represented as a collection of pairs $(x, f(x))$ of real numbers. If we plot these pairs as points in the Cartesian plane, we obtain the *graph* of the function. In general we have the following definition:

DEFINITION 0.26

The Graph of a Function

The *graph* of a function f is the collection of ordered pairs $(x, f(x))$ for which x is in the domain of f. In set notation we can write

$$\text{Graph}(f) = \{\, (x, f(x)) \mid x \in \text{Domain}(f) \,\}.$$

For example, the graph of $f(x) = x^2$ is the collection of ordered pairs of the form (x, x^2), for $x \in \mathbb{R}$. Since $f(-1) = (-1)^2 = 1$ and $f(2) = 2^2 = 4$, the points $(-1, 1)$ and $(2, 4)$ are on the graph of $f(x) = x^2$. In contrast, the point $(1, 2)$ is *not* a part of the graph, because $f(1) \neq 2$, as shown in the following graph:

Graph of $f(x) = x^2$ and partial table of values

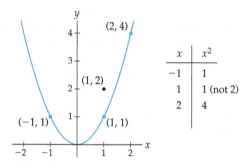

x	x^2
-1	1
1	1 (not 2)
2	4

A function always has exactly one output value for every input in the domain, which means that the graph of a function always passes the following test, which you will prove in Exercise 90:

THEOREM 0.27

The Vertical Line Test

A graph represents a function if and only if every vertical line intersects the graph in at most one point.

For example, consider the three graphs that follow this paragraph. The leftmost graph passes the vertical line test and thus is the graph of a function. The graph in the middle fails the vertical line test because the vertical line $x = 2$ intersects the graph in two points, $(2, 1)$ and $(2, 3)$; therefore the middle graph does not represent a function. The rightmost graph assigns the same output to two distinct inputs, but that is perfectly fine for a function. Because the graph on the right passes the vertical line test, it is the graph of a function.

A graph that is a function

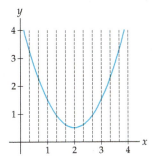

A graph that is not a function

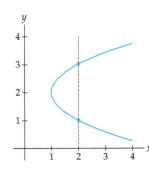

A function, but not one-to-one

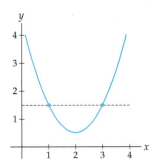

If each element of the range of a function is the output of exactly one element of the domain, then the function is said to be ***one-to-one***. Graphically, we can tell if a function f is one-to-one by checking to see if it passes the ***horizontal line test***: if f is one-to-one, then every horizontal line meets the graph of f at most once; see Exercise 91. Algebraically, this means that a function f is one-to-one if two distinct elements in the domain are always sent to different elements of the range:

DEFINITION 0.28

One-to-One Function

A function f is ***one-to-one*** if, for all a and b in the domain of f,

$$a \neq b \implies f(a) \neq f(b).$$

In this definition the notation $\Rightarrow$ is pronounced "implies," and it means that if the left-hand part of the expression is true, then the right-hand part of the expression is also true. In other words, the statement "$A \Rightarrow B$" is synonymous with the statement "if A, then B."

A logically equivalent form of Definition 0.28 is its so-called ***contrapositive***:

$$f(a) = f(b) \implies a = b.$$

As we will see in Section 0.7, the contrapositive of an implication $A \Rightarrow B$ is the equivalent statement (not B) $\Rightarrow$ (not A). The contrapositive form of Definition 0.28 is often easier to use, because it is an affirmative rather than a negative statement. For example, $f(x) = 3x$ is one-to-one because if $3a = 3b$, then we can guarantee that $a = b$. In contrast, the squaring function $f(x) = x^2$ is not one-to-one, because we cannot guarantee that if $a^2 = b^2$, then $a = b$ (for example, $(-3)^2 = 3^2$, but $-3 \neq 3$).

Properties of Graphs

The table that follows gives us vocabulary and precise mathematical definitions for various types of graphical behavior. Rows 1, 2, 5, and 6 describe behaviors that a function could exhibit at a specific point. The remaining rows describe graphical behaviors that occur over an interval I of real numbers. Much of the material in the early chapters of this book will be dedicated to developing techniques for properly defining and identifying these types of properties of functions. For now we present them just to set terminology and to familiarize ourselves with various types of function behavior.

Vocabulary	Definition	Behavior
f has a **root** at $x = c$	$f(c) = 0$	graph intersects the x-axis at $x = c$
f has a **y-intercept** at $y = b$	$f(0) = b$	graph intersects the y-axis at $y = b$
f is **positive** on I	$f(x) > 0$ for all $x \in I$	graph is above the x-axis on I
f is **increasing** on I	$f(b) > f(a)$ for all $b > a$ in I	graph moves up as we look from left to right on I
f has a **local maximum** at $x = c$	$f(c) \geq f(x)$ for all x near $x = c$	graph has a relative "hilltop" at $x = c$
f has a **global maximum** at $x = c$	$f(c) \geq f(x)$ for all $x \in \text{Domain}(f)$	graph is the highest at $x = c$
f is **concave up** on I	*will state precisely in Section 3.3*	graph curves upwards on I like part of a "U"
f has an **inflection point** at $x = c$	*will state precisely in Section 3.3*	graph of f changes concavity at $x = c$

Of course, there are similar definitions for local and global minima and for negative, decreasing, and concave-down behavior; see Exercises 20 and 21. Notice that we describe **extrema** (maxima and minima) by *where* on the x-axis they occur, since we can always find the corresponding y-values from these x-values. The concept of "near" in the description of a local maximum will be made more precise in Chapters 1 and 2. Inflection points and concavity cannot be precisely defined until we learn about **derivatives** in Chapters 2 and 3. In that chapter we will also learn ways for algebraically calculating the locations of extrema and inflection points. Until then, we will have to be content with examining such things graphically.

For example, the list that follows at the right describes some aspects of the graphical behavior of the graph $y = f(x)$ shown on the left.

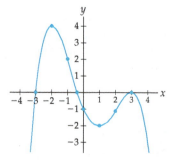

- roots at $x = -3$, $x \approx -0.4$, and $x = 3$
- y-intercept at $y = -1$
- local maxima at $x = -2$ and $x = 3$
- global maximum at $x = -2$
- inflection points at $x = -1$ and $x = 2$
- positive on $(-3, -0.4)$
- increasing on $(-\infty, -2)$ and $(1, 3)$
- concave up on $(-1, 2)$

In fact, technically the function f graphed at the left is increasing on the larger intervals $(-\infty, -2]$ and $[1, 3]$. This is because we do have $f(b) > f(a)$ for all values $b > a$ in these *closed* intervals. Most of the time we will be concerned only with the open intervals on which a function is increasing or decreasing, but there will be a few times in later chapters when we need to consider closed or half-closed intervals. For now, we will use open intervals unless we require otherwise.

The increasing and/or decreasing behavior of a function is related to its ***average rate of change*** on various intervals. The average rate of change of a function f on an interval $[a, b]$ measures how much the output $f(x)$ changes over that interval. Average rates of change will be extremely important in Chapter 2 when we study the derivative.

DEFINITION 0.29

Average Rate of Change

The ***average rate of change*** of a function f on an interval $[a, b]$ is the slope of the line from $(a, f(a))$ to $(b, f(b))$, which is given by the quotient

$$\frac{f(b) - f(a)}{b - a}.$$

For example, the function whose properties we just enumerated is increasing on the interval $(1, 3)$, moving up from $(1, f(1)) = (1, -2)$ to $(3, f(3)) = (3, 0)$. The average rate of change tells us how much the function increased per unit change in the input, on average:

$$\frac{f(3) - f(1)}{3 - 1} = \frac{0 - (-2)}{2} = 1$$

unit up for every unit across. We can also measure average rate of change over intervals where the function both increases and decreases; for example, with the same function, on the interval $[-3, 3]$ there is an average rate of change of

$$\frac{f(3) - f(-3)}{3 - (-3)} = \frac{0 - 0}{6} = 0$$

units up for every unit across; look at the graph to see why this makes sense.

Sometimes a graph gets closer and closer to a horizontal or vertical line, or ***asymptote***. In Chapter 1, we will define asymptotes precisely, using limits. For now, we will use the following definition: A line l is an ***asymptote*** of a function f if the difference between the graph of l and the graph of f gets as small as we want as either x or y increases in magnitude. For example, the following graph of a function f has vertical asymptotes at $x = -2$ and $x = 2$, and a horizontal asymptote at $y = 1$:

A function with three asymptotes

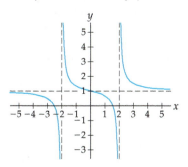

Notice that a graph *can* cross one of its horizontal asymptotes; the preceding graph above does so at the point $(0, 1)$. This is just one of the reasons that we are avoiding using the overly loose definition of asymptote that you may have heard in previous courses ("an asymptote is a line that the graph gets infinitely close to, but never reaches").

Examples and Explorations

EXAMPLE 1   ### Identifying functions and their domains and ranges

Determine whether or not each of the following relationships is a function. For each relationship that is a function, describe its natural domain and range and determine whether or not it is one-to-one. For each relationship that is not a function, describe the parts of the definition of a function that are violated.

(a) The rule $g: \mathbb{R} \to \mathbb{R}$ that assigns each real number x to the numbers whose square is x.

(b) The relationship defined by this table:

x	1	2	3	4	5	6
$P(x)$	5	2	9	−1	0	9

(c) Let P be the set of all living people in the world, and let W be the set of all women that have ever lived. Define $f: P \to W$ so that each person is assigned to his or her biological mother.

(d) $f(x) = 2 - \sqrt{x+5}$

(e) $h(x) = \dfrac{1}{x^2 - 4}$

SOLUTION

(a) This rule is not a function, for two reasons. First of all, negative real numbers do not have real square roots, so g is not defined on the given domain of $\mathbb{R}$. Second, each positive number x has two numbers whose square is x, namely, $\sqrt{x}$ and $-\sqrt{x}$, so this rule would not send each domain element to exactly one output.

(b) The relationship $P(x)$ defined by the table is a function, because the table assigns each value in the domain $\{1, 2, 3, 4, 5, 6\}$ to exactly one element of the range $\{-1, 0, 2, 5, 9\}$. This function is not one-to-one because $P(3)$ and $P(6)$ are both equal to 9.

(c) This relationship is a function because each person has one exactly one woman who is his or her biological mother. No person is without a biological mother, and no person has more than one biological mother. Here the domain is P and the range is the subset of W consisting of women that have had biological children. This function is not one-to-one, since there are examples of different people that have the same biological mother.

(d) This rule is a function because for each value x for which the formula makes sense, there is exactly one real number described by $2 - \sqrt{x+5}$. For x to be in the domain, we must have $x+5 \geq 0$ (since $x+5$ is under a square-root sign), and thus we must have $x \geq -5$. Therefore the domain of f is $[-5, \infty)$. The range of $y = f(x)$ is the set of y-values that can occur as outputs. Since $\sqrt{x+5}$ can take on any value greater than or equal to 0, the expression $2 - \sqrt{x+5}$ can take on any value less than or equal to 2. Therefore the range of f is $(-\infty, 2]$. This function is one-to-one because if $f(a) = f(b)$, then

$$2 - \sqrt{a+5} = 2 - \sqrt{b+5} \implies \sqrt{a+5} = \sqrt{b+5}$$
$$\implies a+5 = b+5$$
$$\implies a = b.$$

(e) The rule $h(x)$ is a function because for each value x at which $\dfrac{1}{x^2-4}$ is defined, there is exactly one real number that h describes. The domain of $h(x)$ is the set of all x-values for which $x^2 - 4 \neq 0$ (since $x^2 - 4$ is in a denominator). Therefore the domain of h is everything except $x = \pm 2$. To find the range of $h(x)$ we must find the y-values that can be expressed in the form $y = h(x)$ for some x. Solving for x in terms of y we obtain $x = \sqrt{\dfrac{1}{y} + 4}$. This means we can find an x that maps via f to y as long as $y \neq 0$ and $\dfrac{1}{y} + 4 \geq 0$. It can be shown that the solution of the latter inequality is

$\left(-\infty, -\frac{1}{4}\right] \cup [0, \infty)$. Therefore the range of $h(x)$ is $\left(-\infty, -\frac{1}{4}\right] \cup (0, \infty)$. This function is not one-to-one because, for example, $h(1)$ and $h(-1)$ are both equal to $-\frac{1}{3}$. ☐

✔ CHECKING THE ANSWER

The following functions f and h have domains marked in blue on the x-axis and ranges marked in red on the y-axis:

f has domain $[-5, \infty)$
and range $(-\infty, 2]$

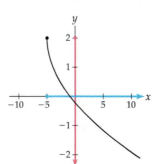

h has domain $x \neq \pm 2$
and range $\left(-\infty, -\frac{1}{4}\right] \cup (0, \infty)$

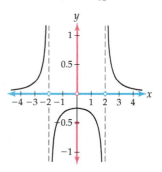

EXAMPLE 2 Evaluating function notation

Given that $f(x) = x\sqrt{3-x}$, evaluate $f(2)$, $f(a)$, $f(x+1)$, and $f(f(x))$.

SOLUTION

To evaluate $f(x) = x\sqrt{3-x}$ at a given input, simply replace x in the formula with whatever the input is:

$$f(2) = 2\sqrt{3-2} = 2\sqrt{1} = 2;$$
$$f(a) = a\sqrt{3-a};$$
$$f(x+1) = (x+1)\sqrt{3-(x+1)};$$
$$f(f(x)) = f(x)\sqrt{3-f(x)} = (x\sqrt{3-x})\sqrt{3-x\sqrt{3-x}}.$$

☐

EXAMPLE 3 Finding a "good" graphing window

Use a graphing utility to find a graphing window that accurately represents the key features of the graph of the function $f(x) = x^3 - 6x^2 - x + 6$.

SOLUTION

The three graphs that follow show $y = f(x)$ in various graphing windows. Each of these windows is "bad" in the sense that the true behavior of the graph of f is not represented.

$f(x) = x^3 - 6x^2 - x + 6$
$x \in [-3, 3], y \in [-10, 10]$

$f(x) = x^3 - 6x^2 - x + 6$
$x \in [-100, 100], y \in [-50, 50]$

$f(x) = x^3 - 6x^2 - x + 6$
$x \in [-20, 20], y \in [-1000, 1000]$

A "good" window (if one exists) is a window in which the local behavior of the graph of f is clear and the global behavior is accurately represented (the "ends" of the graph keep going in the direction indicated). The following figure shows the graph of f in a "good" window:

$$f(x) = x^3 - 6x^2 - x + 6$$
$$x \in [-3, 7], \ y \in [-40, 40]$$

For now we will use trial and error to find an effective graphing window. We will be able to be more systematic after we have learned more about derivatives and function behavior. □

EXAMPLE 4   **Function behavior at points and on intervals**

(a) Sketch the graph of a function that has the following characteristics:
- ▶ roots at $x = -2$, $x = 1$, and $x = 3$ ▶ local minimum at $x = -1$
- ▶ horizontal asymptote at $y = -2$ ▶ local maximum at $x = 2$

(b) Approximate the locations of the inflection points on your graph.

(c) Given the graph that you sketched, find the intervals on which f is
- ▶ positive   ▶ increasing   ▶ decreasing   ▶ concave up

SOLUTION

(a) By plotting the points $(-2, 0)$, $(1, 0)$, and $(3, 0)$, drawing a dashed asymptote at $y = -2$, and plotting some low point for the function at $x = -1$ and some high point at $x = 2$ (the information in the problem does not tell us exactly how high or how low), one might make the following sketch:

One possible graph of f

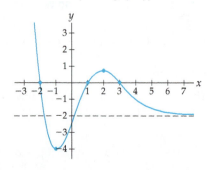

(b) The inflection points on this graph occur where the concavity of the graph changes from a ∪ shape to a ∩ shape, or vice versa. These are the locations where the graph "flexes." In our graph, these points occur at approximately $x = \frac{1}{2}$ and $x = 4$. Note that these points are also the locations of the steepest parts of the graph.

(c) Given our graph, the function f is positive on $(-\infty, -2)$ and $(1, 3)$, increasing on $(-1, 2)$, decreasing on $(-\infty, -1)$ and $(2, \infty)$, and concave up on $\left(-\infty, \frac{1}{2}\right)$ and $(4, \infty)$.

□

EXAMPLE 5

Finding the domain of a function by using equalities and inequalities

Find the domain of the function $f(x) = \sqrt{\dfrac{1 - 2x}{x + 1}}$.

SOLUTION

To find the domain of f, we ask which values of x can be validly plugged into the equation that defines $f(x)$. In order for the value of f to be defined for a real number x, that value of x must make the quotient underneath the square-root sign nonnegative and the denominator nonzero. Thus the domain of $f(x)$ is the set of real numbers that simultaneously satisfy the following:

$$\frac{1 - 2x}{x + 1} \geq 0 \quad \text{and} \quad x + 1 \neq 0.$$

The only x-values at which the quotient in the inequality can change sign are $x = \frac{1}{2}$ and $x = -1$, since those are the values that make either the numerator or denominator equal to zero. To determine the intervals on which the quotient is positive or negative we need only check its sign between the possible change points $x = \frac{1}{2}$ and $x = -1$.

We can record this information on a number line with a ***sign chart*** as follows:

Since the quotient in question is negative on $(-\infty, -1)$ and $\left(\frac{1}{2}, \infty\right)$, the function $f(x)$ is not defined on those intervals. Note that $x = \frac{1}{2}$ is in the domain of f because there is no problem taking the square root of zero. However, since we cannot divide by zero, the function is not defined at $x = -1$. Therefore the domain of f is the half-open interval $\left(-1, \frac{1}{2}\right]$. Note that the inequality we just solved was the same one that we solved in the solution of Example 3 of Section 0.3.

□

EXAMPLE 6

Finding the average rate of change of a function on an interval

Calculate the average rate of change of the function $f(x) = x^2 - x + 1$ (a) on the interval $[0, 3]$ and (b) on the interval $[-1, 1]$. Then (c) illustrate these average rates of change graphically.

SOLUTION

(a) Using the formula from Definition 0.29 we find that the average rate of change of f on $[0, 3]$ is

$$\frac{f(3) - f(0)}{3 - 0} = \frac{(3^2 - 3 + 1) - (0^2 - 0 + 1)}{3 - 0} = \frac{7 - 1}{3 - 0} = \frac{6}{3} = 2.$$

(b) Using the same formula again, we find that the average rate of change of f on $[-1, 1]$ is

$$\frac{f(1) - f(-1)}{1 - (-1)} = \frac{(1^2 - 1 + 1) - ((-1)^2 - (-1) + 1)}{1 - (-1)} = \frac{1 - 3}{1 - (-1)} = \frac{-2}{2} = -1.$$

(g) *True or False:* Average rates of change can be thought of as slopes.

(h) *True or False:* A function can have different average rates of change on different intervals.

2. *Examples:* Construct examples of the thing(s) described in the following. Try to find examples that are different than any in the reading.

(a) A function that is defined with a formula.

(b) A function that is not defined with a formula.

(c) A formula that does not define a function.

3. State the mathematical definition of a function, and describe its meaning in your own words. Support your answer with an example of something that is a function and an example of something that is not.

4. Suppose P is the set of people alive today and C is the set of possible eye colors. Let $f: P \to C$ be the rule that assigns to each person his or her eye color. Is f a function? Why or why not?

5. Use set notation to define the domain of a function. Then use the same notation to express the domain of the function $f(x) = \sqrt{x}$.

6. Use set notation to define the range of a function. Then use the same notation to express the range of the function $f(x) = x^2$.

7. Determine whether the points (a) $(3, 2)$, (b) $(1, 1)$, and (c) $(-5, 2)$ lie on the graph of $f(x) = \sqrt{x + 1}$, without referring to a picture of the graph of f.

8. Describe the graph of the function $f(x) = 3x + 2$ as a set of ordered pairs.

9. Consider the function $f(x) = x^2 + 1$.

(a) Explain why $y = 5$ is in the range of f.

(b) Explain why $y = 0$ is *not* in the range of f.

(c) Argue that the range of $f(x) = x^2 + 1$ is $[1, \infty)$.

10. Determine whether or not each diagram that follows represents a function. If it does, find its domain and range, and determine whether it is one-to-one. If it does not, explain what goes wrong.

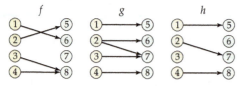

11. Construct a rule $f: \{2, 4, 6, 8, 10\} \to \{1, 2, 3, 4\}$ that is a function. Express this function three ways: as a list, as a table, and as a diagram. Is your function one-to-one? What is its range?

12. Construct a rule $f: \{2, 4, 6, 8, 10\} \to \{1, 2, 3, 4\}$ that is not a function. Justify your answer.

13. If the graph of a rule $y = f(x)$ passes through $(-2, 1)$ and $(2, 1)$, could that rule be a function? Why or why not?

14. A *constant function* is a function $f: A \to B$ with the property that there is some $b \in B$ for which $f(x) = b$ for all $x \in A$. (The output of the function is *constantly* the same.) Describe

(a) a constant function $f: \mathbb{R} \to \mathbb{R}$

(b) a constant function $g: \mathbb{R} \to [-5, -2]$

(c) a constant function $h: \mathbb{R} \to \mathbb{R}^2$

15. The *identity function* for a set A is the function $f: A \to A$ defined by $f(x) = x$ (so called because the output is *identical* to the input). For which of the following domains and ranges is there a well-defined identity function? Why or why not?

(a) $f: \mathbb{R} \to \mathbb{R}$

(b) $g: \mathbb{R}^2 \to \mathbb{R}$

(c) $h: \mathbb{R}^3 \to \mathbb{R}^3$

16. Let P be the set of all people living in the United States. Give examples of each of the following functions and state their ranges:

(a) the identity function $f: P \to P$

(b) two different constant functions $g: P \to P$

(c) a non-constant, non-identity function $g: P \to P$

(d) a constant function $h: P \to \mathbb{R}$

(e) a non-constant function $h: P \to \mathbb{R}$

17. Explain in your own words why the vertical line test determines whether a graph is a function.

18. Explain in your own words why the horizontal line test determines whether a function is one-to-one.

19. Show that $f(x) = x^2 + 1$ is not one-to-one, using values of f (*not* the horizontal line test).

20. Define what it means for a function f with domain $\mathbb{R}$ to have (a) a global minimum at $x = c$ and (b) a local minimum at $x = c$.

21. Define what it means for a function f with domain $\mathbb{R}$ to be (a) negative on an interval I and (b) decreasing on an interval I.

22. Make a labeled graph that illustrates why it makes sense that a function is increasing on an interval I if, for all $b > a$ in I, we have $f(b) > f(a)$. Include labels for a, b, $f(a)$, and $f(b)$, and for the interval I.

23. How is the formula for average rate of change related to the formula for computing slope?

24. Illustrate on a graph of $f(x) = 1 - x^2$ that the average rate of change of f on $[-1, 3]$ is -2.

25. For each local maximum $x = c$ in the following graph, approximate the largest possible $\delta > 0$ so that $f(c) \geq f(x)$ for all $x \in (c - \delta, c + \delta)$. Similarly, for the one local minimum $x = b$, find the largest δ so that $f(b) \leq f(x)$ for all $x \in (b - \delta, b + \delta)$.

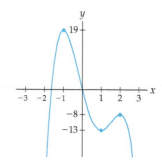

26. Use the definition of a local maximum to explicitly argue why the function graphed in Exercise 25 does *not* have a local maximum at $x = 0$.

Skills

Find the domain and range of each function in Exercises 27–32. Use a graphing utility or plot points to sketch a graph of the function, and illustrate the domain and range on the graph.

27. $f(x) = \sqrt{x-1}$

28. $f(x) = \sqrt{x} - 1$

29. $f(x) = \dfrac{1}{x+2}$

30. $f(x) = \dfrac{1}{\sqrt{5-x}}$

31. $f(x) = \dfrac{1}{x^2+1}$

32. $f(x) = \dfrac{1}{x^2-1}$

Find the domain of each function in Exercises 33–42.

33. $f(x) = \sqrt{x(x-2)}$

34. $f(x) = \dfrac{3x+1}{2x-1}$

35. $f(x) = \sqrt{(x-1)(x+3)}$

36. $f(x) = \dfrac{1}{\sqrt{x^2-4}}$

37. $f(x) = \dfrac{1}{\sqrt{(x-1)(x+3)}}$

38. $f(x) = \sqrt{\dfrac{1}{x^2} - 4}$

39. $f(x) = \dfrac{\sqrt{x^2-1}}{x^2-9}$

40. $f(x) = \dfrac{x^{3/4}}{3x-5}$

41. $f(x) = \dfrac{1}{\sqrt{x-1}} - \dfrac{\sqrt{x}}{x-2}$

42. $f(x) = \dfrac{\sqrt{x^2-1}}{\sqrt{x^2-9}}$

Evaluate each function in Exercises 43–47 at the values indicated. Simplify your answers if possible. *(Note: We will not need to work with such functions until much later in the book, but we present them here to illustrate the variety of functions we will be encountering in this course.)*

43. If $f(x) = x^2 + 1$, find

   (a) $f(-4)$ (b) $f(a^3)$ (c) $f(f(x))$

44. If $k(x) = \dfrac{x^2}{x+1}$, find

   (a) $k(5)$ (b) $k(x+h)$ (c) $k(k(x))$

45. If $l(a, b, c) = \sqrt{a^2 + b^2 + c^2}$, find:

   (a) $l(5, 3, 2)$ (b) $l(3, 0, 4)$ (c) $l(x, y, z)$

46. If $g(v) = (v-1, v, v^2)$, find

   (a) $g(0)$ (b) $g(1)$ (c) $g(x+1)$

47. If $F(u, v, w) = (3u + v, u - w, v + 2w)$, find

   (a) $F(2, 3, 5)$ (b) $F(5, 2, 3)$ (c) $F(a, b, 0)$

For each piecewise-defined function in Exercises 48–50, (a) calculate $f(-1), f(0), f(1)$, and $f(2)$, and (b) sketch a graph of f.

48. $f(x) = \begin{cases} x^2 + 3, & \text{if } x < 0 \\ 3 - x, & \text{if } x \geq 0 \end{cases}$

49. $f(x) = \begin{cases} 3x + 1, & \text{if } x \leq 0 \\ 4, & \text{if } 0 < x \leq 1 \\ x^3, & \text{if } x > 1 \end{cases}$

50. $f(x) = \begin{cases} 4x - 1, & \text{if } x < 0 \\ 2, & \text{if } x = 0 \\ -3x + 5, & \text{if } x > 0 \end{cases}$

Use a graphing utility to sketch a graph of each function in Exercises 51–56. Use trial and error to find a graphing window so that your graph represents the local and global behavior of the function. Include the x and y ranges of your window in your answer.

51. $f(x) = x^2 - 0.1$

52. $f(x) = (x^2 - 5)^7$

53. $f(x) = x^5 - 3x^4 - 7x$

54. $f(x) = x^3 - 11x^2 + 10x$

55. $f(x) = x^2 - 17x - 18$

56. $f(x) = \dfrac{2x^2 - 2}{x^2 - 3x - 5}$

Describe the key properties of each graph in Exercises 57–62, including the following:

▶ domain and range;

▶ locations of roots, intercepts, local and global maxima and minima, and inflection points;

▶ intervals on which the function is positive or negative, increasing or decreasing, and concave up or down;

▶ any horizontal or vertical asymptotes.

57.

58.

59.

60.

61.

62.

Sketch the graph of functions f that satisfy the lists of conditions given in Exercises 63–72, if possible.

63. Domain $\mathbb{R}$, concave up everywhere, and decreasing everywhere.

64. Domain $\mathbb{R}$, concave down everywhere, and decreasing everywhere.

65. Domain $\mathbb{R}$, concave up everywhere, increasing everywhere, and negative everywhere.

66. Domain $\mathbb{R}$, concave down everywhere, increasing everywhere, and negative everywhere.

67. Always increasing, with two horizontal asymptotes, one at $y = -2$ and one at $y = 2$.

68. Domain $(0, \infty)$, always negative, and always increasing.

69. Four roots but no y-intercept.

70. Concave down on $(-\infty, 2)$, concave up on $(2, \infty)$, and always increasing.

71. Concave down on $(-\infty, 0)$ and concave up on $(0, \infty)$ but *without* an inflection point at $x = 0$.

72. Average rate of change of 3 on $[0, 2]$, average rate of change of -1 on $[0, 1]$, and average rate of change of 0 on $[-2, 2]$.

In Exercises 73–78, find the average rate of change of the function f on the interval $[a, b]$.

73. $f(x) = -0.5 + 4.2x, \quad [a, b] = [1, 3.5]$

74. $f(x) = 3, \quad [a, b] = [-100, 100]$

75. $f(x) = \sqrt{x + 1}, \quad [a, b] = [1, 9]$

76. $f(x) = \dfrac{1 - x}{1 + x^3}, \quad [a, b] = [0, 0.5]$

77. $f(x) = \dfrac{1}{x}, \quad [a, b] = [0.9, 1.1]$

78. $f(x) = (x - 2)^2 + \dfrac{3}{x}, \quad [a, b] = [-2, 2]$

Applications

In Exercises 79–82, sketch and label the graph of a function that describes the given situation.

79. An island warthog population initially grows quickly, but as space and food become sparse on the island, the population growth slows down. Eventually the population of the island levels off at 512 warthogs.

80. Susie is late for calculus class and leaves her dorm in a panic. She hurries towards the math building, but about halfway there, she realizes she has left her notebook in her room. She sprints back to her dorm and gets her notebook. Coming out of the dorm, she sprains her ankle, so the best she can do is limp as fast as she can to her classroom.

81. Suppose that after you drink a cup of coffee the amount of caffeine in your body rises sharply and then decreases by half every hour. You have one cup of coffee in the morning and then no more.

82. On a dare, you go skydiving. Gravity causes you to fall faster and faster as you plummet towards the ground. When you open your parachute, your speed is drastically reduced. After opening your parachute you approach the ground at a constant speed.

(a) Graph your distance from the ground as a function of time.

(b) Graph your velocity as a function of time.

83. For each situation described, identify any independent and dependent variables, and express their relationship as an equation in multivariable function notation (see Example 8):

(a) H is the length of the hypotenuse of the right triangle with legs of length a and b.

(b) V is the volume of a rectangular prism ("box") with dimensions x, y, and z.

84. If your rain-catching bucket starts empty and collects 3 inches of rain during a 6-hour rainstorm, what is the average rate of change of the level of rainwater in the bucket over the 6 hours that it rained? Did the rain necessarily collect in the bucket at a constant rate?

85. A disgruntled pet store owner abandoned an unknown number of groundhogs on a small island in 1996. Since then it has been determined that the average rate of change of the groundhog population was 4 groundhogs per year and that the groundhog population was a linear function of time. When the abandoned groundhogs were discovered in 2001, there were 376 groundhogs on the island. How many groundhogs did the disgruntled pet store owner originally leave on the island?

86. The number N of operating drive-in movie theatres in the state of Virginia in various years y is given in the table below.

y	1958	1967	1972	1977	1982	1999
N	143	90	102	87	56	9

(a) Find the average rate of change in the number of drive-in movie theatres in Virginia over each time interval between table entries.

(b) Describe the units and real-world significance of these average rates of change.

(c) Over which time period was the average rate of change the most drastic? On average, assuming that no new theatres were built, how many drive-ins closed per year during that period?

87. In your first job after graduating from college you make \$36,000 a year before taxes. After four years you get a raise of \$2,500. Two years after that you change jobs and go to work for a company that pays you \$49,000 a year.

(a) Construct a piecewise-defined function that describes your pretax income in the year that is t years after you graduate from college.

(b) Write down a function that describes the total amount of money you will have earned t years after graduating from college.

(c) How many years after graduating from college will you have earned a total of one million pre-tax dollars?

88. The following table shows the year 2000 Federal Tax Rate Schedule for single filers:

Taxable income:		The federal tax owed is:		
Over	Not over	Amount	Plus %	Of amt. over
$0	$26,250	$0	15%	$0
$26,250	$63,550	$3,937	28%	$26,250
$63,550	$132,600	$14,381	31%	$63,550
$132,600	$288,350	$35,787	36%	$132,600
$288,350	——	$91,857	39.6%	$288,350

(a) How much tax would you owe if you made $18,000 of taxable income? What if you made $180,000?

(b) What percentage of your taxable income did you owe in taxes if your taxable income was $18,000? What if your taxable income was $180,000?

(c) Construct a piecewise-defined function describing the dollar amount of tax T owed by a single person with m dollars of taxable yearly income. Each piece of your function will be linear. Do the pieces "match up"? Does this make financial sense?

Proofs

89. Use Definition 0.25 to prove that the range of the function $f(x) = 3x - 1$ is $\mathbb{R}$.

90. Use Definition 0.24 to prove that a graph represents a function if and only if it passes the vertical line test.

91. Use Definition 0.28 to prove that a function is one-to-one if and only if its graph passes the horizontal line test.

92. Use the contrapositive form of Definition 0.28 to prove that the function $f(x) = 3x + 1$ is one-to-one.

93. Use the definition of decreasing to prove that the function $f(x) = 1 - 3x$ is decreasing on $(-\infty, \infty)$.

94. Use the definition of increasing to prove that the function $f(x) = \dfrac{1}{3-x}$ is increasing on $(-\infty, 3)$.

95. Prove that the average rate of change of the linear function $f(x) = -2x + 4$ on *any* interval I is always equal to -2.

Thinking Forward

Evaluations for slopes and derivatives: Evaluate each function at the values indicated. Simplify your answers if possible.

▶ If $f(x) = 4 - x^2$, find

(a) $\dfrac{f(1+0.1) - f(1)}{0.1}$ (b) $\dfrac{f(1+0.001) - f(1)}{0.001}$

▶ If $f(x) = \sqrt{x}$, find

(a) $\dfrac{f(1+h) - f(1)}{h}$ (b) $\dfrac{f(x) - f(1)}{x - 1}$

▶ If $q(x, h) = \dfrac{(x+h)^2 - x^2}{h}$, find

(a) $q(3, 0.5)$ (b) $q(3, h)$ (c) $q(x, 0.5)$

Evaluations for series: Evaluate each function at the values indicated. Simplify your answers if possible.

▶ If $S(x, n) = x - \dfrac{x^2}{2} + \dfrac{x^3}{3} - \dfrac{x^4}{4} + \cdots + (-1)^{n+1}\dfrac{x^n}{n}$, find

(a) $S\left(\dfrac{1}{2}, 5\right)$ (b) $S\left(\dfrac{1}{2}, n\right)$ (c) $S(x, 5)$

▶ If $c(x, n) = 1 - \dfrac{x^2}{2!} + \dfrac{x^4}{4!} - \dfrac{x^6}{6!} + \cdots + (-1)^n\dfrac{x^{2n}}{(2n)!}$, find

(a) $c(\pi, 3)$ (b) $c(\pi, n)$ (c) $c(x, 3)$

0.5   A BASIC LIBRARY OF FUNCTIONS

▶ Types of algebraic and transcendental functions

▶ An in-depth study of linear functions and their equations

▶ Piecewise-defined functions and absolute values

Algebraic Functions

A function is **algebraic** if it can be expressed in terms of constants and a variable x by using only arithmetic operations ($+$, $-$, $\times$, and $\div$) and rational constant powers of the variable. For example,

$$f(x) = \frac{3x^2 - 5}{1 - x} \quad \text{and} \quad g(x) = (\sqrt{x} + \sqrt[3]{x})^{7/2}$$

are algebraic functions, but $h(x) = 2^x$ and $k(x) = \sin x$ are not.

There are four basic types of algebraic functions: linear functions, power functions, polynomial functions, and rational functions. More complicated algebraic functions can be constructed by using combinations of these four basic types. The following table gives a general form and examples for each type:

Type	General Form	Examples
Linear	$f(x) = mx + b$, where m and b are any real numbers	$f(x) = 2x - 1$, $f(x) = 1.4x$
Power	$f(x) = Ax^k$, where $A \neq 0$ is real and k is rational	$f(x) = 3x^3$, $f(x) = 1.7x^{-1/5}$
Polynomial	$f(x) = a_n x^n + a_{n-1} x^{n-1} + \ldots + a_1 x + a_0$, where n is a nonnegative integer and each a_i is a real number	$f(x) = 3x^5 - 2x^3 + x - 6$
Rational	$f(x) = \frac{p(x)}{q(x)}$, where $p(x)$ and $q(x)$ are polynomials	$f(x) = \frac{3x^2 - 1}{x^5 + x^3 + 1}$
Other	any other algebraic function that is not one of the types listed	$f(x) = \frac{1 + \sqrt{x}}{1 + x}$, $f(x) = x^{2/3} + 5$

There is a certain amount of overlap among the types listed in the table; for example, $f(x) = x^2$ is both a power function and a polynomial function, and $f(x) = \frac{1}{x} = x^{-1}$ is both a power function and a rational function.

We will examine power functions, polynomial functions, rational functions, and more complicated algebraic functions in detail in Chapter 4. Until that point we will keep our algebraic calculations relatively simple so that we can focus on the key concepts of calculus. In those later chapters we will learn how to sketch the graphs of many algebraic functions by hand; for now we will be content with knowing a few simple graphs, such as graphs of the following four simple power functions:

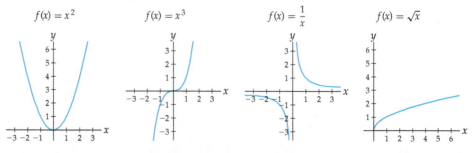

$$f(x) = x^2 \qquad\qquad f(x) = x^3 \qquad\qquad f(x) = \frac{1}{x} \qquad\qquad f(x) = \sqrt{x}$$

Linear Functions

Since linear functions will be central to our study of derivatives, we will spend some time studying them in depth right away. As we will do for all of our investigations of functions in this book, we begin with the algebraic definition:

DEFINITION 0.30   **Linear Functions**

A **linear function** is a function that can be written in the form $f(x) = mx + b$ for some real numbers m and b.

For example, each of the following equations describes y as a linear function of x, since each can be rewritten in the form $y = mx + b$:

$$y = 3x + 1, \quad y = 1.2, \quad y = x, \quad y = \frac{3x + 1}{2}, \quad \text{and} \quad 2y - 1 = 6x.$$

Note that constant functions such as $y = 1.2$ and the identity function $y = x$ are special cases of linear functions.

You may already know many properties of linear functions; for example, they have constant rates of change and their graphs are lines, as their name would suggest. All of these properties follow directly from the algebraic definition that we began with in this section, and we collect and prove them here:

THEOREM 0.31   **Properties of Linear Functions**

Suppose $f(x) = mx + b$ is a linear function.

(a) The domain of $f(x) = mx + b$ is $\mathbb{R}$.

(b) The range of $f(x) = mx + b$ is $\mathbb{R}$ if $m \neq 0$ and is $\{b\}$ if $m = 0$.

(c) The average rate of change of $f(x) = mx + b$ on any interval is m.

(d) The graph of $f(x) = mx + b$ is a line with slope m and y-intercept b.

Proof. The domain of $f(x) = mx + b$ is all of $\mathbb{R}$ because the expression $mx + b$ makes sense and produces a real number for all inputs x, regardless of the values of m and b. As x varies among all possible real numbers, the expression $mx + b$ takes on all possible real values, provided that $m \neq 0$; this means that when $m \neq 0$, the range of $f(x) = mx + b$ is $\mathbb{R}$. However, if $m = 0$, then $f(x) = 0x + b = b$ and the only possible output of f is the number b. Thus when $m = 0$, the range of $f(x) = mx + b$ is the singleton set $\{b\}$.

To prove (c) we need only compute the average rate of change of f on an arbitrary interval. To avoid confusion, we will use the notation $[c, d]$ for this interval instead of our usual notation $[a, b]$, since we are already using the letter b in the description of $f(x)$. The average rate of change of $f(x) = mx + b$ on the interval $[c, d]$ is

$$\frac{f(d) - f(c)}{d - c} = \frac{(m(d) + b) - (m(c) + b)}{d - c} = \frac{md + b - mc - b}{d - c} = \frac{m(d - c)}{d - c} = m.$$

Part (d) now immediately follows, since lines are the only graphs that have constant average rates of change on every interval, and the y-intercept of f is $f(0) = m(0) + b = b$.   ∎

The graph of a linear function with slope m moves a **rise** of m units up or down for each **run** of 1 unit across. On a general interval $[a, b]$, if we let $\Delta x = b - a$ be the change in x and

$\Delta y = f(b) - f(a)$ be the change in y, then the following quantities are equivalent:

$$\frac{\Delta y}{\Delta x} = \frac{\text{rise}}{\text{run}} = \frac{f(b) - f(a)}{b - a} = \quad \begin{array}{c} \text{Average rate of change} \\ \text{of } f \text{ on } [a, b]. \end{array}$$

For example, the graph of $f(x) = 3x - 5$ is shown next. On the interval $[1, 3]$, we have a run of $3 - 1 = 2$ units. From the point $(1, f(1)) = (1, -2)$ to the point $(3, f(3)) = (3, 4)$, we have a corresponding rise of $4 - (-2) = 6$ units. Thus the average rate of change of f on the interval $[1, 3]$ is $\frac{\Delta y}{\Delta x} = \frac{f(3) - f(1)}{3 - 1} = \frac{4 - (-2)}{3 - 1} = \frac{6}{2} = 3$.

$$f(x) = 3x - 5 \text{ has } \frac{\text{rise}}{\text{run}} = 3 \text{ everywhere}$$

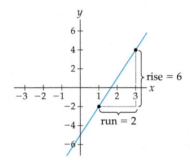

A line in the plane is completely described by its slope m and y-intercept b. You could tell someone the values of m and b over the telephone and he or she would know exactly which line you were referring to. We can also describe a line in the plane with its slope and the coordinates of any point on the line, or with the coordinates of any two points on the line. These three ways of describing lines give us the following three forms of equations for lines:

Name of form	Information given	Equation of line
slope-intercept form	slope m, y-intercept b	$y = mx + b$
point-slope form	slope m and any point (x_0, y_0) on the line	$y - y_0 = m(x - x_0)$
two-point form	any two points (x_0, y_0) and (x_1, y_1) on the line	$y - y_0 = \left(\dfrac{y_1 - y_0}{x_1 - x_0} \right)(x - x_0)$

In Exercises 75 and 77, you will prove that these ways of describing lines yield equations of lines.

One important special case of linear functions involves **proportional functions**, which are functions that can be written in the form $y = kx$ for some real number k called the **proportionality constant**. Proportional functions are simply linear functions that pass through the origin; they are often used to describe how two real-world quantities compare to each other. In word problems we will often say that one variable or quantity y is "proportional to" another variable or quantity x. When we say this we mean that y is a proportional function of x (i.e., that y is some constant multiple of x).

Piecewise-Defined Functions and Absolute Values

Recall that the absolute value $|x|$ of a real number x is equal to x if x is positive (or zero) and is equal to $-x$ if x is negative. For example, $|2| = 2$ while $|-2| = -(-2) = 2$. Thus we can write the function $f(x) = |x|$ as a piecewise-defined function by splitting the definition of $|x|$ into two cases: $x \geq 0$ and $x < 0$.

DEFINITION 0.32

The Absolute Value Function

The **absolute value function** is the piecewise-defined function $|x| = \begin{cases} x, & \text{if } x \geq 0 \\ -x, & \text{if } x < 0. \end{cases}$

The graph of $y = |x|$ is a combination of the graph of $y = x$ on $[0, \infty)$ and the graph of $y = -x$ on $(-\infty, 0)$:

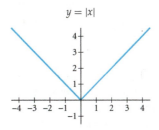

$y = |x|$

Of course, in general we might wish to take the absolute value of a more complicated expression. In the more general case we do exactly the same thing: The absolute value will leave positive quantities untouched, but flip the sign of negative quantities.

DEFINITION 0.33

The Absolute Value of a Function

The **absolute value of a function** $g(x)$ is $|g(x)| = \begin{cases} g(x), & \text{for all } x \text{ with } g(x) \geq 0 \\ -g(x), & \text{for all } x \text{ with } g(x) < 0. \end{cases}$

For example, consider the function $g(x) = 2x - 1$ and its absolute value $f(x) = |g(x)| = |2x - 1|$. When $2x - 1$ is positive or zero (i.e., when $x \geq 1/2$) the absolute value remains $2x - 1$. But when $2x - 1$ is negative (i.e., when $x < 1/2$) the absolute value of g is $-(2x - 1)$. The graph of $y = |2x - 1|$ is a combination of the graphs of $y = 2x - 1$ and $y = -(2x - 1)$, switching between graphs at $x = 1/2$, as shown in the following figures:

$$y = 2x - 1 \qquad y = -(2x - 1) \qquad y = |2x - 1| = \begin{cases} 2x - 1, & \text{if } x \geq 1/2 \\ -(2x - 1), & \text{if } x < 1/2. \end{cases}$$

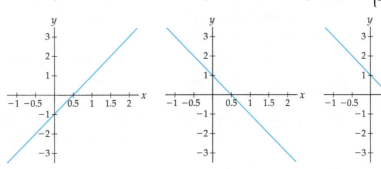

Transcendental Functions

Functions that are not algebraic are called **transcendental** functions. In this book we will investigate four basic types of transcendental functions: exponential, logarithmic, trigonometric, and inverse trigonometric functions. We will not go into their definitions or general forms right now; however, the following table provides some examples:

Type	Examples
Exponential	$f(x) = 2^x$, $f(x) = 3e^{4x}$, $f(x) = 1.2(3.4)^x$
Logarithmic	$f(x) = \log_{10} x$, $f(x) = \ln x$, $f(x) = \log_2 x$
Trigonometric	$f(x) = \sin x$, $f(x) = \cos x$, $f(x) = \cot x$
Inverse Trigonometric	$f(x) = \arcsin x$, $f(x) = \cos^{-1} x$, $f(x) = \arctan x$
Other	$f(x) = x + \sin x$, $f(x) = \ln(\sqrt{x} + 12)$, $f(x) = 2^x \arctan x$

Although we will not study such transcendental functions until Chapters 5 and 6, there will be a few instances before that where it will be necessary to use transcendental functions as examples. When we do so, we will rely on a graphing utility to provide us with graphs and values. In other words, right now we will think of functions like $\ln x$ and $\sin x$ simply as functions that our calculators can calculate and graph.

Examples and Explorations

EXAMPLE 1

Identifying algebraic function types

For each function listed, determine whether it is a linear, power, polynomial, rational, algebraic, or transcendental function. Some functions may satisfy more than one classification.

(a) $f(x) = x^{-2}$

(b) $g(x) = 3x^2 + 2^x + 1$

(c) $h(x) = \dfrac{\sqrt{x}}{1+x}$

(d) $k(x) = (x-1)^2 - x^2$

SOLUTION

(a) $f(x) = x^{-2}$ is a power function of the form Ax^k with real-number coefficient $A = 1$ and rational-number exponent $k = -2$. Since $x^{-2} = \frac{1}{x^2}$, f is also a rational function of the form $\frac{p(x)}{q(x)}$ with the polynomials $p(x) = 1$ and $q(x) = x^2$. Thus f is an algebraic function that is both a power function and a rational function.

(b) The function 2^x is not algebraic, because it involves a nonconstant exponent. Therefore the function $g(x) = 3x^2 + 2^x + 1$ also fails to be algebraic and thus is transcendental.

(c) $h(x) = \frac{\sqrt{x}}{1+x}$ a quotient of algebraic functions, but not a quotient of *polynomials*, since $\sqrt{x}$ is not a polynomial. Therefore $h(x)$ is an algebraic function, but not a rational function.

(d) This function can be simplified: $k(x) = (x-1)^2 - x^2 = x^2 - 2x + 1 - x^2 = -2x + 1$. After the last simplification it is easy to see that $k(x)$ is a linear function, with $m = -2$ and $b = 1$. The function $k(x)$ is also a polynomial function, with $n = 1$, $a_1 = -2$, and $a_0 = 1$. Furthermore, since we can write $-2x + 1 = \frac{-2x+1}{1}$, $k(x)$ is a rational function as well, with $p(x) = -2x + 1$ and $q(x) = 1$. Therefore $k(x)$ is an algebraic function that is linear, polynomial, and rational.

EXAMPLE 2   Finding equations of lines

Find equations for each of the linear functions described. Two lines are said to be **parallel** if they have the same slope, and **perpendicular** if their slopes are negative reciprocals of each other (meaning that $m_2 = -1/m_1$).

(a)   The line with slope -2 that passes through the point $(3, 1)$, in slope-intercept form.

(b)   The line that passes through the points $(-1, 2)$ and $(3 - 4)$, in slope-intercept form.

(c)   The line that passes through the point $(2, 0)$ and is perpendicular to the line $y = \frac{1}{3}x + 1$, in slope-intercept form.

SOLUTION

(a)   Using the point-slope formula with slope $m = -2$ and point $(x_0, y_0) = (3, 1)$, we get the equation $y - 1 = -2(x - 3)$. To put this equation into slope-intercept form, we just solve for y:

$$y - 1 = -2(x - 3) \implies y - 1 = -2x + 6 \implies y = -2x + 7.$$

(b)   The slope of the line through the points $(x_0, y_0) = (-1, 2)$ and $(x_1, y_1) = (3, 4)$ is $\frac{y_1 - y_0}{x_1 - x_0} = \frac{4 - 2}{3 - (-1)} = \frac{2}{4} = \frac{1}{2}$. Using this slope, the point $(-1, 2)$, and the two-point formula, we get

$$y - 2 = \frac{1}{2}(x - (-1)) \implies y - 2 = \frac{1}{2}(x + 1) \implies y = \frac{1}{2}x + \frac{5}{2}.$$

(c)   The line we are looking for is perpendicular to $y = \frac{1}{3}x + 1$ and thus has slope $m = -\frac{1}{1/3} = -3$. Using the point-slope form with slope $m = -3$ and point $(x_0, y_0) = (2, 0)$, we obtain $y - 0 = -3(x - 2)$. In slope-intercept form the equation is written $y = -3x + 6$.

☐

✔ CHECKING THE ANSWER

To verify that the three linear functions we just found are reasonable, we can graph them with a graphing utility (or by hand):

$y = -2x + 7$

$y = \frac{1}{2}x + \frac{5}{2}$

$y = -3x + 6 \text{ and } y = \frac{1}{3}x + 1$

Notice that the two lines in the rightmost graph don't look exactly perpendicular, because the scales on the x- and y-axes are different.

EXAMPLE 3   A real-world example of proportionality

Some people would argue that your expected grade point average G in a calculus course is proportional to the number of hours r spent reading the textbook each week.

(a)   Would you expect the proportionality constant k to be positive or negative in this situation, and why?

TEST YOUR UNDERSTANDING

▶ Why does the func

▶ Why does the func

▶ Which linear funct

▶ In part (b) of Exam to get the point-sl

▶ From memory and $y = x^3$, $y = \frac{1}{x}$, and

EXERCISES 0.5

Thinking Back

Solving inequalities: Find the solution sets of the foll equalities in two ways: first, by hand, using a sign ond, by using a graphing utility to examine the grap

▶ $x^2 - 4 \geq 0$ ▶ $x^2 - 5x + 6$

▶ $\frac{x}{x - 1} < 0$ ▶ $3 - \frac{1}{x} \geq 0$

Concepts

0. *Problem Zero:* Read the section and make your mary of the material.

1. *True/False:* Determine whether each of the stater follow is true or false. If a statement is true, ex If a statement is false, provide a counterexampl

  (a) *True or False:* Every linear function is a p function.

  (b) *True or False:* Every linear function is function.

  (c) *True or False:* Every constant function is function.

  (d) *True or False:* Every power function is a p function.

  (e) *True or False:* Every polynomial function is function.

  (f) *True or False:* Every constant function is function.

  (g) *True or False:* Every rational function is an function.

(b) If $k = \frac{2}{3}$, what wou
each week reading
the textbook to get

(c) For what values of

SOLUTION

(a) Presumably the cc
textbook would res

(b) If $k = \frac{2}{3}$, then the p
per week, we wou
average in this mo
textbook reading p

(c) Although the prop
and r, it makes rea
seen, $G = 4$ will oc

EXAMPLE 4 **Writing an absolute v**

Write the function $f(x)$
defined function to calc

SOLUTION

When $x^2 - 1$ is positive
ative, its absolute value

$$f(x) =$$

Although this is one wa
work with. For example
$x^2 - 1 \le 0$ when $x = -$
the conditions as inter
$x^2 - 1 < 0$ when $-1 <$

$$f(x$$

Now we can easily

$$f(-2) = (-2)$$
$$f(0) = -((0$$
$$f(1) = (1)^2 \cdot$$

Of course, we could ha
we are practicing writin
a skill that will come in
involve absolute values

The graph of $y = |$
$(-\infty, -1]$ and $[1, \infty)$.
site that of $x^2 - 1$, so i
x-axis, as the following

5. Suppose $f(x) = -3x + 1$ and that over a certain interval you have $\Delta x = 4$. What is the corresponding value of Δy? (*Hint: Think about the relationship between Δy, Δx, and the slope of a linear function.*)

6. Illustrate, on a graph of $f(x) = -5x + 1$, that the average rate of change of f on $[-1, 3]$ is -5. (Think about "rise over run.")

7. What is the slope of a horizontal line? What is the slope of a vertical line? Explain your answers in terms of average rates of change.

8. Label each of the six lines in the following figure with an approximate slope value (think about "rise over run"):

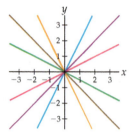

9. Without a calculator, and on the same set of axes, sketch graphs of the functions $f(x) = 3x - 1$, $g(x) = \frac{1}{10x+2}$, and $h(x) = 10x$. Label the coordinates of three points on each graph.

10. Without a calculator, and on the same set of axes, sketch graphs of the functions $f(x) = 1 - 2x$, $g(x) = \sqrt{x}$, and $h(x) = \frac{1}{x}$. Label the coordinates of three points on each graph.

11. Assuming that f is a linear function, use the values given in the following table to fill in the missing values:

x	1	3		7		10
$f(x)$	1	−5	−8		−23	

12. If y is proportional to x with proportionality constant k, is x necessarily proportional to y? If so, what is the proportionality constant? What if $k = 0$?

13. If $f(0) = 1$, $f(1) = 2$, and $f(4) = 5$, explain why f could not be a proportional function. Could f be a linear function? If not, explain why not. If so, find a linear equation for $f(x)$.

14. If $f(1) = 4$, $f(2) = 5$, and $f(4) = 6$, could f be a linear function? If not, explain why not. If so, find a linear equation for $f(x)$.

For each graph $y = f(x)$ shown, sketch the graph of $y = |f(x)|$.

15.

16.

17.

18.
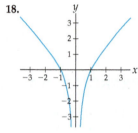

19. An alternative definition for the absolute value function is $|x| = \sqrt{x^2}$. Explain in your own words why this definition is equivalent to that of the absolute value function given in Definition 0.32. This exercise shows that the absolute value function is algebraic, not just piecewise algebraic.

20. Suppose that f is a function defined in pieces, using some functions g and h, as follows:

$$f(x) = \begin{cases} g(x), & \text{if } x \le 2 \\ h(x), & \text{if } x > 2. \end{cases}$$

Will the piece of the graph on $(-\infty, 2]$ always "match up" with the piece of the graph on $(2, \infty)$? Why or why not? Use graphs to illustrate your answer.

Skills

Determine whether each function f in Exercises 21–26 is a linear, power, polynomial, or rational function, or none of these. Some functions may satisfy more than one classification. For each function type that f belongs to, write $f(x)$ explicitly in general form as given in the table at the start of this section.

21. $f(x) = \frac{x^2}{5}$

22. $f(x) = x(x + 1)(x + 2)$

23. $f(x) = x^{-2} + x^{-3}$

24. $f(x) = x^2 - (x - 3)^2$

25. $f(x) = \frac{x^2 - 1}{x^3 + x + 2}$

26. $f(x) = \frac{x^3 + x}{x^2 + 1}$

Determine whether or not each of the functions in Exercises 27–32 are linear. For each linear function, find the slope and y-intercept.

27. $f(x) = \frac{3 - x}{8}$

28. $f(x) = -2(1 + x) + 7$

29. $f(x) = 3x(2x + 1)$

30. $x - 2y = \frac{1}{2}$

31. $y - 3 = 2(x + 5)$

32. $\frac{4 + x}{2y} = 3$

Find equations for each of the functions described in Exercises 33–44.

33. The linear function whose graph has slope -1 and passes through the point $(3, -2)$.

34. The linear function whose graph passes through the points $(-2, 1)$ and $(3, -4)$.

35. The linear function whose graph has slope 6 and y-intercept -3.

36. The linear function whose graph has y-intercept 5 and passes through $(1, 2)$.

37. The linear function whose graph is perpendicular to $y = -0.5x + 2$ and passes through the point $(1, 0)$.

38. The linear function whose graph is parallel to $y = 15x - 7$ and passes through the point $(-1, 8)$.

39. The linear function whose graph is parallel to $y = 2x + 1$ and passes through the point $(-1, 4)$.

40. A power function whose graph passes through $(0, 0)$ and $(1, 3)$.

41. A power function whose graph passes through $(0, 0)$ and $(2, 8)$.

42. A polynomial function whose graph passes through $(-2, 0)$, $(1, 0)$, and $(3, 0)$.

43. A polynomial function whose graph passes through $(0, 0)$, $(2, 0)$, and $(4, 0)$.

44. A polynomial function whose graph passes through $(0, 0)$, $(2, 0)$, $(4, 0)$, and $(1, 2)$.

Use a graphing utility to graph each of the functions in Exercises 45–52. Choose your graphing window so that the key features of the graph are represented, and label the coordinates of three points on each graph.

45. $f(x) = 3e^{0.5x}$

46. $f(x) = 2^x$

47. $f(x) = \log_2 x$

48. $f(x) = \ln x$

49. $f(x) = \sin x$

50. $f(x) = \cos x$

51. $f(x) = \tan^{-1} x$

52. $f(x) = \sin^{-1} x$

Write each function in Exercises 53–60 as a piecewise-defined function where each piece is defined on an interval of x-values. Then sketch a labeled graph of each piecewise-defined function.

53. $f(x) = |5 - 3x|$

54. $f(x) = |1.5x + 2.3|$

55. $f(x) = |x^2 + 1|$

56. $f(x) = |x^2 - 4|$

57. $f(x) = |9 - x^2|$

58. $f(x) = |3 - 4x + x^2|$

59. $f(x) = |x^2 - 3x - 4|$

60. $f(x) = |x^3 + x^2 - 2x|$

Applications

For each scenario described in Exercises 61–64, identify and name the dependent and independent variables and use function notation to describe the situation.

61. The number of hours per week that Tim spends watching television is proportional to the amount of credit card debt that he has.

62. A grapefruit is dropped from the top of a tall building. The velocity of the grapefruit (in meters per second) is proportional to the number of seconds it has been falling, with proportionality constant 9.8. What is the velocity of the grapefruit at the moment it is released? After 2 seconds?

63. The population on an island can be expressed as a linear function of time. In 1985 there were 35 people on the island. In the year 2000 the population had increased to 91. According to this model, in what year will there be 200 people on the island?

64. Calvin captured 15 fireflies on Tuesday and put them in a jar. Since then there have been 3 new fireflies in the jar every day. Assuming this growth rate continues, how many fireflies will Calvin have on Sunday?

65. A man who lives on your street insists that the number of Twinkies produced each year is proportional to the amount that the U.S. government spends on national security. If 500 million Twinkies were produced last year and the U.S. government spent 25 billion dollars on national security, and if this man is correct about this proportional relationship, then what is the proportionality constant that relates Twinkies to the national security budget?

66. The table that follows describes the monthly rent for Linda's New York City apartment in the years 1995–2000. Could a linear function go through all of these data points? If not, why not? If so, use this linear function to predict the price of Linda's apartment this year.

Year	1995	1996	1997	1998	2000
Rent	$1,475	$1,537	$1,599	$1,661	$1,785

67. The table below shows the number of compact discs that Alina owned at the end of certain months last year. Could a linear function go through all of these data points? If not, why not? If so, use this linear function to determine how many more months after December it will take before Alina has a collection of 200 CDs.

Month	Jan.	April	June	July	Sept.	Dec.
CDs	51	60	66	69	75	84

68. A department store is having trouble selling a particular jacket. In desperation the store reduces the price of the jacket each Friday in the hopes that someone will eventually find it a bargain and purchase it. The table that follows shows the price of the jacket t weeks after the store starts to mark it down. Could a linear function go through all of these data points? If not, why not? If so, use this linear function to determine when the jacket will cost only $50.

Week	1	2	4	5	7	8
Price	$150	$135	$120	$105	$90	$75

For example, if $f(4) = 6$ and $g(10) = 4$, then $(f \circ g)(10) = f(g(10)) = f(4) = 6$. You should think of compositions as nestings of functions. The notation $(f \circ g)$ is pronounced "f composed with g" or sometimes "f circle g." The notation $f(g(x))$ is pronounced "f of g of x."

If $g: X \to Y$ and $f: Y \to Z$, then their composition is a function $(f \circ g): X \to Z$ that takes an input x first to $g(x)$ and then to $f(g(x))$. For example, if $f(x) = x^2$ and $g(x) = 3x + 1$ then

$$(f \circ g)(x) = f(g(x)) = f(3x + 1) = (3x + 1)^2.$$

Notice that although the function f appears first (i.e., on the left) in the notation, it is the function g that gets applied to the input x first. Composition is not a **commutative** operation, which means that $f \circ g$ is not necessarily the same function as $g \circ f$. With the same example of $f(x) = x^2$ and $g(x) = 3x + 1$, if we compose in the other order, we get

$$(g \circ f)(x) = g(f(x)) = g(x^2) = 3(x^2) + 1.$$

 CAUTION | You may have noticed that the notation for composition looks a bit like multiplication notation, but there is a key difference. When we want to denote multiplication we will use a small closed dot or no dot at all. To denote composition of functions we will always use an open circle.

Transformations and Symmetry

Another way we can obtain new functions from old is through **transformations**. Given a function $f(x)$ and constants C and k, we could consider such modifications as $f(x) + C$, $f(x + C)$, $kf(x)$, and $f(kx)$. Each of these transformations changes $f(x)$ graphically and algebraically.

For example, transforming $f(x)$ to $f(x) \pm C$ clearly adds C units to every output of $f(x)$. This means that the graph of $y = f(x)$ shifts up or down vertically C units everywhere, to become the graph of $y = f(x) + C$ or $y = f(x) - C$, as illustrated by the red and green graphs, respectively, shown in the figure next at the left. If we instead add a constant to the independent variable and transform $f(x)$ to $f(x \pm C)$, the graph shifts left or right horizontally by C units, as illustrated in the green and red graphs shown in the figure at the right. (Note that the shift to the left for $f(x + 2)$ and to the right for $f(x - 2)$ might be the opposite of what we might initially expect.) These additive transformations are called **translations**.

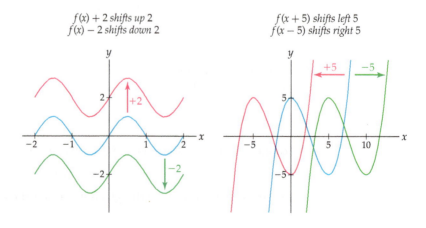

$f(x) + 2$ *shifts up 2*
$f(x) - 2$ *shifts down 2*

$f(x + 5)$ *shifts left 5*
$f(x - 5)$ *shifts right 5*

If we instead transform $f(x)$ by multiplication to $kf(x)$, then the graph of $y = f(x)$ expands or contracts vertically by a factor of k to become the graph of $y = kf(x)$, as shown in the red and green graphs next at the left. In contrast, if we do the same transformation to the independent variable and transform $f(x)$ to $f(kx)$, this contracts or expands the graph of $y = f(x)$ by a factor of k in the horizontal direction, as illustrated in the red and green graphs next at the right.

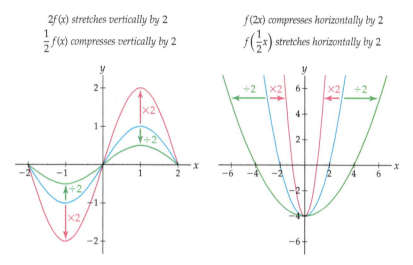

$2f(x)$ *stretches vertically by* 2

$\frac{1}{2}f(x)$ *compresses vertically by* 2

$f(2x)$ *compresses horizontally by* 2

$f\left(\frac{1}{2}x\right)$ *stretches horizontally by* 2

What happens if we multiply x or $f(x)$ by a negative number? We can answer that question by just looking at what happens when we multiply by -1. Changing $f(x)$ to $-f(x)$ transforms all positive outputs into negative outputs, and vice versa. The graph of $y = f(x)$ is then reflected across the x-axis to become the graph of $y = -f(x)$, as shown in the red graph in the figure that follows. If we instead multiply the independent variable by -1, then we obtain a reflection across the y-axis, as shown in the green graph.

$-f(x)$ *reflects across the x-axis*

$f(-x)$ *reflects across the y-axis*

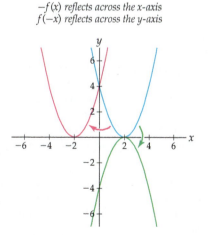

Now if we want to transform $f(x)$ to $f(-2x)$, for example, we can transform $f(x)$ first to $f(2x)$ and then by reflection to $f(-2x)$.

The table that follows summarizes the graphical and algebraic effects of the transformations just discussed. You will prove these general results in Exercises 86–88.

Transformation	Graphical Result	Algebraic Result
$f(x) + C$	shifts up C units if $C > 0$ shifts down C units if $C < 0$	$(x, y) \rightarrow (x, y + C)$
$f(x + C)$	shifts left C units if $C > 0$ shifts right C units if $C < 0$	$(x, y) \rightarrow (x - C, y)$
$k f(x)$	vertical stretch by k if $k > 1$ vertical compression by k if $0 < k < 1$	$(x, y) \rightarrow (x, ky)$
$f(kx)$	horizontal compression by k if $k > 1$ horizontal stretch by k if $0 < k < 1$	$(x, y) \rightarrow \left(\frac{1}{k}x, y \right)$
$-f(x)$	graph reflects across the x-axis	$(x, y) \rightarrow (x, -y)$
$f(-x)$	graph reflects across the y-axis	$(x, y) \rightarrow (-x, y)$

Some graphs do not change under certain transformations. For example, the graph of $f(x) = x^2$ shown next at the left remains the same if we reflect it across the y-axis. We say that this function has **y-axis symmetry**. As another example, the graph of $g(x) = x^3$ shown at the right remains the same if we reflect it first across the y-axis and then across the x-axis.

$f(x) = x^2$ *preserved under y-axis reflection* $g(x) = x^3$ *preserved under 180° rotation*

 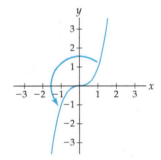

It turns out that the double-reflection we just described for $g(x) = x^3$ is equivalent to rotation around the origin by 180°. You can try this equivalence out for yourself by physically double-reflecting and rotating the book while looking at the preceding graph of $g(x) = x^3$. You can also see the equivalence by using a piece of paper with a smiley-face drawn on the front: Flipping the paper vertically and then horizontally is equivalent to rotating the paper by 180 degrees. A function that is preserved under the transformation of 180° rotation is said to have 180° **rotational symmetry**.

These types of symmetries are also called **even symmetry** and **odd symmetry**, since power functions with even powers all have y-axis symmetry and power functions with odd powers all have rotational symmetry. Because graphical reflections correspond to multiplication by -1, we can describe functions with even and odd symmetry algebraically as follows:

DEFINITION 0.36   **Even and Odd Functions**

A function f is an **even function** if $f(-x) = f(x)$ for all x in the domain of f.

A function f is an **odd function** if $f(-x) = -f(x)$ for all x in the domain of f.

For example, the function $f(x) = x^2$ is even because for all x we have

$$f(-x) = (-x)^2 = x^2 = f(x).$$

In contrast, the function $g(x) = x^3$ is odd because for all x we have

$$g(-x) = (-x)^3 = -(x^3) = -g(x).$$

Note that some functions are neither even nor odd; for example $h(x) = x^2 + x$ is one such function, because $h(-x) = (-x)^2 + (-x) = x^2 - x$, which is equal neither to $h(x)$ nor to $-h(x)$. Consequently, the function $h(x) = x^2 + x$ has neither y-axis symmetry nor rotational symmetry.

Inverse Functions

The ***inverse*** of a function f is a function that undoes the action of f. For example, the function that adds 1 to each real number can be undone by subtracting 1 from each real number. If two functions undo each other, then composing them results in the identity function. This property suggests the following definition:

DEFINITION 0.37

The Inverse of a Function

If f and g are functions such that

$$g(f(x)) = x, \quad \text{for all } x \text{ in the domain of } f$$
$$f(g(x)) = x, \quad \text{for all } x \text{ in the domain of } g$$

then g is the ***inverse*** of f and we denote g by f^{-1}.

Note that the two conditions in this definition guarantee that if a function g is the inverse of a function f, then f is the inverse of the function g.

For example, the functions $f(x) = x^3$ and $g(x) = \sqrt[3]{x}$ are inverses of each other. Intuitively, taking the cube of a number is undone by taking the cube root, and vice versa. It is easy to verify that these two functions satisfy the condition in Definition 0.37:

$$g(f(x)) = g(x^3) = \sqrt[3]{x^3} = x,$$
$$f(g(x)) = f(\sqrt[3]{x}) = (\sqrt[3]{x})^3 = x.$$

⊗ CAUTION

It is important to note that although we use the notation x^{-1} to denote the reciprocal $\frac{1}{x}$, the notation f^{-1} does *not* stand for the reciprocal $\frac{1}{f}$ of f. The notation f^{-1} used in Definition 0.37 is pronounced "f inverse." We are now using the same notation for two very different things, but it should be clear from the context which one we mean.

Not all functions have inverses. For example, consider the squaring function $f(x) = x^2$ on the domain $\mathbb{R}$. Since $f(2) = 4$ and $f(-2) = 4$, an inverse of f would have to send the input 4 to both 2 and -2; this is clearly not a function. As the following theorem asserts, a function has an inverse only if no two inputs are ever sent to the same output:

THEOREM 0.38

Invertible Functions are One-to-One

A function f has an inverse if and only if f is one-to-one.

Since a function is one-to-one if and only if it passes the horizontal line test, this theorem implies that a function has an inverse if and only if it passes the horizontal line test. The proof follows directly from the properties of inverse functions:

Proof. We first prove that if a function is invertible, then it must be one-to-one. If f is an invertible function, then it has an inverse f^{-1}. Suppose two domain values a and b are sent by f to the same output $f(a) = f(b)$. Applying f^{-1} to both sides, we have

$$f(a) = f(b) \implies f^{-1}(f(a)) = f^{-1}(f(b)) \implies a = b.$$

For the converse, suppose f is one-to-one. Then each element b in the range of f is the output of *exactly one* element a from the domain. We can define $f^{-1}(b)$ to be this element a. Since f is one-to-one, this new relationship f^{-1} will be a function and we will have $f^{-1}(b) = a$ if and only if $f(a) = b$. Therefore if f is one-to-one, then f has an inverse. ∎

The properties of inverses given in the next theorem follow directly from our definition of an inverse function, that is, from the fact that f^{-1} undoes the function f. Functions that have inverses are said to be ***invertible***.

THEOREM 0.39

Properties of Inverses

If f is an invertible function with inverse f^{-1}, then the following statements hold.

(a) Domain(f^{-1}) = Range(f) and Range(f^{-1}) = Domain(f).

(b) $f^{-1}(b) = a$ if and only if $f(a) = b$.

(c) The graph of $y = f^{-1}(x)$ is the graph of $y = f(x)$ reflected across the line $y = x$.

Proof. The proofs of parts (a) and (c) are left to Exercises 91 and 92. To prove part (b), suppose $f^{-1}(b) = a$. Applying f to both sides, we have $f(f^{-1}(b)) = f(a)$. Since f and f^{-1} are inverses, their composition is the identity function. Therefore we have $b = f(a)$, as desired. With an entirely similar argument we can show that if $f(a) = b$, then $f^{-1}(b) = a$; see Exercise 93. ∎

For example, consider the one-to-one function $f(x) = \sqrt{x+1}$. To find a function that undoes $f(x)$ we solve $y = \sqrt{x+1}$ for y, obtaining $x = y^2 - 1$. Changing notation so that x is again the independent variable, we see that the inverse of $f(x)$ is $f^{-1}(x) = x^2 - 1$. The following three figures illustrate the properties from Theorem 0.39 for $f(x) = \sqrt{x+1}$.

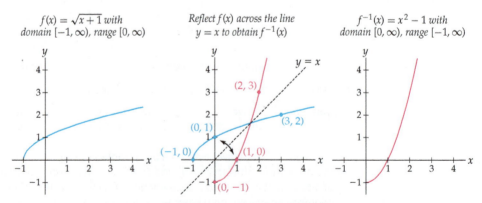

Sometimes a function is not invertible on its largest domain, but *is* invertible on some smaller, "restricted" domain. For example, the function $g(x) = x^2 - 1$ is not invertible on its usual domain $\mathbb{R}$, but it is invertible on the restricted domain $[0, \infty)$. In general, to find a restricted domain on which a function is one-to-one, we choose a smaller domain on which the graph passes the horizontal line test; see Example 4.

Examples and Explorations

EXAMPLE 1

Combinations of functions and their domains

Describe the domains of each of the combinations of $f(x) = \frac{1}{x}$ and $g(x) = \sqrt{x+1}$ that follow. Then find an expression for the combination function.

(a) $3f$ (b) $f + g$ (c) $\frac{f}{g}$ (d) $f \circ g$ (e) $g \circ f$

SOLUTION

(a) The domain of $3f$ is the same as the domain of f, which is $x \neq 0$. We have $(3f)(x) = 3f(x) = 3\left(\frac{1}{x}\right) = \frac{3}{x}$.

(b) The domain of f is $x \neq 0$, and the domain of g is $x \geq -1$. The domain of their sum $f + g$ is the intersection of these domains, or $[-1, 0) \cup (0, \infty)$. For values of x in this domain we have $(f + g)(x) = f(x) + g(x) = \frac{1}{x} + \sqrt{x+1}$.

(c) We have $g(x) = 0$ only when $x = -1$, so the quotient $\frac{f}{g}$ is defined on the intersection of the domains of f and g with the point $x = -1$ removed, or $(-1, 0) \cup (0, \infty)$. On this domain we have

$$\left(\frac{f}{g}\right)(x) = \frac{f(x)}{g(x)} = \frac{1/x}{\sqrt{x+1}} = \frac{1}{x\sqrt{x+1}}.$$

(d) For a value x to be in the domain of $f \circ g$, it must first be in the domain $[-1, \infty)$ of g. Then the value of $g(x) = \sqrt{x+1}$ must be in the domain of f, so we must have $\sqrt{x+1} \neq 0$, or in other words, $x \neq -1$. Therefore the domain of $f \circ g$ is $(-1, \infty)$. For values of x in this domain we have $(f \circ g)(x) = f(g(x)) = f(\sqrt{x+1}) = \frac{1}{\sqrt{x+1}}$. Notice that this equation is consistent with our calculation of the domain.

(e) For a value x to be in the domain of $g \circ f$, it must first be in the domain $(-\infty, 0) \cup (0, \infty)$ of f. Then the value of $f(x) = \frac{1}{x}$ must be in the domain of g, so we must have $\frac{1}{x} \in [-1, \infty)$. Since $\frac{1}{x} \geq -1$ when $x \leq -1$ or $x > 0$, the domain of $g \circ f$ is $(-\infty, -1] \cup (0, \infty)$. For values of x in this domain we have $(g \circ f)(x) = g(f(x)) = g\left(\frac{1}{x}\right) = \sqrt{\frac{1}{x} + 1}$. Again notice that the domain that we found does make sense with this equation.   □

EXAMPLE 2

Vertical and horizontal translations, stretches, and reflections

The figure that follows shows a piece of the graph of $f(x) = 3 + 2x - x^2$ with five marked points. Find equations and graphs for each of the given transformations. On each graph, mark the new coordinates of the five marked points.

(a) $f(x) + 3, f(x) - 3, f(x + 3),$ and $f(x - 3)$

(b) $2f(x), \frac{1}{2}f(x), f(2x),$ and $f\left(\frac{1}{2}x\right)$

(c) $-f(x)$ and $f(-x)$

SOLUTION

(a) The equations for the four transformations are

$$f(x) + 3 = (3 + 2x - x^2) + 3 = 6 + 2x - x^2,$$
$$f(x) - 3 = (3 + 2x - x^2) - 3 = 2x - x^2,$$
$$f(x + 3) = 3 + 2(x + 3) - (x + 3)^2 = -4x - x^2, \text{ and}$$
$$f(x - 3) = 3 + 2(x - 3) - (x - 3)^2 = -12 + 8x - x^2.$$

The graphs of the transformations are shifts up, down, left, and right, respectively, of the original graph by 3 units, as shown in the red graphs in the following figures:

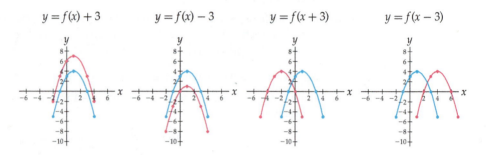

$$y = f(x) + 3 \qquad y = f(x) - 3 \qquad y = f(x + 3) \qquad y = f(x - 3)$$

In each case, every point on the original graph is shifted in some direction by 3 units. For example, the point $(1, 4)$ on the original graph becomes $(1, 7)$ on the graph of $f(x) + 3$, $(1, 1)$ on the graph of $f(x) - 3$, $(-2, 4)$ on the graph of $f(x + 3)$, and $(4, 4)$ on the graph of $f(x - 3)$. The other four marked points move in a similar fashion.

(b) Algebraically, the four transformations are given by the equations

$$2f(x) = 2(3 + 2x - x^2) = 6 + 4x - 2x^2,$$
$$\frac{1}{2}f(x) = \frac{1}{2}(3 + 2x - x^2) = \frac{3}{2} + x - \frac{1}{2}x^2,$$
$$f(2x) = 3 + 2(2x) - (2x)^2 = 3 + 4x - 4x^2, \text{ and}$$
$$f\left(\frac{1}{2}x\right) = 3 + 2\left(\frac{1}{2}x\right) - \left(\frac{1}{2}x\right)^2 = 3 + x - \frac{1}{4}x^2.$$

The first two transformations cause the graph of f to stretch or compress vertically, and the last two cause the graph to compress or stretch horizontally, as shown in the red graphs in the next four figures.

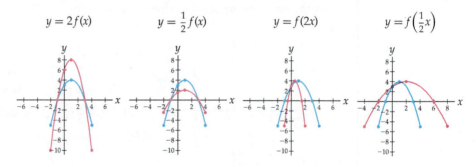

$$y = 2f(x) \qquad y = \frac{1}{2}f(x) \qquad y = f(2x) \qquad y = f\left(\frac{1}{2}x\right)$$

In the first two transformations it is the y-coordinate that changes; for example, the point $(1, 4)$ on the original graph moves to $(1, 8)$ for the first transformation and to $(1, 2)$ for the second. For the last two transformations, it is the x-coordinate that changes; for example, $(1, 4)$ moves to $\left(\frac{1}{2}, 4\right)$ for the third transformation and to $(2, 4)$ for the fourth. The other marked points move in a similar fashion, with either the x- or the y-coordinate being multiplied or divided by 2.

(c) The equations for the two transformations are

$$-f(x) = -(3 + 2x - x^2) = -3 - 2x + x^2 \text{ and}$$
$$f(-x) = 3 + 2(-x) - (-x)^2 = 3 - 2x - x^2.$$

The first transformation gives a vertical reflection across the x-axis, with each marked point (x, y) moving to the point $(x, -y)$, as shown next at the left. The second transformation causes a horizontal reflection across the y-axis, with each marked point (x, y) moving to the point $(-x, y)$, as shown at the right.

$y = -f(x)$

$y = f(-x)$

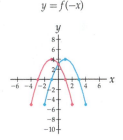

EXAMPLE 3

Testing if functions are even or odd

Determine whether each of the following functions is even, odd, or neither:

(a) $f(x) = \dfrac{1}{x}$ **(b)** $g(x) = x^4 - x^2$ **(c)** $h(x) = \dfrac{2 + x}{1 + x^2}$

SOLUTION

(a) To determine whether f is even or odd (or neither) we must calculate $f(-x)$ and determine if it is equal to $f(x)$, $-f(x)$, or neither. We have

$$f(-x) = \frac{1}{-x} = -\frac{1}{x} = -f(x),$$

so $f(x) = \dfrac{1}{x}$ is an odd function.

(b) The function $g(x)$ is even because

$$g(-x) = (-x)^4 - (-x)^2 = x^4 - x^2 = g(x).$$

(c) Since we have

$$h(-x) = \frac{2 + (-x)}{1 + (-x)^2} = \frac{2 - x}{1 + x^2},$$

which is equal neither to $h(x)$ nor to $-h(x)$, this function is neither even nor odd.

The graphs of the functions f, g, and h are shown next. Note that f has rotational symmetry about the origin, g has y-axis reflectional symmetry, and h does not have either type of symmetry.

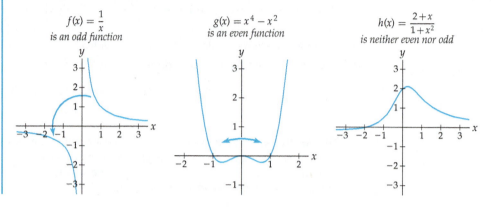

$f(x) = \dfrac{1}{x}$
is an odd function

$g(x) = x^4 - x^2$
is an even function

$h(x) = \dfrac{2+x}{1+x^2}$
is neither even nor odd

EXAMPLE 4

Graphically finding a restricted domain on which a function is invertible

Explain why the function f graphed here is not invertible on its domain. Then find three restricted domains on which the function *does* have an inverse.

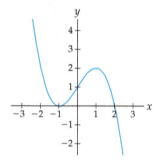

SOLUTION

This function is not invertible on the domain $\mathbb{R}$, because it does not pass the horizontal line test and therefore fails to be a one-to-one function. However, small enough pieces of the graph of f do pass the horizontal line test and therefore have an inverse on that restricted domain. The three graphs that follow show the graph of f on the restricted-domain domains $[-1, 1]$, $(-\infty, -1]$, and $[1, \infty)$, respectively. All three of these restricted graphs pass the horizontal line test and thus are invertible.

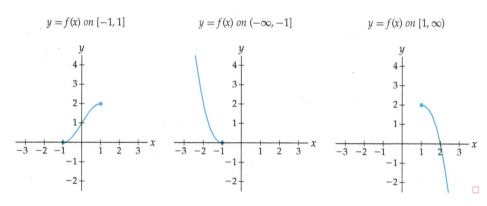

$y = f(x)$ on $[-1, 1]$

$y = f(x)$ on $(-\infty, -1]$

$y = f(x)$ on $[1, \infty)$

| EXAMPLE 5 | **The graphical and algebraic relationships of inverse functions** |

Consider the function $f(x) = 2x - 2$.

(a) Explain why f must have an inverse. Then sketch graphs of $y = f(x)$ and $y = f^{-1}(x)$ and explain the relationship between these two graphs.

(b) Find a formula for $f^{-1}(x)$.

(c) Algebraically verify that $f(f^{-1}(x)) = x$ and $f^{-1}(f(x)) = x$.

SOLUTION

(a) $f(x) = 2x - 2$ has an inverse because it is a non-constant linear function and thus passes the horizontal line test and is one-to-one. By reflecting the graph of $y = f(x)$ across the line $y = x$, we obtain the graph of this inverse; see the following figures:

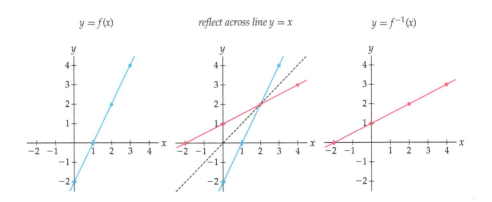

Whenever (x, y) is a point on the graph of f, the point (y, x) is on the graph of f^{-1}. For example, in the preceding graphs we see that $(2, 2)$ and $(3, 4)$ are on the graph of f while $(2, 2)$ and $(4, 3)$ are on the graph of f^{-1}.

(b) We could find a formula for f^{-1} by using the two-point form of a line and the two points shown in the rightmost figure. However, we will find it by solving the equation $y = 2x - 2$ for x:

$$y = 2x - 2 \implies y + 2 = 2x \implies x = \frac{1}{2}(y + 2) = \frac{1}{2}y + 1.$$

If $y = f(x)$, then $x = f^{-1}(y)$; thus we have shown that $f^{-1}(y) = \frac{1}{2}y + 1$. Since we would rather represent the independent variable of f^{-1} by the traditional letter x, replace the y's in the equation with x's to get $f^{-1}(x) = \frac{1}{2}x + 1$. Notice that this equation does appear reasonable, given the slope and y-intercept in the rightmost graph.

(c) For $f(x) = 2x - 2$ and $f^{-1}(x) = \frac{1}{2}x + 1$, we have

$$f(f^{-1}(x)) = f\left(\frac{1}{2}x + 1\right) = 2\left(\frac{1}{2}x + 1\right) - 2 = x + 2 - 2 = x,$$

$$f^{-1}(f(x)) = f^{-1}(2x - 2) = \frac{1}{2}(2x - 2) + 1 = x - 1 + 1 = x.$$

17. If an invertible function f has domain $[-1, 1)$ and range $(-\infty, 3]$, what are the domain and range of f^{-1}? Sketch the graph of a function f with the domain and range given, and on the same set of axes sketch the graph of its inverse f^{-1}.

18. Consider the function $f(x) = x^5 + 4x^3 + 2x + 1$. Alina says that f is invertible, because its graph appears to be one-to-one. Linda says that f is not invertible, because she cannot figure out how to solve $y = x^5 + 4x^3 + 2x + 1$ for x in terms of y. Who is correct, and why?

19. Use the values given in the table to fill in the missing values. There is only one correct way to fill in the table.

x	0	1	2
$f(x)$	1		
$g(x)$	0		-2
$h(x)$		1	
$(-2f)(x)$		4	
$(f + 2g - h)(x)$	4		
$(fg)(x)$			2
$\left(\frac{g}{h}\right)(x)$		3	-1

20. Use the values given in the table to fill in the missing values. There is only one correct way to fill in the table.

x	1	2	3	4
$f(x)$	4	3	1	2
$g(x)$	3		1	2
$h(x)$		2	4	
$(f \circ g)(x)$		4		
$(h \circ f)(x)$	3			2
$(g \circ g)(x)$				
$(g \circ f \circ h)(x)$	2			

Skills

Given that $f(x) = x^2 + 1$, $g(x) = \dfrac{1}{x-2}$, and $h(x) = \sqrt{x}$, find the domain of each function in Exercises 21–29. Then find an equation for the function and calculate its value at $x = 1$.

21. $(f + g)(x)$ **22.** $(3fh)(x)$ **23.** $\left(\dfrac{g}{h}\right)(x)$

24. $(g \circ f)(x)$ **25.** $(g \circ g)(x)$ **26.** $(g \circ h \circ f)(x)$

27. $g(x - 5)$ **28.** $h(3x) + 1$ **29.** $h(3x + 1)$

The table that follows defines three functions f, g, and h. Create additional rows for the table for each function in Exercises 30–38. (For some transformations, you may have to use different x-values than the ones in the table.)

x	0	1	2	3	4	5	6
$f(x)$	0	1	3	2	3	0	2
$g(x)$	1	0	1	1	0	1	0
$h(x)$	3	2	0	3	2	3	1

30. $(f - g)(x)$ **31.** $2f(x) + 3$ **32.** $(gh)(x)$

33. $(h \circ g)(x)$ **34.** $(g \circ h)(x)$ **35.** $(f \circ f \circ f)(x)$

36. $-h(-x)$ **37.** $g(x - 1)$ **38.** $f(2x)$

Use the graphs of f and g given here to sketch the graphs of the functions in Exercises 39–50. Label at least four points on each graph. Don't find or use equations for the given graphs.

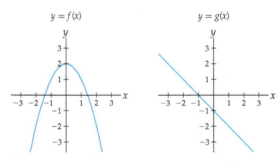

39. $(f + g)(x)$ **40.** $2 - 3f(x)$ **41.** $g\left(\dfrac{1}{2}x\right)$

42. $g(x - 2) + 1$ **43.** $(-.5f)(x)$ **44.** $(fg)(x)$

45. $(f \circ g)(x)$ **46.** $(g \circ f)(x)$ **47.** $(f \circ f)(x)$

48. $g^{-1}(x)$ **49.** $\left(\dfrac{1}{f}\right)(x)$ **50.** $\left(\dfrac{1}{g}\right)(x)$

Suppose f and g are the piecewise-defined functions defined here. For each combination of functions in Exercises 51–56, (a) find its values at $x = -1$, $x = 0$, $x = 1$, $x = 2$, and $x = 3$, (b) sketch its graph, and (c) write the combination as a piecewise-defined function.

$$f(x) = \begin{cases} 2x + 1, & \text{if } x \le 0 \\ x^2, & \text{if } x > 0 \end{cases} \qquad g(x) = \begin{cases} -x, & \text{if } x < 2 \\ 5, & \text{if } x \ge 2 \end{cases}$$

51. $(f + g)(x)$ **52.** $3f(x)$ **53.** $(g \circ f)(x)$

54. $f(x) - 1$ **55.** $f(x - 1)$ **56.** $g(3x)$

Find two nontrivial ways to write each of the functions in Exercises 57–60 as a composition $f = g \circ h$. "Nontrivial" means that you should *not* choose $g(x) = x$ or $h(x) = x$.

57. $f(x) = 3x^2 + 5$ **58.** $f(x) = (3x + 5)^2$

59. $f(x) = \dfrac{6}{x + 1}$ **60.** $f(x) = \dfrac{x^2}{\sqrt{1 + x^2}}$

Determine algebraically whether the functions in Exercises 61–66 are even, odd, or neither. Afterwards, verify your answers by inspecting the symmetries of the graphs.

61. $f(x) = x^4 + 1$ **62.** $f(x) = 1 - 2x$

63. $f(x) = x^3 + x^2$ **64.** $f(x) = -\dfrac{2}{5x^3}$

65. $f(x) = \dfrac{x^3}{x^2 + 1}$ **66.** $f(x) = \dfrac{1}{x + 1}$

Use Definition 0.37 to show that each pair of functions in Exercises 67–70 are inverses of each other.

67. $f(x) = 2 - 3x$ and $g(x) = -\dfrac{1}{3}x + \dfrac{2}{3}$

68. $f(x) = x^2$ restricted to $[0, \infty)$ and $g(x) = \sqrt{x}$

69. $f(x) = \dfrac{x}{1 - x}$ and $g(x) = \dfrac{x}{1 + x}$

70. $f(x) = \dfrac{1}{2x}$ and $g(x) = \dfrac{1}{2x}$

For each invertible function in Exercises 71–74, sketch the graph of f^{-1} and label three points on its graph. If a function is not invertible, then find a restricted domain on which it is invertible and sketch a graph of the restricted inverse. Don't find or use equations for the given graphs.

71.

72.

73.

74.
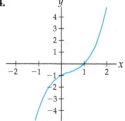

Find the inverse of each function in Exercises 75–80. Then check your answers by graphing f and f^{-1}.

75. $f(x) = \dfrac{1 - 5x}{2}$ **76.** $f(x) = 1.2 - 3.5x$

77. $f(x) = 1 + \dfrac{1}{x}$ **78.** $f(x) = \dfrac{2}{x + 1}$

79. $f(x) = \dfrac{x - 1}{x + 1}$ **80.** $f(x) = \dfrac{x - 2}{x}$

Applications

81. CarpetKing charges \$4.25 per square foot for its Deluxe ThriftySoft carpet, plus a flat fee of \$200.00 for delivery and installation.

(a) If a square room measures x feet on a side, and S is the number of square feet of floor in the room, write down S as a function of x.

(b) Write down a function that describes the cost C of carpeting a room enclosing S square feet.

(c) Write down a function that describes the cost C of carpeting a square room that measures x feet on a side. Explain how this function is a composition.

82. The first table that follows shows the number of deer in Happyland Forest Park during 1990–1995. The number of deer seems to affect the number of ticks in the park; the second table shows the number of ticks that can be expected for various numbers of deer in the park.

Year	1990	1991	1992	1993	1994	1995
Deer	183	180	177	179	184	181

Deer	177	178	179	180	181	182	183	184
Ticks	850	855	860	865	870	875	880	885

(a) Use the tables to estimate the number of ticks that were in the park in 1995.

(b) Make a table that predicts the number of ticks in the park during 1990–1995.

(c) Explain how your table represents a composition of the deer and tick tables.

83. Juri's custom printing t-shirt shop charges \$12.00 per t-shirt plus a \$20.00 setup fee.

(a) Write a function that describes the cost C of printing n t-shirts at Juri's store.

(b) Show that $C(n)$ is an invertible function.

(c) Find the inverse of $C(n)$. What does this inverse function $n(C)$ represent in terms of C and n?

(d) Use the function you found in part (c) to calculate the number of shirts you can produce with \$150.

84. The math teachers at Pinnacle High School have discovered a relationship between the number of hours their students watch television each day and their students' scores on math tests. The following table shows this relationship:

Hours of T.V.	0	1	2	3	4	5	6
Grade on test	92	86	80	74	68	62	56

(a) Find a linear function that describes the grade g on a student's math test as a function of the number of hours t of television that the student watches every day.

(b) Show that your function $g(t)$ is invertible.
(c) Find the inverse of $g(t)$. What does this function $t(g)$ represent in terms of t and g?
(d) Use $t(g)$ to predict the number of hours that Calvin watched television each day given that he scored a 40 on his math test. How much television can he be allowed to watch each day if his mother wants him to get an 85 on his next test?

Proofs

85. Prove algebraically that if $f(x) = x^k$, where k is a positive integer, then the graphs of $y = f(2x)$ and $y = 2^k f(x)$ are the same.

86. Prove that if (x, y) is a point on the graph of $y = f(x)$ and C is a real number, then

(a) $(x, y + C)$ is a point on the graph of $y = f(x) + C$.
(b) $(x - C, y)$ is a point on the graph of $y = f(x + C)$.

87. Prove that if (x, y) is a point on the graph of $y = f(x)$ and $k \neq 0$, then

(a) (x, ky) is a point on the graph of $y = kf(x)$.
(b) $\left(\frac{1}{k}x, y\right)$ is a point on the graph of $y = f(kx)$.

88. Prove that if (x, y) is a point on the graph of $y = f(x)$, then

(a) $(x, -y)$ is a point on the graph of $y = -f(x)$.
(b) $(-x, y)$ is a point on the graph of $y = f(-x)$.

89. Prove that every odd function that is defined at $x = 0$ must pass through the origin.

90. Prove that the algebraic definitions of even and odd functions imply even and odd graphical symmetry, by showing that:

(a) If (x, y) is a point on the graph of an even function $y = f(x)$, then $(-x, y)$ is also on the graph.
(b) If (x, y) is a point on the graph of an odd function $y = f(x)$, then $(-x, -y)$ is also on the graph.

91. Prove Theorem 0.39 (a) by using Definition 0.37 to argue that the domain of f^{-1} is the range of f and that the range of f^{-1} is the domain of f.

92. Prove Theorem 0.39 (c) by showing that if f is a function with an inverse f^{-1} and (x, y) is on the graph of $y = f(x)$, then (y, x) is on the graph of $y = f^{-1}(x)$. Why does this conclusion imply that the graph of $f^{-1}(x)$ is the reflection of the graph of $f(x)$ across the line $y = x$?

93. Suppose f is invertible with inverse f^{-1}. Prove that if $f(a) = b$, then $f^{-1}(b) = a$.

94. Prove that an invertible function f can have only *one* inverse. (*Hint: Suppose g and h are both inverses of a function f, and suppose also that there is some real number x_0 for which $g(x_0) \neq h(x_0)$. Show that these suppositions together produce a contradiction.*)

Thinking Forward

Transformations of trigonometric graphs: Consider the function $f(x) = \sin x$ shown in the following figure.

$$f(x) = \sin x$$

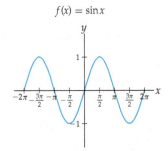

▶ Use transformations to sketch the graph of the function $f\left(x + \frac{\pi}{2}\right)$. (The resulting graph will be the graph of $\cos x$, so this problem illustrates the identity $\sin\left(x + \frac{\pi}{2}\right) = \cos x$.)

▶ Use transformations to sketch the graph of the function $f(x + 2\pi)$. What do you notice about this new graph? What does this mean about the function $f(x) = \sin x$?

▶ Sketch the graph of the reciprocal $\frac{1}{f(x)}$ of $f(x)$. Be especially careful when considering what happens at the places where $f(x)$ has zeros. (The resulting graph will be the graph of $\csc x$, so this problem illustrates that $\frac{1}{\sin x} = \csc x$.)

▶ The function $f(x) = \sin x$ is not one-to-one and therefore is not invertible. However, $f(x) = \sin x$ is invertible on the restricted domain $\left[-\frac{\pi}{2}, \frac{\pi}{2}\right]$. Sketch this restricted function, and then sketch its inverse $f^{-1}(x)$. (The graph of $f^{-1}(x)$ is the graph of the function known as $\sin^{-1} x$.)

0.7   LOGIC AND MATHEMATICAL THINKING*

▶ Logical statements that involve quantifiers or implications

▶ Using a counterexample to show that a statement is false

▶ Mathematical proof techniques, including direct proof and proof by contradiction

From Definitions to Theorems

Throughout this chapter we have learned a lot of the mathematical language that we will use throughout the book. It is now time to start thinking about how to build on our basic definitions and develop the theory of calculus. Developing such a theory is like building a skyscraper. With our library of mathematical definitions and language we have set the foundation. Throughout the rest of this book we will build on that foundation by using logical, mathematical arguments to develop new theorems. Each new theorem will be the springboard to a new set of definitions and theorems and will form a new level of our skyscraper. In this section we present an overview of the logic and proof techniques needed for our construction of calculus.

In what follows we will keep the definitions simple so that we can focus on the logic and proofs. Most of the mathematical statements we will discuss will concern real numbers and whether they are positive, negative, or zero, or will build upon the following five definitions:

▶ An *integer* is a whole number that is positive, negative, or zero. The set of all integers is

$$\{\ldots, -5, -4, -3, -2, -1, 0, 1, 2, 3, 4, 5, \ldots\}.$$

▶ An integer n is *divisible* by an integer m if we can write $n = km$ for some integer k.

▶ An integer n is *even* if we can write $n = 2k$ for some integer k.

▶ An integer n is *odd* if we can write $n = 2k + 1$ for some integer k.

▶ A *rational number* is a real number that can be written in the form $\frac{p}{q}$ for some integers p and q, where $q \neq 0$.

Quantifiers

We will often be interested in stating that a property is true all of the time, some of the time, at least once, or none of the time. Logical *quantifiers* are the key to making such statements precisely.

DEFINITION 0.40

The Quantifiers "For All" and "There Exists"

Suppose P is a property that depends on a value x.

(a) *For all x, property P* means that property P holds for all possible values of x.

(b) *There exists x such that property P* means that property P is true for at least one value of x.

For example, consider the following four quantified statements:

For all real numbers x, $2x + 2 = 2(x + 1)$.

For all integers x, $\frac{1}{x}$ is a rational number.

There exists an integer x such that x is even.

There exists a real number x such that $\frac{1}{x} = 0$.

The first statement is true because its claim that $2x + 2 = 2(x + 1)$ is indeed true for *every* real number x. The second statement is false because its claim that $\frac{1}{x}$ is rational fails for $x = 0$. The third statement is true because many even integers exist; for example, $x = 2$. The fourth statement is false because there is not even one real number x for which $\frac{1}{x} = 0$.

For statements with more than one quantifier, the order in which the quantifiers appear can make a big difference. For example, consider the following two statements, which differ only in the order that their quantifiers are listed:

"For all $x > 0$, there exists $y > 0$ such that $y > x$."

"There exists $y > 0$ such that for all $x > 0$, $y > x$."

The first statement claims that given any positive number x, we can find some number y that is greater than x. This is clearly true, since we can always choose y as large as we like to ensure that it is larger than the given number x. The second statment claims that there is some positive number y with the property that y is greater than *every* number x. This is clearly false, since no real number is greater than all other real numbers.

Implications

Statements of the form "If A, then B" are called **implications.** Most of the theorems in this book have the form of an implication. The statement A is called the **hypothesis,** and the statement B is called the **conclusion.** We will use implications in this book so often that we have a shorthand notation for them, namely, "$A \Rightarrow B$" (pronounced "A implies B").

DEFINITION 0.41	**Implications**
	An *implication* is a statement of the form **if A, then B** (also written $A \Rightarrow B$). Such an implication is true if, whenever statement A is true, statement B must also be true.

For example, the statement "If $x > 2$, then $x > 0$" is an implication. In the arrow notation we would write this as "$x > 2 \Rightarrow x > 0$," and as a quantified statement we could equivalently write "For all $x > 2$, $x > 0$." The hypothesis of the statement is "$x > 2$," and the conclusion is "$x > 0$." Thus, if we know that x is greater than 2, then we can conclude that x must also be greater than 0.

Suppose we have statements A and B, and the statement that A implies B. Does this mean that B also implies A? Not necessarily; for example, it is true that $x > 2$ implies $x > 0$, but the reverse implication is not true: $x > 0$ does not imply $x > 2$. When we switch the roles of the hypothesis and the conclusion of an implication, we have a new implication called the *converse* of the original.

DEFINITION 0.42	**The Converse of an Implication**
	The *converse* of the implication $A \Rightarrow B$ is the implication $B \Rightarrow A$.

If an implication $A \Rightarrow B$ and its converse $B \Rightarrow A$ happen to both be true, then we have a two-way implication. We write this as $A \Longleftrightarrow B$, which is pronounced **A if and only if B**. For example, the statements "x is even" and "x is divisible by 2" are equivalent; each implies the other. Thus we can say that "x is even if and only if x is divisible by 2."

The converse of an implication is obtained when the hypothesis and conclusion switch places. This results in a very different statement from the original. However, if the hypothesis and conclusion switch places *and* are negated, then surprisingly, we end up with a statement that is equivalent to the original.

DEFINITION 0.43

The Contrapositive of an Implication

The **contrapositive** of the implication $A \Rightarrow B$ is the statement (Not B) $\Rightarrow$ (Not A).

The contrapositive of an implication is *always* logically equivalent to the original implication. For example, the contrapositive of the statement "If x is an integer, then x is a rational number" is the statement "If x is not a rational number, then x is not an integer." These two statements are logically equivalent (and happen to be true).

Counterexamples

A statement of the form "for all x, property P" is false if there is even one instance x for which P is false. Such an instance is called a **counterexample**. For example, the statement "all cats are grey" means that *every* cat is grey. If there is one cat that is not grey, then *some* cats are *not* grey, and this shows that the statement "all cats are grey" is false.

DEFINITION 0.44

Counterexamples

A **counterexample** is an example of a value that makes a statement false.

THEOREM 0.45

Counterexamples to "For All" Statements

Suppose P is a property that depends on a value x. Then the statement "For all x, we have P" is false if and only if there is a counterexample for which P is false.

For example, consider the statement "For all integers x, $\frac{1}{x}$ is a rational number." This statement is false, because we can find a counterexample: when $x = 0$, $\frac{1}{x}$ is not a rational number. In contrast, the statement "For all real numbers x, $2x + 2 = 2(x + 1)$" is true, since there are no counterexamples x that do not have the property $2x + 2 = 2(x + 1)$. Finding a counterexample is a fast and easy way to prove that a "for all" statement is false. To show that a "for all" statement is true, however, requires substantially more work, as we will see shortly in this section.

What about counterexamples to implications? An implication "$A \Rightarrow B$" implies nothing about the truth or falsehood of A. Moreover, if A is false, then the implication "$A \Rightarrow B$" does not imply anything about statement B. For example, consider the statement "If I am elected, then I will lower your taxes." A politician who says this, gets elected, but then does not lower taxes is a liar. But that same politician is *not* a liar if he fails to get elected, whether he raises taxes or not. A statement of the form "If A, then B" is false only if there is an instance when its hypothesis A is true and yet its conclusion B is false.

(b) The converse of the given statement is "If $x \geq 0$, then x is even." This statement is false because not all nonnegative integers are even. One counterexample is $x = 3$, since $3 \geq 0$ but 3 is not even. In this example the original statement in part (a) and the converse here in part (b) both happened to be false. In general, however, a statement and its converse may or may not have the same truth value, since they are logically different statements.

(c) The contrapositive of the statement in part (a) is "If $x < 0$, then x is odd." Notice that the negation of the statement $x \geq 0$ is *not* the statement $x \leq 0$; why? The contrapositive statement is false, because not all negative numbers are odd. One counterexample is $x = -2$, since $-2 < 0$ but -2 is not odd. Notice that we can use the same counterexample for the contrapositive as for the statement in part (a), since the contrapositive is logically equivalent to the original statement. ☐

EXAMPLE 3   A simple calculational proof

Prove that for all real numbers a and b, $a^3 - b^3 = (a - b)(a^2 + ab + b^2)$.

SOLUTION

Sometimes a proof is nothing more than a calculation, written out with justifications for the steps:

Proof. For any real numbers a and b,

$$(a - b)(a^2 + ab + b^2) = a^3 + a^2 b + ab^2 - a^2 b - ab^2 - b^3 \qquad \leftarrow \text{multiply out}$$
$$= a^3 - b^3. \qquad \leftarrow \text{simplify}$$

Therefore, $a^3 - b^3 = (a - b)(a^2 + ab + b^2)$. ∎

EXAMPLE 4   A direct proof

Prove that the sum of any two rational numbers is a rational number.

SOLUTION

Before writing a proof, it is helpful to write down the hypothesis that you are given and the conclusion that you are trying to show. While doing this, give names to the variables involved.

> *Given:* a and b are any rational numbers.

> *Show:* $a + b$ is a rational number.

The first line of your proof will be the "given," the last will be the "show," and the middle will be an argument that uses definitions and logical inferences to connect the two. Since rational numbers are involved, we should remind ourselves of their definition: Recall that a number is **rational** if it can be written as the quotient of two integers. This gives us an updated and more descriptive version of our "given" and "show":

> *Given:* $a = \frac{p}{q}$ and $b = \frac{r}{s}$ for some integers p, q, r, and s with $q \neq 0$ and $s \neq 0$.

> *Show:* $a + b$ can be written in the form $\frac{u}{v}$ for some integers u and v with $v \neq 0$.

We now have a very clear road map that indicates how we should prove the implication; we just have to get from the "given" to the "show" by calculating $a + b$:

Proof. Suppose a and b are rational numbers. Then we can write a and b as quotients of integers, say, $a = \frac{p}{q}$ and $b = \frac{r}{s}$, for some integers p, q, r, and s, where q and s are nonzero. With this notation, the sum of a and b is

$$a + b = \frac{p}{q} + \frac{r}{s} = \frac{ps + qr}{qs}.$$

Since p, q, r, and s are integers, so are $ps + qr$ and qs. Moreover, since q and s are nonzero, so is qs. Therefore we have written $a + b$ as a valid quotient of two integers, and thus $a + b$ is a rational number.   ■

EXAMPLE 5   **A proof by contradiction**

Prove that the sum of a rational number and an irrational number is irrational.

SOLUTION

Let's begin by writing out the "given" and "show:"

> **Given:** r is a rational number and x is an irrational number.

> **Show:** $r + x$ is irrational.

A number is irrational if it cannot be written as the quotient of two integers. It can be difficult to show that we *can't* write a number as a quotient of two integers, so instead of using a direct proof we will use the method of **proof by contradiction**. This means that we will show that $r + x$ is irrational by showing that it cannot possibly be rational. More precisely, we will suppose that $r + x$ is rational and then show that this supposition causes a logical contradiction.

Proof. Suppose r is a rational number and x is an irrational number. Seeking a contradiction, suppose that the sum $r + x$ is rational. In the previous example we proved that the sum of two rational numbers is rational. In addition, if a number r is rational, then so is $-r$. This means that the sum of the rational numbers $r + x$ and $-r$ must also be rational, so $(r + x) + (-r) = x$ must be rational. But this conclusion contradicts our assumption that x is irrational. If $r + x$ is rational, we are led to a contradiction; therefore $r + x$ must be an irrational number.   ■

TEST YOUR UNDERSTANDING

▶ What does it mean to say that a statement A "implies" a statement B? How is it different than asserting that A and B are true?

▶ What can you say about an implication if its hypothesis is always false? What about if its conclusion is always true?

▶ Why is $x = 3$ *not* a counterexample to the implication "If x is even, then $x \geq 0$"?

▶ What does it mean to say that an implication and its contrapositive are logically equivalent statements?

▶ Define each of the following: counterexample, implication, converse, contrapositive.

EXERCISES 0.7

Thinking Back

Basic definitions: Recall the definitions of each of the following terms or quantities.

▶ nonnegative

▶ rational number

▶ $|x|$

▶ integer

▶ irrational number

▶ dist(a, b)

▶ *Types of numbers:* Are all integers rational? Are all rational numbers integers? Is 0 a rational number? Why or why not?

▶ *Inequality opposites:* If $x > 9$ is false for a particular value of x, does this necessarily mean that $x < 9$? Why or why not? What is the logical opposite of the statement "$x > 9$"?

Concepts

0. *Problem Zero:* Read the section and make your own summary of the material.

1. *True/False:* Determine whether each of the statements that follow is true or false. If a statement is true, explain why. If a statement is false, provide a counterexample.

   (a) *True or False:* You can show that "For all x, P" is true by exhibiting just one value of x that makes P true.

   (b) *True or False:* You can show that "For all x, P" is false by exhibiting just one value of x that makes P false.

   (c) *True or False:* You can show that "There exists x such that we have P" is true by exhibiting just one value of x that makes P true.

   (d) *True or False:* You can show that "There exists x such that we have P" is false by exhibiting just one value of x that makes P false.

   (e) *True or False:* The converse of an implication is also an implication.

   (f) *True or False:* When A is true and B is false, the implication $A \Rightarrow B$ is false.

   (g) *True or False:* When A is false and B is true, the implication $A \Rightarrow B$ is false.

   (h) *True or False:* When A is false and B is false, the implication $A \Rightarrow B$ is false.

2. *Examples:* Construct examples of the thing(s) described in the following. Try to find examples that are different than any in the reading.

   (a) A true statement involving two quantifiers, and a false statement involving two quantifiers.

   (b) A statement of the form "For all x, $A \Rightarrow B$" that has just one counterexample, and another that has many counterexamples.

   (c) An implication whose converse is false, and an implication whose converse is true.

3. Suppose the implication "$C \Rightarrow D$" is true. If C is true, what can you say about D? If C is false, what can you say about D?

4. Suppose the implication "$R \Rightarrow S$" is false. What does this mean about statements R and S?

5. Consider the statement "Every positive real number is greater than -2." Write this statement using the quantifier "for all." Then write a statement that is logically equivalent but uses "if ..., then ..." instead of quantifiers.

6. Consider the statement "The square of any real number is nonnegative." Write this statement using the quantifier "for all." Then write a statement that is logically equivalent but uses "if ..., then ..." instead of quantifiers.

7. Consider the statement "Every square is a rectangle." Is this statement true? Write down the converse and the contrapositive of the statement, and determine whether they are true or false.

8. What is the converse of the statement $C \Rightarrow D$? Is the converse logically equivalent to the original statement? Why or why not?

9. What is the contrapositive of the statement $P \Rightarrow Q$? Is the contrapositive logically equivalent to the original statement? Why or why not?

10. Prove, by exhibiting examples, that the sum of two irrational numbers can be either rational or irrational. Why is it okay to prove "by example" here, whereas it is not okay to prove "by example" in general?

For Exercises 11–16, suppose you know the following (no more and no less) about a function $f(x)$ at values of x:

$$\text{If } 0 < |x - 3| < 0.1, \text{ then } |f(x) + 5| < 0.2.$$

Note that this implication means that *if* we know that x is a solution of the double inequality $0 < |x - 3| < 0.1$, *then* we can conclude that $f(x)$ is a solution of the inequality $|f(x) + 5| < 0.2$. Use the meaning of the given implication to determine whether or not each of the following related implications is guaranteed to be true. (*Hint: You may have to solve inequalities or sketch number lines and think about distances to determine the meanings of the hypotheses and conclusions of the implications.*)

11. If $0 < |x - 3| < 0.1$, then $|f(x) + 5| < 0.1$.

12. If $0 < |x - 3| < 0.1$, then $|f(x) + 5| < 0.3$.

13. If $0 < |x - 3| < 0.05$, then $|f(x) + 5| < 0.2$.

14. If $0 < |x - 3| < 0.05$, then $|f(x) + 5| < 0.1$.

15. If $0 < |x - 3| < 0.05$, then $|f(x) + 5| < 0.4$.

16. If $0 < |x - 3| < 0.2$, then $|f(x) + 5| < 0.4$.

Skills

Determine whether each statement in Exercises 17–46 is true or false. Justify your answers with reasoning, examples, or counterexamples, as appropriate.

17. There is some real number between 2 and 3.

18. Every real number is a rational number.

19. No real number is both rational and irrational.

20. Every real number is either rational or irrational.

21. For all real numbers x, there is some real number y with $y = x^2$.

22. For all real numbers x, there is some real number y with $x = y^2$.

23. If x is an integer greater than 1, then $x \geq 2$.

24. All real numbers are either greater than zero or less than zero.

25. For all real numbers x, either x is even or x is odd.

26. For all real numbers x, if $x < -2$, then $x^2 > 4$.

27. There exists a real number x such that $x \leq 1$ and $x \geq 2$.

28. There exist real numbers x and y such that $x + y = 4$.

29. For all real numbers x, if x is negative, then x is irrational.

30. For all real numbers x, if x is an integer, then x is rational.

31. There exists a real number x such that $x \leq 1$ or $x \geq 2$.

32. There exists a real number x such that $x > 0$ and $x^2 > 10$.

33. For all real numbers a and b, if $a < b$, then $3a + 1 < 3b + 1$.

34. No rational number is both less than $\frac{1}{2}$ and greater than $\frac{1}{3}$.

35. For all real numbers x, $x^2 \geq 0$ and $|x| \geq 0$.

36. For all real numbers x, either $x \geq 2$ or $x \leq 1$.

37. There exist real numbers $x < 0$ and $y < 0$ such that $xy < 0$.

38. There exist real numbers $x > 0$ and $y > 0$ such that $xy = 0$.

39. For all real numbers x and y, if $x < y$, then $2x - 1 < 2y - 1$.

40. For all real numbers x and y, if $x < y$, then $x^2 < y^2$.

41. There exists a real number x such that for all real numbers y, $|y| > x$.

42. For all real numbers x and y, $xy = 0$ if and only if $x = 0$ or $y = 0$.

43. For all real numbers x, there exists some y such that $x < y$.

44. For all real numbers x, there exists some y such that $x = y^2$.

45. For all real numbers x and y, $\frac{x}{y} = 0$ if and only if $x = 0$.

46. There exists a real number x such that for all real numbers y, $y > x$.

Suppose A and B represent logical statements. In Exercises 47–54, write (a) the converse and (b) the contrapositive of the given statement. Simplify each of your statements if possible.

47. (Not A) $\Rightarrow B$

48. $A \Rightarrow$ (Not B)

49. (Not B) $\Rightarrow$ (Not A)

50. (Not A) $\Rightarrow$ (Not B)

51. (A and B) $\Rightarrow C$

52. (A or B) $\Rightarrow C$

53. $A \Rightarrow$ (B and C)

54. $A \Rightarrow$ (B or C)

In Exercises 55–66, write (a) the converse and (b) the contrapositive of each statement. Simplify your statements as much as possible. (c) Provide counterexamples if the original, the converse, and/or the contrapositive statements are false.

55. If x is a real number, then x is rational.

56. If $x \geq 2$, then $x \geq 3$.

57. If $x > 2$, then $x \geq 3$.

58. If $x \geq 2$, then $x \geq 1$.

59. If x is negative, then $\sqrt{x}$ is not a real number.

60. If x is rational, then x is not irrational.

61. If $x \leq 0$, then $|x| = -x$.

62. If $x < -2$, then $|x| = -x$.

63. If x is not zero, then $x^2 > x$.

64. If x is positive and rational, then $x - 1$ is positive and rational.

65. If x is odd, then there is some integer n such that $x = 2n + 1$.

66. If x is even, then there is some integer n such that $x = 2n$.

Applications

Use logic to solve the puzzles in Exercises 67 and 68. Then write proofs to argue that your solutions are correct.

67. Linda, Alina, Phil, and Stuart are wearing different-colored hats: either red, yellow, green, or blue. From the following statements, determine which hat each person is wearing:

▶ Neither boy wears a red hat.

▶ The oldest person wears a green hat.

▶ Linda is older than Alina.

▶ Alina never wears yellow or red.

▶ Stuart is the youngest and hates blue.

68. Xena, Yolanda, and Zeke each have different favorite fruits: either apples, bananas, or cantaloupes. Use the statements that follow to determine which person prefers which fruit. Then write a proof that argues that your solution is correct.

▶ Xena likes bananas better than apples.

▶ Zeke is allergic to cantaloupes.

▶ Bananas are the favorite fruit of one of the girls.

▶ Yolanda likes bananas better than cantaloupes.

Consider an Island X where there are exactly two types of people: those who always tell the truth ("truth-tellers") and those who always lie ("liars"). Given the statements in Exercises 69–71, determine which people are truth-tellers and which are liars. Then write a proof which argues that your solution is correct. (*These types of puzzles, based on those made popular by Raymond Smullyan, are a great avenue to developing your logical thinking skills and your proof-writing abilities.*)

69. You meet Liz, Rein, and Zubin, who say the following:

> **Liz:** "We all tell the truth."
> **Rein:** "Exactly two of us tell the truth."
> **Zubin:** "Liz and Rein always lie."

70. You meet Anita, Bill, and Chris, who say the following:

> **Anita:** "At least one of Bill or Chris tells the truth."
> **Bill:** "Anita is a liar."
> **Chris:** "Bill is a liar."

71. You meet Hyun, Jaan, and Kate, and only Kate and Hyun have something to say:

> **Kate:** "Hyun and Jaan are liars."
> **Hyun:** "Kate always tells the truth."

Proofs

72. Prove that $x^2 - y^2 = (x - y)(x + y)$ for all real numbers x and y.

73. Prove that if x is an irrational number and r is a rational number, then the difference $x - r$ must be an irrational number. You may assume that the sum of two rational numbers is a rational number.

74. Prove that for any real numbers a and b, $|b - a| = |a - b|$.

75. Prove that the sum of an even integer and an odd integer is odd.

76. Prove that the sum of an odd integer and an odd integer is even.

77. Prove that the product of an even integer and an odd integer is even.

78. The Pythagorean theorem states that if a right triangle has legs of length a and b, and a hypotenuse of length c, then $a^2 + b^2 = c^2$. Use the Pythagorean theorem and the definition of the distance between two real numbers to prove that the distance between any two points $P = (x_1, y_1)$ and $Q = (x_2, y_2)$ in the plane is given by the "distance formula" $\sqrt{(x_2 - x_1)^2 + (y_2 - y_1)^2}$. (*Hint: Draw an example of two points P and Q in the plane, label their coordinates, and use an appropriate right triangle.*)

79. Prove that the midpoint $\left(\frac{x_1 + x_2}{2}, \frac{y_1 + y_2}{2}\right)$ between the points $P = (x_1, y_1)$ and $Q = (x_2, y_2)$ is equidistant from P and Q.

80. Prove that the numbers

$$x = \frac{-b + \sqrt{b^2 - 4ac}}{2a} \quad \text{and} \quad x = \frac{-b - \sqrt{b^2 - 4ac}}{2a}$$

found by the quadratic formula are solutions of the quadratic equation $ax^2 + bx + c = 0$.

81. Use the fact that $\frac{(a/b)}{(c/d)} = \frac{ad}{bc}$ to prove that $\frac{(a/b)}{c} = \frac{a}{bc}$. You may assume that all denominators are nonzero.

82. Use the fact that $\left(\frac{a}{b}\right)\left(\frac{c}{d}\right) = \frac{ac}{bd}$ to prove that $c\left(\frac{a}{b}\right) = \frac{ac}{b}$. You may assume that any denominators are nonzero.

83. Follow the steps outlined here to prove the **triangle inequality**: $|a + b| \leq |a| + |b|$ for any real numbers a and b:

(a) Argue that for any real number x, $|x| = \sqrt{x^2}$.

(b) Show that $(a + b)^2 \leq (|a| + |b|)^2$. (*Hint: Start on the left-hand side, multiply out the expression, and use the fact that $a \leq |a|$ and $b \leq |b|$.*)

(c) Take the square root of both sides of the inequality from part (b) (this is valid since both sides are positive), and use part (a) to show that $|a + b| \leq |a| + |b|$.

Use the triangle inequality to prove the following two inequalities, for any real numbers a and b:

84. $|a - b| \leq |a| + |b|$ **85.** $|a - b| \geq |a| - |b|$

For Exercises 86 and 87, use the definition of absolute value and systems of inequalities to prove that for any real numbers x and c, and any positive real number δ, the given statement is true:

86. $|x - c| < \delta \iff x \in (c - \delta, c + \delta)$

87. $|x - c| > \delta \iff x \in (-\infty, c - \delta) \cup (c + \delta, \infty)$

88. The "Monty Hall Problem" is a good example of a problem about which people's initial intuition is often incorrect. On the game show *Let's Make a Deal*, the host, Monty Hall, presents you with a choice of three doors. Behind one door is a lot of money. Behind the other two doors are worthless gag prizes. You pick a door and point at it. Monty Hall knows which door conceals the prize, and he opens one of the two doors you didn't pick to show you a gag prize. Then he gives you the option of keeping your original choice or switching your choice to the remaining door.

(a) Is it better to switch or to stick with your original choice? Or are both choices equally likely to lead to the money? Think about this problem for awhile and convince yourself of an answer before you go on to the next part.

(b) Come up with a proof or argument that would convince another person of the correct answer.

Thinking Forward

Quantified statements about distance: Show that each of the statements that follow is true by exhibiting a value of δ that satisfies each statement. These types of statements will be the backbone of our study of limits in Chapter 1. (Note: The symbols ϵ and δ are Greek letters that represent real numbers that are usually positive and quite small. In the third statement you will need to write δ in terms of ϵ.)

▶ There exists a $\delta > 0$ such that for all x, if x is within distance δ of 2, then $3x + 1$ is within distance 1 of 7.

▶ There exists a $\delta > 0$ such that for all x, if $|x - 2| < \delta$, then $|3x - 6| < 0.3$.

▶ For all $\epsilon > 0$, there exists a $\delta > 0$ such that if x is within distance δ of 3, then $2x$ is within distance ϵ of 6.

Negating quantified statements about distance: Write down the negation of each statement that follows. We will see these types of statements again when we formally define the concept of a limit in Chapter 1.

▶ For all $M > 0$, there exists an $N > 0$ such that for all x, if $x > N$, then $x^2 > M$.

▶ For all $\epsilon > 0$, there exists a $\delta > 0$ such that for all x, if $0 < |x - 2| < \delta$, then $|x^2 - 4| < \epsilon$.

▶ For all $\epsilon > 0$, there exists a $\delta > 0$ such that for all x, if $0 < |x - 4| < \delta$, then $|\sqrt{x} - 2| < \epsilon$.

CHAPTER REVIEW, SELF-TEST, AND CAPSTONES

Before you progress to the next chapter, be sure you are familiar with the definitions, concepts, and basic skills outlined here. The capstone exercises at the end bring together ideas from this chapter and look forward to future chapters.

Definitions

Give precise mathematical definitions or descriptions of each of the concepts that follow. Then illustrate the definition with a graph or algebraic example, if possible.

▶ a *nonnegative* real number, a *complex* number, an *irrational number*, and a *rational number* (without referring to decimals)

▶ an *open interval* and a *closed interval* of real numbers

▶ the *distance* and the *midpoint* between two real numbers a and b

▶ the *distance* and the *midpoint* between two points P and Q in the plane

▶ the *union* $A \cup B$ and the *intersection* $A \cap B$ of two sets A and B

▶ the *solution set* of an equation

▶ an equation that is an *identity*

▶ the *discriminant* of a quadratic equation

▶ a *function* from a set A to a set B

▶ the *independent variable* and the *dependent variable* of a function

▶ the *domain* and *range* of a function, in set notation

▶ the *graph* of a function, in set notation and as a picture

▶ a *one-to-one* function

▶ a *root* of a function and a *y-intercept* of a function

▶ a *local maximum*, *local minimum*, *global maximum*, or *global minimum* of a function

▶ an *inflection point* of a function (in rough terms)

▶ what it means for a function f to be *positive*, *negative*, *increasing*, or *decreasing* on an interval I

▶ what it means (roughly) for a function f to be *concave up* or *concave down* on an interval I

▶ the *average rate of change* of a function f on an interval $[a, b]$

▶ a *piecewise-defined function*

▶ the *slope–intercept form*, *point–slope form*, and *two–point form* of a linear function

▶ the form of a *proportional function*, and its *proportionality constant*

▶ the form of a *power function* and the form of a *polynomial function*

▶ the form of a *rational function*

▶ what it means for a function to be *algebraic*

▶ the piecewise-defined form of $f(x) = |x|$, and more generally of a function $f(x) = |g(x)|$

▶ the algebraic definitions of *even functions* and *odd functions*, and the graphical meaning of *y-axis symmetry* and *rotational symmetry*

▶ what it means for functions f and g to be *inverses* of each other

▶ the meaning of the quantified statement *"For all x, P"*

▶ the meaning of the quantified statement *"There exists x such that P"*

▶ the meaning of the implication *"If A, then B"*

▶ the *hypothesis* and the *conclusion* of an implication

▶ the *converse* and the *contrapositive* of an implication $A \Rightarrow B$

▶ a *counterexample* to a statement

Theorems

Fill in the blanks to complete each of the following statements of theorems:

▶ For any real numbers a and b, $|ab| =$ _____ and $\left|\dfrac{a}{b}\right| =$ _____ .

▶ *The Quadratic Formula:* If a, b, and c are any _____ , then the _____ of the quadratic equation _____ are of the form $x =$ _____ .

▶ For any real expressions A and B, $AB = 0$ if and only if _____ .

▶ For any real expressions A and B, $\dfrac{A}{B} = 0$ if and only if _____ and _____ .

▶ For any real expression A, the expression $\dfrac{A}{A}$ is equal to 1 if _____ , but is undefined if _____ .

▶ Fill in the square with the appropriate inequality: If $a < b$ and $c \,\square\, 0$, then $ac < bc$.

▶ Fill in the square with the appropriate inequality: If $a < b$ and $c \,\square\, 0$, then $ac > bc$.

▶ Fill in the square with the appropriate inequality: If $0 < a < b$, then $\dfrac{1}{a} \,\square\, \dfrac{1}{b}$.

▶ Fill in the square with the appropriate inequality: If $a < b$, then $a + c \,\square\, b + c$.

▶ For any real expressions A and B, $AB > 0$ if and only if (____ and ____) or (____ and ____).

▶ For any real expressions A and B, $AB < 0$ if and only if (____ and ____) or (____ and ____).

▶ A product or quotient of nonzero real numbers is positive if and only if ____ and is negative if and only if ____ .

▶ A polynomial can change sign only at values where it is ____ .

▶ A quotient of polynomials can change sign only at values where its ____ or ____ is equal to 0.

▶ The solutions of $|x - c| < \delta$ can be expressed in interval notation as ____ .

▶ The solutions of $|x - c| > \delta$ can be expressed in interval notation as ____ .

▶ Select one: If A is a real expression, then $|A| < \delta$ if and only if $A < \delta$ (and)/(or) $A > -\delta$.

▶ Select one: If A is a real expression, then $|A| > \delta$ if and only if $A < \delta$ (and)/(or) $A > -\delta$.

▶ *The Vertical Line Test:* A graph represents a function if and only if every vertical line has the property that it ____ .

▶ If $f(x) = mx + b$ is a linear function, then its domain is ____ , its range is ____ (if $m \neq 0$), its average rate of change on any interval is ____ , and its graph is a ____ .

▶ $y = f(x) + C$ is shifted C units ____ from $y = f(x)$ if $C > 0$ and C units ____ from $y = f(x)$ if $C < 0$.

▶ $y = f(x + C)$ is shifted C units ____ from $y = f(x)$ if $C > 0$ and C units ____ from $y = f(x)$ if $C < 0$.

▶ $y = kf(x)$ is vertically stretched from $y = f(x)$ by a factor of k if ____ and vertically compressed by a factor of k if ____ .

▶ $y = f(kx)$ is horizontally stretched from $y = f(x)$ by a factor of k if ____ and horizontally compressed by a factor of k if ____ .

▶ If f is an invertible function with inverse f^{-1}, then the domain of f^{-1} is ____ and the range of f^{-1} is ____ .

▶ If f is an invertible function with inverse f^{-1}, then $f^{-1}(b) = a$ if and only if ____ , and the graph of $y = f^{-1}(x)$ can be obtained from the graph of $y = f(x)$ by ____ .

▶ A function has an inverse if and only if the function is ____ .

▶ A statement of the form "For all x, P" is false if and only if there is a counterexample in which ____ .

▶ A statement of the form "For all x, if A, then B" is false if and only if there is a counterexample in which ____ but ____ .

Notation and Algebraic Rules

Notation: Describe the meanings of each of the following mathematical expressions.

▶ $\{x \mid x^2 = 1\}$

▶ $\{x \mid x \geq -1\}$

▶ $x \in [0, 5)$

▶ $(-\infty, 3]$

▶ $\varnothing$

▶ $\mathbb{R}$

▶ $|a|$

▶ $|b - a|$

▶ $|x - c| < \delta$

▶ $|x - c| > \delta$

▶ $|A| < \delta$

▶ $|A| > \delta$

▶ $f : A \to B$

▶ $y = f(x)$

▶ $f(x) = |x|$

▶ $(kf)(x)$

▶ $(f + g)(x)$

▶ $(f \cdot g)(x)$

▶ $\left(\dfrac{f}{g}\right)(x)$

▶ $(f \circ g)(x)$

▶ $(f \circ g \circ h)(x)$

▶ $A \Rightarrow B$

▶ $A \Longleftrightarrow B$

▶ $\blacksquare$

Factoring: Fill in the blanks to complete each of the factorizations that follow. You may assume that a and b are real numbers and n is a positive integer.

▶ $a^2 - b^2 = (a - b)(\underline{})$

▶ $a^3 - b^3 = (a - b)(\underline{})$

▶ $a^n - b^n = (a - b)(\underline{})$

▶ $a^n - 1 = (a - 1)(\underline{})$

Fractions: Fill in the blanks to complete each of the algebraic rules that follow. You may assume that a, b, c, and d are real numbers and that all denominators are nonzero.

▶ $\dfrac{a}{b} + \dfrac{c}{d} = \underline{}$

▶ $\dfrac{a}{b} - \dfrac{c}{d} = \underline{}$

▶ $\left(\dfrac{a}{b}\right)\left(\dfrac{c}{d}\right) = \underline{}$

▶ $\dfrac{(a/b)}{(c/d)} = \underline{}$

▶ $\dfrac{(a/b)}{c} = \underline{}$

▶ $\dfrac{a}{(b/c)} = \underline{}$

Skill Certification: Basic Algebra and Simple Functions

Simplifying expressions: Simplify each expression as much as possible.

1. $\dfrac{x^3 - 2^3}{x - 2}$

2. $\dfrac{\frac{1}{2+h} - \frac{1}{2}}{h}$

3. $\dfrac{x - 2}{x^3 - x^2 - 4x + 4}$

4. $\dfrac{x^4 + 27x}{x^2 + 5x + 6}$

5. $|-2(x^2 + 1)|$

6. $\dfrac{|4 - 2x|}{x - 2}$

Solving inequalities: Solve each of the inequalities, and express each solution set in interval notation.

7. $2x^2 - 7x + 3 > 0$

8. $\dfrac{x^2 - 9}{x - 1} \le 0$

9. $\dfrac{3}{x - 2} < 1$

10. $|3x - 4| < \dfrac{1}{2}$

11. $|5x - 2| > 1$

12. $|x^2 - 4| < 2$

Finding zeros and undefined values: Determine the x-values for which each function is zero, and the x-values for which each function does not exist.

13. $f(x) = \dfrac{2x^2 - 5x + 3}{x}$

14. $f(x) = 3x^4 - 6x^3 + x^2$

15. $f(x) = |x - 2| - 5$

16. $f(x) = \dfrac{1}{x} - \dfrac{1}{1 - x}$

17. $f(x) = \dfrac{2(x - 1)^2 - 4}{x - 1}$

18. $f(x) = \dfrac{2x^3 - 1}{2x^2 - 5x - 3}$

Finding domains: Find the domain of each function, and express the domain in interval notation.

19. $f(x) = \dfrac{\sqrt{x + 2}}{x}$

20. $f(x) = \sqrt{x^2 - 2x - 3}$

21. $f(x) = \dfrac{x + 2}{x^2 - x - 6}$

22. $f(x) = \dfrac{1}{\sqrt{x^2 - 3x}}$

23. $f(x) = \sqrt{x} + \sqrt{2 - x}$

24. $f(x) = \sqrt{\dfrac{x - 3}{1 - x}}$

Graphs of simple functions: Sketch the graph of each function by hand, using your knowledge of simple graphs and transformations. Label any important points or features.

25. $f(x) = 3 - 2x$

26. $f(x) = 4(x - 1) + 2$

27. $f(x) = \sqrt{x - 3}$

28. $f(x) = (x - 3)^2 + 1$

29. $f(x) = 2x^3 - 1$

30. $f(x) = \sqrt{1 - x}$

31. $f(x) = \dfrac{1}{x} - 2$

32. $f(x) = \dfrac{1}{x - 2}$

33. $f(x) = |4x - 3|$

34. $f(x) = |x^2 - 9|$

Capstone Problems

A. *Transformations of linear functions:* Prove algebraically that a vertical or horizontal shift or stretch of a linear function is also a linear function. That is, prove that if $f(x) = mx + b$ is a linear function, then so are $f(x) + C$, $f(x + C)$, $kf(x)$, and $f(kx)$.

B. *Peeking forward to derivatives:* Suppose $f(x) = x^2$, and consider the two-variable function

$$q(x, h) = \frac{f(x + h) - f(x)}{h}.$$

(a) Simplify $q(3, h)$ as much as possible, and argue that it approaches 6 as h gets closer to 0.

(b) Simplify $q(x, h)$ as much as possible, and argue that it approaches $2x$ as h gets closer to 0.

C. *Approximating the area of a region in the plane:* Sketch the region that lies between the graph of $f(x) = \sqrt{x - 2}$, the x-axis, and the lines $y = 2$ and $y = 4$. Use geometric figures to approximate upper and lower bounds for the area of this region.

D. *Optimizing the area of a region given its perimeter:* Elizabeth wants to build a rectangular pen for her dogs with 100 feet of spare fencing.

(a) Write down an equation in terms of length l and width w for the perimeter P of the pen.

(b) Use the perimeter equation and the constraint on fencing material to construct a one-variable equation for the area A of the enclosure.

(c) Sketch a graph of the one-variable function A, and use the graph to argue that its maximum occurs when $l = w = 25$ and the enclosure is square.

Limits

1.1   AN INTUITIVE INTRODUCTION TO LIMITS

▶ Examples of the types of limits that are important in calculus

▶ Intuitive descriptions of two-sided and one-sided limits

▶ Infinite limits, limits at infinity, and horizontal and vertical asymptotes

Examples of Limits

Limits are the backbone of calculus. Limits are the key to investigating the local behavior of functions, giving meaning to the slope of a curve, finding areas under curves and volumes inside surfaces, and determining the long-term behavior of infinite sequences and sums. In the next section we will formally and mathematically define limits; in this section we focus on intuitive examples. Let's start with three examples that illustrate how limits can arise.

Limits of sequences: As a starting point for thinking about the concept of a limit, consider the following **sequence** of numbers:

$$\frac{1}{2}, \frac{1}{4}, \frac{1}{8}, \frac{1}{16}, \frac{1}{32}, \frac{1}{64}, \cdots, \frac{1}{2^k}, \cdots$$

If the pattern of this sequence continues, then the numbers will continue to get smaller and smaller, approaching zero. We say that 0 is the **limit** of the sequence $\left\{\frac{1}{2^k}\right\}$ as k approaches infinity. We can never actually let k be equal to infinity, because infinity is not a real number. However, we can let k get as large as we like. For each large value of k, the value of $\frac{1}{2^k}$ is very small, but not actually zero. When we "take the limit," we make an important theoretical transition: Instead of evaluating $\frac{1}{2^k}$ at a particular value of k, we think about the behavior of the sequence as k gets larger and larger. We think about what the sequence *approaches*, even if it never actually gets there for any real number k.

Limits of sequences of sums: Now consider the sequence defined by adding up more and more terms from the previous sequence:

$$\frac{1}{2}, \quad \frac{1}{2}+\frac{1}{4}, \quad \frac{1}{2}+\frac{1}{4}+\frac{1}{8}, \quad \frac{1}{2}+\frac{1}{4}+\frac{1}{8}+\frac{1}{16}, \quad \frac{1}{2}+\frac{1}{4}+\frac{1}{8}+\frac{1}{16}+\frac{1}{32}, \quad \cdots$$

After computing the sums, this sequence is equal to

$$\frac{1}{2}, \frac{3}{4}, \frac{7}{8}, \frac{15}{16}, \frac{31}{32}, \frac{63}{64}, \cdots$$

The terms get closer and closer to 1 as we go further and further out in this **sequence of sums**. You may have noticed a pattern in the sequence: It turns out that for any given k, the value of $\frac{1}{2} + \frac{1}{4} + \cdots + \frac{1}{2^k}$ will be equal to the quantity $\frac{2^k - 1}{2^k}$. The larger the value of k, the closer this quantity gets to 1. Moreover, we can get the sum to be as close as we like to 1 by choosing a sufficiently large value of k. Again, we can't plug in infinity for k, since we can't in real life add up infinitely many numbers, even if those numbers are getting infinitesimally small, as they are here. However, mathematically we can use the concept of a limit to notice that this sequence of sums *approaches* the quantity 1. We will study limits of sequences of sums in depth in Chapter 7; for now we present this type of limit only as an interesting example to consider.

Limits of average rates of change: Now let's switch gears and think of something a little more practical. Suppose you drop a bowling ball from the top of a 100-foot parking deck and want to know how fast it is falling when it hits the ground. Suppose that the height of the ball t seconds after it is dropped is given by $s(t) = 100 - 16t^2$ feet and thus that the ball hits the ground after 2.5 seconds. By the old formula "distance equals rate times time," the average rate that the ball falls over any time period Δt will be $r = \frac{\Delta s}{\Delta t}$, where Δs is the elapsed distance over the period Δt. A good approximation of the final speed of the bowling ball is the average rate in the last half-second, which is

$$r = \frac{s(2.5) - s(2)}{2.5 - 2} = \frac{0 - 36}{0.5} = -72 \text{ feet per second}$$

(negative since the ball is falling downwards). To get a better approximation for the ball's final **instantaneous velocity**, we could calculate the rate over the last quarter-second. The following table records a sequence of better and better approximations for the final velocity of the bowling ball:

Time interval	$[2, 2.5]$	$[2.25, 2.5]$	$[2.4, 2.5]$	$[2.49, 2.5]$	$[2.4999, 2.5]$
Average rate	$-72 \dfrac{\text{ft}}{\text{sec}}$	$-76 \dfrac{\text{ft}}{\text{sec}}$	$-78.4 \dfrac{\text{ft}}{\text{sec}}$	$-79.84 \dfrac{\text{ft}}{\text{sec}}$	$-79.9984 \dfrac{\text{ft}}{\text{sec}}$

We can't compute the actual final velocity of the ball this way, because we can't use $r = \frac{\Delta s}{\Delta t}$ when Δt is zero. But the average rates seem to *approach* -80 feet per second as Δt gets smaller and smaller. This sounds like another limit, and it is. In fact it is a very famous and useful type of limit called a **derivative** that we will introduce in Chapter 2. In general, limits help us discuss what happens when we let things get infinitesimally small, infinitely large, or arbitrarily close to some number.

Limits of Functions

Intuitively, a limit is what the output of a function approaches as we let the input of that function approach some value. In the previous examples, we saw that:

▶ As k approaches infinity, the quantity $\frac{1}{2^k}$ approaches 0.

▶ As k approaches infinity, the sum $\frac{1}{2} + \frac{1}{4} + \cdots + \frac{1}{2^k}$ approaches 1.

▶ As t approaches 2.5, the average rate $\frac{s(2.5) - s(t)}{2.5 - t}$ approaches -80 feet per second.

If the values of a function $f(x)$ approach some number L as x gets closer and closer to some value $x = c$, we will write

$$\lim_{x \to c} f(x) = L.$$

We can also consider limits of functions as $x \to \infty$, that is, as x grows without bound. For example, in this notation we have

$$\lim_{k \to \infty} \frac{1}{2^k} = 0 \quad \text{and} \quad \lim_{t \to 2.5} \frac{s(2.5) - s(t)}{2.5 - t} = -80.$$

When considering a limit, it only matters what happens as x gets closer and closer to c, not what happens when it actually gets there. This means that $\lim_{x \to c} f(x)$ may or may not in general be the same as the value $f(c)$ of the function at $x = c$. For example, the functions $f(x) = x + 1$ and $g(x) = \frac{x^2 - 1}{x - 1}$ shown in the following figures are not equal *at* the point $x = 1$, but they do *approach* the same value as $x \to 1$:

$\lim_{x \to 1} f(x) = 2$ *and* $f(1) = 2$

$\lim_{x \to 1} g(x) = 2$ *but* $g(1)$ *is undefined*

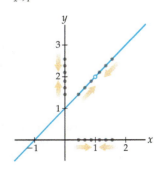

The dots on the x-axis represent a sequence of values of x that approach $x = 1$. When we evaluate the functions f and g at these values, in both cases we get a sequence of values of y that gets closer and closer to $y = 2$. Although $f(1) \neq g(1)$, we do have $\lim\limits_{x \to 1} f(x) = \lim\limits_{x \to 1} g(x) = 2$.

We can also consider limits as x grows without bound, and/or as $f(x)$ grows without bound. The following definition summarizes the notation we will use in each case:

DEFINITION 1.1

Intuitive Description of and Notation for Limits

Suppose f is a function and L and c are real numbers.

(a) *Limit:* If the values of a function $f(x)$ approach L as x approaches c, then we say that L is the *limit* of $f(x)$ as x approaches c and we write

$$\lim_{x \to c} f(x) = L.$$

(b) *Limit at Infinity:* If the values of a function $f(x)$ approach L as x grows without bound, then we say that L is the *limit* of $f(x)$ as x approaches ∞ and we write

$$\lim_{x \to \infty} f(x) = L.$$

(c) *Infinite Limit:* If the values of a function $f(x)$ grow without bound as x approaches c, then we say that ∞ is the *limit* of $f(x)$ as x approaches c and we write

$$\lim_{x \to c} f(x) = \infty.$$

(d) *Infinite Limit at Infinity:* If the values of a function $f(x)$ grow without bound as x grows without bound, then we say that ∞ is the *limit* of $f(x)$ as x approaches ∞ and we write

$$\lim_{x \to \infty} f(x) = \infty.$$

When a limit approaches a real number, we say that the limit ***exists***. When a limit approaches ∞ or $-\infty$ we say that the limit ***does not exist*** (because ∞ and $-\infty$ are not real numbers), but we will always be as specific as possible and describe the sign of infinity in such cases. Later we will see more pathological limits that "do not exist" in a way other than being infinite.

When we say that x approaches a real number c, we implicitly mean to consider values of x that are close to c from either the right or the left. In other words, when trying to find $\lim\limits_{x \to c} f(x)$, we consider both values of x that are slightly less than c as well as values of x that are slightly greater than c. Sometimes it is convenient to consider these two cases separately:

DEFINITION 1.2

Intuitive Description of One-Sided Limits

If the values of a function $f(x)$ approach a value L as x approaches c from the left, we say that L is the *left-hand limit* of $f(x)$ as x approaches c and we write

$$\lim_{x \to c^-} f(x) = L.$$

If the values of a function $f(x)$ approach a value R as x approaches c from the right, we say that R is the *right-hand limit* of $f(x)$ as x approaches c and we write

$$\lim_{x \to c^+} f(x) = R.$$

Note that the notation $x \to c^-$ does not mean anything about whether c is a positive or negative number, only that x approaches c from the left. The two-sided limit of $f(x)$ as $x \to c$

exists if and only if the left and right limits as x approaches c exist and are equal. This means that both the left and right limits approach the same real number.

For example, the function graphed here has a different limit from the left than from the right as x approaches 1:

$$\lim_{x \to 1^-} f(x) = 2 \textit{ but } \lim_{x \to 1^+} f(x) = 3$$

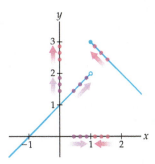

The purple sequence of values of x that approach $x = 1$ from the left determines a sequence of values of $f(x)$ that approach $y = 2$, while the red sequence determines values that approach $y = 3$. The value of the function at $x = 1$ happens to be $f(1) = 3$, but that is not relevant to either limit calculation. Since the limits from the left and right are not the same, there is no one real number that the function approaches as $x \to c$ and we say the two-sided limit **does not exist**.

Infinite Limits, Limits at Infinity, and Asymptotes

Armed with the concept of limits, we can now give proper definitions for horizontal and vertical asymptotes. If a function f increases or decreases without bound as x approaches a real number c from either the right or the left, then f has a vertical asymptote at $x = c$:

DEFINITION 1.3

Vertical Asymptotes

A function f has a **vertical asymptote** at $x = c$ if one or more of the following are true:

$$\lim_{x \to c^+} f(x) = \infty, \quad \lim_{x \to c^-} f(x) = \infty, \quad \lim_{x \to c^+} f(x) = -\infty, \quad \text{or} \quad \lim_{x \to c^-} f(x) = -\infty.$$

If $f(x)$ approaches ∞ from both the left and the right as $x \to c$, then we say that $\lim_{x \to c} f(x) = \infty$, as happens in the leftmost graph that follows. If $f(x)$ approaches $-\infty$ from both the left and the right, then we say that $\lim_{x \to c} f(x) = -\infty$. If $f(x)$ approaches different signs of infinity from the left and the right, then the two-sided limit $\lim_{x \to c} f(x)$ does not exist, as happens in the middle graph.

$$\lim_{x \to 1} f(x) = \infty$$

$$\lim_{x \to 1^-} f(x) = -\infty, \lim_{x \to 1^+} f(x) = \infty$$

$$\lim_{x \to \infty} f(x) = 10$$

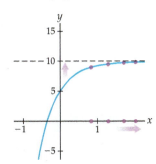

If the values of a function $f(x)$ approach a real-number value as x increases or decreases without bound, then f has a horizontal asymptote. For example, the rightmost graph of the three just presented shows a function with a horizontal asymptote and its corresponding limit. In general, we have the following definition:

DEFINITION 1.4   **Horizontal Asymptotes**

A nonconstant function f has a **horizontal asymptote** at $y = L$ if one or both of the following are true:

$$\lim_{x \to \infty} f(x) = L, \quad \text{or} \quad \lim_{x \to -\infty} f(x) = L.$$

Note that by convention, if $f(x)$ is actually *equal* to L as $x \to \infty$ or as $x \to -\infty$, then we do not consider f to have a horizontal asymptote at $y = L$. For example, the constant function $f(x) = 2$ has $\lim_{x \to \infty} f(x) = 2$ and yet does *not* have a horizontal asymptote at $y = 2$, since $f(x)$ is constantly equal to 2 as $x \to \infty$.

Examples and Explorations

EXAMPLE 1   Determining limits with tables of values

Use tables of values to find (a) $\lim_{x \to 1}(x + 1)$ and (b) $\lim_{x \to \infty} \dfrac{x}{x - 1}$.

SOLUTION

(a) To see what happens to $x + 1$ as $x \to 1$, we choose a sequence of values approaching $x = 1$ from the left and a sequence approaching $x = 1$ from the right, and record the corresponding values of $x + 1$:

x	.9	.99	.999	1	1.001	1.01	1.1
$x + 1$	1.9	1.99	1.999	*	2.001	2.01	2.1

From both the left and the right, the values of of $x + 1$ approach 2. Assuming that this pattern continues for values of x that are even closer to 1, we have $\lim_{x \to 1}(x + 1) = 2$.

(b) To see what happens to $\dfrac{x}{x-1}$ as $x \to \infty$, we choose a sequence of values of x that gets larger and larger, and record the corresponding (rounded) values of $\dfrac{x}{x-1}$:

x	25	50	100	1000	10,000
$\dfrac{x}{x-1}$	1.04167	1.02041	1.0101	1.001	1.0001

As x grows larger, the quantity $\dfrac{x}{x-1}$ approaches 1, so, assuming that the pattern in the table continues for even larger values of x, we have $\lim_{x \to \infty} \dfrac{x}{x-1} = 1$. $\square$

EXAMPLE 2   Graphically identifying limits

Determine the limits at any holes, corners, or asymptotes on the graphs of the functions (a) f, (b) g, and (c) h:

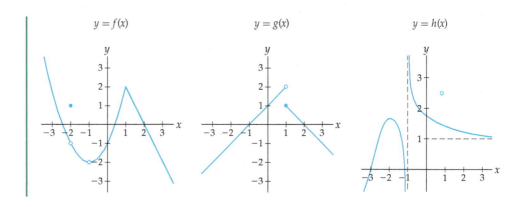

SOLUTION

(a) Observe that the graph in question has holes at $x = -2$ and $x = -1$ and a corner at $x = 1$, so we will examine limits at those three points. As x approaches -2, the height of the graph of $y = f(x)$ approaches -1. The value of $f(x)$ at $x = -2$ is $f(-2) = 1$, but this is not relevant to what $f(x)$ *approaches* as $x \to -2$ and thus does not affect the limit. Therefore $\lim_{x \to -2} f(x) = -1$.

As x approaches -1, the height of the graph approaches -2. The value $f(-1)$ is not defined, but that is not relevant to the limit as $x \to -1$. Thus $\lim_{x \to -1} f(x) = -2$.

As x approaches 1, the height of the graph approaches 2. The value of $f(x)$ at $x = 1$ happens to also equal 2, although this is irrelevant to the value of the limit as $x \to 1$. We have $\lim_{x \to 1} f(x) = 2$.

(b) The function $g(x)$ approaches different values as we approach $x = 1$ from the left and the right. As x approaches 1 from the left, the height of the graph approaches $y = 2$. As x approaches 1 from the right, the height of the graph approaches $y = 1$. The value of $g(x)$ at $x = 1$ happens to be $g(1) = 1$, but that is not relevant to either limit. Therefore we have $\lim_{x \to 1^-} g(x) = 2$ and $\lim_{x \to 1^+} g(x) = 1$, but the two-sided limit $\lim_{x \to 1} g(x)$ does not exist.

(c) The function $h(x)$ has a vertical asymptote at $x = -1$, with the height of the function decreasing without bound as we approach from the left and increasing without bound as we approach from the right. Therefore we have $\lim_{x \to -1^-} h(x) = -\infty$ and $\lim_{x \to -1^+} h(x) = \infty$, but the limit $\lim_{x \to -1} h(x)$ does not exist.

Let's also investigate the limits at the ends of the graph of $h(x)$. On the left side, as $x \to -\infty$, the graph decreases without bound; therefore $\lim_{x \to -\infty} h(x) = -\infty$. On the right side, as $x \to \infty$, the height of the function approaches $y = 1$. Therefore $h(x)$ has a horizontal asymptote on the right, and $\lim_{x \to \infty} h(x) = 1$.   ☐

EXAMPLE 3   A function with infinitely many oscillations as x approaches 0

Use a table of values and various graphing windows on a calculator or other graphing utility to investigate the limit of the function $f(x) = \sin\left(\frac{1}{x}\right)$ as x approaches 0.

SOLUTION

Be sure that your calculator or other graphing utility is set to radian mode, rather than degree mode, for this example. To see what happens to the quantity $\sin\left(\frac{1}{x}\right)$ as x approaches 0, we choose points progressively closer to $x = 0$ from both the left and the right and record the corresponding (rounded) values of $\sin\left(\frac{1}{x}\right)$ in a table:

x	−0.001	−0.0001	−0.00000001	0	0.00000001	0.0001	0.001
$\sin\left(\frac{1}{x}\right)$	−0.827	0.306	−0.932	??	0.932	−0.306	0.827

From the table we see that as x approaches 0, the values of $f(x)$ seem to jump around! It is not clear whether or not the function $f(x) = \sin\left(\frac{1}{x}\right)$ will eventually approach any particular value as $x \to 0$. It is impossible to make an educated guess for $\lim\limits_{x\to 0} \sin\left(\frac{1}{x}\right)$ from this table.

The reason that this is happening is that as $x \to 0$ the function $f(x) = \sin\left(\frac{1}{x}\right)$ oscillates faster and faster between −1 and 1, and never settles down. The graph on the left that follows shows this function on $[-3, 3]$, and the graph on the right shows the same function after reducing by a factor of 10 on the x-scale (but keeping the y-scale the same). No matter how much we "zoom in" towards $x = 0$, this function will keep oscillating. Therefore $\lim\limits_{x\to 0} \sin\left(\frac{1}{x}\right)$ does not exist.

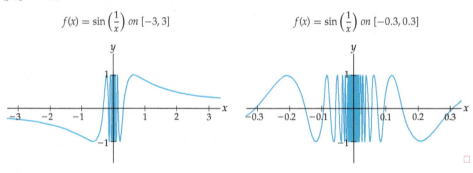

$f(x) = \sin\left(\frac{1}{x}\right)$ on $[-3, 3]$ $f(x) = \sin\left(\frac{1}{x}\right)$ on $[-0.3, 0.3]$

EXAMPLE 4   Areas under curves as another application of limits

Consider the area between the graph of $f(x) = x^2$ and the x-axis from $x = 0$ to $x = 1$, as shown in the leftmost figure.

(a) Find the value of the four-rectangle approximation shown in the middle figure.

(b) Find the value of the eight-rectangle approximation shown in the rightmost figure.

(c) Describe what would happen if we were to do similar approximations with more and more rectangles.

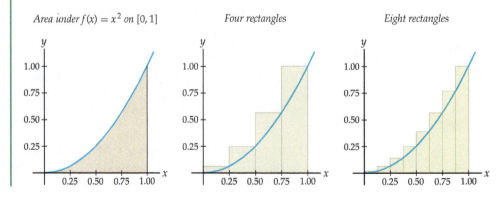

Area under $f(x) = x^2$ on $[0, 1]$ *Four rectangles*   *Eight rectangles*

SOLUTION

(a) The four rectangles in the given middle graph each have width $\frac{1}{4}$, and their heights are given by $f\left(\frac{1}{4}\right), f\left(\frac{2}{4}\right), f\left(\frac{3}{4}\right),$ and $f\left(\frac{4}{4}\right)$, from left to right. Therefore, since $f(x) = x^2$, the sum of the areas of the four rectangles is

$$\left(\frac{1}{4}\right)^2\left(\frac{1}{4}\right) + \left(\frac{2}{4}\right)^2\left(\frac{1}{4}\right) + \left(\frac{3}{4}\right)^2\left(\frac{1}{4}\right) + \left(\frac{4}{4}\right)^2\left(\frac{1}{4}\right) = \frac{15}{32} \approx 0.4688.$$

(b) Similarly, the eight rectangles in the given rightmost graph each have width $\frac{1}{8}$, and their heights are given by $f\left(\frac{1}{8}\right), f\left(\frac{2}{8}\right), f\left(\frac{3}{8}\right), \ldots, f\left(\frac{8}{8}\right)$, from left to right. Therefore the sum of their eight areas is

$$\left(\frac{1}{8}\right)^2\left(\frac{1}{8}\right) + \left(\frac{2}{8}\right)^2\left(\frac{1}{8}\right) + \left(\frac{3}{8}\right)^2\left(\frac{1}{8}\right) + \cdots + \left(\frac{8}{8}\right)^2\left(\frac{1}{8}\right) = \frac{51}{128} \approx 0.3984.$$

(c) So what happens when we do similar approximations with more rectangles? Consider the following three figures, where we consider more and more rectangles:

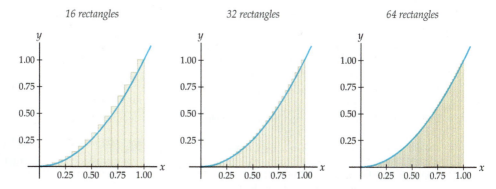

Clearly as we let the number N of rectangles get larger and larger, the sum of the areas of the rectangles gets closer and closer to a particular real number, namely, the actual area under the graph of $f(x) = x^2$ on $[0, 1]$. In limit notation we have

$$\lim_{N \to \infty} (N\text{-rectangle area approximation}) = (\text{actual area under graph}).$$

It turns out that the area approximations corresponding to the 16-, 32-, and 64-rectangle figures are 0.3652, 0.3491, and 0.3412, respectively. In Chapter 7 we will develop theory that will allow us to show that these area approximations have a limit of $\frac{1}{3}$ as $N \to \infty$.   □

TEST YOUR UNDERSTANDING

▶ What is the difference between taking a limit of some quantity as, say, x approaches 2 and actually computing the value of the quantity at $x = 2$?

▶ What can you say about the two-sided limit of a function as $x \to c$ if the left and right limits as $x \to c$ both exist but are not equal to each other?

▶ Can a function have more than one horizontal asymptote? More than two? Use Definition 1.4 to support your answer.

▶ Explain why the table in Example 1 cannot guarantee that $\lim_{x \to 1}(x+1)$ is actually 2 rather than some other number that is close to 2, such as 2.000035.

▶ Why does the table in Example 3 suggest that $\lim_{x \to 0}\sin\left(\frac{1}{x}\right)$ does not exist?

EXERCISES 1.1

Thinking Back

Finding the pattern in a sequence: For each sequence shown, find the next two terms. Then write a general form for the kth term of the sequence.

▶ $2, 6, 10, 14, 18, 22, \ldots$ ▶ $1, 8, 27, 64, 125, 216, \ldots$

▶ $3, \dfrac{3}{4}, \dfrac{3}{9}, \dfrac{3}{16}, \dfrac{3}{25}, \dfrac{3}{36}, \cdots$ ▶ $1, \dfrac{1}{3}, \dfrac{1}{9}, \dfrac{1}{27}, \dfrac{1}{81}, \dfrac{1}{243}, \cdots$

▶ $\dfrac{3}{5}, \dfrac{4}{7}, \dfrac{5}{9}, \dfrac{6}{11}, \dfrac{7}{13}, \dfrac{8}{15}, \cdots$ ▶ $\dfrac{3}{2}, \dfrac{5}{5}, \dfrac{7}{10}, \dfrac{9}{17}, \dfrac{11}{26}, \dfrac{13}{37}, \cdots$

Distance, rate, and time: A watermelon dropped from the top of a 50-foot building has height given by $s(t) = 50 - 16t^2$ feet after t seconds. Calculate each of the following:

▶ The average rate of change of the watermelon over its entire fall, over the first half of its fall, and over the second half of its fall.

▶ The average rate of change over the last second, the last half-second, and the last quarter-second of its fall.

Concepts

0. *Problem Zero:* Read the section and make your own summary of the material.

1. *True/False:* Determine whether each of the statements that follow is true or false. If a statement is true, explain why. If a statement is false, provide a counterexample.

(a) *True or False:* A limit exists if there is some real number that it is equal to.

(b) *True or False:* The limit of $f(x)$ as $x \to c$ is the value $f(c)$.

(c) *True or False:* The limit of $f(x)$ as $x \to c$ might exist even if the value $f(c)$ does not.

(d) *True or False:* The two-sided limit of $f(x)$ as $x \to c$ exists if and only if the left and right limits of $f(x)$ exist as $x \to c$.

(e) *True or False:* If the graph of f has a vertical asymptote at $x = 5$, then $\lim\limits_{x \to 5} f(x) = \infty$.

(f) *True or False:* If $\lim\limits_{x \to 5} f(x) = \infty$, then the graph of f has a vertical asymptote at $x = 5$.

(g) *True or False:* If $\lim\limits_{x \to 2} f(x) = \infty$, then the graph of f has a horizontal asymptote at $x = 2$.

(h) *True or False:* If $\lim\limits_{x \to -\infty} f(x) = 2$, then the graph of f has a horizontal asymptote at $y = 2$.

2. *Examples:* Construct examples of the thing(s) described in the following. Try to find examples that are different than any in the reading.

(a) The graph of a function f for which $f(2)$ does not exist but $\lim\limits_{x \to 2} f(x)$ does exist.

(b) The graph of a function f for which $f(2)$ exists and $\lim\limits_{x \to 2} f(x)$ exists, but the two are not equal.

(c) The graph of a function f for which neither $f(2)$ nor $\lim\limits_{x \to 2} f(x)$ exist.

3. If $\lim\limits_{x \to 1^-} f(x) = 5$ and $\lim\limits_{x \to 1^+} f(x) = 5$, what can you say about $\lim\limits_{x \to 1} f(x)$? What can you say about $f(1)$?

4. If $\lim\limits_{x \to 0^+} f(x) = -2$, $\lim\limits_{x \to 0^-} f(x) = 3$, and $f(0) = -2$, what can you say about $\lim\limits_{x \to 0} f(x)$?

5. If $\lim\limits_{x \to 2^+} f(x) = 8$ but $\lim\limits_{x \to 2} f(x)$ does not exist, what can you say about $\lim\limits_{x \to 2^-} f(x)$?

6. If $\lim\limits_{x \to -1^+} f(x) = -\infty$ and $\lim\limits_{x \to -1^-} f(x) = -\infty$, what can you say about $\lim\limits_{x \to -1} f(x)$?

7. If $\lim\limits_{x \to -\infty} f(x) = \infty$, $\lim\limits_{x \to \infty} f(x) = 3$, and $\lim\limits_{x \to 1^+} f(x) = \infty$, what can you say about any horizontal and vertical asymptotes of f?

8. Consider the sequence $\dfrac{1}{2}, \dfrac{2}{3}, \dfrac{3}{4}, \dfrac{4}{5}, \cdots, \dfrac{k}{k+1}, \cdots$.

(a) What happens to the terms of this sequence as k gets larger and larger? Express your answer in limit notation.

(b) Use a calculator to find a sufficiently large value of k so that every term past the kth term of this sequence will be within 0.01 unit of 1.

9. Consider the sequence $\dfrac{1}{3}, \dfrac{1}{9}, \dfrac{1}{27}, \dfrac{1}{81}, \cdots, \dfrac{1}{3^k}, \cdots$.

(a) What happens to the terms of this sequence as k gets larger and larger? Express your answer in limit notation.

(b) Find a sufficiently large value of k so that every term past the kth term of this sequence will be less than 0.0001.

10. Consider the sequence of sums $\dfrac{1}{3}, \dfrac{1}{3} + \dfrac{1}{9}, \dfrac{1}{3} + \dfrac{1}{9} + \dfrac{1}{27}, \dfrac{1}{3} + \dfrac{1}{9} + \dfrac{1}{27} + \dfrac{1}{81}, \cdots$.

(a) What happens to the terms of this sequence of sums as k gets larger and larger?

(b) Find a sufficiently large value of k which will guarantee that every term past the kth term of this sequence of sums is in the interval $(0.49999, 0.5)$.

11. Consider the sequence of sums 1, 1 + 2, 1 + 2 + 3, 1 + 2 + 3 + 4, 1 + 2 + 3 + 4 + 5, ...

(a) What happens to the terms of this sequence of sums as k gets larger and larger?

(b) Find a sufficiently large value of k that will guarantee that every term past the kth term of this sequence of sums is greater than 1000.

12. An orange falling from 20 feet has a height of $s(t) = 20 - 16t^2$ feet when it has fallen for t seconds.

(a) Graph the position function $s(t)$ and find the time that the orange will hit the ground.

(b) Make a table to record the average rates that the orange is falling during the last second, half-second, quarter-second, and eighth-of-a-second of its fall.

(c) From the data in your table, make a guess for the instantaneous final velocity of the orange at the moment it hits the ground.

13. If you are on the moon, then an orange falling from 20 feet has a height of $s(t) = 20 - 2.65t^2$ feet when it has fallen for t seconds.

(a) Graph the position function $s(t)$ and find the time that the orange will hit the surface of the moon.

(b) Make a table to record the average rates that the orange is falling during the last second, half-second, quarter-second, and eighth-of-a-second of its fall on the moon.

(c) From the data in your table, make a guess for the instantaneous final velocity of the orange at the moment it hits the surface of the moon.

14. Consider the area between the graph of $f(x) = \sqrt{x}$ and the x-axis on $[0, 4]$.

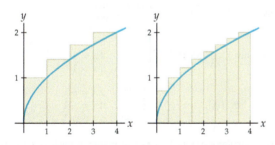

(a) Use the four rectangles shown on the left to approximate the given area, and then use the eight rectangles shown on the right to obtain another approximation of that area. Be sure to use the fact that the graph shown is that of the function $f(x) = \sqrt{x}$ in your calculations.

(b) Describe what would happen if we did similar approximations with more and more rectangles, and make a guess for the resulting limit.

15. Consider the area between the graph of $f(x) = 4 - x^2$ and the x-axis on $[0, 2]$.

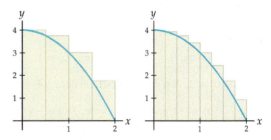

(a) Use the four rectangles shown on the left to approximate the given area, and then use the eight rectangles shown on the right to obtain another approximation of that area. Be sure to use the fact that the graph shown is that of the function $f(x) = 4 - x^2$ in your calculations.

(b) Describe what would happen if we did similar approximations with more and more rectangles, and make a guess for the resulting limit.

16. Sketch a function that has the following table of values, but whose limit as $x \to \infty$ is equal to $-\infty$:

x	100	200	500	1,000	10,000
$f(x)$	50	55	56.2	56.89	56.99

17. Sketch a function that has the following table of values, but whose limit as $x \to 2$ does not exist:

x	1.9	1.99	1.999	2	2.001	2.01	2.1
$f(x)$	3.12	3.09	3.01	-	2.99	2.92	2.87

18. Use a calculator or other graphing utility to graph the function $f(x) = \dfrac{x-2}{x^2 - x - 2}$.

(a) Show that $f(x)$ is not defined at $x = 2$. How is this reflected in your calculator graph?

(b) Use the graph to argue that even though $f(2)$ is undefined, we have $\lim\limits_{x \to 2} f(x) = \dfrac{1}{3}$.

19. Use a calculator or other graphing utility to graph the function $g(x) = \dfrac{x^2 - 2x + 1}{x - 1}$.

(a) Show that $g(x)$ is not defined at $x = 1$. How is this reflected in your calculator graph?

(b) Use the graph to argue that even though $g(1)$ is undefined, we have $\lim\limits_{x \to 1} g(x) = 0$.

20. Use a calculator or other graphing utility to investigate the graph $f(x) = x \sin\left(\dfrac{1}{x}\right)$ near $x = 0$. Be sure to have your calculator set to radian mode. Use the graphs to make an educated guess for $\lim\limits_{x \to 0} f(x)$.

Skills

Sketch the graphs of functions that have the given limits and values in Exercises 21–30. (There are multiple correct answers.)

21. $\lim\limits_{x \to -\infty} f(x) = 3$ and $\lim\limits_{x \to \infty} f(x) = -\infty$

22. $\lim\limits_{x \to 2} f(x) = -4$ and $\lim\limits_{x \to -\infty} f(x) = -\infty$

23. $\lim\limits_{x \to 0^+} f(x) = \infty$ and $\lim\limits_{x \to 0^-} f(x) = \infty$

24. $\lim\limits_{x \to 5^-} f(x) = 3$ and $\lim\limits_{x \to 5^+} f(x) = 1$

25. $\lim\limits_{x \to -\infty} f(x) = 2$ and $\lim\limits_{x \to \infty} f(x) = 2$

26. $\lim\limits_{x \to 2^-} f(x) = \infty$, $\lim\limits_{x \to 2^+} f(x) = -\infty$, and $f(2) = 1$

27. $\lim\limits_{x \to 3^-} f(x) = 2$, $\lim\limits_{x \to 3^+} f(x) = 2$, but $f(3)$ does not exist

28. $\lim\limits_{x \to -\infty} f(x) = -2$ and $\lim\limits_{x \to 3} f(x) = \infty$, $f(0) = -5$

29. $\lim\limits_{x \to 2^-} f(x) = 2$, $\lim\limits_{x \to 2^+} f(x) = -1$, and $f(2) = 2$

30. $\lim\limits_{x \to 2^-} f(x) = 3$, $\lim\limits_{x \to 2^+} f(x) = 3$, and $f(2) = 0$

For the function f graphed as follows, approximate each of the limits and values in Exercises 31–34:

31. $\lim\limits_{x \to -2^-} f(x)$, $\lim\limits_{x \to -2^+} f(x)$, $\lim\limits_{x \to -2} f(x)$, and $f(-2)$.

32. $\lim\limits_{x \to -1^-} f(x)$, $\lim\limits_{x \to -1^+} f(x)$, $\lim\limits_{x \to -1} f(x)$, and $f(-1)$.

33. $\lim\limits_{x \to 2^-} f(x)$, $\lim\limits_{x \to 2^+} f(x)$, $\lim\limits_{x \to 2} f(x)$, and $f(2)$.

34. $\lim\limits_{x \to 0} f(x)$, $\lim\limits_{x \to 1} f(x)$, $\lim\limits_{x \to -\infty} f(x)$, and $\lim\limits_{x \to \infty} f(x)$.

For the function $g(x)$ graphed as follows, approximate each of the limits and values in Exercises 35–38:

35. $\lim\limits_{x \to -1^-} g(x)$, $\lim\limits_{x \to -1^+} g(x)$, $\lim\limits_{x \to -1} g(x)$, and $g(-1)$.

36. $\lim\limits_{x \to 1^-} g(x)$, $\lim\limits_{x \to 1^+} g(x)$, $\lim\limits_{x \to 1} g(x)$, and $g(1)$.

37. $\lim\limits_{x \to 2^-} g(x)$, $\lim\limits_{x \to 2^+} g(x)$, $\lim\limits_{x \to 2} g(x)$, and $g(2)$.

38. $\lim\limits_{x \to 0} g(x)$, $\lim\limits_{x \to 3} g(x)$, $\lim\limits_{x \to -\infty} g(x)$, and $\lim\limits_{x \to \infty} g(x)$.

Use tables of values to make educated guesses for each of the limits in Exercises 39–50.

39. $\lim\limits_{x \to 2^-} (x^2 + x + 1)$

40. $\lim\limits_{x \to 3^+} (1 - 3x + x^2)$

41. $\lim\limits_{x \to 2} \dfrac{1}{x^2 - 4}$

42. $\lim\limits_{x \to 1} \dfrac{1}{1 - x}$

43. $\lim\limits_{x \to 3} \dfrac{x - 3}{(x^2 - 2)(x - 3)}$

44. $\lim\limits_{x \to 5} \dfrac{x - 5}{x^2 - 25}$

45. $\lim\limits_{x \to -\infty} \dfrac{3x + 1}{1 - x}$

46. $\lim\limits_{x \to \infty} \dfrac{1 + 2x}{x - 1}$

47. $\lim\limits_{x \to \infty} \dfrac{x + 1}{x^2 - 1}$

48. $\lim\limits_{x \to \infty} \left(1 + \dfrac{1}{2x + 1} + \dfrac{1}{x^2}\right)$

49. $\lim\limits_{x \to \infty} \sin x$

50. $\lim\limits_{x \to \infty} \sin\left(\dfrac{1}{x}\right)$

Sketch graphs by hand and use them to make approximations for each of the limits in Exercises 51–62. If a two-sided limit does not exist, describe the one-sided limits.

51. $\lim\limits_{x \to 2}(3x + 1)$

52. $\lim\limits_{x \to \infty} (1 - 2x)$

53. $\lim\limits_{x \to 2} |x - 2|$

54. $\lim\limits_{x \to \infty} |5x - 4|$

55. $\lim\limits_{x \to 0} \dfrac{1}{x}$

56. $\lim\limits_{x \to -1} (x^3 - 2)$

57. $\lim\limits_{x \to 1} \dfrac{x^2 - 1}{x - 1}$

58. $\lim\limits_{x \to -2} \dfrac{x^2 + x - 2}{x + 2}$

59. $\lim\limits_{x \to 2} f(x)$, for $f(x) = \begin{cases} x^2, & \text{if } x < 2 \\ 1 - 3x, & \text{if } x \geq 2 \end{cases}$

60. $\lim\limits_{x \to 0} f(x)$, for $f(x) = \begin{cases} 2x + 1, & \text{if } x \leq 0 \\ 2x - 1, & \text{if } x > 0 \end{cases}$

61. $\lim\limits_{x \to 1} f(x)$, for $f(x) = \begin{cases} x^2 + 1, & \text{if } x < 1 \\ 3, & \text{if } x = 1 \\ 3 - x, & \text{if } x > 1 \end{cases}$

62. $\lim\limits_{x \to -1} f(x)$, for $f(x) = \begin{cases} x + 1, & \text{if } x < -1 \\ 2, & \text{if } x = -1 \\ -x^2, & \text{if } x > -1 \end{cases}$

Use calculator graphs to make approximations for each of the limits in Exercises 63–70.

63. $\lim\limits_{x \to 4}(3 - 4x - 5x^2)$

64. $\lim\limits_{x \to \infty} (-0.2x^5 + 100x)$

65. $\lim\limits_{x \to 1} \dfrac{3 - x}{x - 1}$

66. $\lim\limits_{x \to 2} \dfrac{x + 1}{x - 2}$

67. $\lim\limits_{x \to \infty} \dfrac{x^{100}}{2^x}$

68. $\lim\limits_{x \to \infty} \dfrac{\ln x}{x}$

69. $\lim\limits_{x \to 0} \dfrac{\sin x}{x}$

70. $\lim\limits_{x \to 0} \dfrac{1 - \cos x}{x}$

Applications

71. There are four squirrels currently living in Linda's attic. If she does nothing to evict these squirrels, the number of squirrels in her attic after t days will be given by the formula $S(t) = \dfrac{12 + 5.5t}{3 + 0.25t}$.

(a) Verify that there are four squirrels in Linda's attic at time $t = 0$.

(b) Determine the number of squirrels in Linda's attic after 30 days, 60 days, and one year.

(c) Approximate $\lim\limits_{t \to \infty} S(t)$ with a table of values. What does this limit mean in real-world terms?

(d) Graph $S(t)$ with a graphing utility, and use the graph to verify your answer to part (c).

72. The following graph describes the temperature $T(t)$ of a yam in an oven, where temperature T is measured in degrees Fahrenheit and time t is measured in minutes:

Temperature of yam

(a) Use the graph to approximate the temperature of the yam when it is first put in the oven.

(b) Use the graph to approximate $\lim\limits_{t \to \infty} T(t)$.

(c) What is the temperature of the oven, and why?

73. In 1960, H. von Foerster suggested that the human population could be measured by the function

$$P(t) = \frac{179 \times 10^9}{(2027 - t)^{0.99}}.$$

Here P is the size of the human population. The time t is measured in years, where $t = 1$ corresponds to the year 1 A.D., time $t = 1973$ corresponds to the year 1973 A.D., and so on.

(a) Use a graphing utility to graph this function. You will have to be very careful when choosing a graphing window!

(b) Use the graph you found in part (a) to approximate $\lim\limits_{t \to 2027^-} P(t)$.

(c) This population model is sometimes called the ***doomsday model.*** Why do you think this is? What year is doomsday, and why?

(d) In part (b), we considered only the *left* limit of $P(t)$ as $x \to 2027$. Why? What is the real-world meaning of the part of the graph that is to the right of $t = 2027$?

Proofs

74. Prove that for all $k > 100$, the quantity $\dfrac{1}{k^2}$ is in the interval $(0, 0.0001)$. What does this have to do with the limit of the sequence $\left\{ \dfrac{1}{k^2} \right\}$ as $k \to \infty$?

75. For any positive integer k, the following equation holds: $1 + 2 + 3 + \cdots + k = \dfrac{k(k+1)}{2}$. Use this fact to prove that for all $k > 100$, the value of the sum of the first k integers is greater than 5000. What does this have to do with the limit of a sequence of sums as $k \to \infty$?

76. Prove that for all x within 0.01 of the value $x = 1$, the quantity $(x - 1)^2$ is within the interval $(0, 0.0001)$. What does this have to do with $\lim\limits_{x \to 1}(x - 1)^2$?

77. Prove that for all x within 0.01 of the value $x = 1$, the quantity $\dfrac{1}{(x-1)^2}$ is greater than 10,000. What does this have to do with $\lim\limits_{x \to 1} \dfrac{1}{(x-1)^2}$?

Thinking Forward

Convergence and divergence of sequences: If a sequence $a_1, a_2, a_3, \ldots, a_k, \ldots$ approaches a real-number limit as $k \to \infty$, then we say that the sequence $\{a_k\}$ ***converges.*** If the terms of the sequence do not get arbitrarily close to some real number, then we say that the sequence ***diverges.*** Write out enough terms of each sequence to make an educated guess as to whether it converges or diverges.

▶ $\left\{ \left(\dfrac{1}{4} \right)^k \right\}$ ▶ $\left\{ \left(\dfrac{5}{4} \right)^k \right\}$ ▶ $\left\{ \dfrac{k}{k+2} \right\}$ ▶ $\left\{ \dfrac{k+1}{k} \right\}$

Convergence and divergence of series: A ***series*** can be thought of as an infinite sum $a_1 + a_2 + a_3 + a_4 + \cdots + a_k + \cdots$. A series ***converges*** if this sum gets closer and closer to some real number limit as we add up more and more terms. Otherwise, the series is said to ***diverge.***

▶ In Calculus 2, you will learn that the series $1 + \dfrac{1}{4} + \dfrac{1}{9} + \dfrac{1}{16} + \cdots + \dfrac{1}{k^2} + \cdots$ converges. Calculate partial sums including more and more terms until you are convinced that the sum eventually approaches a real-number limit and does not grow without bound.

▶ Although you might think that the series $1 + \dfrac{1}{2} + \dfrac{1}{3} + \dfrac{1}{4} + \cdots + \dfrac{1}{k} + \cdots$ converges because its terms get smaller and smaller, it in fact does not. Calculate partial sums including more and more terms until you are convinced that this sum diverges and in fact grows without bound, never approaching a real-number limit. (*A calculator will come in handy here!*)

1.2   FORMAL DEFINITION OF LIMIT

▶ Moving from an intuitive concept of limit to a formal mathematical definition

▶ Uniqueness and existence of limits

▶ Limits from the left and right, limits at infinity, and infinite limits

Formalizing the Intuitive Definition of Limit

In the previous section we gave an intuitive description of limits. Now that we understand the basic concept, we are ready to give a precise, rigorous mathematical definition. Let's start with our intuitive description: For real numbers c and L and a function f, we have $\lim_{x \to c} f(x) = L$ if the values of $f(x)$ get closer and closer to L as x gets closer and closer to c. For example, $\lim_{x \to 2} x^2 = 4$ because the values of $f(x) = x^2$ approach 4 as x approaches 2. From the left, $f(1.9) = (1.9)^2 = 3.61$, $f(1.99) = (1.99)^2 = 3.9601$, $f(1.999) = (1.999)^2 \approx 3.996$, and so on, getting closer and closer to 4. A similar thing happens as x approaches 2 from the right.

Note that to be able to discuss $\lim_{x \to c} f(x)$, we must know how to calculate $f(x)$ near, but not necessarily at, the point $x = c$. Throughout this section we will assume that $f(x)$ is defined on a **punctured interval** $(c - \delta, c) \cup (c, c + \delta)$, where $\delta > 0$ represents a small distance to the left and right of $x = c$, as shown on the number line that follows. Notice that in our discussion of limits we will never be concerned with what happens *at* the point $x = c$, only *near* the point $x = c$.

Punctured δ-interval around c

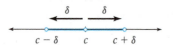

To make the definition of limit precise, we have to be very clear about what we mean when we say that $f(x)$ "approaches" L. We want to capture the idea that we can make the values of $f(x)$ not just close to L, but *as close as we like* to L if only we choose values of x that are sufficiently close to c. For example, we can guarantee that $f(x) = x^2$ is within 0.05 unit of 4 if we choose values of x that are within 0.01 unit of 2. Note that $f(2.01) = (2.01)^2 = 4.0401$ and $f(1.99) = (1.99)^2 = 3.9601$ are both within 0.05 unit of 4, and values of x that are closer to 2 will result in values of $f(x)$ that are even closer to 4. If we want values of $f(x)$ that are even closer to $L = 4$, then we can just choose values of x that are even closer to $c = 2$.

In general, suppose we want to guarantee that the values of $f(x)$ are within some very small distance ϵ above or below limit value L, as shown at the left in the graphs that follow. To do this we must choose values of x that are sufficiently close to c, say, some distance $\delta > 0$ left or right of c, as shown in the middle graph. The Greek letters **delta** (δ) and **epsilon** (ϵ) are the traditional letters used for these small distances. The figure at the right illustrates that a choice of x-value inside the blue punctured δ-interval $(c - \delta, c) \cup (c, c + \delta)$ determines an $f(x)$-value within the beige ϵ-interval $(L - \epsilon, L + \epsilon)$. In these figures we have omitted the point at $x = c$ to emphasize that we are not concerned with the actual value of $f(x)$ at the point $x = c$, only with the behavior of $f(x)$ at points near $x = c$.

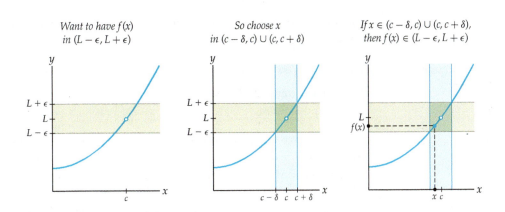

If the values of $f(x)$ get arbitrarily close to L as x approaches c, then we can choose smaller and smaller beige ϵ-intervals $(L - \epsilon, L + \epsilon)$ and in each case always find some blue punctured δ-interval $(c - \delta, c) \cup (c, c + \delta)$ that determines values of $f(x)$ which are within ϵ of the limit L. The following three figures illustrate this idea:

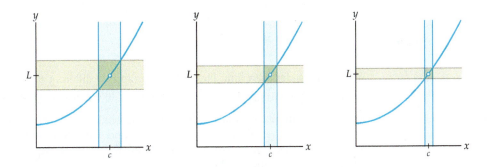

We want the limit statement $\lim\limits_{x \to c} f(x) = L$ to mean that no matter how small a distance we choose for ϵ, we can find some δ so that values of x that are within δ of $x = c$ will yield values of $f(x)$ that are within ϵ of $y = L$. Writing this in terms of intervals gives us the following definition:

DEFINITION 1.5   **Formal Definition of Limit**

The ***limit*** $\lim\limits_{x \to c} f(x) = L$ means that for all $\epsilon > 0$, there exists $\delta > 0$ such that

$$\text{if } x \in (c - \delta, c) \cup (c, c + \delta), \text{ then } f(x) \in (L - \epsilon, L + \epsilon).$$

Stop and think about that for a minute until it makes sense. Understanding this definition is the key to understanding limits, and limits are the foundation of everything in calculus. So take a few minutes, have some tea, and get everything straight in your head before you continue reading.

Uniqueness of Limits

A limit $\lim\limits_{x \to c} f(x)$ ***exists*** if it is equal to some real number L. If a limit exists, then it can be equal to one and only one number. That sounds obvious, since the values of a function $f(x)$ cannot approach two different values L and M as x approaches c. However, as we are about to see, to prove uniqueness of limits we must carefully apply the formal definition of limits.

THEOREM 1.6   **Uniqueness of Limits**

If $\lim_{x \to c} f(x) = L$ and $\lim_{x \to c} f(x) = M$, then $L = M$.

Proof. Suppose to the contrary that somehow $\lim_{x \to c} f(x) = L$ and $\lim_{x \to c} f(x) = M$, with $L \neq M$. Let's suppose that $L > M$, since, if not, then we can just reverse the roles of L and M. If $L > M$, then we must have $L = M + k$ for some positive real number k. Now consider $\epsilon = \dfrac{k}{2}$; note that with this choice of ϵ, the intervals $(L - \epsilon, L + \epsilon)$ and $(M - \epsilon, M + \epsilon)$ do not overlap.

Since $\lim_{x \to c} f(x) = L$, we can find $\delta_1 > 0$ such that for all $x \in (c - \delta_1, c) \cup (c, c + \delta_1)$, we have $f(x) \in (L - \epsilon, L + \epsilon)$. Similarly, since $\lim_{x \to c} f(x) = M$, we can find $\delta_2 > 0$ such that for all $x \in (c - \delta_2, c) \cup (c, c + \delta_2)$, we have $f(x) \in (M - \epsilon, M + \epsilon)$. Now if we let δ be the smaller of δ_1 and δ_2, we can say that for any $x \in (c - \delta, c) \cup (c, c + \delta)$, we can guarantee that both $f(x) \in (L - \epsilon, L + \epsilon)$ and $f(x) \in (M - \epsilon, M + \epsilon)$. But this cannot be, since the intervals $(L - \epsilon, L + \epsilon)$ and $(M - \epsilon, M + \epsilon)$ do not overlap. Therefore we could not have initially had $f(x)$ approaching two different limits L and M; we must have $L = M$. ∎

One-Sided Limits

We can consider each limit $\lim_{x \to c} f(x) = L$ from two different directions: from the left and from the right. We say that we have a left limit $\lim_{x \to c^-} f(x) = L$ if, given an ϵ-interval $(L - \epsilon, L + \epsilon)$, we can always find a sufficiently small half-neighborhood $(c - \delta, c)$ to the left of $x = c$ so that values of x that are in that left hand δ-interval yield values of $f(x)$ that are in the ϵ-interval, as shown in the left-hand graph that follows. We define right limits similarly, as shown in the right-hand graph:

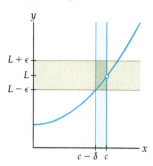

Can get $f(x)$ in $(L - \epsilon, L + \epsilon)$ by choosing x in $(c - \delta, c)$

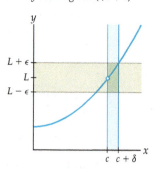

Can get $f(x)$ in $(L - \epsilon, L + \epsilon)$ by choosing x in $(c, c + \delta)$

DEFINITION 1.7   **One-Sided Limits**

The **left limit** $\lim_{x \to c^-} f(x) = L$ means that for all $\epsilon > 0$, there exists $\delta > 0$ such that

$$\text{if } x \in (c - \delta, c), \text{ then } f(x) \in (L - \epsilon, L + \epsilon).$$

The **right limit** $\lim_{x \to c^+} f(x) = L$ means that for all $\epsilon > 0$, there exists $\delta > 0$ such that

$$\text{if } x \in (c, c + \delta), \text{ then } f(x) \in (L - \epsilon, L + \epsilon).$$

A two-sided limit $\lim_{x \to c} f(x)$ is equal to some real number L if and only if the corresponding left and right limits exist and are also equal to that same number L.

THEOREM 1.8   **For a Limit to Exist, the Left and Right Limits Must Exist and Be Equal**

$\lim_{x \to c} f(x) = L$ if and only if $\lim_{x \to c^-} f(x) = L$ and $\lim_{x \to c^+} f(x) = L$.

The proof of this theorem is a straightforward application of the definitions of two-sided and one-sided limits.

> **Proof.** Suppose $\lim_{x \to c} f(x) = L$. Then for all $\epsilon > 0$, there is some $\delta > 0$ such that if $x \in (c - \delta, c) \cup (c, c + \delta)$, then $f(x) \in (L - \epsilon, L + \epsilon)$. In particular this means that if $x \in (c - \delta, c)$ or if $x \in (c, c + \delta)$, then we will have $f(x) \in (L - \epsilon, L + \epsilon)$. Therefore $\lim_{x \to c^-} f(x) = L$ and $\lim_{x \to c^+} f(x) = L$.
>
> For the converse, suppose $\lim_{x \to c^-} f(x) = L$ and $\lim_{x \to c^+} f(x) = L$. Then for all $\epsilon > 0$, there exist numbers $\delta_1 > 0$ and $\delta_2 > 0$ such that for either $x \in (c - \delta_1, c)$ or $x \in (c, c + \delta_2)$, we have $f(x) \in (L - \epsilon, L + \epsilon)$. If we let δ be the smaller of δ_1 and δ_2, then we can say that for $x \in (c - \delta, c) \cup (c, c + \delta)$, we can guarantee that $f(x) \in (L - \epsilon, L + \epsilon)$. Therefore $\lim_{x \to c} f(x) = L$. ∎

Infinite Limits and Limits at Infinity

So far we have formalized the definition of limit only in the case where both x and $f(x)$ are approaching real numbers. Now we consider what happens if one or both of x and $f(x)$ approach $\pm\infty$. For example, we want $\lim_{x \to c} f(x) = \infty$ to capture the idea that as x approaches c, the values of $f(x)$ grow without bound. In other words, $\lim_{x \to c} f(x) = \infty$ should guarantee that values of $f(x)$ will lie above any given large number M as long as we choose values of x that are sufficiently close to c; see the figure that follows at the left.

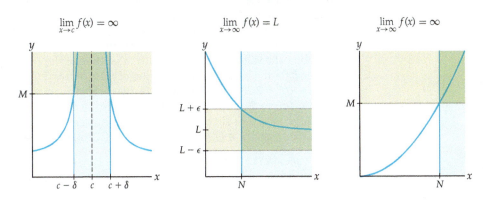

$$\lim_{x \to c} f(x) = \infty \qquad \lim_{x \to \infty} f(x) = L \qquad \lim_{x \to \infty} f(x) = \infty$$

Similarly, we want the limit statement $\lim_{x \to \infty} f(x) = L$ to indicate that given any ϵ-interval around L, we can choose values of x sufficiently large so that $f(x)$ is in the ϵ-interval; see the middle graph. And $\lim_{x \to \infty} f(x) = \infty$ should mean that given any large number M, we can get values of $f(x)$ that are greater than M if we choose sufficiently large values of x, say, larger than some big number N, as in the graph at the right. The definition that follows expresses these three limits in terms of intervals; compare the three parts with the preceding three figures. You will define other related types of limits in Exercises 37–42.

DEFINITION 1.9   **Limits Involving Infinity**

The **infinite limit** $\lim\limits_{x \to c} f(x) = \infty$ means that for all $M > 0$, there exists $\delta > 0$ such that

$$\text{if } x \in (c - \delta, c) \cup (c, c + \delta), \text{ then } f(x) \in (M, \infty).$$

The **limit at infinity** $\lim\limits_{x \to \infty} f(x) = L$ means that for all $\epsilon > 0$, there exists $N > 0$ such that

$$\text{if } x \in (N, \infty), \text{ then } f(x) \in (L - \epsilon, L + \epsilon).$$

The **infinite limit at infinity** $\lim\limits_{x \to \infty} f(x) = \infty$ means that for all $M > 0$, there exists $N > 0$ such that

$$\text{if } x \in (N, \infty), \text{ then } f(x) \in (M, \infty).$$

Examples and Explorations

EXAMPLE 1   Approximating δ given ϵ for a limit

Use a graph to illustrate and approximate

(a) the largest δ which guarantees that if $x \in (2 - \delta, 2) \cup (2, 2 + \delta)$, then $x^2 \in (3, 5)$.

(b) the largest δ which guarantees that if $x \in (2 - \delta, 2) \cup (2, 2 + \delta)$, then $x^2 \in (3.5, 4.5)$.

What limit statement do these problems have to do with, and why?

SOLUTION

These problems concern the limit statement $\lim\limits_{x \to 2} x^2 = 4$, which by definition means that for all $\epsilon > 0$, there exists $\delta > 0$ such that if $x \in (2 - \delta, 2) \cup (2, 2 + \delta)$, then $x^2 \in (4 - \epsilon, 4 + \epsilon)$. Therefore parts (a) and (b) of this example ask us to find corresponding values of δ for $\epsilon = 1$ and $\epsilon = 0.5$, respectively.

(a) To find the largest δ corresponding to $\epsilon = 1$, we begin by drawing $f(x) = x^2$ and the beige ϵ-interval of width 1 around $y = 4$, as shown in the graph that follows at the left. We then draw the vertical blue band shown in the figure, to represent the range of values of x which determine values of $f(x)$ that are within the horizontal beige band. The leftmost x-value a in the blue band is a solution of $f(a) = a^2 = 3$, and the rightmost x-value b for the blue band is a solution of $f(b) = b^2 = 5$. Therefore $a = \sqrt{3} \approx 1.732$ and $b = \sqrt{5} \approx 2.236$. Now, what is δ in this case? We can move $2 - 1.732 = 0.268$ unit to the left of $x = 2$ and $2.236 - 2 = 0.236$ unit to the right of $x = 2$. We need the largest δ that will work in *both* directions, which is the smaller of the two distances we just found: $\delta = 0.236$. Then if $x \in (2 - 0.236, 2) \cup (2, 2 + 0.236)$, we can guarantee that $x^2 \in (3, 5)$.

If $x \in (1.732, 2.236)$, then $x^2 \in (3, 5)$ *If $x \in (1.87, 2.12)$, then $x^2 \in (3.5, 4.5)$*

(b) We now repeat the problem with a smaller value of ϵ. For $\epsilon = 0.5$ we draw the smaller horizontal beige ϵ-bar shown in the preceding graph at the right, and the corresponding vertical blue bar. Solving $f(a) = a^2 = 3.5$ and $f(b) = b^2 = 4.5$, we get $a \approx 1.87$ and $b \approx 2.12$ as the leftmost and rightmost values of x contained in the vertical blue bar. Therefore values of x that are at most $2 - 1.87 = 0.13$ unit to the left of $x = 2$ or at most $2.12 - 2 = 0.12$ unit to the right of $x = 1$ will determine values of $f(x)$ that are within 0.5 unit of $y = 4$. The smaller of these two distances is the largest δ that will work in both directions, namely, $\delta = 0.12$. If $x \in (2 - 0.12, 2) \cup (2, 2 + 0.12)$, then we can guarantee that $x^2 \in (3.5, 4.5)$.   □

EXAMPLE 2

Approximating *N* given ϵ for a limit at infinity

Use a graph to illustrate and approximate

(a) the smallest N which guarantees that if $x \in (N, \infty)$, then $\frac{x+1}{x} \in (0.75, 1.25)$.

(b) the smallest N which guarantees that if $x \in (N, \infty)$, then $\frac{x+1}{x} \in (0.9, 1.1)$.

What limit statement do these problems have to do with, and why?

SOLUTION

These problems are about the limit statement $\lim\limits_{x \to \infty} \frac{x+1}{x} = 1$, which by definition means that for all $\epsilon > 0$, there is some $N > 0$ such that if $x \in (N, \infty)$, then $\frac{x+1}{x} \in (1 - \epsilon, 1 + \epsilon)$. Therefore parts (a) and (b) of this example ask us to find the corresponding values of N for $\epsilon = 0.25$ and $\epsilon = 0.1$, respectively.

(a) The figure that follows at the left shows $f(x) = \frac{x+1}{x}$ and a beige bar representing all the heights within 0.25 unit of $y = 1$. The blue area shows all of the values of x for which the corresponding values of $f(x)$ lie within the beige bar. According to this graph, to find the leftmost point $x = a$ of the blue area we must solve $f(a) = 1.25$:

$$f(a) = 1.25 \implies \frac{a+1}{a} = 1.25 \implies a + 1 = 1.25a \implies 1 = 0.25a \implies a = 4.$$

Therefore if $x \in (4, \infty)$, then we can guarantee that $\frac{x+1}{x} \in (0.75, 1.25)$.

If $x \in (4, \infty)$, then $\frac{x+1}{x} \in (0.75, 1.25)$ *If $x \in (10, \infty)$, then $\frac{x+1}{x} \in (0.9, 1.1)$*

    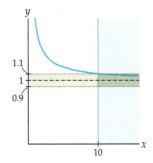

(b) We now do the same thing but for $\epsilon = 0.1$. For this smaller value of ϵ we must draw a smaller beige bar around $y = 1$, which in turn requires a different blue area of values of x for which the corresponding values of $f(x)$ lie within the ϵ-bar, as shown in the preceding graph at the right. To find the leftmost point $x = a$ of the blue area, we solve $f(a) = 1.1$:

$$f(a) = 1.1 \implies \frac{a+1}{a} = 1.1 \implies a + 1 = 1.1a \implies 1 = 0.1a \implies a = 10.$$

Thus for $x \in (10, \infty)$ we can guarantee that $\frac{x+1}{x} \in (0.9, 1.1)$. ▫

EXAMPLE 3   A real–world example of finding δ given ϵ

Fuel efficiency depends on driving speed. A typical car runs at 100% fuel efficiency when driven at 55 miles per hour. Suppose that the fuel efficiency percentage at speed s (in MPH) is given by $E(s) = -0.033058(s^2 - 110s)$. If you want your car to run with at least 95% fuel efficiency, how close to 55 miles per hour do you have to drive?

SOLUTION

In the language of limits, we are asking: For the limit $\lim\limits_{s \to 55} E(s) = 100$, if $\epsilon = 5$, what is δ? The corresponding ϵ-bar and δ-bar are shown below in the following graph of $E(s)$:

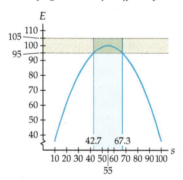

Staying within 5% fuel efficiency

The leftmost and rightmost values $x = a$ and $x = b$ of the blue δ-interval shown can be found by using the quadratic formula to find the two solutions of the equation $E(s) = 95$ or by using a calculator graph to approximate values. In either case, we find $a \approx 42.7$ and $b \approx 67.3$. Therefore, according to the graph you can drive anywhere between 42.7 and 67.3 miles per hour and get at least 95% fuel efficiency. ▫

TEST YOUR UNDERSTANDING

▶ When discussing limits as $x \to c$, why do we consider *punctured* intervals around $x = c$?

▶ In the definition of limit, why do we need the statement to be true for *all* values $\epsilon > 0$?

▶ In the proof of Theorem 1.6 we had $L = M + k$ and $\epsilon = \frac{k}{2}$. Why does this mean that $(L - \epsilon, L + \epsilon)$ and $(M - \epsilon, M + \epsilon)$ do not overlap?

▶ How are left limits and right limits related to two-sided limits?

▶ How is a limit at infinity different from an infinite limit?

EXERCISES 1.2

Thinking Back

Logical quantifiers: Determine whether each of the following statements about real numbers is true or false, and why.

- ▶ For all a, there exists some b such that $b = a^2$.
- ▶ For all a, there exists some b such that $a = b^2$.
- ▶ For all a, there exists some b such that $b = a + 1$.
- ▶ For all integers a, there exists some integer b such that if $x \geq a$, then $x > b$.
- ▶ For all integers a, there exists some integer b such that if $x > a$, then $x = b$.

Solving function equations: Solve each of the following equations for x, and illustrate these solutions on a graph of $y = f(x)$.

- ▶ If $f(x) = x^3$, solve $f(x) = 7.5$ and $f(x) = 8.5$.
- ▶ If $f(x) = \sqrt{x-1}$, solve $f(x) = 1.8$ and $f(x) = 2.2$.
- ▶ If $f(x) = -0.033058(x^2 - 110x)$, solve $f(x) = 90$.
- ▶ If $f(x) = 2 - x^2$ and $x > 0$, solve $f(x) = -7.01$ and $f(x) = -6.99$.
- ▶ If $f(x) = \dfrac{x^2 - 2x - 3}{x - 1}$ and $x < 0$, solve $f(x) = -2$ and $f(x) = 0$.

Concepts

0. *Problem Zero:* Read the section and make your own summary of the material.

1. *True/False:* Determine whether each of the statements that follow is true or false. If a statement is true, explain why. If a statement is false, provide a counterexample.

  (a) *True or False:* For $\lim_{x \to c} f(x)$ to be defined, the function f must be defined at $x = c$.

  (b) *True or False:* We can calculate a limit of the form $\lim_{x \to c} f(x)$ simply by finding $f(c)$.

  (c) *True or False:* If $\lim_{x \to c} f(x) = 10$, then $f(c) = 10$.

  (d) *True or False:* If $f(c) = 10$, then $\lim_{x \to c} f(x) = 10$.

  (e) *True or False:* A function can approach more than one limit as x approaches c.

  (f) *True or False:* If $\lim_{x \to 4} f(x) = 10$, then we can make $f(x)$ as close to 4 as we like by choosing values of x sufficiently close to 10.

  (g) *True or False:* If $\lim_{x \to 6} f(x) = \infty$, then we can make $f(x)$ as large as we like by choosing values of x sufficiently close to 6.

  (h) *True or False:* If $\lim_{x \to \infty} f(x) = 100$, then we can find values of $f(x)$ between 99.9 and 100.1 by choosing values of x that are sufficiently large.

2. *Examples:* Construct examples of the thing(s) described in the following. Try to find examples that are different than any in the reading.

  (a) A function f and a value c such that $\lim_{x \to c} f(x)$ happens to be equal to $f(c)$.

  (b) A function f and a value c such that $\lim_{x \to c} f(x)$ is not equal to $f(c)$.

  (c) A function f and a value c such that $\lim_{x \to c} f(x)$ exists but $f(c)$ does not exist.

3. What are punctured intervals, and why do we need to use them when discussing limits?

4. Describe the punctured interval around $x = 2$ that has a radius of 3 and the punctured interval around $x = 4$ that has a radius of 0.25.

5. Find punctured intervals on which the function $f(x) = \dfrac{1}{x^2 - x}$ is defined, centered around

  (a) $x = 1.5$ (b) $x = 0.25$ (c) $x = 1$

6. Find punctured intervals on which the function $f(x) = \dfrac{1}{x\sqrt{x+2}}$ is defined, centered around

  (a) $x = 0$ (b) $x = -1$ (c) $x = -1.5$

Use interval notation to fill in the blanks that follow. Your answers will involve δ, ϵ, N, and/or M.

7. If $\lim_{x \to 2} f(x) = 5$, then for all $\epsilon > 0$, there is some $\delta > 0$ such that if $x \in$ _____ , then $f(x) \in$ _____ .

8. If $\lim_{x \to 3^-} f(x) = 1$, then for all $\epsilon > 0$, there is some $\delta > 0$ such that if $x \in$ _____ , then $f(x) \in$ _____ .

9. If $\lim_{x \to \infty} f(x) = 2$, then for all $\epsilon > 0$, there is some $N > 0$ such that if $x \in$ _____ , then $f(x) \in$ _____ .

10. If $\lim_{x \to \infty} f(x) = -\infty$, then for all $M > 0$, there is some $N > 0$ such that if $x \in$ _____ , then $f(x) \in$ _____ .

11. If $\lim_{x \to 1^+} f(x) = \infty$, then for all $M > 0$, there is some $\delta > 0$ such that if $x \in$ _____ , then $f(x) \in$ _____ .

12. Sketch a labeled graph that illustrates what is going on in the proof of Theorem 1.6 in the reading. Your graph should include two different ϵ-bars and a graphical reason that they cannot overlap.

13. Sketch a labeled graph that illustrates what is going on in the proof of Theorem 1.8 in the reading. Your graph should include two different δ-bars and a graphical reason why they combine to make a punctured delta-interval.

14. Suppose f is a function with $f(2) = 5$ where for all $\epsilon > 0$, there is some $\delta > 0$ such that if $x \in (2 - \delta, 2) \cup (2, 2 + \delta)$, then $f(x) \in (3 - \epsilon, 3 + \epsilon)$. Sketch a possible graph of f.

15. If $x \in (1.5, 2.5)$, what is the largest interval $I = (4 - \epsilon, 4 + \epsilon)$ for which we can guarantee that $x^2 \in I$?

16. It is *false* that $\lim_{x \to 1}(x + 1.01) = 2$. Express this fact in a mathematical sentence involving δ and ϵ, to show how the formal definition of limit fails in this case.

17. It is *false* that $\lim\limits_{x \to \infty} \dfrac{1000}{x} = \infty$. Express this fact in a mathematical sentence involving M and N, to show how the formal definition of limit fails in this case.

18. Show that the limit as $x \to 2$ of $f(x) = \sqrt{x - 1.1}$ is *not* equal to 1, by finding an $\epsilon > 0$ for which there is no corresponding $\delta > 0$ satisfying the formal definition of limit.

Skills

Write each limit in Exercises 19–42 as a formal statement involving δ, ϵ, N, and/or M, and sketch a graph that illustrates the roles of these constants. In the last six exercises you may take f to be any appropriate graph.

19. $\lim\limits_{x \to -3} \sqrt{x + 7} = 2$

20. $\lim\limits_{x \to 2} \dfrac{x^2 - 4}{x + 2} = 0$

21. $\lim\limits_{x \to -1} (x^3 - 2) = -3$

22. $\lim\limits_{x \to 1^-} \sqrt{1 - x} = 0$

23. $\lim\limits_{x \to 2} \dfrac{x^2 - 4}{x - 2} = 4$

24. $\lim\limits_{x \to 1} (x^2 - 3) = -2$

25. $\lim\limits_{x \to 0^+} \sqrt{x} = 0$

26. $\lim\limits_{x \to 3^-} (4 - x^2) = -5$

27. $\lim\limits_{x \to -2^+} \dfrac{1}{x + 2} = \infty$

28. $\lim\limits_{x \to 0^-} \dfrac{1}{x} = -\infty$

29. $\lim\limits_{x \to 3^-} \dfrac{1}{3 - x} = \infty$

30. $\lim\limits_{x \to 1^+} \dfrac{1}{1 - x} = -\infty$

31. $\lim\limits_{x \to \infty} \dfrac{x}{1 - 2x} = -.5$

32. $\lim\limits_{x \to -\infty} \dfrac{1}{x^2 + 1} = 0$

33. $\lim\limits_{x \to \infty} (x^3 + x + 1) = \infty$

34. $\lim\limits_{x \to -\infty} (1 - 3x) = \infty$

35. $\lim\limits_{h \to 0} \dfrac{(2 + h)^2 - 4}{h} = 4$

36. $\lim\limits_{h \to 0} \dfrac{\frac{1}{2 + h} - \frac{1}{2}}{h} = -\dfrac{1}{4}$

37. $\lim\limits_{x \to c^+} f(x) = -\infty$

38. $\lim\limits_{x \to c^-} f(x) = \infty$

39. $\lim\limits_{x \to -\infty} f(x) = L$

40. $\lim\limits_{x \to -\infty} f(x) = -\infty$

41. $\lim\limits_{x \to -\infty} f(x) = \infty$

42. $\lim\limits_{x \to \infty} f(x) = -\infty$

For each limit $\lim\limits_{x \to c} f(x) = L$ in Exercises 43–54, use graphs and algebra to approximate the largest value of δ such that if $x \in (c - \delta, c) \cup (c, c + \delta)$, then $f(x) \in (L - \epsilon, L + \epsilon)$.

43. $\lim\limits_{x \to 2} x^3 = 8, \ \epsilon = 0.5$

44. $\lim\limits_{x \to 2} x^3 = 8, \ \epsilon = 0.25$

45. $\lim\limits_{x \to 5} \sqrt{x - 1} = 2, \ \epsilon = 1$

46. $\lim\limits_{x \to 5} \sqrt{x - 1} = 2, \ \epsilon = 0.2$

47. $\lim\limits_{x \to -2} x^2 = 4, \ \epsilon = 0.5$

48. $\lim\limits_{x \to -2} x^2 = 4, \ \epsilon = 0.1$

49. $\lim\limits_{x \to 1} \dfrac{2x^2 - 2x}{x - 1} = 2, \ \epsilon = 0.5$

50. $\lim\limits_{x \to 1} \dfrac{2x^2 - 2x}{x - 1} = 2, \ \epsilon = 0.25$

51. $\lim\limits_{x \to 3} (2 - x^2) = -7, \ \epsilon = 0.01$

52. $\lim\limits_{x \to 3} (2 - x^2) = -7, \ \epsilon = 0.001$

53. $\lim\limits_{x \to -1} \dfrac{x^2 - 2x - 3}{x + 1} = -4, \ \epsilon = 1$

54. $\lim\limits_{x \to -1} \dfrac{x^2 - 2x - 3}{x + 1} = -4, \ \epsilon = 0.1$

For each limit in Exercises 55–64, use graphs and algebra to approximate the largest δ or smallest-magnitude N that corresponds to the given value of ϵ or M, according to the appropriate formal limit definition.

55. $\lim\limits_{x \to 1^+} \dfrac{1}{x^2 - 1} = \infty, \ M = 1000, \ \text{find largest } \delta > 0$

56. $\lim\limits_{x \to 1^+} \dfrac{1}{x^2 - 1} = \infty, \ M = 10{,}000, \ \text{find largest } \delta > 0$

57. $\lim\limits_{x \to \infty} \dfrac{3x}{x + 1} = 3, \ \epsilon = 0.5, \ \text{find smallest } N > 0$

58. $\lim\limits_{x \to \infty} \dfrac{3x}{x + 1} = 3, \ \epsilon = 0.1, \ \text{find smallest } N > 0$

59. $\lim\limits_{x \to \infty} \sqrt{x - 2} = \infty, \ M = 500, \ \text{find smallest } N > 0$

60. $\lim\limits_{x \to \infty} \sqrt{x - 2} = \infty, \ M = 5000, \ \text{find smallest } N > 0$

61. $\lim\limits_{x \to -\infty} \dfrac{1}{x} = 0, \ \epsilon = 0.5, \ \text{find smallest-magnitude } N < 0$

62. $\lim\limits_{x \to -\infty} \dfrac{1}{x} = 0, \ \epsilon = 0.25, \ \text{find smallest-magnitude } N < 0$

63. $\lim\limits_{x \to \infty} (4 - x^2) = -\infty, \ M = -100, \ \text{find smallest } N > 0$

64. $\lim\limits_{x \to \infty} (4 - x^2) = -\infty, \ M = -10{,}000, \ \text{find smallest } N > 0$

Applications

65. Every month, Jack hides $50 under a broken floorboard to save up for a new boat. After t months of saving, he will have $F(t) = 50t$ dollars.
   (a) The boat Jack wants costs at least $7,465. How many months does Jack have to save money before he will have enough to pay for the boat? Illustrate this information on a graph of $F(t)$.
   (b) Suppose a different boat costs M dollars. Will there be a value $t = N$ for which $F(N) > M$? What does this mean in real-world terms? Illustrate the roles of M and N on a graph of $F(t)$.

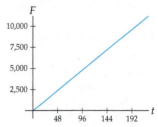

Money saved for Jack's boat

66. Len's company produces different-sized cylindrical cans that are each 6 inches tall. The cost to produce a can with radius r is $C(r) = 10\pi r^2 + 24\pi r$ cents.

Different cans with radius 6 inches   *Cost of producing a can*

(a) Len's boss wants him to construct the cans so that the cost of each can is within 25 cents of $4.00. Given these cost requirements, what is the acceptable range of values for r?

(b) Len's boss now says that he wants the cans to cost within 10 cents of $4.00. Under these new cost requirements, what is the acceptable range of values for r?

(c) Interpret this problem in terms of δ and ϵ ranges. Specifically, what is c? What is L? What is ϵ for part (a) and part (b)? What are the corresponding values of δ? Illustrate these values of c, L, ϵ, and δ on a graph of $C(r)$.

67. You work for a company that sells velvet Elvis paintings. The function $N(p) = 9.2p^2 - 725p + 16,333$ predicts the number N of velvet Elvis paintings that your company will sell if they are priced at p dollars each, and is shown in the following graph. The Presley estate does not allow you to charge more than $50 per painting.

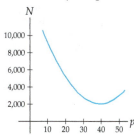

Velvet Elvis paintings sold

(a) Use a graphing utility to estimate the price your company should charge per painting if it wishes to sell 6000 velvet Elvis paintings.

(b) Find the range of prices that would enable your company to sell between 5000 and 7000 velvet Elvis paintings.

(c) Interpret this problem in terms of δ and ϵ ranges. Specifically, what is c? What is L? What is ϵ? What is the corresponding value of δ? Illustrate these values of c, L, ϵ, and δ on a graph of $N(p)$.

Proofs

Prove the four limit statements in Exercises 68–71. In the next section we will present a systematic method for such proofs.

68. Prove that $\lim_{x \to 1} 3x = 3$, with these steps:

(a) What is the δ–ϵ statement that must be shown to prove that $\lim_{x \to 1} 3x = 3$?

(b) Argue that $x \in (1 - \delta, 1) \cup (1, 1 + \delta)$ if and only if $-\delta < x - 1 < \delta$, with $x \neq 1$. Then use algebra to show that this means that $0 < |x - 1| < \delta$.

(c) Argue that $3x \in (3 - \epsilon, 3 + \epsilon)$ if and only if $-\epsilon < 3(x - 1) < \epsilon$. Then use algebra to show that this means that $3|x - 1| < \epsilon$.

(d) Given any particular $\epsilon > 0$, what value of δ would guarantee that if $0 < |x - 1| < \delta$, then $3|x - 1| < \epsilon$? Your answer will depend on ϵ.

(e) Put the previous four parts together to prove the limit statement.

69. Prove that $\lim_{x \to 2}(7 - x) = 5$, with these steps:

(a) What is the δ–ϵ statement that must be shown to prove that $\lim_{x \to 2} 7 - x = 5$?

(b) Argue that $x \in (2 - \delta, 2) \cup (2, 2 + \delta)$ if and only if $-\delta < x - 2 < \delta$. Then use algebra to show that this means that $0 < |x - 2| < \delta$.

(c) Argue that $7 - x \in (5 - \epsilon, 5 + \epsilon)$ if and only if $-\epsilon < 2 - x < \epsilon$. Then use algebra to show that this means that $|x - 2| < \epsilon$.

(d) Given any particular $\epsilon > 0$, what value of δ would guarantee that if $0 < |x - 2| < \delta$, then $|x - 2| < \epsilon$? Your answer will depend on ϵ.

(e) Put the previous four parts together to prove the limit statement.

70. Prove that $\lim_{x \to 0^+} \dfrac{1}{x} = \infty$, with these steps:

(a) What is the M–δ statement that must be shown to prove that $\lim_{x \to 0^+} \dfrac{1}{x} = \infty$?

(b) Argue that $x \in (0, 0 + \delta)$ if and only if $0 < x < \delta$.

(c) Argue that $\dfrac{1}{x} \in (M, \infty)$ if and only if $x < \dfrac{1}{M}$. You may assume that $M > 0$.

(d) Given any particular $M > 0$, what value of δ would guarantee that if $0 < x < \delta$, then $x < \dfrac{1}{M}$? Your answer will depend on M.

(e) Put the previous four parts together to prove the limit statement.

71. Prove that $\lim_{x \to \infty} \dfrac{1}{x} = 0$, with these steps:

(a) What is the ϵ–N statement that must be shown to prove that $\lim_{x \to \infty} \dfrac{1}{x} = 0$?

(b) Argue that $x \in (N, \infty)$ if and only if $x > N$.

(c) Argue that $\dfrac{1}{x} \in (0 - \epsilon, 0 + \epsilon)$ if and only if $-\epsilon < \dfrac{1}{x} < \epsilon$. Then argue that for this limit, it suffices to consider $0 < \dfrac{1}{x} < \epsilon$.

(d) Given any particular ϵ, what value of $N > 0$ would guarantee that if $x > N$, then $0 < \dfrac{1}{x} < \epsilon$? Your answer will depend on ϵ.

(e) Put the previous four parts together to prove the limit statement.

Thinking Forward

Continuity: As you have already seen, sometimes $\lim_{x \to c} f(x)$ is equal to $f(c)$, and sometimes it is not. As we will see in Section 1.4, when the limit of a function f as $x \to c$ does happen to be equal to the value of $f(x)$ at $x = c$, we say that the function f is **continuous** at $x = c$.

▶ You may have heard the following loose, only partially accurate definition of continuity in a previous class: A function is continuous if you "can draw it without picking up your pencil." Why does it make sense that this would be related to the definition just presented of continuity in terms of limits?

▶ State the limit-definition of continuity with a formal δ–ϵ statement.

▶ The function $f(x) = x^2$ is continuous at every point. Use this fact and the formal definition of continuity to calculate $\lim_{x \to 2} x^2$, $\lim_{x \to 5} x^2$, and $\lim_{x \to -4} x^2$.

Limits of Sequences: We say that an infinite sequence of real numbers $a_1, a_2, a_3, \ldots, a_k, \ldots$ **converges to a limit** L, and we write $\lim_{k \to \infty} s_k = L$, if for all $\epsilon > 0$, there exists some $N > 0$ such that if $k > N$, then $|a_k - L| < \epsilon$.

▶ Use algebra to solve the inequality $|a_k - L| < \epsilon$ for a_k. Your answer should be in the form $a_k \in \underline{\quad}$, where the blank is filled in with interval notation.

▶ Relate the definition of the convergence of a sequence to the definition of a limit at infinity.

1.3   DELTA-EPSILON PROOFS*

▶ Developing an equivalent algebraic definition of limits from our geometric definition

▶ Finding delta in terms of epsilon so that we can prove a limit statement

▶ The formal logic of writing delta–epsilon proofs

Describing Limits with Absolute Value Inequalities

In Definition 1.5, we formally defined the limit statement $\lim_{x \to c} f(x) = L$ to mean that for all $\epsilon > 0$, there exists $\delta > 0$ such that whenever $x \in (c - \delta, c) \cup (c, c + \delta)$, we can guarantee that $f(x) \in (L - \epsilon, L + \epsilon)$. This definition of limit has a very geometric flavor, since it is stated in the language of ϵ-intervals and punctured δ-intervals. That kind of language is useful when looking at specific values of ϵ or δ, but not as useful when trying to prove that *every* value of ϵ has a corresponding value of δ. For the purposes of proving limit statements, we give the following algebraic definition of limit and prove that it is equivalent to our previous geometric definition:

DEFINITION 1.10

Algebraic Definition of Limit

The **limit** $\lim_{x \to c} f(x) = L$ means that for all $\epsilon > 0$, there exists $\delta > 0$ such that

$$\text{if } 0 < |x - c| < \delta, \text{ then } |f(x) - L| < \epsilon.$$

THEOREM 1.11

Equivalence of Geometric and Algebraic Definitions of Limit

The two definitions of limit in Definition 1.5 and 1.10 are equivalent. Specifically,

(a) $x \in (c - \delta, c) \cup (c, c + \delta)$ if and only if $0 < |x - c| < \delta$;

(b) $f(x) \in (L - \epsilon, L + \epsilon)$ if and only if $|f(x) - L| < \epsilon$.

Proof. We will begin by proving part (b), since it is slightly easier. The statement $f(x) \in (L - \epsilon, L + \epsilon)$ means that $L - \epsilon < f(x) < L + \epsilon$. Subtracting L from all three parts of this double inequality, we get $-\epsilon < f(x) - L < \epsilon$. This is precisely the solution of the inequality $|f(x) - L| < \epsilon$.

For part (a) we have a similar situation, except that we must deal with a punctured interval. The statement $x \in (c - \delta, c) \cup (c, c + \delta)$ means that $c - \delta < x < c + \delta$ and $x \neq c$. Subtracting c from all three parts of the double inequality, we get $-\delta < x - c < \delta$, which is the solution set for the inequality $|x - c| < \delta$. The fact that $x \neq c$ means that $x - c \neq 0$. This is equivalent to saying that $|x - c| \neq 0$, and since the absolute value of a number is always positive or zero, it is also equivalent to saying that $|x - c| > 0$. Therefore $x \in (c - \delta, c) \cup (c, c + \delta)$ if and only if $0 < |x - c| < \delta$.   ◼

Finding a Delta for Every Epsilon

With our new algebraic definition of limit, we have the final tool we need to be able to effectively *prove* limit statements. This is the first step towards being able to *calculate* limits, something that, perhaps surprisingly, we do not yet know how to do.

Consider for example the limit statement $\lim_{x \to 2}(3x - 1) = 5$. We can examine this limit with a table of values, noticing that as the values of x approach $x = 2$ from the left and the right, the corresponding values of $3x - 1$ approach 5:

x	1.9	1.99	1.999	2	2.001	2.01	2.1
$3x - 1$	4.7	4.97	4.997	-	5.003	5.03	5.3

We can also investigate the limit statement $\lim_{x \to 2}(3x - 1) = 5$ with the following graph of $f(x) = 3x - 1$ at the left, noticing that a sequence of values of x approaching $x = 2$ from either the left or the right determines a sequence of values of $f(x)$ that approach $y = 5$:

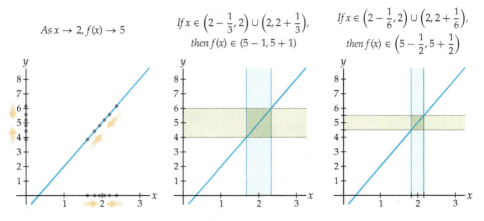

As $x \to 2$, $f(x) \to 5$

If $x \in \left(2 - \frac{1}{3}, 2\right) \cup \left(2, 2 + \frac{1}{3}\right)$, then $f(x) \in (5 - 1, 5 + 1)$

If $x \in \left(2 - \frac{1}{6}, 2\right) \cup \left(2, 2 + \frac{1}{6}\right)$, then $f(x) \in \left(5 - \frac{1}{2}, 5 + \frac{1}{2}\right)$

We can even show that the formal geometric definition of limit holds for a particular ϵ, by finding some δ for which $x \in (2 - \delta, 2) \cup (2, 2 + \delta)$ guarantees that $3x - 1 \in (5 - \epsilon, 5 + \epsilon)$. For example, it turns out that if $\epsilon = 1$, then $\delta = \frac{1}{3}$ will work, as illustrated in the middle figure shown, and if $\epsilon = \frac{1}{2}$, then $\delta = \frac{1}{6}$ will work, as in the rightmost figure.

Although the preceding investigations are encouraging evidence, none of them *prove* that $\lim_{x \to 2}(3x - 1)$ is equal to 5. In the table, perhaps we did not consider values of x close enough to 2 to see the true behavior of $f(x) = 3x - 1$, or perhaps the limit is actually 5.0001 and not 5. The three graphs shown each have the same possible problems. To prove that $\lim_{x \to 2}(3x - 1) = 5$ we would need to know for certain that we could get values of $f(x)$ *arbitrarily close* to $y = 5$ by choosing values of x that are sufficiently close to $x = 2$. For *every* epsilon we would have to be able to find a delta.

Specifically, in this example, we need to show that for any choice of $\epsilon > 0$, we can find some $\delta > 0$ such that whenever a real number x satisfies $0 < |x - 2| < \delta$, we can also guarantee that x satisfies $|(3x - 1) - 5| < \epsilon$. Notice that $|(3x - 1) - 5| = |3x - 6| = 3|x - 2|$, so what we really need is to conclude that $3|x - 2| < \epsilon$, i.e., that $|x - 2| < \frac{\epsilon}{3}$. But that is easy if we are allowed to choose a different δ for each ϵ: Given any $\epsilon > 0$, simply choose $\delta = \frac{\epsilon}{3}$. Then whenever $0 < |x - 2| < \delta$, we also have $|x - 2| < \frac{\epsilon}{3}$, as desired. Notice that in both the second and third of the graphs shown previously, we did in fact choose δ to be one-third of ϵ. We have just *proved* the limit statement $\lim_{x \to 2}(3x - 1) = 5$ by showing that *every* value of ϵ has a corresponding value of $\delta = \frac{\epsilon}{3}$ for which the formal algebraic definition of limit holds.

Writing Delta-Epsilon Proofs

We have just proved that $\lim_{x \to 2}(3x - 1) = 5$, but our proof meandered about in a paragraph of discussion. We now present a systematic way to write up delta-epsilon proofs for limit statements. Remember that in our example what we must show is the following doubly quantified logical implication:

"For all $\epsilon > 0$, there exists a $\delta > 0$ such that if $0 < |x - 2| < \delta$, then $|(3x - 1) - 5| < \epsilon$."

Logically, to prove such a statement we must first let ϵ be an arbitrary positive number and then choose a value for δ in terms of ϵ. We must then show that all values of x with $0 < |x - 2| < \delta$, we can also say that $|(3x - 1) - 5| < \epsilon$.

In our example, we already know that for any $\epsilon > 0$ we should choose $\delta = \frac{\epsilon}{3}$. Using this fact, we could arrange our proof very concisely as follows:

Proof. Given $\epsilon > 0$, choose $\delta = \frac{\epsilon}{3}$. For all x with $0 < |x - 2| < \delta$, we have

$$|(3x - 1) - 5| = |3x - 6| = |3(x - 2)| = 3|x - 2| < 3\delta \overset{*}{=} 3\left(\frac{\epsilon}{3}\right) = \epsilon.$$

Therefore whenever $0 < |x - 2| < \delta$, we also have $|(3x - 1) - 5| < \epsilon$.   ■

The first three equals signs in the proof are simple algebra. The less-than step followed from our assumption that $|x - 2| < \delta$. The starred equality followed from the fact that $\delta = \frac{\epsilon}{3}$.

In general we will not know at the outset what to choose for δ, as we did just now. In those cases we can either do a side-calculation to find δ in advance or just leave a blank space for δ and continue with the proof. When we get to the starred equality, after using the assumption $0 < |x - c| < \delta$, we will be able to see what to choose for δ and can fill in the blank as if we knew it all along; see Example 2. Throughout the examples we will try our hand at proving all kinds of limit statements, including one-sided limits, infinite limits, and limits at infinity.

Examples and Explorations

EXAMPLE 1   Finding delta as a function of epsilon

Find formulas for δ in terms of ϵ for each of the following limit statements:

(a) $\lim_{x \to 4}(2x + 1) = 9$ **(b)** $\lim_{x \to 2}(x^2 - 4x + 5) = 1$

Then use those formulas to find punctured δ-intervals for $\epsilon = 1$ and $\epsilon = 0.5$.

SOLUTION

(a) The limit statement $\lim_{x \to 4}(2x + 1) = 9$ means that given any $\epsilon > 0$, there is some $\delta > 0$ so that if $0 < |x - 4| < \delta$, then we can conclude that $|(2x + 1) - 9| < \epsilon$. We have

$$|(2x + 1) - 9| = |2x - 8| = |2(x - 4)| = 2|x - 4|,$$

so $|(2x + 1) - 9| < \epsilon$ when $2|x - 4| < \epsilon$, i.e., when $|x - 4| < \frac{\epsilon}{2}$. Therefore we should choose $\delta = \frac{\epsilon}{2}$. In particular, when $\epsilon = 1$, we have $\delta = \frac{1}{2} = 0.5$ and punctured δ-interval $(3.5, 4) \cup (4, 4.5)$. When $\epsilon = 0.5$, we choose $\delta = \frac{0.5}{2} = 0.25$, which gives us the punctured δ-interval $(3.75, 4) \cup (4, 4.25)$.

(b) The limit statement $\lim_{x \to 2}(x^2 - 4x + 5) = 1$ means that for any $\epsilon > 0$, we can find $\delta > 0$ so that whenever $0 < |x - 2| < \delta$, we can conclude that $|(x^2 - 4x + 5) - 1| < \epsilon$. Notice that

$$|(x^2 - 4x + 5) - 1| = |x^2 - 4x + 4| = |(x - 2)^2| = |x - 2|^2,$$

which is clearly very closely related to our δ-inequality $0 < |x - 2| < \delta$. In fact, the inequality $|x - 2|^2 < \epsilon$ is equivalent to the inequality $|x - 2| < \sqrt{\epsilon}$. Therefore we should choose $\delta = \sqrt{\epsilon}$. In particular, when $\epsilon = 1$, we should choose $\delta = \sqrt{1} = 1$ and thus punctured δ-interval $(1, 2) \cup (2, 3)$. When $\epsilon = 0.5$, we should choose $\delta = \sqrt{0.5} \approx 0.707$ and thus punctured δ-interval $(1.293, 2.707)$.

 CHECKING THE ANSWER

To check that our formulas for δ in terms of ϵ are reasonable, we can graph the functions and the punctured δ-intervals that we found in each case. For the limit $\lim_{x \to 4}(2x + 1) = 9$, $\delta = 0.5$ looks right for $\epsilon = 1$ and $\delta = 0.25$ looks right for $\epsilon = 0.5$:

If $x \in (4 - 0.5, 4) \cup (4, 4 + 0.5)$, then $f(x) \in (9 - 1, 9 + 1)$

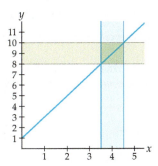

If $x \in (4 - 0.25, 4) \cup (4, 4 + 0.25)$, then $f(x) \in (9 - 0.5, 9 + 0.5)$

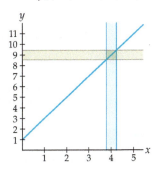

For $\lim_{x \to 2}(x^2 - 4x + 5) = 1$, $\delta = 1$ looks right for $\epsilon = 1$ and $\delta \approx 0.707$ looks right for $\epsilon = 0.5$:

If $x \in (2 - 1, 2) \cup (2, 2 + 1)$, then $f(x) \in (1 - 1, 1 + 1)$

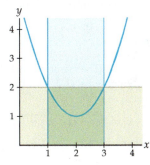

If $x \in (2 - 0.707, 2) \cup (2, 2 + 0.707)$, then $f(x) \in (1 - 0.5, 1 + 0.5)$

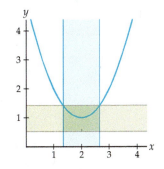

EXAMPLE 2 | Writing basic delta–epsilon proofs

Write delta–epsilon proofs for each of the following limit statements:

(a) $\lim_{x \to 4} (2x + 1) = 9$ (b) $\lim_{x \to 2} (x^2 - 4x + 5) = 1$

SOLUTION

(a) To show that $\lim_{x \to 4}(2x + 1) = 9$ we must start with an arbitrary $\epsilon > 0$, choose some $\delta > 0$, and then show that for all x with $0 < |x - 4| < \delta$, we also have $|(2x+1) - 9| < \epsilon$. From the previous example we know that we should choose $\delta = \frac{\epsilon}{2}$.

> **Proof.** Given $\epsilon > 0$, choose $\delta = \frac{\epsilon}{2}$. For all x with $0 < |x - 4| < \delta$, we have
>
> $$|(2x + 1) - 9| = |2x - 8| = |2(x - 4)| = 2|x - 4| < 2\delta \overset{\star}{=} 2\left(\frac{\epsilon}{2}\right) = \epsilon.$$
>
> Therefore whenever $0 < |x - 4| < \delta$, we also have $|(2x + 1) - 9| < \epsilon$. ■

If we had not already known that we should choose $\delta = \frac{\epsilon}{2}$, we could have determined that choice when we reached the starred equality. At that point we had shown that $|(2x + 1) - 9|$ was less than 2δ. What we were trying to show was that $|(2x + 1) - 9|$ was less than ϵ. If 2δ were equal to ϵ, then we would be done; solving $2\delta = \epsilon$ for δ, we arrive at the choice $\delta = \frac{\epsilon}{2}$.

(b) To show that $\lim_{x \to 2}(x^2 - 4x + 5) = 1$ we must start with an arbitrary $\epsilon > 0$, choose some $\delta > 0$, and then show that for all x with $0 < |x - 2| < \delta$, we also have $|(x^2 - 4x + 5) - 1| < \epsilon$. From the previous example we know that we should choose $\delta = \sqrt{\epsilon}$.

> **Proof.** Given $\epsilon > 0$, choose $\delta = \sqrt{\epsilon}$. For all x with $0 < |x - 2| < \delta$, we have
>
> $$|(x^2 - 4x + 5) - 1| = |x^2 - 4x + 4| = |(x - 2)^2| = |x - 2|^2 < \delta^2 \overset{\star}{=} (\sqrt{\epsilon})^2 = \epsilon.$$
>
> Thus we can conclude that whenever $0 < |x - 2| < \delta$, we also have $|(x^2 - 4x + 5) - 1| < \epsilon$. ■

Again, if we had not already known that we should choose $\delta = \sqrt{\epsilon}$, then at the starred equality we would have δ^2 where we wish to have ϵ. Solving $\delta^2 = \epsilon$ for δ, we get our choice of $\delta = \sqrt{\epsilon}$. □

EXAMPLE 3 | Proofs for one-sided and infinite limits

Give formal proofs for each of the following limit statements:

(a) $\lim_{x \to 1^+} \sqrt{x - 1} = 0$ (b) $\lim_{x \to \infty} \frac{1}{x} = 0$ (c) $\lim_{x \to 3^-} \frac{1}{3 - x} = \infty$

SOLUTION

(a) The limit statement $\lim_{x \to 1^+} \sqrt{x - 1} = 0$ means that for all $\epsilon > 0$, there exists $\delta > 0$ such that if $x \in (1, 1 + \delta)$, then $|\sqrt{x - 1} - 0| < \epsilon$. Other than a small bit of extra work to translate the meaning of $x \in (1, 1 + \delta)$, this is a standard delta-epsilon proof.

> **Proof.** Given $\epsilon > 0$, choose $\delta = \epsilon^2$. (The reason for that choice of δ will become clear at the starred equality that follows.) Suppose $x \in (1, 1 + \delta)$. Then $1 < x < 1 + \delta$, which means that $0 < x - 1 < \delta$. We then have
>
> $$|\sqrt{x - 1} - 0| = |\sqrt{x - 1}| = \sqrt{x - 1} < \sqrt{\delta} = \sqrt{\epsilon^2} = \epsilon.$$
>
> Thus we can conclude that whenever $x \in (1, 1 + \delta)$, we also have $|\sqrt{x - 1} - 0| < \epsilon$. ■

(b) Geometrically, the limit statement $\lim_{x\to\infty} \frac{1}{x} = 0$ means that for all $\epsilon > 0$, there exists $N > 0$ such that if $x \in (N, \infty)$, then $\frac{1}{x} \in (0 - \epsilon, 0 + \epsilon)$. Algebraically, the implication after the quantifiers can be rewritten as follows: If $x > N$, then $\left|\frac{1}{x} - 0\right| < \epsilon$. This is what we shall prove.

> **Proof.** Given $\epsilon > 0$, choose $N = \frac{1}{\epsilon}$. (The reason for that choice of N will become clear at the starred equality that follows.) If $x > N$, then $\frac{1}{x} < \frac{1}{N}$ and x is positive, and therefore
>
> $$\left|\frac{1}{x} - 0\right| = \left|\frac{1}{x}\right| = \frac{1}{x} < \frac{1}{N} = \frac{1}{\frac{1}{\epsilon}} = \epsilon.$$
>
> Thus we can conclude that whenever $x > N$, we also have $\left|\frac{1}{x} - 0\right| < \epsilon$.   ∎

(c) Geometrically, the limit $\lim_{x\to 3^-} \frac{1}{3-x} = \infty$ means that for all $M > 0$, there exists $\delta > 0$ such that if $x \in (3 - \delta, 3)$, then $\frac{1}{3-x} \in (M, \infty)$. Note that $x \in (3 - \delta, 3)$ means that $3 - \delta < x < 3$ and therefore that $-\delta < x - 3 < 0$. Multiplying by -1 and flipping inequalities, this becomes $\delta > -(x - 3) > 0$, or equivalently, $0 < 3 - x < \delta$. Hence the implication in our limit statement can be expressed as follows: If $0 < 3 - x < \delta$, then $\frac{1}{3-x} > M$. This is what we will prove:

> **Proof.** Given $M > 0$, choose $\delta = \frac{1}{M}$. (As usual, the reason for this choice will become clear at the starred equality that follows.) For all x with $0 < 3 - x < \delta$, we have $\frac{1}{3-x} > \frac{1}{\delta}$, and therefore
>
> $$\frac{1}{3-x} > \frac{1}{\delta} = \frac{1}{\frac{1}{M}} = M.$$
>
> Thus we can conclude that whenever $0 < 3 - x < \delta$, we also have $\frac{1}{3-x} > M$.   ∎

EXAMPLE 4   **A delta–epsilon proof where it is necessary to bound delta from above**

Write a delta-epsilon proof for the limit statement $\lim_{x\to 2} 5x^4 = 80$.

SOLUTION

To prove the limit statement $\lim_{x\to 2} 5x^4 = 80$, we must show that for all $\epsilon > 0$, there exists a choice of $\delta > 0$ such that whenever $0 < |x - 2| < \delta$, we also have $|5x^4 - 80| < \epsilon$. In our previous delta-epsilon proofs, the algebra has always magically worked out nicely. In this example there will be a point at which we get stuck. What will get us out of the jam will be assuming that δ is no larger than 1. This assumption will allow us to put a bound on an expression that would otherwise be in our way. For that reason, our choice of δ in this proof will be the minimum of 1 and an expression that depends on epsilon.

> **Proof.** Given $\epsilon > 0$, choose $\delta = \min\left(1, \frac{\epsilon}{325}\right)$, i.e., the smaller of 1 and $\frac{\epsilon}{325}$. (The reason for this elaborate choice of δ will be made clear at the two starred inequalities that follow.) If x is a real number with $0 < |x - 2| < \delta$, then we have
>
> $$\begin{aligned}
> |5x^4 - 80| &= |5(x^4 - 16)| = 5|(x^2 - 4)(x^2 + 4)| \\
> &= 5|(x - 2)(x + 2)(x^2 + 4)| \\
> &= 5|x - 2| \cdot |(x + 2)(x^2 + 4)| \\
> &< 5\delta \cdot |(x + 2)(x^2 + 4)|.
> \end{aligned}$$

Because our choice of δ ensures that δ is at most 1, we can say that $0 < |x-2| \leq 1$ and therefore that x is between 1 and 3. This means that we can bound the troublesome quantity $|(x+2)(x^2+4)|$ from above; it is largest when $x = 3$, so we know that $|(x+2)(x^2+4)|$ is at most $|(3+2)(3^2+4)| = 65$. Combining this result with the work we just did, we have (using our choice of $\delta = \min\left(1, \frac{\epsilon}{325}\right)$ at the two starred inequalities)

$$|5x^4 - 80| \overset{\star}{<} 5\delta \cdot 65 = 325\,\delta \overset{\star}{\leq} 325\left(\frac{\epsilon}{325}\right) = \epsilon.$$

Thus we can conclude that whenever $0 < |x-2| < \delta$, we also have $|5x^4 - 80| < \epsilon$. ◼

TEST YOUR UNDERSTANDING

▶ Why do we say that Definition 1.5 from the previous section is geometric in nature while we say that Definition 1.10 is algebraic? In what situations might one definition be more useful than another?

▶ How can we express the formal definitions for one-sided limits, infinite limits, and limits at infinity "algebraically," in the spirit of Definition 1.10?

▶ Suppose that we have a limit of the form $\lim_{x \to c} f(x) = L$ and that we can find a value of δ for some given value of ϵ. Why does this not prove definitively that $\lim_{x \to c} f(x) = L$?

▶ In a delta-epsilon proof, how do we come up with a choice for δ in terms of ϵ?

▶ Why is it sometimes necessary to require that $\delta \leq 1$ in a delta-epsilon proof?

EXERCISES 1.3

Thinking Back

Inequalities: Find the solution sets of each of the following inequalities.

▶ $0 < |x - 2| < 0.5$

▶ $0 < |x + 5| < 0.1$

▶ $|x^2 - 4| < 0.5$

▶ $|(3x + 1) - 10| < 1$

▶ $\dfrac{1}{x^2} < 0.01$

▶ $\dfrac{1}{x^2} > 1000$

Logical implications: Suppose A and B are statements and that the implication "If A, then B" holds.

▶ If A is true, then what, if anything, can you conclude about B, and why?

▶ If A is false, then what, if anything, can you conclude about B, and why?

▶ If B is true, then what, if anything, can you conclude about A, and why?

Concepts

0. *Problem Zero:* Read the section and make your own summary of the material.

1. *True/False:* Determine whether each of the statements that follow is true or false. If a statement is true, explain why. If a statement is false, provide a counterexample.

   (a) *True or False:* If $x \neq c$, then $|x - c|$ is strictly greater than zero.

   (b) *True or False:* If $|x-c|$ is strictly greater than zero, then $x \neq c$.

   (c) *True or False:* x is a solution of $0 < |x - c| < \delta$ if and only if $c - \delta < x < c + \delta$.

   (d) *True or False:* If $0 < |x - c| < \delta$, then $x \in (c - \delta, c) \cup (c, c + \delta)$.

   (e) *True or False:* If $|f(x)-L| < \epsilon$, then $L-\epsilon < f(x) < L+\epsilon$.

   (f) *True or False:* If $f(x) \in (L - \epsilon, L + \epsilon)$, then $0 < f(x) < |L + \epsilon|$.

   (g) *True or False:* The fact that $0 < |x - 3| < 0.25$ guarantees that $|(2x - 1) - 5| < 0.5$ proves that $\lim_{x \to 3}(2x - 1) = 5$.

   (h) *True or False:* $\lim_{x \to 3}(2x - 1) = 5$ means that for all $\delta > 0$ there is some $\epsilon > 0$ such that if $0 < |x - c| < \delta$, then $|(2x - 1) - 5| < \epsilon$.

2. *Examples:* Construct examples of the thing(s) described in the following. Try to find examples that are different than any in the reading.

   (a) A function f with values given in the following table but whose limit as $x \to 2$ is *not* equal to 5:

x	1.9	1.99	1.999	2	2.001	2.01	2.1
$f(x)$	4.7	4.97	4.997	-	5.003	5.03	5.3

   (b) An inequality involving absolute values whose solution set is $(2.75, 3) \cup (3, 3.25)$.

   (c) An inequality involving absolute values whose solution set is $(-0.01, 0) \cup (0, 0.01)$.

3. Write down the formal delta-epsilon statement you would have to prove in order to prove the limit statement $\lim_{x \to -2} \frac{3}{x+1} = -3$.

4. Suppose you show that $|(1 - 2x) - (-5)| < 0.05$ for all x with $0 < |x - 3| < 0.025$. Explain why this does *not* prove that $\lim_{x \to 3}(1 - 2x) = -5$.

5. Write down a mathematical equation that expresses the sentence "x is not equal to 5, and the distance between x and 5 is less than 0.01." Then write an equation that means "the distance between $f(x)$ and -2 is less than 0.5."

6. Why do we have $0 < |x - c| < \delta$ instead of just $|x - c| < \delta$ in Definition 1.10?

Write each of the following inequalities in interval notation:

7. $0 < |x - 2| < 0.1$

8. $0 < |x + 3| < 0.05$

9. $|(x^2 - 1) + 3| < 0.5$

10. $|(3x + 1) - 2| < 0.1$

11. $|f(x) - L| < \epsilon$

12. $0 < |x - c| < \delta$

Determine whether each implication that follows is true or false. Use graphs to justify any implications that are true, and counterexamples for any implications that are false.

13. If $0 < |x - 2| < 1$, then $|x^2 - 4| < 0.5$.

14. If $0 < |x - 2| < 0.2$, then $|x^2 - 4| < 1$.

15. If $0 < |x - 0| < 0.75$, then $|x^2 - 0| < 0.5$.

16. If $0 < |x + 2| < 0.1$ then $|x^2 - 4| < 0.4$.

17. If $0 < |x + 2| < 0.075$, then $|x^2 - 4| < 0.4$.

18. In Example 2 we proved that $\lim_{x \to 2}(x^2 - 4x + 5) = 1$. Use the proof to find values of δ corresponding to (a) $\epsilon = 1$, (b) $\epsilon = 0.1$, and (c) $\epsilon = 0.01$. Illustrate that your choices of δ work by examining a graph of $f(x) = x^2 - 4x + 5$ and sketching appropriate ϵ and δ intervals.

19. In Example 4 we proved that $\lim_{x \to 2} 5x^4 = 80$. Use the proof to find values of δ corresponding to (a) $\epsilon = 5$, (b) $\epsilon = 0.01$, and (c) $\epsilon = 350$. Illustrate that your choices of δ work by examining a graph of $f(x) = 5x^4$ and sketching appropriate ϵ and δ intervals.

20. Use algebra to solve the inequality $0 < |x - c| < \delta$ and show that its solution set is $x \in (c - \delta, c) \cup (c, c + \delta)$.

21. Use algebra to solve the inequality $|f(x) - L| < \epsilon$ and show that its solution set is $f(x) \in (L - \epsilon, L + \epsilon)$.

22. Suppose $f(x) = mx + b$ is a linear function with $m \neq 0$, and let c be any real number.

   (a) Show that for all $\epsilon > 0$, if $0 < |x - c| < \frac{\epsilon}{|m|}$, then $|f(x) - f(c)| < \epsilon$.

   (b) What does the implication in part (a) have to do with limits?

   (c) Illustrate the implication in part (a) with a labeled graph. Explain in terms of slopes why it makes sense that given $\epsilon > 0$, the corresponding $\delta > 0$ is $\delta = \frac{\epsilon}{|m|}$.

Skills

Use algebra to find the largest possible value of δ or smallest possible value of N that makes each implication in Exercises 23–28 true. Then verify and support your answers with labeled graphs.

23. If $0 < |x - 2| < \delta$, then $|(3x - 1) - 5| < 0.25$.

24. If $0 < |x - 3| < \delta$, then $\left| \frac{1}{x} - \frac{1}{3} \right| < 0.2$.

25. If $x \in (1, 1 + \delta)$, then $|\sqrt{x - 1} - 0| < 0.5$.

26. If $x \in (3 - \delta, 3)$, then $\frac{1}{3 - x} > 1000$.

27. If $x > N$, then $\left| \frac{1}{x^2} - 0 \right| < 0.001$.

28. If $x > N$, then $1 - 2x < -500$.

For each limit statement $\lim_{x \to c} f(x) = L$ in Exercises 29–40, use algebra to find $\delta > 0$ in terms of $\epsilon > 0$ so that if $0 < |x - c| < \delta$, then $|f(x) - L| < \epsilon$.

29. $\lim_{x \to 3}(x + 5) = 8$

30. $\lim_{x \to -2}(4 - 2x) = 8$

31. $\lim_{x \to 0}(3 - 4x) = 3$

32. $\lim_{x \to 1}(3x + 8) = 11$

33. $\lim_{x \to 0}(5x^2 - 1) = -1$

34. $\lim_{x \to 3}(x^2 - 6x + 5) = -4$

35. $\lim_{x \to 2}(x^2 - 4x + 6) = 2$

36. $\lim_{x \to 0}(x^3 + 1) = 1$

37. $\lim_{x \to 2} \frac{1}{x} = \frac{1}{2}$; you may assume $\delta \leq 1$

38. $\lim_{x \to 3} \frac{1}{x} = \frac{1}{3}$; you may assume $\delta \leq 1$

39. $\lim_{x \to 3}(x^2 - 2x - 3) = 0$; you may assume $\delta \leq 1$

40. $\lim_{x \to 1} 2x^4 = 2$; you may assume $\delta \leq 1$

For each limit statement in Exercises 41–44, use algebra to find δ or N in terms of ϵ or M, according to the appropriate formal limit definition.

41. $\lim_{x \to -2^+}(1 + \sqrt{x + 2}) = 1$, find δ in terms of ϵ

42. $\lim_{x \to \infty} \frac{x - 1}{x} = 1$, find N in terms of ϵ

43. $\lim_{x \to 1^-} \frac{1}{1 - x} = \infty$, find δ in terms of M

44. $\lim_{x \to \infty}(x^2 + 2) = \infty$, find N in terms of M

Applications

For Exercises 45 and 46, suppose you work for a company that manufactures gourmet soup cans. The material for the curved sides of the cans costs 0.25 cent (a quarter of a cent) per square

Cost of materials for producing a soup can

0.1 cents per linear inch

5 in.

0.25 cents per square inch

0.5 cents per square inch

inch, the material for the top and bottom costs 0.5 cent per square inch, and the reinforcing weld around the seams costs 0.1 cent per linear inch. The seams run around the edges of the top and bottom and also in a straight line from the top to the bottom of the curved side.

45. Find a formula for the cost $C(r)$ of producing a gourmet soup can with radius r and height 5 inches, and answer the following questions:
  (a) What is the radius of a can that is 5 inches tall and costs 30 cents to produce?
  (b) Your manager wants you to produce 5-inch-tall cans that cost between 20 and 40 cents. Write this requirement as an absolute value inequality.
  (c) What range of radii would satisfy your manager? Write an absolute value inequality whose solution set lies inside this range of radii.

46. Find a formula for the cost $C(h)$ of producing a gourmet soup can with height h and radius 2 inches, and answer the following questions:
  (a) What is the height of a can that has radius 2 inches and costs 45 cents to produce?
  (b) Your manager wants you to produce 2-inch-radius cans that cost between 40 and 50 cents. Write this requirement as an absolute value inequality.
  (c) What range of heights would satisfy your manager? Write an absolute value inequality whose solution set lies within this range of heights.

Proofs

Write delta-epsilon proofs for each of the limit statements $\lim_{x \to c} f(x) = L$ in Exercises 47–60.

47. $\lim_{x \to 1}(2x + 4) = 6$

48. $\lim_{x \to 2}(3 - 4x) = -5$

49. $\lim_{x \to -6}(x + 2) = -4$

50. $\lim_{x \to -3}(1 - x) = 4$

51. $\lim_{x \to 4}(6x - 1) = 23$

52. $\lim_{x \to 8}(3x - 11) = 13$

53. $\lim_{x \to 0}(3x^2 + 1) = 1$

54. $\lim_{x \to 3}(x^2 - 6x + 11) = 2$

55. $\lim_{x \to 1}(2x^2 - 4x + 3) = 1$

56. $\lim_{x \to 2}(3x^2 - 12x + 15) = 3$

57. $\lim_{x \to 1}\dfrac{x^2 - 1}{x - 1} = 2$

58. $\lim_{x \to 2}\dfrac{x^2 - 3x + 2}{x - 2} = 1$

59. $\lim_{x \to 5^+}\sqrt{x - 5} = 0$

60. $\lim_{x \to 2^+}3\sqrt{2x - 4} = 0$

For each of the limit statements in Exercises 61–66, write a δ–M, N–ϵ, or N–M proof, according to the type of limit statement.

61. $\lim_{x \to -2^+}\dfrac{1}{x + 2} = \infty$

62. $\lim_{x \to -2^-}\dfrac{1}{x + 2} = -\infty$

63. $\lim_{x \to \infty}\dfrac{2x - 1}{x} = 2$

64. $\lim_{x \to -\infty}\dfrac{2x - 1}{x} = 2$

65. $\lim_{x \to \infty}(3x - 5) = \infty$

66. $\lim_{x \to -\infty}(3x - 5) = -\infty$

Prove each of the limit statements in Exercises 67–72. You will have to bound δ.

67. $\lim_{x \to 3}(x^2 - 2x - 3) = 0$

68. $\lim_{x \to -1}(x^2 - 2x - 3) = 0$

69. $\lim_{x \to 5}(x^2 - 6x + 7) = 2$

70. $\lim_{x \to 1}(x^2 - 6x + 7) = 2$

71. $\lim_{x \to 2}\dfrac{4}{x^2} = 1$

72. $\lim_{x \to 3}\dfrac{18}{x^2} = 2$

Thinking Forward

Calculating limits: We still do not have a way to calculate limits easily. In the following problems you will develop rules for calculating limits of some very simple functions.

▶ Explain why it makes intuitive sense that $\lim_{x \to c} x = c$ for any real number c. Then use a delta–epsilon argument to prove it.

▶ Explain why it makes intuitive sense that $\lim_{x \to c} x^2 = c^2$ for any real number c. Then use a delta–epsilon argument to prove it. (*Hint: You will need to assume that* $\delta \le 1$.)

▶ Use the preceding two problems and the result of Exercise 22 to calculate the following limits:
  • $\lim_{x \to -1} x$
  • $\lim_{x \to 4} x$
  • $\lim_{x \to \pi} x$
  • $\lim_{x \to 0} x^2$
  • $\lim_{x \to 5} x^2$
  • $\lim_{x \to -2} x^2$
  • $\lim_{x \to 0}(2x - 3)$
  • $\lim_{x \to 1}(1 - x)$
  • $\lim_{x \to 3}(3x + 1)$

▶ When calculating each of these limits $\lim_{x \to c} f(x)$, you simply used the value of $f(c)$. Will that method always work for any limit? Why or why not?

1.4   CONTINUITY AND ITS CONSEQUENCES

▶   Continuity of functions at points and on intervals, and basic types of discontinuities

▶   Simple functions that are continuous on their domains

▶   The Extreme Value Theorem and the Intermediate Value Theorem

Defining Continuity with Limits

Intuitively, a function is ***continuous*** if its graph has no breaks, jumps, or holes. Loosely speaking, you can sketch the graph of a continuous function "without picking up your pencil." We can make the notion of continuity more precise by using limits. For example, consider the following four graphs:

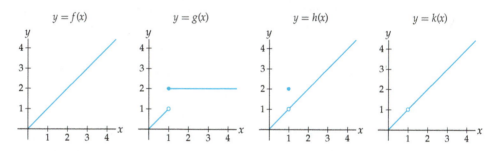

While the first graph has no breaks or holes, the remaining three graphs all have some sort of bad behavior at $x = 1$. It turns out that limits as $x \to 1$ detect exactly this bad behavior: In each case, the limit as $x \to 1$ is not the same as the value at $x = 1$. For example, $\lim\limits_{x \to 1} g(x)$ does not exist, but $g(1) = 2$. For $h(x)$, $\lim\limits_{x \to 1} h(x)$ is equal to 1, while the value $h(1)$ is equal to 2. Finally, for $k(x)$, the limit is $\lim\limits_{x \to 1} k(x) = 1$ but the value $k(1)$ does not exist. On the other hand, for $f(x)$, we have both $\lim\limits_{x \to 1} f(x) = 1$ and $f(1) = 1$.

The preceding examples suggest the following definition: A function is continuous at a point $x = c$ if its limit as $x \to c$ is equal to a real number that is the same as the value of the function at $x = c$.

DEFINITION 1.12   **Continuity of a Function at a Point**

A function f is ***continuous at*** $x = c$ if $\lim\limits_{x \to c} f(x) = f(c)$.

By considering one-sided limits, we can get a more detailed picture of continuity. For example, with the previous function $g(x)$, the left limit as $x \to 1$ is not equal to the value $g(1)$ but the right limit is. We say that $g(x)$ is ***right continuous*** at $x = 1$ but not ***left continuous***.

DEFINITION 1.13   **Left and Right Continuity at a Point**

A function f is ***left continuous at*** $x = c$ if $\lim\limits_{x \to c^-} f(x) = f(c)$ and is ***right continuous at*** $x = c$ if $\lim\limits_{x \to c^+} f(x) = f(c)$.

Sometimes it is convenient to talk about continuity of a function on an interval. We say that a function is continuous on an open interval if it is continuous at each point in the interval. For non-open intervals we also require one-sided continuity as we approach any closed endpoints.

DEFINITION 1.14   **Continuity of a Function on an Interval**

A function f is **continuous on an interval** I if it is continuous at every point in the interior of I, right continuous at any closed left endpoint, and left continuous at any closed right endpoint.

The graphs that follow provide examples of continuity on the four possible types of bounded intervals. For example, a function f is continuous on $I = (1, 3]$ if it is continuous at every point in the interior $(1, 3)$ and left continuous at the right endpoint $x = 3$, as shown in the third figure. In terms of limits this means that $\lim_{x \to c} f(x) = f(c)$ for all $c \in (1, 3)$ and $\lim_{x \to 3^-} f(x) = f(3)$.

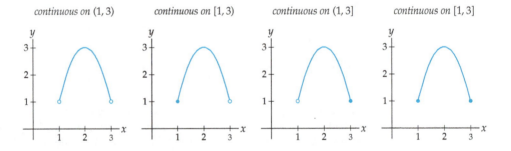

continuous on $(1, 3)$ *continuous on $[1, 3)$*   *continuous on $(1, 3]$*   *continuous on $[1, 3]$*

Types of Discontinuities

When a function is not continuous at a point $x = c$, we say that it is **discontinuous** at $x = c$. In terms of limits this means that the limit $\lim_{x \to c} f(x)$ is not equal to the value $f(c)$. The three most basic types of discontinuities that a function can have are illustrated as follows:

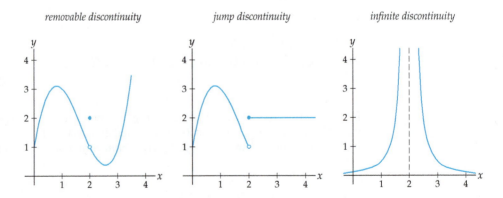

removable discontinuity   *jump discontinuity*   *infinite discontinuity*

Intuitively, we say that a discontinuity is **removable** if we could remove it just by changing one function value. At a **jump discontinuity**, the function jumps from one value to another, and at an **infinite discontinuity** the function has a vertical asymptote.

These types of discontinuities can be described precisely in terms of limits as follows:

DEFINITION 1.15

Removable, Jump, and Infinite Discontinuities

Suppose f is discontinuous at $x = c$. We say that $x = c$ is a

(a) *removable discontinuity* if $\lim_{x \to c} f(x)$ exists but is not equal to $f(c)$;

(b) *jump discontinuity* if $\lim_{x \to c^-} f(x)$ and $\lim_{x \to c^+} f(x)$ both exist but are not equal;

(c) *infinite discontinuity* if one or both of $\lim_{x \to c^-} f(x)$ and $\lim_{x \to c^+} f(x)$ is infinite.

For example, in the first figure shown, the limit $\lim_{x \to 2} f(x) = 1$ exists but is not equal to $f(2) = 2$, and therefore $f(x)$ has a removable discontinuity at $x = 2$. The function $g(x)$ in the second figure has left and right limits $\lim_{x \to 2^-} g(x) = 1$ and $\lim_{x \to 2^+} g(x) = 2$, respectively; the limits from both sides exist, but they are not equal to each other, and therefore $h(x)$ has a jump discontinuity at $x = 2$. Finally, the function $h(x)$ in the third graph has an infinite limit from both the left and the right at $x = 2$, and therefore has an infinite discontinuity at that point.

Continuity of Very Basic Functions

We say that a function is ***continuous on its domain*** if it is continuous on every interval on which it is defined. The following theorem proves that, unsurprisingly, our simplest examples of functions are continuous on their domains:

THEOREM 1.16

Continuity of Simple Functions

(a) Constant, identity, and linear functions are continuous everywhere. In terms of limits, for every k, c, m, and b in $\mathbb{R}$ we have

$$\lim_{x \to c} k = k, \quad \lim_{x \to c} x = c, \quad \text{and} \quad \lim_{x \to c}(mx + b) = mc + b.$$

(b) Power functions are continuous on their domains. In terms of limits, if A is real and k is rational, then for all values $x = c$ at which x^k is defined we have

$$\lim_{x \to c} Ax^k = Ac^k.$$

This is a powerful theorem, because it tells us that we can calculate limits of certain simple functions at domain points just by evaluating the functions at those points. For example, $f(x) = \dfrac{1}{x} = x^{-1}$ is a power function, so by the preceding theorem, it is continuous on its domain $(-\infty, 0) \cup (0, \infty)$. This means that at any point $c \neq 0$ we can calculate $\lim_{x \to c} \dfrac{1}{x}$ by simply calculating the value $f(c) = \dfrac{1}{c}$. However, the theorem does not tell us how to calculate $\lim_{x \to 0} \dfrac{1}{x}$; we will discuss such limits in a later section.

Technical point: In part (b) of the theorem the limit may sometimes be only one-sided. For example, $f(x) = x^{1/2}$ is defined at $x = 0$ and to the right of $x = 0$, but not for $x < 0$. Therefore the corresponding limit statement is one-sided: $\lim_{x \to 0^+} x^{1/2} = 0$.

Although it seems graphically obvious that the simple types of functions described in Theorem 1.16 are continuous everywhere they are defined, to actually prove continuity we need to appeal to the definition of limit. We will prove part (a) here and discuss the proof of part (b) after we learn about limit rules in the next section. You will prove that $f(x) = x^k$ is continuous on its domain for $k = 2$ and $k = -1$ in Exercises 88 and 89.

Proof. *(This proof requires material covered in optional Section 1.3.)*

The limit $\lim_{x \to c} k = k$ makes intuitive sense because as x approaches c, the number k should simply remain k; there is no x involved. To prove this limit statement we must show that for all $\epsilon > 0$, there is some $\delta > 0$ such that if $x \in (c - \delta, c) \cup (c, c + \delta)$, then $k \in (k - \epsilon, k + \epsilon)$. But k is *always* in the interval $(k - \epsilon, k + \epsilon)$, so the implication is trivially true for all values of ϵ and δ. This is illustrated in the leftmost figure that follows.

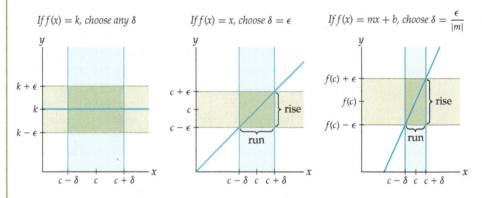

To prove that $\lim_{x \to c} x = c$ we must show that for all $\epsilon > 0$, there exists $\delta > 0$ such that if $x \in (c - \delta, c) \cup (c, c + \delta)$, then $x \in (L - \epsilon, L + \epsilon)$. This clearly holds if we choose δ to be equal to ϵ, as shown in the middle figure.

To prove that $\lim_{x \to c}(mx + b) = (mc + b)$ we will use the definition of limit in terms of absolute value inequalities from Definition 1.10. In the case when $m = 0$, the linear function $f(x) = mx + b$ is the constant function $f(x) = b$; we have already proved that case. If $m \neq 0$, then given $\epsilon > 0$, choose $\delta = \dfrac{\epsilon}{|m|}$. Then for all x satisfying $0 < |x - c| < \delta$ we also have

$$|(mx + b) - (mc + b)| = |mx - mc| = |m||x - c| < |m|\delta = |m|\left(\frac{\epsilon}{|m|}\right) = \epsilon.$$

It makes intuitive sense that δ should be $\dfrac{\epsilon}{m}$, since the ratio $\dfrac{\epsilon}{\delta}$ is equal to the ratio $\dfrac{\text{rise}}{\text{run}}$, or the slope of $f(x) = mx + b$, as shown in the rightmost figure. ∎

Extreme and Intermediate Values of Continuous Functions

In this section we examine two important consequences of continuity. First, a continuous function on a closed interval must be bounded and attain its upper and lower bounds. Second, if f is continuous between two values $x = a$ and $x = b$, then the corresponding values of $f(x)$ go through every possible intermediate value between the y-values $f(a)$ and $f(b)$. Both of these consequences are intuitively obvious if we think of continuous functions as having "unbroken" graphs.

For example, consider the function f in the first figure that follows. This function is continuous on $[a, b]$. In the second graph we see that f attains its maximum on that interval at $x = M$ and its minimum at $x = m$. In the third graph we see that the function attains every intermediate value K between $f(a)$ and $f(b)$.

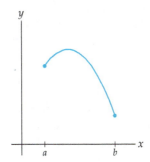

f is continuous on [a, b]

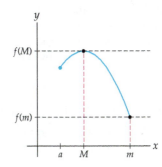

Maximum at M, minimum at m

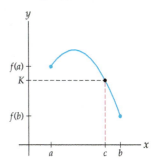

Height of y = K at x = c

In general, attaining extreme values and passing through all intermediate values are properties that hold for every function that is continuous on a closed interval [a, b]:

THEOREM 1.17

The Extreme Value Theorem

If f is continuous on a closed interval [a, b], then there exist values M and m in the interval [a, b] such that f(M) is the maximum value of f(x) on [a, b] and f(m) is the minimum value of f(x) on [a, b].

THEOREM 1.18

The Intermediate Value Theorem

If f is continuous on a closed interval [a, b], then for any K strictly between f(a) and f(b), there exists at least one c ∈ (a, b) such that f(c) = K.

These two important consequences of continuity may seem obvious, but in fact they rely on a subtle mathematical property of the real numbers called the ***Least Upper Bound Axiom***. Properly explaining the proofs of these theorems is outside of the scope of this book.

In the Extreme and Intermediate Value Theorems, the hypothesis that f be continuous on a closed interval [a, b] is essential. If f either fails to be continuous on the interior of the interval or fails to be continuous at a closed endpoint, then the conclusions of these theorems do not necessarily hold. For example, each of the following three functions fails to be continuous on [a, b] and also fails to satisfy at least one of the conclusions of the two theorems.

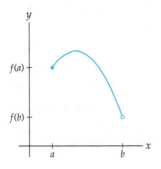

No minimum value on [a, b]

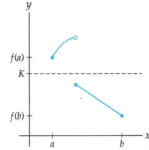

No c with f(c) = K

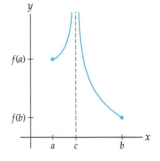

No maximum value on [a, b]

An extremely useful special case of the Intermediate Value Theorem is the case when we consider the intermediate value $K = 0$. In this case, the Intermediate Value Theorem says that if f is continuous on $[a, b]$ and $f(a)$ and $f(b)$ have opposite signs, then there exists at least one $c \in (a, b)$ where $f(c) = 0$. We use an equivalent variant of this special case every time we solve an inequality by checking signs between roots and discontinuities:

THEOREM 1.19

A Function Can Change Sign Only at Roots and Discontinuities

A function f can change sign (from positive to negative or vice versa) at a point $x = c$ only if $f(x)$ is zero, undefined, or discontinuous at $x = c$.

The graph that follows at the left shows a function f that is continuous on $[a, b]$, and changes sign only at its roots c_1, c_2, and c_3. The graph at the right is discontinuous somewhere in $[a, b]$ and therefore can change sign as we move from left to right without ever crossing the y-axis.

f(x) can change sign at roots

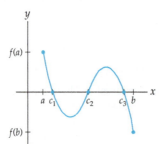

f(x) can change sign at a discontinuity

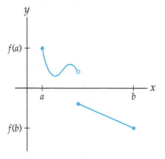

Examples and Explorations

EXAMPLE 1 **Limits and continuity of piecewise-defined functions**

Describe the continuity or discontinuity of each piecewise-defined function that follows at $x = 1$ by using graphs to determine the left, right, and two-sided limits at $x = 1$. Then describe the intervals on which each function is continuous.

(a) $f(x) = \begin{cases} x + 1, & \text{if } x < 1 \\ 3 - x^2, & \text{if } x \geq 1 \end{cases}$ **(b)** $g(x) = \begin{cases} 4 - x^2, & \text{if } x \leq 1 \\ x - 1, & \text{if } x > 1 \end{cases}$

SOLUTION

(a) The graph of f looks like $y = x + 1$ to the left of $x = 1$ and like $y = 3 - x^2$ to the right of $x = 1$, as shown next at the left. From the graph we see that $\lim\limits_{x \to 1^-} f(x) = 2$ and $\lim\limits_{x \to 1^+} f(x) = 2$, and thus $\lim\limits_{x \to 1} f(x) = 2$. We also have $f(1) = 3 - 1^2 = 2$. Since the limit of $f(x)$ as $x \to 1$ is equal to the value of $f(x)$ at $x = 1$, we can conclude that f is continuous at $x = 1$. In fact, according to the graph, the function f is continuous on all of $(-\infty, \infty)$.

Left and right limits both equal f(1)   *Only the left limit equals g(1)*

                                        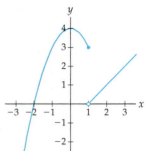

(b) The graph of $g(x)$ looks like $4 - x^2$ to the left of $x = 1$ and like $x - 1$ to the right of $x = 1$, with value $g(1) = 4 - 1^2 = 3$, as shown previously at the right. From the graph we see that $\lim_{x \to 1^-} g(x) = 3$ while $\lim_{x \to 1^+} g(x) = 0$. The left and right limits both exist, but they are not equal; thus $g(x)$ has a jump discontinuity at $x = 1$. Since $\lim_{x \to 1^-} g(x) = 3 = g(1)$ but $\lim_{x \to 1^+} g(x) = 0 \neq g(1)$, we can also say that $g(x)$ is left continuous, but not right continuous, at $x = 1$. According to the graph, the function $g(x)$ is continuous on $(-\infty, 1]$ and on $(1, \infty)$. □

EXAMPLE 2   **Continuity of a function that is defined separately for rationals and irrationals**

Determine graphically whether or not the rather exotic functions that follow are continuous at $x = 0$. Although these types of functions are not going to be a major focus in this course, this example helps get at the root of what continuity really means. You might be surprised by the solution!

(a) $f(x) = \begin{cases} 1, & \text{if } x \text{ is rational} \\ -1, & \text{if } x \text{ is irrational} \end{cases}$ **(b)** $g(x) = \begin{cases} 1, & \text{if } x \text{ is rational} \\ x + 1, & \text{if } x \text{ is irrational} \end{cases}$

SOLUTION

In the graphs of f and g that follow, the lighter dotted line represents the values of the function at rational-number inputs and the darker dotted line represents the values at irrational-number inputs. Note that the graphs of both f and g pass the vertical line test, since every input x is either rational or irrational and never both.

Overall limit does not exist as x → 0   *Limit as x → 0 approaches 1 in both cases*

                                  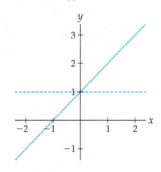

(a) We must consider the limit of f as $x \to 0$ separately for rational and irrational values of x. From the graph at the left, we see that for rational values of x we have $\lim_{x \to 0} f(x) = 1$

while for irrational values of x we have $\lim_{x \to 0} f(x) = -1$. Note that every punctured interval $(-\delta, 0) \cup (0, \delta)$ around $x = 0$ contains both rational and irrational numbers. Since the limit of $f(x)$ as $x \to 0$ is different depending on whether we choose rational or irrational values of x, the overall limit does not exist. Therefore the function f is not continuous at $x = 0$.

(b) On the other hand, looking at the graph of g at the right, we see that for rational values of x we have $\lim_{x \to 0} g(x) = 1$ and for irrational values of x we also have $\lim_{x \to 0} g(x) = 1$. Therefore $\lim_{x \to 0} g(x)$ exists and is equal to 1. Since $g(0)$ is also equal to 1, the function g is in fact continuous at $x = 0$, as strange as that may seem.   □

EXAMPLE 3   Calculating limits of very basic functions

Use continuity to calculate each of the limits that follow, if possible. If we do not yet have enough information to calculate a limit, explain why not.

(a) $\lim_{x \to 3} 2$ **(b)** $\lim_{x \to 3} \dfrac{1}{x}$ **(c)** $\lim_{x \to 0} \dfrac{1}{x}$ **(d)** $\lim_{x \to 4} \sqrt{x - 1}$

SOLUTION

By Theorem 1.16 we know that the constant function $f(x) = 2$ and the power function $g(x) = \dfrac{1}{x}$ are continuous on their domains. Therefore for parts (a) and (b) we can calculate the limits just by evaluating at $x = 3$: $\lim_{x \to 3} 2 = 2$ and $\lim_{x \to 3} \dfrac{1}{x} = \dfrac{1}{3}$.

We do not yet know how to calculate the remaining two limits algebraically. In part (c), the point $x = 0$ is not in the domain of $g(x) = \dfrac{1}{x}$, so we cannot apply Theorem 1.16. In part (d), the function $h(x) = \sqrt{x - 1}$ is not a constant, identity, linear, or power function, and thus at this point we cannot conclude anything about its continuity or its limits.   □

EXAMPLE 4   A real-world illustration of the Extreme and Intermediate Value Theorems

Consider the function $w(t)$ that describes a particular person's weight at t years of age between the ages of 18 and 45. Why does it make sense that this function is continuous on $[18, 45]$? What do the Extreme Value Theorem and the Intermediate Value Theorem say about $w(t)$?

SOLUTION

The weight function $w(t)$ should be continuous on $[18, 45]$ because a person's weight changes continuously over time and cannot jump from one value to another. (We are assuming typical circumstances, so that a person does not get a serious haircut, lose a limb, or somehow otherwise get their weight to change drastically in an instant.)

The Extreme Value Theorem tells us that there is some time $M \in [18, 45]$ at which the person's weight was greatest and some time $m \in [18, 45]$ at which that person weighed the least. In other words, at some time between 18 and 45 years of age, the person must have had a maximum weight and a minimum weight.

The Intermediate Value Theorem tells us that for every weight K between $w(18)$ and $w(45)$, there is some time $c \in (18, 45)$ for which $w(c) = K$. For example, if the person weighed $w(18) = 130$ pounds at age 18 and $w(45) = 163$ pounds at age 45, then there must be some age between 18 and 45 at which the person weighed, say, exactly 144 pounds.   □

EXAMPLE 5

Applying the Intermediate Value Theorem to a Continuous Function

The function $f(x) = x^3 - 3x + 1$ is continuous everywhere. (We will see this later in Section 1.5.) Use the Intermediate Value Theorem to conclude that there is some point c for which $f(c) = 2$. Then use a graph of f to approximate at least one such value of c.

SOLUTION

To show that there is some c with $f(c) = 2$ we need to find values a and b such that $K = 2$ is between $f(a)$ and $f(b)$, and apply the Intermediate Value Theorem. By trial and error we can find such values a and b, by testing different values of $f(x)$ until we find one that is less than and one that is greater than 2. For example,

$$f(0) = 0^3 - 3(0) + 1 = 1 < 2,$$

$$f(2) = 2^3 - 3(2) + 1 = 3 > 2.$$

Since f is continuous on $[0, 2]$ and $f(0) < 2 < f(2)$, by the Intermediate Value Theorem there is some value $c \in (0, 2)$ for which $f(c) = 2$. Note that the Intermediate Value Theorem doesn't tell us where c is, only that such a c exists somewhere in the interval $(0, 2)$.

We can approximate some values of c for which $f(c) = 2$ by approximating the values of x for which the graph of $f(x) = x^3 - 3x + 1$ intersects the line $y = 2$:

From this graph we can conclude that $f(c) = 2$ at $c \approx -1.5$, $c \approx -0.4$, and $c \approx 1.9$. To get better approximations we could trace along the graph on a calculator or other graphing utility.

EXAMPLE 6

Determining sign information between zeroes and discontinuities

Determine the intervals on which the functions that follow are positive or negative. You may assume that the function in part (a) is continuous everywhere and the function in part (b) is continuous on each piece.

(a) $f(x) = 3x^3 + 3x^2 - 6x$ **(b)** $g(x) = \begin{cases} x + 1, & \text{if } x < -2 \\ (x + 1)^2, & \text{if } -2 \leq x \leq 1 \\ 2 - x, & \text{if } x > 1 \end{cases}$

SOLUTION

(a) The roots of $f(x) = 3x^3 + 3x^2 - 6x = 3x(x^2 + x - 2) = 3x(x - 1)(x + 2)$ are $x = 0$, $x = 1$, and $x = -2$. By Theorem 1.19, in each of the intervals between these points the function f is either always positive or always negative. We need to test the sign of

$f(x)$ at one point in each interval. For example, $f(-3) = -36 < 0$, $f(-1) = 6 > 0$, $f(0.5) = -1.875 < 0$, and $f(2) = 24 > 0$. Reading from the resulting sign chart shown at the left, we can see that $f(x) = 3x^3 + 3x^2 - 6x$ is negative on $(-\infty, -2) \cup (0, 1)$ and positive on $(-2, 0) \cup (1, \infty)$:

(b) The piecewise-defined function g can be discontinuous only at its break points $x = -2$ and $x = 1$. Furthermore, its first component $x+1$ is never zero on $(-\infty, -2)$, its second component $(x + 1)^2$ is zero only at $x = -1$, and its third component $2 - x$ is zero when $x = 2$. By Theorem 1.19 the function g can change sign only at the roots and discontinuities at $-2, -1, 1$, and 2. All that now remains is to check the sign of $g(x)$ one time between each of these points; the results are recorded on the preceding sign chart at the right. We marked the discontinuous points with "dc" to distinguish them from the zeros. Reading from this sign chart and keeping careful track of the sign of $g(x)$ at the break points, we see that $g(x)$ is negative on $(-\infty, -2) \cup (2, \infty)$ and positive on $[-2, -1) \cup (-1, 2]$. □

CHECKING THE ANSWER

We can graph f and g with a calculator or other graphing utility to verify that the sign charts we found are reasonable. Notice that the intervals where $f(x)$ or $g(x)$ are positive are the intervals on which their graphs are above the x-axis.

$f(x)$ is above the x-axis on $(-2, 0) \cup (1, \infty)$

$g(x)$ is above the x-axis on $[-2, -1) \cup (-1, 2)$

TEST YOUR UNDERSTANDING

▶ Use limits to give definitions of each of the following: continuity at a point, continuity on an interval, left and right continuity.

▶ Use limits to give definitions of each of the following: removable discontinuity, jump discontinuity, infinite discontinuity.

▶ In the reading is a graph of a function with a removable discontinuity. We could make this graph continuous if we could change just one function value. What value?

▶ Why would it be difficult to prove at this point that all power functions, no matter what kind of power, are continuous on their domains?

▶ The conclusion of the Intermediate Value Theorem tells us of the existence of a point $c \in (a, b)$ with $f(c) = K$. Why do we need only the open interval (a, b), and not the closed interval $[a, b]$, in this conclusion?

EXERCISES 1.4

Thinking Back

Finding roots of piecewise-defined functions: For each function f that follows, find all values $x = c$ for which $f(c) = 0$. Check your answers by sketching a graph of f.

▶ $f(x) = \begin{cases} 4 - x^2, & \text{if } x < 0 \\ x + 1, & \text{if } x \geq 0 \end{cases}$

▶ $f(x) = \begin{cases} x + 1, & \text{if } x < 0 \\ 4 - x^2, & \text{if } x \geq 0 \end{cases}$

▶ $f(x) = \begin{cases} 2x - 1, & \text{if } x \leq 1 \\ 2x^2 + x - 3, & \text{if } x > 1 \end{cases}$

Logical existence statements: Determine whether each of the statements that follow are true or false. Justify your answers.

▶ If x is an integer, then there exists some positive integer y such that $|y| = x$.

▶ If x is a positive integer, then there exists some negative integer y such that $|y| = x$.

▶ If $x \in [-2, 2]$, then there exists some $y \in (0, 4)$ such that $y = x^2$.

▶ If $x \in [0, 100]$, then there exists some $y \in [-10, 10]$ such that $x = y^2$.

Concepts

0. *Problem Zero:* Read the section and make your own summary of the material.

1. *True/False:* Determine whether each of the statements that follow is true or false. If a statement is true, explain why. If a statement is false, provide a counterexample.

   (a) *True or False:* If f is both left and right continuous at $x = c$, then f is continuous at $x = c$.

   (b) *True or False:* If f is continuous on the open interval $(0, 5)$, then f is continuous at every point in $(0, 5)$.

   (c) *True or False:* If f is continuous on the closed interval $[0, 5]$, then f is continuous at every point in $[0, 5]$.

   (d) *True or False:* If f is continuous on the interval $(2, 4)$, then f must have a maximum value and a minimum value on $(2, 4)$.

   (e) *True or False:* If $f(3) = -5$ and $f(9) = -2$, then there must be a value c at which $f(c) = -3$.

   (f) *True or False:* If f is continuous everywhere, and if $f(-2) = 3$ and $f(1) = 2$, then $f(x)$ must have a root somewhere in $(-2, 1)$.

   (g) *True or False:* If f is continuous everywhere, and if $f(0) = -2$ and $f(4) = 3$, then $f(x)$ must have a root somewhere in $(0, 4)$.

   (h) *True or False:* If $f(0) = f(6) = 0$ and $f(2) > 0$, then $f(x)$ is positive on the entire interval $(0, 6)$.

2. *Examples:* Construct examples of the thing(s) described in the following. Try to find examples that are different than any in the reading.

   (a) The graph of a function with $f(4) = 2$ that has a removable discontinuity at $x = 4$.

   (b) The graph of a function that is continuous on its domain but not continuous at $x = 0$.

   (c) The graph of a function that is continuous on $(0, 2]$ and $(2, 3)$ but not on $(0, 3)$.

3. If f is a continuous function, what can you say about $\lim_{x \to 1} f(x)$?

4. Explain what it means for a function f to be continuous at a point $x = c$, with a sentence that includes the words "approaches" and "value."

5. In our proof that constant functions are continuous, we used the fact that given any $\epsilon > 0$, a choice of *any* $\delta > 0$ will work in the formal definition of limit. Use a graph to explain why this makes intuitive sense. (*This exercise depends on Section 1.3.*)

6. In our proof that linear functions are continuous, we used the fact that given any $\epsilon > 0$, the choice of $\delta = \dfrac{\epsilon}{|m|}$ will work in the formal definition of limit. Use a graph to explain why this makes intuitive sense. (*This exercise depends on Section 1.3.*)

7. Given the following function f, define $f(1)$ so that f is continuous at $x = 1$, if possible:

$$f(x) = \frac{x^2 - 2x + 1}{x^2 - 6x + 5}.$$

8. Given the following function f, define $f(1)$ so that f is continuous at $x = 1$, if possible:

$$f(x) = \begin{cases} 3x - 1, & \text{if } x < 1 \\ x^2 + 1, & \text{if } x > 1. \end{cases}$$

Each function in Exercises 9–12 is discontinuous at some value $x = c$. Describe the type of discontinuity and any one-sided continuity at $x = c$, and sketch a possible graph of f.

9. $\lim\limits_{x \to -1^-} f(x) = 2$, $\lim\limits_{x \to -1^+} f(x) = 2$, $f(-1) = 1$.

10. $\lim\limits_{x \to 2^-} f(x) = 2$, $\lim\limits_{x \to 2^+} f(x) = 1$, $f(2) = 1$.

11. $\lim\limits_{x \to 0^-} f(x) = -1$, $\lim\limits_{x \to 0^+} f(x) = 1$, $f(0) = 0$.

12. $\lim\limits_{x \to 2^-} f(x) = -\infty$, $\lim\limits_{x \to 2^+} f(x) = \infty$, $f(2) = 3$.

13. State what it means for a function f to be continuous at a point $x = c$, in terms of the delta–epsilon definition of limit. (*This exercise depends on Section 1.3.*)

14. State what it means for a function f to be left continuous at a point $x = c$, in terms of the delta–epsilon definition of limit. (*This exercise depends on Section 1.3.*)

15. State what it means for a function f to be right continuous at a point $x = c$, in terms of the delta–epsilon definition of limit. (*This exercise depends on Section 1.3.*)

16. Sketch a labeled graph of a function that satisfies the hypothesis of the Extreme Value Theorem, and illustrate on your graph that the conclusion of the Extreme Value Theorem follows.

17. Sketch a labeled graph of a function that satisfies the hypothesis of the Intermediate Value Theorem, and illustrate on your graph that the conclusion of the Intermediate Value Theorem follows.

18. Sketch a labeled graph of a function that fails to satisfy the hypothesis of the Intermediate Value Theorem, and illustrate on your graph that the conclusion of the Intermediate Value Theorem does not necessarily hold.

19. Sketch a labeled graph of a function that fails to satisfy the hypothesis of the Extreme Value Theorem, and illustrate on your graph that the conclusion of the Extreme Value Theorem does not necessarily hold.

20. Explain why the Intermediate Value Theorem allows us to say that a function can change sign only at discontinuities and zeroes.

For each of the following sign charts, sketch the graph of a function f that has the indicated signs, zeros, and discontinuities:

21.

22.

Skills

For each function f graphed in Exercises 23–26, describe the intervals on which f is continuous. For each discontinuity of f, describe the type of discontinuity and any one-sided continuity. Justify your answers about discontinuities with limit statements.

23.

24.

25.

26.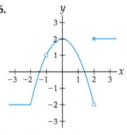

Sketch the graph of a function f described in Exercises 27–32, if possible. If it is not possible, explain why not.

27. f is left continuous at $x = 1$ and right continuous at $x = 1$, but is not continuous at $x = 1$, and $f(1) = -2$.

28. f is left continuous at $x = 2$ but not continuous at $x = 2$, and $f(2) = 3$.

29. f has a jump discontinuity at $x = -1$ and is left continuous at $x = -1$, and $f(-1) = 2$.

30. f has an infinite discontinuity at 0 but is right continuous at 0, and $f(0) = 1$.

31. f has a removable discontinuity at $x = -2$ and is right continuous at $x = -2$, and $f(-2) = 0$.

32. f is continuous on $[0, 2)$ but not on $[0, 2]$.

For each limit in Exercises 33–38, either use continuity to calculate the limit or explain why Theorem 1.16 does not apply.

33. $\lim\limits_{x \to -1} 6$

34. $\lim\limits_{x \to -1} x$

35. $\lim\limits_{x \to -5} (3x - 2)$

36. $\lim\limits_{x \to 3} x^4$

37. $\lim\limits_{x \to 0} x^{-3}$

38. $\lim\limits_{x \to -5} \sqrt{x}$

In Exercises 39–44, use Theorem 1.16 and left and right limits to determine whether each function f is continuous at its break point(s). For each discontinuity of f, describe the type of discontinuity and any one-sided discontinuity.

39. $f(x) = \begin{cases} x - 3, & \text{if } x < 3 \\ -(x - 3), & \text{if } x \geq 3 \end{cases}$

40. $f(x) = \begin{cases} x - 3, & \text{if } x < 0 \\ -(x - 3), & \text{if } x \geq 0 \end{cases}$

41. $f(x) = \begin{cases} x^2, & \text{if } x < 2 \\ 4, & \text{if } x = 2 \\ 2x + 1, & \text{if } x > 2 \end{cases}$

42. $f(x) = \begin{cases} \sqrt{-x}, & \text{if } x < 0 \\ 2, & \text{if } x = 0 \\ \sqrt{x}, & \text{if } x > 0 \end{cases}$

43. $f(x) = \begin{cases} x + 1, & \text{if } x < 1 \\ 3x - 1, & \text{if } 1 \leq x < 2 \\ x + 2, & \text{if } x \geq 2 \end{cases}$

44. $f(x) = \begin{cases} x^3, & \text{if } x \leq 0 \\ 1 - x, & \text{if } 0 < x < 3 \\ x - 5, & \text{if } x \geq 3 \end{cases}$

Use graphs to determine if each function f in Exercises 45–48 is continuous at the given point $x = c$.

45. $f(x) = \begin{cases} 2 - x, & \text{if } x \text{ rational} \\ x^2, & \text{if } x \text{ irrational,} \end{cases} \quad c = 2$

46. $f(x) = \begin{cases} x^2 - 3, & \text{if } x \text{ rational} \\ 3x + 1, & \text{if } x \text{ irrational,} \end{cases} \quad c = 0$

47. $f(x) = \begin{cases} 2 - x, & \text{if } x \text{ rational} \\ x^2, & \text{if } x \text{ irrational,} \end{cases} \quad c = 1$

48. $f(x) = \begin{cases} x^2 - 3, & \text{if } x \text{ rational} \\ 3x + 1, & \text{if } x \text{ irrational,} \end{cases} \quad c = 4$

Use the Extreme Value Theorem to show that each function f in Exercises 49–54 has both a maximum and a minimum value on $[a, b]$. Then use a graphing utility to approximate values M and m in $[a, b]$ at which f has a maximum and a minimum, respectively. You may assume that these functions are continuous everywhere.

49. $f(x) = x^4 - 3x^2 - 2$, $[a, b] = [-2, 2]$

50. $f(x) = x^4 - 3x^2 - 2$, $[a, b] = [0, 2]$

51. $f(x) = x^4 - 3x^2 - 2$, $[a, b] = [-1, 1]$

52. $f(x) = 3 - 2x^2 + x^3$, $[a, b] = [-1, 2]$

53. $f(x) = 3 - 2x^2 + x^3$, $[a, b] = [0, 2]$

54. $f(x) = 3 - 2x^2 + x^3$, $[a, b] = [-1, 1]$

Use the Intermediate Value Theorem to show that for each function f, interval $[a, b]$, and value K in Exercises 55–60, there is some $c \in (a, b)$ for which $f(c) = K$. Then use a graphing utility to approximate all such values c. You may assume that these functions are continuous everywhere.

55. $f(x) = 5 - x^4$, $[a, b] = [0, 2]$, $K = 0$

56. $f(x) = 5 - x^4$, $[a, b] = [-2, -1]$, $K = 0$

57. $f(x) = x^3 - 3x^2 - 2$, $[a, b] = [-2, 4]$, $K = -4$

58. $f(x) = x^3 - 3x^2 - 2$, $[a, b] = [0, 2]$, $K = -4$

59. $f(x) = x^3 - 3x^2 - 2$, $[a, b] = [2, 4]$, $K = -4$

60. $f(x) = 2 + x + x^3$, $[a, b] = [-1, 2]$, $K = 3$

Use the Intermediate Value Theorem to show that for each function f and value K in Exercises 61–66, there must be some $c \in \mathbb{R}$ for which $f(c) = K$. You will have to select an appropriate interval $[a, b]$ to work with. Then find or approximate one such value of c. You may assume that these functions are continuous everywhere.

61. $f(x) = x^3 + 2$, $K = -15$

62. $f(x) = -2x^2 + 4$, $K = 0$

63. $f(x) = (x - 3)^4 - 2$, $K = -1$

64. $f(x) = (x - 3)^4 - 2$, $K = 100$

65. $f(x) = |3x + 1|$, $K = 1$

66. $f(x) = |2 - 3x|$, $K = 2$

Find the intervals on which each function in Exercises 67–74 is positive or negative. Make clear how your work uses the Intermediate Value Theorem and continuity. You may assume that polynomials and their quotients are continuous on the intervals on which they are defined.

67. $f(x) = 2 + 5x + 2x^2$

68. $f(x) = x^3 - 2x^2 - 3x$

69. $f(x) = \dfrac{x^2 - 4}{x^2 - 1}$

70. $f(x) = \dfrac{(x + 4)(x - 1)}{2x + 3}$

71. $f(x) = \begin{cases} x - 4, & \text{if } x \leq 1 \\ x^2 - 4, & \text{if } x > 1 \end{cases}$

72. $f(x) = \begin{cases} 3x + 1, & \text{if } x < 0 \\ x, & \text{if } x \geq 0 \end{cases}$

73. $f(x) = \begin{cases} x^3, & \text{if } x \leq 2 \\ 4x - x^3, & \text{if } x > 2 \end{cases}$

74. $f(x) = \begin{cases} x^2 - 9, & \text{if } x \leq -2 \\ x^2 + x - 2, & \text{if } x > -2 \end{cases}$

Applications

Explain in practical terms what the Extreme Value Theorem says about each continuous function defined in Exercises 75–77. Then explain in practical terms what the Intermediate Value Theorem says in each situation.

75. Alina hasn't cut her hair for six years. Six years ago her hair was just 2 inches long. Now her hair is 42 inches long. Let $H(t)$ be the function that describes the length, in inches, of Alina's hair t years after she stopped cutting it.

76. Linda collects rain in a bucket outside her back door. Since the first day of April she has been keeping track of how the amount of water in the bucket changes as it fills with rain and evaporates. On April 1 the bucket was empty, and today it contains 4 inches of water. Let $w(t)$ be the height, in inches, of rainwater in the bucket t days after the first day of April.

77. The number of gallons of gas in Phil's new station wagon t days after he bought it is given by the function $g(t)$. When

he purchased the station wagon one year ago, the tank had 19 gallons of gas in it. Today he ran out of gas.

78. Lars was 20 inches tall when he was born, and six foot one when he died at age 83. Use the Intermediate Value Theorem to show that there must have been some point in Lars's life at which his height in inches was equal to his age in years. (*Hint: Think about when the difference between his height and age is zero.*)

79. As a vacuum cleaner salesman, Alex earns a salary of $8,500 a year, whether he sells any vacuum cleaners or not. In addition, for every 30 vacuum cleaners he sells, he earns a $1,500 commission.

  (a) Construct a piecewise-defined function $M(v)$ that describes the amount of money M that Alex will make in a year if he sells v vacuum cleaners over the course of the year. Assume he sells between 0 and 90 vacuum cleaners in a year.

(b) Check that your function makes sense by using it to calculate $M(0)$, $M(30)$, $M(59)$, $M(61)$, and $M(90)$. Then sketch a graph of $M(v)$ on the interval $0 \leq v \leq 90$.

(c) The piecewise-defined function $M(v)$ is *not* continuous. List all the values at which $M(v)$ fails to be continuous, and use the definition of continuity to support your answers.

80. One immediate application of the Intermediate Value Theorem is the method of finding roots called the **Bisection Method**. In this problem you will develop this method and then use it to approximate the square root of 2.

(a) Suppose f is continuous on $\mathbb{R}$ and that a and b are some real numbers for which $f(a)$ is negative and $f(b)$ is positive. Explain why the Intermediate Value Theorem guarantees that there must be some point in (a, b) where $f(x)$ has a root.

(b) Consider the function $f(x) = x^2 - 2$. Show that $f(0)$ is negative and $f(2)$ is positive. What conclusion can we draw from the Intermediate Value Theorem?

(c) We can bisect the interval $(0, 2)$ by finding the midpoint of the interval, which in this case is $x = 1$. Is $f(1)$ positive or negative? Does the Intermediate Value Theorem say anything about $f(x) = x^2 - 2$ on the interval $(0, 1)$? What about on the interval $(1, 2)$?

(d) Your answer to part (b) tells you that $f(x) = x^2 - 2$ must have a root somewhere in the interval $(0, 2)$ of length 2. Your answer to part (c) tells you that $f(x) = x^2 - 2$ must have a root in a shorter interval of length 1. Now repeat! Bisect the interval of length 1 to find an interval of length $\frac{1}{2}$ on which $f(x)$ must have a root.

(e) Describe why this Bisection Method will in general give better and better approximations for finding a root of a given function. In this particular example, with $f(x) = x^2 - 2$, why does the Bisection Method give us an approximation for $\sqrt{2}$?

Proofs

81. Write a delta–epsilon proof that shows that the function $f(x) = 3x - 5$ is continuous at $x = 2$. *(This exercise depends on Section 1.3.)*

82. Write a delta–epsilon proof that shows that the function $f(x) = 2x + 1$ is continuous at $x = 5$. *(This exercise depends on Section 1.3.)*

83. Write a delta–epsilon proof that shows that the function $f(x) = |x|$ is continuous. You may find the following inequality useful: For any real numbers a and b, $||a| - |b|| \leq |a - b|$. *(This exercise depends on Section 1.3.)*

84. Use what you know about one-sided limits to prove that a function f is continuous at a point $x = c$ if and only if it is both left and right continuous at $x = c$.

For each function f in Exercises 85–86, use the delta–epsilon definition of continuity to argue that f is or is not continuous at the indicated point $x = c$.

85. $f(x) = \begin{cases} 2 - x, & \text{if } x \text{ rational} \\ x^2, & \text{if } x \text{ irrational,} \end{cases} \quad c = 2$

86. $f(x) = \begin{cases} 2 - x, & \text{if } x \text{ rational} \\ x^2, & \text{if } x \text{ irrational,} \end{cases} \quad c = 1$

87. Use the Intermediate Value Theorem to prove that every cubic function $f(x) = Ax^3 + Bx^2 + Cx + D$ has at least one real root. You will have to first argue that you can find real numbers a and b so that $f(a)$ is negative and $f(b)$ is positive.

For each power function f in Exercises 88–89, write a delta–epsilon proof which proves that f is continuous on its domain. In each case you will need to assume that δ is less than or equal to 1. *(These exercises depend on Section 1.3.)*

88. $f(x) = x^{-1}$

89. $f(x) = x^2$

Thinking Forward

Interesting trigonometric limits: For each of the functions that follow, use a calculator or other graphing utility to examine the graph of f near $x = 0$. Does it appear that f is continuous at $x = 0$? Make sure your calculator is set to radian mode.

▶ $f(x) = \begin{cases} \dfrac{1}{x} \sin(x), & \text{if } x \neq 0 \\ 1, & \text{if } x = 0 \end{cases}$

▶ $f(x) = \begin{cases} \sin\left(\dfrac{1}{x}\right), & \text{if } x \neq 0 \\ 0, & \text{if } x = 0 \end{cases}$

▶ $f(x) = \begin{cases} x \sin\left(\dfrac{1}{x}\right), & \text{if } x \neq 0 \\ 0, & \text{if } x = 0 \end{cases}$

▶ $f(x) = \begin{cases} x^2 \sin\left(\dfrac{1}{x}\right), & \text{if } x \neq 0 \\ 0, & \text{if } x = 0 \end{cases}$

1.5   LIMIT RULES AND CALCULATING BASIC LIMITS

▶ Rules for calculating limits of arithmetic combinations and compositions of functions

▶ Continuity of algebraic and transcendental functions at domain points

▶ The Cancellation Theorem for calculating limits

Limits of Combinations of Functions

Although we now understand in depth what limit statements *mean*, at this point we do not have many tools for calculating limits. We can calculate limits of continuous functions at domain points by evaluation, and we know that very simple functions, such as constant and linear functions, are continuous. What about more complicated functions? For example, we already know from the continuity of power functions that

$$\lim_{x \to 2} x^2 = 2^2 = 4 \quad \text{and} \quad \lim_{x \to 2} x^3 = 2^3 = 8.$$

Can we use these results to say something about $\lim_{x \to 2}(x^2 + x^3)$ of the sum of these functions?

The key theorem that follows will help us answer this question; it says that limits behave well with respect to all of the arithmetic operations and even with respect to composition.

THEOREM 1.20

Rules for Calculating Limits of Combinations

If $\lim_{x \to c} f(x)$ and $\lim_{x \to c} g(x)$ exist, then the following rules hold for their combinations:

Constant Multiple Rule: $\lim_{x \to c} kf(x) = k \lim_{x \to c} f(x)$, for any real number k.

Sum Rule: $\lim_{x \to c}(f(x) + g(x)) = \lim_{x \to c} f(x) + \lim_{x \to c} g(x)$

Difference Rule: $\lim_{x \to c}(f(x) - g(x)) = \lim_{x \to c} f(x) - \lim_{x \to c} g(x)$

Product Rule: $\lim_{x \to c}(f(x)g(x)) = \left(\lim_{x \to c} f(x)\right)\left(\lim_{x \to c} g(x)\right)$

Quotient Rule: $\lim_{x \to c} \dfrac{f(x)}{g(x)} = \dfrac{\lim_{x \to c} f(x)}{\lim_{x \to c} g(x)}$, if $\lim_{x \to c} g(x) \neq 0$

Composition Rule: $\lim_{x \to c} f(g(x)) = f(\lim_{x \to c} g(x))$, if f is continuous at $\lim_{x \to c} g(x)$

This theorem is a powerful tool for calculating limits, since it tells us how to find limits of compound functions in terms of the limits of their components. For example, we can calculate the limit of the sum $x^2 + x^3$ as $x \to 2$ by taking the sum of the limits of x^2 and x^3 as $x \to 2$:

$$\lim_{x \to 2}(x^2 + x^3) = \lim_{x \to 2} x^2 + \lim_{x \to 2} x^3 = 4 + 8 = 12.$$

We will postpone the proofs of the limit rules in Theorem 1.20 until the end of this section so that we can first explore their consequences and practical uses. For example, an immediate consequence of Theorem 1.20 is that constant multiples, sums, differences, products, quotients, and compositions of continuous functions are continuous:

THEOREM 1.21

Combinations of Continuous Functions Are Continuous

If f and g are continuous at $x = c$ and k is any constant, then the functions kf, $f + g$, $f - g$, and fg are also continuous at $x = c$.

Moreover, if $g(c) \neq 0$, then $\frac{f}{g}$ is continuous at $x = c$, and if f is also continuous at $g(c)$, then $f \circ g$ is continuous at $x = c$.

For example, since $f(x) = x^2$ and $g(x) = x^3$ are continuous at $x = 2$, Theorem 1.21 tells us that $(f + g)(x) = x^2 + x^3$ must be also be continuous at $x = 2$. This makes sense given Theorem 1.20 because

$$\lim_{x \to 2}(f + g)(x) = \lim_{x \to 2}(x^2 + x^3) = \lim_{x \to 2}x^2 + \lim_{x \to 2}x^3 = 2^2 + 2^3 = 4 + 8 = 12 = (f + g)(2).$$

Limits of Algebraic Functions

With the limit rules we can now prove that most of the functions we will use in this book are continuous on their domains. We will start with the algebraic functions. Recall that a function is **algebraic** if it can be expressed with the use of only arithmetic operations ($+$, $-$, $\times$, and $\div$) and rational constant powers. Power functions, polynomial functions, and rational functions are all examples of algebraic functions.

THEOREM 1.22   **Continuity of Algebraic Functions**

All algebraic functions are continuous on their domains. In particular, if $x = c$ is in the domain of an algebraic function f, then we can calculate $\lim_{x \to c} f(x)$ by evaluating $f(c)$.

With this theorem we can do lots of basic limit calculations. For example, $\lim_{x \to 4} x^{1/2} = \sqrt{4} = 2$, $\lim_{x \to 1}(3x^4 - 2x) = 3(1)^4 - 2(1) = 1$, and $\lim_{x \to 2}\dfrac{1+x}{3-x} = \dfrac{1+2}{3-2} = 3$. For certain special cases the limits are only one-sided; for example, $\lim_{x \to 2^+} \sqrt{x - 2} = 0$. Note that the theorem does not tell us how to calculate limits at non-domain points; for example, we still do not know how to calculate $\lim_{x \to 2}\dfrac{1}{x - 2}$.

Proof. Algebraic functions are by definition built out of rational powers and arithmetic combinations of real numbers and the variable x. We already know how to handle limits of constant multiples, sums, products, quotients, and compositions by using the limit rules. We also know from Theorem 1.16 that $\lim_{x \to c} k = k$ and $\lim_{x \to c} x = c$. Therefore to show that every algebraic function is continuous on its domain, it suffices to show that every function of the form $f(x) = x^k$ is continuous on its domain.

We must show that for any rational number k, if $x = c$ is in the domain of x^k, then $\lim_{x \to c} x^k = c^k$. There are a few cases to consider. If k is a positive integer, then we just repeatedly apply the product rule for limits so that we can use the known limit $\lim_{x \to c} x = c$:

$$\lim_{x \to c} x^k = \underbrace{(\lim_{x \to c}x)(\lim_{x \to c}x)\cdots(\lim_{x \to c}x)}_{k \text{ times}} = \underbrace{(c)(c)\cdots(c)}_{k \text{ times}} = c^k.$$

For negative integer powers, we apply the quotient rule for limits and the result for positive integer powers. In this case we must require $c \neq 0$ so that c will be in the domain of x^{-k}, and we obtain the following limit:

$$\lim_{x \to c} x^{-k} = \lim_{x \to c}\frac{1}{x^k} = \frac{\lim\limits_{x \to c} 1}{\lim\limits_{x \to c} x^k} = \frac{1}{c^k} = c^{-k}.$$

Although we will not prove so here, it can be shown that $\lim_{x \to c} x^{1/q} = c^{1/q}$ when c is in the domain of $x^{1/q}$. Given these facts, the composition rule for limits allows us to prove that $x^{p/q}$ is continuous at domain points for any rational power $\dfrac{p}{q}$:

$$\lim_{x \to c} x^{p/q} = \lim_{x \to c}\sqrt[q]{x^p} = \sqrt[q]{\lim_{x \to c} x^p} = \sqrt[q]{c^p} = c^{p/q}.$$
∎

Finding Limits by Cancelling

The continuity of algebraic functions and the limit rules can help us calculate a great many limits, but only at domain points. One thing that can help us at non-domain points is the cancellation of common factors. For example, consider the limit $\lim\limits_{x \to 1} \frac{x^2 - 1}{x - 1}$.
At $x = 1$ we have $x^2 - 1 = 0$ and $x - 1 = 0$, and therefore the limit is of the form $\frac{0}{0}$. Limits of this form are said to be **indeterminate**, which means that they may or may not exist, depending on the situation. We will examine indeterminate forms in depth in Section 1.6. In the example we are considering, we can determine the limit by simple cancellation:

$$\lim_{x \to 1} \frac{x^2 - 1}{x - 1} = \lim_{x \to 1} \frac{(x - 1)(x + 1)}{x - 1} = \lim_{x \to 1} (x + 1) = 2.$$

The cancellation of the common factor $x - 1$ is valid because $\frac{(x - 1)(x + 1)}{x - 1}$ and $x + 1$ differ only when $x = 1$, and when we take the limit, we are not concerned with what happens *at* the point $x = 1$. In general, by definition, limits as $x \to c$ never have anything to do with what happens at $x = c$, which proves the following theorem:

THEOREM 1.23

The Cancellation Theorem for Limits

If $\lim\limits_{x \to c} g(x)$ exists, and f is a function that is equal to g for all x sufficiently close to c except possibly at c itself, then $\lim\limits_{x \to c} f(x) = \lim\limits_{x \to c} g(x)$.

Delta-Epsilon Proofs of the Limit Rules

The limit rules seem almost obvious; for example, if $f(x)$ approaches L and $g(x)$ approaches M as $x \to c$, it is reasonable to expect that $f(x) + g(x)$ approaches $L + M$ as $x \to c$. To *prove* the limit rules in Theorem 1.20, however, we must appeal to the delta–epsilon definition of limit.

Proof. *(This proof requires material covered in optional Section 1.3.)*
We will prove the (a) sum, (b) product, and (c) composition rules and leave the proofs of the remaining rules to Exercises 69, 70, and 71.

(a) To prove the sum rule for limits, we must show that we can get $f(x) + g(x)$ as close as we like to $L + M$ by choosing δ so that $f(x)$ and $g(x)$ are each half of that distance from L and M, as illustrated in the graph that follows at the left. Given $\epsilon > 0$, choose δ_1 to get $f(x)$ within $\frac{\epsilon}{2}$ of L and choose δ_2 to get $g(x)$ within $\frac{\epsilon}{2}$ of M. Then for $\delta = \min(\delta_1, \delta_2)$ and $x \in (c - \delta, c) \cup (c, c + \delta)$, we have

$$L - \frac{\epsilon}{2} < f(x) < L + \frac{\epsilon}{2} \quad \text{and} \quad M - \frac{\epsilon}{2} < g(x) < M + \frac{\epsilon}{2}.$$

Adding these two double inequalities together, we get our desired conclusion:

$$(L + M) - \epsilon < f(x) + g(x) < (L + M) + \epsilon.$$

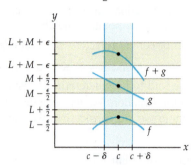

*Given ϵ for $f+g$,
choose δ to get $\dfrac{\epsilon}{2}$ for f and g*

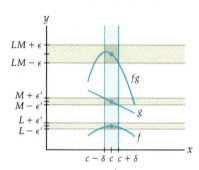

*Given ϵ for fg,
choose δ to get ϵ' for f and g*

(b) The proof for the product rule for limits is similar: We will get $f(x)g(x)$ as close as we like to LM by choosing δ so that $f(x)$ and $g(x)$ are sufficiently close enough to L and M, respectively, as illustrated in the preceding graph at the right. We will assume here that both L and M are positive; the other cases are similar. Given $\epsilon > 0$, choose $\epsilon' > 0$ sufficiently small so that $\epsilon' < L$, $\epsilon' < M$, and $(L + M)\epsilon' + (\epsilon')^2 < \epsilon$. The reason for this odd choice of ϵ' will become clear in a moment. Now choose δ_1 and δ_2 to get $f(x)$ and $g(x)$ within ϵ' of L and M, respectively. Then for $\delta = \min(\delta_1, \delta_2)$ and $x \in (c - \delta, c) \cup (c, c + \delta)$, we have

$$L - \epsilon' < f(x) < L + \epsilon' \quad \text{and} \quad M - \epsilon' < g(x) < M + \epsilon'.$$

Since L and M are assumed to be positive, we can assume that δ has been chosen small enough so that for $x \in (c - \delta, c) \cup (c, c + \delta)$ the values of $f(x)$ and $g(x)$ are also positive. Therefore, multiplying our two double inequalities together gives

$$(L - \epsilon')(M - \epsilon') < f(x)g(x) < (L + \epsilon')(M + \epsilon')$$

By our choice of ϵ' it is easy to show that $(L + \epsilon')(M + \epsilon') < LM + \epsilon$ and that $LM - \epsilon < (L - \epsilon')(M - \epsilon')$. Putting these two inequalities together with our double inequality for $f(x)g(x)$, we can say that for $x \in (c - \delta, c) \cup (c, c + \delta)$ we have

$$LM - \epsilon < f(x)g(x) < LM + \epsilon.$$

(c) To prove the composition rule for limits, let $\lim_{x \to c} g(x) = L$. We will show that we can get $f(g(x))$ as close as we like to $f(L)$ by choosing δ so that $g(x)$ is sufficiently close to L. Since f is continuous at L, we know that $\lim_{u \to L} f(u) = L$. Thus given $\epsilon > 0$, we can choose $\delta' > 0$ so that whenever $u \in (L - \delta', L) \cup (L, L + \delta')$, we also have $f(u) \in (f(L) - \epsilon, f(L) + \epsilon)$. In fact, if $u = L$, then $f(u) = f(L)$, so we can say a little bit more:

$$\text{if } u \in (L - \delta', L + \delta'), \text{ then } f(u) \in (f(L) - \epsilon, f(L) + \epsilon).$$

Now, $\lim_{x \to c} g(x) = L$ allows us to choose $\delta > 0$ so that

$$\text{if } x \in (c - \delta, c) \cup (c, c + \delta), \text{ then } g(x) \in (L - \delta', L + \delta').$$

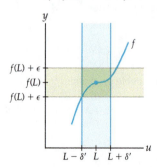

Given ϵ for $f(u)$, choose δ' for u

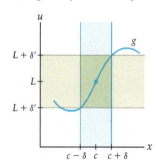

Then given δ' for u, choose δ for x

The two figures above illustrate our choices for δ' and δ. Now given our choice of δ we can let $g(x) = u$ and string together the last two displayed implications above to conclude that

$$\text{if } x \in (c - \delta, c) \cup (c, c + \delta), \text{ then } f(g(x)) \in (f(L) - \epsilon, f(L) + \epsilon).$$ ∎

Examples and Explorations

EXAMPLE 1   **Calculating limits by using continuity and limit rules**

Calculate the limits that follow, using only the continuity of linear and power functions and the limit rules in Theorem 1.20. Cite each limit rule that you apply.

(a) $\lim\limits_{x \to 3}(3x^2 - 2x + 1)$ **(b)** $\lim\limits_{x \to 1}(3x - 1)^{12}$

SOLUTION

(a) We apply the rules for limits of combinations of functions until we have reduced the problem to limits of functions that we know to be continuous:

$$\lim_{x \to 3}(3x^2 - 2x + 1) = \lim_{x \to 3}3x^2 - \lim_{x \to 3}2x + \lim_{x \to 3}1 \quad \leftarrow \text{ sum and difference rules}$$

$$= 3\lim_{x \to 3}x^2 - 2\lim_{x \to 3}x + \lim_{x \to 3}1 \quad \leftarrow \text{ constant multiple rule}$$

$$= 3(3^2) - 2(3) + 1 = 22. \quad \leftarrow \text{ limits of continuous functions}$$

(b) We could find this limit by multiplying out $(3x - 1)^{12}$ and then applying the sum and constant multiple rules as we did in part (a), but it is much faster to instead apply the composition rule for limits. We know that the power function x^{12} is continuous everywhere and, in particular, at $\lim\limits_{x \to 1}(3x - 1) = 2$. Therefore we have

$$\lim_{x \to 1}(3x - 1)^{12} = (\lim_{x \to 1}(3x - 1))^{12} \quad \leftarrow \text{ composition rule for limits}$$

$$= (3(1) - 1)^{12} = 2^{12} = 4096. \quad \leftarrow \text{ linear functions are continuous}$$ ☐

EXAMPLE 2   **Limits and continuity of piecewise-defined functions**

Describe the continuity or discontinuity of each of the following piecewise-defined functions at $x = 1$ by algebraically calculating the left, right, and two-sided limits at $x = 1$:

(a) $f(x) = \begin{cases} x + 1, & \text{if } x < 1 \\ 3 - x^2, & \text{if } x \geq 1 \end{cases}$ **(b)** $g(x) = \begin{cases} 4 - x^2, & \text{if } x \leq 1 \\ x - 1, & \text{if } x > 1 \end{cases}$

SOLUTION

(a) These two functions are the same ones we investigated graphically in Example 1 of Section 1.4. Here, for each function we must compute the limit as $x \to 1$ and compare it with the value at $x = 1$. Since the piecewise-defined function f has a break point at $x = 1$ we must compute its left and right limits separately. Because the component functions $x + 1$ and $3 - x^2$ are continuous, we can calculate the left and right limits by evaluation:

$$\lim_{x\to 1^-} f(x) = \lim_{x\to 1^-} (x+1) = 1 + 1 = 2 \quad \text{and}$$

$$\lim_{x\to 1^+} f(x) = \lim_{x\to 1^+} (3 - x^2) = 3 - (1)^2 = 2.$$

Since the left and right limits both exist and are equal to 2, we have $\lim_{x\to 1} f(x) = 2$, which is equal to the value $f(1) = 2$. Therefore, f is continuous at $x = 2$.

(b) Once again we must calculate the left and right limits of $g(x)$ separately. By continuity of the component functions we have

$$\lim_{x\to 1^-} g(x) = \lim_{x\to 1^-} (4 - x^2) = 4 - (1)^2 = 3 \quad \text{and}$$

$$\lim_{x\to 1^+} g(x) = \lim_{x\to 1^+} (x - 1) = 1 - 1 = 0.$$

Since the left and right limits both exist but are not equal as $x \to 1$, the function $g(x)$ has a jump discontinuity at $x = 1$. Because $\lim_{x\to 1^-} g(x) = 3 = g(1)$ but $\lim_{x\to 1^+} g(x) = 0 \ne g(1)$, we can also say that $g(x)$ is left continuous, but not right continuous, at $x = 1$.   ☐

EXAMPLE 3   Calculating limits

Calculate each of the following limits:

(a) $\displaystyle\lim_{x\to 1} \frac{1}{\sqrt{x^3 + 3}}$ **(b)** $\displaystyle\lim_{x\to 2} \frac{2 - x}{4 - x^2}$ **(c)** $\displaystyle\lim_{x\to 1} \frac{2x^2 + x - 3}{x - 1}$

SOLUTION

(a) $f(x) = \dfrac{1}{\sqrt{x^3 + 3}}$ is an algebraic function and thus is continuous on its domain by Theorem 1.22. Since $x = 1$ is in the domain of f, we can solve this limit by simple evaluation:

$$\lim_{x\to 1} \frac{1}{\sqrt{x^3 + 3}} = \frac{1}{\sqrt{1^3 + 3}} = \frac{1}{2}.$$

(b) The function $f(x) = \dfrac{2 - x}{4 - x^2}$ is algebraic and thus continuous on its domain, but unfortunately $x = 2$ is not in that domain. As $x \to 2$, both the numerator and the denominator approach 0, and therefore this limit is of the form $\dfrac{0}{0}$, which is indeterminate.

This means that we don't know at this point whether or not the limit exists or not, or if it does, what it might be equal to. We can solve this indeterminacy by doing some preliminary algebra; after cancellation we get a limit that is no longer indeterminate and, in fact, that we can find by evaluation at $x = 2$:

$$\lim_{x\to 2} \frac{2 - x}{4 - x^2} = \lim_{x\to 2} \frac{2 - x}{(2 - x)(2 + x)} = \lim_{x\to 2} \frac{1}{2 + x} = \frac{1}{4}.$$

(c) As $x \to 1$, both the numerator and the denominator approach zero, so this limit is of the form $\dfrac{0}{0}$, which is indeterminate. After factoring and cancelling, we can solve the problem:

$$\lim_{x\to 1} \frac{2x^2 + x - 3}{x - 1} = \lim_{x\to 1} \frac{(2x + 3)(x - 1)}{x - 1} = \lim_{x\to 1} (2x + 3) = 2(1) + 3 = 5.$$ ☐

TEST YOUR UNDERSTANDING

▶ The sum rule for limits says that the limit of a sum is the sum of the limits. In English, what do the other limit rules say?

▶ In the proof of the sum rule for limits, in order to get $f(x) + g(x)$ within ϵ of $L + M$, how close do we have to get $f(x)$ and $g(x)$ to L and M, respectively?

▶ What is the rule for the limit of a constant? What is the rule for the limit of a constant multiple of a function? How are these two rules different?

▶ Why does it make sense that cancellation would be a valid operation when dealing with a limit as $x \to c$, even if what is being cancelled approaches zero as $x \to c$?

▶ Given that x^2 and $\sin x$ are functions that are continuous everywhere, what can you say about the function $x^2 + \sin x$, and why?

EXERCISES 1.5

Thinking Back

Values of algebraic functions: Without a calculator, find each of the function values that follow. For some values the answer may be undefined.

▶ If $f(x) = \sqrt{x^3 - 2}$, find $f(3)$ and $f(0)$.

▶ If $f(x) = (x+1)^{-1}$, find $f(1)$ and $f(-1)$.

▶ If $f(x) = \dfrac{1}{x^2 - 4x - 5}$, find $f(4)$ and $f(5)$.

▶ If $f(x) = \dfrac{x^2 - 4}{x+2}$, find $f(2)$ and $f(-2)$.

The δ–ϵ definition of limit: Write each limit statement that follows in terms of the δ–ϵ definition of limit. Then approximate the largest value of δ corresponding to $\epsilon = 0.5$, and illustrate this choice of δ on a graph of f.

▶ $\displaystyle\lim_{x \to 2}(3x - 2) = 4$

▶ $\displaystyle\lim_{x \to 1}(x^3 - 1) = 0$

▶ $\displaystyle\lim_{x \to 1}\dfrac{x^2 - 1}{x + 3} = 0$

▶ $\displaystyle\lim_{x \to 0^+}\sqrt{x + 4} = 2$

Concepts

0. *Problem Zero:* Read the section and make your own summary of the material.

1. *True/False:* Determine whether each of the statements that follow is true or false. If a statement is true, explain why. If a statement is false, provide a counterexample.

(a) *True or False:* The limit of a difference of functions as $x \to c$ is equal to the difference of the limits of those functions as $x \to c$, provided that all limits involved exist.

(b) *True or False:* If $f(x)$ is within 0.25 unit of 7 and $g(x)$ is within 0.25 unit of 2, then $f(x) + g(x)$ is within 0.5 unit of 9.

(c) *True or False:* If $f(x)$ is within 0.25 unit of 7 and $g(x)$ is within 0.25 unit of 2, then $f(x)g(x)$ is within 0.5 unit of 9.

(d) *True or False:* Every algebraic function f is continuous at every real number $x = c$.

(e) *True or False:* Every power function $f(x) = Ax^k$ is continuous at the point $x = 2$.

(f) *True or False:* The function $f(x) = \dfrac{1}{x-1}$ is continuous at $x = 1$.

(g) *True or False:* The value of $\dfrac{(x-c)f(x)}{(x-c)g(x)}$ at $x = c$ is equal to the limit of $\dfrac{f(x)}{g(x)}$ at $x = c$.

(h) *True or False:* The limit of $\dfrac{(x-c)f(x)}{(x-c)g(x)}$ as $x \to c$ is equal to the limit of $\dfrac{f(x)}{g(x)}$ as $x \to c$.

2. *Examples:* Construct examples of the thing(s) described in the following. Try to find examples that are different than any in the reading.

(a) Two limits that are initially in an indeterminate form but can be solved with the Cancellation Theorem.

(b) Two limits that can be solved by simple evaluation.

(c) Two limits that we do not yet know how to calculate.

3. State the constant multiple rule, sum rule, product rule, quotient rule, and composition rule for limits.

4. Explain in your own words the types of functions whose limits we can calculate with the limit rules in this section.

5. Explain why we can't calculate every limit $\displaystyle\lim_{x \to c}f(x)$ just by evaluating $f(x)$ at $x = c$. Support your argument with the graph of a function f for which $\displaystyle\lim_{x \to c}f(x) \neq f(c)$.

6. Find functions f and g and a real number c such that $\displaystyle\lim_{x \to c}f(x) + \lim_{x \to c}g(x) \neq \lim_{x \to c}(f(x) + g(x))$. Does this example contradict the sum rule for limits? Why or why not?

7. Find functions f and g and a real number c such that $(\displaystyle\lim_{x \to c}f(x))(\lim_{x \to c}g(x)) \neq \lim_{x \to c}(f(x)g(x))$. Does this example contradict the product rule for limits? Why or why not?

8. Write the constant multiple rule for limits in terms of delta–epsilon statements.

9. Write the difference rule for limits in terms of delta–epsilon statements.

10. Write the product rule for limits in terms of delta–epsilon statements.

11. Explain how the algebraic function

$$f(x) = (\sqrt{x} + 1)^3$$

is a combination of identity, constant, and power functions. Why does this mean that we can calculate limits of this function at domain points by evaluation?

12. Explain how the algebraic function

$$f(x) = \frac{(x^2 + 1)(4 - 3x)}{3x^2}$$

is a combination of identity, constant, and power functions. Why does this mean that we can calculate limits of this function at domain points by evaluation?

Suppose f and g are functions such that $\lim_{x\to3} f(x) = 5$, $\lim_{x\to4} f(x) = 2$, and $\lim_{x\to3} g(x) = 4$. Given this information, calculate the limits that follow, if possible. If it is not possible with the given information, explain why.

13. $\lim_{x\to3}(2f(x) - 3g(x))$

14. $\lim_{x\to3} -2f(x)$

15. $\lim_{x\to7} f(x)$

16. $\lim_{x\to4} f(x)g(x)$

17. $\lim_{x\to3} \dfrac{f(x) - 3}{g(x)}$

18. $\lim_{x\to3} f(g(x))$

19. Graph the functions $f(x) = x + 1$ and $g(x) = \dfrac{x^2 - 1}{x - 1}$, and show that they are equal everywhere except at one point. Then show that $f(x)$ and $g(x)$ have different values, but the same limit, at this point.

20. Graph the functions $f(x) = 2 - x$ and $g(x) = \dfrac{4 - x^2}{x + 2}$, and show that they are equal everywhere except at one point. Then show that $f(x)$ and $g(x)$ have different values, but the same limit, at this point.

Skills

Calculate the limits in Exercises 21–24, using only the continuity of linear and power functions and the limit rules. Cite each limit rule that you apply.

21. $\lim_{x\to1} 15(3 - 2x)$

22. $\lim_{x\to-1} \dfrac{x - 1}{(x + 4)(x + 2)}$

23. $\lim_{x\to3}(3x + x^2(2x + 1))$

24. $\lim_{x\to0} \dfrac{3}{2x^2 - 4x + 1}$

Calculate each of the limits in Exercises 25–56.

25. $\lim_{x\to0}(x^2 - 1)$

26. $\lim_{x\to2}(x - 1)(x + 1)(x + 5)$

27. $\lim_{x\to-1}\left(x^2 - \dfrac{3x}{x + 2}\right)$

28. $\lim_{x\to1.7}(3.1x^2 - 4x + 0.8)$

29. $\lim_{x\to2} \dfrac{4 + 2x}{x + 2}$

30. $\lim_{x\to-2} \dfrac{4 + 2x}{x + 2}$

31. $\lim_{x\to1} \dfrac{x^2 - 1}{x - 1}$

32. $\lim_{x\to0} \dfrac{x^2 - 1}{x - 1}$

33. $\lim_{x\to-3} \dfrac{x + 3}{3x^2 + 8x - 3}$

34. $\lim_{x\to-3} \dfrac{3x^2 + 8x - 3}{x + 3}$

35. $\lim_{x\to2} \dfrac{x^2(x - 5)}{3(1 - 2x)(2 - 3x)}$

36. $\lim_{x\to0} \dfrac{x^2(x - 5)}{3(1 - 2x)(2 - 3x)}$

37. $\lim_{x\to-3} \dfrac{2}{3x + 1}$

38. $\lim_{x\to4} \dfrac{2 - 3x}{x^3}$

39. $\lim_{x\to0} \dfrac{6x + 3x^2}{x + 2}$

40. $\lim_{x\to-2} \dfrac{6x + 3x^2}{x + 2}$

41. $\lim_{x\to1/2} \dfrac{4x - 2}{6x^2 + x - 2}$

42. $\lim_{x\to-2/3} \dfrac{6x^2 + x - 2}{3x + 2}$

43. $\lim_{x\to1^+} \dfrac{x - 1}{\sqrt{x - 1}}$

44. $\lim_{x\to1^+} \dfrac{1 - \sqrt{x}}{1 - x}$

45. $\lim_{x\to1} \dfrac{x + x^2 - 2x^3}{x - x^2}$

46. $\lim_{x\to0} \dfrac{x + x^2 - 2x^3}{x - x^2}$

47. $\lim_{h\to0} \dfrac{(1 + h)^2 - 1}{h}$

48. $\lim_{h\to0} \dfrac{(-1 + h)^2 - 1}{h}$

49. $\lim_{h\to0} \dfrac{(3 + h)^2 - 3^2}{h}$

50. $\lim_{h\to0} \dfrac{(2 + h)^2 - 2^2}{h}$

51. $\lim_{h\to0} \dfrac{(1 + h)^3 - 1^3}{h}$

52. $\lim_{h\to0} \dfrac{(-1 + h)^3 - (-1)^3}{h}$

53. $\lim_{h\to0} \dfrac{\frac{1}{1+h} - 1}{h}$

54. $\lim_{h\to0} \dfrac{\frac{1}{2+h} - \frac{1}{2}}{h}$

55. $\lim_{h\to0} \dfrac{\frac{4}{(2+h)^2} - 1}{h}$

56. $\lim_{h\to0} \dfrac{\frac{4}{(1+h)^2} - 4}{h}$

Describe the intervals on which each function f in Exercises 57–64 is continuous. At each point where f fails to be continuous, use limits to determine the type of discontinuity and any left- or right-continuity.

57. $f(x) = \begin{cases} x^2 + 1, & \text{if } x \le 0 \\ 1 - x, & \text{if } x > 0 \end{cases}$

58. $f(x) = \begin{cases} 3x + 2, & \text{if } x < -1 \\ 5 + 4x^3, & \text{if } x \ge -1 \end{cases}$

59. $f(x) = \begin{cases} x^2 - 3x - 1, & \text{if } x \ne -2 \\ 3, & \text{if } x = -2 \end{cases}$

60. $f(x) = \begin{cases} \dfrac{3x^2}{4 + x}, & \text{if } x < 2 \\ x^2 - 2, & \text{if } x \ge 2 \end{cases}$

61. $f(x) = \begin{cases} \dfrac{x^2 - 1}{x - 1}, & \text{if } x < 1 \\ 0, & \text{if } x = 1 \\ 3x - 1, & \text{if } x > 1 \end{cases}$

62. $f(x) = \begin{cases} x + 1, & \text{if } x < 3 \\ 2, & \text{if } x = 3 \\ x^2 - 9, & \text{if } x > 3 \end{cases}$

63. $f(x) = \begin{cases} \dfrac{2 - x}{x^2 - 4}, & \text{if } x \le 2 \\ 1 - \sqrt{2x}, & \text{if } x > 2 \end{cases}$

64. $f(x) = \begin{cases} \dfrac{1}{x}, & \text{if } x < 1 \\ x, & \text{if } x \ge 1 \end{cases}$

Applications

In Exercise 88 in Section 0.4, you constructed a piecewise-defined function from the 2000 Federal Tax Rate Schedule that you will use in the next two problems. Specifically, you found that a person who makes m dollars a year will pay $T(m)$ dollars in tax, given by the function

$$0.15m, \text{ if } 0 \leq m \leq 26{,}250$$
$$3{,}937 + 0.28(m - 26{,}250), \text{ if } 26{,}250 < m \leq 63{,}550$$
$$14{,}381 + 0.31(m - 63{,}550), \text{ if } 63{,}550 < m \leq 132{,}600$$
$$35{,}787 + 0.36(m - 132{,}600), \text{ if } 132{,}600 < m \leq 288{,}350$$
$$91{,}857 + 0.396(m - 288{,}350), \text{ if } m > 288{,}350.$$

65. Suppose you make \$63,550 a year and pay taxes according to the given formula.

(a) Calculate the value of $T(63{,}550)$ and the limit of $T(m)$ as m approaches 63,550 from the left and from the right.

(b) Use part (a) to argue that the function $T(m)$ is continuous at $m = 63{,}550$. What does this mean in real-world terms?

66. Suppose you make \$288,350 a year and pay taxes according to the given formula.

(a) Calculate the value of $T(288{,}350)$ and the limit of $T(m)$ as m approaches 288,350 from the left and from the right.

(b) Use part (a) to argue that the function $T(m)$ is continuous at $m = 288{,}350$. What does this mean in real-world terms?

Proofs

67. Use limit rules and the continuity of power functions to prove that every polynomial function is continuous everywhere.

68. Use limit rules and the continuity of polynomial functions to prove that every rational function is continuous on its domain.

69. Prove the constant multiple rule for limits: If $\lim_{x \to c} f(x) = L$ and $k \in \mathbb{R}$, then $\lim_{x \to c} kf(x) = kL$.

70. Prove the difference rule for limits by applying the sum and constant multiple rules for limits.

71. Suppose that we know the reciprocal rule for limits: If $\lim_{x \to c} g(x) = M$ exists and is nonzero, then $\lim_{x \to c} \dfrac{1}{g(x)} = \dfrac{1}{M}$. This limit rule is tedious to prove and we do not include it here. Use the reciprocal rule and the product rule for limits to prove the quotient rule for limits.

Thinking Forward

Limits for Derivatives: In Chapter 2 we will define the **derivative** of a function f at a point $x = c$ to be the slope of the line that points in the direction of the graph of f at $x = c$. Algebraically, the derivative of f at c is given by the following limit:

$$\lim_{h \to 0} \frac{f(c+h) - f(c)}{h}.$$

Such limits are always of the indeterminate form $\dfrac{0}{0}$, so we must do algebra before we can resolve the limit.

▶ Calculate the derivative of $f(x) = x^2$ at $c = 0$.

▶ Calculate the derivative of $f(x) = x^2$ at $c = 2$.

▶ Calculate the derivative of $f(x) = x^2$ at $c = 4$.

▶ Sketch a graph of $f(x) = x^2$, and sketch the lines that point in the direction of the curve at $(0, f(0))$, $(2, f(2))$, and $(4, f(4))$. Relate the slopes of these lines to the answers to the last three exercises.

1.6   INFINITE LIMITS AND INDETERMINATE FORMS

▶ Calculating limits at infinity and infinite limits

▶ Recognizing non-indeterminate forms and dealing with indeterminate forms

▶ The Squeeze Theorem for calculating limits

Infinite Limits

The utility of continuity is that it enables us to calculate limits by evaluation. However, we can't solve limits by evaluation if continuity fails. For example, consider the limit $\lim\limits_{x\to 1}\dfrac{1}{x-1}$.

The function $\dfrac{1}{x-1}$ is neither defined nor continuous at $x=1$, so we cannot find its limit by evaluation. If we try to evaluate $\dfrac{1}{x-1}$ at $x=1$, we would not get a real number, because the denominator would be 0. In terms of limits, we say that

$$\lim_{x\to 1}\frac{1}{x-1} \text{ is of the form } \frac{1}{0}, \text{ and thus } \lim_{x\to 1}\frac{1}{x-1} \text{ is not a real number.}$$

Can we be more specific than just pointing out that limits of type $\dfrac{1}{0}$ "do not exist" (i.e., are not real numbers)? Consider the behavior of $\dfrac{1}{x-1}$ as x approaches 1 from the right, as shown in the table below. As $x\to 1^+$ (see the first row of the table) the values of $x-1$ approach 0^+ (see the second row). The reciprocals of these values (see the third row) then approach ∞:

x	1.1	1.01	1.001	1.0001	1.00001	$\to$	1^+
$x-1$	0.1	0.01	0.001	0.0001	0.00001	$\to$	0^+
$\dfrac{1}{x-1}$	10	100	1000	10,000	100,000	$\to$	∞

In symbols, this means that $\dfrac{1}{x-1}\to\infty$ as $x\to 1^+$. In terms of limits we can express this behavior from the right by saying that

$$\lim_{x\to 1^+}\frac{1}{x-1} \text{ is of the form } \frac{1}{0^+}, \text{ and thus } \lim_{x\to 1^+}\frac{1}{x-1}=\infty.$$

We will prove a more general version of this statement in Theorem 1.24.

From the left, we have a similar situation: As $x\to 1^-$, we have $x-1\to 0^-$, so the quantity $\dfrac{1}{x-1}$ will remain negative as $x\to 1^-$, and as the magnitude of the denominator $x-1$ gets smaller and smaller, the magnitude of $\dfrac{1}{x-1}$ gets larger and larger. In terms of limits we can express this behavior from the left by saying that

$$\lim_{x\to 1^-}\frac{1}{x-1} \text{ is of the form } \frac{1}{0^-}, \text{ and thus } \lim_{x\to 1^-}\frac{1}{x-1}=-\infty.$$

We have now shown that $\lim\limits_{x\to 1}\dfrac{1}{x-1}$ does not exist, and more specifically, that the limit is ∞ from the right and $-\infty$ from the left.

If an expression approaches ∞ from both the right and the left, then we say that the two-sided limit is ∞. For example, since $\dfrac{1}{(x-1)^2}$ approaches ∞ from both the right and the left as $x\to 1$, we would write $\lim\limits_{x\to 1}\dfrac{1}{(x-1)^2}$ is of the form $\dfrac{1}{0^+}$, and thus $\lim\limits_{x\to 1}\dfrac{1}{(x-1)^2}=\infty$.

The theoretical basis for the discussion above is summarized in the following theorem:

THEOREM 1.24

Limits Whose Denominators Approach Zero from the Right or the Left

(a) If $\lim\limits_{x \to c} \dfrac{f(x)}{g(x)}$ is of the form $\dfrac{1}{0^+}$, then $\lim\limits_{x \to c} \dfrac{f(x)}{g(x)} = \infty$.

(b) If $\lim\limits_{x \to c} \dfrac{f(x)}{g(x)}$ is of the form $\dfrac{1}{0^-}$, then $\lim\limits_{x \to c} \dfrac{f(x)}{g(x)} = -\infty$.

Theorem 1.24 also applies to one-sided limits and to limits as $x \to \infty$ or as $x \to -\infty$. We will prove only the case for limits from the right:

Proof. We will prove the case where $\lim\limits_{x \to c^+} \dfrac{f(x)}{g(x)}$ is of the form $\dfrac{1}{0^+}$. The other cases are similar; you will handle another in Exercise 63. Since $f(x) \to 1$ as $x \to 0$, it follows that for any $\epsilon_1 > 0$, we can find $\delta_1 > 0$ such that

$$\text{if } c < x < c + \delta_1, \text{ then } 1 - \epsilon_1 < f(x) < 1 + \epsilon_1.$$

Similarly, since $g(x) \to 0^+$ as $x \to 0$, it follows that for any $\epsilon_2 > 0$, we can find $\delta_2 > 0$ such that

$$\text{if } c < x < c + \delta_2, \text{ then } 0 < g(x) < 0 + \epsilon_2.$$

Now to prove that $\lim\limits_{x \to c^+} \dfrac{f(x)}{g(x)} = \infty$, take any $M > 0$. Choose δ to be the minimum of the δ_1 corresponding to $\epsilon_1 = \dfrac{1}{2}$ and the δ_2 corresponding to $\epsilon_2 = \dfrac{1}{2M}$. With this choice of δ, we have

$$\text{if } c < x < c + \delta, \text{ then } \frac{f(x)}{g(x)} > \frac{1 - \epsilon_1}{\epsilon_2} = \frac{\frac{1}{2}}{\frac{1}{2M}} = M.$$

∎

Limits at Infinity

We have just seen that limits of the form $\dfrac{1}{0^+}$ and $\dfrac{1}{0^-}$ are always infinite. A sort of reverse of this is also true, and is the subject of our next theorem: Limits of the form $\dfrac{1}{\infty}$ and $\dfrac{1}{-\infty}$ are always zero. This makes intuitive sense because the reciprocal of a number of large magnitude is a number of small magnitude. Note that even though the expression "$\dfrac{1}{\infty}$" does not represent a real number, a limit of that form *will* be equal to a real number, namely, 0.

THEOREM 1.25

Limits Whose Denominators Become Infinite Approach Zero

(a) If $\lim\limits_{x \to \infty} \dfrac{f(x)}{g(x)}$ is of the form $\dfrac{1}{\infty}$, then $\lim\limits_{x \to \infty} \dfrac{f(x)}{g(x)} = 0$.

(b) If $\lim\limits_{x \to \infty} \dfrac{f(x)}{g(x)}$ is of the form $\dfrac{1}{-\infty}$, then $\lim\limits_{x \to \infty} \dfrac{f(x)}{g(x)} = 0$.

For example, as $x \to \infty$ we have $x + 1 \to \infty$ and thus $\dfrac{1}{x+1} \to 0$; therefore $\lim\limits_{x \to \infty} \dfrac{1}{x+1} = 0$.

Similarly, as $x \to -\infty$ we have $x + 1 \to -\infty$ and thus $\dfrac{1}{x+1} \to 0$; therefore $\lim\limits_{x \to -\infty} \dfrac{1}{x+1} = 0$.

Theorem 1.25 also applies for limits as $x \to -\infty$ and as $x \to c$, although we will not prove that here.

Proof. We prove the first part and leave the second part to Exercise 64. Since $f(x) \to 1$, it follows that for any $\epsilon_1 > 0$, we can find $N_1 > 0$ such that

$$\text{if } x > N_1, \text{ then } 1 - \epsilon < f(x) < 1 + \epsilon.$$

Similarly, since $g(x) \to \infty$, it follows that for any $M > 0$, we can find $N_2 > 0$ such that

$$\text{if } x > N_2, \text{ then } g(x) > M.$$

Now to prove that $\lim_{x \to \infty} \dfrac{f(x)}{g(x)} = 0$, take any $\epsilon > 0$. Choose N to be the maximum of the N_1 corresponding to $\epsilon_1 = 1$ and the N_2 corresponding to $M = \dfrac{2}{\epsilon}$. With this choice of N, we have

$$\text{if } x > N, \text{ then } 0 < \frac{f(x)}{g(x)} < \frac{1 + \epsilon_1}{M} = \frac{2}{\left(\dfrac{2}{\epsilon}\right)} = \epsilon.$$

∎

The next theorem describes the limits at infinity of power functions. You will prove this theorem in Exercises 66 and 67.

THEOREM 1.26   **Limits of Power Functions at Infinity**

If $k > 0$, then $\lim_{x \to \infty} x^k = \infty$ and $\lim_{x \to \infty} x^{-k} = 0$.

Rather than memorizing these limits, it is better to remember the behavior on the right side of the graphs of power functions as $x \to \infty$, as in the following examples:

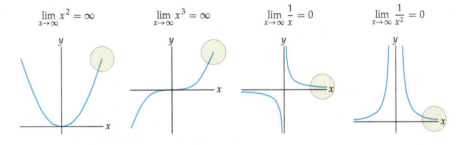

$$\lim_{x \to \infty} x^2 = \infty \qquad \lim_{x \to \infty} x^3 = \infty \qquad \lim_{x \to \infty} \frac{1}{x} = 0 \qquad \lim_{x \to \infty} \frac{1}{x^2} = 0$$

We can say similar things about limits as $x \to -\infty$. In fact, in most cases limits as $x \to -\infty$ can be rewritten with algebra as limits as $x \to \infty$. It is also helpful to remember the behavior on the left side of common graphs to determine limits as $x \to -\infty$. Although we will not prove so in general here, the limit rules from Theorem 1.20 also apply when $x \to \infty$ and when $x \to -\infty$. You will prove this for the sum rule for limits in Exercise 65. For example, given that $\lim_{x \to \infty} x^{-2} = 0$ and that $\lim_{x \to \infty} x^{-3} = 0$, we can conclude that $\lim_{x \to \infty} (x^{-2} + x^{-3}) = 0 + 0 = 0$.

Indeterminate and Non-Indeterminate Forms

If a limit is *indeterminate*, then we cannot initially say whether or not it exists—or if it exists, what real number it is equal to. Many indeterminate limits can be resolved with algebra such as cancellation or factoring. For example, the four limits that follow are all initially of the indeterminate form $\dfrac{0}{0}$. After some simple algebra, we see that three limits exist and one limit becomes infinite. The three that exist are each equal to different real numbers.

$$\lim_{x \to 1} \frac{(x-1)^2}{x-1} = \lim_{x \to 1} \frac{x-1}{1} = \frac{0}{1} = 0 \qquad \lim_{x \to 1} \frac{x-1}{(x-1)^3} = \lim_{x \to 1} \frac{1}{(x-1)^2} = \infty$$

$$\lim_{x \to 1} \frac{x-1}{x-1} = \lim_{x \to 1} 1 = 1 \qquad \lim_{x \to 1} \frac{x-1}{3(x-1)} = \lim_{x \to 1} \frac{1}{3} = \frac{1}{3}$$

When a limit is indeterminate, it is essentially because there is a "fight" going on between two parts of the limit. For example, limits of the form $\frac{0}{0}$ approach different things depending on whether the numerator or the denominator approaches 0 faster. If the numerator does, then it "wins" the fight and the limit is equal to zero. If the denominator approaches 0 faster, then the limit will become infinite. If the numerator and denominator are balanced appropriately, then they cancel each other out and the limit will approach a nonzero real number. Over time you will develop an intuition for what types of expressions are likely to win such fights, and the examples at the end of this section illustrate algebraic techniques for resolving such indeterminacies.

The following theorem identifies seven common indeterminate forms for limits:

THEOREM 1.27

Indeterminate Forms for Limits

Each of the following is an ***indeterminate form***, meaning that a limit in one of these forms may or may not exist, depending on the situation:

$$\frac{0}{0} \qquad \frac{\infty}{\infty} \qquad 0 \cdot \infty \qquad \infty - \infty \qquad 0^0 \qquad 1^\infty \qquad \infty^0$$

To prove this theorem, for each indeterminate form we need only exhibit an example of a limit of that form that exists and an example of a limit of that form that does not, as we did before for limits of the indeterminate form $\frac{0}{0}$. Identifying such examples for the first four forms is left to you in Exercises 15–18; such forms can often be solved after simple factoring or cancelling. Limits of the remaining three types pose more of a challenge; see Section 5.5.

 CAUTION

Of course, the indeterminate expression "$\frac{0}{0}$" is not a real number and we cannot actually divide the number 0 by the number 0. Theorem 1.27 tells us that limits of this form are indeterminate, and thus cannot be determined until we somehow rewrite or re-examine the limit, perhaps by factoring, cancelling, or some other method. Note that this is very different than saying that a limit "does not exist."

We have already seen that a limit of the form $\frac{1}{\infty}$ is equal to zero and a limit of the form $\frac{1}{0^+}$ is infinite. The following limit forms also always either approach 0 or become infinite and thus are not indeterminate:

THEOREM 1.28

Non-Indeterminate Forms for Limits

(a) A limit in any of these forms must be equal to 0:

$$\frac{1}{\infty} \qquad \frac{0}{\infty} \qquad \frac{0}{1} \qquad 0^\infty \qquad 0^1$$

(b) A limit in any of these forms must be ∞:

$$\frac{1}{0^+} \qquad \frac{\infty}{0^+} \qquad \frac{\infty}{1} \qquad \infty + \infty \qquad \infty \cdot \infty \qquad \infty^\infty \qquad \infty^1$$

We are *not* suggesting that you should memorize Theorem 1.28. Each of the limit forms in the theorem can be easily determined by investigation. We will not give a formal proof of this theorem, but rather, we can argue that in each case there is no "fight"; it is clear what limits of each form must approach. For example, in a limit of the form $\dfrac{0}{\infty}$, the numerator approaches zero, making the quotient smaller and smaller; at the same time the denominator grows without bound, which also makes the quotient smaller and smaller. Thus for the form $\dfrac{0}{\infty}$, the behavior of both the numerator and the denominator causes the limit to approach 0. You will see some of the other non-indeterminate forms in Exercises 11–14.

The Squeeze Theorem for Limits

Another useful tool for calculating new types of limits is the Squeeze Theorem. This theorem says that if the output of a function $f(x)$ is always bounded between a lower function $l(x)$ and an upper function $u(x)$, and if the lower and upper functions approach the same value L as $x \to c$, then $f(x)$ gets squeezed between the lower and upper functions and also approaches L as $x \to c$.

THEOREM 1.29

The Squeeze Theorem for Limits

If $l(x) \le f(x) \le u(x)$ for all x sufficiently close to c, but not necessarily at $x = c$, and if $\lim\limits_{x \to c} l(x)$ and $\lim\limits_{x \to c} u(x)$ are both equal to L, then $\lim\limits_{x \to c} f(x) = L$.

Similar results hold for limits at infinity and one-sided limits.

For example, the figure that follows shows a function f that is sandwiched between two functions u and l as $x \to 0$. Since $u(x)$ and $l(x)$ have the same limit at $x \to 0$ and $f(x)$ is squeezed between them, we know that $f(x)$ must share that same limit as $x \to 0$.

$$l(x) \le f(x) \le u(x)$$

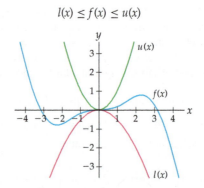

Proof.   Given $\epsilon > 0$, we can choose $\delta_1 > 0$ to get $u(x)$ within ϵ of L and also choose δ_2 to get $l(x)$ within ϵ of L. If $\delta = \min(\delta_1, \delta_2)$, then whenever $x \in (c - \delta, c) \cup (c, c + \delta)$, we also have

$$L - \epsilon < l(x) \le f(x) \le u(x) < L + \epsilon.$$

Similar arguments prove that the Squeeze Theorem holds for limits as $x \to \infty$ and $x \to c^+$ or $x \to c^-$.

The Squeeze Theorem for limits also applies to one-sided limits and to limits at infinity. For example, consider the graph of the function $f(x) = x^2 \sin \frac{1}{x}$ shown next at the left.

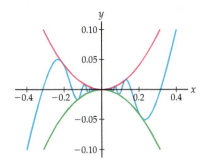

Squeezed between $y = x^2$ and $y = -x^2$

From this graph it seems reasonable that $\lim_{x \to 0} x^2 \sin \frac{1}{x} = 0$. We can use the Squeeze Theorem to find this limit algebraically. The sine function has outputs that are always between -1 and 1. Therefore $-1 < \sin \frac{1}{x} < 1$, which means that $-x^2 < x^2 \sin \frac{1}{x} < x^2$, as shown in the graph at the right. We know that $\lim_{x \to 0} -x^2 = 0$ and $\lim_{x \to 0} x^2 = 0$. Therefore by the Squeeze Theorem we must also have $\lim_{x \to 0} x^2 \sin \frac{1}{x} = 0$.

Examples and Explorations

EXAMPLE 1

Calculating limits to determine horizontal asymptotes

Use limits to find any horizontal asymptotes of the following functions:

(a) $f(x) = \dfrac{x}{x-1}$ **(b)** $g(x) = \dfrac{x^2 - 1}{x^3 - 4x}$ **(c)** $h(x) = \dfrac{\sin x}{x}$

SOLUTION

(a) To find horizontal asymptotes we must examine limits as $x \to \pm\infty$. As $x \to \infty$, we have $x \to \infty$ and $x - 1 \to \infty$, and therefore
$$\lim_{x \to \infty} \frac{x}{x-1} \text{ is of the indeterminate form } \frac{\infty}{\infty}.$$
Limits of the indeterminate form $\frac{\infty}{\infty}$ can often be resolved by dividing the numerator and denominator by the highest power of x that appears, as follows:
$$\lim_{x \to \infty} \frac{x}{x-1} = \lim_{x \to \infty} \frac{x}{x-1}\left(\frac{1/2x}{1/2x}\right) = \lim_{x \to \infty} \frac{1}{1 - (1/x)} = \frac{1}{1 - 0} = 1.$$
In a similar fashion we can show that $\lim_{x \to -\infty} \frac{x}{x-1} = 1$. Therefore $f(x) = \frac{x}{x-1}$ has a two-sided horizontal asymptote at $y = 1$.

(b) As $x \to \infty$ we have $x^2 - 1 \to \infty$ in the numerator and $x^3 \to \infty$ and $4x \to \infty$ in the denominator. Since $\infty - \infty$ is an indeterminate form, we cannot even be certain what the denominator $x^3 - 4x$ approaches at this point. To resolve this limit we will once again divide the numerator and denominator by the highest power of x:
$$\lim_{x \to \infty} \frac{x^2 - 1}{x^3 - 4x} = \lim_{x \to \infty} \frac{x^2 - 1}{x^3 - 4x}\left(\frac{1/x^3}{1/x^3}\right) = \lim_{x \to \infty} \frac{1/x - 1/x^3}{1 - (4/x^2)} = \frac{0 - 0}{1 - 0} = \frac{0}{1} = 0.$$

Similarly, $\lim\limits_{x \to -\infty} \dfrac{x^2 - 1}{x^3 - 4x}$ is equal to 0. Therefore $g(x) = \dfrac{x^2 - 1}{x^3 - 4x}$ has a two-sided horizontal asymptote at $y = 0$.

(c) Looking at $h(x) = \dfrac{\sin x}{x}$ as $x \to \infty$, we see that the numerator $\sin x$ oscillates between -1 and 1 while the denominator x gets infinitely large. A bounded quantity divided by a quantity that increases without bound must approach zero; in other words, as $x \to \infty$ we have

$$\frac{\sin x}{x} \to \frac{\text{bounded}}{\infty} \to 0.$$

The same is true as $x \to -\infty$, and thus $\lim\limits_{x \to \infty} \dfrac{\sin x}{x}$ and $\lim\limits_{x \to -\infty} \dfrac{\sin x}{x}$ are both equal to 0.

Therefore the function $h(x) = \dfrac{\sin x}{x}$ has a two-sided horizontal asymptote at $y = 0$.   □

CHECKING THE ANSWER

We can use calculator graphs to verify the horizontal asymptotes that we just found. These graphs also provide verification for the vertical asymptotes in the next example.

f has horizontal asymptote at $y = 1$ *g has horizontal asymptote at $y = 0$*   *h has horizontal asymptote at $y = 0$*

EXAMPLE 2

Calculating limits to determine vertical asymptotes

Use limits to describe any vertical asymptotes of the following functions, if possible:

(a) $f(x) = \dfrac{x}{x - 1}$ **(b)** $g(x) = \dfrac{x^2 - 1}{x^3 - 4x}$ **(c)** $h(x) = \dfrac{\sin x}{x}$

SOLUTION

(a) From the formula for $f(x)$, we see that $x = 1$ is the only serious candidate for a vertical asymptote. As $x \to 1$ we have $x - 1 \to 0$ and therefore

$$\lim_{x \to 1} \frac{x}{x - 1} \text{ is of the form } \frac{1}{0}.$$

This tells us that $f(x) = \dfrac{x}{x - 1}$ has a vertical asymptote at $x = 1$. If we want to describe the behavior of f near this asymptote more precisely, we can calculate the right and left limits separately. As $x \to 1^-$ we have $x - 1 \to 0^-$, and as $x \to 1^+$ we have $x - 1 \to 0^+$; therefore

$$\lim_{x \to 1^-} \frac{x}{x - 1} = -\infty \text{ and } \lim_{x \to 1^+} \frac{x}{x - 1} = \infty.$$

This means that the vertical asymptote at $x = 1$ is downward-pointing on the left and upward-pointing on the right; see the leftmost graph from the previous "Checking the Answer" figures.

(b) The function $g(x)$ factors as

$$\frac{x^2 - 1}{x^3 - 4x} = \frac{(x - 1)(x + 1)}{x(x - 2)(x + 2)}.$$

The values of x that cause the denominator of this quotient to approach zero are $x = 0$, $x = 2$, and $x = -2$. None of these values cause the numerator to approach zero, so in each case we will get a limit that is either ∞ or $-\infty$ from the left and/or the right. In any case, we know that $g(x)$ has vertical asymptotes at $x = 0$, $x = -2$, and $x = 2$. If we want to know the precise behavior of $g(x)$ at one of these vertical asymptotes, we can look from the left and the right. For example, as $x \to 2^-$ we have $x(x-2)(x+2) \to 2(0^-)(4) \to 0^-$, and as $x \to 2^+$ we have $x(x-2)(x+2) \to 2(0^+)(4) \to 0^+$. Therefore the left and right limits at $x = 2$ are

$$\lim_{x \to 2^-} \frac{x^2 - 1}{x^3 - 4x} = \lim_{x \to 2^-} \frac{(x-1)(x+1)}{x(x-2)(x+2)} = -\infty$$

and

$$\lim_{x \to 2^+} \frac{x^2 - 1}{x^3 - 4x} = \lim_{x \to 2^+} \frac{(x-1)(x+1)}{x(x-2)(x+2)} = \infty.$$

Notice that in these calculations we kept track only of the left/right "$\pm$" directions when we encountered 0. This is because whether a multiplicative factor in the limit is approaching, say, 1^+ or 1^-, will not affect the overall sign. On the other hand, the difference between a factor of 0^+ and 0^- does affect the sign, which in turn will determine whether the limit approaches ∞ or $-\infty$. Notice also that we used the factored form of $g(x)$ to determine the sign of infinity in each case, since it would have been difficult to determine the sign of the denominator as $x \to 2^+$ and as $x \to 2^-$ in the unfactored expression for $g(x)$.

(c) The denominator of $h(x) = \frac{\sin x}{x}$ is zero only when $x = 0$, so $x = 0$ is the only candidate for a vertical asymptote. Unfortunately, $x = 0$ also makes the numerator of $h(x)$ equal to zero, since $\sin 0 = 0$. This means that as $x \to 0$ the limit is indeterminate:

$$\lim_{x \to 0} \frac{\sin x}{x} \;\to\; \frac{\sin 0}{0} \;\to\; \frac{0}{0}.$$

This limit is not one that we know how to solve. In Chapter 6 we will see that this limit is actually equal to 1. Therefore $h(x)$ does not have a vertical asymptote at $x = 0$; see the third graph in the Checking the Answer discussion before this example.   ☐

EXAMPLE 3	**Indeterminate and non-indeterminate forms**

Determine whether each of the limits that follow is initially in indeterminate form or non-indeterminate form. Then calculate each limit.

(a) $\lim_{x \to \infty} x^{-1} \sqrt{x}$ 　　　　 **(b)** $\lim_{x \to \infty} (x^2 - x^3)$ 　　　　 **(c)** $\lim_{x \to 0^+} x^{1/x}$

SOLUTION

(a) Since -1 is a negative power and $\frac{1}{2}$ is a positive power, Theorem 1.26 tells us that as $x \to \infty$ we have $x^{-1} \to 0$ and $\sqrt{x} \to \infty$. Therefore this limit is of the indeterminate form $0 \cdot \infty$. One way to use algebra to resolve this indeterminate form is as follows:

$$\lim_{x \to \infty} x^{-1} \sqrt{x} = \lim_{x \to \infty} \frac{\sqrt{x}}{x} = \lim_{x \to \infty} \frac{1}{\sqrt{x}} = 0.$$

The last step in the calculation above follows from Theorem 1.25, since the limit is of the form $\frac{1}{\infty}$.

(b) Since 2 and 3 are positive powers, as $x \to \infty$ we have $x^2 \to \infty$ and $x^3 \to \infty$. Therefore this limit is of the indeterminate form $\infty - \infty$. Limits of this form can often be resolved by factoring. We have

$$x^2 - x^3 = x^2(1 - x),$$

and as $x \to \infty$ we have $x^2 \to \infty$ and $1 - x \to -\infty$. Since the form $\infty(-\infty)$ is the same as $-\infty$, we have

$$\lim_{x \to \infty} (x^2 - x^3) = \lim_{x \to \infty} x^2(1 - x) = -\infty.$$

(c) As $x \to \infty$, this limit approaches a non-indeterminate form that we can solve immediately by using Theorem 1.28. Specifically, as $x \to \infty$, the base x approaches 0 and the exponent $\dfrac{1}{x}$ approaches ∞. Therefore this limit is of the form 0^∞, which is always equal to 0. Therefore we have

$$\lim_{x \to \infty} x^{1/x} = 0.$$

We can use intuition to verify that these answers seem reasonable. Remember that each time a limit has an indeterminate form, two parts of the limit are "fighting" against each other. In part (a) of the preceding example, as $x \to \infty$ we have $x^{-1}\sqrt{x} \to 0 \cdot \infty$. While this form is indeterminate, it makes intuitive sense that it would approach 0, since -1 has a larger magnitude than $\frac{1}{2}$, and thus it is reasonable to believe that x^{-1} should approach 0 faster than $\sqrt{x} = x^{1/2}$ approaches ∞.

For part (b), we have $x^2 - x^3 \to \infty - \infty$ as $x \to \infty$. Since x^3 should approach ∞ much faster than x^2 does, it makes sense to expect that x^3 should win the battle and $x^2 - x^3$ should eventually approach $-\infty$.

In part (c), as $x \to \infty$ the base of $x^{1/x}$ approaches 0, making the overall limit smaller. At the same time, the exponent approaches ∞, and large powers of numbers that are close to zero are smaller still; for example, $0.01^2 = 0.0001$. Both the base and the exponent are working together to make the overall limit approach 0.

EXAMPLE 4   **The global behavior of a polynomial is determined by its leading term**

Use limits to show that the polynomial $f(x) = x^4 - x^3 - 11x^2 + 9x + 18$ behaves like its leading term x^4 as $x \to \infty$ and $x \to -\infty$. Then use graphs to compare the graph of f with the graph of $y = x^4$ in different graphing windows.

SOLUTION

It is not immediately obvious how to calculate this limit, because, as $x \to \infty$, the terms x^4 and $9x$ approach ∞ while the terms $-x^3$ and $-11x^2$ approach $-\infty$. Therefore $\lim\limits_{x \to \infty} x^4 - x^3 - 11x^2 + 9x + 18$ is indeterminate.

However, with some simple algebra we can change this sum and difference of infinities into a product that is easier to work with. Specifically, we can factor out the largest power of x:

$$\lim_{x \to \infty} (x^4 - x^3 - 11x^2 + 9x + 18) = \lim_{x \to \infty} x^4 \left(1 - \frac{1}{x} - \frac{11}{x^2} + \frac{9}{x^3} + \frac{18}{x^4}\right).$$

Since as $x \to \infty$ we have $x^4 \to \infty$ and the remainder of the expression approaching $1 - 0 - 0 + 0 + 0 = 1$, we can say that the limit is equal to ∞.

Similarly, the limit as $x \to -\infty$ is also ∞. Notice that the only term which ended up being relevant in the limit calculation was the leading term. The figures that follow show the function $f(x) = x^4 - x^3 - 11x^2 + 9x + 18$ in blue and its leading term $y = x^4$ in red, in three different viewing windows. The more we enlarge the graphing window, the more the graph of the function $y = f(x)$ looks like the graph of $y = x^4$.

$f(x)$ and x^4 on $[-3, 3]$

$f(x)$ and x^4 on $[-6, 6]$

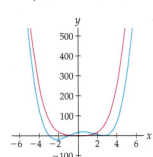

$f(x)$ and x^4 on $[-8, 8]$

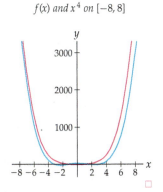

? TEST YOUR UNDERSTANDING

▶ In terms of large and small numbers, why does it make intuitive sense that limits of the form $\frac{1}{0^+}$ must always equal ∞?

▶ If a limit is of the form $\frac{1}{0}$ as $x \to c$, why should we examine the corresponding left and right limits separately?

▶ Why does it make sense that limits of the form $\frac{1}{\infty}$ and of the form $\frac{1}{-\infty}$ are always equal to zero?

▶ Why does it make intuitive sense that limits of the form $\frac{0}{0}$ are indeterminate? What "fight" is happening between the numerator and the denominator? What will happen if the numerator "wins"? The denominator? If there is a tie?

▶ In the Squeeze Theorem for limits, why do we require that the upper and lower functions $u(x)$ and $l(x)$ have the same limit as $x \to c$?

EXERCISES 1.6

Thinking Back

Behavior of algebraic functions: Determine whether each function approaches 0, approaches a nonzero real number, or becomes infinite as x approaches each indicated value.

▶ $f(x) = \sqrt{x^3 - 2}$, with $x \to 2$ and $x \to -2$

▶ $f(x) = (x + 1)^{-1}$, with $x \to 0$ and $x \to 1$

▶ $f(x) = \dfrac{1}{x^2 - 4x - 5}$, with $x \to 0$ and $x \to 5$

▶ $f(x) = \dfrac{x^2 - 4}{x + 2}$, with $x \to 2$ and $x \to -2$

The definition of infinite limits and limits at infinity: Write each limit statement that follows in terms of the formal definition of limit. Then approximate the largest value of δ or N corresponding to $\epsilon = 0.5$ or $M = 100$, as appropriate, and illustrate this choice of δ or N on a graph of f.

▶ $\displaystyle\lim_{x \to \infty} \frac{2x}{x - 1} = 2$

▶ $\displaystyle\lim_{x \to 2^+} \frac{1}{x^2 - 4} = \infty$

▶ $\displaystyle\lim_{x \to \infty} \sqrt{x - 1} = \infty$

▶ $\displaystyle\lim_{x \to -\infty} \frac{1}{x} = 0$

Concepts

0. *Problem Zero:* Read the section and make your own summary of the material.

1. *True/False:* Determine whether each of the statements that follow is true or false. If a statement is true, explain why. If a statement is false, provide a counterexample.

  (a) *True or False:* If $f(x) \to 0^+$, then $\dfrac{1}{f(x)} \to \infty$.

  (b) *True or False:* If $f(x) \to \infty^+$, then $\dfrac{1}{f(x)} \to 0^+$.

  (c) *True or False:* If a limit initially has an indeterminate form, then it can never be solved.

  (d) *True or False:* A limit "does not exist" if there is no real number that it approaches.

  (e) *True or False:* As limit forms, $\infty^2 \to \infty$.

  (f) *True or False:* As limit forms, $2^\infty \to \infty$.

  (g) *True or False:* As limit forms, $\infty - \infty \to 0$.

  (h) *True or False:* The limit of a function f as $x \to c$ is always equal to the value $f(c)$, provided that $f(c)$ exists.

2. *Examples:* Construct examples of the thing(s) described in the following. Try to find examples that are different than any in the reading.

  (a) A limit of the form $\dfrac{1}{0}$ that approaches ∞ as $x \to c^+$ and $-\infty$ as $x \to c^-$.

  (b) Two limits that can be solved with the Squeeze Theorem.

  (c) Formulas for three functions that are discontinuous at $x = 3$: one removable, one jump, and one infinite discontinuity.

In Exercises 3–6, $\lim\limits_{x \to c} f(x) = L$ and $\lim\limits_{x \to c} g(x) = M$ for some real numbers L and M. What, if anything, can you say about $\lim\limits_{x \to c} \dfrac{f(x)}{g(x)}$ in each case?

3. $L \neq 0$ and $M \neq 0$ **4.** $L = 0$ and $M \neq 0$

5. $L \neq 0$ and $M = 0$ **6.** $L = 0$ and $M = 0$

7. Determine which of the given forms are indeterminate. For each form that is not indeterminate, describe the behavior of a limit of that form.

$$\infty + \infty \qquad \infty - \infty \qquad \infty + 1 \qquad 0 + \infty$$

8. Determine which of the given forms are indeterminate. For each form that is not indeterminate, describe the behavior of a limit of that form.

$$\infty \cdot \infty \qquad 0 \cdot \infty \qquad 5 \cdot \infty \qquad 5 \cdot 0 \qquad 0 \cdot 0$$

9. Determine which of the given forms are indeterminate. For each form that is not indeterminate, describe the behavior of a limit of that form.

$$\frac{0}{0} \qquad \frac{0}{\infty} \qquad \frac{\infty}{0} \qquad \frac{1}{0} \qquad \frac{0}{1} \qquad \frac{\infty}{1} \qquad \frac{1}{\infty} \qquad \frac{\infty}{\infty}$$

10. Determine which of the given forms are indeterminate. For each form that is not indeterminate, describe the behavior of a limit of that form.

$$0^1 \qquad 0^0 \qquad 0^\infty \qquad 1^\infty \qquad \infty^1 \qquad \infty^0 \qquad \infty^\infty$$

11. Describe in terms of large and small numbers why it makes intuitive sense that limits of the form (a) $\dfrac{1}{\infty}$, (b) $\dfrac{0}{\infty}$, and (c) $\dfrac{0}{1}$ must equal 0.

12. Describe in terms of large and small numbers why it makes intuitive sense that limits of the form (a) 0^∞ and (b) 0^1 must equal 0.

13. Describe in terms of large and small numbers why it makes intuitive sense that limits of the form (a) $\dfrac{1}{0^+}$, (b) $\dfrac{\infty}{0^+}$, and (c) $\dfrac{\infty}{1}$ must be infinite.

14. Describe in terms of large and small numbers why it makes intuitive sense that limits of the form (a) $\infty + \infty$, (b) $\infty \cdot \infty$, (c) ∞^∞, and (d) ∞^1 must be infinite.

To prove that the limit forms in Theorem 1.27 are indeterminate, we need only list explicit examples of limits that do and do not exist for each form. Do so for each of the limit forms from Exercises 15–18.

15. $\dfrac{0}{0}$ that approaches (a) 0, (b) 2, (c) ∞.

16. $\dfrac{\infty}{\infty}$ that approaches (a) 1, (b) 6, (c) ∞.

17. $0 \cdot \infty$ that approaches (a) 0 (b) 1, (c) ∞.

18. $\infty - \infty$ that approaches (a) 0, (b) 5, (c) ∞.

Skills

Find the roots, discontinuities, and horizontal and vertical asymptotes of the functions in Exercises 19–24. Support your answers by explicitly computing any relevant limits.

19. $f(x) = \dfrac{-3x^2}{(x+3)^2}$ **20.** $f(x) = \dfrac{x+2}{x^2 + 4x + 4}$

21. $f(x) = \dfrac{x^3}{4(x-2)}$ **22.** $f(x) = x + \dfrac{x}{2x+1}$

23. $f(x) = \dfrac{x^2 - 2x - 3}{x - 3}$ **24.** $f(x) = \dfrac{2x^2 - 1}{x^2 - 2x + 1}$

Calculate each limit in Exercises 25–52.

25. $\lim\limits_{x \to 0} -4x^{-3}$ **26.** $\lim\limits_{x \to 0} 2x^{-3/4}$

27. $\lim\limits_{x \to \infty} 2x^{-4/3}$ **28.** $\lim\limits_{x \to -\infty} -5x^{3/5}$

29. $\lim\limits_{x \to \infty} (\sqrt{x} - x)$ **30.** $\lim\limits_{x \to -\infty} (x^4 - x^5)$

31. $\lim\limits_{x \to \infty} (-3x^5 + 4x + 11)$ **32.** $\lim\limits_{x \to -\infty} (5 - 2x + 3x^3)$

33. $\lim\limits_{x\to 2} \dfrac{4-2x}{x-2}$

34. $\lim\limits_{x\to -2} \dfrac{4-2x}{x+2}$

35. $\lim\limits_{x\to 0}(x^{-2}+1)$

36. $\lim\limits_{x\to 0}(3x^{-1}-2x^{-2})$

37. $\lim\limits_{x\to 1} \dfrac{1}{1-x}$

38. $\lim\limits_{x\to 1} \dfrac{1}{x^2-1}$

39. $\lim\limits_{x\to -4} \dfrac{x^2+8x+16}{(x+4)^2(x+1)}$

40. $\lim\limits_{x\to 2} \dfrac{x+1}{(x-2)^2}$

41. $\lim\limits_{x\to 0} \dfrac{x^2+1}{x(x-1)}$

42. $\lim\limits_{x\to -4} \dfrac{x+4}{x^2+8x+16}$

43. $\lim\limits_{x\to 0} \dfrac{x}{x^2-x}$

44. $\lim\limits_{x\to 1} \dfrac{x-1}{x^2-2x+1}$

45. $\lim\limits_{x\to \infty} \dfrac{(3x+1)^2(x-1)}{1-x^3}$

46. $\lim\limits_{x\to -\infty} \dfrac{1-2x^2}{(3-x)(3+4x)}$

47. $\lim\limits_{x\to 0^+}(x^{-1/3}-x^{-1/2})$

48. $\lim\limits_{x\to \infty}(x^{-1/3}-x^{-1/2})$

49. $\lim\limits_{x\to \infty} \dfrac{x^{-3}}{x^2-x^{-1}}$

50. $\lim\limits_{x\to 0^+} \dfrac{x^{-3}}{x^2-x^{-1}}$

51. $\lim\limits_{x\to 0^+} \dfrac{x^{7/2}-x^{8/3}}{x^2}$

52. $\lim\limits_{x\to \infty} \dfrac{x^{7/2}-x^{8/3}}{x^2}$

Use the Squeeze Theorem to solve each of the limits in Exercises 53–58. Explain exactly how the Squeeze Theorem applies in each case. You will need to think about the graphs of sine and cosine, including their long-term behavior and the fact that for any real number u, we have $-1 \le \sin u \le 1$ and $-1 \le \cos u \le 1$.

53. $\lim\limits_{x\to 0} x \sin \dfrac{1}{x}$

54. $\lim\limits_{x\to 0} x \sin \dfrac{1}{x^2}$

55. $\lim\limits_{x\to 0}(e^x-1) \sin \dfrac{1}{x}$

56. $\lim\limits_{x\to 0} \sin x \sin \dfrac{1}{x}$

57. $\lim\limits_{x\to 1}(x-1) \cos \dfrac{1}{x-1}$

58. $\lim\limits_{x\to 1}(x-1)^2 \cos \dfrac{1}{x-1}$

Applications

59. In 1960, H. von Foerster suggested that the human population could be measured by the function

$$P(t) = \frac{179 \times 10^9}{(2027-t)^{0.99}}.$$

The time t is measured in years, where $t=1$ corresponds to the year 1 A.D., $t=1973$ corresponds to the year 1973 A.D., and so on. (We saw this "doomsday model" for population in Problem 73 of Section 1.1, on page 105.) Use limit techniques to calculate $\lim\limits_{t\to 2027^-} P(t)$. What does this limit mean in real-world terms?

60. Suppose instead we consider the population model

$$Q(t) = \frac{44 \times 10^{10}}{1+(2027-t)^{4/3}},$$

with t measured in years as in the previous problem.
   (a) Use limit techniques to calculate $\lim\limits_{t\to \infty} Q(t)$. What does this limit mean in real–world terms? What happens in this model in the year 2027?
   (b) Use calculator graphs to compare the population models in this exercise with those in the previous exercise. Describe the long–term population growth scenarios that are suggested by these models.

61. Consider a mass hanging from the ceiling at the end of a spring. If you pull down on the mass and let go, it will oscillate up and down according to the equation

$$s(t) = A \sin\left(\sqrt{\dfrac{k}{m}}\, t\right) + B \cos\left(\sqrt{\dfrac{k}{m}}\, t\right),$$

where $s(t)$ is the distance of the mass from its equilibrium position, m is the mass of the bob on the end of the spring, and k is a "spring coefficient" that measures how tight or stiff the spring is. The constants A and B depend on initial conditions—specifically, how far you pull down the mass (s_0) and the velocity at which you release the mass (v_0).

This equation does *not* take into effect any friction due to air resistance.

spring coefficient, k

$s(t)$

v_0 mass, m

s_0

   (a) Determine whether or not the limit of $s(t)$ as $t\to\infty$ exists. What does this say about the long-term behavior of the mass on the end of the spring?
   (b) Explain how this limit relates to the fact that the equation for $s(t)$ does *not* take friction due to air resistance into account.
   (c) Suppose the bob at the end of the spring has a mass of 2 grams and that the coefficient for the spring is $k=9$. Suppose also that the spring is released in such a way that $A=\sqrt{2}$ and $B=2$. Use a graphing utility to graph the function $s(t)$ that describes the distance of the mass from its equilibrium position. Use your graph to support your answer to part (a).

62. In the previous exercise we gave an equation describing spring motion without air resistance. If we take into account friction due to air resistance, the mass will oscillate up and down according to the equation

$$s(t) = e^{(-f/2m)t}\left(A \sin\left(\dfrac{\sqrt{4km-f^2}}{2m}\, t\right)\right.$$
$$\left. + B \cos\left(\dfrac{\sqrt{4km-f^2}}{2m}\, t\right)\right),$$

where m, k, A, and B are the constants described in Problem D and f is a positive "friction coefficient" that measures the amount of friction due to air resistance.

(a) Find the limit of $s(t)$ as $t \to \infty$. What does this say about the long-term behavior of the mass on the end of the spring?

(b) Explain how this limit relates to the fact that the new equation for $s(t)$ *does* take friction due to air resistance into account.

(c) Suppose the bob at the end of the spring has a mass of 2 grams, the coefficient for the spring is $k = 9$, and the friction coefficient is $f = 6$. Suppose also that the spring is released in such a way that $A = 4$ and $B = 2$. Use a graphing utility to graph the function $s(t)$ that describes the distance of the mass from its equilibrium position. Use your graph to support your answer to part (a).

Proofs

63. Prove the second part of Theorem 1.24: If $\lim\limits_{x \to c} \dfrac{f(x)}{g(x)}$ is of the form $\dfrac{1}{0^-}$, then $\lim\limits_{x \to c} \dfrac{f(x)}{g(x)} = -\infty$.

64. Prove the second part of Theorem 1.25: If $\lim\limits_{x \to \infty} \dfrac{f(x)}{g(x)}$ is of the form $\dfrac{1}{-\infty}$, then $\lim\limits_{x \to \infty} \dfrac{f(x)}{g(x)} = 0$.

65. Prove that the sum rule for limits also applies for limits as $x \to \infty$: If $\lim\limits_{x \to \infty} f(x) = L$ and $\lim\limits_{x \to \infty} g(x) = M$, then $\lim\limits_{x \to \infty} (f(x) + g(x)) = L + M$.

66. Prove the first part of Theorem 1.26: If $k > 0$, then $\lim\limits_{x \to \infty} x^k = \infty$. (*Hint: Given $M > 0$, choose $N = M^{1/k}$. Then show that for $x > N$ it must follow that $x^k > M$.*)

67. Prove the second part of Theorem 1.26: If $k > 0$, then $\lim\limits_{x \to \infty} x^{-k} = 0$.

Thinking Forward

A limit representing an instantaneous rate of change: After t seconds, a bowling ball dropped from 350 feet has height $h(t) = 350 - 16t^2$, measured in feet.

350 ft

h

▶ Calculate the average rate of change of the height of the bowling ball from $t = 3$ to $t = 3 + h$ seconds in the cases where h is equal to 0.5, 0.25, 0.1, and 0.01.

▶ Write down a formula for the average rate of change of the height of the bowling ball from time $t = 3$ to time $t = 3 + h$, assuming that $h > 0$. The only letter in your formula should be h.

▶ Take the limit as $h \to 0^+$ of the formula you found for average rate of change in the previous problem. What does this limit represent in real–world terms?

Taylor Series: In this section we learned that e can be thought of as the following limit:

$$\lim_{h \to 0} (1 + h)^{1/h} = e.$$

CHAPTER REVIEW, SELF-TEST, AND CAPSTONES

Before you progress to the next chapter, be sure you are familiar with the definitions, concepts, and basic skills outlined here. The capstone exercises at the end bring together ideas from this chapter and look forward to future chapters.

Definitions

Give precise mathematical definitions or descriptions of each of the concepts that follow. Then illustrate the definition with a graph or algebraic example, if possible.

▶ the intuitive meaning of the limit statements $\lim_{x \to c} f(x) = L$, $\lim_{x \to c^-} f(x) = L$, and $\lim_{x \to c^+} f(x) = L$

▶ the intuitive meaning of the limit statements $\lim_{x \to c} f(x) = \infty$ and $\lim_{x \to \infty} f(x) = L$

▶ the formal δ–ϵ definition of the limit statements $\lim_{x \to c} f(x) = L$, $\lim_{x \to c^-} f(x) = L$, and $\lim_{x \to c^+} f(x) = L$

▶ the formal δ–M, N–ϵ, and N–M definitions of the limit statements $\lim_{x \to c} f(x) = \infty$, $\lim_{x \to \infty} f(x) = L$, and $\lim_{x \to \infty} f(x) = \infty$, respectively

▶ what we mean when we say that a limit *exists*, or that a limit *does not exist*

▶ what it means, in terms of limits, for a function f to have a *vertical asymptote* at $x = c$ or a *horizontal asymptote* at $y = L$

▶ what it means, in terms of limits, for a function f to be *continuous at a point* $x = c$, *left continuous* at $x = c$, and *right continuous* at $x = c$

▶ what it means for a function f to be *continuous on a closed interval* $[a, b]$, or *continuous on an open interval* (a, b), or *continuous on a half-closed interval* $[a, b)$

▶ what it means, in terms of limits, for a function to have a *removable discontinuity*, a *jump discontinuity*, or an *infinite discontinuity* at $x = c$

Theorems

Fill in the blanks to complete each of the following theorem statements:

▶ If $\lim_{x \to c} f(x) = L$ and $\lim_{x \to c} f(x) = M$, then _____ .

▶ $\lim_{x \to c} f(x) = L$ if and only if $\lim_{x \to c^-} f(x) = $ _____ and $\lim_{x \to c^+} f(x) = $ _____ .

▶ For $\delta > 0$, $x \in (c - \delta, c) \cup (c, c + \delta)$ if and only if $0 < $ _____ $< \delta$.

▶ For $\epsilon > 0$, $f(x) \in (L - \epsilon, L + \epsilon)$ if and only if _____ $< \epsilon$.

▶ *The Extreme Value Theorem:* If f is _____ on a closed interval $[a, b]$, then there exist values M and m in the interval $[a, b]$ such that $f(M)$ is _____ and $f(m)$ is _____ .

▶ *The Intermediate Value Theorem:* If f is _____ on a closed interval $[a, b]$, then for any K strictly between _____ and _____ , there exists at least one $c \in (a, b)$ such that _____ .

▶ A function f can change sign from positive to negative, or vice versa, at $x = c$ only if $f(x)$ is _____ , _____ , or _____ at $x = c$.

▶ Constant, identity, and linear functions are continuous everywhere, which means in terms of limits that _____ , _____ , and _____ .

▶ Power functions are continuous everywhere, which means in terms of limits that _____ .

▶ All algebraic functions are _____ on their domains, which means in terms of limits that if $x = c$ is in the domain of an algebraic function f, then _____ .

▶ If $\lim_{x \to c} g(x)$ exists and $f(x)$ is a function that is _____ to $g(x)$ for all x sufficiently close to _____ , but not necessarily at _____ , then _____ .

▶ *The Squeeze Theorem for Limits:* If $l(x) \le f(x) \le u(x)$ for all x sufficiently close to _____ , but not necessarily at _____ , and if $\lim_{x \to c} l(x)$ and $\lim_{x \to c} u(x)$ are both equal to L, then _____ .

Limit Rules and Indeterminate Forms

Limits of basic functions: Fill in the blanks to complete the limit rules that follow. You may assume that k is positive.

▶ $\lim_{x \to c} k = $ _____

▶ $\lim_{x \to c} (mx + b) = $ _____

▶ $\lim_{x \to \infty} x^k = $ _____

▶ $\lim_{x \to c} x = $ _____

▶ $\lim_{x \to c} Ax^n = $ _____ .

▶ $\lim_{x \to \infty} x^{-k} = $ _____

Limits of combinations: Fill in the blanks to complete the limit rules that follow. You may assume that k and c are any real numbers and that both $\lim_{x \to c} f(x)$ and $\lim_{x \to c} g(x)$ exist.

▶ $\lim_{x \to c} kf(x) = $ _____

▶ $\lim_{x \to c} (f(x) + g(x)) = $ _____

▶ $\lim_{x \to c} (f(x) - g(x)) = $ _____

▶ $\lim_{x \to c} (f(x)g(x)) = $ _____

▶ $\lim_{x \to c} \dfrac{f(x)}{g(x)} = $ ____ , provided that ____

▶ $\lim_{x \to c} f(g(x)) = $ ____ , provided that ____

Indeterminate forms: Identify which of the limit forms listed here are indeterminate. For each form that is not indeterminate, describe the behavior of a limit of that form.

▶ $\dfrac{1}{0^+}$　　▶ $\dfrac{1}{0^-}$　　▶ $\dfrac{1}{\infty}$　　▶ $\dfrac{1}{-\infty}$

▶ $\dfrac{0}{1}$　　▶ $\dfrac{0}{0}$　　▶ $\dfrac{\infty}{1}$　　▶ $\dfrac{0}{\infty}$

▶ $\dfrac{\infty}{\infty}$　　▶ $\dfrac{\infty}{0^+}$　　▶ $0 \cdot \infty$　　▶ $\infty \cdot \infty$

▶ $\infty(-\infty)$　　▶ $\infty + \infty$　　▶ $\infty - \infty$　　▶ 0^0

▶ 0^1　　▶ 0^∞　　▶ 1^∞　　▶ ∞^0

▶ ∞^∞　　▶ ∞^1　　▶ $0^{-\infty}$　　▶ $\infty^{-\infty}$

Skill Certification: Basic Limits

Calculating limits: Find each limit by hand.

1. $\lim_{x \to 0} 3x^{-4}$

2. $\lim_{x \to \infty} -2x^{-1/2}$

3. $\lim_{x \to 2} \dfrac{1}{2 - x}$

4. $\lim_{x \to 1} \dfrac{1}{x^2 - 1}$

5. $\lim_{x \to 1} \dfrac{2x^3 - x^2 - 2x + 1}{x^2 - 2x + 1}$

6. $\lim_{x \to -\infty} \dfrac{x^3 + 2x - 1}{1 - x^4}$

7. $\lim_{x \to \infty} (\sqrt{x} - x)$

8. $\lim_{x \to 0^+} \dfrac{\sqrt{x} - x^3}{x^2}$

9. $\lim_{x \to 3} \dfrac{\frac{1}{x-3} - \frac{1}{x}}{x - 3}$

10. $\lim_{x \to 4} \dfrac{2 - \sqrt{x}}{4 - x}$

11. $\lim_{x \to \infty} (-2x^3 + x^2 - 10)$

12. $\lim_{x \to 0} (x^{-3} - 2x^{-1})$

13. $\lim_{x \to \infty} \dfrac{(2x - 1)(x^2 + 1)}{x^2 - 4}$

14. $\lim_{x \to \infty} \dfrac{(x - 1)(3x + 1)^3}{(x - 2)^4}$

15. $\lim_{x \to \infty} \dfrac{\sqrt{x}}{1 - \sqrt{x}}$

16. $\lim_{x \to -\infty} e^x \tan^{-1} x$

17. $\lim_{x \to \infty} \sin x$

18. $\lim_{x \to \infty} \sin \dfrac{1}{x}$

19. $\lim_{x \to \infty} \dfrac{1}{x} \sin x$

20. $\lim_{x \to 0} x \sin \dfrac{1}{x}$

Capstone Problems

A. *Continuity of piecewise-defined functions:* For each given function f, find a real number a that makes f continuous at $x = 0$, if possible.

(a) $f(x) = \begin{cases} 3x + 1, & \text{if } x < 0 \\ 2x + a, & \text{if } x \geq 0 \end{cases}$

(b) $f(x) = \begin{cases} \dfrac{a}{x+2}, & \text{if } x < 0 \\ 3, & \text{if } x = 0 \\ ax + 1, & \text{if } x > 0 \end{cases}$

B. *Limits that define derivatives:* In the next chapter we will be interested in **derivatives**, which we will define as limits of the form

$$\lim_{h \to 0} \frac{f(c + h) - f(c)}{h}.$$

(a) Calculate this limit for $f(x) = x^3$ and $c = 0$.
(b) Calculate this limit for $f(x) = x^3$ and $c = 2$.
(c) Calculate this limit for $f(x) = x^3$ and general $c = x$. This time your answer will be a function of x instead of a number.

C. *The limit of a model at infinity:* Leila is interested in the effect of a stabilized wolf population on the eventual population of beavers in Idaho. The following table gives estimated beaver populations $B(t)$ for $t = 0, 1, 2, 3, 4,$ and 5 years after 2005:

t	0	1	2	3	4	5
$B(t)$	48,112	42,256	47,088	43,684	46,320	44,704

(a) Leila makes a plot of these values of $B(t)$ and notes that the population of beavers is cyclical with diminishing amplitude. She finds that the quadratic function

$$M(t) = -51x^2 + 918x + 41,389$$

is a good model for the relative maximum data points at $t = 0, 2,$ and 4, and that

$$m(t) = 33.25x^2 - 583.5x + 48,122$$

is a good model for the relative minimum data points at $t = 1, 3,$ and 5. Verify that these functions do in fact pass through the relevant data points, and graph the data for $B(t)$ along with the two functions.

(b) Do the two quadratics $M(t)$ and $m(t)$ ever meet? If so, where? What conclusion could Leila make concerning the eventual steady population $\lim_{t \to \infty} B(t)$ of beavers in Idaho?

D. *The limit of a rational function model at infinity:* Upon further reflection, Leila decides that the quadratics used in the previous problem are unreasonable, since the quadratic model for the relative maximum values could be interpreted as indicating that the eventual number of beavers would be unbounded. She decides to change her model for the relative maximum beaver populations to

$$M(t) = \frac{40944\, t^2 + 454512\, t - 1732032}{t^2 + 9\, t - 36}.$$

(a) Verify that this function does pass through the data points at $t = 0, 2,$ and 4. Is this function continuous everywhere? (*Hint: Consider* $\lim_{t \to 3} M(t)$.)

(b) Compute $\lim_{t \to \infty} M(t)$. What is the significance of this number?

Derivatives

2.1   AN INTUITIVE INTRODUCTION TO DERIVATIVES

▶ Associated slope functions, tangent lines, and secant lines

▶ Velocity and other instantaneous rates of change

▶ Approximating slopes of tangent lines and instantaneous rates of change

Slope Functions

We begin our study of the derivative with an intuitive introduction in terms of slopes and rates of change. We also start thinking about how one might calculate, or at least approximate, derivatives. In Section 2.2 we will give a formal mathematical definition of the derivative in terms of limits.

Intuitively speaking, if the graph of a function f is **smooth** on an interval (a, b)—meaning that it does not have any corners, cusps, jumps, or holes—then at every point $(x, f(x))$ on the the graph of f on the interval (a, b) we can consider the direction, or **slope**, of the function at that point. For example, in the figure that follows at the left, the **tangent line** drawn at $x = -1$ points in the same direction as the function f at the point $(-1, f(-1))$. More precisely, if you imagine yourself in a tiny car driving along the graph of f with your headlights on, then that line represents the direction that your headlights are pointing when you reach the point $(-1, f(-1))$ from the right or the left. Similarly, the line drawn at $x = 4$ represents the direction of the graph of f at the point $(4, f(4))$.

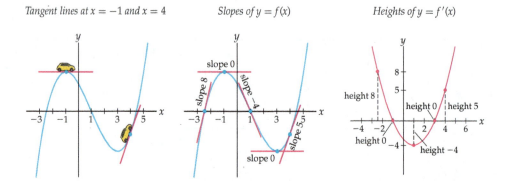

Tangent lines at $x = -1$ and $x = 4$ *Slopes of $y = f(x)$*   *Heights of $y = f'(x)$*

We can use the graph of a smooth function f to define a new function whose output at each point x is the slope of the tangent line at x. For example, the slopes shown on the graph of f in the middle figure are used to define heights on the graph of the **associated slope function** shown at the right. This associated slope function is what we will define in Section 2.2 as the **derivative** of $f(x)$ and denote as $f'(x)$ (pronounced "f prime of x"). Notice the following relationships between a function f and its derivative f':

▶ For each x, the slope of $f(x)$ is the height of $f'(x)$.

▶ Where f has a horizontal tangent line, the derivative f' has a root.

▶ Where the graph of f is increasing, the derivative f' is above the x-axis.

▶ Where the graph of f is decreasing, the derivative f' is below the x-axis.

▶ Where f has steep slope, the derivative f' has large magnitude.

▶ Where f has shallow slope, the derivative f' has small magnitude.

Position and Velocity

Suppose an object moves in a straight path so that after t seconds it is a distance of $s(t)$ units from its starting point. We will call the function $s(t)$ describing the motion of the object a **position function**. The moving object has a speed and a direction at any time t, and the combination of these two measurements defines a **velocity function** for the object. Specifically, if we consider one direction on the straight path as the "positive" direction and the other as the "negative" direction, then the velocity of the object at time t is the speed of the object times either $+1$ or -1, depending on the direction in which the object is moving.

The velocity $v(t)$ of such a moving object is a measurement of how the position function of the object is changing over time. Intuitively, because the way position changes at a particular moment in time is measured by the slope of its graph, velocity is the associated slope function for position. In other words, velocity is the derivative of position, or in symbols, $v(t) = s'(t)$. Similarly, acceleration $a(t)$ measures how velocity changes, and thus $a(t) = v'(t)$. We will examine these relationships more precisely in Section 2.2.

For example, suppose you throw a grapefruit straight up into the air, releasing it at a height of 4 feet and an upwards velocity of 32 feet per second, as illustrated in the figure that follows on the left. On the right is a plot of the height of the grapefruit over time. The points A, B, C, D, and E show the height $s(t)$, in feet, of the grapefruit at $t = 0$, $t = 0.6$, $t = 1$, $t = 1.75$, and $t = 2.118$ seconds. At A, the grapefruit is moving upwards quickly. Because of the downwards pull of gravity, the grapefruit is moving upwards more slowly at B. At C the grapefruit is at the top of its flight and about to fall to the ground. Gravity then causes the grapefruit to fall faster and faster through D and then finally E when it hits the ground.

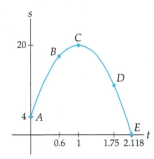

Position increases, then decreases

The next leftmost figure shows the velocity $v(t)$ of the grapefruit. Notice that at times A and B, when the grapefruit is moving upwards, its velocity is positive; at C, when the grapefruit turns around at the top of its flight, its velocity is zero; and at D and E, when the grapefruit is falling to the ground, its velocity is negative. The rightmost figure shows the constant acceleration of the grapefruit due to gravity.

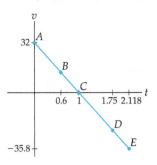

Velocity is positive, then negative

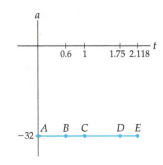

Acceleration is constant

In general, our earlier list of relationships between a function $f(x)$ and its derivative $f'(x)$ translates into a list of relationships between position $s(t)$ and velocity $v(t) = s'(t)$ as follows:

▶ For each t, the way position $s(t)$ is changing is measured by velocity $v(t)$.
▶ When position $s(t)$ is not changing, velocity $v(t)$ is zero.
▶ When position $s(t)$ is increasing, velocity $v(t)$ is positive.
▶ When position $s(t)$ is decreasing, velocity $v(t)$ is negative.
▶ When position $s(t)$ is changing rapidly, velocity $v(t)$ has large magnitude.
▶ When position $s(t)$ is changing slowly, velocity $v(t)$ has small magnitude.

Approximating the Slope of a Tangent Line

Usually, calculating the slope of a line is a simple matter: Simply take two points (x_0, y_0) and (x_1, y_1) on the line and calculate the "rise over run," which is equal to the average rate of change $\frac{\Delta y}{\Delta x} = \frac{y_1 - y_0}{x_1 - x_0}$. With tangent lines the situation is more complicated, because we know only *one* point on a tangent line, namely, the point $(c, f(c))$ where it touches the function. The slope of the tangent line measures the "direction" of the function, but how do we calculate that from only one point? The key will be to use nearby points on the function to approximate nearby slopes.

The ***secant line*** from a to b for a function f is the line that passes through the points $(a, f(a))$ and $(b, f(b))$. If f is a smooth function and $x = z$ is a point that is close to $x = c$, then the slope of the secant line from $x = c$ to $x = z$ will be close to the slope of the tangent line to f at $x = c$, as shown in the middle graph that follows:

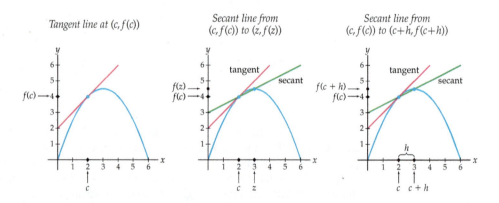

If we choose points z that are closer and closer to the point $x = c$, we will get secant lines that get closer and closer to the tangent line we are interested in. Equivalently, we could think of the second point z as "c plus a little bit," where the little bit is called h. In other words, $z = c + h$, as in the rightmost graph shown.

Since we know two points on a secant line, we can easily calculate its slope. The slope $f'(c)$ of the tangent line to f at $x = c$ can be approximated by the slope of a nearby secant line from $x = c$ to $x = z$, or equivalently, from $x = c$ to $x = c + h$:

$$f'(c) \approx \frac{f(z) - f(c)}{z - c}, \quad \text{or equivalently,} \quad f'(c) \approx \frac{f(c + h) - f(c)}{h}.$$

The preceding expressions are often called *difference quotients*, and when we find the derivative of a function f, we say that we are *differentiating* the function. If the graph of f is smooth, then as z gets closer to c, or as h gets closer to 0, these approximations get closer and closer to $f'(c)$. In Section 2.2 we will in fact take the limit as $z \to c$, or, equivalently, as $h \to 0$, to define the derivative exactly.

Approximating an Instantaneous Rate of Change

You may have noticed that the approximations we have been using for $f'(c)$ are closely related to the formula for average rate of change. This is no coincidence, since average rates of change are in fact the same as slopes of secant lines. As we choose points $z = c + h$ closer and closer to $x = c$, these average rates of change approach the *instantaneous rate of change* of the function at $x = c$. For a general function f, this instantaneous rate of change at $x = c$ is the derivative, that is, the slope $f'(c)$ of the tangent line to f at $x = c$. In the case of a position function $s(t)$, the instantaneous rate of change is the velocity $v(c) = s'(c)$.

We can approximate instantaneous rates of change in a position or velocity context in much the same way as we approximated slopes of tangent lines in the previous discussion. Suppose an object is moving along a straight path. The *distance formula* says that for such a moving object, the distance travelled, average rate, and time elapsed are related by the formula $d = rt$ ("distance equals rate times time"). We can also write this formula as $r = \frac{d}{t}$, or more accurately, as $r = \frac{\Delta d}{\Delta t}$, since we want to consider the change in distance over a corresponding change in time. If an object starts at position s_0 at time t_0 and ends at position s_1 at time t_1, then we have

$$\begin{array}{c} \text{average velocity} \\ \text{from } t_0 \text{ to } t_1 \end{array} = \begin{array}{c} \text{average rate of change of} \\ \text{position from } t_0 \text{ to } t_1 \end{array} = \frac{\Delta d}{\Delta t} = \frac{s_1 - s_0}{t_1 - t_0}.$$

Now suppose $s(t)$ describes the position of the object at time t and we are interested in finding the velocity $v(c)$ at some time $t = c$. This instantaneous velocity can be approximated by the average velocity over a small time interval $[c, z]$, or equivalently, $[c, c + h]$:

$$v(c) \approx \frac{s(z) - s(c)}{z - c}, \quad \text{or equivalently,} \quad v(c) \approx \frac{s(c + h) - s(c)}{h}.$$

Notice that this is just a special case of what we did earlier for a general function $f(x)$ and its derivative $f'(x)$ at a point $x = c$.

We can use derivatives to examine instantaneous rates of change in many contexts. In general, the derivative of a function $y(x)$ represents the instantaneous rate of change of the variable y as the variable x varies. The units for the derivative $y'(x)$ are the units for the variable y divided by the units for the variable x. For example, if time t is measured in hours and position $s(t)$ is measured in miles, then the velocity $v(t) = s'(t)$ is measured in miles per hour. As another example, if $Q(t)$ is the amount of money in a savings account after t years, measured in dollars, then $Q'(t)$ is the rate at which the savings balance changes over time, with units measured in dollars per year.

Examples and Explorations

EXAMPLE 1   **Sketching the graph of an associated slope function**

Given the following graph of the smooth function f, sketch the graph of its associated slope function f':

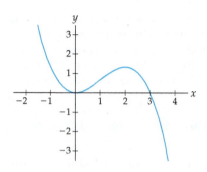

SOLUTION

A good place to begin is by marking all the locations on the graph of f where the tangent line is horizontal and thus has slope 0. In this case that happens at $x = 0$ and at $x = 2$, as shown next at the left. Thus the associated slope function f' has zeroes at $x = 0$ and $x = 2$, as shown next at the right.

$f(x)$ with slopes marked $f'(x)$ heights are $f(x)$ slopes

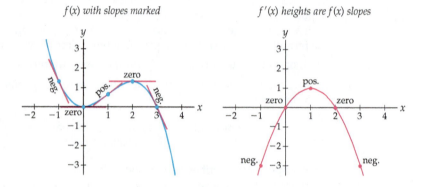

Looking again at the graph of f, we see that its tangent lines have positive slope between $x = 0$ and $x = 2$ (see, for example, the positive slope marked at $x = 1$). This means that, in the graph of f', the heights will be positive between $x = 0$ and $x = 2$. Similarly, the negative slopes on the graph of f to the left of $x = 0$ and to the right of $x = 2$ correspond to negative heights on the graph of f'.   ☐

EXAMPLE 2   Graphing velocity from the graph of position

Suppose the graph that follows describes your distance from home one morning as you drive back and forth from your sister's house. Describe a possible scenario for your travels that morning. Then sketch the corresponding graph of your velocity.

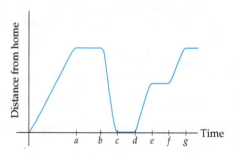

SOLUTION

One possible scenario is this: You drive to your sister's house for a visit. After talking to her for a few minutes, you realize you forgot something at home and race back to get it. In the middle of returning to your sister's house, you have to stop at a red light for a couple of minutes. Following the times marked on the time axis, we see that from 0 to a you drive to your sister's house, you talk until b, race home from b to c, leave your house at d and get stopped at the light at e, and move on at f until you get back to your sister's house at g.

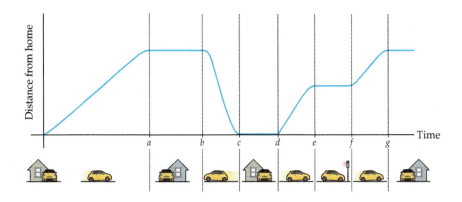

The graph of your velocity that morning is the graph of the associated slope function for the given position graph. The slope of the position graph is positive from 0 to a, zero from a to b, negative and steep from b to c, zero from c to d, positive and steep from d to e, zero from e to f, positive and steep from f to g, and finally zero again after g. The previous sentence also describes the height of the corresponding velocity graph, where steep slope values correspond to large magnitudes of velocity:

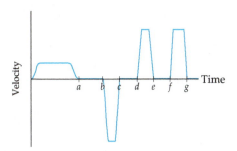

□

EXAMPLE 3

Estimating the slope of a tangent line with a sequence of secant lines

Estimate the slope of the line tangent to the graph of $f(x) = -\frac{1}{2}x^2 + 3x$ at the point $(2, f(2))$ by calculating a sequence of slopes of secant lines.

SOLUTION

The tangent line passes through the point $(2, f(2)) = \left(2, -\frac{1}{2}(2)^2 + 3(2)\right) = (2, 4)$ and is shown in red in each of the graphs that follow. Since we only know one point on this line, we cannot compute its slope directly. We will approximate its slope by considering a sequence the slopes of secant lines on smaller and smaller intervals, namely, [2, 3], [2, 2.5], [2, 2.25], and [2, 2.1], as shown in the following four graphs below:

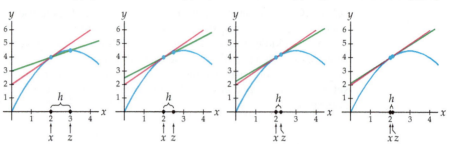

Secant line on [2, 3]   Secant line on [2, 2.5]   Secant line on [2, 2.25]   Secant line on [2, 2.1]

In our difference quotient notation, these intervals correspond to a sequence of points $z = 3$, $z = 2.5$, $z = 2.25$, and $z = 2.1$ that approach $c = 2$. Equivalently, we can think of this sequence as a series of h-values $h = 1$, $h = 0.5$, $h = 0.25$, and $h = 0.1$ approaching zero.

The slope of the secant line from $x = 2$ to $x = 3$ in the leftmost graph is given by the difference quotient:

$$\frac{f(3) - f(2)}{3 - 2} = \frac{\left(-\frac{1}{2}(3)^2 + 3(3)\right) - \left(-\frac{1}{2}(2)^2 + 3(2)\right)}{3 - 2} = \frac{4.5 - 4}{3 - 2} = 0.5.$$

Similarly, the slopes of the remaining three secant lines are given by the difference quotients:

$$\frac{f(2.5) - f(2)}{2.5 - 2} = 0.75, \quad \frac{f(2.25) - f(2)}{2.25 - 2} = 0.875, \quad \text{and} \quad \frac{f(2.1) - f(2)}{2.1 - 2} = 0.95.$$

Each of these slopes is an approximation to the slope of the red tangent line. As the graphs shown suggest, we would expect this sequence of slopes to be getting closer and closer to the actual slope of the red tangent line; notice for example that, in the last figure shown, the green secant line is almost indistinguishable from the red tangent line.

In a similar fashion we can calculate the slopes of secant lines from the left of $x = 2$. For example, the slope of the secant line from $x = 1$ to $x = 2$ is given by the difference quotient:

$$\frac{f(1) - f(2)}{1 - 2} = \frac{\left(-\frac{1}{2}(1)^2 + 3(1)\right) - \left(-\frac{1}{2}(2)^2 + 3(2)\right)}{1 - 2} = \frac{2.5 - 4}{1 - 2} = 1.5.$$

Over the smaller intervals $[1.5, 2]$, $[1.75, 2]$, and $[1.9, 2]$ we have secant lines with slopes given by

$$\frac{f(1.5) - f(2)}{1.5 - 2} = 1.25, \quad \frac{f(1.75) - f(2)}{1.7 - 2} = 1.125, \quad \text{and} \quad \frac{f(1.9) - f(2)}{1.9 - 2} = 1.05.$$

Putting all this information together, we obtain the following table:

Interval	[1, 2]	[1.5, 2]	[1.75, 2]	[1.9, 2]	*	[2, 2.1]	[2, 2.25]	[2, 2.5]	[2, 3]
Slope	1.5	1.25	1.125	1.05	*	0.95	0.875	0.75	0.5

From this table, we might guess that the slope of the tangent line is 1. This guess is only an estimate; the slope of the tangent line might instead be something like 0.97 or 1.02, but we don't have enough information to say otherwise at this point. □

EXAMPLE 4   Estimating instantaneous velocity with a sequence of average velocities

It can be shown that a watermelon dropped from a height of 100 feet will be $s(t) = -16t^2 + 100$ feet off the ground t seconds after it is dropped. Approximate the instantaneous velocity of the watermelon at time $t = 1$ by calculating a sequence of average velocities. Then interpret these average velocities graphically as slopes of secant lines.

SOLUTION

To estimate the instantaneous velocity at $t=1$ we will look at a sequence of small time intervals near $t = 1$ and consider the corresponding average velocities. The time intervals we choose to consider are $[1, 2]$, $[1, 1.5]$, $[1, 1.25]$, and $[1, 1.1]$. These intervals correspond to $z = 2, 1.5, 1.25$, and 1.1, or equivalently, to $h = 1, 0.5, 0.25$, and 0.1.

Let's look first at the interval $[1, 2]$. At $t=1$ the watermelon is $s(1) = 84$ feet from the ground, and at $t = 2$ the watermelon is $s(2) = 36$ feet from the ground, as illustrated here:

The average velocity over the first interval is therefore given by the difference quotient:

$$\frac{s(2) - s(1)}{2 - 1} = \frac{(-16(2)^2 + 100) - (-16(1)^2 + 100)}{1} = 36 - 84$$

$$= -48 \text{ feet per second.}$$

Similarly, the average velocities over the remaining three time intervals, in feet per second, are

$$\frac{s(1.5) - s(1)}{1.5 - 1} = -40, \quad \frac{s(1.25) - s(1)}{1.25 - 1} = -36, \quad \text{and} \quad \frac{s(1.1) - s(1)}{1.1 - 1} = -33.6.$$

Each of these average velocities is an approximation to the instantaneous velocity of the watermelon at time $t = 1$. Since the approximations should be improving as z gets closer to 1 (or, equivalently, as h gets closer to 0), we might guess that the instantaneous velocity of the watermelon at time $t = 1$ is some value greater than, but close to, -33.6 feet per second. For example, we might estimate that the instantaneous velocity at time $t = 1$ is -33 feet per second.

Each average velocity just calculated is an average rate of change of position, and thus can be thought of as the slope of a secant line, as in the four graphs shown next. As we consider smaller and smaller time intervals, we see that the slopes corresponding to these average velocities approach the slope of the red tangent line to $s(t)$ at $t = 1$, which in turn represents the instantaneous velocity at $t = 1$.

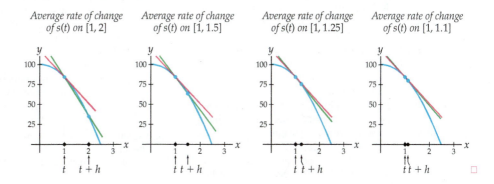

Average rate of change of $s(t)$ on $[1, 2]$ Average rate of change of $s(t)$ on $[1, 1.5]$ Average rate of change of $s(t)$ on $[1, 1.25]$ Average rate of change of $s(t)$ on $[1, 1.1]$

▶ Why does it make intuitive sense that when the graph of a smooth function f has a horizontal tangent line, the graph of its associated slope function will have a root?

▶ Why does it make intuitive sense that when the graph of a smooth function f is increasing, the graph of its associated slope function will be above the x-axis?

▶ In our initial discussion of position and velocity in this section, we showed graphs for the position, velocity, and acceleration of a grapefruit thrown into the air. Why does it make sense that the graph of acceleration is constant, given that the graph of velocity is a straight line?

▶ How can we use a sequence of slopes of secant lines to estimate the slope of a tangent line? Why is considering a sequence of secant lines better than considering just one secant line?

▶ In real–world examples, how are the units of a derivative of a function related to the units of the independent and dependent variables of that function?

EXERCISES 2.1

Thinking Back

▶ *Slope and linear functions:* If f is a linear function with slope -3 such that $f(2) = 1$, find the following, *without* first finding an equation for $f(x)$.
- $f(4)$
- $f(7)$
- $f(-2)$

▶ *Approximating limits:* Use sequences of approximations to estimate the values of
- $\lim\limits_{x \to 2} \dfrac{4 - x^2}{2 - x}$
- $\lim\limits_{z \to 3} \dfrac{z^3 - 27}{z - 3}$

▶ *Identifying increasing and decreasing behavior:* Use a graphing utility to determine the intervals on which $f(x) = -4x^5 + 25x^4 - 40x^3$ is increasing or decreasing.

▶ *Interpreting distance graphically:* When flying home for the holidays, Eva often flies between Denver International Airport (DIA) and Chicago O'Hare (ORD). Suppose Eva's plane takes off from DIA and 50 miles from ORD the plane has to circle the airport because of snow. The plane circles ORD four times and then lands.

(a) Draw a graph depicting the distance from DIA to Eva's plane as a function of time.

(b) Draw a graph depicting the distance from ORD to Eva's plane as a function of time.

Concepts

0. *Problem Zero:* Read the section and make your own summary of the material.

1. *True/False:* Determine whether each of the statements that follow is true or false. If a statement is true, explain why. If a statement is false, provide a counterexample.

(a) *True or False:* The slope of the tangent line to a function f at the point $x = 4$ is given by $f'(4)$.

(b) *True or False:* The instantaneous rate of change of a function f at the point $x = -3$ is given by $f'(-3)$.

(c) *True or False:* The instantaneous rate of change of a function f at a point $x = a$ can be represented as the slope of a secant line.

(d) *True or False:* Where a function f is positive, its associated slope function f' is increasing.

(e) *True or False:* Where a function f is decreasing, its associated slope function f' is negative.

(f) *True or False:* When a function f has a steep slope at a point on its graph, its instantaneous rate of change at that point will have a large magnitude.

(g) *True or False:* When the graph of a function f is decreasing with a steep slope, the graph of the associated slope function f' is negative with a large magnitude.

(h) *True or False:* Suppose an object is moving in a straight path with position function $s(t)$. If $s(t)$ is positive and decreasing, then the velocity $v(t)$ is negative.

2. *Examples:* Construct examples of the thing(s) described in the following. Try to find examples that are different than any in the reading.

(a) The graph of a function whose associated slope function f' is positive on $(-\infty, 2)$ and negative on $(2, \infty)$.

(b) The graph of a function with the following three properties: The average rate of change of f on $[0, 2]$ is 3, the average rate of change of f on $[0, 1]$ is -1, and the average rate of change of f on $[-2, 2]$ is 0.

(c) The graph of a function f with the following three properties: The instantaneous rate of change of f at $x = 2$ is zero, the average rate of change of f on $[1, 2]$ is 2, and the average rate of change of f on $[2, 4]$ is 1.

3. Explain why it is not a simple task to calculate the slope of the tangent line to a function f at a point $x = c$. Shouldn't calculating the slope of a line be really easy? What goes wrong here?

4. Let l be the line connecting two points $(a, f(a))$ and $(b, f(b))$ on the graph of a function f. What does this line l have to do with the average rate of change of f on the interval $[a, b]$, and why?

5. Given that $s(t)$ measures the distance an object has travelled over time, explain what the expression $\frac{s(b) - s(a)}{b - a}$ has to do with the distance formula $d = rt$.

6. How is velocity different from speed? What does it mean if velocity is negative?

7. What is the relationship between the derivative of a function f at a point $x = c$, the slope of the tangent line to the graph of f at $x = c$, and the instantaneous rate of change of f at $x = c$?

8. On a graph of $f(x) = x^2$,

(a) draw the tangent line to the graph of f at the point $(2, f(2))$;

(b) draw the secant line from $(2, f(2))$ to $(2.75, f(2.75))$;

(c) draw the secant line from $(1.75, f(1.75))$ to $(2, f(2))$.

(d) Which secant line is a better approximation to the tangent line, and why?

9. In Example 3 we estimated the slope of the tangent line to $f(x) = -\frac{1}{2}x^2 + 3x$ at $x = 2$. Get a better estimate by calculating the slopes of secant lines with values of z even closer to $x = 2$—for example, $z = 2.01$, $z = 2.001$, and $z = 2.0001$.

10. In Example 3 we estimated the slope of the tangent line to $f(x) = -\frac{1}{2}x^2 + 3x$ at $x = 2$ by finding slopes of secant lines from $x = 2$ to various points $x = z$ with $z > 2$. Draw a sequence of graphs that illustrates how to do this for $z < 2$, and then make specific calculations for $z = 1$, $z = 1.5$, $z = 1.75$, and $z = 1.9$. What are the corresponding values of h in this example?

11. For the graph of f appearing next at the left, label each of the following quantities to illustrate that $f'(c) \approx \frac{f(c + h) - f(c)}{h}$:

(a) the locations c, $c + h$, $f(c)$, and $f(c + h)$

(b) the distances h and $f(c + h) - f(c)$

(c) the slopes $\frac{f(c + h) - f(c)}{h}$ and $f'(c)$

12. For the graph of $g(x)$ appearing next at the right, label each of the following quantities to illustrate that $g'(c) \approx \frac{g(c + h) - g(c)}{h}$:

(a) the locations c, $c + h$, $g(c)$, and $g(c + h)$

(b) the distances h and $g(c + h) - g(c)$

(c) the slopes $\frac{g(c + h) - g(c)}{h}$ and $g'(c)$

f(x), Exercises 11 and 13   *g(x), Exercises 12 and 14*

          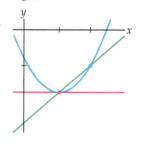

13. Consider again the graph of f at the left. Label each of the following quantities to illustrate that $f'(c) \approx \frac{f(z) - f(c)}{z - c}$:

(a) the locations c, z, $f(c)$, and $f(z)$

(b) the distances $z - c$ and $f(z) - f(c)$

(c) the slopes $\frac{f(z) - f(c)}{z - c}$ and $f'(c)$

14. Consider again the graph of $g(x)$ at the right. Label each of the following quantities to illustrate that $g'(c) \approx \frac{g(z) - g(c)}{z - c}$:

(a) the locations c, z, $g(c)$, and $g(z)$

(b) the distances $z - c$ and $g(z) - g(c)$

(c) the slopes $\frac{g(z) - g(c)}{z - c}$ and $g'(c)$

15. For the graph of f shown next at the left, list the following quantities in order from least to greatest:

(a) the average rate of change of f on $[-1, 1]$

(b) the instantaneous rate of change of f at $x = 1$

(c) $f'(-1)$

(d) $\frac{f(2) - f(-1)}{2 - (-1)}$

16. For the graph of $g(x)$ shown next at the right, list the following quantities in order from least to greatest:

(a) the average rate of change of g on $[0, 1]$

(b) the instantaneous rate of change of g at $x = 1$

(c) $\frac{g(-1 + 0.1) - g(-1)}{0.1}$

(d) $\frac{g(1) - g(-1)}{1 - (-1)}$

f(x), Exercises 15 and 17   *g(x), Exercises 16 and 18*

          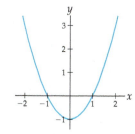

17. Consider again the function f graphed at the left. At which values of x does f have the greatest instantaneous rate of change? The least? At which values of x is the instantaneous rate of change of f equal to zero?

18. Consider again the function $g(x)$ graphed at the right. For which values of x does $g(x)$ have a positive instantaneous rate of change? Negative? Zero?

19. Make a copy of the graph of f used in Exercises 11 and 13, and sketch additional secant lines to illustrate that as $h \to 0$ (or equivalently, as $z \to c$) the slopes of the secant line get closer and closer to the slope of the tangent line to f at $x = c$.

20. The derivative of a smooth function f at a point $x = c$ can also be approximated with a **symmetric difference quotient**:

$$f'(c) \approx \frac{f(c + h) - f(c - h)}{2h}.$$

(a) Use a graph to illustrate what the symmetric difference measures. Why would it be reasonable to use the two-sided symmetric difference to approximate $f'(c)$? (*Hint: Your answer should involve a certain kind of secant line and a discussion of what happens as h gets close to 0.*)

(b) Use a sequence of symmetric difference approximations to estimate the derivative of $f(x) = x^2$ at $x = 3$. Illustrate your answer with a sequence of graphs.

Skills

In Exercises 21–24, sketch the graph of a function f that has the listed characteristics.

21. $f(1) = 2$, $f'(1) = 0$, $f'(3) = 2$

22. $f'(-3) = 0$, $f'(-1) = 0$, $f'(2) = 0$

23. $f(-1) = 2$, $f'(-1) = 3$, $f(1) = -2$, $f'(1) = 3$

24. $f'(-2) = 2$, $f'(0) = 1$, $f(1) = -5$

Sketch a graph of the associated slope function f' for each function f in Exercises 25–30.

25.

26.

27.

28.

29.

30.
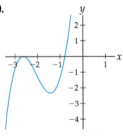

Now go the other way! Each graph in Exercises 31–34 can be thought of as the associated slope function f' for some unknown function f. In each case sketch a possible graph of f.

31.

32.

33.

34.
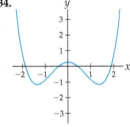

For each function f and value $x = c$ in Exercises 35–44, use a sequence of approximations to estimate $f'(c)$. Illustrate your work with an appropriate sequence of graphs of secant lines.

35. $f(x) = 4 - x^2$, $c = 1$

36. $f(x) = 4 - x^2$, $c = 0$

37. $f(x) = x + x^3$, $c = 0$

38. $f(x) = x + x^3$, $c = 1$

39. $f(x) = 4$, $c = -5$

40. $f(x) = 3x + 1$, $c = 10$

41. $f(x) = x^2$, $c = 3$

42. $f(x) = \dfrac{1}{x+1}$, $c = -2$

43. $f(x) = |x - 1|$, $c = 3$

44. $f(x) = |x^2 - 4|$, $c = 1$

Applications

A bowling ball dropped from a height of 400 feet will be $s(t) = 400 - 16t^2$ feet from the ground after t seconds. Use a sequence of average velocities to estimate the instantaneous velocities described in Exercises 45–48.

45. When the bowling ball is first dropped, with $h = 0.5$, $h = 0.25$, and $h = 0.1$

46. After $t = 1$ seconds, with $h = 0.5$, $h = 0.25$, $h = -0.5$, and $h = -0.2$

47. After $t = 2$ seconds, with $h = 0.1$, $h = 0.01$, $h = -0.1$, and $h = -0.01$

48. When the bowling ball hits the ground, with $h = -0.5$, $h = -0.2$, and $h = -0.1$

49. Think about what you did today and how far north you were from your house or dorm throughout the day. Sketch a graph that represents your distance north from your house or dorm over the course of the day, and explain how the graph reflects what you did today. Then sketch a graph of your velocity.

50. Stuart left his house at noon and walked north on Pine Street for 20 minutes. At that point he realized he was late for an appointment at the dentist, whose office was located *south* of Stuart's house on Pine Street; fearing he would be late, Stuart sprinted south on Pine Street, past his house, and on to the dentist's office. When he got there, he found the office closed for lunch; he was 10 minutes early for his 12:40 appointment. Stuart waited at the office for 10 minutes and then found out that his appointment was actually for the next day, so he walked back to his house. Sketch a graph that describes Stuart's position over time. Then sketch a graph that describes Stuart's velocity over time.

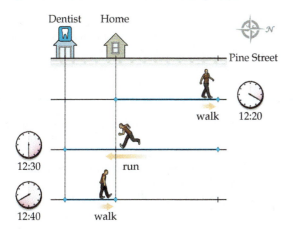

51. Every morning Linda takes a thirty-minute jog in Central Park. Suppose her distance *s* in feet from the oak tree on the north side of the park *t* minutes after she begins her jog is given by the function $s(t)$ shown that follows at the left, and suppose she jogs on a straight path leading into the park from the oak tree.

(a) What was the average rate of change of Linda's distance from the oak tree over the entire thirty-minute jog? What does this mean in real-world terms?

(b) On which ten-minute interval was the average rate of change of Linda's distance from the oak tree the greatest: the first 10 minutes, the second 10 minutes, or the last 10 minutes?

(c) Use the graph of $s(t)$ to estimate Linda's average velocity during the 5-minute interval from $t = 5$ to $t = 10$. What does the sign of this average velocity tell you in real-world terms?

(d) Approximate the times at which Linda's (instantaneous) velocity was equal to zero. What is the physical significance of these times?

(e) Approximate the time intervals during Linda's jog that her (instantaneous) velocity was negative. What does a negative velocity mean in terms of this physical example?

52. Last night Phil went jogging along Main Street. His distance from the post office *t* minutes after 6:00 P.M. is shown in the preceding graph at the right.

(a) Give a narrative (that matches the graph) of what Phil did on his jog.

(b) Sketch a graph that represents Phil's instantaneous velocity *t* minutes after 6:00 P.M. Make sure you label the tick marks on the vertical axis as accurately as you can.

(c) When was Phil jogging the fastest? The slowest? When was he the farthest away from the post office? The closest to the post office?

53. Suppose $h(t)$ represents the average height, in feet, of a person who is *t* years old.

(a) In real-world terms, what does $h(12)$ represent and what are its units? What does $h'(12)$ represent, and what are its units?

(b) Is $h(12)$ positive or negative, and why? Is $h'(12)$ positive or negative, and why?

(c) At approximately what value of *t* would $h(t)$ have a maximum, and why? At approximately what value of *t* would $h'(t)$ have a maximum, and why?

54. A tomato plant given *x* ounces of fertilizer will successfully bear $T(x)$ pounds of tomatoes in a growing season.

(a) In real-world terms, what does $T(5)$ represent and what are its units? What does $T'(5)$ represent and what are its units?

(b) A study has shown that this fertilizer encourages tomato production when less than 20 ounces are used, but inhibits production when more than 20 ounces are used. When is $T(x)$ positive and when is $T(x)$ negative? When is $T'(x)$ positive and when is $T'(x)$ negative?

55. If Katie walked at 3 miles per hour for 20 minutes and then sprinted at 10 miles an hour for 8 minutes, how fast would Dave have to walk or run to go the same distance as Katie did in the same time while moving at a constant speed? Sketch a graph of Katie's position over time and a graph of Dave's position over time on the same set of axes.

Katie

t, minutes

d, distance

Dave

d, distance

56. Velocity $v(t)$ is the derivative of position $s(t)$. It is also true that acceleration $a(t)$ (the rate of change of velocity) is the derivative of velocity. If a race car's position in miles t hours after the start of a race is given by the function $s(t)$, what are the units of $s(1.2)$? What are the units and real-world interpretation of $v(1.2)$? What are the units and real-world interpretation of $a(1.2)$?

57. The total yearly expenditures by public colleges and universities from 1990 to 2000 can be modeled by the function $E(t) = 123(1.025)^t$, where expenditures are measured in billions of dollars and time is measured in years since 1990.

(a) Estimate the total yearly expenditures by these colleges and universities in 1995.

(b) Compute the average rate of change in yearly expenditures between 1990 and 2000.

(c) Compute the average rate of change in yearly expenditures between 1995 and 1996.

(d) Estimate the rate at which yearly expenditures of public colleges and universities were increasing in 1995.

Proofs

58. Show that if f is a function and $z = x + h$, then

$$\frac{f(z) - f(x)}{z - x} = \frac{f(x + h) - f(x)}{h}.$$

59. Suppose f is a linear function with positive slope. Show that the average rate of change of f on any interval $[a, b]$ is positive, and then use this fact to show that f is always increasing.

Thinking Forward

Taking the limit: We have seen that if f is a smooth function, then $f'(c) \approx \frac{f(c+h) - f(c)}{h}$. This approximation should get better as h gets closer to zero. In fact, in the next section we will define the derivative in terms of such a limit.

$$f'(c) = \lim_{h \to 0} \frac{f(c + h) - f(c)}{h}.$$

▶ Use the limit just defined to calculate the exact slope of the tangent line to $f(x) = x^2$ at $x = 4$.

▶ Instead of choosing small values of h, we could have chosen values of z close to c. What limit involving z instead of h is equivalent to the one involving h?

▶ Use the limit you just found to calculate the exact slope of the tangent line to $f(x) = x^2$ at $x = 4$. Obviously you should get the same final answer as you did earlier.

2.2   FORMAL DEFINITION OF THE DERIVATIVE

▶ Using limits to define derivatives, tangent lines, and instantaneous rates of change

▶ Differentiability at a point, from one side, and on intervals

▶ Leibniz notation and multiple derivatives

The Derivative at a Point

In the previous section we examined derivatives intuitively, by discussing tangent lines and instantaneous rates of change. We will now use limits to make these ideas precise. We have seen that the slope $f'(c)$ of the tangent line to f at $x = c$ can be approximated by the slope of a nearby secant line from $x = c$ to $x = c + h$, or equivalently, from $x = c$ to $x = z$:

$$f'(c) \approx \frac{f(c + h) - f(c)}{h}, \quad \text{or equivalently,} \quad f'(c) \approx \frac{f(z) - f(c)}{z - c}.$$

If the limit of these quantities approaches a real number as $h \to 0$, or as $z \to c$, then we will define that real number to be the **derivative** of f at the point $x = c$.

DEFINITION 2.1

The Derivative of a Function at a Point

The **derivative at $x = c$** of a function f is the number

$$f'(c) = \lim_{h \to 0} \frac{f(c + h) - f(c)}{h}, \quad \text{or equivalently,} \quad f'(c) = \lim_{z \to c} \frac{f(z) - f(c)}{z - c},$$

provided that this limit exists.

The derivative of a function f at a point $x = c$ measures the **instantaneous rate of change** of the function at that point. Notice that this instantaneous rate of change is a limit of average rates of change.

For example, consider the function $f(x) = x^2$. We can calculate the derivative of this function at the point $x = 3$ with a limit as $h \to 0$ or with a limit as $z \to 3$. Using the $h \to 0$ definition of the derivative, we have

$$f'(3) = \lim_{h \to 0} \frac{(3 + h)^2 - 3^2}{h} = \lim_{h \to 0} \frac{6h + h^2}{h} = \lim_{h \to 0} \frac{h(6 + h)}{h} = \lim_{h \to 0} (6 + h) = 6.$$

Using the $z \to c$ definition of derivative we obtain the same answer:

$$f'(3) = \lim_{z \to c} \frac{z^2 - 3^2}{z - 3} = \lim_{z \to c} \frac{(z + 3)(z - 3)}{z - 3} = \lim_{z \to c} (z + 3) = 6.$$

In both calculations above we have shown the instantaneous rate of change of $f(x) = x^2$ at $x = 3$ is equal to 6. At the instant that we have $x = 3$, the function $f(x) = x^2$ is changing at a rate of 6 vertical units for each horizontal unit. This is equivalent to saying that the slope of the tangent line is equal to 6, as shown here:

Tangent line to $f(x) = x^2$ at $x = 3$ has slope $f'(3) = 6$

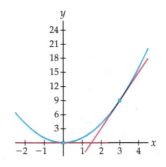

The Derivative as a Function

By putting all of the point-derivatives of a function together, we can define a function f' whose output at any value of x is defined to be the derivative, or instantaneous rate of change, of f at that point.

DEFINITION 2.2

The Derivative of a Function

The **derivative** of a function f is the function f' defined by

$$f'(x) = \lim_{h \to 0} \frac{f(x+h) - f(x)}{h}, \quad \text{or equivalently,} \quad f'(x) = \lim_{z \to x} \frac{f(z) - f(x)}{z - x}.$$

The domain of f' is the set of values x for which the defining limit of f' exists.

The function f' is the **associated slope function** that we investigated in the previous section, since at each point x its value is the slope of the graph of f. In addition, the function f' represents the instantaneous rate of change at every point x. In particular, if $s(t)$ is a position function, then its instantaneous rate of change is the velocity function $v(t) = s'(t)$. Similarly, the instantaneous rate of change of velocity $v(t)$ is the the acceleration function $a(t) = v'(t)$.

For example, we can calculate the derivative of $f(x) = x^2$ for *all* values of x with either a limit as $h \to 0$ or a limit as $z \to x$; again we choose the first method:

$$f'(x) = \lim_{h \to 0} \frac{(x+h)^2 - x^2}{h} = \lim_{h \to 0} \frac{2xh + h^2}{h} = \lim_{h \to 0} \frac{h(2x + h)}{h} = \lim_{h \to 0} (2x + h) = 2x.$$

Finding $f'(x)$ for general x is like calculating $f'(c)$ for all possible values $x = c$ at the same time. Once we have a formula for $f'(x)$, we can easily calculate any particular value $f'(c)$. For example, evaluating $f'(x) = 2x$ at $x = 3$ does give us $f'(3) = 2(3) = 6$, as we calculated before. The following figures show slopes on the graph of $f(x) = x^2$ together with heights on the graph of $f'(x) = 2x$, for $x = -2$, $x = 0$, and $x = 3$.

Slopes of $f(x) = x^2$

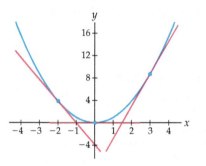

Heights of $f'(x) = 2x$

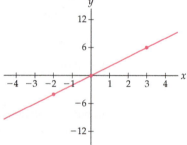

In this book we will most often use the $h \to 0$ version of the derivative, but will use the $z \to x$ version when it suits our purposes or makes a calculation easier. You will verify that these two versions of the derivative are equivalent in Exercise 5.

Differentiability

Given a function f and a value $x = c$, there may or may not be a well-defined tangent line to the graph of f at $(c, f(c))$. When there is a well-defined tangent line with a finite slope, we say that f is **differentiable** at $x = c$:

DEFINITION 2.3

Differentiability at a Point

A function f is **differentiable at $x = c$** if $\lim_{h \to 0} \dfrac{f(c+h) - f(c)}{h}$ exists.

We used the $h \to 0$ definition of the derivative in Definition 2.3, but everything would work equally well with the equivalent $z \to c$ definition of the derivative. If a point-derivative $f'(c)$ is infinite, then we say that f has a ***vertical tangent line*** at $x = c$. In this case a line exists that is tangent to the graph of the function, but since that line is vertical, its slope is undefined and the function fails to be differentiable at that point.

Like continuity, differentiability can be considered from the left or from the right. A function is ***left differentiable at*** $x = c$ if its left derivative exists, and ***right differentiable at*** $x = c$ if its right derivative exists, where the left and right derivatives are defined with left and right limits as follows:

DEFINITION 2.4

One-sided Differentiability at a Point

The ***left derivative*** and ***right derivative*** of a function f at a point $x = c$ are, respectively, equal to the following, if they exist:

$$f'_-(c) = \lim_{h \to 0^-} \frac{f(c + h) - f(c)}{h}, \qquad f'_+(c) = \lim_{h \to 0^+} \frac{f(c + h) - f(c)}{h}.$$

We could also use the $z \to c$ definition of the derivative to define left and right derivatives, by considering limits of difference quotients as $z \to c^-$ and as $z \to c^+$.

For example, consider the function $f(x) = |x|$. This function has a sharp corner at $x = 0$, and therefore we would not expect it to have a well-defined tangent line at that point. Indeed, when we try to calculate $f'(0)$, we encounter the following limit:

$$f'(0) = \lim_{h \to 0} \frac{|0 + h| - |0|}{h} = \lim_{h \to 0} \frac{|h|}{h}.$$

This limit is initially in the indeterminate form $\frac{0}{0}$, but we cannot cancel anything in it until we get rid of the absolute value. Recall that if $h \geq 0$, then $|h| = h$, but if $h < 0$, then $|h| = -h$. Looking from the left and the right, we have the following limits:

$$f'_-(0) = \lim_{h \to 0^-} \frac{|h|}{h} = \lim_{h \to 0^-} \frac{-h}{h} = \lim_{h \to 0^-} -1 = -1,$$

$$f'_+(0) = \lim_{h \to 0^+} \frac{|h|}{h} = \lim_{h \to 0^+} \frac{h}{h} = \lim_{h \to 0^+} 1 = 1.$$

Since $f'_-(0)$ and $f'_+(0)$ exist but are not equal, $f'(0)$ does not exist. The first two graphs shown next illustrate the left and right derivatives at 0, for small negative h and small positive h, respectively. The third graph shows the graph of the derivative function f'; note that this function has a jump discontinuity and is not defined at $x = 0$.

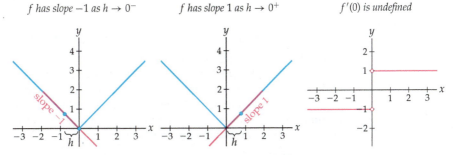

f has slope -1 as $h \to 0^-$ f has slope 1 as $h \to 0^+$ $f'(0)$ is undefined

As with continuity, we say that a function is ***differentiable on an interval*** I if it is differentiable at every point in the interior of I, right differentiable at any closed left endpoint, and left differentiable at any closed right endpoint. For example, the first graph shown next is differentiable on $[2, 3]$, since it is right differentiable at $x = 2$, even though it is not differentiable at $x = 2$. The second and third graphs show two more ways that a function could fail to be differentiable.

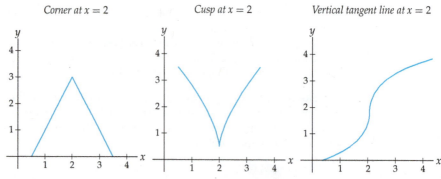

In the first and second graphs, the left and right derivatives exist at $x = 2$ but the two-sided derivative at $x = 2$ does not. (If the cusp in the second graph were "steeper," then we could also have a situation where the left and right derivatives at $x = 2$ were infinite, with opposite signs.) In the third graph, the left, right, and two-sided derivatives at $x = 2$ are all infinite, because the tangent line is vertical.

Another way that a function f can fail to be differentiable at a point $x = c$ is if f fails to be continuous at $x = c$. Intuitively, if a function is not continuous, then it has absolutely no chance of being differentiable; think for a minute about secant lines from the left and the right at a jump or removable discontinuity, for example. (See also Example 6.) What this means is that although not every continuous function is differentiable, every differentiable function is continuous:

THEOREM 2.5   **Differentiability Implies Continuity**

If f is differentiable at $x = c$, then f is continuous at $x = c$.

Proof. If f is differentiable at $x = c$, then $\lim\limits_{x \to c} \dfrac{f(x) - f(c)}{x - c} = f'(c)$ exists; that is, the limit, and thus $f'(c)$, is equal to some real number. We will use this fact to show that f is continuous at $x = c$, by showing that $\lim\limits_{x \to c} f(x) = f(c)$. By the sum rule for limits, it is equivalent to show that $\lim\limits_{x \to c} (f(x) - f(c)) = 0$, which we can do by using the expression for the derivative:

$$\lim_{x \to c} (f(x) - f(c)) = \lim_{x \to c} \left(\frac{f(x) - f(c)}{x - c} (x - c) \right) \qquad \leftarrow \frac{x - c}{x - c} = 1 \text{ if } x \neq c$$

$$= \left(\lim_{x \to c} \frac{f(x) - f(c)}{x - c} \right) \left(\lim_{x \to c} (x - c) \right) \qquad \leftarrow \text{product rule for limits}$$

$$= f'(c) \lim_{x \to c} (x - c) \qquad \leftarrow \text{definition of the derivative}$$

$$= f'(c)(0) = 0. \qquad \leftarrow \text{limit rules} \qquad \blacksquare$$

Tangent Lines and Local Linearity

Although we have an intuitive sense of the tangent line to a graph at a point, up until now we did not have a formal mathematical definition for this tangent line. Our geometrical intuition helped us arrive at an algebraic definition for the derivative, but it is the algebraic definition that now allows us to define the tangent line precisely. Specifically, we define the **tangent line** through a point on the graph of a function to be the line whose slope is given by the derivative of the function at that point.

THEOREM 2.6   **Equation of the Tangent Line to a Function at a Point**

The **tangent line** to the graph of a function f at a point $x = c$ is defined to be the line passing through $(c, f(c))$ with slope $f'(c)$, provided that the derivative $f'(c)$ exists. This line has equation

$$y = f(c) + f'(c)(x - c).$$

Proof. We need only calculate the form of the line that passes through the point $(c, f(c))$ and has slope $f'(c)$. Using the point-slope form $y - y_0 = m(x - x_0)$, we see that this line has equation

$$y - f(c) = f'(c)(x - c),$$

and solving for y, we obtain the desired equation $y = f(c) + f'(c)(x - c)$.   ■

For example, we saw earlier that the derivative of $f(x) = x^2$ at $x = 3$ is equal to $f'(3) = 6$. This means that the tangent line to the graph of $f(x) = x^2$ at $x = 3$ has slope $f'(3) = 6$ and passes through $(3, f(3)) = (3, 9)$. This line has equation $y = 9 + 6(x - 3)$, which, in slope-intercept form, is the equation $y = 6x - 9$.

The tangent line to a function f at a point $x = c$ is the unique line that "agrees with" both the height of the function and the slope of the function at $x = c$. This means that near $x = c$ the graph of a function is very close to the graph of its tangent line. Therefore we can use the tangent line as a rough approximation to the function f itself near $x = c$.

DEFINITION 2.7   **Local Linearity**

If f has a well-defined derivative $f'(c)$ at the point $x = c$, then, for values of x near c, the function $f(x)$ can be approximated by the tangent line to f at $x = c$ with the **linearization of f around $x = c$** given by

$$f(x) \approx f(c) + f'(c)(x - c).$$

Note that this definition does not assert how good an approximation one can make by using the tangent line. What constitutes "near" and what constitutes "good" will be determined by the context of the problem at hand.

For example, since the tangent line to $f(x) = x^2$ at $x = 3$ is the line $y = 6x - 9$, the line $y = 6x - 9$ can be used as a rough approximation to the graph of the function $f(x) = x^2$, at least for values of x close to 3. This approximation will be better the closer we are to $x = 3$. As the following figure shows, the two graphs have nearly the same height at the point $x = 3.5$ and are still relatively close even at $x = 4$.

$y = 6x - 9$ is close to $f(x) = x^2$ near $x = 3$

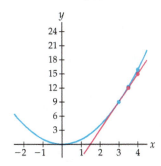

The concept of local linearity will be particularly helpful for approximating roots with Newton's method in Example 7 and Exercises 71–76, as well as for approximating solutions of differential equations with Euler's method in Section 7.5.

Leibniz Notation and Differentials

Derivatives are used in so many different fields of study that they are represented with a wide variety of notations. In this book we will focus on two types: first, the "prime" notation $f'(x)$ that we have already established, and second, the **Leibniz notation** $\frac{df}{dx}$. If $y = f(x)$, then we can also write $y'(x)$ or $\frac{dy}{dx}$ to represent the derivative.

Leibniz notation is named for one of the founders of calculus and is intentionally structured to remind us of the connection between derivatives and average rates of change. (The "prime" notation we have been using so far is due to Lagrange.) Intuitively, the expression dx represents an infinitesimally small change in x, just as Δx represents a small finite change in x. In Leibniz notation the definition of the derivative as a limit of average rates of change is strikingly clear:

$$\frac{dy}{dx} = \lim_{\Delta x \to 0} \frac{\Delta y}{\Delta x}.$$

It is sometimes convenient to think of the expressions dy and dx as **differentials** in a functional relationship, with dy as a function of dx, in the following sense: We know that $\frac{\Delta y}{\Delta x}$ represents the slope of a line with vertical change of Δy for a given horizontal change Δx, as in the figure shown next at the left. In the same way we might try to think of $\frac{dy}{dx}$ as the slope of a line with a vertical change of dy for a given horizontal change dx, as in the figure at the right.

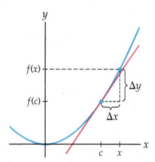

Δy depends on Δx
and gives the height of the function

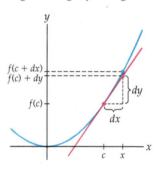

As differentials, dy depends on dx
and give the height of the tangent line

With this interpretation of dy as a differential we can recast local linearity (Definition 2.7) as saying that when x is sufficiently close to a point c, we can approximate $f(x)$ by adding dy to $f(c)$. If $x = c + dx$ as in the rightmost figure, this approximation can be expressed as

$$f(c + dx) \approx f(c) + dy.$$

 CAUTION

It is important to note that we will *not* be thinking of dx and dy as numbers, but rather thinking of $\frac{dy}{dx}$ as a formal symbol that represents the derivative $f'(x)$. This formal symbol suggests meanings such as those given to the differentials in the preceding figure, but $\frac{dy}{dx}$ is not an actual quotient of numbers; it is a limit of slopes of secant lines as defined in Definition 2.2.

We can also write Leibniz notation in a slightly different way, called **operator notation**, as follows:

$$\frac{dy}{dx} = \frac{d}{dx}(y(x)).$$

Here we are thinking of $\frac{d}{dx}$ as a sort of metafunction that operates on functions instead of numbers: It takes functions as inputs and returns the derivatives of those functions as outputs. Using operator notation, we can express the statement "if $y = x^2$, then $\frac{dy}{dx} = 2x$" in the more compact form "$\frac{d}{dx}(x^2) = 2x$."

Although Leibniz notation is sometimes more convenient or informative than the usual "prime" notation, it has one drawback: It is cumbersome to write down the point-derivative of a function in Leibniz notation. For example, as we showed earlier, the derivative of $f(x) = x^2$ is $f'(x) = 2x$. This derivative is easy to express in Leibniz notation as $\frac{d}{dx}(x^2) = 2x$.

Now suppose we wish to consider the same derivative at the point $x = 3$. In "prime notation" it is easily expressed as $f'(3) = 2(3) = 6$. In Leibniz notation it is more difficult to work the $x = 3$ evaluation into the notation. We could write any of the following:

$$\frac{df}{dx}\Big|_3 = 6, \quad \frac{d}{dx}(x^2)\Big|_{x=3} = 6, \quad \text{or} \quad \frac{d}{dx}(x^2)\Big|_3 = 6.$$

If f is a function, then its derivative f' is a function as well. This means that we can also consider *its* derivative, which we call f'', the **second derivative** of f. In Leibniz notation we write the second derivative as $\frac{d^2f}{dx^2}$. This notation is supposed to suggest the fact that we are differentiating twice, that is, finding $\frac{d}{dx}\left(\frac{d}{dx}(f(x))\right)$. Similarly, we could find the third, fourth, or fifth derivatives of a function f, and so on. For example, the third derivative of $f(x)$ can be written as $f'''(x)$. For larger values of n we will replace the primes with a parenthetical notation: for example, $f''''''(x) = f^{(6)}(x)$. In general, the *nth* **derivative** of a function f is denoted by:

$$f^{(n)}(x) = \frac{d^nf}{dx^n} = \underbrace{\frac{d}{dx}\left(\frac{d}{dx}\left(\frac{d}{dx}\cdots\left(\frac{d}{dx}(f(x))\right)\cdots\right)\right).}_{n \text{ times}}$$

Examples and Explorations

> **EXAMPLE 1**
>
> ### The derivative of a linear function
>
> Consider the linear function $f(x) = 3x + 1$.
>
> **(a)** Use the definition of the derivative to show that $f'(1) = 3$.
>
> **(b)** Use the definition of the derivative to show that $f'(x) = 3$ for all x.
>
> **(c)** Why does it make intuitive, graphical sense that the derivative of the linear function $f(x) = 3x + 1$ is constantly equal to 3?

SOLUTION

(a) If $f(x) = 3x + 1$, then by the $h \to 0$ definition of the derivative, the derivative of f at $x = 1$ is

$$f'(1) = \lim_{h \to 0}\frac{f(1+h) - f(1)}{h} \qquad \leftarrow \text{definition of the derivative}$$

$$= \lim_{h \to 0}\frac{(3(1+h) + 1) - (3(1) + 1)}{h} \qquad \leftarrow \text{use formula } f(x) = 3x + 1$$

$$= \lim_{h \to 0}\frac{3 + 3h + 1 - 3 - 1}{h} = \lim_{h \to 0}\frac{3h}{h} \qquad \leftarrow \text{algebra}$$

$$= \lim_{h \to 0}(3) = 3. \qquad \leftarrow \text{cancel } h\text{'s and evaluate limit}$$

(b) The calculation for general values of x is nearly identical:

$$f'(x) = \lim_{h \to 0}\frac{f(x+h) - f(x)}{h} \qquad \leftarrow \text{definition of the derivative}$$

$$= \lim_{h \to 0}\frac{(3(x+h) + 1) - (3x + 1)}{h} \qquad \leftarrow \text{use formula } f(x) = 3x + 1)$$

$$= \lim_{h \to 0}\frac{3x + 3h + 1 - 3x - 1}{h} = \lim_{h \to 0}\frac{3h}{h} \qquad \leftarrow \text{algebra}$$

$$= \lim_{h \to 0}(3) = 3. \qquad \leftarrow \text{cancel } h\text{'s and evaluate limit}$$

(c) The graph of the linear function $f(x) = 3x + 1$ is a line with slope 3, so it makes sense that for each value of x, the line tangent to $f(x) = 3x + 1$ at x has slope 3. In this case the line that points "in the direction of the function" at any point is in fact the line $f(x) = 3x + 1$ itself. $\square$

EXAMPLE 2   Calculating the derivative at a point

Consider the function $f(x) = x^3$.

(a) Use the $h \to 0$ definition of the derivative to find $f'(2)$.

(b) Use the $z \to x$ definition of the derivative to find $f'(2)$.

(c) Find the equation of the tangent line to $f(x) = x^3$ at $x = 2$, and graph $f(x) = x^3$ and this line on the same set of axes.

SOLUTION

(a) Using the $h \to 0$ definition of the derivative, we have

$$f'(2) = \lim_{h \to 0} \frac{f(2+h) - f(2)}{h} = \lim_{h \to 0} \frac{(2+h)^3 - 2^3}{h} \quad \leftarrow \text{derivative with } x = 2$$

$$= \lim_{h \to 0} \frac{(8 + 12h + 6h^2 + h^3) - 8}{h} \quad \leftarrow \text{multiply out } (2+h)^3$$

$$= \lim_{h \to 0} \frac{12h + 6h^2 + h^3}{h} = \lim_{h \to 0} \frac{h(12 + 6h + h^2)}{h} \quad \leftarrow \text{algebra}$$

$$= \lim_{h \to 0} (12 + 6h + h^2) = 12 + 6(0) + (0)^2 = 12. \quad \leftarrow \text{cancellation, limit rules}$$

(b) With the $z \to 2$ definition of the derivative the algebra is different, but the final answer will be the same. Along the way we will need to use the factoring formula $a^3 - b^3 = (a - b)(a^2 + ab + b^2)$, as follows:

$$f'(x) = \lim_{z \to 2} \frac{f(z) - f(2)}{z - 2} = \lim_{z \to 2} \frac{z^3 - 2^3}{z - 2} \quad \leftarrow \text{derivative with } x = 2$$

$$= \lim_{z \to 2} \frac{(z - 2)(z^2 + 2z + 4)}{z - 2} \quad \leftarrow \text{factoring formula}$$

$$= \lim_{z \to 2} (z^2 + 2z + 4) = 2^2 + 2(2) + 4 = 12. \quad \leftarrow \text{cancellation, limit rule}$$

(c) We have just shown two ways that if $f(x) = x^3$, then $f'(2) = 12$. Thus the tangent line to $f(x)$ at $x = 2$ has slope 12 and passes through the point $(2, f(2)) = (2, 8)$. The equation of this line is

$$y - 8 = 12(x - 2) \implies y = 12x - 24 + 8 \implies y = 12x - 16.$$

When we graph the line along with the original function, we see that indeed the line $y = 12x - 16$ is the tangent line to $f(x) = x^3$ at $x = 2$:

$y = 12x - 16$ is the tangent line to $f(x) = x^2$ at $x = 2$

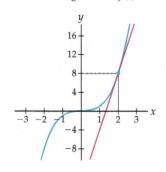

$\square$

EXAMPLE 3

Calculating the derivative of a function

Consider the function $f(x) = x^3$.

(a) Use the $h \to 0$ definition of the derivative to find $f'(x)$.

(b) Use the $z \to x$ definition of the derivative to find $f'(x)$.

(c) Graph the function $f(x) = x^3$ and the derivative $f'(x)$ you found in parts (a) and (b), and argue that one graph is the slope function of the other.

SOLUTION

(a) The calculations in this example will be similar to those in the previous example, except that we will not specify a specific value of x here. Using the $h \to 0$ definition of the derivative, we have

$$f'(x) = \lim_{h \to 0} \frac{f(x+h) - f(x)}{h} = \lim_{h \to 0} \frac{(x+h)^3 - x^3}{h} \qquad \leftarrow \text{derivative}$$

$$= \lim_{h \to 0} \frac{(x^3 + 3x^2 h + 3xh^2 + h^3) - x^3}{h} \qquad \leftarrow \text{multiply out } (x+h)^3$$

$$= \lim_{h \to 0} \frac{3x^2 h + 3xh^2 + h^3}{h} = \lim_{h \to 0} \frac{h(3x^2 + 3xh + h^2)}{h} \qquad \leftarrow \text{algebra}$$

$$= \lim_{h \to 0} (3x^2 + 3xh + h^2) = 3x^2 + 3x(0) + (0)^2 = 3x^2. \qquad \leftarrow \text{cancellation, limit rules}$$

Notice that before we did any algebra, the limit was of the indeterminate form $\frac{0}{0}$. For derivative calculations with the $h \to 0$ definition, the goal is often to expand and simplify until a common factor of h can be cancelled, as we did here.

(b) With the $z \to x$ definition of the derivative the algebra is different, but the final answer is the same:

$$f'(x) = \lim_{z \to x} \frac{f(z) - f(x)}{z - x} = \lim_{z \to x} \frac{z^3 - x^3}{z - x} \qquad \leftarrow \text{derivative}$$

$$= \lim_{z \to x} \frac{(z - x)(z^2 + zx + x^2)}{z - x} \qquad \leftarrow \text{factoring formula}$$

$$= \lim_{z \to x} (z^2 + zx + x^2) = x^2 + x^2 + x^2 = 3x^2. \qquad \leftarrow \text{cancellation, limit rules}$$

Again, notice that the limit was initially of the indeterminate form $\frac{0}{0}$. For derivative calculations with the $z \to x$ definition, the goal is often to factor and simplify until a common factor such as $z - x$ can be cancelled, as we just did.

(c) We have just shown two ways that if $f(x) = x^3$, then $f'(x) = 3x^2$. When we graph these two functions, we can see that at each value of x, the slope of the graph of $f(x) = x^3$ is equal to the height of the graph of $f'(x) = 3x^2$. In particular notice that the slopes of $f(x) = x^3$ are always positive and the graph of $f'(x) = 3x^2$ is always positive.

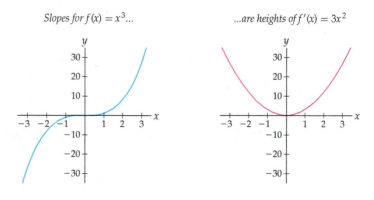

Slopes for $f(x) = x^3$...   *...are heights of $f'(x) = 3x^2$*

EXAMPLE 4   **Finding the equation of a tangent line**

Suppose $f(x) = \dfrac{2}{x}$. Find the equation of the tangent line to f at $x = 1$.

SOLUTION

By definition, the tangent line to f at $x = 1$ has slope $f'(1)$ and passes through the point $(1, f(1)) = \left(1, \dfrac{2}{1}\right) = (1, 2)$. Before we can find the equation of this line, we must calculate the value of $f'(1)$. A lot of algebra will be needed before we can cancel a common factor of h and solve the limit:

$$f'(1) = \lim_{h \to 0} \frac{f(1 + h) - f(1)}{h} = \lim_{h \to 0} \frac{\frac{2}{1+h} - \frac{2}{1}}{h} \qquad \leftarrow \text{derivative with } x = 1$$

$$= \lim_{h \to 0} \frac{\left(\frac{2 - 2(1+h)}{1+h}\right)}{h} = \lim_{h \to 0} \frac{2 - 2 - 2h}{h(1+h)} \qquad \leftarrow \text{algebra}$$

$$= \lim_{h \to 0} \frac{-2h}{h(1+h)} = \lim_{h \to 0} \frac{-2}{1+h} = \frac{-2}{1+0} = -2. \qquad \leftarrow \text{cancellation, limit rules}$$

We now use the point-slope form of a line to find the equation of the line that has slope $f'(1) = -2$ and that passes through the point $(1, f(1)) = (1, 2)$:

$$y - 2 = -2(x - 1) \implies y = -2(x - 1) + 2 \implies y = -2x + 4. \qquad \square$$

 CHECKING THE ANSWER

We can verify the reasonableness of the answer we just found by graphing the function $f(x) = \dfrac{2}{x}$ and the line $y = -2x + 4$ on the same set of axes and checking that the line appears to be tangent to the graph of f at the point $(1, f(1))$:

$$f(x) = \frac{2}{x} \text{ and } y = -2x + 4$$

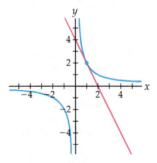

EXAMPLE 5   **The derivative of a piecewise-defined function**

Consider the piecewise-defined function $f(x) = \begin{cases} 3x + 2, & \text{if } x < 1 \\ 6 - x, & \text{if } x \geq 1. \end{cases}$

(a) Calculate $f'_+(1)$ and $f'_-(1)$. What can you say about $f'(1)$?

(b) Write down a formula for $f'(x)$ as a piecewise-defined function.

SOLUTION

(a) We start by finding the right derivative of f at $x = 1$. In this case we examine $h \to 0^+$, which means that $h > 0$, and thus $1 + h > 1$. Therefore we will use the second part of the piecewise-defined function f to evaluate $f(1 + h)$ in this case:

$$f'_+(1) = \lim_{h \to 0^+} \frac{f(1 + h) - f(1)}{h} = \lim_{h \to 0^+} \frac{(6 - (1 + h)) - (6 - 1)}{h}$$

$$= \lim_{h \to 0^+} \frac{6 - 1 - h - 6 + 1}{h} = \lim_{h \to 0^+} \frac{-h}{h} = \lim_{h \to 0^+} (-1) = -1.$$

In contrast, when we calculate the left derivative of f at $x = 1$, we will have $h \to 0^-$, and thus $h < 0$. This means that $1 + h < 1$, so we will use the first part of the piecewise-defined function f to evaluate $f(1 + h)$. Of course we still have $1 \geq 1$, so we still use the *second* part of f to evaluate $f(1)$:

$$f'_-(1) = \lim_{h \to 0^-} \frac{f(1 + h) - f(1)}{h} = \lim_{h \to 0^-} \frac{(3(1 + h) + 2) - (6 - 1)}{h}$$

$$= \lim_{h \to 0^-} \frac{3 + 3h + 2 - 6 + 1}{h} = \lim_{h \to 0^-} \frac{3h}{h} = \lim_{h \to 0^-} 3 = 3.$$

Since the left and right derivatives of f at $x = 1$ are not equal to each other, the derivative $f'(1)$ of f at $x = 1$ is undefined. Note that f is left differentiable and right differentiable at $x = 1$, but *not* differentiable at $x = 1$.

(b) We have just calculated that $f'(1)$ does not exist. It now remains to determine $f'(x)$ for values of x that are less than or greater than 1. For $x < 1$ the value of $f(x)$ is equal to $3x + 2$. Using the definition of the derivative, we see that for $x < 1$ we have

$$f'(x) = \lim_{h \to 0} \frac{(3(x + h) + 2) - (3x + 2)}{h} = \lim_{h \to 0} \frac{3x + 3h + 2 - 3x - 2}{h}$$

$$= \lim_{h \to 0} \frac{3h}{h} = \lim_{h \to 0} 3 = 3.$$

For $x > 1$ the value of $f(x)$ is equal to $6 - x$. A similar calculation shows that for $x > 1$ we have

$$f'(x) = \lim_{h \to 0} \frac{(6 - (x + h)) - (6 - x)}{h} = \lim_{h \to 0} \frac{6 - x - h - 6 + x}{h}$$

$$= \lim_{h \to 0} \frac{-h}{h} = \lim_{h \to 0} -1 = -1.$$

Therefore the derivative of the piecewise-defined function f is

$$f'(x) = \begin{cases} 3, & \text{if } x < 1 \\ \text{undefined}, & \text{if } x = 1 \\ -1, & \text{if } x > 1, \end{cases}$$

where "DNE" stands for "does not exist," representing the fact that the function f' is not defined at $x = 1$ (i.e., that there is no real number assigned to $f'(1)$). □

CHECKING THE ANSWER

The figure that follows at the left shows the graphs of the function f and its derivative f' from the previous example. We can see from the graph of f that the right derivative of the function at $x = 1$ is negative (think about the secant lines when $h > 0$) while the left derivative at $x = 1$ is positive and steeper (think about the secant lines when $h < 0$). In fact, since the pieces of this graph are linear, we can clearly see from the graph that the slopes of the secant lines from the right and the left are -1 and 3, respectively.

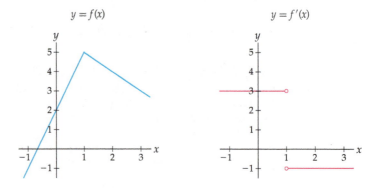

Because the left and right derivatives at $x = 1$ do not agree, there is no well-defined tangent line at $x = 1$; the (two-sided) derivative of f at $x = 1$ does not exist. Graphically, there is no *unique* tangent line that passes through the point $(1, f(1))$. In some sense, the graph of f has *two* directions at $x = 1$, one with slope -1 and one with slope 3. In the graph of f' at the right we see that the slopes of f are recorded correctly, with a slope of 3 for $x < 1$, a slope of -1 for $x > 1$, and an undefined slope for $x = 1$.

EXAMPLE 6   A function that fails to be continuous also fails to be differentiable

Consider the piecewise-defined function f given by the equation

$$f(x) = \begin{cases} x^2, & \text{if } x < 3 \\ 12, & \text{if } x \geq 3. \end{cases}$$

Argue graphically that f is not differentiable at $x = 3$ by examining secant lines on the graph of f from both the left and the right of $x = 3$.

SOLUTION

The graph of f looks like this:

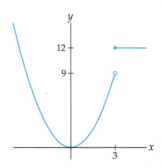

On the one hand, for $h > 0$, the secant lines over intervals of the form $[3, 3 + h]$ all have slope zero, since, on the interval $[3, \infty)$, the graph of f is a horizontal line. Therefore the right derivative of f at $x = 3$ must be $f'_+(3) = 0$. On the other hand, for $h < 0$, the secant lines over intervals of the form $[3 + h, 3]$ behave as shown in the following three figures:

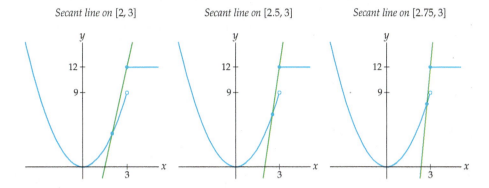

As $h \to 0^-$, we get a sequence of secant lines that are becoming more and more vertical. The slopes of these secant lines are large and positive, and approach ∞ as $h \to 0^-$. Therefore the left derivative $f'_-(3)$ does not exist. This means that f is not differentiable at $x = 3$, since its left derivative does not exist. □

EXAMPLE 7   **Approximating roots with local linearity and Newton's method**

Use local linearity to approximate a root of $f(x) = -x^3 - x + 1$.

SOLUTION

First of all, notice that by the Intermediate Value Theorem and the fact that $f(0) = 1$ is positive and $f(1) = -1$ is negative, the continuous function $f(x) = -x^3 - x + 1$ must have at least one root between $x = 0$ and $x = 1$. However, since we cannot factor $f(x)$, we are unable to solve for the root directly. Our approximation strategy will be to guess a location for a root, find the tangent line to the function at that point, and then look at the root of that tangent line. If the tangent line is a good local approximation for the function, then the root of the tangent line will be close to the root of the function if our initial guess is close enough. We then repeat the process, using the root of the tangent line as our next guess for the root of the function, to get better and better approximations of the root we are looking for. This strategy is known as **Newton's method** for approximating roots.

We'll start with $x_1 = 0$ as our first guess for a root of $f(x)$. Clearly this is not actually a root of the function, because $f(0) = 1$ is not equal to zero. Hopefully the tangent line at $x_1 = 0$ will help us make a better guess for a root. Before we compute the tangent line, we must find the derivative of $f(x) = -x^3 - x + 1$. We will leave the bulk of the computational details to the reader (see Exercise 40):

$$f'(x) = \lim_{h \to 0} \frac{f(x + h) - f(x)}{h}$$

$$= \lim_{h \to 0} \frac{(-(x + h)^3 - (x + h) + 1) - (-x^3 - x + 1)}{h}$$

$$= \cdots = -3x^2 - 1.$$

Therefore the derivative of f at $x_1 = 0$ is $f'(0) = -3(0)^2 - 1 = -1$, and thus the tangent line to $f(x)$ at $(0, f(0)) = (0, 1)$ has equation

$$y = f(0) + f'(0)(x - 0) = 1 + (-1)(x - 0) = 1 - x.$$

Although it is not easy to calculate a root of our original cubic $f(x) = -x^3 - x + 1$, it is easy to calculate that 1 is a root of the linear function $y = 1 - x$. Although $x_2 = 1$ is still not a root of $f(x)$ (since $f(1) = -1$), the hope is that repeating this process will get us closer and closer to a root. The following figure shows our initial guess of $(x_1, f(x_1)) = (0, 1)$, the root at $x = 1$ of the tangent line to $f(x)$ at $x_1 = 0$, and the new guess of $(x_2, f(x_2)) = (1, -1)$.

$f(x) = -x^3 - x + 1$ *with* $x_1 = 0$ *and* $x_2 = 1$

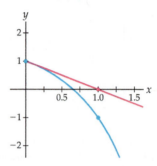

Repeating this process, we can use a root of the tangent line to $f(x)$ at x_2 to obtain a revised guess x_3 for a root. The tangent line passes through $(1, -1)$ and has slope $f'(1) = -3(1) - 1 = -4$, so its equation is

$$y = f(1) + f'(1)(x - 1) = -1 + (-4)(x - 1) = -4x + 3.$$

The root of this line, and our new guess for a root of $f(x)$, is $x_3 = \frac{3}{4}$; see the graph that follows. Note that already our third approximation, x_3, appears quite close to the actual root of $f(x)$:

$f(x) = -x^3 - x + 1$ *with* $x_2 = 1$ *and* $x_3 = \frac{3}{4}$

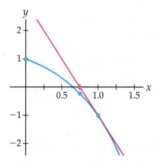

Repeating this process one more time, we obtain $x_4 \approx 0.686047$, which is in fact extremely close to the actual root of $x \approx 0.682328$. In Exercise 22 you will compare the accuracy of this method with that of the Bisection Method of finding roots that we used in Exercise 80 of Section 1.4.   □

TEST YOUR UNDERSTANDING

▶ What is the definition of the derivative of a function f at a point $x = c$? Give two answers, one with a limit that involves h and one with a limit that involves z.

▶ When calculating a derivative with the $h \to 0$ definition of the derivative, why is the limit always initially of the form $\frac{0}{0}$? What about when we use the $z \to x$ definition of the derivative?

▶ Is a function that is continuous at $x = c$ necessarily differentiable at $x = c$? Is a function that is differentiable at $x = c$ necessarily continuous at $x = c$?

▶ What is the connection between the Leibniz notation $\frac{dy}{dx}$ for the derivative and the difference quotient $\frac{\Delta y}{\Delta x}$?

▶ How can we express the fifth derivative of a function f at the point $x = 2$ in Leibniz notation? In "prime" notation?

EXERCISES 2.2

Thinking Back

Simplifying quotients: Simplify and rewrite the following expressions until you can cancel a common factor in the numerator and the denominator.

▶ $\dfrac{(x + h)^4 - x^4}{h}$

▶ $\dfrac{(x + h)^{-2} - x^{-2}}{h}$

▶ $\dfrac{z^4 - x^4}{z - x}$

▶ $\dfrac{z^{-2} - x^{-2}}{z - x}$

Limit calculations: Find each of the following limits.

▶ $\lim\limits_{h \to 0} \dfrac{(1 - h) - 1}{h}$

▶ $\lim\limits_{h \to 0} \dfrac{(3(-1 + h)^2 + 1) - 4}{h}$

▶ $\lim\limits_{h \to 0} \dfrac{\frac{1}{2 + h} - 0.5}{h}$

▶ $\lim\limits_{z \to 2} \dfrac{z^2 - 4}{z - 2}$

▶ $\lim\limits_{z \to 4} \dfrac{(1 - 3z) + 11}{z - 4}$

▶ $\lim\limits_{z \to 1} \dfrac{\frac{1}{z} - 1}{z - 1}$

Concepts

0. *Problem Zero:* Read the section and make your own summary of the material.

1. *True/False:* Determine whether each of the statements that follow is true or false. If a statement is true, explain why. If a statement is false, provide a counterexample.

  (a) *True or False:* $f'(x) = \dfrac{f(x + h) - f(x)}{h}$.

  (b) *True or False:* $f'(x) = \lim\limits_{x \to 0} \dfrac{f(x + h) - f(x)}{h}$.

  (c) *True or False:* $f'(x) = \lim\limits_{z \to 0} \dfrac{f(z) - f(x)}{z - x}$.

  (d) *True or False:* If $f(x) = x^3$, then $f(x + h) = x^3 + h$.

  (e) *True or False:* If $f(x) = x^3$, then $f'(x) = \lim\limits_{h \to 0} \dfrac{f(x^3 + h) - f(x)}{h}$.

  (f) *True or False:* A function f is differentiable at $x = c$ if and only if $f'_-(c)$ and $f'_+(c)$ both exist.

  (g) *True or False:* If f is continuous at $x = c$, then f is differentiable at $x = c$.

  (h) *True or False:* If f is not continuous at $x = c$, then f is not differentiable at $x = c$

2. *Examples:* Construct examples of the thing(s) described in the following. Try to find examples that are different than any in the reading.

  (a) The graph of a function that is continuous, but not differentiable, at $x = 2$.

  (b) The graph of a function that is left and right differentiable, but not differentiable, at $x = 3$.

  (c) The graph of a function that is differentiable on the interval $[-1, 1]$ but not differentiable at the point $x = 1$.

3. Use limits to give mathematical definitions for each of the following derivatives, first with the $h \to 0$ definition of the derivative, and then with the $z \to x$ definition:

  (a) the derivative of a function f at the point $x = 5$

  (b) the derivative of a function f

  (c) the right derivative of a function f at the point $x = -2$

4. Use limits to give mathematical definitions for:

  (a) the slope of the line tangent to the graph of a function f at the point $x = 4$

  (b) the line tangent to the graph of a function f at the point $x = 4$

  (c) the instantaneous rate of change of a function f at the point $x = 1$

  (d) the acceleration at time $t = 1.65$ of an object that moves with position function $s(t)$

5. Explain why the limits $\lim\limits_{h \to 0} \dfrac{f(x + h) - f(x)}{h}$ and $\lim\limits_{z \to x} \dfrac{f(z) - f(x)}{z - x}$ are the same for any function f. (*Hint: Consider the substitution $z = x + h$.*)

6. Explain why the limits $\lim\limits_{h \to 0} \dfrac{f(x + h) - f(x)}{h}$ and $\lim\limits_{z \to x} \dfrac{f(z) - f(x)}{z - x}$ are each initially in the form $\dfrac{0}{0}$. Why would cancelling a common factor of h or $z - x$ be likely to resolve this indeterminate form?

7. The function $f(x) = 4x^3 - 5x + 1$ is both continuous and differentiable at $x = 2$. Write these facts as limit statements.

8. The function $f(x) = 4 - x^2$ is both continuous and differentiable at $x = 1$. Write these facts as limit statements.

9. If $\lim\limits_{x \to c} \dfrac{f(x) - f(c)}{x - c}$ exists, what can you say about the differentiability of f at $x = c$? What can you say about the continuity of f at $x = c$?

10. Suppose $f(0) = 1$, $\lim\limits_{x \to 0^+} f(x) = 1$, $\lim\limits_{x \to 0^-} f(x) = 1$, $\lim\limits_{x \to 0^+} \dfrac{f(x) - f(0)}{x} = 3$, and $\lim\limits_{x \to 0^-} \dfrac{f(x) - f(0)}{x} = -2$.

  (a) Is f continuous and/or differentiable at $x = 0$? What about from the left or right?

  (b) Sketch a possible graph of f.

11. Suppose $f(1) = 3$, $\lim\limits_{x \to 1^-} f(x) = 3$, $\lim\limits_{x \to 1^+} f(x) = 3$, $\lim\limits_{h \to 0^-} \dfrac{f(1 + h) - f(1)}{h} = 2$, and $\lim\limits_{h \to 0^+} \dfrac{f(1 + h) - f(1)}{h} = 0$.

(a) Is f continuous and/or differentiable at $x = 1$? What about from the left or right?

(b) Sketch a possible graph of f.

12. Consider the function f graphed here:

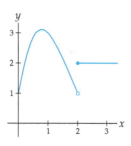

(a) Sketch secant lines from $(2, f(2))$ to $(2 + h, f(2 + h))$ on the graph of f, for the following values of h: $h = 0.5$, $h = 0.25$, $h = 0.1$, $h = -0.5$, $h = -0.25$, and $h = -0.1$.

(b) Use the secant lines you sketched in part (a) to graphically evaluate $\lim\limits_{h \to 0^+} \dfrac{f(2+h) - f(2)}{h}$ and $\lim\limits_{h \to 0^-} \dfrac{f(2+h) - f(2)}{h}$.

(c) Use your answer from part (b) to show that f is not differentiable at $x = 2$.

13. Sketch secant lines on a graph of $f(x) = |x|$, and use them to argue that the absolute value function is not differentiable at $x = 0$.

14. The two-sided symmetric difference approximation for the slope of a tangent line (see Exercise 20 in Section 2.1) can sometimes be misleading. Use a sequence of symmetric difference approximations to estimate the derivative of $f(x) = |x|$ at $x = 0$. What does your sequence of approximations suggest about $f'(0)$? Does this seem right?

15. Express the sentence "the derivative of $f(x) = 3x^2 - 1$ at $x = -4$ is -24" in each of the following notations:

(a) prime   (b) Leibniz   (c) operator

16. Express the sentence "the fourth derivative of $f(x) = 3x^2 - 1$ is equal to 0" in each of the following notations:

(a) prime   (b) Leibniz   (c) operator

Suppose $f(x) = x^3 - 2x + 1$, and let $y = f(x)$. It can be shown that $f'(x) = 3x^2 - 2$. Use this information to determine the expressions in Exercises 17 and 18. (*Note: No differentiation will be necessary, since the derivatives are given. These problems are just a test of your ability to interpret derivative notation.*)

17. $\dfrac{dy}{dx}$, $\dfrac{df}{dx}$, $\dfrac{d}{dx}(f(x))$, and $\dfrac{dy}{dx}(f(x))$.

18. $\dfrac{dy}{dx}\Big|_{x=2}$, $\dfrac{df}{dx}\Big|_{x=2}$, $\dfrac{d}{dx}(f(x))\Big|_{x=2}$, and $\dfrac{dy}{dx}(f(2))$

Express each of the following limit statements as delta–epsilon statements:

19. $f'(3) = \lim\limits_{h \to 0} \dfrac{(3 + h)^2 - 9}{h}$

20. $f'(3) = \lim\limits_{x \to 3} \dfrac{x^2 - 9}{x - 3}$

21. Suppose that you know that the derivative of $f(x) = \sqrt{x}$ is equal to $f'(x) = \dfrac{1}{2\sqrt{x}}$. Use this fact, local linearity, and the fact that $\sqrt{4} = 2$ to approximate the value of $\sqrt{4.1}$. How close is your approximation with the approximation of $\sqrt{4.1}$ that you can find with a calculator? (*Hint: Consider the tangent line to $f(x)$ at $x = 4$, and use it to approximate the function nearby.*)

22. Suppose you wish to find a root of $f(x) = x^3 + x^2 + x + 1$ in the interval $[-3, 2]$. Compare the accuracy of Newton's method as applied in Example 7 with the accuracy of the Bisection Method used in Exercise 80 of Section 1.4 for the same function. Which method gets closest to the root in three iterations?

Skills

Use (a) the $h \to 0$ definition of the derivative and then (b) the $z \to c$ definition of the derivative to find $f'(c)$ for each function f and value $x = c$ in Exercises 23–32.

23. $f(x) = 1 - 5x$, $x = 2$

24. $f(x) = 3$, $x = -1$

25. $f(x) = 4x + 3$, $x = -2$

26. $f(x) = \dfrac{2 - x}{3}$, $x = 2$

27. $f(x) = x^2$, $x = -3$

28. $f(x) = x^3$, $x = 1$

29. $f(x) = \dfrac{1}{x}$, $x = -1$

30. $f(x) = \dfrac{1}{x^2}$, $x = 2$

31. $f(x) = 1 - x^3$, $x = -1$

32. $f(x) = x^4 + 1$, $x = 2$

Use the definition of the derivative to find f' for each function f in Exercises 33–44.

33. $f(x) = 5x - 2$

34. $f(x) = 3 - x$

35. $f(x) = -1$

36. $f(x) = \dfrac{3x + 1}{2}$

37. $f(x) = -2x^2$

38. $f(x) = 4 + x^2$

39. $f(x) = x^3 + 2$

40. $f(x) = -x^3 - x + 1$

41. $f(x) = \dfrac{2}{x + 1}$

42. $f(x) = \dfrac{2}{1 - x}$

43. $f(x) = \dfrac{1}{x^2}$

44. $f(x) = \dfrac{1}{x^3}$

Use the definition of the derivative to find the derivatives described in Exercises 45–48.

45. Find $\dfrac{d}{dx}(2x^3)$, $\dfrac{d^2}{dx^2}(2x^3)$, and $\dfrac{d^3}{dx^3}(2x^3)$.

46. Find $\dfrac{d}{dx}(2x^3)\Big|_3$, $\dfrac{d^2}{dx^2}(2x^3)\Big|_1$, and $\dfrac{d^3}{dx^3}(2x^3)\Big|_{-2}$.

47. Given $f^{(3)}(x) = 3x^2 + 1$, find $\dfrac{d^4 f}{dx^4}$ and $\dfrac{d^4 f}{dx^4}\Big|_2$.

48. Given $\dfrac{d^2 f}{dx^2} = 2 - x^2$, find $f^{(3)}(4)$ and $\dfrac{d^4}{dx^4}(f(x))$.

Use the definition of the derivative to find the equations of the lines described in Exercises 49–54.

49. The tangent line to $f(x) = x^2$ at $x = -3$.

50. The tangent line to $f(x) = x^2$ at $x = 0$.

51. The line tangent to the graph of $y = 1 - x - x^2$ at the point $(1, -1)$.

52. The line tangent to the graph of $y = 4x + 3$ at the point $(-2, -5)$.

53. The line that passes through the point $(3, 2)$ and is parallel to the tangent line to $f(x) = \dfrac{1}{x}$ at $x = -1$.

54. The line that is perpendicular to the tangent line to $f(x) = x^4 + 1$ at $x = 2$ and also passes through the point $(-1, 8)$.

For each function f graphed in Exercises 55–58, determine the values of x at which f fails to be continuous and/or differentiable. At such points, determine any left or right continuity or differentiability. Sketch secant lines supporting your answers.

55.

56.

57.

58.

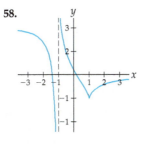

In Exercises 59–70, determine whether or not f is continuous and/or differentiable at the given value of x. If not, determine any left or right continuity or differentiability. For the last four functions, use graphs instead of the definition of the derivative.

59. $f(x) = \dfrac{1}{x}$, $x = 0$

60. $f(x) = -3x^2$, $x = 0$

61. $f(x) = |x^2 - 4|$, $x = 2$

62. $f(x) = |x^2 - 4|$, $x = -2$

63. $f(x) = \begin{cases} x + 4, & \text{if } x < 2 \\ 3x, & \text{if } x \geq 2, \end{cases}$ $x = 2$

64. $f(x) = \begin{cases} x^2 - 3, & \text{if } x < 3 \\ x + 2, & \text{if } x \geq 3, \end{cases}$ $x = 3$

65. $f(x) = \begin{cases} x^2, & \text{if } x \leq 1 \\ 2x - 1, & \text{if } x > 1, \end{cases}$ $c = 1$

66. $f(x) = \begin{cases} x^2, & \text{if } x \leq 1 \\ 2x + 4, & \text{if } x > 1, \end{cases}$ $c = 1$

67. $f(x) = \begin{cases} x \sin\left(\dfrac{1}{x}\right), & \text{if } x \neq 0 \\ 0, & \text{if } x = 0, \end{cases}$ $c = 0$

68. $f(x) = \begin{cases} x^2 \sin\left(\dfrac{1}{x}\right), & \text{if } x \neq 0 \\ 0, & \text{if } x = 0, \end{cases}$ $c = 0$

69. $f(x) = \begin{cases} 1, & \text{if } x \text{ rational} \\ x + 1, & \text{if } x \text{ irrational}, \end{cases}$ $c = 1$

70. $f(x) = \begin{cases} x^2, & \text{if } x \text{ rational} \\ 2x - 1, & \text{if } x \text{ irrational}, \end{cases}$ $c = 1$

For each function $f(x)$ and interval $[a, b]$ in Exercises 71–76, use the Intermediate Value Theorem to argue that the function must have at least one real root on $[a, b]$. Then apply Newton's method to approximate that root.

71. $f(x) = x^2 - 5$, $[a, b] = [1, 3]$

72. $f(x) = x^2 - 2$, $[a, b] = [1, 2]$

73. $f(x) = x^3 - 3x + 1$, $[a, b] = [0, 1]$

74. $f(x) = x^3 - 3x + 1$, $[a, b] = [1, 2]$

75. $f(x) = x^3 + 1$, $[a, b] = [-2, 1]$

76. $f(x) = x^4 - 2$, $[a, b] = [1, 2]$

Applications

77. In Example 4 of Section 2.1, we saw that a watermelon dropped from a height of 100 feet will be $s(t) = -16t^2 + 100$ feet above the ground t seconds after it is dropped. In that example, we approximated the velocity of such a watermelon at time $t = 1$ by calculating a sequence of average rates of change. Now we can calculate this velocity exactly, using the definition of the derivative. Do so, and compare the exact answer to the approximation we found earlier.

78. On a long road trip you are driving along a straight portion of Route 188. Suppose that t hours after entering Nevada your distance from the Donut Hole is $s(t) = -10t^2 - 40t + 120$ miles.

(a) How long will it take you to reach the Donut Hole after entering Nevada?

(b) Find your velocity $v(t)$ as you drive toward the Donut Hole.

(c) Are you accelerating or decelerating as you approach the Donut Hole? At what rate?

79. Suppose your position $s(t)$ as you drive north along a straight highway is as shown in the graph that follows at the left, with t measured in hours and s measured in miles.

(a) Sketch a graph of your velocity $v(t)$, and use the graph to describe your velocity over the course of the two-hour drive.

(b) Sketch a graph of your acceleration $a(t)$, and use the graph to describe your acceleration over the course of the two-hour drive.

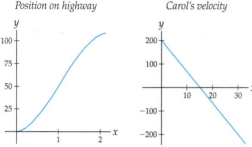

Position on highway   Carol's velocity

80. While Carol is walking along a straight north-south forest path, her velocity, in feet per minute, after t minutes is given by the preceding graph at the right. Suppose the "positive" direction is north.

(a) Describe the sort of walk Carol must have taken to have this velocity graph. Be sure your description explains the physical significance of the fact that her velocity is zero at $t = 15$ minutes and the fact that her velocity is negative for the second half of her walk.

(b) Find Carol's average acceleration over the 30-minute walk. Was her acceleration constant over the duration of her walk? Why or why not?

(c) What was Carol's average velocity over her entire walk? Why does your answer make sense?

81. To save up for a car, you take a job working 10 hours a week at the school library. For the first six weeks the library pays you $8.00 an hour. After that you earn $11.50 an hour. You put all of the money you earn each week into a savings account. On the day you start work your savings account already holds $200.00. Let $S(t)$ be the function that describes the amount in your savings account t weeks after your library job begins.

(a) Find the values of $S(3)$, $S(6)$, $S(8)$, $S'(3)$, $S'(6)$, and $S'(8)$, if possible, and describe their meanings in practical terms. If it is not possible to find one or more of these values, explain why.

(b) Write an equation for the function $S(t)$. (*Hint: $S(t)$ will be a piecewise-defined function.*) Be sure that your equation correctly produces the values you calculated in part (a).

(c) Sketch a labeled graph of $S(t)$. By looking at this graph, determine whether $S(t)$ is continuous and whether $S(t)$ is differentiable. Explain the practical significance of your answers.

(d) Show algebraically that $S(t)$ is a continuous function, but not a differentiable function.

Proofs

82. Use the definition of the derivative to prove that every quadratic function $f(x) = ax^2 + bx + c$ has the property that its graph has a horizontal tangent line at the point $x = -\dfrac{b}{2a}$.

83. Use the definition of the derivative to prove that our concept of slope for linear functions matches the slope that is defined by the derivative. In other words, show that if $f(x) = mx + b$ is any linear function, then $f'(x) = m$.

84. Use Problem 83 to prove that a linear function is its own tangent line at every point. In other words, show that if $f(x) = mx + b$ is any linear function, then the tangent line to f at any point $x = c$ is given by $y = mx + b$.

85. Use the mathematical definition of a tangent line and the point-slope form of a line to show that if f is differentiable at $x = c$, then the tangent line to f at $x = c$ is given by the equation $y = f'(c)(x - c) + f(c)$.

86. Prove that if a function f is differentiable at $x = c$, then f is continuous at $x = c$.

(a) We are given that f is differentiable at $x = c$. Use the alternative definition of the derivative to write down what that statement means.

(b) We want to show that f is continuous at $x = c$. Use the definition of continuity to show that this statement is equivalent to the statement $\lim\limits_{x \to c} (f(x) - f(c)) = 0$.

(c) Now use part (a) to show that $\lim\limits_{x \to c} (f(x) - f(c)) = 0$. (*Hint: Multiply $(f(x) - f(c))$ by $\dfrac{x - c}{x - c}$ and use the product rule for limits.*)

87. Use the definition of two-sided and one-sided derivatives, together with properties of limits, to prove that $f'(c)$ exists if and only if $f'_-(c)$ and $f'_+(c)$ exist and are equal.

88. Show that if a function $y = f(x)$ is differentiable at x_0 and

$$\Delta y = f(x_0 + \Delta x) - f(x_0),$$

then

$$\Delta y = f'(x_0)\Delta x + \epsilon \Delta x,$$

where ϵ is a function satisfying $\lim\limits_{\Delta x \to 0} \epsilon = 0$.

Thinking Forward

Derivatives of power functions: After differentiating a few power functions, you may have noticed a pattern emerging. In the following exercises we will investigate a possible formula for differentiating power functions.

▶ Use the $z \to x$ definition of the derivative to show that $\frac{d}{dx}(x^4) = 4x^3$.

▶ Use the $z \to x$ definition of the derivative to show that $\frac{d}{dx}(x^8) = 8x^7$. *(Hint: The following factoring formula will come in handy: If n is a positive integer, then $z^n - x^n = (z - x)(z^{n-1} + z^{n-2}x + z^{n-3}x^2 + \cdots + z^2 x^{n-3} + zx^{n-2} + x^{n-1})$.)*

▶ Use the preceding two derivative formulas to make a conjecture about a formula for $\frac{d}{dx}(x^n)$, where n is a positive integer.

Derivatives of combinations of functions: We have already seen that the limit of a sum is the sum of the limits and that the limit of a product is the product of the limits. Do derivatives also interact well with sums and products?

▶ Use the definition of the derivative (or exercises done previously in this section) to find (a) $\frac{d}{dx}(3x)$, (b) $\frac{d}{dx}(x^2)$, and (c) $\frac{d}{dx}(3x + x^2)$. Use your answers to make a conjecture as to whether or not $\frac{d}{dx}(f(x) + g(x)) = \frac{df}{dx} + \frac{dg}{dx}$.

▶ Use the definition of the derivative (or exercises done previously in this section) to find (a) $\frac{d}{dx}(x - 3)$, (b) $\frac{d}{dx}(2x + 1)$, and (c) $\frac{d}{dx}((x - 3)(2x + 1))$. Use your answers to make a conjecture as to whether or not $\frac{d}{dx}(f(x)g(x)) = \left(\frac{df}{dx}\right)\left(\frac{dg}{dx}\right)$.

2.3 RULES FOR CALCULATING BASIC DERIVATIVES

▶ Formulas for differentiating constant, identity, linear, and power functions

▶ Rules for differentiating constant multiples, sums, products, and quotients

▶ Using differentiation rules to quickly calculate derivatives and antiderivatives

Derivatives of Linear Functions

You may have noticed by now that using the definition of the derivative to calculate derivatives can be rather tedious. Since derivatives will be used often throughout this course, we need to develop a faster method of calculating them. Let's start with linear functions. If f is a linear function, then it has slope m everywhere, and therefore its derivative is constantly m. Since constant and identity functions are linear functions with slopes 0 and 1, their derivatives are constantly 0 and 1, respectively. These are our first differentiation rules.

THEOREM 2.8   **Derivatives of Constant, Identity, and Linear Functions**

For any real numbers k, m, and b,

(a) $\frac{d}{dx}(k) = 0$ 　　　　　(b) $\frac{d}{dx}(x) = 1$ 　　　　　(c) $\frac{d}{dx}(mx + b) = m$

With this differentiation rule we can find derivatives of linear functions very quickly, without having to consider the definition of the derivative. For example, $\frac{d}{dx}(3) = 0$, $\frac{d}{dx}(3x + 1) = 3$, and $\frac{d}{dx}(2x - 99) = 2$.

Proof. We need only prove the third formula, since first two are special cases of that formula. Our proof is just a general calculation with the definition of the derivative: If m and b are any constants and $f(x) = mx + b$, then

$$f'(x) = \lim_{h \to 0} \frac{f(x+h) - f(x)}{h} \qquad \leftarrow \text{ definition of the derivative}$$

$$= \lim_{h \to 0} \frac{(m(x+h) + b) - (mx + b)}{h} \qquad \leftarrow \text{ use formula for } f(x)$$

$$= \lim_{h \to 0} \frac{mx + mh + b - mx - b}{h} \qquad \leftarrow \text{ algebra}$$

$$= \lim_{h \to 0} \frac{mh}{h} = \lim_{h \to 0} m = m. \qquad \leftarrow \text{ limit of a constant}$$

The Power Rule

You may have already noticed a particular pattern for the derivatives of power functions. In various examples and exercises in the previous section, we have seen that

$$\frac{d}{dx}(x^3) = 3x^2, \quad \frac{d}{dx}(x^4) = 4x^3, \quad \frac{d}{dx}\left(\frac{1}{x^2}\right) = \frac{-2}{x^3}, \quad \text{and} \quad \frac{d}{dx}\left(\frac{1}{x^3}\right) = \frac{-3}{x^4}.$$

The pattern becomes clear if we write these derivative formulas in exponent notation:

$$\frac{d}{dx}(x^3) = 3x^2, \quad \frac{d}{dx}(x^4) = 4x^3, \quad \frac{d}{dx}(x^{-2}) = -2x^{-3}, \quad \text{and} \quad \frac{d}{dx}(x^{-3}) = -3x^{-4}.$$

From these examples it appears that, to take the derivative of x^k, we bring down the exponent k to the front of the expression and then decrease the exponent by one, to get kx^{k-1}. This is in fact the case in general:

THEOREM 2.9

The Power Rule

For any nonzero rational number k, $\frac{d}{dx}(x^k) = kx^{k-1}$.

Although we require that k be a rational number in this formula, the power rule is actually true for any nonzero real number k. If $k = 0$, then x^0 is the constant function 1, which we already know how to differentiate.

Considering the algebra involved in applying the definition of the derivative, it is a relief to have such a simple formula for finding derivatives of power functions! With this formula we can quickly say, for example, that $\frac{d}{dx}(x^{15}) = 15x^{14}$ or that $\frac{d}{dx}(x^{-1000}) = -1000x^{-1001}$, or even that $\frac{d}{dx}(x^{17/12}) = \frac{17}{12}x^{5/12}$. We can also find higher derivatives very easily; for example, if $f(x) = x^4$, then $f'(x) = 4x^3, f''(x) = 12x^2, f'''(x) = 24x, f^{(4)}(x) = 24$, and $f^{(5)}(x) = 0$. These functions are graphed next; each one is the associated slope function for the one before.

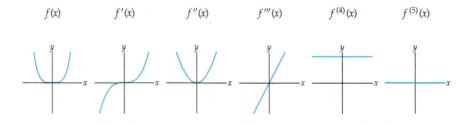

$$f(x) \qquad\qquad f'(x) \qquad\qquad f''(x) \qquad\qquad f'''(x) \qquad\qquad f^{(4)}(x) \qquad\qquad f^{(5)}(x)$$

Proof. We will prove the power rule by cases. The case where k is a positive integer is proved in what follows. The case where k is a negative integer is left for Exercise 86. The case where k is a general rational number $\frac{p}{q}$ must be put off until after we know about implicit differentiation in Section 2.4.

Given any positive integer k, we need to apply the definition of the derivative to find $\frac{d}{dx}(x^k)$. It turns out that the $z \to x$ definition of the derivative is easier to use in this case. Along the way we will require the following factoring formula: If k is a positive integer, then

$$z^k - x^k = (z-x)(z^{k-1} + z^{k-2}x + z^{k-3}x^2 + \cdots + z^2 x^{k-3} + zx^{k-2} + x^{k-1}).$$

(You can verify this formula by simply multiplying out the right-hand side of the equation; everything except the first and last terms will cancel, and you will obtain the left side of the equation.) Now using the definition of the derivative, we have

$$\frac{d}{dx}(x^k) = \lim_{z \to x} \frac{z^k - x^k}{z - x} \qquad \leftarrow \text{definition of the derivative}$$

$$= \lim_{z \to x} \frac{(z-x)(z^{k-1} + z^{k-2}x + z^{k-3}x^2 + \ldots + zx^{k-2} + x^{k-1})}{z - x} \qquad \leftarrow \text{factoring formula}$$

$$= \lim_{z \to x}(z^{k-1} + z^{k-2}x + z^{k-3}x^2 + \ldots + zx^{k-2} + x^{k-1}) \qquad \leftarrow \text{cancellation}$$

$$= x^{k-1} + x^{k-2}x + x^{k-3}x^2 + \ldots + xx^{k-2} + x^{k-1}. \qquad \leftarrow \text{evaluate limit}$$

$$= x^{k-1} + x^{k-1} + x^{k-1} + \ldots + x^{k-1} + x^{k-1} \qquad \leftarrow \text{there are } k \text{ of these}$$

$$= kx^{k-1}.$$

The Constant Multiple and Sum Rules

In Chapter 1 we saw that limits behave well with respect to arithmetic combinations of functions. For example, limits commute with sums: The limit of a sum is the sum of the limits, provided that the limits involved exist. Is the same thing true for derivatives? For constant multiples and sums, the answer is yes:

THEOREM 2.10

Derivatives of Constant Multiples and Sums of Functions

If f and g are functions and k is a constant, then for all x where the functions involved are differentiable, we have the following differentiation formulas:

$$\textbf{\textit{Constant Multiple Rule:}} \quad (kf)'(x) = kf'(x)$$

$$\textbf{\textit{Sum Rule:}} \quad (f+g)'(x) = f'(x) + g'(x)$$

$$\textbf{\textit{Difference Rule:}} \quad (f-g)'(x) = f'(x) - g'(x)$$

The notation $(kf)'(x)$ indicates that we are differentiating the function kf with respect to x. In Leibniz notation we would write this differentiation as $\frac{d}{dx}(kf(x))$. The difference rule is of course just a combination of the sum and constant multiple rules, since $f(x) - g(x) = f(x) + (-g(x))$.

These rules mean that we can factor out constants and split up sums when calculating derivatives. For example, derivatives of sums or constant multiples of the functions $f(x) = x^2$ and $g(x) = x^4$ can be expressed as sums or constant multiples of their derivatives

$f'(x) = 2x$ and $g'(x) = 4x^3$, as illustrated here:

$$\frac{d}{dx}(5x^2 + 2x^4) = \frac{d}{dx}(5x^2) + \frac{d}{dx}(2x^4) \qquad \leftarrow \text{sum rule}$$

$$= 5\frac{d}{dx}(x^2) + 2\frac{d}{dx}(x^4) \qquad \leftarrow \text{constant multiple rule}$$

$$= 5(2x) + 2(4x^3) \qquad \leftarrow \text{power rule}$$

$$= 10x + 8x^3.$$

Proof. We will use the definition of the derivative to prove (a) the constant multiple and (b) sum rules in what follows. The difference rule can also be proved with the definition of the derivative, or by applying the sum and constant multiple rules; see Exercise 87.

(a) The proof of the constant multiple rule uses the definition of the derivative and the constant multiple rule for limits. Given a function $f(x)$ and a constant k, we wish to show that the derivative of the function $kf(x)$ is the same as k times the derivative of $f(x)$. We'll start from the left and work to the right:

$$(kf)'(x) = \lim_{h \to 0} \frac{kf(x+h) - kf(x)}{h} \qquad \leftarrow \text{definition of the derivative for } kf(x)$$

$$= \lim_{h \to 0} \frac{k(f(x+h) - f(x))}{h} \qquad \leftarrow \text{factor out } k \text{ from numerator}$$

$$= \lim_{h \to 0} \left(k \left(\frac{f(x+h) - f(x)}{h} \right) \right) \qquad \leftarrow \text{factor out } k \text{ from quotient}$$

$$= k \left(\lim_{h \to 0} \frac{f(x+h) - f(x)}{h} \right) \qquad \leftarrow \text{constant multiple rule for limits}$$

$$= kf'(x). \qquad \leftarrow \text{definition of the derivative for } f(x)$$

(b) Similarly, the proof of the sum rule uses the definition of the derivative and the sum rule for limits. We wish to show that the derivative of the function $f + g$ is the sum of the derivative of f and the derivative of g. We will work from the left to the right; our goal is to use the definition of the derivative to write the left-hand statement as a limit and then use algebra and the sum rule for limits to split this limit into two limits, one of which will be the derivative of f, and one the derivative of g:

$$(f+g)'(x) = \lim_{h \to 0} \frac{(f(x+h) + g(x+h)) - (f(x) + g(x))}{h} \qquad \leftarrow \text{definition of derivative}$$

$$= \lim_{h \to 0} \frac{f(x+h) + g(x+h) - f(x) - g(x)}{h} \qquad \leftarrow \text{simplify}$$

$$= \lim_{h \to 0} \frac{(f(x+h) - f(x)) + (g(x+h) - g(x))}{h} \qquad \leftarrow \text{reordering terms}$$

$$= \lim_{h \to 0} \left(\frac{f(x+h) - f(x)}{h} + \frac{g(x+h) - g(x)}{h} \right) \qquad \leftarrow \text{algebra}$$

$$= \left(\lim_{h \to 0} \frac{f(x+h) - f(x)}{h} \right) + \left(\lim_{h \to 0} \frac{g(x+h) - g(x)}{h} \right) \qquad \leftarrow \text{sum rule for limits}$$

$$= f'(x) + g'(x). \qquad \leftarrow \text{definition of derivative} \ \blacksquare$$

The Product and Quotient Rules

We now know that derivatives interact nicely with constant multiples and sums; for example, the derivative of a sum is the sum of the derivatives. Do derivatives also commute with products and quotients? Sadly, the answer is no. For example, consider the functions $f(x) = x^2 + 1$ and $g(x) = x^3$, whose derivatives are $f'(x) = 2x$ and $g'(x) = 3x^2$. Now consider the product $f(x)g(x) = (x^2 + 1)(x^3) = x^5 + x^3$; by the power and sum rules, the derivative of that product is $5x^4 + 3x^2$. However, this is clearly not equal to the product $f'(x)g'(x) = (2x)(3x^2) = 6x^3$.

In general, the derivative of a product is not necessarily the product of its component derivatives. Similarly, the derivative of a quotient is not in general the quotient of its component derivatives. However, we can write the derivative of a product fg or quotient $\frac{f}{g}$ in terms of f, g, f', and g', with the following, somewhat surprising, formulas:

THEOREM 2.11

Derivatives of Products and Quotients of Functions

If f and g are functions, then for all x such that both f and g are differentiable, we have the following differentiation formulas:

Product Rule: $(fg)'(x) = f'(x)\, g(x) + f(x)\, g'(x)$

Quotient Rule: $\left(\dfrac{f}{g}\right)'(x) = \dfrac{f'(x)\, g(x) - f(x) g'(x)}{(g(x))^2}$

Some people remember the product rule by remembering the pattern of differentiating one piece and not the other in both possible ways and then taking the sum. Some people remember the quotient rule with a phrase like "lo d-hi minus hi d-lo over lo lo," although it might be easier just to remember the quotient rule than to try and remember that! In any case, you will have to memorize these differentiation formulas, because they will be needed often throughout this book.

Armed with the product and quotient rules, we can now differentiate a lot more functions. For example, with our earlier example of $f(x) = x^2 + 1$ and $g(x) = x^3$, we can differentiate $f(x)g(x)$ without multiplying it out first:

$$\frac{d}{dx}(f(x)g(x)) = f'(x)\, g(x) + f(x)\, g'(x) \qquad \leftarrow \text{product rule}$$

$$= (2x)(x^3) + (x^2 + 1)(3x^2) \qquad \leftarrow f'(x) = 2x,\, g(x) = 3x^2$$

$$= 2x^4 + 3x^4 + 3x^2 \qquad \leftarrow \text{algebra}$$

$$= 5x^4 + 3x^2.$$

This is exactly what we found earlier by multiplying out $f(x)g(x)$ first and then taking the derivative.

Proof. The proof of the product rule is yet another general calculation using the definition of the derivative, with a small twist that we will explain shortly. A similar method of proof works for the quotient rule and is left to Exercise 89. We will have an easier way to prove the quotient rule once we learn the chain rule in Section 2.4.

Suppose f and g are differentiable functions. We wish to show that $(fg)' = f'g + fg'$. We will apply the definition of the derivative to the left side of that equation and then use a long string of algebra and limit rules to rewrite it in terms of f', g, f, and g'. This will be made possible in the second step of the following calculation by adding and subtracting the same expression, $f(x)g(x+h)$, from the numerator.

$$(fg)'(x) = \lim_{h \to 0} \frac{f(x+h)g(x+h) - f(x)g(x)}{h} \qquad \leftarrow \text{derivative}$$

$$= \lim_{h \to 0} \frac{f(x+h)g(x+h) - f(x)g(x+h) + f(x)g(x+h) - f(x)g(x)}{h} \qquad \leftarrow \text{algebra}$$

$$= \lim_{h \to 0} \frac{(f(x+h) - f(x))g(x+h) + f(x)(g(x+h) - g(x))}{h} \qquad \leftarrow \text{factoring}$$

$$= \lim_{h \to 0} \left(\frac{f(x+h) - f(x)}{h} \cdot g(x+h) + f(x) \cdot \frac{g(x+h) - g(x)}{h} \right) \qquad \leftarrow \text{algebra}$$

$$= \left(\lim_{h \to 0} \frac{f(x+h) - f(x)}{h} \right) \lim_{h \to 0} g(x+h) + \lim_{h \to 0} f(x) \left(\lim_{h \to 0} \frac{g(x+h) - g(x)}{h} \right) \qquad \leftarrow \text{limit rules}$$

$$= \left(\lim_{h \to 0} \frac{f(x+h) - f(x)}{h} \right) g(x) + f(x) \left(\lim_{h \to 0} \frac{g(x+h) - g(x)}{h} \right) \qquad \leftarrow \text{continuity}$$

$$= f'(x)g(x) + f(x)g'(x). \qquad \leftarrow \text{derivative}$$

The step labeled "continuity" follows from the fact that $g(x)$ is differentiable and therefore continuous, because that is what allows us to say that $\lim_{h \to 0} g(x+h) = g(x)$. Notice that throughout the entire calculation, our goal was to extract the expressions that represent the derivatives of f and g. The algebra steps were not meant to "simplify"; they were meant to get us closer to the final form of $f'g + fg'$. ∎

Examples and Explorations

EXAMPLE 1

Determining whether the differentiation rules apply

Find the derivatives of each of the functions that follow, using the differentiation rules from this section, if possible. If it is not possible, explain why not.

(a) $f(x) = 2(5^x)$ **(b)** $g(x) = 3^4$ **(c)** $h(x) = (2x+3)^2$ **(d)** $k(x) = \sqrt{3x^2 + 1}$

SOLUTION

(a) We cannot differentiate $f(x) = 2(5^x)$ with any of the rules from this section. In particular, the power rule applies only to power functions of the form x^k, where the variable x is in the base and a constant k is in the exponent. The function 5^x is not a power function, but an exponential function, which we will see how to differentiate in Chapter 5.

(b) Since $g(x) = 3^4 = 81$ is a constant function, $\frac{d}{dx}(3^4) = 0$.

(c) $h(x) = (2x+3)^2$ is a composition of functions, and we do not yet have a rule for differentiating compositions. Therefore we must use algebra to expand the function first. We could write the function as $(2x+3)(2x+3)$ and use the product rule, or we could just expand $(2x+3)^2$ entirely and then need only the sum and constant multiple rules. We present the latter approach here:

$$\frac{d}{dx}((2x+3)^2) = \frac{d}{dx}(4x^2 + 12x + 9) \qquad \leftarrow \text{algebra}$$

$$= \frac{d}{dx}(4x^2) + \frac{d}{dx}(12x) + \frac{d}{dx}(9) \qquad \leftarrow \text{sum rule}$$

$$= 4\frac{d}{dx}(x^2) + 12\frac{d}{dx}(x) + \frac{d}{dx}(9) \qquad \leftarrow \text{constant multiple rule}$$

$$= 4(2x^1) + 12(1) + 0 = 8x + 12. \qquad \leftarrow \text{power rule}$$

(d) The function $k(x) = \sqrt{3x^2 + 1}$ is a composition that cannot be simplified, and we do not yet have a rule for differentiating compositions. (Note in particular that this function is *not* equal to $\sqrt{3x} + 1$, because square roots do not distribute over sums; there is no useful simplification that can be done to this function.) We will see how to differentiate such functions in Section 2.4.   □

EXAMPLE 2

Finding derivatives with and without simplifying first

Differentiate each of the following functions with and without the quotient rule:

(a) $f(x) = \dfrac{4}{3x^2}$ **(b)** $g(x) = \dfrac{x^2}{x^{-3}\sqrt{x}}$ **(c)** $h(x) = \dfrac{x^7 + \sqrt{x}}{x^3}$

SOLUTION

(a) Applying the quotient rule gives us

$$\frac{d}{dx}\left(\frac{4}{3x^2}\right) = \frac{\frac{d}{dx}(4) \cdot (3x^2) - (4) \cdot \frac{d}{dx}(3x^2)}{(3x^2)^2} \qquad \leftarrow \text{quotient rule}$$

$$= \frac{(0)(3x^2) - (4)(6x)}{9x^4} \qquad \leftarrow \text{constant and power rules}$$

$$= \frac{-24x}{9x^4} = -\frac{8}{3}x^{-3}. \qquad \leftarrow \text{algebra}$$

Alternatively, we could simplify f first and then apply the constant multiple and power rules. We of course will get the same final answer:

$$\frac{d}{dx}\left(\frac{4}{3x^2}\right) = \frac{d}{dx}\left(\frac{4}{3}x^{-2}\right) \qquad \leftarrow \text{algebra}$$

$$= \frac{4}{3}\frac{d}{dx}(x^{-2}) \qquad \leftarrow \text{constant multiple rule}$$

$$= \frac{4}{3}(-2x^{-3}) = -\frac{8}{3}x^{-3}. \qquad \leftarrow \text{power rule, algebra}$$

(b) If we apply differentiation rules immediately without simplifying first, we need to use both the quotient and product rules:

$$\frac{d}{dx}\left(\frac{x^2}{x^{-3}\sqrt{x}}\right) = \frac{\frac{d}{dx}(x^2) \cdot (x^{-3}\sqrt{x}) - (x^2) \cdot \frac{d}{dx}(x^{-3}\sqrt{x})}{(x^{-3}\sqrt{x})^2} \qquad \leftarrow \text{quotient rule}$$

$$= \frac{(2x)(x^{-3}\sqrt{x}) - (x^2)\left(-3x^{-4}\sqrt{x} + x^{-3}\left(\frac{1}{2}x^{-1/2}\right)\right)}{(x^{-3}\sqrt{x})^2}. \qquad \leftarrow \text{more rules}$$

However, if we do some preliminary algebra, then the differentiation steps and final simplification will both be much, much simpler:

$$\frac{d}{dx}\left(\frac{x^2}{x^{-3}\sqrt{x}}\right) = \frac{d}{dx}(x^2 x^3 x^{-1/2}) \qquad \leftarrow \text{algebra}$$

$$= \frac{d}{dx}(x^{9/2}) \qquad \leftarrow \text{algebra}$$

$$= \frac{9}{2}x^{7/2}. \qquad \leftarrow \text{power rule}$$

In Exercise 17 you will show that the preceding two answers are equal. Students often want to differentiate as soon as possible, since the differentiation rules can be easier to implement than algebraic simplification. However, in many cases, in the long run, simplifying before differentiating ends up saving a lot of even messier algebra later on. Look at the end result of the two calculations we just did, and ask yourself which one you would rather work with!

(c) Applying the quotient rule first, we have

$$\frac{d}{dx}\left(\frac{x^7 + \sqrt{x}}{x^3}\right) = \frac{\frac{d}{dx}(x^7 + \sqrt{x})\cdot(x^3) - (x^7 + \sqrt{x})\cdot\frac{d}{dx}(x^3)}{(x^3)^2} \qquad \leftarrow \text{quotient rule}$$

$$= \frac{\left(7x^6 + \frac{1}{2}x^{-1/2}\right)(x^3) - (x^7 + \sqrt{x})(3x^2)}{x^6}. \qquad \leftarrow \text{sum, power rules}$$

Alternatively, we could do some algebra before differentiating and then apply the sum and power rules. This gives us a simpler but equivalent answer:

$$\frac{d}{dx}\left(\frac{x^7 + \sqrt{x}}{x^3}\right) = \frac{d}{dx}\left(\frac{x^7}{x^3} + \frac{x^{1/2}}{x^3}\right) \qquad \leftarrow \text{algebra}$$

$$= \frac{d}{dx}(x^4 + x^{-5/2}) \qquad \leftarrow \text{algebra}$$

$$= 4x^3 - \frac{5}{2}x^{-7/2}. \qquad \leftarrow \text{sum, power rules}$$

Doing algebra before differentiating almost always means that there is much less algebra needed to simplify your answer after differentiating. In Exercise 18 you will show that the two answers we just found are in fact equal.   □

CHECKING THE ANSWER | You can always check the reasonableness of the answer to a differentiation problem by graphing both f and f' and verifying that f' appears to be the associated slope function for f. Do this for the following graphs, for part (c) of the example:

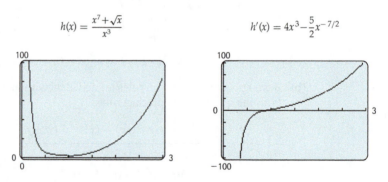

$$h(x) = \frac{x^7 + \sqrt{x}}{x^3} \qquad\qquad h'(x) = 4x^3 - \frac{5}{2}x^{-7/2}$$

EXAMPLE 3   Differentiating a piecewise-defined function

Find the derivative of the function $f(x) = \begin{cases} x^2, & \text{if } x \leq 1 \\ \dfrac{1}{x}, & \text{if } x > 1. \end{cases}$

SOLUTION

For $x < 1$ we use the first case of the function to find the derivative:

$$f'(x) = \frac{d}{dx}(x^2) = 2x.$$

Similarly, for $x > 1$ we use the second case:

$$f'(x) = \frac{d}{dx}\left(\frac{1}{x}\right) = \frac{d}{dx}(x^{-1}) = -x^{-2} = \frac{-1}{x^2}.$$

It now remains only to determine what the derivative is at the breakpoint $x = 1$, if it exists. The function f will be differentiable at $x = 1$ if it is continuous at $x = 1$ and the derivatives of each of its pieces are equal at $x = 1$. Since $(1)^2 = \frac{1}{1}$, f is continuous at $x = 1$. However, since $2(1) \neq \frac{-1}{(1)^2}$, the derivatives of the left and right pieces are not equal at $x = 1$. Therefore f is not differentiable at $x = 1$, that is, $f'(1)$ does not exist. Thus the derivative of f is

$$f'(x) = \begin{cases} 2x, & \text{if } x < 1 \\ \text{DNE}, & \text{if } x = 1 \\ \frac{-1}{x^2}, & \text{if } x > 1. \end{cases}$$

□

Consider the graphs of f and f' that follow. Note that f is indeed continuous, but not differentiable, at $x = 1$. In the graph of f' we have a hole at $x = 1$. We can see in *both* graphs the fact that $f'_-(1) = 2$ but $f'_+(1) = -1$. We can also see that, except at $x = 1$, the slopes of $f(x)$ do seem to be the same as the heights of $f'(x)$.

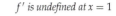

f is not differentiable at x = 1   *f' is undefined at x = 1*

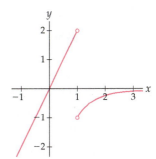

EXAMPLE 4   **Finding antiderivatives**

An ***antiderivative*** of a function $g(x)$ is a function whose derivative is $g(x)$. In this problem you may assume that any two functions with the same derivative must differ by a constant. (We will prove this fact in Section 3.2.)

(a) Find one antiderivative of $g(x) = 8x^3$.

(b) Describe the set of antiderivatives $g(x) = 8x^3$.

(c) Find the one antiderivative of $g(x) = 8x^3$ that passes through $(1, 4)$.

SOLUTION

(a) We can use a targeted guess-and-check method to find an antiderivative of $g(x) = 8x^3$. Since differentiating a power function decreases its power by one, we might start with the function $f(x) = x^4$, whose derivative is $f'(x) = 4x^3$; this is almost what we want, but off by a factor of two. A good second guess would be $f(x) = 2x^4$, and indeed its derivative is $f'(x) = 8x^3$, as desired.

(b) We now have one function whose derivative is $f'(x) = 8x^3$. Since any two functions with that derivative must differ by a constant, the functions whose derivative is $f'(x) = 8x^3$ are all the functions of the form $f(x) = 2x^4 + C$ for some real number

C, as shown in the next figure at the left. The graph corresponding to our previous choice of $f(x) = 2x^4$, with $C = 0$, is shown in black.

One antiderivative has C = 0

One antiderivative has f(1) = 4

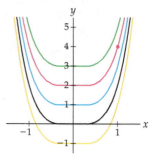

(c) Only one of the antiderivatives of $f'(x) = 8x^3$ passes through the point $(1, 4)$, as shown in red at the right. To find this antiderivative we set $f(1) = 4$ and solve for C:

$$f(1) = 4 \implies 2(1)^4 + C = 4 \implies C = 2.$$

Therefore the only antiderivative of $f'(x) = 8x^3$ that passes through $(1, 4)$ is the function $f(x) = 2x^4 + 2$.

? TEST YOUR UNDERSTANDING

▶ Why are the rules for differentiating constants and the identity function special cases of the rule for differentiating a linear function?

▶ What is the difference between the rule for differentiating a constant function and the constant multiple rule?

▶ Can you find examples of functions f and g such that the derivative of their quotient is not the same as the quotient of their derivatives?

▶ Why do we add and subtract $f(x)g(x + h)$ in the calculation in the proof of the product rule? What does that enable us to do in the calculation?

▶ What sorts of functions can we differentiate with the rules in this section? Are there any functions that we can't differentiate with these rules?

EXERCISES 2.3

Thinking Back

Factoring formulas: Pull out a linear factor from each of the following expressions.

▶ $z^2 - 100$ ▶ $z^3 - 27$ ▶ $z^6 - 64$

Definition-of-derivative calculations: Use the definition of the derivative to find f' for each function.

▶ $f(x) = x^4$ ▶ $f(x) = x^{-2}$

▶ $f(x) = x^2(x + 1)$ ▶ $f(x) = \dfrac{x^2}{x + 1}$

Associated slope functions: For each of the following two function graphs, sketch a careful, labeled graph of its associated slope function.

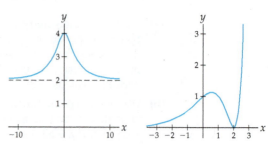

Concepts

0. *Problem Zero:* Read the section and make your own summary of the material.

1. *True/False:* Determine whether each of the statements that follow is true or false. If a statement is true, explain why. If a statement is false, provide a counterexample.

(a) *True or False:* $\dfrac{d}{dx}(5) = 0$.

(b) *True or False:* $\dfrac{d}{dr}(ks + r) = k$.

(c) *True or False:* $\dfrac{d}{ds}(ks + r) = k$.

(d) *True or False:* $\dfrac{d}{dx}(3x + 1)^k = k(3x + 1)^{k-1}$.

(e) *True or False:* $\dfrac{d}{dx}\left(\dfrac{1}{x^3}\right) = \dfrac{1}{3x^2}$.

(f) *True or False:* If f and g are differentiable functions, then $(f(x)g(x))' = g'(x)f(x) + f'(x)g(x)$.

(g) *True or False:* If g and h are differentiable functions, then $\left(\dfrac{g(x)}{h(x)}\right)' = \dfrac{h(x)g'(x) - g(x)h'(x)}{(h(x))^2}$.

(h) *True or False:* Proving the sum rule for differentiation involves the definition of the derivative, a lot of algebraic manipulation, and the sum rule for limits.

2. *Examples:* Construct examples of the thing(s) described in the following. Try to find examples that are different than any in the reading.

(a) Functions f and g, which illustrate that, in general, derivatives and products do not commute.

(b) Functions f and g, which illustrate that, in general, derivatives and quotients do not commute.

(c) Three functions whose derivatives we cannot calculate by using the differentiation rules we have developed so far.

3. Express the constant multiple, sum, and difference rules in Leibniz/operator notation.

4. Express the product and quotient rules in Leibniz/operator notation.

5. Why does it make graphical sense that the derivative of a constant is zero? That the derivative of the identity function is constantly equal to 1? That the derivative of a linear function $f(x) = mx + b$ is equal to m?

6. Why does the proof of the power rule (Theorem 2.9) in this text work only when k is an integer? Also, at which point in the proof do we use the fact that k is positive?

7. Explain why the power rule does *not* say that the derivative of 3^x is $x3^{x-1}$. Specifically, why doesn't 3^x fit the pattern required for the power rule to apply?

8. Explain why the power rule cannot be used to differentiate the function $(2 - x)^{1/3}$.

9. Use the product rule to find and state the rule for differentiating a product of *three* functions f, g, and h. In other words, fill in the blank:$(f(x)g(x)h(x))' = $ _____ .

Then use your rule to differentiate the function $y = (2x - 1)(x^2 + x + 1)(1 - 3x^4)$. Check your answer by differentiating the function $y(x)$ another way.

10. Two operations commute if they can be done in either order. Does multiplying a function by a constant commute with differentiation? Does adding two functions commute with differentiation? What about products and quotients and the operation of differentiation?

Given that f, g, and h are functions with values $f(2) = 1$, $g(2) = -4$, and $h(2) = 3$ and point-derivatives $f'(2) = 3$, $g'(2) = 0$, and $h'(2) = -1$, calculate

11. $(3f + 4h)'(2)$

12. $(2f + 3g - h)'(2)$

13. $(fh)'(2)$

14. $\left(\dfrac{f}{g}\right)'(2)$

15. In the text of this section we displayed graphs of $f(x) = x^4$ and its first five derivatives. Use the slope-height behavior of the graphs to verify that each is the associated slope function of the one before.

16. Sketch graphs of $f(x) = \dfrac{1}{x}$ and its first five derivatives. Then use the slope-height behavior of the graphs to verify that each is the associated slope function of the one before.

17. In Example 2(b) we calculated a derivative two different ways. Use algebra to simplify the first answer and show that it is equal to the second.

18. In Example 2(c) we calculated a derivative two different ways. Use algebra to simplify the first answer and show that it is equal to the second.

19. Suppose f is a piecewise-defined function, equal to $g(x)$ if $x < 2$ and $h(x)$ if $x \geq 2$, where g and h are continuous and differentiable everywhere. If $g(2) = h(2)$, is the function f necessarily differentiable at $x = 2$? Why or why not?

20. Suppose f is a piecewise-defined function, equal to $g(x)$ if $x < 2$ and $h(x)$ if $x \geq 2$, where g and h are continuous and differentiable everywhere. If $g'(2) = h'(2)$, is the function f necessarily differentiable at $x = 2$? Why or why not?

21. If possible, find constants a and b so that the function f that follows is continuous and differentiable everywhere. If it is not possible, explain why not.

$$f(x) = \begin{cases} 3x + a, & \text{if } x < 1 \\ x^{b/2}, & \text{if } x \geq 1. \end{cases}$$

22. If possible, find constants a and b so that the function f that follows is continuous and differentiable everywhere. If it is not possible, explain why not.

$$f(x) = \begin{cases} ax - b, & \text{if } x < 2 \\ bx^2 + 1, & \text{if } x \geq 2. \end{cases}$$

Skills

Suppose $g(x)$, $h(x)$, and $j(x)$ are differentiable functions with the values of the function and its derivative given in the following table:

x	$g(x)$	$h(x)$	$j(x)$	$g'(x)$	$h'(x)$	$j'(x)$
-1	3	0	1	-1	-2	-2
0	2	3	0	-2	3	-2
1	0	-1	-2	-2	-2	-1
2	-2	-2	-3	-1	0	2
3	-3	0	1	0	1	2

Use the table to calculate the values of the derivatives listed in Exercises 23–28.

23. If $f(x) = 3j(x) - 2h(x)$, find $f'(3)$.

24. If $f(x) = g(x)h(x)$, find $f'(0)$.

25. If $f(x) = g(x)(h(x) + j(x))$, find $f'(2)$.

26. If $f(x) = h(x)g(x)j(x)$, find $f'(1)$.

27. If $f(x) = \dfrac{3h(x)}{g(x) + j(x)}$, find $f'(0)$.

28. If $f(x) = \dfrac{g(x)h(x) + j(x)}{h(x)}$, find $f'(-1)$.

Differentiate each of the functions in Exercises 29–34 in two different ways: first with the product and/or quotient rules and then without these rules. Then use algebra to show that your answers are the same.

29. $f(x) = x^2(x + 1)$

30. $f(x) = \dfrac{3x + 1}{x^4}$

31. $f(x) = x^{7/2}(2 - 5x^3)$

32. $f(x) = \sqrt{x}\,(x^{-1} + 1)$

33. $f(x) = \dfrac{x^2 - x^3}{\sqrt{x}}$

34. $f(x) = \dfrac{\sqrt{x}}{x^{-2}x^3}$

Use the differentiation rules developed in this section to find the derivatives of the functions in Exercises 35–64. Note that it may be necessary to do some preliminary algebra before differentiating.

35. $f(x) = 4 - 3x^7$

36. $f(x) = 2x - 3 + 4x^2$

37. $f(x) = 2(1 + 3x^2)$

38. $f(x) = 5x^3 - 2x^2 + 7$

39. $f(x) = 2(3x+1) - 4x^5$

40. $f(x) = x^2 + x(2 - 3x^2)$

41. $f(x) = (x + 2)(x - 1)$

42. $f(x) = (x^2 - 3)^2$

43. $f(x) = (3x + 2)^3$

44. $f(x) = (3 - x)^2 + 5$

45. $f(x) = \dfrac{1 - 6x^3}{3}$

46. $f(x) = \dfrac{x}{x + 1}$

47. $f(x) = \dfrac{x^2 - 1}{x + 1}$

48. $f(x) = \dfrac{x^4 - 7x^3}{2x}$

49. $f(x) = \pi^2$

50. $f(x) = (\sqrt{x} - \sqrt[3]{x})^2$

51. $f(x) = \dfrac{x^2}{\sqrt[5]{x^2}}$

52. $f(x) = \dfrac{x - 1}{(x + 1)(x + 2)}$

53. $f(x) = \dfrac{\sqrt[5]{x^7} - 2x^4}{x^3}$

54. $f(x) = (3x\sqrt{x})^{-2}$

55. $f(x) = \dfrac{x^7 - 3x^5 + 4}{1 - 3x^4}$

56. $f(x) = \dfrac{x^3 + x - 1}{x^2 - 7}$

57. $f(x) = \sqrt{\dfrac{1}{x^3} + \left(\dfrac{1}{\sqrt{x}}\right)^3}$

58. $f(x) = \dfrac{x^2 - 3x}{x^2 - 2x + 1}$

59. $f(x) = \dfrac{2x - 3}{5x + 4}$

60. $f(x) = x^3\sqrt{x}\,(x^{2/3})$

61. $f(x) = \dfrac{1}{(x - 2)(x - 3)}$

62. $f(x) = \dfrac{1}{(x + 1)^3}$

63. $f(x) = \dfrac{x^2}{x^3 + 5x^2 - 3x}$

64. $f(x) = \dfrac{(x - 2)^2}{(x^2 + 1)(x - 3)}$

Find the derivatives of each of the absolute value and piecewise-defined functions in Exercises 65–72.

65. $f(x) = |x|$

66. $f(x) = |3x + 1|$

67. $f(x) = |1 - 2x|$

68. $f(x) = |x^2 - 1|$

69. $f(x) = \begin{cases} x^3, & \text{if } x < 1 \\ x, & \text{if } x \geq 1 \end{cases}$

70. $f(x) = \begin{cases} 1, & \text{if } x \leq -1 \\ x^{2/3}, & \text{if } x > -1 \end{cases}$

71. $f(x) = \begin{cases} -x^2, & \text{if } x \leq 0 \\ x^2, & \text{if } x > 0 \end{cases}$

72. $f(x) = \begin{cases} 3x + 1, & \text{if } x \leq 1 \\ x^3, & \text{if } x > 1 \end{cases}$

In Exercises 73–78, find a function that has the given derivative and value. In each case you can find the answer with an educated guess-and-check process. It may be helpful to do some preliminary algebra.

73. $f'(x) = 3x^5 - 2x^2 + 4$, $f(0) = 1$

74. $f'(x) = 7x^2 + 8x^{11} - 18$, $f(0) = -2$

75. $f'(x) = 1 - 4x^6$, $f(1) = 3$

76. $f'(x) = x(4 - 2x)$, $f(0) = 0$

77. $f'(x) = (x^4 - 8)(1 - 3x^5)$, $f(0) = 2$

78. $f'(x) = (3x + 1)^3$, $f(2) = 1$

Applications

79. A spaceship is moving along a straight path from Venus into the heart of the Sun. The velocity of the spaceship t hours after leaving Venus is $v(t) = 0.012t^2 + 400$ thousands of miles per hour. (To simplify matters we will pretend that Venus is not moving with respect to the sun; you may assume that everything is fixed in place in this exercise.)

(a) Say what you can about the initial values s_0, v_0, and a_0, and then use derivatives and antiderivatives to find equations for the position and acceleration of the spaceship.

(b) Is the spaceship always moving towards the sun? How can you tell?

(c) Is the spaceship travelling at a constant acceleration? Is it speeding up or slowing down, or neither? How can you tell?

(d) The distance between Venus and the sun is about 67 million miles. How long will it take the spaceship to reach the sun? How fast will the spaceship be going when it gets there?

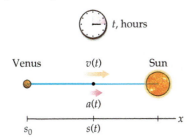

80. A bowling ball is thrown down from a 20th-story window. After 3 seconds, the bowling ball is 26 feet from the ground and falling at a rate of −106 feet per second (downwards). You may assume that gravity causes a constant downward acceleration of −32 feet per second.

(a) If the height $s(t)$ of the bowling ball t seconds after being thrown is given by a quadratic polynomial function, use $s(3)$, $s'(3)$, and $s''(3)$ to find an equation for $s(t)$.

(b) How high is the 20th-story window from which the bowling ball was thrown?

(c) How fast was the bowling ball initially thrown?

81. On Earth, a falling object has a downward acceleration of −32 feet per second per second due to gravity. Suppose an object falls from an initial height of s_0 feet, with an initial velocity of v_0 feet per second. Use antiderivatives to show that the equations for

the position and velocity of the falling object after t seconds are, respectively, $s(t) = -16t^2 + v_0 t + s_0$ and $v(t) = -32t + v_0$.

82. In a science fiction novel, gravity on the planet XV-37 acts very differently than it does in our universe. An object dropped from a 1000-foot building on planet XV-37 will have a downward gravitational acceleration of $a(t) = -6t$ feet per second per second after falling for t seconds. Use antiderivatives to find equations for the position and velocity of such a falling object. What might be the consequences of living on this strange planet?

83. In another science fiction novel, gravity on planet Xillian again acts very strangely. The height $s(t)$ of a falling object on Xillian is always a cubic polynomial function. Suppose a kiwi fruit is dropped (with initial velocity of zero) from the top of a Xillian radio tower. After 5 seconds, the kiwi fruit is 100 feet from the ground and falling at a rate of −200 feet per second. The acceleration of the kiwi fruit at that moment is −46 feet per second per second.

(a) Use the values of $s'(0)$ (note the kiwi initially has velocity zero), $s(5)$, $s'(5)$, and $s''(5)$ given in the preceding description to find a formula for the height $s(t)$ of the kiwi fruit t seconds after being dropped from the Xillian radio tower. Specifically, use these four data points to solve for the coefficients of a cubic polynomial $s(t) = at^3 + bt^2 + ct + d$.

(b) Verify that the function $s(t)$ you just found produces the correct values for $s'(0)$, $s(5)$, $s'(5)$, and $s''(5)$ in this exercise.

(c) How high is the Xillian radio tower from which the kiwi fruit was dropped?

(d) On Earth, acceleration due to gravity is given by a constant −32 feet per second per second, that is, on Earth we always have $a(t) = -32$ for falling objects. What is the function $a(t)$ for acceleration due to Xillian "gravity"? Is this acceleration constant? What are the physical implications of gravity on Xillian?

Proofs

84. Use the definition of the derivative to directly prove the differentiation rules for constant and identity functions.

85. Use the $h \to 0$ definition of the derivative to prove that the power rule holds for positive integer powers.

86. Prove, in two ways, that the power rule holds for negative integer powers:

(a) By using the $z \to x$ definition of the derivative.

(b) By using the $h \to 0$ definition of the derivative.

87. Prove the difference rule in two ways:

(a) Using the definition of the derivative.

(b) Using the sum and constant multiple rules.

88. Use the definition of the derivative to prove the following special case of the product rule:

$$\frac{d}{dx}(x^2 f(x)) = 2x f(x) + x^2 f'(x).$$

89. Use the definition of the derivative to prove the quotient rule. (*Hint: At some point you will have to add and subtract the quantity $f(x)g(x)$ from the numerator.*)

90. The following **reciprocal rule** tells us how to differentiate the reciprocal of a function:

$$\frac{d}{dx}\left(\frac{1}{f(x)}\right) = -\frac{f'(x)}{(f(x))^2}.$$

Prove the reciprocal rule in two ways:

(a) By using the definition of the derivative.

(b) By using the quotient rule.

91. Consider the following formula for antidifferentiating power functions: If $f'(x) = x^k$ and $k \neq -1$, then $f(x) = \frac{1}{k+1}x^{k+1} + C$ for some constant C.

(a) Prove this antidifferentiation formula. You may assume that any two functions with the same derivative differ by a constant, as we will prove in Section 3.2.

(b) What part of your argument from part (a) breaks down when $f(x) = x^{-1}$?

92. Consider the piecewise-defined function

$$f(x) = \begin{cases} g(x), & \text{if } x \leq c \\ h(x), & \text{if } x > c. \end{cases}$$

Prove that if $g(x)$ and $h(x)$ are continuous and differentiable at $x = c$, and if $g(c) = h(c)$ and $g'(c) = h'(c)$, then f is differentiable at $x = c$.

Thinking Forward

For each function f that follows, find all of the x-values in the domain of f for which $f'(x) = 0$ and all values for which $f'(x)$ does not exist. In later sections we will call these values the **critical points** of f.

▶ $f(x) = x^3 - 2x$

▶ $f(x) = \sqrt{x} - x$

▶ $f(x) = \dfrac{1}{1 + \sqrt{x}}$

▶ $f(x) = \dfrac{x^2(x-1)}{(x-2)^2}$

Taylor polynomials: In the exercises that follow, you will investigate the relationship between the coefficients of a polynomial and its higher order derivatives. This same sort of idea can be used to locally approximate differentiable functions with polynomials.

▶ Prove that if f is any quadratic polynomial function $f(x) = ax^2 + bx + c$, then the coefficients of f are completely determined by the values of $f(x)$ and its derivatives at $x = 0$, as follows:

$$a = \frac{f''(0)}{2}, \quad b = f'(0), \quad \text{and} \quad c = f(0).$$

In particular, this means that if f is any quadratic polynomial function, then $f(x) = \frac{f''(0)}{2}x^2 + f'(0)x + f(0)$.

▶ Prove that if f is any cubic polynomial function $f(x) = ax^3 + bx^2 + cx + d$, then the coefficients of f are completely determined by the values of $f(x)$ and its derivatives at $x = 0$, as follows:

$$a = \frac{f'''(0)}{6}, \quad b = \frac{f''(0)}{2}, \quad c = f'(0),$$
$$\text{and} \quad d = f(0).$$

This means that every cubic polynomial function can be written $f(x) = \frac{f'''(0)}{6}x^3 + \frac{f''(0)}{2}x^2 + f'(0)x + f(0)$.

▶ Suppose f is any cubic polynomial function $f(x) = ax^3 + bx^2 + cx + d$. Prove that the coefficients a, b, c, and d of f can be expressed in terms of the values of $f(x)$ and its derivatives at the point $x = 2$. (*Hint: In other words, show that you can write the coefficients a, b, c, and d in terms of $f(2)$, $f'(2)$, $f''(2)$, and $f'''(2)$.*)

▶ Suppose f is a polynomial of degree n, and let k be some integer with $0 \leq k \leq n$. Prove that if $f(x)$ is of the form

$$f(x) = a_n x^n + a_{n-1}x^{n-1} + \cdots + a_k x^k + \cdots + a_1 x + a_0,$$

then $a_k = \dfrac{f^{(k)}(0)}{k!}$, where $f^{(k)}(x)$ is the kth derivative of $f(x)$ and $k! = k(k-1)\cdots(2)(1)$. (*Hint: Find $f^{(k)}(x)$, and use it to show that $f^{(k)}(0) = k!a_k$.*)

2.4   THE CHAIN RULE AND IMPLICIT DIFFERENTIATION

▶ The chain rule for differentiating compositions of functions

▶ Implicit functions and implicit differentiation

▶ Using implicit differentiation to prove derivative formulas

Differentiating Compositions of Functions

With the rules we have developed so far, we can differentiate any arithmetic combination of functions whose components have known derivatives. There is, however, one other way that functions can be combined: by composition. We still don't know how to differentiate, say, the composition $y = (x^2 + 1)^{1/2}$. Is the derivative of a composition $f(g(x))$ somehow related to the derivatives of f and g?

Let's consider a simple example involving rates of change. Suppose you own a small factory that makes 30 widgets an hour and you make a profit of $10.00 for each widget made. This means that you can make a profit of $300.00 an hour by making widgets at your factory. We arrive at this answer by using the product:

$$300 \, \frac{\text{dollars}}{\text{hour}} = \left(10 \, \frac{\text{dollars}}{\text{widget}} \right) \left(30 \, \frac{\text{widgets}}{\text{hour}} \right).$$

What does this have to do with the derivative of a composition? Let $w(t)$ be the number of widgets you have t hours after starting production, and let $p(w)$ be the profit made from producing w widgets. Then the composition $p(w(t))$ gives the profit made after t hours. We are interested in the rate of change of profit per hour, which is the derivative $\frac{dp}{dt}$. The fact that you make a profit of $\frac{dp}{dw} = 10$ dollars per widget and the fact that your factory can make $\frac{dw}{dt} = 30$ widgets per hour are also statements about derivatives. Repeating our previous calculation, we can relate the derivative of the composition $p(w(t))$ to the derivatives of $p(w)$ and $w(t)$:

$$\frac{dp}{dt} = \frac{dp}{dw} \frac{dw}{dt}.$$

This example suggests that the derivative of a composition is the product of the derivatives of the component functions. That is in fact the case in general:

THEOREM 2.12   **The Chain Rule**

Suppose $f(u(x))$ is a composition of functions. Then for all values of x at which u is differentiable at x and f is differentiable at $u(x)$, the derivative of f with respect to x is equal to the product of the derivative of f with respect to u and the derivative of u with respect to x.

In Leibniz notation, we write this as

$$\frac{df}{dx} = \frac{df}{du} \frac{du}{dx}.$$

In "prime" notation, we write it as

$$(f \circ u)'(x) = f'(u(x)) \, u'(x).$$

To use the chain rule, we must first recognize a function as a composition $f(u(x))$ and identify the "outside" and "inside" functions f and u. For example, the function $y = (x^2 + 1)^{1/2}$ can be thought of as a composition $y = f(u(x))$ with inside function $u(x) = x^2 + 1$ and outside function $f(u) = u^{1/2}$. The chain rule says that we should differentiate the outside function f with respect to u and then multiply the result by the derivative of the inside function:

$$(f(u(x))' = f'(u(x))\, u'(x) = \frac{1}{2}(u(x))^{-1/2}u'(x) = \frac{1}{2}(x^2 + 1)^{-1/2}(2x).$$

For more examples of how to apply the chain rule, see Examples 1 and 2.

The "prime" notation in Theorem 2.12 makes clear why we want u to be differentiable at x and f to be differentiable at $u(x)$. When it suits our purposes, we can also write $(f(u(x)))'$ for $(f \circ u)'(x)$. The Leibniz notation suggests why we call it the "chain" rule: We are taking the derivative of a chain of functions by multiplying a chain of derivatives. If we had a longer chain of functions, then we would have a longer chain of derivatives; for example, the derivative of the composition $f(u(v(x)))$ with respect to x is given by the chain of derivatives

$$\frac{df}{dx} = \frac{df}{du}\frac{du}{dv}\frac{dv}{dx}.$$

The chain rule seems sensible if you consider its analog with difference quotients, since we can cancel Δu's to justify the equation $\frac{\Delta f}{\Delta x} = \frac{\Delta f}{\Delta u}\frac{\Delta u}{\Delta x}$. However, this cancellation does not automatically apply to the Leibniz notation, since the differentials df, dx, and du do not represent numerical quantities. They cannot be cancelled just because the notation makes it look tempting to do so. The proof of the chain rule requires more work than simply "cancelling," but except for a certain technical point, is not that difficult.

Proof. Our proof will start with the definition of the derivative for $(f \circ g)$. After some algebra and a limit rule, a change of variables will give us the result we desire. In the middle of the calculation we will make a simplifying substitution:

$$(f \circ g)'(x) = \lim_{h \to 0} \frac{f(g(x+h)) - f(g(x))}{h} \qquad \leftarrow \text{derivative}$$

$$= \lim_{h \to 0}\left(\frac{f(g(x+h)) - f(g(x))}{g(x+h) - g(x)} \frac{g(x+h) - g(x)}{h}\right) \qquad \leftarrow \text{algebra}$$

$$= \left(\lim_{h \to 0}\frac{f(g(x+h)) - f(g(x))}{g(x+h) - g(x)}\right)\left(\lim_{h \to 0}\frac{g(x+h) - g(x)}{h}\right) \qquad \leftarrow \text{product rule for limits}$$

$$= \left(\lim_{k \to 0}\frac{f(g(x) + k) - f(g(x))}{k}\right)(g'(x)) \qquad \leftarrow \text{see below}$$

$$= f'(g(x))\, g'(x) \qquad \leftarrow \text{derivative}$$

There is one technical point to consider in this proof. In the fourth step we applied the substitution $k = g(x+h) - g(x)$. Since $g(x)$ is differentiable, it is also continuous; therefore as $h \to 0$, we also have $k = g(x+h) - g(x) \to 0$. However, the preceding calculation assumes that $k = g(x+h) - g(x)$ is nonzero for small enough h, which in some situations may not be the case. In the case when this happens because $g(x+h) = g(x)$ as $h \to 0$, it is easy to show that $(f \circ g)'(x)$ and $f'(g(x))g'(x)$ are both zero and therefore equal. There are other cases when this can happen, but handling the details for those functions is beyond the scope of this course, so we will assume here that $g(x)$ is nice enough that we can avoid those cases.

Implicit Differentiation

Consider the equation $x^2 + y^2 = 1$ that describes the circle of radius 1 centered at the origin, as shown next at the left. Clearly, this graph does not represent a function, since it does not pass the vertical line test. However, *locally*, that is, in small pieces, the graph does define y as a function of x. For example, the top half of the graph shown in the middle figure does represent a function, as does the graph of the bottom half shown at the right.

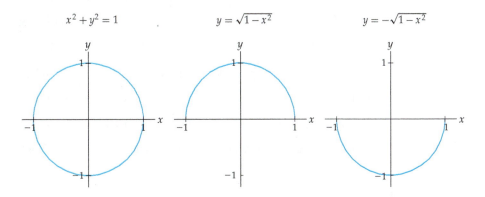

Although we cannot solve the equation $x^2 + y^2 = 1$ for y and obtain a single well-defined function, we can still think of the x-values as inputs and the y-values as outputs; the only difference is that there may be more than one y-value for each x-value. In cases such as these, we say that y is an ***implicit function*** of x.

Thinking locally, we can use our usual differentiation techniques to show that the unit circle has horizontal tangent lines at $(0, 1)$ and at $(0, -1)$. Looking at the top of the circle, we have $y = \sqrt{1 - x^2}$, with derivative

$$y' = \frac{1}{2}(1 - x^2)^{-1/2}(-2x) = \frac{-x}{\sqrt{1 - x^2}},$$

which is clearly zero only when $x = 0$. Therefore the unit circle has a horizontal tangent line at the point $(0, \sqrt{1 - 0^2}) = (0, 1)$. In a similar fashion we could use the equation $y = -\sqrt{1 - x^2}$ for the bottom half of the circle to find the other horizontal tangent line.

We will not always be able to divide an implicit function into pieces given by actual functions whose equations we know. That is, we will not always be able to "solve for y." However, given an equation that defines an implicit function, we can still find information about slopes and derivatives simply by differentiating both sides of the equation with respect to x. This technique is known as ***implicit differentiation***. The key to implementing the technique will be applying the chain rule appropriately.

For example, we can differentiate both sides of the equation $x^2 + y^2 = 1$ from the previous example. Along the way we will have to remember that we are thinking of $y = y(x)$ as a function of x and apply the chain rule:

$$x^2 + y^2 = 1 \qquad \leftarrow \text{given equation}$$

$$\frac{d}{dx}(x^2 + (y(x))^2) = \frac{d}{dx}(1) \qquad \leftarrow \text{differentiate both sides}$$

$$2x + 2y\frac{dy}{dx} = 0 \qquad \leftarrow \text{power and chain rules}$$

$$\frac{dy}{dx} = \frac{-2x}{2y} = -\frac{x}{y}. \qquad \leftarrow \text{solve for } \frac{dy}{dx}$$

The use of the chain rule in the differentiation step is crucial to these types of calculations. Notice that the derivative of x^2 with respect to x is the familiar $2x$, but the derivative of y^2 with respect to x uses the chain rule to arrive at $2y\frac{dy}{dx}$. Here y is thought of as an implicit function of x, so the expression y^2 is really the composition of the implicit function $y(x)$ with the squaring function. That is why the chain rule must come into play when we differentiate y^2 with respect to x.

The fact that $\frac{dy}{dx} = -\frac{x}{y}$ means that for any point (a, b) that lies on the graph of the circle $x^2 + y^2 = 1$, the slope of the tangent line to the circle at (a, b) is given by $-\frac{a}{b}$. For example, the point $\left(\frac{1}{2}, \frac{\sqrt{3}}{2}\right)$ is on the graph of the circle because $\left(\frac{1}{2}\right)^2 + \left(\frac{\sqrt{3}}{2}\right)^2 = 1$ and the slope of the line tangent to the circle at that point is $\frac{-1/2}{\sqrt{3}/2} = -\frac{1}{\sqrt{3}}$. Similarly, at the point $\left(-\frac{1}{2}, -\frac{\sqrt{3}}{2}\right)$ the slope of the tangent line is also $-\frac{1}{\sqrt{3}}$, as shown next at the left. At the point $(0, 1)$ on the circle, the tangent line has slope $-\frac{0}{1} = 0$ and is thus horizontal, and at the point $(1, 0)$ the tangent line has undefined slope $-\frac{1}{0}$ and thus is vertical, as shown at the right.

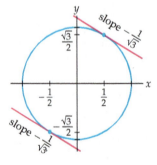

Slope at (a, b) given by $-\frac{a}{b}$

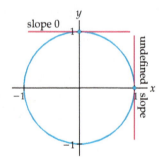

Horizontal and vertical tangent lines

Examples and Explorations

EXAMPLE 1   **Applying the chain rule one step at a time**

Use the chain rule to differentiate $h(x) = \left(\dfrac{x}{1 - 3x^2}\right)^4$, writing out all steps.

SOLUTION

We can think of $h(x)$ as a composition $f(u(x))$ with inside function $u(x) = \dfrac{x}{1 - 3x^2}$ and outside function $f(u) = u^4$. Applying the chain rule, we have

$$h'(x) = f'(u(x))u'(x) \qquad \leftarrow \text{chain rule}$$

$$= 4(u(x))^3 u'(x) \qquad \leftarrow \text{derivative of outside function}$$

$$= 4\left(\frac{x}{1 - 3x^2}\right)^3 u'(x) \qquad \leftarrow \text{evaluate at inside function}$$

$$= 4\left(\frac{x}{1 - 3x^2}\right)^3 \left(\frac{1(1 - 3x^2) - x(-6x)}{(1 - 3x^2)^2}\right). \qquad \leftarrow \text{derivative of inside function}$$

When you become comfortable using the chain rule, you won't write out each step separately as we just did; instead you will simply think something like "the derivative of the outside with the inside plugged in, times the derivative of the inside," and write out the derivative calculation in one step. Note that once again we have not simplified the derivative that we found. In this text we will not simplify such answers unless we have a specific use for the derivative in question.

☐

EXAMPLE 2 **Determining which differentiation rule to apply first**

For each function f, find the derivative f'. What is the *first* differentiation rule that you must use in each case?

(a) $f(x) = \left(\sqrt[3]{x} - \dfrac{1}{x}\right)^{-2}$ **(b)** $g(x) = \dfrac{\sqrt{x^2 - 1}}{1 - \dfrac{1}{x}}$ **(c)** $h(x) = (x^2 - 1)^5 (3x - 1)^2$

SOLUTION

(a) The function f is at its outermost layer a composition of the form $(u(x))^{-2}$, where $u(x) = \sqrt[3]{x} - \dfrac{1}{x}$. Therefore, the first rule to use is the chain rule. The power rule is the *second* differentiation rule we must use, since we need to use the chain rule to even begin differentiating the function. Before applying the chain rule, we use algebra to rewrite the function f in a more convenient exponent form:

$$\frac{d}{dx}\left(\left(\sqrt[3]{x} - \frac{1}{x}\right)^{-2}\right) = \frac{d}{dx}((x^{1/3} - x^{-1})^{-2}) \qquad \leftarrow \text{algebra}$$

$$= -2(x^{1/3} - x^{-1})^{-3} \cdot \frac{d}{dx}(x^{1/3} - x^{-1}) \qquad \leftarrow \text{chain, power rules}$$

$$= -2(x^{1/3} - x^{-1})^{-3}\left(\frac{1}{3}x^{-2/3} - (-1)x^{-2}\right). \qquad \leftarrow \text{difference, power rules}$$

(b) The function $g(x)$ is at its outermost layer a quotient, so we will begin by using the quotient rule. Again we write roots and fractions as exponents first, so that the differentiation steps will be easier:

$$\frac{d}{dx}\left(\frac{\sqrt{x^2 - 1}}{1 - (1/x)}\right) = \frac{d}{dx}\left(\frac{(x^2 - 1)^{1/2}}{1 - x^{-1}}\right) \qquad \leftarrow \text{algebra}$$

$$= \frac{\frac{d}{dx}((x^2 - 1)^{1/2})(1 - x^{-1}) - (x^2 - 1)^{1/2}\frac{d}{dx}(1 - x^{-1})}{(1 - x^{-1})^2} \qquad \leftarrow \text{quotient rule}$$

$$= \frac{\frac{1}{2}(x^2 - 1)^{-1/2}(2x)(1 - x^{-1}) - (x^2 - 1)^{1/2}(0 - (-1)x^{-2})}{(1 - x^{-1})^2}. \qquad \leftarrow \text{more rules}$$

(c) The function $h(x)$ is at its outermost layer a product, so we begin by applying the product rule:

$$\frac{d}{dx}((x^2 - 1)^5(3x - 1)^2) = \frac{d}{dx}((x^2 - 1)^5)(3x - 1)^2 + (x^2 - 1)^5\frac{d}{dx}((3x - 1)^2) \qquad \leftarrow \text{product rule}$$

$$= 5(x^2 - 1)^4(2x)(3x - 1)^2 + (x^2 - 1)^5(2)(3x - 1)(3). \qquad \leftarrow \text{more rules}$$

☐

EXAMPLE 3 **Implicit differentiation and the chain rule**

Each of the equations that follow defines $y = y(x)$ as an implicit function of x. Use implicit differentiation to find $y' = y'(x)$.

(a) $3x^4 + 2y^4 + x + y = 0$ **(b)** $x^2y + xy^2 = 1$

SOLUTION

(a) Before we begin, note that $y = y(x)$ is an implicit function of x, so when we differentiate any expression involving y with respect to x, we will need the chain rule. We will write y and y' to avoid any confusion with multiplication here, but each time we do so we will think of the (implicit) functions $y(x)$ and $y'(x)$. Differentiating both sides of the given equation and then solving for y' gives

$$\frac{d}{dx}(3x^4 + 2y^4 + x + y) = \frac{d}{dx}(0) \qquad \leftarrow \text{differentiate both sides}$$

$$3(4x^3) + 2(4y^3)y' + 1 + y' = 0 \qquad \leftarrow \text{note } x \text{ and } y \text{ are dealt with differently}$$

$$8y^3y' + y' = -12x^3 - 1 \qquad \leftarrow \text{isolate } y' \text{ terms to one side}$$

$$y'(8y^3 + 1) = -12x^3 - 1 \qquad \leftarrow \text{factor out } y' \text{ on the left}$$

$$y' = \frac{-12x^3 - 1}{8y^3 + 1}. \qquad \leftarrow \text{divide to finish solving for } y'$$

(b) This time we will need the product rule, since x^2y is the product of the function x^2 and the implicit function y, and xy^2 is the product of the function x and the implicit function y^2. For the latter product we will also require the chain rule. The overall process is much the same as in the previous calculation:

$$\frac{d}{dx}(x^2y + xy^2) = \frac{d}{dx}(1) \qquad \leftarrow \text{differentiate both sides}$$

$$(2xy + x^2y') + (1y^2 + x(2yy')) = 0 \qquad \leftarrow \text{product and chain rules}$$

$$x^2y' + 2xyy' = -2xy - y^2 \qquad \leftarrow \text{isolate } y' \text{ terms to one side}$$

$$y'(x^2 + 2xy) = -2xy - y^2 \qquad \leftarrow \text{factor out } y' \text{ on the left}$$

$$y' = \frac{-2xy - y^2}{x^2 + 2xy}. \qquad \leftarrow \text{divide to finish solving for } y'$$

EXAMPLE 4   **Finding values and tangent lines on the graph of an implicit function**

Consider the equation $y^3 + xy + 2 = 0$ that defines y as an implicit function of x.

(a) Show that $y^3 - 5y + 2$ factors as $(y - 2)(y^2 + 2y - 1)$.

(b) If $x = -5$, find all the possible values for y.

(c) Find the slope of the line tangent to $y^3 + xy + 2 = 0$ at the point $(-5, 2)$.

SOLUTION

(a) We could factor $y^3 - 5y + 2$ by using something called **synthetic division** that we will see in Section 4.1, but since the factorization is already given in the problem, all we need to do is multiply out and check:

$$(y - 2)(y^2 + 2y - 1) = y^3 + 2y^2 - y - 2y^2 - 4y + 2 = y^3 - 5y + 2.$$

(b) Given the factorization from part (a), we substitute $x = -5$ into the equation $y^3 + xy + 2 = 0$ and solve for all corresponding values of y:

$$y^3 - 5y + 2 = 0 \qquad \leftarrow \text{the equation with } x = -5$$

$$(y - 2)(y^2 + 2y - 1) = 0 \qquad \leftarrow \text{factorization in part (a)}$$

$$y = 2, \ y = -1 + \sqrt{2}, \text{ or } y = -1 - \sqrt{2}. \qquad \leftarrow \text{quadratic formula}$$

Thus $(-5, 2)$, $(-5, -1 + \sqrt{2})$, and $(-5, -1 - \sqrt{2})$ are points on the graph of $y^3 + xy + 2 = 0$.

(c) To find the slope of the tangent line we must first calculate $\frac{dy}{dx}$. Using implicit differentiation and the chain rule, we have

$$y^3 + xy + 2 = 0 \qquad \leftarrow \text{the given equation}$$

$$\frac{d}{dx}\left((y(x))^3 + x \cdot y(x) + 2\right) = \frac{d}{dx}(0) \qquad \leftarrow \text{differentiate both sides}$$

$$3y^2\frac{dy}{dx} + (1)(y) + (x)\left(\frac{dy}{dx}\right) + 0 = 0 \qquad \leftarrow \text{chain and product rules}$$

$$3y^2\frac{dy}{dx} + x\frac{dy}{dx} = -y \qquad \leftarrow \text{start solving for } \frac{dy}{dx}$$

$$\frac{dy}{dx}(3y^2 + x) = -y \qquad \leftarrow \text{algebra}$$

$$\frac{dy}{dx} = \frac{-y}{3y^2 + x}. \qquad \leftarrow \text{algebra}$$

This means that if a point (a, b) is on the graph of $y^3 + xy + 2 = 0$, then the line tangent to the graph of $y^3 + xy + 2 = 0$ at (a, b) has slope

$$\frac{dy}{dx}\Big|_{\substack{x=a \\ y=b}} = \frac{-b}{3b^2 + a}.$$

Therefore the slope of the tangent line at the point $(a, b) = (-5, 2)$ is

$$\frac{dy}{dx}\Big|_{\substack{x=-5 \\ y=2}} = \frac{-2}{3(2)^2 + (-5)} = \frac{-2}{12 - 5} = \frac{-2}{7}. \qquad \square$$

CHECKING THE ANSWER

We can use a graphing utility to sketch the graph of the implicit function from the previous example, as shown next at the left. Notice that for $x = -5$ there are three corresponding y-values, as we showed in part (b). At each of these points $(-5, y)$ there is a tangent line to the graph. From the graph at the right it does appear (taking axes scales into account) that the slope of the tangent line at $(-5, 2)$ could be approximately $-\frac{2}{7}$.

$y^3 + xy + 2 = 0$ *and the points at* $x = -5$ $\qquad$ *Slope at* $(-5, 2)$ *is* $-\frac{2}{7}$

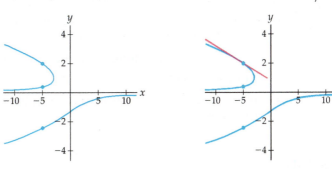

EXAMPLE 5 **Using implicit differentiation to prove the general power rule**

Although we have been using the power rule for general rational powers, we have in fact proved it only for integer powers.

(a) Use implicit differentiation and the power rule for integer powers to prove that $\frac{d}{dx}(x^{5/3}) = \frac{5}{3}x^{2/3}$.

(b) Use implicit differentiation and the power rule for integer powers to prove the power rule for rational powers.

SOLUTION

(a) If $y = x^{5/3}$, then $y^3 = x^5$. With this equation we can use implicit differentiation and the power rule for integer powers to find y':

$$y^3 = x^5 \qquad \leftarrow \text{since } y = x^{5/3}$$

$$\frac{d}{dx}((y(x))^3) = \frac{d}{dx}(x^5) \qquad \leftarrow \text{differentiate both sides}$$

$$3y^2 y' = 5x^4 \qquad \leftarrow \text{chain rule and integer power rule}$$

$$y' = \frac{5x^4}{3y^2} = \frac{5x^4}{3(x^{5/3})^2} \qquad \leftarrow \text{solve for } y', \text{ substitute } y = x^{5/3}$$

$$y' = \frac{5}{3}x^{4-(10/3)} = \frac{5}{3}x^{2/3}. \qquad \leftarrow \text{algebra}$$

(b) Suppose $y = x^{p/q}$, where $\frac{p}{q}$ is a rational number. Generalizing the method used in part (a), we use implicit differentiation on the equation $y^q = x^p$ to solve for y':

$$y^q = x^p \qquad \leftarrow \text{since } y = x^{p/q}$$

$$\frac{d}{dx}((y(x))^q) = \frac{d}{dx}(x^p) \qquad \leftarrow \text{differentiate both sides}$$

$$qy^{q-1}y' = px^{p-1} \qquad \leftarrow \text{chain rule, integer power rule}$$

$$y' = \frac{px^{p-1}}{qy^{q-1}} = \frac{px^{p-1}}{q(x^{p/q})^{q-1}} \qquad \leftarrow \text{solve for } y', \text{ substitute } y = x^{p/q}$$

$$y' = \frac{px^{p-1}}{qx^{p(q-1)/q}} = \frac{p}{q}x^{((p-1)-(p(q-1)/q))} \qquad \leftarrow \text{algebra}$$

$$y' = \frac{p}{q}x^{p/q-1}. \qquad \leftarrow \text{even more algebra}$$

TEST YOUR UNDERSTANDING

- What is the difference between $f'(g(x))$ and $(f(g(x)))'$?

- The chain rule formula $f'(u(x))u'(x)$ is not exactly the same as the product of the derivatives of f and u. What is the difference?

- What is the difference between saying that y is a function of x and saying that y is an implicit function of x?

- Suppose y is an implicit function of x. Would it be correct or incorrect to say that $\frac{d}{dx}(y^3) = 3y^2$, and why?

- If f is an invertible function with inverse f^{-1}, then what is the relationship between the derivatives of f and f^{-1}?

EXERCISES 2.4

Thinking Back

Differentiation review: Without using the chain rule, find the derivatives of each of the functions f that follow. Some algebra may be required before differentiating.

- $f(x) = (3x + 1)^4$

- $f(x) = \left(\frac{1}{x} + \frac{1}{x^2}\right)^2$

- $f(x) = (x + 1)^2\sqrt{x}$

- $f(x) = \frac{(x + 1)^2}{\sqrt{x}}$

Decomposing functions: For each function k that follows, find functions f, g, and h so that $k = f \circ g \circ h$. There may be more than one possible answer.

- $k(x) = (x^2 + 1)^3$

- $k(x) = \sqrt{1 + (x - 2)^2}$

- $k(x) = \sqrt{\dfrac{1}{3x + 1}}$

- $k(x) = \dfrac{1}{\sqrt{3x + 1}}$

Concepts

0. *Problem Zero:* Read the section and make your own summary of the material.

1. *True/False:* Determine whether each of the statements that follow is true or false. If a statement is true, explain why. If a statement is false, provide a counterexample.

   (a) *True or False:* The chain rule is used to differentiate compositions of functions.

   (b) *True or False:* If f and g are differentiable functions, then the derivative of $f \circ g$ is equal to the derivative of $g \circ f$.

   (c) *True or False:* If f and g are differentiable functions, then $\frac{d}{dx}(f(g(x))) = f'(x)g'(x)$.

   (d) *True or False:* If u and v are differentiable functions, then $\frac{d}{dx}(u(v(x))) = u'(v'(x))$.

   (e) *True or False:* If h and k are differentiable functions, then $\frac{d}{dx}(k(h(x))) = k'(h(x))h'(x)$.

   (f) *True or False:* If y is an implicit function of x, then there can be more than one y-value corresponding to a given x-value.

   (g) *True or False:* The graph of an implicit function can have vertical tangent lines.

   (h) *True or False:* If y is an implicit function of x and $\frac{dy}{dx}\big|_{x=2} = 0$, then the graph of the implicit function has a horizontal tangent line at $(2, 0)$.

2. *Examples:* Construct examples of the thing(s) described in the following. Try to find examples that are different than any in the reading.

   (a) Three functions that we could not have differentiated before learning the chain rule, even after algebraic simplification.

   (b) An equation that defines y as an implicit function of x, but not as a function of x.

   (c) The graph of an implicit function with three horizontal tangent lines and two vertical tangent lines.

3. State the chain rule for differentiating a composition $g(h(x))$ of two functions expressed (a) in "prime" notation and (b) in Leibniz notation.

4. In the text we noted that if $f(u(v(x)))$ was a composition of three functions, then its derivative is $\frac{df}{dx} = \frac{df}{du}\frac{du}{dv}\frac{dv}{dx}$. Write this rule in "prime" notation.

5. Write down a rule for differentiating a composition $f(u(v(w(x))))$ of four functions (a) in "prime" notation and (b) in Leibniz notation.

6. Suppose $u(x) = \sqrt{3x^2 + 1}$ and $f(u) = \frac{u^2 + 3u^5}{1 - u}$. Use the chain rule to find $\frac{d}{dx}(f(u(x)))$ without first finding the formula for $f(u(x))$.

7. Differentiate $f(x) = (3x + \sqrt{x})^2$ in three ways. When you have completed all three parts, show that your three answers are the same:

   (a) with the chain rule

   (b) with the product rule but not the chain rule

   (c) without the chain or product rules

8. Differentiate $f(x) = \left(\frac{x^4 - 2}{\sqrt{x}}\right)^3$ in three ways. When you have completed all three parts, show that your three answers are the same:

   (a) with the chain rule

   (b) with the quotient rule but not the chain rule

   (c) without the chain or quotient rules

Suppose g, h, and j are differentiable functions with the values for the function and derivative given in the following table:

x	$g(x)$	$h(x)$	$j(x)$	$g'(x)$	$h'(x)$	$j'(x)$
-3	0	3	1	1	0	2
-2	1	2	3	2	-3	0
-1	3	0	1	-1	-2	-2
0	2	3	0	-2	3	-2
1	0	-1	-2	-2	-2	-1
2	-2	-2	-3	-1	0	2
3	-3	0	1	0	1	2

Use the table to calculate the values of the derivatives listed in Exercises 9–16.

9. If $f(x) = g(h(x))$, find $f'(3)$.

10. If $f(x) = h(g(x))$, find $f'(3)$.

11. If $f(x) = (g(x))^3$, find $f'(-2)$.

12. If $f(x) = g(x^3 - 6)$, find $f'(2)$.

13. If $f(x) = h(g(j(x)))$, find $f'(1)$.

14. If $f(x) = j(2x)$, find $f'(-1)$.

15. If $f(x) = h(g(x)j(x))$, find $f'(0)$.

16. If $f(x) = h(h(h(x)))$, find $f'(1)$

17. If y is a function of x, then how is the chain rule involved in differentiating y^3 with respect to x, and why?

18. Show that, for any integers p and q (with $q \neq 0$),

$$(p - 1) - \frac{p(q - 1)}{q} = \frac{p}{q} - 1.$$

What does this equation have to do with the current section?

19. Match the two graphs shown here to the equations $(x + 1)(y^2 + y - 1) = 1$ and $xy^2 + y = 1$. Explain your choices.

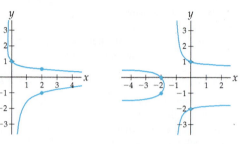

20. Match the two graphs shown here to the equations $xy^2 + x = 1$ and $1 + x + xy^2 = 0$. Explain your choices.

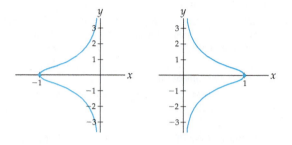

Skills

Find the derivatives of the functions in Exercises 21–46. Keep in mind that it may be convenient to do some preliminary algebra before differentiating.

21. $f(x) = \dfrac{1}{x^3 + 1}$

22. $f(x) = (\sqrt{x} + 4)^5$

23. $f(x) = x(3x^2 + 1)^9$

24. $f(x) = \sqrt{x^2 + 1}$

25. $f(x) = \dfrac{1}{\sqrt{x^2 + 1}}$

26. $f(x) = \dfrac{3x + 1}{\sqrt{x^2 + 1}}$

27. $f(x) = (x\sqrt{x + 1})^{-2}$

28. $f(x) = \dfrac{(1 + \sqrt{x})^2}{3x^2 - 4x + 1}$

29. $f(x) = \dfrac{\dfrac{1}{x} - 3x^2}{x^5 - \dfrac{1}{\sqrt{x}}}$

30. $f(x) = \dfrac{(x + 1)(3x - 4)}{\sqrt{x^3 - 27}}$

31. $f(x) = (x^{1/3} - 2x)^{-1}$

32. $f(x) = \sqrt{2 - \sqrt{3x + 1}}$

33. $f(x) = x^{-1/2}(x^2 - 1)^3$

34. $f(x) = (x^{-1/2}(x^2 - 1))^3$

35. $f(x) = \sqrt{3x - 4(2x + 1)^6}$

36. $f(x) = \dfrac{x^2 - \sqrt[3]{x} + 5x^9}{\sqrt{x^{-1}}}$

37. $f(x) = (5(3x^4 - 1)^3 + 3x - 1)^{100}$

38. $f(x) = (1 - 4x)^2(3x^2 + 1)^9$

39. $f(x) = 3((x^2 + 1)^8 - 7x)^{-2/3}$

40. $f(x) = \dfrac{(3x + 1)(x^4 - 3)^4}{(x + 5)^{-2}(1 + x^2)^5}$

41. $f(x) = (5x^4 - 3x^2)^7(2x^3 + 1)$

42. $f(x) = x\sqrt{3x^2 + 1}\sqrt[3]{2x + 5}$

43. $f(x) = ((2x + 1)^{-5} - 1)^{-9}$

44. $f(x) = x(x^2)(\sqrt{x})(x^{2/3})$

45. $f(x) = (x^4 - \sqrt{3 - 4x})^8 + 5x$

46. $f(x) = \sqrt{1 + \sqrt{1 + \sqrt{1 - 2x}}}$

Calculate each of the derivatives or derivative values in Exercises 47–52.

47. $\dfrac{d^2}{dx^2}((x\sqrt{x + 1})^{-2})$

48. $\dfrac{d^2}{dx^2}\left(\dfrac{\dfrac{1}{x} - 3x^2}{x^5 - \dfrac{1}{\sqrt{x}}}\right)$

49. $\dfrac{d^2}{dx^2}\left(\dfrac{3}{x^{-3/2}\sqrt{x}}\right)\bigg|_{x=2}$

50. $\dfrac{d^2}{dx^2}((5x^4 - 3x^2)^7(2x^3 + 1))$

51. $\dfrac{d^3}{dx^3}(x(3x^2 + 1)^9)|_{x=0}$

52. $\dfrac{d}{dx}\left(\dfrac{x^2 + 1}{(x^2 + 4)(3x - 2)}\right)\bigg|_{x=-1}$

Suppose that r is an independent variable, s is a function of r, and q is a constant. Calculate the derivatives in Exercises 53–58. Your answers may involve r, s, q, or their derivatives.

53. $\dfrac{d}{dr}(s^3)$

54. $\dfrac{d}{dr}(r^3)$

55. $\dfrac{d}{dr}(q^3)$

56. $\dfrac{d}{dr}(sr^2)$

57. $\dfrac{d}{dr}(rs^2)$

58. $\dfrac{d}{dr}(qs^2)$

For each of the equations in Exercises 59–62, y is defined as an implicit function of x. Solve for y, and use what you find to sketch a graph of the equation.

59. $\dfrac{1}{4}x^2 + y^2 = 9$

60. $(x - 1)^2 + (y + 2)^2 = 4$

61. $x^2 - 3y^2 = 16$

62. $4y^2 - x^2 + 25 = 0$

In Exercises 63–68, find a function that has the given derivative and value. In each case you can find the answer with an educated guess-and-check process while thinking about the chain rule.

63. $f'(x) = 5(x^2 + 1)^4(2x)$, $f(0) = 1$

64. $f'(x) = 5(x^2 + 1)^4(2x)$, $f(1) = 25$

65. $f'(x) = x(x^2 + 1)^4$, $f(0) = 0$

66. $f'(x) = -6x(x^2 + 1)^4$, $f(0) = 1$

67. $f'(x) = \sqrt{3x+1}$, $f(0) = 1$

68. $f'(x) = \dfrac{x^2}{\sqrt{x^3+1}}$, $f(1) = 2$

Each of the equations in Exercises 69–80 defines y as an implicit function of x. Use implicit differentiation (*without solving for y first*) to find $\dfrac{dy}{dx}$.

69. $4x^2 - y^2 = 9$

70. $x^2 + y^2 = 4$

71. $xy^2 + 3x^2 = 4$

72. $y^6 - 3x + 4 = 0$

73. $(3x+1)(y^2 - y + 6) = 0$

74. $x^2y - y^2x = x^2 + 3$

75. $\sqrt{3y-1} = 5xy$

76. $(3y^2 + 5xy - 2)^4 = 2$

77. $\dfrac{y^2 + 1}{3y - 1} = x$

78. $3y = 5x^2 + \sqrt[3]{y-2}$

79. $\dfrac{1}{y} - \dfrac{1}{x} = \dfrac{x^2}{y+1}$

80. $\dfrac{x+1}{y^2 - 3} = \dfrac{1}{xy}$

In Exercises 81–84, use implicit differentiation to algebraically find each quantity or location related to the given implicit function.

81. Consider the circle of radius 1 centered at the origin, that is, the solutions of the equation $x^2 + y^2 = 1$.

(a) Find all points on the graph with an x-coordinate of $x = \dfrac{1}{2}$, and then find the slope of the tangent line at each of these points.

(b) Find all points on the graph with a y-coordinate of $y = \dfrac{\sqrt{2}}{2}$, and then find the slope of the tangent line at each of these points.

(c) Find all points on the graph where the tangent line is vertical.

(d) Find all points on the graph where the tangent line has a slope of -1.

82. Consider the graph of the solutions of the equation $4y^2 - x^2 + 2x = 2$.

(a) Find all points on the graph with an x-coordinate of $x = 3$, and then find the slope of the tangent line at each of these points.

(b) Find all points on the graph with a y-coordinate of $y = 3$, and then find the slope of the tangent line at each of these points.

(c) Find all points where the graph has a horizontal tangent line.

(d) Find all points where the graph has a vertical tangent line.

83. Consider the graph of the solutions of the equation $y^3 + xy + 2 = 0$.

(a) Find all points on the graph with an x-coordinate of $x = 1$, and then find the slope of the tangent line at each of these points.

(b) Find all points on the graph with a y-coordinate of $y = 1$, and then find the slope of the tangent line at each of these points.

(c) Find all points where the graph has a horizontal tangent line.

(d) Find all points where the graph has a vertical tangent line.

84. Consider the graph of the solutions of the equation $y^3 - 3y - x = 1$.

(a) Find all points on the graph with an x-coordinate of $x = -1$, and then find the slope of the tangent line at each of these points.

(b) Find all points on the graph with a y-coordinate of $y = 2$, and then find the slope of the tangent line at each of these points.

(c) Find all points where the graph has a horizontal tangent line.

(d) Find all points where the graph has a vertical tangent line.

Applications

85. Linda can sell 12 magazine subscriptions per week and makes $4.00 for each magazine subscription she sells. Obviously this means that Linda will make $(12)(\$4.00) = \$48.00$ per week from magazine subscriptions. Explain this result mathematically, using mathematical notation and the chain rule.

86. If you drop a pebble into a large lake, you will cause a circle of ripples to expand outward. The area $A = A(t)$ and radius $r = r(t)$ are clearly functions of t (they change over time) and are related by the formula $A = \pi r^2$.

Area $A = A(t)$

$r = r(t)$

t, seconds

(a) If r is measured in inches and t is measured in seconds, what are the units of $\dfrac{dA}{dt}$? What are the units of $\dfrac{dA}{dr}$?

(b) Find $\dfrac{dA}{dr}$ and explain the practical meaning of $\dfrac{dA}{dr}\Big|_{r=2}$.

(c) Find $\dfrac{dA}{dt}$ and explain the practical meaning of $\dfrac{dA}{dt}\Big|_{r=2}$.

87. The area of a circle can be written in terms of its radius as $A = \pi r^2$, where both A and r are functions of time. Suppose a circular area from a spotlight on a stage floor is slowly expanding.

(a) Find $\dfrac{dA}{dr}$ and explain its meaning in practical terms.

(b) Does the rate $\dfrac{dA}{dr}$ depend on how fast the radius of the circle is increasing? Does it depend on the radius of the circle? Why or why not?

(c) Find $\dfrac{dA}{dt}$ and explain its meaning in practical terms.

(d) Does the rate $\dfrac{dA}{dt}$ depend on how fast the radius of the circle is increasing? Does it depend on the radius of the circle?

(e) If the radius of the circle of light is increasing at a constant rate of 2 inches per second, how fast is the area of the circle of light increasing at the moment that the spotlight has a radius of 24 inches?

Proofs

88. Use the chain rule twice to prove that $\dfrac{d}{dx}(f(u(v(x)))) = f'(u(v(x)))u'(v(x))v'(x)$.

89. In Exercise 89 of the Section 2.3 you used the definition of derivative to prove the quotient rule. Prove it now another way: by writing a quotient $\dfrac{f}{g}$ as a product and applying the product, power, and chain rules. Point out where you use each rule.

90. Use implicit differentiation and the fact that $\dfrac{d}{dx}(x^4) = 4x^3$ to prove that $\dfrac{d}{dx}(x^{-4}) = -4x^{-5}$.

91. Use implicit differentiation and the fact that $\dfrac{d}{dx}(x^3) = 3x^2$ and $\dfrac{d}{dx}(x^5) = 5x^4$ to prove that $\dfrac{d}{dx}(x^{3/5}) = \dfrac{3}{5}x^{-2/5}$.

92. Use implicit differentiation and the power rule for integer powers (*not* the general power rule) to prove that $\dfrac{d}{dx}(x^{2/3}) = \dfrac{2}{3}x^{-1/3}$.

93. Use implicit differentiation, the product rule, and the power rule for positive integer powers to prove the power rule for negative integer powers.

94. Use implicit differentiation and the power rule for integer powers to prove the power rule for rational powers.

Thinking Forward

Finding critical points: For each of the following functions f, find all of the x-values for which $f'(x) = 0$ and all of the x-values for which $f'(x)$ does not exist.

▶ $f(x) = x^3\sqrt{3x+1}$

▶ $f(x) = (x^2+3)(x-2)^{3/2}$

▶ $f(x) = \dfrac{\sqrt{x}}{x\sqrt{x-1}}$

▶ $f(x) = (1 - x^4)^7$

▶ $f(x) = 3x\left(x + \dfrac{1}{x}\right)$

▶ $f(x) = (x\sqrt{x+1})^{-2}$

Finding antiderivatives by undoing the chain rule: For each function f that follows, find a function F with the property that $F'(x) = f(x)$. You may have to guess and check to find such a function.

▶ $f(x) = x\sqrt{1+x^2}$

▶ $f(x) = \dfrac{1}{(2 - 5x)^3}$

▶ $f(x) = x^2\sqrt{1+x^3}$

▶ $f(x) = \dfrac{1}{\sqrt{1+3x}}$

CHAPTER REVIEW, SELF-TEST, AND CAPSTONES

Before you progress to the next chapter, be sure you are familiar with the definitions, concepts, and basic skills outlined here. The capstone exercises at the end bring together ideas from this chapter and look forward to future chapters.

Definitions

Give precise mathematical definitions or descriptions of each of the concepts that follow. Then illustrate the definition with a graph or algebraic example, if possible.

▶ the graphical interpretations of a *tangent line* and a *secant line* to a graph

▶ the real–world interpretations of *position*, *velocity*, and *acceleration*

▶ the real–world interpretations of *average rate of change* and *instantaneous rate of change*

▶ the formal definition of the *derivative* of a function f at a point $x = c$ (both $z \to x$ form and $h \to 0$ form)

▶ the formal definition of the *derivative* of a function f, as a function (both $z \to x$ form and $h \to 0$ form)

▶ the formal definitions of the *tangent line* and the *instantaneous rate of change* of the graph of a function f at a point $x = c$

▶ what it means for a function f to be *differentiable, left differentiable*, and *right differentiable* at a point $x = c$.

▶ what it means for a function f to be *differentiable* on an open or closed interval I

▶ what it means to say that y is an *implicit function* of x, and the meaning of *implicit differentiation*

Theorems

Fill in the blanks to complete each of the following theorem statements:

▶ If a function f is differentiable at $x = c$, then f is _____ at $x = c$.

▶ If f and f^{-1} are inverse functions, then for all appropriate values of x we can write the derivative of $f^{-1}(x)$ in terms of the derivative of $f'(x)$ as follows: _____ .

Notation and Differentiation Rules

Leibniz notation: Describe the meanings of each of the mathematical expressions that follow. Translate expressions written in Leibniz notation to "prime" notation, and vice versa.

▶ $f'(x)$ ▶ $f'(2)$ ▶ $f''(x)$

▶ $\dfrac{df}{dx}$ ▶ $\dfrac{dy}{dx}$ ▶ $\dfrac{d}{dx}(y(x))$

▶ $\dfrac{dg}{dt}\Big|_{t=3}$ ▶ $\dfrac{d}{dx}(x^2)$ ▶ $\dfrac{d}{dx}\Big|_{-1}(x^2)$

▶ $\dfrac{d^2y}{dx^2}$ ▶ $\dfrac{d}{dx}\left(\dfrac{d}{dx}(x^2)\right)$ ▶ $f^{(5)}(x)$

Derivatives of basic functions: Fill in the blanks to differentiate each of the given basic functions. You may assume that k, m, and b are appropriate constants.

▶ $\dfrac{d}{dx}(k) =$ _____   ▶ $\dfrac{d}{dx}(x) =$ _____

▶ $\dfrac{d}{dx}(mx + b) =$ _____   ▶ $\dfrac{d}{dx}(x^k) =$ _____

▶ $\dfrac{d}{dx}(\sqrt{x}) =$ _____   ▶ $\dfrac{d}{dx}\left(\dfrac{1}{x}\right) =$ _____

Derivatives of combinations: Fill in the blanks to complete each of the given differentiation rules. You may assume that f and g are differentiable everywhere.

▶ $(kf)'(x) =$ _____   ▶ $(f + g)'(x) =$ _____

▶ $(f - g)'(x) =$ _____   ▶ $(fg)'(x) =$ _____

▶ $\left(\dfrac{f}{g}\right)'(x) =$ _____   ▶ $(f \circ g)'(x) =$ _____

Skill Certification: Basic Derivatives

Basic definition-of-derivative calculations: Find the derivatives of the functions that follow, using (a) the $h \to 0$ definition of the derivative and (b) the $z \to x$ definition of the derivative.

1. $f(x) = x^2$ **2.** $f(x) = x^3$

3. $f(x) = \dfrac{1}{x}$ **4.** $f(x) = \dfrac{1}{x^2}$

Calculating basic derivatives: Find the derivatives of the functions that follow, using the differentiation rules developed in this chapter.

5. $f(x) = 4(3x^2 - 7)^3$ **6.** $f(x) = (x^2 + 1)(x - 2)$

7. $f(x) = \pi^3$ **8.** $f(x) = \dfrac{x^3 - 2x^2}{3x}$

9. $f(x) = \dfrac{1}{x^4 - 5x^3 + 2}$ **10.** $f(x) = \dfrac{x - 2}{(x - 1)(x + 3)}$

11. $f(x) = (x\sqrt{2x - 1})^{-3}$ **12.** $f(x) = x^{-1}(4 - x)^2$

13. $f(x) = \dfrac{1 - (1/x)}{\sqrt{x}}$ **14.** $f(x) = \dfrac{3x\sqrt{2x + 1}}{1 - x}$

15. $f(x) = |x|$ **16.** $f(x) = |3x + 1|$

Calculating antiderivatives: For each exercise that follows, find a function f that has the given derivative f' and value $f(c)$. In each case you can find the answer with an educated guess-and-check process.

17. $f'(x) = -32$, $f(0) = 4$

18. $f'(x) = -32x + 4$, $f(0) = 100$

19. $f'(x) = x(3x + 1)$, $f(2) = 4$

20. $f'(x) = \dfrac{3x^2}{\sqrt{x^3 + 1}}$, $f(2) = 6$

Differentiating with respect to different variables: Find each derivative described.

21. If $3v^2 + xv - 1 = 0$, find $\dfrac{dv}{dx}$.

22. If $3v^2 + xv - 1 = 0$, find $\dfrac{dx}{dv}$.

23. If $A = \pi r^2$, find $\dfrac{dA}{dr}$.

24. If $A = \pi r^2$, and A and r are both functions of time t, find $\dfrac{dA}{dt}$.

25. If $y = 3x^2t - t^k$, where x and k are constant, find $\dfrac{dy}{dt}$.

26. If $y = 3x^2t - t^k$, where t and k are constant, find $\dfrac{dy}{dx}$.

Capstone Problems

A. *The sum rule for differentiation:* Use the definition of the derivative to prove that the derivative of a sum of functions $f(x) + g(x)$ is equal to the sum of their derivatives $f'(x) + g'(x)$.

B. *The power rule for differentiation:* Use the definition of the derivative and factoring formulas to prove that for any positive integer k, the derivative of x^k is kx^{k-1}.

C. *Rates of change from data:* The following table lists the consumption of gasoline in billions of gallons in the United States from 1994 to 2000:

Year	1994	1995	1996	1997	1998	1999	2000
Gas	109	111	113	117	118	121	122

(a) Compute the average rate of change in gasoline consumption in the United States for each year from 1994 to 2000.

(b) During which year was gasoline consumption increasing most rapidly? Least rapidly? Estimate the instantaneous rates of change in gasoline consumption during those years.

D. *Derivatives and graphical behavior:* In the next chapter we will see that we can get a lot of information about the graph of a function f by looking at the signs of $f(x)$ and its first and second derivatives. Let's do this for the function $f(x) = x^3 - 3x^2 - 9x + 27$.

(a) Find the roots of f, and then determine the intervals on which f is positive or negative.

(b) Find the roots of f', and then determine the intervals on which f' is positive or negative.

(c) Find the roots of f'', and then determine the intervals on which f'' is positive or negative.

(d) The graph of f will be above the x-axis when $f(x)$ is positive and below the x-axis when $f(x)$ is negative. Moreover, the graph of f will be increasing when f' is positive and decreasing when f' is negative. Finally, the graph of f will be concave up when f'' is positive and concave down when f'' is negative. Given this information and your answers from parts (a)–(c), sketch a careful, labeled graph of f.

Applications of the Derivative

3.1   THE MEAN VALUE THEOREM

▶ Local extrema, critical points, and their relationships

▶ Rolle's Theorem and the Mean Value Theorem

▶ Using critical points to find local extrema

The Derivative at a Local Extremum

Suppose a function f has a local maximum at some point $x = c$. This means that the value $f(c)$ is greater than or equal to all other nearby $f(x)$ values. The following definition makes this notion precise:

DEFINITION 3.1   **Local Extrema of a Function**

(a) f has a *local maximum* at $x = c$ if there exists some $\delta > 0$ such that $f(c) \geq f(x)$ for all $x \in (c - \delta, c + \delta)$.

(b) f has a *local minimum* at $x = c$ if there exists some $\delta > 0$ such that $f(c) \leq f(x)$ for all $x \in (c - \delta, c + \delta)$.

Intuitively, at a local extremum, the tangent line of a function must be either horizontal or undefined; for example, consider the following three graphs:

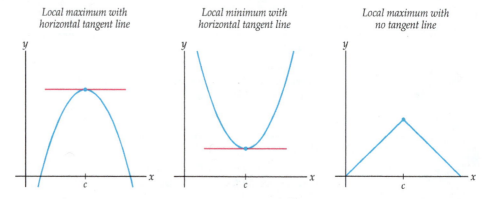

Local maximum with horizontal tangent line   Local minimum with horizontal tangent line   Local maximum with no tangent line

When a function has a horizontal or an undefined tangent line at a point, its derivative at that point is either zero or undefined. We call such points *critical points* of the function:

DEFINITION 3.2   **Critical Points of a Function**

A point $x = c$ in the domain of f is called a *critical point* of f if $f'(c) = 0$ or $f'(c)$ does not exist.

Notice that only points in the domain of f are considered critical points. For example, consider the function $f(x) = \frac{1}{x}$, whose derivative is $f'(x) = -\frac{1}{x^2}$. Clearly $f'(0)$ does not exist; however, since $x = 0$ is not in the domain of f, it is not called a critical point.

The preceding graphs suggest that every local extremum is also a critical point. This seemingly obvious relationship between critical points and extrema turns out to be the foundation on which we will build two more theorems that are key in our development of calculus:

THEOREM 3.3   **Local Extrema are Critical Points**

If $x = c$ is the location of a local extremum of f, then $x = c$ is a critical point of f.

The converse of this theorem is not true. That is, not every critical point is a local extremum of f; see Example 1. Although the implication in Theorem 3.3 is intuitively obvious just by thinking about graphs and the behavior of the derivative at local maxima and minima, actually proving it requires a somewhat subtle argument. The key is to look at the left and right derivatives at a local extremum. Theorem 3.3 is also known as *Fermat's Theorem for Local Extrema*, when formulated equivalently as saying that if $x = c$ is a local extremum and f is differentiable at $x = c$, then $f'(c)$ must be zero.

Proof. We will prove the case for local maxima and leave the similar proof for local minima to Exercise 62. Suppose $x = c$ is the location of a local maximum of f. If $f'(c)$ does not exist, then $x = c$ is a critical point and we are done. It now suffices to show that if $f'(c)$ exists, then it must be equal to 0. We will do so by examining the right and left derivatives at $x = c$.

Since $x = c$ is the location of a local maximum of f, there is some $\delta > 0$ such that for all $x \in (c - \delta, c + \delta)$, $f(c) \geq f(x)$, and thus $f(x) - f(c) \leq 0$. In the case where $x \in (c, c + \delta)$, it follows that $x > c$, which means that $x - c$ is positive. Thus in this case $\frac{f(x) - f(c)}{x - c} \leq 0$, and therefore

$$f'_+(c) = \lim_{x \to c^+} \frac{f(x) - f(c)}{x - c} \leq 0.$$

By a similar argument for $x \in (c - \delta, c)$, we have $x - c < 0$ and $f(x) - f(c) \leq 0$, and therefore

$$f'_-(c) = \lim_{x \to c^-} \frac{f(x) - f(c)}{x - c} \geq 0.$$

Since we are assuming that $f'(c)$ exists, we know that both $f'_+(c)$ and $f'_-(c)$ must exist and be equal to $f'(c)$. We have just shown both that $f'(c) \leq 0$ and that $f'(c) \geq 0$. Therefore, we must have $f'(c) = 0$, as desired.   ☐

Rolle's Theorem

Suppose a differentiable function f has two roots $x = a$ and $x = b$. What can you say about the graph of f between a and b? The three graphs that follow next provide a clue; if the graph of a function is smooth and unbroken, then somewhere between each root of f the function must turn around, and at that turning point it must have a local extremum with a horizontal tangent line:

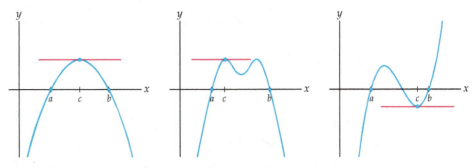

The preceding discussion is a summary of both the statement and the proof of the following key theorem:

THEOREM 3.4

Rolle's Theorem

If f is continuous on $[a, b]$ and differentiable on (a, b), and if $f(a) = f(b) = 0$, then there exists at least one value $c \in (a, b)$ for which $f'(c) = 0$.

Actually, Rolle's Theorem also holds in the more general case where $f(a)$ and $f(b)$ are equal to each other (not necessarily both zero). For example, Rolle's Theorem is also true if $f(a) = f(b) = 5$, or if $f(a) = f(b) = -3$, and so on, because vertically shifting a function by adding a constant term does not change its derivative. However, the classic way to state Rolle's Theorem is with $f(a)$ and $f(b)$ both equal to zero.

> **Proof.** Rolle's Theorem is an immediate consequence of the Extreme Value Theorem from Section 1.4 and the fact that every extremum is a critical point. Suppose f is continuous on the closed interval $[a, b]$ and differentiable on the open interval (a, b), with $f(a) = f(b) = 0$. By the Extreme Value Theorem, we know that f attains both a maximum and a minimum value on $[a, b]$. If one of these extreme values occurs at a point $x = c$ in the interior (a, b) of the interval, then $x = c$ is a local extremum of f. By the previous theorem, this means that $x = c$ is a critical point of f. Since f is assumed to be differentiable at $x = c$, it follows that $f'(c) = 0$ and we are done.
>
> It remains to consider the special case where all of the maximum and minimum values of f on $[a, b]$ occur at the endpoints of the interval (i.e., at $x = a$ or at $x = b$). In this case, since $f(a) = f(b) = 0$, the maximum and minimum values of $f(x)$ must both equal zero. For all x in $[a, b]$ we would have $0 \le f(x) \le 0$, which means that f would have to be the constant function $f(x) = 0$ on $[a, b]$. Since the derivative of a constant function is always zero, in this special case we have $f'(x) = 0$ for *all* values of c in (a, b), and we are done. ∎

Just as the Intermediate Value Theorem and the Extreme Value Theorem illustrate basic properties of continuous functions, Rolle's Theorem illustrates a basic property of functions that are both continuous and differentiable. Like those two theorems before, Rolle's Theorem is a theorem about *existence*, not calculation; it tells you that there must exist some value $c \in (a, b)$ where the derivative of f is zero, but it does not tell you what that value is. It is important to note that the continuity and the differentiability hypotheses of Rolle's Theorem are essential: If a function f fails to be continuous on $[a, b]$ or fails to be differentiable on (a, b), then the conclusion of Rolle's Theorem does not necessarily follow; see Example 2.

The Mean Value Theorem

The Mean Value Theorem is a generalization of Rolle's Theorem to the case where $f(a)$ and $f(b)$ are not necessarily equal. Suppose f is a continuous, differentiable function. What can we say about the derivative of f between two points $x = a$ and $x = b$? The three graphs that follow suggest an answer: Somewhere between a and b the slope of the tangent line must be the same as the slope of the line from $(a, f(a))$ to $(b, f(b))$. If you turn your head so that the green line is horizontal in each figure, you can see that these figures are similar to rotated versions of the earlier figures used in illustrating Rolle's Theorem.

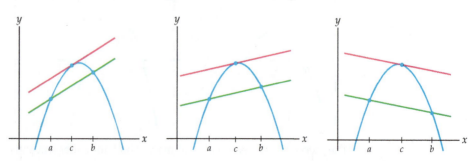

Algebraically, this means that there must be some $c \in (a, b)$ whose derivative value $f'(c)$ is equal to the average rate of change of f on $[a, b]$:

THEOREM 3.5   **The Mean Value Theorem**

If f is continuous on $[a, b]$ and differentiable on (a, b), then there exists at least one value $c \in (a, b)$ such that

$$f'(c) = \frac{f(b) - f(a)}{b - a}.$$

The "mean" in the "Mean Value Theorem" refers to an ***average***. Basically, this theorem says that if a function is continuous on a closed interval and differentiable on its interior, then there is always at least one place in the interval where the instantaneous rate of change of the function is equal to its average rate of change over the whole interval. As a real–world example, suppose you drove at an average speed of 50 miles per hour on a short road trip. The Mean Value Theorem guarantees that at some point along your journey you must have been travelling at exactly 50 miles per hour.

The Mean Value Theorem is intuitively clear if you believe that you can just "turn your head to the side" and see Rolle's Theorem. In fact, the proof of the Mean Value Theorem is based on an algebraic version of this intuition:

Proof.   Suppose f is a function that is continuous on $[a, b]$ and differentiable on (a, b), and let $l(x)$ be the secant line from $(a, f(a))$ to $(b, f(b))$. The idea of the proof is to "turn our heads" algebraically. To rotate so that the secant line $l(x)$ plays the role of the x-axis, we will consider the function $g(x) = f(x) - l(x)$. The graph of this new function $g(x)$ will have roots at $x = a$ and $x = b$, and we will be able to apply Rolle's Theorem.

Since the secant line $l(x)$ has slope $\frac{f(b) - f(a)}{b - a}$ and passes through the point $(a, f(a))$, its equation is

$$l(x) = \frac{f(b) - f(a)}{b - a}(x - a) + f(a).$$

This means that the function $g(x) = f(x) - l(x)$ is equal to

$$g(x) = f(x) - \frac{f(b) - f(a)}{b - a}(x - a) - f(a).$$

If we want to apply Rolle's Theorem to $g(x)$, then we must first verify that $g(x)$ satisfies all the hypotheses of Rolle's Theorem. First, $g(x)$ is continuous on $[a, b]$ because it is a combination of continuous functions. Second, $g(x)$ is differentiable on (a, b) because f is differentiable on (a, b). Finally, $g(a) = 0$ and $g(b) = 0$:

$$g(a) = f(a) - \frac{f(b) - f(a)}{b - a}(a - a) - f(a) = f(a) - 0 - f(a) = 0,$$

$$g(b) = f(b) - \frac{f(b) - f(a)}{b - a}(b - a) - f(a) = f(b) - (f(b) - f(a)) - f(a) = 0.$$

Since Rolle's Theorem applies to the function $g(x)$, we can conclude that there exists some $c \in (a, b)$ for which $g'(c) = 0$. How does this conclusion relate to our original problem? To answer that, we must first calculate $g'(x)$:

$$g'(x) = \frac{d}{dx}\left(f(x) - \frac{f(b) - f(a)}{b - a}(x - a) - f(a)\right) \qquad \leftarrow \text{definition of } g(x)$$

$$= f'(x) - \frac{f(b) - f(a)}{b - a}(1) - 0 \qquad \leftarrow \frac{f(b) - f(a)}{b - a} \text{ and } f(a) \text{ are constants}$$

$$= f'(x) - \frac{f(b) - f(a)}{b - a}. \qquad \leftarrow \text{simplify}$$

Rolle's Theorem now guarantees that there exists a $c \in (a, b)$ for which $g'(c) = 0$. By our previous calculation, for this value of c we have $g'(c) = f'(c) - \frac{f(b) - f(a)}{b - a} = 0$, and therefore $f'(c) = \frac{f(b) - f(a)}{b - a}$, as desired.   ∎

Examples and Explorations

EXAMPLE 1   **Not every critical point is a local extremum**

Show that $x = 1$ is a critical point of $f(x) = x^3 - 3x^2 + 3x$. Then use a graph to show that $x = 1$ is not a local extremum of f.

SOLUTION

If $f(x) = x^3 - 3x^2 + 3x$, then $f'(x) = 3x^2 - 6x + 3$ and thus $f'(1) = 3(1)^2 - 6(1) + 3 = 0$. Therefore $x = 1$ is a critical point of f. However, looking at the following graph of f, we can see that f has neither a local minimum nor a local maximum at $x = 1$:

x = 1 is a critical point but not an extremum

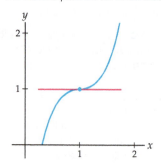

EXAMPLE 2   **The hypotheses of Rolle's Theorem are important**

Sketch graphs of three functions that fail to satisfy the hypotheses of Rolle's Theorem, for which the conclusion of Rolle's Theorem does not follow.

SOLUTION

The function f in the first graph that follows at the right fails to be differentiable on $(1, 3)$, and therefore can "turn around" at $x = 2$ without having a horizontal tangent line. For this function, there is no value of $c \in (a, b)$ with $f'(c) = 0$.

f is not differentiable on (1, 3)   *g is not continuous on [1, 3]*   *h is not continuous on [1, 3]*

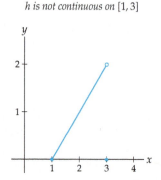

In the second graph, the function $g(x)$ fails to be continuous at the very place where we would have expected its derivative to be zero. Since this function is not continuous at $x = 2$, it is also not differentiable at $x = 2$, and there is no value $c \in (1, 3)$ with $g'(c) = 0$.

The third graph illustrates a function $h(x)$ that fails to be continuous at the right endpoint $x = 3$ of the interval. There is no value $c \in (1, 3)$ with $h'(c) = 0$; the function never has to "turn around," since it just jumps down to the root at $x = 3$.

EXAMPLE 3   Applying Rolle's Theorem

Use Rolle's Theorem to show that there must exist some value of c in $(-2, 5)$ at which the function $f(x) = x^2 - 3x - 10$ has a horizontal tangent line. Then use f' to find such a value c algebraically, and verify your answer with a graph.

SOLUTION

First notice that $f(x) = x^2 - 3x - 10 = (x + 2)(x - 5)$ has roots at $x = -2$ and at $x = 5$. Since f is a polynomial, it is continuous and differentiable. In particular, it is continuous on $[-2, 5]$ and differentiable on $(-2, 5)$. Therefore Rolle's Theorem applies to the function f, and we can conclude that there must exist some value of $c \in (-2, 5)$ for which $f'(c) = 0$. At this value of c the graph of f will have a horizontal tangent line.

Rolle's Theorem tells us that there exists some $c \in (-2, 5)$ where $f'(c) = 0$, but it doesn't tell us exactly where. We can find such a c by solving the equation $f'(x) = 0$. Since $f(x) = x^2 - 3x - 10$, we have $f'(x) = 2x - 3$, which is equal to zero when $x = \frac{3}{2}$. Therefore f has a horizontal tangent line at $c = \frac{3}{2}$, which is in the interval $(-2, 5)$. The following graph illustrates that $f(x) = x^2 - 3x - 10$ does appear to have a horizontal tangent line at $x = \frac{3}{2}$.

Horizontal tangent at $x = \dfrac{3}{2}$

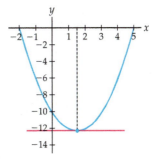

EXAMPLE 4   Using critical points and Rolle's Theorem to find local extrema

The function $f(x) = x(x - 1)(x - 3)$ is a cubic polynomial with one local maximum and one local minimum. Use Rolle's Theorem to identify intervals on which these extrema exist. Then use derivatives to find the exact locations of these extrema.

SOLUTION

The roots of $f(x) = x(x - 1)(x - 3)$ are $x = 0$, $x = 1$, and $x = 3$. Since f is a polynomial, it is continuous and differentiable everywhere. Therefore Rolle's Theorem applies on the intervals $[0, 1]$ and $[1, 3]$, and it tells us that at least one critical point must exist inside each of these intervals.

The critical points of f are the possible locations of the local extrema that we seek. To find the critical points we must solve the equation $f'(x) = 0$. It is simpler to do some algebra before differentiating:

$$f'(x) = \frac{d}{dx}(x(x - 1)(x - 3)) = \frac{d}{dx}(x^3 - 4x^2 + 3x) = 3x^2 - 8x + 3.$$

By the quadratic formula, we have $f'(x) = 0$ at the points

$$x = \frac{-(-8) \pm \sqrt{8^2 - 4(3)(3)}}{2(3)} = \frac{8 \pm \sqrt{28}}{6} = \frac{4 \pm \sqrt{7}}{3}.$$

These x-values are approximately $x \approx 0.451$ and $x \approx 2.215$. If we look at the graph of f, then we can see that the smaller of these two x-values is the location of the local maximum and the larger is the location of the local minimum; see the figure that follows. □

The graph of $f(x) = x(x - 1)(x - 3)$ is shown next. Notice that the local extrema do seem to occur at the values we just found.

Extrema at $x \approx 0.451$ and $x \approx 2.215$

EXAMPLE 5   **Applying the Mean Value Theorem**

Use the Mean Value Theorem to show that there is some value $c \in (0, 2)$ at which the tangent line to the function $f(x) = x^2 - 2$ has slope 2. Then use f' to find such a value c algebraically, and verify your answer with a graph.

SOLUTION

The function $f(x) = x^2 - 2$ is a polynomial and thus is continuous and differentiable; in particular it is continuous on $[0, 2]$ and differentiable on $(0, 2)$. Therefore, the Mean Value Theorem applies to this function on the interval $[0, 2]$. The slope of the line from $(0, f(0))$ to $(2, f(2))$ is

$$\frac{f(2) - f(0)}{2 - 0} = \frac{(2^2 - 2) - (0^2 - 2)}{2} = \frac{2 - (-2)}{2} = \frac{4}{2} = 2.$$

By the Mean Value Theorem, there must exist at least one point $c \in (0, 2)$ with $f'(c) = 2$.

To find such a value of c algebraically, observe that the derivative of $f(x) = x^2 - 2$ is $f'(x) = 2x + 0 = 2x$. We want to find $c \in (0, 2)$ with $f'(c) = 2$, so we solve:

$$f'(c) = 2 \implies 2c = 2 \implies c = 1.$$

The point $c = 1$ is indeed in the interval $(0, 2)$, and $f'(1) = 2$, so we are done. The following figure illustrates that $f(x) = x^2 - 2$ does appear to have the same slope at $x = 1$ as the secant line from $(0, f(0))$ to $(2, f(2))$.

$f'(1)$ equals average rate of change on $[0, 2]$

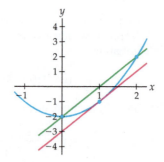

□

? **TEST YOUR** **UNDERSTANDING**

▶ Is every critical point a local extremum? Is every local extremum a critical point?

▶ What is the role of δ in the definitions of local extrema given in Definition 3.1?

▶ Why do you think $f(a) = f(b)$ would be a sufficient hypothesis in the statement of Rolle's Theorem? Think about the situation graphically.

▶ Can Rolle's Theorem tell you the exact location of a root of f'?

▶ How is the Mean Value Theorem related to Rolle's Theorem?

EXERCISES 3.1

Thinking Back

Review of definitions and theorems: State each theorem or definition that follows in precise mathematical language. Then give an illustrative graph or example, as appropriate.

▶ f has a local maximum at $x = c$

▶ f has a local minimum at $x = c$

▶ f is continuous on $[a, b]$

▶ f is differentiable on (a, b)

▶ The secant line from $(a, f(a))$ to $(b, f(b))$

▶ The right derivative $f'_+(c)$ at a point $x = c$

▶ The left derivative $f'_-(c)$ at a point $x = c$

▶ The Extreme Value Theorem

▶ The Intermediate Value Theorem

Concepts

0. *Problem Zero:* Read the section and make your own summary of the material.

1. *True/False:* Determine whether each of the statements that follow is true or false. If a statement is true, explain why. If a statement is false, provide a counterexample.

(a) *True or False:* Rolle's Theorem is a special case of the Mean Value Theorem.

(b) *True or False:* The Mean Value Theorem is so named because it concerns the average (or "mean") rate of change of a function on an interval.

(c) *True or False:* If f is differentiable on $\mathbb{R}$ and has an extremum at $x = -2$, then $f'(-2) = 0$.

(d) *True or False:* If f has a critical point at $x = 1$, then f has a local minimum or maximum at $x = 1$.

(e) *True or False:* If f is any function with $f(2) = 0$ and $f(8) = 0$, then there is some c in the interval $(2, 8)$ such that $f'(c) = 0$.

(f) *True or False:* If f is continuous and differentiable on $[-2, 2]$ with $f(-2) = 4$ and $f(2) = 0$, then there is some $c \in (-2, 2)$ with $f'(c) = -1$.

(g) *True or False:* If f is continuous and differentiable on $[0, 10]$ with $f'(5) = 0$, then f has a local maximum or minimum at $x = 5$.

(h) *True or False:* If f is continuous and differentiable on $[0, 10]$ with $f'(5) = 0$, then there are some values a and b in $(0, 10)$ for which $f(a) = 0$ and $f(b) = 0$.

2. *Examples:* Construct examples of the thing(s) described in the following. Try to find examples that are different than any in the reading.

(a) A function with a local minimum at $x = 3$ that is continuous but not differentiable at $x = 3$.

(b) A function with a local maximum at $x = -2$ that is not differentiable at $x = -2$ because of a removable discontinuity.

(c) A function with a local minimum at $x = 1$ that is not differentiable at $x = 1$ because of a jump discontinuity.

3. If f has a local maximum at $x = 1$, then what can you say about $f'(1)$? What if you also know that f is differentiable at $x = 1$?

4. If f has a local minimum at $x = 0$ and a local maximum at $x = 2$, what can you say about $f'(0)$ and $f'(2)$? Is there anything else you can say about f'?

5. Suppose that f is defined on $(-\infty, \infty)$ and differentiable everywhere except at $x = -2$ and $x = 4$, and that $f'(x) = 0$ only at $x = 0$ and $x = 5$. List all the critical points of f and sketch a possible graph of f.

6. Suppose that f is defined for $x \neq 0$ and differentiable everywhere except at $x = 0$ and $x = 1$, and that $f'(x) = 0$ only at $x = \pm 2$. List all the critical points of f and sketch a possible graph of f.

7. If a continuous, differentiable function f has zeroes at $x = -4$, $x = 1$, and $x = 2$, what can you say about f' on $[-4, 2]$?

8. If a continuous, differentiable function f is equal to 2 at $x = 3$ and at $x = 5$, what can you say about f' on $[3, 5]$?

9. If a continuous, differentiable function f has values $f(-2) = 3$ and $f(4) = 1$, what can you say about f' on $[-2, 4]$?

10. Restate Theorem 3.3 so that its conclusion has to do with tangent lines.

11. Restate Rolle's Theorem so that its conclusion has to do with tangent lines.

12. Restate the Mean Value Theorem so that its conclusion has to do with tangent lines.

In Exercises 13–22, sketch the graph of a function that satisfies the given description. Label or annotate your graph so that it is clear that it satisfies each part of the description.

13. A function that satisfies the hypothesis, and therefore the conclusion, of Rolle's Theorem on $[2, 6]$.

14. A function that satisfies the hypothesis, and therefore the conclusion, of the Mean Value Theorem.

15. A function f that satisfies the hypotheses of Rolle's Theorem on $[-2, 2]$ and for which there are exactly three values $c \in (-2, 2)$ that satisfy the conclusion of the theorem.

16. A function f that satisfies the hypothesis of the Mean Value Theorem on $[0, 4]$ and for which there are exactly three values $c \in (0, 4)$ that satisfy the conclusion of the theorem.

17. A function f that is defined on $[-2, 2]$ with $f(-2) = f(2) = 0$ such that f is continuous everywhere, differentiable everywhere except at $x = -1$, and fails the conclusion of Rolle's Theorem.

18. A function f defined on $[1, 5]$ with $f(1) = f(5) = 0$ such that f is continuous everywhere except for $x = 2$, differentiable everywhere except at $x = 2$, and fails the conclusion of Rolle's Theorem.

19. A function f defined on $[-3, -1]$ with $f(-3) = f(-1) = 0$ such that f is continuous everywhere except at $x = -1$ and differentiable everywhere except at $x = -1$, and fails the conclusion of Rolle's Theorem.

20. A function f defined on $[0, 4]$ such that f is continuous everywhere, differentiable everywhere except at $x = 2$, and fails the conclusion of the Mean Value Theorem with $a = 0$ and $b = 4$.

21. A function f defined on $[-3, 3]$ such that f is continuous everywhere except at $x = 1$, differentiable everywhere except at $x = 1$, and fails the conclusion of the Mean Value Theorem with $a = -3$ and $b = 3$.

22. A function f defined on $[-2, 0]$ such that f is continuous everywhere except at $x = -2$, differentiable everywhere except at $x = -2$, and fails the conclusion of the Mean Value Theorem with $a = -2$ and $b = 0$.

Skills

For each graph of f in Exercises 23–26, approximate all the values $x \in (0, 4)$ for which the derivative of f is zero or does not exist. Indicate whether f has a local maximum, minimum, or neither at each of these critical points.

23.

24.

25.

26.
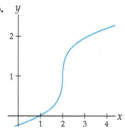

Find the critical points of each function f in Exercises 27–34. Then use a graphing utility to determine whether f has a local minimum, a local maximum, or neither at each of these critical points.

27. $f(x) = x^2 - 3x + 1$

28. $f(x) = 3x + 1$

29. $f(x) = (x - 1)^3$

30. $f(x) = x^3 - 1$

31. $f(x) = (x - 1.7)(x + 3)$

32. $f(x) = x^3 + x^2 + 1$

33. $f(x) = 3x^4 + 8x^3 - 18x^2$

34. $f(x) = (2x - 1)^5$

For each graph of f in Exercises 35–38, explain why f satisfies the hypotheses of Rolle's Theorem on the given interval $[a, b]$. Then approximate any values $c \in (a, b)$ that satisfy the conclusion of Rolle's Theorem.

35. $[a, b] = [-3, 1]$
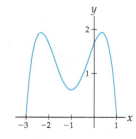

36. $[a, b] = [-3, 3]$
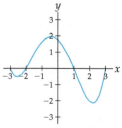

37. $[a, b] = [0, 4]$
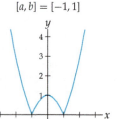

38. $[a, b] = [-1, 1]$

Determine whether or not each function f in Exercises 39–46 satisfies the hypotheses of Rolle's Theorem on the given interval $[a, b]$. For those that do, use derivatives and algebra to find the exact values of all $c \in (a, b)$ that satisfy the conclusion of Rolle's Theorem.

39. $f(x) = x^2 - 3x - 4$, $[a, b] = [-1, 4]$

40. $f(x) = 3 + 2x - x^2$, $[a, b] = [-1, 3]$

41. $f(x) = \dfrac{1}{x-1}$, $[a,b] = [0,3]$

42. $f(x) = \dfrac{1}{x-1}$, $[a,b] = [2,3]$

43. $f(x) = x^3 - 4x^2 + 3x$, $[a,b] = [0,3]$

44. $f(x) = x^3 - 4x^2 + 3x$, $[a,b] = [1,3]$

45. $f(x) = x^4 - 3.24x^2 - 3.04$, $[a,b] = [-2,2]$

46. $f(x) = \dfrac{x^2 - 4x}{x^2 - 4x + 3}$, $[a,b] = [0,4]$

For each graph of f in Exercises 47–50, explain why f satisfies the hypotheses of the Mean Value Theorem on the given interval $[a,b]$ and approximate any values $c \in (a,b)$ that satisfy the conclusion of the Mean Value Theorem.

47. $[a,b] = [0,2]$ **48.** $[a,b] = [-1,3]$

    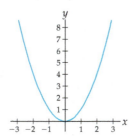

49. $[a,b] = [-3,0]$ **50.** $[a,b] = [0,4]$

Determine whether or not each function f in Exercises 51–58 satisfies the hypotheses of the Mean Value Theorem on the given interval $[a,b]$. For those that do, use derivatives and algebra to find the exact values of all $c \in (a,b)$ that satisfy the conclusion of the Mean Value Theorem.

51. $f(x) = 4 - x^2$, $[a,b] = [-1,3]$

52. $f(x) = \dfrac{1}{3x+1}$, $[a,b] = [-2,2]$

53. $f(x) = (x-1)(x+3)$, $[a,b] = [-3,2]$

54. $f(x) = (x-2)(x+5)$, $[a,b] = [0,4]$

55. $f(x) = x^2 + \dfrac{1}{x}$, $[a,b] = [-3,2]$

56. $f(x) = x^3 - 2x + 1$, $[a,b] = [0,6]$

57. $f(x) = -x^3 + 3x^2 - 7$, $[a,b] = [-2,3]$

58. $f(x) = (x^2 - 1)(x^2 - 4)$, $[a,b] = [-3,3]$

Applications

59. The cost of manufacturing a container for frozen orange juice is $C(h) = h^2 - 7.4h + 13.7$ cents, where h is the height of the container in inches. Your boss claims that the containers will be cheapest to make if they are 4 inches tall. Use Theorem 3.3 to quickly show that he is wrong.

60. Last night at 6 P.M., Linda got up from her blue easy chair. She did not return to her easy chair until she sat down again at 8 P.M. Let $s(t)$ be the distance between Linda and her easy chair t minutes after 6 P.M. last night.

  (a) Sketch a possible graph of $s(t)$, and describe what Linda did between 6 P.M. and 8 P.M. according to your graph. (Questions to think about: Will Linda

necessarily move in a continuous and differentiable way? What are good ranges for t and s?)

  (b) Use Rolle's Theorem to show that at some point between 6 P.M. and 8 P.M., Linda's velocity $v(t)$ with respect to the easy chair was zero. Find such a place on the graph of $s(t)$.

61. It took Alina half an hour to drive to the grocery store that is 20 miles from her house.

  (a) Use the Mean Value Theorem to show that, at some point during her trip, Alina must have been travelling exactly 40 miles per hour.

  (b) Why does what you have shown in part (a) make sense in real-world terms?

Proofs

62. Prove the part of Theorem 3.3 that was not proved in the reading: If a function f has a local minimum at $x = c$, then either $f'(c)$ does not exist or $f'(c) = 0$.

63. Prove Rolle's Theorem: If f is continuous on $[a,b]$ and differentiable on (a,b), and if $f(a) = f(b) = 0$, then there is some value $c \in (a,b)$ with $f'(c) = 0$.

64. Prove the Mean Value Theorem: If f is continuous on $[a,b]$ and differentiable on (a,b), then there is some value $c \in (a,b)$ with $f'(c) = \dfrac{f(b)-f(a)}{b-a}$.

65. Use Rolle's Theorem to prove that if f is continuous and differentiable everywhere and has three roots, then its derivative f' has at least two roots.

66. Follow the method of proof that we used for Rolle's Theorem to prove the following slightly more general theorem: If f is continuous on $[a,b]$ and differentiable on (a,b), and if $f(a) = f(b)$, then there is some value $c \in (a,b)$ with $f'(c) = 0$.

67. Use Rolle's Theorem to prove the slightly more general theorem from Exercise 66: If f is continuous on $[a,b]$ and differentiable on (a,b), and if $f(a) = f(b)$, then there is some value $c \in (a,b)$ with $f'(c) = 0$. (*Hint: Apply Rolle's Theorem to the function $g(x) = f(x) - f(a)$.*)

Thinking Forward

Sign analyses for derivatives: For each function f that follows, find the derivative f'. Then determine the intervals on which the derivative f' is positive and the intervals on which the derivative f' is negative. Record your answers on a sign chart for f', with tick-marks only at the x-values where f' is zero or undefined.

▶ $f(x) = (x-2)^2$

▶ $f(x) = \sqrt{x-2}$

▶ $f(x) = 2x^3 - 3x^2 + 1$

▶ $f(x) = \dfrac{1}{x-1}$

Sign analyses for second derivatives: Repeat the instructions of the previous block of problems, except find sign intervals for the second derivative f'' instead of the first derivative.

▶ $f(x) = (x-2)^2$

▶ $f(x) = \sqrt{x-2}$

▶ $f(x) = 2x^3 - 3x^2 + 1$

▶ $f(x) = \dfrac{1}{x-1}$

3.2   THE FIRST DERIVATIVE AND CURVE SKETCHING

▶ The relationship between the derivative and increasing/decreasing functions

▶ Proving that all antiderivatives of a function differ by a constant

▶ Using the first-derivative test to determine whether critical points are maxima, minima, or neither

Derivatives and Increasing/Decreasing Functions

In Section 0.4 we defined a function f to be **increasing** on an interval if, for all a and b in the interval with $b > a$, $f(b) > f(a)$. In other words, the height of f at points farther to the right are higher. Similarly, f is **decreasing** on an interval if, for all $b > a$ in the interval, $f(b) < f(a)$. These definitions can be difficult to work with if we wish to find the intervals on which a given function is increasing or decreasing. Luckily, the derivative will provide us with an easier method.

We have seen that the first derivative f' in some sense measures the direction of the graph of a function f at each point, since f' is the associated slope function for f. In particular, if f' is positive at a point $x = c$, then the graph of f must be moving in an upwards direction, that is, increasing, as it passes $x = c$. Similarly, if $f'(c)$ is negative, then the graph of f must be decreasing at $x = c$. For example, we can divide the real-number line into intervals according to where the function $f(x) = x^3 - 3x^2 - 9x + 11$ is increasing or decreasing, as shown in the figure that follows at the left. This same division into subintervals describes where the derivative $f'(x) = 3x^2 - 6x - 9$ is positive and negative, as shown at the right.

Intervals where f increases/decreases

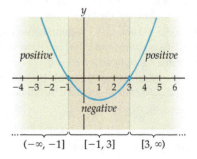

Intervals where f' is positive/negative

This relationship between intervals on which f is increasing or decreasing exactly when f' is positive or negative, respectively, holds in general. The only wrinkle is at the extremum of a function f; for example, in the graph at the left we say that f is increasing on $(-\infty, -1]$, which includes the extremum at $x = -1$, because it is true that for all $b > a$ less than *or equal to* $x = -1$ we do have $f(b) > f(a)$. However, the derivative at $x = -1$ is not positive, but zero. Therefore f' is positive only on the interior $(-\infty, -1)$ of the interval, but the function f is increasing on the entire interval $(-\infty, -1]$.

THEOREM 3.6

The Derivative Measures Where a Function is Increasing or Decreasing

Let f be a function that is differentiable on an interval I.

(a) If f' is positive in the interior of I, then f is increasing on I.

(b) If f' is negative in the interior of I, then f is decreasing on I.

(c) If f' is zero in the interior of I, then f is constant on I.

Theorem 3.6 is intuitively obvious if we consider the slopes of tangent lines on increasing and decreasing graphs. To prove this theorem formally we require the Mean Value Theorem.

> **Proof.** We'll prove part (a) here and leave the similar proofs of parts (b) and (c) to Exercises 71 and 72, respectively. The key to the proof will be the Mean Value Theorem.
>
> Let f be a function that is differentiable on an interval I and whose derivative f' is positive on the interior of that interval. Suppose also that $a, b \in I$ with $b > a$. By the definition of increasing, we must show that $f(b) > f(a)$. Since f is differentiable, and thus also continuous, on the interval I, and since $[a, b]$ is contained in the interval I, f satisfies the hypotheses of the Mean Value Theorem on $[a, b]$. Therefore we can conclude that there exists some $c \in (a, b)$ such that
>
> $$f'(c) = \frac{f(b) - f(a)}{b - a}.$$
>
> To show that $f(b) > f(a)$ it suffices to show that $f(b) - f(a) > 0$; with this in mind we can rearrange the preceding equation as
>
> $$f(b) - f(a) = f'(c)(b - a).$$
>
> Since $c \in (a, b)$, it follows that c is in the interior of I, and thus by hypothesis $f'(c) > 0$. Furthermore, since $b > a$, we have $(b - a) > 0$. Therefore $f(b) - f(a)$ is the product of two positive numbers and must itself be positive, which is what we wanted to show. ∎

Up to this point we could only graphically approximate the intervals on which a function is increasing or decreasing. With Theorem 3.6 we can now find these intervals algebraically, by examining the sign of f'. We have thus reduced the difficult problem of finding the intervals on which a function is increasing or decreasing to the much simpler problem of finding the intervals on which an associated function—the derivative—is positive or negative.

Functions with the Same Derivative

If two functions differ by a constant, then obviously they will have the same derivative, because the derivative of a constant is zero. For example, $f(x) = x^3$ and $g(x) = x^3 + 10$ differ by a constant because their difference $g(x) - f(x)$ is equal to the constant 10, and their derivatives $f'(x) = 3x^2$ and $g'(x) = 3x^2 + 0 = 3x^2$ are equal.

Although it is less obvious, the converse is also true: Any two functions that have the same derivative must differ by a constant. Algebraically, this means that if you find one antiderivative of a function, then all other antiderivatives of that function differ from the one that you found by a constant. For example, one antiderivative of $3x^2$ is x^3, and thus all antiderivatives of $3x^2$ are of the form $x^3 + C$ for some constant C. Graphically, this means that if two functions have the same derivative, then one is a vertical shift of the other. For example, the graph of $3x^2$ is shown next at the left and five of its antiderivatives $x^3 + C$ are shown at the right, for $C = 0$, $C = \pm10$, and $C = \pm20$. The red graph of $y = 3x^2$ yields information about *every* one of the blue graphs $y = x^3 + C$, regardless of the vertical shift C.

Graph of $y = 3x^2$

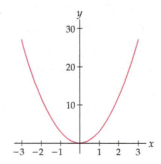

All antiderivatives of $3x^2$ are of the form $x^3 + C$

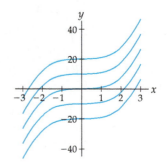

THEOREM 3.7   **Functions with the Same Derivative Differ by a Constant**

If $f'(x) = g'(x)$ for all $x \in [a, b]$, then, for some constant C, $f(x) = g(x) + C$ for all $x \in [a, b]$.

Proof. Suppose $f'(x) = g'(x)$ for all $x \in [a, b]$. Then $f'(x) - g'(x) = 0$, which by the difference rule means that $\frac{d}{dx}(f(x) - g(x)) = 0$ on $[a, b]$. By the third part of Theorem 3.6 this means that the function $f(x) - g(x)$ is constant for all $x \in [a, b]$, say, $f(x) - g(x) = C$ for some real number C. Therefore $f(x) = g(x) + C$ for all $x \in [a, b]$. ■

The First-Derivative Test

In the previous section we saw that the set of critical points of a function, that is, the values of x for which f' is zero or does not exist—is a complete list of all the possible local extrema of f. We now develop a method for using the first derivative to determine which critical points are local maxima, which are local minima, and which are neither.

Suppose $f'(c) = 0$ and that f is differentiable near c. Then if f is not constant, there are four different ways that f can behave near $x = c$:

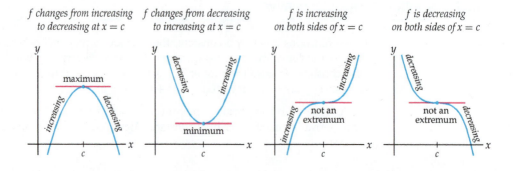

On the one hand, if f changes from increasing to decreasing at $x = c$, then f has a local maximum at $x = c$. If f changes from decreasing to increasing at the point $x = c$, then f has a local minimum at $x = c$. On the other hand, if f does not change direction at $x = c$, then f does not have a local extremum at $x = c$. Since a function f is increasing or decreasing according to whether its derivative is positive or negative, we can record the preceding four situations on sign charts for f', with the corresponding information for f recorded as the following sign analyses:

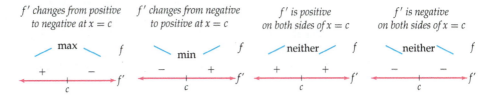

| f' changes from positive to negative at $x = c$ | f' changes from negative to positive at $x = c$ | f' is positive on both sides of $x = c$ | f' is negative on both sides of $x = c$ |

CAUTION | Please recall that, throughout this book, we will make tick-marks on sign charts *only* at the locations where the function is zero or fails to exist. We will mark with "DNE" any locations where the function does not exist, but we will not explicitly mark the zeroes. You should assume that any unlabeled tick-marks on a sign chart are zeroes of the function in question. Do not make additional tick-marks on your sign charts if you follow this convention.

Using a sign chart for the first derivative f' to test whether critical points of f are local maxima, minima, or neither is quite sensibly known as the ***first-derivative test***. The statement of the test is wordy and its proof is somewhat technical, but its meaning is equivalent to information obtained from the previous sign chart analyses.

THEOREM 3.8   **The First-Derivative Test**

Suppose $x = c$ is the location of a critical point of a function f, and let (a, b) be an open interval around c that is contained in the domain of f and does not contain any other critical points of f. If f is continuous on (a, b) and differentiable at every point of (a, b) except possibly at $x = c$, then the following statements hold.

(a) If $f'(x)$ is positive for $x \in (a, c)$ and negative for $x \in (c, b)$, then f has a local maximum at $x = c$.

(b) If $f'(x)$ is negative for $x \in (a, c)$ and positive for $x \in (c, b)$, then f has a local minimum at $x = c$.

(c) If $f'(x)$ is positive for both $x \in (a, c)$ and $x \in (c, b)$, then f does not have a local extremum at $x = c$.

(d) If $f'(x)$ is negative for both $x \in (a, c)$ and $x \in (c, b)$, then f does not have a local extremum at $x = c$.

Proof. We will prove parts (a) and (c) here and leave the similar proofs of parts (b) and (d) to Exercises 73 and 74, respectively. The proof will be an application of Theorem 3.6. Suppose $x = c$ is a critical point of f, and let (a, b) be an interval around $x = c$ satisfying the hypotheses of the theorem.

To prove part (a), suppose $f'(x) > 0$ for $x \in (a, c)$ and $f'(x) < 0$ for $x \in (c, b)$, that is, suppose that f is increasing on $(a, c]$ and decreasing on $[c, b)$. We will show that $f(c) \geq f(x)$ for all $x \in (a, b)$, which will tell us that f has a local maximum at $x = c$. Given any $x \in (a, b)$, there are three cases to consider. First, if $x = c$, then clearly $f(c) = f(x)$. Second, if $a < x < c$, then since f is increasing on $(a, c]$, we have $f(x) < f(c)$. Third, if $c < x < b$, then since f is decreasing on $[c, b)$, we have $f(c) > f(x)$. In all three cases we have $f(c) \geq f(x)$, and therefore f has a local maximum at $x = c$.

To prove part (c), suppose that $f'(x) > 0$ for all $x \in (a, c) \cup (c, b)$. Then by Theorem 3.6, f must be increasing on all of (a, b). Now, the point $x = c$ cannot be the location of a local maximum of f, because, for all $x > c$ in (a, b), $f(x) > f(c)$, since f is increasing on (a, b). But neither can $x = c$ be a local minimum of f, because, for all $x < c$ in (a, b), $f(x) < f(c)$. Therefore f has neither a local minimum nor a local maximum at $x = c$. ∎

Now, to find the local extrema of a function f, we need only find the critical points of f and then test each one with the first-derivative test. In other words, we find the derivative f', determine where f' is zero or does not exist, and then make a sign chart for f' around these critical points to determine whether f has a local maximum, a local minimum, or neither at each critical point. This method will find all local extrema for functions f that are defined on open intervals. For functions defined on closed or half-closed intervals, we will also have to consider endpoint extrema, which we will discuss in Section 3.4.

Examples and Explorations

EXAMPLE 1   Using the derivative to determine where a function is increasing or decreasing

Use Theorem 3.6 to determine the intervals on which each of the following functions are increasing or decreasing:

(a) $f(x) = x^3$ **(b)** $g(x) = x^2 - 2x + 1$ **(c)** $h(x) = \dfrac{x^2 + 10x + 1}{x - 2}$

SOLUTION

(a) If $f(x) = x^3$, then $f'(x) = 3x^2$, which is positive as long as $x \neq 0$. Therefore f' is positive on $(-\infty, 0)$ and $(0, \infty)$. By Theorem 3.6 we can say that f is increasing on the entire half-closed intervals $(-\infty, 0]$ and $[0, \infty)$. Thus f is increasing on all of $\mathbb{R}$. Notice that, as shown in the graph that follows, the tangent line to $f(x) = x^3$ is horizontal at $x = 0$. However, the function $f(x) = x^3$ is still increasing everywhere: For all real numbers $b > a$ we have $b^3 > a^3$, even if one of a or b is zero.

$f(x) = x^3$ *is increasing everywhere*

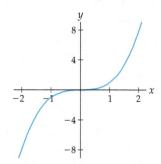

(b) If $g(x) = x^2 - 2x + 1$, then $g'(x) = 2x - 2$. Recall that functions can change signs only at roots, non-domain points, and discontinuities. The derivative $g'(x) = 2x - 2$ is always defined and continuous, and $g'(x) = 0$ when $2x - 2 = 0$ (i.e., when $x = 1$). Therefore we need only check the sign of $g'(x)$ to the left and right of $x = 1$. For example, $g'(0) = 2(0) - 2 = -2$ is negative, and $g'(2) = 2(2) - 2 = 2$ is positive, as the following sign chart shows:

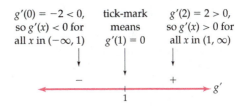

By Theorem 3.6, the information on the sign chart for $g'(x)$ shows that $g(x)$ is decreasing on $(-\infty, 1]$ and increasing on $[1, \infty)$. However, from this point forward we will record the closed-endpoint information only when we need it and instead record only the open-interval information. Therefore here we would say simply that $g(x)$ is decreasing on $(-\infty, 1)$ and increasing on $(1, \infty)$.

(c) By the quotient rule, the derivative of $h(x)$ is

$$h'(x) = \frac{(2x+10)(x-2) - (x^2 + 10x + 1)(1)}{(x-2)^2}$$

$$= \frac{x^2 - 4x - 21}{(x-2)^2} = \frac{(x+3)(x-7)}{(x-2)^2}.$$

Notice that we simplified the preceding equation so that it would be immediately clear that $h'(x) = 0$ when $x = -3$ and $x = 7$ and that $h'(x)$ does not exist when $x = 2$, a point that is not in the domain of $h(x)$ in the first place. To determine the sign chart for $h'(x)$ we need only check the sign of $h'(x)$ one time between each of these critical points and non-domain points. For example, using the factored form of $h'(x)$ we have:

$$h'(-5): \frac{(-)(-)}{(-)^2} \text{ (positive)} \qquad h'(0): \frac{(+)(-)}{(-)^2} \text{ (negative)}$$

$$h'(5): \frac{(+)(-)}{(+)^2} \text{ (negative)} \qquad h'(10): \frac{(+)(+)}{(+)^2} \text{ (postiive)}$$

We can record this information on a sign chart for $h'(x)$ as follows, with annotations above the chart for the corresponding increasing/decreasing behavior of $h(x)$:

Reading off the sign chart, we see that $h(x)$ is increasing on $(-\infty, -3)$ and $(7, \infty)$ and decreasing on $(-3, 2)$ and on $(2, 7)$.   □

To check our answers to parts (b) and (c) of the example, we can simply graph $g(x)$ and $h(x)$. Notice that the graph of $g(x)$ shown next at the left does appear to be decreasing on $(-\infty, 1)$ and increasing on $(1, \infty)$. The graph of $h(x)$ at the right does appear to be increasing on $(-\infty, -3)$ and $(7, \infty)$ and decreasing elsewhere.

$$g(x) = x^2 - 2x + 1 \qquad\qquad h(x) = \frac{x^2 + 10x + 1}{x - 2}$$

EXAMPLE 2   **Interpreting the sign chart for the derivative of a function**

Let f be a function whose derivative f' has the signs indicated on the sign chart shown. Recall our convention that tick-marks on this sign chart indicate the locations where f' is zero (if unlabeled) or does not exist (if marked with "DNE"). Use the sign chart to sketch possible graphs for f and f'.

SOLUTION

This sign chart says that f' is positive on the intervals $(-\infty, -1)$ and $(2, \infty)$, and negative on the interval $(-1, 2)$. It also says that $f'(-1) = 0$ and that $f'(2)$ does not exist. Given this information, one possible sketch of f' is shown next at the left.

We can determine a lot about the graph of a function f from the graph of its derivative f'. The function f must be increasing on $(-\infty, -1)$ and $(2, \infty)$, and decreasing on $(-1, 2)$, with a horizontal tangent line at $x = -1$, and a non-differentiable point at $x = 2$. One possible sketch of a function f that has these characteristics is shown next at the right. Notice that the information about f' does not tell us how high or low to sketch the graph of f. In fact, any vertical shift of f would do just as well, since every antiderivative of f' is of the form $f(x) + C$.

Possible graph of f'

Possible graph of f

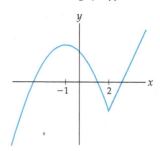

EXAMPLE 3   **Detailed curve-sketching analyses using the first derivative**

Sketch the graph of each of the given functions. Along the way, determine all local extrema and important features of the graph.

(a) $f(x) = x^3 - 3x + 2$ **(b)** $f(x) = \dfrac{x^2}{x^2 - 2x + 1}$

SOLUTION

(a) For $f(x) = x^3 - 3x + 2$ we have $f'(x) = 3x^2 - 3$. This derivative is always defined and continuous, so the critical points of f are just the places where $f'(x) = 0$:

$$3x^2 - 3 = 0 \implies 3x^2 = 3 \implies x^2 = 1 \implies x = \pm 1.$$

These critical points divide the real-number line into three intervals, namely, $(-\infty, -1)$, $(-1, 1)$, and $(1, \infty)$. We'll test the sign of f' on each interval by testing the sign of $f'(x)$ at one point in that interval, say, at $x = -2$, $x = 0$, and $x = 2$:

$$f'(-2) = 3(-2)^2 - 3 = 12 - 3 = 9 > 0,$$
$$f'(0) = 3(0)^2 - 3 = 0 - 3 = -3 < 0,$$
$$f'(2) = 3(2)^2 - 3 = 12 - 3 = 9 > 0.$$

Recording this information on a sign chart for f', with consequences for f illustrated above the chart, we have

The maximum and minimum of f listed on the chart were identified by the first-derivative test: Since f' changes from positive to negative at $x = -1$, f changes from increasing to decreasing there; thus the point $x = -1$ must be a local maximum of f. Since f' changes from negative to positive at $x = 1$, f changes from decreasing to increasing there; thus f must have a local minimum at the point $x = 1$.

With all the first-derivative information we have collected, we can sketch a reasonable graph of f by plotting just a few points. As a general rule it is a good idea to plot the points $(c, f(c))$ for each critical point $x = c$ and then use the information about the derivative to connect the dots accordingly. In this example, the critical points are $x = -1$ and $x = 1$, so we calculate:

$$f(-1) = (-1)^3 - 3(-1) + 2 = -1 + 3 + 2 = 4,$$
$$f(1) = (1)^3 - 3(1) + 2 = 1 - 3 + 2 = 0.$$

Using this information, we can make a sketch of the graph of $f(x) = x^3 - 3x + 2$, as shown next. Compare the features of this graph with the information in the sign chart for f':

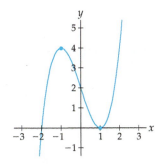

(b) Before we do anything else, we should note that the domain of

$$f(x) = \frac{x^2}{x^2 - 2x + 1} = \frac{x^2}{(x - 1)^2}$$

is all points except $x = 1$. The function f is zero only at $x = 0$, and at all other points in its domain it is positive, since both its numerator and denominator are perfect squares. So far we know that the entire graph of f will be above the x-axis, except at the root $(0, 0)$ and at the non-domain point $x = 1$, where something interesting may occur.

To determine local extrema and increasing/decreasing behavior, we must find the derivative of $f(x) = \frac{x^2}{x^2 - 2x + 1}$. Because we will be interested in where f' is zero or does not exist, we also simplify as much as possible:

$$f'(x) = \frac{(2x)(x^2 - 2x + 1) - (x^2)(2x - 2)}{(x^2 - 2x + 1)^2} = \frac{-2x(x - 1)}{(x - 1)^4} = \frac{-2x}{(x - 1)^3}.$$

This derivative is zero when its numerator is zero but its denominator is not, that is, when $x = 0$. The derivative is undefined when $x = 1$, which makes sense because the original function f is also undefined at $x = 1$. The critical point $x = 0$ and the non-domain point $x = 1$ are the only places at which the sign of f' can change.

Now we make a sign chart representing the sign of the derivative f' between each of these critical points. For example, we can find that $f'(-1) < 0$, $f'(0.5) > 0$, and $f'(2) < 0$:

We can identify the local minimum at $x = 0$ indicated on the chart with the first-derivative test, since $f'(x)$ changes sign from negative to positive at $x = 0$. However, the same type of argument cannot be applied to the point $x = 1$: Even though $f'(x)$ changes sign at $x = 1$, the function f does not have a local extremum at $x = 1$, because it is not defined at that point.

Information about limits can be important to figuring out the key features of a graph, so we will examine limits of $f(x) = \dfrac{x^2}{x^2 - 2x + 1}$ at any "interesting" places. In this case the interesting limits to consider are at the point $x = 1$, where the function does not exist, and at the ends as $x \to \pm\infty$. As $x \to 1$ we have $x^2 \to 1$ and $x^2 - 2x + 1 = (x - 1)^2 \to 0^+$, and therefore the limit in question is of the form $\dfrac{1}{0^+}$. Thus

$$\lim_{x \to 1} \frac{x^2}{x^2 - 2x + 1} = \infty.$$

We can calculate the limit of f as $x \to \infty$ with the familiar strategy of dividing numerator and denominator by the highest power of x:

$$\lim_{x \to \infty} \frac{x^2}{x^2 - 2x + 1} \left(\frac{1/x^2}{1/x^2} \right) = \lim_{x \to \infty} \frac{1}{1 - \dfrac{2}{x} + \dfrac{1}{x^2}} = \frac{1}{1 - 0 + 0} = 1.$$

A similar argument shows that $\lim\limits_{x \to -\infty} \dfrac{x^2}{x^2 - 2x + 1}$ is also equal to 1. Since $\lim\limits_{x \to 1} f(x) = \infty$, we can see that f has a vertical asymptote at the non-domain point $x = 1$, with the function approaching ∞ on both sides of this asymptote. Since $\lim\limits_{x \to \infty} f(x) = 1$ and $\lim\limits_{x \to -\infty} f(x) = 1$, f has a horizontal asymptote on both the left and the right at $y = 1$.

Putting all of this information together, we get the following graph:

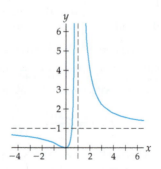

? **TEST YOUR** **UNDERSTANDING**	▶ If f' is positive on $(0, 3)$, why does it make sense that we can say that f is increasing on all of $[0, 3]$?

▶ Can you give an example of a function f for which f' is zero on all of $(-1, 1)$?

▶ If $x^3 - 4x^2 - 2$ is one antiderivative of some function f, what can you say about the other antiderivatives of f?

▶ If f is defined, continuous, and differentiable at $x = 3$, and if f' changes sign at $x = 3$, then what can you say about f at $x = 3$?

▶ Can you sketch an example of a function f with a critical point at $x = 2$ but no local extremum at $x = 2$?

EXERCISES 3.2

Thinking Back

Differentiation: Find the derivative of each function f, and then simplify as much as possible.

▶ $f(x) = (2x - 1)^3(3x + 1)^2$

▶ $f(x) = \dfrac{(2x - 1)^3}{(3x + 1)^2}$

▶ $f(x) = 3x^2\sqrt{x - 1}$

▶ $f(x) = \sqrt{\dfrac{x^2}{1 - x^3}}$

Solving equations: For each of the following functions $g(x)$, find the solutions of $g(x) = 0$ and also find the values of x for which $g(x)$ does not exist.

▶ $g(x) = \dfrac{3(x^2 - 1) - (3x + 1)(2x)}{(x^2 - 1)^2}$

▶ $g(x) = \dfrac{\frac{1}{2}x^{-1/2}(1 + 5x) - \sqrt{x}(5)}{(1 + 5x)^2}$

▶ $g(x) = 2x\sqrt{x - 1} + x^2\left(\frac{1}{2}(x - 1)^{-1/2}\right)$

Sign analyses: For each of the following functions $g(x)$, use algebra and a sign chart to find the intervals on which $g(x)$ is positive and the intervals on which $g(x)$ is negative.

▶ $g(x) = 6x^2 - 18x$

▶ $g(x) = 2 - x - 2x^2 + x^3$

▶ $g(x) = \dfrac{(x + 2)(x - 1)^4}{x - 3}$

▶ $g(x) = \dfrac{3x^2 - 5x - 2}{(3x + 1)^2}$

Concepts

0. *Problem Zero:* Read the section and make your own summary of the material.

1. *True/False:* Determine whether each of the statements that follow is true or false. If a statement is true, explain why. If a statement is false, provide a counterexample.

(a) *True or False:* If $f'(x) < 0$ for all $x \in (0, 3)$, then f is decreasing on $[0, 3]$.

(b) *True or False:* If f is increasing on $(-2, 2)$, then $f'(x) \geq 0$ for all $x \in (-2, 2)$.

(c) *True or False:* If $f'(x) = 2x$, then $f(x) = x^2 + C$ for some constant C.

(d) *True or False:* If f' is continuous on $(1, 8)$ and $f'(3)$ is negative, then f' is negative on all of $(1, 8)$.

(e) *True or False:* If f' changes sign at $x = 3$, then $f'(3) = 0$.

(f) *True or False:* If $f'(-2) = 0$, then f has either a local maximum or a local minimum at $x = -2$.

(g) *True or False:* If $x = 1$ is the only critical point of f and $f'(0)$ is positive, then $f'(2)$ must be negative.

(h) *True or False:* If $f'(1)$ is negative and $f'(3)$ is positive, then f has a local minimum at $x = 2$.

2. *Examples:* Construct examples of the thing(s) described in the following. Try to find examples that are different than any in the reading.

(a) A function that is decreasing on $(-\infty, 0)$, increasing on $(0, \infty)$, and undefined at $x = 0$.

(b) A function that is decreasing on $(-\infty, 0]$ and increasing on $[0, \infty)$.

(c) A function that is always positive and always decreasing, on all of $\mathbb{R}$.

3. State the definition of what it means for a function f to be increasing on an interval I and what it means for a function f to be decreasing on an interval I.

4. Describe what a critical point is, intuitively and in mathematical language. Then describe what a local extremum is. How are these two concepts related?

5. Can a point $x = c$ be both a local extremum and a critical point of a differentiable function f? Both an inflection point and a critical point? Both an inflection point and a local extremum? Sketch examples, or explain why such a point cannot exist.

6. Suppose f is a function that may be non-differentiable at some points. Can a point $x = c$ be both a local extremum and a critical point of such a function f? Both an inflection point and a critical point? Both an inflection point and a local extremum? Sketch examples, or explain why such a point cannot exist.

7. Suppose f is defined and continuous everywhere. Why is testing the sign of the derivative f' at just one point sufficient to determine the sign of f' on the whole interval between critical points of f?

8. Describe what the first-derivative test is for and how to use it. Sketch graphs and sign charts to illustrate your description.

9. Sketch the graph of a function f with the following properties:

   ▶ f is continuous and defined on $\mathbb{R}$;
   ▶ $f(0) = 5$;
   ▶ $f(-2) = -3$ and $f'(-2) = 0$;
   ▶ $f'(1)$ does not exist;
   ▶ f' is positive only on $(-2, 1)$.

10. Sketch the graph of a function f with the following properties:

    ▶ f is continuous and defined on $\mathbb{R}$;
    ▶ f has critical points at $x = -3, 0$, and 5;
    ▶ f has inflection points at $x = -3, -1$, and 2.

11. Use the definitions of *increasing* and *decreasing* to argue that $f(x) = x^4$ is decreasing on $(-\infty, 0]$ and increasing on $[0, \infty)$. Then use derivatives to argue the same thing.

12. Use the definition of *increasing* to argue that $f(x) = x^5$ is increasing on all of $\mathbb{R}$. Then use derivatives to argue the same thing.

13. Suppose f is a function that is continuous and differentiable everywhere and that the derivative of f is

$$f'(x) = \sqrt{1 + x^2} - 4.$$

What are the critical points of f?

14. Suppose f is a function that is continuous and differentiable everywhere and that the derivative of f is

$$f'(x) = \frac{(x - 1)(x - 2)}{x - 3}.$$

What are the critical points of f?

15. If $g(x)$ and $h(x)$ are both antiderivatives of some function $f(x)$, then what can you say about the function $g(x) - h(x)$?

16. If $g(x)$ is an antiderivative of $f(x)$, then what is the relationship between the functions $g(x) + 10$ and $f(x)$?

17. One of the graphs shown is a function f and the other is its derivative f'. Which one is which, and why?

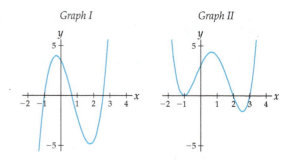

| Graph I | Graph II |

18. One of the graphs shown is a function f and the other is its derivative f'. Which one is which, and why?

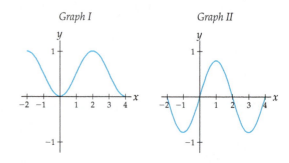

| Graph I | Graph II |

Skills

For each function f graphed in Exercises 19–22, sketch a possible graph of its derivative f'.

19.

20.

21.

22.
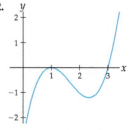

Each graph in Exercises 23–26 represents the derivative f' of some function f. Use the given graph of f' to sketch a possible graph of f.

23.

24.

25.

26.

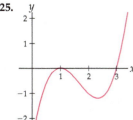

Use a sign chart for f' to determine the intervals on which each function f in Exercises 27–34 is increasing or decreasing. Then verify your algebraic answers with graphs from a calculator or graphing utility.

27. $f(x) = 3x + 1$

28. $f(x) = (x^2 - 1)(x - 3)$

29. $f(x) = x(x + 3)(x - 1)$

30. $f(x) = (x - 1)^3 - 2x^2$

31. $f(x) = 2x^3 - 9x^2 + 1$

32. $f(x) = x^3 + 4x^2 + 4x - 5$

33. $f(x) = \dfrac{x}{x^2 + 4}$

34. $f(x) = \dfrac{3x + 1}{x^2 - 1}$

Use the first-derivative test to determine the local extrema of each function f in Exercises 35–42. Then verify your algebraic answers with graphs from a calculator or graphing utility.

35. $f(x) = 3 - 4x$

36. $f(x) = -x^2 - 10x - 33$

37. $f(x) = x^2 - 5x - 6$

38. $f(x) = 9 - x^4$

39. $f(x) = (x - 2)^2(1 + x)$

40. $f(x) = x^2(x - 1)(x + 1)$

41. $f(x) = \dfrac{1 + x + x^2}{x^2 + x - 2}$

42. $f(x) = \dfrac{(x - 1)^2}{x + 2}$

For each sign chart for f' in Exercises 43–48, sketch possible graphs of both f' and f. On each sign chart, unlabeled tick-marks are locations where $f'(x)$ is zero and x-values where $f'(x)$ does not exist are indicated by tick-marks labeled "DNE."

43.

44.

45.

46.

47.

48.

Sketch careful, labeled graphs of each function f in Exercises 49–66 by hand, without consulting a calculator or graphing utility. As part of your work, make sign charts for the signs, roots, and undefined points of f and f', and examine any relevant limits so that you can describe all key points and behaviors of f.

49. $f(x) = (x - 2)(3x + 1)$

50. $f(x) = x^2 - x + 100$

51. $f(x) = x^3 - x^2 - x + 1$

52. $f(x) = x^3 - 9x + 1$

53. $f(x) = x^3 - 6x^2 + 12x$

54. $f(x) = x(x^2 - 4)$

55. $f(x) = (2x + 11)(x^2 + 10)$

56. $f(x) = 3x^5 - 10x^4$

57. $f(x) = (1 - x^4)^7$

58. $f(x) = \sqrt{x} - \dfrac{1}{\sqrt{x}}$

59. $f(x) = \dfrac{x + 1}{x - 1}$

60. $f(x) = \dfrac{1}{x} + \dfrac{1}{x^2}$

61. $f(x) = \sqrt{x^2 + 1}$

62. $f(x) = \dfrac{1}{\sqrt{x^2 - 1}}$

63. $f(x) = \dfrac{(x - 1)^2}{x^2 + x - 6}$

64. $f(x) = \dfrac{(x - 1)^2}{x^2 - 1}$

65. $f(x) = \dfrac{x^2(x - 1)}{(x - 2)^2}$

66. $f(x) = \dfrac{x^3}{x^2 - 3x + 2}$

Applications

67. Dr. Alina is interested in the behavior of rats trapped in a long tunnel. Her rat Bubbles is released from the left-hand side of the tunnel and runs back and forth in the tunnel for 4 minutes. Bubbles' velocity $v(t)$, in feet per minute, is given by the following graph.

Velocity of rat in a long tunnel

(a) On which time intervals is Bubbles moving towards the right-hand side of the tunnel?

(b) At which point in time is Bubbles farthest away from the left-hand side of the tunnel, and why? Do you think that Bubbles ever comes back to the left-hand side of the tunnel?

(c) On which time intervals does Bubbles have a positive acceleration?

(d) Find an interval on which Bubbles has a negative velocity but a positive acceleration. Describe what Bubbles is doing during this period.

68. Calvin uses a slingshot to launch an orange straight up in the air to see what will happen. The distance in feet between the orange and the ground after t seconds is given by the equation $s(t) = -16t^2 + 90t + 5$. Use this equation to answer the following questions:

(a) What is the initial height of the orange? What is the initial velocity of the orange? What is the initial acceleration of the orange?

(b) What is the maximum height of the orange?

(c) When will the orange hit the ground?

Proofs

69. Prove that every nonconstant linear function is either always increasing or always decreasing.

70. Prove that every quadratic function has exactly one local extremum.

71. Prove part (b) of Theorem 3.6: Suppose f is differentiable on an interval I; if f' is negative on the interior of I, then f is decreasing on I.

72. Prove part (c) of Theorem 3.6: Suppose f is differentiable on an interval I; if f' is zero on the interior of I, then f is constant on I. (*Hint: use the Mean Value Theorem to show that any two numbers a and b in I must be equal.*)

73. Prove part (b) of Theorem 3.8: With hypotheses as stated in the theorem, if $x = c$ is a critical point of f, where $f'(x) < 0$ to the left of c and $f'(x) > 0$ to the right of c, then f has a local minimum at $x = c$.

74. Prove part (d) of Theorem 3.8: With hypotheses as stated in the theorem, if $x = c$ is a critical point of f, where $f'(x) < 0$ to the left and to the right of c, then $x = c$ is not a local extremum of f.

Thinking Forward

▶ *Second-derivative graphs:* The three graphs shown are graphs of a function f and its first and second derivatives f' and f'', in no particular order. Identify which graph is which.

Graph I   *Graph II*

Graph III

▶ *More second-derivative graphs:* The three graphs shown are graphs of a function f and its first and second derivatives f' and f'', in no particular order. Identify which graph is which.

Graph I   *Graph II*

Graph III

3.3   THE SECOND DERIVATIVE AND CURVE SKETCHING

▶ Using first and second derivatives to define and detect concavity

▶ The behavior of the first and second derivatives at inflection points

▶ Using the second-derivative test to determine whether critical points are maxima, minima, or neither

Derivatives and Concavity

In Section 0.4 we gave an informal definition of concavity: The graph of a function is *concave up* if it "curves upward" and *concave down* if it "curves downward." This is equivalent to saying that the graph of a concave-up function lies below its secant lines and above its tangent lines, and the graph of a concave-down function lies above its secant lines and below its tangent lines. Now that we know about derivatives, we are finally able to give a more precise definition of concavity.

DEFINITION 3.9

Formally Defining Concavity

Suppose f and f' are both differentiable on an interval I.

(a) f is *concave up* on I if f' is increasing on I.

(b) f is *concave down* on I if f' is decreasing on I.

How does this formal definition of concavity correspond with our intuitive notion of concavity? Consider the functions graphed next. On each graph four slopes are illustrated and estimated. Notice that when f is concave up, its slopes increase from left to right, and when f is concave down, its slopes decrease from left to right.

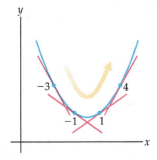

Slopes increase when f is concave up

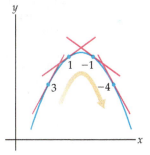

Slopes decrease when f is concave down

As we have already seen, a function increases where its derivative is positive. Taking this up one level, we see that the derivative function f' is increasing where *its* derivative function f'' is positive. Therefore we can check whether a function f is concave up or concave down by looking at the sign of its second derivative:

THEOREM 3.10

The Second Derivative Determines Concavity

Suppose both f and f' are differentiable on an interval I.

(a) If f'' is positive on I, then f is concave up on I.

(b) If f'' is negative on I, then f is concave down on I.

Proof. We will prove part (a) and leave part (b) to Exercise 75. Suppose that f and f' are differentiable on I and that $f''(x) > 0$ for all x in I. Then since the derivative of f' is f'', it follows from Theorem 3.6 that f' is increasing on I. By the definition of concavity this means that f is concave up on the interval I. ∎

For example, we can divide the real-number line into intervals according to where the function $f(x) = x^3$ is concave up or concave down. This same division into subintervals describes where the derivative $f'(x) = 3x^2$ is increasing or decreasing and where the second derivative $f''(x) = 6x$ is positive or negative:

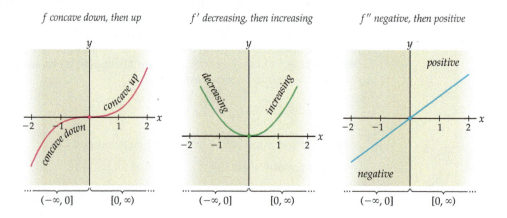

Inflection Points

Recall from Section 0.1 that the **inflection points** of a function f are the points in the domain of f at which its concavity changes. Since the sign of f'' measures the concavity of f, we can find inflection points by looking for the places where f'' changes sign. For example, if $f(x) = x^3$, then $f''(x) = 6x$, which is zero only when $x = 0$. The sign of f'' changes from negative to positive at $x = 0$, and therefore the concavity of f changes from down to up at that inflection point.

If $x = c$ is an inflection point of f and $f''(c) = 0$, then the graph of f could look one of the following four ways near $x = c$, depending on how f changes concavity and whether f is increasing or decreasing:

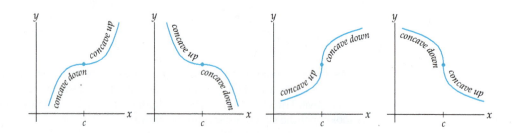

We know that at an inflection point of a function f, the function changes concavity and the second derivative f'' changes sign. What happens to the first derivative f'? The answer lies in the fact that f'' is the derivative of f'. Thus we know that If f'' is positive, then f' is increasing, and If f'' is negative, then f' is decreasing. The four possible scenarios corresponding to the preceding graphs are shown here:

Notice that in each case the sign of f'' changes at $x = c$, causing f to have an inflection point at $x = c$ and in addition causing f' to have a local maximum or minimum at $x = c$. If you sketch tangent lines on the four graphs shown, you should be able to see that the slopes are at their maximum or minimum values at the inflection points. For example, in the first graph, the slopes start out large and positive, decrease to a minimum of zero at $x = c$, and then increase to larger and larger positive slopes as we move from left to right.

The Second-Derivative Test

In the previous section we saw how to apply the first-derivative test to determine whether the critical points of a function were local maxima, local minima, or neither. We can also use the second derivative to test critical points, by examining the concavity of the function at each critical point.

Suppose f is a differentiable function and $x = c$ is a critical point of f with $f'(c) = 0$. Then there are four possible ways that f can behave near $x = c$, as shown in the figures that follow. In each case we can examine the second derivative at the point $x = c$:

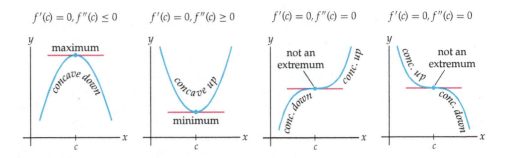

In the first graph, the second derivative is negative at $x = c$, so f curves downwards and has a local maximum at that point. In the second graph, the second derivative is positive at $x = c$, so f curves upwards and has a local minimum at that point.

In general, knowing that $f''(c) = 0$ does not tell us whether f has a maximum, minimum, or neither at $x = c$. The reason is that $f''(c) = 0$ is a possibility in all four of the cases just graphed. It is obvious that the last two graphs must have $f''(c) = 0$ because in those cases the function changes concavity at $x = c$. Perhaps less obvious is that we could have $f''(c) = 0$ in the first two graphs, where no inflection points occur; for example see the function $f(x) = x^4$ that we will examine later in part (a) of Example 1.

What we have just illustrated with the preceding four graphs is the **second-derivative test**. When we applied the first-derivative test to a critical point, we were interested in the sign of $f'(x)$ to the left and the right of the critical point. When we use the second derivative to test a critical point, we will look at the sign of the second derivative *at* the critical point, as follows:

THEOREM 3.11

The Second-Derivative Test

Suppose $x = c$ is the location of a critical point of a function f with $f'(c) = 0$, and suppose both f and f' are differentiable and f'' is continuous on an interval around $x = c$.

(a) If $f''(c)$ is positive, then f has a local minimum at $x = c$.

(b) If $f''(c)$ is negative, then f has a local maximum at $x = c$.

(c) If $f''(c) = 0$, then this test says nothing about whether or not f has an extremum at $x = c$.

Proof. Suppose both f and f' are differentiable in a neighborhood of a critical point $x = c$ with $f'(c) = 0$. To prove part (a), suppose $f''(c) > 0$. Since f'' is assumed to be continuous near c, f'' must be positive in a small neighborhood $(c - \delta, c + \delta)$ of c. Because f'' is the derivative of f', it follows from Theorem 3.6 that f' is increasing on $(c - \delta, c + \delta)$. Since f' is increasing near c, and is zero at c, we must have $f'(x) < 0$ to the immediate left of c and $f'(x) > 0$ to the immediate right of c. Therefore, by the first-derivative test, f has a local minimum at $x = c$. The proof of part (b) is similar and is left to Exercise 76.

To prove part (c), it suffices to exhibit three functions with $f''(c) = 0$ at some point c where one function has a local maximum at $x = c$, one function has a local minimum at $x = c$, and one function does not have a local extremum at $x = c$. You will do this in Exercise 10. ■

Curve-Sketching Strategies

When we graph a function by hand, a good place to start is to use algebra, derivatives, and sign charts to determine the intervals where the function is positive or negative, increasing or decreasing, and concave up or down, as well as the coordinates of any roots, extrema, and inflection points of f. Armed with this information and occasionally a few strategic function values and limits, we can often sketch a fairly accurate graph.

Of course, actually sketching a graph based on the information just described can take a little bit of practice. One useful thing to notice is that a continuous, differentiable function can change sign only at it roots, change direction only at its local extrema, and change concavity only at its inflection points. This means that on the intervals between such points the graph of the function is relatively homogeneous. In fact, if f is a sufficiently well-behaved function, then we can connect each adjacent pair of dots determined by the coordinates of its local extrema and inflection points with one of the following four types of arcs:

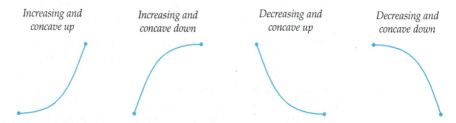

| *Increasing and concave up* | *Increasing and concave down* | *Decreasing and concave up* | *Decreasing and concave down* |

Here is a summary of what you might consider when sketching the graph of a function f by hand:

► Find the domain of f.

► Determine the roots of f and the intervals on which f is positive or negative. Record this information on a sign chart for f.

▶ Find f' and determine the points where f' is zero or does not exist. Each of these points that is in the domain of f is a critical point of f and therefore a possible local extremum of f.

▶ Determine the intervals on which f' is positive or negative. These are the same as the intervals on which f is increasing or decreasing, respectively. Record this information on a sign chart for f'.

▶ Determine whether f has a local minimum, local maximum, or neither at each critical point.

▶ Find f'' and determine the points where f'' is zero or does not exist. Each of these points that is in the domain of f is a critical point of f' and therefore a possible inflection point of f.

▶ Determine the intervals on which f'' is positive or negative. These are the same as the intervals on which f is concave up or concave down, respectively. Record this information on a sign chart for f''.

▶ Determine whether or not each critical point of f' is an inflection point of f.

▶ At each local extremum or inflection point $x = c$, determine the value of $f(c)$. Plot these key points $(c, f(c))$.

▶ At each non-continuous or non-differentiable point $x = c$ in the domain of f, determine the value of $f(c)$. Use limits of f and f' to determine the type of discontinuity or non-differentiable point. Plot these key points $(c, f(c))$.

▶ Calculate limits of f at any non-domain points and as $x \to \pm\infty$. Determine any horizontal or vertical asymptotes and the long-term behavior of the graph.

▶ Between key points and non-domain points, use the sign charts for f' and f'' to determine which of the four types of arc shapes the graph will have.

▶ Graph the function by connecting the key points with arcs.

For some functions you will not be able to obtain all of the information you want, but in every case you should explore as much as you can until you can confidently determine the behavior of the graph of f.

Why would we want to spend time using derivatives and algebra to sketch the graph of a function by hand when we could make essentially the same sketch with a graphing calculator in just a few keystrokes? One reason is that graphing by hand provides more specific information, such as the exact locations and values of the key points on the graph. Moreover, if we do the work by hand, then we can determine a graphing window that captures all of the important features of the graph and possibly also discern features of the graph that are not apparent with a graphing calculator, such as holes and asymptotes. Derivatives, limits, and algebra not only enable us to sketch graphs by hand, but also help us correctly interpret graphs made by calculators.

Examples and Explorations

| EXAMPLE 1 | Using the second derivative to determine concavity and inflection points |

Use second derivatives to determine the intervals on which each of the functions that follow are concave up or concave down. Then determine any inflection points.

(a) $f(x) = x^4$ **(b)** $g(x) = x^3 - 3x + 2$

SOLUTION

(a) We start by calculating the first and second derivatives of f:

$$f(x) = x^4 \implies f'(x) = 4x^3 \implies f''(x) = 12x^2.$$

Therefore $f''(x) = 0$ only when $x = 0$, and f'' is positive to both the left and right of $x = 0$, as shown in the following sign chart:

Since the sign of $f''(x) = 12x^2$ does not change at $x = 0$, the function $f(x) = x^4$ does not have an inflection point at $x = 0$. Therefore $f(x) = x^4$ has no inflection points and is concave up on both $(-\infty, 0)$ and $(0, \infty)$.

(b) Again we start by calculating derivatives:

$$g(x) = x^3 - 3x + 2 \implies g'(x) = 3x^2 - 3 \implies g''(x) = 6x.$$

Therefore $g''(x) = 0$ only when $x = 0$. Testing the sign of $g''(x)$ to the left and right of $x = 0$, we have $g''(-1) = -6 < 0$ and $g''(1) = 6 > 0$. This information about $g''(x)$ is summarized in the following sign chart:

Thus g is concave down on $(-\infty, 0)$, is concave up on $(0, \infty)$, and has an inflection point at $x = 0$.   □

CHECKING THE ANSWER

To verify the preceding calculations, we sketch graphs of f and g. In the graph that follows at the left we see that the curve $f(x) = x^4$ is flat enough to have zero curvature at $x = 0$, but has positive curvature to both the left and the right of the origin. In the graph at the right we see that $g(x) = x^3 - 3x + 2$ does have an inflection point at $x = 0$, where its concavity changes from concave-down to concave-up. Note that $g(x)$ is the same function we examined in Example 3(a) of the previous section.

$y = x^4$ *has no inflection points* $g(x)$ *has an inflection point at $x = 0$*

          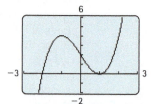

EXAMPLE 2   **Comparing the first- and second-derivative tests**

Determine the local extrema of the function $f(x) = x^3 - x^2 + 2$, using

(a) the first-derivative test   **(b)** the second-derivative test

SOLUTION

(a) The derivative of $f(x) = x^3 - x^2 + 2$ is $f'(x) = 3x^2 - 2x = x(3x - 2)$.

This derivative f' is zero at the points $x = 0$ and $x = \frac{2}{3}$, and always exists. Thus the only critical points of f are $x = 0$ and $x = \frac{2}{3}$. To apply the first-derivative test we must find the sign of the derivative between each of these points. For example, we could calculate:

$$f'(-1) = 5 > 0, \quad f'\left(\tfrac{1}{2}\right) = -\tfrac{1}{4} < 0, \quad \text{and} \quad f'(1) = 1 > 0.$$

The following sign chart summarizes this information about f':

By the first-derivative test, f has a local maximum at $x = 0$ and a local minimum at $x = \tfrac{2}{3}$.

(b) The calculation for the second derivative starts out the same way, by computing the derivative $f'(x) = 3x^2 - 2x$ and then finding the critical points $x = 0$ and $x = \tfrac{2}{3}$. The difference is that we will test the sign of the second derivative $f''(x) = 6x - 2$ at each of these critical points:

$$f''(0) = -2 < 0 \quad \text{and} \quad f''\left(\tfrac{2}{3}\right) = 2 > 0.$$

By the second-derivative test, since f is concave down at the critical point $x = 0$, f has a local maximum at $x = 0$. Similarly, since f is concave up at the critical point $x = \tfrac{2}{3}$, we know that f has a local minimum at $x = \tfrac{2}{3}$. This is of course the same conclusion we reached when we applied the first-derivative test. ☐

EXAMPLE 3 **A detailed curve-sketching analysis**

Use derivatives, algebra, and sign charts to sketch the graph of $f(x) = x^5 - 15x^3$. Identify the coordinates of each root, local extremum, and inflection point.

SOLUTION

We'll start by finding and then simplifying the functions f, f', and f''. We will factor each function as much as possible so that we can easily identify its roots:

$$f(x) = x^5 - 15x^3 = x^3(x^2 - 15) = x^3(x - \sqrt{15})(x + \sqrt{15});$$
$$f'(x) = 5x^4 - 45x^2 = 5x^2(x^2 - 9) = 5x^2(x - 3)(x + 3);$$
$$f''(x) = 20x^3 - 90x = 10x(2x^2 - 9) = 10x(\sqrt{2}x - 3)(\sqrt{2}x + 3).$$

From these factorizations we can see that f has roots at $x = 0$ and $x = \pm\sqrt{15} \approx \pm 3.87$, that the only possible local extrema of f are $x = 0$ and $x = \pm 3$, and that the only possible inflection points of f are $x = 0$ and $x = \pm\dfrac{3}{\sqrt{2}} \approx \pm 2.12$. By checking the signs of $f(x)$, $f'(x)$, and $f''(x)$ at appropriate values, we obtain the following sign charts:

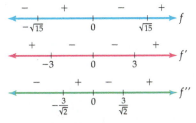

From this information we can see that f has a maximum at $x = -3$, a minimum at $x = 3$, and inflection points at $x = 0$ and $x = \pm\dfrac{3}{\sqrt{2}}$. We can also identify the intervals on which

f is positive or negative, increasing or decreasing, and concave up or concave down, as summarized in the following three sign charts:

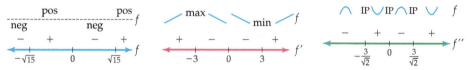

By evaluating $f(x) = x^5 - 15x^3$ at its key points we can obtain the coordinates of its roots, extrema, and inflection points:

▶ roots at $(-\sqrt{15}, 0)$, $(0, 0)$, and $(\sqrt{15}, 0)$;

▶ local maximum at $(-3, 162)$, local minimum at $(3, -162)$;

▶ inflection points at $\left(-\frac{3}{\sqrt{2}}, 100.232\right)$, $(0, 0)$, and $\left(\frac{3}{\sqrt{2}}, -100.232\right)$.

Now we need only plot these points and connect the dots with appropriate arcs according to the sign information in the sign charts. For example, we can see from the sign charts for f, f', and f'' that between $x = -3$ and $x = -\frac{3}{\sqrt{2}}$ the graph of f should be positive, decreasing, and concave down.

We already have all of the information we need about the derivatives, but for some people it helps to collect all this information in one place. The arc shapes on each subinterval between key points are recorded on the combined number-line chart that follows. Note that on this number line the tick-marks represent the locations of all interesting points on the graph of f, meaning that they are a composite of the tick-marks from the three sign charts for f, f', and f''.

Although at this point the shape of the graph is pretty clear, for completeness we should compute limits at any interesting points. The function $f(x) = x^5 - 15x^3$ has no non-domain points, discontinuities, or non-differentiable points, so the only limits to check are those as $x \to \pm\infty$. We can use what we know about the behavior of fifth-degree polynomials, or we can just compute these limits directly:

$$\lim_{x \to \infty} (x^5 - 15x^3) = \lim_{x \to \infty} (x^3)(x^2 - 15) = \infty,$$
$$\lim_{x \to -\infty} (x^5 - 15x^3) = \lim_{x \to -\infty} (x^3)(x^2 - 15) = -\infty.$$

This information tells us that the graph has no horizontal asymptotes and indicates what happens at the "ends" of the graph of f.

Putting all of the information together into a labeled graph, we have

$$f(x) = x^5 - 15x^3$$

We can verify the graph we constructed in the preceding example by using a calculator or graphing utility to graph $f(x) = x^5 - 15x^3$ in a similar graphing window. Notice that we did indeed capture the major features of this graph. In addition, because we did the work by hand, we know the exact values of every key point on the graph of f.

Calculator graph of $f(x) = x^5 - 15x^3$

EXAMPLE 4

A curve-sketching analysis with asymptotes

Sketch an accurate, labeled graph of the function $f(x) = \dfrac{6}{4 - 2^x}$. Include complete sign analyses of f, f', and f'', and calculate any relevant limits.

SOLUTION

Let's begin by finding and simplifying the first and second derivatives of $f(x)$. The first derivative of $f(x) = 6(4 - 2^x)^{-1}$ is

$$f'(x) = 6(-1)(4 - 2^x)^{-2}(-(\ln 2)2^x) = \frac{6(\ln 2)2^x}{(4 - 2^x)^2}.$$

Differentiating that result and then simplifying as much as possible so that we can easily identify roots, we find that the second derivative of f is

$$f''(x) = \frac{6(\ln 2)(\ln 2)2^x(4 - 2^x)^2 - (6(\ln 2)2^x)(2)(4 - 2^x)^1(-(\ln 2)2^x)}{(4 - 2^x)^4}$$

$$= \frac{6(\ln 2)^2 2^x(4 - 2^x)((4 - 2^x) + 2 \cdot 2^x)}{(4 - 2^x)^4} = \frac{6(\ln 2)^2 2^x(4 + 2^x)}{(4 - 2^x)^3}.$$

To determine the intervals on which f, f', and f'' are positive and negative we must first locate the values of x for which these functions are zero or do not exist. The function $f(x) = \dfrac{6}{4 - 2^x}$ is never zero, but is undefined where its denominator is zero, at $x = 2$. Looking at the formula for $f'(x)$, we can easily see that it is never zero (since 2^x is never zero), but is undefined if $x = 2$ (since the denominator $4 - 2^x$ is zero for $x = 2$). Thus $x = 2$ is the only critical point of f and the only point we will mark on the number line for f'. Similarly, from the formula for $f''(x)$, it is clear that $f''(x)$ is never zero (since neither 2^x nor $4 + 2^x$ can ever be zero), but is undefined at $x = 2$. Checking signs on either side of $x = 2$ for each function, we obtain the following set of number lines:

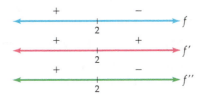

We now know that $f(x)$ is positive, increasing, and concave up on $(-\infty, 2)$ and negative, increasing, and concave down on $(2, \infty)$. The graph has no roots, no extrema, and no inflection points.

It remains to calculate any interesting limits. Since the domain of f is $(-\infty, 2) \cup (2, \infty)$, we must investigate the limits of $f(x) = \frac{6}{4-2^x}$ as $x \to \pm\infty$ and as $x \to 2$ from the left and the right. As $x \to \infty$ the denominator $4 - 2^x$ of $f(x)$ approaches $-\infty$, and thus

$$\lim_{x \to \infty} \frac{6}{4 - 2^x} = 0.$$

As $x \to -\infty$, the denominator $4 - 2^x$ approaches 4, and thus

$$\lim_{x \to -\infty} \frac{6}{4 - 2^x} = \frac{6}{4} = \frac{3}{2}.$$

This information tells us that the graph of f has *two* horizontal asymptotes: at $y = 0$ on the right and at $y = \frac{3}{2}$ on the left.

As $x \to 2^-$ we have $4 - 2^x \to 0^+$ and thus

$$\lim_{x \to 2^-} \frac{6}{4 - 2^x} = \infty,$$

and as $x \to 2^+$ we have $4 - 2^x \to 0^-$ and thus

$$\lim_{x \to 2^+} \frac{6}{4 - 2^x} = -\infty.$$

Thus the graph of f has a vertical asymptote at $x = 2$, where the graph approaches ∞ to the left of 2 and approaches $-\infty$ to the right of 2.

Putting all of this information together, we can now sketch the graph:

$$f(x) = \frac{6}{4 - 2^x}$$

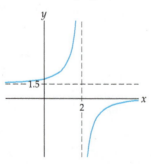

TEST YOUR UNDERSTANDING

▶ Why could we not give a precise mathematical definition of concavity before this section?

▶ The domain points $x = c$ where $f''(c) = 0$ or where $f''(c)$ does not exist are the critical points of the function f'. Why?

▶ Why is it not clear to say a sentence such as "Because it is positive, it is concave up"? How could this information be conveyed more precisely?

▶ Why does it make sense that f' is increasing when f'' is positive?

▶ Suppose $x = c$ is a critical point with $f'(c) = 0$. Why does it make graphical sense that f has a local minimum at $x = c$ when f is concave up in a neighborhood around $x = c$?

EXERCISES 3.3

Thinking Back

Finding the second derivative: For each of the following functions f, calculate and simplify the second derivative f''.

▶ $f(x) = x^4 - 3x^2 + 2x$

▶ $f(x) = \dfrac{1}{3x + 1}$

▶ $f(x) = \dfrac{x - 1}{3x^2 - 4x + 2}$

▶ $f(x) = \dfrac{x}{\sqrt{x^2 + 1}}$

Solving for zeroes and non-domain points: For each of the following expressions, find all values of x for which $g(x)$ is zero or does not exist.

▶ $g(x) = x^3 + 3x^2 + 2x$

▶ $g(x) = \dfrac{x - 2}{x^2 - 4}$

▶ $g(x) = \dfrac{3x^2 - x - 2}{x^4 + 2x^2 - 3}$

▶ $g(x) = \dfrac{1}{x - 2} - \dfrac{3x + 1}{x + 2}$

Concepts

0. *Problem Zero:* Read the section and make your own summary of the material.

1. *True/False:* Determine whether each of the statements that follow is true or false. If a statement is true, explain why. If a statement is false, provide a counterexample.

(a) *True or False:* If $f''(2) = 0$, then $x = 2$ is an inflection point of f.

(b) *True or False:* If f'' is concave up on an interval I, then it is positive on I.

(c) *True or False:* If f is concave up on an interval I, then f'' is positive on I.

(d) *True or False:* If $f''(2)$ does not exist and $x = 2$ is in the domain of f, then $x = 2$ is a critical point of the function f'.

(e) *True or False:* If f has an inflection point at $x = 3$ and f is differentiable at $x = 3$, then the derivative f' has a local minimum or maximum at $x = 3$.

(f) *True or False:* If $f'(1) = 0$ and $f''(1) = -2$, then f has a local minimum at $x = 1$.

(g) *True or False:* The second-derivative test involves checking the sign of the second derivative on each side of every critical point.

(h) *True or False:* The second-derivative test always produces exactly the same information as the first-derivative test.

2. *Examples:* Construct examples of the thing(s) described in the following. Try to find examples that are different than any in the reading.

(a) The graph of a function f for which f' is positive everywhere, $f''(x) > 0$ for $x < -2$, and $f''(x) < 0$ for $x > -2$.

(b) The graph of a function f for which $f(3) = 0$, $f'(3) = 0$, and $f''(3) = 0$.

(c) The graph of a function f for which $f(x)$ is zero at $x = -1$, $x = 2$, and $x = 4$; $f'(x)$ is zero at $x = -1$, $x = 1$, and $x = 3$; and $f''(x)$ is zero at $x = 0$ and $x = 2$.

3. Sketch the graph of a function f that is concave up everywhere. Then draw five tangent lines on the graph, and explain how you can see that the derivative of f is increasing.

4. Sketch the graph of a function f that is concave down everywhere. Then draw five tangent lines on the graph, and explain how you can see that the derivative of f is decreasing.

5. State the converse of Theorem 3.10(a). Is the converse true? If so, explain why; if not, provide a counterexample.

6. State the contrapositive of Theorem 3.10(a). Is the contrapositive true? If so, explain why; if not, provide a counterexample.

7. Sketch the graph of a function f that has an inflection point at $x = c$ in such a way that the derivative f' has a local maximum at $x = c$. Add tangent lines to your sketch to illustrate that f' does have a local maximum at $x = c$.

8. Sketch the graph of a function f that has an inflection point at $x = c$ in such a way that the derivative f' has a local minimum at $x = c$. Add tangent lines to your sketch to illustrate that f' does have a local minimum at $x = c$.

9. Show that for $f(x) = x^6$ we have $f''(0) = 0$ but the point $x = 0$ is not an inflection point of f.

10. In this problem we will verify part (c) of Theorem 3.11.

(a) For $f(x) = x^3$, show that $f'(0) = 0$ and $f''(0) = 0$ while f does not have a local extremum at $x = 0$.

(b) For $g(x) = x^4$, show that $g'(0) = 0$ and $g''(0) = 0$ while g has a local minimum at $x = 0$.

(c) For $h(x) = -x^4$, show that $h'(0) = 0$ and $h''(0) = 0$ while h has a local maximum at $x = 0$.

(d) Explain how the parts (a)–(c) show that the second-derivative test does not tell us anything when the second derivative at a critical point is zero.

11. We could use part (c) from Theorem 3.6 to add a third part to Theorem 3.10 that would tell us what it means when f'' is zero in the interior of an interval I. Fill in the blank accordingly: If f'' is zero on I, then f is _____ on I.

12. Describe what the second-derivative test is for and how to use it. Sketch graphs and sign charts to illustrate your description.

13. If a function f has four critical points, how many calculations after finding derivatives are required in order to apply the first-derivative test? The second-derivative test?

14. Describe in words, and then illustrate in pictures, the four types of arcs that are the building blocks of most continuous, differentiable graphs.

For Exercises 15–20, sketch the graph of a function f that has the indicated characteristics. If a graph is not possible, explain why.

15. f positive, f' negative, and f'' positive on $[0, 3]$.

16. f negative, f' negative, and f'' negative on $[0, 3]$.

17. f negative, f' positive, and f'' positive on $[0, 3]$.

18. f positive, f' negative, and f'' positive on $\mathbb{R}$.

19. f negative, f' negative, and f'' negative on $\mathbb{R}$.

20. f negative, f' positive, and f'' positive on $\mathbb{R}$.

Skills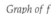

In Exercises 21–28, graphs of f, f', or f'' are given. Whichever is shown, sketch graphs of the remaining two functions. Label the locations of any roots, extrema, and inflection points on each graph.

21. *Graph of f*

22. *Graph of f*

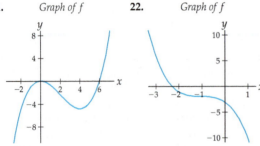

23. *Graph of f*

24. *Graph of f'*

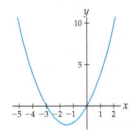

25. *Graph of f'*

26. *Graph of f'*

27. *Graph of f''*

28. *Graph of f''*

Use the second-derivative test to determine the local extrema of each function f in Exercises 29–36. If the second-derivative test fails, you may use the first-derivative test. Then verify your algebraic answers with graphs from a calculator or graphing utility. *(Note: These are the same functions that you examined with the first-derivative test in Exercises 35–42 of Section 3.2.)*

29. $f(x) = 3 - 4x$

30. $f(x) = -x^2 - 10x - 33$

31. $f(x) = x^2 - 5x - 6$

32. $f(x) = 9 - x^4$

33. $f(x) = (x - 2)^2(1 + x)$

34. $f(x) = x^2(x - 1)(x + 1)$

35. $f(x) = \dfrac{1 + x + x^2}{x^2 + x - 2}$

36. $f(x) = \dfrac{(x - 1)^2}{x + 2}$

Use a sign chart for f'' to determine the intervals on which each function f in Exercises 37–46 is concave up or concave down, and identify the locations of any inflection points. Then verify your algebraic answers with graphs from a calculator or graphing utility.

37. $f(x) = 2x^2 - x^3$

38. $f(x) = 3x + 1$

39. $f(x) = x^2(2 - x)^2$

40. $f(x) = x^4 - 16$

41. $f(x) = (x - 2)^4$

42. $f(x) = (x - 3)^3(x - 1)$

43. $f(x) = x^4 - 2x^3 - 5$

44. $f(x) = \dfrac{1}{1 + x^2}$

45. $f(x) = \dfrac{1}{x^2 + x + 1}$

46. $f(x) = \dfrac{\sqrt{x}}{x - 2}$

For each set of sign charts in Exercises 47–56, sketch a possible graph of f.

47.

48.

49.

50.

51.

52.

53.

54.

55.

56.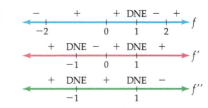

Sketch careful, labeled graphs of each function f in Exercises 57–68 by hand, without consulting a calculator or graphing utility. As part of your work, make sign charts for the signs, roots, and undefined points of f, f', and f'', and examine any relevant limits so that you can describe all key points and behaviors of f.

57. $f(x) = x^2 + 3x$

58. $f(x) = (x-2)(x+2)$

59. $f(x) = x^3 + 3x^2$

60. $f(x) = \dfrac{1}{x^2+3}$

61. $f(x) = x^3 - 2x^2 + x$

62. $f(x) = (1-x)^4 - 2$

63. $f(x) = x^3(x+2)$

64. $f(x) = \dfrac{x}{x^2+1}$

65. $f(x) = \sqrt{x}(4-x)$

66. $f(x) = \dfrac{1}{(x-1)^2(x-2)}$

67. $f(x) = \dfrac{x^2-1}{x^2-5x+4}$

68. $f(x) = \dfrac{x^2-x}{x^2-3x+2}$

In Exercises 69–72, use the given derivative f' to find any local extrema and inflection points of f and sketch a possible graph without first finding an formula for f.

69. $f'(x) = x^2 + x - 2$

70. $f'(x) = 3 - 5x$

71. $f'(x) = x^3 - 3x^2 + 3x$

72. $f'(x) = x^4 - 1$

Applications

73. Jason's distance in miles north from the corner of Main Street and High Street t minutes after noon on Tuesday is given by the following function $s(t)$:

Distance north of Main and High

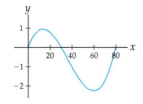

(a) Find an interval on which Jason's velocity is positive and decreasing. Describe what Jason is doing over this time interval.

(b) Find a time interval on which Jason is moving north and his velocity is increasing. Describe what Jason is doing over this time interval.

(c) Find a time interval on which Jason's acceleration and velocity are both negative. Describe what Jason is doing over this time interval.

(d) At which time is Jason's velocity at a minimum? What is he doing at that moment?

74. Suppose Juri drives for two hours and that his distance from home in miles is given by the function $s(t)$ shown in the following figure.

Distance from home

(a) Find a time interval on which Juri's acceleration is positive. Is his velocity positive or negative on this interval? Describe what Juri is doing over this time interval.

(b) The graph $y = s(t)$ of Juri's position has an inflection point at $t = 1$ hour. What does this inflection point say about Juri's velocity at $t = 1$? About his acceleration at $t = 1$?

Proofs

75. Prove part (b) of Theorem 3.10: If both f and f' are differentiable on an interval I, and f'' is negative on I, then f is concave down on I.

76. Prove Theorem 3.11 (b): If $x = c$ is a critical point of f, both f and f' are differentiable near $x = c$, and if $f''(c)$ is negative, then f has a local maximum at $x = c$.

77. Prove that every quadratic function is either always concave up or always concave down.

78. Prove that every cubic function (i.e., every function of the form $f(x) = ax^3 + bx^2 + cx + d$ for some constants a, b, c, and d) has exactly one inflection point. (*Note: It is not enough just to show that the second derivative of any cubic function has exactly one zero; you must also show that the sign of the second derivative changes.*)

79. Prove that if f'' is zero on an interval, then f is linear on that interval.

Thinking Forward

▶ *Global extrema on an interval:* The first-derivative test can be used to show that the function $f(x) = x^2 + 3x$ has a local minimum at $x = -\dfrac{3}{2}$. Is this a global minimum of the function? Is there a global maximum? What are the global extrema (if any) if we consider the function restricted to the interval $[-3, 3]$?

▶ *Global extrema and derivatives:* The first-derivative test can be used to show that the function $f(x) = x^3 - x^2 + x$ has a local maximum at $x = \dfrac{1}{3}$ and a local minimum at $x = 1$. Are either of these local extrema also global extrema? Can the first or second derivative tell you whether or not a local extremum is a global extremum?

3.4   OPTIMIZATION

▶ Comparing local extrema and endpoint behavior to find global extrema

▶ Strategies for translating real–world problems into mathematical problems

▶ Optimizing real–world quantities by solving global extrema problems

Finding Global Extrema

Many real-world problems involve optimization, that is, finding the global maximum and minimum values of a function on an interval. For example, a company might be interested in the maximum amount of profit it can make or in the minimum cost of producing an object. In this section we will learn how to use limits, derivatives, and values to find the global extrema of a function on an interval, so that we can solve such real-world problems.

Recall from Section 0.4 that $x = c$ is the location of a ***global maximum*** of a function f on an interval I if $f(c) \geq f(x)$ for all $x \in I$. Sometimes local extrema are also global extrema, and sometimes they are not. For example, the following four functions each have a local maximum at the critical point $x = 2$:

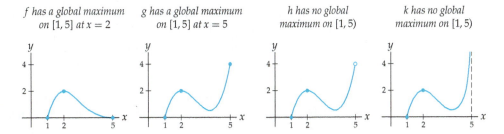

f has a global maximum on $[1, 5]$ at $x = 2$ g has a global maximum on $[1, 5]$ at $x = 5$ h has no global maximum on $[1, 5)$ k has no global maximum on $[1, 5)$

Only in the first graph is the local maximum also a global maximum. In the graph of g the global maximum on $[1, 5]$ is instead at the right endpoint $x = 5$. In the last two graphs there is no global maximum on the interval, because there is no point on either graph that is higher than all other points on its graph.

Although derivatives can help us locate local extrema, they do not always provide enough information to detect global extrema. The previous figures suggest that to find a global extremum of a continuous function we must compare the following:

▶ the values $f(c)$ for each interior local extremum $c \in I$;

▶ the values $f(c)$ at any closed endpoints $x = c$ of I;

▶ the limits $\lim_{x \to c} f(x)$ at any open endpoints or non-domain points $x = c$ of I.

Whichever of these three values is largest determines the location, if any, of the global maximum of f on I. Whichever is smallest determines the location, if any, of the global minimum. For example, in the graph of f that we just looked at, the value at the local extremum is $f(2) = 2$ and at the left and right endpoints we have $f(1) = 0$ and $f(5) = 0$. Since the value is highest at $x = 2$, that point is the global maximum of f on $[1, 5]$. A different thing happens with the graph of h; we again have $h(2) = 2$ at the local extremum and $h(1) = 0$ at the left endpoint, but at the open right endpoint we have $\lim_{x \to 5^-} h(x) = 4$. Since the limit at the open right endpoint is larger than the value at the left endpoint and the values at any local extrema, the function h has no global maximum on $[1, 5)$.

Translating Word Problems into Mathematical Problems

Finding the global extrema of a function f on an interval I is a mathematical problem that we can tackle with function values, limits, and derivatives. Real–world problems are less straightforward, since they are expressed in full sentences in which the variables, constants, functions, and relationships are described in words instead of in mathematical notation. To solve a real–world optimization problem we must first translate it into a mathematical global extrema problem. Once we solve the mathematical problem, we can translate the solution back into the real–world context. This translating procedure is illustrated in the following diagram:

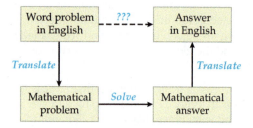

Many students have difficulty with word problems because of the translation step, not because of the mathematical solving step. As a simple example, suppose we wanted to solve the following real–world problem:

> *Calvin throws a baseball straight into the air at 50 feet per second, releasing the ball when his hand is 5 feet above the ground. How high did Calvin throw the baseball?*

To figure out how to answer this problem with calculus, we need to get some equations and variables to work with. A good place to start is to list what you know and set variable names as you go along:

▶ Let $s(t)$ be the height/position of the baseball at time t, in feet.

▶ We know that the position equation will be of the form $s(t) = -16t^2 + v_0 t + s_0$.

▶ $v_0 = 50$ feet per second is the initial velocity of the baseball.

▶ $s_0 = 5$ feet is the initial position of the baseball.

▶ We seek the greatest height of the baseball during the time that it is in the air.

▶ Solving $s(t) = 0$, we see that the ball hits the ground at $t \approx 3.22$ seconds.

At some point while developing this list of facts it is good to collect the information into a figure, such as this one:

Putting this all together, we see that the mathematical problem that we wish to solve is

Find the global maximum value of $s(t) = -16t^2 + 50t + 5$ on the interval $[0, 3.22]$.

Compare this mathematical statement with our earlier real–world statement. After the hard work of translating, we now have a fairly straightforward calculus problem to solve. Here is an outline of the steps we could use to solve this optimization problem:

▶ Take the derivative: $s'(t) = -32t + 50$.

▶ Find critical points: $s'(t) = 0$ when $t = \frac{50}{32} = 1.5625$.

▶ Test critical points: $s''(1.5625) = -32 < 0$, so $s(t)$ has a local maximum at $x = 1.5625$.

▶ Check height at local maximum: $s(1.5625) \approx 44.06$.

▶ Check heights at endpoint values: $s(0) = 0$ and $s(3.22) \approx 0$.

▶ Conclusion: The maximum value of $s(t)$ on $[0, 3.22]$ is approximately 44.06.

Translating back into the real–world context, we can say that Calvin threw the baseball to a height of just over 44 feet.

In general, when translating a word problem we must identify any variables or functions and express them in mathematical notation with letters and symbols. With those same letters and symbols we can construct labeled diagrams, formulas, and relationships in mathematical notation. We must keep translating until we form a well-posed mathematical problem that we know how to solve with the tools and techniques of calculus. In the context of optimization problems, this means that we must identify a specific function and a specific interval on which we wish to find a global maximum or minimum value. At this point we only know how to find global extrema of one-variable functions, so we will sometimes have to use constraint equations to reduce multivariable functions to single-variable functions; see Example 2.

Examples and Explorations

EXAMPLE 1 Finding global extrema of a function on an interval

Find the global maximum and minimum values, if any, of the function $f(x) = 2x^3 - 15x^2 + 24x + 20$ on the interval $(0, 6)$.

SOLUTION

Our goal will be to use the derivative to identify local extrema on the interior of the interval and then compare those extrema with values or limits at the ends of the interval. In this case the interval is open, so we will compare with limits at $x \to 0$ and $x \to 6$. Step by step, we have

▶ Take the derivative: $f'(x) = 6x^2 - 30x + 24 = 6(x^2 - 5x + 4) = 6(x - 1)(x - 4)$.

▶ Find critical points: $f'(x) = 0$ at $x = 1$ and $x = 4$.

▶ Test critical points: $f''(x) = 12x - 30$, so $f''(1) = -18 < 0$ and $f''(4) = 18 > 0$; thus f has a local maximum at $x = 1$ and a local minimum at $x = 4$.

▶ Check height at local extrema: $f(1) = 31$ and $f(4) = 4$.

▶ Check limits at open endpoints:

$$\lim_{x \to 0^+} f(x) = \lim_{x \to 0^+} (2x^3 - 15x^2 + 24x + 20) = 20,$$

$$\lim_{x \to 6^-} f(x) = \lim_{x \to 6^-} (2x^3 - 15x^2 + 24x + 20) = 56.$$

The local maximum value of $f(1) = 31$ is exceeded by the limit of f at the right endpoint, so the function f has no global maximum on $(0, 6)$. The local minimum value of $f(4) = 4$ is less than the limit at either end of the interval and therefore is also a global minimum of f on $(0, 6)$. The graph of f on $(0, 6)$ looks like this:

Global minimum at x = 4,
no global maximum on (0, 6)

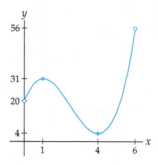

EXAMPLE 2   **Optimizing a function on an interval given a constraint**

Farmer Joe wants to build a rectangular chicken pen for his chickens. He wants to build the pen so that it has the largest area possible, and he has only 100 feet of chicken-wire fencing. What dimensions should he use for the pen?

SOLUTION

Let l and w represent the length and width of the rectangular pen, in feet. We know that the perimeter P of the pen must be 100 feet, since that is how much chicken-wire fencing Farmer Joe has; therefore $P = 2w + 2l = 100$. We are interested in maximizing the area $A = lw$ of the pen. The following diagram summarizes this information:

Constraint: $P = 2w + 2l = 100$

Maximize: $A = lw$

At this point the function A that we wish to maximize is written in terms of two variables, but our calculus techniques work only for functions of one variable. Luckily, we can use the constraint $P = 2w + 2l = 100$ to solve for one variable in terms of the other; this will reduce A to a one-variable function that we can optimize. Solving for l in terms of w, we have

$$P = 2w + 2l = 100 \quad \Longrightarrow \quad 2l = 100 - 2w \quad \Longrightarrow \quad l = 50 - w.$$

Using this equation, we can write the area function entirely in terms of w:

$$A = lw = (50 - w)w = 50w - w^2.$$

We now have a one-variable function $A(w) = 50w - w^2$ to maximize, but on what interval? We need to know which values of w we are willing to consider in this problem. Clearly we must have $w \geq 0$, since the width of the chicken pen cannot be negative. What is the upper bound on w? The key is to realize that the width w is the largest when the length l is the smallest. The smallest value that l could have is $l = 0$, and when $l = 0$, we have $0 = 50 - w$, or $w = 50$. Therefore we must have $w \leq 50$. We have now determined that the underlying mathematical problem for this word problem is

Find the global maximum of $A(w) = 50w - w^2$ on the interval $[0, 50]$.

Now we just have a simple mathematical problem to solve, and we follow the usual procedure for finding a global extremum on an interval:

- ▶ Take the derivative: $A'(w) = 50 - 2w$.
- ▶ Find critical points: $A'(w) = 0$ at $w = 25$.
- ▶ Test critical points: $A''(w) = -2$, so $A''(25) = -2$, and by the second-derivative test, $A(w)$ has a local maximum at $w = 25$.
- ▶ Check height at local extremum: $A(25) = 625$.
- ▶ Check heights at closed endpoints: $A(0) = 0$ and $A(50) = 0$.

From this work we see that $w = 25$ is indeed the global maximum of $A(w)$ on the interval $[0, 50]$. Thus the width of the pen should be $w = 25$ feet long. Since $l = 50 - w$, the length of the pen should also be $l = 25$ feet long, and the pen is square. The final answer to the word problem is therefore that Farmer Joe should build a square chicken pen where each side is 25 feet long.

□

EXAMPLE 3   Modeling and minimizing a cost function

You work for a company that makes jewelry boxes. Your boss tells you that each jewelry box must have a square base and an open top and that you can spend $3.75 on the materials for each box. The people in production tell you that the material for the sides of the box costs 2 cents per square inch while the reinforced material for the base of the box costs 5 cents per square inch. What is the largest volume jewelry box that you can make and still stay within budget?

SOLUTION

From a quick reading of the problem we can see that we wish to maximize the volume of a jewelry box given certain monetary constraints. Let's begin with a simple picture, assign variable names, and list what we know in mathematical language. Suppose x is the length of the sides of the square base and y is the height of the jewelry box, both measured in inches. The material to make each of the four sides of the box will cost $0.02(xy)$ dollars, and the material to make the base of the box will cost $0.05(x^2)$ dollars. Therefore we must have $0.02(4xy) + 0.05(x^2) = 3.75$, since we have $3.75 to spend on materials. We wish to maximize the volume $V = x^2y$ of the box:

x
$0.02(xy)$ dollars
$0.05(x^2)$ dollars
y
x
x
y
x
x

Constraint: $0.02(4xy) + 0.05(x^2) = 3.75$

Maximize: $V = x^2y$

Again, the function to be optimized is written in terms of two variables at this point, but we can use the constraint to solve for one variable in terms of the other. It is easiest to solve for y in terms of x:

$$0.02(4xy) + 0.05(x^2) = 3.75 \implies 0.08xy = 3.75 - 0.05x^2 \implies y = \frac{3.75 - 0.05x^2}{0.08x}.$$

We can use this equation for y to write the volume function V in terms of one variable. Doing this and then simplifying as much as possible, we have

$$V = x^2 y = x^2 \left(\frac{3.75 - 0.05x^2}{0.08x} \right) = \frac{1}{0.08} x(3.75 - 0.05x^2) = 46.875x - 0.625x^3.$$

Now we must determine the appropriate domain for $V(x) = 46.875x - 0.625x^3$ in the context of this problem. Clearly the smallest that the length x can be is zero. To find the largest that the base side length x can be, consider that the smallest possible value of the height of the box is $y = 0$. When $y = 0$, we have

$$0 = \frac{3.75 - 0.05x^2}{0.08x} \implies 0 = 3.75 - 0.05x^2 \implies x = \sqrt{\frac{3.75}{0.05}} = \sqrt{75}.$$

Note that we do not consider the negative square root of 75, since we know that the length x must be nonnegative. We have now completely translated the original word problem into the following mathematical problem:

Find the global maximum value of $V(x) = 46.875x - 0.625x^3$ on the interval $[0, \sqrt{75}]$.

To solve this problem we will find all the local interior extrema of $V(x)$ in the interval $[0, \sqrt{75}]$ and compare their values with the values of V at the endpoints of the interval. The steps are the same as those from previous examples:

▶ Take the derivative: $V'(x) = 46.875 - 3(0.625)x^2$.

▶ Find critical points: $V'(x) = 0$ when $x = \pm 5$, but only $x = 5$ is in the interval $[0, \sqrt{75}]$.

▶ Test critical points: $V''(5) = -2(3)(0.625)(5)$ is negative, so by the second-derivative test, $V(x)$ has a local maximium at $x = 5$.

▶ Check height at local extremum: $V(5) = 156.25$.

▶ Check heights at closed endpoints: $V(0) = 0$ and $V(\sqrt{75}) = 0$.

We now see that $x = 5$ is not only the location of a local maximum, but in fact the location of the *global* maximum of $V(x)$ on $[0, \sqrt{75}]$. Therefore the largest jewelry box that we can make with the given cost restrictions has volume 156.25 cubic inches.   ☐

 CHECKING THE ANSWER

When $x = 5$, we must have $y = \frac{3.75 - 0.05(5)^2}{0.08(5)} = 6.25$. To algebraically check some of the work we have done, we can verify that, with these values for x and y, the cost of producing the jewelry box is indeed $0.02(4xy) + 0.05(x^2) = \$3.75$. We can check our work regarding the optimization of V and the choice of endpoints for the interval by graphing $V(x) = 46.875x - 0.625x^3$, as shown next. From this graph, it does seem reasonable that at $x = 5$ we have a global maximum of 156.25 and that the point $x = \sqrt{75} \approx 8.66$ should be the right end of the interval for the problem.

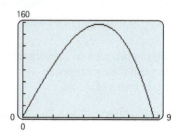

EXAMPLE 4

The fastest way to put out a flaming tent

While you are on a camping trip, your tent accidentally catches fire. Luckily, you happen to be standing right at the edge of a stream with a bucket in your hand. The stream runs east–west, and the tent is 40 feet north of the stream and 100 feet farther east than you are, as shown here:

You can run only half as fast while carrying the full bucket as you can empty handed, and thus any distance travelled with the full bucket is effectively twice as long. What is the fastest way for you to get water to the tent?

SOLUTION

On the one hand, notice that if you were to get water immediately and run diagonally to the tent, you would run the entire distance with a full bucket of water at half your normal speed. On the other hand, if you were to run along the side of the stream until you were directly south of the tent, then get water and run north to the tent, you would have a lot of total distance to run. Clearly, it would be more efficient for you to run along the side of the stream for a while, fill the bucket, and then run diagonally to the tent. The question is, how far should you run along the side of the stream before you stop to fill up the bucket?

We need some mathematical notation to get our heads around this problem. Suppose x is the distance you will run along the stream before filling your bucket. The distance you will run with the full bucket is then the hypotenuse of a right triangle with legs of lengths 40 feet and $100 - x$ feet. By the Pythagorean theorem, you will have to run $\sqrt{40^2 + (100 - x)^2}$ feet with a full bucket of water, as shown here:

Since it takes you twice as long to run with a full bucket of water, the total *effective* distance you will have to run is

$$D(x) = x + 2\sqrt{40^2 + (100 - x)^2}.$$

This is the function we want to minimize. What is the interval of x-values that we are interested in? From the diagram we can see that x must be between 0 and 100 feet, since clearly you wouldn't want to run away from the tent or past the tent. The endpoint cases $x = 0$ and $x = 100$ correspond to the two special cases we discussed earlier: The case when you get water immediately and the case when you run the full 100 feet before getting water. We have now translated the word problem into the following mathematical optimization problem:

Find the global minimum value of $D(x) = x + 2\sqrt{40^2 + (100 - x)^2}$ on the interval $[0, 100]$.

Once again this is a straightforward global extremum problem, and we follow the same steps as previously; but this time the mathematics is a little more involved:

▶ Take the derivative: By applying the chain rule twice, we see that the derivative of $D(x)$ is

$$D'(x) = 1 + 2\left(\frac{1}{2}\right)(40^2 + (100 - x)^2)^{-1/2}(2(100 - x)(-1)).$$

▶ Find critical points: We need to simplify $D'(x)$ before we attempt to find its zeroes or the values where it does not exist. With a bit of algebra we can write $D'(x)$ in the form of a quotient:

$$D'(x) = 1 + \frac{-2(100 - x)}{\sqrt{40^2 + (100 - x)^2}} = \frac{\sqrt{40^2 + (100 - x)^2} - 2(100 - x)}{\sqrt{40^2 + (100 - x)^2}}.$$

From this quotient form it is easy to pick out any places where $D'(x)$ is zero or undefined. The denominator of $D'(x)$ is never zero, and thus $D'(x)$ always exists. To find the values of x for which $D'(x) = 0$, we set the numerator equal to 0 and solve:

$$\sqrt{40^2 + (100 - x)^2} - 2(100 - x) = 0$$

$$\sqrt{40^2 + (100 - x)^2} = 2(100 - x)$$

$$40^2 + (100 - x)^2 = 4(100 - x)^2 \qquad \leftarrow \text{square both sides}$$

$$1600 + 10,000 - 200x + x^2 = 40,000 - 800x + 4x^2$$

$$0 = 3x^2 - 600x + 28400$$

$$x = 100 \pm \frac{40}{3}\sqrt{3}. \qquad \leftarrow \text{quadratic formula}$$

Note that $x = 100 + \frac{40}{3}\sqrt{3} \approx 123.1$ is not in the interval $[0, 100]$, but $x = 100 - \frac{40}{3}\sqrt{3} \approx 76.9$ is. In addition, recall that squaring both sides of an equation sometimes leads to extraneous, or false, solutions; interestingly, $x = 100 + \frac{40}{3}\sqrt{3}$ happens to be one of those false solutions. Therefore the only critical point of $D(x)$ is $x = 100 - \frac{40}{3}\sqrt{3} \approx 76.9$.

▶ Test critical points: The first-derivative test can be applied to show that $D(x)$ has a local minimum at $x = 100 - \frac{40}{3}\sqrt{3} \in [0, 100]$.

▶ Check height at local extremum: $D(100 - \frac{40}{3}\sqrt{3}) \approx 169.282$.

▶ Check heights at closed endpoints: $D(0) \approx 215.407$, $D(100) = 180$.

The work that we just did shows that the global minimum of $D(x)$ on $[0, 100]$ is at $x = 100 - \frac{40}{3}\sqrt{3}$. This means that the quickest way to bring water to the burning tent is to run for $100 - \frac{40}{3}\sqrt{3} \approx 76.9$ feet along the side of the stream, then fill the bucket with water and run diagonally to the tent.

TEST YOUR UNDERSTANDING

▶ Why can't the derivative necessarily detect the global minimum of a function f on an interval $[a, b]$ if that minimum happens to occur at an endpoint of the interval?

▶ Suppose a function f satisfies $\lim_{x \to 3^-} f(x) = -\infty$. Why does this mean that f has no global minimum on $[0, 3]$. What can you say about any global minima of f on $[0, 6]$?

▶ Suppose a function f satisfies $\lim_{x \to 1^+} f(x) = \infty$. What can you say about any global maxima of f on $[-2, 4]$?

▶ What types of real–world problems translate into mathematical problems in which we must find the global maximum or minimum of a function on an interval?

▶ What are the general steps for solving an optimization word problem?

EXERCISES 3.4

Thinking Back

Local and global extrema: Use mathematical notation, including inequalities as used in the definition of local and global extrema, to express each of the following statements.

▶ On the interval $[-3, 5]$, f has a local maximum at $x = 2$.

▶ On the interval $[0, 2]$, f has a global maximum at $x = -1$.

▶ On the interval $[-4, 4]$, f has a global minimum at $x = 0$.

▶ On the interval $[0, 5]$, f has no global minimum.

Critical points: Find the critical points of each of the following functions.

▶ $f(x) = (x + 2)(x - 3)$

▶ $f(x) = -4 + 4x - x^2$

▶ $f(x) = \dfrac{3x - 2}{x - 1}$

▶ $f(x) = \dfrac{1}{1 + \sqrt{x}}$

Concepts

0. *Problem Zero:* Read the section and make your own summary of the material.

1. *True/False:* Determine whether each of the statements that follow is true or false. If a statement is true, explain why. If a statement is false, provide a counterexample.

(a) *True or False:* Every local maximum is a global maximum.

(b) *True or False:* Every global minimum is a local minimum.

(c) *True or False:* If f has a global maximum at $x = 2$ on the interval $(-\infty, \infty)$, then the global maximum of f on the interval $[0, 4]$ must also be at $x = 2$.

(d) *True or False:* If f has a global maximum at $x = 2$ on the interval $[0, 4]$, then the global maximum of f on the interval $(-\infty, \infty)$ must also be at $x = 2$.

(e) *True or False:* If f is continuous on an interval I, then f has both a global maximum and a global minimum on I.

(f) *True or False:* Suppose f has two local minima on the interval $[0, 10]$, one at $x = 2$ with a value of 4 and one at $x = 7$ with a value of 1. Then the global minimum of f on $[0, 10]$ must be at $x = 7$.

(g) *True or False:* If f has no local maxima on $(-\infty, \infty)$, then it will have no global maximum on the interval $[0, 5]$.

(h) *True or False:* If $f'(3) = 0$, then f has either a local minimum or a local maximum at $x = 3$.

2. *Examples:* Construct examples of the thing(s) described in the following. Try to find examples that are different than any in the reading.

(a) The graph of a function with a local minimum at $x = 2$ but no global minimum on $[0, 4]$.

(b) The graph of a function with no local or global extrema on $(-3, 3)$.

(c) The graph of a function whose global maximum on $[2, 6]$ does not occur at a critical point.

3. When you try to find the local extrema of a function f on an interval I, one of the first steps is to find the critical points of f. Explain why these critical points of f won't help you locate any "endpoint" extrema.

4. Explain why you can't find the global maximum of a function f on an interval I just by finding all the local extrema of f and then checking to see which one has the highest value $f(c)$.

5. Suppose f is a function that is defined and continuous on an open interval I. Will the endpoints of I always be local extrema of f? Will f necessarily have a global maximum or minimum in the interval I? Justify your answers.

6. Suppose f is a function that is defined and continuous on a closed interval I. Will the endpoints of I always be local extrema of f? Will f necessarily have a global maximum or minimum in the interval I? Justify your answers.

7. Suppose f is a function that is discontinuous somewhere on an interval I. Explain why comparing the values of any local extrema of f on I and the values or limits of f at the endpoints of I is not in general sufficient to determine the global extrema of f on I.

8. Given the following graph of f, graphically estimate the global extrema of f on each of the six intervals listed:

(a) $[-2, 4]$ (b) $(-2, 4)$ (c) $(-1, 1)$
(d) $(0, 4]$ (e) $[0, 4)$ (f) $(-\infty, \infty)$

9. Given the following graph of f, graphically estimate the global extrema of f on each of the six intervals listed:

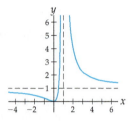

(a) $[0, 4]$ (b) $[2, 5]$ (c) $(-2, 1)$
(d) $[0, \infty)$ (e) $(-\infty, 0]$ (f) $(-\infty, \infty)$

10. Given the following graph of f, graphically estimate the global extrema of f on each of the six intervals listed:

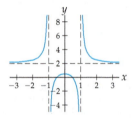

(a) $(-1, 1)$ (b) $[2, \infty)$ (c) $[-2, -1)$
(d) $[0, 2]$ (e) $(1, \infty)$ (f) $(-\infty, \infty)$

Skills

Find the locations and values of any global extrema of each function f in Exercises 11–18 on each of the four given intervals. Do all work by hand by considering local extrema and endpoint behavior. Afterwards, check your answers with graphs.

11. $f(x) = (x + 2)(x - 3)$, on the intervals

   (a) $[-2, 3]$ (b) $[0, 4]$ (c) $[1, 5)$ (d) $(-2, 2)$

12. $f(x) = -4 + 4x - x^2$, on the intervals

   (a) $[-1, 5]$ (b) $[3, 5)$ (c) $(-1, 2]$ (d) $(0, \infty)$

13. $f(x) = 2x^3 - 3x^2 - 12x$, on the intervals

   (a) $[-3, 3]$ (b) $[0, 3]$ (c) $(-1, 2]$ (d) $(-2, 1)$

14. $f(x) = 3x^4 + 4x^3 - 36x^2$, on the intervals

   (a) $[-5, 5]$ (b) $[-2, 2]$ (c) $(-3, 1]$ (d) $(-1, 3)$

15. $f(x) = -12x + 6x^2 + 4x^3 - 3x^4$, on the intervals

   (a) $[-1, 1]$ (b) $(-1, 1)$ (c) $(-3, 0]$ (d) $[0, 3]$

16. $f(x) = \dfrac{3x - 2}{x - 1}$, on the intervals

   (a) $[0, 2]$ (b) $[-2, 0]$ (c) $[-1, 1]$ (d) $(-1, 1)$

17. $f(x) = \dfrac{1}{1 + \sqrt{x}}$, on the intervals

   (a) $[0, 3]$ (b) $(0, 3)$ (c) $[1, 2]$ (d) $[0, \infty)$

18. $f(x) = \dfrac{4}{\sqrt{x^2 + 1}} + 3$, on the intervals

   (a) $(-5, 2]$ (b) $[-5, 2)$ (c) $[1, 10]$ (d) $[0, 20]$

Find dimensions for each shape in Exercises 19–22 so that the total area enclosed is as large as possible, given that the total edge length is 120 inches. The rounded shapes are half-circles, and the triangles are equilateral.

19.   20.

21.   22.

Use optimization techniques to answer the questions in Exercises 23–28.

23. Find two real numbers x and y whose sum is 36 and whose product is as large as possible.

24. Find two real numbers x and y whose sum is 36 and whose product is as small as possible.

25. Find real numbers a and b whose sum is 100 and for which the sum of the squares of a and b is as small as possible.

26. Find the area of the largest rectangle that fits inside a circle of radius 4.

27. Find the area of the largest rectangle that fits inside a circle of radius 10.

28. Find the volume of the largest cylinder that fits inside a sphere of radius 10.

In Exercises 29–32, find the point on the graph of the function f that is closest to the point (a, b) by minimizing the square of the distance from the graph to the point.

29. $f(x) = 3x + 1$ and the point $(-2, 1)$

30. $f(x) = x^2$ and the point $(0, 3)$

31. $f(x) = x^2 - 2x + 1$ and the point $(1, 2)$

32. $f(x) = \sqrt{x^2 + 1}$ and the point $(2, 0)$

Applications

A farmer wants to build four fenced enclosures on his farm-land for his free-range ostriches. To keep costs down, he is always interested in enclosing as much area as possible with a given amount of fence. For the fencing projects in Exercises 33–36, determine how to set up each ostrich pen so that the maximum possible area is enclosed, and find this maximum area.

33. A rectangular ostrich pen built with 350 feet of fencing material.

34. A rectangular ostrich pen built along the side of a river (so that only three sides of fence are needed), with 540 feet of fencing material.

35. A rectangular ostrich pen built with 1000 feet of fencing material, divided into three equal sections by two interior fences that run parallel to the exterior side fences, as shown next at the left.

Ostrich pen with three sections   *Ostrich pen with six sections*

36. A rectangular ostrich pen that is divided into six equal sections by two interior fences that run parallel to the east and west fences, and another interior fence running parallel to the north and south fences, as shown previously at the right. The farmer has allotted 2400 feet of fencing material for this important project.

You are in charge of constructing a zoo habitat for prairie dogs, with the requirement that the habitat must enclose 2500 square feet of area and use as little border fencing as possible. For each of the habitat designs described in Exercises 37–38, find the amount of border fencing that the project will require.

37. A rectangular habitat with a 20-foot-wide nesting hutch along one side (so that fencing is not needed along those 20 feet), as shown next at the left.

Rectangular habitat with hutch   *Arena-style habitat*

38. An arena-style habitat whose front area is a semicircle and whose back area is rectangular, as shown previously at the right.

Alina wants to make keepsake boxes for her two best friends. She doesn't have a lot of money, so she wants to make each box described in Exercises 39–40 so that it holds as much as possible with a limited amount of material.

39. For Jen, Alina wants to make a box with a square base whose sides and base are made of wood and whose top is made of metal. The wood she wants to use costs 5 cents per square inch, while the material for the metal top costs 12 cents per square inch. What is the largest possible box that Alina can make for Jen if she only has $20.00 to spend on materials?

Wood and metal box   *Velvet-lined box*

0.12 per sq. in.

0.05 per sq. in. $l = 2w$ w

40. For Eliza, Alina wants to make a rectangular box whose base is twice as long as it is wide. This box will be lined on the entire inside with velvet and in addition the outside of the top of the box is to be lined in velvet. If Alina has 240 square inches of velvet, how can she make Eliza's box so that it holds as many keepsakes as possible?

The U.S. Postal Service ships a package under large-package rates if the sum of the length and the girth of the package is greater than 84 inches and less than or equal to 108 inches. The *length* of a package is considered to be the length of its longest side, and the *girth* of the package is the distance around the package perpendicular to its length. In each of Exercises 41–43, Linda wants to ship packages under the USPS large-package rates.

41. Linda needs to mail a rectangular package with square ends (in other words, with equal width and height). What is the largest volume that her package can have? What is the largest surface area that her package can have?

42. Linda's second package must be rectangular and 40 inches in length. What is the largest volume that her package can have? What is the largest surface area that her package can have?

43. Linda also needs to mail some architectural plans, which must be shipped in a cylindrical container. What is the largest volume that her package can have? What is the largest surface area that her package can have? (*Hints: The volume of a right circular cylinder with radius r and height h is $V = \pi r^2 h$; the total surface area of such a cylinder is $SA = 2\pi rh + 2\pi r^2$.*)

For Exercises 44–46, consider a toy car that moves back and forth on a long, straight track for 4 minutes in such a way that the function $s(t) = 96t - 84t^2 + 28t^3 - 3t^4$ describes how far the car is to the right of the starting point, in centimeters, after t minutes.

44. When is the toy car farthest away from the starting point? Does it ever return to the starting point? For how long does this model make sense?

45. Within the domain of your model, when is the toy car moving fastest to the right? When is the toy car moving fastest to the left?

46. When is the toy car accelerating the fastest to the right? When is the toy car accelerating fastest to the left?

Suppose you have a 10-inch length of wire that you wish to cut and form into shapes. In each of Exercises 47–49 you will

determine how to cut the wire to minimize or maximize the area of the resulting shapes.

47. Suppose you wish to make one cut in the wire and use the two pieces to form a square and a circle. Determine how to cut the wire so that the combined area enclosed by the square and the circle is (a) as small as possible and (b) as large as possible.

48. Suppose you wish to make one cut in the wire and use the two pieces to form a square and an equilateral triangle. Determine how to cut the wire so that the combined area is (a) as small as possible and (b) as large as possible.

49. Suppose you wish to make one cut in the wire and use the two pieces to form a circle and an equilateral triangle. Determine how to cut the wire so that the combined area of these two shapes is (a) as small as possible and (b) as large as possible.

In each situation described in Exercises 50–58, set up and solve a global extrema problem that solves the given real–world optimization problem.

50. Alina needs to make a flyer for her band's concert. The flyer must contain 20 square inches of printed material and for design purposes should have side margins of 1 inch and top and bottom margins of 2 inches. What size paper should Alina use in order to use the least amount of paper per flyer as possible?

51. Your company produces cylindrical metal oil drums that must each hold 40 cubic feet of oil. How should the oil drums be constructed so that they use as little metal as possible? Can they be constructed to use as *much* metal as possible?

52. An airplane leaves Chicago at noon and travels south at 500 miles per hour. Another airplane is travelling east towards Chicago at 650 miles per hour and arrives at 2:00 P.M.. When were these two airplanes closest to each other, and how far apart were they at that time?

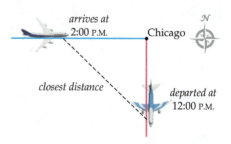

53. The cost of the material for the top and bottom of a cylindrical can is 5 cents per square inch. The material for the rest of the can costs only 2 cents per square inch. If the can must hold 400 cubic inches of liquid, what is the cheapest way to make the can? What is the most expensive way?

54. Consider the can-making situation in the previous exercise, but suppose that the cans are made with open tops.

If each can must hold 400 cubic inches of liquid, what is the cheapest way to make the cans? What is the most expensive way?

55. A steam pipe must be buried underground to reach from one corner of a rectangular parking lot to the diagonally opposite corner. The dimensions of the parking lot are 500 feet by 800 feet. It costs 5 dollars per foot to lay steam pipe under the pavement but only 3 dollars per foot to lay the pipe along one of the long edges of the parking lot. Because of nearby sidewalks, the pipe cannot be laid along the 500-foot sides of the parking lot. How should the steam pipe be buried so as to cost as little as possible?

56. Suppose you want to make an open-topped box out of a 4×6 index card by cutting a square out of each corner and then folding up the edges, as shown in the figure. How large a square should you cut out of each corner in order to maximize the volume of the resulting box?

57. Your family makes and sells velvet Elvis paintings. After many years of research you have found a function that predicts how many paintings you will sell in a year, based on the price that you charge per painting. You always charge between \$5.00 and \$55.00 per painting. Specifically, if you charge c dollars per painting, then you can sell $N(c) = 0.6c^2 - 54c + 1230$ paintings in a year.

(a) What price should you charge to sell the *greatest* number of velvet Elvis paintings, and how many could you sell at that price? For what price would you sell the *least* number of paintings, and how many would you sell?

(b) Write down a function that predicts the revenue $R(c)$, in dollars, that you will earn in a year if you charge c dollars per painting. (*Hint: Try some examples first; for example, what would your yearly revenue be if you charged \$10.00 per painting? What about \$50.00? Then write down a function that works for all values of c.*)

(c) What price should you charge to earn the *most* money, and how much money would you earn? What price per painting would cause you to make the *least* amount of money in a year, and how much money would you make in that case?

(d) Explain why you do not make the most money at the same price per painting for which you sell the most paintings.

58. While you are on a camping trip, your tent accidentally catches fire. At the time, you and the tent are both 50 feet from a stream and you are 200 feet away from the tent, as shown in the diagram. You have a bucket with you, and need to run to the stream, fill the bucket, and run to the tent as fast as possible. You can run only half as fast while carrying the full bucket as you can empty handed, and thus any distance travelled with the full bucket is effectively twice as long. Complete parts (a)–(f) to determine how you can put out the fire as quickly as possible.

(a) Let x represent the distance from the point on the stream directly "below" you to the point on the stream that you run to. Sketch the path that you would follow to run from your starting position, to the point x along the stream, to the tent.

(b) Let $D(x)$ represent the effective distance (counting twice any distance travelled while carrying a full bucket) you have to run in order to collect water and get to the tent. Write a formula for $D(x)$.

(c) Determine the interval I of x-values on which $D(x)$ should be minimized. Explain in practical terms what happens at the endpoints of this interval, and calculate the value of $D(x)$ at these endpoints.

(d) Find $D'(x)$, and simplify as much as possible. Are there any points (in the interval I) at which $D'(x)$ is undefined?

(e) It is difficult to find the zeroes of $D'(x)$ by hand. Use a graphing utility to approximate any zeroes of $D'(x)$ in the interval I, and test these zeroes by evaluating D' at each one.

(f) Use the preceding information to determine the minimum value of $D(x)$, and then use this value to answer the original word problem.

Proofs

59. Prove that the rectangle with the largest possible area given a fixed perimeter P is always a square.

60. Prove that the most efficient way to build a rectangular fenced area along a river—so that only three sides of fencing are needed—is to make the side parallel to the river twice as long as the other sides. You may assume that you have a fixed amount of fencing material.

Thinking Forward

Consider the graph of the function f shown next. Define $A(x)$ to be the area of the region between the graph of f and the x-axis from 0 to x. We will count areas of regions above the x-axis positively and areas of regions below the x-axis negatively.

A(x) is area under this graph on $[0, x]$

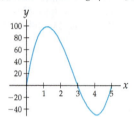

▶ Use the graph to approximate the values of $A(0)$, $A(1)$, $A(2)$, $A(3)$, $A(4)$, and $A(5)$.

▶ From the graph of f, estimate all local maxima and minima of $A(x)$.

▶ From the graph of f, estimate all global maxima and minima of $A(x)$, if any.

▶ It turns out that the function f whose graph is shown is given by the formula $f(x) = 12x^3 - 96x^2 + 180x$ and that the area function A is given by the formula $A(x) = 3x^4 - 32x^3 + 90x^2$. What surprising relationship do f and A have?

▶ Show that your answers for the local and global extrema of $A(x)$ are reasonable by using optimization techniques on the area function $A(x) = 3x^4 - 32x^3 + 90x^2$.

3.5   RELATED RATES

▶ Using implicit differentiation to obtain relationships between rates

▶ Formulas for volume, surface area, and relationships between side lengths of triangles

▶ Techniques for solving related-rates problems

Related Quantities Have Related Rates

If two quantities that change over time are related to each other, then their rates of change over time will also be related to each other. For example, consider an expanding circle. Clearly the radius $r = r(t)$ of the circle and the area $A = A(t)$ of the circle are related: If you know one of these quantities at some time t, then you also know the other, via the formula $A = \pi r^2$. As the circle expands over time, the rate $\frac{dr}{dt}$ at which its radius increases is related to the rate $\frac{dA}{dt}$ at which its area increases. We can find an equation that relates these rates by applying implicit differentiation to the formula that relates the quantities r and A:

$$A = \pi r^2 \qquad \leftarrow \text{relationship between } A \text{ and } r$$

$$\frac{d}{dt}(A(t)) = \frac{d}{dt}(\pi (r(t))^2) \qquad \leftarrow \text{differentiate both sides with respect to } t$$

$$\frac{dA}{dt} = \pi \left(2r\frac{dr}{dt}\right) \qquad \leftarrow \text{chain rule}$$

$$\frac{dA}{dt} = 2\pi r\frac{dr}{dt}. \qquad \leftarrow \text{relationship between } \frac{dA}{dt} \text{ and } \frac{dr}{dt}$$

Notice that in this calculation we use A interchangeably with $A(t)$ and r interchangeably with $r(t)$, employing the explicit function notation only when we need to be reminded that A and r are functions of t.

According to the formula we just found, the rate $\frac{dA}{dt}$ depends on both the size of the radius r at time t and the rate $\frac{dr}{dt}$ at which the radius is increasing. For example, suppose that our expanding circle has an initial radius of $r(0) = 2$ inches at time $t = 0$, and that the radius then increases at a constant rate of $\frac{dr}{dt} = 3$ inches per second. Then the formula we found tells us that at the instant the circle has a radius of 4 inches, the area of the circle is increasing at a rate of

$$\frac{dA}{dt}\Big|_{r=4} = 2\pi(4)(3) = 24\pi \text{ square inches per second.}$$

In contrast, when the radius of the circle is 20 inches, the area of the circle is increasing at the faster rate of

$$\frac{dA}{dt}\Big|_{r=20} = 2\pi(20)(3) = 120\pi \text{ square inches per second.}$$

Notice that although the rate of change of the radius of the circle is constant, the rate at which the area is changing is not: As the circle gets larger and larger, changes to the radius create larger and larger changes to the area. We could also determine the rate of change of the area at a particular time, say, after $t = 3$ seconds. Since $r(t) = 2 + 3t$ in our example, at time $t = 3$ seconds, we have a radius of $r(3) = 2 + 3(3) = 11$ inches. The rate of change of the area at this moment is

$$\frac{dA}{dt}\Big|_{t=3} = \frac{dA}{dt}\Big|_{r=11} = 2\pi(11)(3) = 66\pi \text{ square inches per second.}$$

Most related-rates problems involve two rates, one of which is known and one of which you are asked to find. These rates will be related in some way that is determined by the way the corresponding quantities are related. Translating a related-rates problem should result in an equation that relates the two quantities whose rates you are interested in. You can then implicitly differentiate the equation relating the quantities to get an equation relating the rates, as we did at the outset of this section.

Volumes and Surface Areas of Geometric Objects

Many related-rates problems involve geometric quantities such as volume, area, and surface area. Many of the formulas for these quantities should already be familiar to you. For example, here are some formulas for two-dimensional objects: A circle with radius r has area $A = \pi r^2$ and circumference $C = 2\pi r$, a rectangle with length l and width w has area $A = lw$ and perimeter $P = 2l + 2w$, and any triangle with base b and height h has area $A = \frac{1}{2}bh$.

We now gather some three-dimensional geometric formulas for reference, although we will not have the tools to prove them until much later in the book.

THEOREM 3.12   **Volume and Surface Area Formulas**

The formulas that follow describe the volume V and surface area S of a rectangular box, sphere, right circular cylinder, and right circular cone. The lateral (side) surface area L is also given for the cylinder and cone.

(a) The volume and surface area of a **rectangular box** of length x, width y, and height z are

$$V = xyz$$
$$S = 2xy + 2yz + 2xz$$

(b) The volume and surface area of a **sphere** of radius r are

$$V = \frac{4}{3}\pi r^3$$
$$S = 4\pi r^2$$

(c) The volume, surface area, and lateral surface area of a **right circular cylinder** of radius r and height h are

$$V = \pi r^2 h$$
$$S = 2\pi rh + 2\pi r^2$$
$$L = 2\pi rh$$

(d) The volume, surface area, and lateral surface area of a **right circular cone** of radius r and height h are

$$V = \frac{1}{3}\pi r^2 h$$
$$S = \pi r\sqrt{r^2 + h^2} + \pi r^2$$
$$L = \pi r\sqrt{r^2 + h^2}$$

As always, whenever possible you should *think* rather than memorize. For example, let's walk through why the volume and surface area formulas for the cylinder make sense. The volume of a cylinder is the product of the area πr^2 of its top and its height h. The surface area of a cylinder is the sum of two things: the area of its curvy side and the areas of its top and bottom circles. Notice that if we unrolled the curvy side, it would be a rectangle whose width is the circumference $2\pi r$ of the top and bottom circles and whose height is h. Therefore the lateral, or side, surface area of the cylinder is the product $2\pi rh$. Of course, the area of the top and bottom circles each have area πr^2, so the top and bottom together have area $2\pi r^2$. Thus the total surface area of the cylinder is $2\pi rh + 2\pi r^2$.

Similar Triangles

It is also common for related-rates word problems to involve the following two well-known theorems concerning right triangles, which we present here for reference, without proof:

THEOREM 3.13   ## Two Theorems About Right Triangles

The following two theorems describe well-known relationships between the side lengths of right triangles:

(a) The ***Pythagorean theorem*** states that if a right triangle has legs of lengths a and b and hypotenuse of length c, then:
$$a^2 + b^2 = c^2$$

(b) The ***law of similar triangles*** states that if two right triangles have the same three angle measures, so that one is just a scaled-up version of the other, then the ratios of side lengths on one triangle are equal to the ratios of corresponding side lengths on the other. Specifically, with the side lengths shown in the diagram at the right, we have

$$\frac{h}{b} = \frac{H}{B}, \qquad \frac{d}{b} = \frac{D}{B}, \qquad \frac{d}{h} = \frac{D}{H}$$

The reason these theorems about triangles arise in related-rates problems is that both theorems give us ways to relate quantities that might change together over time. Finding an equation that relates two quantities is often the first step in finding an equation that relates the rates of change of those quantities.

Examples and Explorations

EXAMPLE 1   ### Relating quantities and rates

In each part that follows, write down an equation that relates the two given quantities. Then use implicit differentiation to obtain a relationship between the rates at which the following quantities change over time:

(a) the circumference C and the area A of a circle;

(b) the surface area S and the radius r of a cylinder with a fixed height of 4 units;

(c) the lengths a and b of the legs of a right triangle with hypotenuse of fixed length 7 units.

SOLUTION

(a) We know that a circle of radius r has circumference $C = 2\pi r$ and area $A = \pi r^2$. We need to find an equation relating C and A. Since $C = 2\pi r$, we have $r = \frac{C}{2\pi}$; substituting the right-hand side into $A = \pi r^2$ gives

$$A = \pi \left(\frac{C}{2\pi}\right)^2 = \frac{1}{4\pi}C^2.$$

Now suppose that the circle is expanding or contracting, so that its area $A = A(t)$ and circumference $C = C(t)$ are changing over time. By implicitly differentiating the preceding equation with respect to t we have

$$\frac{dA}{dt} = \frac{1}{4\pi} 2C \frac{dC}{dt} = \frac{1}{2\pi} C \frac{dC}{dt}.$$

(b) The formula for the surface area of a cylinder with radius r and height 4 is

$$S = 2\pi r(4) + 2\pi r^2 = 8\pi r + 2\pi r^2.$$

By differentiating both sides we can express the relationship between the radius $r = r(t)$ and surface area $S = S(t)$ if the cylinder changes size over time:

$$\frac{d(S)}{dt} = 8\pi \frac{dr}{dt} + 2\pi \, 2r \frac{dr}{dt} = (8\pi + 4\pi r)\frac{dr}{dt}.$$

(c) By the Pythagorean theorem, if a right triangle has legs of length a and b and hypotenuse of length 7, then

$$a^2 + b^2 = 7^2.$$

If the triangle is changing shape or size over time in such a way that the hypotenuse remains 7 units in length, then by implicit differentiation the rates of change of the leg lengths $a = a(t)$ and $b = b(t)$ are related as follows:

$$2a \frac{da}{dt} + 2b \frac{db}{dt} = 0. \qquad \qquad \square$$

EXAMPLE 2   **Relating the changing radius and area of an expanding circle**

Suppose a rock dropped into a pond causes a circular wavefront of ripples whose radius increases at 3 inches per second. How fast is the area of the circle of ripples expanding at the instant that the circle has a radius of 12 inches?

SOLUTION

Like many related-rates problems, this situation involves (1) two quantities that are related and (2) known information about the rate of change of one of these quantities; we are then asked to find information about the rate of change of the other quantity. In this case the related quantities are the radius $r = r(t)$ and area $A = A(t)$ of the circle, which are related by the formula $A = \pi r^2$. Both of these quantities change over time as the circle expands. We are given that the radius changes constantly at a rate of $\frac{dr}{dt} = 3$ inches per second and asked to find the rate of change $\frac{dA}{dt}$ of the area at a particular moment. This information is summarized as follows:

$r(t) = $ radius

$A(t) = $ area

Relationship: $A = \pi r^2$

Given: $\dfrac{dr}{dt} = 3$

Find: $\dfrac{dA}{dt}\Big|_{r=12}$

At this point we are done translating the original real–world problem into a mathematical related-rates problem that we know how to solve. To get a formula relating the rates of change of $r(t)$ and $A(t)$ we differentiate both sides of the area formula, yielding $\frac{dA}{dt} = 2\pi r \frac{dr}{dt}$. By evaluating this equation for $\frac{dA}{dt}$ at $r = 12$ and using the given information that $\frac{dr}{dt} = 3$, we see that at the moment that the circle has a 12-inch radius, it is expanding at a rate of

$$\frac{dA}{dt}\Big|_{r=12} = 2\pi(12)\frac{dr}{dt} = 2\pi(12)(3) = 72\,\pi \text{ square inches per second.} \qquad \square$$

EXAMPLE 3   Relating the changing volume and radius of an inflating balloon

Suppose a pink spherical party balloon is being inflated at a constant rate of 44 cubic inches per second.

(a) How fast is the radius of the balloon increasing at the instant that the balloon has a radius of 4 inches?

(b) How fast is the radius of the balloon increasing at the instant that the balloon contains 100 cubic inches of air?

SOLUTION

This is a related-rates problem because it involves two rates, namely, the rate at which the balloon is being inflated and the rate of change of the radius of the balloon. We know something about the first rate and wish to say something about the second. Suppose $r = r(t)$ is the radius of the balloon in inches after t seconds and $V = V(t)$ is the volume of the balloon in cubic inches after t seconds. The quantities r and V are related by the volume equation $V = \frac{4}{3}\pi r^3$. We are given that the rate of change of the volume is constantly $\frac{dV}{dt} = 44$, and we want to find the rate of change $\frac{dr}{dt}$ of the radius when $r = 4$ and when $V = 100$. The following diagram summarizes this translation of the problem into mathematical notation:

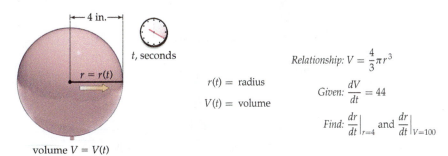

$$\text{Relationship: } V = \frac{4}{3}\pi r^3$$

t, seconds

$r(t) = $ radius

$V(t) = $ volume

$$\text{Given: } \frac{dV}{dt} = 44$$

$$\text{Find: } \frac{dr}{dt}\Big|_{r=4} \text{ and } \frac{dr}{dt}\Big|_{V=100}$$

$r = r(t)$

volume $V = V(t)$

To get an equation relating the rates $\frac{dV}{dt}$ and $\frac{dr}{dt}$ we differentiate both sides of the equation that relates V and r with respect to t:

$$\frac{dV}{dt} = \frac{4}{3}\pi\,3r^2\frac{dr}{dt} = 4\pi r^2\frac{dr}{dt}.$$

We know from the given information that $\frac{dV}{dt}$ is always equal to 44. Therefore we have the following formula concerning the rate of change of the radius of the balloon:

$$44 = 4\pi r^2\frac{dr}{dt}.$$

(a) To find the rate of change of the radius at the instant that the ballon has a 4-inch radius, we evaluate the preceding equation at $r = 4$ and then solve for $\frac{dr}{dt}\Big|_{r=4}$:

$$44 = 4\pi(4^2)\frac{dr}{dt}\Big|_{r=4} \quad\Longrightarrow\quad \frac{dr}{dt}\Big|_{r=4} = \frac{44}{4\pi(16)} \approx 0.2188 \text{ inch per second.}$$

(b) For the second part we must find the rate at which the radius is changing at the moment that the volume of the balloon is 100 cubic inches. Our formula for the rate of change of the radius of the balloon depends on r, not V, so we must first use the equation $V = \frac{4}{3}\pi r^3$ to determine the value of r when $V = 100$:

$$100 = \frac{4}{3}\pi r^3 \quad\Longrightarrow\quad r^3 = \frac{100}{(4/3)\pi} \quad\Longrightarrow\quad r = \left(\frac{100}{(4/3)\pi}\right)^{1/3} \approx 2.879.$$

Using our previously developed formula, we find that when the volume of the balloon is $V = 100$ cubic inches and thus the radius is approximately $r \approx 2.879$ inches, the rate of change of the radius of the balloon is

$$44 = 4\pi(2.879)^2 \left.\frac{dr}{dt}\right|_{V=100} \quad\Longrightarrow\quad \left.\frac{dr}{dt}\right|_{V=100} = \frac{44}{4\pi(2.879)^2} \approx 0.422 \text{ inch per second.} \quad\square$$

EXAMPLE 4   **The shadow of a person walking away from a streetlight**

Matt is 6 feet tall and is walking away from a 10-foot streetlight at a rate of 3 feet per second. As he walks away from the streetlight, his shadow gets longer. How fast is the length of Matt's shadow increasing when he is 8 feet from the streetlight?

SOLUTION

We are given the rate at which Matt walks away from the streetlight, and we wish to find the rate of change of the length of Matt's shadow. To find a relationship between these two rates we will find a relationship between their underlying variables: the distance s between Matt and the streetlight and the length l of Matt's shadow. By the law of similar triangles, s and l are related by the equation $\frac{10}{s+l} = \frac{6}{l}$, as shown in the following diagram:

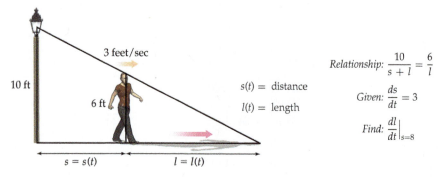

Relationship: $\dfrac{10}{s+l} = \dfrac{6}{l}$

Given: $\dfrac{ds}{dt} = 3$

Find: $\left.\dfrac{dl}{dt}\right|_{s=8}$

To find the relationship between $\frac{ds}{dt}$ and $\frac{dl}{dt}$ we must implicitly differentiate the equation relating s and l. We will simplify the equation first to make our lives easier:

$$\frac{10}{s+l} = \frac{6}{l} \quad\Longrightarrow\quad 10l = 6(s+l) \quad\Longrightarrow\quad 4l = 6s.$$

Differentiating both sides and then using the fact that $\frac{ds}{dt} = 3$ gives us

$$4\frac{dl}{dt} = 6\frac{ds}{dt} \quad\Longrightarrow\quad 4\frac{dl}{dt} = 6(3) \quad\Longrightarrow\quad \frac{dl}{dt} = 4.5.$$

Interestingly, we have just discovered that Matt's shadow is increasing at a *constant* rate of 4.5 feet per second. In particular, when Matt is 8 feet from the streetlight, the length of his shadow is increasing at a rate of $\left.\frac{dl}{dt}\right|_{s=8} = 4.5$ feet per second. $\quad\square$

TEST YOUR
UNDERSTANDING

▶ Suppose $r = r(t)$ is a function of t. Why is $\frac{d}{dt}(r^3)$ not equal to $3r^2$?

▶ What are the formulas for the volumes of spheres, cylinders, and cones? What about the formulas for surface area?

▶ What do we mean when we say that two triangles are similar? What does the law of similar triangles say?

▶ In Example 2, why did we not label the diagram of the circle with the number 12? Why did we use the variable r instead?

▶ In Example 3 the radius of the balloon increases at a faster rate when the balloon is smaller. Why does this make sense?

EXERCISES 3.5

Thinking Back

Using the chain rule: Given that $r = r(t)$, $s = s(t)$, and $u = u(t)$ are functions of t and that c and k are constants, find each of the following derivatives.

▶ $\frac{d}{dt}(\pi u^2)$

▶ $\frac{d}{dt}(3r + 2s)$

▶ $\frac{d}{dt}(cu + rs)$

▶ $\frac{d}{dt}(k + cu^3)$

▶ $\frac{d}{dt}(cr^2 u)$

▶ $\frac{d}{dt}\left(\frac{c + s}{k + u}\right)$

Evaluation in Leibniz notation: Given that $r = r(t)$, $s = s(t)$, and $u = u(t)$ are functions of t, answer each of the following.

▶ If $\frac{ds}{dt} = 3s^2 - 4$, find $\frac{ds}{dt}\big|_{s=2}$.

▶ If $r^2 \frac{dr}{dt} - 2r = 0$, find $\frac{dr}{dt}\big|_{r=3}$.

▶ If $4 = 2u\frac{du}{dt}$ and $u = 2 + 3t$, find $\frac{du}{dt}\big|_{t=4}$.

Concepts

0. *Problem Zero:* Read the section and make your own summary of the material.

1. *True/False:* Determine whether each of the statements that follow is true or false. If a statement is true, explain why. If a statement is false, provide a counterexample.

  (a) *True or False:* If a square grows larger, so that its side length increases at a constant rate, then its area will also increase at a constant rate.

  (b) *True or False:* If a square grows larger, so that its side length increases at a constant rate, then its perimeter will also increase at a constant rate.

  (c) *True or False:* If a circle grows larger, so that its radius increases at a constant rate, then its circumference will also increase at a constant rate.

  (d) *True or False:* If a sphere grows larger, so that its radius increases at a constant rate, then its volume will also increase at a constant rate.

  (e) *True or False:* The volume of a right circular cone is one-third of the volume of the right circular cylinder with the same radius and height.

  (f) *True or False:* If $V(r)$ is the volume of a sphere as a function of its radius, and $S(r)$ is the surface area of a sphere as a function of its radius, then $V'(r) = S(r)$.

  (g) *True or False:* If you unroll the side of a right circular cylinder with radius r and height h, you get a flat rectangle with height h and width $2\pi r$.

  (h) *True or False:* Given a right triangle with side lengths a and c and hypotenuse of length b, we must have $a^2 + b^2 = c^2$.

2. *Examples:* Construct examples of the thing(s) described in the following. Try to find examples that are different than any in the reading.

  (a) Two triangles that are similar to the right triangle with legs of length 1 and hypotenuse $\sqrt{2}$.

  (b) Two triangles with a hypotenuse of length 5.

  (c) Two cylinders with a volume of 100 cubic units.

3. Give formulas for the volume and the surface area of a cylinder with radius y and height s.

4. Give formulas for the volume and the surface area of a cone with radius u and height w.

5. Give formulas for the volume and the surface area of a cylinder whose radius r is half of its height h.

6. Give formulas for the volume and the surface area of a cone whose height h is three times its radius r.

7. State the Pythagorean theorem and give an example of a triangle that illustrates the theorem.

8. State the law of similar triangles and give an example of a pair of triangles that illustrate this law.

9. If the volume and radius of a sphere are functions of time, what is the relationship between the rate of change of the volume of the sphere and the rate of change of the radius of the sphere?

10. If the volume and radius of a cone are functions of time, what is the relationship between the rate of change of the volume of the cone and the rate of change of the radius of the cone?

11. If the volume and equator of a sphere are functions of time, what is the relationship between the rate of change of the volume of the sphere and the rate of change of the equator of the sphere?

12. Suppose the side lengths x, y, and z of a rectangular box are each functions of time.

    (a) How is the rate of change of the volume of the box related to the rates of change of x, y, and z?

    (b) How is the rate of change of the surface area of the box related to the rates of change of x, y, and z?

13. Suppose the radius r, volume V, and surface area S of a sphere are functions of time t.

    (a) How are $\dfrac{dV}{dt}$ and $\dfrac{dr}{dt}$ related?

    (b) How are $\dfrac{dS}{dt}$ and $\dfrac{dr}{dt}$ related?

14. Suppose the radius r, volume V, and surface area S of a sphere are functions of time t. How are $\dfrac{dV}{dt}$ and $\dfrac{dS}{dt}$ related?

15. Suppose the radius r, height h, and volume V of a cylinder are functions of time t. How is $\dfrac{dV}{dt}$ related to $\dfrac{dr}{dt}$ if the height of the cylinder is constant?

16. Suppose the radius r, height h, and volume V of a cylinder are functions of time t. How is $\dfrac{dV}{dt}$ related to $\dfrac{dh}{dt}$ if the radius of the cylinder is constant?

17. Suppose the radius r, height h, and volume V of a cylinder are functions of time t, and further suppose that the height of the cylinder is always twice its radius. Write $\dfrac{dV}{dt}$ in terms of h and $\dfrac{dh}{dt}$.

18. Suppose the radius r, height h, and volume V of a cylinder are functions of time t, and further suppose that the volume of the cylinder is always constant. Write $\dfrac{dr}{dt}$ in terms of r, h, and $\dfrac{dh}{dt}$.

Skills

In Exercises 19–26, write down an equation that relates the two quantities described. Then use implicit differentiation to obtain a relationship between the rates at which the quantities change over time.

19. The area A and perimeter P of a square.

20. The area A and perimeter P of an equilateral triangle.

21. The surface area S and height h of a cylinder with a fixed radius of 2 units.

22. The volume V and radius r of a cylinder with a fixed height of 10 units.

23. The surface area S and radius r of a cone with a fixed height of 5 units.

24. The volume V and height h of a cone with a fixed radius of 3 units.

25. The area A and hypotenuse c of an isosceles right triangle.

26. The area A and hypotenuse c of a triangle that is similar to a right triangle with legs of lengths 3 and 4 units and hypotenuse of length 5 units.

Given that $u = u(t)$, $v = v(t)$, and $w = w(t)$ are functions of t and that k is a constant, calculate the derivative $\dfrac{df}{dt}$ of each function $f(t)$ in Exercises 27–36. Your answers may involve u, v, w, $\dfrac{du}{dt}$, $\dfrac{dv}{dt}$, $\dfrac{dw}{dt}$, k, and/or t.

27. $f(t) = u^2 + kv$

28. $f(t) = u + v + w$

29. $f(t) = tv + kv$

30. $f(t) = kuvw$

31. $f(t) = 2v\sqrt{u + w}$

32. $f(t) = 3u^2v + vt$

33. $f(t) = w(u + t)^2$

34. $f(t) = \dfrac{w}{uv}$

35. $f(t) = \dfrac{ut + w}{k}$

36. $f(t) = \dfrac{k}{u^2w}$

Applications

37. A rock dropped into a pond causes a circular wave of ripples whose radius increases at 4 inches per second. How fast is the area of the circle of ripples expanding at the instant that the radius of the circle is 12 inches? 24 inches? 100 inches? Explain why it makes sense that the rate of change of the area increases as the radius increases.

38. A rock dropped into a pond causes a circular wave of ripples whose radius increases at 6 inches per second. How fast is the area of the circle of ripples expanding at the instant that the area of the circle is 100 square inches? 200 square inches? 1000 square inches? Explain why it makes sense that the rate of change of the area increases as the area increases.

In Exercises 39–42, suppose the sides of a cube are expanding at a rate of 2 inches per minute.

39. How fast is the volume of the cube changing at the moment that the cube has a side length of 8 inches?

40. How fast is the volume of the cube changing at the moment that the cube has a side length of 20 inches?

41. How fast is the volume of the cube changing at the moment that the cube's volume is 55 cubic inches?

42. How fast is the volume of the cube changing at the moment that the area of the cube's base is 10 square inches?

In Exercises 43–45, consider a large helium balloon that is being inflated at the rate of 120 cubic inches per second.

43. How fast is the radius of the balloon increasing at the instant that the balloon has a radius of 12 inches?

44. How fast is the radius of the balloon increasing at the instant that the balloon contains 300 cubic inches of air?

45. How fast is the surface area of the balloon increasing at the instant that the radius of the balloon is 15 inches?

In Exercises 46–48, suppose that Stuart is 6 feet tall and is walking towards a 20-foot streetlight at a rate of 4 feet per second. As he walks towards the streetlight, his shadow gets shorter.

46. How fast is the length of Stuart's shadow changing? Does it depend on how far Stuart is from the streetlight?

47. How fast is the tip of Stuart's shadow moving? Does it depend on how far Stuart is from the streetlight?

48. How fast is the area of the triangle made up of Stuart's legs and his shadow changing? Is it increasing or decreasing as Stuart walks towards the streetlight?

In Exercises 49–51, Alina props a 12-foot ladder against the side of her house so that she can sneak into her upstairs bedroom window. Unfortunately, the ground is muddy because of a recent rainstorm, and the base of the ladder slides away from the house at a rate of half a foot per second.

12 ft

4 feet

$\frac{1}{2}$ foot per second

49. How fast is the top of the ladder moving down the side of the house when the base of the ladder is 4 feet from the house?

50. How fast is the top of the ladder moving down the side of the house when the base of the ladder is 10 feet from the house?

51. How fast is the area of the triangle formed by the ladder, the house, and the ground changing when the top of the ladder is 6 feet from the ground?

In Exercises 52–55, Linda is bored and decides to pour an entire container of salt into a pile on the kitchen floor. She pours 3 cubic inches of salt per second into a conical pile whose height is always two-thirds of its radius.

52. How fast is the radius of the conical salt pile changing when the radius of the pile is 2 inches?

53. How fast is the radius of the conical salt pile changing when the height of the pile is 4 inches?

54. How fast is the height of the conical salt pile changing when the radius of the pile is 2 inches?

55. How fast is the height of the conical salt pile changing when the height of the pile is 4 inches?

56. Linda is still bored and is now pouring sugar onto the floor. The poured-out sugar forms a conical pile whose height is three-quarters of its radius and whose height is growing at a rate of 1.5 inches per second. How fast is Linda pouring the sugar at the instant that the pile of sugar is 3 inches high?

57. Riley is holding an ice cream cone on a hot summer day. As usual, the cone has a small hole at the bottom, and ice cream is melting and dripping through the hole at a rate of half a cubic inch per minute. The cone has a radius of 2 inches and a height of 5 inches. How fast is the height of the ice cream changing when the height of the ice cream in the cone is 3 inches?

2 in.

r

5 in.

h

58. Suppose the width w of a rectangle is decreasing at a rate of 3 inches per second while the height h of the rectangle is increasing at a rate of 3 inches per second. The rectangle initially has a width of 100 inches and a height of 75 inches.

(a) Find the rate of change of the area of the rectangle in terms of its width and height.

(b) On what intervals do the variables w and h make sense in this problem? On what time interval does the problem make sense?

(c) When will the area of the rectangle be increasing, and when will it be decreasing? Answer these questions both in terms of the width and height of the rectangle and in terms of time.

59. Suppose the length a of one leg of a right triangle is increasing at a rate of 4 inches per second while the length b of its other leg is decreasing at a rate of 2 inches per second. The triangle initially has legs of width $a = 1$ inch and $b = 10$ inches.

(a) Find the rate of change of the area of the triangle over time, in terms of its width and height.

(b) On what intervals do the variables a and b make sense in this problem? On what time interval does the problem make sense?

(c) When will the area of the triangle be increasing, and when will it be decreasing? Answer these questions both in terms of the width and height of the triangle and in terms of time.

60. Annie is paddling her kayak through the San Juan Islands and is a quarter of a mile away from where she wants to cross a channel. She sees a ferry in the channel approaching fast from her left, about 2 miles away. The ferry travels at about 20 mph, while Annie can do about 3 mph if she jams.

(a) To set up a model for this problem, suppose Annie is travelling on the x-axis and is approaching the origin from the right. Suppose also that the ferry is travelling on the y-axis and is approaching the origin from above. Then $x = x(t)$ represents Annie's position at time t and $y = y(t)$ represents the ferry's position. Given what you know about this problem, what are $x'(t)$ and $y'(t)$?

(b) Construct an equation in terms of $x = x(t)$ and $y = y(t)$ that describes the distance between Annie and the ferry at time t.

(c) Use implicit differentiation to determine how fast the distance between Annie and the ferry is decreasing when she first sees the ferry.

(d) If Annie decides to jam across the channel, will the ferry hit her?

61. The sun goes down at a rate of about 11 degrees per hour in Colorado in the middle of summer. Ian finds himself contemplating this fact one evening while sitting at Chasm Lake, below Long's Peak in Colorado, watching the sun descend behind the peak. The point on the ridge where the sun is descending is at 13,200 feet. The lake is at 11,710 feet. Ian is sitting 3,100 horizontal feet from the ridge.

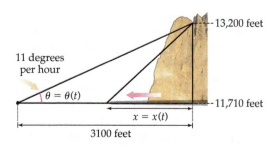

(a) Suppose $x = x(t)$ is the distance of the edge of the shadow from the ridge at time t. This distance is related to the angle $\theta = \theta(t)$ shown in the figure. Find a formula for the speed $x'(t)$ of the shadow.

(b) Use the model from part (a) to determine how fast the shadow of the ridge is moving when it reaches Ian.

Proofs

62. Prove that the lateral surface area of a right circular cone is equal to $\pi r l$, where r is the radius of the cone and l is the length of the diagonal of the cone, that is, the distance from the vertex of the cone to a point on its circumference.

63. Prove that the rate of change of the volume of a sphere with respect to its radius r is equal to the surface area of the sphere. Why does it make geometric sense that the surface area would be related to this rate of change?

64. Prove that the rate of change of the volume of a cylinder with fixed height with respect to its radius r is equal to the lateral surface area of the cylinder. Why does it make geometric sense that the lateral surface area would be related to this rate of change?

65. Suppose a right triangle has legs of lengths a and b and a hypotenuse of length h and that this triangle is changing size, so that the length of its hypotenuse does not change. Prove that the ratio of the rates of change $\dfrac{da}{dt}$ and $\dfrac{db}{dt}$ is equal to $-\dfrac{b}{a}$.

Thinking Forward

Parametric curves: Imagine the curve traced in the xy-plane by the coordinates $(x, y) = (3z + 1, z^2 - 4)$ as z varies, where the parameter z is a function of time t.

▶ Plot the points (x, y) in the plane that correspond to $z = -3, -2, -1, 0, 1, 2,$ and 3.

▶ If the parameter z moves at 3 units per second and $z = 0$ when $t = 0$, plot the points (x, y) in the plane that correspond to $t = 0, 1, 2, 3,$ and 4.

▶ If the parameter z moves at 5 units per second, find the instantaneous rate of change of the x- and y-coordinates as the curve passes through the point $(7, 0)$.

▶ If the x-coordinate moves at 5 units per second, find the instantaneous rate of change of the y-coordinate as the curve passes through the point $(7, 0)$.

CHAPTER REVIEW, SELF-TEST, AND CAPSTONES

Before you progress to the next chapter, be sure you are familiar with the definitions, concepts, and basic skills outlined here. The capstone exercises at the end bring together ideas from this chapter and look forward to future chapters.

Definitions

Give precise mathematical definitions or descriptions of each of the concepts that follow. Then illustrate the definition with a graph or algebraic example, if possible.

▶ f has a *local maximum* or a *local minimum* at $x = c$

▶ f has a *critical point* at $x = c$

▶ f is *increasing* or *decreasing* on an interval I

▶ f is *concave up* or *concave down* on an interval I

▶ f has a *global maximum* or a *global minimum* at $x = c$

▶ f does not have any *global maximum* or does not have any *global minimum*

Theorems

Fill in the blanks to complete each of the following theorem statements:

▶ If $x = c$ is a local extremum of f, then $f'(c)$ is either _____ or _____ .

▶ *Rolle's Theorem:* If f is _____ on $[a, b]$ and _____ on (a, b), and if _____ , then there exists at least one value $c \in (a, b)$ for which $f'(c) = $ _____ .

▶ *The Mean Value Theorem:* If f is _____ on $[a, b]$ and _____ on (a, b), then there exists at least one value $c \in (a, b)$ for which $f'(c) = $ _____ .

▶ If f is differentiable on an interval I and f' is positive in the interior of I, then f is _____ on I.

▶ If f is differentiable on an interval I and f' is _____ in the interior of I, then f is decreasing on I.

▶ If f is differentiable on an interval I and f' is zero in the interior of I, then f is _____ on I.

▶ If $f'(x) = g'(x)$ for all $x \in [a, b]$, then for some constant C, $f(x) = $ _____ for all $x \in [a, b]$.

▶ *The first-derivative test:* Suppose $x = c$ is a _____ of a differentiable function f. If _____ , then f has a local maximum at $x = c$. If _____ , then f has a local minimum at $x = c$. If _____ , then f has neither a local maximum nor a local minimum at $x = c$.

▶ Suppose f and f' are differentiable on an interval I. If _____ is positive on I, then f is concave up on I. If _____ is _____ on I, then f is concave down on I.

▶ *The second-derivative test:* Suppose $x = c$ is a _____ of a twice-differentiable function f. If _____ , then f has a local maximum at $x = c$. If _____ , then f has a local minimum at $x = c$. If _____ , then this test is inconclusive.

Geometric Formulas and Theorems

Volume and Surface Area Formulas: Write a formula for (a) the volume and (b) the surface area of each solid described below.

▶ A rectangular box of width w, length l, and height h

▶ A box of height h with a square base of area A

▶ A sphere of radius r

▶ A sphere of circumference C

▶ A right circular cylinder of radius R and height y

▶ A right circular cylinder whose radius r is half of its height

▶ A cone of height y and circular base of area A

Right-Triangle Theorems: Write an equation that describes the relationships between the variables given for each theorem that follows. Draw a picture to illustrate the theorem and the roles of the variables in your equation.

▶ *The Pythagorean Theorem:* If a right triangle has legs of lengths x and y and a hypotenuse of length h, then _____ .

▶ *The Law of Similar Triangles:* Suppose two right triangles have the same angle measures as each other (i.e., they are similar), where the first has legs of lengths x_1 and y_1 and a hypotenuse of length h_1 and the second has corresponding legs of lengths x_2 and y_2 and a hypotenuse of length h_2. Then we have the following three equations involving ratios: _____ , _____ , and _____ .

Skill Certification: Derivatives and Curve Sketching

Intervals of behavior: For each of the following functions f, determine the intervals on which f is positive, negative, increasing, decreasing, concave up, and concave down.

1. $f(x) = x^2 - 3x + 7$

2. $f(x) = x(x-1)(x+1)$

3. $f(x) = x^3 + 3x^2 - 9x - 27$

4. $f(x) = x^{4/3}$

5. $f(x) = \dfrac{x-1}{x+3}$

6. $f(x) = \dfrac{1}{(x-1)(x+2)}$

Important points: Find all roots, local maxima and minima, and inflection points of each function f. In addition, determine whether any local extrema are also global extrema on the domain of f.

7. $f(x) = (x+1)(x-3)$

8. $f(x) = \dfrac{1}{x}$

9. $f(x) = 3x^4 - 8x^3$

10. $f(x) = x^3 - 15x^2 - 33x$

11. $f(x) = \dfrac{x}{x^2+1}$

12. $f(x) = \dfrac{1}{1+\sqrt{x}}$

Curve sketching: For each function f that follows, construct sign charts for f, f', and f'', if possible. Examine function values or limits at any interesting values and at $\pm\infty$. Then interpret this information to sketch a labeled graph of f.

13. $f(x) = x^2 - x + 1$

14. $f(x) = 1 + \dfrac{1}{x}$

15. $f(x) = x^3 - 2x^2 - 4x + 8$

16. $f(x) = \dfrac{1}{3x+1}$

17. $f(x) = x\sqrt{x+1}$

18. $f(x) = \sqrt{x^2 + 2x + 10}$

Capstone Problems

A. *Critical points, extrema, and inflection points:* Find examples of differentiable functions which illustrate that not every critical point is an extremum and that not every zero of the second derivative is an inflection point. More specifically, find the following:

  (a) A function f with $f'(2) = 0$ and an extremum at $x = 2$, and a function g with $g'(2) = 0$ but no extremum at $x = 2$.

  (b) A function k with $k''(2) = 0$ and an inflection point at $x = 2$, and a function h with $h''(2) = 0$ but no inflection point at $x = 2$.

B. *Optimizing perimeter, given area:* Suppose that you want to cut a rectangular shape with a particular area A from a sheet of material, and that you want the perimeter of the shape to be as small as possible. Use techniques of optimization to argue that the smallest possible perimeter will be achieved if the rectangular shape that you cut out is a square.

C. *The Mean Value Theorem:* Recap the development of the Mean Value Theorem as follows:

  (a) Prove that if f is a differentiable function, then every extremum $x = c$ of f is also a critical point of f. (*Hint: Show that $f'(c) = 0$ by proving that $f'_+(c) \le 0$ and $f'_-(c) \ge 0$.*)

  (b) Use part (a) and the Extreme Value Theorem to prove Rolle's Theorem. (*Hint: Consider the case where f has an extremum on the interior of the interval first.*)

  (c) Explain how the Mean Value Theorem is essentially a rotated version of Rolle's Theorem and how the proof in the reading makes use of that fact.

D. *Area accumulation functions:* Suppose f is the function pictured here, and $A(x)$ is the associated function whose value at any $x \ge 0$ is equal to the area between the graph of f and the x-axis from 0 to x. The quantity $A(2.5)$ is shaded in the figure. We will count area below the x-axis negatively, so that in this example $A(5)$ is less than $A(4)$.

Area function $A(x)$ is defined by the shaded region as x varies

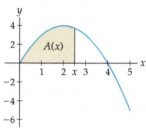

  (a) On what interval of x-values is $A(x)$ an increasing function? On what interval is $A(x)$ decreasing?

  (b) On what interval of x-values is the function f positive? On what interval is f negative?

  (c) Here is a surprising fact: One of these functions is the derivative of the other! Use your answers to parts (a) and (b) to determine whether $A' = f$ or $f' = A$.

CHAPTER 4

Calculus with Power, Polynomial, and Rational Functions

4.1   ADVANCED ALGEBRAIC TECHNIQUES

▶ Simplifying expressions that involve exponents

▶ Factoring and root-guessing techniques

▶ Synthetic division and polynomial long division

Working with Exponents

Recall that an ***algebraic function*** is a function that can be expressed as an arithmetic combination of expressions of the form x^k where x is a real variable and k is a rational number. ***Power functions*** like $3x^{4/3}$, ***polynomial functions*** such as $-5x^4 + 3x^2 + 6x + 2$, and ***rational functions*** like $\frac{1-3x^2}{2+4x+x^2}$ are all algebraic, as are more complicated functions such as $\sqrt{1-x^2}$ and $5x^{3/2}+3x^{-2}$. We've already studied simple examples of these functions and their limits, derivatives, and graphs. In this chapter we will apply our toolbox of calculus techniques to a more in-depth study of algebraic functions.

We have been working with simple functions involving expressions of the form x^k for some time now. Before we consider more complicated calculations, let's make sure we understand the basics completely. When we take a real number and raise it to a power, what are we really talking about? The answer is that it depends on what sort of power we are considering, and writing it out precisely may at first appear surprisingly complex:

DEFINITION 4.1   **The Meaning of Exponents**

Suppose x is a real number, k is a positive integer, $\frac{p}{q}$ is a positive rational number, and r is a rational number. Then

(a) $x^k = \underbrace{x \cdot x \cdot x \cdots x}_{k \text{ times}}$ **(b)** $x^{-k} = \dfrac{1}{x^k}$ **(c)** $x^0 = \begin{cases} 1, & \text{if } x \neq 0 \\ \text{undefined,} & \text{if } x = 0 \end{cases}$

(d) $x^{1/k} = \sqrt[k]{x} = \begin{cases} \text{the number whose } k\text{th power is } x, \text{ if } k \text{ is odd} \\ \text{the nonnegative number whose } k\text{th power is } x, \text{ if } k \text{ is even and } x \geq 0 \\ \text{not a real number, if } k \text{ is even and } x < 0 \end{cases}$

(e) $x^{p/q} = (\sqrt[q]{x})^p$ **(f)** $x^{-p/q} = \dfrac{1}{(\sqrt[q]{x})^p}$ **(g)** $0^r = \begin{cases} 0, & \text{if } r > 0 \\ \text{undefined,} & \text{if } r \leq 0 \end{cases}$

For example, $8^3 = 8 \cdot 8 \cdot 8 = 512$ and $8^{-3} = \frac{1}{8 \cdot 8 \cdot 8} = \frac{1}{512}$. Also, since $2^3 = 8$, we have $8^{1/3} = 2$, and since $(-2)^3 = -8$, it follows that $(-8)^{1/3} = \sqrt[3]{-8} = -2$. In contrast, $(-16)^{1/2}$ is undefined, because there is no real number x such that $x^2 = -16$.

From the preceding definitions, we can derive various algebraic rules for simplifying expressions with exponents. These rules should already be familiar to you, but to strengthen understanding we present them here and prove them in some simple cases.

THEOREM 4.2   **Algebraic Rules for Exponents**

For any numbers x, y, a, and b for which the expressions that follow are defined,

(a) $x^a x^b = x^{a+b}$ **(b)** $(xy)^a = x^a y^a$ **(c)** $(x^a)^b = (x^b)^a = x^{ab}$

(d) $\dfrac{x^a}{x^b} = x^{a-b}$ **(e)** $\left(\dfrac{x}{y}\right)^a = \dfrac{x^a}{y^a}$ **(f)** $(\sqrt[b]{x})^a = \sqrt[b]{x^a} = x^{a/b}$

Proof. We will prove rules (a) and (b) in the case where the powers are positive integers and then prove rule (d) from rule (a). Proofs of other rules are left to Exercises 89–92.

(a) When a and b are positive integers, the quantity $x^a x^b$ is the product of a copies of x with b copies of x, and thus

$$x^a x^b = (\underbrace{x \cdot x \cdot x \cdots x}_{a \text{ times}})(\underbrace{x \cdot x \cdot x \cdots x}_{b \text{ times}}) = (\underbrace{x \cdot x \cdot x \cdots x}_{a+b \text{ times}}) = x^{a+b}.$$

(b) When a and b are positive integers, the quantity $(xy)^a$ is a copies of xy multiplied together. Therefore, rearranging terms, we have

$$(xy)^a = \underbrace{xy \cdot xy \cdot xy \cdots xy}_{a \text{ times}} = (\underbrace{x \cdot x \cdot x \cdots x}_{a \text{ times}})(\underbrace{y \cdot y \cdot y \cdots y}_{a \text{ times}}) = x^a y^a.$$

(d) This rule follows directly from rule (a) and the definition of negative exponents:

$$\frac{x^a}{x^b} = (x^a)\left(\frac{1}{x^b}\right) = (x^a)(x^{-b}) = x^{a+(-b)} = x^{a-b}.$$

Factoring and Roots

Writing an expression as the product of two simpler expressions is called *factoring*. Factoring is a key skill in calculus because we are often interested in where a function or its derivative is zero and it is usually easier to see where an expression is zero if that expression is factored. For example, one way to factor the expression $\sqrt{x} - x^3$ is to pull out the common factor $\sqrt{x}$:

$$\sqrt{x} - x^3 = x^{1/2} - x^3 = x^{1/2}(1 - x^{5/2}).$$

For polynomial functions, we are usually interested in factoring into a product of polynomials of lower degree. For example, with the familiar reverse-FOIL method we can factor this quadratic polynomial:

$$3x^2 - x - 2 = (3x + ?)(x + ?) = (3x + 1)(x - 2).$$

And using the method of factoring by grouping, we can factor this cubic polynomial:

$$x^3 - 3x^2 + 5x - 15 = (x^3 - 3x^2) + (5x - 15) = x^2(x - 3) + 5(x - 3) = (x^2 + 5)(x - 3).$$

These simple factoring methods should already be familiar to you. But what do we do in more complicated situations when such methods do not apply?

For example, consider the polynomial $f(x) = x^3 - 3x^2 + 2$. We can't factor it by grouping, and it has no common factors. It would be nice if we could factor this cubic function into the product of a linear function and a quadratic function, such as

$$f(x) = x^3 - 3x^2 + 2 = (x - c)g(x),$$

for some real number c and quadratic polynomial $g(x)$. Notice that if there were some such number c, it would be a root of f, since we'd have $f(c) = (c - c)g(c) = 0$. Our factoring strategy will be to try to guess a root $x = c$ and then use some clever algebra to pull out the linear factor $(x - c)$ from $f(x)$.

So how do we guess a root of $f(x) = x^3 - 3x^2 + 2$? The following theorem provides a method for making intelligent guesses at roots of a polynomial function that happen to be integers:

THEOREM 4.3 **The Integer Root Theorem**

Let $f(x) = a_n x^n + a_{n-1} x^{n-1} + \cdots + a_1 x + a_0$ be a polynomial function such that each a_i is an integer and $a_0 \neq 0$. Then every integer root of f is a positive or negative divisor of the constant term a_0.

We won't present a formal proof of the Integer Root Theorem here, but the basic idea is that the constant term of a polynomial is always the product of the constant terms in the factors of that polynomial and each root $x = c$ corresponds to a factor $(x - c)$. This theorem allows us to make a list of all the possible integer roots of f. Not everything in the list will be a root, and non-integer roots won't be in the list, but if we are lucky, we will be able to identify a root with this method that we can then attempt to factor out.

For example, our function $f(x) = x^3 - 3x^2 + 2$ has constant term 2, and therefore if it has any integer roots, then they are in the list $\pm 1, \pm 2$. By evaluating $f(1) = 1^3 - 3(1^2) + 2 = 0$, we see that $x = 1$ is a root. Now how do we pull out the factor $(x - 1)$? Since we want to find $g(x)$ so that $f(x) = (x - 1)g(x)$, we essentially want to solve the division problem $g(x) = \dfrac{f(x)}{(x-1)}$.

Synthetic Division

The next theorem provides a method for dividing a polynomial $f(x)$ by a linear function $(x - c)$. If that linear function is a factor of the polynomial, then this method will give us a partial factorization $f(x) = (x - c)g(x)$, and if that linear function is not a factor, then the method will give us a remainder.

THEOREM 4.4   **Synthetic Division**

Let $f(x) = a_n x^n + a_{n-1} x^{n-1} + \cdots + a_1 x + a_0$ be any polynomial function, and let $x = c$ be any real number. To divide $f(x)$ by $(x - c)$, make a diagram like the one shown here with the coefficients a_i of the polynomial $f(x)$ and the number $x = c$. We will find the new coefficients $b_{n-1}, b_{n-2}, \ldots b_1, b_0$ by using the following diagram:

Bring the first coefficient a_n down to become the new coefficient b_{n-1}. Each diagonal arrow indicates multiplication by c, and each vertical arrow represents addition (of the term a_i and the term in the circle). Follow these arrows to find the remaining coefficients b_i and the remainder R. Let $g(x) = b_{n-1} x^{n-1} + b_{n-2} x^{n-2} + \cdots + b_1 x + b_0$.

If $x = c$ is a root of $f(x)$, then the remainder R will be zero, and it follows that

$$\frac{f(x)}{x - c} = g(x), \quad \text{i.e., } f(x) = (x - c)g(x).$$

If $x = c$ is not a root of $f(x)$, then the remainder R will be nonzero, and it follows that

$$\frac{f(x)}{x - c} = g(x) + \frac{R}{x - c}, \quad \text{i.e., } f(x) = (x - c)g(x) + R.$$

We will not provide a formal proof of this method, although in the next subsection we will see that synthetic division is a special case of the method of polynomial long division.

To apply the method of synthetic division to divide our function $f(x) = x^3 - 3x^2 + 2$ by its factor $(x - 1)$, we set up the diagram next at the left. After completing the synthetic division algorithm, we have the diagram at the right.

  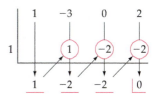

Notice that the last entry is zero, which indicates that the remainder is zero and thus $(x-1)$ divides evenly into $f(x)$. We can read the coefficients of the remaining factor $g(x)$ from the bottom row of our diagram to get $g(x) = 1x^2 - 2x - 2$. The function $g(x)$ happens to be an **irreducible quadratic**, which means that we cannot factor it any further. Therefore the complete factorization of $f(x)$ is

$$f(x) = x^3 - 3x^2 + 2 = (x - 1)(x^2 - 2x - 2).$$

Polynomial Long Division

We can use synthetic division when we want to divide a polynomial by a linear function of the form $(x - c)$. The more general algorithm known as polynomial long division will allow us to divide any polynomial by any other polynomial of equal or lower degree. Polynomial long division works almost like the long-division algorithm that we use to divide one integer by another, except that it involves polynomials instead of positive integers.

Let's begin by reviewing the long-division algorithm for dividing positive integers, with the example of dividing 131 by 4:

$$
\begin{array}{r}
3\,2 \\
4\,\overline{)1\,3\,1} \\
1\,2 \\
\hline
1\,1 \\
8 \\
\hline
3
\end{array}
$$

The work shown can be explained as follows: The first step is to figure out how many times 4 goes into 13, which is 3 times. We then multiply 3 by 4 and subtract the resulting product from 13, giving us a remainder of 1. We bring down the remaining 1 from the number 131. Now we repeat the process and determine that 4 goes into 11 twice. This gives us the "2" in the "32" at the top of the diagram. If we multiply 2 by 4 and subtract the resulting product from 11, we are left with a remainder of 3. Since 4 does not go into 3, we are done.

We have shown that 4 goes into 131 thirty-two times with a remainder of 3, and thus that

$$131 = 4(32) + 3.$$

We can rewrite this equality by dividing both sides by 4:

$$\frac{131}{4} = 32 + \frac{3}{4}.$$

The algorithm for long division of polynomials is exactly the same, but with powers of x determining place value instead of powers of 10:

THEOREM 4.5

Polynomial Long Division

To use polynomial long division to divide a polynomial $p(x)$ by a polynomial $q(x)$ of equal or lower degree, we fill in the following diagram as described next:

$$q(x) \overline{\smash{\big)}\, p(x)}^{\textstyle m(x)}$$
$$\vdots$$
$$R(x)$$

▶ Find an expression ax^k whose product with the leading term of $q(x)$ is equal to the leading term of $p(x)$. Record this expression above the line.

▶ Multiply $q(x)$ by the expression ax^k you found in the first step, and *subtract* the product from $p(x)$. For the purposes of the next step, call the resulting polynomial the "remainder polynomial."

▶ Repeat, with $p(x)$ replaced by the remainder polynomial, until the degree of the final remainder polynomial $R(x)$ is less than the degree of $q(x)$.

At the end of the algorithm, we will have found $m(x)$ and $R(x)$ so that

$$p(x) = q(x)\, m(x) + R(x).$$

In other words,

$$\frac{p(x)}{q(x)} = m(x) + \frac{R(x)}{q(x)}.$$

We won't present a formal proof that the polynomial long-division algorithm always works. However, given that this algorithm allows you to divide one polynomial by another, we can use it to explain why *synthetic* division works; you will do so in Exercise 22. (See Example 6 for an example of the implementation of polynomial long division.)

Examples and Explorations

EXAMPLE 1

Working with different kinds of exponents

Calculate each of the following quantities by hand, without a calculator:

(a) 10^{-34} **(b)** 0^{-3} **(c)** $(-8)^{-2/3}$ **(d)** $(-9)^{3/2}$ **(e)** $1^{-17/5}$

SOLUTION

(a) $10^{-34} = \frac{1}{10^{34}} = 0.0000000000000000000000000000000001$, with 33 zeroes between the decimal place and the final 1. In scientific notation, this number is written as 1×10^{-34}.

(b) The expression $0^{-3} = \frac{1}{0^3} = \frac{1}{0}$ is undefined.

(c) $(-8)^{-2/3} = (\sqrt[3]{-8})^{-2} = (-2)^{-2} = \frac{1}{(-2)^2} = \frac{1}{4}$.

(d) $(-9)^{3/2} = (\sqrt{-9})^3$. Since we cannot take an even root of a negative number, $(\sqrt{-9})^3$ is not a real number.

(e) $1^{-17/5} = \frac{1}{1^{17/5}} = \frac{1}{(\sqrt[5]{1})^{17}} = \frac{1}{1^{17}} = \frac{1}{1} = 1$. In fact, 1 raised to *any* power is equal to 1. ☐

EXAMPLE 2   **Simplifying expressions that involve multiple exponents**

Simplify each of the expressions that follow until no more than one exponent is involved. Cite all of the rules you use from Theorem 4.2.

(a) $\dfrac{x^6 + (x^2)^3}{x^4}$ **(b)** $(16x^5x^7)^{1/4}(\sqrt{x})^{-6}$ **(c)** $\dfrac{\sqrt[3]{r^{12}}}{(2s)^4}$

SOLUTION

(a) By rules (c) and (d),

$$\frac{x^6 + (x^2)^3}{x^4} = \frac{x^6 + x^6}{x^4} = \frac{2x^6}{x^4} = 2x^2,$$

where the final step is valid only for $x \neq 0$, since when $x = 0$, the expression $\frac{2x^6}{x^4}$ is undefined but the expression $2x^2$ is equal to zero.

(b) By rules (a), (b), and (c), we have

$$(16x^5x^7)^{1/4}(\sqrt{x})^{-6} = (16x^{12})^{1/4}(x^{1/2})^{-6} = (16)^{1/4}(x^{12})^{1/4}(x^{1/2})^{-6} = 2x^3x^{-3}$$

$$= 2x^{3-3} = 2x^0 = 2(1) = 2.$$

This computation holds only for $x > 0$, since if $x \leq 0$, then the original expression is undefined.

(c) Using Rules (f), (b), and (e) from Theorem 4.2, we have

$$\frac{\sqrt[3]{r^{12}}}{(2s)^4} = \frac{r^{12/3}}{16s^4} = \frac{1}{16} \cdot \frac{r^4}{s^4} = \frac{1}{16}\left(\frac{r}{s}\right)^4.$$ ☐

 CHECKING THE ANSWER

One way to check each of the preceding answers is to verify that the graph of the original expression is the same as the graph of the final expression. Alternatively, choose a value for x and substitute it into the original and final expressions; if you get two different answers, then you've probably made an algebra mistake. Neither of these methods absolutely ensures that your answer is correct; however, if your calculation fails either of these tests, then you must have done something wrong.

EXAMPLE 3   **Applying standard factorization techniques**

Find the roots of the polynomial function $f(x) = x^4 - 2x^3 - 4x^2 + 8x$ by factoring as much as possible.

SOLUTION

Luckily, after taking out the common factor x, we can find a grouping that works for the factoring-by-grouping technique:

$$
\begin{aligned}
x^4 - 2x^3 - 4x^2 + 8x &= x(x^3 - 2x^2 - 4x + 8) && \leftarrow \text{take out common factor} \\
&= x((x^3 - 2x^2) + (-4x + 8)) && \leftarrow \text{group terms together} \\
&= x(x^2(x - 2) - 4(x - 2)) && \leftarrow \text{find common factor in each group} \\
&= x(x^2 - 4)(x - 2) && \leftarrow \text{distributive property} \\
&= x(x + 2)(x - 2)(x - 2) && \leftarrow \text{difference of squares} \\
&= x(x + 2)(x - 2)^2. && \leftarrow \text{combine factors}
\end{aligned}
$$

Since the factors of $f(x)$ are x, $(x + 2)$, and $(x - 2)$, the roots of the polynomial function $f(x) = x^4 - 2x^3 - 4x^2 + 8x$ are $x = 0$, $x = -2$, and $x = 2$ (repeated twice).   ☐

CHECKING THE ANSWER

You can algebraically check the roots we just found either by multiplying out the factorization or by checking that $x = 0$, $x = -2$, and $x = 2$ are all roots of the original function. Graphically, we could look at a graph of $y = f(x)$ and check that it has roots only at $x = 0$, $x = -2$, and $x = 2$, as shown here:

$$f(x) = x^4 - 2x^3 - 4x^2 + 8x$$

Note that the doubled root at $x = 2$ looks different than the other roots of f; we'll discuss how to graphically recognize various types of repeated roots in Section 4.3.

EXAMPLE 4

Factoring when all the roots of a polynomial are known

Every root of the polynomial function $f(x) = 2x^3 + 2x^2 - 8x - 8$ is an integer. Find these roots, and use them to factor $f(x)$.

SOLUTION

Since the constant term of $f(x) = 2x^3 + 2x^2 - 8x - 8$ is $a_0 = -8$, the possible integer roots of $f(x)$ are ± 1, ± 2, ± 4, and ± 8. By evaluating $f(x)$ at each of these values, one can easily see that $x = -1$, $x = 2$, and $x = -2$ are roots of $f(x)$. Since $f(x)$ is a cubic polynomial, it can have at most three roots, so we have found *all* the roots of $f(x)$. Because every root $x = c$ of $f(x)$ corresponds to a linear factor $(x - c)$ of $f(x)$, we can write

$$f(x) = A(x + 1)(x - 2)(x + 2)$$

for some constant A. Since $(x + 1)(x - 2)(x + 2) = x^3 + x^2 - 4x - 4$ differs from $f(x) = 2x^3 + 2x^2 - 8x - 8$ by a factor of 2, it follows that $A = 2$. Thus we can write $f(x)$ in factored form as

$$f(x) = 2(x + 1)(x - 2)(x + 2).$$

In this particular example, we could have also factored $f(x)$ by grouping; however, the method applied here will be useful for many examples of polynomial functions that would otherwise be difficult to factor. $\square$

EXAMPLE 5

Factoring by using the Integer Root Theorem and synthetic division

Factor the polynomial function $f(x) = 2x^4 + 7x^3 + x^2 - 7x - 3$ as much as possible.

SOLUTION

By the Integer Root Theorem, the only possible integer roots of this function are ± 1 and ± 3, the integer divisors of the constant term -3. Evaluating $f(x)$ at these values, we see that $x = \pm 1$ and $x = -3$ are roots of $f(x)$ and thus that $(x - 1)$, $(x + 1)$, and $(x + 3)$ are factors of $f(x)$. It does not matter which root we use for the first step; we will begin by using synthetic division with the root $x = -3$:

$$
\begin{array}{r|rrrrr}
     & 2 & 7 & 1 & -7 & -3 \\
-3   &   & -6 & -3 & 6 & 3 \\
\hline
     & 2 & 1 & -2 & -1 & 0 \\
\end{array}
$$

Notice that we left out the arrows and circles and included just the numbers; this is the most typical way to write synthetic division computations. Our calculation shows that

$$ f(x) = 2x^4 + 7x^3 + x^2 - 7x - 3 = (x + 3)(2x^3 + x^2 - 2x - 1). $$

Now that we have managed to extract one linear factor from $f(x)$, we concentrate on what is left to factor, namely, $2x^3 + x^2 - 2x - 1$. We can do another synthetic division calculation to divide $2x^3 + x^2 - 2x - 1$ by $x - 1$, since $x = 1$ is one of the other roots of $f(x)$:

$$
\begin{array}{r|rrrr}
    & 2 & 1 & -2 & -1 \\
1   &   & 2 & 3 & 1 \\
\hline
    & 2 & 3 & 1 & 0 \\
\end{array}
$$

Now we know that $2x^3 + x^2 - 2x - 1 = (x - 1)(2x^2 + 3x + 1)$. Therefore we can factor $f(x)$ completely as

$$ f(x) = (x + 3)(x - 1)(2x^2 + 3x + 1) = (x + 3)(x - 1)(2x + 1)(x + 1). $$ □

EXAMPLE 6   Dividing one polynomial by another

Divide the polynomial $4x^2 + 2x + 3$ by the polynomial $x - 2$ in two different ways: (a) with polynomial long division and (b) with synthetic division.

SOLUTION

(a) The leading term of $x - 2$ is x, and the leading term of $4x^2 + 2x + 3$ is $4x^2$. The first step is to determine how many times x goes into $4x^2$, which is $4x$. Therefore the first entry on top of the line is $4x$. Now we multiply $x - 2$ by $4x$ to get $4x^2 - 8x$ and then subtract this, term-by-term, from the polynomial $4x^2 + 2x + 3$. At this point we have the diagram at the left:

$$
\begin{array}{r}
\;4x \\
x - 2 \,\overline{)\,4x^2 + 2x + 3} \\
\underline{4x^2 - 8x} \\
10x + 3
\end{array}
\qquad\qquad
\begin{array}{r}
\;4x + 10 \\
x - 2 \,\overline{)\,4x^2 + 2x + 3} \\
\underline{4x^2 - 8x} \\
10x + 3 \\
\underline{10x - 20} \\
23
\end{array}
$$

Now we repeat the process. The leading term of $x - 2$ goes into the leading term of $10x + 3$ ten times, since $10(x) = 10x$, so we put a 10 above the top line. Now we multiply 10 by $x - 2$ and subtract the product $10x - 20$, term-by-term, from $10x + 3$. When we do so, we get a remainder of 23. Because 23 has a degree less than $x - 2$, the process is over, and we can say that

$$ 4x^2 + 2x + 3 = (x - 2)(4x + 10) + 23. $$

(b) Since we are dividing by a linear function of the form $x - c$, we can also perform the calculation by synthetic division:

$$
\begin{array}{c|ccc}
 & 4 & 2 & 3 \\
2 & & 8 & 20 \\
\hline
 & 4 & 10 & \boxed{23}
\end{array}
$$

Once again we see that $4x^2 + 2x + 3 = (x - 2)(4x + 10) + 23$. Compare this calculation to the polynomial long-division calculation we did before. In particular, notice where the numbers 4, 10, and 23 appear in both calculations and how they are found. Where do the numbers 8 and 20 appear in the polynomial long-division algorithm? □

EXAMPLE 7   Using polynomial long division to rewrite an improper rational function

A rational function is said to be **_proper_** if the degree of its numerator is strictly less than the degree of its denominator; otherwise it is said to be **_improper_**. Use polynomial long division to write the improper rational function $f(x) = \dfrac{x^4 - 3x^3 + x - 1}{x^2 + x - 3}$ as a sum of a polynomial and a proper rational function.

SOLUTION

Following the algorithm for polynomial long division to divide $p(x) = x^4 - 3x^3 + x - 1$ by $q(x) = x^2 + x - 3$, we have

$$
\require{enclose}
\begin{array}{r}
x^2 - 4x + 7 \\
x^2 + x - 3 \enclose{longdiv}{x^4 - 3x^3 - 0x^2 + x - 1} \\
\underline{x^4 + x^3 - 3x^2 } \\
-4x^3 + 3x^2 + x - 1 \\
\underline{-4x^3 - 4x^2 + 12x } \\
7x^2 - 11x - 1 \\
\underline{7x^2 + 7x - 21} \\
-18x + 20
\end{array}
$$

Note that we added "$0x^2$" to the function $p(x)$ so that the terms of all our polynomials would line up by degree. Note also that we stopped when the degree of the remainder polynomial $-18x + 20$ became smaller than the degree of $q(x) = x^2 + x - 3$. We now have

$$x^4 - 3x^3 + x - 1 = (x^2 + x - 3)(x^2 - 4x + 7) + (-18x + 20).$$

Dividing both sides of the preceding equality by $x^2 + x - 3$, we get

$$f(x) = \frac{x^4 - 3x^3 + x - 1}{x^2 + x - 3} = (x^2 - 4x + 7) + \frac{-18x + 20}{x^2 + x - 3}.$$

We have now written the original improper rational function as a sum of a polynomial and a proper rational function, as desired. □

CHECKING THE ANSWER

To check our division, we could multiply out $(x^2 + x - 3)(x^2 - 4x + 7) + (-18x + 20)$ and see that it is equal to $x^4 - 3x^3 + x - 1$. Alternatively, we could combine the terms of $(x^2 - 4x + 7) + \frac{-18x + 20}{x^2 + x - 3}$ by finding a common denominator and then show that the result is equal to the original improper rational function $\frac{x^4 - 3x^3 + x - 1}{x^2 + x - 3}$.

TEST YOUR UNDERSTANDING

▶ What can you say about the values of 0^2, 2^0, and 0^0?

▶ Suppose a is a real number. What does a^r represent if r is a positive integer? A negative integer? A rational number?

▶ Can all the roots of the polynomial $f(x) = 3x^2 + x - 2$ be found by using the Integer Root Theorem? Why or why not?

▶ Could synthetic division be used to divide $x^3 + 4x^2 - x + 1$ by $3x + 1$? If so, how? If not, why not?

▶ Polynomial long division can be used to divide any polynomial $f(x)$ by any polynomial $g(x)$ of equal or lower degree. Sometimes synthetic division can also be used for this purpose, and sometimes not. What conditions on $f(x)$ and/or $g(x)$ must be satisfied for synthetic division to be a viable method of division?

EXERCISES 4.1

Thinking Back

Factoring formulas: Write down factoring formulas for taking out a linear factor $(a - b)$ from each of the following expressions.

▶ $a^2 - b^2$

▶ $a^3 - b^3$

▶ $a^4 - b^4$

▶ $a^n - b^n$

Each of the statements that follow are *false* and represent common algebra mistakes. Find a counterexample for each statement; that is, a value of x and/or y for which the statement is false.

▶ False: $(3x)^2 = 3x^2$

▶ False: $x^2 x^3 = x^6$

▶ False: $\dfrac{x^{12}}{x^3} = x^4$

▶ False: $x^2 + x^3 = x^5$

▶ False: $\dfrac{1 + x^{-2}}{y} = \dfrac{1}{y + x^2}$

▶ False: $(x + y)^{-1} = \dfrac{1}{x} + \dfrac{1}{y}$

▶ False: $\dfrac{x^{-1} + y^{-1}}{z} = \dfrac{1}{(x + y)z}$

▶ False: $\sqrt{x^2} = x$

▶ False: $\sqrt{x^2 + y^2} = x + y$

Concepts

0. *Problem Zero:* Read the section and make your own summary of the material.

1. *True/False:* Determine whether each of the statements that follow is true or false. If a statement is true, explain why. If a statement is false, provide a counterexample.

(a) *True or False:* The sum of any two algebraic functions is itself an algebraic function.

(b) *True or False:* The product of any two algebraic functions is itself an algebraic function.

(c) *True or False:* The quotient of any two algebraic functions is itself an algebraic function.

(d) *True or False:* If f is a polynomial function with constant term $a_0 = 0$, then $x = 0$ is a root of f.

(e) *True or False:* If f is a polynomial function with constant term $a_0 = 3$, then $x = 3$ is a root of f.

(f) *True or False:* The Integer Root Theorem can be used to find all the roots of any polynomial function f.

(g) *True or False:* Synthetic division can always be used in place of polynomial long division.

(h) *True or False:* Polynomial long division can be used to write any improper rational function as the sum of a polynomial function and a proper rational function.

2. *Examples:* Construct examples of the thing(s) described in the following. Try to find examples that are different than any in the reading.

(a) A polynomial with four real roots, only two of which are integers.

(b) A polynomial whose roots are all rational, but non-integer, numbers.

(c) A polynomial whose roots are all irrational numbers.

3. What is an *algebraic function*? Give three examples of algebraic functions and three examples of functions that are not algebraic.

4. Explain why the functions $f(x) = 3^x$ and $g(x) = x^x$ are not algebraic functions.

5. The function $f(x) = |x|$ doesn't appear at first glance to be algebraic, but it is! Show that $f(x) = |x|$ can be written in terms of arithmetic operations, integer powers, and roots of the independent variable x.

6. Suppose A is any real number and k is any rational number.

(a) Why is $f(x) = Ax^k$ a function? Justify your answer with the formal definition of function.

(b) When $k = \frac{1}{2}$, $f(x) = A\sqrt{x}$. What is the domain of this function? Is it really true that for every x in the domain of f we have only *one* output $A\sqrt{x}$? Why might someone incorrectly think otherwise?

7. By convention, for nonzero x we set $x^0 = 1$. Why couldn't we instead adopt the convention that $x^0 = 0$? Let's find out.

   (a) Explain why it is not immediately obvious what the number x^0 should represent, if $x \neq 0$. What does it mean to say that setting $x^0 = 1$ is a *convention*?

   (b) Suppose we were to adopt the alternative convention that $x^0 = 0$. With this convention, what would rule (d) say about $\frac{x^2}{x^2}$? What happens? Why is it bad?

   (c) Suppose again that we were to adopt the alternative convention that $x^0 = 0$. With this convention, use rule (a) and the fact that $x^2 = x^{2+0}$ to show that x^2 is equal to zero for all x. Why is this bad?

   (d) You've just seen that if we set $x^0 = 0$, then the rules in Theorem 4.2 no longer work. So what? Why do you think we want those particular rules to work?

8. Rule (f) of Theorem 4.2 states that $x^{a/b} = \sqrt[b]{x^a} = (\sqrt[b]{x})^a$. Suppose we have $a = 6$, $b = 4$, and $x = -2$. Show that rule (f) does not hold in this case, and explain why this is not a contradiction to the rule.

9. In this problem you will examine the indeterminate expression 0^0.

   (a) We'll first examine what happens when $k = 0$ and x approaches zero. Approximate the value of $\lim_{x \to 0} x^0$ by evaluating it at small nonzero values of x.

   (b) Now suppose that $x = 0$ and k approaches zero. Use small nonzero values of k to approximate the value of $\lim_{k \to 0} 0^k$.

   (c) Explain why your answers from parts (a) and (b) show that there is no reasonable convention or definition for the expression 0^0.

10. Consider the polynomial $f(x) = 4x^3 + 11x^2 - 3x$.

    (a) Find the roots of this polynomial.

    (b) For each root $x = c$ you found in part (a), verify that $(x - c)$ is a factor of f by writing $f(x)$ in the form $f(x) = (x - c)g(x)$ for some polynomial function $g(x)$.

    (c) Write $f(x)$ in the form $f(x) = A(x - b)(x - c)(x - d)$ for some real numbers A, b, c, and d.

11. Consider the polynomial $f(x) = 2x^3 - 3x^2 + 8x$.

    (a) Factor this polynomial as much as possible. You should find that there is one linear factor and one irreducible quadratic factor. Explain how you know that the quadratic factor is irreducible.

(b) Write $f(x)$ in the form $f(x) = A(x - b)(x^2 + cx + d)$ for some real numbers A, b, c, and d.

12. What is the distributive property, and what does it have to do with the method of factoring by grouping?

13. Verify that $x = -1$, $x = 1$, and $x = \frac{1}{2}$ are roots of the polynomial function $f(x) = 2x^4 - 3x^3 - x^2 + 3x - 1$. Also show that $x = 2$ and $x = -3$ are *not* roots of this function.

14. Suppose f is a polynomial with constant term a_0. If the only possible integer roots of f are ±1, ±2, ±3, and ±6, what is a_0? Is there only one possible answer?

15. Suppose f is a polynomial with constant term $a_0 = -12$. List all the possible integer roots of f.

16. What is synthetic division? Specifically, what is synthetic division used for, and how do you "do it"?

17. Fill in the blank: If there is no remainder after using synthetic division to divide a polynomial function $f(x)$ by a linear expression $x - c$, then $f(x) =$ _____ for some polynomial function $g(x)$.

18. Fill in the blank: If there is a nonzero remainder R after using synthetic division to divide a polynomial function $f(x)$ by a linear expression $x - c$, then $f(x) =$ _____ for some polynomial function $g(x)$.

19. What is polynomial long division used for? Give three examples of ways that polynomial long division is like the long division algorithm for numbers, and give two examples of ways that it is different.

20. What would happen if you tried to use polynomial long division to divide a two-degree polynomial by a three-degree polynomial? Identify the part of the algorithm that justifies your answer. (*Hint: It may help to try an example first.*)

21. Suppose you use polynomial long division to divide $p(x)$ by $q(x)$, and after doing your calculations you end up with the polynomial $x^2 - x + 3$ as the quotient above the top line, and the polynomial $3x - 1$ at the bottom as the remainder. Then $p(x) =$ _____ and $\frac{p(x)}{q(x)} =$ _____ .

22. Use polynomial long division to divide the polynomial $x^3 - 2x + 3$ by the polynomial $x - 2$. Then use synthetic division to perform the same calculation. Compare the two calculations, and determine where each number that appears in the synthetic division algorithm appears in the polynomial long division algorithm. Explain how the synthetic division algorithm is really the same as the polynomial long division algorithm, but with abbreviated notation.

Skills

23. Evaluate, by hand and without a calculator, each of the expressions that follow. Some of the expressions may be undefined.

    (a) $\left(\frac{2}{3}\right)^{-3}$ (b) $(-4)^{5/2}$ (c) $\left(\frac{8}{27}\right)^{2/3}$ (d) $0^{1/100}$

24. Evaluate, by hand and without a calculator, each of the expressions that follow. Some of the expressions may be undefined.

    (a) $\left(\frac{1}{2}\right)^{-1}$ (b) $64^{5/6}$ (c) $1^{4/9}$ (d) 0^{-3}

Solve each of the inequalities in Exercises 25–30.

25. $x^3 \le x^5$

26. $x^2 > x^4$

27. $x^{2/3}(x-2)^{1/5} \ge 0$

28. $x^{3/2} + x^{1/2} < 0$

29. $\sqrt[4]{x} + x^8 \ge 0$

30. $\sqrt[3]{x}\,(x-1)^{-2} \le 0$

Find the domain and the zeroes of each function in Exercises 31–40. Check your answers afterwards with graphs.

31. $f(x) = -\dfrac{1}{4}x^{2/3}$

32. $f(x) = \dfrac{-4}{\sqrt[6]{x^{7/3}}}$

33. $f(x) = 2x^{-7/4}$

34. $f(x) = \dfrac{x^3 + x^2 - 2x}{2x^3 - x^2 - 6x}$

35. $f(x) = (x^2 - 1)^{-1/4}$

36. $f(x) = |x^2 - 9|^{-3/4}$

37. $f(x) = \dfrac{\sqrt{3-x}}{\sqrt[3]{x^2 - 3x - 4}}$

38. $f(x) = \sqrt[4]{x^{-2} + 1}$

39. $f(x) = \sqrt{\dfrac{x^2 - 1}{x^3 - 7x + 6}}$

40. $f(x) = \dfrac{3x^{-3}}{x^{-1/4} - x^{3/4}}$

Factor each of the polynomials in Exercises 41–60 as much as possible. Some factorizations require that you first guess a root with the Integer Root Theorem and/or use synthetic division.

41. $f(x) = x^3 - 3x^2 - x + 3$

42. $f(x) = x^3 - 5x^2 - 2x + 24$

43. $f(x) = x^4 + 7x^3 + 9x^2 - 7x - 10$

44. $f(x) = x(x+3) + 2$

45. $f(x) = x^3 - 5x^2 + 3x + 9$

46. $f(x) = x^4 - 6x^3 + 13x^2 - 12x + 4$

47. $f(x) = x^3 - 2x^2 + x - 2$

48. $f(x) = x^3 + 8x^2 - 4x - 32$

49. $f(x) = -8x^2 + 14x - 3$

50. $f(x) = 5 - x + 5x^4 - x^5$

51. $f(x) = 2x^3 + x^2 - 32x - 16$

52. $f(x) = 3(x^4 + 2) - x(3 + 6x^2)$

53. $f(x) = x^3 - 2x^2 - 5x + 6$

54. $f(x) = 3x^3 - 8x^2 + 5x - 2$

55. $f(x) = x^3 + 4x^2 - 11x + 6$

56. $f(x) = x^4 - 8x^3 + 24x^2 - 32x + 16$

57. $f(x) = 2x^4 + 6x^2 - 8$

58. $f(x) = 3x^4 + 11x^3 + 15x^2 + 9x + 2$

59. $f(x) = 2x^6 - 2x^5 - 10x^4 + 2x^3 + 16x^2 + 8x$

60. $f(x) = 3x^4 + 7x^3 - 21x^2 - 6x - 8$

There is an analog of the Integer Root Theorem called the **Rational Root Theorem** that finds rational roots of polynomials:

> If $f(x) = a_n x^n + a_{n-1}x^{n-1} + \cdots + a_1 x + a_0$, where each a_i is an integer and $a_0 \ne 0$, then every reduced rational root $\dfrac{p}{q}$ has the property that p is a positive or negative divisor of the constant term a_0 and q is a positive or negative divisor of the leading coefficient a_n.

Use this more general theorem to guess the rational roots of each of the polynomials in Exercises 61–64, and then use synthetic division to completely factor them.

61. $f(x) = 2x^3 + x^2 + 8x + 4$

62. $f(x) = 6x^4 - 7x^3 + 8x^2 - 7x + 2$

63. $f(x) = 16x^3 - 12x^2 + 1$

64. $f(x) = 27x^4 - 18x^2 - 8x - 1$

Use either synthetic division or polynomial long division to write each of the improper rational functions $f(x) = \dfrac{p(x)}{q(x)}$ in Exercises 65–80 as the sum of a polynomial function $m(x)$ and a proper rational function $\dfrac{R(x)}{q(x)}$. Check each of your answers by multiplying out $q(x)m(x) + R(x)$.

65. $f(x) = \dfrac{x^3 + 2x^2 - x - 5}{x - 3}$

66. $f(x) = \dfrac{2x^3 - 3x^2 + 8}{x - 2}$

67. $f(x) = \dfrac{1 - x^4}{x + 3}$

68. $f(x) = \dfrac{(x-2)(x^5 - 3x^2) + 4}{x - 2}$

69. $f(x) = \dfrac{5x^4 - 3x + 2}{x + 1}$

70. $f(x) = \dfrac{x^6 + x^5 - 3x^3 - 2x^2 - 7x - 8}{x + 1}$

71. $f(x) = \dfrac{x^3 - 2x^2 + 5}{x - 2}$

72. $f(x) = \dfrac{x^2 + 3x - 2}{x^2 + x - 1}$

73. $f(x) = \dfrac{x^5 - 5x^3 - 4x^2 + 6x + 12}{x^2 - 3}$

74. $f(x) = \dfrac{x^4 - 2x^3 - 3x^2 + 8x - 4}{x^2 - 2x + 1}$

75. $f(x) = \dfrac{1 - x^4}{x^2 + 2x - 1}$

76. $f(x) = \dfrac{x^4 + 12}{x^3 - 3}$

77. $f(x) = \dfrac{2 - 3x^2 + x^4}{2x^2 + 1}$

78. $f(x) = \dfrac{x^5 - 3x^2 + x - 1}{2x + 5}$

79. $f(x) = \dfrac{x^7 - 1}{1 - 2x + x^2}$

80. $f(x) = \dfrac{x^5 - 3x^4 - 3x^3 - 3x^2 - 4x}{x^2 - 3x - 4}$

Combine your new algebra skills with your knowledge of derivatives and number lines to find any local extrema and inflection points of the functions in Exercsies 81–86.

81. $f(x) = x^{1/4} - 2x^{-1/2}$

82. $f(x) = x^{-2/3}(x^2 - x)$

83. $f(x) = x^4 - 4x^3 + 6x^2 - 36x$

84. $f(x) = x^4 - 2x^3 - 8x^2 - 6x$

85. $f(x) = \dfrac{2x^3 + 3x^2 - 2x - 3}{x^2 - 2x - 3}$

86. $f(x) = \dfrac{x^3 - 3x^2 - x + 3}{x^3 - 6x^2 + 12x - 8}$

Applications

Suppose an object in a wind tunnel is moving left and right along a straight path over a 6-second interval. By collecting data with a video camera and using a computer program, you determine that the following function models the position of the object, in centimeters:

$$s(t) = t^4 - 11t^3 + 37t^2 - 45t + 18.$$

In this model, distances to the right of the center point of the tunnel are positive and distances to the left are negative. Use the model to answer the questions in Exercises 87–88.

87. The object is in the center point of the tunnel when $s = 0$. According to this model, are there any times in $[0, 6]$ for which the object is at the center point?

88. Use what you know about derivatives and factoring to determine the times in $[0, 6]$ at which the object is changing direction.

Proofs

In Exercises 89–92, prove each rule from Theorem 4.2 in the manner specified.

89. Prove that if a and b are positive integers, then $(x^a)^b = (x^b)^a = x^{ab}$.

90. Prove that if a and b are positive integers, then $\frac{x^a}{x^b} = x^{a-b}$.

91. Use rules (b) and (c) to prove that $\left(\frac{x}{y}\right)^a = \frac{x^a}{y^a}$.

92. Use rule (c) to prove that $(\sqrt[b]{x})^a = \sqrt[b]{x^a} = x^{a/b}$.

93. Prove that if rule (a) of Theorem 4.2 is true, then $x^0 = 1$ for all nonzero values of x. Make sure to point out where you use the fact that x is nonzero. *(Hint: Look at Exercise 7(c).)*

Thinking Forward

Multiplying by the conjugate: Complete each of the following calculations by using the definition of the derivative.

▶ Use the definition of the derivative to calculate the derivative of $f(x) = \sqrt{x}$ at $c = 4$. At some point you will need to multiply numerator and denominator by the **conjugate** of $\sqrt{4+h} - 2$, which is $\sqrt{4+h} + 2$.

▶ Use the definition of the derivative to calculate the derivative of $f(x) = x^{-1/2}$ at $c = 4$. As in the previous calculation, you will need to multiply numerator and denominator by a conjugate at some point.

▶ Use the definition of the derivative to calculate the derivative of $f(x) = \sqrt{x}$. You will need to multiply by a conjugate at some point in your calculation.

4.2   POWER FUNCTIONS

▶ Properties and graphs of power functions with integer powers

▶ Properties and graphs of power functions with rational powers

▶ Applying limit and derivative techniques to functions that involve power functions

Power Functions

Power functions are the building blocks of algebraic functions; in fact, all algebraic functions are combinations and/or compositions of power functions. In this section we will use what we know about algebra, limits, and derivatives to investigate general power functions and their graphical properties. Such an investigation will not only allow us to understand and classify power functions; it will also give us a chance to review everything in our calculus toolbox.

DEFINITION 4.6   **Power Functions**

A **power function** is a function that can be written in the form

$$f(x) = Ax^k$$

for some nonzero real number A and some rational number k.

In this definition, the constant k is called the **power** or **exponent**, and the constant A is called the **coefficient**. Note that the exponent k must be a rational number and the variable must be in the base; this means that, for example, $f(x) = x^\pi$ and $f(x) = 10^x$ are not considered power functions.

Notice that we say a function f is a power function if it *can* be written in the form $f(x) = Ax^k$ for some constants A and k; it may not start out written that way. Sometimes it can take a great deal of algebra to determine whether or not a function is a power function "in disguise." For example, is the function

$$f(x) = \frac{(2x)^3 + 8x^{17/5}}{x^{2/5} + 1}$$

a power function? If so, what are the numbers A and k? Surprisingly, the answer is yes; see Example 2.

Integer Power Functions

The simplest power functions are those whose powers are integers. It turns out that we can classify such power functions into four types. The next theorem illustrates these types for $f(x) = Ax^k$ when $A = 1$. The graphs for $A > 0$ are similar, and the graphs for $A < 0$ are reflected over the x-axis.

THEOREM 4.7   **Integer Power Functions**

If k is an integer with $k \neq 1$ and $k \neq 0$, then the graph of $f(x) = x^k$ has the following shape, domain, range, roots, and asymptotes, depending on whether k is even or odd, and positive or negative:

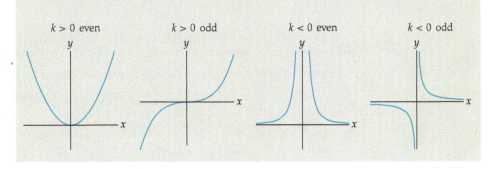

For example, the first graph could be the graph of $f(x) = x^2$ or $f(x) = x^6$. The second graph could be the graph of $f(x) = x^5$ or $f(x) = x^{13}$. The third and fourth graphs could be the graphs of $f(x) = x^{-4}$ and $f(x) = x^{-7}$, respectively.

Proof. We will prove this theorem in the case where k is a positive even integer and leave the remaining three cases to Exercises 87–89. Our proof will basically be a curve-sketching analysis, but for a general even positive integer power k instead of for a specific power. There are quite a few things that we must show about $f(x) = x^k$ when k is even and positive, including that it

▶ has domain $(-\infty, \infty)$ and range $[0, \infty)$;

▶ is continuous and differentiable everywhere;

▶ is an even function; that is, has y-axis symmetry;

▶ has a root and a global minimum at $x = 0$ and has no inflection points;

▶ is positive on $(-\infty, 0) \cup (0, \infty)$;

▶ is decreasing on $(-\infty, 0)$ and increasing on $(0, \infty)$;

▶ is concave up on $(-\infty, \infty)$; and

▶ increases without bound as $x \to \infty$ and as $x \to -\infty$.

We'll handle these bullet points in order. Since k is positive, $f(x) = x^k$ is defined everywhere. Because k is even, $f(x) = x^k$ will be positive for all x, except at $x = 0$, where $f(0) = 0^k = 0$. Therefore f has domain $(-\infty, \infty)$ and range $[0, \infty)$.

Since power functions are continuous on their domains, f is continuous everywhere. Because k is a positive even integer, it follows that $k \geq 2$ and therefore $k - 1 \geq 1$. This means that $f'(x) = kx^{k-1}$ is defined everywhere and thus that f is differentiable everywhere.

Since k is an even integer, $f(-x) = (-x)^k = x^k$ and therefore f is an even function and has y-axis symmetry.

A number line analysis will take care of all but the last remaining bullet point. We have $f(x) = x^k$, $f'(x) = kx^{k-1}$, and $f''(x) = k(k-1)x^{k-2}$. Since all three of these functions are power functions with positive integer powers, they are all defined everywhere and are zero at exactly one point, namely, $x = 0$. (There is one exception: If $k = 2$, then $f''(0) = 2$ is not zero. That case is simple and can be handled separately by the reader.) Using the fact that k is even, $k - 1$ is odd, and $k - 2$ is even, we obtain the following number lines:

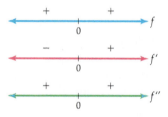

From these number lines we can see that $f(x) = x^k$ has the listed root, minimum, inflection point, and positive/negative, decreasing/increasing, and concave up/down behavior.

For the last bullet point, we need to compute some limits. Since k is positive and even,

$$\lim_{x \to \infty} x^k \to \infty^k \to \infty \quad \text{and} \quad \lim_{x \to -\infty} x^k \to (-\infty)^k \to \infty.$$

Therefore $f(x) = x^k$ increases without bound at both ends, as desired.   ▪

Graphical Effects of Exponents and Coefficients

The sign and the parity of the integer power k together determine the general shape of a power function $f(x) = x^k$. In addition, the *size* of k determines the steepness of the curve.

For example, in the first figure that follows, $f(x) = x^2$, $f(x) = x^4$, and $f(x) = x^6$ all have the same basic "U" shape, but the red graph of $f(x) = x^4$ is steeper on the ends and flatter in the middle than the blue graph of $f(x) = x^2$, and the green graph of $f(x) = x^6$ is steeper and flatter still. In each of the following figures, the largest value of $|k|$ has the steepest and flattest curve:

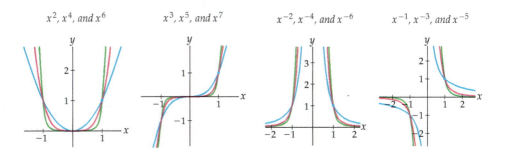

$x^2, x^4,$ and x^6 $x^3, x^5,$ and x^7 $x^{-2}, x^{-4},$ and x^{-6} $x^{-1}, x^{-3},$ and x^{-5}

Note that each of the four graphs shown goes through the point $(1, 1)$, because if $f(x) = x^k$, then $f(1) = 1^k = 1$ regardless of the sign or the parity of the power k. This means that none of the graphs are a vertical or horizontal stretch of one another. For example, the graph of $y = x^4$ is not a vertical or horizontal stretch of the graph of $y = x^2$.

Multiplying $f(x) = x^k$ by a positive constant $A > 1$ produces a vertical stretch, multiplying by $0 < A < 1$ causes a vertical compression of x^k, and multiplying by a negative number A will in addition reflect the graph over the x-axis. In each figure that follows we see the graphs of $f(x) = x^k$, $f(x) = 0.5x^k$, and $f(x) = 2x^k$ for some value of k. In each case the graph of $f(x) = x^k$ is the blue middle graph, the graph of $f(x) = 2x^k$ is the green graph with heights doubled, and the graph of $f(x) = 0.5x^k$ is the red graph whose heights are halved.

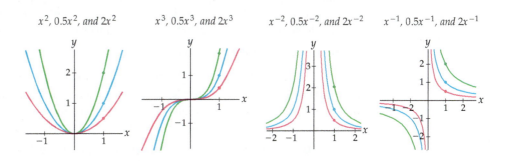

$x^2, 0.5x^2,$ and $2x^2$ $x^3, 0.5x^3,$ and $2x^3$ $x^{-2}, 0.5x^{-2},$ and $2x^{-2}$ $x^{-1}, 0.5x^{-1},$ and $2x^{-1}$

Power Functions with Rational Powers

Things get a little more complicated when we consider graphs of power functions $f(x) = x^k$ where k is a reduced rational number $\frac{p}{q}$. For example, when q is even, $x^{p/q}$ will be undefined for negative values of x, since we cannot take an even root of a negative number. For positive rational powers $\frac{p}{q}$ there are six possible types of graphs that a power function $x^{p/q}$ can have:

THEOREM 4.8

Power Functions with Positive Rational Powers

If $\frac{p}{q}$ is a positive reduced rational number, then the graph of $f(x) = x^{p/q} = (\sqrt[q]{x})^p$ has the following shape, domain, range, roots, and asymptotes, depending on the parity of p and q and whether $\frac{p}{q}$ is greater than or less than 1:

(a) If $0 < \frac{p}{q} < 1$, then $f(x) = x^{p/q}$ has one of the following three types of graphs:

p even, q odd p odd, q odd q even

(b) If $\frac{p}{q} > 1$, then $f(x) = x^{p/q}$ has one of the following three types of graphs:

p even, q odd p odd, q odd q even

For example, the first row of graphs could be the graphs of $y = x^{2/3}$, $y = x^{1/3}$, and $y = x^{1/2}$, from left to right. The second row of graphs could be $y = x^{4/3}$, $y = x^{5/3}$, and $y = x^{3/2}$. We can prove these graph shapes with case-by-case curve-sketching analyses, just as we did for Theorem 4.7. We leave the details to Exercises 90–92. In Examples 4 and 5 we will investigate limit and derivative calculations that involve rational non-integer powers.

So far, we have examined power functions $f(x) = x^{p/q}$ where $\frac{p}{q}$ is positive. What happens when the power is negative? By properties of exponents we have $f(x) = x^{-p/q} = \frac{1}{x^{p/q}}$, and thus the function $f(x) = x^{-p/q}$ is the reciprocal of the function $g(x) = x^{p/q}$. This means that the height of $f(x)$ at any point x is simply the reciprocal of the height of $g(x)$ at that point x. For example, wherever $g(x)$ is equal to 1, $f(x)$ will also equal $\frac{1}{1} = 1$. Whenever $g(x)$ is large, $f(x)$ will be small; and whenever $g(x)$ is small, $f(x)$ will be large. Finally, if $g(x)$ is zero, then $f(x)$ will have a vertical asymptote, and vice versa, as illustrated in the following three figures:

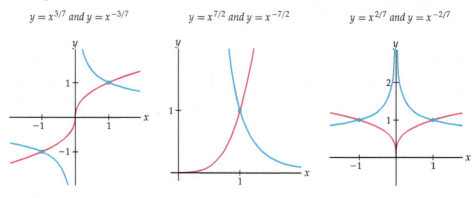

$y = x^{3/7}$ and $y = x^{-3/7}$ $y = x^{7/2}$ and $y = x^{-7/2}$ $y = x^{2/7}$ and $y = x^{-2/7}$

Therefore just as there are six possible types of graphs for positive rational power functions, there are six possible types of graphs for negative rational power functions, obtained by taking reciprocals of graphs.

Examples and Explorations

EXAMPLE 1

Identifying power functions

Which of the following are power functions?

(a) $3x^2$　　　**(b)** $4x^{\sqrt{2}}$　　　**(c)** $5x^{3/2}$　　　**(d)** πx^8　　　**(e)** $3(5^x)$

SOLUTION

All of the given functions are power functions, except for (b) and (e). The function $4x^{\sqrt{2}}$ is not considered a power function because its exponent is an irrational number. The function $3(5^x)$ cannot be written in the form Ax^k because the variable x is in the exponent, rather than the base, of the expression. The remaining expressions are written in the form Ax^k for some real number A and rational number k.　□

EXAMPLE 2

Writing an expression in the form of a power function

Write each of the following functions in the form Ax^k:

(a) $f(x) = \dfrac{(2x)^3 + 8x^{17/5}}{x^{2/5} + 1}$　　　　**(b)** $g(x) = \dfrac{x^{1/6}x^{4/3}}{(x^4)^{1/3}}$

SOLUTION

(a) We can expand, factor, and then make a lucky cancellation to write this expression in the form Ax^k:

$$f(x) = \frac{(2x)^3 + 8x^{17/5}}{x^{2/5} + 1} = \frac{8x^3 + 8x^{17/5}}{x^{2/5} + 1} = \frac{8x^3(1 + x^{(17/5)-3})}{x^{2/5} + 1} = \frac{8x^3(1 + x^{2/5})}{x^{2/5} + 1} = 8x^3.$$

Since $1 + x^{2/5}$ is never zero, the cancellation in the last step is valid for all x. Therefore f is a power function with $A = 8$ and $k = 3$.

(b) Simplifying with the rules of exponents, we have

$$g(x) = \frac{x^{1/6}x^{4/3}}{(x^4)^{1/3}} = \frac{x^{(1/6+4/3)}}{x^{4/3}} = \frac{x^{3/2}}{x^{4/3}} = x^{(3/2-4/3)} = x^{1/6}.$$

The final expression is defined for $x = 0$, but the initial expression is not. For $x > 0$, the function $g(x)$ is the same as the power function with $A = 1$ and $k = \dfrac{1}{6}$.　□

EXAMPLE 3

Finding domains of power functions

Find the domains of the following power functions:

(a) $f(x) = 3x^{-2}$　**(b)** $g(x) = \dfrac{1}{2}x^{3/4}$　**(c)** $h(x) = 2x^{4/5}$　**(d)** $k(x) = 8x^{-1/2}$

SOLUTION

(a) If we rewrite the power function so that any roots or denominators are visible, the calculation of the domain is obvious: $f(x) = 3x^{-2} = \dfrac{3}{x^2}$ is defined everywhere but at $x = 0$, so its domain is $(-\infty, 0) \cup (0, \infty)$.

(b) The function $g(x) = \dfrac{1}{2}x^{3/4} = \dfrac{(\sqrt[4]{x})^3}{2}$ is defined everywhere except $x < 0$, so its domain is $[0, \infty)$.

(c) The function $h(x) = 2x^{-4/5} = \dfrac{2}{(\sqrt[5]{x})^4}$ is defined for all $x \neq 0$, so its domain is $(-\infty, 0) \cup (0, \infty)$.

(d) The function $k(x) = 8x^{-1/2} = \dfrac{8}{\sqrt{x}}$ fails to be defined for $x = 0$ and for $x < 0$, so its domain is $(0, \infty)$.   ☐

CHECKING THE ANSWER

We can use calculator graphs to verify the domains we just found, as shown in the diagrams that follow. In each case compare the domain we found with the values on which the graph is defined.

$f(x) = 3x^{-2}$

$g(x) = \dfrac{1}{2}x^{-3/4}$

$h(x) = 2x^{-4/5}$

$k(x) = 8x^{-1/2}$

EXAMPLE 4

Limits of power functions and combinations of power functions

Calculate each of the following limits:

(a) $\displaystyle\lim_{x \to 0} x^{-1/2}$ (b) $\displaystyle\lim_{x \to -\infty} x^{1/3}$ (c) $\displaystyle\lim_{x \to \infty} (x^{2/3} - x^{3/4})$

SOLUTION

(a) As $x \to 0$, $x^{-1/2} = \dfrac{1}{\sqrt{x}} \to \dfrac{1}{0}$, so this limit does not exist and in fact must be infinite. Investigating the limit from the right and the left, we have

$$\lim_{x \to 0^+} x^{-1/2} = \lim_{x \to 0^+} \frac{1}{\sqrt{x}} \to \frac{1}{0^+} \to \infty,$$

$$\lim_{x \to 0^-} x^{-1/2} = \lim_{x \to 0^-} \frac{1}{\sqrt{x}} \text{ does not exist, since } \sqrt{x} \text{ is not defined for } x < 0.$$

Therefore $\displaystyle\lim_{x \to 0} x^{-1/2}$ does not exist. The graph of $y = x^{-1/2}$ approaches ∞ as x approaches zero from the right. The graph is not defined for $x < 0$.

(b) Since the cube root of a negative number is a negative number, it follows that

$$\lim_{x \to -\infty} x^{1/3} = \lim_{x \to -\infty} \sqrt[3]{x} \to \sqrt[3]{-\infty} \to -\infty.$$

Therefore the graph of $y = x^{1/3}$ decreases without bound as $x \to -\infty$.

(c) Since $\dfrac{2}{3}$ and $\dfrac{3}{4}$ are positive powers, as $x \to \infty$ we have $x^{2/3} \to \infty$ and $x^{3/4} \to \infty$. Therefore the limit in question is of the indeterminate form $\infty - \infty$. Limits of this form can often be resolved by factoring. We have

$$x^{2/3} - x^{3/4} = x^{2/3}(1 - x^{1/12}),$$

and as $x \to \infty$ we have $x^{2/3} \to \infty$ and $(1 - x^{1/12}) \to -\infty$. Therefore the limit in question is now of the form $(\infty)(-\infty)$, which means that

$$\lim_{x \to \infty} (x^{2/3} - x^{3/4}) = \lim_{x \to \infty} x^{2/3}(1 - x^{1/12}) = -\infty$$

This answer makes intuitive sense because the original limit $\displaystyle\lim_{x \to \infty} (x^{2/3} - x^{3/4})$ was of the indeterminate form $\infty - \infty$, and since $3/4 > 2/3$ we might expect that $x^{3/4}$ would approach ∞ faster than $x^{2/3}$ would.   ☐

EXAMPLE 5 Using the definition of the derivative to examine a non-differentiable point

Investigate the derivative of $f(x) = x^{2/3}$ at $x = 0$.

SOLUTION

If $f(x) = x^{2/3}$, then by the definition of derivative,

$$f'(0) = \lim_{h \to 0} \frac{(0+h)^{2/3} - 0^{2/3}}{h} = \lim_{h \to 0} \frac{h^{2/3}}{h} = \lim_{h \to 0} \frac{1}{h^{1/3}} \to \frac{1}{0},$$

which does not exist. At this point we already know that f fails to be differentiable at $x = 0$, but we can say more. Specifically, from the right and the left we have

$$f'_+(0) = \lim_{h \to 0^+} \frac{(0+h)^{2/3} - 0^{2/3}}{h} = \lim_{h \to 0^+} \frac{1}{h^{1/3}} \to \frac{1}{0^+} \to \infty,$$

$$f'_-(0) = \lim_{h \to 0^-} \frac{(0+h)^{2/3} - 0^{2/3}}{h} = \lim_{h \to 0^-} \frac{1}{h^{1/3}} \to \frac{1}{0^-} \to -\infty.$$

The graph of f must therefore have a vertical cusp at $x = 0$.   □

 CHECKING THE ANSWER

A graph of $f(x) = x^{2/3}$ is shown next. To verify the behavior of the left and right derivatives, look at the slopes of the secant lines from $(0, f(0))$ to $(0+h, f(0+h))$ for various small positive and negative values of h.

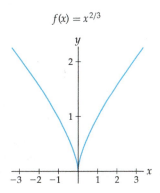

$$f(x) = x^{2/3}$$

EXAMPLE 6 Distinguishing between graphs with different sizes of exponents

The four graphs shown in the two figures that follow are the graphs of $y = x^{1/4}$, $y = x^{3/4}$, $y = x^{5/4}$, and $y = x^{11/4}$. The dotted line in each figure is the line $y = x$. Determine which of the four graphs is which, without using a calculator.

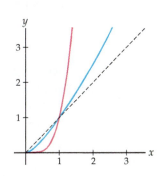

SOLUTION

Since the graphs in the first figure are concave down, they must be the graphs of power functions whose exponents are less than 1: $y = x^{1/4}$ and $y = x^{3/4}$. Because 3/4 is closer to 1 than 1/4 is, the graph of $y = x^{3/4}$ is the blue graph that is straighter and closer to the graph of $y = x$.

Now we know that the graphs in the second figure must be $y = x^{5/4}$ and $y = x^{11/4}$. Since 5/4 is closer to 1 than 11/4 is, the graph of $y = x^{5/4}$ must be the straighter blue graph that is closest to the line $y = x$.

$\square$

EXAMPLE 7   Modeling graphs with power functions

Find power functions that could have the following graphs:

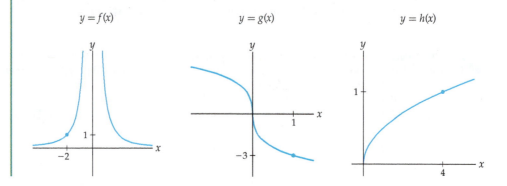

SOLUTION

(a) From the shape of the first graph $y = f(x)$, we know that if $f(x) = Ax^k$ is a power function, then k must be an even negative integer. Let's try the simplest possible case: $k = -2$. Then $f(x) = Ax^{-2}$ and $f(-2) = 1$ imply that

$$1 = A(-2)^{-2} = \frac{A}{(-2)^2} = \frac{A}{4} \quad \Longrightarrow \quad A = 4.$$

Thus $f(x) = 4x^{-2}$ is a power function that could have the first graph shown.

(b) For the middle graph $y = g(x)$, we need a power function of the form $g(x) = Ax^{p/q}$ where q is odd, p is odd, A is negative, and $\frac{p}{q} < 1$. The simplest power $k = \frac{p}{q}$ that satisfies these conditions is $\frac{p}{q} = \frac{1}{3}$. Since $g(1) = -3$, we have

$$-3 = A(1)^{1/3} = A \quad \Longrightarrow \quad A = -3.$$

Therefore $g(x) = -3x^{1/3}$ is a power function whose graph could be the middle graph.

(c) From the shape of the last graph $y = h(x)$, we see that we need a rational power function $h(x) = Ax^{p/q}$ where q is even, p is odd, and $0 < \frac{p}{q} < 1$. The simplest rational number satisfying these conditions is $\frac{p}{q} = \frac{1}{2}$. Assuming that $k = \frac{1}{2}$ and letting $h(4) = 1$, we find that:

$$1 = A(4)^{1/2} = A\sqrt{4} = 2A \implies A = \frac{1}{2}.$$

Thus $h(x) = \frac{1}{2}x^{1/2}$ is a function that could have the final graph. $\square$

? **TEST YOUR UNDERSTANDING**

▶ What is the definition of a power function?

▶ What are the four types of graphs that a power function with an integer power can have if that power is not 1 or 0? What happens if the power is 1 or 0?

▶ Why does it make sense that power functions with positive even powers are special cases of the first graph in part (b) of Theorem 4.8? Why does it make sense that power functions with positive odd powers are special cases of the second graph in that part?

▶ In Theorem 4.8, why do we never have to consider the case where p and q are both even? Can $\frac{p}{q}$ be reduced if both p and q are even?

▶ How many possible types of graphs are there for power functions of the form $f(x) = x^{-p/q}$ where $-\frac{p}{q}$ is a negative rational number, and why?

EXERCISES 4.2

Thinking Back

Toolbox review: Each of the following problems reviews a calculus concept that we will revisit in our study of functions in this and the next few chapters.

▶ What is the delta–epsilon definition of the limit statement $\lim_{x \to c} f(x) = L$? Illustrate the graphical meaning of this definition with the function $f(x) = 4 - x^2$ at the point $x = 1$.

▶ What is the N–epsilon definition of the limit statement $\lim_{x \to \infty} f(x) = L$? Illustrate the graphical meaning of this definition with the function $f(x) = \frac{2x+1}{x+1}$ as $x \to \infty$.

▶ What is the $h \to 0$ definition of the derivative of a function f at a point $x = c$? Illustrate the graphical meaning of this definition with the function $f(x) = \sqrt{x}$ at $x = 4$.

▶ What is the $z \to x$ definition of the derivative of a function f at a point $x = c$? Illustrate the graphical meaning of this definition with the function $f(x) = \frac{1}{x}$ at $x = 2$.

▶ What can number lines for f, f', and f'' tell us about the graphical properties of a function f? Illustrate how such number lines describe the graph of $f(x) = x^3 - x$.

Concepts

0. *Problem Zero:* Read the section and make your own summary of the material.

1. *True/False:* Determine whether each of the statements that follow is true or false. If a statement is true, explain why. If a statement is false, provide a counterexample.

(a) *True or False:* The product of two power functions is a power function.

(b) *True or False:* The sum of two power functions is a power function.

(c) *True or False:* Every power function is a polynomial function.

(d) *True or False:* The derivative of a power function is always a power function.

(e) *True or False:* The antiderivative of a power function is always a power function.

(f) *True or False:* Power functions are continuous everywhere on their domains.

(g) *True or False:* Power functions are differentiable everywhere on their domains.

(h) *True or False:* Graphs of power functions always pass through the origin.

2. *Examples:* Construct examples of the thing(s) described in the following. Try to find examples that are different than any in the reading.

   (a) Two graphs of functions of the form $f(x) = Ax^k$ where $k > 0$ is odd and A is a negative number.

   (b) Two graphs of functions of the form $f(x) = Ax^{p/q}$ where $p > 0$ is odd, $q > 0$ is even, $A < 0$, and $p > q$.

   (c) Two graphs of functions of the form $f(x) = Ax^{-p/q}$ where $p > 0$ and $q > 0$ are odd, $A > 0$, and $p < q$.

3. Explain why none of $f(x) = x^2 + 1$, $g(x) = (x+1)^2$, or $h(x) = x^2 + x^3$ are power functions.

4. Explain why $f(x) = \sqrt[7]{3x^2}$ and $g(x) = \dfrac{1}{2x^{17}}$ are power functions.

5. Sketch graphs of $y = -2x^3$ and $y = 3x^{-2}$ by hand, without using a calculator. Label three points on each graph.

6. Sketch graphs of $y = -3x^{1/2}$ and $y = x^{2/3}$ by hand, without using a calculator. Label three points on each graph.

7. Sketch graphs of $y = x^2$, $y = x^4$, and $y = x^6$ by hand, without a calculator, on the same set of axes. Label three points on each graph.

8. Sketch graphs of $y = x^3$, $y = x^5$, and $y = x^7$ by hand, without a calculator, on the same set of axes. Label three points on each graph.

9. If $f(x) = Ax^k$ is a power function, is its reciprocal $\dfrac{1}{f(x)}$ also a power function? If so, write the reciprocal in the form Cx^r for some real numbers C and r.

10. Suppose $f(x) = Ax^k$ is a power function, where k is an integer.

    (a) What must be true about k for f to be one-to-one, and why?

    (b) If f is one-to-one, is its inverse f^{-1} also a power function? If so, write the inverse in the form $f^{-1}(x) = Cx^r$ for some real numbers C and r.

11. For which values of x is the function $f(x) = x^{1/2}$ continuous, and why? In what sense is $x^{1/2}$ continuous at $x = 0$? In what sense is it not?

12. The function $f(x) = 2x^{-4}$ is continuous at $x = 3$. Write this fact as (a) a limit statement and (b) a δ-ϵ statement.

13. The function $f(x) = 3x^{1/3}$ is continuous at $x = 0$. Write this fact as (a) a limit statement and (b) a δ-ϵ statement.

14. The function $f(x) = \sqrt[6]{x}$ is not continuous at $x = -1$. Write this fact as (a) a limit statement and (b) a δ-ϵ statement.

15. Is the derivative of a power function always a power function? Is the antiderivative of a power function always a power function? Justify your answer.

16. Let $f(x) = 3\sqrt{x}$. In this exercise you will approximate the slope of the tangent line to f at $x = 4$ by calculating the slopes of approaching secant lines.

    (a) Calculate the slope of the secant lines from $(4, f(4))$ to $(4+h, f(4+h))$ for the following values of h: $h = 0.5$, $h = 0.25$, $h = 0.1$, $h = 0.01$.

    (b) What value do these slopes seem to be approaching, and what does this mean about the slope of the tangent line to $f(x) = 3\sqrt{x}$ at $x = 4$?

   (c) Sketch a graph of f together with each of the secant lines whose slopes you computed in part (a), and illustrate the fact that the slopes of these secant lines are indeed approaching the slope of the tangent line to f at $x = 4$.

   (d) Use the power rule to find the exact value of $f'(4)$, and compare this exact answer with your approximations in part (a).

17. Find a function whose derivative is $4\sqrt{x}$. Then find two more functions whose derivative is $4\sqrt{x}$. Graph all of these functions on the same set of axes. What do you notice about the graphs?

Use mathematical language to list all the graphical properties of the following two graphs:

18.   19.

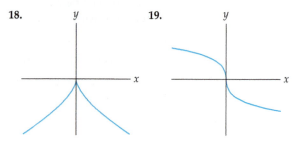

20. Give brief, intuitive explanations for the following statements concerning a power function $f(x) = Ax^{p/q}$ where $\dfrac{p}{q}$ is a positive, reduced, rational number:

    (a) If q is even, then $f(x)$ will be defined only for $x \geq 0$.

    (b) If p is even, then the range of f will have only non-negative values.

    (c) If q is odd and p is even, then f will be an even function.

    (d) If both p and q are odd, then f will be an odd function.

21. If $A > 0$, then $y = Ax^k$ is a vertical stretch or shrink of $y = x^k$ by a factor of A. In the case of power functions this vertical stretch or shrink can also be thought of as a *horizontal* stretch or shrink. You may need to review Section 0.6 to do this problem.

    (a) If k is a positive integer, then $y = Ax^k$ is a horizontal stretch or shrink of $y = x^k$ by what factor?

    (b) If k is a positive integer, then $y = Ax^{-k}$ is a horizontal stretch or shrink of $y = x^{-k}$ by what factor?

22. Sometimes, knowing the general shape and two points other than the origin on the graph of a power function completely determines the function. However, in certain special cases this is not true. Sketch graphs with labeled points satisfying the following criteria:

    (a) The graph of a power function $f(x) = Ax^k$, where k is even and positive and two points other than the origin are known, but there is not enough information to determine both A and k.

    (b) The graph of a power function $f(x) = Ax^k$, where k is odd and negative and two points other than the origin are known, but there is not enough information to determine both A and k.

Skills

Which of the functions in Exercises 23–30 are power functions? Write each power function in the form $f(x) = Ax^k$.

23. $f(x) = \dfrac{1}{\sqrt{x^{3/5}}}$

24. $f(x) = x^{3/2} + x^{5/3}$

25. $f(x) = \dfrac{x^{-3} - x^{-2}}{x^{-1} - 1}$

26. $f(x) = \left(\dfrac{1}{x^{-1}}\right)^{-2}$

27. $f(x) = 3\sqrt{2^x}$

28. $f(x) = \dfrac{x^3 - x^2}{x - 1}$

29. $f(x) = \dfrac{x^{-2}x^3}{x^5(4x^{-3})^2}$

30. $f(x) = 5\sqrt[3]{x} + 3\sqrt[5]{x}$

Sketch rough graphs of the functions in Exercises 31–42 by hand. Identify any roots, holes, or asymptotes, and label the coordinates $(1, f(1))$ and $(-1, f(-1))$ on each graph.

31. $f(x) = -3x^{5/3}$

32. $f(x) = -5x^{-3}$

33. $f(x) = 0.5x^{4/5}$

34. $f(x) = x^{-2/3}$

35. $f(x) = 2x^{4/3}$

36. $f(x) = -1.3x^{-0.25}$

37. $f(x) = -\dfrac{1}{3}x^{-3/7}$

38. $f(x) = -\dfrac{1}{3}x^{3/7}$

39. $f(x) = \sqrt[3]{x+1}$

40. $f(x) = (x+3)^{2/3} - 2$

41. $f(x) = \dfrac{1}{16 - x^4}$

42. $f(x) = x^4 - 6x^2 + 9$

In Exercises 43–46, list three power functions that could have the graphs shown. Keep close attention to the relative shape and positions of the graphs.

43.

44.

45.

46.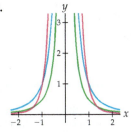

For each graph in Exercises 47–52, find a function whose graph looks like the one shown and passes through any marked points.

47.

48.

49.

50.

51.

52.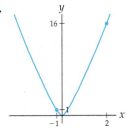

Calculate each of the limits in Exercises 53–62. If a limit does not exist, examine the left and right limits and describe the behavior of the graph of the function.

53. $\displaystyle\lim_{x \to -\infty} -3x^{4/3}$

54. $\displaystyle\lim_{x \to 0} 2x^{-3/4}$

55. $\displaystyle\lim_{x \to \infty} -2x^{-3/4}$

56. $\displaystyle\lim_{x \to -\infty} 3x^{-3/4}$

57. $\displaystyle\lim_{x \to -\infty} -5x^{3/4}$

58. $\displaystyle\lim_{x \to 0}(x^{-2} - x^{-1})$

59. $\displaystyle\lim_{x \to -\infty}(2x^{2/3} - 3x^2)$

60. $\displaystyle\lim_{x \to \infty}(3x^{3/2} - 2x^{5/3})$

61. $\displaystyle\lim_{x \to \infty} \dfrac{x^{-3}}{x^2 - x^{-1}}$

62. $\displaystyle\lim_{x \to -\infty}(\sqrt{-3x} + 5x^3)$

Use (a) the $h \to 0$ definition of the derivative and then (b) the $z \to c$ definition of the derivative to find $f'(c)$ for each function f and value $x = c$ in Exercises 63–68. If a function is not differentiable at $x = c$, then examine the left and right derivatives and describe the behavior of the graph at $x = c$.

63. $f(x) = x^{1/2},\ x = 9$

64. $f(x) = x^{-1/2},\ x = 9$

65. $f(x) = x^{6/5},\ x = 0$

66. $f(x) = x^{1/3},\ x = 0$

67. $f(x) = x^{1/3},\ x = 0$

68. $f(x) = x^{2/5},\ x = 0$

Use the definition of the derivative to find f' for each function f in Exercises 69–72.

69. $f(x) = 3\sqrt{x}$

70. $f(x) = \dfrac{3}{\sqrt{x}}$

71. $f(x) = \sqrt{2x+1}$

72. $f(x) = \dfrac{1}{\sqrt{2x+1}}$

Find the locations and values of any global extrema of each function f in Exercises 73–75 on each of the four given intervals. Do all work by hand by considering local extrema and endpoint behavior. Afterwards, check your answers with graphs.

73. $f(x) = x^{3/2}(3x - 5)$, on the intervals

   (a) $[0, 4]$ (b) $[0, 4)$ (c) $[0, 1)$ (d) $(0, 1)$

74. $f(x) = x^{2/3} - x$, on the intervals

   (a) $[0, 1]$ (b) $(0, 1)$ (c) $\left(0, \dfrac{1}{2}\right)$ (d) $[0, \infty)$

75. $f(x) = (x^2 - 4x + 3)^{-1/2}$, on the intervals

   (a) $[0, 4]$ (b) $[0, 10]$ (c) $[0, 3.5]$ (d) $(3, \infty)$

In Exercises 76–79, find a function f that has the given derivative and value. Check your answers by verifying that your function f indeed has the derivative and value given.

76. $f'(x) = \dfrac{1}{x^3}$, $f(1) = 2$

77. $f'(x) = \dfrac{1}{\sqrt{x^3}}$, $f(9) = 1$

78. $f'(x) = \sqrt{x}(1 + x^{3/2})$, $f(0) = 4$

79. $f'(x) = \dfrac{3}{2}x^{-4/5}$, $f(1) = 2$

Applications

Suppose an object moves in a straight line in such a way that its position from a marking flag t seconds after starting is $s(t)$ meters.

80. If the position of the object at time t is $s(t) = \dfrac{3.2}{x^3}$, find equations for (a) the velocity $v(t)$ and (b) the acceleration $a(t)$ of the object.

81. If the velocity of the object at time t is $v(t) = 3t^{-1/2}$ meters per second and the initial position of the object is 100 meters from the marking flag, find equations for (a) the position $s(t)$ and (b) the acceleration $a(t)$ of the object.

82. If the acceleration of the object at time t is $a(t) = 1.3t^{0.24}$, and the initial position of the object is 20 meters from the marking flag, and the initial velocity of the object is 2.5 meters per second, find equations for (a) the velocity $v(t)$ and (b) the position $s(t)$ of the object.

Proofs

83. Prove that the composition of two power functions is a power function.

84. Prove that the product of two power functions is a power function.

85. Prove that the sum of two power functions is not always a power function.

86. Prove (with a, M, N argument) that $\lim\limits_{x \to \infty} \sqrt[3]{x} = \infty$. You may assume that $b > a \Rightarrow \sqrt[3]{b} > \sqrt[3]{a}$ for all real numbers a and b.

For each type of power function in Exercises 87–92, do a general curve-sketching analysis to prove that the graph of that type of power function always has the shape and features described in Theorem 4.7 or Theorem 4.8.

87. $f(x) = x^k$, where k is a positive odd integer.

88. $f(x) = x^{-k}$, where $-k$ is a negative even integer.

89. $f(x) = x^{-k}$, where $-k$ is a negative odd integer.

90. $f(x) = x^{p/q}$, where p is an odd positive integer, q is an odd positive integer, and $0 < \dfrac{p}{q} < 1$.

91. $f(x) = x^{p/q}$, where p is an odd positive integer, q is an even positive integer, and $0 < \dfrac{p}{q} < 1$.

92. $f(x) = x^{p/q}$, where p is an even positive integer, q is an odd positive integer, and $\dfrac{p}{q} > 1$.

For each power function f in Exercises 93–96, write a delta–epsilon proof which shows that f is continuous on its domain. You may need to assume that δ is less than or equal to 1 and/or multiply by a conjugate.

93. $f(x) = x^{-2}$

94. $f(x) = x^3$

95. $f(x) = x^{1/2}$

96. $f(x) = x^{-1/2}$

Thinking Forward

Separable differential equations: Some differential equations can be solved by **separating** them so that they are of the form $g(y)y' = h(x)$ and then antidifferentiating both sides of that equation with respect to x. Notice that the form of the left-hand side of the equation allows a reverse chain rule calculation.

▶ Use the technique of separation of variables to find a function $y(x)$ that satisfies the differential equation $\sqrt[3]{x}\,y' = 2$.

▶ Use the technique of separation of variables to find a function $y(x)$ that satisfies the differential equation $y' = \sqrt{\dfrac{x}{y}}$.

▶ Use the technique of separation of variables to find a function $y(x)$ that satisfies the differential equation $2xyy' = 1$ and the initial condition $y(1) = 2$.

▶ Use the technique of separation of variables to find a function $y(x)$ that satisfies the differential equation $(xy)^{2/3}y' = x$ and the initial condition $y(1) = 1$.

4.3   POLYNOMIAL FUNCTIONS

▶ Polynomials and their coefficients, leading terms, and degrees

▶ Local and global properties of polynomial functions

▶ Graphing polynomial functions and modeling graphs with polynomials

Polynomial Functions

A ***polynomial function*** is a finite sum of power functions that have nonnegative integer powers. The following definition provides a general notation:

DEFINITION 4.9

Polynomial Functions

A ***polynomial function*** of degree n is a function that can be written in the form

$$f(x) = a_n x^n + a_{n-1} x^{n-1} + a_{n-2} x^{n-2} + \cdots + a_2 x^2 + a_1 x + a_0$$

for some integer $n \geq 0$ and some real numbers $a_0, a_1, \ldots a_n$ with $a_n \neq 0$.

As a matter of convention, we also say that the constant zero function $f(x) = 0$ is a polynomial function (and that its degree is undefined).

The numbers a_i (for $i = 0, 1, 2, \ldots, n$) are called the ***coefficients*** of the polynomial. Note that the coefficient belonging to the term containing the power x^i is conveniently named a_i; for example, the coefficient of the x^2 term is called a_2. The coefficient a_n belonging to the highest power of x is called the ***leading coefficient***, and the term $a_n x^n$ containing the highest power of x with $a_n \neq 0$ is called the ***leading term***. The coefficient a_0 is called the ***constant term***.

We have special names for some polynomials according to their degrees. For example, polynomials of degrees 0, 1, 2, 3, 4, and 5 are called ***constant***, ***linear***, ***quadratic***, ***cubic***, ***quartic***, and ***quintic*** polynomials, respectively. (Note: We will require non-zero leading coefficients for quadratics and higher degrees, but not for linear functions; in other words, we will consider constant functions to be linear.) Higher degrees can sometimes result in more roots and more turning points in the graph of a polynomial; for example, examine the following cubic, quartic, and quintic polynomials:

$$f(x) = x^3 + x^2 - 2x$$
$$= x(x-1)(x+2)$$

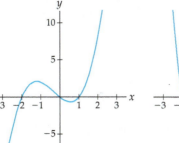

$$g(x) = x^4 + x^3 - 2x^2$$
$$= x^2(x-1)(x+2)$$

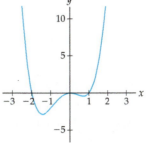

$$h(x) = x^5 - x^4 - 4x^3 + 4x^2$$
$$= x^2(x-1)(x+2)(x-2)$$

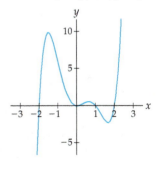

A quadratic expression is **_irreducible_** if it cannot be factored with real-number coefficients; that is, $ax^2 + bx + c$ is irreducible if it cannot be written in the form $a(x - r_1)(x - r_2)$ for some real numbers r_1 and r_2. A quadratic expression $ax^2 + bx + c$ is irreducible if and only if its **_discriminant_** $b^2 - 4ac$ is negative; think about the quadratic formula to see why. For example, the quadratic expressions $x^2 + 5$ and $x^2 + x + 7$ are irreducible.

Note that every linear factor $(x - r)$ of a polynomial corresponds to a root $x = r$ of that polynomial, since if a polynomial $f(x)$ can be factored as $f(x) = (x - r)g(x)$ for some other polynomial $g(x)$, then $f(r) = (r - r)g(r) = 0$ and thus $x = r$ is a root of $f(x)$. It happens that every polynomial function can be factored into linear factors (which correspond to real-number roots) and/or irreducible quadratic factors (which do not correspond to real-number roots). However, just because a polynomial _has_ a factorization doesn't mean that we have an easy way to actually factor that polynomial!

Properties of Polynomial Functions

The next theorem describes four key graphical properties of polynomial functions. The first part of Theorem 4.10 is related to the **_Fundamental Theorem of Algebra_**, and the proof of this deep theorem is beyond the scope of this course.

THEOREM 4.10

Graphical Properties of Polynomial Functions

If f is a polynomial function of degree n, then the graph of f

(a) has at most n real roots;

(b) has at most $n - 1$ local extrema; which we will call **_turning points_**

(c) is continuous and differentiable on $\mathbb{R}$ and has no asymptotes;

(d) behaves like the graph of its leading term at the "ends" of the graph.

The last part tells us that a polynomial function $f(x) = a_n x^n + a_{n-1} x^{n-1} + \cdots + a_1 x + a_0$ with $a_n \neq 0$ will behave like the power function $a_n x^n$ at its "ends." This means that the "ends" of the graph of a polynomial always looks like one of the four graphs that follow, depending on whether the degree n is even or odd and whether the leading coefficient a_n is positive or negative. The dashed part of each graph indicates that this part of the theorem does not tell us the behavior in the middle of the graph, only at the ends.

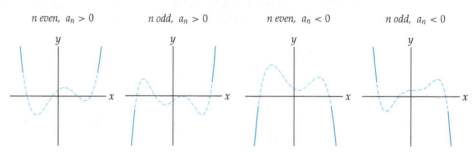

n even, $a_n > 0$ n odd, $a_n > 0$ n even, $a_n < 0$ n odd, $a_n < 0$

Proof. Part (a) follows from the fact that all polynomials factor into linear and/or irreducible quadratic factors and the fact that each linear factor contributes one real root.

The proof of part (b) is really an application of part (a). If f is a polynomial of degree $n > 0$, then its derivative f' will be a polynomial of degree $n - 1$ and thus by part (a) can have at most

$n - 1$ real roots. Since polynomials are continuous and differentiable everywhere, the only possible turning points of f are at the roots of f'; therefore there can be at most $n - 1$ turning points of f.

Part (c) follows directly from the fact that polynomials are sums of power functions with non-negative integer powers and the fact that continuity and differentiability are preserved by sums. Since nonnegative integer power functions are continuous and differentiable on $\mathbb{R}$, polynomial functions are as well. Because nonnegative power functions have no vertical asymptotes, neither does a polynomial sum of such functions. The fact that polynomials have no horizontal asymptotes follows from the last part of the theorem.

To prove part (d) we must show that the limits of a polynomial $f(x) = a_n x^n + a_{n-1} x^{n-1} + \cdots + a_0$ as $x \to \pm\infty$ are the same as the limits of its leading term $a_n x^n$ as $x \to \pm\infty$. The proof will be a general version of the limit calculation in Example 4 of Section 1.6. The key is to factor out the leading term:

$$\lim_{x \to \infty} (a_n x^n + a_{n-1} x^{n-1} + \cdots + a_0) = \lim_{x \to \infty} \left(a_n x^n \left(1 + \frac{a_{n-1}}{a_n x} + \cdots + \frac{a_0}{a_n x^n} \right) \right)$$
$$= (\lim_{x \to \infty} a_n x^n)(1 + 0 + \cdots + 0) = \lim_{x \to \infty} a_n x^n. \qquad \blacksquare$$

A polynomial of degree n might have fewer than n real roots and $n - 1$ turning points. For example, all three polynomials shown next have degree 5, but none of them have the maximum number of five real roots. If we shifted the first graph down about two units, then we would have an example with five real roots. As for turning points, the first graph has the maximum of $5 - 1 = 4$ turning points, but the second and third graphs have fewer turning points.

$$f(x) = x^5 - 5x^3 + 4x + 2 \qquad\qquad g(x) = x^5 - 4x^3 + 1 \qquad\qquad h(x) = x^5 - 5x^4 + 10x^3 - 10x^2 + 5x$$

  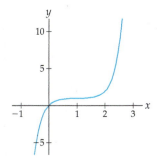

We have just seen that the degree of a polynomial determines an upper bound for the number of its turning points. The roots of the polynomial can give us a lower bound. By Rolle's Theorem, between every two roots of a polynomial f must lie at least one root of its derivative f'. In fact, from the proof of Rolle's Theorem in Section 3.1 we can guarantee even more: Between each adjacent pair of roots of f there must be some root of f' at which the graph of f actually turns around and has a local maximum or minimum value. This property proves the following useful theorem:

THEOREM 4.11   **Turning Points Between Roots of Polynomials**

If f is a polynomial function of degree n with k real roots, then f has at least $k - 1$ turning points, with at least one turning point between each root.

As a corollary to this theorem, notice that if a polynomial of degree n splits into n distinct linear factors, then it will have the maximum number n of roots and also the maximum number $n - 1$ of turning points, one between and/or at each root. If the polynomial splits into k linear factors, some of which are repeated, then it will have exactly $k - 1$ turning points, one between each root. Knowing how many turning points can exist makes graphing such polynomials easy; see Example 2.

Repeated Roots

When a linear factor $x - c$ occurs more than once in the factorization of a polynomial f, we say that f has a ***repeated root*** at $x = c$. We say that repeated roots are ***double roots***, ***triple roots***, and so on, according to the number of times they are repeated. For example, the function $f(x) = (x + 1)(x - 2)^2(x - 6)$ has a double root at $x = 2$ and the polynomial $g(x) = (x + 1)^3(x - 2)^4$ has a triple root at $x = -1$ and a quadruple root at $x = 2$. Notice in the graphs of these functions that there is some interesting behavior at the repeated roots:

$$f(x) = (x + 1)(x - 2)^2(x - 6)$$

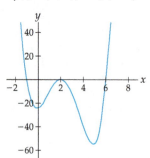

$$g(x) = (x + 1)^3(x - 2)^4$$

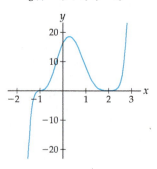

In the first figure, the graph of f does not go straight through the x-axis at the double root $x = 2$; instead it bounces off, much like an upside-down graph of x^2. In the second graph, the triple root at $x = -1$ behaves like x^3 and the quadruple root behaves like x^4. This phenomenon happens in general:

THEOREM 4.12

Repeated Roots of Polynomials

If f is a polynomial function with a k-times repeated root at $x = c$, then the graph of f behaves like the graph of $\pm(x - c)^k$ near $x = c$. Specifically, $f(c), f'(c), f''(c), \ldots f^{[k-1]}(c)$ are all zero, but $f^{[k]}(c)$ is not zero.

Proof. Suppose f is a polynomial with a k-times repeated root at $x = c$. Then we can write

$$f(x) = (x - c)^k g_0(x)$$

for some polynomial $g_0(x)$ that is not divisible by $x - c$. By the product rule, the derivative of this function is

$$f'(x) = k(x - c)^{k-1} g_0(x) + (x - c)^k g_0'(x)$$
$$= (x - c)^{k-1}(k g_0(x) + (x - c)g_0'(x)) = (x - c)^{k-1} g_1(x),$$

where again $g_1(x) = k g_0(x) + (x - c)g_0'(x)$ is not divisible by $x - c$. Since $(x - c)$ is a factor of $f'(x)$, it follows that $f'(c) = 0$. Repeating this process we can differentiate and simplify to write

$$f''(x) = (x - c)^{k-2} g_2(x),$$
$$f^{[3]}(x) = (x - c)^{k-3} g_3(x),$$

$$\vdots$$

$$f^{[k-1]}(x) = (x - c)g_{k-1}(x),$$

where all of the $g_i(x)$ are functions that are not zero at $x = c$. All of these derivatives of f, however, will be zero at $x = c$. The kth derivative will not be zero at $x = c$, since

$$f^{[k]}(x) = g_{k-1}(x) + (x - c)g'_{k-1}(x).$$

We have just shown that $f(x)$ and $(x - c)^k$ have the same values and the same first through $(k - 1)$th derivatives, which is why the graph of $y = f(x)$ behaves like the graph of $y = (x - c)^k$ near $x = c$.   ■

Examples and Explorations

EXAMPLE 1   Constructing polynomials that have different numbers of roots

Give formulas and graphs for three cubic polynomials: one with three distinct real roots, one with two distinct real roots, and one with only one real, non-repeated, root.

SOLUTION

To construct a cubic polynomial with three distinct real roots we simply multiply together three distinct linear factors, as shown next at the left. To obtain exactly two distinct roots we can make one of our factors repeated, as in the middle figure. Finally, to have only one root that is not repeated, we can multiply a linear factor and an irreducible quadratic factor, since irreducible quadratic factors do not have any real roots; see the rightmost figure.

$f(x) = (x - 2)(x - 3)(x - 4)$ $f(x) = (x - 2)(x - 3)^2$ $f(x) = (x - 2)(x^2 + 1)$

      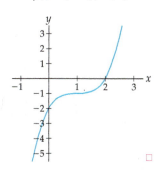

EXAMPLE 2   Quickly graphing a polynomial function that splits into linear factors

Sketch a rough graph of the polynomial function $f(x) = -2(x + 1)^2(x - 2)(x - 3)$.

SOLUTION

The polynomial function f is completely split into linear factors, with no irreducible quadratic factors. Therefore f has the maximum number of roots. More precisely, f is a fourth-degree polynomial, and we have four roots if we count repeated roots separately: $x = -1$ twice, $x = 2$, and $x = 3$.

By Theorem 4.11, we know that there is exactly one turning point between the roots $x = -1$ and $x = 2$ and exactly one turning point between the roots $x = 2$ and $x = 3$. Using this fact, together with the fact that $x = -1$ is a double root and the fact that f is a fourth-degree polynomial with a negative leading coefficient, we can quickly sketch a graph of f, as follows: First, we use the fact that the degree of f is even to determine that the "ends" of the graph of f will either both point up or both point down. Since the leading coefficient $a_n = -2$ is negative, we know that these "ends" must point down. Because f is the product of linear factors, there is exactly one turning point between $x = -1$ and $x = 2$ and exactly one turning point between $x = 2$ and $x = 3$. Finally, we evaluate $f(0) = -2(0+1)^2 (0-2)(0-3) = -2(1)(-2)(-3) = -12$ so that we can mark the height of one point on the graph.

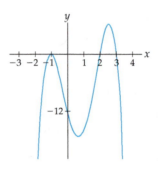

<table>
<tr><td>**EXAMPLE 3**</td><td>**Using derivatives to determine the local behavior of a polynomial function**</td></tr>
</table>

Find all the local extrema and inflection points of the function $f(x) = x^4 + 4x^3 - 16x$.

SOLUTION

To find the local extrema of f we must first find its critical points. The derivative of f is $f'(x) = 4x^3 + 12x^2 - 16$, which is a polynomial and thus always defined; we need only find its zeroes. First we pull out the common factor of 4:

$$f'(x) = 4x^3 + 12x^2 - 16 = 4(x^3 + 3x^2 - 4).$$

The polynomial $x^3 + 3x^2 - 4$ has no convenient groupings or common factors, so we'll try to guess a root with the Integer Root Theorem. Since the constant term is -4, the only possible integer roots are ± 1, ± 2, and ± 4. We need only to find one root in order to use synthetic division and reduce the problem of factoring a cubic to a problem of factoring a quadratic. It turns out that $x = 1$ is a root, since $(1)^3 + 3(1)^2 - 4 = 1 + 3 - 4 = 0$. Now a quick synthetic division calculation will enable us to take out a factor of $x - 1$ from $x^3 + 3x^2 - 4$:

$$
\begin{array}{r|rrrr}
 & 1 & 3 & 0 & -4 \\
1 & & 1 & 4 & 4 \\
\hline
 & 1 & 4 & 4 & 0
\end{array}
$$

Thus $x^3 + 3x^2 - 4 = (x-1)(x^2 + 4x + 4)$, so we have

$$f'(x) = 4x^3 + 12x^2 - 16 = 4(x-1)(x^2 + 4x + 4) = 4(x-1)(x+2)^2.$$

Therefore the only critical points of f are $x = 1$ and $x = -2$. The second derivative is easy to factor:

$$f''(x) = 12x^2 + 24x = 12x(x+2).$$

Sign analyses of f' and f'' give us the following number lines:

From these number lines we see that f has a local minimum at $x = 1$ and inflection points at $x = -2$ and $x = 0$.

THE ANSWER

A calculator graph of $f(x) = x^4 + 4x^3 - 16x$ is shown next. Note that $x = -2$ is a critical point that is not a local extremum and $x = 1$ is a critical point that is a local minimum. Moreover, both $x = -2$ and $x = 0$ are inflection points of f.

$$f(x) = x^4 + 4x^3 - 16x$$

EXAMPLE 4

Modeling a graph with a polynomial

Explain why the graph shown here could be modeled with a polynomial function f. Then say what you can about the degree and leading coefficient of f, and find a possible equation for $f(x)$.

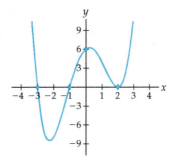

SOLUTION

This graph is defined on all of $\mathbb{R}$, is smooth and unbroken everywhere, and has no asymptotes, so it could be part of the graph of a polynomial function f. We will now find a polynomial whose graph is like the one pictured. Since both ends of the graph point upwards, we know that the degree of f must be even and the leading coefficient of f must be positive. The fact that the graph has three roots means that degree of f is at least 3. The fact that the graph has three local extrema means that the degree of f is in fact at least 4.

The roots of f are $x = -3$, $x = -1$, and $x = 2$. Note that the function f behaves differently at the root $x = 2$ than at the other roots. Near $x = 2$ the graph appears to have a quadratic type of shape. This means that $x = 2$ is a repeated root of f. Given all of this information, we see that one possible form for $f(x)$ is

$$f(x) = A(x + 3)(x + 1)(x - 2)^2,$$

where A is some positive constant. The y-intercept $(0, 6)$ marked on the graph can help us determine A. Since $f(0) = 6$, we have

$$6 = A(0 + 3)(0 + 1)(0 - 2)^2 \quad\Longrightarrow\quad 6 = A(3)(1)(4) \quad\Longrightarrow\quad 6 = 12A \quad\Longrightarrow\quad A = \frac{1}{2}.$$

Therefore a function that could have the given graph is the quartic polynomial function $f(x) = \frac{1}{2}(x + 3)(x - 1)(x - 2)^2$.

EXAMPLE 5   **Using a polynomial to model a graph with an unknown root**

Find a polynomial function f that could have the following graph:

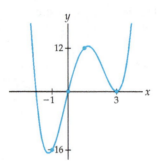

SOLUTION

If we were given the coordinates of the leftmost root on the graph, then this problem would be easy; unfortunately we are not given this root value. We can tell from the graph that f could be a fourth-degree polynomial with positive leading coefficient and that f has three distinct roots: $x = 0$, $x = 3$ (twice), and one that we do not know. Thus f could have the form

$$f(x) = Ax(x - 3)^2(x - c)$$

for some positive real number A and some unidentified root $x = c$.

Also, since $y = f(x)$ goes through the points $(-1, -16)$ and $(1, 12)$, it follows that

$$f(-1) = -16 \implies A(-1)(-1 - 3)^2(-1 - c) = -16 \implies A(1 + c) = -1,$$
$$f(1) = 12 \implies A(1)(1 - 3)^2(1 - c) = 12 \implies A(1 - c) = 3.$$

We can solve this system of equations to find A and c. We have $A = \frac{-1}{1+c}$ from the first equation and $A = \frac{3}{1-c}$ from the second equation. Notice that $c \neq \pm 1$, since we can see from the graph that $x = 1$ and $x = -1$ are not roots of $f(x)$. By setting the two expressions of A equal to each other, we can solve for c:

$$\frac{-1}{1+c} = \frac{3}{1-c} \implies -1 + c = 3(1 + c) \implies -1 + c = 3 + 3c \implies c = -2.$$

Since $c = -2$, $A = \frac{-1}{1+c} = \frac{-1}{1-2} = 1$. Therefore the graph can be modeled by the polynomial function

$$f(x) = x(x - 3)^2(x + 2). \qquad \square$$

EXAMPLE 6   **Using information about derivatives to find a polynomial model for a graph**

Find a polynomial function that could have the following graph:

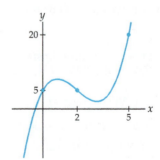

SOLUTION

We can see that f has one root just to the left of the origin, but we are not told what that value is. What we know from the given graph is the three points $f(0) = 5$, $f(2) = 5$, and $f(5) = 20$. In addition, since $x = 2$ seems to be an inflection point of f, we will suppose that $f''(2) = 0$. We can also see from the global behavior of the graph that f must be an odd-degree polynomial function with a positive leading coefficient. Because there are two turning points in the graph, we know that the degree of f must be at least three; that's the lowest possible degree, so let's assume that f is a cubic polynomial of the form

$$f(x) = ax^3 + bx^2 + cx + d$$

for some real numbers a, b, c, and d. Then the first and second derivatives of f are

$$f'(x) = 3ax^2 + 2bx + c,$$
$$f''(x) = 6ax + 2b.$$

The values of $f(x)$ and $f''(x)$ that we got from the graph now result in the following system of equations:

$$\begin{cases} a(0)^3 + b(0)^2 + c(0) + d = 5 \\ a(2)^3 + b(2)^2 + c(2) + d = 5 \\ a(5)^3 + b(5)^2 + c(5) + d = 20 \\ 6a(2) + 2b = 0 \end{cases} \implies \begin{cases} d = 5 \\ 8a + 4b + 2c + d = 5 \\ 125a + 25b + 5c + d = 20 \\ 12a + 2b = 0. \end{cases}$$

A simple but tedious calculation shows that the solution of this system of equations is $a = 1$, $b = -6$, $c = 8$, and $d = 5$. Therefore the graph can be modeled by the polynomial

$$f(x) = x^3 - 6x^2 + 8x + 5.$$

Notice that we had four coefficients to solve for, but only three given points on the graph of f. The extra, fourth piece of information that made it possible to solve for a, b, c, and d was our assumption that $f''(2) = 0$ at the inflection point of the graph.   □

? TEST YOUR UNDERSTANDING

▶ How can you tell if a quadratic polynomial is irreducible?

▶ Given that the degree of a polynomial function is the integer n that represents the highest power of x with a nonzero coefficient in the polynomial, why do you think we say that the degree of the zero polynomial (the function f defined by $f(x) = 0$ for *all* values of x) is undefined?

▶ How can you tell how a polynomial function will behave as $x \to \infty$ and as $x \to -\infty$?

▶ Can an even-degree polynomial have an even number of turning points? Why or why not? What about an odd-degree polynomial?

▶ What does the graph of a polynomial function look like at a repeated root?

EXERCISES 4.3

Thinking Back

▶ *Rolle's Theorem:* State Rolle's Theorem, and give a general outline of its proof.

▶ *Discriminants:* Use the quadratic formula to argue that a quadratic $ax^2 + bx + c$ is irreducible if and only if its discriminant $b^2 - 4ac$ is negative.

Limits of positive integer power functions: Calculate the given limits. Sketch graphs of relevant power functions to support your answers.

▶ $\lim\limits_{x \to \infty} -3x^4$

▶ $\lim\limits_{x \to \infty} \frac{1}{2}x^5$

▶ $\lim\limits_{x \to -\infty} \frac{1}{2}x^5$

▶ $\lim\limits_{x \to -\infty} 5x^7$

▶ $\lim\limits_{x \to -\infty} -3x^8$

▶ $\lim\limits_{x \to \infty} -0.25x^5$

Concepts

0. *Problem Zero:* Read the section and make your own summary of the material.

1. *True/False:* Determine whether each of the statements that follow is true or false. If a statement is true, explain why. If a statement is false, provide a counterexample.

(a) *True or False:* Every polynomial function is a sum of power functions.

(b) *True or False:* If f is a polynomial function of degree n, then the graph of f must have $n - 1$ turning points.

(c) *True or False:* If f is a polynomial function with k roots, then f must have at least $k - 1$ turning points.

(d) *True or False:* If f is a polynomial function with odd degree, then f must have at least one real root.

(e) *True or False:* If f is a polynomial function with odd degree, then f is an odd function.

(f) *True or False:* If f is a polynomial function with k real roots, then the degree of f must be at most k.

(g) *True or False:* If f is a polynomial function with k turning points, then the degree of f must be at least $k + 1$.

(h) *True or False:* For any real numbers a and b, the polynomial function $f(x) = x^4 - ax^3 + bx + 7$ has exactly four real roots.

2. *Examples:* Construct examples of the thing(s) described in the following. Try to find examples that are different than any in the reading.

(a) Three polynomial functions with roots at $x = 0$, $x = 2$, and $x = -5$.

(b) Three irreducible quadratics.

(c) A quadratic polynomial whose values are always negative.

3. For the polynomial $f(x) = x(3x + 1)(x - 2)^2$, determine the leading coefficient, leading term, degree, constant term, and coefficients a_1 and a_3.

4. For the polynomial $f(x) = x^5(1 - 2x)(1 - x^3)$, determine the leading coefficient, leading term, degree, constant term, and coefficients a_1 and a_5.

5. Use the quadratic formula to explain why a quadratic polynomial function $f(x) = ax^2 + bx + c$ is irreducible if, and only if, the discriminant $b^2 - 4ac$ is negative. Then use the discriminant to show that $f(x) = 3x^2 + 2x + 6$ is irreducible.

6. Give an example of each of the following types of polynomials:

(a) A quintic polynomial with only one real root.

(b) A polynomial, all of whose roots are rational, but non-integer, numbers.

(c) A polynomial with integer coefficients that has four real roots, only two of which are integers.

7. Fill in the blank: If $x = c$ is a double root of a polynomial function f, then f can be written in the form _____ for some polynomial function g.

8. Fill in the blank: If $x = c$ is a triple root of a polynomial function f, then f can be written in the form _____ for some polynomial function g.

9. Suppose f is the polynomial function graphed next at the left. Approximate values k so that the translation $f(x) + k$ has (a) no roots, (b) one root, (c) two roots, (d) three roots, and (e) four roots. Sketch the graph of $f(x) + k$ for each value of k.

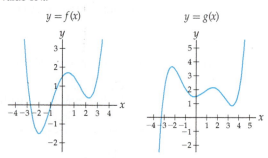

$y = f(x)$

$y = g(x)$

10. Suppose $g(x)$ is the polynomial function graphed previously at the right. Approximate values k so that the translation $g(x) + k$ has (a) one root, (b) two roots, (c) three roots, (d) four roots, and (e) five roots. Sketch the graph of $g(x) + k$ for each value of k.

In Exercises 11–16, either sketch a graph representing a polynomial function with the given conditions or explain why such a polynomial function cannot exist.

11. A polynomial function with odd degree and negative leading coefficient.

12. A polynomial function of degree 4 with five turning points.

13. A polynomial function with even degree, three turning points, and one root.

14. A polynomial function of degree 5 with three turning points.

15. A polynomial function of degree 7 that has seven roots and four turning points.

16. A polynomial function of degree 6 that has five turning points and three roots.

In Exercises 17–22, either write down an equation for a polynomial function with the given properties or explain why such a polynomial function cannot exist.

17. A seven-degree polynomial with five real roots, but only four *distinct* real roots.

18. A polynomial with four distinct real roots, one of which is a triple root.

19. A six-degree polynomial with exactly two distinct real roots.

20. A polynomial with five turning points but only one root, which is a double root.

21. A polynomial with one real root and two turning points.

22. A seven-degree polynomial with four distinct roots, one of which is a double root, and with more than four turning points.

Skills

Calculate each of the limits in Exercises 23–26.

23. $\lim_{x \to \infty} (-2x^5 + 8x^4 - 6)$

24. $\lim_{x \to \infty} (-5x^4 + 3x^2 - 10)$

25. $\lim_{x \to -\infty} (2x^7 - 8x^3 + 16)$

26. $\lim_{x \to -\infty} (3x^3 - 1)(5 - 7x^4)$

Sketch rough graphs of the polynomial functions in Exercises 27–34, without using a graphing calculator and without taking any derivatives.

27. $f(x) = 2x^4 - x^3 - x^2$

28. $f(x) = x^4 - 2x^2 + 1$

29. $f(x) = x^3 - 2x^2 - 4x + 8$

30. $f(x) = x(x+2)(x-3)^2$

31. $f(x) = -2(x+2)^4(x-1)(x-3)^2$

32. $f(x) = x^5 + x^4 - 2x^3$

33. $f(x) = |(x-2)^3(x+1)^2|$

34. $f(x) = |x^2(x-1)(x+2)|$

In Exercises 35–42, find an equation for a polynomial function f with the given characteristics. Do not use a calculator; instead, use known values of f and its derivatives to determine conditions on the coefficients of f. There may be more than one possible answer for some problems.

35. A quartic polynomial function f with two double roots and a y-intercept of 5.

36. A quadratic polynomial function f with no roots and a local maximum at $x = -1$.

37. A cubic polynomial function f with an inflection point at $x = 2$.

38. A cubic polynomial function f with $f(0) = 2$, $f'(0) = 3$, $f''(0) = -4$, and $f'''(0) = 12$.

39. A cubic polynomial function f with $f(0) = -5$, $f'(0) = -3$, $f''(0) = -2$, and $f'''(0) = 6$.

40. A quartic polynomial function f with $f(1) = 4$, $f'(1) = 0$, $f''(1) = 0$, $f'''(1) = -24$, and $f^{(4)}(1) = -48$.

41. A quadratic polynomial function f with $f(2) = 5$, $f'(2) = 7$, and $f''(2) = 2$.

42. A quintic polynomial function for which $f(1) = 2$ and $f^{(k)}(1) = 0$ for all integers $k \geq 1$.

Use derivatives, limits, and number lines to make labeled, accurate graphs for each polynomial function f in Exercises 43–54.

43. $f(x) = x^4 - 6x^2 + 8x$

44. $f(x) = 2x^3 - 3x^2 - 12x$

45. $f(x) = x^4 + 4x^3 + 6x^2 + 8x$

46. $f(x) = 2x^5 - 15x^4 - 30x^3 + 70x^2$

47. $f(x) = (x^2 - 2x)(x^2 - 2x - 6)$

48. $f(x) = x^3(1 + 3x)$

49. $f(x) = x^3 - 2x^2 + x - 2$

50. $f(x) = x^3 - 12x^2 - 45x$

51. $f(x) = 6x^5 - 15x^4 + 10x^3$

52. $f(x) = (1 + x^2)(x - 3)$

53. $f(x) = x^4 + 4x^2 + 4$

54. $f(x) = 8x^3 + 3x^4$

For each graph in Exercises 55–62, find a polynomial function whose graph looks like the one shown.

55.

56.

57.

58.

59.

60.

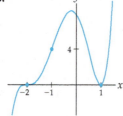

(b) Use your answer to part (a) to write down a possible polynomial expression for f'.

(c) Find an antiderivative f of f'. (You should multiply out the expression for f' before antidifferentiating.)

(d) Use the one known value of f to determine the value of the antidifferentiation constant C.

(e) Use a graphing calculator to graph the function f that results from parts (a)–(c), and check that it does have the properties of the graph of f shown.

61.

62.

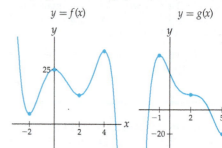

$$y = f(x) \qquad\qquad y = g(x)$$

63. In this exercise you will find a polynomial function f that could have the graph shown next at the left, by finding f' and then antidifferentiating, as follows:

(a) List all of the zeroes of the *derivative* f' of the function f. Assuming that f' is a polynomial function, determine a possible value for the degree of f' and the sign of its leading coefficient.

64. Find a polynomial function that could have the graph $y = g(x)$ shown before this exercise at the right, by the method used in Problem 63. Be aware that the derivative f' has a repeated root; you'll have to determine the location of this root and how many times it is repeated.

Applications

65. Alina is driving from New York to San Francisco. At 12:00 noon she is 550 miles into her trip. Her velocity t hours after noon is $v(t) = 40t^3 - 100t^2 + 50t + 45$ miles per hour. Find equations for Alina's position and acceleration over time.

66. After you drink a cup of coffee, the concentration of caffeine in your bloodstream increases at first and then decreases steadily. Let $C(t)$ represent the amount of caffeine, in grams, in your bloodstream t hours after drinking a cup of coffee. Explain why $C(t)$ cannot be a polynomial function. (*Hint: Try drawing a possible graph of $C(t)$.*)

The following table for Exercises 67 and 68 describes the number of cars that were on a particular 1-mile stretch of Route 97 at t hours after 6:00 A.M. Monday morning:

t (hours after 6:00 A.M.)	0	1	2	3
N (cars on 1 mile of road)	0	28	12	18

67. Make a plot of the data given in the preceding table. Suppose you wanted to find a polynomial function $N(t)$ that passes through each of the four data points. What is the minimum degree that this polynomial function could be, and why?

68. Given the same table of car densities,

(a) Find a polynomial function $N(t)$ that goes through all four data points in the table. (*Hint: Use the data to solve for the coefficients.*)

(b) Overlay a plot of the function $N(t)$ you found with the plot of the data. Does $N(t)$ look the way you expected it to?

(c) Use your function $N(t)$ to predict the number of cars that were on that 1-mile stretch at 7:30 A.M. How many cars does your function predict will be there at 3 P.M.? On what time interval does your model function make practical sense?

Proofs

69. Prove that the composition of two linear functions is also a linear function.

70. Prove that the sum of two polynomial functions is a polynomial.

71. Prove that the product of two cubic polynomials is a polynomial of degree six.

72. Suppose f is a polynomial function that factors as $f(x) = 3(x - a)(x^2 + cx + d)$, where $c^2 < 4d$. Prove that f has only one real root.

73. Use limits to prove that the limits of a polynomial $f(x) = a_n x^n + a_{n-1}x^{n-1} + a_1 x + a_0$ are the same as the limits of its leading term $a_n x^n$ as $x \to \infty$ and as $x \to -\infty$. (*Hint: Show that $\lim\limits_{x\to\infty} f(x)$ is equal to $\lim\limits_{x\to\infty} a_n x^n$ by factoring out $a_n x^n$ from $f(x)$.*)

74. Suppose f is a polynomial function of odd degree. Use the global behavior of f and the Intermediate Value Theorem to prove that f must have at least one real root.

75. Prove that every polynomial function is defined, continuous, and differentiable on all of $(-\infty, \infty)$. (*Hint: Use what you know about power functions.*)

76. Suppose $g(x) = a_n x^n + a_{n-1} x^{n-1} + a_{n-2} x^{n-2} + \cdots + a_2 x^2 + a_1 x + a_0$ is any polynomial function. Prove that an antiderivative of g is

$$f(x) = \frac{a_n}{n+1}x^{n+1} + \frac{a_{n-1}}{n}x^n + \cdots + \frac{a_1}{2}x^2 + a_0 x + C.$$

77. Suppose f is a polynomial function with k roots. Use Rolle's Theorem to prove that f must have at least $k - 1$ turning points.

78. Prove that the $(n + 1)$st derivative of any n-degree polynomial function is zero.

79. Prove that the nth derivative of an n-degree polynomial function with leading coefficient a_n is equal to $n! a_n$. (Recall that $n!$ stands for n **factorial**, which is equal to the product $n(n - 1)(n - 2) \cdots (2)(1)$ of all integers from 1 to n.)

Thinking Forward

Taylor series: Not all functions are polynomials, but sufficiently well behaved functions can be expressed as infinitely long polynomials, at least in some neighborhood. In this problem you will examine the stipulation that for $-1 < x < 1$ the function $\frac{1}{1-x}$ is equal to the infinite *Taylor series*

$$\frac{1}{1-x} = 1 + x + x^2 + x^3 + x^4 + x^5 + \cdots .$$

The partial sums of this infinite polynomial are approximations for $\frac{1}{1-x}$ and are called *Taylor polynomials*:

$$T_k(x) = 1 + x + x^2 + x^3 + x^4 + \cdots + x^k.$$

▶ Clearly $f(x) = \frac{1}{1-x}$ and $T_k(x)$ have the same value at $x = 0$ for all possible k. For other values of x close to 0, the values of $f(x)$ and $T_k(x)$ should be close. Verify that this is the case by comparing these values at $x = -0.5$ and $x = 0.25$ for $0 \le k \le 8$.

▶ Use a calculator to graph $f(x) = \frac{1}{1-x}$ with $T_k(x)$ in the window $-1 < x < 1$, for $0 \le k \le 8$. What happens as k gets larger?

▶ Show that $f(x) = \frac{1}{1-x}$ and its kth Taylor polynomial $T_k(x)$ have the same value and the same first through kth derivatives at $x = 0$. Why would this mean that the graph of $T_k(x)$ is a better and better approximation for the graph of f near $x = 0$ as k gets larger and larger?

▶ As we will see in later chapters, if $f(x) = a_0 + a_1 x + a_2 x^2 + a_3 x^3 + \cdots$, then $a_k = \frac{f^{(k)}(0)}{k!}$. Show that this is indeed the case for $f(x) = \frac{1}{1-x}$ and the given Taylor series.

4.4   RATIONAL FUNCTIONS

▶   Rational functions and their roots, holes, and domains

▶   Horizontal, vertical, slant, and curve asymptotes of rational functions

▶   Graphing rational functions and modeling graphs with rational functions

Rational Functions

A rational number is a number that can be written as a quotient of the simplest possible numbers you can imagine, namely, integers. Similarly, a rational *function* is a quotient of very simple functions, namely, polynomials.

DEFINITION 4.13   **Rational Functions Are Quotients of Polynomials**

A **rational function** is a function that can be written as the quotient of two polynomial functions

$$f(x) = \frac{p(x)}{q(x)} = \frac{a_n x^n + a_{n-1} x^{n-1} + \cdots + a_1 x + a_0}{b_m x^m + b_{m-1} x^{m-1} + \cdots + b_1 x + b_0},$$

for any x such that $q(x) \neq 0$.

Graphs of rational functions are highly dependent on small changes in their numerators and denominators; for example, examine the three following rational functions:

$$f(x) = \frac{(x-3)}{(x-3)(x-1)}$$

$$g(x) = \frac{(x-3)^2}{(x-3)(x-1)}$$

$$h(x) = \frac{(x-3)^2}{(x-3)^3(x-1)}$$

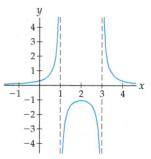

Roots, Holes, and Vertical Asymptotes of Rational Functions

As the preceding examples suggest, the domain, roots, and holes in the graph of a rational function are determined by the roots of its numerator and denominator.

THEOREM 4.14   **Graphical Properties of Rational Functions**

If $f(x) = \frac{p(x)}{q(x)}$ is a rational function, then the following are true.

(a) f is not defined at the roots of $q(x)$.

(b) f has roots at the points that are roots of $p(x)$ but not roots of $q(x)$.

(c) f has holes at the points that are roots of both $p(x)$ and $q(x)$, provided that the roots have higher or equal multiplicity in $p(x)$ than in $q(x)$.

(d) f has vertical asymptotes at the points that are roots of $q(x)$, provided that the roots have higher multiplicity in $q(x)$ than in $p(x)$.

The first two parts of this theorem follow directly from properties of polynomials, roots, and quotients; see Exercises 73 and 74. Proofs for the last two parts are general limit calculations:

Proof. To prove part (c), suppose that $f(x) = \frac{p(x)}{q(x)}$ is a rational function and that $x = c$ is a root of $p(x)$ with multiplicity k and a root of $q(x)$ with multiplicity l, with $k \geq l$. Then we can write $p(x) = (x - c)^k r(x)$ and $q(x) = (x - c)^l s(x)$ for some polynomials $r(x)$ and $s(x)$ that are not divisible by $x - c$. As $x \to c$,

$$\lim_{x \to c} f(x) = \lim_{x \to c} \frac{(x - c)^k r(x)}{(x - c)^l s(x)} = \lim_{x \to c} \frac{(x - c)^{k-l} r(x)}{s(x)}.$$

If $k = l$, then this limit exists and is equal to $\frac{r(c)}{s(c)}$. If $k > l$, then the limit exists and is equal to $\frac{(0) r(c)}{s(c)} = 0$. In any case, the limit as $x \to c$ exists. However, the function $f(x)$ is undefined at $x = c$. Since its limit as $x \to c$ exists, but its value at $x = c$ does not, f must have a hole at $x = c$.

Similarly, for part (d) suppose that $f(x) = \frac{(x-c)^k r(x)}{(x-c)^l s(x)}$ for some polynomials $r(x)$ and $s(x)$ that are not divisible by $x - c$, but this time with $l > k$. This means that $l - k > 0$, so, as $x \to c$,

$$\lim_{x \to c} f(x) = \lim_{x \to c} \frac{(x - c)^k r(x)}{(x - c)^l s(x)} = \lim_{x \to c} \frac{r(x)}{(x - c)^{l-k} s(x)} \to \frac{r(c)}{0},$$

which is infinite because $x - c$ is not a factor of $r(x)$ and therefore $r(c) \neq 0$. This shows that f has a vertical asymptote at $x = c$, as desired.   ∎

Horizontal Asymptotes of Rational Functions

As we saw in the previous section, polynomials behave like their leading terms as $x \to \infty$ and as $x \to -\infty$. For example, the function $f(x) = 2x^3 - 5x - 1$ will be dominated by its leading term $2x^3$ as x takes on larger and larger magnitudes. Therefore f approaches ∞ as $x \to \infty$, and approaches $-\infty$ as $x \to -\infty$. Since rational functions are by definition quotients of polynomial functions, we can use what we know about the global behavior of polynomials to determine the global behavior of rational functions:

THEOREM 4.15

Horizontal Asymptote Theorem for Rational Functions

If $f(x) = \frac{p(x)}{q(x)}$ is a rational function in which the polynomials $p(x)$ and $q(x)$ have leading terms $a_n x^n$ and $b_m x^m$, respectively, then

(a) if $n < m$, then the graph of $y = f(x)$ has a horizontal asymptote at $y = 0$.

(b) if $n = m$, then the graph of $y = f(x)$ has a horizontal asymptote at $y = \frac{a_n}{b_m}$.

(c) if $n > m$, then the graph of $y = f(x)$ does not have a horizontal asymptote.

Proof. In the proof of the last part of Theorem 4.10 we saw that the limit of a polynomial as $x \to \pm\infty$ is equal to the limit of its leading term as $x \to \pm\infty$. We will do all proofs assuming that $x \to \infty$; the proofs assuming that $x \to -\infty$ are similar. To prove part (a), suppose that $f(x) = \frac{p(x)}{q(x)}$ is a rational function where $p(x)$ has leading term $a_n x^n$ and $q(x)$ has leading term $b_m x^m$ and that $n < m$. Then we have $m - n > 0$, and therefore

$$\lim_{x \to \infty} \frac{p(x)}{q(x)} = \lim_{x \to \infty} \frac{a_n x^n}{b_m x^m} = \lim_{x \to \infty} \frac{a_n}{b_m x^{m-n}} \to \frac{a_n}{\infty} \to 0,$$

since the leading coefficient a_n of $p(x)$ cannot be 0. Thus f has a horizontal asymptote at $y = 0$ in this case.

To prove part (b) we do the same calculation but with $n = m$. We have

$$\lim_{x \to \infty} \frac{p(x)}{q(x)} = \lim_{x \to \infty} \frac{a_n x^n}{b_m x^m} = \lim_{x \to \infty} \frac{a_n}{b_m},$$

which means that f has a horizontal asymptote at $y = \frac{a_n}{b_m}$ when $n = m$.

Finally, for part (c) we have $n > m$, and therefore $n - m > 0$. Thus

$$\lim_{x \to \infty} \frac{p(x)}{q(x)} = \lim_{x \to \infty} \frac{a_n x^n}{b_m x^m} = \lim_{x \to \infty} \frac{a_n x^{n-m}}{b_m} \to \frac{\infty}{b_m} \to \pm\infty,$$

since the leading coefficient b_m of $q(x)$ cannot be zero. In this case f has no horizontal asymptote. ∎

Slant and Curve Asymptotes of Rational Functions

Recall that a rational function is ***improper*** if the degree of is numerator is greater than or equal to the degree of its denominator. We have seen that with polynomial long division we can write any improper rational function $f(x) = \frac{p(x)}{q(x)}$ as the sum of a polynomial and a proper rational function:

$$\frac{p(x)}{q(x)} = a(x) + \frac{R(x)}{q(x)}.$$

Since the degree of $R(x)$ must be strictly less than the degree of $q(x)$, the proper rational function $\frac{R(x)}{q(x)}$ must go to zero as $x \to \pm\infty$. This means that the difference between $\frac{p(x)}{q(x)}$ and $a(x)$ gets smaller and smaller as $x \to \pm\infty$, so we know exactly how the function $f(x) = \frac{p(x)}{q(x)}$ behaves as $x \to \pm\infty$:

THEOREM 4.16

Slant and Curve Asymptotes

Suppose $f(x)$ is an improper rational function as just described, which after polynomial long division can be written as

$$f(x) = \frac{p(x)}{q(x)} = a(x) + \frac{R(x)}{q(x)}.$$

Then as $x \to \pm\infty$, the graph of f gets infinitely close to the graph of $y = a(x)$, and

(a) If $\deg p(x) = \deg q(x) + 1$, then $a(x)$ will be a linear function and we say that it is a ***slant asymptote*** of f.

(b) If $\deg p(x) \geq \deg q(x) + 2$, then $a(x)$ will be a polynomial function of degree $p(x) - q(x)$ and we say that it is a ***curve asymptote*** of f.

Theorem 4.16 does not mention the case where $\deg p(x) = \deg q(x)$, because we already handled it in Theorem 4.15. In this case $a(x)$ is the constant function $y = \frac{a_n}{b_m}$ that is the ratio of the leading coefficients of $p(x)$ and $q(x)$, and it is a horizontal asymptote of f.

For example, with polynomial long division we can write the improper rational function $f(x) = \frac{x^3 - 9x}{x^2 - 4}$ as the sum $x + \frac{-5x}{x^2 - 4}$, and therefore f has a slant asymptote of $y = x$, as shown next at the left. As another example using polynomial long division or synthetic division, we can write the improper rational function $g(x) = \frac{x^3 - 4x}{x - 1}$ as $(x^2 + x - 3) + \frac{-3}{x - 1}$, and therefore $g(x)$ has a quadratic curve asymptote of $y = x^2 + x - 3$, as shown in the figure at the right.

$$f(x) = \frac{x^3 - 9x}{x^2 - 4} \quad and \quad y = x \qquad\qquad g(x) = \frac{x^3 - 4x}{x - 1} \quad and \quad y = x^2 + x - 3$$

 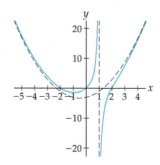

Examples and Explorations

EXAMPLE 1

Finding domains, roots, holes, and asymptotes

Find the domain, roots, holes, vertical asymptotes, and horizontal asymptotes of the following rational functions:

(a) $f(x) = \dfrac{2x^3 - 8x}{x^3 + 2x^2 - x - 2}$ **(b)** $g(x) = \dfrac{x^2 - 1}{x^3 - 2x^2 - 5x + 6}$

SOLUTION

(a) Since the numerator of f has the same degree as the denominator of $g(x)$, we know that f will have a horizontal asymptote at the ratio $y = \frac{2}{1} = 2$ of its leading coefficients.

Finding vertical asymptotes takes a little more work, since we need to factor the numerator and denominator first:

$$f(x) = \frac{2x(x^2 - 4)}{x^2(x + 2) - 1(x + 2)} = \frac{2x(x - 2)(x + 2)}{(x^2 - 1)(x + 2)} = \frac{2x(x - 2)(x + 2)}{(x - 1)(x + 1)(x + 2)}.$$

From this calculation we see that f has a hole at $x = -2$, roots at $x = 0$ and $x = 2$, and vertical asymptotes at $x = 1$ and $x = -1$.

(b) Clearly the numerator of $g(x)$ factors as $p(x) = x^2 - 1 = (x - 1)(x + 1)$. To factor the denominator we need to use the Integer Root Theorem to guess a root. The only integers that could be roots of $q(x) = x^3 - 2x^2 - 5x + 6$ are the divisors of the constant term: $\pm 1, \pm 2, \pm 3,$ and ± 6. A quick check shows that $x = 1$ is a root of $q(x)$. We can now use synthetic division to pull out a factor of $x - 1$ from $q(x)$:

$$
\begin{array}{r|rrrr}
 & 1 & -2 & -5 & 6 \\
1 & & 1 & -1 & -6 \\
\hline
 & 1 & -1 & -6 & \boxed{0}
\end{array}
$$

Therefore $q(x) = (x - 1)(x^2 - x - 6) = (x - 1)(x + 2)(x - 3)$, and thus the rational function $g(x)$ can be written in factored form as

$$g(x) = \frac{x^2 - 1}{x^3 - 2x^2 - 5x + 6} = \frac{(x - 1)(x + 1)}{(x - 1)(x + 2)(x - 3)}.$$

This function $g(x)$ is defined everywhere except at $x = 1$, at $x = -2$, and at $x = 3$, so its domain is $(-\infty, -2) \cup (-2, 1) \cup (1, 3) \cup (3, \infty)$. Because $x = -1$ is a root of the numerator but not the denominator, it is a root of $g(x)$. Since $x = 1$ is a root of both the numerator and the denominator, with the same multiplicity, it is a hole of $g(x)$. Since $x = -2$ and $x = 3$ are roots of the denominator but not the numerator, they are vertical asymptotes of $g(x)$. Finally, because the degree of the numerator is less than the degree of the denominator of $g(x)$, $g(x)$ has a horizontal asymptote at $y = 0$.   □

CHECKING THE ANSWER   Calculator graphs for the preceding examples above are shown next and do have the listed features. Note that some calculators may draw asymptotes as solid lines that are connected to your graphs and often won't show holes explicitly unless you trace along the graph and look for them.

$$f(x) = \frac{2x^3 - 8x}{x^3 + 2x^2 - x - 2}$$

$$g(x) = \frac{x^2 - 1}{x^3 - 2x^2 - 5x + 6}$$

EXAMPLE 2   **Common factors and multiplicities**

Sketch quick graphs of the following rational functions:

(a) $f(x) = \dfrac{(x-1)^2}{x-1}$ **(b)** $g(x) = \dfrac{x-1}{x-1}$ **(c)** $h(x) = \dfrac{x-1}{(x-1)^2}$

SOLUTION

(a) Since $f(x) = \frac{(x-1)^2}{x-1} = x - 1$ except at $x = 1$, where it is undefined, the graph of f is the same as the graph of $y = x - 1$ but with a hole at $x = 1$, as shown in the figure next at the left.

(b) Since $g(x) = \frac{x-1}{x-1} = 1$ except at $x = 1$, where it is undefined, the graph of $g(x)$ is the same as the graph of $y = 1$ but with a hole at $x = 1$, as shown in the middle figure.

(c) Since $h(x) = \frac{x-1}{(x-1)^2}$ and $y = \frac{1}{x-1}$ are equal everywhere they are defined, the graph of $h(x)$ is the same as the graph of $y = \frac{1}{x-1}$, as shown in the right-hand figure.

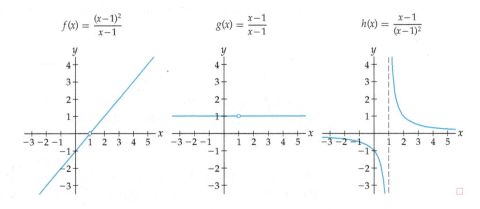

□

<table>
<tr><td>EXAMPLE 3</td><td></td></tr>
</table>

Making a rough graph of a rational function

Without a calculator, sketch a rough graph of the function $f(x) = \dfrac{2(x-1)^2(x+2)}{(x-1)(x+1)(x-2)}$.

SOLUTION

By Theorem 4.14, we can see immediately from the factors in the numerator and denominator that the graph of f will have

▶ a hole at $x = 1$;

▶ a root at $x = -2$;

▶ vertical asymptotes at $x = -1$ and $x = 2$;

▶ a horizontal asymptote at $y = \dfrac{a_n}{b_n} = \dfrac{1}{2} = 2.$

A quick sign analysis tells us where the graph of f is above or below the x-axis. We must check the sign of f on each subinterval between the roots and non-domain points:

Note that on this number line we include tick-marks only at the locations where $f(x)$ is zero or does not exist. The unlabeled tick-marks are the locations where $f(x)$ is zero, and the ones labeled "DNE" are the locations where $f(x)$ is not defined.

Plotting a few key points will help us make a more accurate graph. The y-intercept of the graph is $f(0) = 2$. The height of the hole at $x = 1$ will be the value of f at $x = 1$ after cancelling common factors:

$$\frac{2(1-1)(1+2)}{(1+1)(1-2)} = \frac{2(0)(3)}{(2)(-1)} = 0.$$

Finally, just to get a value to the right side of the graph, we also calculate $f(3) = 5$. The figure that follows at the left shows some of the information we have collected. The figure at the right uses this information and the sign analysis to fill in a rough sketch of the graph.

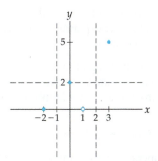

Some information about the graph

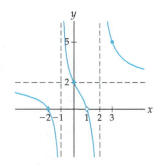

A rough sketch of f

EXAMPLE 4   **Modeling a graph with a slant asymptote by a rational function**

Find a rational function f that could have this graph:

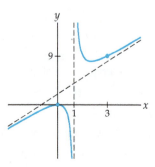

SOLUTION

Suppose the given graph is that of a rational function $f(x) = \frac{p(x)}{q(x)}$. The graph has one root and no holes, and therefore $p(x)$ must have only one root, at $x = 0$. Since there is a vertical asymptote at $x = 1$, the denominator $q(x)$ must have a root at $x = 1$. Because the graph has a slant asymptote, the degree of $p(x)$ must be exactly one higher than the degree of $q(x)$. Perhaps the simplest type of rational function having all these characteristics is

$$f(x) = \frac{Ax^2}{x - 1}.$$

We put a repeated linear factor x^2 in the numerator to satisfy our degree requirement without adding any more roots. We included a constant A in the numerator to determine the extent to which the graph of f is stretched vertically. Since the graph passes through the point $(3, 9)$, we have

$$f(3) = 9 \quad \Longrightarrow \quad \frac{A(3)^2}{3 - 1} = 9 \quad \Longrightarrow \quad \frac{9A}{2} = 9 \quad \Longrightarrow \quad A = 2.$$

The function $f(x) = \frac{2x^2}{x-1}$ is currently our best guess for the given graph. Let's verify that this function seems to have the sort of slant asymptote shown in the graph. Using synthetic division, we can write $f(x) = (2x + 2) + \frac{2}{x-1}$, which gives us a slant asymptote equation of $y = 2x + 2$. This equation is appropriate for the slope and y-intercept of the slant asymptote in the figure. Therefore $f(x) = \frac{2x^2}{x-1}$ is a good model for the graph.   ☐

TEST YOUR UNDERSTANDING

▶   How are rational functions analogous to rational numbers? How are they different?

▶   If you wanted to find the roots and holes of a rational function, would you prefer that it were written in factored form or expanded form? What if you were looking for vertical asymptotes? Horizontal asymptotes?

▶   Calculators are notoriously bad at graphing rational functions. Sometimes they connect a graph over its vertical asymptotes, and most times the holes of a rational function are not immediately clear from a calculator graph. What do you think causes calculators to make these errors?

▶   What types of rational functions will have a horizontal asymptote at $y = 0$, and why?

▶   When will a rational function have a slant asymptote? What about a cubic curve asymptote?

EXERCISES 4.4

Thinking Back

Factoring after root-guessing: Factor as much as possible each of the polynomial and rational functions that follow. In these exercises it is necessary to guess a root and use synthetic division to get started with factoring.

▶ $f(x) = 2x^4 + 6x^2 - 8$

▶ $f(x) = x^3 + 4x^2 - 11x + 6$

▶ $f(x) = \dfrac{x^3 + 6x^2 + 3x - 10}{x^5 + 3x^4 - x - 3}$

Polynomial long division: Write each of the following improper rational functions as a sum of a polynomial function and a proper rational function.

▶ $f(x) = \dfrac{x^2 - 1}{x^2 - x - 2}$

▶ $f(x) = \dfrac{(x+1)(x-3)(x+2)}{(x+2)(x-3)}$

▶ $f(x) = \dfrac{2x^3 + 3x^2 - 2x - 3}{x^2 - 2x - 3}$

Concepts

0. *Problem Zero:* Read the section and make your own summary of the material.

1. *True/False:* Determine whether each of the statements that follow is true or false. If a statement is true, explain why. If a statement is false, provide a counterexample.

(a) *True or False:* Every polynomial function is a rational function.

(b) *True or False:* Rational functions are quotients of integers.

(c) *True or False:* If $f(x) = \dfrac{p(x)}{q(x)}$ is a rational function with $p(3) = 0$, then $x = 3$ is a root of f.

(d) *True or False:* If $f(x) = \dfrac{p(x)}{q(x)}$ is a rational function with $q(3) = 0$, then $x = 3$ is not in the domain of f.

(e) *True or False:* If $f(x) = \dfrac{p(x)}{q(x)}$ is a rational function with $p(3) = 0$ and $q(3) = 0$, then $x = 3$ is a hole of f.

(f) *True or False:* If $p(x)$ and $q(x)$ are both cubic polynomials, then $f(x) = \dfrac{p(x)}{q(x)}$ is an improper rational function.

(g) *True or False:* A rational function f has a horizontal asymptote if and only if it is a proper rational function.

(h) *True or False:* Every rational function has either a horizontal asymptote, a slant asymptote, or a curve asymptote.

2. *Examples:* Construct examples of the thing(s) described in the following. Try to find examples that are different than any in the reading.

(a) Two rational functions and two functions that involve quotients but are not rational.

(b) Two reduced rational functions and two rational functions that are not reduced.

(c) Two improper rational functions and two proper rational functions.

3. What is a rational function?

4. Consider the rational functions

$$f(x) = \frac{x^2 - 1}{x^2 + 4x - 5} \quad \text{and} \quad g(x) = \frac{x+1}{x+5}.$$

For which values of x are these two functions equal? For which values are they different?

5. Explain why the graph of $f(x) = \dfrac{(x-1)(x+2)}{(x+2)^2}$ does not have a hole in it at $x = -2$.

6. Use a calculator to graph the function $f(x) = \dfrac{x^2 + x - 2}{x^2 - 4}$. This graph has a hole, where the function is not defined; determine the location of the hole, and trace along the graph on your calculator until you find it there.

7. Construct an equation of a rational function whose graph has no roots, no holes, vertical asymptotes at $x = \pm 2$, and a horizontal asymptote at $y = 3$.

8. Construct an equation of a rational function whose graph has a hole at the coordinates $(-2, 0)$, vertical asymptotes at $x = 1$ and $x = -3$, and a horizontal asymptote at $y = -5$.

9. Suppose f is a rational function with roots at $x = 1$ and $x = 3$, a hole at $x = -1$, a vertical asymptote at $x = 2$, and a horizontal asymptote at $y = -1$.

(a) Sketch three possible graphs of f. Make the graphs as different as you can while still having the given characteristics.

(b) Write down the equations of three functions f that have the given properties. (Your equations do not have to match your graphs from part (a).)

10. Suppose f is a rational function with root at $x = -2$, holes at $x = 0$ and $x = 3$, a vertical asymptote at $x = 1$, and no horizontal asymptote.

(a) Sketch three possible graphs of f. Make the graphs as different as you can while still having the given characteristics.

(b) Write down the equations of three functions f that have the given properties. (Your equations do not have to match your graphs from part (a).)

11. Suppose $f(x) = \dfrac{p(x)}{q(x)}$ is a rational function and that $p(x)$ has degree n while $q(x)$ has degree m. Fill in the blanks in the following statements:

(a) The graph of f will have a horizontal asymptote if _____ .

(b) The graph of f will have a horizontal asymptote at $y = 0$ if _____

(c) The graph of f will have a slant asymptote if _____ .

(d) The graph of f will have a curve asymptote of degree greater than or equal to 2 if _____ .

(e) The graph of f will have a curve asymptote of degree 4 if _____ .

12. The graph that follows shows asymptotes, roots, and holes that a rational function f could have. Assuming that the only asymptotes, roots, and holes are those shown in the graph, sketch three possible graphs with the given characteristics. What additional information would you need to know in order to completely determine the graph of f?

13. Given the number lines that follow, sketch a possible graph of f. Make sure that your graph reflects all of the information given in the number lines. Identify all roots, holes, and asymptotes of your graph. What additional information would you need in order to sketch a completely accurate graph of f?

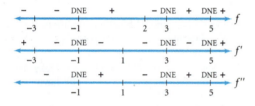

14. Find an equation for a rational function that would have the number line from Exercise 13. You may have to do some educated guessing (and use a graphing calculator to check and refine your guesses).

15. We have seen that a common linear factor $x - c$ of a rational function f can cause the graph of f to have a "hole" at $x = c$. What happens if a rational function f has an irreducible quadratic factor that is common to the numerator and the denominator, and why? Give an example to support your argument.

16. The rational function $f(x) = \dfrac{1}{x-1}$ has a vertical asymptote at $x = 1$. Determine the behavior of the graph of f on either side of $x = 1$ in two ways:

(a) By examining $\lim\limits_{x \to 1^+} f(x)$ and $\lim\limits_{x \to 1^-} f(x)$.

(b) By creating and using a sign chart for the derivative f'.

17. Suppose $f(x) = \dfrac{p(x)}{q(x)}$ is a rational function whose numerator $p(x)$ is of degree n and whose denominator $q(x)$ is of degree m. When we use the quotient rule to differentiate f, the resulting function f' will also be a rational function.

(a) What can you say about the degree of the numerator of f'? What can you say about the degree of the denominator of f'? Be careful when examining the numerator of f'; is it possible that there will be some cancellation that lowers the degree of the numerator of f' more than you would expect?

(b) If f is a *proper* rational function, is f' necessarily a proper rational function? Use your answer part (a) to explain your answer to this question.

(c) Suppose f has a horizontal asymptote at $y = 2$. Does f' necessarily have a horizontal asymptote, and if so, can you tell where it is? Explain your answer in two ways: first, in terms of the degrees of the numerator and denominator of f and f'; second, in terms of the graph of f and its behavior at the horizontal asymptote $y = 2$.

18. If you can factor the denominator of a rational function $f(x) = \dfrac{p(x)}{q(x)}$, will you also necessarily be able to factor the denominators of f' and f''? Why or why not? What if you can factor the numerator of f?

Skills

Find the roots or holes, and any horizontal, vertical, slant, or curve asymptotes, of the rational functions in Exercises 19–34.

19. $f(x) = \dfrac{(x+1)(x-2)}{(x-2)(x+2)}$

20. $f(x) = \dfrac{(x+1)(x-2)^2}{(x-2)(x+2)}$

21. $f(x) = \dfrac{(x+1)(x-2)}{(x-2)^2(x+2)}$

22. $f(x) = \dfrac{(x+1)(x-2)}{(x-2)(x+2)^2}$

23. $f(x) = \dfrac{16 - x^4}{x^4 - 4x^2}$

24. $f(x) = \dfrac{9x^4 + 12x^2 + 4}{x^4 + 6x^2 + 9}$

25. $f(x) = \dfrac{2x^2 + x - 3}{x + 1}$

26. $f(x) = \dfrac{x^3 + x^2 - 6x}{x^2 - 4x + 4}$

27. $f(x) = \dfrac{16x^4 - 81}{3 - 2x}$

28. $f(x) = \dfrac{x^4 + 2x^2 + 1}{x^3 - 1}$

29. $f(x) = \dfrac{x^3 + 2x^2 - 4x - 8}{2x^3 - x^2 - 1}$

30. $f(x) = \dfrac{x^3 - x^2 - 4x + 4}{x^3 - 7x - 6}$

31. $f(x) = \dfrac{(x-1)(3x+2)(x+1)}{x^2 + x + 1}$

32. $f(x) = \dfrac{(x+3)(x-2)^2}{(x-2)(x+1)}$

33. $f(x) = \dfrac{x^5 - 2x^4 + x^3}{x^2 - 1}$

34. $f(x) = \dfrac{x^4 - 2x^3 - 3x^2 + 4x + 4}{x^2 - x - 2}$

Sketch rough graphs of the rational functions in Exercises 35–46, without using a graphing calculator. Be as accurate as you can, and identify any roots, holes, or asymptotes.

35. $f(x) = \dfrac{x^2 + x - 6}{x - 2}$

36. $f(x) = \dfrac{x^2 - x}{(x + 1)(x - 1)}$

37. $f(x) = \dfrac{3x(x + 1)}{(x - 5)^2(x + 2)}$

38. $f(x) = \dfrac{x^2 + 1}{x - 1}$

39. $f(x) = \dfrac{x^2 - 1}{2x^2 - 3x - 2}$

40. $f(x) = \dfrac{(x^2 - 1)^2}{2x^2 - 3x - 2}$

41. $f(x) = \dfrac{2x^3 + 4x^2 - 6x}{x^2 - 4}$

42. $f(x) = \dfrac{(x^2 - 4)^2}{2x^2 - 3x - 2}$

43. $f(x) = \dfrac{(x + 1)(x - 3)(x + 2)}{(x + 2)(x - 3)}$

44. $f(x) = \dfrac{(x - 1)(x + 2)}{(x + 1)(x - 1)}$

45. $f(x) = \dfrac{2x^3 + 3x^2 - 2x - 3}{x^2 - 2x - 3}$

46. $f(x) = \left| \dfrac{x^2(x - 3)(x + 2)}{x - 1} \right|$

Use derivatives, limits, and number lines to sketch accurate, labeled graphs of the rational functions in Exercises 47–54. Give the equations of any slant or curve asymptotes.

47. $f(x) = \dfrac{x - 2}{(x - 1)^2}$

48. $f(x) = \dfrac{x^2 - 1}{x^2 - x - 2}$

49. $f(x) = \dfrac{3x^2 - 2x + 4}{x^2 - 4x + 4}$

50. $f(x) = \dfrac{x^2}{x - 1}$

51. $f(x) = \dfrac{x(x + 1)(x - 1)}{(x - 2)(x + 1)}$

52. $f(x) = \dfrac{x^3 - 5x}{x - 1}$

53. $f(x) = \dfrac{(x^2 - 1)^2}{3x^2 - x - 2}$

54. $f(x) = \dfrac{x^4}{x - 2}$

For each graph in Exercises 56–66, find a function whose graph looks like the one shown. When you are finished, use a graphing calculator to check that your function f has the properties and features of the given graph.

55.

56.

57.

58.

59.

60.

61.

62.

63.

64.

65.

66.

Applications

67. A new energy-efficient clothes dryer costs \$800 to purchase and \$55 each year in electrical costs.

(a) If this clothes dryer lasts 10 years, what is the overall cost of owning and operating the dryer over those 10 years? What is the yearly cost? What if the dryer lasts 20 years?

(b) Express the yearly cost $Y(n)$ of owning and operating the dryer as a rational function of the number of years n that the dryer lasts.

(c) What is $\lim\limits_{n \to \infty} Y(n)$, and what is its real-world significance? Why does this limit make sense in terms of yearly costs?

(d) Write $Y(n)$ as the sum of a (constant) polynomial and a proper rational function. Why does this formula make real-world sense?

68. Suppose the population of rats on a desert island t days after a crate of them washes ashore is given by the rational function $P(t) = \dfrac{300 + 4t}{5 + 0.1t}$.

   (a) How many rats were in the crate that washed up on shore? Does the population of rats increase or decrease over time?

   (b) Does the rat population eventually level off to an equilibrium? Why or why not?

   (c) Suppose a scientist drops a similar crate of rats on another island and this time the population of rats is given by $A(t) = \dfrac{300 + 4t}{5 + 0.1t^2}$. Is this new island more or less hospitable to the rats than the first, and why? What happens to the rats on this new island after a long time?

For Exercises 69 and 70, suppose that Emmy is investigating a release of toxins from a tank farm on the Hanford nuclear reservation into groundwater. The groundwater eventually forms a spring that runs into the Columbia River. The following table describes the amount of toxins in the spring:

Years after 2006	0	1	2
Concentration of toxin (ppm)	0.53	0.65	0.74

69. Emmy believes that the leak is not getting larger, so that the concentration of toxin in the spring will approach some steady constant value. She wants to make estimates of the date on which the leak started and of the eventual steady concentration of toxin. Is a polynomial the best choice of a function to fit to the data for this purpose? Why or why not?

70. Emmy wants to fit a rational function of the form $T(t) = \dfrac{at + b}{t + d}$ to the data, where t is the number of years after 2006.

   (a) What values should Emmy use for the coefficients a, b and d?

   (b) Use your rational function model to estimate the date that the leak started.

   (c) What asymptotic value will the concentration approach as time t increases without bound?

   (d) What are the potential problems in using this model?

Proofs

71. Prove that the sum of two rational functions is a rational function.

72. Prove that the product of two rational functions is a rational function.

73. Prove that the domain of a rational function $f(x) = \dfrac{p(x)}{q(x)}$ is the set $\{x \mid q(x) \neq 0\}$.

74. Prove that the graph of a rational function $f(x) = \dfrac{p(x)}{q(x)}$ has a root at $x = c$ if and only if $p(c) = 0$ but $q(c) \neq 0$.

75. Prove that a rational function of the form

$$f(x) = \frac{(x - c)^k\, r(x)}{(x - c)^l\, s(x)},$$

where $(x - c)$ is not a factor of $r(x)$ or $s(x)$, has a vertical asymptote at $x = c$ if $k < l$, but has a hole at $x = c$ if $k \geq l$.

76. Use limit techniques to prove that a rational function $f(x) = \dfrac{p(x)}{q(x)}$ will have

   (a) a horizontal asymptote at $y = 0$, if the degree of $p(x)$ is less than the degree of $q(x)$;

   (b) a horizontal asymptote at $y = \dfrac{a_n}{b_m}$, where a_n and b_m are the leading terms of $p(x)$ and $q(x)$, respectively, if $p(x)$ and $q(x)$ have the same degree;

   (c) no horizontal asymptote, if the degree of $p(x)$ is greater than the degree of $q(x)$.

77. Prove that if f is a rational function with a horizontal asymptote, then its derivative f' also has a horizontal asymptote.

78. Prove that if f is a rational function with a slant asymptote, then its derivative f' has a horizontal asymptote.

Thinking Forward

Polynomial long division and antidifferentiating: Rational functions can be difficult to antidifferentiate, but polynomial long division sometimes comes to the rescue. Write each of the given improper rational functions as the sum of a polynomial and a proper rational function. Then use a guess-and-check strategy to find an antiderivative.

▶ $f(x) = \dfrac{x^3 + 3x}{x^2 + 1}$

▶ $f(x) = \dfrac{-3x^3 + 3x^2 - 9x}{x^3 + 3x + 1}$

▶ $f(x) = \dfrac{3x^7 - 2x^5 + x^4 + 3x^2 - 2}{x^5 + 1}$

Partial fractions before antidifferentiating: Another way to make rational functions easier to antidifferentiate is by splitting them into what is known as **partial fractions**. Use algebra to solve for A, B, and C, and then write each of the given rational functions in its given partial-fraction form. Then use a guess-and-check strategy to find an antiderivative.

▶ $f(x) = \dfrac{8}{(x - 1)^2(3x + 1)} = \dfrac{A}{x - 1} + \dfrac{B}{(x - 1)^2} + \dfrac{C}{3x + 1}$

▶ $f(x) = \dfrac{2x + 1}{(x - 5)^3} = \dfrac{A}{x - 5} + \dfrac{B}{(x - 5)^2} + \dfrac{C}{(x - 5)^3}$

▶ $f(x) = \dfrac{3x - 2}{x(x^2 + 1)} = \dfrac{A}{x} + \dfrac{Bx + C}{1 + x^2}$

CHAPTER REVIEW, SELF-TEST, AND CAPSTONES

Before you progress to the next chapter, be sure you are familiar with the definitions, concepts, and basic skills outlined here. The capstone exercises at the end bring together ideas from this chapter and look forward to future chapters.

Definitions

Give precise mathematical definitions or descriptions of each of the concepts that follow. Then illustrate the definition with a graph or an algebraic example.

▶ the form of a *power function*

▶ the form of a *polynomial function*

▶ the *leading coefficient*, the *leading term*, and the *constant term* of a polynomial

▶ the forms of *constant*, *linear*, *quadratic*, *cubic*, *quartic*, and *quintic* polynomials

▶ what it means for a quadratic polynomial to be *irreducible*, and how this is related to the *discriminant*

▶ what it means to *factor* an expression

▶ the description of the algorithm known as *synthetic division*, and the type of situation in which one would apply synthetic division

▶ the description of the algorithm known as *polynomial long division*, and the type of situation in which one would apply polynomial long division

▶ what it means for a polynomial to have a *repeated root* at $x = c$

▶ the form of a *rational function*

▶ what it means for a rational function to be *proper* or *improper*

Theorems

Fill in the blanks, or circle as appropriate, to complete each of the following theorem statements:

▶ *The Integer Root Theorem*: If $f(x) = a_n x^n + a_{n-1} x^{n-1} + \cdots + a_1 x + a_0$ is a polynomial function with coefficients a_i that are _____ and with a constant term that is _____, then every integer root of f is a _____ of the constant term a_0.

▶ Suppose that we use synthetic division to divide a polynomial f by $x - c$, and obtain a polynomial $g(x)$ and a remainder $R = 0$. Then $x = c$ (is)/(is not) a root of f, and

$$\frac{f(x)}{x - c} = \underline{\quad}.$$

▶ Suppose that we use synthetic division to divide a polynomial f by $x - c$, and obtain a polynomial $g(x)$ and a remainder $R \neq 0$. Then $x = c$ (is)/(is not) a root of f, and

$$\frac{f(x)}{x - c} = \underline{\quad} + \underline{\quad}.$$

▶ Suppose that we divide a polynomial $p(x)$ by another polynomial $g(x)$ of equal or lower degree and obtain a polynomial $m(x)$ and a remainder $R(x)$. Then the polynomials $p(x)$, $q(x)$, $m(x)$, and $R(x)$ are related in the following way:

$$p(x) = \underline{\quad} + \underline{\quad}.$$

Rephrasing this, we can write

$$\frac{p(x)}{q(x)} = \underline{\quad} + \underline{\quad}.$$

▶ If k is an integer such that $k \neq 1$ and $k \neq 0$, then the graph of $f(x) = x^k$ has one of the following four shapes: _____, _____, _____, or _____ . (In each case, provide a graph and indicate the necessary conditions on the power k.)

▶ If p/q is a positive reduced rational number and $0 < p/q < 1$, then the graph of $f(x) = x^{p/q}$ has one of the following three shapes: _____, _____, or _____ . (In each case, provide a graph and indicate the necessary conditions on p and q.)

▶ If p/q is a positive reduced rational number and $p/q > 1$, then the graph of $f(x) = x^{p/q}$ has one of the following three shapes: _____, _____, or _____ . (In each case, provide a graph and indicate the necessary conditions on p and q.)

▶ If f is a polynomial function of degree n, then the graph of f has at most _____ real roots and at most _____ local extrema.

▶ If f is a polynomial function, then the graph of f behaves like the graph of _____ at its "ends."

▶ If f is a polynomial function of degree n with k real roots, then f has at least _____ turning points, with at least one turning point between each _____ .

▶ If f is a polynomial function with a k-times repeated root at $x = c$, then the graph of f behaves like the graph of _____ near $x = c$. Specifically, the values _____, _____, $\ldots$, _____ are all zero, but the value _____ is not zero.

▶ If $f(x) = \frac{p(x)}{q(x)}$ is a rational function, then f is not defined at the roots of _____, and f has roots at the points that are roots of _____ but not roots of _____ .

▶ If $f(x) = \frac{p(x)}{q(x)}$ is a rational function, then f has holes at the points that are roots of _____, provided that _____ .

▶ If $f(x) = \frac{p(x)}{q(x)}$ is a rational function, then f has vertical asymptotes at the points that are roots of _____, provided that _____ .

▶ Suppose $f(x) = \frac{p(x)}{q(x)}$ is a rational function with $\deg(p(x)) = n$ and $\deg(q(x)) = m$. If $n < m$, then f has a horizontal asymptote at _____; if $n = m$, then f has a horizontal asymptote at _____; and if $n > m$, then f _____ .

▶ Suppose $f(x) = \frac{p(x)}{q(x)} = a(x) + \frac{R(x)}{q(x)}$ is an improper rational function. If $\deg p(x) = \deg q(x) + 1$, then $a(x)$ will be a _____ function that is a _____ asymptote of f.

▶ Suppose $f(x) = \frac{p(x)}{q(x)} = a(x) + \frac{R(x)}{q(x)}$ is an improper rational function. If $\deg p(x) \geq \deg q(x) + 2$, then $a(x)$ will be a _____ function of degree _____ that is a _____ asymptote of f.

Notation and Algebraic Rules

Exponents: What does it mean to raise a number to a power? Define each expression that follows in simpler terms. You may assume that x is a real number, k is a positive integer, $\frac{p}{q}$ is a positive rational number, and r is a rational number.

▶ x^k ▶ x^{-k} ▶ $x^{1/k}$

▶ $\sqrt[k]{x}$ ▶ $x^{p/q}$ ▶ $x^{-p/q}$

▶ 0^r ▶ x^0 ▶ 0^0

Rules of Exponents: Fill in the blanks to rewrite each expression according to the algebraic rules of exponentiation. You may assume that x, y, a, and b are numbers for which each expression is defined.

▶ $x^a x^b =$ _____   ▶ $(xy)^a =$ _____

▶ $(x^a)^b =$ _____   ▶ $\frac{x^a}{x^b} =$ _____

▶ $\frac{x^a}{y^a} =$ _____   ▶ $(\sqrt[b]{x})^a =$ _____

Skill Certification: Algebra Techniques, Limits, and Basic Algebraic Graphs

Simplifying expressions: Use rules of exponents to write each function f in the form of a power function or a sum of power functions, if possible.

1. $f(x) = \frac{2x^4 - x^3}{2x - 1}$

2. $f(x) = \frac{x^{-1} + x^{-2}}{x^{-3}}$

3. $f(x) = \frac{x^{1/4} + x^{1/3}}{x^2}$

4. $f(x) = \frac{x^{-2/5}\sqrt{4x}}{\sqrt[3]{x}}$

Factoring polynomials: Factor each of the given polynomial functions as much as possible. Some factorizations require root-guessing and/or synthetic division.

5. $f(x) = x^4 - 8x^2 + 16$

6. $f(x) = 3x^3 - 2x^2 - 7x - 2$

7. $f(x) = x^3 - 6x^2 + 12x - 8$

8. $f(x) = 2x^3 - 25x^2 + 84x - 36$

Dividing polynomials: Use either synthetic division or polynomial long division to write each of the given improper rational functions $f(x) = \frac{p(x)}{q(x)}$ as the sum of a polynomial $m(x)$ and a proper rational function $\frac{R(x)}{q(x)}$.

9. $f(x) = \frac{3x^3 - 6x^2 + x}{x - 2}$

10. $f(x) = \frac{2x^3 + 7x - 1}{x^3 + 2x - 1}$

11. $f(x) = \frac{(3x + 1)^3}{(x - 1)(x - 2)}$

12. $f(x) = \frac{x^4 + 2x^2 - x - 7}{2 - x^2}$

Limits of algebraic functions: Calculate each limit by hand, showing all work. Verify your answers afterwards with a graphing utility.

13. $\lim\limits_{x \to 0} x^{-1/5}$

14. $\lim\limits_{x \to \infty} (\sqrt{x} - \sqrt[3]{x})$

15. $\lim\limits_{x \to \infty} (x - 2)^2(3 - 5x)$

16. $\lim\limits_{x \to \infty} \frac{x(1 - 3x)}{2x^2 + 1}$

17. $\lim\limits_{x \to 1} \frac{x^2 + 1}{x^2 - 1}$

18. $\lim\limits_{x \to 2} \frac{x^3 - x^2 - x - 2}{x^2 - 3x + 2}$

Graphing: For each given function f, (a) sketch a quick graph of f by hand, using your knowledge of simple graphs, transformations, and/or properties of polynomials or rational functions. Then (b) use derivatives, number lines, and limits to sketch a more detailed graph of the function. Label any important points or features.

19. $f(x) = -x^6$

20. $f(x) = \frac{1}{2}x^7$

21. $f(x) = \frac{1}{x^5}$

22. $f(x) = x^{6/5}$

23. $f(x) = -2x^{1/3}$

24. $f(x) = 2x^{-1/3}$

25. $f(x) = -x(x + 1)^2$

26. $f(x) = 3x^3 + x^2 - 3x - 1$

27. $f(x) = \frac{(x - 1)(x + 2)}{(x - 1)^2}$

28. $f(x) = \frac{x^3 + x^2 + x + 1}{2x^2 + x - 1}$

Modeling a graph with a function: For each of the given graphs, use what you know about transformations and graphs of power, polynomial, and rational functions to find the equation of a function that has the given graph. Be sure that your function has the same intercepts, marked points, and other features of the graph.

29.

30.

31.

32.

33.

34.

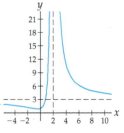

Capstone Problems

A. *Transformations of rational functions:* Prove algebraically that any vertical or horizontal shift or stretch of a rational function is also a rational function. That is, prove that if $f(x) = \frac{p(x)}{q(x)}$ is a rational function, then so are $f(x) + C$, $f(x + C)$, $kf(x)$, and $f(kx)$.

B. *Limits of power, polynomial, and rational functions at infinity:* The limit of a polynomial as $x \to \infty$ depends on the leading term of the polynomial. The limit of a rational function as $x \to \infty$ depends on the ratio of the numerator and denominator polynomials of the rational function. This means that the limit of a rational function as $x \to \infty$ depends on the ratio of the leading terms of the numerator and denominator polynomials of the rational function. Write these statements in terms of limits, and then prove that they are true.

C. *Derivatives of power functions:* Given the fact that the power rule $\frac{d}{dx}(x^k) = kx^{k-1}$ holds when k is a positive integer, use implicit differentiation to prove the following statements:

(a) The power rule $\frac{d}{dx}(x^{-k}) = -kx^{-k-1}$ holds for negative integers $-k$.

(b) The power rule $\frac{d}{dx}(x^{p/q}) = \frac{p}{q}x^{(p/q)-1}$ holds for positive rational numbers $\frac{p}{q}$.

(c) The power rule $\frac{d}{dx}(x^{-p/q}) = -\frac{p}{q}x^{-(p/q)-1}$ holds for negative rational numbers $-\frac{p}{q}$.

D. *Horizontal, slant, and curve asymptotes:* What are the circumstances under which a rational function will have a horizontal asymptote? A slant asymptote? A curve asymptote that is a polynomial of degree k? Are there any other possibilities for the behavior of a rational function as $x \to \pm\infty$, and why or why not?

Calculus with Exponential and Logarithmic Functions

5.1 DEFINING EXPONENTIAL AND LOGARITHMIC FUNCTIONS

▶ Definition, properties, and algebra of exponential and logarithmic functions

▶ The natural exponential function and the natural logarithmic function

▶ The relationship between exponential and logarithmic functions

Exponential Functions

Functions that are not algebraic are called **transcendental** functions. In this book we will investigate four basic types of transcendental functions: exponential, logarithmic, trigonometric, and inverse trigonometric functions. Exponential functions are similar to power functions, but with the roles of constant and variable reversed in the base and exponent:

DEFINITION 5.1   **Exponential Functions**

An **exponential function** is a function that can be written in the form

$$f(x) = Ab^x$$

for some real numbers A and b such that $A \neq 0$, $b > 0$, and $b \neq 1$.

There is an important technical problem with this definition: We know what it means to raise a number to a rational power by using integer roots and powers, but we don't know what it means to raise a number to an irrational power. We need to be able to raise numbers to irrational powers to talk about exponential functions; for example, if $f(x) = 2^x$, then we need to be able to compute $f(\pi) = 2^\pi$. One way to think of b^x where x is irrational is as a limit:

$$b^x = \lim_{\substack{r \to x \\ r \text{ rational}}} b^r.$$

For example, 2^π can be approximated by 2^r for rational numbers r that are close to π:

$$2^\pi \approx 2^{3.14} = 2^{314/100} = \sqrt[100]{2^{314}}.$$

As we consider rational numbers r that are closer and closer to π, the expression 2^r will get closer and closer to 2^π; see Exercise 4. In Chapter 7 we will give a more rigorous definition of exponential functions as the inverses of certain accumulation integrals.

We will assume that you are familiar with the basic algebraic rules of exponents, for example that $b^{x+y} = b^x b^y$, that $b^0 = 1$ for any nonzero b, and that $(b^x)^y = b^{xy}$. Proving those rules requires the more rigorous definition of exponential functions that we will see in Section 7.7 so for the moment we will take these algebraic rules as given. From those basic rules it follows that an exponential function $f(x) = b^x$ is one-to-one, and that b^x is never zero for any value of x. (See Exercises 85 and 86.)

Since we require that $b > 0$, we can see intuitively that for any value of x, the quantity b^x will be defined and in fact must be positive. To prove this we would have to use the more rigorous definition of exponential functions in Chapter 7, but it is true that every exponential function of the form b^x has domain $\mathbb{R}$ and range $(0, \infty)$.

In Theorem 4.2 we summarized basic rules for manipulating expressions with exponents. The following theorem rephrases these rules in notation more compatible with exponential functions, with the variables in the exponents:

THEOREM 5.2   **Rules for Exponents**

For any numbers b, c, x, and y for which the expressions that follow are defined,

(a) $b^{x+y} = b^x b^y$ **(b)** $(bc)^x = b^x c^x$ **(c)** $b^{xy} = (b^x)^y = (b^y)^x$

(d) $b^{x-y} = \dfrac{b^x}{b^y}$ **(e)** $\left(\dfrac{b}{c}\right)^x = \dfrac{b^x}{c^x}$ **(f)** $b^{x/y} = (b^x)^{1/y} = (b^{1/y})^x$

From these rules we can prove two basic facts about exponential functions:

THEOREM 5.3   **Exponential Functions Are One-To-One and Never Zero**

Suppose b is a real number with $b > 0$ and $b \neq 1$.

(a) The value of b^x is never zero for any $x \in \mathbb{R}$.

(b) The function given by $y = b^x$ is one-to-one and therefore invertible.

Proof. Suppose $b > 0$ and $b \neq 1$. To prove part (a) suppose, seeking a contradiction, that $b^x = 0$ for some value of x. Clearly we could not have $x = 0$, because we know that $b^0 = 1$, which is not equal to 0. If $x \neq 0$, then we can use the rules of exponents to conclude that

$$b^x = 0 \implies (b^x)^{1/x} = 0^{1/x} \implies b = 0.$$

But by assumption $b \neq 0$. We now have a contradiction, and thus there is no number x such that $b^x = 0$.

For part (b), to show that $y = b^x$ is one-to-one, we must show that whenever $b^x = b^y$, it follows that $x = y$. Using the rules of exponents, we have

$$b^x = b^y \implies \frac{b^x}{b^y} = 1 \implies b^{x-y} = 1.$$

It now suffices to show that $b^z = 1$ only when $z = 0$. Seeking a contradiction, suppose $z \neq 0$ and $b^z = 1$. Then we could write $(b^z)^{1/z} = 1^{1/z} = 1$ and conclude that $b = 1$, which violates our assumption that $b \neq 1$. Therefore $b^z = 1$ only when $z = 0$, and thus since $b^{x-y} = 1$, it follows that $x - y = 0$. This means that $x = y$, as desired.   ■

Natural Exponential Functions

Interestingly, the most natural base b to use for an exponential function isn't a simple integer, like $b = 2$ or $b = 3$. Instead, for reasons that will become clear later in this chapter, the most natural base is the irrational number known as e, and the function e^x is therefore called the ***natural exponential function***. An approximation of the number e to 65 digits is:

2.71828182845904523536028747135266249775724709369995749669676277 . . . .

Of course, since e is an irrational number, we cannot define it just by writing an approximation of e in decimal notation; we will define e properly with a limit in the next section.

It turns out that every exponential function can be written so that its base is the natural number e, as the next theorem states:

THEOREM 5.4   **Natural Exponential Functions**

Every exponential function can be written in the form

$$f(x) = Ae^{kx}$$

for some real number A and some nonzero real number k.

Proving this is so requires us to develop notation for the inverses of exponential functions. That is where logarithms come in; see the discussion after Theorem 5.6.

As we will prove with limits and derivatives later in this chapter, every exponential function has a graph similar to either the **exponential growth** graph that follows at the left or the **exponential decay** graph at the right, depending on the values of k and b. Of course, if the coefficient A is negative, then the graph of $f(x) = Ae^{kx}$ or $f(x) = Ab^x$ will be a reflection of one of these two graphs over the x-axis.

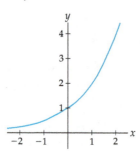

$$f(x) = e^{kx} \text{ with } k > 0,$$
$$f(x) = b^x \text{ with } b > 1$$

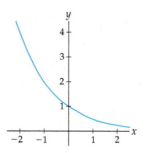

$$f(x) = e^{kx} \text{ with } k < 0,$$
$$f(x) = b^x \text{ with } 0 < b < 1$$

Logarithmic Functions

Since every exponential function b^x is one-to-one, every exponential function has an inverse. These inverses are what we call the **logarithmic functions**:

DEFINITION 5.5

Logarithmic Functions as Inverses of Exponential Functions

The inverse of the exponential function b^x is the **logarithmic function**

$$\log_b x.$$

As a special case, the inverse of the natural exponential function e^x is the **natural logarithmic function**

$$\log_e x = \ln x.$$

We require that the base b satisfy $b > 0$ and $b \neq 1$, because these are exactly the conditions we must have for $y = b^x$ to be an exponential function. In Section 7.7 we will define logarithms another way, in terms of integrals and accumulation functions.

We can use properties of inverse functions to write down some relationships between logarithmic and exponential functions:

THEOREM 5.6

Relationships Between $\log_b x$ and b^x

Since exponential functions and logarithmic functions are inverses of each other, for $b > 0$ and $b \neq 1$ we have the following relationships:

(a) $\log_b(b^x) = x$ for all x, and $b^{\log_b x} = x$ for all $x > 0$;

(b) $\log_b x$ has domain $(0, \infty)$ and range $\mathbb{R}$;

(c) $\log_b y = x$ if and only if $y = b^x$;

(d) The graph of $y = \log_b x$ is the graph of $y = b^x$ reflected over the line $y = x$.

These relationships are just the statements about inverse functions from Definition 0.37 and Theorem 0.39 of Chapter 0, with $f(x) = b^x$ and $f^{-1}(x) = \log_b x$. They are of course also true when $b = e$; that is, for $\ln x$ and e^x.

Part (a) of Theorem 5.6 is what allows us to prove that every exponential function can be written so that its base is e. Suppose $f(x) = Ab^x$ is an exponential function. Then $b = e^{\ln b}$, and thus, if $k = \ln b$ we have

$$f(x) = Ab^x = A(e^{\ln b})^x = Ae^{(\ln b)x} = Ae^{kx}.$$

Part (d) of the theorem tells us that the graphs of logarithmic functions can be obtained easily from the graphs of exponential functions by reflection over the line $y = x$, resulting in the following graphs:

$y = b^x$ and $y = \log_b x$ with $b > 1$ $y = b^x$ and $y = \log_b x$ with $0 < b < 1$

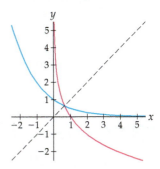

The rules for manipulating logarithmic expressions follow directly from the rules for manipulating exponential expressions; see Exercises 92–96 for proofs.

THEOREM 5.7

Algebraic Rules for Logarithmic Functions

For all values of x, y, b, and a for which these expressions are defined, we have

(a) $\log_b x = y$ if and only if $b^y = x$

(b) $\log_b(b^x) = x$

(c) $b^{\log_b x} = x$

(d) $\log_b(x^a) = a \log_b x$

(e) $\log_b(xy) = \log_b x + \log_b y$

(f) $\log_b\left(\frac{1}{x}\right) = -\log_b x$

(g) $\log_b\left(\frac{x}{y}\right) = \log_b x - \log_b y$

(h) $\log_b x = \dfrac{\log_a x}{\log_a b}$

The first three properties follow from properties of inverse functions, and tell us that $\log_b x$ is the exponent to which you have to raise b in order to get x. For example, $\log_2 8$ is the power to which you have to raise 2 to get 8; since $2^3 = 8$, we have $\log_2 8 = 3$. All of these rules also apply to the natural exponential function, because $\ln x$ is just $\log_b x$ with base $b = e$.

Properties (d) and (e) follow from the algebraic rules of exponents, and properties (f) and (g) are their immediate consequences. The final property in Theorem 5.7 is called the **base conversion formula**, because it allows us to translate from one logarithmic base to another. The base conversion formula is especially helpful for converting to base e or base 10 so that we can calculate logarithms on a calculator. For example, $\log_7 2$ is equal to $\frac{\ln 2}{\ln 7}$, which we can approximate using the built-in "ln" key on a calculator.

Examples and Explorations

EXAMPLE 1

Identifying exponential functions

Determine which of the given functions are exponential. Write each exponential function in the form $f(x) = Ab^x$ for some real numbers A and b.

(a) $f(x) = 4(1.2)^x - 5$　　　**(b)** $f(x) = 3.1x^{7.4}$　　　**(c)** $f(x) = 2^x 3^{1-x}$

SOLUTION

The functions in parts (a) and (b) are not exponential functions, because they cannot be written in the form $f(x) = Ab^x$. The function in part (a) is a translation of an exponential function, but is not itself exponential. The function in part (b) is not an exponential function because the variable x appears in the base of the expression.

The function in part (c) can be written in the form $f(x) = Ab^x$ after some algebra:

$$f(x) = 2^x 3^{1-x} = 2^x 3^1 3^{-x} = 2^x (3)\frac{1}{3^x} = 3(2)^x \left(\frac{1}{3}\right)^x = 3\left((2)\left(\frac{1}{3}\right)\right)^x = 3\left(\frac{2}{3}\right)^x.$$

Thus $f(x) = Ab^x$ with $A = 3$ and $b = \frac{2}{3}$. □

EXAMPLE 2

Using properties of inverse functions to simplify expressions

Use the fact that $y = e^x$ and $y = \ln x$ are inverses to simplify the following expressions:

(a) $e^{\ln 3}$　　　　　　　**(b)** $e^{3 \ln x}$　　　　　　　**(c)** $\ln(e^{3x})$

SOLUTION

(a) Since $e^{\ln x} = x$ for all $x > 0$, it follows that $e^{\ln 3} = 3$.

(b) We know that $e^{\ln x} = x$ for all $x > 0$, but the expression $e^{3 \ln x}$ is not quite in this form. We must first do some algebra to take care of the extra "3" in the expression:

$$e^{3 \ln x} = (e^{\ln x})^3 = x^3.$$

(c) Since $\ln(e^x) = x$ for any value of x, it follows that $\ln(e^{3x}) = 3x$ for all values of x. □

EXAMPLE 3

Converting between forms of exponential functions

Write the function $f(x) = 2^x$ in the form $f(x) = e^{kx}$, and write the function $g(x) = e^{2x}$ in the form $g(x) = b^x$.

SOLUTION

For the first part, we need to find a number k so that $2^x = e^{kx}$. Taking the natural logarithm of both sides of this equation, we have

$$2^x = e^{kx} \implies \ln(2^x) = \ln(e^{kx}) \implies x \ln 2 = kx \implies k = \ln 2 \approx 0.693147.$$

Therefore

$$f(x) = 2^x = e^{(\ln 2)x} \approx e^{0.693147x}.$$

For the second part, we need to find a number b so that $e^{2x} = b^x$. We could solve for b by taking logarithms as we just did, but for variety we present another way here. With a little algebra we can see that b must be e^2:

$$g(x) = e^{2x} = (e^2)^x \approx (7.38906)^x.$$ □

| EXAMPLE 4 | **Evaluating logarithmic expressions** |

Calculate each of the following by hand, without a calculator.

(a) $\log_2 64$ **(b)** $\log_6 3 + \log_6 12$ **(c)** $2\log_3 6 - \log_3 4$

SOLUTION

(a) The number $\log_2 64$ is the exponent to which you must raise 2 in order to get 64. Since $2^6 = 64$, we have $\log_2 64 = 6$.

(b) $\log_6 3$ is the exponent to which we would have to raise 6 to get 3; think $6^? = 3$. It is not immediately apparent what this exponent is. Similarly, it is not clear how to calculate $\log_6 12$ without a calculator. However, using the additive property of logarithms we can write

$$\log_6 3 + \log_6 12 = \log_6(3 \cdot 12) = \log_6 36 = 2.$$

The final equality holds because $6^2 = 36$.

(c) Calculating $\log_3 6$ and $\log_3 4$ by hand is not possible, since we know that there are no obvious numbers that can fill in the boxes in the expressions $3^\square = 6$ and $3^\square = 4$. However, if we apply the algebraic rules for logarithms, it turns out that this is an easy logarithm in disguise:

$$2\log_3 6 - \log_3 4 = \log_3(6^2) - \log_3 4 = \log_3\left(\frac{6^2}{4}\right) = \log_3\left(\frac{36}{4}\right) = \log_3 9 = 2.$$

| EXAMPLE 5 | **Solving equations that involve exponential functions** |

Solve each of the following equations:

(a) $3.25(1.72)^x = 1000$ **(b)** $3^x = 9^{x-2}$ **(c)** $9(9^x) - 10(3^x) + 1 = 0$

SOLUTION

(a) To solve for x we will isolate the expression $(1.72)^x$ and then apply the natural logarithm so that we can get x out of the exponent:

$$3.25(1.72)^x = 1000 \implies \ln((1.72)^x) = \ln\left(\frac{1000}{3.25}\right) \implies x\ln(1.72) = \ln\left(\frac{1000}{3.25}\right).$$

It is now a simple matter to solve for

$$x = \frac{\ln(1000/3.25)}{\ln(1.72)} \approx 10.564.$$

(b) We will use algebra to write each side of the equation with the same base and then use the fact that exponential functions are one-to-one to set the exponents equal to each other:

$$3^x = 9^{x-2} \implies 3^x = (3^2)^{x-2} \implies 3^x = 3^{2(x-2)} \implies x = 2(x-2) \implies x = 4.$$

One way to check this answer is to verify that $x = 4$ satisfies the original equation. It does, since $3^4 = 81$ and $9^{4-2} = 9^2 = 81$. Of course, this won't tell you if you missed any solutions, but it does tell you that $x = 4$ is a valid solution.

(c) Since $9 = 3^2$, we can rewrite this equation as

$$9(3^x)^2 - 10(3^x) + 1 = 0.$$

This equation is in the form $A(3^x)^2 + B(3^x) + C = 0$, which is a quadratic equation in which the role of x is played by 3^x. We can factor it as follows:

$$(9(3^x) - 1)(3^x - 1) = 0.$$

To find the solutions, we set each factor equal to zero and solve. When $9(3^x) - 1 = 0$, we have $3^x = \frac{1}{9}$ and thus $x = -2$. When $3^x - 1 = 0$, we have $3^x = 1$ and thus $x = 0$. Therefore the solutions of the equation are $x = -2$ and $x = 0$.   □

EXAMPLE 6   **Domains and graphs of exponential and logarithmic functions**

Find the domain of each of the functions that follow. Then use transformations to sketch careful graphs of each function by hand, without a graphing calculator.

(a) $f(x) = 5 - 3e^{1.7x}$ **(b)** $g(x) = \dfrac{1}{\ln(x-2)}$

SOLUTION

(a) The domain of $f(x) = 5 - 3e^{1.7x}$ is $\mathbb{R}$, and the graph of f is a transformation of the exponential growth function $e^{1.7x}$ shown in the left-hand figure that follows. $y = -3e^{1.7x}$ can be obtained by reflecting the leftmost graph over the x-axis and then stretching vertically by a factor of 3, as shown in the middle figure. The graph of $f(x) = 5 - 3e^{1.7x}$ can now be obtained by shifting the middle graph up five units, as shown at the right.

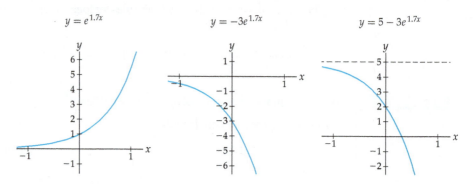

(b) For the function $g(x) = \dfrac{1}{\ln(x-2)}$ to be defined at a value x, we must have $x - 2 > 0$, and thus $x > 2$. We must also have $\ln(x-2) \neq 0$, which means that $x - 2 \neq 1$, and thus $x \neq 3$. Therefore the domain of $g(x)$ is $(2, 3) \cup (3, \infty)$. To sketch the graph of $g(x) = \dfrac{1}{\ln(x-2)}$ we start with the graph of $y = \ln x$ in the left-hand figure that follows, translate to the right two units as shown in the middle figure, and then sketch the reciprocal as shown at the right.

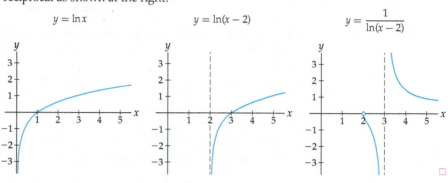

? **TEST YOUR UNDERSTANDING**

▶ Why do we require that $A \neq 0$ and $b > 0$, $b \neq 1$ in the definition of exponential functions? (Think about what the graph of $y = Ab^x$ would look like if $A = 0$, $b \leq 0$, or $b = 1$ and whether these have typical exponential behavior.)

▶ Is $f(x) = x^x$ an exponential function? Is it a power function? Why or why not?

▶ In the reading we calculated $\log_7 2$ by finding $\frac{\ln 2}{\ln 7}$ with a calculator. Would we get the same answer if we computed $\frac{\log_{10} 2}{\log_{10} 7}$?

▶ Where are the functions $f(x) = b^x$ and $g(x) = \log_b x$ defined? Where are they zero?

▶ List ten algebraic rules for manipulating exponential and/or logarithmic expressions.

EXERCISES 5.1

Thinking Back

Working with exponents: Use algebra to simplify each of the following expressions until just one exponent appears in the expression.

▶

▶ $a^c b^{2c}$

▶

Inverse functions: Suppose f and g are inverses of each other.

▶ What can you say about $f(g(x))$ and $g(f(x))$?

▶ If f has a horizontal asymptote at $y = 0$, what can you say about g?

▶ If f has a y-intercept at $y = 1$, what can you say about g?

Concepts

0. *Problem Zero:* Read the section and make your own summary of the material.

1. *True/False:* Determine whether each of the statements that follow is true or false. If a statement is true, explain why. If a statement is false, provide a counterexample.

(a) *True or False:* The function $f(x) = 3e^{0.5x} - 2$ is an exponential function.

(b) *True or False:* The function $f(x) = 3e^x + 2e^x$ is an exponential function.

(c) *True or False:* For all $x > 0$, $\ln(x^3) = 3 \ln x$.

(d) *True or False:* For all $x > 0$, $\dfrac{\log_2 x}{\log_2 3} = \dfrac{\log_6 x}{\log_6 3}$.

(e) *True or False:* If $b > 0$ and $b \neq 1$, and if x and y are any real numbers, then $b^x = b^y$ if and only if $x = y$.

(f) *True or False:* $3(2^x)$ is always equal to 6^x.

(g) *True or False:* Every exponential function has a horizontal asymptote at $y = 0$.

(h) *True or False:* The function $f(x) = 50 - 2^x$ has a horizontal asymptote at $y = 50$.

2. *Examples:* Construct examples of the thing(s) described in the following. Try to find examples that are different than any in the reading.

(a) Two exponential functions and their inverses.

(b) Three exponential functions of the form $f(x) = Ae^{kx}$ that represent exponential growth.

(c) Three exponential functions of the form $f(x) = Ab^x$ that represent exponential decay.

3. What is the definition of an exponential function, and how is such a function different from a power function? Is the function $f(x) = x^x$ a power function, an exponential function, or neither, and why?

4. In this exercise we will examine two ways to think of a^b when b is an irrational number, and in particular, we will consider what the quantity 2^π represents.

(a) One way to define 2^π is to think of it as a limit. If we take a sequence $a_1, a_2, a_3, \ldots$ of rational numbers that approaches π, then the sequence $2^{a_1}, 2^{a_2}, 2^{a_3}, \ldots$ should approach 2^π. Said in terms of limits, this means that

$$2^\pi = \lim_{a \to \pi} 2^a,$$

where each a is assumed to be a rational number. Can you think of a sequence of rational numbers that gets closer and closer to π? (*Hint: Think about the decimal expansion of π.*)

(b) Another way to consider 2^π is to write it as an infinite product:

$$2^\pi = 2^3 \, 2^{1/10} \, 2^{4/100} \, 2^{1/1000} \, 2^{5/10000} \, 2^{9/100000} \cdots.$$

What will the next term in the product be? How could 2^π equal the product of infinitely many numbers? Wouldn't that make 2^π infinitely large? Calculate some of the later terms in the product (for example, $2^{5/10000}$ or $2^{9/100000}$), and use these calculations to argue that even though 2^π can be written as a product of infinitely many numbers, it is not necessarily infinitely large.

5. Approximate $2^{\sqrt{3}}$ by calculating 2^r for rational values r that get closer and closer to $\sqrt{3}$. (*Hint: You can use the decimal expansion of $\sqrt{3}$ to get a sequence of rational numbers that approaches $\sqrt{3}$.*)

6. Why can't we define the number e just by writing it down in decimal notation to lots of decimal places?

7. Write the exponential function $f(x) = 3e^{-2x}$ in the form Ab^x for some real numbers A and b. Then write the exponential function $g(x) = -2(3^x)$ in the form Ae^{kx} for some real numbers A and k.

8. Fill in each blank with an interval of real numbers.

   (a) An exponential function $f(x) = Ab^x$ represents exponential growth if $b \in$ _____ and exponential decay if $b \in$ _____ .

   (b) An exponential function $f(x) = Ae^{kx}$ represents exponential growth if $k \in$ _____ and exponential decay if $k \in$ _____ .

   (c) Suppose that $e^{kx} = b^x$ for some real numbers k and b. Then $k \in (0, \infty)$ if and only if $b \in$ _____ .

   (d) Suppose that $e^{kx} = b^x$ for some real numbers k and b. Then $k \in (-\infty, 0)$ if and only if $b \in$ _____ .

9. The functions $f(x) = 2^{(x^2)}$ and $g(x) = (2^x)^2$ look similar, but are very different functions. (This is why we try to avoid using the ambiguous notation 2^{x^2}.) Calculate $f(3)$ and $g(3)$, and show that they are not the same (and thus that $f(x)$ and $g(x)$ are not the same function). Then find all the values for which $f(x) = g(x)$.

10. Every exponential function of the form $f(x) = b^x$ (with $b > 0$ and $b \neq 1$) is one-to-one. Explain why this fact implies that $b^x = b^y$ if and only if $x = y$.

11. In the definition of the logarithmic function $\log_b x$, what are the allowable values for the base b, and why?

12. Fill in the blanks in each of the following statements.

   (a) For all $x \in$ _____ , $\log_2 x = y$ if and only if $x =$ _____ .

   (b) For all $x \in$ _____ , $3^{\log_3 x} =$ _____ .

   (c) For all $x \in$ _____ , $\log_4(4^x) =$ _____ .

   (d) $\log_2 3$ is the exponent to which you have to raise _____ to get _____ .

13. The graphs of $y = \log_2 x$ and $y = \log_4 x$ are shown here. Determine which graph is which, without using a calculator. (Hint: Think about the graphs $y = 2^x$ and $y = 4^x$, and then reflect those graphs over the line $y = x$.)

$y = \log_2 x$ and $y = \log_4 x$

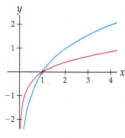

14. State the algebraic properties of the natural logarithm function that correspond to the eight properties of logarithmic functions in Theorem 5.7.

15. Use algebraic properties of logarithms, the graph of $y = \ln x$, and your knowledge of transformations to sketch graphs of $f(x) = \ln(x^2)$ and $g(x) = \ln\left(\dfrac{1}{x}\right)$.

16. Determine whether each of the given statements is true or false. Explain your answers.

   (a) *True or False:* $(\log_2 5)^3 = 3\log_2 5$.

   (b) *True or False:* $\dfrac{\log_3 7}{\log_3 2} = \log_3 5$.

   (c) *True or False:* $\dfrac{\log_5 2}{\log_5 7} = \dfrac{2}{7}$.

   (d) *True or False:* $(\log_{10} 3)(\log_{10} 4) = \log_{10} 7$.

   (e) *True or False:* $(\log_{10} 3)(\log_{10} 4) = \log_{10} 12$.

   (f) *True or False:* $\log_4 5 + \log_3 5 = \log_7 5$.

17. Use the base conversion formula for logarithms to argue that the function $f(x) = \log_2 x$ is equal to the function $g(x) = \log_3 x$ only when $x = 1$.

18. Explain how we know that logarithmic functions are one-to-one. Why does this mean that $A = B$ if and only if $\log_b A = \log_b B$ (assuming that A and B are positive)?

19. Solve the equation $3^{x-1} = 2$ in two ways: (a) by using the natural logarithm function and (b) by using the base-3 logarithm function. Show that the answers you get are the same.

20. Solve the equation $4(2^x) = 3$ in two ways: (a) by immediately applying the natural logarithm function to both sides and then solving and (b) by isolating the expression 2^x and *then* applying the natural logarithm function to both sides. Do you get the same answer? Is one way easier than the other?

21. If $e^{kx} = b^x$ for all values of x, how must the constants k and b be related? Express your answer in two ways: with k in terms of b as well as with b in terms of k.

22. If $e^{kx} = b^x$ and $k > 0$, what can you say about b? If $e^{kx} = b^x$ and $k < 0$, what can you say about b?

Skills

Determine whether or not each function in Exercises 23–28 is exponential. For those which are, use algebra to write them (a) in the form $f(x) = Ab^x$ and (b) in the form $f(x) = Ae^{kx}$.

23. $f(x) = \dfrac{3}{2^x}$

24. $f(x) = \dfrac{3^x}{2^x}$

25. $f(x) = 2e^{3x-4}$

26. $f(x) = 3e^{x+1}$

27. $f(x) = 5^x 2^{3-x}$

28. $f(x) = 3^{x^2+1}$

Find the exact values of each of the quantities in Exercises 29–36. Do not use a calculator.

29. $e^0 + \ln 1 + \ln e$

30. $e^{2\ln 3}$

31. $e^{\ln 3 + \ln 2}$

32. $2^{1+3\log_2 1}$

33. $\ln\left(\dfrac{1}{e^2}\right)$

34. $\log_{1/2} 4$

35. $4\log_2 6 - 2\log_2 9$

36. $\dfrac{\log_7 9}{\log_7 1/3} + \log_3 1$

Find the domains of the functions in Exercises 37–48.

37. $f(x) = \frac{1}{3}(3^{-2x+1})$

38. $f(x) = \frac{x-1}{2(4^{3x})}$

39. $f(x) = \frac{1}{3^x - 1}$

40. $f(x) = \sqrt{\left(\frac{1}{2}\right)^x}$

41. $f(x) = \frac{\ln(x+1)}{\ln(x-2)}$

42. $f(x) = \frac{1}{e^x - e^{2x}}$

43. $f(x) = \frac{1}{\sqrt{\ln(x-1)}}$

44. $f(x) = \frac{1}{1 - e^{\ln x}}$

45. $f(x) = \ln\left(\frac{x+1}{x-2}\right)$

46. $f(x) = \frac{\ln(x^2-1)}{\ln(\sqrt{3x+1})}$

47. $f(x) = \frac{\sqrt{x+1}}{3^x - 2^x}$

48. $f(x) = \frac{1}{9^x - 2(3^x) + 1}$

Solve the equations in Exercises 49–62 by hand. When you are finished, check your answers either by testing your solutions or by graphing.

49. $\ln(x^2 + x - 5) = 0$

50. $1.76(0.24)^x = 500$

51. $x3^{x+1} = x^2 3^x$

52. $\frac{2^x + 1}{3^x - 5} = 0$

53. $2\ln(3x^2 - 1) = 4$

54. $10^x = e^x$

55. $2^x = 3^{x-1}$

56. $2 = 10\left(1 + \frac{0.19}{12}\right)^{12x}$

57. $\log_2\left(\frac{x-1}{x+1}\right) = 4$

58. $25^x - 26(5^x) + 25 = 0$

59. $\frac{1}{4^x} - \frac{5}{2^x} + 4 = 0$

60. $4(2^{2x}) = 1 - 5(2^x)$

61. $\frac{3^x + 1}{3^x - 1} = 4$

62. $\frac{3^{x-1}}{3^{2x+1}} = 4$

Sketch graphs of the functions in Exercises 63–70 by hand, without using a calculator. Indicate any roots, intercepts, and asymptotes on your graphs.

63. $f(x) = 2(3^x)$

64. $f(x) = -2e^{-3x}$

65. $f(x) = -\left(\frac{1}{2}\right)^x + 10$

66. $f(x) = -0.25(3^{x-2})$

67. $f(x) = 20 - 5e^{-2x}$

68. $f(x) = \log_{1/2} x$

69. $f(x) = -\log_2(x-3)$

70. $f(x) = 3\log_4 x - 8$

For each graph in Exercises 71–74, find a function whose graph looks like the one shown. When you are finished, use a graphing calculator to check that your function f has the properties and features of the given graph.

71.

72.

73.

74.

For each pair of values $f(a)$ and $f(b)$ in Exercises 75–80, find an exponential function $f(x) = Ae^{kx}$ that has those given values. (*Hint: Set up equations to solve for the constants A and k.*)

75. $f(0) = 9, f(2) = 6$

76. $f(-4) = 10, f(0) = 2$

77. $f(0) = -3, f(5) = -7$

78. $f(2) = 18, f(3) = 54$

79. $f(-2) = 8, f(3) = \frac{1}{4}$

80. $f(-1) = 1, f(1) = 3$

Applications

81. The population of fish in a stream t years after 1990 is given by the exponential function $P(t) = 750(1.08)^t$. In what year will the fish population have doubled?

82. In January you borrowed $500 from LoanSharks, Incorporated. LoanSharks charges interest on your debt in such a way that after t months you will owe the company $L(t) = 500(1.17)^t$ dollars. How much will you owe them in April? How much will you owe LoanSharks in November?

83. Ten years ago, Jenny deposited $10,000 into an investment account. Her investment account now holds $22,609.80. Her accountant tells her that her investment account balance $I(t)$ is an exponential function.

(a) Find an exponential function of the form $I(t) = Ae^{kt}$ to model Jenny's investment account balance.

(b) Find an exponential function of the form $I(t) = Ab^t$ to model Jenny's investment account balance.

84. Suppose there were 500 rats on a certain island in 1973 and 1,697 rats on the same island 10 years later. Assume that the number $R(t)$ of rats on the island t years after 1973 is an exponential function.

(a) Find an equation for the exponential function $R(t)$ that describes the number of rats on the island. Let $t = 0$ represent the year 1973.

(b) According to your function $R(t)$, how many rats will be on the island in 2020?

(c) How long did it take for the population of rats to double from its 1973 amount? How long did it take for it to double again? And again?

85. Suppose a rock sample initially contains 250 grams of the radioactive substance unobtainium, and that the amount of unobtainium after t years is given by an exponential function of the form $S(t) = Ae^{kt}$. The half-life of unobtainium is 29 years, which means that it takes 29 years for the amount of the substance to decrease by half.

(a) Find an equation for the exponential function $S(t)$.

(b) What percentage of unobtainium decays each year?

(c) How long will it be before the rock sample contains only 6 grams of unobtainium?

86. Again considering the rock sample described in Exercise 85, answer the following questions:

(a) At one point the rock sample contained 900 grams of unobtainium; how long ago?

(b) What percentage of the unobtainium will be left in 300 years?

(c) How long will it be before 95% of the unobtainium has decayed?

Proofs

87. Prove by contradiction that every exponential function $f(x) = Ab^x$ has the property that $f(x)$ is never zero. *(Hint: Use what you know about the algebraic properties of exponential functions, and the fact that if $f(x) = Ab^x$ is an exponential function, then neither A nor b is zero.)*

88. Use the definition of a one-to-one function to prove that every exponential function $f(x) = Ab^x$ is one-to-one. *(Hint: Use the fact that $b^x = 1$ only when $x = 0$.)*

89. Use the base conversion formula for logarithms to prove that the function $f(x) = \log_2 x$ is equal to the function $g(x) = \log_3 x$ only when $x = 1$.

90. Use logarithms to prove that every exponential function of the form $f(x) = Ab^x$ can be written in the form $f(x) = Ae^{kx}$, and vice versa.

91. Use the definition of a logarithmic function $y = \log_b x$ to prove that for any $b > 0$ with $b \neq 1$, the quantity $\log_b 1$ is equal to zero.

In Exercises 92–96, assume that x, y, a, and b are values which make sense in the expressions involved.

92. Use the fact that logarithmic functions are the inverses of exponential functions to prove that:

(a) $\log_b x = y$ if and only if $b^y = x$

(b) $\log_b (b^x) = x$

(c) $b^{\log_b x} = x$

93. Prove that $\log_b (x^a) = a \log_b x$. *(Hint: Start with $\log_b (x^a)$ and replace x with $b^{\log_b x}$.)*

94. Prove that $\log_b (xy) = \log_b x + \log_b y$. *(Hint: Show that this statement is equivalent to the statement $xy = b^{\log_b x + \log_b y}$, and prove the new statement instead.)*

95. Use the results of the two exericses above to prove that:

(a) $\log_b \left(\dfrac{1}{x}\right) = -\log_b x$

(b) $\log_b \left(\dfrac{x}{y}\right) = \log_b x - \log_b y$

96. Prove the base conversion formula $\log_b x = \dfrac{\log_a x}{\log_a b}$. *(Hint: Set $y = \log_b x$ and then show that $b^y = x$.)*

Thinking Forward

▶ *A special exponential limit:* Use a calculator to approximate $\dfrac{e^h - 1}{h}$ for the following values of h: (a) $h = 0.1$; (b) $h = 0.01$; (c) $h = 0.001$. As h gets closer to zero, what number does your approximations seem to approach?

▶ *Logarithms with absolute values:* Sketch a graph of the function $f(x) = \ln |x|$. What is the domain of this function? Is the function even, odd, or neither, and why?

5.2   LIMITS OF EXPONENTIAL AND LOGARITHMIC FUNCTIONS

▶   Giving a proper definition for the irrational number e

▶   Continuity of exponential and logarithmic functions on their domains

▶   Calculating limits of exponential and logarithmic functions

Defining the Number e

Before we can discuss continuity and limits of exponential functions, we must have a proper definition for the irrational number e that we have so far been approximating as $e \approx 2.71828$. It turns out that this definition itself involves a limit.

DEFINITION 5.8

The Number e

We define e to be the number that $(1 + h)^{1/h}$ approaches as h approaches 0:

$$e = \lim_{h \to 0} (1 + h)^{1/h}.$$

It is important to note that this is a weak definition, because we have not proven that the limit in this definition exists. That is, we have not shown that $(1 + h)^{1/h}$ converges to a real number as $h \to 0$.

For small values of h it is easy to see that the quantity $(1 + h)^{1/h}$ is close to the approximation $e \approx 2.71828$ we have been using so far. For example, when $h = 0.0001$, we have

$$(1 + 0.0001)^{1/0.0001} \approx 2.71815.$$

Proving that the limit in Definition 5.8 does converge to a real number is beyond the scope of this chapter. In Example 1 and Exercise 12 we will use tables of values to show that this limit is reasonable. We will use accumulation integrals to give another definition of the number e in Section 7.7.

As we will see in the next section, our definition of the number e is partially motivated by derivatives. Specifically, the reason that e is the natural base for exponential functions has to do with the slope of the graph of $y = e^x$ at $x = 0$. In the next section we will see that the slope of the graph of an exponential function $y = b^x$ at $x = 0$ is given by the limit $\lim_{h \to 0} \frac{b^h - 1}{h}$. We can use our definition of e to show that when $b = e$, this slope is equal to 1:

THEOREM 5.9

Another Characterization of the Number e

The number e satisfies the following limit statement:

$$\lim_{h \to 0} \frac{e^h - 1}{h} = 1.$$

Proof.   We will give here only a convincing argument that uses approximations. Given that $e = \lim_{h \to 0} (1 + h)^{1/h}$ as in Definition 5.8, for sufficiently small values of h we have

$$e \approx (1 + h)^{1/h} \quad \Longleftrightarrow \quad e^h \approx 1 + h \quad \Longleftrightarrow \quad e^h - 1 \approx h \quad \Longleftrightarrow \quad \frac{e^h - 1}{h} \approx 1$$

Since the preceding approximations get better as $h \to 0$, it is reasonable that $\lim_{h \to 0} \frac{e^h - 1}{h} = 1$.   ■

Continuity of Exponential and Logarithmic Functions

We can use the definition of e to prove that exponential and logarithmic functions are continuous everywhere they are defined. This allows us to calculate limits such as $\lim_{x \to 2} 3^x = 3^2 = 9$ and $\lim_{x \to 1} \ln x = \ln 1 = 0$ by simple evaluation.

THEOREM 5.10 | **Continuity of Exponential and Logarithmic Functions**

All exponential and logarithmic functions are continuous on their domains.

Proof. The proof hinges entirely on algebra, limit rules, and the definition of e in Definition 5.8. We will prove continuity for (a) the natural exponential function and (b) the natural logarithmic function here, and use limit rules to extend these results to general exponential and logarithmic functions in Exercises 84–86. The proofs are a little technical, but without them we would not be able to compute even the simplest limits of exponential and logarithmic functions!

(a) To prove that e^x is continuous on its domain $\mathbb{R}$ we must show that for all $c \in \mathbb{R}$ we have $\lim_{x \to c} e^x = e^c$. If we want to use the definition of e, then we need to have a limit as $h \to 0$, so we define $h = x - c$. Then $x = c + h$, and as $x \to c$, we have $h \to 0$. This makes our limit equal to

$$\lim_{x \to c} e^x = \lim_{h \to 0} e^{c+h} = \lim_{h \to 0} e^c e^h = e^c \lim_{h \to 0} e^h.$$

The last step follows from the constant multiple rule for limits, since e^c is a constant. At this point we would be done if we could show that $\lim_{h \to 0} e^h = e^0 = 1$. In other words, the proof that e^x is continuous for all x essentially boils down to showing that it is continuous at one point, namely, 0. To finish the proof we employ a series of algebraic manipulations followed by some limit rules and the definition of e:

$$\lim_{h \to 0} e^h = \lim_{h \to 0} (e^h - 1 + 1) = \lim_{h \to 0} \left(\frac{e^h - 1}{h} \cdot h + 1 \right) \qquad \leftarrow \text{algebra}$$

$$= \left(\lim_{h \to 0} \frac{e^h - 1}{h} \right) (\lim_{h \to 0} h) + (\lim_{h \to 0} 1) \qquad \leftarrow \text{limit rules}$$

$$= (1)(0) + (1) = 1. \qquad \leftarrow \text{Theorem 5.9}$$

One technical point: In the second line of the above calculation we applied the product rule for limits, which is only valid when the limits involved are known to exist. Therefore this proof hinges on knowing that $\lim_{h \to 0} \frac{e^h - 1}{h}$ exists, which is a nontrivial fact that we will not prove here.

(b) To show that $\ln x$ is continuous on its domain, we must show that for all $c \in (0, \infty)$ we have $\lim_{x \to c} \ln x = \ln c$. We will use what we just proved about the continuity of e^x. By the composition rule for limits and the fact that e^x is continuous everywhere, for $c > 0$ we have

$$\lim_{x \to c} e^{\ln x} = e^{\lim_{x \to c} \ln x}.$$

Since $\ln x$ is the inverse of e^x, we know that $e^{\ln x} = x$. Therefore for $c > 0$ we also have

$$\lim_{x \to c} e^{\ln x} = \lim_{x \to c} x = c = e^{\ln c}.$$

Since e^x is a one-to-one-function, putting these two calculations together we see that $\lim_{x \to c} \ln x$ must be equal to $\ln c$. (Note that once again, we are making an important assumption here, that $\lim_{x \to c} \ln x$ is equal to some real number. If that number does not exist then we cannot apply the composition rule for limits here.) ∎

Limits at Zero and Infinity

We now know that we can calculate limits of exponential and logarithmic functions at domain points simply by evaluation. What about at non-domain points? Exponential functions are defined everywhere, so the only thing that remains is to determine their limits as $x \to \infty$ and as $x \to -\infty$:

THEOREM 5.11   **Limits of Exponential Functions at Infinity and Negative Infinity**

If k is a positive real number, then

(a) $\lim\limits_{x \to \infty} e^{kx} = \infty$ and $\lim\limits_{x \to -\infty} e^{kx} = 0$.

(b) $\lim\limits_{x \to \infty} e^{-kx} = 0$ and $\lim\limits_{x \to -\infty} e^{-kx} = \infty$.

For example, $f(x) = e^{2x}$ approaches ∞ on the right and approaches 0 on the left. In contrast, $f(x) = e^{-2x}$ does the opposite, approaching 0 on the right and ∞ on the left. We can say a similar thing for exponential functions of the form b^x, with the first case of the theorem occurring when $b > 1$ and the second case when $0 < b < 1$.

Logarithmic functions have domain $(0, \infty)$, so we need to examine their limits as $x \to 0^+$ and as $x \to \infty$. Again there are two cases:

THEOREM 5.12   **Limits of Logarithmic Functions at Infinity and Zero**

(a) If $b > 1$, then $\lim\limits_{x \to \infty} \log_b x = \infty$ and $\lim\limits_{x \to 0^+} \log_b x = -\infty$.

(b) If $0 < b < 1$, then $\lim\limits_{x \to \infty} \log_b x = -\infty$ and $\lim\limits_{x \to 0^+} \log_b x = \infty$.

For example, $f(x) = \log_2 x$ approaches ∞ as $x \to \infty$, and approaches $-\infty$ as $x \to 0^+$. In contrast, $g(x) = \log_{1/2} x$ is flipped over, approaching $-\infty$ as $x \to \infty$ and approaching ∞ as $x \to 0^+$. The important case, in which $b = e$, says that $\lim\limits_{x \to \infty} \ln x = \infty$ and $\lim\limits_{x \to 0^+} \ln x = -\infty$.

These theorems are easier to remember in pictures than in symbols. What they say is that the "ends" of the graphs of exponential and logarithmic functions behave like this:

e^{kx} for $k > 0$
b^x for $b > 1$

e^{-kx} for $k > 0$
b^x for $0 < b < 1$

$\log_b x$ for $b > 1$

$\log_b x$ for $0 < b < 1$

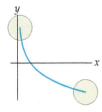

The proofs of these theorems are a bit technical, but don't be put off by that. In practice you need only remember the pictures when evaluating limits of exponential and logarithmic functions at the ends of their domains.

Proof.   We will prove that $\lim\limits_{x \to \infty} \ln x = \infty$ just to show the kind of technical argument that is required. Proofs of some of the other parts of the theorems are in Exercises 87 and 88.

Our proof will require that we return to the definition of a limit at infinity. We wish to prove that $\lim\limits_{x \to \infty} \ln x = \infty$, which by definition means we need to show that, for all $M > 0$, there is some $N > 0$ such that $x > N$ guarantees that $\ln x > M$.

Here is the technical argument: Suppose that we are given some $M > 0$, and consider the number $N = e^M$. If $x > N$, then $x = kN$ for some number $k > 1$. For such values of x,

$$\ln x = \ln(kN) = \ln k + \ln N > \ln N = \ln(e^M) = M.$$

The inequality follows from the fact that $k > 1$ and therefore $\ln k > 0$. The equality after that follows from our choice of $N = e^M$. We have now shown that for $x > N$ we have $\ln x > M$, which is what we wanted to prove.   ◼

As a consequence of the limits in Theorems 5.11 and 5.12, we can say that exponential functions always have a one-sided horizontal asymptote at $y = 0$ and that logarithmic functions always have a vertical asymptote at $x = 0$.

Examples and Explorations

EXAMPLE 1   **Using tables of values to approximate limits related to e**

Use tables of values to approximate each of the limits that follow. In the second limit you will have to use a calculator approximation of e to perform the calculations.

(a) $\displaystyle\lim_{h \to 0} \frac{3^h - 1}{h}$ (b) $\displaystyle\lim_{h \to 0} \frac{e^h - 1}{h}$ (c) $\displaystyle\lim_{h \to 0} (1 + h)^{1/h}$

SOLUTION

(a) This limit is similar to the one in Theorem 5.9, but with the base e replaced by 3. Limits of the form $\displaystyle\lim_{h \to 0} \frac{b^h - 1}{h}$ converge to different numbers, depending on the base b, but such a limit converges to 1 only for the number $b = e$. Thus we would expect that $\displaystyle\lim_{h \to 0} \frac{3^h - 1}{h}$ should converge to some number other than 1. The following table of values of $\frac{3^h - 1}{h}$ suggests that this is indeed the case:

h	-0.1	-0.01	-0.001	0	0.001	0.01	0.1
$\dfrac{3^h - 1}{h}$	1.040415	1.092600	1.098009	?	1.099216	1.104669	1.161232

If the pattern in the table continues, then we would expect this limit to converge to some number between 1.098009 and 1.099216. The numbers in that range are close to 1, but none of them are equal to 1. This is not unexpected, because the base in our limit, which is 3, is close to the base $e \approx 2.71828$ that would cause the limit we seek to approach exactly 1.

(b) From Theorem 1.26 we know that $\displaystyle\lim_{h \to 0} \frac{e^h - 1}{h}$ should approach 1. Let's see if that is the case. For small values of h approaching 0 we have

h	-0.1	-0.01	-0.001	0	0.001	0.01	0.1
$\dfrac{e^h - 1}{h}$	0.951626	0.995017	0.999500	?	1.000500	1.005017	1.051709

As expected, this limit does seem to be approaching 1 as h approaches 0.

(c) We can approximate the limit in Definition 5.8 (and thus the value of e) by using a table of approximate values of $(1 + h)^{1/h}$ for small values of h:

h	−0.1	−0.01	−0.001	0	0.001	0.01	0.1
$(1+h)^{1/h}$	2.867972	2.731999	2.719642	?	2.716924	2.704814	2.593742

Although $(1 + h)^{1/h}$ does not seem to approach a very *nice* number as $h \to 0$, it appears that the limit does exist and is somewhere between 2.719642 and 2.716923. If we evaluate $(1 + h)^{1/h}$ at an extremely small value of h, say, $h = 0.000001$, then we can get a relatively accurate approximation for the limit in Definition 5.8 and thus for the number e:

$$e \approx (1 + 0.000001)^{1/0.000001} \approx 2.718280.$$

EXAMPLE 2   **Calculating limits that involve exponential functions**

Calculate each of the following limits:

(a) $\displaystyle\lim_{x \to 0} \frac{e^{2x} + e^x - 2}{e^x - 1}$ **(b)** $\displaystyle\lim_{x \to \infty} \frac{3^x - 2^x}{1 + 4^x}$ **(c)** $\displaystyle\lim_{x \to \infty} \left(1 + \frac{3}{x}\right)^{2x}$

SOLUTION

(a) All of our limit-solving techniques can be brought to bear on limits that involve exponential functions. In general, we first try evaluating the limit. If we find that it is indeterminate, then we use our usual arsenal of "tricks" to rewrite the limit until it can be evaluated. As $x \to 0$ both the numerator and the denominator approach zero, so this limit is of the form $\frac{0}{0}$, which is indeterminate. After factoring and cancelling we can resolve this problem:

$$\lim_{x \to 0} \frac{e^{2x} + e^x - 2}{e^x - 1} = \lim_{x \to 0} \frac{(e^x - 1)(e^x + 2)}{e^x - 1} = \lim_{x \to 0}(e^x + 2) = e^0 + 2 = 3.$$

(b) By Theorem 6.11, as $x \to \infty$ the expressions 3^x, 2^x, and 4^x all approach ∞. Therefore the limit in question is an indeterminate form. An analog of the method of dividing by the highest power works in this case, but this time we divide numerator and denominator by the exponential function with the largest base:

$$\lim_{x \to \infty} \frac{3^x - 2^x}{1 + 4^x} = \lim_{x \to \infty} \frac{3^x - 2^x}{1 + 4^x}\left(\frac{1/4^x}{1/4^x}\right) = \lim_{x \to \infty} \frac{(3/4)^x - (1/2)^x}{(1/4)^x + 1} = \frac{0 - 0}{0 + 1} = \frac{0}{1} = 0.$$

(c) As $x \to \infty$, the base $1 + \frac{3}{x}$ approaches $1 + 0 = 1$ and the exponent $2x$ approaches ∞. Therefore this limit is of the form 1^∞, which is indeterminate. Fortunately, with a substitution we can rewrite the limit in such a way that allows us to apply the definition of e from Definition 5.8. Let $h = \frac{3}{x}$. Then as $x \to \infty$, we have $h \to 0^+$. Using this relationship and the fact that $x = \frac{3}{h}$, we have

$$\lim_{x \to \infty} \left(1 + \frac{3}{x}\right)^{2x} = \lim_{h \to 0^+} (1 + h)^{2(3/h)} = \left(\lim_{h \to 0^+} (1 + h)^{1/h}\right)^6 = e^6.$$

EXAMPLE 3   **Calculating limits that involve logarithmic functions**

Calculate each of the following limits:

(a) $\displaystyle\lim_{x \to 0^+} \frac{x^2}{\log_3 x}$

(b) $\displaystyle\lim_{x \to \infty} (\log_2(x+1) - \log_2(x^2+1))$

SOLUTION

(a) From the shape of the graph of $y = \log_3 x$ we know that as $x \to 0^+$, the quantity $\log_3 x$ in the denominator approaches $-\infty$. At the same time the numerator x^2 approaches 0, and therefore the limit in question is of the form $\dfrac{0}{-\infty}$. This limit is not indeterminate; limits of this form are always equal to zero. As the numerator approaches zero, the quotient gets smaller. Similarly, as the magnitude of the denominator gets larger, the quotient gets smaller. Both the numerator and the denominator contribute to the quotient approaching 0. Therefore

$$\lim_{x \to 0^+} \frac{x^2}{\log_3 x} = 0.$$

(b) This limit is of the indeterminate form $\infty - \infty$, since as $x \to \infty$ both $\log_2(x+1)$ and $\log_2(x^2+1)$ approach ∞. Luckily, using the algebraic rules of logarithms, we can rewrite that limit in the alternative form, as follows:

$$\lim_{x \to \infty} (\log_2(x+1) - \log_2(x^2+1)) = \lim_{x \to \infty} \log_2\left(\frac{x+1}{x^2+1}\right).$$

As $x \to \infty$, the proper rational function $\frac{x+1}{x^2+1}$ approaches 0, which in turn means that the quantity $\log_2\left(\frac{x+1}{x^2+1}\right)$ approaches $-\infty$. Therefore we have

$$\lim_{x \to \infty} (\log_2(x+1) - \log_2(x^2+1)) = -\infty.$$

$\square$

EXAMPLE 4   **A function with two different horizontal asymptotes**

Find any horizontal asymptotes of the function $f(x) = \dfrac{3}{2+e^{-4x}}$.

SOLUTION

To see if $f(x)$ has any horizontal asymptotes we must examine the limit of $f(x)$ as $x \to \infty$ and as $x \to -\infty$. As $x \to \infty$ we have

$$\lim_{x \to \infty} f(x) = \lim_{x \to \infty} \frac{3}{2+e^{-4x}} \to \frac{3}{2+0} = \frac{3}{2}.$$

As $x \to -\infty$, the quantity $2 + e^{-4x}$ in the denominator approaches ∞, and thus $\frac{3}{2+e^{-4x}}$ approaches 0. Thus we have

$$\lim_{x \to -\infty} f(x) = \lim_{x \to -\infty} \frac{3}{2+e^{-4x}} = 0.$$

The graph of $f(x) = \dfrac{3}{2+e^{-4x}}$ therefore has two different horizontal asymptotes. To the right there is an asymptote at $y = \frac{3}{2}$, and to the left there is an asymptote at $y = 0$.

$\square$

CHECKING THE ANSWER | We can check the answers with a graph, as shown next. Notice that the graph does indeed have horizontal asymptotes at both $y = 0$ and $y = \dfrac{3}{2}$.

$$f(x) = \frac{3}{2+e^{-4x}} \text{ has two horizontal asymptotes}$$

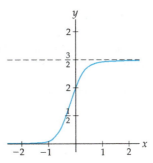

TEST YOUR UNDERSTANDING

▶ What is the definition of the number e? What is an alternative, but equivalent, definition of the number e?

▶ Why is it important for us to know that exponential and logarithmic functions are continuous on their domains? What calculations are made possible by this knowledge?

▶ What can you say about the graph of an exponential function $f(x) = e^{kx}$ or $f(x) = b^x$ as $x \to \infty$? How does it depend of the value of k or b?

▶ What can you say about the graph of a logarithmic function $g(x) = \log_b x$ as $x \to \infty$? How does it depend on the value of b?

▶ Do all exponential functions have an asymptote? What about logarithmic functions?

EXERCISES 5.2

Thinking Back

Asymptotes and Limits: Think back to Chapter 1 to answer each of the following in formal mathematical notation.

▶ By definition, a function $f(x)$ has a horizontal asymptote at $y = 2$ if what limit statement is true?

▶ By definition, a function $f(x)$ has a vertical asymptote at $x = 3$ if what limit statement is true?

▶ Write $\lim\limits_{x \to \infty} f(x) = 5$ as a formal statement involving the quantities ϵ and N.

▶ Write $\lim\limits_{x \to 2^+} f(x) = \infty$ as a formal statement involving the quantities M and δ.

Review of Limit Techniques: Calculate the following limits by hand, and verify your answers with a graphing calculator.

▶ $\lim\limits_{x \to 0} \dfrac{x^2 - x^3}{x^4}$

▶ $\lim\limits_{x \to \infty} \dfrac{x^2 - x^3}{x^4}$

▶ $\lim\limits_{x \to 2} \dfrac{3x^2 + 7x + 2}{x + 2}$

▶ $\lim\limits_{x \to \infty} \dfrac{3x^2 + 7x + 2}{x + 2}$

▶ $\lim\limits_{x \to \infty} (\sqrt[5]{x} - \sqrt[3]{x})$

▶ $\lim\limits_{x \to -\infty} (\sqrt[5]{x} - \sqrt[3]{x})$

Concepts

0. *Problem Zero:* Read the section and make your own summary of the material.

1. *True/False:* Determine whether each of the statements that follow is true or false. If a statement is true, explain why. If a statement is false, provide a counterexample.

(a) *True or False:* The number e is irrational and therefore cannot be defined by listing its digits.

(b) *True or False:* The number e is approximately equal to $(1 + 0.001)^{1000}$.

(c) *True or False:* If $\lim\limits_{h \to 0} \dfrac{b^h - 1}{h} = 1$, then $b = e$.

(d) *True or False:* $f(x) = 3(2^x)$ is continuous on $(-\infty, \infty)$.

(e) *True or False:* $f(x) = 3\log_2 x$ is continuous on $(-\infty, \infty)$.

(f) *True or False:* For all real numbers c, $\lim_{x \to c} e^{2x} = e^{2c}$.

(g) *True or False:* Every exponential function has a horizontal asymptote.

(h) *True or False:* Every logarithmic function has a vertical asymptote.

2. *Examples:* Construct examples of the thing(s) described in the following. Try to find examples that are different than any in the reading.

(a) Three exponential functions that approach 0 as $x \to \infty$.

(b) Three logarithmic functions that approach ∞ as $x \to \infty$.

(c) Two limits that define the irrational number e.

3. Write each of the following as a limit statement:

(a) $f(x) = 2e^{3x}$ is continuous at $x = 1$.

(b) $f(x) = 2e^{3x}$ grows without bound as x approaches ∞.

(c) $f(x) = 2e^{3x}$ has a horizontal asymptote at $y = 0$ on the left.

4. Write each of the following as a limit statement:

(a) $f(x) = 3\log_2 x$ is continuous at $x = 1$.

(b) $f(x) = 3\log_2 x$ grows without bound as x approaches ∞.

(c) $f(x) = 3\log_2 x$ has a vertical asymptote at $x = 0$.

5. Fill in the blanks:

(a) If $k < 0$, then $\lim_{x \to \infty} e^{kx} = $ _____ .

(b) If $b > 1$, then $\lim_{x \to -\infty} b^x = $ _____ .

(c) If $k \in$ _____ , then $\lim_{x \to -\infty} e^{kx} = 0$.

(d) If $b \in$ _____ , then $\lim_{x \to \infty} b^x = 0$.

6. Fill in the blanks:

(a) If $0 < b < 1$, then $\lim_{x \to 0^+} \log_b x = $ _____ .

(b) If $b > 1$, then $\lim_{x \to \infty} \log_b x = $ _____ .

(c) If $b \in$ _____ , then $\lim_{x \to \infty} \log_b x = -\infty$.

(d) If $b \in$ _____ , then $\lim_{x \to 0^+} \log_b x = -\infty$.

Use an appropriate formal definition of limit to express each of the limits in Exercises 7–10. According to the type of limit, your answer may involve δ, ϵ, N, or M. Also, draw a picture that illustrates your answer.

7. $\lim_{x \to 2} e^x = e^2$

8. $\lim_{x \to \infty} (0.5)^x = 0$

9. $\lim_{x \to 1} \ln x = 0$

10. $\lim_{x \to \infty} \log_2 x = \infty$

11. Using properties of inverse functions, explain how the horizontal asymptote of $f(x) = 2^x$ is related to the vertical asymptote of $g(x) = \log_2 x$.

12. In this exercise you will use a calculator to investigate the number e.

(a) Make a table of values that describes the behavior of the quantity $(1 + h)^{1/h}$ as $h \to 0$.

(b) Make a table of values that describes the behavior of the quantity $\dfrac{e^h - 1}{h}$ as $h \to 0$.

(c) What do your tables of values have to do with Definition 5.8 and Theorem 5.9?

13. Argue that the characterizations of the number e in Definition 5.8 and Theorem 5.9 are equivalent, by showing that $e \approx (1 + h)^{1/h}$ for small h if and only if $\dfrac{e^h - 1}{h} \approx 1$ for small h.

14. Consider the limit $\lim_{x \to \infty} \dfrac{2^x - 4^x}{3^x}$.

(a) Solve the limit by using the method of dividing numerator and denominator by the "largest" exponential function that appears in the expression.

(b) Solve the limit by using algebra to simplify $\dfrac{2^x - 4^x}{3^x}$.

15. Consider the limit $\lim_{x \to -\infty} \dfrac{2^x - 4^x}{3^x}$.

(a) Explain why the method of dividing numerator and denominator by the "largest" exponential function does *not* work for this limit.

(b) Solve this limit by using algebra to simplify $\dfrac{2^x - 4^x}{3^x}$.

16. Describe the intervals on which the function

$$f(x) = \begin{cases} 2^x - 1, & \text{if } x \le 0 \\ \dfrac{4^x - 1}{2^x - 1}, & \text{if } x > 0 \end{cases}$$

is continuous. At each point where $f(x)$ fails to be continuous, use limits to determine the type of discontinuity and any left- or right-continuity.

Skills

Use tables of values to estimate each of the limits in Exercises 17–22.

17. $\lim_{h \to 0} \dfrac{1 - e^h}{h}$

18. $\lim_{h \to 0} (1 + h)^{2/h}$

19. $\lim_{x \to \infty} \ln x$

20. $\lim_{x \to 0^+} \ln x$

21. $\lim_{x \to 2} \dfrac{3}{4 - 2^x}$

22. $\lim_{x \to \infty} (3e^{-2x} + 1)$

Calculate each of the limits in Exercises 23–68.

23. $\lim_{x \to 0} \dfrac{2 - 3e^x}{x + 4}$

24. $\lim_{x \to -3} 2(4 - 3e^{4x})$

25. $\lim_{x \to 3} \dfrac{2^x - 1}{3^x}$

26. $\lim_{x \to \infty} -2e^{4x}$

27. $\lim_{x \to \infty} (1 - e^{-x})$

28. $\lim_{x \to -\infty} (3e^{4x} + 1)$

29. $\lim_{x \to 5} \log_3(x^2 - 4)$

30. $\lim_{x \to 2^+} \log_3(x^2 - 4)$

31. $\lim_{x \to 0^+} (2 - \log_3 x)$

32. $\lim_{x \to \infty} \ln(x - 5)$

33. $\lim_{x \to -\infty} \dfrac{1}{2 + 3^x}$

34. $\lim_{x \to \infty} \dfrac{1}{2 + 3^x}$

35. $\lim_{x \to 0} \dfrac{2}{4 + e^{-2x}}$

36. $\lim_{x \to \infty} \dfrac{2}{4 + e^{-2x}}$

37. $\lim_{x \to 0^+} \log_{1/2}\left(\dfrac{1}{x}\right)$

38. $\lim_{x \to \infty} x^{\ln x}$

39. $\lim_{x \to 0} \dfrac{2^x - 3^x}{4^x}$

40. $\lim_{x \to 2^+} \sqrt{2^x - 4}$

41. $\lim_{x \to 4} (3e^{1.7x} + 1)$

42. $\lim_{x \to 0^+} \ln(1 + \sqrt{x})$

43. $\lim_{x \to 0} \dfrac{e^x - 1}{e^{2x} + 2e^x - 3}$

44. $\lim_{x \to 1} \dfrac{e^x - 1}{e^{2x} + 2e^x - 3}$

45. $\lim_{x \to 0} \dfrac{2x}{e^x - 1}$

46. $\lim_{x \to 0} \dfrac{e^x - 1}{x}$

47. $\lim_{x \to 0} \dfrac{3^{2x} + 3^x - 2}{1 - 3^x}$

48. $\lim_{x \to \infty} \dfrac{3^x - 2^x}{1 + 2(3^x)}$

49. $\lim_{x \to \infty} \dfrac{4^x - 3^x}{5^x}$

50. $\lim_{x \to \infty} \dfrac{2^x - 4^{-x}}{3^x}$

51. $\lim_{x \to -\infty} \dfrac{3^x - 5^x}{4^x}$

52. $\lim_{x \to \infty} \dfrac{4(3^x)}{2 + 3^x}$

53. $\lim_{x \to \infty} \dfrac{2e^{1.5x}}{3e^{2x} + e^{1.5x}}$

54. $\lim_{x \to \infty} \dfrac{1 - 5e^{2x}}{3e^x + 4e^{2x}}$

55. $\lim_{x \to 3^+} \ln(x^2 - 9)$

56. $\lim_{x \to 0^+} \ln\left(\dfrac{1}{x}\right)$

57. $\lim_{x \to \infty} (\ln x^2 - \ln(2x + 1))$

58. $\lim_{x \to \infty} (\ln 3x - \ln 2x)$

59. $\lim_{x \to 0} \dfrac{4^x - 6(2^x) + 5}{1 - 2^x}$

60. $\lim_{x \to 2} \dfrac{4^x - 3(2^x) - 4}{2^x - 4}$

61. $\lim_{x \to 0^+} \dfrac{x}{\log_2 x}$

62. $\lim_{x \to \infty} \dfrac{e^{2x} - e^{4x}}{e^{3x}}$

63. $\lim_{x \to 0} (1 + x)^{2/x}$

64. $\lim_{x \to 0} (1 + 2x)^{3/x}$

65. $\lim_{x \to \infty} \left(1 + \dfrac{1}{x}\right)^{3x}$

66. $\lim_{x \to \infty} \left(1 - \dfrac{5}{x}\right)^x$

67. $\lim_{x \to \infty} \ln\left(\dfrac{e^x}{1 + e^x}\right)$

68. $\lim_{x \to \infty} \log_3\left(\dfrac{x^2 - 1}{2 - 3x^2}\right)$

Find the roots, discontinuities, and horizontal and vertical asymptotes of the functions in Exercises 69–76. Support your answers by explicitly computing any relevant limits.

69. $f(x) = 25 - (0.5)^x$

70. $f(x) = 3e^{-2x} + 4$

71. $f(x) = \dfrac{2}{4 + e^{-2x}}$

72. $f(x) = \dfrac{1}{2 + 3^x}$

73. $f(x) = \dfrac{2^x - 4^x}{3^x}$

74. $f(x) = \dfrac{4^x - 6(2^x) + 5}{1 - 2^x}$

75. $f(x) = 3 - 4\ln(x + 1)$

76. $f(x) = \dfrac{1}{1 + \log_2 x}$

To support each limit statement in Exercises 77–80, use graphs and algebra to approximate the smallest magnitude value of N that corresponds to the given value of ϵ or M, according to the appropriate formal limit definition.

77. $\lim_{x \to \infty} \ln x = \infty$, $M = 100$, find smallest $N > 0$

78. $\lim_{x \to \infty} \ln x = \infty$, $M = 100{,}000$, find smallest $N > 0$

79. $\lim_{x \to -\infty} 3^x = 0$, $\epsilon = \dfrac{1}{2}$, find smallest-magnitude $N < 0$

80. $\lim_{x \to -\infty} 3^x = 0$, $\epsilon = \dfrac{1}{4}$, find smallest-magnitude $N < 0$

Applications

81. The temperature T, in degrees Fahrenheit, of a yam that has been sitting in a hot oven for t minutes is given by the function

$$T(t) = 350 - 280e^{-0.2t}.$$

(a) What is the initial temperature of the yam, before it is put in the oven?

(b) Given that over time the temperature of the yam will approach the temperature inside the oven, use a limit to determine the temperature of the oven.

(c) How long will it take for the yam to be within 5 degrees Fahrenheit of the temperature of the oven?

82. A political candidate starts an advertising campaign in Hamtown, Virginia. The number of people in Hamtown that have heard of him t days after the start of his advertising campaign is given by

$$P(t) = \dfrac{45{,}000}{1 + 35e^{-0.12t}}.$$

(a) How many people knew about the candidate before the start of his advertising campaign?

(b) Given that over time the advertising will eventually reach everyone in the town, use a limit to determine the population of Hamtown.

(c) How many days will it take for all but one person in Hamtown to have heard of the candidate?

Proofs

83. Use approximations to argue that if $\lim\limits_{h \to 0}(1 + h)^{1/h} = e$, then $\lim\limits_{h \to 0} \dfrac{e^h - 1}{h} = 1$.

84. Use algebra, limit rules, and the continuity of e^x to prove that every exponential function of the form $f(x) = Ae^{kx}$ is continuous everywhere.

85. Use algebra, limit rules, and the continuity of e^x to prove that every exponential function of the form $f(x) = Ab^x$ is continuous everywhere.

86. Use algebra, limit rules, and the continuity of $\ln x$ on $(0, \infty)$ to prove that every logarithmic function of the form $f(x) = \log_b x$ is continuous on $(0, \infty)$.

87. Use limit rules, the base conversion formula for logarithms, and the fact that $\lim\limits_{x \to \infty} \ln x = \infty$ to prove that $\lim\limits_{x \to \infty} \log_b x$ is equal to ∞ if $b > 1$ and is equal to $-\infty$ if $0 < b < 1$.

88. Prove that $\lim\limits_{x \to \infty} e^x = \infty$. (*Hint: Given $M > 0$, choose $N = \ln M$. Then if $x > N = \ln M$, we must have $x = \ln M + c$ for some positive number c. Use this to show that $e^x > M$.*)

89. Prove that $\lim\limits_{x \to \infty} e^{-x} = 0$.

Thinking Forward

Differentiating e^x: Now that we know how to calculate limits involving exponential functions, we can apply that knowledge to the limits that define derivatives.

▶ Calculate $\lim\limits_{h \to 0} \dfrac{e^{x+h} - e^x}{h}$. At some point you should need the definition of e.

▶ What does your calculation say about the derivative of $f(x) = e^x$ at $c = 0$?

▶ Describe what your calculation means in terms of graphs, heights, and slopes.

Taylor Series: In this section we learned that e can be thought of as the following limit:

$$\lim_{h \to 0}(1 + h)^{1/h} = e.$$

In the following exercise you will investigate the convergence of this limit and also get a preview of **Taylor series**, which we will see in Chapter 8:

▶ Use the substitution $n = \dfrac{1}{h}$ to show that the preceding limit statement is equivalent to the limit statement

$$\lim_{n \to \infty}\left(1 + \frac{1}{n}\right)^n = e.$$

▶ The **Binomial Theorem** says that an expression of the form $(a + b)^n$ can be expanded to

$$\binom{n}{0}a^n b^0 + \binom{n}{1}a^{n-1}b^1 + \binom{n}{2}a^{n-1}b^2 + \cdots + \binom{n}{n}x^0 y^n,$$

where, for any $0 \leq k \leq n$, the symbol $\binom{n}{k}$ is equal to $\dfrac{n!}{k!(n-k)!}$. Here $n!$ is n **factorial**, the product of the integers from 1 to n. By convention we set $0! = 1$. Apply this expansion to the expression $\left(1 + \dfrac{1}{n}\right)^n$.

▶ Show that as $n \to \infty$ we would expect the preceding expansion to approach

$$1 + \frac{1}{1!} + \frac{1}{2!} + \frac{1}{3!} + \frac{1}{4!} + \frac{1}{5!} + \cdots.$$

(*Hint: Think about limits of rational functions and ratios of leading coefficients.*)

▶ Use a calculator to find the sum of the first six terms of the sum from the previous problem, and compare this sum with your calculator's best approximation of the number e.

5.3   DERIVATIVES OF EXPONENTIAL AND LOGARITHMIC FUNCTIONS

▶ Formulas for differentiating exponential and logarithmic functions

▶ Rates of growth of exponential functions

▶ Proving graph characteristics of exponential and logarithmic functions

Derivatives of Exponential Functions

The power rule tells us that the derivative of x^k is kx^{k-1}. This rule works only for power functions, where the base is the variable x and the power is a constant k; it does not tell us how to differentiate an exponential function like 2^x or e^{3x}, where the variable is in the exponent. To determine such derivatives we must return to the definition of the derivative:

$$\frac{d}{dx}(b^x) = \lim_{h \to 0} \frac{b^{x+h} - b^x}{h} = \lim_{h \to 0} \frac{b^x b^h - b^x}{h} = \lim_{h \to 0} \frac{b^x(b^h - 1)}{h} = b^x \left(\lim_{h \to 0} \frac{b^h - 1}{h} \right).$$

Although we have simplified as much as possible, this is a limit that we do not yet know how to calculate.

In one special case we do already know how to evaluate this limit. In the previous section we defined e to be the unique number such that $\lim_{h \to 0} \frac{e^h - 1}{h} = 1$. Therefore when $b = e$, the preceding calculation looks like this:

$$\frac{d}{dx}(e^x) = \lim_{h \to 0} \frac{e^{x+h} - e^x}{h} = \cdots = e^x \left(\lim_{h \to 0} \frac{e^h - 1}{h} \right) = e^x(1) = e^x.$$

We have just shown that the function $f(x) = e^x$ is its own derivative! This is in fact the exact reason that we defined the number e the way that we did. As illustrated here, $y = e^x$ is its own associated slope function:

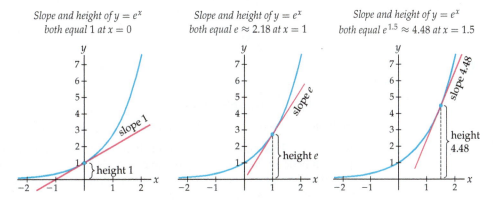

Slope and height of $y = e^x$ both equal 1 at $x = 0$

Slope and height of $y = e^x$ both equal $e \approx 2.18$ at $x = 1$

Slope and height of $y = e^x$ both equal $e^{1.5} \approx 4.48$ at $x = 1.5$

Other exponential functions of the form e^{kx} or b^x have graphs similar to the graph of e^x, but only e^x is scaled in exactly the right way to be its own derivative. With the chain rule, we can use the derivative of e^x to find the derivatives of general exponential functions:

THEOREM 5.13   **Derivatives of Exponential Functions**

For any constant k, any constant $b > 0$ with $b \neq 1$, and all $x \in \mathbb{R}$,

(a) $\dfrac{d}{dx}(e^x) = e^x$ **(b)** $\dfrac{d}{dx}(b^x) = (\ln b)\, b^x$ **(c)** $\dfrac{d}{dx}(e^{kx}) = ke^{kx}$

For example, we can use these rules to quickly calculate $\dfrac{d}{dx}(2e^{3x}) = 2(3)e^{3x} = 6e^{3x}$ and $\dfrac{d}{dx}(5(2^x)) = 5(\ln 2)2^x$.

> **Proof.** The previous discussion proves that $\dfrac{d}{dx}(e^x) = e^x$. To prove the second rule we rewrite b^x as $(e^{\ln b})^x$ and then apply the chain rule:
>
> $$\frac{d}{dx}(b^x) = \frac{d}{dx}((e^{\ln b})^x) \qquad\qquad \leftarrow b = e^{\ln b}$$
>
> $$= \frac{d}{dx}(e^{(\ln b)x}) \qquad\qquad \leftarrow \text{algebra of exponents}$$
>
> $$= e^{(\ln b)x}\frac{d}{dx}((\ln b)x) \qquad \leftarrow \text{chain rule and derivative of } e^x$$
>
> $$= (e^{\ln b})^x(\ln b) \qquad\qquad \leftarrow \text{algebra and derivative of linear function}$$
>
> $$= (\ln b)\, b^x. \qquad\qquad\qquad \leftarrow \text{algebra}$$
>
> The proof for e^{kx} is similar, with a simpler application of the chain rule, and is left for Exercise 91.   ∎

By combining our new differentiation rules for exponential functions with the chain rule, we obtain the following rules for exponential compositions:

$$\frac{d}{dx}(e^{u(x)}) = e^{u(x)}u'(x),$$

$$\frac{d}{dx}(b^{u(x)}) = (\ln b)b^{u(x)}u'(x).$$

For example, we have

$$\frac{d}{dx}(e^{x^2+1}) = e^{x^2+1} \cdot \frac{d}{dx}(x^2 + 1) = e^{x^2+1}(2x),$$

$$\frac{d}{dx}(2^{x^2+1}) = (\ln 2)2^{x^2+1} \cdot \frac{d}{dx}(x^2 + 1) = (\ln 2)2^{x^2+1}(2x).$$

Exponential Functions Grow Proportionally to Themselves

Notice that all exponential functions have the property that their derivatives are constant multiples of the original function; for example, $\dfrac{d}{dx}(e^{2x}) = 2e^{2x}$ is just 2 times the original function e^{2x}. This means that the rate of change of an exponential function f is proportional to the function f itself. In fact, the converse is also true, although proving that it is true is outside the scope of this course. Together, both proportionalities give us the following two-sided statement, which will prove useful in many application problems:

THEOREM 5.14   **Rates of Change and Exponential Functions**

$f'(x) = kf(x)$ for some constant k if and only if f is an exponential function of the form $f(x) = Ae^{kx}$.

Recall that one quantity y is **proportional** to another quantity x if y is a constant multiple of x. Therefore the preceding theorem gives us a nice characterization of exponential functions: *All* exponential functions f, and *only* exponential functions, have the property that f' is proportional to f.

If a word problem states that the rate of change of a function is constant, we immediately know that the function is linear. Similarly, by Theorem 5.14, if a word problem states that the rate of change of a function is proportional to the function itself, we immediately know that that function is exponential. Then it is just a question of finding values A and k to determine the function $f(x) = Ae^{kx}$ that models the situation. Exponential functions often arise when modeling populations, for example, since a larger population can produce greater numbers of offspring than a smaller population and can therefore grow at a faster rate.

Derivatives of Logarithmic Functions

Up to this point, the derivatives that we have seen are similar to their original functions. For example, derivatives of power functions are power functions, derivatives of polynomials are polynomials, and derivatives of exponential functions are exponential. Now something surprising happens: Derivatives of logarithmic functions are not logarithmic. Even more surprisingly, logarithmic functions are transcendental but their derivatives are algebraic!

THEOREM 5.15   **Derivatives of Logarithmic Functions**

For any constant $b > 0$ with $b \neq 1$, and all appropriate values of x,

(a) $\dfrac{d}{dx}(\log_b x) = \dfrac{1}{(\ln b)x}$ **(b)** $\dfrac{d}{dx}(\ln x) = \dfrac{1}{x}$ **(c)** $\dfrac{d}{dx}(\ln |x|) = \dfrac{1}{x}$

The second rule is a special case of the first, with $b = e$. Because $\ln x$ has domain $(0, \infty)$, when we say that $\dfrac{d}{dx}(\ln x) = \dfrac{1}{x}$, we are also restricting $\dfrac{1}{x}$ to the domain $(0, \infty)$. In the third rule we generalize the second rule so that we are considering the full domain $(-\infty, 0) \cup (0, \infty)$ of $\dfrac{1}{x}$.

> **Proof.** We will prove the second rule and leave the proofs of the first and third rules to Exercises 94 and 95, respectively. Since $y = \ln x$ and $y = e^x$ are inverses, $e^{\ln x} = x$ for all x in the domain $(0, \infty)$ of $y = \ln x$. Differentiating both sides of the equation gives
>
> $$e^{\ln x} = x \qquad\qquad \leftarrow \text{property of inverses}$$
>
> $$\frac{d}{dx}(e^{\ln x}) = \frac{d}{dx}(x) \qquad\qquad \leftarrow \text{differentiate both sides}$$
>
> $$e^{\ln x}\frac{d}{dx}(\ln x) = 1 \qquad\qquad \leftarrow \text{exponential and chain rules}$$
>
> $$\frac{d}{dx}(\ln x) = \frac{1}{e^{\ln x}} \qquad\qquad \leftarrow \text{solve for } \frac{d}{dx}(\ln x)$$
>
> $$\frac{d}{dx}(\ln x) = \frac{1}{x}. \qquad\qquad \leftarrow \text{since } e^{\ln x} = x$$

The two graphs that follow illustrate that the associated slope function for $\ln |x|$ is the function $\dfrac{1}{x}$. For negative values of x, as we move from left to right, the slopes of $\ln |x|$ are negative with larger and larger magnitude while the heights of $\dfrac{1}{x}$ behave the same way. For positive values of x, as we move from left to right, the slopes of $\ln x$ are positive but getting smaller while the heights of $\dfrac{1}{x}$ do the same.

Slopes of $f(x) = \ln|x|$

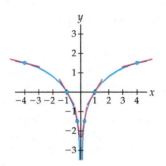

Heights of $f'(x) = \dfrac{1}{x}$

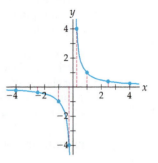

The rule for differentiating $\ln|x|$ will come in handy for calculating certain derivatives. Functions that involve many products or quotients or that have variables in both a base and an exponent can be difficult to differentiate. One strategy for differentiating such functions is to apply $\ln|x|$ to both sides of the equation $y = f(x)$ and then differentiate both sides. This process is called **logarithmic differentiation** and is illustrated in Examples 5 and 6.

By combining our new differentiation rules for logarithmic functions with the chain rule, we obtain the following rules for logarithmic compositions:

$$\frac{d}{dx}(\ln u(x)) = \frac{1}{u(x)} \cdot u'(x) = \frac{u'(x)}{u(x)},$$

$$\frac{d}{dx}(\log_b u(x)) = \frac{1}{(\ln b)u(x)} \cdot u'(x) = \frac{u'(x)}{(\ln b)u(x)}.$$

For example, we have

$$\frac{d}{dx}(\ln(x^2 + 1)) = \frac{1}{x^2 + 1} \cdot 2x = \frac{2x}{x^2 + 1},$$

$$\frac{d}{dx}(\log_b(x^2 + 1)) = \frac{1}{(\ln b)(x^2 + 1)} \cdot 2x = \frac{2x}{(\ln b)(x^2 + 1)}.$$

Exponential and Logarithmic Graphs

Now that we understand both limits and derivatives of exponential and logarithmic functions, we can *prove* that those types of functions must always have the types of graphs we have grown to expect. For exponential functions we have two types of graphs:

THEOREM 5.16   **Graphs of Basic Exponential Functions**

Every exponential function $f(x) = e^{kx}$ or $f(x) = b^x$ has a graph with one of the following two shapes:

$k > 0$ or $b > 1$

$k < 0$ or $0 < b < 1$

To prove this theorem we need only do a curve-sketching analysis. For example, in the first case we need to show that $f(x) = e^{kx}$ has a y-intercept of $y = 1$; is always positive, increasing, and concave up; increases without bound as $x \to \infty$; and has a horizontal asymptote at $y = 0$ as $x \to -\infty$. The details are left to Exercises 97 and 98.

We can similarly classify the graphs of logarithmic functions, simply by reflecting the graphs in the previous theorem across the line $y = x$:

THEOREM 5.17   **Graphs of Basic Logarithmic Functions**

Every logarithmic function $g(x) = \log_b x$ has a graph with one of the following two shapes:

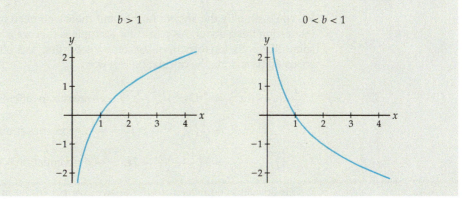

We could do a limit and derivative analysis to prove this theorem, but it is much easier to simply note that since logarithmic functions are inverses of exponential functions, we can obtain the graphs in Theorem 5.17 by reflecting the graphs in Theorem 5.16 over the line $y = x$.

Examples and Explorations

EXAMPLE 1   **Differentiating simple combinations of exponential and logarithmic functions**

Find the derivatives of each of the following functions:

(a) $f(x) = \dfrac{5^{3x}}{1+x}$ **(b)** $f(x) = e^{x^2} \ln x$ **(c)** $f(x) = 3\sqrt{e^x} - x^2$

SOLUTION

(a) Starting with the quotient rule, we have

$$\frac{d}{dx}\left(\frac{5^{3x}}{1+x}\right) = \frac{\frac{d}{dx}(5^{3x}) \cdot (1+x) - 5^{3x} \cdot \frac{d}{dx}(1+x)}{(1+x)^2} \qquad \leftarrow \text{quotient rule}$$

$$= \frac{(\ln 5)5^{3x}(3)(1+x) - 5^{3x}(1)}{(1+x)^2} \qquad \leftarrow \text{chain, exponential rules}$$

$$= \frac{5^{3x}(3(\ln 5)(1+x) - 1)}{(1+x)^2}. \qquad \leftarrow \text{simplify a bit}$$

(b) Using the product and chain rules, we have

$$\frac{d}{dx}\left(e^{x^2}\ln x\right) = \frac{d}{dx}\left(e^{x^2}\right) \cdot \ln x + e^{x^2} \cdot \frac{d}{dx}(\ln x) = 2xe^{x^2}\ln x + e^{x^2}\left(\frac{1}{x}\right).$$

(c) In this example, an exponential function is on the *inside* of a composition, since e^x appears inside the square root. Thus we have

$$\frac{d}{dx}(3\sqrt{e^x} - x^2) = \frac{d}{dx}(3(e^x)^{1/2} - x^2) \qquad \leftarrow \text{rewrite so differentiation will be easier}$$

$$= 3\left(\frac{1}{2}\right)(e^x)^{-1/2}(e^x) - 2x \quad \leftarrow \text{don't forget the chain rule!}$$

$$= \frac{3}{2}e^x(e^x)^{-1/2} - 2x. \qquad \leftarrow \text{some algebra}$$

We could simplify the answer further, but there's no need to do that right now. One interesting thing about this particular example is that we could have saved ourselves some differentiation work by doing some algebra first, as is often the case. In particular, we could have done the calculation this way:

$$\frac{d}{dx}(3\sqrt{e^x} - x^2) = \frac{d}{dx}(3(e^x)^{1/2} - x^2) \quad \leftarrow \text{rewrite so differentiation will be easier}$$

$$= \frac{d}{dx}(3e^{1/2x} - x^2) \qquad \leftarrow \text{algebra makes differentiation easier still}$$

$$= 3\left(\frac{1}{2}\right)e^{1/2x} - 2x. \qquad \leftarrow \text{differentiation is easy now}$$

This answer is equivalent to the answer we got in the first calculation. (Do some algebra to see why.) Notice that the differentiation steps used in the second calculation were *much* easier than the differentiation steps used in the first calculation!   □

EXAMPLE 2   **More complicated differentiation problems**

Find the derivatives of each of the following functions:

(a) $f(x) = \dfrac{7e^{3x} - x^2 2^x}{\log_5 x}$ **(b)** $f(x) = \ln\left(\left(\dfrac{x^2 - 1}{1 - 2x}\right)^2\right)$

SOLUTION

(a) The function $f(x)$ is a quotient, so we begin by applying the quotient rule:

$$f'(x) = \frac{\frac{d}{dx}(7e^{3x} - x^2 2^x) \cdot \log_5 x - (7e^{3x} - x^2 2^x) \cdot \frac{d}{dx}(\log_5 x)}{(\log_5 x)^2} \quad \leftarrow \text{quotient rule}$$

$$= \frac{\left(21e^{3x} - \frac{d}{dx}(x^2 2^x)\right)(\log_5 x) - (7e^{3x} - x^2 2^x)\left(\frac{1}{(\ln 5)x}\right)}{(\log_5 x)^2} \quad \leftarrow \text{other rules}$$

$$= \frac{(21e^{3x} - (2x2^x + x^2(\ln 2)2^x))(\log_5 x) - (7e^{3x} - x^2 2^x)\left(\frac{1}{(\ln 5)x}\right)}{(\log_5 x)^2} \quad \leftarrow \text{product rule}$$

In the preceding calculation, we saved the product rule calculation for the last step. For lengthy derivative problems it is often helpful to postpone some calculations to later steps.

(b) There are a number of ways we could find $f'(x)$. One way is to jump right in and start differentiating; although it is possible to do this all in one step, we'll do the calculation in a few separate steps to be as clear as possible:

$$\frac{d}{dx}\left(\ln\left(\left(\frac{x^2-1}{1-2x}\right)^2\right)\right) = \frac{1}{\left(\frac{x^2-1}{1-2x}\right)^2}\cdot\frac{d}{dx}\left(\left(\frac{x^2-1}{1-2x}\right)^2\right) \qquad \leftarrow \text{chain rule}$$

$$= \frac{1}{\left(\frac{x^2-1}{1-2x}\right)^2}\,(2)\left(\frac{x^2-1}{1-2x}\right)^1\cdot\frac{d}{dx}\left(\frac{x^2-1}{1-2x}\right) \qquad \leftarrow \text{chain rule}$$

$$= \frac{1}{\left(\frac{x^2-1}{1-2x}\right)^2}\,(2)\left(\frac{x^2-1}{1-2x}\right)^1\left(\frac{(2x)(1-2x)-(x^2-1)(-2)}{(1-2x)^2}\right). \qquad \leftarrow \text{quotient rule}$$

The first step used the chain rule and the logarithmic rule, the second step used the chain rule and the power rule, and the last step used the quotient rule.

If we do a little bit of algebra first, the differentiation step becomes much easier:

$$\frac{d}{dx}\left(\ln\left(\left(\frac{x^2-1}{1-2x}\right)^2\right)\right) = \frac{d}{dx}\left(2\ln\left(\frac{x^2-1}{1-2x}\right)\right) \qquad \leftarrow \text{algebra}$$

$$= \frac{d}{dx}(2(\ln(x^2-1)-\ln(1-2x))) \qquad \leftarrow \text{algebra}$$

$$= 2\left(\frac{1}{x^2-1}(2x)-\frac{1}{1-2x}(-2)\right). \qquad \leftarrow \text{differentiate}$$

In the preceding calculation, the first two steps were algebra; only the final step involved differentiation. Although the two answers obtained look very different, they are in fact the same. (As an exercise, use algebra to show this.) ☐

EXAMPLE 3

A real-world problem that can be modeled with an exponential function

Suppose a population of wombats on a small island is growing at a rate proportional to the number of wombats on the island. If there were 12 wombats on the island in 1990 and 37 wombats on the island in 1998, how many wombats were on the island in the year 2010?

SOLUTION

We are given that the rate of change $W'(t)$ of the population of wombats is proportional to the population $W(t)$ of wombats at time t; in other words, $W'(t) = kW(t)$ for some real number k. By Theorem 5.14, this means that the population $W(t)$ of wombats on the island must be an exponential function $W(t) = Ae^{kt}$. We now only need to find constants A and k that match the information given in the problem.

If we let $t = 0$ represent 1990 (so that $t = 8$ will represent 1998), then from the information given in the problem, we have $W(0) = 12$ and $W(8) = 37$. We will use these two data points to find A and k. Using the first data point, we have

$$W(0) = 12 \implies Ae^{k(0)} = 12 \implies Ae^0 = 12 \implies A(1) = 12 \implies A = 12,$$

so $W(t) = 12e^{kt}$ for some k. Using the second data point, we can now solve for the value of k:

$$W(8) = 37 \implies 12e^{k(8)} = 37 \implies e^{8k} = \frac{37}{12} \implies 8k = \ln\left(\frac{37}{12}\right) \implies k = \frac{\ln\left(\frac{37}{12}\right)}{8} \approx 0.14.$$

Thus $W(t) = 12e^{0.14t}$. Using this function, we can now easily calculate the number of wombats that were on the island in the year 2010. Since 2010 is 20 years after 1990, we need to find

$$W(20) = 12e^{0.14(20)} \approx 197.34.$$

In the year 2010, there were approximately 197 wombats on the island, assuming of course that we cannot have "parts" of wombats.   ☐

EXAMPLE 4

A curve-sketching analysis with asymptotes

Sketch an accurate, labeled graph of the function $f(x) = \frac{6}{4 - 2^x}$. Include complete sign analyses of f, f', and f'', and calculate any relevant limits.

SOLUTION

Let's begin by finding and simplifying the first and second derivatives of $f(x)$. The first derivative of $f(x) = 6(4 - 2^x)^{-1}$ is

$$f'(x) = 6(-1)(4 - 2^x)^{-2}(-(\ln 2)2^x) = \frac{6(\ln 2)2^x}{(4 - 2^x)^2}.$$

Differentiating that result and then simplifying as much as possible so that we can easily identify roots, we find that the second derivative of f is

$$f''(x) = \frac{6(\ln 2)(\ln 2)2^x(4 - 2^x)^2 - (6(\ln 2)2^x)(2)(4 - 2^x)^1(-(\ln 2)2^x)}{(4 - 2^x)^4}$$

$$= \frac{6(\ln 2)^2 2^x(4 - 2^x)((4 - 2^x) + 2 \cdot 2^x)}{(4 - 2^x)^4} = \frac{6(\ln 2)^2 2^x(4 + 2^x)}{(4 - 2^x)^3}.$$

To determine the intervals on which f, f', and f'' are positive and negative we must first locate the values of x for which these functions are zero or do not exist. The function $f(x) = \frac{6}{4 - 2^x}$ is never zero, but is undefined where its denominator is zero, at $x = 2$. Looking at the formula for $f'(x)$, we can easily see that it is never zero (since 2^x is never zero), but is undefined if $x = 2$ (since the denominator $4 - 2^x$ is zero for $x = 2$). Thus $x = 2$ is the only critical point of f and the only point we will mark on the number line for f'. Similarly, from the formula for $f''(x)$, it is clear that $f''(x)$ is never zero (since neither 2^x nor $4 + 2^x$ can ever be zero), but is undefined at $x = 2$. Checking signs on either side of $x = 2$ for each function, we obtain the following set of number lines:

We now know that $f(x)$ is positive, increasing, and concave up on $(-\infty, 2)$ and negative, increasing, and concave down on $(2, \infty)$. The graph has no roots, no extrema, and no inflection points.

It remains to calculate any interesting limits. Since the domain of f is $(-\infty, 2) \cup (2, \infty)$, we must investigate the limits of $f(x) = \frac{6}{4-2^x}$ as $x \to \pm\infty$ and as $x \to 2$ from the left and the right. As $x \to \infty$ the denominator $4 - 2^x$ of $f(x)$ approaches $-\infty$, and thus

$$\lim_{x \to \infty} \frac{6}{4 - 2^x} = 0.$$

As $x \to -\infty$, the denominator $4 - 2^x$ approaches 4, and thus

$$\lim_{x \to -\infty} \frac{6}{4 - 2^x} = \frac{6}{4} = \frac{3}{2}.$$

This information tells us that the graph of f has *two* horizontal asymptotes: at $y = 0$ on the right and at $y = \frac{3}{2}$ on the left.

As $x \to 2^-$ we have $4 - 2^x \to 0^+$ and thus

$$\lim_{x \to 2^-} \frac{6}{4 - 2^x} = \infty,$$

and as $x \to 2^+$ we have $4 - 2^x \to 0^-$ and thus

$$\lim_{x \to 2^+} \frac{6}{4 - 2^x} = -\infty.$$

Thus the graph of f has a vertical asymptote at $x = 2$, where the graph approaches ∞ to the left of 2 and approaches $-\infty$ to the right of 2.

Putting all of this information together, we can now sketch the graph:

$$f(x) = \frac{6}{4 - 2^x}$$

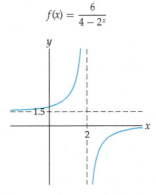

EXAMPLE 5   **Using logarithmic differentiation when products and quotients are involved**

Use logarithmic differentiation to calculate the derivative of the function

$$f(x) = \frac{\sqrt{x}\,(x^2 - 1)^5}{(x + 2)(x - 4)^3}.$$

SOLUTION

Obviously it would take a great deal of work to differentiate this function as it is currently written. We would have to apply the quotient rule once and the product rule twice, among other things. We could do some algebra and multiply out some of the factors in the numerator and the denominator of $f(x)$ to make our job easier, but that is in itself a pretty nasty calculation.

However, it happens to be not nearly as difficult to differentiate the related function $\ln|f(x)|$. Taking the logarithm of both sides and then differentiating both sides, we have

$$y = \frac{\sqrt{x}\,(x^2-1)^5}{(x+2)(x-4)^3} \qquad\qquad \leftarrow \text{set } y = f(x)$$

$$\ln|y| = \ln\left|\frac{\sqrt{x}(x^2-1)^5}{(x+2)(x-4)^3}\right| \qquad\qquad \leftarrow \text{apply } \ln|x|$$

$$\ln|y| = \ln|\sqrt{x}| + \ln|(x^2-1)^5| - \ln|x+2| - \ln|(x-4)^3| \qquad \leftarrow \text{algebra}$$

$$\ln|y| = \frac{1}{2}\ln|x| + 5\ln|x^2-1| - \ln|x+2| - 3\ln|x-4| \qquad\qquad \leftarrow \text{algebra}$$

$$\frac{d}{dx}(\ln|y|) = \frac{d}{dx}\left(\frac{1}{2}\ln|x| + 5\ln|x^2-1| - \ln|x+2| - 3\ln|x-4|\right) \leftarrow \text{differentiate}$$

$$\frac{1}{y}y' = \frac{1}{2x} + \frac{5(2x)}{x^2-1} - \frac{1}{x+2} - \frac{3}{x-4} \qquad\qquad \leftarrow \text{derivative rules}$$

$$y' = y\left(\frac{1}{2x} + \frac{10x}{x^2-1} - \frac{1}{x+2} - \frac{3}{x-4}\right) \qquad\qquad \leftarrow \text{solve for } y'$$

$$y' = \frac{\sqrt{x}(x^2-1)^5}{(x+2)(x-4)^3}\left(\frac{1}{2x} + \frac{10x}{x^2-1} - \frac{1}{x+2} - \frac{3}{x-4}\right). \qquad \leftarrow \text{definition of } y$$

Although the calculation as a whole is algebraically complicated, the differentiation step was simple. The most difficult part was to remember to use the chain rule when we differentiated $\frac{d}{dx}(\ln y) = \frac{1}{y}y'$. Note also that we used the fact that $|ab| = |a||b|$ and the fact that $|a^b| = |a|^b$ in our calculations. □

EXAMPLE 6   **Using logarithms to differentiate a function with variables in both base and exponent**

Use logarithmic differentiation to find $\frac{d}{dx}((x^2-3)^x)$.

SOLUTION

In this example using logarithmic differentiation is not a choice, but a necessity: Neither the product rule nor the exponential rule applies to this function. When taking the derivative of a function that involves the variable in both the base and the exponent, we must use logarithmic differentiation. After setting $y = (x^2-3)^x$ and then applying the natural logarithm to both sides, we will be able to use algebra to remove the variable x from the exponent and then differentiate both sides:

$$y = (x^2-3)^x \qquad\qquad \leftarrow \text{set } y = f(x)$$

$$\ln y = \ln((x^2-3)^x) \qquad\qquad \leftarrow \text{apply } \ln(x) \text{ to both sides}$$

$$\ln y = x\ln(x^2-3) \qquad\qquad \leftarrow \text{algebra}$$

$$\frac{d}{dx}(\ln y) = \frac{d}{dx}(x\ln(x^2-3)) \qquad\qquad \leftarrow \text{differentiate both sides}$$

$$\frac{1}{y}y' = (1)\ln(x^2-3) + (x)\frac{2x}{x^2-3} \qquad\qquad \leftarrow \text{chain, product rules}$$

$$y' = y\left(\ln(x^2-3) + \frac{2x^2}{x^2-3}\right) \qquad\qquad \leftarrow \text{solve for } y'$$

$$y' = (x^2-3)^x\left(\ln(x^2-3) + \frac{2x^2}{x^2-3}\right). \qquad \leftarrow \text{since } y = (x^2-3)^x$$

Note that we did not need any absolute values in this calculation, because $(x^2 - 3)^x$ is always positive where it is defined.   □

▶ Why is it so convenient for us to define the number e to be the unique number with $\lim_{h \to 0} \frac{e^h - 1}{h} = 1$? What differentiation fact is the consequence of this definition?

▶ Why does the power rule not apply to exponential functions? That is, why is the derivative of $f(x) = e^x$ *not* equal to $f'(x) = xe^{x-1}$?

▶ In the proof that $\frac{d}{dx}(\ln x) = x$, we used the fact that $e^{\ln x} = x$. It is also true that $\ln(e^x) = x$; could we have started with this equality instead? Why or why not?

▶ Although there are absolute values in the fifth line of the calculation in Example 5, there are no absolute values in the sixth line; what happened to the absolute values and why?

▶ Why was logarithmic differentiation necessary in Example 6? In particular, why did neither the power nor the exponential rule apply?

EXERCISES 5.3

Thinking Back

Curve sketching: Express each of the following properties in mathematical language, using derivatives, limits, or values.

▶ $f(x)$ has roots at $x = 2$ and $x = 3$

▶ $f(x)$ is increasing on $(0, \infty)$

▶ $f(x)$ is concave up on $(-2, 2)$

▶ $f(x)$ has a local maximum at $x = 5$

▶ $f(x)$ has a global maximum at $x = 1$

▶ $f(x)$ has a horizontal asymptote at $y = -3$

▶ $f(x)$ has a vertical asymptote at $x = 2$

Compositions: For each function k, find functions f, g, and h such that $k = f \circ g \circ h$. There may be more than one possible answer.

▶ $k(x) = \ln(\sqrt{x^2 + 5})$ ▶ $k(x) = e^{1/\sqrt{x+1}}$

▶ $k(x) = (\ln 3x)^2$ ▶ $k(x) = 3\ln(5^x + 2)$

Concepts

0. *Problem Zero:* Read the section and make your own summary of the material.

1. *True/False:* Determine whether each of the statements that follow is true or false. If a statement is true, explain why. If a statement is false, provide a counterexample.

  (a) *True or False:* $\frac{d}{dx}(e^\pi) = 0$.

  (b) *True or False:* $\frac{d}{dz}(e^z) = e^z$.

  (c) *True or False:* $\frac{d}{dx}\left(\frac{1}{x}\right) = \ln x$.

  (d) *True or False:* $\frac{d}{dx}(\ln|x|) = \frac{1}{|x|}$.

  (e) *True or False:* If $f'(x)$ is a constant multiple of $f(x)$, then $f(x)$ is an exponential function.

  (f) *True or False:* If $f(x)$ is an exponential function, then $f'(x)$ is a constant multiple of $f(x)$.

  (g) *True or False:* Graphs of basic exponential functions are always concave up everywhere.

  (h) *True or False:* Graphs of basic logarithmic functions are always concave down everywhere.

2. *Examples:* Construct examples of the thing(s) described in the following. Try to find examples that are different than any in the reading.

  (a) Three functions $f(x)$ whose derivatives are just constant multiples of $f(x)$.

  (b) Three functions that are transcendental, but whose derivatives are algebraic.

  (c) A function whose derivative would be difficult to find without the method of logarithmic differentiation.

3. Does the exponential rule apply to the function $f(x) = x^x$? What about the power rule? Explain your answers.

4. The natural exponential function is its own derivative. Explain what this means graphically. (Use words like "height" and "slope.")

5. Explain how the formula for differentiating the natural exponential function is a special case of the formula for differentiating exponential functions of the form e^{kx}. Then explain why it is a special case of the formula for differentiating functions of the form b^x.

6. The function $f(x) = e^x$ is its own derivative. Are there other functions with this property? If not, explain why not. If so, give three examples.

7. Explain how the formula for differentiating the natural logarithm function is a special case of the formula for differentiating logarithmic functions of the form $\log_b x$.

8. When we say that $\dfrac{d}{dx}(\ln x) = \dfrac{1}{x}$, we really mean to consider the function $\dfrac{1}{x}$ on the restricted domain $(0, \infty)$. Why?

9. The graphs of the exponential functions $y = 2^x$, $y = 4^x$, and $y = 2(2^x)$ are shown in the figure at the left. Use your knowledge of transformations to determine which graph is which without using a graphing calculator.

$$y = 2^x, \; y = 4^x, \; y = 2(2^x) \qquad y = e^x, \; y = e^{3x}, \; y = e^{-2x}$$

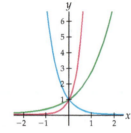

10. The graphs of the exponential functions $y = e^x$, $y = e^{3x}$, and $y = e^{-2x}$ are shown in the preceding graph at the right. Use your knowledge of transformations to determine which graph is which without using a graphing calculator.

11. Describe the process called logarithmic differentiation. What types of differentiation problems is logarithmic differentiation useful for?

12. Why do we need to consider absolute values when we apply logarithmic differentiation to $f(x) = xe^x \sin x$? In contrast, why do we not need to consider absolute values when we apply logarithmic differentiation to $f(x) = x^x$?

For each function $f(x)$ and value $x = c$ in Exercises 13–16, use a sequence of approximations and the definition of the derivative to estimate $f'(c)$. Illustrate your work with an appropriate sequence of graphs of secant lines.

13. $f(x) = \ln(x^2 + 1), c = 0$

14. $f(x) = e^x, c = 0$

15. $f(x) = 2^x, c = 3$

16. $f(x) = \log_2 x, c = 1$

Skills

Find the derivatives of each of the functions in Exercises 17–44. In some cases it may be convenient to do some preliminary algebra.

17. $f(x) = \dfrac{1}{2 - e^{5x}}$

18. $f(x) = \log_2(3x^2 - 5)$

19. $f(x) = 3x^2 e^{-4x}$

20. $f(x) = e^{3x} \ln(x^2 + 1)$

21. $f(x) = \dfrac{1 - x}{e^x}$

22. $f(x) = \ln\left(\dfrac{x^3}{x^2 + x + 1}\right)$

23. $f(x) = e^x(x^2 + 3x - 1)$

24. $f(x) = \dfrac{e^x \ln x}{x^2 - 1}$

25. $f(x) = e^{3 \ln x}$

26. $f(x) = 3^x + \log_3 x$

27. $f(x) = e^{(e^x)}$

28. $f(x) = e^{(x^e)}$

29. $f(x) = (e^x)^e$

30. $f(x) = (x^e)^e$

31. $f(x) = \dfrac{\ln(x^5)}{\ln(x^4)}$

32. $f(x) = 11 + e^\pi - \ln 2$

33. $f(x) = x^{-3} e^{2x}$

34. $f(x) = \ln(x^2 2^x)$

35. $f(x) = x^2 \log_2(x 2^x)$

36. $f(x) = \sqrt{\ln(x^2 + 1)}$

37. $f(x) = \ln(x^2 + e^{\sqrt{x}})$

38. $f(x) = \dfrac{\sqrt{e^x}}{\ln \sqrt{x}}$

39. $f(x) = \sqrt{\log_2(3^x - 5)}$

40. $f(x) = \ln(x^2 + 1)(e^x)^{-1/3}$

41. $f(x) = x^2 \ln(\ln x)$

42. $f(x) = \ln(x^x)$

43. $f(x) = (2^x)^x$

44. $f(x) = 2^{1-3^x}$

Use the first-derivative test to find all the local extrema of each function $f(x)$ in Exercises 45–48.

45. $f(x) = 3x - 2e^x$

46. $f(x) = 3^x - 2^x$

47. $f(x) = \dfrac{\ln 2x}{x}$

48. $f(x) = 2^{1 - \ln x}$

Use the second-derivative test to find all the local extrema of each function $f(x)$ in Exercises 49–52.

49. $f(x) = \dfrac{1 - x}{e^x}$

50. $f(x) = \ln(x^2 + 1)$

51. $f(x) = \dfrac{e^x}{x}$

52. $f(x) = e^{3x} - e^{2x}$

Determine the intervals on which each function $f(x)$ in Exercises 53–56 is increasing or decreasing.

53. $f(x) = e^x(x - 2)$

54. $f(x) = \dfrac{e^x}{1 + e^x}$

55. $f(x) = \ln(x^2 + 1)$

56. $f(x) = \ln(\ln x)$

For each function $f(x)$ in Exercises 57–60, find the intervals on which $f(x)$ is concave up or concave down, as well as the locations of any inflection points.

57. $f(x) = e^{3x}(1 - e^x)$

58. $f(x) = \dfrac{x}{\ln x}$

59. $f(x) = e^{1+2x-x^2}$

60. $f(x) = \dfrac{2^x}{1 - 2^x}$

Sketch an accurate, labeled graph of each function $f(x)$ in Exercises 61–68. Your work should include finding the domain; doing a complete sign analysis of f, f', and f''; and calculating any relevant limits.

61. $f(x) = 400 - 30e^{-2x}$

62. $f(x) = \dfrac{3}{9 - 3^x}$

63. $f(x) = \dfrac{10}{2e^{-3x} + 5}$

64. $f(x) = \ln(\ln x)$

65. $f(x) = x \log_2 x$

66. $f(x) = \dfrac{\ln x}{x}$

67. $f(x) = \sqrt{1 - 2^x}$

68. $f(x) = \dfrac{e^{3x}}{1 - e^{3x}}$

Use logarithmic differentiation to find the derivatives of each of the functions in Exercises 69–78.

69. $f(x) = \sqrt{x \ln |2^x + 1|}$

70. $f(x) = 12x\sqrt[3]{1 - x}\sqrt{x + 1}$

71. $f(x) = \dfrac{2^x \sqrt{x^3 - 1}}{\sqrt{x}(2x - 1)}$

72. $f(x) = \dfrac{e^{2x}(x^3 - 2)^4}{x(3e^{5x} + 1)}$

73. $f(x) = x^{\ln x}$

74. $f(x) = (2x + 1)^{3x}$

75. $f(x) = \left(\dfrac{x}{x - 1}\right)^x$

76. $f(x) = (\ln x)^x$

77. $f(x) = (\ln x)^{\ln x}$

78. $f(x) = \left(\dfrac{1}{x + 1}\right)^x$

In Exercises 79–84, find a function $f(x)$ with the given derivative $f'(x)$ and value $f(c)$.

79. $f'(x) = 2e^{3x}$ and $f(0) = 1$

80. $f'(x) = 2^x$ and $f(2) = 5$

81. $f'(x) = \dfrac{1}{x}$ and $f(1) = 4$

82. $f'(x) = \dfrac{3}{2x + 1}$ and $f(2) = 6$

83. $f'(x) = \dfrac{2x}{x^2 + 1}$ and $f(0) = -2$

84. $f'(x) = \dfrac{3x^2}{x^3 + 2}$ and $f(-1) = 3$

Applications

85. A fruit fly population is being cultivated in Dr. Drosophila's laboratory. Dr. Drosophila wonders if the population $P(t)$ of fruit flies after t days could be modeled by an exponential function. His laboratory assistant correctly says, "The population of fruit flies is an exponential function if the rate of change of the population of fruit flies is proportional to _____ ." Fill in the blank.

86. An abandoned building contained 45 rats on the first day of the year and 53 rats 30 days later. Let $r(t)$ be the function that describes the number of rats in the building t days after the first of the year.

 (a) Find a formula for $r(t)$ given that the rate of change of the rat population is constant, and use this formula to predict the number of rats in the building on the 100th day of the year ($t = 99$).

 (b) Find a formula for $r(t)$ given that the rate of change of the rat population is proportional to the number of rats in the building, and use this formula to predict the number of rats in the building on the 100th day of the year ($t = 99$).

87. Alina started an investment account with an initial deposit of one thousand dollars. Following the initial deposit, the amount of money increased at a rate proportional to her investment account balance. After three years her balance was $1,260.

 (a) Write down a function $A(t)$ that describes the amount of money in Alina's investment account t years after her initial deposit.

 (b) How much money will Alina have in her investment account after 30 years?

 (c) How long will it be before Alina's initial investment quadruples?

88. Continuing Exercise 81 from Section 5.2, suppose that the temperature T, in degrees Fahrenheit, of a yam that has been sitting in an oven for t minutes is given by

$$T(t) = 350 - 280e^{-0.2t}.$$

 (a) Use a derivative to show that the temperature of the yam is always increasing.

 (b) Use a second derivative to show that the temperature of the yam increases at a decreasing rate. This relationship is related to the odd saying "Cold water boils faster." How?

89. Continuing Exercise 82 from Section 5.2, suppose that a political candidate in Hamtown, Virginia runs an advertising campaign that increases his name recognition so that, after t days, the number of people in Hamtown that have heard of him is given by

$$P(t) = \frac{45,000}{1 + 35e^{-0.12t}}.$$

 (a) Use derivatives to show that the number of people who have heard of the candidate increases over time, first slowly, then quickly, and then slowly again.

 (b) After how many days is the name recognition of the candidate increasing the fastest?

Proofs

90. Use the definition of the derivative and the definition of the number e to prove that $f(x) = e^x$ is its own derivative.

91. Use the chain rule to prove that $\frac{d}{dx}(e^{kx}) = ke^{kx}$.

92. Use the fact that $\frac{d}{dx}(b^x) = (\ln b)\, b^x$ to prove that $\frac{d}{dx}(e^{kx}) = ke^{kx}$.

93. Prove that if f is an exponential function, then $f'(x)$ is proportional to $f(x)$.

94. Use implicit differentiation and the fact that $\log_b x$ is the inverse of b^x to prove that $\frac{d}{dx}(\log_b x) = \frac{1}{(\ln b)x}$.

95. Use the definition of $|x|$, the chain rule, and the fact that $\frac{d}{dx}(\ln x) = \frac{1}{x}$ for $x > 0$ to prove that $\frac{d}{dx}(\ln |x|) = \frac{1}{x}$ for all $x \neq 0$.

96. Use a direct application of the fact that $\frac{d}{dx}(f^{-1}(x)) = \frac{1}{f'(f^{-1}(x))}$ to prove that $\frac{d}{dx}(\ln x) = \frac{1}{x}$.

97. Prove that every exponential function $f(x) = e^{kx}$ with $k > 0$ has a graph whose shape is as given in the first part of Theorem 5.16.

98. Prove that every exponential function $f(x) = e^{kx}$ with $k < 0$ has a graph whose shape is as given in the second part of Theorem 5.16.

Thinking Forward

L'Hôpital's rule: At the end of this chapter we will see that, under certain conditions, the limit of a quotient of functions is equal to the limit of the quotient of the derivatives of those functions. Specifically, if $f(x)$ and $g(x)$ both approach zero as $x \to c$, then

$$\lim_{x \to c} \frac{f(x)}{g(x)} = \lim_{x \to c} \frac{f'(x)}{g'(x)}.$$

Show that each of the following limits is of the indeterminate form $\frac{0}{0}$ and then use L'Hôpital's rule to calculate the limit:

▶ $\lim_{x \to 0} \dfrac{x^3}{1 - 2^x}$

▶ $\lim_{x \to 1} \dfrac{\ln x}{x - 1}$

▶ $\lim_{x \to 1} \dfrac{3^x - 3}{1 - x^2}$

▶ $\lim_{x \to 3} \dfrac{(x - 3)^2}{1 - e^{x-3}}$

5.4   APPLICATIONS OF EXPONENTIAL FUNCTIONS

▶ Modeling quantities with constant percentage growth with exponential functions

▶ Realizing continuous growth as a limit of multiple compoundings of percentage growth

▶ Doubling time and half-life of quantities modeled by exponential functions

Percentage Growth

If a quantity changes by a fixed amount over repeated periods of the same duration, then that quantity is a linear function of time. For example, if you start with 8 pounds of potatoes on day zero and then add 2 more pounds of potatoes to your pile each subsequent day, then on day t you will have $8 + 2t$ pounds of potatoes, as shown next at the left.

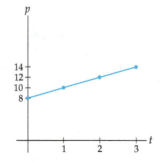

Linear growth of potato pile

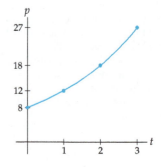

Percentage growth of potato pile

If instead the quantity changes by a fixed *percentage* each such period, then as we will soon see, that quantity will be an exponential function of time. For example, what happens if you start with 8 pounds of potatoes and then increase the number of pounds by 50 percent each subsequent day? Starting with 8 pounds, on day one you add 4 more pounds, for a total of 12 pounds. On day two you add 6 more pounds, for a total of 18 pounds. On day three you add 9 pounds, for a total of 27 pounds, and so on, as shown in the preceding right-hand figure. In the linear example, the rate of growth is constant, but in the percent growth example, the rate of growth gets larger as the number of potatoes gets larger.

As another example, consider a savings account with an initial balance of $1000.00 that earns 3% interest each month. Let $B(t)$ represent the balance of the savings account t months after the initial deposit. After one month, the amount in the savings account will be $B(1) = 1000 + 0.03(1000) = \$1030.00$. Since $1000 + 0.03(1000) = 1.03(1000)$, we could also write this equation:

$$B(1) = 1000(1.03) = \$1030.00.$$

To find the balance after two months, we need to multiply $B(1)$ by 1.03 (since the balance should grow by 3 percent more). The balance of the savings account after two months will be $B(1)(1.03) = (1000(1.03))(1.03) = 1000(1.03)^2$:

$$B(2) = 1000(1.03)^2 = \$1060.90.$$

Similarly, after three months the savings account balance will be

$$B(3) = 1000(1.03)^3 = \$1092.73.$$

Each month the balance in the savings account will be 1.03 times the amount in the savings account the previous month. The savings account balance $B(t)$ after t months is the function $B(t) = 1000(1.03)(1.03)\cdots(1.03)(1.03)$, with t copies of 1.03. This is an exponential function:

$$B(t) = 1000(1.03)^t.$$

In general, if a quantity $Q(t)$ increases by a fixed percentage r each period of a given duration, where t is measured in terms of that period and r is written in decimal form, then $Q(t)$ is an exponential function with base $1 + r$ and exponent t and with coefficient given by the initial amount Q_0 of the quantity.

THEOREM 5.18   **Percentage Growth Formula**

If $Q(t)$ with an initial quantity of $Q(0) = Q_0$ grows at a fixed decimal percentage r over every period t, then $Q(t)$ is modeled by the exponential function

$$Q(t) = Q_0(1 + r)^t.$$

This formula works for any real value of t, not just for integer values. For example, in the context of the savings balance problem, this formula says that you will have $B(3.5) = 1000(1.03)^{3.5} \approx \$1109.00$ in your savings account after three-and-a-half months. However, in real life your savings account balance might grow only at the beginning of each month, and thus $B(3.5)$ in real life would be the same as $B(3)$. The percentage growth formula is a continuous model for what in real life might be a discrete phenomenon.

Multiple Compounding

Sometimes a quantity increases by a fixed percentage each period, but we wish to measure the time in terms of a larger period. For example, some credit card companies that offer an annual percentage rate actually compound the interest monthly. For instance, if a credit card has an annual percentage rate of 14.99% compounded monthly, then your balance increases by $\frac{14.99}{12}$% each month. The general formula for such multiple compounding of percentage growth is given in the following theorem:

THEOREM 5.19

Percentage Growth with Multiple Compounding

If $Q(t)$ grows at r percent compounded n times per period t and has an initial value of $Q(0) = Q_0$, then the function:

$$Q(t) = Q_0 \left(1 + \frac{r}{n}\right)^{nt}.$$

For example, suppose you have a $3000.00 debt on a credit card with a 14.99% annual percentage rate that is compounded monthly, and suppose also that you leave that debt on the card without making any payments or accruing any new charges or fees. Then the size of your debt $D(t)$ in dollars after t years is given by the formula

$$D(t) = 3000 \left(1 + \frac{0.1499}{12}\right)^{12t}.$$

After five years your debt will be $D(5) = 3000 \left(1 + \frac{0.1499}{12}\right)^{12(5)} \approx \$6318.42$. Notice that this is more than the amount $3000(1 + 0.1499)^5 \approx \$6031.45$ that you would pay if the credit card charged 14.99% interest all at once at the end of each year, instead of monthly. The monthly compounding formula increases your debt faster, since interest is applied each month not only to the original balance, but also to the interest accrued during previous months.

The more times interest is applied in a period, the faster the balance will grow. For this reason, many credit card companies now apply interest *daily*, not monthly. For example, a $3000.00 debt at 14.99% interest compounded daily will increase by $\frac{14.99}{365}$% each month. In this case your debt $D(t)$ after t years would be

$$D(t) = 3000 \left(1 + \frac{0.1499}{365}\right)^{365t}.$$

After five years with this type of interest your debt to the credit card company would be

$$D(5) = 3000 \left(1 + \frac{0.1499}{365}\right)^{365(5)} \approx \$6346.85.$$

Imagine now that you are an enterprising young credit card executive. If monthly compounding makes your company more money than yearly compounding, and if daily is even better than monthly, then what would happen if interest were compounded even more frequently, say, every hour? Or even better, every minute? The following table compares these and our previous compounding intervals:

Compounding	Times compounded	Balance after five years	
Yearly	1	$3000(1 + 0.1499)^5$	$= \$6031.45$
Monthly	12	$3000 \left(1 + \dfrac{0.1499}{12}\right)^{12(5)}$	$= \$6318.42$
Daily	365	$3000 \left(1 + \dfrac{0.1499}{365}\right)^{365(5)}$	$= \$6346.85$
Hourly	8760	$3000 \left(1 + \dfrac{0.1499}{8760}\right)^{8760(5)}$	$= \$6347.78$
Every minute	1,314,000	$3000 \left(1 + \dfrac{0.1499}{525,600}\right)^{525,600(5)}$	$= \$6347.82$

Notice that your debt balance is higher when the interest is compounded more often, but that the difference between compounding every day, hour, or minute is not that significant. There seems to be some sort of limit to how much money the credit card companies can make with multiple compounding.

Wait; did we say *limit*? If n is the number of times that the interest is compounded, what happens as $n \to \infty$?

Continuous Growth

In the credit card example we just presented, compounding n times resulted in the formula $D(5) = 3000 \left(1 + \frac{0.1499}{n}\right)^{n(5)}$ for the amount of debt accumulated after five years. As $n \to \infty$, $1 + \frac{0.1499}{n}$ approaches 1 and $n(5)$ approaches ∞, and therefore

$$\lim_{n \to \infty} 3000 \left(1 + \frac{0.1499}{n}\right)^{n(5)}$$

is of the intedeterminate form 1^∞. At this point we cannot be sure whether or not the limit exists, although from the preceding table of credit card balances it would seem reasonable to imagine that it does exist and in fact probably is just less than \$6,348.00.

However, recall from Definition 5.9 the following special limit:

$$\lim_{h \to 0} (1 + h)^{1/h} = e.$$

If we substitute $n = \frac{1}{h}$ in this limit, then as $h \to 0$, we have $n \to \infty$, and thus the special limit becomes

$$\lim_{n \to \infty} \left(1 + \frac{1}{n}\right)^n = e.$$

This expression for e has some similarities with our indeterminate multiple compounding limit and in fact is the key to its solution. The following theorem will help us make the connection:

THEOREM 5.20   **Expressing e^r as a Limit**

For any real number r, $\displaystyle\lim_{n \to \infty} \left(1 + \frac{r}{n}\right)^n = e^r$.

Proof. The key to the proof will be a change of variables. Let $m = \frac{n}{r}$, so that $n = rm$. As $n \to \infty$, we also have $m \to \infty$, and thus

$$\lim_{n \to \infty} \left(1 + \frac{r}{n}\right)^n = \lim_{m \to \infty} \left(1 + \frac{r}{rm}\right)^{rm} \qquad \leftarrow \text{ change of variables } m = \frac{n}{r}$$

$$= \lim_{m \to \infty} \left(\left(1 + \frac{1}{m}\right)^m\right)^r \qquad \leftarrow \text{ algebra}$$

$$= \left(\lim_{m \to \infty} \left(1 + \frac{1}{m}\right)^m\right)^r \qquad \leftarrow \text{ continuity and limit rules}$$

$$= e^r. \qquad \leftarrow \text{ special limit for } e$$

Using this limit expression for e^r, we can now calculate our multiple compounding limit:

$$\lim_{n \to \infty} 3000 \left(1 + \frac{0.1499}{n}\right)^{n(5)} = 3000 \left(\lim_{n \to \infty} \left(1 + \frac{0.1499}{n}\right)^n\right)^5 = 3000(e^{0.1499})^5 = 3000 e^{0.1499(5)}.$$

Therefore if the credit card company could somehow compound infinitely many times each year, at the end of five years you would owe $3000e^{0.08(5)} \approx 6347.83$. At first glance this number looks identical to the every-minute compounding amount that we found earlier, but in fact they differ by a fraction of a cent.

In general, we call compounding infinitely many times *continuous compounding*. Continuous compounding is seen often in biological growth and radioactive decay situations, in which growth happens all the time in small increments rather than all at once at the end of each month, day, or minute. The analysis we just completed suggests that the general formula for continuous growth is as follows:

THEOREM 5.21   **Continuous Growth Formula**

If $Q(t)$ has a *continuous growth rate* k and an initial value $Q_0 = Q(0)$, then

$$Q(t) = Q_0 e^{kt}.$$

Proof. We want to prove that the continuous growth formula is the limit, as $n \to \infty$, of the formula for percentage growth with multiple compounding. By Theorem 5.20, we have

$$\lim_{n \to \infty} Q_0 \left(1 + \frac{r}{n}\right)^{nt} = Q_0 \left(\lim_{n \to \infty} \left(1 + \frac{r}{n}\right)^n\right)^t = Q_0(e^r)^t = Q_0 e^{rt}. \qquad \blacksquare$$

Doubling Time and Half-Life

We have already seen two important characteristics of exponential functions. Earlier in this section we saw that quantities modeled by exponential functions grow at a fixed percentage each period. We also saw that quantities modeled by exponential functions grow at a rate proportional to the quantity itself. Now we examine a third characteristic: Quantities modeled by exponential functions take a fixed amount of time to double in size or to decrease by half.

The amount of time that it takes an increasing quantity to double is called a *doubling time*. Most functions have varying doubling times depending on the starting point, but as we will see in a moment, an exponential function will have the same doubling time no matter what the starting point. Consider for example the following table of values for the linear function $P(t) = 2t$ and the exponential function $Q(t) = 2^t$:

t	0	1	2	3	4	5	6
$P(t) = 2t$	0	2	4	6	8	10	12
$Q(t) = 2^t$	1	2	4	8	16	32	64

By looking at the second row of the table, we can see that the exponential function $Q(t) = 2^t$ doubles for every unit increase in time. Its doubling time is a constant value of one time unit. This is not the case for the linear function $P(t) = 2t$ in the first row of the table. For example, it takes one time unit for $P(t)$ to double from $P(1) = 2$ to $P(2) = 4$, but *two* time units for $P(t)$ to double again from $P(2) = 4$ to $P(4) = 8$.

THEOREM 5.22	**Exponential Functions Have a Constant Doubling Time or Half-Life**

Every exponential function has a constant doubling time or a constant half-life. More precisely,

(a) If $Q(t)$ is an exponential growth function, then there is some positive constant α for which $Q(t + \alpha) = 2Q(t)$ for all t.

(b) If $Q(t)$ is an exponential decay function, then there is some positive constant α for which $Q(t + \alpha) = \frac{1}{2}Q(t)$ for all t.

As we will see in Example 3, knowing the doubling time or half-life of an exponential function is equivalent to knowing the continuous growth value k; if you know one of these values, then you can always find the other. It also happens that this theorem is true for tripling times, quadrupling times, one-third-lives, and so on.

Proof. Suppose $Q(t) = Q_0 e^{kt}$ is an exponential growth function with $k > 0$. You will prove the exponential decay case in Exercise 64. Since exponential quantities increase without bound, we know that there must be some positive real number α so that $Q(\alpha)$ is twice the initial value $Q(0) = Q_0$. For this value of α we have:

$$Q(\alpha) = 2Q_0 \implies Q_0 e^{k\alpha} = 2Q_0 \implies e^{k\alpha} = 2.$$

We want to show that, regardless of the starting point, the quantity $Q(t)$ will *always* take the time α to double. In other words, we must show that $Q(t + \alpha) = 2Q(t)$ for all values of t. Using the fact that $e^{k\alpha} = 2$, we have

$$Q(t + \alpha) = Q_0 e^{k(t+\alpha)} = Q_0 e^{kt+k\alpha} = Q_0 e^{kt} e^{k\alpha} = Q_0 e^{kt}(2) = 2Q_0 e^{kt} = 2Q(t). \qquad \blacksquare$$

Examples and Explorations

EXAMPLE 1	**A population that is reduced by a fixed percentage each week**

A termite colony is slowly dying off. Three weeks ago there were 2500 termites in the colony; now there are only 1250. Assuming that the population of termites is decreasing by a fixed percentage each week, determine when there will be only 300 termites in the colony.

SOLUTION

Let $P(t)$ be the number of termites after t weeks, where $t = 0$ represents three weeks ago. Then $P_0 = P(0) = 2500$, and $P(3) = 1250$. Since the population of termites decreases by a fixed percentage each week, $P(t)$ is an exponential function of the form

$$P(t) = 2500(1 - r)^t.$$

Notice that we chose to write $1 - r$ instead of $1 + r$ because we know that the size of the termite colony is decreasing and thus the percentage rate will be negative. We can use the fact that $P(3) = 1250$ to solve for the percentage decay rate r:

$$P(3) = 1250 \implies 2500(1-r)^3 = 1250 \implies (1-r)^3 = \frac{1}{2} \implies r = 1 - \left(\frac{1}{2}\right)^{1/3} \approx 0.206.$$

This calculation tells us that the population of termites must be decreasing by approximately 20.6% each week and that the number of termites after t weeks is given by the formula

$$P(t) = 2500(1 - 0.206)^t = 2500(0.794)^t.$$

Now that we have a formula for $P(t)$, we can find the time at which there are only 300 termites left in the colony by solving the equation $P(t) = 300$:

$$P(t) = 300 \implies 2500(0.794)^t = 300 \qquad \leftarrow \text{formula for } P(t)$$

$$\implies (0.794)^t = \frac{300}{2500} = 0.12 \qquad \leftarrow \text{algebra}$$

$$\implies \ln((0.794)^t) = \ln 0.12 \qquad \leftarrow \text{take logarithm of both sides}$$

$$\implies t \ln 0.794 = \ln 0.12 \qquad \leftarrow \text{property of logarithms}$$

$$\implies t = \frac{\ln 0.12}{\ln 0.794} \approx 9.19. \qquad \leftarrow \text{solve for } t$$

Thus there will be 300 termites in the colony just after nine weeks from $t = 0$ and therefore just after six weeks from "now" in the problem.   □

CHECKING THE ANSWER

We can do a reality check on our previous answer by using the fact that exponential functions must change by a fixed percentage over periods of the same duration. We were given that the population $P(t)$ of termites decreases by half in the first three weeks, from 2500 at time $t = 0$ to 1250 at time $t = 3$. Since the population of termites is modeled by exponential decay, it should decrease by half again in the three weeks from $t = 3$ to $t = 6$, from 1250 to 625. Then, the next three weeks from $t = 6$ to $t = 9$, it should decrease by half yet again, from 625 to 312.5. This is nearly as low as the population of 300 termites that we seek in the example; therefore it makes sense that it would take just over 9 weeks for the population to decrease from 2500 termites to 300 termites.

EXAMPLE 2

Converting from a percentage growth rate to a continuous growth rate

The number of worms in a backyard garden increases every year by a percentage growth rate of 7%. What is the equivalent continuous growth rate of the worm population?

SOLUTION

The information given tells us that the number of worms in the garden after t years will be

$$W(t) = W_0(1 + 0.07)^t.$$

The equivalent continuous growth rate is the number k such that the population formula $W_0(1 + 0.07)^t$ is equal to W_0e^{kt} for all times t. There must be such a number k because we can always convert from exponential functions in the form b^t to exponential functions in the form e^{kt}. This is nothing more than an algebra problem:

$$W_0(1 + 0.07)^t = W_0e^{kt} \implies (1.07)^t = (e^k)^t \qquad \leftarrow \text{cancel } W_0$$

$$\implies 1.07 = e^k \qquad \leftarrow \text{exponential functions are one-to-one}$$

$$\implies k = \ln(1.07) \approx 0.06766. \qquad \leftarrow \text{solve for } k$$

Therefore the continuous growth rate of the worms is $k = 0.06766$, or 6.766%, which is just under the yearly percentage growth rate of 7%. As a reality check, note that it does make sense that compounding 6.766% continuously would yield an effective yearly percentage growth rate of something a little higher, such as 7%.   □

| **EXAMPLE 3** | ### Using doubling time to determine continuous growth rate |

Suppose a population of bacteria is growing in a petri dish in such a way that the number of bacteria doubles every 3 hours. If there are initially 1000 bacteria in the petri dish, how long will it take for there to be 1 million bacteria?

SOLUTION

Let $B(t)$ be the number of bacteria in the dish after t hours. Since $B(t)$ has a constant doubling time, we will model it with an exponential function. Moreover, we know that $B(0) = 1000$; therefore $B(t) = 1000e^{kt}$ for some value of k. Since the bacteria population doubles every three hours, $B(3) = 2B(0) = 2(1000) = 2000$. We can use this fact to solve for the continuous growth rate k:

$$B(3) = 2000 \implies 1000e^{k(3)} = 2000 \implies e^{3k} = 2 \implies 3k = \ln 2 \implies k = \frac{\ln 2}{3}.$$

We could make a numerical approximation of k at this point, but we will wait until the end of the problem to make any approximations; this approach will make our final answer much more accurate.

Now that we know that the number of bacteria in the petri dish after t hours is given by $B(t) = 1000e^{(\ln 2/3)t}$, we can find out when there will be 1 million bacteria by solving the equation $B(t) = 1,000,000$ for t:

$$
\begin{aligned}
B(t) = 1,000,000 \implies 1000e^{(\ln 2/3)t} &= 1,000,000 & &\leftarrow \text{formula for } B(t) \\
\implies e^{(\ln 2/3)t} &= 1000 & &\leftarrow \text{algebra} \\
\implies \left(\frac{\ln 2}{3}\right)t &= \ln 1000 & &\leftarrow \text{take logarithm of both sides} \\
\implies t &= \frac{3\ln 1000}{\ln 2} \approx 29.9. & &\leftarrow \text{solve for } t
\end{aligned}
$$

Therefore it will be approximately 29.9 hours before there are 1 million bacteria in the petri dish.   □

| **EXAMPLE 4** | ### Using half-life to determine percentage growth rate |

Radioactive substances have constant half-lives. This means that a quantity of a radioactive substance always takes the same amount of time to decrease by half, no matter how much or how little of that substance you start with. For example, suppose that the half-life of the fictional radioactive substance unobtainium is 29 years. A sample of rock is found with 80 grams of unobtainium. What is the percentage by which the amount of unobtainium in the rock decreases each year?

SOLUTION

Let $S(t)$ represent the number of grams of unobtainium that are left after t years. Since $S(t)$ has a constant half-life, we will model $S(t)$ with an exponential function. We also know that its initial value is $S_0 = S(0) = 80$ grams. Therefore $S(t) = 80e^{kt}$ for some value k.

We begin by using the given half-life to solve for the value of the continuous growth rate k. Since unobtainium has a half-life of 29 years and we began with 80 grams of it, we know that $S(29) = \frac{S_0}{2} = \frac{80}{2} = 40$ grams. We can use this fact to solve for k:

$$S(29) = 40 \implies 80e^{k(29)} = 40 \implies e^{29k} = \frac{1}{2} \implies k = \frac{\ln(1/2)}{29}.$$

This means that the number of grams of unobtainium in the rock after t years is given by the exponential function $S(t) = 80e^{(\ln(0.5)/29)t}$.

Since we were asked by what percentage the amount of unobtainium decreased each year, we need to convert $S(t)$ into an exponential function of the form $S(t) = S_0(1-r)^t$. The value of r is the percentage decay rate. We have

$$S(t) = 80e^{(\ln(0.5)/29)t} = 80(e^{\ln(0.5)/29})^t \approx 80(0.97638)^t = 80(1 - 0.02362)^t.$$

Therefore the amount of unobtainium in the rock decreases by approximately 2.362% each year.   □

CHECKING THE ANSWER

To check the yearly percentage growth rate we just found, we can verify that decreasing the initial amount of 80 grams by 2.362% a year for 29 years will bring us down to half the amount of unobtainium, or 40 grams. Indeed, we have

$$S(29) = 80(1 - 0.02362)^{29} \approx 39.9979.$$

The only reason we did not end up with exactly 40 grams is that we had done some rounding previously.

? TEST YOUR UNDERSTANDING

▶ In the potato pile example at the start of this section, does the linear model grow by the same percentage in each period? Why or why not?

▶ Why does it make sense that a $1000 savings account balance that earns 3% interest a year would be worth $1000(1.03)^3$ dollars after three years?

▶ Would it be to a credit card bank's advantage to compile interest every day instead of every month? What about every second instead of every hour? Why or why not?

▶ Why might it make more sense to use a continuous growth model rather than a multiple-compounding growth model to discuss the growth of a biological population?

▶ What does it mean for a quantity to have a constant doubling time?

EXERCISES 5.4

Thinking Back

Exponential functions: Express each of the following in the form $f(x) = Ae^{kx}$ for some constants A and k.

▶ $f(x) = 3(2^x)$ ▶ $f(x) = \dfrac{-5}{4^x}$ ▶ $f(x) = 3^{-2x}$

Exponential functions, part 2: Express each of the following in the form $f(x) = C(b^x)$ for some constants C and b.

▶ $f(x) = 2e^{-2x}$ ▶ $f(x) = 5e^{1.7x}$ ▶ $f(x) = 4e^{3x+1}$

▶ *Exponential growth and decay:* For which values of k does an exponential function of the form $f(x) = Ae^{kx}$ exhibit exponential growth? Exponential decay? Draw a sketch of each type of graph.

▶ *Exponential growth and decay, part 2:* For which values of b does an exponential function of the form $f(x) = C(b^x)$ exhibit exponential growth? Exponential decay? Draw a sketch of each type of graph.

Concepts

0. *Problem Zero:* Read the section and make your own summary of the material.

1. *True/False:* Determine whether each of the statements that follow is true or false. If a statement is true, explain why. If a statement is false, provide a counterexample.

(a) *True or False:* If a quantity $Q(t)$ starts with $Q(0) = 5$ and then grows by 17% a year, then $Q(t) = 5(1+17)^t$.

(b) *True or False:* If a credit card balance has an annual percentage rate of 11.99% with interest compounded monthly, then the outstanding balance will be increased by 11.99% each month.

(c) *True or False:* A savings account that earns a 2.5% annual interest rate, compounded monthly, will earn less money over five years than a savings account that earns a 2.5% annual interest rate compounded once a year.

(d) *True or False:* The number e is equal to $\lim_{h \to 0}(1+h)^{1/h}$.

(e) *True or False:* If $Q(t)$ is exponential with percentage growth rate r, then $Q'(t) = rQ(t)$.

(f) *True or False:* If $Q(t)$ is exponential with continuous growth rate k, then $Q'(t) = kQ(t)$.

(g) *True or False:* All exponential functions have a constant doubling time.

(h) *True or False:* All exponential growth functions have a constant doubling time.

2. *Examples:* Construct examples of the thing(s) described in the following. Try to find examples that are different than any in the reading.

(a) Two real-world quantities that might exhibit percentage growth.

(b) Two real-world quantities that might exhibit percentage growth with multiple compounding.

(c) Two real-world quantities that might exhibit continuous exponential growth.

3. Fill in the blank: If a quantity $Q(t)$ grows at a rate proportional to _____ , then $Q(t)$ is an exponential function.

4. Fill in the blank: If a quantity $Q(t)$ grows or decreases by a fixed percentage each period, then $Q(t)$ is a _____ function.

5. Fill in the blanks: A quantity $Q(t)$ that has a _____ doubling time or half-life can be modeled by a _____ function.

6. Show that the percentage growth formula is just a special case of the formula for percentage growth with multiple compounding, where the compounding occurs only once per period.

7. If a credit card company charges an annual percentage rate of 13.99% and compounds interest monthly, what percentage is charged each month as interest on any outstanding balances?

8. Suppose you owe money to a loan company that charges 15% interest annually. Would you rather have them com-

pound your interest monthly, daily, or continuously, and why?

9. Suppose you wish to put some money in an investment account. Which is better, to have 16% annual interest compounded yearly or to have 15% annual interest compounded daily?

10. State the formula for the exponential function $Q(t)$ that models a percentage growth rate r over each period t from an initial amount Q_0.

11. State the formula for the exponential function $Q(t)$ that models a percentage growth rate r compounded n times over each period t from an initial amount Q_0.

12. State the formula for the exponential function $Q(t)$ that models a continuous growth rate k over time t from an initial amount Q_0.

13. What limit statement allows us to show that the formula for multiple compounding of percentage growth becomes the formula for continuous growth as we let the number of compoundings per period approach infinity?

14. Suppose a quantity $Q(t)$ is an exponential function that increases with time (measured in days). If the quantity takes 7 days to double in size, how long will it take for that quantity to increase eightfold?

15. Suppose a quantity $Q(t)$, with time measured in years, has a half-life of 17 years and an initial amount Q_0. Without finding an equation for $Q(t)$, fill in the blanks: $Q(17) = $ _____ , $Q(51) = $ _____ , and $Q(-34) = $ _____ .

16. Suppose a quantity $Q(t) = Q_0 e^{kt}$, with time measured in days, has a doubling time of five days. Without finding a value for k, fill in the blanks: $e^{5k} = $ _____ and $e^{10k} = $ _____ .

17. Suppose a growing population $W(t)$ of warthogs on an island grows at a rate proportional to the number of warthogs on the island. Will the function $W(t)$ have a constant doubling time? Why or why not?

18. Show that the function $P(t) = t^2 + 1$ does not have a constant doubling time.

Skills

Calculate each of the limits in Exercises 19–24 by using the fact that $\lim_{x \to \infty}\left(1+\dfrac{r}{x}\right)^x = e^r$.

19. $\lim_{x \to \infty}\left(1+\dfrac{3}{x}\right)^{2x}$

20. $\lim_{x \to \infty}(1+7x^{-1})^{-3x}$

21. $\lim_{x \to \infty}\left(1-\dfrac{1}{x}\right)^x$

22. $\lim_{x \to \infty}\left(1+\dfrac{1}{3x}\right)^{4x+1}$

23. $\lim_{x \to 0^+}\left(1-\dfrac{x}{2}\right)^{3/x}$

24. $\lim_{x \to 0^+}(1+5x)^{1/x}$

In Exercises 25–37, $Q(t)$ is a quantity that can be modeled by an exponential function, with t measured in years.

25. If $Q(t)$ has a yearly percentage growth rate of 24%, what is its continuous growth rate?

26. If $Q(t)$ has a continuous growth rate of 0.17, what is its yearly percentage growth rate?

27. If $Q(t)$ has a monthly percentage growth rate of 2.3%, what is its yearly percentage growth rate?

28. If $Q(t)$ has a monthly percentage growth rate of 3.1%, what is its continuous growth rate?

29. If $Q(t)$ has a half-life of 32 years, what is its continuous growth rate?

30. If $Q(t)$ triples every 17 years, what is its yearly percentage growth rate?

31. If $Q(t)$ has a yearly percentage growth rate of 13%, what is its doubling time?

32. If $Q(t)$ has a doubling time of 6.5 years, what is the percentage by which it increases each year?

33. If $Q(t)$ has a doubling time of 6.5 years, how long will it take for it to triple?

34. If $Q(t)$ has a doubling time of 6.5 years, what is its continuous growth rate?

35. If $Q(t)$ decreases by 4% a year, how long will it take for it to decrease by half?

36. If $Q(t)$ decreases by 4% a year, how long will it take for it to decrease by a third?

37. If $Q(t)$ decreases by 4% a year, what is the continuous decay rate?

Applications

38. Suppose you earn 4% annual interest, compounded monthly, in a savings account. If you wanted to change to an account on which interest is compounded yearly, what annual percentage rate would earn you the same total amount of interest over the course of the year?

39. Suppose you owe 18% annual interest, compounded monthly, on a credit card. By what percentage does your account increase each year? What would that effective yearly percentage be if the credit card decided to compound your interest daily instead?

40. A television commercial for Quickbuck claims that investing your money in the company's fund will double your money in 10 years. Is this likely? What annual interest rate would correspond to such growth?

You owe $5,000 to Transcam, a loan company that charges 22% interest annually. Answer the questions in Exercises 41–47 under the assumption that you do not make any payments on, or add to, this loan.

41. How much would you owe Transcam in 15 years if the company compounded the interest only once a year? What if it compounded the interest monthly?

42. How much would you owe Transcam in 15 years if the company compounded the interest daily? Four times a day?

43. How much would you owe Transcam in 15 years if the company compounded the interest continuously?

44. By what percentage would your balance with Transcam increase each year if the company compounded the interest yearly? Monthly?

45. By what percentage would your balance with Transcam increase each year if the company compounded the interest daily? Continuously?

46. How many years will it be before you owe Transcam 1 million dollars, assuming that the company compounds interest yearly? Monthly?

47. How many years will it be before you owe Transcam 1 million dollars, assuming that the company compounds interest daily? Continuously?

Dr. Drosophila is breeding fruit flies in his laboratory. He started with 600 fruit flies, and after three days there were already 840 fruit flies in Dr. Drosophila's laboratory. Answer the questions in Exercises 48–51, assuming that the fruit fly population grows at a rate proportional to the number of fruit flies in the laboratory.

48. By what percentage does the fruit fly population increase each day?

49. What is the continuous growth rate of the fruit fly population?

50. How long will it take for the fruit fly population to quadruple?

51. How long will it be before there are 1 million fruit flies packed into Dr. Drosophila's laboratory?

Calvin starts a rumor that peanut butter makes your hair fall out. Answer the questions in Exercises 52–55 given that the number of people who have heard this rumor triples every four hours.

52. How long will it be before 700 people have heard the rumor?

53. How long will it take for the number of people who have heard the rumor to double?

54. By what percentage does the number of people who have heard the rumor increase each hour?

55. By what percentage does the number of people who have heard the rumor increase each *day*? (Careful; this one is harder!)

Suppose that the half-life of the fictional radioactive substance unobtainium is 29 years. Answer the questions in Exercises 56–61 for a rock sample that initially contains 250 grams of unobtainium.

56. What percentage of unobtainium decays each year?

57. What is the continuous decay rate of unobtainium?

58. How long will it be before the rock sample contains only 6 grams of unobtainium?

59. At one point in the past, the rock sample contained 900 grams of unobtainium; how long ago?

60. What percentage of the unobtainium will be left in 300 years?

61. How long will it be before 95% of the unobtainium has decayed?

Proofs

62. Given that $\lim\limits_{n \to \infty} \left(1 + \frac{r}{n}\right)^n = e^r$, prove that the limit of the multiple-compounding formula $Q_0 \left(1 + \frac{r}{n}\right)^{nt}$ as $n \to \infty$ is the continuous growth formula $Q_0 e^{rt}$.

63. Suppose that $Q(t)$ is an exponential growth function and that it takes 12 time units for the quantity $Q(t)$ to increase from its initial amount, Q_0, to twice that amount, $2Q_0$. Prove that $Q(t)$ has a *constant* doubling time of 12 time units. (In other words, prove that, for *any* time t, $Q(t + 12) = 2Q(t)$.)

64. Prove that every decreasing exponential function has a constant half-life, as follows: Suppose $Q(t)$ is a decreasing exponential function with $Q_0 = Q(0)$, and suppose that it takes α time units for the function $Q(t)$ to decrease by half from Q_0 to $\frac{Q_0}{2}$. Show that it always takes $Q(t)$ a period of α time units to decrease by half, regardless of the starting point.

In Exercises 65–68, suppose that a quantity $Q(t)$, with t measured in years, has a yearly percentage growth rate r, a continuous growth rate k, a doubling time α, and a tripling time β.

65. Find and prove a formula that expresses the continuous growth rate k in terms of the percentage growth rate r, and prove this formula.

66. Find and prove a formula that expresses the percentage growth rate r in terms of the doubling time α.

67. Find and prove a formula that expresses the doubling time α in terms of the continuous growth rate k.

68. Find and prove a formula that expresses the tripling time β in terms of the doubling time α.

Thinking Forward

Differential equations: A function $y(x)$ is exponential if and only if its derivative is proportional to itself. This means that exponential functions are solutions of **differential equations** of the form $\frac{dy}{dx} = ky$. A solution of a differential equation is a function $y(x)$ that makes the equation true.

▶ Show that $y(x) = 4e^{3x}$ is a solution of the differential equation $\frac{dy}{dx} = 3y$.

▶ Show that $y(x) = 1.7e^{-2.1x}$ is a solution of the differential equation $\frac{dy}{dx} = -2.1y$.

▶ Show that $y(x) = 3(2^x)$ is a solution of the differential equation $\frac{dy}{dx} = (\ln 2)y$.

▶ Describe all of the solutions of the differential equation $\frac{dy}{dx} = 3y$.

▶ Describe all of the solutions of the differential equation $\frac{dy}{dx} = 3y$ that satisfy $y(0) = 2$.

5.5   L'HÔPITAL'S RULE

▶ L'Hôpital's rule for calculating limits of the indeterminate forms $\frac{0}{0}$ and $\frac{\infty}{\infty}$

▶ Rewriting limits of the indeterminate form $0 \cdot \infty$ so that L'Hôpital's rule applies

▶ Using logarithms to calculate limits of the indeterminate forms 0^0, 1^∞, and ∞^0

Geometrical Motivation for L'Hôpital's Rule

As we have already seen, limits of the form $\frac{0}{0}$ are indeterminate. At first glance it is not clear whether such a limit exists or what it might be equal to. In some cases we can resolve the indeterminate form $\frac{0}{0}$ with some algebra, such as in this example:

$$\lim_{x \to 0} \frac{x^2}{x^3 - x} = \lim_{x \to 0} \frac{x}{x^2 - 1} = \frac{0}{0^2 - 1} = 0.$$

Other limits of the indeterminate form $\frac{0}{0}$ are not so easy to simplify. In particular, limits of quotients that involve a mixture of different types of functions are usually more resistant to algebra. For example, consider the limit

$$\lim_{x \to 0} \frac{x^2}{2^x - 1}.$$

As $x \to 0$ we have $x^2 \to 0$ and $2^x - 1 \to 1 - 1 = 0$, and thus this limit is of the indeterminate form $\frac{0}{0}$. This indeterminate limit cannot be simplified with algebra. So what can we do?

Let's approach the problem graphically. The graphs of $f(x) = x^2$ and $g(x) = 2^x - 1$ are shown next at the left. Since we are interested in a limit as $x \to 0$, we should focus on what happens as we look at smaller and smaller graphing windows around $x = 0$, as shown in the second and third graphs.

$f(x) = x^2$ and $g(x) = 2^x - 1$ *Same graph but in smaller window*   *Graphs are almost linear here*

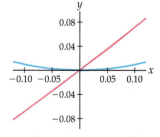

Near $x = 0$, the graph of $f(x) = x^2$ looks a lot like its horizontal tangent line $y = 0$, and the graph of $g(x) = 2^x - 1$ looks a lot like its tangent line $y = x$. Thus we would anticipate the behavior of the quotient $\frac{f(x)}{g(x)} = \frac{x^2}{2^x - 1}$ as $x \to 0$ to be similar to that of the quotient $\frac{0}{x}$ of the corresponding tangent lines at $x = 0$. From this information it would be reasonable to guess that $\lim_{x \to 0} \frac{x^2}{2^x - 1} = 0$. As we are about to see, that is in fact the case. Indeed, we will soon see that, in general, limits of the indeterminate form $\frac{0}{0}$ are related to the limit of the quotient of the slopes, or derivatives, of those numerator and denominator functions.

We can also use tangent lines to examine limits of the indeterminate form $\frac{\infty}{\infty}$. For example, consider the limit

$$\lim_{x \to \infty} \frac{x^2}{2^x - 1}.$$

This is the same quotient $\frac{x^2}{2^x - 1}$ of functions as before, but with a limit as $x \to \infty$ instead of as $x \to 0$. As $x \to \infty$ we have $x^2 \to \infty$ and $2^x - 1 \to \infty$, and therefore this limit is of the indeterminate form $\frac{\infty}{\infty}$.

Again, we cannot simplify this expression with algebra. Let's examine what happens as we look at larger graphing windows to see the behavior of $f(x) = x^2$ and $g(x) = 2^x - 1$ as $x \to \infty$:

$f(x) = x^2$ and $g(x) = 2^x - 1$	*Same graph but in larger window*	*Graphs are much different out here*

  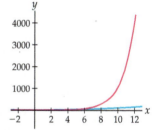

From the rightmost graph we can see that the heights on the graph of $g(x) = 2^x - 1$ are in some sense approaching ∞ in a fundamentally faster way than the heights on the graph of $f(x) = x^2$. Specifically, the slopes of $g(x) = 2^x - 1$ are significantly steeper than the slopes of $f(x) = x^2$ for large values of x. Accordingly, we would anticipate that as $x \to \infty$ the values of $2^x - 1$ would win the race to ∞ and that the limit $\lim\limits_{x \to \infty} \frac{x^2}{2^x - 1}$ would therefore be equal to zero. Once again a ratio of slopes has helped us to guess the value of the limit of a quotient. This geometric intuition is the basis for the powerful limit technique known as **L'Hôpital's rule**.

L'Hôpital's Rule for the Indeterminate Forms $\frac{0}{0}$ and $\frac{\infty}{\infty}$

The ideas we have developed regarding limits of quotients suggest the theorem that follows. This key theorem will allow us to solve some limits of the indeterminate form $\frac{0}{0}$ or $\frac{\infty}{\infty}$ by relating them to the quotients of the corresponding derivatives:

THEOREM 5.23   **L'Hôpital's Rule**

Suppose f and g are differentiable functions on some punctured interval around $x = c$ on which $g(x)$ is nonzero. If $\lim\limits_{x \to c} \frac{f(x)}{g(x)}$ is of the indeterminate form $\frac{0}{0}$ or $\frac{\infty}{\infty}$, then

$$\lim_{x \to c} \frac{f(x)}{g(x)} = \lim_{x \to c} \frac{f'(x)}{g'(x)},$$

as long as the second limit exists or is infinite.

The conclusion holds also if $x \to \infty$ (or $x \to -\infty$), as long as f and g are differentiable on some interval (N, ∞) (or $(-\infty, N)$) on which $g(x)$ is nonzero.

There are a few technical points to notice about the hypotheses of this theorem. First, notice that the functions f and g must be differentiable with $g(x)$ nonzero for x-values near, but not necessarily at, $x = c$. Second, notice that the conclusion does not necessarily hold if the limit on the right does not exist for some reason other than being infinite. In practice, most of the functions we look at will satisfy the two conditions we just mentioned. The vital thing to check when applying L'Hôpital's rule is that the limit is of the indeterminate form $\frac{0}{0}$ or $\frac{\infty}{\infty}$. If the limit is not in one of those two indeterminate forms, then L'Hôpital's rule cannot be applied.

The proof of L'Hôpital's rule requires a more general version of the Mean Value Theorem called the **Cauchy Mean Value Theorem**. This theorem—and the resulting proof of L'Hôpital's rule—are not that difficult, but proper proofs would take us too far afield for our purposes, so we leave that for future mathematics courses. To simplify matters we will prove L'Hôpital's rule only in the special case where f and g are very well behaved and the limit is of the indeterminate form $\frac{0}{0}$:

Proof. We prove the theorem in the special case where f and g are continuous and differentiable on an interval around and including $x = c$, and $f(c) = g(c) = 0$ but $g'(c) \neq 0$. We wish to show that

$$\lim_{x \to c} \frac{f(x)}{g(x)} = \frac{f'(c)}{g'(c)}.$$

We will work backwards from the right-hand side of the equality to the left-hand side. Applying the definition of the derivative to the numerator and the denominator, we have

$$\frac{f'(c)}{g'(c)} = \frac{\lim\limits_{x \to c} \dfrac{f(x)-f(c)}{x-c}}{\lim\limits_{x \to c} \dfrac{g(x)-g(c)}{x-c}} = \lim_{x \to c} \frac{f(x)-f(c)}{g(x)-g(c)} = \lim_{x \to c} \frac{f(x)-0}{g(x)-0} = \lim_{x \to c} \frac{f(x)}{g(x)}.$$

L'Hôpital's rule can be an extremely powerful tool for resolving indeterminate limits of the indeterminate form $\frac{0}{0}$ or $\frac{\infty}{\infty}$. In Example 2 we will use the rule to find the limits that we examined graphically at the start of this section. L'Hôpital's rule can also be useful for resolving other indeterminate forms, provided that we can rewrite them so that they are of the form $\frac{0}{0}$ or $\frac{\infty}{\infty}$; see Example 3.

Using Logarithms for the Indeterminate Forms 0^0, 1^∞, and ∞^0

Recall from Section 1.6 that limits of the form 0^0, 1^∞, and ∞^0 are indeterminate. For example, all three of the following limits are of the indeterminate form 1^∞, but each of them approaches something different.

$$\lim_{x \to \infty} \left(\frac{x}{x-1}\right)^{x^2} = \infty; \qquad \lim_{x \to 1^+} x^{1/(x^2-1)} = \sqrt{e}; \qquad \lim_{x \to \infty} \left(1 + \frac{1}{x}\right)^x = e.$$

In each of these limits there is a race between how fast the base approaches 1 and how fast the exponent approaches ∞, and in some sense the winner of that race determines the limit. But how can we determine who wins this race?

One difficulty with limits of the indeterminate forms 0^0, 1^∞, and ∞^0 is that such limits involve a variable in *both* the base *and* the exponent. Fortunately, logarithms have the power to change exponentiation into multiplication, in the sense that $\ln(a^b) = b \ln a$. The key to using logarithms to calculate limits is the following theorem:

THEOREM 5.24   **Relating the Limit of a Function to the Limit of Its Logarithm**

(a) If $\lim_{x \to c} \ln(f(x)) = L$, then $\lim_{x \to c} f(x) = e^L$.

(b) If $\lim_{x \to c} \ln(f(x)) = \infty$, then $\lim_{x \to c} f(x) = \infty$.

(c) If $\lim_{x \to c} \ln(f(x)) = -\infty$, then $\lim_{x \to c} f(x) = 0$.

To use this theorem to calculate a limit of the form $\lim_{x \to c} u(x)^{v(x)}$, we consider instead the limit

$$\lim_{x \to c} \ln(u(x)^{v(x)}) = \lim_{x \to c} v(x)\, \ln(u(x)).$$

Notice that the logarithm allows us to consider the limit of a product rather than a limit involving an exponent. Once we find this limit L, the theorem tells us that the answer to our original limit $\lim_{x \to c} u(x)^{v(x)}$ must be e^L. If instead of L we get $\pm\infty$, our original limit must be equal to $e^{\pm\infty}$.

> **_Proof._** We will prove only the first part of the theorem. The proof will follow directly from the composition rule for limits of continuous functions from Section 1.5. Suppose f is a function that is positive as x approaches c, so that $\ln(f(x))$ is defined near $x = c$. For functions f with an exponent and base both involving the variable x, domain restrictions will ensure that this will always be the case. Since the function $f(x) = \ln x$ is continuous on $(0, \infty)$, by the rule for limits of compositions of continuous functions, we have
>
> $$L = \lim_{x \to c} \ln(f(x)) = \ln(\lim_{x \to c} f(x)).$$
>
> Since $L = \ln(A)$ if and only if $A = e^L$, this equation implies that, as desired,
>
> $$\lim_{x \to c} f(x) = e^L.$$
> ■

Examples and Explorations

EXAMPLE 1   **Checking to see if L'Hôpital's rule applies**

Determine whether or not L'Hôpital's rule applies to each of the following limits as they are written here (without any preliminary algebra or simplification):

(a) $\lim_{x \to 1} \dfrac{(x-1)^2}{x-1}$ **(b)** $\lim_{x \to 0} \dfrac{x^2}{1+2^x}$ **(c)** $\lim_{x \to \infty} (2^x - x^3)$ **(d)** $\lim_{x \to \infty} \dfrac{2^x}{1-3^x}$

SOLUTION

(a) L'Hôpital's rule does apply, since as $x \to 1$, both the numerator $(x-1)^2$ and the denominator $x - 1$ approach zero, and therefore the limit $\lim_{x \to 1} \dfrac{(x-1)^2}{x-1}$ is of the indeterminate form $\dfrac{0}{0}$. The more technical hypotheses of L'Hôpital's rule are also satisfied by the numerator and denominator, as they are in most common examples. It is also possible to use algebra to solve this limit.

(b) L'Hôpital's rule does not apply, since $\dfrac{x^2}{1+2^x}$ approaches $\dfrac{0^2}{1+2^0} = \dfrac{0}{2} = 0$ as $x \to 0$.

(c) L'Hôpital's rule does not apply, since $2^x - x^3$ is not a quotient.

(d) L'Hôpital's rule does apply, since as $x \to \infty$ we have $2^x \to \infty$ and $1 - 3^x \to -\infty$ and therefore this limit is of the indeterminate form $\frac{\infty}{-\infty}$. The negative sign in the denominator could be factored out, but L'Hôpital's rule will work even if we do not extract the negative sign.   □

EXAMPLE 2 **Applying L'Hôpital's rule**

Use L'Hôpital's rule to calculate **(a)** $\lim\limits_{x \to 0} \dfrac{x^2}{2^x - 1}$ and **(b)** $\lim\limits_{x \to \infty} \dfrac{x^2}{2^x - 1}$.

SOLUTION

(a) Since the limit $\lim\limits_{x \to 0} \dfrac{x^2}{2^x-1}$ is of the indeterminate form $\frac{0}{0}$, L'Hôpital's rule applies and says that we can calculate it by considering instead the limit of the quotient of the derivatives of the numerator and denominator:

$$\lim_{x \to 0} \frac{x^2}{2^x - 1} \overset{\text{L'H}}{=} \lim_{x \to 0} \frac{\frac{d}{dx}(x^2)}{\frac{d}{dx}(2^x - 1)} = \lim_{x \to 0} \frac{2x}{(\ln 2)2^x} = \frac{2(0)}{(\ln 2)2^0} = \frac{0}{\ln 2} = 0.$$

Notice that we wrote "L'H" above the equals sign where we applied L'Hôpital's rule, to indicate our reasoning in that step. Notice also that we did not apply the quotient rule to differentiate the quotient $\frac{f(x)}{g(x)}$, because that is not the way that L'Hopital's rule works. Instead, following L'Hôpital's rule, we differentiated the numerator and denominator individually.

(b) The limit $\lim\limits_{x \to \infty} \dfrac{x^2}{2^x - 1}$ is of the indeterminate form $\frac{\infty}{\infty}$, so L'Hôpital's rule applies. Again we replace the numerator and denominator of our quotient with their derivatives:

$$\lim_{x \to \infty} \frac{x^2}{2^x - 1} \overset{\text{L'H}}{=} \lim_{x \to \infty} \frac{\frac{d}{dx}(x^2)}{\frac{d}{dx}(2^x - 1)} = \lim_{x \to \infty} \frac{2x}{(\ln 2)2^x}.$$

Unfortunately, our application of L'Hôpital's rule was not sufficient to resolve the limit here, because if we let $x \to \infty$ we have $2x \to \infty$ and $(\ln 2)2^x \to \infty$, so our limit is still in the indeterminate form $\frac{\infty}{\infty}$. However, this means that we can apply L'Hôpital's rule again!

$$\lim_{x \to \infty} \frac{x^2}{2^x - 1} \overset{\text{L'H}}{=} \lim_{x \to \infty} \frac{2x}{(\ln 2)2^x} \overset{\text{L'H}}{=} \lim_{x \to \infty} \frac{2}{(\ln 2)(\ln 2)2^x}$$

As $x \to \infty$ the denominator of this limit approaches ∞, while the numerator is equal to 2. Therefore we have

$$\lim_{x \to \infty} \frac{2}{(\ln 2)(\ln 2)2^x} = 0.$$

Notice that both of the preceding answers agree with what we guessed from a graphical analysis at the start of this section.   □

EXAMPLE 3 **Rewriting limits in the form $\frac{0}{0}$ or $\frac{\infty}{\infty}$ so that L'Hôpital's rule applies**

Use L'Hôpital's rule to calculate each of the following limits:

(a) $\lim\limits_{x \to \infty} x^2 e^{-3x}$ 　　　　　　**(b)** $\lim\limits_{x \to 0^+} x \ln x$ 　　　　　　**(c)** $\lim\limits_{x \to 0}\left(\dfrac{1}{x} - \dfrac{1}{e^x - 1}\right)$

SOLUTION

(a) As $x \to \infty$ we have $x^2 \to \infty$ and $e^{-3x} \to 0$, so the limit $\lim\limits_{x \to \infty} x^2 e^{-3x}$ is in the indeterminate form $\infty \cdot 0$. Therefore, this limit is not yet in a form to which we can apply L'Hôpital's rule. Luckily, limits of the indeterminate form $\infty \cdot 0$ can always be rewritten as a quotient of the form $\frac{\infty}{\infty}$ or as a quotient of the form $\frac{0}{0}$, simply by inverting one of the factors and placing it in the denominator. We can then choose whichever of these two indeterminate forms we prefer and apply L'Hôpital's rule. One way we could rewrite the limit is

$$\lim_{x \to \infty} x^2 e^{-3x} = \lim_{x \to \infty} \frac{x^2}{1/e^{-3x}} = \lim_{x \to \infty} \frac{x^2}{e^{3x}},$$

which is of the indeterminate form $\frac{\infty}{\infty}$. Another way we could write the limit is

$$\lim_{x \to \infty} x^2 e^{-3x} = \lim_{x \to \infty} \frac{e^{-3x}}{1/x^2},$$

which is of the indeterminate form $\frac{0}{0}$.

The first way of rewriting seems like it would be easier to deal with, so we apply L'Hôpital's rule to that version:

$$\lim_{x \to \infty} x^2 e^{-3x} = \lim_{x \to \infty} \frac{x^2}{e^{3x}} \qquad \leftarrow \text{rewrite to form } \frac{\infty}{\infty}$$

$$\overset{\text{L'H}}{=} \lim_{x \to \infty} \frac{2x}{3e^{3x}} \qquad \leftarrow \text{apply L'Hôpital's rule; still of form } \frac{\infty}{\infty}$$

$$\overset{\text{L'H}}{=} \lim_{x \to \infty} \frac{2}{9e^{3x}} \qquad \leftarrow \text{apply L'Hôpital's rule}$$

$$= 0. \qquad \leftarrow \text{since } 9e^{3x} \to \infty \text{ as } x \to \infty$$

(b) Note that we consider only the limit from the right, since $\ln x$ is not defined for negative numbers. As $x \to 0^+$ we have $x \to 0$ and $\ln x \to -\infty$, and therefore the limit $\lim\limits_{x \to 0^+} x \ln x$ is in the indeterminate form $0(-\infty)$. In order to apply L'Hôpital's rule we must rewrite the limit in the form $\frac{0}{0}$ or $\frac{\infty}{\infty}$. In this case it is easier to leave $\ln x$ in the numerator; as $x \to 0^+$ we have

$$\lim_{x \to 0^+} x \ln x = \lim_{x \to 0^+} \frac{\ln x}{1/x} \qquad \leftarrow \text{limit is now in the form } \frac{-\infty}{\infty}$$

$$\overset{\text{L'H}}{=} \lim_{x \to 0^+} \frac{1/x}{-1/x^2} \qquad \leftarrow \text{apply L'Hôpital's rule}$$

$$= \lim_{x \to 0^+} \frac{-x^2}{x} \qquad \leftarrow \text{use algebra to simplify}$$

$$= \lim_{x \to 0^+} (-x) = 0. \qquad \leftarrow \text{simplify more and evaluate limit}$$

Note that we could have applied L'Hôpital's rule a second time in this problem, since immediately after the first application of the rule the limit was again in the indeterminate form $\frac{\infty}{\infty}$. However, simplifying instead resulted in a very simple limit that we could easily evaluate.

(c) As $x \to 0$, both $\frac{1}{x}$ and $\frac{1}{e^x - 1}$ become infinite, so the limit $\lim_{x \to 0} \left(\frac{1}{x} - \frac{1}{e^x - 1} \right)$ is potentially of the indeterminate form $\infty - \infty$.

To be honest, we are playing pretty fast and loose here: Specifically, we are not bothering to examine whether the terms $\frac{1}{x}$ and $\frac{1}{e^x - 1}$ approach ∞ or $-\infty$. Without knowing that, we don't know for sure whether this is a limit of the indeterminate form $\infty - \infty$ or the non-indeterminate form $\infty + \infty$. We could look from the left and right and do this more precisely if we cared to, but it is easier to instead do some algebra so that we can apply L'Hôpital's rule:

$$\lim_{x \to 0} \left(\frac{1}{x} - \frac{1}{e^x - 1} \right) = \lim_{x \to 0} \frac{(e^x - 1) - x}{x(e^x - 1)} \quad \leftarrow \text{combine fractions; form is now } \frac{0}{0}$$

$$\overset{\text{L'H}}{=} \lim_{x \to 0} \frac{e^x - 1}{(e^x - 1) + xe^x} \quad \leftarrow \text{apply L'Hôpital's rule; still } \frac{0}{0}$$

$$\overset{\text{L'H}}{=} \lim_{x \to 0} \frac{e^x}{e^x + e^x + xe^x} \quad \leftarrow \text{apply L'Hôpital's rule again}$$

$$= \frac{e^0}{e^0 + e^0 + 0e^0} = \frac{1}{2}. \quad \leftarrow \text{evaluate the limit}$$

✔ CHECKING THE ANSWER We can check these limits with calculator graphs:

$$\lim_{x \to \infty} x^2 e^{-3x} = 0$$

$$\lim_{x \to 0^+} x \ln x = 0$$

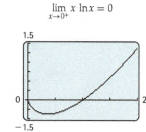

$$\lim_{x \to 0} \left(\frac{1}{x} - \frac{1}{e^x - 1} \right) = \frac{1}{2}$$

EXAMPLE 4

Using logarithms to calculate a limit

Use logarithms and L'Hôpital's rule to calculate each of the following limits:

(a) $\lim_{x \to \infty} x^{1/x}$ **(b)** $\lim_{x \to 0^+} (\sin x)^x$

SOLUTION

(a) Since $x \to \infty$ and $\frac{1}{x} \to 0$ as x approaches ∞, this limit is of the indeterminate form ∞^0. Let's calculate the related, but *different*, limit $\lim_{x \to \infty} \ln(x^{1/x})$ and see what we get:

$$\lim_{x \to \infty} \ln(x^{1/x}) = \lim_{x \to \infty} \frac{1}{x} \ln x \quad \leftarrow \text{algebra; now the limit is of the form } 0 \cdot \infty$$

$$= \lim_{x \to \infty} \frac{\ln x}{x} \quad \leftarrow \text{algebra; now the limit is of the form } \frac{\infty}{\infty}$$

$$\overset{\text{L'H}}{=} \lim_{x \to \infty} \frac{1/x}{1} \quad \leftarrow \text{apply L'Hôpital's rule}$$

$$= \lim_{x \to \infty} \frac{1}{x} = 0. \quad \leftarrow \text{simplify and evaluate the limit}$$

But wait! This is not the answer to our *original* limit. Since this limit is equal to 0, we now know by Theorem 5.24 that our original limit must be equal to $e^0 = 1$.

(b) Once again we have a limit that involves a variable in both the base and the exponent, and we will need to use logarithms to resolve this problem. As $x \to 0^+$ we have $\sin x \to 0$ and $x \to 0$, so $\lim\limits_{x \to 0^+} (\sin x)^x$ is in the indeterminate form 0^0. Let's look at the related limit obtained by first taking the logarithm:

$$\lim_{x \to 0^+} \ln((\sin x)^x) = \lim_{x \to 0^+} x \ln(\sin x) \quad \leftarrow \text{algebra; limit is now of form } 0(-\infty)$$

$$= \lim_{x \to 0^+} \frac{\ln(\sin x)}{1/x} \quad \leftarrow \text{algebra; limit is now of form } \frac{-\infty}{\infty}$$

$$\overset{\text{L'H}}{=} \lim_{x \to 0^+} \frac{(\cos x)/(\sin x)}{-1/x^2} \quad \leftarrow \text{apply L'Hôpital's rule}$$

$$= \lim_{x \to 0^+} \frac{-x^2 \cos x}{\sin x} \quad \leftarrow \text{algebra; limit is now of form } \frac{0}{0}$$

$$\overset{\text{L'H}}{=} \lim_{x \to 0^+} \frac{-2x \cos x + x^2 \sin x}{\cos x} \quad \leftarrow \text{apply L'Hôpital's rule again}$$

$$= \frac{0 + 0}{1} = 0. \quad \leftarrow \text{evaluate the limit}$$

Since this limit is equal to 0, our original limit $\lim\limits_{x \to 0^+} (\sin x)^x$ must be equal to $e^0 = 1$. □

TEST YOUR UNDERSTANDING

▶ What sorts of limits does L'Hôpital's rule help us calculate?

▶ How do you explain in words what L'Hôpital's rule says we can do to solve limits?

▶ When calculating a limit by using L'Hôpital's rule multiple times, how do you know when to stop applying L'Hôpital's rule and evaluate the limit?

▶ How can L'Hôpital's rule sometimes be used to solve limits of the indeterminate form $0 \cdot \infty$?

▶ How can we use logarithms to solve limits of the indeterminate forms 0^0, 1^∞, and ∞^0?

EXERCISES 5.5

Thinking Back

Indeterminate forms: Determine which of the given forms are indeterminate. For each form that is not indeterminate, describe the behavior of a limit of that form.

▶ $\infty \cdot \infty$ ▶ $\dfrac{\infty}{\infty}$ ▶ 0^0 ▶ ∞^∞

▶ $\dfrac{0}{0}$ ▶ 1^∞ ▶ $\infty - \infty$ ▶ $\dfrac{0}{\infty}$

▶ 0^∞ ▶ $\infty \cdot 0$ ▶ $\dfrac{\infty}{0}$ ▶ ∞^0

Simple limit calculations: Determine each of the limits that follow. You should be able to solve all of these very quickly by thinking about the graphs of the functions.

▶ $\lim\limits_{x \to \infty} 2^x$ ▶ $\lim\limits_{x \to \infty} x^{-5}$ ▶ $\lim\limits_{x \to -\infty} e^{3x}$

▶ $\lim\limits_{x \to 0} \dfrac{1}{x^2}$ ▶ $\lim\limits_{x \to 0^+} \ln x$ ▶ $\lim\limits_{x \to \infty} \left(\dfrac{1}{2}\right)^x$

Concepts

0. *Problem Zero:* Read the section and make your own summary of the material.

1. *True/False:* Determine whether each of the statements that follow is true or false. If a statement is true, explain why. If a statement is false, provide a counterexample.

  (a) *True or False:* If a limit has an indeterminate form, then that limit does not have a real number as its solution.

  (b) *True or False:* L'Hôpital's rule can be used to find the limit of any quotient $\dfrac{f(x)}{g(x)}$ as $x \to c$.

  (c) *True or False:* When using L'Hôpital's rule, you need to apply the quotient rule in the differentiation step.

  (d) *True or False:* L'Hôpital's rule applies only to limits as $x \to 0$ or as $x \to \infty$.

  (e) *True or False:* L'Hôpital's rule applies only to limits of the indeterminate form $\dfrac{0}{0}$ or $\dfrac{\infty}{\infty}$.

  (f) *True or False:* If $\lim\limits_{x \to 2} \ln(f(x)) = 4$, then $\lim\limits_{x \to 2} f(x) = \ln 4$.

  (g) *True or False:* If $\lim\limits_{x \to 2} \ln(f(x)) = \infty$, then $\lim\limits_{x \to 2} f(x) = \infty$.

  (h) *True or False:* If $\lim\limits_{x \to 2} \ln(f(x)) = -\infty$, then $\lim\limits_{x \to 2} f(x) = -\infty$.

2. *Examples:* Construct examples of the thing(s) described in the following. Try to find examples that are different than any in the reading.

  (a) Three limits of the indeterminate form $\dfrac{0}{0}$, one that approaches ∞, one that equals 0, and one that equals 3.

  (b) Three limits of the indeterminate form $\dfrac{\infty}{\infty}$, one that approaches ∞, one that equals 0, and one that equals 3.

  (c) Three limits of the indeterminate form $0 \cdot \infty$, one that approaches ∞, one that equals 0, and one that equals 3.

3. Explain graphically why it makes sense that a limit $\lim\limits_{x \to c} \dfrac{f(x)}{g(x)}$ of the indeterminate form $\dfrac{0}{0}$ would be related to $\lim\limits_{x \to c} \dfrac{f'(x)}{g'(x)}$.

4. Explain graphically why it might make sense that a limit $\lim\limits_{x \to \infty} \dfrac{f(x)}{g(x)}$ of the indeterminate form $\dfrac{\infty}{\infty}$ would be related to $\lim\limits_{x \to \infty} \dfrac{f'(x)}{g'(x)}$.

5. Suppose $f(x) = x^2 - 1$ and $g(x) = \ln x$. Find the equations of the tangent lines to these functions at $x = 1$. Then argue graphically that it would be reasonable to think that the limit of the quotient $\dfrac{f(x)}{g(x)}$ as $x \to 1$ might be equal to the limit of the quotient of these tangent lines as $x \to 1$.

6. Suppose $f(x) = 2^x - 4$ and $g(x) = x - 2$. Find the equations of the tangent lines to these functions at $x = 2$. Then argue graphically that it would be reasonable to think that the limit of the quotient $\dfrac{f(x)}{g(x)}$ as $x \to 2$ might be equal to the limit of the quotient of these tangent lines as $x \to 2$.

Each of the limits in Exercises 7–12 is of the indeterminate form $0 \cdot \infty$ or $\infty \cdot 0$. Rewrite each limit so that it is (a) in the form $\dfrac{0}{0}$ and then (b) in the form $\dfrac{\infty}{\infty}$. Then (c) determine which of these indeterminate forms would be easier to work with when applying L'Hôpital's rule.

7. $\lim\limits_{x \to \infty} 2^{-x} x$

8. $\lim\limits_{x \to 0} (2^x - 1) x^{-2}$

9. $\lim\limits_{x \to \infty} x^{-2} \ln x$

10. $\lim\limits_{x \to 0} x^3 \ln x$

11. $\lim\limits_{x \to \infty} e^{-x} \ln x$

12. $\lim\limits_{x \to 1} \sqrt{x - 1}\, \ln(x - 1)$

13. Find the error in the following *incorrect* calculation, and then calculate the limit correctly:

$$\lim_{x \to 0} \frac{x^2 - x}{2^x - 1} \overset{\text{L'H}}{=} \lim_{x \to 0} \frac{2x - 1}{(\ln 2)2^x} \overset{\text{L'H}}{=} \lim_{x \to 0} \frac{2}{(\ln 2)^2 2^x}$$
$$= \frac{2}{(\ln 2)^2 2^0} = \frac{2}{(\ln 2)^2}.$$

14. Find the error in the following *incorrect* calculation. Then calculate the limit correctly.

$$\lim_{x \to \infty} \frac{e^{-x}}{x^2} \overset{\text{L'H}}{=} \lim_{x \to \infty} \frac{-e^{-x}}{2x} \overset{\text{L'H}}{=} \lim_{x \to \infty} \frac{e^{-x}}{2} = \frac{0}{2} = 0.$$

Skills

Calculate each of the limits in Exercises 15–20 (a) using L'Hôpital's rule and (b) without using L'Hôpital's rule.

15. $\lim\limits_{x \to 1} \dfrac{x^2 + x - 2}{x - 1}$

16. $\lim\limits_{x \to 2} \dfrac{x^2 - 4x + 4}{x - 2}$

17. $\lim\limits_{x \to \infty} \dfrac{x - 1}{2 - 3x^2}$

18. $\lim\limits_{x \to \infty} \dfrac{3^x}{1 - 4^x}$

19. $\lim\limits_{x \to \infty} \dfrac{e^{3x}}{1 - e^{2x}}$

20. $\lim\limits_{x \to 0} \dfrac{2^x - 1}{4^x - 1}$

Calculate each of the limits in Exercises 21–40. Some of these limits are made easier by L'Hôpital's rule, and some are not.

21. $\lim\limits_{x \to 3} \dfrac{2^x - 8}{3 - x}$

22. $\lim\limits_{x \to 1} \dfrac{x - 1}{e^{x-1} - 1}$

23. $\lim\limits_{x \to 2} \dfrac{e^{2x-4} - 1}{x^2 - 4}$

24. $\lim\limits_{x \to \infty} \dfrac{e^x - 7}{8x^2 + 12x + 5}$

25. $\lim\limits_{x \to \infty} \dfrac{2^{-x}}{x^2 + 1}$

26. $\lim\limits_{x \to \infty} \dfrac{x + 3^x}{3 - 2^x}$

27. $\lim\limits_{x \to 0} \dfrac{x^{-2}}{x^{-3}+1}$

28. $\lim\limits_{x \to \infty} x^2 e^{-x}$

29. $\lim\limits_{x \to \infty} \left(e^x - \dfrac{e^x}{x+1} \right)$

30. $\lim\limits_{x \to 0^+} \left(x^{-1} - \dfrac{1}{2^x - 1} \right)$

31. $\lim\limits_{x \to \infty} x \left(\dfrac{1}{2} \right)^x$

32. $\lim\limits_{x \to \infty} \dfrac{xe^{-3x}}{x^2 + 3x + 1}$

33. $\lim\limits_{x \to \infty} \dfrac{xe^x}{e^{2x} + 1}$

34. $\lim\limits_{x \to 0^+} \left(\dfrac{2^x - 1}{x^2} - \dfrac{1}{1 - e^x} \right)$

35. $\lim\limits_{x \to \infty} \dfrac{e^{-3x}}{x^2 + 3x + 1}$

36. $\lim\limits_{x \to \infty} \dfrac{\ln x}{\ln(2x+1)}$

37. $\lim\limits_{x \to 0^+} \dfrac{\log_2 x}{\log_2 3x}$

38. $\lim\limits_{x \to 2^+} \ln \left(\dfrac{2^x - 4}{x - 2} \right)$

39. $\lim\limits_{x \to 2^+} \dfrac{\ln(x-2)}{\ln(x^2 - 4)}$

40. $\lim\limits_{x \to 0^+} \dfrac{\ln x}{\ln x - x + 1}$

Calculate each of the limits in Exercises 41–52. Some of these limits are made easier by considering the logarithm of the limit first, and some are not.

41. $\lim\limits_{x \to 0^+} x^{\ln x}$

42. $\lim\limits_{x \to \infty} x^{\ln x}$

43. $\lim\limits_{x \to 2^+} (x - 2)^{x^2 - 4}$

44. $\lim\limits_{x \to 0^+} (x^2 + 1)^x$

45. $\lim\limits_{x \to 1^+} x^{1/(x-1)}$

46. $\lim\limits_{x \to 0^+} x^{2x}$

47. $\lim\limits_{x \to \infty} x^{1/x}$

48. $\lim\limits_{x \to \infty} \left(\dfrac{1}{x} \right)^x$

49. $\lim\limits_{x \to \infty} \left(\dfrac{1}{x+1} \right)^x$

50. $\lim\limits_{x \to \infty} \left(\dfrac{x}{x-1} \right)^x$

51. $\lim\limits_{x \to 1^+} (x - 1)^{\ln x}$

52. $\lim\limits_{x \to 1^+} (\ln x)^{x-1}$

A function f **dominates** another function g as $x \to \infty$ if $f(x)$ and $g(x)$ both grow without bound as $x \to \infty$ and if

$$\lim\limits_{x \to \infty} \dfrac{f(x)}{g(x)} = \infty.$$

Intuitively, f dominates g as $x \to \infty$ if $f(x)$ is very much larger than $g(x)$ for very large values of x. In Exercises 53–62, use limits to determine whether $u(x)$ dominates $v(x)$, or $v(x)$ dominates $u(x)$, or neither.

53. $u(x) = x + 100,\ v(x) = x$

54. $u(x) = 5x^2 + 1,\ v(x) = x^3$

55. $u(x) = 100x^2,\ v(x) = 2x^{100}$

56. $u(x) = x^2,\ v(x) = 2^x$

57. $u(x) = 2^x,\ v(x) = e^x$

58. $u(x) = x^{10} + 3,\ v(x) = 10^x + 3$

59. $u(x) = \log_2 x,\ v(x) = \log_{30} x$

60. $u(x) = \ln(x^2 + 1),\ v(x) = x^2 + 1$

61. $u(x) = 0.001e^{0.001x},\ v(x) = 100x^{100}$

62. $u(x) = 0.001x^2 - 100x,\ v(x) = 100 \log_3 x$

As you will prove in Exercises 77 and 78, exponential growth functions e^{kx} always dominate power functions x^r, and power functions x^r with positive powers always dominate logarithmic functions $\log_b x$. Use these facts to quickly determine each of the limits in Exercises 63–68.

63. $\lim\limits_{x \to \infty} \dfrac{x^{101} + 500}{e^x}$

64. $\lim\limits_{x \to \infty} \dfrac{e^x}{x^{101} + 500}$

65. $\lim\limits_{x \to \infty} \dfrac{\sqrt{x}}{300 \ln x}$

66. $\lim\limits_{x \to \infty} \dfrac{\ln x^{100}}{x^8}$

67. $\lim\limits_{x \to \infty} 2^x x^{-100}$

68. $\lim\limits_{x \to \infty} e^{-x} \ln x$

Now that we know L'Hôpital's rule, we can apply it to solve more sophisticated global optimization problems. Consider domains, limits, derivatives, and values to determine the global extrema of each function f in Exercises 69–72 on the given intervals I and J.

69. $f(x) = x \ln x,\ I = (0, 1],\ J = (0, \infty)$

70. $f(x) = x^2 \ln(0.2x),\ I = (0, 4],\ J = (0, \infty)$

71. $f(x) = x^3 e^{-x},\ I = [0, \infty),\ J = (-\infty, \infty)$

72. $f(x) = \dfrac{\ln x}{\ln 2x},\ I = [0, 1],\ J = [1, \infty)$

Applications

In Exercises 73–74, suppose that Leila is a population biologist with the Idaho Fish and Game Service. Wolves were introduced formally into Idaho in 1994, but there were some wolves in the state before that. Leila has been assigned the task of estimating the rate at which the number of wolves in Idaho increased naturally, before the animals were introduced. The only information she has is a population model, which indicates that the wolf population currently satisfies the formula

$$w(t) = 835(1 - e^{-.006t}),$$

where t is the number of years since 1994.

73. Leila reasons that the average rate of change of wolves from 1994 to a given time t is given by $\dfrac{w(t)}{t}$. Why does her reasoning make sense?

74. To approximate the rate of increase of wolves per year at the beginning of 1994, Leila decides to take the limit of $\dfrac{w(t)}{t}$ as $t \to 0$. Why does this approach make sense, and what is the value of that limit? Is there another way she could find the same number?

Proofs

Exercises 75–78 concern dominance of functions as defined earlier in Exercises 53–62.

75. Use L'Hôpital's rule to prove that every exponential growth function dominates the power function $g(x) = x^2$.

76. Use L'Hôpital's rule to prove that every power function with a positive power dominates the logarithmic function $g(x) = \ln x$.

77. Use L'Hôpital's rule to prove that exponential growth functions always dominate power functions.

78. Use L'Hôpital's rule to prove that power functions with positive powers always dominate logarithmic functions.

79. Use logarithms to prove that for any real number r,
$$\lim_{x \to \infty} \left(1 + \frac{r}{x}\right)^x = e^r.$$

Thinking Forward

Convergence and divergence of sequences: If a sequence $a_1, a_2, a_3, \ldots, a_k, \ldots$ approaches a real-number limit as $k \to \infty$, then the sequence $\{a_k\}$ **converges**. If the terms of the sequence do not get arbitrarily close to some real number, then the sequence **diverges**. Determine the general form $\{a_k\}$ for each of the following sequences, and then use L'Hôpital's rule to determine whether that sequence converges or diverges.

▶ $\dfrac{1}{2}, \dfrac{4}{4}, \dfrac{9}{8}, \dfrac{16}{16}, \dfrac{25}{32}, \dfrac{36}{64} \cdots$

▶ $\dfrac{\ln 1}{\ln 2}, \dfrac{\ln 2}{\ln 3}, \dfrac{\ln 3}{\ln 4}, \dfrac{\ln 4}{\ln 5}, \dfrac{\ln 5}{\ln 6}, \dfrac{\ln 6}{\ln 7} \cdots$

▶ $\dfrac{1}{10}, \dfrac{3}{100}, \dfrac{7}{1000}, \dfrac{15}{10,000}, \dfrac{31}{100,000}, \dfrac{63}{1,000,000} \cdots$

▶ $\dfrac{1}{301}, \dfrac{8}{304}, \dfrac{27}{309}, \dfrac{64}{316}, \dfrac{125}{325}, \dfrac{216}{336} \cdots$

CHAPTER REVIEW, SELF-TEST, AND CAPSTONES

Before you progress to the next chapter, be sure you are familiar with the definitions, concepts, and basic skills outlined here. The capstone exercises at the end bring together ideas from this chapter and look forward to future chapters.

Definitions

Give precise mathematical definitions or descriptions of each of the concepts that follow. Then illustrate the definition with a graph or an algebraic example.

▶ what it means for a function to be *transcendental*

▶ the form of an *exponential function* and of the *natural exponential function*

▶ the form of a *logarithmic function* and of the *natural logarithmic function*

▶ the definition of the number e as a limit

▶ the definition of the number e as a number that makes a particular limit equal to 1

▶ what it means for one quantity to be *proportional* to another

▶ the method of *logarithmic differentiation*, including when it is useful and how it is applied

▶ the mathematical meaning of *doubling time* and *half-life*

▶ what it means for one function to *dominate* another as $x \to \infty$

Theorems

Fill in the blanks to complete each of the following theorem statements:

▶ If b is a real number with $b > 0$ and $b \neq 1$, then b^x is never equal to _____ for any value of x.

▶ If b is a real number with _____ and _____ , then the function $y = b^x$ is _____ and is therefore invertible.

▶ Since exponential functions and logarithmic functions are inverses of each other, for $b > 0$ and $b \neq 1$ we have _____ $= x$ for all x and _____ $= x$ for all $x > 0$.

▶ For $b > 0$ and $b \neq 1$, the function $y = \log_b x$ has domain _____ and range _____ .

▶ For $b > 0$ and $b \neq 1$, $\log_b y = x$ if and only if $y = $ _____ .

▶ For $b > 0$ and $b \neq 1$, the graph of $y = \log_b x$ can be obtained from the graph of _____ by reflecting over the line _____ .

▶ All exponential and logarithmic functions are _____ on their domains.

▶ $f'(x) = kf(x)$ for some constant k if and only if f is a ____ function.

▶ If $f(x) = e^{kx}$ or $f(x) = b^x$ is an exponential growth function, then k is in the interval ____ , b is in the interval ____ , and the graph of f has the following shape: ____ (provide a graph).

▶ If $f(x) = \log_b x$ is a logarithmic growth function, then b is in the interval ____ and the graph of f has the following shape: ____ (provide a graph).

▶ If $f(x) = \log_b x$ is a logarithmic decay function, then b is in the interval ____ and the graph of f has the following shape: ____ (provide a graph).

▶ If a quantity $Q(t)$ with initial amount $Q(0) = Q_0$ grows at a fixed decimal percentage r over every period t, then $Q(t) = $ ____ .

▶ If a quantity $Q(t)$ with initial amount $Q(0) = Q_0$ grows at r percent compounded n times per period t, then $Q(t) = $ ____ .

▶ If a quantity $Q(t)$ with initial amount $Q(0) = Q_0$ grows with a continuous growth rate k, then $Q(t) = $ ____ .

▶ Every exponential growth function has a ____ doubling time, in the sense that there is some positive constant α for which $Q(t + \alpha) = $ ____ for all t.

▶ Every exponential decay function has a constant ____ , in the sense that there is some positive constant α for which ____ $= \frac{1}{2} Q(t)$ for all t.

▶ *L'Hôpital's Rule*: Suppose f and g are ____ functions on some punctured interval around $x = c$ on which $g(x)$ is ____ . If $\lim\limits_{x \to c} \frac{f(x)}{g(x)}$ is of the indeterminate form ____ or ____ , then

$$\lim_{x \to c} \frac{f(x)}{g(x)} = \_\_\_\_ ,$$

as long as the second limit exists or is infinite.

Rules of Algebra, Limits, and Derivatives

The algebra of exponents: Fill in the blanks to rewrite each expression according to the algebraic rules of exponentiation. You may assume that x, y, b, and c are numbers for which each expression is defined.

▶ $b^x b^y = $ ____

▶ $(b^x)^y = $ ____

▶ $\dfrac{b^x}{c^x} = $ ____

▶ $(bc)^x = $ ____

▶ $\dfrac{b^x}{b^y} = $ ____

▶ $b^{x/y} = $ ____

The algebra of logarithms: Fill in the blanks to rewrite each expression according to the algebraic rules of logarithms. You may assume that x, y, a, and b are numbers for which each expression is defined.

▶ $\log_b (b^x) = $ ____

▶ $\log_b (x^a) = $ ____

▶ $\log_b \left(\dfrac{1}{x} \right) = $ ____

▶ $\dfrac{\log_a x}{\log_a b} = $ ____

▶ $b^{\log_b x} = $ ____

▶ $\log_b (xy) = $ ____

▶ $\log_b \left(\dfrac{x}{y} \right) = $ ____

Limits of exponential and logarithmic functions: Fill in the blanks to complete each basic limit statement. You should assume that k and b are real numbers with the given properties, and that L is any real number.

▶ If $k > 0$, then $\lim\limits_{x \to \infty} e^{kx} = $ ____ .

▶ If $k > 0$, then $\lim\limits_{x \to -\infty} e^{kx} = $ ____ .

▶ If $k > 0$, then $\lim\limits_{x \to \infty} e^{-kx} = $ ____ .

▶ If $k > 0$, then $\lim\limits_{x \to -\infty} e^{-kx} = $ ____ .

▶ If $b > 1$, then $\lim\limits_{x \to \infty} \log_b x = $ ____ .

▶ If $b > 1$, then $\lim\limits_{x \to 0^+} \log_b x = $ ____ .

▶ If $0 < b < 1$, then $\lim\limits_{x \to \infty} \log_b x = $ ____ .

▶ If $0 < b < 1$, then $\lim\limits_{x \to 0^+} \log_b x = $ ____ .

▶ $\lim\limits_{h \to 0} (1 + h)^{1/h} = $ ____ .

▶ $\lim\limits_{h \to 0} \dfrac{e^h - 1}{h} = $ ____ .

▶ For any real number r, $\lim\limits_{n \to \infty} \left(1 + \dfrac{r}{n} \right)^n = $ ____ .

▶ If $\lim\limits_{x \to c} \ln(f(x)) = L$, then $\lim\limits_{x \to c} f(x) = $ ____ .

▶ If $\lim\limits_{x \to c} \ln(f(x)) = \infty$, then $\lim\limits_{x \to c} f(x) = $ ____ .

▶ If $\lim\limits_{x \to c} \ln(f(x)) = -\infty$, then $\lim\limits_{x \to c} f(x) = $ ____ .

Derivatives of exponential and logarithmic functions: Fill in the blanks to complete each basic statement about derivatives. You may assume that k and b are real numbers with $b > 0$ and $b \neq 1$ and that x is any real number for which each expression is defined.

▶ $\dfrac{d}{dx} (e^x) = $ ____

▶ $\dfrac{d}{dx} (e^{kx}) = $ ____

▶ $\dfrac{d}{dx} (\ln x) = $ ____

▶ $\dfrac{d}{dx} (b^x) = $ ____

▶ $\dfrac{d}{dx} (\log_b x) = $ ____

▶ $\dfrac{d}{dx} (\ln |x|) = $ ____

Skill Certification: Working with Exponential and Logarithmic Functions

Solving equations and inequalities: Use rules of algebra for exponential and logarithmic functions to solve each of the following equations and inequalities.

0. $\log_x 8 = 3$

1. $e^{2\ln x + 4} = 6$

2. $\dfrac{4^{x-1}}{2^x} = 6$

3. $2e^{3x+1} - 5 = 0$

4. $\ln\left(\dfrac{x+1}{x-1}\right) \geq 0$

5. $\dfrac{1}{e^{3x} - e^x} < 0$

Basic graphs: Use your knowledge of transformations to sketch quick graphs of each of the given functions. Label any roots, intercepts, or asymptotes.

6. $f(x) = 3 - e^{2x}$

7. $f(x) = e^{-x} - 4$

8. $f(x) = 2\log_{1/2} x$

9. $f(x) = \ln|x - 2| + 1$

Limits: Calculate each of the given limits. Some limits may need L'Hôpital's rule, logarithms, or expressions for e or e^r.

10. $\lim\limits_{x\to\infty} (2 - e^{-x})$

11. $\lim\limits_{x\to 0^+} (1 - \ln x)$

12. $\lim\limits_{x\to\infty} \dfrac{2^x}{1 + 3^x}$

13. $\lim\limits_{x\to\infty} x^3 e^{-x}$

14. $\lim\limits_{x\to\infty} \dfrac{2x}{4 + e^{-2x}}$

15. $\lim\limits_{x\to\infty} \dfrac{xe^x}{1 - e^x}$

16. $\lim\limits_{x\to\infty} \dfrac{\ln(3x + 1)}{\ln x}$

17. $\lim\limits_{x\to 0} \dfrac{x^{-1}}{x^{-2} + 3}$

18. $\lim\limits_{x\to\infty} \left(1 + \dfrac{2}{x}\right)^{3x}$

19. $\lim\limits_{x\to 0^+} (1 + 5x)^{1/x}$

20. $\lim\limits_{x\to 0^+} x^{3x}$

21. $\lim\limits_{x\to\infty} \left(\dfrac{x}{x-3}\right)^x$

Derivatives: Calculate the derivative of each of the given functions. Some problems may require or benefit from logarithmic differentiation.

22. $f(x) = 2^x \ln(x^2)$

23. $f(x) = \dfrac{xe^x}{\log_3 x}$

24. $f(x) = \ln(\ln(2 + e^{3x}))$

25. $f(x) = x^2 e^{4x^3 + 5}$

26. $f(x) = x^{3x}$

27. $f(x) = \dfrac{x^2(2^x)\sqrt{x^2 + 1}}{(3x + 1)e^x}$

Curve sketching: Use algebra, derivatives, limits, and number lines to sketch careful, labeled graphs of each of the following functions.

28. $f(x) = \dfrac{x}{\ln x}$

29. $f(x) = \dfrac{2^x}{1 - 2^x}$

30. $f(x) = \dfrac{1}{1 - e^x}$

31. $f(x) = x^2 \ln x$

Capstone Problems

A. *Forms of exponential functions:* Every exponential function f can be written both as $f(x) = Ae^{kx}$ and as $f(x) = Ab^x$ for some real numbers A, k, and b.

   (a) Solve the equation $Ae^{kx} = Ab^x$ for k to express k in terms of b.

   (b) Solve the equation $Ae^{kx} = Ab^x$ for b to express b in terms of k.

B. *The definition of e:* Discuss the two definitions of e in this chapter (see Definition 5.8 and Theorem 5.9). Why are these "weak" definitions of the irrational number we call e? Why are these definitions reasonable? Look ahead to Definition 7.34 to see how we will define $\ln x$ in the future, and read the discussion after that definition to see how e^x will be properly defined in that chapter.

C. *Compounding interest:* Investigate the result of a starting investment of $500.00 over the next 30 years with a 5 percent interest rate, under each of the following scenarios.

   (a) Compounded yearly.

   (b) Compounded monthly.

   (c) Compounded daily.

   (d) Compounded continuously.

   (e) What yearly interest rate would give the same result as the continuously compounded interest rate scenario in part (d) of this problem?

   (f) What continuous interest rate would give the same result as the yearly interest rate scenario in part (a) of this problem?

D. *Dominance of functions:* Which types of functions dominate other types of functions, and why? Consider linear, power, polynomial, rational, exponential, and logarithmic functions, as well as more exotic functions like x^x, and determine which ones dominate others as $x \to \infty$. Prove your answers with limits.

CHAPTER 6

Calculus with Trigonometric and Inverse Trigonometric Functions

6.1   DEFINING THE TRIGONOMETRIC FUNCTIONS

▶ Using ratios of side lengths of right triangles to define trigonometric functions for acute angles

▶ Using ratios of coordinates on the unit circle to define trigonometric functions of any angle

▶ Side lengths of 30–60–90 and 45–45–90 triangles

Right-Triangle Trigonometry

For **acute** angles that measure less than 90° we can define trigonometric functions in terms of side lengths of triangles. Every acute angle θ can be realized as part of a right triangle with sides labeled as shown here at the left:

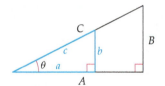

For any given acute angle there are many possible right triangles that we could draw, but as illustrated at the right, all of the possibilities will have the same three angles and thus be similar triangles. This relationship is key, since the **Law of Similar Triangles** tells us that similar triangles may have different side lengths, but always have the same *ratios* of side lengths:

$$\frac{a}{b} = \frac{A}{B}, \qquad \frac{a}{c} = \frac{A}{C}, \qquad \frac{b}{c} = \frac{B}{C}.$$

In all, there are six ratios of side lengths for any given triangle: the three ratios we just listed and their reciprocals. Whatever size of right triangle we choose for our acute angle θ, these ratios will always be the same. The six ratios are the trigonometric functions of θ. In the following definition, "opp," "adj,", and "hyp" denote the lengths of the opposite side, adjacent side, and hypotenuse, respectively:

DEFINITION 6.1   **Trigonometric Functions for Acute Angles**

Given an acute angle θ, the six **trigonometric functions** are defined as the following six ratios of side lengths:

(a) $\sin \theta = \dfrac{\text{opp}}{\text{hyp}}$ **(b)** $\cos \theta = \dfrac{\text{adj}}{\text{hyp}}$ **(c)** $\tan \theta = \dfrac{\text{opp}}{\text{adj}}$

(d) $\csc \theta = \dfrac{\text{hyp}}{\text{opp}}$ **(e)** $\sec \theta = \dfrac{\text{hyp}}{\text{adj}}$ **(f)** $\cot \theta = \dfrac{\text{adj}}{\text{opp}}$

The function $\sin \theta$ is pronounced as "sign theta" but its full name is the **sine function**. The full names of the remaining trigonometric functions are **cosine, tangent, cosecant, secant,** and **cotangent**.

For example, consider the angle θ shown in the 3–4–5 right triangle shown here at the left:

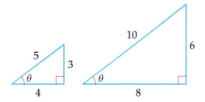

The side lengths here are opp $= 3$, adj $= 4$, and hyp $= 5$. The ratios of these side lengths give us

$$\sin\theta = \frac{3}{5}, \quad \cos\theta = \frac{4}{5}, \quad \tan\theta = \frac{3}{4}, \quad \csc\theta = \frac{5}{3}, \quad \sec\theta = \frac{5}{4}, \quad \text{and} \quad \cot\theta = \frac{4}{3}.$$

The same angle θ can be thought of inside the 6–8–10 triangle just displayed at the right, and although all the side lengths are twice as large, their ratios will be exactly the same.

Throughout our study of trigonometric functions we will often use the fact that by definition, all of the trigonometric functions can be expressed in terms of the sine and cosine:

THEOREM 6.2

Sine and Cosine Determine All of the Trigonometric Functions

Given any acute angle θ,

(a) $\tan\theta = \dfrac{\sin\theta}{\cos\theta}$ **(b)** $\csc\theta = \dfrac{1}{\sin\theta}$ **(c)** $\sec\theta = \dfrac{1}{\cos\theta}$ **(d)** $\cot\theta = \dfrac{1}{\tan\theta}$

Two Special Triangles

In order to calculate the trigonometric functions of an acute angle θ, we must know the ratios of the lengths of the sides of a right triangle that contains the angle θ. Usually, finding these ratios is difficult; for example, it is difficult to determine the possible side lengths of a right triangle that contains an angle of 12 degrees.

However, there are two special right triangles for which we can determine the ratios of the side lengths by hand: the 30–60–90 and 45–45–90 triangles. The following theorem describes the lengths of the legs of these triangles when we assume that the hypotenuse is 1 unit in length:

THEOREM 6.3

Side Lengths of Two Important Right Triangles

The 45–45–90 right triangle and 30–60–90 right triangle with hypotenuse of length 1 unit have side lengths as shown in the following figures:

If you have seen something like Theorem 6.3 before, you may remember it a bit differently, scaled up by a factor of two:

We will most often use the versions from Theorem 6.3, since we will often be interested in situations in which the hypotenuse of a triangle is 1 unit in length.

Proof. The proof of Theorem 6.3 is just an application of the Pythagorean theorem. Let's start with the 45–45–90 triangle. Suppose a right triangle with a hypotenuse of length 1 has two angles of 45 degrees. Then the triangle is an isosceles triangle; that is, its two legs have the same length a as shown next at the left. It remains to show that $a = \frac{\sqrt{2}}{2}$. By the Pythagorean theorem:

$$a^2 + a^2 = 1 \implies 2a^2 = 1 \implies a^2 = \frac{1}{2} \implies a = \sqrt{\frac{1}{2}} = \frac{1}{\sqrt{2}} = \frac{1}{\sqrt{2}}\left(\frac{\sqrt{2}}{\sqrt{2}}\right) = \frac{\sqrt{2}}{2}.$$

Notice that we considered only the positive square root of $\frac{1}{2}$, since we know that a is a length and thus must be positive.

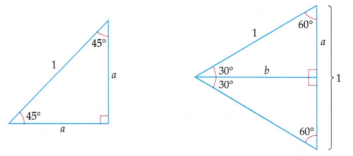

Now suppose that a right triangle has a hypotenuse of length 1, and angles of 30, 60, and 90 degrees. Let a be the length of the leg across from the 30-degree angle, and let b be the length of the leg across from the 60-degree angle. Two of these triangles together make the equilateral triangle shown in the right-hand figure. Since this triangle is equilateral, it follows that $2a = 1$ and thus $a = \frac{1}{2}$. It now remains to show that $b = \frac{\sqrt{3}}{2}$. Again we use the Pythagorean theorem:

$$a^2 + b^2 = 1^2 \implies \frac{1}{4} + b^2 = 1 \implies b^2 = \frac{3}{4} \implies b = \sqrt{\frac{3}{4}} = \frac{\sqrt{3}}{2}.$$

Note that we considered only the positive square root in this calculation, since b is a length and must be positive.   ■

Now we can easily calculate the trigonometric functions of the angles $\theta = 45°$, $\theta = 30°$, and $\theta = 60°$. For example, using the second triangle in Theorem 6.3, we can see that

$$\cos 30° = \frac{\text{adj}}{\text{hyp}} = \frac{\sqrt{3}/2}{1} = \frac{\sqrt{3}}{2}.$$

To calculate trigonometric values of other angles we can use a calculator, set to **degree mode** if we are working with degrees. For example, with a calculator we can approximate that $\cos 12° \approx 0.978$ and that $\csc 75° = \frac{1}{\sin 75°} \approx \frac{1}{0.966} \approx 1.035$.

Radians and Standard Position

Right triangles were sufficient to define trigonometric functions for acute angles. But what about other angles? To answer that question we first need to think about angles in a more general context.

So far we have been measuring angles by thinking of a circle divided into 360 pieces called degrees. But why would we divide a circle into 360 pieces? Why not 100 pieces? Or 1000 pieces? There is no intrinsic mathematical reason that we should measure angles in terms of units that represent $\frac{1}{360}$ of a circle. We now develop a much more natural way to represent angle measure by using the intrinsic property of the circumference of a circle of radius 1.

We can place any angle in a ***standard position*** in the xy-plane by placing its vertex at the origin and its ***initial edge*** along the positive side of the x-axis, as shown next at the left. The angle then opens up in either a counterclockwise or clockwise direction until it reaches its ***terminal edge***. A ***positive angle*** is measured counterclockwise from its initial edge, while a ***negative angle*** is measured clockwise from its initial edge.

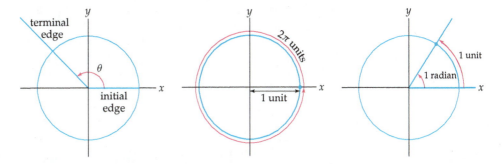

Now consider the ***unit circle*** shown in the center diagram. Since the unit circle has radius $r = 1$ unit, its circumference is $C = 2\pi r = 2\pi(1) = 2\pi$ units. That's a circumference of approximately 6.283185 units, which certainly is not as nice as a number like 360 that we can easily divide into integer-sized pieces. However, we can still measure everything in terms of this circumference by defining a new unit of angle measure called a ***radian*** that represents the size of an angle in standard position whose terminal edge intersects the unit circle after an arc length of 1 unit, as shown in the diagram at the right.

Since the distance all the way around the circle is 2π units, the distance halfway around is π units and the distance one-quarter of the way around the circle is $\frac{\pi}{2}$ units. This means that an angle of 90° measures $\frac{\pi}{2}$ radians, an angle of 180° measures π radians, and an angle of 360° measures, of course, 2π radians. The following three diagrams illustrate some common positive angles in radian measure around the unit circle:

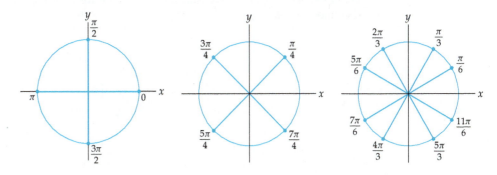

Of course we can also consider negative angles; for example, the angle that opens up in the clockwise direction for one quarter of the distance around the bottom half of the circle has radian measure $-\frac{\pi}{4}$. Its terminal edge intersects the unit circle in the same location as the angle $\frac{7\pi}{4}$ shown in the middle figure. We can also consider angles that go more than once around the circle; for example, the angle $\frac{5\pi}{2} = 2\pi + \frac{\pi}{2}$ intersects the unit circle at the same point as the angle $\frac{\pi}{2}$ in the diagram at the left.

Unit-Circle Trigonometry

Using the unit circle and standard position, we can now define trigonometric functions for all angles, not just acute angles. Instead of ratios of side lengths we will consider coordinates on the unit circle. Given any angle θ in standard position, the terminal edge of θ intersects the unit circle at some point (x, y) in the xy-plane. We will define the height y of that point to be the **sine** of θ, while the **cosine** of θ will be defined as the x-coordinate of that point.

DEFINITION 6.4

Trigonometric Functions for Any Angle

Given any angle θ measured in radians and in standard position, let (x, y) be the point where the terminal edge of θ intersects the unit circle. The six **trigonometric functions** of an angle θ are the six possible ratios of the coordinates x and y for θ:

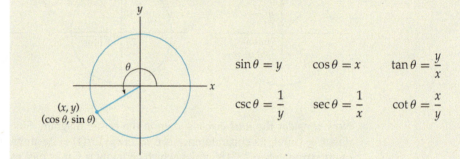

$$\sin\theta = y \qquad \cos\theta = x \qquad \tan\theta = \frac{y}{x}$$

$$\csc\theta = \frac{1}{y} \qquad \sec\theta = \frac{1}{x} \qquad \cot\theta = \frac{x}{y}$$

Notice that once again the sine and cosine functions determine the remaining four trigonometric functions, since $\tan\theta$ is the ratio $\frac{\sin\theta}{\cos\theta}$, and the last three trigonometric functions are the reciprocals of the first three.

Is Definition 6.4 compatible with our previous definition of trigonometric functions for acute angles? In other words, do the two definitions agree in the case where θ is an acute angle? Fortunately, yes. Acute angles in standard position must terminate in the first quadrant, as illustrated next. The ratios of the side lengths of the right triangle in the figure are the same as the ratios of the coordinates on the unit circle that we used in the unit-circle definition of the trigonometric functions. Therefore, for acute angles, the two trigonometric definitions agree.

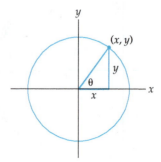

It is important to note that the values of the trigonometric functions depend only on the terminal edge of θ in standard position, and not on how many times or in what direction the angle rotates around the unit circle. Therefore, angles whose terminal edges intersect the unit circle at the same point will have the same trigonometric function values. For example, the angles $\frac{5\pi}{6}$, $\frac{17\pi}{6}$, and $-\frac{7\pi}{6}$ all have the same terminal edge and thus all have the same trigonometric values.

Now to actually calculate, say, the sine and cosine of the angle $\frac{5\pi}{6}$, we have to know the coordinates (x, y) at which the terminal edge of $\frac{5\pi}{6}$ meets the unit circle. As shown in the circle that follows at the left, we can make a ***reference triangle*** by dropping down a line from the point where the terminal edge of $\frac{5\pi}{6}$ meets the unit circle. Luckily for us, this reference triangle is a 30–60–90 triangle and therefore we already know its side lengths, as shown in the circle that follows at the right.

Reference angle is 30°

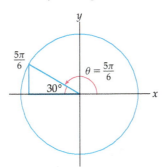

Known side lengths give coordinates

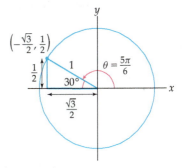

The x-coordinate of the point where the terminal edge meets the unit circle tells us that $\cos\left(\frac{5\pi}{6}\right) = -\frac{\sqrt{3}}{2}$, and the y-coordinate of that point tells us that $\sin\left(\frac{5\pi}{6}\right) = \frac{1}{2}$.

For angles whose reference triangles are not our well-known 30–60–90 or 45–45–90 triangles, we can calculate trigonometric values with a calculator. Now that we are working in radians rather than degrees, it is vital that these calculations be done in ***radians mode***. For example, with a calculator in radians mode we can approximate $\cos\left(-\frac{\pi}{5}\right) \approx 0.809$. We can verify that this makes sense with a picture. As shown in the next figure, it does seem realistic that the angle $-\frac{\pi}{5}$ (which is one-fifth of the way clockwise on the bottom half of the unit circle) meets the unit circle at a point whose x-coordinate is approximately 0.809:

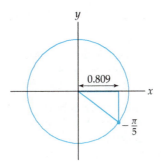

Trigonometric Functions as Graphs

Each trigonometric function takes an angle as an input and returns a ratio of coordinates on the unit circle as an output. Each possible angle gets assigned to exactly one ratio of coordinates. That is precisely the definition of a function. By shifting perspective we can think of these functions as graphs in the xy-plane, where the angles are represented by the independent variable x and the ratios are represented by the dependent variable y. For example, as an angle x changes, the height at which its terminal edge meets the unit circle oscillates between 1 and -1. Since $\sin x$ is the height of the point where the terminal edge of angle x meets the unit circle, this means that the graph of $\sin x$ should oscillate up and down between 1 and -1 as x varies.

The graphs of the six trigonometric functions are shown next. Each of the graphs in the second row is the reciprocal of the graph immediately above it. Remember that you can use the graph of a function f to sketch the graph of its reciprocal $\frac{1}{f}$. In particular, the zeroes of f will be vertical asymptotes of $\frac{1}{f}$, large heights on the graph of f will become small heights on the graph of $\frac{1}{f}$, and vice versa.

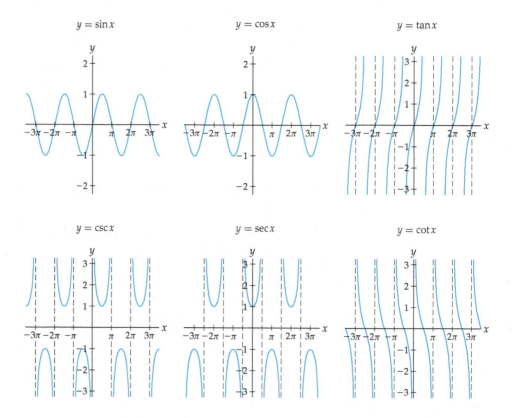

From the unit-circle definitions it is not hard to show that the trigonometric functions have the domains, ranges, roots, and asymptotes shown in the preceding graphs. You will do so in the exercises of this section and future sections.

Notice from the graphs that each of the trigonometric functions repeats a certain pattern over and over. For example, on the graph of $f(x) = \sin x$ the piece of the graph on $[0, 2\pi]$ is repeated over and over across the graph. In general, a function has a ***period*** p if p is the smallest positive real number such that, for all x in the domain of $f(x)$, $f(x+p) = f(x)$.

A function with a period is said to be **periodic**. All of the trigonometric functions are periodic with period either 2π or π; see Exercise 24.

Examples and Explorations

EXAMPLE 1   Scaling up all side lengths to obtain a similar triangle

Find the lengths of the sides of a 30–60–90 triangle whose shortest side is 5 units long.

SOLUTION

Our favorite 30–60–90 triangle is shown next at the left, with hypotenuse of length 1 unit. The shortest side is the one across from the 30° angle, which has length $\frac{1}{2}$ unit. To get a similar triangle with the corresponding side 5 units long, we must scale up all the sides by a factor of 10, as shown in the triangle at the right.

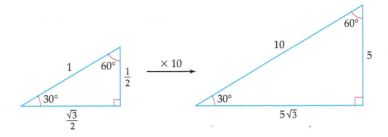

From this diagram, we see that the length of the hypotenuse of the new triangle is $1(10) = 10$ and the length of the remaining side is $\frac{\sqrt{3}}{2}(10) = 5\sqrt{3}$.   □

EXAMPLE 2   Converting between radians and degrees

Express 60° in radians, and express 3 radians in degrees.

SOLUTION

Since an angle of 180° is the same as an angle of π radians, we can convert from degrees to radians by multiplying by $\frac{\pi}{180°}$:

$$60° = 60° \cdot \frac{\pi}{180°} = \frac{\pi}{3}.$$

Similarly, we can convert from radians to degrees by multiplying by the conversion factor $\frac{180°}{\pi}$:

$$3 = 3 \cdot \frac{180°}{\pi} \approx 171.89°.$$

It is customary not to list radians explicitly as units, which is why we wrote simply "3" rather than "3 radians."   □

EXAMPLE 3   Calculating trigonometric values by hand

Sketch the angle $-\frac{11\pi}{4}$ in standard position, and then use the unit circle to find the values of all six trigonometric functions of that angle.

SOLUTION

Note that $-\frac{11\pi}{4} = -\frac{11}{4}\pi = -2\pi - \frac{3}{4}\pi$. Since 2π radians is one full revolution around the unit circle, the angle $-\frac{11\pi}{4}$ opens up in the clockwise direction for one full revolution and then an additional three-quarters of the bottom half of the unit circle, as shown at the left:

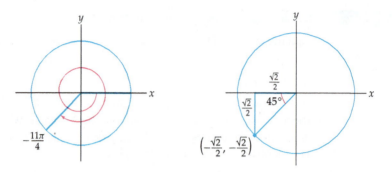

By the unit-circle definitions of the sine and cosine, the coordinates $(x, y) = \left(-\frac{\sqrt{2}}{2}, -\frac{\sqrt{2}}{2}\right)$ are equal to $\left(\cos\left(-\frac{11\pi}{4}\right), \sin\left(-\frac{11\pi}{4}\right)\right)$. Therefore

$$\cos\left(-\frac{11\pi}{4}\right) = -\frac{\sqrt{2}}{2}, \quad \sin\left(-\frac{11\pi}{4}\right) = -\frac{\sqrt{2}}{2}.$$

The remaining trigonometric functions can be obtained by taking reciprocals or quotients of the sine and cosine:

$$\sec\left(-\frac{11\pi}{4}\right) = \frac{1}{\cos(-11\pi/4)} = -\frac{2}{\sqrt{2}} = -\sqrt{2}, \quad \csc\left(-\frac{11\pi}{4}\right) = \frac{1}{\sin(-11\pi/4)} = -\frac{2}{\sqrt{2}} = -\sqrt{2},$$

$$\tan\left(-\frac{11\pi}{4}\right) = \frac{\sin(-11\pi/4)}{\cos(-11\pi/4)} = \frac{-\sqrt{2}/2}{-\sqrt{2}/2} = 1, \quad \cot\left(-\frac{11\pi}{4}\right) = \frac{\cos(-11\pi/4)}{\sin(-11\pi/4)} = \frac{-\sqrt{2}/2}{-\sqrt{2}/2} = 1.$$

☐

EXAMPLE 4

Using one trigonometric value to find another

Given that $\cos\theta = -\frac{1}{3}$ and that θ is an angle that terminates in the third quadrant, find $\tan\theta$.

SOLUTION

Since $\cos\theta = -\frac{1}{3}$, the angle θ must intersect the unit circle at a point whose x-coordinate is $-\frac{1}{3}$, as shown next. Only one of these points is in the third quadrant, and it is of the form $\left(-\frac{1}{3}, y\right)$ where y is some negative number.

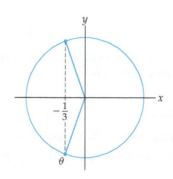

To find the tangent of θ we need to know the y-coordinate of the point we have identified in the third quadrant. By the Pythagorean theorem, it follows that

$$a^2 + \left(\frac{1}{3}\right)^2 = 1^2 \quad \Longrightarrow \quad a^2 + \frac{1}{9} = 1 \quad \Longrightarrow \quad a = \pm\sqrt{1 - \frac{1}{9}} = \pm\sqrt{\frac{8}{9}} = \pm\frac{2\sqrt{2}}{3} \approx \pm 0.9428.$$

Since θ terminates in the third quadrant, the y-coordinate corresponding to θ is the negative choice: $-\frac{2\sqrt{2}}{3} \approx -0.9428$. As a reality check, notice that -0.9428 is a reasonable answer, since the point in question has a y-coordinate very close to -1. We can now say that

$$\tan\theta = \frac{y}{x} = \frac{2\sqrt{2}/3}{-1/3} \approx \frac{-0.9428}{-1/3} = 3(0.9428) = 2.8284.$$

It is interesting to note that we never actually had to determine the angle measure of θ; we only needed to know the (x, y) coordinates of the corresponding point on the unit circle.   □

EXAMPLE 5   **Solving an equation that involves transcendental functions**

Find all of the solutions of the equation $\sin\theta = \cos\theta$.

SOLUTION

If $\sin\theta = \cos\theta$, then θ is an angle whose terminal edge intersects the unit circle at a point (x, y) with $x = y$. The only such points on the unit circle are $(\sqrt{2}/2, \sqrt{2}/2)$ and $(-\sqrt{2}/2, -\sqrt{2}/2)$, as shown in the circle that follows. The angles that end at these points are all of the form $\theta = \frac{\pi}{4} + \pi k$ for some integer k. Thus the solution set for the equation is

$$\left\{\ldots, -\frac{3\pi}{4}, \frac{\pi}{4}, \frac{5\pi}{4}, \frac{9\pi}{4}, \ldots\right\}.$$

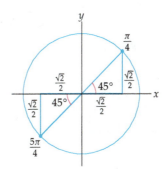

□

EXAMPLE 6   **Using trigonometry to find the height of a building**

Suppose you are 500 feet from the base of an office building, and suppose that the angle of elevation of your line of sight (i.e., the angle between the horizontal ground and your diagonal line of sight) to the top of the building is 36 degrees. How tall is the building?

SOLUTION

Suppose h is the height of the building. You are standing 500 feet away from the building as shown in the following diagram:

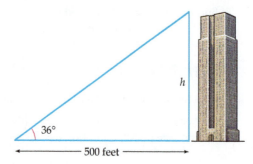

We know the length of the side of the triangle adjacent to the 36° angle, and we would like to find the length of the side that is opposite that angle. The tangent function relates those two sides; specifically, $\tan 36°$ is the ratio of the height of the building to the distance you are from the building:

$$\tan 36° = \frac{\text{opp}}{\text{adj}} = \frac{h}{500} \implies h = 500 \tan 36° \implies h \approx 363.27 \text{ feet.}$$

In the last step we used a calculator to approximate $\tan 36° \approx 0.7265$. □

EXAMPLE 7 Domains and graphs of trigonometric functions

Find the domain of $h(x) = 3 \sec 2x$. Then use transformations to sketch a careful graph of $h(x)$, without using a graphing calculator.

SOLUTION

The function $h(x) = 3 \sec 2x = \frac{3}{\cos 2x}$ is defined when $\cos 2x \neq 0$. The latter condition occurs when $2x$ is not an odd multiple of $\frac{\pi}{2}$ and thus when x is not an odd multiple of $\frac{\pi}{4}$. Therefore the domain of $h(x)$ is $x \neq \frac{\pi}{4}(2k+1)$ for positive integers k. To sketch the graph of $h(x)$, we start with the graph of $y = \sec x$ as follows at the left, stretch vertically by a factor of 3 as shown in the middle figure, and then compress horizontally by a factor of 2 as shown at the right. (See Section 0.6 for a review of transformations of graphs.)

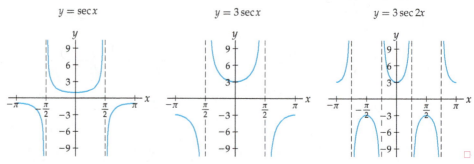

□

? TEST YOUR UNDERSTANDING

▶ How are the unit-circle definitions of the trigonometric functions related to the right-triangle definitions of the trigonometric functions?

▶ How do you convert from radians to degrees and vice versa?

▶ What are the side lengths of a 45–45–90 triangle whose hypotenuse is 1 unit in length? How about a 30–60–90 triangle whose hypotenuse is 1 unit in length?

▶ How is the graph of the reciprocal of a function related to the graph of that function? How can this information be useful for remembering the graphs of $y = \csc x$, $y = \sec x$, and $y = \tan x$?

▶ Why does it make sense that the graph of $y = \csc x$ has vertical asymptotes at multiples of π?

EXERCISES 6.1

Thinking Back

The Pythagorean theorem: State the Pythagorean theorem, and use it to answer each of the following questions.

▶ If a right triangle has legs of lengths 10 and 5, what is the length of its hypotenuse?

▶ If a right triangle has hypotenuse of length 8 and one leg of length 6, what is the length of the remaining leg?

The Law of Similar Triangles: State the Law of Similar Triangles, and use it to answer each of the following questions.

▶ Suppose that a right triangle is similar to the 3–4–5 right triangle and that its hypotenuse has length 100. What are the lengths of the legs of the triangle?

▶ Suppose that a right triangle is similar to the 5–12–13 right triangle and that its shortest leg has length 1. What are the lengths of the other leg and the hypotenuse of this triangle?

Concepts

0. *Problem Zero:* Read the section and make your own summary of the material.

1. *True/False:* Determine whether each of the statements that follow is true or false. If a statement is true, explain why. If a statement is false, provide a counterexample.

(a) *True or False:* If θ is any angle in standard position, then θ and $2\pi + \theta$ have the same terminal edge.

(b) *True or False:* There are only two angles whose sine is equal to $-\frac{1}{4}$.

(c) *True or False:* If (x, y) is the point on the unit circle corresponding to the angle $-\frac{7\pi}{3}$, then x is positive and y is negative.

(d) *True or False:* The sine of an angle θ is always equal to the sine of the reference angle for θ.

(e) *True or False:* If θ is an angle in standard position whose terminal edge is in the first quadrant, then the values of all six trigonometric functions of θ are positive.

(f) *True or False:* If θ is an angle in standard position whose terminal edge is in the third quadrant, then the values of all six trigonometric functions of θ are negative.

(g) *True or False:* If $\tan \theta = 1$, then the sine and cosine of θ must be equal.

(h) *True or False:* For every angle θ, $-1 \leq \cos \theta \leq 1$.

2. *Examples:* Construct examples of the thing(s) described in the following. Try to find examples that are different than any in the reading.

(a) Two positive angles and two negative angles with the same terminal edge as the angle $\frac{4\pi}{3}$.

(b) Two positive angles and two negative angles whose sine is $-\frac{1}{2}$.

(c) Two positive angles and two negative angles whose cosine is $-\frac{4}{5}$.

3. Use a right triangle to argue that, for any acute angle, $\sin \theta$ is always less than 1. (*Hint: Use the fact that the longest side of a right triangle is its hypotenuse.*)

4. Why is the Law of Similar Triangles important for defining the trigonometric functions of acute angles?

5. Show that the side lengths of the 30–60–90 and 45–45–90 triangles with hypotenuse 1 satisfy the Pythagorean theorem.

6. Given the angle β shown in the following right triangle, find $\sin \beta$, $\cos \beta$, $\tan \beta$, $\sec \beta$, $\csc \beta$, and $\cot \beta$:

7. Given the angle α shown in the following right triangle, find $\sin \alpha$, $\cos \alpha$, $\tan \alpha$, $\sec \alpha$, $\csc \alpha$, and $\cot \alpha$:

8. Suppose θ is an acute angle. As θ increases, does $\sin \theta$ increase or decrease, and why? What about $\cos \theta$? What about $\tan \theta$? Use pictures of triangles to support your arguments.

9. Given the angle θ in the right triangle that follows, which of the following are true, which are false, and why?

(a) *True or False:* $\sin \theta < \cos \theta$.

(b) *True or False:* $\tan \theta > 1$.

(c) *True or False:* $\sec \theta < \csc \theta$.

10. Use a calculator in radians mode to approximate each of the following trigonometric values:

(a) $\cos \frac{\pi}{7}$ (b) $\sec \frac{3\pi}{10}$ (c) $\tan 1$

11. Use a calculator in degree mode to approximate each of the following trigonometric values:
    (a) $\cos 11°$ (b) $\sec 72°$ (c) $\tan 1°$

12. What is a radian? Is it larger or smaller than a degree? Show an angle of 3 degrees in standard position and an angle of 3 radians in standard position.

13. Give a mathematical definition of $\sin\theta$ for any angle θ. Your definition should include the words "unit circle," "standard position," "terminal," and "coordinate."

14. Give a mathematical definition of $\tan\theta$ for any angle θ. Your definition should include the words "unit circle," "standard position," "terminal," and "coordinate."

15. Use the definition of the sine function to explain why $\sin\left(\frac{\pi}{4}\right)$ is equal to both $\sin\left(\frac{9\pi}{4}\right)$ and $\sin\left(-\frac{7\pi}{4}\right)$.

16. Fill in each blank with an interval of real numbers.
    (a) The function $f(x) = \cos x$ has domain _____ and range _____ .
    (b) The function $f(x) = \csc x$ has domain _____ and range _____ .
    (c) The angle $-\frac{5\pi}{6}$ is in the _____ quadrant and has a reference angle of _____ degrees.
    (d) For any angle θ, the cosecant of θ is the _____ of the sine of θ.

17. Suppose θ is an angle in standard position whose terminal edge intersects the unit circle at the point (x, y). If $y = -\frac{1}{3}$, what are the possible values of $\cos\theta$? If you know that the terminal edge of θ is in the third quadrant, what can you say about $\cos\theta$? What if the terminal edge of θ is in the fourth quadrant? Could the terminal edge of θ be in the first or second quadrant?

18. Show that $-\sqrt{3}$ is in the range of the tangent function by finding an angle θ for which $\tan\theta = -\sqrt{3}$.

19. Determine whether or not each of the following is equal to $\cot\theta$, and why:
    (a) $\dfrac{\cos\theta}{\sin\theta}$ (b) $\csc\theta\cos\theta$
    (c) $\dfrac{1}{\tan\theta}$ (d) $\dfrac{\csc\theta}{\sec\theta}$

20. Use geometric arguments (i.e., pictures rather than calculations) to support each of the following statements:
    (a) If $\theta = -\frac{6\pi}{7}$, then the sine of θ is between $-\frac{1}{2}$ and 0.
    (b) If $\theta = \frac{\pi}{5}$, then the tangent of θ is less than 1.

21. Suppose θ is an angle in standard position whose terminal edge intersects the unit circle at the point $\left(-\frac{1}{4}, \frac{\sqrt{15}}{4}\right)$. Verify that this is in fact a point that lies on the unit circle. Then use it to find $\sin\theta$, $\cos\theta$, and $\tan\theta$.

22. Use the unit-circle definition of the cosine to explain why the graph of $y = \cos x$ looks the way that it does. In particular, why are the zeroes of the graph where they are? Why is the height of the graph always between -1 and 1? Why does the graph keep repeating itself every 2π units?

23. Use the unit circle to explain why $\cos x$ has domain $(-\infty, \infty)$ and range $[-1, 1]$.

24. Use the unit circle to explain why $\sin x$, $\cos x$, $\csc x$, and $\sec x$ have period 2π and why $\tan x$ and $\cot x$ have period π.

Skills

Find the exact values of each of the expressions in Exercises 25–33 by hand. Support your work with a labeled 30–60–90 or 45–45–90 triangle.

25. $\sin 30°$ 26. $\sin 60°$ 27. $\csc 30°$
28. $\cos 45°$ 29. $\cot 45°$ 30. $\sec 45°$
31. $\csc 45°$ 32. $\sec 60°$ 33. $\tan 30°$

For each triangle in Exercises 35–38, use the information given to find the lengths of the remaining sides. Find your answers by hand, using similar triangles whenever possible; otherwise use a calculator.

34.

35.

36.

37.

38.

39.

Convert each of the radian angles in Exercises 40–42 to degree measure.

40. $-\dfrac{3\pi}{4}$ 41. $\dfrac{5\pi}{6}$ 42. 4

Sketch each angle θ in Exercises 43–45 in standard position on the unit circle, and identify the reference angle.

43. $\theta = \dfrac{7\pi}{6}$ 44. $\theta = \dfrac{37\pi}{3}$ 45. $\theta = -\dfrac{11\pi}{4}$

Sketch each angle θ in Exercises 46–48 in standard position on the unit circle, and find the coordinates of the point where the terminal edge of θ meets the unit circle.

46. $\theta = \dfrac{2\pi}{3}$ 47. $\theta = -\dfrac{3\pi}{4}$ 48. $\theta = \dfrac{11\pi}{6}$

Find the exact values of each of the quantities in Exercises 49–66. Do not use a calculator.

49. $\sin 0$

50. $\cos \pi$

51. $\cot \dfrac{\pi}{2}$

52. $\cos \dfrac{3\pi}{2}$

53. $\tan\left(-\dfrac{\pi}{4}\right)$

54. $\cos\left(\dfrac{48\pi}{3}\right)$

55. $\csc\left(-\dfrac{5\pi}{4}\right)$

56. $\sin(201\pi)$

57. $\sin\left(-\dfrac{\pi}{3}\right)$

58. $\sec\left(-\dfrac{\pi}{6}\right)$

59. $\csc \dfrac{42\pi}{3}$

60. $\tan\left(-\dfrac{15\pi}{6}\right)$

61. $\sec\left(-\dfrac{\pi}{2}\right)$

62. $\sin\left(-\dfrac{7\pi}{6}\right)$

63. $\cos\left(-\dfrac{9\pi}{4}\right)$

64. $\cot(35\pi)$

65. $\tan \dfrac{11\pi}{6}$

66. $\sec(102\pi)$

Use the information given in Exercises 67–70 to calculate the exact value of the given trigonometric functions of θ. Show your work, using a sketch of the unit circle and a triangle.

67. Given that $\sin \theta = \dfrac{1}{5}$ and $\dfrac{\pi}{2} < \theta < \pi$, find $\cos \theta$.

68. Given that $\cos \theta = -\dfrac{1}{4}$ and θ is in the second quadrant, find $\sin \theta$.

69. Given that $\csc \theta = -3$ and that $\sec \theta$ is positive, find $\cot \theta$.

70. Given that $\tan \theta = 1$ and $\sin \theta$ is negative, find $\sec \theta$.

Find all of the solutions of the equations in Exercises 71–76 by hand. Verify your answers by examining information about the unit circle.

71. $\cos 2x = 1$

72. $\sin x = \dfrac{1}{2}$

73. $\sin\left(\theta - \dfrac{\pi}{3}\right) = \dfrac{1}{2}$

74. $\tan(\pi - \theta) = 1$

75. $\sec \theta = 0$

76. $\sin \theta = -\cos \theta$

Find the domains of the functions in Exercises 77–82.

77. $f(x) = \tan(2x)$

78. $f(x) = 3\cos(5x - 2) + 1$

79. $f(x) = \dfrac{\sec x}{\tan x}$

80. $f(x) = \csc x \cot x$

81. $f(x) = \sqrt{\sec x}$

82. $f(x) = \dfrac{1}{1 - \tan x}$

Identify the period of each function $f(x)$ in Exercises 83–88. Then sketch a careful graph of $f(x)$ by hand, using transformations.

83. $f(x) = \cos\left(-\dfrac{\pi}{2}(x+1)\right)$

84. $f(x) = 2\sin(\pi x) + 8$

85. $f(x) = \sin(5 - 4x) - 1$

86. $f(x) = \sec(2(x - \pi))$

87. $f(x) = \tan(\pi(x - 1))$

88. $f(x) = -2\cot(1-x)+3$

Applications

89. Find the area of an isosceles triangle with two 30-degree angles and two sides of length 3. (*Hint: Use the fact that a triangle with base b and height h has area* $A = \dfrac{1}{2}bh$.)

90. Suppose you are standing on the ground looking up to the top of a 2500-foot building, and suppose that the angle of elevation of your line of sight is 18 degrees. How far away are you from the building? (You may assume that your eyes are level with the ground, or, if you prefer a slightly more challenging problem, then you can take the height of your eyes into account as well.)

91. Alina is flying a kite and has managed to get her kite so high in the air that she has let out 400 feet of kite string. If the angle made by the ground and the line of kite string is 32 degrees, how high is the kite?

92. Suppose two stars are each 60 light-years away from Earth. The angle between the line of sight to the first star and the line of sight to the second star is 2 degrees. In other words, if you look at the first star, then turn your head to look at the second star, your head will move through an angle of 2 degrees. How far apart are the stars?

You are in charge of constructing a zoo habitat for prairie dogs, with the requirement that the habitat must enclose 2500 square feet of area and use as little border fencing as possible. For each of the habitat designs described in Exercises 93-94, find the amount of border fencing that the project will require.

93. A trapezoid-shaped habitat whose angled side is an enclosed walkway for zoo patrons (so that no fencing is needed along the walkway), where the walkway makes an angle of 60° with the right fence, as shown next at the left.

Habitat along walkway   *Habitat with mural*

94. An arena-style trapezoid-shaped habitat whose long back side is a wall with a landscape mural (so no fencing is needed along the back wall), where the back wall makes an angle of 60° with the slanted side fences, as shown previously at the right.

Proofs

95. Prove that a 30–60–90 triangle with a hypotenuse of length 1 must have legs of lengths $\frac{\sqrt{3}}{2}$ and $\frac{1}{2}$.

96. Prove that a 45–45–90 triangle with a hypotenuse of length 1 must have two legs of length $\frac{\sqrt{2}}{2}$.

97. Prove that if θ is an acute angle, then the unit-circle definition of $\cos \theta$ is equivalent to the right-triangle definition of $\cos \theta$.

98. Use the unit-circle definition of the trigonometric functions to prove that $\csc \theta = \frac{1}{\sin \theta}$ and $\tan \theta = \frac{\sin \theta}{\cos \theta}$.

99. Use a unit-circle argument to prove that $\sin \theta = 0$ if and only if $\theta = \pi k$ for some integer k.

100. Prove that $\cos \theta = 0$ if and only if $\theta = \frac{\pi}{2} + \pi k$ for some integer k.

Thinking Forward

Pythagorean identities: In each of the following problems, suppose θ is an acute angle in a right triangle with hypotenuse 1.

▶ Use the Pythagorean theorem and the right-triangle definitions of the sine and cosine to show that, for any acute angle θ,

$$(\sin \theta)^2 + (\cos \theta)^2 = 1.$$

▶ Now suppose θ is any angle in standard position. Use the unit-circle definitions of the sine and cosine, together with the fact that the equation of the unit circle is $x^2 + y^2 = 1$, to argue that, for every angle θ,

$$(\sin \theta)^2 + (\cos \theta)^2 = 1.$$

▶ Divide both sides of the preceding equation by $(\cos \theta)^2$, and use properties of trigonometric functions, to show that, for every angle θ,

$$(\tan \theta)^2 + 1 = (\sec \theta)^2.$$

6.2   TRIGONOMETRIC IDENTITIES

▶ Basic trigonometric identities that follow from the unit-circle definitions

▶ The sum and difference identities for sine and cosine

▶ Double-angle identities

Pythagorean Identities

An equation that is always true is called an ***identity***. For example, the equation $x^2 = 4$ is not an identity, because it is true only for the values $x = 2$ and $x = -2$. By contrast, the equation $3^{2x} = 9^x$ is true for *all* values of x and thus is an identity. Trigonometric functions have a lot of structure and are closely related to each other and thus there are a lot of trigonometric identities, where two seemingly different trigonometric expressions are in fact equal for all relevant values of x. These identities will be the key to rewriting and simplifying trigonometric expressions.

For instance, the sine and cosine of an angle θ are the y-coordinate and x-coordinates, respectively, of the point where the terminal edge of θ meets the unit circle. Since the unit circle has equation $x^2 + y^2 = 1$, the quantities $x = \cos \theta$ and $y = \sin \theta$ are related by the trigonometric identity

$$(\cos \theta)^2 + (\sin \theta)^2 = 1.$$

This equation is known as a ***Pythagorean identity***, since, for acute angles, it follows from the Pythagorean theorem. If θ is an angle in a right triangle with hypotenuse 1, then $\sin \theta = \frac{\text{opp}}{1}$ is the length of the side opposite θ and $\cos \theta = \frac{\text{adj}}{1}$ is the length of the side adjacent to θ. By the Pythagorean theorem, the sum of the squares of those lengths must be equal to the square of the hypotenuse, again giving us the identity just listed. The following pictures illustrate these two ways of obtaining that identity:

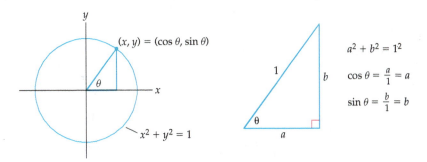

For notational convenience, from now on we will use the abbreviation $\sin^2 \theta$ to represent $(\sin \theta)^2$. We will use exponents similarly for the other trigonometric functions and for higher integer powers. From the preceding discussion we can derive three trigonometric identities:

THEOREM 6.5

Pythagorean Identities

For every angle θ for which the expressions that follow are defined,

(a) $\sin^2 \theta + \cos^2 \theta = 1$ **(b)** $\tan^2 \theta + 1 = \sec^2 \theta$ **(c)** $1 + \cot^2 \theta = \csc^2 \theta$

Proof. We have already proved the first identity. To prove the second, we simply divide both sides of the first identity by $\cos^2 \theta$ and then simplify:

$$\frac{\sin^2 \theta + \cos^2 \theta}{\cos^2 \theta} = \frac{1}{\cos^2 \theta} \implies \frac{\sin^2 \theta}{\cos^2 \theta} + \frac{\cos^2 \theta}{\cos^2 \theta} = \frac{1}{\cos^2 \theta} \implies \tan^2 \theta + 1 = \sec^2 \theta.$$

The proof of the third identity is similar and is left to Exercise 52.   ∎

Even and Odd Identities

We can quickly obtain a few more simple identities by further examining the unit-circle definition of the trigonometric functions. For example, for every angle θ we clearly have $\sin(\theta + 2\pi) = \sin \theta$, since $\theta + 2\pi$ and θ have the same terminal edge and thus intersect the unit circle at the same point.

As another example, from the unit circle we can determine that each of the six trigonometric functions is either an even function or an odd function. Recall from Section 0.6 that a function $f(x)$ is **even** if $f(-x) = f(x)$ for all x in its domain and **odd** if $f(-x) = -f(x)$. Graphs of odd functions have $180°$ rotational symmetry, while graphs of even functions have y-axis reflection symmetry. The following theorem tells us that the sine and tangent are odd functions while the cosine is an even function.

THEOREM 6.6

Even and Odd Identities

For every angle θ,

(a) $\sin(-\theta) = -\sin \theta$ **(b)** $\cos(-\theta) = \cos \theta$ **(c)** $\tan(-\theta) = -\tan \theta$

We can easily see from the graphs of $\sin x$, $\cos x$, and $\tan x$ that the sine and the tangent have odd (rotational) symmetry while the cosine has even (reflection) symmetry, as shown in the next three figures. Of course, to prove these symmetries we will appeal to the algebraic definitions of those functions.

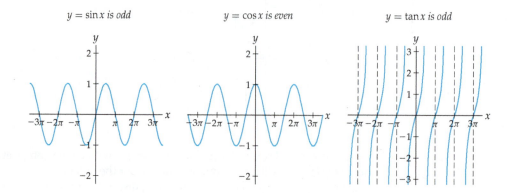

$y = \sin x$ is odd $y = \cos x$ is even $y = \tan x$ is odd

Proof. We'll prove part (a) and leave the proofs of parts (b) and (c) to Exercises 53 and 54, respectively. Given any angle θ in standard position, the angle $-\theta$ opens up by the same angle measure, but in the opposite direction, as shown next at the left. This means that if (x, y) is the point on the unit circle corresponding to θ, then the point $(x, -y)$ is the point on the unit circle corresponding to $-\theta$.

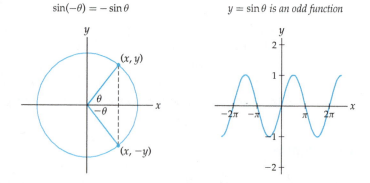

$\sin(-\theta) = -\sin\theta$ $y = \sin\theta$ is an odd function

Since $\sin\theta$ is the y-coordinate of the point at which the terminal edge of θ meets the unit circle, and $\sin(-\theta)$ is the corresponding y-coordinate for $-\theta$, we have shown that $\sin(-\theta) = -\sin\theta$. This identity means that $y = \sin\theta$ is an odd function, which is reflected in the 180° rotational symmetry of the graph of $y = \sin\theta$ in the right-hand figure.

Sum and Difference Identities

Given two angles α and β, is $\sin(\alpha + \beta)$ the same thing as $\sin\alpha + \sin\beta$? That would be very nice, but alas, it is not so. For example, if $\alpha = \frac{\pi}{4}$ and $\beta = \frac{\pi}{2}$, then $\sin\left(\frac{\pi}{4} + \frac{\pi}{2}\right) = \sin\frac{3\pi}{4} = \frac{\sqrt{2}}{2}$, which is different from $\sin\frac{\pi}{4} + \sin\frac{\pi}{2} = \frac{\sqrt{2}}{2} + 1$. We can, however, relate the sine of the sum of two angles to the sine and cosine of those angles:

THEOREM 6.7 **Sum and Difference Identities**

For every angle θ,

(a) $\sin(\alpha + \beta) = \sin\alpha\cos\beta + \sin\beta\cos\alpha$ **(c)** $\sin(\alpha - \beta) = \sin\alpha\cos\beta - \sin\beta\cos\alpha$

(b) $\cos(\alpha + \beta) = \cos\alpha\cos\beta - \sin\alpha\sin\beta$ **(d)** $\cos(\alpha - \beta) = \cos\alpha\cos\beta + \sin\alpha\sin\beta$

We can verify the first identity in the example we just discussed. In that example, the left side of the identity is $\sin\left(\dfrac{\pi}{4} + \dfrac{\pi}{2}\right) = \dfrac{\sqrt{2}}{2}$ and the right side is

$$\sin\frac{\pi}{4}\cos\frac{\pi}{2} + \sin\frac{\pi}{2}\cos\frac{\pi}{4} = \frac{\sqrt{2}}{2}\cdot 0 + 1\cdot\frac{\sqrt{2}}{2} = \frac{\sqrt{2}}{2}.$$

Proof. The proofs of the two sum identities are lengthy, and we do not have time to include them here. The proofs of the difference identities follow easily from the sum identities, using the fact that sine is an odd function and cosine is an even function. We prove part (c) in this way and leave the proof of part (d) to Exercise 55:

$$\sin(\alpha - \beta) = \sin(\alpha + (-\beta)) \qquad \leftarrow \text{rewrite difference as a sum}$$
$$= \sin(\alpha)\cos(-\beta) + \sin(-\beta)\cos(\alpha) \qquad \leftarrow \text{sum identity for sine}$$
$$= \sin\alpha\cos\beta - \sin\beta\cos\alpha. \qquad \leftarrow \text{cosine is even, sine is odd} \qquad ■$$

Previously we could calculate the exact values of trigonometric functions only for angles that were halves, thirds, quarters, or sixths of π. Now we can use the sum and difference identities to expand this ability to sums or differences of these angles. For example, we know how to calculate $\sin 45°$ and $\sin 30°$ by hand, so we can use the difference identity for the sine to find $\sin(45° - 30°) = \sin(15°)$ by hand. In the language of radians, this means that we can use $\sin\left(\dfrac{\pi}{4}\right)$ and $\sin\left(\dfrac{\pi}{6}\right)$ to calculate $\sin\left(\dfrac{\pi}{4} - \dfrac{\pi}{6}\right) = \sin\left(\dfrac{\pi}{12}\right)$; see Example 1.

Double-Angle Identities

Many more trigonometric identities can be obtained from the sum and difference identities. For example, we can write down identities for rewriting expressions like $\sin\left(\theta + \dfrac{\pi}{2}\right)$ or $\sin(\theta + \pi)$; see Exercises 17 and 18. We can also obtain identities for the sine and cosine of "double angles" that tell us how to rewrite the sine and cosine of 2θ in terms of the sine and cosine of θ:

THEOREM 6.8 **Double-Angle Identities**

For every angle θ,

(a) $\sin 2\theta = 2\sin\theta\cos\theta$ **(c)** $\cos 2\theta = 1 - 2\sin^2\theta$

(b) $\cos 2\theta = \cos^2\theta - \sin^2\theta$ **(d)** $\cos 2\theta = 2\cos^2\theta - 1$

Note that we have three different ways to express the cosine of a double angle.

Proof. To prove the double-angle identity for the sine, we write 2θ as the sum $\theta + \theta$ of two angles and apply the sum identity for the sine:

$$\sin 2\theta = \sin(\theta + \theta) \qquad \leftarrow \text{rewrite } 2\theta \text{ as a sum}$$
$$= \sin\theta\cos\theta + \sin\theta\cos\theta \qquad \leftarrow \text{sum identity for sine}$$
$$= 2\sin\theta\cos\theta.$$

The proof for the first double-angle identity for the cosine in part (b) is similar and is left to Exercise 56.

We can prove the alternative double-angle identities for the cosine by applying Pythagorean identities. For example, to prove part (c) we can use the fact that since $\sin^2\theta + \cos^2\theta = 1$, we have $\cos^2\theta = 1 - \sin^2\theta$:

$$\cos 2\theta = \cos^2\theta - \sin^2\theta \qquad \leftarrow \text{first double-angle identity for cosine}$$
$$\cos 2\theta = (1 - \sin^2\theta) - \sin^2\theta \qquad \leftarrow \text{Pythagorean identity}$$
$$\cos 2\theta = 1 - 2\sin^2\theta.$$

The proof of the remaining identity is similar and is left to Exercise 57. ∎

The double-angle identities from Theorem 6.8 can be rewritten by using simple algebra to obtain the following identities, which will prove particularly useful in Section 8.4:

THEOREM 6.9

More Double-Angle Identities

For every angle θ,

(a) $\sin^2\theta = \dfrac{1 - \cos 2\theta}{2}$ **(b)** $\cos^2\theta = \dfrac{1 + \cos 2\theta}{2}$

The first identity in particular is very useful when we want to write sines in terms of cosines or vice versa. Both are useful for reducing exponents in trigonometric expressions, as we will see in Section 8.4.

Proof. We will prove the first identity and leave the second to Exercise 58. We start with a known double-angle identity for cosine and rewrite it in the desired form:

$$\cos 2\theta = 1 - 2\sin^2\theta \implies 2\sin^2\theta = 1 - \cos 2\theta \implies \sin^2\theta = \frac{1 - \cos 2\theta}{2}.$$
∎

There are many other trigonometric identities, as you will see in the exercises. For example, there are sum, difference, and double-angle identities for the tangent and the other trigonometric functions. You should memorize the basic trigonometric identities in this section and know where to find formulas for more complicated trigonometric identities when you need them.

Examples and Explorations

EXAMPLE 1 Using sum or difference identities to calculate a trigonometric value

Calculate the exact value of $\sin\dfrac{\pi}{12}$ by hand.

SOLUTION

The reference triangle for the angle $\frac{\pi}{12}$ is not a 30–60–90 or a 45–45–90 triangle, so we cannot calculate the value of $\sin\frac{\pi}{12}$ directly. However, we can write $\frac{\pi}{12}$ as a difference of angles that have the desired reference triangles:

$$\frac{\pi}{12} = \frac{3\pi}{12} - \frac{2\pi}{12} = \frac{\pi}{4} - \frac{\pi}{6}.$$

By the difference identity for the sine, we have

$$\sin\frac{\pi}{12} = \sin\left(\frac{\pi}{4} - \frac{\pi}{6}\right)$$

$$= \sin\frac{\pi}{4}\cos\frac{\pi}{6} - \sin\frac{\pi}{6}\cos\frac{\pi}{4} \qquad \leftarrow \text{difference identity for sine}$$

$$= \frac{\sqrt{2}}{2}\cdot\frac{\sqrt{3}}{2} - \frac{1}{2}\cdot\frac{\sqrt{2}}{2} \qquad \leftarrow \text{using reference triangles}$$

$$= \frac{\sqrt{6}}{4} - \frac{\sqrt{2}}{4} = \frac{\sqrt{6}-\sqrt{2}}{4}. \qquad \square$$

 CHECKING THE ANSWER

You can check the answer to the preceding example by using a calculator in radian mode to approximate $\sin\frac{\pi}{12} \approx 0.2588$ and then comparing your answer to an approximation of $\frac{\sqrt{6}-\sqrt{2}}{4}$.

EXAMPLE 2   Even and odd trigonometric functions

Determine algebraically whether the function $f(x) = \frac{\tan x}{1+\cos x}$ is even, odd, or neither.

SOLUTION

Recall from Section 0.6 that a function is even if $f(-x) = f(x)$ for all x and is odd if $f(-x) = -f(x)$ for all x. Therefore we must calculate $f(-x)$ and determine whether it is equal to $f(x)$, $-f(x)$, or neither. We already know from the odd and even identities that $\tan(-x) = -\tan x$ and $\cos(-x) = \cos x$; therefore

$$f(-x) = \frac{\tan(-x)}{1+\cos(-x)} = \frac{-\tan x}{1+\cos x} = -\frac{\tan x}{1+\cos x} = -f(x).$$

Since $f(-x) = -f(x)$ for all values of x in its domain, it follows that $f(x)$ is odd. $\square$

EXAMPLE 3   Identifying trigonometric identities

For each equation that follows, if the equation is an identity, then prove that it holds for every angle θ. If the equation is not an identity, then find an angle θ for which the equation is false and then solve for all angles θ that make the equation true.

(a) $\sin 2\theta = 2\sin\theta$ **(b)** $\sin(\theta + \pi) = -\sin\theta$

SOLUTION

(a) This is not a trigonometric identity because it does not hold for *all* angles. For example, if $\theta = \frac{\pi}{2}$, then $\sin 2\theta = \sin\pi = 0$ but $2\sin\theta = 2\sin\frac{\pi}{2} = 2(1) = 2$. Since we know from one of the double-angle identities that $\sin 2\theta = 2\sin\theta\cos\theta$, we have

$$\sin 2\theta = 2\sin\theta \implies 2\sin\theta\cos\theta = 2\sin\theta \implies 2\sin\theta(\cos\theta - 1) = 0.$$

This is true when $\sin\theta = 0$ or when $\cos\theta = 1$, which happens exactly when θ is a multiple of π. Therefore the solutions of the equation $\sin 2\theta = 2\sin\theta$ are of the form $\theta = \pi k$, where k is any integer.

(b) Thinking about transformations of the graph of the sine should convince you that this equation is an identity, since shifting the graph of the sine π units to the left has the same effect as reflecting the graph of the sine over the x-axis. To prove the given identity we will apply the sum identity for the sine:

$$\sin(\theta + \pi) = \sin\theta\cos\pi + \sin\pi\cos\theta = (\sin\theta)(-1) + (0)(\cos\theta) = -\sin\theta. \quad \square$$

EXAMPLE 4   **Proving a trigonometric identity by simplifying both sides**

Prove the following double-angle identity for the tangent:

$$\tan 2\theta = \frac{2\tan\theta}{1 - \tan^2\theta}.$$

SOLUTION

Since most of the identities we know at this point involve sines and cosines, it is a good idea to recast this identity in terms of those functions. We begin by simplifying the right side of the identity:

$$\frac{2\tan\theta}{1 - \tan^2\theta} = \frac{2(\sin\theta/\cos\theta)}{1 - (\sin^2\theta/\cos^2\theta)} = \frac{2\sin\theta\cos\theta}{\cos^2\theta - \sin^2\theta}.$$

Now we simplify the left side:

$$\tan 2\theta = \frac{\sin 2\theta}{\cos 2\theta} = \frac{2\sin\theta\cos\theta}{\cos^2\theta - \sin^2\theta}.$$

Combining these two calculations, it follows that $\tan 2\theta = \frac{2\tan\theta}{1 - \tan^2\theta}$. $\quad \square$

TEST YOUR UNDERSTANDING

▶ What is the difference between an identity and an equation?

▶ List two trigonometric identities that follow directly from the unit-circle definitions of the sine and/or cosine.

▶ List two trigonometric identities that follow directly from the sum or difference identities for the sine and/or cosine.

▶ List at least four double-angle identities.

▶ Using the sum and/or difference identities, for what types of angles do we now know how to calculate trigonometric values by hand?

EXERCISES 6.2

Thinking Back

Trigonometric calculations: Calculate each of the following by hand, using the unit circle and reference triangles.

▶ $\sin\dfrac{\pi}{3}$

▶ $\tan\dfrac{\pi}{4}$

▶ $\cos\left(-\dfrac{13\pi}{2}\right)$

▶ $\cot\left(-\dfrac{5\pi}{3}\right)$

▶ $\sec\dfrac{3\pi}{4}$

▶ $\csc\dfrac{7\pi}{6}$

Even and odd functions: Determine whether each of the following functions is even, odd, or neither.

▶ $f(x) = x^3$

▶ $f(x) = 3^x$

▶ $f(x) = \dfrac{x^3 - x}{x^4 + 1}$

▶ $f(x) = \dfrac{x^2 - 1}{x^3 - 1}$

Concepts

0. *Problem Zero:* Read the section and make your own summary of the material.

1. *True/False:* Determine whether each of the statements that follow is true or false. If a statement is true, explain why. If a statement is false, provide a counterexample.

(a) *True or False:* Every equation is an identity.

(b) *True or False:* For every x, $\cos^2 3x + \sin^2 3x = 1$.

(c) *True or False:* For every x, $1 - \cos^2(5x^3) = \sin^2(5x^3)$.

(d) *True or False:* Every point (x, y) on the unit circle satisfies the equation $x^2 + y^2 = 1$.

(e) *True or False:* For any angle θ, $(\cos\theta, \sin\theta)$ is a point on the unit circle.

(f) *True or False:* The sine of a sum of angles is equal to the sum of the sines of the angles.

(g) *True or False:* The sine of a product of angles is equal to the product of the sines of the angles.

(h) *True or False:* $\cos^2 \theta$ is the same as $(\cos \theta)^2$.

2. *Examples:* Construct examples of the thing(s) described in the following. Try to find examples that are different than any in the reading.

(a) Two trigonometric identities that are related to the Pythagorean theorem.

(b) Two trigonometric equations that are not trigonometric identities.

(c) Two trigonometric identities that are consequences of the sum formula for the sine.

3. Which of the six trigonometric functions are even functions? Which are odd functions?

4. List as many trigonometric identities as you can, from memory.

Show by providing specific counterexamples that each equation in Exercises 5–8 is not a trigonometric identity.

5. $\cos 2\theta = 2\cos \theta$

6. $\cos(\alpha+\beta) = \cos \alpha + \cos \beta$

7. $\sin \theta \cos \theta = 1 + \cos \theta$

8. $\sin^2 \theta + \sec^2 \theta = 1$

Use unit-circle calculations to show that each of the trigonometric identities in Exercises 9–12 holds for the angle $\theta = \frac{4\pi}{3}$.

9. $\sin^2 \theta + \cos^2 \theta = 1$

10. $\tan(-\theta) = -\tan \theta$

11. $\cos(2\theta) = 1 - 2\sin^2 \theta$

12. $\cos^2 \theta = \frac{1+\cos 2\theta}{2}$

Use unit-circle calculations to show that each of the trigonometric identities in Exercises 13–16 holds for the angles $\alpha = \frac{3\pi}{2}$ and $\beta = -\frac{\pi}{4}$.

13. $\sin(\alpha + \beta) = \sin \alpha \cos \beta + \sin \beta \cos \alpha$

14. $\sin(\alpha - \beta) = \sin \alpha \cos \beta - \sin \beta \cos \alpha$

15. $\cos(\alpha + \beta) = \cos \alpha \cos \beta - \sin \alpha \sin \beta$

16. $\cos(\alpha - \beta) = \cos \alpha \cos \beta + \sin \alpha \sin \beta$

17. Use graphs to determine trigonometric identities for shifting the sine function. Specifically, find identities that involve $\sin(\theta + 2\pi)$, $\sin(\theta + \pi)$, and $\sin\left(\theta + \frac{\pi}{2}\right)$.

18. Use graphs to determine trigonometric identities for shifting the cosine function. Specifically, find identities that involve $\cos(\theta + 2\pi)$, $\cos(\theta + \pi)$, and $\cos\left(\theta - \frac{\pi}{2}\right)$.

19. The following "proof" of the trigonometric identity $\sin 2\theta = 2 \sin \theta \cos \theta$ uses bad style and bad logic. What is wrong with the way this proof is written? Rewrite the proof so that it is correct.

$$\sin 2\theta = 2 \sin \theta \cos \theta$$
$$\sin(\theta + \theta) = 2 \sin \theta \cos \theta$$
$$\sin \theta \cos \theta + \sin \theta \cos \theta = 2 \sin \theta \cos \theta$$
$$2 \sin \theta \cos \theta = 2 \sin \theta \cos \theta$$

20. Use a right triangle and the Pythagorean theorem to argue that $\sin^2 \theta + \cos^2 \theta = 1$ for every acute angle θ.

21. Use the equation for the unit circle to argue that $\sin^2 \theta + \cos^2 \theta = 1$ for every angle θ.

22. Use the unit-circle definition of the sine to argue that $\sin(-\theta) = -\sin \theta$ for every angle θ.

Skills

Suppose that $\cos(\theta) = \frac{1}{6}$, $\sin(\theta) > 0$, $\sin(\phi) = \frac{3}{5}$, and $\cos(\phi) < 0$. Use trigonometric identities to identify the quantities in Exercises 23–28.

23. $\sin(\theta)$

24. $\sin(-\phi)$

25. $\cos(2\theta)$

26. $\sin\left(\theta + \frac{\pi}{2}\right)$

27. the sign of $\cos(\theta + \phi)$

28. the sign of $\tan(\theta + \pi)$

Calculate each of the trigonometric values in Exercises 29–37 by hand, without a calculator. You will need to use the unit circle and sum or difference identities.

29. $\sin\left(-\frac{\pi}{12}\right)$

30. $\cos\left(-\frac{\pi}{12}\right)$

31. $\tan\left(-\frac{\pi}{12}\right)$

32. $\cos\left(\frac{7\pi}{12}\right)$

33. $\sec\left(\frac{7\pi}{12}\right)$

34. $\cot\left(\frac{7\pi}{12}\right)$

35. $\csc\left(\frac{100\pi}{12}\right)$

36. $\sin\left(-\frac{29\pi}{12}\right)$

37. $\tan\left(-\frac{11\pi}{12}\right)$

Determine algebraically whether each of the functions in Exercises 38–43 is odd, even, or neither.

38. $f(\theta) = \frac{\cos \theta}{1+\sin \theta}$

39. $f(\theta) = \sin \theta + \cos \theta$

40. $f(\theta) = \sin^2 \theta - \sec \theta$

41. $f(\theta) = \frac{\sin \theta}{1+\cos \theta}$

42. $f(\theta) = \cos 2\theta \sin 2\theta$

43. $f(\theta) = \frac{\tan \theta}{\csc \theta + \cot \theta}$

For each equation in Exercises 44–49, if the equation is an identity, then show why it holds for every angle θ. If the equation is not an identity, then find an angle θ for which the equation is false and solve for all angles θ that make the equation true.

44. $\tan \theta = \sin \theta \cos \theta$

45. $\tan \theta = \sin \theta \sec \theta$

46. $\sec(-\theta) = -\sec \theta$

47. $\csc(-\theta) = -\csc \theta$

48. $2\sin^2 \theta - \cos 2\theta = 1$

49. $2\cos^2 \theta - \cos 2\theta = 1$

Applications

The following two exercises concern triangles without right angles. Suppose a triangle has sides and angles labeled as follows (not to scale):

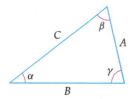

50. The *Law of Cosines* says that regardless of the angle measures or side lengths, it is always true that

$$A^2 = B^2 + C^2 - 2BC \cos \alpha.$$

Suppose two flights leave O'Hare Airport at the same time, one headed to Boston at 500 miles per hour and the other headed to Atlanta at 400 miles per hour. Their initial flight paths take them each in a straight line, separated by an angle of 39 degrees. Use the Law of Cosines to determine how far apart the planes are 30 minutes into their flights.

51. The *Law of Sines* says that regardless of the angle measures or side lengths, it is always true that

$$\frac{\sin \alpha}{A} = \frac{\sin \beta}{B} = \frac{\sin \gamma}{C}.$$

The Leaning Tower of Pisa leans at an angle of 3.99° from the vertical. If you stand 100 feet from the base of the short side with the tower leaning towards you, the angle between the ground and the top of that side is about 64.55°. Use the Law of Sines to determine the "height" of the tower in the direction that it leans.

Proofs

Prove the trigonometric identities in Exercises 52–58 in the manners specified, to complete the proofs of the theorems in the reading.

52. Use the fact that $\sin^2 \theta + \cos^2 \theta = 1$ to prove the Pythagorean identity for the cotangent and cosecant: $1 + \cot^2 \theta = \csc^2 \theta$.

53. Use the unit circle to prove the even identity for the cosine: $\cos(-\theta) = \cos \theta$.

54. Use the fact that the sine is an odd function and the cosine is an even function to prove the odd identity for the tangent: $\tan(-\theta) = -\tan \theta$.

55. Use a sum identity to prove the difference identity for the cosine: $\cos(\alpha - \beta) = \cos \alpha \cos \beta + \sin \alpha \sin \beta$.

56. Use a sum identity to prove the following double-angle identity: $\cos 2\theta = \cos^2 \theta - \sin^2 \theta$.

57. Use the identity in Exercise 56 and a Pythagorean identity to prove the double-angle identity $\cos 2\theta = 2\cos^2 \theta - 1$.

58. Use a double-angle identity from Theorem 6.8 to prove the identity $\cos^2 \theta = \frac{1+\cos 2\theta}{2}$.

Prove each of the trigonometric identities in Exercises 59–63.

59. $(\sin \theta + \cos \theta)^2 = \sin 2\theta + 1$

60. $\csc 2\theta = \dfrac{1}{2} \sec \theta \csc \theta$

61. $\sec \theta \cot \theta = \dfrac{\sin \theta}{1 - \cos^2 \theta}$

62. $\tan(\alpha + \beta) = \dfrac{\tan \alpha + \tan \beta}{1 - \tan \alpha \tan \beta}$

63. $\tan(\alpha - \beta) = \dfrac{\tan \alpha - \tan \beta}{1 + \tan \alpha \tan \beta}$

Thinking Forward

Rewriting trigonometric expressions: In later chapters we will be concerned with rewriting trigonometric expressions in particular ways. Rewrite each of the given expressions as specified. You may have to multiply out some expressions and/or use identities more than once.

▶ Use the Pythagorean identity $\sin^2 x + \cos^2 x = 1$ to rewrite the expression $\cos^5 x$ in the form $g(x) \cos x$, where $g(x)$ is an expression that involves sines but not cosines.

▶ Use the Pythagorean identity $\tan^2 x + 1 = \sec^2 x$ to rewrite the expression $\sec^6 x \tan^5 x$ in the form $h(x) \sec^2 x$, where $h(x)$ is an expression that involves tangents but not secants.

▶ Use the Pythagorean identity $\tan^2 x + 1 = \sec^2 x$ to rewrite the expression $\sec^6 x \tan^5 x$ in the form $k(x) \sec x \tan x$, where $k(x)$ is an expression that involves secants but not tangents.

▶ Use the double-angle identity $\sin^2 x = \dfrac{1 - \cos 2x}{2}$ to rewrite the expression $\sin^4 x \cos^2 x$ in terms of a sum of expressions of the form $A \cos kx$.

▶ Use the double-angle identity $\cos^2 x = \dfrac{1 + \cos 2x}{2}$ to rewrite the expression $\cos^6 x$ in terms of a sum of expressions of the form $A \cos^n kx$.

6.3   LIMITS AND DERIVATIVES OF TRIGONOMETRIC FUNCTIONS

▶   Continuity and basic limits of trigonometric functions

▶   Two important limits involving the sine and cosine

▶   Formulas for differentiating the trigonometric functions

Continuity of Trigonometric Functions

As usual, our first step in learning to calculate limits of a new type of function is to determine when taking a limit happens to be the same as evaluating a function at a point. Fortunately, as with all the previous types of functions we have studied, limits of trigonometric functions are equal to their values everywhere that they happen to be defined:

THEOREM 6.10   **Continuity of Trigonometric Functions**

All six trigonometric functions are continuous on their domains.

For example, $\cos x$ is defined at $x = \pi$, and this theorem says that the limit $\lim_{x \to \pi} \cos x$ is equal to the value $\cos(\pi) = -1$. The theorem does not tell us how to calculate $\lim_{x \to \pi} \sec x$, however, because π is not in the domain of $\sec x$.

> **Proof.** * *(The details of this proof can be safely omitted in some courses)*
> Our strategy will be (a) to show that the sine and cosine are continuous at zero and then (b) to use this fact together with a sum identity to show that the sine is continuous everywhere. Proving that the cosine is continuous everywhere is similar and is left to Exercise 95.
>
> **(a)**   We can use the Squeeze Theorem from Section 1.6 to calculate the limits of $\sin \theta$ and $\cos \theta$ at $x = 0$. What we need to show is that $\lim_{\theta \to 0} \sin \theta = \sin 0 = 0$ and that $\lim_{\theta \to 0} \cos \theta = \cos 0 = 1$. We will consider the case where $\theta \to 0^+$; the proof for limits as $\theta \to 0^-$ are similar. Consider the following diagrams of portions of the unit circle, with angles measured in radians and $0 < \theta < \dfrac{\pi}{4}$:

$$0 < \sin \theta < \theta \qquad\qquad\qquad 0 < 1 - \cos \theta < \theta$$

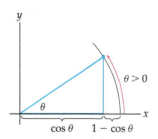

> According to the leftmost figure, we clearly have $0 < \sin \theta < \theta$. Since $\lim_{\theta \to 0^+} 0 = 0$ and $\lim_{\theta \to 0^+} \theta = 0$, by the Squeeze Theorem we must also have $\lim_{\theta \to 0^+} \sin \theta = 0$. In the second figure we can see that $0 < 1 - \cos \theta < \theta$. Thus again by the Squeeze Theorem we have $\lim_{\theta \to 0^+} (1 - \cos \theta) = 0$. Rewriting this limit with some simple limit rules, we see that $\lim_{x \to 0^+} \cos \theta = 1$. We can illustrate similar limits as $\theta \to 0^-$ by drawing pictures of small negative values of θ in the fourth quadrant (see Exercises 10 and 11).

(b) To show that $\sin x$ is continuous on its domain, we must show that $\lim\limits_{x \to c} \sin x = \sin c$ for all $c \in \mathbb{R}$. Our strategy will be to change variables with $h = x - c$, and relate to some limits that we already know. By applying the sum identity for the sine and the results of part (a) of this proof, we have

$$\lim_{x \to c} \sin x = \lim_{h \to 0} \sin(c + h) \qquad \leftarrow \text{change variables}$$

$$= \lim_{h \to 0} (\sin c \cos h + \sin h \cos c) \qquad \leftarrow \text{sum identity for sine}$$

$$= \sin c \left(\lim_{h \to 0} \cos h \right) + \cos c \left(\lim_{h \to 0} \sin h \right) \qquad \leftarrow \text{sum and constant multiple rules}$$

$$= (\sin c)(1) + (\cos c)(0) = \sin c. \qquad \leftarrow \text{known limits}$$

Infinite Limits and Limits at Infinity

Recall that the graphs of the six trigonometric functions look like this:

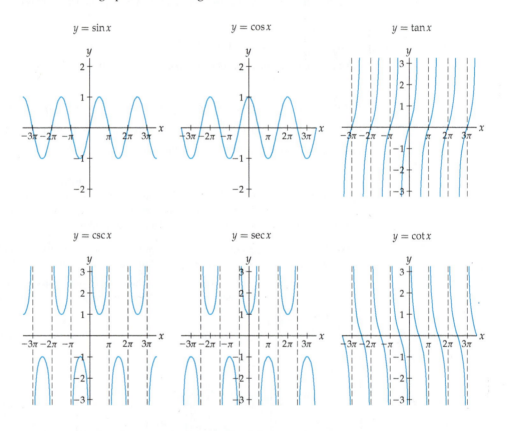

In terms of limits, these graphs suggest the following theorem:

THEOREM 6.11	**Asymptotes and Long-Term Behavior of Trigonometric Functions**

(a) All six trigonometric functions exhibit periodic behavior as $x \to \infty$, and thus their limits as $x \to \pm\infty$ do not exist.

(b) $\tan x$ and $\sec x$ have vertical asymptotes at $x = \frac{\pi}{2} + \pi k$, for any integer k.

(c) $\csc x$ and $\cot x$ have vertical asymptotes at $x = \pi k$, for any integer k.

As usual, this theorem is easy to remember if you think of the graphs of the six trigonometric functions, and that is what you should do in practice. Proving the theorem guarantees that the six trigonometric functions have the graphs that they do; see Exercises 98–100.

Special Limits Involving Sine and Cosine

Our ultimate goal in this section is to determine the derivatives of the trigonometric functions. Finding the derivatives of the sine and cosine requires knowing the following two special limits:

THEOREM 6.12   **Two Important Trigonometric Limits**

(a) $\displaystyle\lim_{\theta \to 0} \frac{\sin \theta}{\theta} = 1$ **(b)** $\displaystyle\lim_{\theta \to 0} \frac{1 - \cos \theta}{\theta} = 0$

It is not immediately obvious how to calculate these limits. For example, as $\theta \to 0$, the first limit approaches $\frac{\sin 0}{0} \to \frac{0}{0}$, which is indeterminate. At this point we might be tempted to apply L'Hôpital's Rule, but to do that we would need to know the derivative of the sine, and we do not yet know that. Similar problems occur if we attempt to calculate the second limit.

Proof. We can intuitively see that the two special trigonometric limits make sense based on the following figures of portions of the unit circle, with angles measured in radians:

$\sin \theta \approx \theta$ $1 - \cos \theta$ *much smaller than* θ

    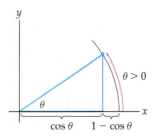

For small positive values of θ, the picture on the left suggests that $\sin \theta \approx \theta$ and therefore that $\frac{\sin\theta}{\theta} \approx 1$. Similarly, for small positive θ, the picture on the right suggests that $1 - \cos \theta$ approaches 0 much faster than θ and therefore that $\frac{1 - \cos\theta}{\theta} \approx 0$.

More formally, from the earlier figure on the left, we can see that $\sin \theta \leq \theta \leq \tan \theta$. Dividing all expressions in this chain of inequalitites by $\sin \theta$ and using the fact that $\tan \theta = \frac{\sin\theta}{\cos\theta}$, we have

$$1 \leq \frac{\theta}{\sin \theta} \leq \frac{1}{\cos \theta}.$$

Since all of the expressions in this chain of inequalities are positive, we can take reciprocals to obtain:

$$\cos \theta \leq \frac{\sin \theta}{\theta} \leq 1.$$

Now by applying the Squeeze Theorem to these two inequalities we have $\displaystyle\lim_{\theta \to 0^+} \frac{\sin\theta}{\theta} = 1$. The argument for $\theta \to 0^-$ is similar, with a picture in the fourth quadrant.

The second trigonometric limit can be proved from the first by using the double-angle formula $\cos 2\theta = 1 - 2\sin^2 \theta$; see Exercise 106.   ◼

The next two figures show the functions $y = \frac{\sin x}{x}$ and $y = \frac{1 - \cos x}{x}$. Notice that neither function is defined at $x = 0$, but both approach a specific real-number value as $x \to 0$.

$$\lim_{x \to 0} \frac{\sin x}{x} = 1$$

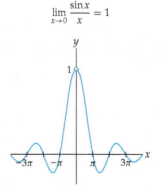

$$\lim_{x \to 0} \frac{1 - \cos x}{x} = 0$$

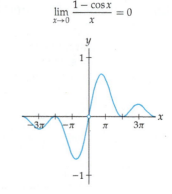

We can use the limits in Theorem 6.12 to calculate limits such as

$$\lim_{x \to 1} \frac{\sin(x - 1)}{x - 1}, \quad \lim_{x \to 0} \frac{1 - \cos 4x}{4x}, \quad \text{and} \quad \lim_{x \to -2} \frac{1 - \cos(x + 2)}{x + 2}.$$

Each of these limits is equivalent to one of the two special trigonometric limits; the first limit is equal to 1, while the last two limits are equal to 0.

Derivatives of the Trigonometric Functions

We are now in a position to develop formulas for differentiating the trigonometric functions. Because trigonometric functions have periodic oscillating behavior, and their slopes also have periodic oscillating behavior, it would make sense if the derivatives of trigonometric functions were trigonometric. For example, the two graphs that follow show the function $f(x) = \sin x$ and its derivative $f'(x) = \cos x$. As we will prove in Theorem 6.13, it turns out that, at each value of x, the slope of the graph of $f(x) = \sin x$ is given by the height of the graph of $f'(x) = \cos x$. Before we tackle this fact algebraically, take a minute to verify that it is the case with these graphs for the values $x = -5.2$, $x = \frac{\pi}{2}$, and $x = 4$, as shown in the following figures:

Slopes of $f(x) = \sin x$ at three points

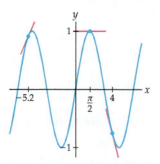

Heights of $f'(x) = \cos x$ at three points

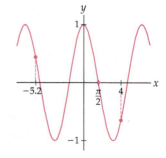

The six trigonometric functions have the following derivatives:

| THEOREM 6.13 | **Derivatives of the Trigonometric Functions** |

For all values of x at which the following functions below are defined,

(a) $\dfrac{d}{dx}(\sin x) = \cos x$ **(c)** $\dfrac{d}{dx}(\tan x) = \sec^2 x$ **(e)** $\dfrac{d}{dx}(\cot x) = -\csc^2 x$

(b) $\dfrac{d}{dx}(\cos x) = -\sin x$ **(d)** $\dfrac{d}{dx}(\sec x) = \sec x \tan x$ **(f)** $\dfrac{d}{dx}(\csc x) = -\csc x \cot x$

It is important to note that these formulas for derivatives are true only if angles are measured in radians; see Exercise 19.

Proof. We will prove the formulas for $\sin x$ and $\tan x$ from parts (a) and (c) and leave the proofs of the remaining four formulas to Exercises 101–104.

(a) The proof of the first formula is nothing more than an annotated calculation using the definition of the derivative. To simplify the limit we obtain, we will rewrite $\sin(x + h)$ with a trigonometric identity. Our goal after that will be to rewrite the limit so that we can apply the two trigonometric limits from Theorem 1.35 in Section 1.6. The calculation is as follows:

$$\frac{d}{dx}(\sin x) = \lim_{h \to 0} \frac{\sin(x + h) - \sin x}{h} \qquad \leftarrow \text{ definition of the derivative}$$

$$= \lim_{h \to 0} \frac{(\sin x \cos h + \sin h \cos x) - \sin x}{h} \qquad \leftarrow \text{ sum identity for sine}$$

$$= \lim_{h \to 0} \frac{\sin x(\cos h - 1) + \sin h \cos x}{h} \qquad \leftarrow \text{ algebra}$$

$$= \lim_{h \to 0} \left(\sin x \frac{\cos h - 1}{h} + \cos x \frac{\sin h}{h} \right) \qquad \leftarrow \text{ algebra}$$

$$= \sin x \left(\lim_{h \to 0} \frac{\cos h - 1}{h} \right) + \cos x \left(\lim_{h \to 0} \frac{\sin h}{h} \right) \qquad \leftarrow \text{ limit rules}$$

$$= (\sin x)(0) + (\cos x)(1) = \cos x. \qquad \leftarrow \text{ trigonometric limits}$$

(c) We do not have to resort to the definition of the derivative in order to prove the formula for differentiating $\tan x$. Instead we can use the quotient rule, the fact that $\tan x = \dfrac{\sin x}{\cos x}$, and the formulas for differentiating $\sin x$ and $\cos x$:

$$\frac{d}{dx}(\tan x) = \frac{d}{dx}\left(\frac{\sin x}{\cos x}\right)$$

$$= \frac{\dfrac{d}{dx}(\sin x) \cdot (\cos x) - (\sin x) \cdot \dfrac{d}{dx}(\cos x)}{(\cos x)^2} \qquad \leftarrow \text{ quotient rule}$$

$$= \frac{(\cos x)(\cos x) - (\sin x)(-\sin x)}{\cos^2 x} \qquad \leftarrow \text{ derivatives of } \sin x \text{ and } \cos x$$

$$= \frac{\cos^2 x + \sin^2 x}{\cos^2 x} = \frac{1}{\cos^2 x} = \sec^2 x. \qquad \leftarrow \text{ algebra and identities}$$

Examples and Explorations

| EXAMPLE 1 | **Basic trigonometric limits** |

Calculate each of the following limits:

(a) $\displaystyle\lim_{x \to 0} \sqrt{\sin\left(x + \frac{\pi}{2}\right)}$ **(b)** $\displaystyle\lim_{x \to -\pi} \csc x$

SOLUTION

(a) $f(x) = \sqrt{\sin\left(x + \frac{\pi}{2}\right)}$ is a combination of functions that are continuous on their domains, and thus is continuous on its domain. Since $x = 0$ is in the domain of f we can solve this limit by simple evaluation:

$$\lim_{x \to 0} \sqrt{\sin\left(x + \frac{\pi}{2}\right)} = \sqrt{\sin\left(0 + \frac{\pi}{2}\right)} = \sqrt{1} = 1.$$

(b) Since $\csc x = \frac{1}{\sin x}$ and $\sin(-\pi) = 0$, $\lim_{x \to -\pi} \csc x$ is of the form $\frac{1}{0}$ and therefore the function $y = \csc x$ will have a vertical asymptote at $x = -\pi$. However, we don't yet know if the function approaches $-\infty$ or ∞ from the left and the right. From the left, as $x \to -\pi^-$, we are considering angles slightly less than $-\pi$ (for instance $-\pi - 0.1$, which terminates in the second quadrant). For such angles, the sine function is negative, and therefore as $x \to -\pi^-$ we have $\sin x \to 0^+$ and thus

$$\lim_{x \to -\pi^-} \csc x = \infty.$$

Similarly, from the right, as $x \to -\pi^+$, we have $\sin x \to 0^-$ and thus

$$\lim_{x \to -\pi^+} \csc x = -\infty.$$

This means that $\lim_{x \to -\pi} \csc x$ does not exist. □

EXAMPLE 2 | **Limits of trigonometric expressions at infinity**

Investigate the following two limits:

(a) $\lim_{x \to \infty} x \sin x$ **(b)** $\lim_{x \to \infty} \frac{\sin x}{x}$

SOLUTION

(a) As $x \to \infty$, $\sin x$ oscillates between -1 and 1. Your first thought might be that since x is approaching ∞, the function $x \sin x$ will also approach infinity. However, because $\sin x$ oscillates between negative and positive numbers, so will the function $x \sin x$. Note that as x increases in magnitude, so does $x \sin x$. Therefore as $x \to \infty$, the quantity $x \sin x$ oscillates without converging to any real number, and therefore $\lim_{x \to \infty} x \sin x$ does not exist. The right side of the graph of $y = x \sin x$ is shown in the graph here at the left:

$y = x \sin x$

$y = \dfrac{\sin x}{x}$

(b) Again, as $x \to \infty$, $\sin x$ oscillates between -1 and 1. Therefore, the numerator $\sin x$ stays bounded while the denominator x becomes infinitely large. A bounded quantity divided by an infinite quantity that increases without bound is zero; in other words, loosely speaking,

$$\frac{\text{bounded}}{\infty} \to 0.$$

Therefore $\lim\limits_{x \to \infty} \frac{\sin x}{x} = 0$. The right side of the graph of $y = \frac{\sin x}{x}$ is shown in the right-hand figure.   □

EXAMPLE 3   **Calculating limits without L'Hôpital's Rule**

Calculate the following limits, using the special trigonometric limits from Theorem 6.12 instead of L'Hôpital's Rule:

(a) $\lim\limits_{x \to 0} \dfrac{\sin 3x}{2x}$ **(b)** $\lim\limits_{x \to 0} \dfrac{2x}{\tan^2 x}$

SOLUTION

(a) If we can write this limit in the form $\frac{\sin u}{u}$, then we will be able to evaluate the limit with Theorem 6.12. We need the expression inside the sine function to match the expression in the denominator. We can rewrite the $2x$ in the denominator as $2x = \frac{2}{3}(3x)$. We then have

$$\lim_{x \to 0} \frac{\sin 3x}{2x} = \lim_{x \to 0} \frac{\sin 3x}{(2/3)(3x)} = \frac{3}{2} \lim_{x \to 0} \frac{\sin 3x}{3x} = \frac{3}{2}(1) = \frac{3}{2}.$$

(b) Since $\tan^2 0 = 0$, this limit is of the indeterminate form $\frac{0}{0}$. We will write the limit in terms of sines and cosines and then apply Theorem 6.12:

$$\lim_{x \to 0} \frac{2x}{\tan^2 x} = \lim_{x \to 0} \frac{2x \cos^2 x}{\sin^2 x} \qquad\qquad \leftarrow \tan x = \frac{\sin x}{\cos x}$$

$$= \lim_{x \to 0} \left(\frac{x}{\sin x} \cdot \frac{2 \cos^2 x}{\sin x} \right) \qquad\qquad \leftarrow \text{algebra}$$

$$= \left(\lim_{x \to 0} \frac{x}{\sin x} \right) \left(\lim_{x \to 0} \frac{2 \cos^2 x}{\sin x} \right) \qquad \leftarrow \text{product rule for limits}$$

$$= (1) \left(\lim_{x \to 0} \frac{2 \cos^2 x}{\sin x} \right) \qquad\qquad \leftarrow \text{Theorem 6.12}$$

Note that we used the fact that since $\frac{\sin x}{x}$ approaches 1 as $x \to 0$, its reciprocal $\frac{x}{\sin x}$ approaches $\frac{1}{1} = 1$ as $x \to 0$.

Now as $x \to 0$, the quantity $\sin x$ in the denominator of the remaining expression approaches 0, while the quantity $2 \cos^2 x$ in the numerator approaches 2. Thus the remaining limit is of the form $\frac{2}{0}$. We know that this limit is infinite, but we need to examine both sides of the limit to see if they approach ∞ or $-\infty$. From the left, as $x \to 0^-$, we are considering angles that terminate in the fourth quadrant, where $\sin x < 0$. Therefore as $x \to 0^-$, $\sin x \to 0^-$ and thus

$$\lim_{x \to 0^-} \frac{2 \cos^2 x}{\sin x} = -\infty.$$

Similarly, as $x \to 0^+$, $\sin x \to 0^+$ and thus

$$\lim_{x \to 0^+} \frac{2\cos^2 x}{\sin x} = \infty.$$

Combining these results, we see that the original limit does not exist as $x \to 0$; it approaches ∞ from the right and $-\infty$ from the left.   □

EXAMPLE 4

Calculating derivatives that involve trigonometric functions

Calculate each of the following derivatives:

(a) $f(x) = \dfrac{\tan x}{x^3 - 2}$ **(b)** $f(x) = \sec^2 e^x$

SOLUTION

(a) The function $f(x) = \frac{\tan x}{x^3 - 2}$ is a quotient of two functions. By the quotient rule and the rule for differentiating tangent, we have

$$\frac{d}{dx}\left(\frac{\tan x}{x^3 - 2}\right) = \frac{\frac{d}{dx}(\tan x) \cdot (x^3 - 2) - (\tan x) \cdot \frac{d}{dx}(x^3 - 2)}{(x^3 - 2)^2}$$

$$= \frac{(\sec^2 x)(x^3 - 2) - (\tan x)(3x^2)}{(x^3 - 2)^2}.$$

(b) The function $f(x) = \sec^2 e^x$ is a composition of three functions, and thus we need to apply the chain rule twice:

$$\frac{d}{dx}(\sec^2 e^x) = \frac{d}{dx}((\sec(e^x))^2) \qquad \leftarrow \text{rewrite so compositions are clear}$$

$$= 2(\sec e^x)^1 \cdot \frac{d}{dx}(\sec e^x) \qquad \leftarrow \text{first application of chain rule}$$

$$= 2(\sec e^x)(\sec e^x)(\tan e^x) \cdot \frac{d}{dx}(e^x) \qquad \leftarrow \text{second application of chain rule}$$

$$= 2(\sec e^x)(\sec e^x)(\tan e^x)e^x \qquad \leftarrow \text{derivative of } e^x$$

Perhaps the most difficult part of the preceding calculation is that the derivative of $\sec x$ has *two* instances of the independent variable: $\frac{d}{dx}(\sec x) = \sec \underline{x} \tan \underline{x}$. This means that we needed to put the "inside" function e^x into *both* of the slots for variables.   □

EXAMPLE 5

Calculating trigonometric limits with L'Hôpital's Rule

Calculate each of the following limits:

(a) $\displaystyle\lim_{x \to 0^+} \frac{\csc x}{\ln x}$ **(b)** $\displaystyle\lim_{x \to \pi/2} (\tan x - \sec x)$

SOLUTION

(a) As $x \to 0^+$, $\csc x$ approaches ∞ and $\ln x$ approaches $-\infty$; therefore this is a limit of the form $\frac{\infty}{\infty}$, which is indeterminate. L'Hôpital's Rule applies and gives us

$$\lim_{x \to 0^+} \frac{\csc x}{\ln x} \overset{\text{L'H}}{=} \lim_{x \to 0^+} \frac{-\csc x \cot x}{1/x} = \lim_{x \to 0^+} (-x \csc x \cot x).$$

As $x \to 0^+$ we have $-x \to 0$, $\csc x \to \infty$, and $\cot x \to \infty$, and therefore this limit is of the indeterminate form $0 \cdot \infty$. After converting to sines and cosines and simplifying, we will be able to apply L'Hôpital's Rule again:

$$\lim_{x \to 0^+} (-x \csc x \cot x) = \lim_{x \to 0^+} (-x) \left(\frac{1}{\sin x}\right)\left(\frac{\cos x}{\sin x}\right) \qquad \leftarrow \text{trigonometric identities}$$

$$= \lim_{x \to 0^+} \frac{-x \cos x}{\sin^2 x} \qquad \leftarrow \text{algebra; limit now of}$$
$$\text{form } \frac{0}{0}$$

$$\overset{\text{L'H}}{=} \lim_{x \to 0^+} \frac{(-1)\cos x + (-x)(-\sin x)}{2 \sin x \cos x} \qquad \leftarrow \text{note use of product and}$$
$$\text{chain rules}$$

$$= \lim_{x \to 0^+} \frac{-\cos x + x \sin x}{2 \sin x \cos x}. \qquad \leftarrow \text{simplify}$$

Now as $x \to 0^+$ we have $\sin x \to 0+$ and $\cos 0 \to 1$, and therefore the denominator in the remaining expression approaches 0^+. Since the numerator approaches $-\cos 0 + 0(\sin 0) = -1$, the limit in question is of the form $\frac{-1}{0^+}$ and therefore

$$\lim_{x \to 0^+} \frac{-\cos x + x \sin x}{2 \sin x \cos x} = -\infty.$$

(b) At first glance this does not look like a L'Hôpital's Rule problem, since $\tan x - \sec x$ is not in the form of a quotient. Worse, the limit is of the indeterminate form $\infty - \infty$, so we can't solve it simply by evaluating the function at $x = \frac{\pi}{2}$. After using a bit of algebra, we will see that L'Hôpital's Rule does apply here:

$$\lim_{x \to \pi/2} (\tan x - \sec x) = \lim_{x \to \pi/2} \left(\frac{\sin x}{\cos x} - \frac{1}{\cos x}\right) \qquad \leftarrow \text{trigonometric identities}$$

$$= \lim_{x \to \pi/2} \frac{\sin x - 1}{\cos x} \qquad \leftarrow \text{this is now in the form } \frac{0}{0}$$

$$\overset{\text{L'H}}{=} \lim_{x \to \pi/2} \frac{\cos x}{-\sin x} \qquad \leftarrow \text{L'Hôpital's Rule}$$

$$= \frac{\cos \pi/2}{-\sin \pi/2} = \frac{0}{-1} = 0. \qquad \leftarrow \text{evaluate limit}$$

CHECKING THE ANSWER

After all that work it would be wise to check our answers with some graphs. Notice that the graph on the left does approach $-\infty$ as x approaches 0 from the right and the graph on the right does approach 0 as $x \to \frac{\pi}{2}$:

$$f(x) = \frac{\csc x}{\ln x}$$

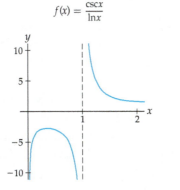

$$f(x) = \tan x - \sec x$$

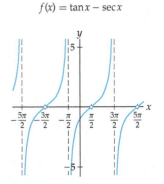

? **TEST YOUR**
UNDERSTANDING

▶ Why is it important that the trigonometric functions are continuous on their domains?

▶ What trigonometric identity helped us prove that $\sin x$ is continuous on its domain?

▶ Which trigonometric functions have vertical asymptotes, and where?

▶ What two special trigonometric limits were used to find the derivative of $\sin x$?

▶ How can we obtain the derivative of $\sec x$ from the derivative of $\cos x$?

EXERCISES 6.3

Thinking Back

Compositions: For each function k that follows, find functions f, g, and h so that $k = f \circ g \circ h$. There may be more than one possible answer.

▶ $k(x) = \dfrac{1}{\sin(x^3)}$

▶ $k(x) = \sin^{-1}(\cos^2 x)$

L'Hôpital's Rule: Calculate each of the following limits.

▶ $\displaystyle\lim_{x\to\infty} \dfrac{e^x}{e^x - 1}$

▶ $\displaystyle\lim_{x\to\infty} \dfrac{x^3}{e^x - 1}$

▶ $\displaystyle\lim_{x\to 0^+} x \ln x$

▶ $\displaystyle\lim_{x\to 0^+} \dfrac{\ln x}{\ln 3x}$

Concepts

0. *Problem Zero:* Read the section and make your own summary of the material.

1. *True/False:* Determine whether each of the statements that follow is true or false. If a statement is true, explain why. If a statement is false, provide a counterexample.

(a) *True or False:* When θ is a small angle, the quantities $\sin\theta$ and θ are equal.

(b) *True or False:* The function $f(x) = \sec x$ is continuous at $x = \dfrac{\pi}{2}$.

(c) *True or False:* The function $f(x) = \csc x$ is continuous at $x = \dfrac{\pi}{2}$.

(d) *True or False:* The graph of $y = \csc x$ has vertical asymptotes at $x = k\pi$, for any integer k.

(e) *True or False:* To find the derivative of $\sin x$ we had to use the definition of the derivative.

(f) *True or False:* To find the derivative of $\tan x$ we had to use the definition of the derivative.

(g) *True or False:* If $f(x)$ is a trigonometric function, then $f'(x)$ is also a trigonometric function.

(h) *True or False:* The derivative of $\dfrac{x^4}{\sin x}$ is $\dfrac{4x^3}{\cos x}$.

2. *Examples:* Construct examples of the thing(s) described in the following. Try to find examples that are different than any in the reading.

(a) Two limits that can be solved with the Squeeze Theorem.

(b) Two limits that can be solved with the special trigonometric limits from Theorem 6.12.

(c) Two limits involving trigonometric expressions that can be solved with L'Hôpital's Rule.

3. Suppose $x = c$ is a real number in the domain of $f(x) = \sec x$. How can we calculate $\displaystyle\lim_{x\to c} \sec x$? What does your answer have to do with the continuity of the secant function?

4. Suppose you know that $\displaystyle\lim_{x\to 0} \sin x = 0$; explain why this means that $f(x) = \sin x$ is continuous at $x = 0$.

5. Suppose you know that $\displaystyle\lim_{x\to 0}(\cos x - 1) = 0$; explain why this means that $f(x) = \cos x$ is continuous at $x = 0$.

6. Explain why $\displaystyle\lim_{x\to c} \cos x$ is equivalent to $\displaystyle\lim_{h\to 0} \cos(c + h)$.

7. Use a table of values to estimate $\displaystyle\lim_{\theta\to 0} \dfrac{\sin 2\theta}{\theta}$. Then check your approximation by examining a graph of $f(\theta) = \dfrac{\sin 2\theta}{\theta}$ near $\theta = 0$.

8. Use a table of values to estimate $\displaystyle\lim_{\theta\to 0} \dfrac{1 - \cos 2\theta}{\theta}$. Then check your approximation by examining a graph of $f(\theta) = \dfrac{1 - \cos 2\theta}{\theta}$ near $\theta = 0$.

9. In the Squeeze Theorem for limits, we require that $l(x) \le f(x) \le u(x)$ for all x sufficiently close to c, but we do not require this inequality to hold *at* the point $x = c$. Why not?

10. Use a geometric argument and the Squeeze Theorem for limits to argue that

$$\lim_{\theta\to 0^-} \sin\theta = 0$$

for sufficiently small negative angles θ.

11. Use a geometric argument and the Squeeze Theorem for limits to argue that

$$\lim_{\theta\to 0^-} \cos\theta = 1$$

for sufficiently small negative angles θ.

12. What does the height of the graph of $y = \cos x$ at $x = \dfrac{\pi}{2}$ tell you about the graph of $y = \sin x$ at $x = \dfrac{\pi}{2}$? What does the height of the graph of $y = \sin x$ at $x = \dfrac{\pi}{2}$ tell you about the graph of $y = \cos x$ at $x = \dfrac{\pi}{2}$?

13. Find all of the errors in the following calculation, and then calculate the limit correctly:

$$\lim_{x\to\pi/2}\frac{\cos x}{1-\sin 2x}\overset{\text{L'H}}{=}\lim_{x\to\pi/2}\frac{-\sin x}{-\cos 2x}=\frac{-\sin\pi/2}{-\cos\pi}=1.$$

14. Find all of the errors in the following calculation, and then calculate the limit correctly:

$$\lim_{x\to0}\frac{\sin x}{x^2-x}\overset{\text{L'H}}{=}\lim_{x\to0}\frac{\cos x}{2x-1}\overset{\text{L'H}}{=}\lim_{x\to0}\frac{\sin x}{2}=0.$$

15. For $f(x)=\sin x$ and $c=\dfrac{\pi}{2}$, use a sequence of approximations to approximate $f'(c)$. Illustrate your work with an appropriate sequence of graphs of secant lines.

16. List the derivatives of the six trigonometric functions from memory.

17. What limit facts and trigonometric identities are used in the proof that $\dfrac{d}{dx}(\sin x)=\cos x$?

18. Sketch graphs of $\sin x$ and $\cos x$ on $[-2\pi,2\pi]$.

(a) Use the graph of $\sin x$ to determine where $\sin x$ is increasing and and where it is decreasing.

(b) Use the graph of $\cos x$ to determine where $\cos x$ is positive and where it is negative.

(c) Explain why your answers to parts (a) and (b) suggest that $\cos x$ is the derivative of $\sin x$.

19. The differentiation formula $\dfrac{d}{dx}(\sin x)=\cos x$ is valid only if x is measured in radians. In this exercise you will explore why this derivative relationship does not hold if x is measured in degrees.

(a) Set your calculator to degree mode, and sketch a graph of $\sin x$ that shows at least two periods. If the derivative of $\sin x$ is $\cos x$, then the slope of your graph at $x=0$ should be equal to $\cos 0=1$. Use your graph to explain why this is not the case when we use degrees. *(Hint: Think about your graphing window scale.)*

(b) Now set your calculator back to radians mode!

20. Suppose you wish to differentiate $g(x)=\sin^2(x)+\cos^2(x)$. What is the fastest way to do this, and why?

21. The derivatives of the function $f(x)=\cos(3x^2)$ that follow are incorrect. What misconception occurs in each case?

(a) *Incorrect:* $f'(x)=(-\sin x)(3x^2)+(\cos x)(6x)$.

(b) *Incorrect:* $f'(x)=-\sin(6x)$.

22. The derivatives of the function $f(x)=\cos(3x^2)$ that follow are incorrect. What misconception occurs in each case?

(a) *Incorrect:* $f'(x)=-\sin(3x^2)$.

(b) *Incorrect:* $f'(x)=-\sin(3x^2)(6x)(6)$.

Skills

Calculate each of the limits in Exercises 23–42. Do your calculations two ways, both with and without L'Hôpital's rule, if possible.

23. $\displaystyle\lim_{x\to\pi/2}\tan x$

24. $\displaystyle\lim_{x\to\pi}\csc x$

25. $\displaystyle\lim_{x\to\pi}\frac{1}{\csc(x-\pi)}$

26. $\displaystyle\lim_{x\to1}\frac{1-\cos(x-1)}{x}$

27. $\displaystyle\lim_{x\to0}\frac{\sin x}{\tan x}$

28. $\displaystyle\lim_{x\to\pi/2}\frac{\cot x}{\cos x}$

29. $\displaystyle\lim_{x\to0}\frac{1-\cos 2x}{7x}$

30. $\displaystyle\lim_{x\to0}\frac{\sin 3x}{5x}$

31. $\displaystyle\lim_{x\to0}\frac{x}{1-\cos x}$

32. $\displaystyle\lim_{x\to0}\frac{3\sin x+x}{x}$

33. $\displaystyle\lim_{x\to0}\frac{\sin^2 3x}{x^3-x}$

34. $\displaystyle\lim_{x\to0}\frac{\sin(3x^2)}{x^3-x}$

35. $\displaystyle\lim_{x\to0^+}\frac{x^2\csc 3x}{1-\cos 2x}$

36. $\displaystyle\lim_{x\to0}\frac{x^2\cot x}{\sin x}$

37. $\displaystyle\lim_{x\to0}\frac{\sec x\tan x}{x}$

38. $\displaystyle\lim_{x\to0}3x^2\cot^2 x$

39. $\displaystyle\lim_{x\to\infty}\frac{\sin x}{e^x}$

40. $\displaystyle\lim_{x\to\infty}e^x\sin x$

41. $\displaystyle\lim_{x\to\infty}\frac{e^x}{\sin x}$

42. $\displaystyle\lim_{x\to\infty}\frac{e^x}{\sin^2 x}$

Find the derivatives of each of the functions in Exercises 43–68. In some cases it may be convenient to do some preliminary algebra.

43. $f(x)=\sin(3x)$ **44.** $f(x)=\sec(1-x)$

45. $f(x)=x^3\cos x$

46. $f(x)=2x\csc x$

47. $f(x)=\cot(3x+1)$

48. $f(x)=\dfrac{\tan x}{x}$

49. $f(x)=\dfrac{x^2+1}{\cos x}$

50. $f(x)=2\cos(x^3)$

51. $f(x)=\cot x-\csc x$

52. $f(x)=\tan^2(3x+1)$

53. $f(x)=4\sin^2 x+4\cos^2 x$

54. $f(x)=\sec^2 x^{-1}$

55. $f(x)=3\sec x\tan x$

56. $f(x)=3^x\sec x+17$

57. $f(x)=\sin(\cos(\sec(x)))$

58. $f(x)=\csc^2(e^x)$

59. $f(x)=e^{\csc^2 x}$

60. $f(x)=e^x\csc^2 x$

61. $f(x)=\dfrac{-2^x}{5x\sin x}$

62. $f(x)=\dfrac{\log_3(3^x)}{\sin^2 x+\cos^2 x}$

63. $f(x)=x\sqrt{\sin x\cos x}$

64. $f(x)=\dfrac{\sin x\csc x}{\cot x\cos x}$

65. $f(x)=\dfrac{3x^2\ln x}{\tan x}$

66. $f(x)=\dfrac{\ln(3x^2)}{\tan x}$

67. $f(x)=\sin(\ln x)$

68. $f(x)=\ln(x\sin x)$

Find the critical points of each function in Exercises 69–72, and use the first-derivative or second-derivative test to determine whether each critical point is a local minimum, a local maximum, or neither.

69. $f(x)=\cos x$

70. $f(x)=\sec x$

71. $f(x)=\cos(\pi(x+1))$

72. $f(x)=\cos(\pi x)$

Determine the intervals on which each function $f(x)$ in Exercises 73–76 is increasing or decreasing.

73. $f(x) = \sin\left(\frac{\pi}{2}x\right)$

74. $f(x) = \cos^2(x)$

75. $f(x) = \sin x \cos x$

76. $f(x) = \dfrac{1}{\sin x}$

For each function $f(x)$ in Exercises 77–80, find the intervals on which $f(x)$ is concave up or concave down and find the locations of any inflection points.

77. $f(x) = \sin\left(x - \dfrac{\pi}{4}\right)$

78. $f(x) = 3\cos\left(\dfrac{\pi}{2}x\right) + 5$

79. $f(x) = \sin x \cos x$

80. $f(x) = \cos^2 x$

To support each limit statement $\lim\limits_{x \to c} f(x) = L$ in Exercises 81–84, use graphs and algebra to approximate the largest value of δ such that if $x \in (c - \delta, c) \cup (c, c + \delta)$, then $f(x) \in (L - \epsilon, L + \epsilon)$.

81. $\lim\limits_{x \to \pi} \sin x = 0$, $\epsilon = \dfrac{\sqrt{2}}{2}$

82. $\lim\limits_{x \to \pi} \sin x = 0$, $\epsilon = \dfrac{1}{2}$

83. $\lim\limits_{x \to 0} \dfrac{1 - \cos x}{x} = 0$, $\epsilon = \dfrac{1}{2}$

84. $\lim\limits_{x \to 0} \dfrac{1 - \cos x}{x} = 0$, $\epsilon = \dfrac{1}{4}$

Use (a) the $h \to 0$ definition of the derivative and then (b) the $z \to c$ definition of the derivative to find $f'(c)$ for each function $f(x)$ and value $x = c$ in Exercises 85–88.

85. $f(x) = \sin x$, $x = 0$

86. $f(x) = \cos x$, $x = 0$

87. $f(x) = \tan x$, $x = 0$

88. $f(x) = \sec x$, $x = 0$

Use logarithmic differentiation to find the derivatives of each of the functions in Exercises 89–91.

89. $(\sin x)^x$

90. $(\sec x)^x$

91. $(\sin x)^{\cos x}$

Applications

92. The oscillating position of a mass hanging from the end of a spring, neglecting air resistance, is given by the following equation, where A, B, k, and m are constants:

$$s(t) = A\sin\left(\sqrt{\dfrac{k}{m}}\,t\right) + B\cos\left(\sqrt{\dfrac{k}{m}}\,t\right).$$

spring coefficient, k

$s(t)$

v_0 mass, m

s_0

(a) Show that the function $s(t)$ has the property that $s''(t) + \dfrac{k}{m}s(t) = 0$. This is the differential equation for the spring motion, an equation involving derivatives

that describes the motion of the bob on the end of the spring.

(b) Suppose the spring is released from an initial position of s_0 and with an initial velocity of v_0. Show that $A = v_0\sqrt{\dfrac{m}{k}}$ and $B = s_0$.

93. In Exercise D from Section 1.6 we learned that the oscillating position of a mass hanging from the end of a spring, taking air resistance into account, is given by the following equation, where A, B, k, f, and m are constants:

$$s(t) = e^{-f/2mt}\left(A\sin\left(\dfrac{\sqrt{4km - f^2}}{2m}t\right) + B\cos\left(\dfrac{\sqrt{4km - f^2}}{2m}t\right)\right).$$

(a) Show that the function $s(t)$ has the property that $s''(t) + \dfrac{f}{m}s'(t) + \dfrac{k}{m}s(t) = 0$ for some constant f. This is the differential equation for spring motion, taking air resistance into account. (*Hint: Find the first and second derivatives of $s(t)$ first, and then show that $s(t)$, $s'(t)$, and $s''(t)$ have the given relationship.*)

(b) Suppose the spring is released from an initial position of s_0 with an initial velocity of v_0. Show that $A = \dfrac{2mv_0 + fs_0}{\sqrt{4km - f^2}}$ and $B = s_0$.

Proofs

94. Use the Squeeze Theorem to prove that $\lim\limits_{x \to 0} \sin x = 0$ and $\lim\limits_{x \to 0} \cos x = 1$.

95. In the reading we used the Squeeze Theorem to prove that $\lim\limits_{h \to 0} \sin h = 0$ and $\lim\limits_{h \to 0} \cos h = 1$. Use these facts, the sum identity for cosine, and limit rules to prove that $f(x) = \cos x$ is continuous everywhere.

96. Use the quotient rule for limits and the continuity of $\sin x$ and $\cos x$ to prove that $f(x) = \tan x$ is continuous on its domain.

97. Use the quotient rule for limits and the continuity of $\cos x$ to prove that $f(x) = \sec x$ is continuous on its domain.

98. Use the unit circle to argue that the graph of $y = \sin x$ oscillates between -1 and 1 as $x \to \infty$ and therefore that $\lim\limits_{x \to \infty} \sin x$ is bounded but does not exist.

99. Prove that $\tan x$ has vertical asymptotes at $x = \dfrac{\pi}{2} + \pi k$, for any integer k.

100. Prove that $\csc x$ has vertical asymptotes at $x = \pi k$, for any integer k.

101. Use the definition of the derivative, a trigonometric identity, and known trigonometric limits to prove that $\frac{d}{dx}(\cos x) = -\sin x$.

102. Use the quotient rule and the derivative of the cosine function to prove that $\frac{d}{dx}(\sec x) = \sec x \tan x$.

103. Use the quotient rule and the derivative of the sine function to prove that $\frac{d}{dx}(\csc x) = -\csc x \cot x$.

104. Use the quotient rule and the derivatives of the sine and cosine functions to prove that $\frac{d}{dx}(\cot x) = -\csc^2 x$.

105. In Exercise 101 we used the definition of the derivative to find the derivative of the cosine function. Give an alternative proof that $\frac{d}{dx}(\cos x) = -\sin x$ which does *not* use the definition of the derivative, but instead uses the derivative of the sine function and the fact that $\cos x = \sin\left(x + \frac{\pi}{2}\right)$.

106. Prove that $\lim\limits_{\theta \to 0} \frac{1-\cos\theta}{\theta} = 0$ by using the double-angle identity $\cos 2\theta = 1 - 2\sin^2\theta$ and the other special trigonometric limit $\lim\limits_{\theta \to 0} \frac{\sin\theta}{\theta} = 1$.

Thinking Forward

Antidifferentiating trigonometric expressions: For each problem that follows, suppose $f(x)$ is a function with the given derivative $f'(x)$ and value $f(c)$. Find an equation for $f(x)$.

▸ $f'(x) = \sin x, \ f(0) = 0$

▸ $f'(x) = \sec^2 x, \ f\left(\frac{\pi}{2}\right) = 2$

▸ $f'(x) = \csc x \cot x, \ f\left(\frac{\pi}{2}\right) = 1$

▸ $f'(x) = 2\cos(3x), \ f(0) = 4$

▸ $f'(x) = 2x\cos(x^2), \ f(0) = 1$

▸ $f'(x) = \sec 2x \tan 2x, \ f(\pi) = 0$

6.4   INVERSE TRIGONOMETRIC FUNCTIONS

▸ Finding inverses for the sine, tangent, and secant on restricted domains

▸ Limits and derivatives of inverse trigonometric functions

▸ Writing certain inverse trigonometric compositions as algebraic functions

Defining the Inverse Trigonometric Functions

Recall from Section 0.6 that a function has an inverse only if the function is one–to–one, meaning that its input values are in a one–to–one correspondence with its output values. This correspondence allows us to "reverse" a function $y = f(x)$ to obtain a function $f^{-1}(y) = x$. For example, if $f(x) = x^3$, then $f^{-1}(y) = \sqrt[3]{x}$. Given a one–to–one function $f(x)$, its inverse is by definition the function $f^{-1}(x)$ with the property that $f^{-1}(f(x)) = x$ and $f(f^{-1}(y)) = y$ for all x in the domain of f and for all y in the domain of f^{-1}. For $f(x) = x^3$ we have

$$f^{-1}(f(x)) = \sqrt[3]{x^3} = x, \quad \text{for all } x \in \mathbb{R}$$
$$f(f^{-1}(y)) = (\sqrt[3]{y})^3 = y, \quad \text{for all } y \in \mathbb{R}.$$

As we have already seen, if a function fails to be one–to–one, then we can find a restricted domain on which it is one–to–one and construct an inverse on that restricted domain. For example, the function $f(x) = x^2 + 1$ fails the horizontal line test and therefore is not one–to–one on $\mathbb{R}$. But on the restricted domain $[0, \infty)$, the function $f(x) = x^2 + 1$ is one–to–one and its inverse is $f^{-1}(y) = \sqrt{y - 1}$:

$$f^{-1}(f(x)) = \sqrt{(x^2 + 1) - 1} = x, \quad \text{for all } x \in [0, \infty)$$
$$f(f^{-1}(y)) = \left(\sqrt{y - 1}\right)^2 + 1 = y, \quad \text{for all } y \in [1, \infty).$$

Now consider the function $\sin\theta = y$ that takes angles as inputs and returns y-coordinates on the unit circle as outputs. Is there a way to reverse this function to an inverse function $\sin^{-1} y = \theta$ that takes y-coordinates on the unit circle as inputs and returns angles as outputs? Not immediately, since $\sin\theta$ is not one–to–one; many angles correspond to any given y-coordinate on the unit circle.

In fact, none of the six trigonometric functions are one-to-one, but after restricting domains we can construct the so-called inverse trigonometric functions. In this section we will focus on the inverses of only three of the six inverse trigonometric functions: Those for sine, tangent, and secant. (The inverses of these functions will be particularly useful to us in Chapter 8 when we study integration techniques, and the inverses of the remaining three trigonometric functions would add no more to that discussion.) There are many different restricted domains that we could use to obtain partial inverses to these three functions. We need to pick one restricted domain for each function and stick with it. In this text we will use the restricted domains shown below. For the remainder of this section we will use x as the name of our independent variable for both functions and their inverses. You will have to know from the context whether x is an angle, a coordinate on the unit circle, or a ratio of such coordinates.

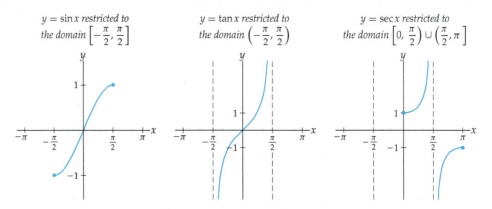

Each of these restricted functions is one-to-one and thus invertible. The inverses of these restricted functions, respectively, are the inverse sine, inverse tangent, and inverse secant functions.

DEFINITION 6.14

The Inverse Trigonometric Functions

(a) The ***inverse sine function*** $\sin^{-1}x$ is the inverse of the restriction of $\sin x$ to the interval $\left[-\frac{\pi}{2}, \frac{\pi}{2}\right]$.

(b) The ***inverse tangent function*** $\tan^{-1}x$ is the inverse of the restriction of $\tan x$ to the interval $\left(-\frac{\pi}{2}, \frac{\pi}{2}\right)$.

(c) The ***inverse secant function*** $\sec^{-1}x$ is the inverse of the restriction of $\sec x$ to the interval $\left[0, \frac{\pi}{2}\right) \cup \left(\frac{\pi}{2}, \pi\right]$.

Notice that since the inputs to the trigonometric functions are angles, it is the *outputs* of the inverse trigonometric functions that are angles. For example, $\sin^{-1}x$ is the unique angle in $\left[-\frac{\pi}{2}, \frac{\pi}{2}\right]$ whose sine is x. We will interchangeably use the alternative notations $\arcsin x$, $\arctan x$, and $\operatorname{arcsec} x$ for these inverse trigonometric functions.

⊗ CAUTION   Although we use the notation $\sin^2 x$ to represent $(\sin x)^2$ and the notation x^{-1} to represent $\frac{1}{x}$, the notation $\sin^{-1}x$ does *not* represent $\frac{1}{\sin x}$. Inverse functions in general have nothing to do with reciprocals, despite what one might imagine from the notation.

All of the properties of $\sin^{-1} x$, $\tan^{-1} x$, and $\sec^{-1} x$ come from the fact that they are the inverses of the restricted functions $\sin x$, $\tan x$, and $\sec x$. For example, we can graph the inverse trigonometric functions simply by reflecting the graphs of the restricted trigonometric functions over the line $y = x$, as follows:

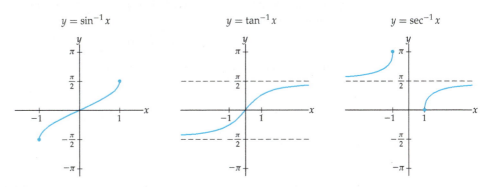

From these graphs we can immediately identify the domains and ranges of the three inverse trigonometric functions:

$$\sin^{-1} x \text{ has domain } [-1, 1] \text{ and range } \left[-\frac{\pi}{2}, \frac{\pi}{2}\right];$$

$$\tan^{-1} x \text{ has domain } \mathbb{R} \text{ and range } \left(-\frac{\pi}{2}, \frac{\pi}{2}\right);$$

$$\sec^{-1} x \text{ has domain } (-\infty, -1] \cup [1, \infty) \text{ and range } \left[0, \frac{\pi}{2}\right) \cup \left(\frac{\pi}{2}, \pi\right].$$

These domains and ranges should make sense to you when you think about angles, coordinates, and ratios on the unit circle; see Exercises 13 and 14.

Limits of Inverse Trigonometric Functions

Since the inverse trigonometric functions are nothing more than the inverses of restricted trigonometric functions, they are continuous where they are defined:

THEOREM 6.15   **Continuity of Inverse Trigonometric Functions**

The three inverse trigonometric functions $\sin^{-1} x$, $\tan^{-1} x$, and $\sec^{-1} x$ are each continuous on their domains.

This continuity property means that we can calculate limits of inverse trigonometric functions by evaluation as long as we are approaching domain values. For example,

$$\lim_{x \to 1} \tan^{-1} x = \tan^{-1} 1 = \frac{\pi}{4} \quad \text{and} \quad \lim_{x \to 1^-} \sin^{-1} x = \sin^{-1} 1 = \frac{\pi}{2}.$$

Notice that the final values in these calculations follow from the fact that $\frac{\pi}{4}$ is the unique angle in $\left(-\frac{\pi}{2}, \frac{\pi}{2}\right)$ whose tangent is 1 and $\frac{\pi}{2}$ is the unique angle in $\left[-\frac{\pi}{2}, \frac{\pi}{2}\right]$ whose sine is 1. You can also verify these answers with the graphs of the inverse sine and inverse tangent functions.

Proof. We will prove that $\sin^{-1} x$ is continuous on its domain and leave to Exercise 93 the proof that the inverse tangent function is domain-continuous. The proof for the inverse secant function is similar. To show that $\sin^{-1} x$ is continuous on its domain $[-1, 1]$ we use the fact that it is the inverse of $\sin x$ restricted to $\left[-\frac{\pi}{2}, \frac{\pi}{2}\right]$, with the same method we used for $\ln x$. Before we do this, a technical point: this proof will show that if $\lim\limits_{x \to c} \sin^{-1} x$ is equal to some real number, then that real number must be $\sin^{-1} c$. We will assume that $\lim\limits_{x \to c} \sin^{-1} x$ exists, a fact that is necessary for the application of limit rules. By the composition rule for limits and the fact that $\sin x$ is continuous everywhere, for $c \in [-1, 1]$ we have

$$\lim_{x \to c} \sin(\sin^{-1} x) = \sin(\lim_{x \to c} \sin^{-1} x).$$

On the other hand, by properties of inverses, for $c \in [-1, 1]$ we also have

$$\lim_{x \to c} \sin(\sin^{-1} x) = \lim_{x \to c} x = c = \sin(\sin^{-1} c).$$

Because $\sin x$ is a one-to-one function on $\left[-\frac{\pi}{2}, \frac{\pi}{2}\right]$, we can put the preceding two calculations together to conclude that $\lim\limits_{x \to c} \sin^{-1} x = \sin^{-1} c$. ∎

What about limits at infinity? From the preceding graphs of the inverse tangent and inverse secant functions, we can see that we have the following:

THEOREM 6.16   **Limits of Inverse Tangent and Inverse Secant at Infinity**

(a) $\lim\limits_{x \to \infty} \tan^{-1} x = \frac{\pi}{2}$ and $\lim\limits_{x \to -\infty} \tan^{-1} x = -\frac{\pi}{2}$.

(b) $\lim\limits_{x \to \infty} \sec^{-1} x = \frac{\pi}{2}$ and $\lim\limits_{x \to -\infty} \sec^{-1} x = \frac{\pi}{2}$.

We don't consider the limit of $\sin^{-1} x$ at infinity, because that function is not defined beyond $[-1, 1]$. Note that each of these limits can be read directly from the graphs of the inverse trigonometric functions.

Derivatives of the Inverse Trigonometric Functions

We can use the formulas for the derivatives of the trigonometric functions to prove the following formulas for the derivatives of the inverse trigonometric functions:

THEOREM 6.17   **Derivatives of Inverse Trigonometric Functions**

For all values of x at which the following functions are defined,

(a) $\dfrac{d}{dx}(\sin^{-1} x) = \dfrac{1}{\sqrt{1 - x^2}}$ **(b)** $\dfrac{d}{dx}(\tan^{-1} x) = \dfrac{1}{1 + x^2}$ **(c)** $\dfrac{d}{dx}(\sec^{-1} x) = \dfrac{1}{|x|\sqrt{x^2 - 1}}$

It is extremely important and surprising to note that although inverse trigonometric functions are transcendental, their derivatives are algebraic! This property makes these derivative formulas particularly useful for finding certain antiderivatives, and in Chapter 8 they will be part of our arsenal of integration techniques. Of course, all of these rules can be used in combination with the sum, product, quotient, and chain rules. For example,

$$\frac{d}{dx}(\sin^{-1}(3x^2 - 1)) = \frac{1}{\sqrt{1 - (3x^2 - 1)^2}}(6x).$$

Proof. We will prove the rule for $\sin^{-1} x$ and leave the remaining two rules to Exercises 95 and 94. Since $\sin(\sin^{-1} x) = x$ for all x in the domain of $\sin^{-1} x$, we have

$$\sin(\sin^{-1} x) = x \qquad\qquad \leftarrow \sin^{-1} x \text{ is the inverse of } \sin x$$

$$\frac{d}{dx}(\sin(\sin^{-1} x)) = \frac{d}{dx}(x) \qquad\qquad \leftarrow \text{differentiate both sides}$$

$$\cos(\sin^{-1} x) \cdot \frac{d}{dx}(\sin^{-1} x) = 1 \qquad\qquad \leftarrow \text{chain rule}$$

$$\frac{d}{dx}(\sin^{-1} x) = \frac{1}{\cos(\sin^{-1} x)} \qquad\qquad \leftarrow \text{algebra}$$

$$\frac{d}{dx}(\sin^{-1} x) = \frac{1}{\sqrt{1 - \sin^2(\sin^{-1} x)}} \qquad\qquad \leftarrow \text{since } \sin^2 x + \cos^2 x = 1$$

$$\frac{d}{dx}(\sin^{-1} x) = \frac{1}{\sqrt{1 - x^2}}. \qquad\qquad \leftarrow \sin x \text{ is the inverse of } \sin^{-1} x$$

We could also have used triangles and the unit circle to show that the composition $\cos(\sin^{-1} x)$ is equal to the algebraic expression $\sqrt{1 - x^2}$, as in Example 6.   ∎

An interesting fact about the derivatives of the inverse sine and inverse secant functions is that their domains are slightly smaller than the domains of the original functions. The graphs of the inverse trigonometric functions are as follows (note their domains):

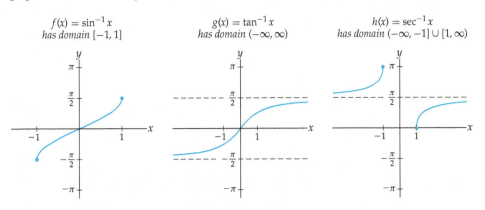

$f(x) = \sin^{-1} x$
has domain $[-1, 1]$

$g(x) = \tan^{-1} x$
has domain $(-\infty, \infty)$

$h(x) = \sec^{-1} x$
has domain $(-\infty, -1] \cup [1, \infty)$

If you look closely at the first and third graphs, you should notice that at the ends of the domains the tangent lines will be vertical. Since a vertical line has undefined slope, the derivative does not exist at these points. This means that the derivatives of $\sin^{-1} x$ and $\sec^{-1} x$ are not defined at $x = 1$ or $x = -1$; see the first and third graphs shown next:

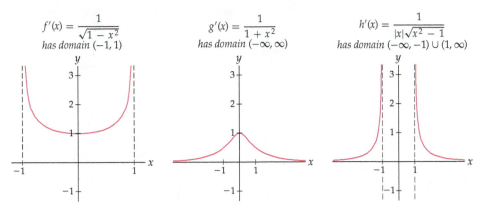

$f'(x) = \dfrac{1}{\sqrt{1 - x^2}}$
has domain $(-1, 1)$

$g'(x) = \dfrac{1}{1 + x^2}$
has domain $(-\infty, \infty)$

$h'(x) = \dfrac{1}{|x|\sqrt{x^2 - 1}}$
has domain $(-\infty, -1) \cup (1, \infty)$

Examples and Explorations

EXAMPLE 1   **Finding values of inverse trigonometric functions by hand**

Calculate each of the following by hand, without a calculator:

(a) $\sin^{-1}\frac{1}{2}$ **(b)** $\tan^{-1}(-\sqrt{3})$ **(c)** $\sec^{-1}\left(-\frac{2}{\sqrt{3}}\right)$

SOLUTION

(a) If $\theta = \sin^{-1}\frac{1}{2}$, then we must have $\sin\theta = \frac{1}{2}$. There are infinitely many angles whose sine is $\frac{1}{2}$, but only one of those angles is in the restricted domain $\left[-\frac{\pi}{2}, \frac{\pi}{2}\right]$. Thus $\theta = \sin^{-1}\left(\frac{1}{2}\right)$ is the unique angle in $\left[-\frac{\pi}{2}, \frac{\pi}{2}\right]$ whose sine is $\frac{1}{2}$, as shown in the following figure. Notice that the triangle must be a 30–60–90 triangle (since its height is $\frac{1}{2}$), and therefore the angle θ we are looking for must be 30° (i.e., $\frac{\pi}{6}$ radians). Therefore $\sin^{-1}\frac{1}{2} = \frac{\pi}{6}$.

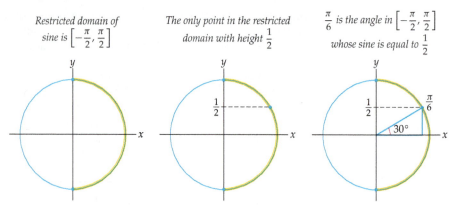

Restricted domain of
sine is $\left[-\frac{\pi}{2}, \frac{\pi}{2}\right]$

The only point in the restricted
domain with height $\frac{1}{2}$

$\frac{\pi}{6}$ is the angle in $\left[-\frac{\pi}{2}, \frac{\pi}{2}\right]$
whose sine is equal to $\frac{1}{2}$

(b) Since $\tan^{-1}x$ is the inverse of the tangent function, it follows that $\theta = \tan^{-1}(-\sqrt{3})$ if and only if $\tan(\theta) = -\sqrt{3}$. More specifically, $\tan^{-1}(-\sqrt{3})$ is the unique angle θ in $\left(-\frac{\pi}{2}, \frac{\pi}{2}\right)$ whose tangent is $-\sqrt{3}$. In terms of sine and cosine this means that

$$\tan\theta = -\sqrt{3} \implies \frac{\sin\theta}{\cos\theta} = -\sqrt{3} \implies \sin\theta = -\sqrt{3}\cos\theta.$$

We now have to think of an angle θ in the first or fourth quadrant of the unit circle whose sine is $-\sqrt{3}$ times its cosine. Since $\tan^{-1}x$ is a function, there must be only one such angle; we will try to "guess" it. Since $\tan\theta$ is negative, the angle θ must terminate in the fourth quadrant. Also, the reference triangle must have a vertical side length that is $\sqrt{3}$ times its horizontal side length; a 30–60–90 triangle has this property:

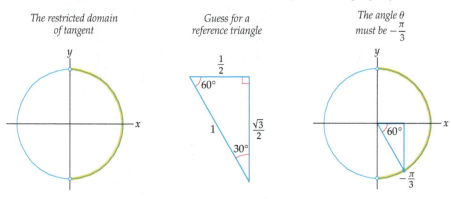

The restricted domain
of tangent

Guess for a
reference triangle

The angle θ
must be $-\frac{\pi}{3}$

The angle in $\left(-\frac{\pi}{2}, \frac{\pi}{2}\right)$ that has this reference triangle is the angle $-\frac{\pi}{3}$. We have now shown that $\tan^{-1}(-\sqrt{3}) = -\frac{\pi}{3}$.

(c) First notice that $x = -\frac{2}{\sqrt{3}}$ is less than -1, since 2 is greater than $\sqrt{3}$. Thus our x is in the domain $(-\infty, -1] \cup [1, \infty)$ of $\sec^{-1}(x)$. We are looking for an angle $\theta = \sec^{-1}\left(-\frac{2}{\sqrt{3}}\right)$; that is, an angle θ such that $\sec\theta = -\frac{2}{\sqrt{3}}$. Since the secant function is the reciprocal of the cosine function, this means that

$$\sec\theta = -\frac{2}{\sqrt{3}} \implies \frac{1}{\cos\theta} = -\frac{2}{\sqrt{3}} \implies \cos\theta = -\frac{\sqrt{3}}{2}.$$

There are infinitely many angles θ with a cosine of $-\frac{\sqrt{3}}{2}$, but we want the *one* angle with this property that is in the range $\left[0, \frac{\pi}{2}\right) \cup \left(\frac{\pi}{2}, \pi\right]$ of the inverse secant function. This means that θ must be in the first or second quadrant. Because $\cos\theta$ is negative, θ must be in the second quadrant. Since $\frac{\sqrt{3}}{2}$ is the length of the long leg of a 30–60–90 triangle, the reference triangle for θ is a 30–60–90 triangle, as shown here:

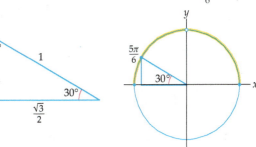

The restricted domain of secant *Guess for a reference triangle* *The angle θ must be $\frac{5\pi}{6}$*

Since $\cos\frac{5\pi}{6} = -\frac{\sqrt{3}}{2}$, we have $\sec\frac{5\pi}{6} = \frac{1}{-\sqrt{3}/2} = -\frac{2}{\sqrt{3}}$, and thus $\sec^{-1}\left(-\frac{2}{\sqrt{3}}\right) = \frac{5\pi}{6}$.

✔ CHECKING
THE ANSWER We can check the preceding calculations with calculator graphs. For example, in part (a) we can use a graph to verify that the point $(x, y) = \left(\frac{1}{2}, \frac{\pi}{6}\right)$ is on the graph of $y = \sin^{-1} x$:

$$\sin^{-1}\frac{1}{2} = \frac{\pi}{6}$$

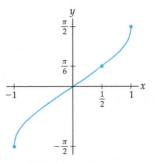

EXAMPLE 2

Compositions of sine and inverse sine

Calculate the exact values of

(a) $\sin^{-1}\left(\sin\left(-\frac{\pi}{4}\right)\right)$ **(b)** $\sin^{-1}\left(\sin\left(\frac{3\pi}{4}\right)\right)$

SOLUTION

(a) Since $-\frac{\pi}{4}$ is in the domain $\left[-\frac{\pi}{2}, \frac{\pi}{2}\right]$ of the restricted sine function, we can use the fact that $\sin^{-1}(\sin x) = x$ for all x in the restricted domain of the sine. Therefore $\sin^{-1}\left(\sin\left(-\frac{\pi}{4}\right)\right) = -\frac{\pi}{4}$.

(b) You may be tempted to say that $\sin^{-1}\left(\sin\left(\frac{3\pi}{4}\right)\right)$ is equal to $\frac{3\pi}{4}$. However, $\frac{3\pi}{4}$ is not in the domain of the restricted sine function. We must calculate $\sin\left(\frac{3\pi}{4}\right)$ and then evaluate the inverse sine function at that value. We know that $\sin\left(\frac{3\pi}{4}\right) = \frac{\sqrt{2}}{2}$, as shown next at the left. To find the value of $\sin^{-1}\left(\frac{\sqrt{2}}{2}\right)$, we must find the angle in $\left[-\frac{\pi}{2}, \frac{\pi}{2}\right]$ whose sine is $\frac{\sqrt{2}}{2}$. This angle is $\frac{\pi}{4}$, as shown at the right.

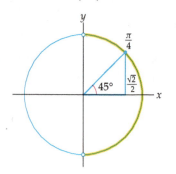

Therefore $\sin^{-1}\left(\sin\left(\frac{3\pi}{4}\right)\right) = \sin^{-1}\left(\frac{\sqrt{2}}{2}\right) = \frac{\pi}{4}$. □

EXAMPLE 3   **Calculating limits that involve inverse trigonometric functions**

Calculate each of the following limits:

(a) $\lim\limits_{x \to 1^-} \dfrac{\sin^{-1} x}{\tan^{-1} x}$ **(b)** $\lim\limits_{x \to \infty} \dfrac{\ln x}{\tan^{-1} x}$ **(c)** $\lim\limits_{x \to 0} \dfrac{\sin^{-1} x}{x^2}$

SOLUTION

(a) Both $\sin^{-1} x$ and $\tan^{-1} x$ are defined to the left of and at the point $x = 1$, so we can solve this limit by simple evaluation. We have $\sin^{-1} 1 = \frac{\pi}{2}$, since $\frac{\pi}{2}$ is the unique angle in $\left[-\frac{\pi}{2}, \frac{\pi}{2}\right]$ whose sine is equal to 1. We also have $\tan^{-1} 1 = \frac{\pi}{4}$, since $\frac{\pi}{4}$ is the unique angle in $\left(-\frac{\pi}{2}, \frac{\pi}{2}\right)$ whose tangent is equal to 1. Therefore

$$\lim_{x \to 1^-} \frac{\sin^{-1} x}{\tan^{-1} x} = \frac{\pi/2}{\pi/4} = 2.$$

(b) As $x \to \infty$, $\ln x \to \infty$ and $\tan^{-1} x \to \frac{\pi}{2}$. Therefore the limit in question has the form $\frac{\infty}{\pi/2}$, which is infinite. Thus

$$\lim_{x \to \infty} \frac{\ln x}{\tan^{-1} x} = \infty.$$

(c) As $x \to 0$, the numerator $\sin^{-1} x$ and the denominator x^2 both approach zero. Therefore this limit is initially in the indeterminate form $\frac{0}{0}$, and L'Hôpital's Rule applies. This gives us

$$\lim_{x \to 0} \frac{\sin^{-1} x}{x^2} \overset{\text{L'H}}{=} \lim_{x \to 0} \frac{1/\sqrt{1-x^2}}{2x} = \lim_{x \to 0} \frac{1}{2x\sqrt{1-x^2}}.$$

Since $2x\sqrt{1-x^2}$ approaches 0 as $x \to 0$, the preceding limit is of the form $\frac{1}{0}$. From the left, as $x \to 0^-$, the denominator approaches 0^-, and thus the limit from the left is of the form $\frac{1}{0^-}$. Similarly, from the right, as $x \to 0^+$, the limit is of the form $\frac{1}{0^+}$. Therefore

$$\lim_{x \to 0^-} \frac{\sin^{-1} x}{x^2} = -\infty \quad \text{and} \quad \lim_{x \to 0^+} \frac{\sin^{-1} x}{x^2} = \infty.$$

Thus the left and right limits as $x \to 0$ are not equal to each other, and the limit as $x \to 0$ does not exist.   □

EXAMPLE 4   Calculating derivatives that involve inverse trigonometric functions

Calculate each of the following derivatives:

(a) $f(x) = (\sec^{-1} x)^2$ **(b)** $f(x) = x \sin^{-1}(3x+1)$ **(c)** $f(x) = \sin^2(\tan^{-1} e^{3x})$

SOLUTION

(a) The function $f(x) = (\sec^{-1} x)^2$ is a composition with an inverse trigonometric function on the inside. By the chain rule,

$$\frac{d}{dx}((\sec^{-1} x)^2) = 2(\sec^{-1} x)^1 \left(\frac{1}{|x|\sqrt{x^2-1}} \right) = \frac{2 \sec^{-1} x}{|x|\sqrt{x^2-1}}.$$

(b) The function $f(x) = x \sin^{-1}(3x+1)$ is a product of two functions, and thus we begin with the product rule. We will also need the chain rule to differentiate the composition $\sin^{-1}(3x+1)$:

$$f'(x) = (1) \cdot \sin^{-1}(3x+1) + x \cdot \frac{1}{\sqrt{1-(3x+1)^2}} \quad (3)$$

$$= \sin^{-1}(3x+1) + \frac{3x}{\sqrt{1-(3x+1)^2}}.$$

(c) We'll begin by rewriting our shortcut notation $\sin^2 u$ in longhand as $(\sin u)^2$ so we can see the correct order for the differentiation rules. The function $f(x)$ is actually a composition of *five* functions: x^2, $\sin x$, $\tan^{-1} x$, e^x, and $3x$. We will have to apply the chain rule quite a few times and thus will proceed carefully, writing several intermediate steps:

$$f'(x) = \frac{d}{dx}(\sin^2(\tan^{-1} e^{3x}))$$

$$= \frac{d}{dx}((\sin(\tan^{-1} e^{3x}))^2) \qquad\qquad \leftarrow \text{rewrite function}$$

$$= 2(\sin(\tan^{-1} e^{3x})) \frac{d}{dx}(\sin(\tan^{-1} e^{3x})) \qquad\qquad \leftarrow \text{chain, power rules}$$

$$= 2(\sin(\tan^{-1} e^{3x})) \cos(\tan^{-1} e^{3x}) \frac{d}{dx}(\tan^{-1} e^{3x}) \qquad \leftarrow \text{chain rule again}$$

$$= 2(\sin(\tan^{-1} e^{3x})) \cos(\tan^{-1} e^{3x}) \left(\frac{1}{1+(e^{3x})^2} \right) \frac{d}{dx}(e^{3x}) \leftarrow \text{chain rule again}$$

$$= 2(\sin(\tan^{-1} e^{3x})) \cos(\tan^{-1} e^{3x}) \left(\frac{1}{1+(e^{3x})^2} \right) (3e^{3x}). \leftarrow \text{chain rule again} \quad □$$

EXAMPLE 5   Finding antiderivatives that involve inverse trigonometric functions

Find a function f whose derivative is $f'(x) = \dfrac{1}{1+4x^2}$.

SOLUTION

Since the derivative of $\tan^{-1} x$ is $\frac{1}{1+x^2}$, we might suspect that the function we are looking for is related to the inverse tangent function. We will use an intelligent guess-and-check method to find f. Clearly $f(x) = \tan^{-1} x$ isn't exactly right, since its derivative is missing the "4." A good guess might be $f(x) = \tan^{-1}(4x)$; let's try that:

$$\frac{d}{dx}(\tan^{-1}(4x)) = \frac{1}{1+(4x)^2}\,(4) = \frac{4}{1+16x^2}.$$

Obviously that wasn't quite right either; but by examining the results we can make a new guess. We might try $\tan^{-1}(2x)$, since the "$2x$" will be squared in the derivative and become the "$4x^2$" we are looking for in the denominator:

$$\frac{d}{dx}(\tan^{-1}(2x)) = \frac{1}{1+(2x)^2}\,(2) = \frac{2}{1+4x^2}.$$

Now we are getting somewhere; this result differs by a multiplicative constant from the derivative $f'(x)$ we are looking for, and that is easy to fix. We need only divide our guess by that constant. Try the function $f(x) = \frac{1}{2}\tan^{-1}(2x)$:

$$\frac{d}{dx}\left(\frac{1}{2}\tan^{-1}(2x)\right) = \left(\frac{1}{2}\right)\frac{1}{1+(2x)^2}\,(2) = \frac{1}{1+4x^2}.$$

We now know that $f(x) = \frac{1}{2}\tan^{-1}(2x)$ is a function whose derivative is $f'(x) = \frac{1}{1+4x^2}$. Of course, we could also add any constant to $f(x)$ and not change its derivative; for example, $f(x) = \frac{1}{2}\tan^{-1}(2x) + 5$ would work as well. In fact, any function of the form $f(x) = \frac{1}{2}\tan^{-1}(2x) + C$ will have $f'(x) = \frac{1}{1+4x^2}$. □

EXAMPLE 6   **Simplifying compositions of inverse trigonometric and trigonometric functions**

Write $\cos(\sin^{-1} x)$ as an algebraic function, that is, a function that involves only arithmetic operations and rational powers.

SOLUTION

If we define $\theta = \sin^{-1} x$, then $\sin\theta = x$ and θ must be in the interval $\left[-\frac{\pi}{2}, \frac{\pi}{2}\right]$. Let's first consider the case where θ is in the first quadrant $\left[0, \frac{\pi}{2}\right]$; the reference triangle for such a θ is shown next at the left. If we wish θ to have a sine of x, then the length of the vertical leg of the triangle must be x. The hypotenuse of the triangle is length 1, since we are on the *unit* circle. We could also have considered that the sine of θ is "opposite over hypotenuse"; thus one triangle involving our angle θ could have an opposite side of length x and a hypotenuse of length 1. Using the Pythagorean theorem, we find that the length of the remaining leg of the triangle is $\sqrt{1-x^2}$, as shown at the right:

Reference triangle for
an angle θ in $\left[0, \dfrac{\pi}{2}\right]$

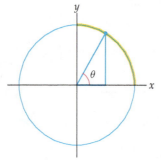

Use Pythagorean theorem to
determine length of remaining leg

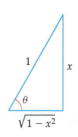

Now $\cos\theta$ is the horizontal coordinate of the point on the unit circle corresponding to θ, or, in terms of "adjacent over hypotenuse," we have

$$\cos\theta = \frac{\sqrt{1-x^2}}{1} = \sqrt{1-x^2}.$$

The case where θ is in the fourth quadrant, that is, where $\theta \in \left[-\frac{\pi}{2}, 0\right]$, is similar and also shows that $\cos\theta = \sqrt{1-x^2}$. Therefore we have shown that $\cos(\sin^{-1}x)$ is equal to the algebraic function $\sqrt{1-x^2}$.   □

 CHECKING THE ANSWER

To verify the strange fact that $\cos(\sin^{-1}x) = \sqrt{1-x^2}$, try evaluating both sides at some simple x-values. While looking at just a few x-values will not prove that the two expressions are equal for all x, it will at least give us some evidence that the equality is reasonable. For example, at $x = 0$ we have

$$\cos(\sin^{-1}0) = \cos 0 = 1 \quad \text{and} \quad \sqrt{1-0^2} = \sqrt{1} = 1,$$

and at $x = 1$ we have

$$\cos(\sin^{-1}1) = \cos\left(\frac{\pi}{2}\right) = 0 \quad \text{and} \quad \sqrt{1-1^2} = \sqrt{0} = 0.$$

As a less trivial example, consider $x = \frac{1}{2}$. At this value we have

$$\cos\left(\sin^{-1}\left(\frac{1}{2}\right)\right) = \cos\left(\frac{\pi}{6}\right) = \frac{\sqrt{3}}{2} \quad \text{and} \quad \sqrt{1-\left(\frac{1}{2}\right)^2} = \sqrt{1-\frac{1}{4}} = \sqrt{\frac{3}{4}} = \frac{\sqrt{3}}{2}.$$

? TEST YOUR UNDERSTANDING

▶ Why is $f(x) = \sin x$ not an invertible function? Why does restricting the domain of the sine to $\left[-\frac{\pi}{2}, \frac{\pi}{2}\right]$ solve this problem?

▶ What are the domains and ranges of $\sin^{-1}x$, $\tan^{-1}x$, and $\sec^{-1}x$?

▶ What are the limits of $\sin^{-1}x$, $\tan^{-1}x$, and $\sec^{-1}x$ as x approaches infinity?

▶ What is the graphical reason that the domains of the derivatives of $\sin^{-1}x$ and $\sec^{-1}x$ are slightly smaller than the domains of the functions themselves?

▶ If $u(x)$ is a function of x, what are the rules for differentiating the composite functions $\sin^{-1}u(x)$, $\tan^{-1}u(x)$, and $\sec^{-1}u(x)$?

EXERCISES 6.4

Thinking Back

Famous triangles, degrees, and radians: The following exercises will help you review and recall basic trigonometry.

▶ Suppose a right triangle has angles 30°, 60°, and 90° and a hypotenuse of length 1. What are the lengths of the remaining legs of the triangle?

▶ Suppose a right triangle has angles 45°, 45°, and 90° and a hypotenuse of length 1. What are the lengths of the remaining legs of the triangle?

▶ What is a radian? Is it larger or smaller than a degree? Compare an angle of 1 degree with an angle of 1 radian, with both angles in standard position.

▶ Show each of the following angles in standard position on the unit circle, in radians:

▶ $\dfrac{3\pi}{4}$ ▶ $-\dfrac{4\pi}{3}$ ▶ $\dfrac{17\pi}{6}$ ▶ 21π

Values of trigonometric functions: Find, without a calculator, each of the given function values. In some instances the answer may be undefined.

▶ If $f(x) = \csc x$, find $f(\pi)$ and $f\left(\dfrac{\pi}{2}\right)$.

▶ If $f(x) = \tan^2 x$, find $f(\pi)$ and $f\left(\dfrac{\pi}{2}\right)$.

Concepts

0. *Problem Zero:* Read the section and make your own summary of the material.

1. *True/False:* Determine whether each of the statements that follow is true or false. If a statement is true, explain why. If a statement is false, provide a counterexample.

(a) *True or False:* $\tan^{-1} x = \arctan x$ for all values of x.

(b) *True or False:* $\sec^{-1}(\sec x) = x$ for all values of x.

(c) *True or False:* $\sec^{-1} x = \dfrac{1}{\cos^{-1} x}$.

(d) *True or False:* $\tan^{-1} x = \dfrac{1}{\tan x}$.

(e) *True or False:* If $\sin \theta = \dfrac{1}{3}$, then $\arcsin\left(\dfrac{1}{3}\right) = \theta$.

(f) *True or False:* If $\arcsin\left(\dfrac{1}{3}\right) = \theta$, then $\sin \theta = \dfrac{1}{3}$.

(g) *True or False:* If $\arcsin x = \theta$, then $\sin \theta$ is greater than or equal to zero.

(h) *True or False:* If $f(x)$ is an inverse trigonometric function, then $f'(x)$ is also an inverse trigonometric function.

2. *Examples:* Construct examples of the thing(s) described in the following. Try to find examples that are different than any in the reading.

(a) Three restricted domains on which the function $f(x) = \sin x$ is invertible.

(b) A function whose domain is larger than the domain of its derivative.

(c) Four functions that are transcendental, but whose derivatives are algebraic.

3. Describe restricted domains for $\sin x$, $\tan x$, and $\sec x$ on which each function is invertible. Then describe the corresponding domains and ranges for $\arcsin x$, $\arctan x$, and $\operatorname{arcsec} x$.

4. Fill in the blanks:

(a) $\sin^{-1} x$ is the angle in the interval _____ whose _____ is x.

(b) $y = \arcsin x$ if and only if $\sin y = $ _____, for all $x \in$ _____ and $y \in$ _____ .

(c) If $\tan^{-1} x = \theta$ and $\tan \theta$ is positive, then θ is in the _____ quadrant.

(d) If $\arctan x = \theta$ and $\sin \theta = \dfrac{1}{3}$, then $\cos \theta = $ _____ .

5. Which of the following expressions are defined? Why or why not?

(a) $\sin^{-1}\left(-\dfrac{1}{25}\right)$

(b) $\sin^{-1}\dfrac{3}{2}$

(c) $\tan^{-1} 100$

(d) $\sec^{-1}\dfrac{\pi}{4}$

6. Sketch a graph of the restricted cosine function on the domain $[0, \pi]$, and argue that this restricted function is one-to-one. Then sketch a graph of $\cos^{-1} x$, and list the domains and ranges of the inverse $\cos^{-1} x$ of this restricted function.

7. Without calculating the exact or approximate values of the following quantities, use the unit circle to determine whether each of those quantities is positive or negative:

(a) $\sin^{-1}\left(-\dfrac{1}{5}\right)$

(b) $\sin^{-1}\left(-\dfrac{2}{3}\right)$

(c) $\tan^{-1} 2$

(d) $\sec^{-1}(-5)$

8. Find all angles whose secant is 2, and then find $\sec^{-1}(2)$.

9. Suppose $\arcsin x = y$. Which variable (x or y) represents an angle, and why? What does the other variable represent (in terms of the unit circle)?

10. Suppose $\tan^{-1} x = y$. Which variable (x or y) represents an angle, and why? What does the other variable represent (in terms of the unit circle)?

11. Explain why $\sin^{-1}\left(\dfrac{1}{2}\right)$ is *not* equal to $\dfrac{5\pi}{6}$, even though $\sin\left(\dfrac{5\pi}{6}\right) = \dfrac{1}{2}$.

12. List nine values of x for which you would be able to calculate the exact value of $\sin^{-1} x$ by hand. (*Hint: Think about the two right triangles whose side lengths we know, and think about the unit circle.*) Could you also calculate $\cos^{-1} x$, $\tan^{-1} x$, and $\sec^{-1} x$ by hand for these values?

13. Illustrate the restricted domain of the sine function on the unit circle, and explain in terms of the unit circle why it makes sense that the domain of the inverse sine function is $[-1, 1]$.

14. Illustrate the restricted domain of the tangent function on the unit circle, and explain in terms of the unit circle why it makes sense that the domain of the inverse tangent function is $\mathbb{R}$.

15. Illustrate the restricted domain of the secant function on the unit circle, and explain in terms of the unit circle why it makes sense that the domain of the inverse secant function is $\left[0, \dfrac{\pi}{2}\right) \cup \left(\dfrac{\pi}{2}, \pi\right]$.

16. For $f(x) = \arctan x$ and $c = 0$, use a sequence of approximations to approximate $f'(c)$. Illustrate your work with an appropriate sequence of graphs of secant lines.

17. In the proof that $\dfrac{d}{dx}(\sin^{-1} x) = \dfrac{1}{\sqrt{1-x^2}}$, we used the fact that $\sin(\sin^{-1} x) = x$. It is also true that $\sin^{-1}(\sin x) = x$; could we have started with that inequality instead? Why or why not?

18. Both of the following equations are true: $\tan(\tan^{-1} x) = x$ and $\tan^{-1}(\tan x) = x$. We can find the derivative of $\tan^{-1} x$ by differentiating both sides of one of these equations and solving for $\dfrac{d}{dx}(\tan^{-1} x)$. Which one of the equations should we use, and why?

19. How can the derivative of $\sin^{-1} x$ be equal to *both* $\dfrac{1}{\sqrt{1-x^2}}$ and $\dfrac{1}{\cos(\sin^{-1} x)}$? Which expression is easier to use, and why?

20. The function $\sin^{-1} x$ is defined on $[-1, 1]$, but its derivative $\dfrac{1}{\sqrt{1-x^2}}$ is defined only on $(-1, 1)$. Explain why the tangent lines to the graph of $y = \sin^{-1} x$ do not exist at $x = \pm 1$. (*Hint: Think about the corresponding tangent lines on the graph of the restricted sine function.*)

21. The function $\sin^{-1} x$ is defined on $[-1, 1]$, but its derivative $\dfrac{1}{\sqrt{1-x^2}}$ is defined only on $(-1, 1)$. Explain why the tangent lines to the graph of $y = \sin^{-1} x$ at $x = 1$ and $x = -1$ do not exist.

22. How can the derivative of $\sin^{-1} x$ be equal to both $\dfrac{1}{\sqrt{1-x^2}}$ and $\dfrac{1}{\cos(\sin^{-1} x)}$? Why might we prefer the algebraic form?

Skills

Find the exact values of each of the quantities in Exercises 23–38. Do not use a calculator.

23. $\sin^{-1}(0)$

24. $\sin^{-1}\left(\dfrac{1}{2}\right)$

25. $\arcsin(1)$

26. $\sin^{-1}\left(-\dfrac{\sqrt{2}}{2}\right)$

27. $\arcsin\left(-\dfrac{\sqrt{3}}{2}\right)$

28. $\sec^{-1}\left(\dfrac{2}{\sqrt{3}}\right)$

29. $\cos^{-1}(-1)$

30. $\sin^{-1}(-1)$

31. $\operatorname{arcsec}\left(-\dfrac{2}{\sqrt{2}}\right)$

32. $\arctan\left(-\dfrac{1}{\sqrt{3}}\right)$

33. $\operatorname{arcsec}(-2)$

34. $\sec^{-1}(-1)$

35. $\tan^{-1}(\sqrt{3})$

36. $\tan^{-1}(-\sqrt{3})$

37. $\sec^{-1}\left(\sec\left(-\dfrac{\pi}{5}\right)\right)$

38. $\sin^{-1}\left(\sin\dfrac{5\pi}{4}\right)$

Use a calculator to find approximate numerical values for each of the expressions in Exercises 39–44, if they exist. If an expression does not exist, use the unit circle to explain why.

39. $\sin^{-1}\left(\dfrac{1}{5}\right)$

40. $\operatorname{arcsec}(0)$

41. $\sin^{-1}\left(\dfrac{3}{2}\right)$

42. $\tan^{-1}(3)$

43. $\sec^{-1}(-5)$

44. $\operatorname{arcsec}\left(\dfrac{\sqrt{2}}{2}\right)$

Calculate each of the limits in Exercises 45–52.

45. $\displaystyle\lim_{x\to 1} x\sin^{-1}\dfrac{x}{2}$

46. $\displaystyle\lim_{x\to 3}\dfrac{\sqrt{x}}{\tan^{-1}\sqrt{x}}$

47. $\displaystyle\lim_{x\to 1^-}\dfrac{1}{\sin^{-1} x}$

48. $\displaystyle\lim_{x\to 0}\dfrac{1}{\sec^{-1} x}$

49. $\displaystyle\lim_{x\to\infty}\dfrac{x^3}{\tan^{-1} x}$

50. $\displaystyle\lim_{x\to 0}\dfrac{x^3}{\tan^{-1} x}$

51. $\displaystyle\lim_{x\to 0}\dfrac{\tan^{-1} x}{\sin^{-1} x}$

52. $\displaystyle\lim_{x\to\infty}\dfrac{\tan^{-1} x}{\sec^{-1} x}$

Find the derivatives of each of the functions in Exercises 53–66.

53. $f(x) = \sin^{-1}(3x^2)$

54. $f(x) = 3(\sin^{-1} x)^2$

55. $f(x) = x^2 \arctan x^2$

56. $f(x) = \tan^{-1}(\ln x)$

57. $f(x) = \sec^{-1} x^2$

58. $f(x) = \sin(\sin^{-1} x)$

59. $f(x) = \sin^{-1}(\sec^2 x)$

60. $f(x) = \sin^2(\sec^{-1} x)$

61. $f(x) = \dfrac{\sin^{-1} x}{\tan^{-1} x}$

62. $f(x) = \dfrac{\sin^{-1} x}{\sec^{-1} x}$

63. $f(x) = \ln(\operatorname{arcsec}(\sin^2 x))$

64. $f(x) = x^{-2} e^{4x} \sin^{-1} x$

65. $f(x) = \sec(1 + \tan^{-1} x)$

66. $f(x) = \dfrac{\sin(\arcsin x)}{\arctan x}$

Write each of the expressions in Exercises 67–74 as an algebraic expression that does not involve trigonometric or inverse trigonometric functions.

67. $\cos(\sin^{-1} x)$

68. $\cot(\tan^{-1} x)$

69. $\tan(\sec^{-1} x)$

70. $\sin(\sec^{-1} x)$

71. $\sin(\cos^{-1} x)$

72. $\tan(\tan^{-1} 2x)$

73. $\sec^2(\tan^{-1} x)$

74. $\sin^2(\tan^{-1} x)$

In Exercises 76–82, find a function $f(x)$ that has the given derivative $f'(x)$.

75. $f'(x) = \dfrac{2}{\sqrt{1-4x^2}}$

76. $f'(x) = \dfrac{2x}{\sqrt{1-4x^2}}$

77. $f'(x) = \dfrac{3x}{1+9x^2}$

78. $f'(x) = \dfrac{1}{1+9x^2}$

79. $f'(x) = \dfrac{3}{\sqrt{4-9x^2}}$

80. $f'(x) = \dfrac{1}{9+x^2}$

81. $f'(x) = \dfrac{2x}{1+x^2}$

82. $f'(x) = \dfrac{2x}{1+x^4}$

Find the critical points of each function in Exercises 83–86, and use the first-derivative or the second-derivative test to determine whether each is a local minimum, a local maximum, or neither.

83. $f(x) = \arctan x$

84. $f(x) = \sin^{-1} x^2$

85. $f(x) = \sin(\cos^{-1} x)$

86. $f(x) = \cos(\sin^{-1} x)$

Use the Squeeze Theorem to solve each of the limits in Exercises 87–88. Explain exactly how the Squeeze Theorem applies in each case.

87. $\displaystyle\lim_{x\to 0} x\tan^{-1}\dfrac{1}{x}$

88. $\displaystyle\lim_{x\to 0} x^2\tan^{-1}\dfrac{1}{x}$

Applications

Suppose your friend Max drops a penny from the top floor of the Empire State Building, 1250 feet from the ground. You are standing about a block away, 250 feet from the base of the building.

89. Find a formula for the angle of elevation $\alpha(t)$ from the ground at your feet to the height of the penny t seconds

after Max drops it. Multiply by an appropriate constant so that $\alpha(t)$ is measured in degrees.

90. Find a formula for the rate at which the angle of elevation $\alpha(t)$ in the previous problem is changing at time t, and use the formula to determine the rate of change of the angle of elevation at the time the penny hits the ground.

Proofs

91. Prove that, for all x in the domain of the inverse secant function, $\sec^{-1} x = \cos^{-1}\left(\frac{1}{x}\right)$.

92. Prove that, for all x in the domain of the inverse tangent function, $\cot^{-1} x = \tan^{-1}\left(\frac{1}{x}\right)$.

93. Use the composition rule for limits and the fact that $\tan x$ is continuous on its domain to prove that $\tan^{-1} x$ is continuous everywhere.

94. Use implicit differentiation and the fact that $\sec(\sec^{-1} x) = x$ for all x in the domain of $\sec^{-1} x$ to prove that $\frac{d}{dx}(\sec^{-1} x) = \frac{1}{|x|\sqrt{x^2 - 1}}$. You will have to consider the cases $x > 1$ and $x < -1$ separately.

95. Use implicit differentiation and the fact that $\tan(\tan^{-1} x) = x$ for all x in the domain of $\tan^{-1} x$ to prove that $\frac{d}{dx}(\tan^{-1} x) = \frac{1}{1 + x^2}$.

Thinking Forward

Trigonometric substitution: In later sections we will use inverse trigonometric functions to find antiderivatives to functions like

$$f(x) = \frac{1}{x^2\sqrt{1 - x^2}}.$$

In the problems that follow you will show, in two ways, that one antiderivative of $f(x)$ is the function $F(x) = -\cot(\sin^{-1} x)$.

▶ Show that $F'(x) = f(x)$ by first differentiating $F'(x)$ and then converting the inverse trigonometric composition to an algebraic function.

▶ Show that $F'(x) = f(x)$ by first converting $F(x)$ to an algebraic function and then differentiating and using algebra to simplify the resulting expression.

CHAPTER REVIEW, SELF-TEST, AND CAPSTONES

Before you progress to the next chapter, be sure you are familiar with the definitions, concepts, and basic skills outlined here. The capstone exercises at the end bring together ideas from this chapter and look forward to future chapters.

Definitions

Give precise mathematical definitions or descriptions of each of the concepts that follow. Then illustrate the definition with a graph or an algebraic example.

▶ an *acute* angle

▶ the definitions of the six trigonometric functions *sine, cosine, tangent, cosecant, secant,* and *cotangent,* for an acute angle, as ratios of side lengths of triangles.

▶ what it means for an angle to be in *standard position,* the corresponding *initial edge* and *terminal edge* of such an angle, and what it means for such an angle to be *positive* or *negative*

▶ the *unit circle*

▶ the unit of angle measurement called a *radian*

▶ the definitions of the six trigonometric functions *sine, cosine, tangent, cosecant, secant,* and *cotangent,* for any angle in standard position, as ratios of coordinates on the unit circle

▶ the *reference triangle* and *reference angle* for an angle in standard position

▶ what it means for a function to be *periodic*

▶ what it means for an equation to be an *identity*

▶ the concept of a *restricted domain*

▶ the definitions of the inverse trigonometric functions *arcsine, arctangent,* and *arcsecant,* and the restricted domains of their associated trigonometric functions

Theorems

Fill in the blanks or circle as appropriate to complete each of the following theorem statements:

▶ *The Law of Similar Triangles:* If a triangle with side lengths a, b, and c is similar to (that is, has the same angle measures as) another triangle with corresponding side lengths A, B, and C, then we have the following three equalities of ratios of side lengths: _____ , _____ , and _____ .

▶ We can express each of the following four trigonometric functions in terms of sine and cosine as follows: $\tan\theta =$ _____ , $\csc\theta =$ _____ , $\sec\theta =$ _____ , and $\cot\theta =$ _____ .

▶ The 45–45–90 right triangle, and the 30–60–90 right triangle, with hypotenuse of length 1 unit have side lengths as shown in the following figures: _____ and _____ (provide labeled figures).

▶ All six trigonometric functions are _____ on their domains.

▶ The inverse trigonometric functions $\sin^{-1} x$, $\tan^{-1} x$, and $\sec^{-1} x$ are _____ on their domains.

Trigonometric Algebra, Graphs, Limits, and Derivatives

Domains and ranges: List the domains and ranges of the six trigonometric funtions. Illustrate your answers with well-labeled graphs of the functions.

▶ $f(x) = \sin x$

▶ $f(x) = \cos x$

▶ $f(x) = \tan x$

▶ $f(x) = \sec x$

▶ $f(x) = \csc x$

▶ $f(x) = \cot x$

▶ $f(x) = \sin^{-1} x$

▶ $f(x) = \tan^{-1} x$

▶ $f(x) = \sec^{-1} x$

Trigonometric identities: Complete each of the given trigonometric identities. You may assume that θ, α, and β are angles for which the expressions are defined.

▶ $\sin^2 \theta + \cos^2 \theta = $ ____

▶ $\sec^2 \theta - 1 = $ ____

▶ $1 + \cot^2 \theta = $ ____

▶ $\sin(-\theta) = $ ____

▶ $\cos(-\theta) = $ ____

▶ $\tan(-\theta) = $ ____

▶ $\sin(\alpha + \beta) = $ ____

▶ $\sin(\alpha - \beta) = $ ____

▶ $\cos(\alpha + \beta) = $ ____

▶ $\cos(\alpha - \beta) = $ ____

▶ $\sin 2\theta = $ ____

▶ $1 - 2\sin \theta = $ ____

▶ $\cos^2 \theta - \sin^2 \theta = $ ____

▶ $2\cos^2 \theta - 1 = $ ____

▶ $\dfrac{1 - \cos 2\theta}{2} = $ ____

▶ $\dfrac{1 + \cos 2\theta}{2} = $ ____

Basic trigonometric and inverse trigonometric limits: Fill in the blanks to complete each of the following statements about limits.

▶ $\displaystyle\lim_{x\to\infty} \sin x = $ ____

▶ $\displaystyle\lim_{x\to\infty} \tan x = $ ____

▶ $\displaystyle\lim_{x\to(\pi/2)^+} \sec x = $ ____

▶ $\displaystyle\lim_{x\to(\pi/2)^-} \sec x = $ ____

▶ $\displaystyle\lim_{x\to(\pi/2)^+} \tan x = $ ____

▶ $\displaystyle\lim_{x\to(\pi/2)^-} \tan x = $ ____

▶ $\displaystyle\lim_{x\to0^+} \cot x = $ ____

▶ $\displaystyle\lim_{x\to0^-} \cot x = $ ____

▶ $\displaystyle\lim_{x\to0} \dfrac{\sin x}{x} = $ ____

▶ $\displaystyle\lim_{x\to0} \dfrac{1 - \cos x}{x} = $ ____

▶ $\displaystyle\lim_{x\to\infty} \tan^{-1} x = $ ____

▶ $\displaystyle\lim_{x\to-\infty} \tan^{-1} x = $ ____

▶ $\displaystyle\lim_{x\to\infty} \sec^{-1} x = $ ____

▶ $\displaystyle\lim_{x\to-\infty} \sec^{-1} x = $ ____

▶ $\displaystyle\lim_{x\to1^-} \sin^{-1} x = $ ____

▶ $\displaystyle\lim_{x\to1^+} \sec^{-1} x = $ ____

Derivatives of trigonometric and inverse trigonometric functions: Fill in the blanks to complete each of the following statements about derivatives.

▶ $\dfrac{d}{dx}(\sin x) = $ ____

▶ $\dfrac{d}{dx}(\cos x) = $ ____

▶ $\dfrac{d}{dx}(\tan x) = $ ____

▶ $\dfrac{d}{dx}(\sec x) = $ ____

▶ $\dfrac{d}{dx}(\csc x) = $ ____

▶ $\dfrac{d}{dx}(\cot x) = $ ____

▶ $\dfrac{d}{dx}(\sin^{-1} x) = $ ____

▶ $\dfrac{d}{dx}(\tan^{-1} x) = $ ____

▶ $\dfrac{d}{dx}(\sec^{-1} x) = $ ____

Skill Certification: Working with Trigonometric and Inverse Trigonometric Functions

Calculating trigonometric values by hand: Use unit-circle arguments to calculate the exact values of each of the following quantities without a calculator or memorized table of values.

1. $\cos \dfrac{2\pi}{3}$

2. $\tan\left(-\dfrac{7\pi}{6}\right)$

3. $\csc\left(-\dfrac{5\pi}{4}\right)$

4. $\sin^{-1}(-1)$

5. $\sec^{-1} \dfrac{1}{\sqrt{3}}$

6. $\tan^{-1}\left(-\dfrac{2}{\sqrt{2}}\right)$

Limits: Calculate each of the given limits. Some limits may need L'Hôpital's rule or logarithms.

7. $\displaystyle\lim_{x\to0} \dfrac{1 + \sin x}{1 - \cos x}$

8. $\displaystyle\lim_{x\to0} \dfrac{\sin x}{\cos x - 1}$

9. $\displaystyle\lim_{x\to-\pi/4} \dfrac{1}{\sin x + \cos x}$

10. $\displaystyle\lim_{x\to0} \dfrac{\sin 3x}{5x}$

11. $\displaystyle\lim_{x\to0} \dfrac{1 - \cos 2x}{7x}$

12. $\displaystyle\lim_{x\to0} \dfrac{\sin^2 3x}{x^3 - x}$

13. $\displaystyle\lim_{x\to1^+} \dfrac{1 - \cos\sqrt{x-1}}{\sqrt{x-1}}$

14. $\displaystyle\lim_{x\to0^+} \dfrac{x^2 \csc 3x}{1 - \cos 2x}$

15. $\displaystyle\lim_{x\to\pi/2} \dfrac{\sin(\cos x)}{\cos x}$

16. $\displaystyle\lim_{x\to1} \dfrac{\sin(\ln x)}{x - 1}$

17. $\displaystyle\lim_{x\to0} \dfrac{x\cos x}{1 - e^x}$

18. $\displaystyle\lim_{x\to0} \dfrac{\tan^{-1} x}{\sin^{-1} x}$

19. $\displaystyle\lim_{x\to0} \dfrac{\tan^{-1} x}{1 - \cos x}$

20. $\displaystyle\lim_{x\to0^+} x^{1 - \cos x}$

Derivatives: Calculate the derivatives of each of the given functions. The last two problems require logarithmic differentiation.

21. $f(x) = e^x \sin x$

22. $f(x) = \dfrac{e^x}{\sin x}$

23. $f(x) = \sin(e^x)$

24. $f(x) = e^{\sin x}$

25. $f(x) = \ln(\tan^2 x)$

26. $f(x) = x^3 \sec x$

27. $f(x) = \sqrt{\sin^{-1}(x^2)}$

28. $f(x) = \dfrac{\sin^2 x + \cos^2 x}{\csc x}$

29. $f(x) = \sin^2(\cos^3 x^4)$

30. $f(x) = \ln((\sin^{-1} x)^2)$

31. $f(x) = x^{\sin x}$

32. $f(x) = (\cos x)^{\tan x}$

Inverse trigonometric compositions: Write each given expression as an algebraic expression that does not involve trigonometric or inverse trigonometric values. Use triangles and a unit circle to support your work.

33. $\csc(\tan^{-1} x)$

34. $\sin(\sec^{-1} x)$

35. $\cos^2(\sin^{-1} x)$

36. $\tan\left(2\tan^{-1}\dfrac{x}{3}\right)$

Capstone Problems

A. *Continuity of three interesting functions:* Use limits and the definition of continuity to determine whether each of the given functions is continuous at $x = 0$. Then use graphs to support your answers.

(a) $f(x) = \begin{cases} \sin\left(\dfrac{1}{x}\right), & \text{if } x \neq 0 \\ 0, & \text{if } x = 0 \end{cases}$

(b) $f(x) = \begin{cases} x\sin\left(\dfrac{1}{x}\right), & \text{if } x \neq 0 \\ 0, & \text{if } x = 0 \end{cases}$

(c) $f(x) = \begin{cases} x^2 \sin\left(\dfrac{1}{x}\right), & \text{if } x \neq 0 \\ 0, & \text{if } x = 0 \end{cases}$

B. *Differentiability of three interesting functions:* Determine whether or not each of the three functions in the previous exercise are differentiable at $x = 0$. Again, follow this work by using graphs to support your answers.

C. *A mass on a spring:* Consider a mass hanging from the ceiling at the end of a spring. If you pull down on the mass and let go, it will oscillate up and down according to the equation

$$s(t) = A\sin\left(\sqrt{\dfrac{k}{m}}\, t\right) + B\cos\left(\sqrt{\dfrac{k}{m}}\, t\right),$$

where $s(t)$ is the distance of the mass from its equilibrium position, m is the mass of the bob on the end of the spring, and k is a "spring coefficient" that measures how tight or stiff the spring is. The constants A and B depend on initial conditions—specifically, how far you pull down the mass (s_0) and the velocity at which you release the mass (v_0). This equation does *not* take into effect any friction due to air resistance.

spring coefficient, k
$s(t)$
v_0 mass, m
s_0

(a) Determine whether or not the limit of $s(t)$ as $t \to \infty$ exists. What does this say about the long-term behavior of the mass on the end of the spring?

(b) Explain how this limit relates to the fact that the equation for $s(t)$ does *not* take friction due to air resistance into account.

(c) Suppose the bob at the end of the spring has a mass of 2 grams and that the coefficient for the spring is $k = 9$. Suppose also that the spring is released in such a way that $A = \sqrt{2}$ and $B = 2$. Use a graphing utility to graph the function $s(t)$ that describes the distance of the mass from its equilibrium position. Use your graph to support your answer to part (a).

D. *A mass on a spring, with air resistance:* In the previous exercise we gave an equation describing spring motion without air resistance. If we take into account friction due to air resistance, the mass will oscillate up and down according to the equation

$$s(t) = e^{(-f/2m)t}\left(A\sin\left(\dfrac{\sqrt{4km - f^2}}{2m}\, t\right) \right.$$
$$\left. + B\cos\left(\dfrac{\sqrt{4km - f^2}}{2m}\, t\right)\right),$$

where m, k, A, and B are the constants described in Problem D and f is a positive "friction coefficient" that measures the amount of friction due to air resistance.

(a) Find the limit of $s(t)$ as $t \to \infty$. What does this say about the long-term behavior of the mass on the end of the spring?

(b) Explain how this limit relates to the fact that the new equation for $s(t)$ *does* take friction due to air resistance into account.

(c) Suppose the bob at the end of the spring has a mass of 2 grams, the coefficient for the spring is $k = 9$, and the friction coefficient is $f = 6$. Suppose also that the spring is released in such a way that $A = 4$ and $B = 2$. Use a graphing utility to graph the function $s(t)$ that describes the distance of the mass from its equilibrium position. Use your graph to support your answer to part (a).

Definite Integrals

7.7 Functions Defined by Integrals

Area Accumulation Functions
The Second Fundamental Theorem of Calculus
Defining the Natural Logarithm Function with an Integral
The Proof of the Second Fundamental Theorem
Examples and Explorations

Chapter Review, Self-Test, and Capstones

7.1   ADDITION AND ACCUMULATION

▶ An introduction to accumulation functions and the area under a curve

▶ Sigma notation and properties of sums

▶ A preview of limits of sums, the backbone of the definition of the definite integral

Accumulation Functions

With this chapter we begin the study of integrals. While derivatives describe the rates at which functions change, integrals can describe how functions accumulate. As we will see throughout this chapter, the concepts of accumulation, area, and differentiation are fundamentally intertwined.

For a simple example of how these concepts are intertwined, imagine that you are driving down a straight road that has stoplights. Suppose that you start from a full stop at one stoplight, increase your velocity for 20 seconds to reach 60 miles per hour (88 feet per second), then decrease your velocity for 20 seconds until you come to a full stop at the next stoplight, as illustrated next at the left. Your speedometer works, so you can tell how fast you are going at any time. However, your odometer is broken, so you have no idea how far you have travelled. Using data from your speedometer, you can show that your velocity on this trip is given by $v(t) = -0.22t^2 + 8.8t$ feet per second, from $t = 0$ to $t = 40$ seconds, as illustrated in the graph shown at the right.

$$v(t) = -0.22t^2 + 8.8t$$

So, how far was it between the two stoplights? If you had travelled at a constant velocity, this would be an easy application of the "distance equals average rate times time" formula. For example, driving at exactly 30 miles per hour (44 feet per second) for 40 seconds would accumulate a distance of $d = (44)(40) = 1760$ feet.

Unfortunately in our example, velocity varies. However, we can approximate the distance travelled by assuming a constant velocity over small chunks of time. For example, we could use $v(5)$, $v(15)$, $v(25)$, and $v(35)$ as constant velocities over the intervals $[0, 10]$, $[10, 20]$, $[20, 30]$, and $[30, 40]$, respectively. Using the $d = rt$ formula over each of the four time intervals would give us an approximate distance travelled of

$$d \approx d_1 + d_2 + d_3 + d_4 = r_1 t_1 + r_2 t_2 + r_3 t_3 + r_4 t_4$$
$$= v(5)(10) + v(15)(10) + v(25)(10) + v(35)(10)$$
$$= (38.5)(10) + (82.5)(10) + (82.5)(10) + (38.5)(10) = 2420 \text{ feet.}$$

Notice that we have just estimated the distance between the two stoplights by means of only the readings on your speedometer at $t = 5$, $t = 15$, $t = 25$, and $t = 35$. Despite having used only this small amount of information, our estimate is fairly close to the actual distance, which in this example happens to be just over 2346 feet.

We can think of the preceding four 10-second distance approximations as areas of four rectangles, as shown next at the left. For example, over the first time interval we have a rectangle of width $t_1 = 10$, height $r_1 = v(5) = 38.5$, and area $d_1 = (t_1)(r_1) = (38.5)(10) = 385$.

Distance was approximated with a sum of areas of rectangles

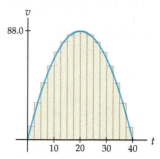

Thinner rectangles give a better approximation

Notice that the sum of the areas of these rectangles is a pretty good approximation for the distance between the stoplights and also a pretty good approximation for the area under the velocity curve in the figure at the left. Using smaller time intervals would give us thinner rectangles (as shown at the right), whose combined areas would be a better approximation for the distance travelled, as well as a better approximation for the area under $v(t)$. These figures suggest that the exact distance travelled might equal the exact area under the velocity curve. Furthermore, since we already know that velocity is the derivative of position, the figures also suggest connections between accumulation, area, and derivatives. Over the rest of this chapter we will make these notions and connections more precise.

Sigma Notation

From the stoplight example we just discussed, it seems that we will have to do a lot of adding in order to investigate area and accumulation functions. Our better approximation with smaller, but more, rectangles involved 16 rectangles with areas to calculate and add up—and for an even better approximation we might consider 100 or more rectangles. We now develop a compact notation to represent the sum of a sequence of numbers—in particular, a sequence of numbers that has a recognizable pattern. This notation is called **sigma notation,** since it uses the letter "sigma" (written Σ), which is the Greek counterpart of the letter "S" (for "sum").

DEFINITION 7.1   **Sigma Notation**

If a_k is a function of k, and m and n are nonnegative integers with $m \leq n$, then

$$\sum_{k=m}^{n} a_k = a_m + a_{m+1} + a_{m+2} + \cdots + a_{n-1} + a_n.$$

As an initial example, consider our earlier sum of four distances. This sum can be written in sigma notation as follows:

$$d \approx \sum_{k=1}^{4} d_k = \sum_{k=1}^{4} r_k t_k = 2420 \text{ feet.}$$

The expression in Definition 7.1 is pronounced "the sum from $k = m$ to n of a_k." The "$k = m$" below the sigma shows where we should begin the sum, and it also specifies that the **index**, or stepping variable, for the sum is k. With sigma notation, the step from one value of k to the next is always equal to 1. If we start with $k = 1$, then the next value of k will be $k = 2$, the next will be $k = 3$, and so on. We end at the value $k = n$, where n is the number appearing above the sigma in the notation. The terms a_k represent the pattern of the items to be summed. When we say that a_k is a function of k, we mean that, for every integer index k, the expression for a_k determines a unique real number. Since the index k can have only integer values, it is traditional to use a_k rather than the usual function notation $a(k)$.

When printed in a line of text, sigma notation is more compact and looks like this: $\sum_{k=m}^{n} a_k$. Sigma notation always follows the general pattern

$$\overset{\text{ending value}}{\underset{k=\text{starting value}}{\sum}} (\text{function of } k).$$

To find the value of a sum in sigma notation $\sum_{k=m}^{n} a_k$ you would evaluate the function a_k at $k = m$, $k = m + 1$, $k = m + 2$, and so on until $k = n$, and then add up all of these values. To put a given sum into sigma notation you must identify a pattern for the function a_k, as well as starting and ending values for k. For example, we could represent the sum

$$1 + \frac{1}{2} + \frac{1}{3} + \frac{1}{4} + \frac{1}{5} + \frac{1}{6} + \frac{1}{7} + \frac{1}{8} + \frac{1}{9} + \frac{1}{10}$$

with the sigma notation $\sum_{k=1}^{10} \frac{1}{k}$, since each of the numbers in our sum is of the form $\frac{1}{k}$ for some integer k, with $k = 1$ for the first term and $k = 10$ for the last term.

The Algebra of Sums in Sigma Notation

The next theorem expresses two common properties of sums in sigma notation. For the moment we will restrict our attention to sums whose index k starts at $k = 1$. However, Theorem 7.2 is true for sums that begin at any integer $k = m$.

THEOREM 7.2

Constant-Multiple and Sum Rules for Sums

If a_k and b_k are functions defined for nonnegative integers k, and c is any real number, then

(a) $\displaystyle\sum_{k=1}^{n} ca_k = c\sum_{k=1}^{n} a_k.$ **(b)** $\displaystyle\sum_{k=1}^{n}(a_k + b_k) = \sum_{k=1}^{n} a_k + \sum_{k=1}^{n} b_k.$

Part (a) of this theorem is a general version of the distributive property $c(x + y) = cx + cy$. Part (b) is a general version of the associative and commutative properties of addition (i.e., the fact that we can reorder and regroup the numbers in a sum). A simple example is the fact that $(a_1 + b_1) + (a_2 + b_2) = (a_1 + a_2) + (b_1 + b_2)$. The proof of Theorem 7.2 consists mostly of translating the sigma notation.

Proof. To prove part (a), we translate the sigma notation to its expanded form and then apply the distributive property:

$$\sum_{k=1}^{n} ca_k = ca_1 + ca_2 + ca_3 + \cdots + ca_{n-1} + ca_n \qquad \leftarrow \text{write out the sum}$$

$$= c(a_1 + a_2 + a_3 + \cdots + a_{n-1} + a_n) \qquad \leftarrow \text{factor out } c$$

$$= c\sum_{k=1}^{n} a_k. \qquad \leftarrow \text{write in sigma notation}$$

The proof of part (b) is similar; we will write the sigma notation in expanded form, reorder and regroup the terms in the sum, and then write the reordered sum in sigma notation:

$$\sum_{k=1}^{n}(a_k + b_k) = (a_1 + b_1) + (a_2 + b_2) + (a_3 + b_3) + \cdots (a_{n-1} + b_{n-1}) + (a_n + b_n)$$

$$= (a_1 + a_2 + a_3 + \cdots + a_{n-1} + a_n) + (b_1 + b_2 + b_3 + \cdots + b_{n-1} + b_n)$$

$$= \sum_{k=1}^{n} a_k + \sum_{k=1}^{n} b_k.$$

We can strip off terms from the beginning or end of a sum, or split a sum into two pieces, by applying the following theorem:

THEOREM 7.3

Splitting a Sum

Given any function a_k defined for nonnegative integers k, and given any integers m, n, and p such that $0 \le m < p < n$,

$$\sum_{k=m}^{n} a_k = \sum_{k=m}^{p-1} a_k + \sum_{k=p}^{n} a_k.$$

Proof. The proof of this theorem is fairly simple; we need only write out the terms of the sum. If $m < p < n$, we have

$$\sum_{k=m}^{n} a_k = a_m + a_{m+1} + \cdots + a_{p-1} + a_p + a_{p+1} + \cdots + a_{n-1} + a_n \qquad \leftarrow \text{expand}$$

$$= (a_m + a_{m+1} + \cdots + a_{p-1}) + (a_p + a_{p+1} + \cdots + a_{n-1} + a_n) \qquad \leftarrow \text{regroup}$$

$$= \sum_{k=m}^{p-1} a_k + \sum_{k=p}^{n} a_k. \qquad \leftarrow \text{sigma notation}$$

Formulas for Common Sums

It can be tedious to calculate a long sum by hand; for example, consider the sum $\sum_{k=1}^{1000} k^2$. To calculate this sum directly, we would have to write out all 1000 of its terms and then add them together. Luckily, the next theorem provides formulas for the values of a few simple sums. Using these formulas, we can quickly calculate any sum whose general term is a constant, linear, quadratic, or cubic polynomial in k.

THEOREM 7.4

Sum Formulas

If n is a positive integer, then

(a) $\displaystyle\sum_{k=1}^{n} 1 = n$

(b) $\displaystyle\sum_{k=1}^{n} k = \frac{n(n+1)}{2}$

(c) $\displaystyle\sum_{k=1}^{n} k^2 = \frac{n(n+1)(2n+1)}{6}$

(d) $\displaystyle\sum_{k=1}^{n} k^3 = \frac{n^2(n+1)^2}{4}$

Proof. To prove the first formula we only have to write out the sum

$$\sum_{k=1}^{n} 1 = \underbrace{1 + 1 + 1 + \cdots + 1}_{n \text{ times}} = n.$$

You will prove the second formula by using two cases in Exercises 61 and 62. The third and fourth formulas are simple to prove by a method called ***induction***, but that method of proof is best left for a future mathematics course, so we do not include proofs of the third and fourth formulas at this time.

Examples and Explorations

EXAMPLE 1   **Converting from sigma notation to an expanded sum**

Write the sum represented by the sigma notation $\sum_{k=2}^{8} 2^k$ in expanded form, and find the value of the sum.

SOLUTION

The given sigma notation is a compact way of writing "the sum of all numbers of the form 2^k, where k is an integer greater than or equal to 2 and less than or equal to 8." This means we must find the values of 2^k for $k = 2$, $k = 3$, and so on until $k = 8$, and then add all of these values:

$$\sum_{k=2}^{8} 2^k = 2^2 + 2^3 + 2^4 + 2^5 + 2^6 + 2^7 + 2^8 = 508.$$

EXAMPLE 2   **Converting from an expanded sum to sigma notation**

Write the following sum in sigma notation:

$$\frac{3}{4} + \frac{4}{5} + \frac{5}{6} + \frac{6}{7} + \frac{7}{8} + \frac{8}{9} + \frac{9}{10}.$$

SOLUTION

First, we look for a pattern in the numbers of the sum, so we can determine the function a_k. One possible pattern is that each number in the sum is of the form $\frac{k}{k+1}$ for some non-negative integer k. Therefore, we can set $a_k = \frac{k}{k+1}$. The first number, $\frac{3}{4}$, corresponds to the value $k = 3$. The last number, $\frac{9}{10}$, corresponds to the value $k = 9$ $\left(\text{when } k = 9, \text{ we have } \frac{k}{k+1} = \frac{9}{10}\right)$. The sum can be written in sigma notation as

$$\frac{3}{4} + \frac{4}{5} + \frac{5}{6} + \frac{6}{7} + \frac{7}{8} + \frac{8}{9} + \frac{9}{10} = \sum_{k=3}^{9} \frac{k}{k+1}.$$

EXAMPLE 3   **Algebraically manipulating sums in sigma notation**

Given that $\sum_{k=1}^{4} a_k = 7$ and $\sum_{k=1}^{4} b_k = 10$, find $\sum_{k=1}^{4} (a_k + 3b_k)$.

SOLUTION

Notice that we do not know what the functions a_k and b_k are in this problem, and thus we cannot directly compute $\sum_{k=1}^{4}(a_k + 3b_k)$. However, using Theorem 7.2, we have

$$\sum_{k=1}^{4}(a_k + 3b_k) = \sum_{k=1}^{4}a_k + \sum_{k=1}^{4}3b_k \qquad \leftarrow \text{Part (a) of Theorem 7.2}$$

$$= \sum_{k=1}^{4}a_k + 3\sum_{k=1}^{4}b_k \qquad \leftarrow \text{Part (b) of Theorem 7.2}$$

$$= 7 + 3(10) = 37. \qquad \leftarrow \text{using the values of the sums given} \quad \square$$

EXAMPLE 4   Sums can be combined only if they start and end at the same value

Use the fact that $\sum_{k=2}^{20}\frac{1}{k} \approx 2.5977$ and $\sum_{k=0}^{19}\sqrt{k} \approx 57.1938$ to estimate $\sum_{k=2}^{19}\left(\frac{2}{k} + \sqrt{k}\right)$.

SOLUTION

We could of course just directly calculate the desired sum by adding up terms, but instead we will combine the two given sums to obtain the information we need. The key is to rewrite the two sums we were given so that they each begin and end and the same value; we will rewrite so that they begin at $k = 2$ and end at $k = 19$.

We are given that $\sum_{k=2}^{20}\frac{1}{k} = \frac{1}{2} + \frac{1}{3} + \cdots + \frac{1}{19} + \frac{1}{20}$ is approximately 2.5977. To get the sum of the $k = 2$ through $k = 19$ terms, all we have to do is subtract the extra term $\frac{1}{20}$:

$$\sum_{k=2}^{19}\frac{1}{k} = \left(\sum_{k=2}^{20}\frac{1}{k}\right) - \frac{1}{20} \approx 2.5977 - \frac{1}{20} = 2.5477.$$

Similarly, we can find the sum from $k = 2$ to $k = 19$ of $\sqrt{k}$ by subtracting the two extra terms (at $k = 0$ and $k = 1$) at the beginning of the sum $\sum_{k=0}^{19}\sqrt{k}$:

$$\sum_{k=2}^{19}\sqrt{k} = \left(\sum_{k=0}^{19}\sqrt{k}\right) - \sqrt{0} - \sqrt{1} \approx 57.1938 - 0 - 1 = 56.1938.$$

We can now use Theorem 7.2 to combine these sums and find the value of the desired sum:

$$\sum_{k=2}^{19}\left(\frac{2}{k} + \sqrt{k}\right) = \sum_{k=2}^{19}\frac{2}{k} + \sum_{k=2}^{19}\sqrt{k} \qquad \leftarrow \text{sum of sums}$$

$$= 2\sum_{k=2}^{19}\frac{1}{k} + \sum_{k=2}^{19}\sqrt{k} \qquad \leftarrow \text{constant times a sum}$$

$$\approx 2(2.5477) + 56.1938 = 61.2892. \quad \leftarrow \text{add previously computed values} \quad \square$$

EXAMPLE 5   Using sum formulas to calculate the value of a sum

Find the value of the sum $\sum_{k=1}^{300}(k^3 - 4k^2 + 2)$.

SOLUTION

By writing the sum as three simple sums and using the formulas in Theorem 7.4, we have

$$\sum_{k=1}^{300}(k^3 - 4k^2 + 2) = \left(\sum_{k=1}^{300} k^3\right) - 4\left(\sum_{k=1}^{300} k^2\right) + 2\left(\sum_{k=1}^{300} 1\right) \qquad \leftarrow \text{properties of sums}$$

$$= \frac{300^2(301)^2}{4} - 4\frac{300(301)(601)}{6} + 2(300) \qquad \leftarrow \text{Theorem 7.4}$$

$$= 2,002,342,900.$$

Of course that was a whole lot easier than expanding the sum and then adding all 300 terms! ☐

EXAMPLE 6   **Using a sum formula to calculate the limit of a sum as $n \to \infty$**

Find the values of the sum $\sum_{k=1}^{n} \frac{k^2}{n^3}$ for $n = 3$, $n = 4$, $n = 100$, and $n = 1000$. Then investigate what happens as n approaches infinity.

SOLUTION

To find the sums for $n = 3$ and $n = 4$, it is easy to expand and calculate the sums directly:

$$\sum_{k=1}^{3} \frac{k^2}{n^3} = \frac{1^2}{3^3} + \frac{2^2}{3^3} + \frac{3^2}{3^3} = \frac{1}{27} + \frac{4}{27} + \frac{9}{27} = \frac{14}{27};$$

$$\sum_{k=1}^{4} \frac{k^2}{n^3} = \frac{1^2}{4^3} + \frac{2^2}{4^3} + \frac{3^2}{4^3} + \frac{4^2}{4^3} = \frac{1}{64} + \frac{4}{64} + \frac{9}{64} + \frac{16}{64} = \frac{30}{64} = \frac{15}{32}.$$

Notice that the denominator of each term in the first sum is constantly equal to n^3 and does not change as k changes. Notice also that the sum from $n = 1$ to $n = 4$ *cannot* be obtained from the sum from $n = 1$ to $n = 3$ just by adding a fourth term; *all* the terms in the second sum are different from the terms in the first sum. In fact, the second sum is actually *smaller* than the first sum, since $\frac{14}{27} \approx 0.5185$ and $\frac{15}{32} = 0.46875$. Although we are adding more terms in the second sum, each of those terms is smaller than the terms in the first sum (since their denominators are larger).

To find the sum for $n = 100$, we will use properties of sums and one of the sum formulas from Theorem 7.4:

$$\sum_{k=1}^{100} \frac{k^2}{n^3} = \sum_{k=1}^{100} \frac{k^2}{100^3} = \frac{1}{100^3} \sum_{k=1}^{100} k^2 \qquad \leftarrow \text{pull out the constant } \frac{1}{100^3}$$

$$= \frac{1}{100^3} \frac{100(101)(201)}{6} \qquad \leftarrow \text{sum formula from Theorem 7.4}$$

$$= \frac{2,030,100}{6,000,000} = 0.33835.$$

Similarly, the sum from $k = 1$ to $k = 1000$ is

$$\sum_{k=1}^{1000} \frac{k^2}{n^3} = \sum_{k=1}^{1000} \frac{k^2}{1000^3} = \frac{1}{1000^3} \sum_{k=1}^{1000} k^2$$

$$= \frac{1}{1000^3} \left(\frac{1000(1001)(2001)}{6}\right) = \frac{2,003,001,000}{6,000,000,000} \approx 0.33383.$$

Notice that even though we just summed up a thousand terms, the sum is a very small number. In fact, the more terms we sum up, the smaller that number seems to get! It looks like if we let n continue to grow, the sum from $k = 1$ to $k = n$ will approach a number that is approximately 0.33 or so. We can show that this is indeed the case by taking a limit of the formula we used to compute the two previous sums:

$$\lim_{n \to \infty} \sum_{k=1}^{n} \frac{k^2}{n^3} = \lim_{n \to \infty} \frac{1}{n^3} \sum_{k=1}^{n} k^2 \qquad \leftarrow n \text{ is constant with respect to } k$$

$$= \lim_{n \to \infty} \frac{1}{n^3} \left(\frac{n(n+1)(2n+1)}{6} \right) \qquad \leftarrow \text{sum formula from Theorem 7.4}$$

$$= \lim_{n \to \infty} \frac{(n+1)(2n+1)}{6n^2} \qquad \leftarrow \text{algebra}$$

$$= \lim_{n \to \infty} \frac{2n^2 + 3n + 1}{6n^2} = \frac{2}{6} = \frac{1}{3}. \qquad \leftarrow \text{take the limit} \qquad \square$$

? TEST YOUR UNDERSTANDING

▶ Considering the stoplight example at the beginning of this section, how can the approximate distances travelled be interpreted as rectangles? How is exact distance travelled related to the area under a velocity curve?

▶ Can you find an example in which $\sum_{k=1}^{n}(a_k b_k)$ is *not* equal to the product of $\sum_{k=1}^{n} a_k$ and $\sum_{k=1}^{n} b_k$?

▶ Do two sums in sigma notation have to start and end at the same index value in order to be combined into one sum? Why or why not?

▶ How can we express the sum in Example 2 by using sigma notation that starts at $k = 4$ and ends at $k = 10$?

▶ How could a sum of infinitely many things ever be a small finite number, as happened in Example 6?

EXERCISES 7.1

Thinking Back

▶ *Approximations and limits:* Describe in your own words how the slope of a tangent line can be approximated by the slope of a nearby secant line. Then describe how the derivative of a function at a point is defined as a limit of slopes of secant lines. What is the approximation/limit situation described in this section?

▶ *Properties of addition:* State the associative law for addition, the commutative law for addition, and the distributive law for multiplication over addition of real numbers. (You may have to think back to a previous algebra course.)

▶ *Sum and constant-multiple rules:* State the sum and constant-multiple rules for (a) derivatives and (b) limits.

Concepts

0. *Problem Zero:* Read the section and make your own summary of the material.

1. *True/False:* Determine whether each of the statements that follow is true or false. If a statement is true, explain why. If a statement is false, provide a counterexample.

(a) *True or False:* The sum formulas in Theorem 7.4 can be applied only to sums whose starting index value is $k = 1$.

(b) *True or False:* $\sum_{k=0}^{n} \frac{1}{k+1} + \sum_{k=1}^{n} k^2$ is equal to $\sum_{k=0}^{n} \frac{k^3 + k^2 + 1}{k+1}$.

(c) *True or False:* $\sum_{k=1}^{n} \frac{1}{k+1} + \sum_{k=0}^{n} k^2$ is equal to $\sum_{k=1}^{n} \frac{k^3 + k^2 + 1}{k+1}$.

(d) *True or False:* $\left(\sum_{k=1}^{n} \frac{1}{k+1} \right)\left(\sum_{k=1}^{n} k^2 \right)$ is equal to $\sum_{k=1}^{n} \frac{k^2}{k+1}$.

(e) *True or False:* $\sum_{k=0}^{m} \sqrt{k} + \sum_{k=m}^{n} \sqrt{k}$ is equal to $\sum_{k=0}^{n} \sqrt{k}$.

(f) *True or False:* $\sum_{k=0}^{n} a_k = -a_0 - a_n + \sum_{k=1}^{n-1} a_k$.

(g) *True or False:* $\left(\sum_{k=1}^{10} a_k\right)^2 = \sum_{k=1}^{10} a_k^2$.

(h) *True or False:* $\sum_{k=1}^{n}(e^x)^2 = \dfrac{e^x(e^x+1)(2e^x+1)}{6}$.

2. *Examples:* Construct examples of the thing(s) described in the following. Try to find examples that are different than any in the reading.

   (a) A sum that would not be suitable for expressing in sigma notation.
   (b) Two different sigma notation expressions of the same sum.
   (c) A sum from $k = 1$ to $k = n$ that is smaller for $n = 10$ than it is for $n = 5$.

3. Explain why it would be difficult to write the sum $\dfrac{1}{3} + \dfrac{1}{4} + \dfrac{1}{5} + \dfrac{1}{8} + \dfrac{1}{11} + \dfrac{1}{12} + \dfrac{1}{13}$ in sigma notation.

4. Use a sentence to describe what the notation $\sum_{k=2}^{100} \sqrt{k}$ means. (*Hint: Start with "The sum of...."*)

5. Use a sentence to describe what the notation $\sum_{k=3}^{87} k^2$ means. (*Hint: Start with "The sum of...."*)

6. Consider the general sigma notation $\sum_{k=m}^{n} a_k$. What do we mean when we say that a_k is a function of k?

7. Consider the sum $\sum_{i=p}^{q} b_i$.

   (a) Write out this sum in expanded form (i.e., without sigma notation).
   (b) What is the index of the sum? What is the starting value? What is the ending value? Which part of the notation describes the form of each of the terms in the sum?
   (c) Do p and q have to be integers? Can they be negative? What about b_i? What else can you say about p and q?

8. Consider the sum $\sum_{k=2}^{5} \dfrac{k}{1-k}$. Identify the terms a_2, a_3, a_4, and a_5.

9. Consider the sum $\sum_{k=m}^{n} a_k = 9 + 16 + 25 + 36 + 49$. What is a_k? What is m? What is n?

10. Show that $\sum_{k=3}^{9} \dfrac{k}{k+1}$ is equal to $\sum_{k=4}^{10} \dfrac{k-1}{k}$ by writing out the terms in each sum.

11. Show that $\sum_{k=0}^{8} \dfrac{1}{k^2+1}$ is equal to $2\sum_{k=0}^{8} \dfrac{1}{2k^2+2}$ by writing out the terms in each sum.

12. Write the sum $\dfrac{4}{7} + \dfrac{5}{8} + \dfrac{6}{9} + \dfrac{7}{10} + \dfrac{8}{11}$ in sigma notation in three ways: with a starting value of (a) $k = 4$, (b) $k = 7$, and (c) $k = 5$.

13. Write the sum $2 + \dfrac{2}{4} + \dfrac{2}{9} + \dfrac{2}{16} + \dfrac{2}{25}$ in sigma notation in three ways: with a starting value of (a) $k = 1$, (b) $k = 2$, and (c) $k = 0$.

14. Split the sum $\sum_{k=4}^{11} \sqrt{k}$ into three sums, each in sigma notation, where the first sum has two terms and the last two sums each have three terms.

15. Verify that $\sum_{k=1}^{n} k$ is equal to $\dfrac{n(n+1)}{2}$ for the cases (a) $n = 2$, (b) $n = 8$, and (c) $n = 9$.

16. Verify that $\sum_{k=1}^{n} k^2$ is equal to $\dfrac{n(n+1)(2n+1)}{6}$ for the cases (a) $n = 1$, (b) $n = 5$, and (c) $n = 10$.

17. State algebraic formulas that express the following sums, where n is a positive integer:

   (a) $\displaystyle\sum_{k=1}^{n} 1$ (b) $\displaystyle\sum_{k=1}^{n} k$ (c) $\displaystyle\sum_{k=1}^{n} k^2$ (d) $\displaystyle\sum_{k=1}^{n} k^3$

18. Explain why terms in the sum in Example 6 with $n = 4$ are completely different from the terms in the sum when $n = 3$. How can the sum from $k = 1$ to $k = 4$ be *smaller* than the sum from $k = 1$ to $k = 3$? What will happen as n gets larger in this example?

19. Considering the discussion at the end of the stoplight example in the reading, would you expect that the area under the graph of a function f is related to the derivative f'? Or would you expect that the area under the graph of a derivative function f' is related to the function f?

20. Consider again the stoplight example from the reading. In making an approximation for distance travelled, why do we assume that velocity is constant on small subintervals? What are some different ways that we could choose which velocity to use on each subinterval? Illustrate a couple of these ways with graphs that involve rectangles.

Skills

Write each of the sums in Exercises 21–28 in sigma notation. Identify m, n, and a_k in each problem.

21. $3 + 3 + 3 + 3 + 3 + 3 + 3 + 3$

22. $\dfrac{4}{3} + \dfrac{5}{4} + \dfrac{6}{5} + \dfrac{7}{6} + \dfrac{8}{7} + \dfrac{9}{8} + \dfrac{10}{9} + \dfrac{11}{10}$

23. $3 + \dfrac{4}{8} + \dfrac{5}{27} + \dfrac{6}{64} + \dfrac{7}{125}$

24. $\dfrac{1}{4} + \dfrac{1}{9} + \dfrac{1}{16} + \dfrac{1}{25} + \dfrac{1}{36} + \dfrac{1}{49} + \dfrac{1}{64}$

25. $5 + 10 + 17 + 26 + 37 + 50 + 65 + 82 + 101$

26. $9 + 12 + 15 + 18 + 21 + 24 + 27$

27. $\dfrac{1}{n} + \dfrac{2}{n} + \dfrac{3}{n} + \cdots + \dfrac{n}{n}$

28. $-2^n - 1^n + 0^n + 1^n + \cdots + n^n$

Write out each sum in Exercises 29–34 in expanded form, and then calculate the value of the sum.

29. $\displaystyle\sum_{k=4}^{9} k^2$

30. $\displaystyle\sum_{k=0}^{6} \dfrac{2}{k+1}$

31. $\displaystyle\sum_{k=0}^{5} \left(\dfrac{1}{2}k\right)^2 \left(\dfrac{1}{2}\right)$

32. $\displaystyle\sum_{k=3}^{10} \ln k$

33. $\displaystyle\sum_{k=0}^{9} \sqrt{3 + \dfrac{1}{10}k}\left(\dfrac{1}{10}\right)$

34. $\displaystyle\sum_{k=1}^{4} ((2+k)^2 + 1)$

Find a formula for each of the sums in Exercises 35–40, and then use these formulas to calculate each sum for $n = 100$, $n = 500$, and $n = 1000$.

35. $\displaystyle\sum_{k=1}^{n}(3-k)$

36. $\displaystyle\sum_{k=1}^{n}(k^3 - 10k^2 + 2)$

37. $\displaystyle\sum_{k=3}^{n}(k+1)^2$

38. $\displaystyle\sum_{k=1}^{n}\frac{k^3 - 1}{4}$

39. $\displaystyle\sum_{k=1}^{n}\frac{k^3 - 1}{n^4}$

40. $\displaystyle\sum_{k=1}^{n}\frac{k^2 + k + 1}{n^3}$

Write each expression in Exercises 41–43 in one sigma notation (with some extra terms added to or subtracted from the sum, as necessary).

41. $\displaystyle 2\sum_{k=0}^{100} a_k - \sum_{k=3}^{101} a_k$

42. $\displaystyle\sum_{k=1}^{40}\frac{1}{k} - \sum_{k=0}^{39}\frac{1}{k+1}$

43. $\displaystyle 3\sum_{k=2}^{25} k^2 + 2\sum_{k=2}^{24} k - \sum_{k=0}^{25} 1$

In Exercises 44–46, find the sum or quantity without completely expanding or calculating any sums.

44. Given $\sum_{k=3}^{10} a_k = 12$ and $\sum_{k=2}^{10} a_k = 23$, find a_2.

45. Given $\sum_{k=1}^{4} a_k = 7$, $\sum_{k=0}^{4} b_k = 10$, and $a_0 = 2$, find the value of $\sum_{k=0}^{4}(2a_k + 3b_k)$.

46. Given $\sum_{k=0}^{25} k = 325$ and $\sum_{k=3}^{28}(k-3)^2 = 14{,}910$, find the value of $\sum_{k=3}^{25}(k^2 - 5k + 9)$.

Determine which of the limit of sums in Exercises 47–52 are infinite and which are finite. For each limit of sums that is finite, compute its value.

47. $\displaystyle\lim_{n\to\infty}\sum_{k=1}^{n}\frac{k^2 + k + 1}{n^3}$

48. $\displaystyle\lim_{n\to\infty}\sum_{k=1}^{n}(k^2 + k + 1)$

49. $\displaystyle\lim_{n\to\infty}\sum_{k=1}^{n}\frac{(k+1)^2}{n^3 - 1}$

50. $\displaystyle\lim_{n\to\infty}\sum_{k=1}^{n}\frac{k^2 + k + 1}{n^2}$

51. $\displaystyle\lim_{n\to\infty}\sum_{k=1}^{n}\left(1+\frac{k}{n}\right)^2\cdot\frac{1}{n}$

52. $\displaystyle\lim_{n\to\infty}\sum_{k=1}^{n}\frac{k^3}{n^4 + n + 1}$

Applications

53. Considering the stoplight example in the reading with velocity $v(t) = -0.22t^2 + 8.8t$ as shown next at the left, approximate the distance travelled by dividing the time interval $[0, 40]$ into eight pieces and assuming constant velocity on each piece. Interpret this distance in terms of rectangles on the graph of $v(t)$.

Velocity of car
$v(t) = -0.22t^2 + 8.8t$

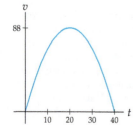

Piecewise approximation of velocity $v(t)$ of race car

54. Suppose you drive on a racetrack for 10 minutes with velocity as shown in the graph at the right.

(a) Describe in words the behavior of your race car over the 10 minutes as shown in the graph.

(b) Find a piecewise-defined formula for your velocity $v(t)$, in miles per hour, t hours after you start from rest. (Note that 1 minute is $\frac{1}{60}$ of an hour.)

(c) Approximate the distance you travelled over the 10 minutes by using 10 subintervals of 1 minute over which you assume a constant velocity. Illustrate this approximation by showing rectangles on the graph of $v(t)$.

(d) Given that distance travelled is the area under the velocity graph, use triangles and squares to calculate the exact distance travelled.

55. The table that follows describes the activity in a college tuition savings account over four years. Notice that 2008 was a particularly bad year for investing! Let $I(t)$ be the amount by which your account increased or decreased in year t, and let $B(t)$ be the balance of your account at the end of year t.

Activity in tuition savings account

Year	2005	2006	2007	2008
Deposited	$ 600	$ 1200	$ 1200	$ 1200
Earnings	$ 10	$ 183	$ 317	$ −1650
Increase	$ 610	$ 1383	$ 1517	$ −450
Balance	$ 610	$ 1994	$ 3512	$ 3061

(a) Describe in your own words how $B(t)$ is the accumulation function for $I(t)$.

(b) Plot a step-function graph of $I(t)$, and describe how $B(t)$ relates to the area under that graph.

(c) What, if anything, can you say about $B(t)$ when $I(t)$ is positive? Negative? If you had to guess that one of these functions was related to the derivative of the other, which one would it be?

56. Suppose 100 mg of a drug is administered to a patient each morning in pill form and it is known that after 24 hours the body processes 80% of the drug from such a pill, leaving 20% of the drug in the body. The amount of the drug in the body right after the first pill is taken is $A(1) = 100$ mg. 24 hours later, after the second pill has been taken, the amount in the body is $A(2) = 100(0.2) + 100 = 120$ mg. 48 hours later, the amount in the body right after taking the third pill is $A(3) = 100(0.2)(0.2) + 100(0.2) + 100 = 124$ mg.

Repeated doses of a drug

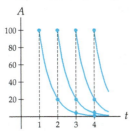

(a) Explain the calculations for $A(1)$, $A(2)$, and $A(3)$ described in the exercise. Which term in $A(3)$ corresponds to the drug left from the first pill?

(b) Interpret the given graph in the context of this problem. What do the marked points represent?

(c) Express $A(n)$ in sigma notation.

(d) Calculate the amount of drug in the body after the 4th through 10th pills. Do you notice anything special about $A(n)$ as n gets larger?

57. Suppose the government enacts a $10 billion tax cut and that the people who save money from this tax cut will spend 70% of it and save the rest. This generates $10(0.7) = \$7$ billion of extra income for other people. Assume these people also spend 70% of their extra

income, and that these transactions continue. Let $A(n)$ be the accumulated amount of spending, in billions of dollars, that has occurred after n such transactions. For example, $A(1)$ is the amount of spending that has occurred after the first group of people has spent its money, so $A(1) = \$10(0.7) = \$7$ billion. $A(2)$ is the amount of spending that has occurred after the first and second groups of people have spent their money, so $A(2) = \$10(0.7) + \$7(0.7) = \$11.9$ billion, as shown in the following graph:

Accumulation of tax cut spending

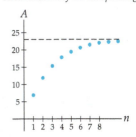

(a) Express $A(n)$ in sigma notation.

(b) Calculate $A(3)$, $A(4)$, and $A(5)$.

(c) Estimate the total spending created by this tax cut by calculating the accumulated spending for at least 10 of these transactions. Interpret your answer in terms of the given graph.

Proofs

58. Give a simple proof that $\sum_{k=5}^{n}(a_k + b_k) = \sum_{k=5}^{n} a_k + \sum_{k=5}^{n} b_k$.

59. Give a simple proof that $\sum_{k=0}^{n} 3a_k = 3\sum_{k=0}^{n} a_k$.

60. Give a simple proof that if n is a positive integer and c is any real number, then $\sum_{k=1}^{n} c = cn$.

61. Prove part (b) of Theorem 7.4 in the case when n is even: If n is a positive even integer, then $\sum_{k=1}^{n} k = \frac{n(n+1)}{2}$. (Hint:

Try some examples first, such as $n = 6$ and $n = 10$, and think about how to group the terms to get the sum quickly.)

62. Prove part (b) of Theorem 7.4 in the case when n is odd: If n is a positive odd integer, then $\sum_{k=1}^{n} k = \frac{n(n+1)}{2}$. (Hint: *Use a method similar to the one for the previous exercise, but take note of what happens with the extra middle term of the sum.)*

Thinking Forward

Functions defined by area accumulation: Let f be the function that is shown here at the left, and define a new function A so that for every $c \geq 1$, $A(c)$ is the area of the region between the graph of f and the x-axis over the interval $[1, c]$. For example, $A(2)$ is the area of the shaded region in the graph at the right.

▶ Use the graph of f to estimate the values of $A(1)$, $A(2)$, and $A(3)$. (Hint: *Consider the grid lines in the graph shown at the right.)*

▶ Describe the intervals on which the function f is positive, negative, increasing, and decreasing. Then describe the intervals on which the function A is positive, negative, increasing, and decreasing.

▶ From the figures, we can see that f is increasing and positive on $[1, \infty)$ and A is also increasing and positive on $[1, \infty)$. What would you be able to say about the area accumulation function A if f were instead

decreasing and positive? Or increasing and negative? Draw some pictures in your investigation.

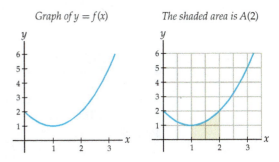

Graph of $y = f(x)$ *The shaded area is $A(2)$*

Approximating the area under a curve with rectangles: Suppose you want to find the area between the graph of a positive function f and the x-axis from $x = a$ to $x = b$. We can approximate such an area by using a sum of areas of small rectangles whose heights depend on the height $f(x)$ at various points. For the problems that follow you should choose rectangles so that each rectangle has the same width and the top left corner of each rectangle intersects the graph of f.

▶ Approximate the area between the graph of $f(x) = x^2$ and the x-axis from $x = 0$ to $x = 4$, by using four rectangles. Include a picture of the rectangles that you are using.

▶ Approximate the same area as earlier, but this time with eight rectangles. Is this an over-approximation or an under-approximation of the exact area under the graph?

Sequences of partial sums: In Exercise 57 we saw a function $A(n)$ that was defined as a sum of n terms, $A(n) = \sum_{k=1}^{n} 10(0.7)^k$. What happens as n approaches infinity? The sum $A(n)$ is called a *partial sum* because it represents part of the sum that accumulates if you let n approach infinity.

▶ Consider the sequence $A(1)$, $A(2)$, $A(3)$, ..., $A(n)$. Write out this sequence up to $n = 10$. What do you notice?

▶ As n approaches infinity, this sequence of partial sums could either *converge*, meaning that the terms eventually approach some finite limit, or it could *diverge* to infinity, meaning that the terms eventually grow without bound. Which do you think is the case here, and why?

7.2 RIEMANN SUMS

▶ Geometric approximation by the process of subdividing, approximating, and adding up

▶ Using rectangles to approximate the area under a curve

▶ Definition and types of Riemann sums in formal mathematical notation

Subdivide, Approximate, and Add Up

As you well know, the formula for finding the area of a circle of radius r is $A = \pi r^2$. In particular, a circle of radius 2 units has area $A = \pi 2^2 = 4\pi$. But wait a moment; where does this area formula come from? Why is it true? Suppose for a moment that we don't know the area formula. How could we find, or at least approximate, the area of a circle of radius 2? The three diagrams that follow suggest an answer.

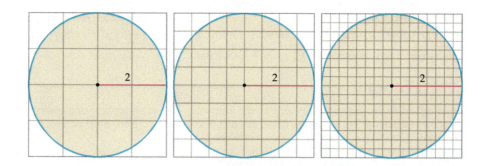

In the figure at the left, a circle of radius 2 units is shown with a grid where each square has side length 1 unit and thus area 1 square unit. We need only count up the approximate number of squares to approximate the area of the circle. The circle encloses four full squares, and 12 partial squares. We will approximate by counting each partial square as *half* of a square. This is just one of many approximation methods we could use. This method produces the approximation

$$A \approx 4(1) + 12\left(\frac{1}{2}\right) = 10 \text{ square units.}$$

As we know, the actual area of the circle is $4\pi \approx 12.5664$, so the approximation we found is not very accurate. If instead we use the grid in the second figure shown previously, we can obtain a better approximation. In this case, the circle covers 32 full squares (each of area $\left(\frac{1}{2}\right)^2 = \frac{1}{4}$) and 28 partial squares (which we will count as half-squares, or $\frac{1}{8}$ square unit each). We now have the approximation

$$A \approx 32\left(\frac{1}{4}\right) + 28\left(\frac{1}{8}\right) = 11.5 \text{ square units.}$$

This approximation is closer to the area we expected, since the squares we used were smaller. An even better approximation can be obtained by using the rightmost figure shown earlier, in which there are 164 full squares of area $\left(\frac{1}{4}\right)^2 = \frac{1}{16}$ and 60 partial squares that we'll count as having an area of $\frac{1}{32}$ square units each. This gives us the even better approximation

$$A \approx 164\left(\frac{1}{16}\right) + 60\left(\frac{1}{32}\right) = 12.125 \text{ square units.}$$

We just used a grid to subdivide a circle into a number of smaller pieces, approximated the area of each of those pieces (by counting each piece as either a full square or a half-square), and then added up each of the small approximated areas. This process of "subdividing, approximating, and adding up" is the cornerstone of the ***definite integral***, which we will introduce in Section 7.3. After we have learned the theory of integrals, we will be able to *prove* that the area formula $A = \pi r^2$ for a circle of radius r is correct. (See Section 8.5.)

Before moving on, notice that we could think of each one of our approximations of the area of a circle as the output of a sort of approximation function whose inputs are the possible grid sizes. Smaller grid sizes should produce more accurate approximations. As the grid sizes get smaller, if those approximations somehow stabilize at a real number, then we will say that the area of the object is the number to which the approximations stabilize. This is yet another place where our study of limits will pay off! We will make the ideas in this paragraph more precise in Section 7.3.

Approximating Area with Rectangles

In the rest of this chapter we will be concerned primarily with finding or approximating the area under a curve, that is, the area enclosed between the graph of some function and the x-axis on some interval $[a, b]$. To keep things simple we will begin by restricting our attention to positive functions. For example, we might be interested in calculating the area between the graph of the function f and the x-axis from $x = 0$ to $x = 2$ as shown here:

Area between f and the x-axis on [0, 2]

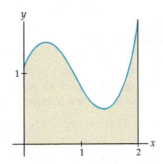

We could use a grid of squares to approximate this area, but it is more efficient to use a set of rectangles whose heights depend on the height of the function f. For example, we could subdivide the interval $[0, 2]$ into four smaller intervals of width $\frac{1}{2}$ as shown next at the left. Then we could use the heights of f at the leftmost points of the subintervals to define four rectangles, as shown in the middle and rightmost figures.

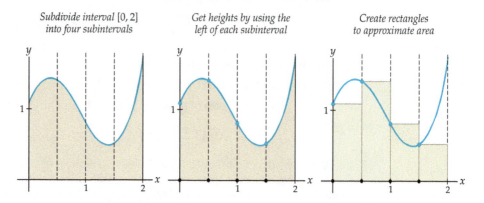

Subdivide interval $[0, 2]$ into four subintervals   *Get heights by using the left of each subinterval*   *Create rectangles to approximate area*

In the third figure, the first rectangle has height $f(0)$ and width $\frac{1}{2}$. Therefore, the area of the first rectangle is $f(0)\left(\frac{1}{2}\right)$. The remaining three rectangles have heights $f(0.5)$, $f(1)$, and $f(1.5)$, respectively, all with width $\frac{1}{2}$. Therefore the sum of the areas of these four rectangles is

$$f(0)\left(\frac{1}{2}\right) + f(0.5)\left(\frac{1}{2}\right) + f(1)\left(\frac{1}{2}\right) + f(1.5)\left(\frac{1}{2}\right).$$

The function f we have been working with happens to have the unwieldy equation

$$f(x) = \frac{1}{80}(125x^3 - 325x^2 + 175x + 89).$$

Evaluating $f(x)$ at $x = 0$, $x = 0.5$, $x = 1$, and $x = 1.5$, we see that the sum of the areas of the four rectangles is approximately

$$(1.11)\left(\frac{1}{2}\right) + (1.39)\left(\frac{1}{2}\right) + (0.80)\left(\frac{1}{2}\right) + (0.53)\left(\frac{1}{2}\right) \approx 1.915.$$

The combined area of the rectangles represents a rough approximation of the area under the graph of f on $[0, 2]$. Notice that this answer is reasonable, since the region whose area we are approximating looks to be about half of the size of the square defined by $0 \le x \le 2$ and $0 \le y \le 2$, a square with area 4.

Riemann Sums

We will now develop mathematical notation to formalize the method of using rectangles to approximate the area under a curve. This formalization will eventually enable us to apply limits so that we can find the *actual* area under a curve. Suppose f is a function that is nonnegative on an interval $[a, b]$. Consider the area under the graph of f on this interval, that is, the area of the region bounded above by the graph of f, below by the x-axis, and to the left and right by the lines $x = a$ and $x = b$. We will use the *subdivide, approximate, and add up* process to define a **Riemann sum** that will help us to approximate this area.

Subdivide: First, we subdivide the interval $[a, b]$ into n subintervals of equal width Δx. This means that $\Delta x = \frac{b - a}{n}$. We will give names to the subdivision points between the subintervals as shown next at the left. Notice that $x_0 = a$, $x_1 = a + \Delta x$, $x_2 = a + 2\Delta x$, and so on, until we end at $x_n = b$. In general, the kth subdivision point in the subdivision x_k is equal to $a + k\Delta x$.

Subdividing [a, b] into n subintervals   *Height and width of kth rectangle*

Approximate: We now need to define, for each subinterval $[x_{k-1}, x_k]$, a rectangle that approximates the area under the graph of f on that subinterval. We need to choose a height for the rectangle so that the height is somehow related to the height of the graph of f. In our previous example we chose the leftmost point of each subinterval, but in general we can choose *any* point x_k^* in $[x_{k-1}, x_k]$. The star in the x_k^* is supposed to suggest that we are choosing any point we like in that subinterval. Now we can define a rectangle over the kth subinterval $[x_{k-1}, x_k]$ such that the height of the rectangle is $f(x_k^*)$, as shown previously at the right.

Add up: If we choose n large enough—or equivalently, make Δx small enough—then the combined areas of the rectangles we define with the points x_k^* will be a decent approximation for the area under the curve. It now remains only to add up the areas of these rectangles. The area of the kth rectangle is its height times its width, or $f(x_k^*)\Delta x$, so the total area under the curve is approximated by the sum

$$f(x_1^*)\Delta x + f(x_2^*)\Delta x + f(x_3^*)\Delta x + \cdots + f(x_k^*)\Delta x + \cdots + f(x_n^*)\Delta x.$$

This sum can be expressed very compactly in sigma notation:

$$\sum_{k=1}^{n} f(x_k^*)\Delta x.$$

Note that to find this sum, we followed a "subdivide, approximate, and add" process. Sums of this form are called **Riemann sums**, named for the prolific mathematician Bernhard Riemann, who developed the notion. The following definition summarizes the notation that we have developed:

DEFINITION 7.5   **Riemann Sums**

A **Riemann sum** for a function f on an interval $[a, b]$ is a sum of the form

$$\sum_{k=1}^{n} f(x_k^*)\,\Delta x,$$

where $\Delta x = \dfrac{b-a}{n}$, $x_k = a + k\Delta x$, and x_k^* is some point in the interval $[x_{k-1}, x_k]$.

As we have seen, a Riemann sum for a function f on an interval $[a, b]$ approximates the area under the graph of f between $x = a$ and $x = b$. If f is continuous on $[a, b]$, then as the number of rectangles n approaches infinity, the approximation will get better and better and approach the actual area under the graph of f on $[a, b]$. We will make this idea mathematically precise in Section 7.3.

It is possible to consider Riemann sums for which the interval $[a, b]$ is partitioned into n subintervals by points $a = x_0, x_1, x_2, \ldots, x_{n-1}, x_n = b$ that are not equal distances apart. In such subdivisions we would have a different width $(\Delta x)_k$ for each subinterval $[x_{k-1}, x_k]$. This more general theory of Riemann sums defined with arbitrary partitions is important in later mathematics courses, but is not needed for our study of calculus here.

Signed Area

When the graph of a function lies below the x-axis, we can still use rectangles to approximate area. For example, consider the following Riemann sum with $n = 4$ rectangles that approximates the area of the region between the graph of $f(x) = x^2 - 1$ and the x-axis on $[0, 1]$:

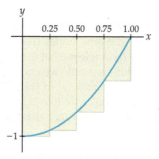

A Riemann sum with $n = 4$ for
$f(x) = x^2 - 1$ *on* $[0, 1]$

Note that the "height" of each of the rectangles in this Riemann sum is *negative*. For example, the first rectangle has a height of $f(0) = 0^2 - 1 = -1$ and is counted as having an "area" of -0.25. The left sum in this case is the sum of the areas of these four "negative" rectangles, $-0.25 - 0.234375 - 0.1875 - 0.109375 = -0.78125$.

Since, by construction, Riemann sums automatically count the area between a negative function and the x-axis as a negative number and the area between a positive function and the x-axis as a positive number, we will say that Riemann sums measure ***signed area***, also known as ***net area***. For example, if f is the following function graphed on $[0, 5]$, then the right sum, with 25 rectangles, will have some rectangles counting area positively and some rectangles counting area negatively, depending on whether f is above or below the x-axis:

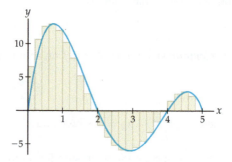

Area is counted positively on $[0, 2]$ and $[4, 5]$, and negatively on $[2, 4]$

Types of Riemann Sums

Depending on how we choose the point x_k^* in each subinterval $[x_{k-1}, x_k]$, we get different types of Riemann sums. For example, we could choose each x_k^* to be the leftmost point of the kth subinterval, as shown in the first graph that follows; this is called a ***left sum***. Alternatively, we could construct a ***right sum*** by choosing x_k^* to be the rightmost point of the kth subinterval, or a ***midpoint sum*** by choosing x_k^* to be the midpoint, as shown in the following middle and rightmost graphs, respectively:

Leftmost point of each subinterval determines heights

Rightmost point of each subinterval determines heights

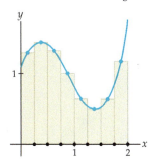

Midpoint of each subinterval determines heights

Algebraically, these three ways of choosing x_k^* from each subinterval $[x_{k-1}, x_k]$ yield the following three types of Riemann sums:

DEFINITION 7.6

Left, Right, and Midpoint Sums

Suppose f is a function defined on the interval $[a, b]$. Given a positive integer n, let $\Delta x = \frac{b-a}{n}$ and $x_k = a + k\Delta x$. Then

(a) The n-rectangle **left sum** for f on $[a, b]$ is $\sum_{k=1}^{n} f(x_{k-1}) \, \Delta x$.

(b) The n-rectangle **right sum** for f on $[a, b]$ is $\sum_{k=1}^{n} f(x_k) \, \Delta x$.

(c) The n-rectangle **midpoint sum** for f on $[a, b]$ is $\sum_{k=1}^{n} f\left(\frac{x_{k-1}+x_k}{2}\right) \Delta x$.

Note that given any interval $[a, b]$ and number n of rectangles, we can write Δx and x_k in terms of a, b, and n. In practice, we will always need to use the explicit expressions $\Delta x = \frac{b-a}{n}$ and $x_k = a + k\Delta x$ (as well as using the definition of the function f) when evaluating a Riemann sum. For example, the right sum expressed earlier is equal to

$$\sum_{k=1}^{n} f\left(a + k\left(\frac{b-a}{n}\right)\right)\left(\frac{b-a}{n}\right).$$

We can also set up Riemann sums that we can guarantee will be over-approximations or under-approximations of the actual area under the graph of a continuous function. The first figure that follows shows the **upper sum,** where each x_k^* is chosen so that $f(x_k^*)$ is the maximum value of f on $[x_{k-1}, x_k]$. The upper sum is always greater than or equal to the actual signed area. Similarly, in the **lower sum** shown in the second figure we choose x_k^* so that $f(x_k^*)$ is the minimum value of f on $[x_{k-1}, x_k]$. The lower sum is always less than or equal to the actual signed area.

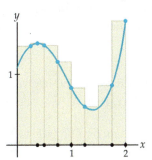

Maximum value of f on each subinterval determines heights

Minimum value of f on each subinterval determines heights

Algebraically, we can express these two types of Riemann sums as follows:

DEFINITION 7.7   **Upper and Lower Sums**

Suppose f is a function that is continuous on the interval $[a, b]$. Given a positive integer n, let $\Delta x = \frac{b-a}{n}$ and $x_k = a + k\Delta x$. Then

(a) The n-rectangle **upper sum** for f on $[a, b]$ is $\sum_{k=1}^{n} f(M_k) \, \Delta x$, where each M_k is chosen so that $f(M_k)$ is the maximum value of f on $[x_{k-1}, x_k]$.

(b) The n-rectangle **lower sum** for f on $[a, b]$ is $\sum_{k=1}^{n} f(m_k) \, \Delta x$, where each m_k is chosen so that $f(m_k)$ is the minimum value of f on $[x_{k-1}, x_k]$.

The reason we require that f be continuous on $[a, b]$ is that this is necessary in order for the Extreme Value Theorem to guarantee that f attains maximum and minimum values on each subinterval.

The upper sum and the lower sum can be more complicated to calculate than the left, right, or midpoint sums, but they can be useful if we wish to not only approximate the area under a curve, but get a bound on how much error is involved in our approximation. For example, if an upper sum approximation for the area under a curve was calculated to be 4.4 and a lower sum approximation for the same area was calculated to be 4.1, then we would know that the actual area was in the interval $[4.1, 4.4]$. In particular, this would mean that both the upper sum and the lower sum were within 0.3 square unit of the actual area under the curve.

We also don't necessarily have to use rectangles to approximate the area under a curve. For example, we could use trapezoids instead, as shown here:

Trapezoids on each subinterval approximate area under a curve

The **trapezoid sum** uses a trapezoid with width Δx, left height $f(x_{k-1})$, and right height $f(x_k)$ to approximate the slice of area in the subinterval $[x_{k-1}, x_k]$. Recall that the area of a trapezoid with base b and heights h_1 and h_2 is $\left(\frac{h_1 + h_2}{2} \right) b$. This formula for the area suggests the following algebraic definition:

DEFINITION 7.8   **Trapezoid Sums**

Suppose f is a function defined on the interval $[a, b]$. Given a positive integer n, let $\Delta x = \frac{b-a}{n}$ and $x_k = a + k\Delta x$. Then

The n-rectangle **trapezoid sum** for f on $[a, b]$ is $\sum_{k=1}^{n} \frac{f(x_{k-1}) + f(x_k)}{2} \, \Delta x$.

TECHNICAL POINT To be honest, in the form of Definition 7.8, the trapezoid sum is not technically a Riemann sum for f; recall that the terms in a Riemann sum must be of the form $f(x_k^*)\Delta x$ for some $x_k^* \in [x_{k-1}, x_k]$. However, if f is a continuous function, then in each subinterval we can use the Intermediate Value Theorem to find some x_k^* in for which $f(x_k^*)$ is equal to the average of $f(x_{k-1})$ and $f(x_k)$. This means that when f is continuous, the trapezoid sum really is a Riemann sum "in disguise."

Examples and Explorations

EXAMPLE 1

Approximating area with a right sum

Approximate the area between the graph of $f(x) = x^2 - 2x + 2$ and the x-axis from $x = 1$ to $x = 3$, using a right sum with four rectangles.

SOLUTION

We begin by subdividing the interval $[1, 3]$ into four subintervals: $[1, 1.5]$, $[1.5, 2]$, $[2, 2.5]$, and $[2.5, 3]$, each of width $\frac{1}{2}$. The rightmost points of these four subintervals are, respectively, 1.5, 2, 2.5, and 3. The values of f at these points will be the heights of our four rectangles, as shown here:

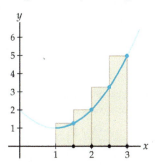

Right sum with four rectangles

It is now a simple matter to add up the areas of the four rectangles:

$$\text{Area} \approx f(1.5)\left(\frac{1}{2}\right) + f(2)\left(\frac{1}{2}\right) + f(2.5)\left(\frac{1}{2}\right) + f(3)\left(\frac{1}{2}\right)$$
$$= (1.25)\left(\frac{1}{2}\right) + (2)\left(\frac{1}{2}\right) + (3.25)\left(\frac{1}{2}\right) + (5)\left(\frac{1}{2}\right) \quad \leftarrow \text{plug into } f(x) = x^2 - 2x + 2$$
$$= 5.75.$$

EXAMPLE 2

Understanding the notation used in Riemann sums

Looking back at Example 1, identify each of the following:

$$a, b; \quad n; \quad \Delta x; \quad x_0, x_1, x_2, x_3, x_4; \quad x_1^*, x_2^*, x_3^*, x_4^*.$$

SOLUTION

We have $a = 1$, $b = 3$, $n = 4$, and $\Delta x = \frac{1}{2}$. The subdivision points are $x_0 = a = 1$, $x_1 = 1.5$, $x_2 = 2$, $x_3 = 2.5$, and $x_4 = b = 3$. The chosen points in each subinterval $[x_{k-1}, x_k]$ are taken to be the rightmost points: $x_1^* = 1.5$, $x_2^* = 2$, $x_3^* = 2.5$, and $x_4^* = 3$. Notice that since this is a right sum, each x_k^* is just equal to x_k.

EXAMPLE 3 **Expressing a right sum in sigma notation**

Consider again the area between the graph of $f(x) = x^2 - 2x + 2$ and the x-axis from $x = 1$ to $x = 3$. Express the n-rectangle right-sum approximation in sigma notation. Simplify until k and n are the only variables that appear in your expression. Then use your expression to calculate the right-sum approximations with $n = 4$ and $n = 8$ rectangles.

SOLUTION

We first subdivide $[1, 3]$ into n subintervals, each of width $\Delta x = \frac{b-a}{n} = \frac{3-1}{n} = \frac{2}{n}$. The subdivision points x_k are of the form $x_k = a + k\Delta x = 1 + \frac{2k}{n}$. Since we are constructing a right sum, in each subinterval $[x_{k-1}, x_k]$ we will choose x_k^* to be the rightmost point x_k. The kth rectangle in our approximation will have height $f(x_k^*)$ and width Δx. Using the notation we have developed, we find that the sum of the areas of the n rectangles is

$$\sum_{k=1}^{n} f(x_k^*) \Delta x = \sum_{k=1}^{n} f\left(1 + \frac{2k}{n}\right)\left(\frac{2}{n}\right)$$

$$= \sum_{k=1}^{n} \left(\left(1 + \frac{2k}{n}\right)^2 - 2\left(1 + \frac{2k}{n}\right) + 2\right)\left(\frac{2}{n}\right) = \sum_{k=1}^{n} \left(1 + \frac{4k^2}{n^2}\right)\left(\frac{2}{n}\right).$$

Substituting $n = 4$ into this expression and then writing out the sum gives us the same four-rectangle right-sum approximation that we found in Example 1:

$$\sum_{k=1}^{4} \left(1 + \frac{4k^2}{4^2}\right)\left(\frac{2}{4}\right) = \left(1 + \frac{4(1)^2}{16}\right)\left(\frac{1}{2}\right) + \left(1 + \frac{4(2)^2}{16}\right)\left(\frac{1}{2}\right)$$

$$+ \left(1 + \frac{4(3)^2}{16}\right)\left(\frac{1}{2}\right) + \left(1 + \frac{4(4)^2}{16}\right)\left(\frac{1}{2}\right)$$

$$= (1.25)\left(\frac{1}{2}\right) + (2)\left(\frac{1}{2}\right) + (3.25)\left(\frac{1}{2}\right) + (5)\left(\frac{1}{2}\right)$$

$$= 5.75.$$

Similarly, we can substitute $n = 8$ and then expand the sum to calculate the eight-rectangle right-sum approximation:

$$\sum_{k=1}^{8} \left(1 + \frac{4k^2}{(8)^2}\right)\left(\frac{2}{8}\right) = \left(1 + \frac{4(1)^2}{64}\right)\left(\frac{1}{4}\right) + \left(1 + \frac{4(2)^2}{64}\right)\left(\frac{1}{4}\right) + \left(1 + \frac{4(3)^2}{64}\right)\left(\frac{1}{4}\right)$$

$$+ \left(1 + \frac{4(4)^2}{64}\right)\left(\frac{1}{4}\right) + \left(1 + \frac{4(5)^2}{64}\right)\left(\frac{1}{4}\right) + \left(1 + \frac{4(6)^2}{64}\right)\left(\frac{1}{4}\right)$$

$$+ \left(1 + \frac{4(7)^2}{64}\right)\left(\frac{1}{4}\right) + \left(1 + \frac{4(8)^2}{64}\right)\left(\frac{1}{4}\right)$$

$$= (1.0625)\left(\frac{1}{4}\right) + (1.25)\left(\frac{1}{4}\right) + (1.5625)\left(\frac{1}{4}\right) + (2)\left(\frac{1}{4}\right)$$

$$+ (2.5625)\left(\frac{1}{4}\right) + (3.25)\left(\frac{1}{4}\right) + (4.0625)\left(\frac{1}{4}\right) + (5)\left(\frac{1}{4}\right)$$

$$= 5.1875.$$

Notice that sigma notation does not in general make it easier to calculate the areas of the rectangles in our approximations. However, the general n-rectangle sigma notation for a Riemann sum will be very important when we discuss the theory of definite integrals in the next section. It would also be very useful if you wanted to write a computer program to approximate the area under a curve, especially with a large number of rectangles. ☐

EXAMPLE 4 **Using the left, midpoint, and trapezoid sums to approximate area**

Approximate the area between the graph of $f(x) = x^2 - 2x + 2$ and the x-axis on $[1, 3]$, using $n = 4$ subintervals with (a) the left sum; (b) the midpoint sum; and (c) the trapezoid sum.

SOLUTION

(a) As in Example 1, we have $\Delta x = \frac{1}{2}$ and $x_k = 1 + k\Delta x = 1 + \frac{k}{2}$. For a left sum we choose x_k^* to be the leftmost point in the interval $[x_{k-1}, x_k]$, and therefore we choose $x_k^* = x_{k-1} = 1 + \frac{k-1}{2}$. The left sum with four rectangles is therefore equal to

$$\sum_{k=1}^{4} f\left(1 + \frac{k-1}{2}\right)\left(\frac{1}{2}\right) = f(1)\left(\frac{1}{2}\right) + f(1.5)\left(\frac{1}{2}\right) + f(2)\left(\frac{1}{2}\right) + f(2.5)\left(\frac{1}{2}\right) = 3.75.$$

(b) For a midpoint sum we choose x_k^* to be the midpoint in the interval $[x_{k-1}, x_k]$, and therefore we choose $x_k^* = \frac{x_{k-1} + x_k}{2} = \frac{(1 + \frac{k-1}{2}) + (1 + \frac{k}{2})}{2} = \frac{3 + 2k}{4}$. The midpoint sum with four rectangles is therefore equal to

$$\sum_{k=1}^{4} f\left(\frac{3 + 2k}{4}\right)\left(\frac{1}{2}\right) = f(1.25)\left(\frac{1}{2}\right) + f(1.75)\left(\frac{1}{2}\right) + f(2.25)\left(\frac{1}{2}\right) + f(2.75)\left(\frac{1}{2}\right) = 4.625.$$

(c) For the trapezoid sum we do something completely different and use the heights at both $x_{k-1} = 1 + \frac{k-1}{2}$ and $x_k = 1 + \frac{k}{2}$ to calculate the area of the kth trapezoid. The trapezoid sum with four trapezoids is equal to

$$\sum_{k=1}^{4} \frac{f\left(1 + \frac{k-1}{2}\right) + f\left(1 + \frac{k}{2}\right)}{2}\left(\frac{1}{2}\right) = \frac{f(1) + f(1.5)}{2}\left(\frac{1}{2}\right) + \frac{f(1.5) + f(2)}{2}\left(\frac{1}{2}\right)$$
$$+ \frac{f(2) + f(2.5)}{2}\left(\frac{1}{2}\right) + \frac{f(2.5) + f(3)}{2}\left(\frac{1}{2}\right) = 4.75.$$

The three figures that follow show the three area approximations we just calculated. Which one do you think is the most accurate, and why? (See Exercise 5.)

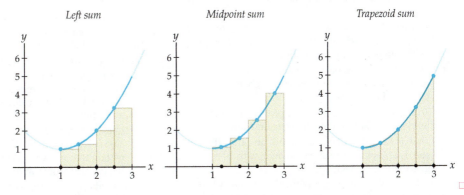

Left sum   *Midpoint sum*   *Trapezoid sum*

TEST YOUR UNDERSTANDING

▶ Recalling the discussion of the circle at the start of this section, can you think of a different way of approximating the area of the circle?

▶ The Riemann sums presented in this section are not the only sums we could use to approximate the area under a curve. For example, we could use a "one-third" sum, which would define x_k^* to be the point one-third of the way across the interval $[x_{k-1}, x_k]$. How would you write x_k^* in terms of x_{k-1} and x_k for the one-third sum? What other methods for choosing x_k^* can you think of?

▶ Both the midpoint sum and the trapezoid sum use an average somewhere in their formula. What is the average that is being taken in each case?

▶ Can you label x_2, x_2^*, and $f(x_2^*)$ in each of the Riemann sum figures in this section?

▶ In general, the left sum will always underestimate the area under an increasing function and overestimate the area under a decreasing function; why? What about the right sum?

EXERCISES 7.2

Thinking Back

Simple area formulas: Give formulas for the areas of each of the following geometric figures.

▶ A circle of radius r

▶ A semicircle of radius r

▶ A right triangle with legs of lengths a and b

▶ A triangle with base b and altitude h

▶ A rectangle with sides of lengths w and l

▶ A trapezoid with width w and heights h_1 and h_2

Concepts

0. *Problem Zero:* Read the section and make your own summary of the material.

1. *True/False:* Determine whether each of the statements that follow is true or false. If a statement is true, explain why. If a statement is false, provide a counterexample.

(a) *True or False:* The n-rectangle lower sum for f on $[a, b]$ is always equal to the corresponding right sum.

(b) *True or False:* If f is positive and increasing on $[a, b]$, then any left sum for f on $[a, b]$ will be an under-approximation.

(c) *True or False:* If $f(a) > f(b)$, then any right-sum approximation for f on $[a, b]$ will be an under-approximation.

(d) *True or False:* A midpoint sum is always a better approximation than a left sum.

(e) *True or False:* If f is positive and concave up on all of $[a, b]$, then any left-sum approximation for f on $[a, b]$ will be an under-approximation.

(f) *True or False:* If f is positive and concave up on all of $[a, b]$, then any trapezoid sum approximation for f on $[a, b]$ will be an over-approximation.

(g) *True or False:* An upper sum approximation for f on $[a, b]$ can never be an under-approximation.

(h) *True or False:* For every function f on $[a, b]$, the left sum is always a better approximation with 10 rectangles than with 5 rectangles.

2. *Examples:* Construct examples of the thing(s) described in the following. Try to find examples that are different than any in the reading.

(a) A graph of a function f on an interval $[a, b]$ for which the left sum with four rectangles finds the *exact* area under the graph of f from $x = a$ to $x = b$.

(b) A graph of a function f on an interval $[a, b]$ for which *all* trapezoid sums (regardless of the size of n) will find the *exact* area under the graph of f from $x = a$ to $x = b$.

(c) A graph of a function f on an interval $[a, b]$ for which the upper sum with four rectangles is a much better approximation than the lower sum with four rectangles.

3. In the reading we used objects whose area we knew (squares) to approximate the area of a more complicated object (a circle). The same kind of technique can be used to approximate the area of the blob pictured here.

(a) Given that each of the squares in the grid has a side length of $\frac{1}{2}$ square unit, approximate the area of the blob.

(b) How could you get a better approximation?

4. Consider the area between the graph of a function f and the x-axis from $x = 0$ to $x = 2$, as shown in each of the two figures that follow.

(a) Use the grid on the left and whatever method you like to approximate this area. Then use the grid on the right and the same method to make another approximation. Which approximation is likely more accurate, and why?

(b) Use the grid at the left to get an upper bound on the area of the region. In other words, make an approximation that you *know* is greater than the actual area. Repeat for the grid at the right.

(c) Use the grid at the left to get a lower bound on the area of the region. In other words, make an approximation that you *know* is less than the actual area. Repeat for the grid at the right.

(d) Use your answers to parts (a)–(c) to come up with your best possible guess for the actual area under the curve.

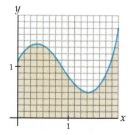

5. In Examples 1 and 4 we found four different approximations for the area between the graph of $f(x) = x^2 - 2x + 2$ and the x-axis on $[1, 3]$.

   (a) Based on the pictures of these approximations given in the reading, which are over-approximations? Which are under-approximations? Which approximation looks like it might be closest to the actual area under the curve?

   (b) The actual area (which we won't know how to calculate until Section 7.5) under the graph of $f(x) = x^2 - 2x + 2$ on $[1, 3]$ is equal to $\frac{14}{3} \approx 4.6667$. Given this information, which of the approximations in Examples 1 and 4 was the most accurate? Was that what you expected?

6. Do you think that one type of Riemann sum (right, left, midpoint, upper, lower, trapezoid) is usually more accurate than the others? Why or why not?

7. Suppose that the left-sum approximation with eight rectangles for the area between the graph of a function f and the x-axis from $x = a$ to $x = b$ is equal to 8.2 and that the corresponding right sum approximation is equal to 7.5.

   (a) What is the corresponding trapezoid sum approximation for this area?

   (b) Is the corresponding midpoint sum for this area necessarily between 7.5 and 8.2? If so, explain why. If not, sketch an example of a function f on an interval $[a, b]$ whose midpoint sum is *not* between the left sum and the right sum.

   (c) What can you say about the corresponding upper sum for this area? The corresponding lower sum?

   (d) Is it necessarily true that f is decreasing on the entire interval $[a, b]$? If so, explain why. If not, sketch an counterexample in which the left sum is greater than the right sum but f is *not* decreasing on all of $[a, b]$.

   (e) Could the function f be increasing on the entire interval $[a, b]$? If not, explain why not. If so, sketch a possible example in which the left sum is greater than the right sum and f is increasing on all of $[a, b]$.

8. In the reading we mentioned that the trapezoid sum is the average of the left sum and the right sum. Use the solutions of Examples 1 and 4 to show that for $f(x) = x^2 - 2x + 2$, $[a, b] = [1, 3]$, and $n = 4$, the trapezoid sum is indeed the average of the left sum and the right sum.

9. Explain why the upper sum approximation for the area between the graph of a function f and the x-axis on $[a, b]$ must always be larger than or equal to any other type of Riemann sum approximation with the same number n of rectangles.

10. Consider the area between the graph of a positive function f and the x-axis on an interval $[a, b]$. Explain why the upper sum approximation for this area with $n = 8$ boxes must be smaller than or equal to the upper sum approximation with $n = 4$ boxes. It may help to sketch some examples.

11. Suppose you wanted to calculate the upper sum approximation for the area between the graph of $f(x) = (x - 1)^2$ and the x-axis from $x = 0$ to $x = 2$. List all of the values M_k used for (a) $n = 2$ rectangles, (b) $n = 3$ rectangles, and (c) $n = 4$ rectangles. Sketch graphs of your rectangles to illustrate your answers.

12. Repeat Exercise 11, using the lower sum approximation and the values m_k.

13. Suppose $v(t)$ is the velocity of a particle moving on a straight path, where v is measured in meters per second and t is measured in seconds. The particle starts moving at time t_0 and moves for Δt seconds.

   (a) What are the units of $v(t_0)\Delta t$?

   (b) Geometrically, what does $v(t_0)\Delta t$ represent?

   (c) What do these questions have to do with this section?

14. Explain in no more than three sentences how we can approximate the derivative of a function f at a point c if we know the graph of f. Then, in no more than three additional sentences, discuss how the method for approximating area is similar to, and how is it different from, approximating the derivative. (Both descriptions should involve *multiple* approximations, each better than the last.)

15. Explain why the sum $\sum_{k=1}^{20} f\left(-3 + \frac{k}{2}\right)(0.25)$ can't be a right sum for f on $[a, b] = [-3, 2]$.

16. Explain why the sum $\sum_{k=1}^{100} f(2 + 0.1(k - 1))(0.1)$ can't be a left sum for f on $[a, b] = [2, 5]$.

Each of the sums in Exercises 17–20 approximates the area between the graph of some function f and the x-axis from $x = a$ to $x = b$. Do some "reverse engineering" to determine the type of approximation (left sum, midpoint sum, etc.) and identify $f(x)$, a, b, n, Δx, and x_k. Then sketch the approximation described.

17. $\displaystyle\sum_{k=1}^{4} \left(1 + \frac{k}{2}\right)^2 \left(\frac{1}{2}\right)$

18. $\displaystyle\sum_{k=0}^{2} \ln\left(2 + \frac{k}{3}\right)\left(\frac{1}{3}\right)$

19. $\displaystyle\sum_{k=1}^{100} \frac{\sin(0.05(k - 1)) + \sin(0.05k)}{2}(0.05)$

20. $\displaystyle\sum_{k=1}^{9} \sqrt{\frac{\left(1 + \frac{k-1}{3}\right) + \left(1 + \frac{k}{3}\right)}{2}} \left(\frac{1}{3}\right)$

Skills

Your calculator should be able to approximate the area between a graph and the x-axis. Determine how to do this on your particular calculator, and then, in Exercises 21–26, use the method to approximate the signed area between the graph of each function f and the x-axis on the given interval $[a, b]$.

21. $f(x) = \sqrt{x - 1}$, $[a, b] = [2, 3]$

22. $f(x) = x^2$, $[a, b] = [0, 3]$

23. $f(x) = e^x$, $[a, b] = [1, 4]$

24. $f(x) = \sin x$, $[a, b] = [0, \pi]$

25. $f(x) = (x-2)^2 + 1$, $[a, b] = [1, 3]$

26. $f(x) = 1 - 2^x$, $[a, b] = [-3, 1]$

For each function f and interval $[a, b]$ in Exercises 27–33, use the given approximation method to approximate the signed area between the graph of f and the x-axis on $[a, b]$. Determine whether each of your approximations is likely to be an over-approximation or an under-approximation of the actual area.

27. $f(x) = x^2$, $[a, b] = [0, 3]$, left sum with
    (a) $n = 3$ (b) $n = 6$

28. $f(x) = \sin x$, $[a, b] = [0, \pi]$, $n = 3$, with
    (a) trapezoid sum   (b) upper sum

29. $f(x) = \sqrt{x-1}$, $[a, b] = [2, 3]$, $n = 4$, with
    (a) left sum   (b) right sum

30. $f(x) = 1 - 2^x$, $[a, b] = [-3, 1]$, $n = 8$, with
    (a) left sum   (b) right sum

31. $f(x) = e^x$, $[a, b] = [1, 4]$, $n = 6$, with
    (a) midpoint sum   (b) trapezoid sum

32. $f(x) = 9 - x^2$, $[a, b] = [0, 5]$, $n = 5$, with
    (a) midpoint sum   (b) lower sum

33. $f(x) = (x - 2)^2 + 1$, $[a, b] = [1, 3]$, lower sum with
    (a) $n = 2$ (b) $n = 3$ (c) $n = 4$

For each function f and interval $[a, b]$ in Exercises 34–38, it is possible to find the *exact* signed area between the graph of f and the x-axis on $[a, b]$ geometrically by using the areas of circles, triangles, and rectangles. Find this exact area, and then calculate the left, right, midpoint, upper, lower, and trapezoid sums with $n = 4$. Which approximation rule is most accurate?

34. $f(x) = 5$, $[a, b] = [-2, 2]$

35. $f(x) = 3x + 1$, $[a, b] = [3, 5]$

36. $f(x) = 4 - x$, $[a, b] = [0, 6]$

37. $f(x) = \sqrt{1 - x^2}$, $[a, b] = [-1, 1]$

38. $f(x) = 3 + \sqrt{4 - x^2}$, $[a, b] = [-2, 2]$

In Exercises 39–44, write out the sigma notation for the Riemann sum described in such a way that the only letter which appears in the general term of the sum is k. Don't calculate the value of the sum; just write it down in sigma notation.

39. $f(x) = \sqrt{x - 1}$, $[a, b] = [2, 3]$, right sum, $n = 4$.

40. $f(x) = x^2$, $[a, b] = [0, 3]$, left sum, $n = 3$.

41. $f(x) = e^x$, $[a, b] = [1, 4]$, midpoint sum, $n = 6$.

42. $f(x) = \ln x$, $[a, b] = [2, 5]$, left sum, $n = 100$.

43. $f(x) = \sin x$, $[a, b] = [0, \pi]$, trapezoid sum, $n = 4$.

44. $f(x) = \sqrt{1 - x^2}$, $[a, b] = [-1, 1]$, midpoint sum, $n = 20$.

Applications

45. Suppose that, as in Section 7.1, you drive in a car for 40 seconds with velocity $v(t) = -0.22t^2 + 8.8t$ feet per second, as shown in the graph that follows. If your total distance travelled is equal to the area under the velocity curve on $[0, 40]$, then find lower and upper bounds for your distance travelled by using

    (a) the lower sum with $n = 4$ rectangles;
    (b) the upper sum with $n = 4$ rectangles.

Velocity of car
$$v(t) = -0.22t^2 + 8.8t$$

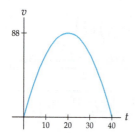

46. Repeat Exercise 45, using

    (a) the midpoint sum with $n = 4$ rectangles;
    (b) the trapezoid sum with $n = 4$ trapezoids.

    If the exact distance travelled is just over 2,346 feet, then which of these approximations is the most accurate?

47. Dad's casserole surprise is hot out of the oven, and its temperature after t minutes is given by the function $T(t)$, measured in degrees. The casserole cools by changing at a rate of $T'(t) = -15e^{-0.5t}$ degrees per minute.

Rate of change of casserole temperature
$$T'(t) = -15e^{-0.5t}$$

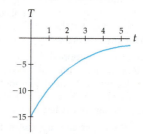

    (a) Discuss what the graph of $T'(t)$ says about the behavior of the casserole's temperature after the casserole is taken out of the oven.
    (b) Just as we can approximate distance travelled by approximating the area under the corresponding velocity curve, we can approximate the change in temperature T of the casserole by approximating the area under the graph of T'. Why?
    (c) Estimate the change in temperature of the casserole over the first 5 minutes it is out of the oven, using any Riemann sum you like, with $n = 10$.

48. The National Oceanic and Atmospheric Administration tabulates flow data from many American rivers. From this data they compute and plot median annual flows. The flows are given by functions whose closed form is

not known, but for which we can read off values for any day of the year we like. The following table describes the flow $f(t)$ in cubic feet per second for Idaho's Lochsa River, t days after January 1:

t	0	60	120	180	240	300	360
$f(t)$	700	1000	6300	4000	500	650	700

(a) Keeping in mind that the data are periodic with period 365, compute the left Riemann sum for the function that these data have sampled.

(b) What is the relation between the numbers you computed and the total amount of water that flows down the Lochsa annually? Estimate the total amount of water that flows down the Lochsa each year.

(c) Most of the flow down the river takes place from April to July. We can get a better idea of the total flow if we add a few data points. Recompute the left Riemann sum, adding the data points $(90, 2100)$, $(150, 11000)$, and $(210, 1000)$.

(d) What would you need to do to get an even better estimate of the total flow?

49. To approximate the flow $f(t)$ of the Lochsa River in its flood stage, we can use a function of the form

$$g(t) = c_1 + c_2 \left(\sin\left(\frac{(t - 90)\pi}{105} \right) - \frac{2}{\pi} \right),$$

where the coefficients c_1 and c_2 are found by evaluating the following two integrals:

$$c_1 = \frac{1}{105} \int_{90}^{195} f(t) \, dt,$$

$$c_2 = \frac{1}{9.95} \int_{90}^{195} f(t) \left(\sin\left(\frac{(t - 90)\pi}{105} \right) - \frac{2}{\pi} \right) dt.$$

(a) Use the data points $(t, f(t)) = (90, 2100)$, $(120, 6300)$, $(150, 11000)$, and $(180, 4000)$, and left Riemann sums to approximate the values of the integrals for c_1 and c_2.

(b) Now that you have found c_1 and c_2, plot the resulting function $g(t)$ against the data points from Exercise 48.

Proofs

50. Use Definition 7.7 to prove that for any function f and interval $[a, b]$, the upper sum with n rectangles is greater than or equal to the lower sum with n rectangles.

51. Use Definition 7.6 and the definition of *increasing* to prove that if a function f is positive and increasing on $[a, b]$, then the left sum with n rectangles is less than or equal to the right sum with n rectangles.

52. Use Definitions 7.6 and 7.8 to prove that for any function f and interval $[a, b]$, the trapezoid sum with n trapezoids is always the average of the left sum and the right sum with n rectangles.

53. Use Definition 7.8 to prove that if a function f is positive and concave up on $[a, b]$, then the trapezoid sum with n trapezoids is an always an over-approximation for the actual area.

Thinking Forward

Approximating the length of a curve: Suppose you want to calculate the driving distance between New York City and Dallas, Texas.

▶ Print out a highway map of the United States, and highlight a route, snaking along with the paths of the major highways.

▶ How could we use the method of "subdivide, approximate, and add up" to approximate this driving distance?

▶ Make an actual approximation of the driving distance from New York City to Dallas, Texas, using the route and method you just described.

▶ How accurate do you think your approximation is? Is it an over-approximation or an under-approximation?

Limits of Riemann sums: In the reading we saw that the area between the graph of $f(x) = x^2 - 2x + 2$ and the x-axis on $[1, 3]$ could be approximated with the right sum $\sum_{k=1}^{n} \left(1 + \frac{4k^2}{n^2} \right) \left(\frac{2}{n} \right)$. Let $A(n)$ be equal to this n-rectangle right-sum approximation. The following table describes various values of $A(n)$:

n	10	100	1000	10,000
$A(n)$	5.08	4.7068	4.6707	4.66707

▶ Describe the meaning of the entries in this table, and verify that the entry for $A(10)$ is correct.

▶ Use the table to make a graph of $A(n)$, and discuss what happens to this graph as n approaches infinity.

▶ What does your graph tell you about the right-sum approximations of the area under the graph of f as n approaches infinity?

Approximations and error: In Section 7.5 we will see that definite integrals can be computed by taking differences of antiderivatives; in particular, the *Fundamental Theorem of Calculus* will reveal that if f is continuous on $[a, b]$, then $\int_a^b f(x) \, dx = F(b) - F(a)$, where F is any antiderivative of f. Armed with this fact, we can check the exact error of Riemann sum approximations for integrals of functions that we can antidifferentiate.

▶ Use the given antiderivative fact to find the exact value of $\int_1^4 \frac{1}{x} \, dx$. (Hint: What is an antiderivative of $\frac{1}{x}$?

In other words, what is a function F whose derivative is
$f(x) = \frac{1}{x}$?)

▶ What is the actual error that results from a right-sum approximation with $n = 4$ for $\int_1^4 \frac{1}{x} dx$?

▶ By trial and error with smaller values of n, find the smallest value of n for which a right sum will approximate $\int_1^4 \frac{1}{x} dx$ to within 0.25.

▶ Repeat the preceding steps, but with the midpoint sum in place of the right sum.

7.3   DEFINITE INTEGRALS

▶ The exact area under the graph of a function f on $[a,b]$

▶ Defining the definite integral as a limit of Riemann sums

▶ Properties of the definite integral

Defining the Area Under a Curve

We have seen how to approximate areas with Riemann sums, but how do we find the *exact* area between the graph of a function f and the x-axis on an interval $[a, b]$? Consider the area of the region bounded by the curve $f(x) = x^2$, the x-axis, and the line $x = 1$ as shown here:

Area under $f(x) = x^2$ on $[0, 1]$

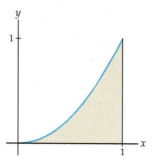

We don't yet have a rigorous mathematical method for finding this area exactly. Using Riemann sums, we can get approximations for the area. For example, the three figures that follow illustrate upper-sum approximations using $n = 10$, $n = 50$, and $n = 100$. Notice that in this case, since $f(x) = x^2$ is increasing on $[0, 1]$, the upper sum happens to be the same as the right sum.

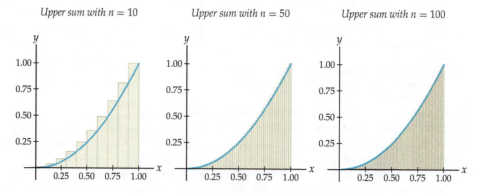

Upper sum with $n = 10$ *Upper sum with $n = 50$*   *Upper sum with $n = 100$*

As the number n of rectangles gets larger, the upper sum approximation gets closer to the area under $f(x) = x^2$ on $[0, 1]$. The same thing happens if we consider any other Riemann sum, say, the lower sum, as pictured here for $n = 10$, $n = 50$, and $n = 100$:

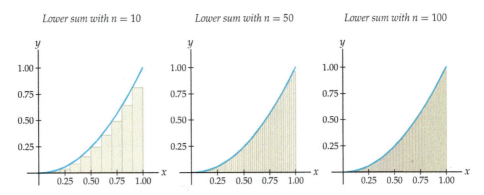

Lower sum with $n = 10$ Lower sum with $n = 50$ Lower sum with $n = 100$

If a sequence of Riemann sum approximations converges to a real number when we take the limit as n approaches infinity, then we call this number the "area" under the graph. Notice that up until this point we did not actually have a truly rigorous definition for even the *concept* of such an area! The following definition describes this limit and the notation we will use to refer to it.

DEFINITION 7.9

The Definite Integral of a Function on an Interval

Let f be a function defined on an interval $[a, b]$. The **definite integral** of f from $x = a$ to $x = b$ is defined to be the number

$$\int_a^b f(x)\,dx = \lim_{n \to \infty} \sum_{k=1}^n f(x_k^*)\,\Delta x,$$

if this limit exists, where $\Delta x = \frac{b-a}{n}$, $x_k = a + k\,\Delta x$, and x_k^* is any choice of point in $[x_{k-1}, x_k]$.

The "$\int$" symbol should remind you of the letter "S" for "Sum." A Riemann sum describes a discrete sum of areas of rectangles, while in a loose sense an integral represents a continuous sum of the areas of infinitely many rectangles, each of which is infinitely thin. The a and b below and above the integral symbol are called the **limits of integration**, where here "limits" is used in the sense of "ends," not in the sense of limits of functions. It may help you remember this definition if you think of it in the following way: As $n \to \infty$, the finite sum ($\sum$) of n things becomes an integral ($\int$) that accumulates everything from $x = a$ to $x = b$. The discrete list of values $f(x_k^*)$ becomes the continuous function $f(x)$, and the small change Δx becomes an "infinitesimal" change dx.

⊗ CAUTION

The "dx" portion of the definite integral is not optional; it must always be included with the notation of the definite integral. Just as

$$\frac{d}{dx}(\underline{\hspace{2em}})$$

is how we denote "the derivative of ____," the notation

$$\int_a^b \underline{\hspace{2em}} dx$$

is how we will denote "the definite integral of ____ on $[a, b]$," which represents "the signed area between the graph of ____ and the x-axis on $[a, b]$." Note that in both cases there is the presence of a "dx" that represents the result of a limiting process on a change Δx in the independent variable with respect to which the differentiation or integration/accumulation/area process is taking place.

Consider again the example with $f(x) = x^2$ and $[a, b] = [0, 1]$. For any given n, we have

$$\Delta x = \frac{b-a}{n} = \frac{1}{n} \quad \text{and} \quad x_k = a + k\Delta x = \frac{k}{n}.$$

We will use the right sum and choose each x_k^* to be the rightmost point $x_k = \frac{k}{n}$ of the subinterval $[x_{k-1}, x_k]$. When we take the limit as the number n of subdivisions approaches infinity, we obtain

$$\int_0^1 x^2 \, dx = \lim_{n \to \infty} \sum_{k=1}^n \left(\frac{k}{n}\right)^2 \left(\frac{1}{n}\right).$$

If, for a given function f and interval $[a, b]$, the limit defining the integral exists for any choice of points x_k^*, then we say that f is **integrable** on $[a, b]$. It turns out that every continuous function is integrable on $[a, b]$. (The proof of this theorem is beyond the scope of the text and will not be presented here.) However, *some* discontinuous functions are still integrable; for example, functions with removable or jump discontinuities are still integrable, but some functions with vertical asymptotes may not be. You will investigate various examples of discontinuous functions in Exercises 19 and 20. In Section 8.6 you will see the somewhat surprising spectacle of certain functions that have vertical asymptotes and yet are integrable.

Properties of Definite Integrals

Recall from the previous section that Riemann sums measure *signed* areas: Areas of regions above the x-axis are counted positively, and areas of regions below the x-axis are counted negatively. Because definite integrals are defined to be limits of Riemann sums, this means that definite integrals also automatically count signed areas:

THEOREM 7.10

Definite Integrals Measure Signed Areas

Definite integrals count areas above the x-axis positively and areas below the x-axis negatively. Algebraically, this means that

(a) If $f(x) \geq 0$ on all of an interval $[a, b]$, then $\int_a^b f(x) \, dx \geq 0$.

(b) If $f(x) \leq 0$ on all of an interval $[c, d]$, then $\int_c^d f(x) \, dx \leq 0$.

For example, if f is the function graphed next, then the definite integral $\int_0^5 f(x) \, dx$ will count the areas of the regions marked A and C positively and count the area of the region marked B negatively:

$$\int_0^5 f(x) \, dx \text{ measures the signed area } A - B + C$$

The fact that definite integrals are defined in terms of sums and limits allows us to state many nice properties of definite integrals. In particular, because definite integrals are just limits of sums, they behave well with sums and constant multiples, as follows:

THEOREM 7.11   **Sum and Constant-Multiple Rules for Definite Integrals**

For any functions f and g that are integrable on $[a, b]$ and any real number k,

(a) $\displaystyle \int_a^b (f(x) + g(x))\, dx = \int_a^b f(x)\, dx + \int_a^b g(x)\, dx.$

(b) $\displaystyle \int_a^b kf(x)\, dx = k \int_a^b f(x)\, dx.$

Before we present a formal proof of this theorem, let's think about the first part of it graphically. The first graph that follows shows a left sum for the area between the graph of $f(x) = x^2$ and the x-axis on $[0, 2]$. The second graph shows a left sum (with the same n) for the area between $g(x) = x$ and the x-axis on the same interval. In the third graph we see that the sum of these two left sums is itself a left sum for the graph of $f(x) + g(x) = x^2 + x$ on $[0, 2]$. As n approaches infinity, the area of the beige rectangles will approach the area under f, the area of the blue rectangles will approach the area under g, and the sum of the areas of the beige and blue rectangles will approach the area under $f + g$.

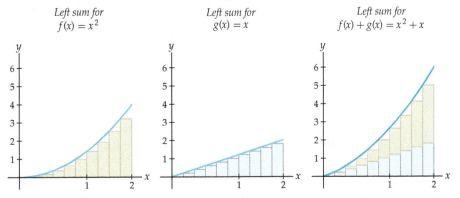

Left sum for
$f(x) = x^2$

Left sum for
$g(x) = x$

Left sum for
$f(x) + g(x) = x^2 + x$

The key to the algebraic proof of Theorem 7.11 is that sums and constant multiples commute with limits and with Riemann sums. Since the definite integral is a limit of Riemann sums, it must be that sums and constant multiples commute with definite integrals.

Proof. Given a positive integer n, define $\Delta x = \dfrac{b - a}{n}$ and $x_k = a + k\Delta x$, and let x_k^* be any point in the subinterval $[x_{k-1}, x_k]$. Then by the definition of the definite integral, we have

$$\int_a^b (f(x) + g(x))\, dx = \lim_{n \to \infty} \sum_{k=1}^n (f(x_k^*) + g(x_k^*))\, \Delta x \qquad \leftarrow \text{definite integral definition}$$

$$= \lim_{n \to \infty} \sum_{k=1}^n (f(x_k^*)\, \Delta x + g(x_k^*)\, \Delta x) \qquad \leftarrow \text{algebra}$$

$$= \lim_{n \to \infty} \left(\sum_{k=1}^n f(x_k^*)\, \Delta x + \sum_{k=1}^n g(x_k^*)\, \Delta x \right) \qquad \leftarrow \text{split into two sums}$$

$$= \lim_{n \to \infty} \sum_{k=1}^n f(x_k^*)\, \Delta x + \lim_{n \to \infty} \sum_{k=1}^n g(x_k^*)\, \Delta x \qquad \leftarrow \text{sum rule for limits}$$

$$= \int_a^b f(x)\, dx + \int_a^b g(x)\, dx. \qquad \leftarrow \text{definite integral definition}$$

The proof of part (b) is similar: Write the definite integral of $y = kf(x)$ as a limit of Riemann sums, use properties of sums and limits to pull out the constant k, and then use the definition of the definite integral again to get k times the definite integral of f. You will write out the details in Exercise 55. ■

The following theorem describes what happens if we take a definite integral over a zero-length interval, or from right to left, or in two pieces:

THEOREM 7.12

Properties Concerning Limits of Integration

If $a < b$ are real numbers and f is integrable on $[a, b]$, then

(a) $\displaystyle\int_a^a f(x)\,dx = 0$ **(b)** $\displaystyle\int_b^a f(x)\,dx = -\int_a^b f(x)\,dx$

Moreover, if c is any real number in $[a, b]$, then

(c) $\displaystyle\int_a^b f(x)\,dx = \int_a^c f(x)\,dx + \int_c^b f(x)\,dx$

Again, the properties stated in this theorem follow directly from the definition of the definite integral as a limit of Riemann sums. The first property makes sense graphically because the area under the graph of a function f from $x = a$ to $x = a$ would have a width of zero. You will prove this property algebraically in Exercise 56. The second property enables us to consider definite integrals like $\int_4^1 x^2\,dx$, in which the starting x-value is greater than the ending x-value. If we integrate "backwards," say, from $x = 4$ to $x = 1$, then we count the area given by the definite integral negatively. This property can also be proved from the definition of the definite integral, where in this situation we would have a *negative* change $\Delta x = \frac{a-b}{n}$. You will prove the property in Exercise 57.

The third part of Theorem 7.12 can also be proved by using the definition of the definite integral, but instead we present a "convincing argument by picture" in the three figures shown next. In this picture we have $a = 0$, $b = 3$, and $c = 1$. The area from $x = 0$ to $x = 3$ is clearly the sum of the area from $x = 0$ to $x = 1$ and the area from $x = 1$ to $x = 3$.

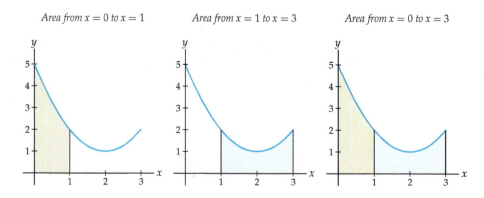

Formulas for Three Simple Definite Integrals

In Section 7.5 we will learn a method for quickly calculating many definite integrals. In the meantime, it will be useful to have a small set of integration formulas that we can use to solve simple problems. The next theorem describes definite integrals of three common functions: constant functions, the identity function, and the squaring function. Using these formulas, together with the sum and constant-multiple rules for definite integrals, we will be able to quickly evaluate definite integrals for any quadradic functions.

THEOREM 7.13

Definite Integral Formulas

For any real numbers a, b, and c,

(a) $\displaystyle\int_a^b c\,dx = c(b-a)$ **(b)** $\displaystyle\int_a^b x\,dx = \frac{1}{2}(b^2 - a^2)$ **(c)** $\displaystyle\int_a^b x^2\,dx = \frac{1}{3}(b^3 - a^3)$

These formulas are just the tip of the iceberg; in Section 7.5 we will show that we can solve many definite integrals by using antiderivatives, and in Chapter 8 we will develop methods for quickly evaluating many more types of definite integrals.

TECHNICAL POINT The definite integral formulas in Theorem 7.13 are true even when the definite integral in question turns out to be a negative number. For example,

$$\int_{-2}^1 x\,dx = \frac{1}{2}(1^2 - (-2)^2) = \frac{1}{2}(1 - 4) = -\frac{3}{2}.$$

In Section 7.6 we will discuss what it means when a definite integral on an interval $[a, b]$ is negative.

Proof. We will use the definition of the definite integral to prove part (a) of the theorem. The proofs of parts (b) and (c) are similar and are left to Exercises 60 and 61. The first two parts of the theorem can also be argued geometrically; see Exercises 58 and 59. Suppose $f(x) = c$ is a constant function on $[a, b]$. Then by the definition of the definite integral, we have

$$\int_a^b f(x)\,dx = \lim_{n\to\infty} \sum_{k=1}^n f(x_k^*)\,\Delta x \qquad \leftarrow \text{definition of definite integral}$$

$$= \lim_{n\to\infty} \sum_{k=1}^n c\left(\frac{b-a}{n}\right) \qquad \leftarrow f(x) = c \text{ and } \Delta x = \frac{b-a}{n}$$

$$= \lim_{n\to\infty} c\left(\frac{b-a}{n}\right) \sum_{k=1}^n 1 \qquad \leftarrow \text{pull constants out of sum}$$

$$= \lim_{n\to\infty} c\left(\frac{b-a}{n}\right)(n) \qquad \leftarrow \text{sum formula}$$

$$= \lim_{n\to\infty} c(b-a) = c(b-a) \qquad \leftarrow \text{limit of a constant} \qquad \blacksquare$$

Examples and Explorations

EXAMPLE 1

Using a sum formula to approximate an area with a large number of rectangles

Find the n-rectangle right-sum approximation for $\int_2^4 (x^2 + 1)\,dx$, and then use it to calculate approximations with $n = 100$ rectangles and with $n = 1000$ rectangles.

SOLUTION

We'll find a formula for the n-rectangle approximation and then evaluate at $n = 100$ and $n = 1000$ at the end. Since $a = 2$ and $b = 4$, we have $\Delta x = \frac{b-a}{n} = \frac{2}{n}$, and therefore $x_k = a + k\Delta x = 2 + \frac{2k}{n}$. For the right sum with $x_k^* = x_k$, we have

$$\sum_{k=1}^{n} f(x_k^*)\Delta x = \sum_{k=1}^{n} \left(\left(2 + \frac{2k}{n}\right)^2 + 1 \right)\left(\frac{2}{n}\right)$$

$$= \sum_{k=1}^{n} \left(\frac{8k^2}{n^3} + \frac{16k}{n^2} + \frac{10}{n} \right) \qquad \leftarrow \text{simple algebra}$$

$$= \frac{8}{n^3}\sum_{k=1}^{n} k^2 + \frac{16}{n^2}\sum_{k=1}^{n} k + \frac{10}{n}\sum_{k=1}^{n} 1 \qquad \leftarrow \text{separate the sums}$$

$$= \frac{8}{n^3}\left(\frac{n(n+1)(2n+1)}{6} \right) + \frac{16}{n^2}\left(\frac{n(n+1)}{2} \right) + \frac{10}{n}(n) \qquad \leftarrow \text{sum formulas}$$

$$= \frac{8n(n+1)(2n+1)}{6n^3} + \frac{16n(n+1)}{2n^2} + 10. \qquad \leftarrow \text{simplify}$$

Notice that our expression for the n-rectangle right sum no longer has any sigma notation in it. This means that it is now a simple matter to evaluate the expression at $n = 100$ and at $n = 1000$. At $n = 100$ we have

$$\frac{800(101)(201)}{6,000,000} + \frac{1600(101)}{20,000} + 10 = 20.7868,$$

and at $n = 1000$ we have

$$\frac{8000(1001)(2001)}{6,000,000,000} + \frac{16000(1001)}{2,000,000} + 10 = 20.6787.$$

 CHECKING THE ANSWER

After all that work it is good to do a reality check. The figure that follows shows a graphing calculator plot of the area that we just approximated. The area we are considering should be just a little less than the area of the trapezoid with base from $(2, 0)$ to $(4, 0)$ and top side connecting $(2, 5)$ and $(4, 17)$. This trapezoid has area $\frac{5+17}{2}(2) = 21$ square units, just slightly more than our earlier approximations.

Area under $f(x) = x^2 + 1$ on $[2, 4]$

EXAMPLE 2   Calculating a definite integral exactly

Take the limit as $n \to \infty$ of an n-rectangle right sum to calculate $\int_2^4 (x^2 + 1) \, dx$ exactly.

SOLUTION

We already did the hard work in the previous example, and all that remains is to take the limit as $n \to \infty$. The exact area between the graph of $f(x) = x^2 + 1$ and the x-axis on $[2, 4]$ is

$$\int_2^4 (x^2 + 1) \, dx = \lim_{n \to \infty} \sum_{k=1}^{n} f(x_k^*) \Delta x \qquad \leftarrow \text{definite integral definition}$$

$$= \lim_{n \to \infty} \left(\frac{8n(n+1)(2n+1)}{6n^3} + \frac{16n(n+1)}{2n^2} + 10 \right) \qquad \leftarrow \text{Example 1}$$

$$= \frac{16}{6} + \frac{16}{2} + 10 = \frac{62}{3}. \qquad \leftarrow \text{ratios of leading coefficients}$$

In the step where we evaluated the limit, we used the fact that the limit of a "balanced" rational function with the same degree in the numerator and denominator is equal to the ratio of leading coefficients.

Notice that we would not have been able to calculate this limit if we did not have a *formula* for the Riemann sum from Example 1 that was expressed in terms of n, but with no "Σ". Note also that our previous approximation with $n = 1000$ was quite accurate; it was only approximately 0.012 square unit larger than the actual area, $\frac{62}{3} \approx 20.6667$.   □

EXAMPLE 3   Using definite integral formulas

Use properties of definite integrals and the definite integral formulas in Theorem 7.13 to verify the calculation of $\int_2^4 (x^2 + 1) \, dx$ from the previous example.

SOLUTION

Using the sum rule for definite integrals from Theorem 7.11, we can indeed calculate the same answer that we did in Example 2:

$$\int_2^4 (x^2 + 1) \, dx = \int_2^4 x^2 \, dx + \int_2^4 1 \, dx \qquad \leftarrow \text{sum rule for definite integrals}$$

$$= \tfrac{1}{3}(4^3 - 2^3) + 1(4 - 2) = \frac{62}{3}. \qquad \leftarrow \text{definite integral formulas} \qquad □$$

EXAMPLE 4   Using the algebraic properties of definite integrals

Given that $\int_1^3 f(x) \, dx = 4$ and $\int_5^3 2f(x) \, dx = -3$, find $\int_1^5 f(x) \, dx$. Identify the theorems or properties that allow each of your steps.

SOLUTION

We pick apart the definite integral we are seeking until we can express it in terms of the definite integrals that we are given:

$$\int_1^5 f(x) \, dx = \int_1^3 f(x) \, dx + \int_3^5 f(x) \, dx \qquad \leftarrow \text{Theorem 7.12(c)}$$

$$= \int_1^3 f(x) \, dx - \int_5^3 f(x) \, dx \qquad \leftarrow \text{Theorem 7.12(b)}$$

$$= \int_1^3 f(x) \, dx - \frac{1}{2} \int_5^3 2f(x) \, dx \qquad \leftarrow \text{Theorem 7.11(b)}$$

$$= 4 - \frac{1}{2}(-3) = \frac{11}{2}. \qquad \leftarrow \text{using what was given} \qquad □$$

| EXAMPLE 5 | Interpreting definite integrals as signed areas |

Use a graph and properties of definite integrals to argue that $\int_1^5 (2-x)\,dx$ is negative. Then find a number a so that $\int_a^5 (2-x)\,dx$ is exactly zero.

SOLUTION

The region between the graph of $f(x) = 2-x$ and the x-axis on $[1, 5]$ consists of one small triangular region above the x-axis and one larger triangular region below the x-axis:

The region between $f(x) = 2 - x$ and the x-axis on $[1, 5]$

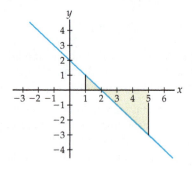

The definite integral will count the region above the x-axis positively and the region below the x-axis negatively; that is,

$$\int_1^2 (2-x)\,dx > 0 \quad \text{and} \quad \int_2^5 (2-x)\,dx < 0.$$

Since the region above the x-axis is smaller than the region below the x-axis, we must have

$$\int_1^5 (2-x)\,dx = \int_1^2 (2-x)\,dx + \int_2^5 (2-x)\,dx < 0.$$

We will have $\int_a^5 (2-x)\,dx$ equal to zero when $a = -1$, because this is the value for which the triangle above the x-axis on the interval $[a, 1]$ is the same size as the triangle below the x-axis on the interval $[2, 5]$. We could compute the areas of the triangles in question exactly from the formula for the area of a triangle, but it is sufficient in this case to see the graph:

The region between $f(x) = 2 - x$ and the x-axis on $[-1, 5]$

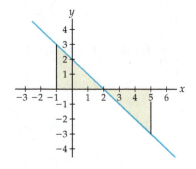

TEST YOUR UNDERSTANDING

▶ How is a definite integral different than a Riemann sum?

▶ What do definite integrals have to do with areas under curves?

▶ What types of functions are integrable?

▶ Why does it make sense that if $a < c < b$, then $\int_a^b f(x)\,dx = \int_a^c f(x)\,dx + \int_c^b f(x)\,dx$?

▶ Why is it not surprising that definite integrals behave well with respect to sums and constant multiples?

EXERCISES 7.3

Thinking Back

▶ *Commuting with sums:* What does it mean to say that derivatives commute with sums? That limits commute with sums? That sums written in sigma notation commute with sums? Express your answers in words and algebraically.

▶ *Commuting with constant multiples:* What does it mean to say that derivatives commute with constant multiples? That limits commute with constant multiples? That sums written in sigma notation commute with constant multiples? Express your answers in words and algebraically.

Concepts

0. *Problem Zero:* Read the section and make your own summary of the material.

1. *True/False:* Determine whether each of the statements that follow is true or false. If a statement is true, explain why. If a statement is false, provide a counterexample.

(a) *True or False:* The left-sum and right-sum approximations are the same if the number n of rectangles is very large.

(b) *True or False:* $\int_{-2}^{5}(x+2)^3\,dx$ is a real number.

(c) *True or False:* $\int(5x^2 - 3x + 2)\,dx$ is exactly 26.167.

(d) *True or False:* $\int_{-3}^{-2} f(x)\,dx = -\int_{2}^{3} f(x)\,dx$.

(e) *True or False:* If $\int_0^2 f(x)\,dx = 3$ and $\int_0^2 g(x)\,dx = 2$, then $\int_0^2 f(g(x))\,dx = 6$.

(f) *True or False:* If $\int_0^2 f(x)\,dx = 3$ and $\int_0^2 g(x)\,dx = 2$, then $\int_0^2 f(x)g(x)\,dx = 6$.

(g) *True or False:* If $\int_0^1 f(x)\,dx = 3$ and $\int_{-1}^0 f(x)\,dx = 4$, then $\int_{-1}^1 f(x)\,dx = 7$.

(h) *True or False:* If $\int_0^2 f(x)\,dx = 3$ and $\int_2^4 g(x)\,dx = 4$, then $\int_0^4 (f(x) + g(x))\,dx = 7$.

2. *Examples:* Construct examples of the thing(s) described in the following. Try to find examples that are different than any in the reading.

(a) A function that is not integrable on $[2, 10]$.

(b) A function f for which we currently know how to calculate $\int_{-2}^2 f(x)\,dx$ exactly.

(c) A function f for which we currently do not know how to calculate $\int_{-2}^2 f(x)\,dx$ exactly.

3. Fill in the blanks: The signed area between the graph of a continuous function f and the x-axis on $[a, b]$ is

represented by the notation _____ and is called the _____ .

4. Explain why it makes sense that *every* Riemann sum for a continuous function f on an interval $[a, b]$ approaches the same number as the number n of rectangles approaches infinity. Illustrate your argument with graphs.

5. Fill in the blanks: The definite integral of an integrable function f from $x = a$ to $x = b$ is defined to be

$$\int_a^b f(x)\,dx = \lim_{\square} \sum_{\square}^{\square} \boxed{},$$

where $\Delta x = $ _____ , $x_k = $ _____ , and x_k^* is _____ .

6. Explain geometrically what the definition of the definite integral as a limit of Riemann sums represents. Include a labeled picture of a Riemann sum (for a particular n) that illustrates the roles of n, Δx, x_k, x_k^*, and $f(x_k^*)$. What happens in the picture as $n \to \infty$?

7. If $f(x)$ is defined at $x = a$, then $\int_a^a f(x)\,dx = 0$. Explain why this makes sense in terms of area.

8. Draw pictures illustrating the fact that if $a \le c \le b$, then

$$\int_a^c f(x)\,dx + \int_c^b f(x)\,dx = \int_a^b f(x)\,dx.$$

Use graphs to determine whether each of the following definite integrals is equal to a positive number, a negative number, or zero:

9. $\int_{-3}^3 (x^2 - 4)\,dx$

10. $\int_{-3}^3 |x^2 - 4|\,dx$

11. $\int_0^{2\pi} \cos x\,dx$

12. $\int_{-3\pi/4}^{3\pi/4} \cos x\,dx$

13. Consider the function f graphed here. Shade in the regions between f and the x-axis on (a) $[-2, 6]$ and (b) $[-4, 2]$. Are the signed areas on these intervals positive or negative, and why?

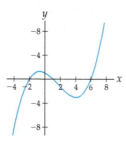

14. In a right sum with $n = 4$ rectangles for the area between the graph of $f(x) = x^2 - 1$ and the x-axis on $[0, 2]$, how many rectangles have negative heights? What if $n = 7$? In each case, is the estimate for the signed area positive or negative? Is the exact value of the signed area positive or negative?

15. Although the definite integral of a sum of functions is equal to the sum of the definite integrals of those functions, the definite integral of a product of functions is *not* the product of two definite integrals.

(a) Use mathematical notation to write the preceding sentence in this form:

_____ = _____,   but   _____ ≠ _____ .

(b) Choose two simple functions f and g so that you can calculate the definite integrals of f, g, and $f + g$ on $[0, 1]$, and show that the sum of the first two definite integrals is equal to the third.

(c) Find two simple functions f and g such that $\int_0^1 f(x)g(x)\,dx$ is not equal to the product of $\int_0^1 f(x)\,dx$ and $\int_0^1 g(x)\,dx$. (*Hint: Choose f and g so that you can calculate the definite integrals involved.*)

16. Suppose f is an integrable function $[a, b]$ and k is a real number. Use pictures of Riemann sums to illustrate that the right sum for the function $kf(x)$ on $[a, b]$ is k times the value of the right sum (with the same n) for f on $[a, b]$. What happens as $n \to \infty$? What does this exercise say about the definite integrals $\int_a^b f(x)\,dx$ and $\int_a^b kf(x)\,dx$?

17. Consider the definite integral $\int_0^3 x^2\,dx$.

(a) Write down an n-rectangle right sum for $\int_0^3 x^2\,dx$, and use algebra and a sum formula to write this sum as a formula in terms of n.

(b) Write down an n-rectangle left sum for $\int_0^3 x^2\,dx$, and use algebra and a sum formula to write this sum as a formula in terms of n.

(c) Use your answers to (a) and (b) to show that the right sum and the left sum for $\int_0^3 x^2\,dx$ are different for $n = 100$ and $n = 1000$.

(d) Use your answers to (a) and (b) to show that the right sum and the left sum for $\int_0^3 x^2\,dx$ approach the same quantity as $n \to \infty$. What does this quantity represent geometrically?

18. The definite integral of a function f on an interval $[a, b]$ is defined as a limit of Riemann sums. How can it be that the sum of the areas of infinitely many rectangles that are each "infinitely thin" is a finite number? On the one hand, shouldn't it be infinite, since we are adding up infinitely many rectangles? On the other hand, shouldn't it always be zero, since the width of each of the rectangles is approaching zero as $n \to \infty$?

19. Some discontinuous functions are not integrable. For example, consider the function $f(x) = \dfrac{1}{x^2}$ on $[-2, 2]$.

(a) Sketch a graph of f on $[-2, 2]$. What kind of discontinuity does f have, and where?

(b) Why might it be reasonable to think that f is not integrable on $[-2, 2]$ because of this discontinuity?

(c) What do you think happens to a Riemann sum approximation for the area between f and the x-axis on $[-2, 2]$ as $n \to \infty$?

(d) Although your intuition probably told you that the area between the graph of $f(x) = \dfrac{1}{x^2}$ and the x-axis on $[-2, 2]$ was likely to be infinite, this is not always the case for functions with vertical asymptotes. Surprisingly, as we will see in Section 8.6, the function $g(x) = \dfrac{1}{x^{2/3}}$ has an asymptote at $x = 0$ and yet its area on $[-2, 2]$ is actually finite! Compare the graphs of $f(x) = \dfrac{1}{x^2}$ and $g(x) = \dfrac{1}{x^{2/3}}$, and speculate as to why this might be the case.

20. Some discontinuous functions are still "nice" enough to be integrable. For example, consider the function

$$f(x) = \begin{cases} x, & \text{if } x \le 2 \\ x + 3, & \text{if } x > 2. \end{cases}$$

(a) Sketch a graph of f on $[0, 5]$. What kind of discontinuity does f have, and where?

(b) Why is it reasonable that f is integrable on $[0, 5]$ despite this discontinuity?

(c) What happens to a Riemann sum approximation for the area between a function f and the x-axis on $[0, 5]$ as $n \to \infty$?

Skills

Use geometry (i.e., areas of triangles, rectangles, and circles) to find the exact values of each of the definite integrals in Exercises 21–28.

21. $\displaystyle\int_0^2 (2 - x)\,dx$

22. $\displaystyle\int_2^1 0(4x - 3)\,dx$

23. $\displaystyle\int_{-3}^8 24\,dx$

24. $\displaystyle\int_{-2}^4 |3x + 1|\,dx$

25. $\displaystyle\int_{-1}^1 \sqrt{1 - x^2}\,dx$

26. $\displaystyle\int_{-r}^r \sqrt{r^2 - x^2}\,dx$

27. $\int_0^3 (1 - |x - 1|)\, dx$ **28.** $\int_{-2}^2 (3 + \sqrt{4 - x^2})\, dx$

If $\int_{-2}^3 f(x)\, dx = 4$, $\int_{-2}^6 f(x)\, dx = 9$, $\int_{-2}^3 g(x)\, dx = 2$, and $\int_3^6 g(x)\, dx = 3$, then find the values of each definite integral in Exercises 29–40. If there is not enough information, explain why.

29. $\int_3^6 f(x)\, dx$ **30.** $\int_{-2}^6 g(x)\, dx$

31. $\int_{-2}^6 (f(x) + g(x))\, dx$ **32.** $\int_3^6 (2f(x) - g(x))\, dx$

33. $\int_{-2}^3 f(x)g(x)\, dx$ **34.** $\int_{-2}^6 (g(x) + x)\, dx$

35. $\int_3^6 (g(x))^2\, dx$ **36.** $\int_{-2}^3 x f(x)\, dx$

37. $\int_6^3 (f(x) + g(x))\, dx$ **38.** $\int_{-2}^6 (4f(x) - 2)\, dx$

39. $\int_3^{-2} (2x^2 - 3g(x))\, dx$ **40.** $\int_{-2}^{-2} x(f(x) + 3)^2\, dx$

For each definite integral in Exercises 41–46, (a) find the general n-rectangle right sum and simplify your answer with sum formulas. Then (b) use your answer to approximate the definite integral with $n = 100$ and $n = 1000$. Finally, (c) take the limit as $n \to \infty$ to find the exact value.

41. $\int_2^5 (5 - x)\, dx$ **42.** $\int_{-3}^3 (2x + 1)\, dx$

43. $\int_0^1 2x^2\, dx$ **44.** $\int_{-3}^2 x^2\, dx$

45. $\int_2^3 (x + 1)^2\, dx$ **46.** $\int_{-1}^2 (1 - x^2)\, dx$

Calculate the exact value of each definite integral in Exercises 47–52 by using properties of definite integrals and the formulas in Theorem 7.13.

47. $\int_2^4 (x^2 + 1)\, dx$ **48.** $\int_0^6 (3x + 2)\, dx$

49. $\int_5^2 (9 + 10x - x^2)\, dx$ **50.** $\int_1^3 (x + 1)^2\, dx$

51. $\int_0^4 ((2x - 3)^2 + 5)\, dx$ **52.** $\int_6^1 (3(1 - 2x)^2 + 4x)\, dx$

Applications

53. Suppose that once again you drive in a car for 40 seconds with velocity $v(t) = -0.22t^2 + 8.8t$ feet per second, as shown in the graph that follows. Suppose also that your total distance travelled is equal to the area under the velocity curve on $[0, 40]$.

Velocity of car
$v(t) = -0.22t^2 + 8.8t$

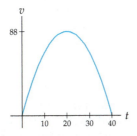

(a) What definite integral would you have to compute in order to find your exact distance travelled over the 40 seconds of your trip?

(b) Find the exact value of that definite integral by taking a limit of Riemann sums.

54. The function for the *standard normal distribution* is

$$f(x) = \frac{1}{\sqrt{2\pi}} e^{-x^2/2}$$

Its graph is that of the **bell curve**. Probability and statistics books often have tables like the one following, which lists some approximate areas under the bell curve:

Areas under the bell curve

b	$\dfrac{1}{\sqrt{2\pi}} \int_{-b}^b e^{-x^2/2}\, dx$
0.5	0.3829
1	0.6827
1.5	0.8664
2	0.9545
2.5	0.9876

Use the information given in the table, properties of definite integrals, and symmetry to find

(a) $\dfrac{1}{\sqrt{2\pi}} \int_{-0.5}^{1.5} e^{-x^2/2}\, dx$ (b) $\dfrac{1}{\sqrt{2\pi}} \int_{1.5}^2 e^{-x^2/2}\, dx$

Proofs

55. Use the definition of the definite integral as a limit of Riemann sums to prove Theorem 7.11(b): For any function f that is integrable on $[a, b]$ and any real number c, $\int_a^b c f(x)\, dx = c \int_a^b f(x)\, dx$.

56. Use the definition of the definite integral as a limit of Riemann sums to prove Theorem 7.12(a): For any function f and real number a, $\int_a^a f(x)\, dx = 0$.

57. Use the definition of the definite integral as a limit of Riemann sums to prove Theorem 7.12(b): For any function f that is integrable on $[a, b]$, $\int_b^a f(x)\, dx = -\int_a^b f(x)\, dx$.

58. Give a geometric argument to prove Theorem 7.13(a): For any real numbers a, b, and c, $\int_a^b c\, dx = c(b - a)$. (*Hint: Use a rectangle.*)

59. Give a geometric argument to prove Theorem 7.13(b): For any real numbers $0 < a < b$, $\int_a^b x\,dx = \frac{1}{2}(b^2 - a^2)$. (*Hint: Use a trapezoid.*)

60. Prove Theorem 7.13(b): For any real numbers a and b, we have $\int_a^b x\,dx = \frac{1}{2}(b^2 - a^2)$. Use the proof of Theorem 7.13(a) as a guide.

61. Prove Theorem 7.13(c): For any real numbers a and b, $\int_a^b x^2\,dx = \frac{1}{3}(b^3 - a^3)$. Use the proof of Theorem 7.13(a) as a guide.

62. Prove that $\int_1^3 (3x + 4)\,dx = 20$ in three different ways:

(a) algebraically, by calculating a limit of Riemann sums;

(b) geometrically, by recognizing the region in question as a trapezoid and calculating its area;

(c) with formulas, by using properties and formulas of definite integrals.

Thinking Forward

Functions defined by area accumulation: We know that for fixed real numbers a and b and an integrable function f, the definite integral $\int_a^b f(x)\,dx$ is a real number. For different real values of b, we get (potentially) different values for the integral $\int_a^b f(x)\,dx$.

▶ Make a table of the values of the integral $\int_0^b 2x\,dx$ corresponding to the values $-3, -2, -1, 0, 1, 2,$ and 3 for b. Conjecture a formula for the relationship between the values of b and the corresponding value of the integral.

▶ What is the word that describes the kind of relationship that exists between the values of b and the corresponding value of the integral?

▶ Now make a table of the values of the integral $\int_1^b 2x\,dx$ corresponding to the values $-3, -2, -1, 0, 1, 2,$ and 3 for b. Conjecture a formula for the relationship between the values of b and the corresponding value of the integral.

▶ What is the relationship between the formula that describes $\int_0^b 2x\,dx$ and the formula that describes $\int_1^b 2x\,dx$?

7.4   INDEFINITE INTEGRALS

▶ The definition of the indefinite integral as a notation for antidifferentiation

▶ Formulas for integrating some basic functions

▶ Guessing and checking to solve integrals of combinations of functions

Antiderivatives and Indefinite Integrals

As we have seen throughout the previous chapters, an **antiderivative** of a function f is a function F whose derivative is f. For example, an antiderivative of $f(x) = 2x$ is the function $F(x) = x^2$. Another antiderivative of $f(x) = 2x$ is the function $G(x) = x^2 + 3$. In fact, any function of the form $x^2 + C$ is an antiderivative of $f(x) = 2x$, and these are the only possible antiderivatives of $f(x) = 2x$. In general, the antiderivatives of a given function are all related to each other, so we call the set of all antiderivatives of f the **family of antiderivatives** of f. As we showed in Theorem 3.7 of Section 3.2, any two antiderivatives of a function must differ by a constant. For convenience we restate that theorem now:

THEOREM 7.14   **Functions with the Same Derivative Differ by a Constant**

If F and G are differentiable functions, then $F'(x) = G'(x)$ if and only if $G(x) = F(x) + C$ for some constant C.

In the previous section we defined the **definite integral** of a function f on an interval $[a, b]$ as a limit of Riemann sums used in calculating the signed area between the graph of a function f and the x-axis on an interval $[a, b]$. We will now define a completely different object called the ***indefinite integral*** of a function f. This new object will be a family of functions, not a number, but we will see in Section 7.5 that these two types of integrals are related, which is why we'll call them by similar names.

The indefinite integral of a function is a notation for expressing the collection of all possible antiderivatives of that function:

DEFINITION 7.15

The Indefinite Integral of a Function

The ***indefinite integral*** of a continuous function f is defined to be the family of antiderivatives

$$\int f(x)\, dx = F(x) + C,$$

where F is an antiderivative of f, that is, a function for which $F' = f$.

For example, since an antiderivative of $f(x) = 2x$ is $F(x) = x^2$, it follows that all antiderivatives of $f(x) = 2x$ are of the form $x^2 + C$, and therefore that $\int 2x\, dx = x^2 + C$. Note that it would be equally accurate to use a different antiderivative, such as $G(x) = x^2 + 3$, and say that $\int 2x\, dx = (x^2 + 3) + C$.

⊗ CAUTION   Although the notation and terminology used for indefinite integrals in Definition 7.15 is similar to what we used for definite integrals in Section 7.3, it is important to note that at this point we have no proof that the two types of integrals are related. When we see the Fundamental Theorem of Calculus in Section 7.5, we will make the surprising discovery that the area under a curve is in fact related to families of antiderivatives, and this relationship will justify why we use such similar notation for two different kinds of objects.

The "dx" in the notation of Definition 7.15 represents the fact that we are antidifferentiating with respect to the variable x. The constant C represents an arbitrary constant. The function f inside the integral notation is called the ***integrand,*** and when we find $\int f(x)\, dx$, we say that we are ***integrating*** the function f. The indefinite integral of a function will often be called simply the ***integral*** of that function. The continuity hypothesis is important (see Exercises 18–20), and we will assume throughout this section that we are working with intervals where our functions are continuous.

Antidifferentiation Formulas

All of the rules that we have developed for differentiating functions can be used to develop *anti*differentiation rules, which in turn will give us formulas for some common indefinite integrals. For example, the rule for differentiating power functions says that for any constant k, $\frac{d}{dx}(x^k) = kx^{k-1}$. The rule for antidifferentiating a power function should "undo" this process and is given in the following theorem:

THEOREM 7.16

Integrals of Power Functions

(a) If $k \neq -1$, then $\displaystyle\int x^k\, dx = \frac{1}{k+1}x^{k+1} + C.$

(b) $\displaystyle\int \frac{1}{x}\, dx = \ln|x| + C.$ *(See Exercise 18 for a technical point.)*

Proof. To prove the first part of the theorem, it suffices to show that if $k \neq -1$, then $\frac{1}{k+1}x^{k+1}$ is an antiderivative of x^k. (Note that if $k = -1$, then $\frac{1}{k+1}$ is not even defined.) In other words, we need only show that the derivative of $\frac{1}{k+1}x^{k+1}$ is x^k. This is a simple application of the power and constant-multiple rules of differentiation:

$$\frac{d}{dx}\left(\frac{1}{k+1}x^{k+1}\right) = \frac{1}{k+1}(k+1)x^k = x^k.$$

The second integration formula in the theorem covers the case when $k = -1$, since $x^{-1} = \frac{1}{x}$. This formula follows immediately from the fact that $\frac{d}{dx}(\ln|x|) = \frac{1}{x}$. ∎

The next three theorems describe formulas for antidifferentiating–and thus integrating–other common types of functions. Each of these formulas can be proved by differentiating; you will do so in Exercises 69–71.

THEOREM 7.17

Integrals of Exponential Functions

(a) If $k \neq 0$, then $\displaystyle\int e^{kx}\,dx = \frac{1}{k}e^{kx} + C.$

(b) If $b > 0$ and $b \neq 1$, then $\displaystyle\int b^x\,dx = \frac{1}{\ln b}b^x + C.$

For example, $\int 2^x\,dx = \frac{1}{\ln 2}2^x$ because $\frac{d}{dx}\left(\frac{1}{\ln 2}2^x\right) = \frac{1}{\ln 2}(\ln 2)2^x = 2^x$. Notice that both of the rules in Theorem 7.17 imply that the integral of e^x is itself, that is, that $\int e^x\,dx = e^x + C$.

THEOREM 7.18

Integrals of Certain Trigonometric Expressions

(a) $\displaystyle\int \sin x\,dx = -\cos x + C$ (b) $\displaystyle\int \cos x\,dx = \sin x + C$

(c) $\displaystyle\int \sec^2 x\,dx = \tan x + C$ (d) $\displaystyle\int \csc^2 x\,dx = -\cot x + C$

(e) $\displaystyle\int \sec x \tan x\,dx = \sec x + C$ (f) $\displaystyle\int \csc x \cot x\,dx = -\csc x + C$

THEOREM 7.19

Integrals Whose Solutions Are Inverse Trigonometric Functions

(a) $\displaystyle\int \frac{1}{\sqrt{1-x^2}}\,dx = \sin^{-1}x + C.$

(b) $\displaystyle\int \frac{1}{1+x^2}\,dx = \tan^{-1}x + C.$

(c) $\displaystyle\int \frac{1}{|x|\sqrt{x^2-1}}\,dx = \sec^{-1}x + C.$

Antidifferentiating Combinations of Functions

We have rules for differentiating constant multiples, sums, products, quotients, and compositions of functions. Only the constant-multiple rule and the sum rule translate directly into antidifferentiation rules:

THEOREM 7.20

Constant Multiple and Sum Rules for Indefinite Integrals

(a) $\displaystyle \int kf(x)\,dx = k \int f(x)\,dx.$

(b) $\displaystyle \int (f(x) + g(x))\,dx = \int f(x)\,dx + \int g(x)\,dx.$

Proof. Suppose F is any antiderivative of f, that is, $F'(x) = f(x)$. Then, by the constant-multiple rule, $kF'(x) = kf(x)$ for any constant k. Furthermore, for any constant D, $F(x) + D$ is also an antiderivative of f, so $(F(x) + D)' = f(x)$. Therefore

$$\int kf(x)\,dx = kF'(x) + C = k\left(F'(x) + \frac{C}{k}\right) = k(F'(x) + D) = k\int f(x)\,dx.$$

Note that in the calculation we just did, $D = \dfrac{C}{k}$ is just a constant.

Similarly, if $F'(x) = f(x)$ and $G'(x) = g(x)$, then, by the sum rule, $(F(x) + G(x))' = f(x) + g(x)$, and therefore

$$\int f(x)\,dx + \int g(x)\,dx = (F'(x) + C_1) + (G'(x) + C_2) = (F'(x) + G'(x)) + C = \int (f(x) + g(x))\,dx,$$

where $C = C_1 + C_2$.   ■

There are no general product, quotient, and chain rules for antidifferentiation. However, if we think of these differentiation rules "backwards," then we can say something about certain types of integrands:

THEOREM 7.21

Reversing the Product, Quotient, and Chain Rules

(a) $\displaystyle \int (f'(x)\,g(x) + f(x)\,g'(x))\,dx = f(x)\,g(x) + C$

(b) $\displaystyle \int \frac{f'(x)\,g(x) - f(x)\,g'(x)}{(g(x))^2}\,dx = \frac{f(x)}{g(x)} + C$

(c) $\displaystyle \int f'(g(x))\,g'(x)\,dx = f(g(x)) + C$

Of course, the trouble is recognizing when an integrand is in one of these special forms. If we are lucky enough to recognize an integrand as the result of a product, quotient, or chain rule calculation, then we can make an educated guess at the integral and check our answer by differentiating. Repeated guess-and-check is at this point our best strategy for integrating combinations of functions. The first and third parts of Theorem 7.21 will form the basis for the methods of integration by parts and integration by substitution that we will see in Chapter 8.

In general, antidifferentiation is much more difficult than differentiation. We can differentiate *every* function that we currently know how to write down; however, at this point we cannot integrate even very simple functions like $\ln x$ or $\sec x$. You should think of integration as a puzzle, not a procedure. Unlike differentiation, where it is always clear which rules to apply, and in which order, it is *not* always immediately clear how to find a given integral. We will learn some more methods for calculating integrals in Chapter 8, but even then we will not be able to calculate all integrals. In fact, as we will see in Section 7.7, some functions have no elementary antiderivative, which means that their antiderivatives cannot even be written down in terms of the functions we now know.

Examples and Explorations

EXAMPLE 1

Identifying antiderivatives

Which of the following are antiderivatives of $f(x) = x^4$, and why?

(a) $4x^3$ **(b)** $\frac{1}{5}x^5$ **(c)** $\frac{1}{5}x^5 - 2$ **(d)** $\frac{1}{5}(x^5 - 2)$

SOLUTION

An antiderivative of $f(x) = x^4$ is a function whose derivative is x^4. Choice (a) is clearly the derivative, not an antiderivative, of $f(x) = x^4$. The remaining three choices are *all* antiderivatives of $f(x) = x^4$, since each of those functions has derivative x^4. □

EXAMPLE 2

Using algebra to identify and then integrate power functions

Find $\displaystyle \int \frac{1}{x^2}\, dx$ and $\displaystyle \int \sqrt{x^3}\, dx$.

SOLUTION

Solving these integrals is an easy application of Theorem 7.16, once we write the integrands in the form x^k:

$$\int \frac{1}{x^2}\, dx = \int x^{-2}\, dx = \frac{1}{-2+1}x^{-2+1} = \frac{1}{-1}x^{-1} + C = -\frac{1}{x} + C;$$

$$\int \sqrt{x^3}\, dx = \int x^{3/2}\, dx = \frac{1}{\frac{3}{2}+1}x^{3/2+1} = \frac{1}{\frac{5}{2}}x^{5/2} + C = \frac{2}{5}\sqrt{x^5} + C.$$

□

EXAMPLE 3

Using the sum and constant-multiple rules for indefinite integrals

Find $\int (5x^3 - \sin x)\, dx$, and show explicitly how the sum and constant-multiple rules for indefinite integrals are applied.

SOLUTION

The function $f(x) = 5x^3 - \sin x$ is a sum of constant multiples of functions whose integrals we know, namely, the functions x^3 and $\sin x$. By Theorem 7.20, we have

$$\int (5x^3 - \sin x)\, dx = \int (5x^3 + (-\sin x))\, dx \qquad \leftarrow \text{write the difference as a sum}$$

$$= \int 5x^3\, dx + \int (-\sin x)\, dx \qquad \leftarrow \text{sum rule}$$

$$= 5 \int x^3\, dx - \int \sin x\, dx \qquad \leftarrow \text{constant-multiple rule}$$

$$= 5 \left(\tfrac{1}{4} x^4 \right) - (-\cos x) + C \qquad \leftarrow \text{antidifferentiation formulas}$$

$$= \tfrac{5}{4} x^4 + \cos x + C.$$

Notice that we added only *one* constant C in this calculation. Technically there are *two* such constants, since the family of antiderivatives for x^3 is $\frac{1}{4}x^4 + C_1$, and the family of antiderivatives for $\sin x$ is $-\cos x + C_2$. The C we are using is really the sum $C_1 + C_2$. There is no need to write the two constants C_1 and C_2 separately, since the sum of two arbitrary constants is also an arbitrary constant.   □

EXAMPLE 4   Recognizing integrands in product, quotient, and chain rule forms

Solve the following indefinite integrals:

(a) $\displaystyle \int (e^x + xe^x)\, dx$ **(b)** $\displaystyle \int 5(x^2 + 1)^4 (2x)\, dx$ **(c)** $\displaystyle \int \frac{2x \tan x - x^2 \sec^2 x}{\tan^2 x}\, dx$

SOLUTION

(a) This integrand looks like it could be the result of a product rule calculation where x and e^x are the factors in the product. We check by differentiating, and indeed $\frac{d}{dx}(xe^x) = 1e^x + xe^x$. Therefore $\int (e^x + xe^x)\, dx = xe^x + C$.

(b) The integrand $5(x^2 + 1)^4 (2x)$ looks like it could be the result of a chain rule calculation, with outside function x^5 and inside function $x^2 + 1$. We can check whether this is the case by differentiating to find $\frac{d}{dx}((x^2 + 1)^5) = 5(x^2 + 1)^4(2x)$, which is what we had hoped for. Therefore $\int 5(x^2 + 1)^4 (2x)\, dx = (x^2 + 1)^5 + C$.

(c) Finally, this integrand could be the result of differentiating a quotient $\frac{f(x)}{g(x)}$ where the denominator is $g(x) = \tan x$. Looking at the numerator of the integrand, we might guess that $f(x) = x^2$. We now check: $\frac{d}{dx}\left(\frac{x^2}{\tan x} \right) = \frac{(2x)\tan x - x^2 \sec^2 x}{(\tan x)^2}$, which is what we wanted for the integrand. Therefore $\int \frac{2x\tan x - x^2 \sec^2 x}{\tan^2 x}\, dx = \frac{x^2}{\tan x} + C$.   □

EXAMPLE 5   Integrating by educated guess-and-check

Solve the following indefinite integrals:

(a) $\displaystyle \int \sin 3x\, dx$ **(b)** $\displaystyle \int xe^{x^2 + 1}\, dx$ **(c)** $\displaystyle \int \frac{1}{1 + 4x^2}\, dx$

SOLUTION

(a) At first glance the function $\sin(3x)$ does not appear to be of the form $f'(g(x))g'(x)$. Specifically, if we choose $f'(x) = \sin x$, so that the inside function is $g(x) = 3x$, then the derivative $g'(x) = 3$ does not appear in the integrand. However, since an antiderivative of $\sin x$ is $-\cos x$, we might try $F(x) = -\cos 3x$ as a first guess for an antiderivative of $\sin 3x$. Testing this guess, we find that $\frac{d}{dx}(-\cos 3x) = -(-\sin 3x)(3) = 3\sin 3x$. This is *almost* the integrand we are looking for, but with an extra constant multiple of 3. Accordingly, we update our guess to $F(x) = -\frac{1}{3}\cos 3x$, so that the $\frac{1}{3}$ at the front will cancel out the factor of 3 that appears when we apply the chain rule. We now have

$$\frac{d}{dx}\left(-\frac{1}{3}\cos 3x\right) = -\frac{1}{3}(-\sin 3x)(3) = \sin 3x,$$

and therefore $\int \sin 3x\, dx = -\frac{1}{3}\cos 3x + C$.

(b) The first thing to recognize about this problem is that the function $x^2 + 1$ is an inside function in the integrand while its derivative $2x$ is *almost* part of the integrand: We have an x instead of a $2x$, so we're missing only a constant multiple. Since the function e^x is its own antiderivative, a good first guess for an antiderivative of xe^{x^2+1} might be $F(x) = e^{x^2+1}$. Checking this guess, we find that $\frac{d}{dx}(e^{x^2+1}) = e^{x^2+1}(2x)$, which is *almost* what we want. To get rid of the extra 2, we update our guess to $F(x) = \frac{1}{2}e^{x^2+1}$; with this guess we have

$$\frac{d}{dx}\left(\frac{1}{2}e^{x^2+1}\right) = \frac{1}{2}e^{x^2+1}(2x) = xe^{x^2+1}.$$

It then follows that $\int xe^{x^2+1}\, dx = \frac{1}{2}e^{x^2+1} + C$. It is important to note that the method used in this example, whereby we update our guess by dividing by what is "missing," works only if what is missing is a *constant*.

(c) In this example we are searching for a function whose derivative is $\frac{1}{1+4x^2}$. Our integrand is similar in structure to the second integrand in Theorem 7.19, so we will start the guessing process with $\tan^{-1}(x)$. We guess, check, and re-guess until we get the derivative we seek:

$$\frac{d}{dx}(\tan^{-1}x) = \frac{1}{1+x^2};$$

$$\frac{d}{dx}(\tan^{-1}(4x)) = \frac{1}{1+(4x)^2}(4) = \frac{4}{1+16x^2};$$

$$\frac{d}{dx}(\tan^{-1}(2x)) = \frac{1}{1+(2x)^2}(2) = \frac{2}{1+4x^2};$$

$$\frac{d}{dx}\left(\frac{1}{2}\tan^{-1}(2x)\right) = \left(\frac{1}{2}\right)\frac{1}{1+(2x)^2}(2) = \frac{1}{1+4x^2}.$$

Therefore $\int \frac{1}{1+4x^2}\, dx = \frac{1}{2}\tan^{-1}(2x) + C$. □

TEST YOUR UNDERSTANDING

▶ How is an indefinite integral related to antiderivatives? In what way is a definite integral different from an antiderivative?

▶ Which of the six trigonometric functions do we know how to integrate at this point, and why?

▶ Is the integral of a difference of two functions always equal to the difference of the integrals of those two functions? Why or why not?

▶ Is the integral of a product of two functions equal to the product of the integrals of those two functions? Why or why not?

▶ How can we solve integrals of combinations of functions by guessing and checking? Why is this a valid method?

EXERCISES 7.4

Thinking Back

▶ *Antiderivatives:* If $g'(x) = h(x)$, then is g an antiderivative of h or is h an antiderivative of g?

▶ *Definite integrals:* State the definition of the definite integral of a function f on an interval $[a, b]$.

▶ *The Mean Value Theorem:* State the Mean Value Theorem.

Derivatives and antiderivatives: Find the derivative and an antiderivative of each of the following functions:

▶ $f(x) = \pi^2$

▶ $f(x) = \dfrac{1}{5x}$

▶ $f(x) = (\sqrt[11]{x})^{-2}$

▶ $f(x) = \sec^2(3x + 1)$

Concepts

0. *Problem Zero:* Read the section and make your own summary of the material.

1. *True/False:* Determine whether each of the statements that follow is true or false. If a statement is true, explain why. If a statement is false, provide a counterexample.

(a) *True or False:* If $f(x) - g(x) = 2$ for all x, then f and g have the same derivative.

(b) *True or False:* If $f(x) - g(x) = 2$ for all x, then f and g have the same antiderivative.

(c) *True or False:* If $f(x) = 2x$, then $F(x) = x^2$.

(d) *True or False:* $\int (x^3 + 1)\,dx = \left(\frac{1}{4}x^4 + x + 2\right) + C$.

(e) *True or False:* $\int \sin(x^2)\,dx = -\cos(x^2) + C$.

(f) *True or False:* $\int e^x \cos x\,dx = e^x \sin x + C$.

(g) *True or False:* $\int \dfrac{1}{x^2 + 1}\,dx = \ln|x^2 + 1| + C$.

(h) *True or False:* $\int \dfrac{1}{x^2 + 1}\,dx = \dfrac{1}{2x}\ln|x^2 + 1| + C$.

2. *Examples:* Construct examples of the thing(s) described in the following. Try to find examples that are different than any in the reading.

(a) An integral that can be solved by recognizing the integrand as the result of a product rule calculation.

(b) An integral that can be solved by recognizing the integrand as the result of a chain rule calculation.

(c) Six functions that we do not currently know how to integrate (and why we cannot integrate each of them given what we currently know).

3. What is the difference between an antiderivative of a function and the indefinite integral of a function?

4. Explain why we call the collection of antiderivatives of a function f a family. How are the antiderivatives of a function related?

5. Compare the definitions of the definite and indefinite integrals. List at least three things that are different about these mathematical objects.

6. Fill in each of the blanks:

(a) $\int$ _____ $dx = x^6 + C$.

(b) x^6 is an antiderivative of _____ .

(c) The derivative of x^6 is _____ .

7. Fill in each of the blanks:

(a) $\int x^6\,dx =$ _____ $+ C$.

(b) _____ is an antiderivative of x^6.

(c) The derivative of _____ is x^6.

8. Explain why the formula for the integral of x^k does not apply when $k = -1$. What *is* the integral of x^{-1}?

9. Explain why at this point we don't have an integration formula for the function $f(x) = \sec x$ whereas we *do* have an integration formula for $f(x) = \sin x$.

10. Why don't we bother to state an integration formula that has to do with $\cos^{-1} x$? (*Hint: Think about the derivatives of* $\cos^{-1} x$ *and* $\sin^{-1} x$.) What would the integration formula be? Why is it "redundant," given the integration formula that has to do with $\sin^{-1} x$?

11. Write out all the integration formulas and rules that we know at this point.

12. Show that $F(x) = \sin x - x \cos x + 2$ is an antiderivative of $f(x) = x \sin x$.

13. Show that $F(x) = \frac{1}{2} e^{(x^2)}(x^2 - 1)$ is an antiderivative of $f(x) = x^3 e^{(x^2)}$.

14. Verify that $\int \cot x \, dx = \ln(\sin x) + C$. (Do not try to solve the integral from scratch.)

15. Verify that $\int \ln x \, dx = x(\ln x - 1) + C$. (Do not try to solve the integral from scratch.)

16. Show by exhibiting a counterexample that, in general, $\int \frac{f(x)}{g(x)} \, dx \neq \frac{\int f(x)\,dx}{\int g(x)\,dx}$. In other words, find two functions f and g such that the integral of their quotient is not equal to the quotient of their integrals.

17. Show by exhibiting a counterexample that, in general, $\int f(x)g(x)\,dx \neq (\int f(x)\,dx)(\int g(x)\,dx)$. In other words, find two functions f and g so that the integral of their product is not equal to the product of their integrals.

In the definition of the definite integral we required integrands to be continuous. If an integrand fails to be continuous everywhere, then a kind of branching can occur in its antiderivatives. In Exercises 18–19 we investigate some discontinuous examples where we choose nonstandard antiderivatives. Outside of this set of examples we will restrict our attention to the standard antiderivatives in Theorems 7.16–7.19.

18. Consider the function
$$F(x) = \begin{cases} \ln |x|, & \text{if } x < 0 \\ \ln |x| + 4, & \text{if } x > 0. \end{cases}$$

Show that the derivative of this function is the function $f(x) = \frac{1}{x}$. Compare the graphs of $F(x)$ and $\ln |x|$, and discuss how this exercise relates to the second part of Theorem 7.16.

19. Consider the function
$$F(x) = \begin{cases} -\cot x, & \text{if } x < 0 \\ -\cot x + 100, & \text{if } x > 0. \end{cases}$$

Show that the derivative of this function is the function $f(x) = \csc^2 x$. Compare the graphs of $F(x)$ and $-\cot x$, and discuss how this exercise relates to the fourth part of Theorem 7.18.

20. Consider the function
$$F(x) = \begin{cases} \sec^{-1} x, -\pi & \text{if } x < -1 \\ \sec^{-1} x + \pi, & \text{if } x > 1. \end{cases}$$

Show that the derivative of this function is the function $f(x) = \frac{1}{|x|\sqrt{x^2 - 1}}$. Compare the graphs of $F(x)$ and $\sec^{-1} x$, and discuss how this exercise relates to the second part of Theorem 7.19.

Skills

Use integration formulas to solve each integral in Exercises 21–58. You may have to use algebra, educated guess-and-check, and/or recognize an integrand as the result of a product, quotient, or chain rule calculation. Check each of your answers by differentiating. (*Hint for Exercise 54:* $\tan x = \frac{\sin x}{\cos x}$).

21. $\int (x^2 - 3x^5 - 7)\,dx$

22. $\int (x^3 + 4)^2 \, dx$

23. $\int (x^2 - 1)(3x + 5)\,dx$

24. $\int \left(\frac{2}{3x} + 1\right) dx$

25. $\int (x^2 + 2^x + 2^2)\,dx$

26. $\int \frac{3}{x^2 + 1}\,dx$

27. $\int \frac{x + 1}{\sqrt{x}}\,dx$

28. $\int \left(\frac{3}{x^2} - 4\right) dx$

29. $\int (3.2(1.43)^x - 50)\,dx$

30. $\int 5 \sin(2x + 1)\,dx$

31. $\int 4(e^{x-3})^2\,dx$

32. $\int 4e^{2x-3}\,dx$

33. $\int \sec(3x)\tan(3x)\,dx$

34. $\int 3 \csc(\pi x) \cot(\pi x)\,dx$

35. $\int 3 \tan^2 x \, dx$

36. $\int (\sin^2 x + \cos^2 x)\,dx$

37. $\int \pi \, dx$

38. $\int \frac{x^{-1}}{1 - x^{-1}}\,dx$

39. $\int \frac{7}{\sqrt{1 - (2x)^2}}\,dx$

40. $\int \frac{-3}{1 + 16x^2}\,dx$

41. $\int \frac{1}{|3x|\sqrt{9x^2 - 1}}\,dx$

42. $\int \frac{1}{4 + x^2}\,dx$

43. $\int \frac{2x}{1 + x^2}\,dx$

44. $\int 2 \sin x \cos x \, dx$

45. $\int (x^2 e^x + 2xe^x)\,dx$

46. $\int 3x^2 \sin(x^3 + 1)\,dx$

47. $\int \frac{6x}{3x^2 + 1}\,dx$

48. $\int x^3 e^{3x^4 - 2}\,dx$

49. $\int \frac{e^{3x} - 2e^{4x}}{e^{2x}}\,dx$

50. $\int \sec^2 x + \csc^2 x \, dx$

51. $\int x^2(x^3 + 1)^5\,dx$

52. $\int (\tan x + x \sec^2 x)\,dx$

53. $\int \frac{2x \ln x - x}{(\ln x)^2}\,dx$

54. $\int \tan x \, dx$

55. $\int \left(e^x \sqrt{x} + \dfrac{e^x}{2\sqrt{x}} \right) dx$ **56.** $\int \dfrac{1}{2\sqrt{x}} e^{\sqrt{x}} \, dx$

57. $\int \left(\dfrac{1}{x} \cos x - \sin x \ln x \right) dx$

58. $\int \dfrac{2x \sin x - (x^2 + 1) \cos x}{\sin^2 x} \, dx$

Solve each of the integrals in Exercises 59–64, where a, b, and c are real numbers with $a \neq 0$, $b \neq 0$, $c > 0$, and $c \neq 1$.

59. $\int a e^{bx+c} \, dx$ **60.** $\int \dfrac{a}{bx^c} \, dx$

61. $\int \dfrac{\sec(ax) \tan(ax)}{b} \, dx$ **62.** $\int (b(c^x) + a) \, dx$

63. $\int \dfrac{a}{1 + (bx)^2} \, dx$ **64.** $\int (a \sin(bx) - c) \, dx$

Applications

65. Suppose you throw a ball straight up into the air at a velocity of $v_0 = 42$ feet per second, initially releasing the ball at a height of 4 feet. Acceleration due to gravity is constantly $a(t) = -32$ feet per second squared (negative since it pulls downward).

Position, velocity, and acceleration of a ball thrown upwards

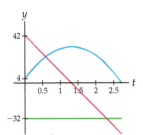

(a) Calculate $\int a(t) \, dt$. Then, given that acceleration is the derivative of velocity and that the initial velocity was $v_0 = 42$, find a formula for the velocity $v(t)$ of the ball after t seconds.

(b) Calculate $\int v(t) \, dt$. Then, given that velocity is the derivative of position and that the initial position was $s_0 = 4$ feet, find a formula for the velocity $s(t)$ of the ball after t seconds.

(c) Discuss the relationships among the graphs of $s(t)$, $v(t)$, and $a(t)$ shown in this exercise.

66. Dad's casserole surprise is for dinner again, and from previous experiments we know that after the casserole is taken out of the oven, the temperature of the casserole will cool by changing at a rate of $T'(t) = -15e^{-0.5t}$ degrees Fahrenheit per minute.

Rate of change of casserole temperature
$T'(t) = -15e^{-0.5t}$

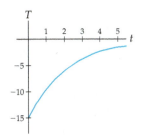

(a) Calculate $\int T'(t) \, dt$. Then, given that the casserole was $T_0 = 350$ degrees when it was removed from the oven, find a formula for the temperature $T(t)$ of the casserole after t minutes.

(b) Calculate $T(5) - T(0)$. Compare your answer with the answer to the last part of Exercise 47 from Section 7.2. Write a few sentences about the connection that this comparison suggests between Riemann sums and antidifferentiation.

Proofs

67. Use differentiation to prove the two formulas from Theorem 7.16 for integrating power functions.

68. Prove that $\int e^x \, dx = e^x + C$ in two ways: (a) by using the first part of Theorem 7.17 and (b) by using the second part of Theorem 7.17.

69. Use differentiation to prove the two formulas from Theorem 7.17 for integrating exponential functions.

70. Use differentiation to prove the six formulas from Theorem 7.18 for integrating certain trigonometric functions.

71. Use differentiation to prove the three formulas from Theorem 7.19 for integrals whose solutions are inverse trigonometric functions.

72. Prove the first part of Theorem 7.20 in your own words: $\int k f(x), dx = k \int f(x) \, dx$.

73. Prove the second part of Theorem 7.20 in your own words: $\int (f(x) + g(x)) \, dx = \int f(x) \, dx + \int g(x) \, dx$.

74. Prove the "forward" direction of Theorem 7.14: If F and G are differentiable and $F'(x) = G'(x)$, then $G(x) = F(x) + C$ for some constant C. (*Hint: Show that if $F'(x) = G'(x)$, then the function $F(x) - G(x)$ is constant. Recall that if the derivative of a function is zero, then that function must be constant.*)

75. Prove the "backwards" direction of Theorem 7.14: If F and G are differentiable and $G(x) = F(x) + C$ for some constant C, then $F'(x) = G'(x)$. (*Hint: Differentiate both sides of the equality $G(x) = F(x) + C$.*)

Thinking Forward

Initial-Value Problems: Find a function f that has the given derivative f' and value $f(c)$, if possible.

▶ $f'(x) = x^8 - x^2 + 1$, $f(0) = -1$

▶ $f'(x) = \dfrac{2}{\sqrt[3]{x}}$, $f(1) = 2$

▶ $f'(x) = 3e^{2x} + 1$, $f(0) = 2$

▶ $f'(x) = 3\left(\dfrac{4}{5}\right)^x$, $f(1) = 0$

▶ $f'(x) = \dfrac{5}{3x - 2}$, $f(1) = 8$

▶ $f'(x) = 3\sin(-2x) - 5\cos(3x)$, $f(0) = 0$

Preview of Differential Equations: Given each of the following equations involving a function f, find a possible formula for $f(x)$.

▶ $f'(x) = f(x)$

▶ $f'(t) = \sin(bt)$

▶ $f'(x) = 2f(x)$

▶ $f''(t) = \cos(t)$

7.5   THE FUNDAMENTAL THEOREM OF CALCULUS

▶ The statement, meaning, and proof of the Fundamental Theorem of Calculus

▶ Other equivalent formulations of the Fundamental Theorem

▶ Using the Fundamental Theorem of Calculus to calculate definite integrals exactly

The Fundamental Theorem

So far we have two very different mathematical objects that we call integrals, namely, the definite integral $\int_a^b f(x)\,dx$ and the indefinite integral $\int f(x)\,dx$. The reason these objects have such similar names and notation is that they are related by two very important theorems collectively called the ***Fundamental Theorem of Calculus***.

We have already seen that our intuition about average velocity suggests that the area under a velocity curve is related to the distance travelled. In other words, we suspect that

$$\int_a^b v(t)\,dt = s(b) - s(a).$$

We have seen hints of this on multiple occasions in previous sections: The signed area under the graph of a function such as $v(t)$ on an interval is related to a difference of values of its antiderivative $s(t)$. This incredible relationship between signed area and antiderivatives is so important, in fact so *fundamental*, to the study of calculus that we call it the Fundamental Theorem of the subject:

THEOREM 7.22   **The Fundamental Theorem of Calculus**

If f is continuous on $[a, b]$ and F is any antiderivative of f, then

$$\int_a^b f(x)\,dx = F(b) - F(a).$$

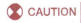 CAUTION

If f is not continuous on $[a, b]$, then the Fundamental Theorem may not apply to f on that interval, because we can't even be sure that its signed area on the interval $[a, b]$ is well defined. For example, consider the function $f(x) = \dfrac{1}{(x-1)^2}$ on the interval $[0, 3]$. The discontinuity of f at $x = 1$ is a problem that we won't learn how to handle until Section 8.6.

The Fundamental Theorem of Calculus expresses a deep fact about the relationship between antiderivatives and signed area and moreover provides a very powerful tool for calculating definite integrals. We will prove the Fundamental Theorem of Calculus at the end of this section, after we have discussed the various ways that we can use this powerful tool. In Section 7.7 we will discuss a related theorem called the Second Fundamental Theorem of Calculus.

Using the Fundamental Theorem of Calculus

When we apply the Fundamental Theorem in practice, we need to identify an antiderivative of the integrand, evaluate that antiderivative at two points, and then subtract. For example, since an antiderivative of $f(x) = x^2$ is $F(x) = \frac{1}{3}x^3$, the Fundamental Theorem tells us that

$$\int_1^4 x^2\, dx = F(4) - F(1) = \frac{1}{3}(4)^3 - \frac{1}{3}(1)^3 = 21.$$

The following notation gives us a more efficient way of working with the Fundamental Theorem in practice:

DEFINITION 7.23   **Evaluation Notation**

For any function F on an interval $[a, b]$, the difference $F(b) - F(a)$ will be called **the evaluation of $F(x)$ on $[a, b]$** and will be denoted by

$$\left[F(x)\right]_a^b = F(b) - F(a).$$

This notation is also sometimes written as $F(x)\,|_a^b$. We call this the "evaluation" of $F(x)$ on $[a, b]$ because $[F(x)]_a^b$ is the difference of $F(x)$ evaluated at $x = b$ and $F(x)$ evaluated at $x = a$. With this new notation, there is no longer a need to give a letter name to the antiderivative if we want to write the antidifferentiation step separately from the evaluation step:

$$\int_1^4 x^2\, dx = \left[\frac{1}{3}x^3\right]_1^4 = \frac{1}{3}(4)^3 - \frac{1}{3}(1)^3 = 21.$$

If F and G are two antiderivatives of a function $f(x)$ on $[a, b]$, then by Theorem 7.14, for some constant C we have $G(x) = F(x) + C$ for all values of x. This means that the evaluations of F and G on $[a, b]$ will be equal:

$$\left[G(x)\right]_a^b = \left[F(x) + C\right]_a^b = (F(b) + C) - (F(a) + C) = F(b) - F(a) = \left[F(x)\right]_a^b.$$

For an integrable function f, the indefinite integral $\int f(x)\,dx$ is the family $F(x) + C$ of antiderivatives of $f(x)$. Using evaluation notation, we can reword the Fundamental Theorem of Calculus to see clearly how definite and indefinite integrals are related:

THEOREM 7.24

The Fundamental Theorem of Calculus in Evaluation Notation

If f is continuous on $[a, b]$, then

$$\int_a^b f(x)\,dx = \left[\int f(x)\,dx\right]_a^b$$

In other words, the definite integral of a function f on an interval $[a, b]$ is equal to the indefinite integral of f evaluated on that interval. The proof of this theorem follows directly from the Fundamental Theorem, evaluation notation, and the definition of the indefinite integral, and is left to Exercise 80.

The Net Change Theorem

We can rephrase the Fundamental Theorem of Calculus in terms of a function and its derivative, rather than in terms of a function and its antiderivative. In this alternative form of the Fundamental Theorem (which you will prove in Exercise 78), the function f plays the role of the antiderivative while its derivative f' plays the role of the integrand.

THEOREM 7.25

The Net Change Theorem

If f is differentiable on $[a, b]$, then

$$\int_a^b f'(x)\,dx = f(b) - f(a).$$

In other words, the *net change* $f(b) - f(a)$ of a function f on $[a, b]$ is equal to the signed area between the graph of its derivative f' and the x-axis on that interval.

In this form, the Fundamental Theorem of Calculus says that in some sense definite integration "undoes" differentiation, in that if you integrate the derivative of a function from $x = a$ to $x = b$, you get something that has to do with the original function (namely, $f(b) - f(a)$). This form of the Fundamental Theorem can also be interpreted as saying that if you accumulate all the instantaneous rates of change of $f(x)$ by integrating its derivative $f'(x)$ on the interval $[a, b]$, you get the total change $f(b) - f(a)$ of the values of $f(x)$ on that interval.

The Net Change Theorem is useful in a variety of practical applications. For example, at the start of this section we saw that if $s(t)$ and $v(t)$ are the position and velocity functions of an object travelling in a straight line, then

$$\int_a^b v(t)\,dt = s(b) - s(a).$$

Since $v(t) = s'(t)$, we can rephrase this as

$$\int_a^b s'(t)\,dt = s(b) - s(a).$$

This formula says that we can find the net change in distance $s(t)$ on $[a, b]$ by finding the net area between the graph of $v(t) = s'(t)$ and the x-axis on $[a, b]$. In any situation where

$Q(t)$ is a quantity that varies over time and whose rate of change is given by the derivative $r(t) = Q'(t)$, we can say that the net change in Q on an interval is the area under r on that interval:

$$\int_a^b r(t)\, dt = \int_a^b Q'(t)\, dt = Q(b) - Q(a)$$

The Proof of the Fundamental Theorem

To prove the Fundamental Theorem of Calculus, we must show that the Riemann sums that define the definite integral of a function f on an interval can be written in terms of an antiderivative F of f. The key to doing this will be applying the Mean Value Theorem to the antiderivative F. This will allow us to relate the average rate of change of F to the instantaneous rate of change–or derivative–of F. Before reading this proof, you may want to review the definition of, and notation for, the definite integral of a function (Section 7.3), as well as the Mean Value Theorem (Section 3.1).

Proof. Suppose f is a continuous function on $[a, b]$, and let F be any antiderivative of f, so that $F' = f$. Since f is continuous, it must be integrable; in other words, its definite integral from $x = a$ to $x = b$ is well defined by a limit of Riemann sums, namely,

$$\int_a^b f(x)\, dx = \lim_{n \to \infty} \sum_{k=1}^n f(x_k^*)\, \Delta x = \lim_{n \to \infty} \sum_{k=1}^n F'(x_k^*)\, \Delta x,$$

where, for each n, we define $\Delta x = \dfrac{b-a}{n}$ and $x_k = a + k\Delta x$, and where each x_k^* is a point in the subinterval $[x_{k-1}, x_k]$. Note that we could write $f(x_k^*) = F'(x_k^*)$, because $f = F'$ by hypothesis.

We can choose each x_k^* to be *any* point we like in the subinterval $[x_{k-1}, x_k]$. The key to this proof is to use the Mean Value Theorem to choose the points x_k^* in a very special way. Since, by hypothesis, F is differentiable on $[x_{k-1}, x_k]$, it is also continuous on that subinterval. Therefore the Mean Value Theorem applies to F on $[a, b] = [x_{k-1}, x_k]$ and says that there exists some point $c_k \in [x_{k-1}, x_k]$ such that

$$F'(c_k) = \frac{F(x_k) - F(x_{k-1})}{x_k - x_{k-1}}.$$

Because we can choose x_k^* to be any point we like in $[x_{k-1}, x_k]$, we will choose it to be the point c_k that we defined for that subinterval. Since $x_k - x_{k-1} = \Delta x$, when we substitute this expression for $F'(x_k^*)$ into our Riemann sum expression for $\int_a^b f(x)\, dx$, we have

$$\int_a^b f(x)\, dx = \lim_{n \to \infty} \sum_{k=1}^n \frac{F(x_k) - F(x_{k-1})}{\Delta x}\, \Delta x = \lim_{n \to \infty} \sum_{k=1}^n (F(x_k) - F(x_{k-1})).$$

For every n, this sum is what is known as a ***telescoping sum***; this means that if we write out the terms of this sum, then, after rearranging terms, we find that most will cancel each other out and, luckily, the sum will collapse like a folding telescope:

$$\sum_{k=1}^n (F(x_k) - F(x_{k-1})) = (F(x_1) - F(x_0)) + (F(x_2) - F(x_1)) + (F(x_3) - F(x_2))$$
$$+ \cdots + (F(x_{n-1}) - F(x_{n-2})) + (F(x_n) - F(x_{n-1}))$$
$$= -F(x_0) + (F(x_1) - F(x_1)) + (F(x_2) - F(x_2))$$
$$+ \cdots + (F(x_{n-1}) - F(x_{n-1})) + F(x_n)$$
$$= -F(x_0) + 0 + 0 + \cdots + 0 + F(x_n) = F(b) - F(a).$$

Putting this computation together with our Riemann sum calculation of the definite integral and taking what is now a very simple limit, we have the equality that we seek:

$$\int_a^b f(x)\, dx = \lim_{n \to \infty} \sum_{k=1}^n (F(x_k) - F(x_{k-1})) = \lim_{n \to \infty} (F(b) - F(a)) = F(b) - F(a). \quad \blacksquare$$

Examples and Explorations

EXAMPLE 1

Distance travelled is the definite integral of velocity

Suppose a car travelled 100 miles along a straight path in two hours. If $s(t)$ describes the car's position in feet after t hours and $v(t) = s'(t)$ describes its velocity in miles per hour after t hours, find the signed area between the graph of $v(t)$ and the t-axis from $t = 0$ to $t = 2$.

SOLUTION

If $s(0) = 0$ miles, then $s(2) = 100$ miles. By the Fundamental Theorem of Calculus and the fact that $s(t)$ is an antiderivative of $v(t)$, the signed area under the graph of $v(t)$ on $[0, 2]$ is

$$\int_0^2 v(t)\, dt = s(2) - s(0) = 100 - 0 = 100.$$

Note that we didn't need to know anything about the velocity function to do this calculation; all we had to know was the change in the position of the car over the time interval $[0, 2]$. Neither did we have to calculate any limits or Riemann sums to find the definite integral $\int_0^2 v(t)\, dt$.   ☐

EXAMPLE 2

Calculating definite integrals with the Fundamental Theorem of Calculus

Use the Fundamental Theorem of Calculus to calculate the exact values of the following definite integrals:

(a) $\displaystyle\int_2^4 (x^2 + 1)\, dx$ **(b)** $\displaystyle\int_{-\pi}^{\pi} \sin x\, dx$ **(c)** $\displaystyle\int_0^2 \frac{1}{x-1}\, dx$

SOLUTION

(a) Calculating this definite integral exactly with the Fundamental Theorem of Calculus is as easy as antidifferentiating $x^2 + 1$ and then evaluating from 2 to 4:

$$\int_2^4 (x^2 + 1)\, dx = \left[\tfrac{1}{3}x^3 + x\right]_2^4 = \left(\tfrac{1}{3}(4)^3 + 4\right) - \left(\tfrac{1}{3}(2)^3 + 2\right) = \tfrac{62}{3}.$$

Notice that the Fundamental Theorem just allowed us to calculate a definite integral quickly without having to make a lengthy computation of limits of Riemann sums. Compare this solution with the work we did to investigate the same definite integral in Examples 1 and 2 from Section 7.3.

(b) Since an antiderivative of $\sin x$ is $-\cos x$ (check this by differentiating), $\cos(\pi) = -1$, and $\cos(-\pi) = -1$, we have

$$\int_{-\pi}^{\pi} \sin x\, dx = \left[-\cos x\right]_{-\pi}^{\pi} = -\cos(\pi) - (-\cos(-\pi)) = -(-1) - (-(-1)) = 0.$$

(c) At first glance we might be tempted to apply the Fundamental Theorem and say that $\int_0^2 \frac{1}{x-1}\, dx$ is equal to $\left[\ln|x - 1|\right]_0^2 = \ln|1| - \ln|-1| = 0$. But this is not the correct answer, because $\frac{1}{x-1}$ is not continuous at $x = 1$, which is part of the interval $[0, 2]$.

The Fundamental Theorem of Calculus can be used *only* if a function is continuous on the entire interval $[a, b]$. At this point we do not have the tools to discuss the area under the graph of a function with a vertical asymptote in the interval (although we will in Section 8.6).   ☐

CHECKING
THE ANSWER

We can use calculator graphs to check that our definite integral calculations are reasonable. The graphs shown next illustrate the three areas we discussed in the previous example. In the example, we found that the first area was $\frac{62}{3}$, which is just over 20 units; this seems reasonable given the leftmost graph. The middle graph clearly shows an area of 0, since the regions below and above the x-axis are the same size; this answer matches with the calculation we made. The rightmost graph shows that the function $\frac{1}{x-1}$ has an asymptote at $x = 1$ over which we cannot integrate.

$$\int_2^4 (x^2 + 1)\,dx \qquad\qquad \int_{-\pi}^{\pi} \sin x\,dx \qquad\qquad \textit{Can't talk about this area yet}$$

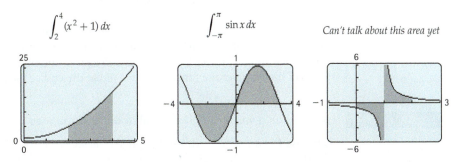

EXAMPLE 3   Using the Fundamental Theorem of Calculus to find a signed area

Find the signed area between the graph of $f(x) = 3e^{2x-5} - 3$ and the x-axis from $x = 1$ to $x = 3$. Use a graph to verify that the sign of your answer is reasonable.

SOLUTION

By definition, the signed area is the definite integral of f on $[1, 3]$. Using integration formulas and a little guess and check, we find that $F(x) = \frac{3}{2}e^{2x-5} - 3x$ is an antiderivative of $f(x) = 3e^{2x-5} - 3$. By the Fundamental Theorem of Calculus, the signed area is then

$$\int_1^3 (3e^{2x-5} - 3)\,dx = \left[\frac{3}{2}e^{2x-5} - 3x \right]_1^3$$
$$= \left(\frac{3}{2}e^{2(3)-5} - 3(3) \right) - \left(\frac{3}{2}e^{2(1)-5} - 3(1) \right)$$
$$= \frac{3}{2}e - \frac{3}{2}e^{-3} - 6 \approx -1.99726.$$

To determine whether it is reasonable that our answer is negative, we must first know where the function $f(x) = 3e^{2x-5} - 3$ is positive and where it is negative. The graph of f crosses the x-axis when $f(x) = 0$. Solving $3e^{2x-5} - 3 = 0$, we get $x = \frac{5}{2} = 2.5$. As you can see in the following graph of f, $f(x)$ is negative on $[1, 2.5]$ and positive on $[2.5, 3]$:

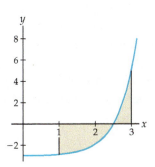

Since the part of the region in question that is below the x-axis is larger than the part of the region that is above the x-axis, it does make sense that the signed area of the region is negative.   ☐

There are many ways that we can check calculations that use the Fundamental Theorem of Calculus. First, we can check the antidifferentiation step by differentiating. For instance, for the previous example we can verify that

$$\frac{d}{dx}\left(\frac{3}{2}e^{2x-5} - 3x\right) = \frac{3}{2}(2)e^{2x-5} - 3 = 3e^{2x-5} - 3.$$

Second, we can check the numerical answer of a definite integral calculation by approximating the value of the definite integral with a graphing calculator or other graphing utility. We can also quickly check that the size and sign of the numerical answer make sense in terms of the graph of the function, as we did at the end of the previous example.

EXAMPLE 4

Computing a value of f given a value of f and the area under f'

Suppose f is a function whose *derivative* f' is shown next. Given that $f(1) = 3$, use the Fundamental Theorem of Calculus and the graph of f' to approximate $f(4)$.

$$\text{Graph of } f' \text{ and the area } \int_{1}^{4} f'(x)\,dx$$

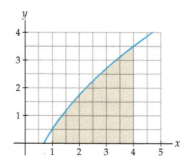

SOLUTION

By the Fundamental Theorem of Calculus, we know that the value $f(1)$, the value $f(4)$, and the area under the graph of f' on $[1, 4]$ are related. Specifically, we know that

$$\int_{1}^{4} f'(x)\,dx = f(4) - f(1).$$

Solving for $f(4)$ gives us

$$f(4) = f(1) + \int_{1}^{4} f'(x)\,dx.$$

We are given that $f(1) = 3$, so all that remains is to approximate the value of the definite integral. We can do this by using the grid in the given figure. Each box in the grid has a length and height of half a unit, and therefore each box has an area of $\left(\frac{1}{2}\right)\left(\frac{1}{2}\right) = \frac{1}{4}$ square units. By counting boxes (and estimating "partial" boxes), we see that the area under the graph of f' from $x = 1$ to $x = 4$ covers approximately 26.5 boxes. Therefore the area in question is approximately $26.5\left(\frac{1}{4}\right) \approx 6.6$. This means that the value of $f(4)$ must be approximately

$$f(4) = f(1) + \int_{1}^{4} f'(x)\,dx \approx 3 + 6.6 = 9.6.$$

Notice that we were able to find an approximate value of $f(4)$ even though we were never given a graph or equation for the function f.   □

Finding net change with a definite integral

A termite colony is slowly dying off in such a way that the number of termites $P(t)$ after t weeks is changing at a rate of $P'(t) = -576.68(0.794)^t$. Determine the number of termites that die in the first three weeks, and interpret your answer as an area under a curve.

SOLUTION

By the Net Change Theorem, we can find the net change in the number of termites over the first three weeks by calculating a definite integral:

$$\int_0^3 -576.68(0.794)^t = \left[\frac{-576.68}{\ln 0.794}(0.794)^t \right]_0^3$$

$$= \frac{-576.68}{\ln 0.794}(0.794)^3 - \frac{-576.68}{\ln 0.794}(0.794)^0$$

$$\approx -1248.58.$$

Therefore about 1249 termites in the colony die in the first three weeks. Compare this example with Example 1 from Section 5.4. To visualize this answer as the area under a curve, consider the graph of the rate $P'(t)$ shown next at the left. Notice that this rate is negative, since the termites are dying, and that the rate has a large magnitude at first and then tapers towards zero. The net change in termites is shown at the right, as the signed area between the graph of $P'(t)$ and the t-axis on the interval $[0, 3]$.

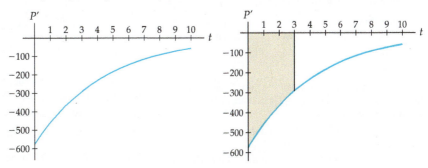

Rate of change $P'(t)$ of the termite population   Net change in the termite population on $[0, 3]$

TEST YOUR UNDERSTANDING

▶ How does the Fundamental Theorem of Calculus connect definite and indefinite integrals?

▶ What important theorem is the key to proving the Fundamental Theorem of Calculus?

▶ What is the identifying characteristic of a telescoping sum?

▶ Does the Fundamental Theorem of Calculus necessarily apply to a discontinuous function? Why or why not?

▶ How can you check your answer after applying the Fundamental Theorem of Calculus to calculate a definite integral?

EXERCISES 7.5

Thinking Back

▶ *Definite integrals:* State the definition of the definite integral of an integrable function f on $[a, b]$.

▶ *Indefinite integrals:* State the definition of the indefinite integral of an integrable function f.

▶ *Approximating with a Riemann sum:* Use a right sum with 10 rectangles to approximate the area under the graph of $f(x) = x^2$ on $[0, 5]$.

▶ *Limits of Riemann sums:* Use a limit of Riemann sums to calculate the exact area under the graph of $f(x) = x^2$ on $[0, 5]$.

Concepts

0. *Problem Zero:* Read the section and make your own summary of the material.

1. *True/False:* Determine whether each of the statements that follow is true or false. If a statement is true, explain why. If a statement is false, provide a counterexample.

   (a) *True or False:* Definite integrals and indefinite integrals are the same.

   (b) *True or False:* If f is integrable on $[a, b]$ and F is any other function, then $\int_a^b f(x)\, dx = F(b) - F(a)$.

   (c) *True or False:* The Fundamental Theorem of Calculus applies to $f(x) = \sin x$ on $[0, \pi]$.

   (d) *True or False:* The Fundamental Theorem of Calculus applies to $f(x) = \tan x$ on $[0, \pi]$.

   (e) *True or False:* If f is an even function, a is a real number, and f is integrable on $[-a, a]$, then $\int_{-a}^a f(x)\, dx = 0$.

   (f) *True or False:* If f is an odd function, a is a real number, and f is integrable on $[-a, a]$, then $\int_{-a}^a f(x)\, dx = 0$.

   (g) *True or False:* If f is an even function, a is a real number, and f is integrable on $[-a, a]$, then $\int_{-a}^0 f(x)\, dx = \int_0^a f(x)\, dx$.

   (h) *True or False:* If f is an odd function, a is a real number, and f is integrable on $[-a, a]$, then $\int_{-a}^0 f(x)\, dx = \int_0^a f(x)\, dx$.

2. *Examples:* Construct examples of the thing(s) described in the following. Try to find examples that are different than any in the reading.

   (a) A function f for which the Fundamental Theorem of Calculus does not apply on $[0, 10]$.

   (b) Three functions f that we do not yet know how to antidifferentiate.

   (c) A non-constant function f for which $\int_0^4 f(x)\, dx = 0$.

3. Why is the Fundamental Theorem of Calculus such an incredible, *fundamental* theorem?

4. State the Fundamental Theorem of Calculus (a) in its original form, (b) in its alternative form, and (c) by using an indefinite integral and evaluation notation.

5. What important theorem is the key to proving the Fundamental Theorem of Calculus?

6. Why must the integrand be an integrable function in the Fundamental Theorem of Calculus?

7. Intuitively, the average value of a velocity function should be the same as the average rate of change of the corresponding position function. Explain why this intuition suggests that the signed area under the velocity graph on an interval is equal to the difference in the position function on that interval, and tell what this has to do with the Fundamental Theorem of Calculus.

8. Why is it not necessary to write down an antiderivative family when using the Fundamental Theorem of Calculus to calculate definite integrals? In other words, why don't we have to use "$+C$"?

Suppose f is a function whose *derivative* f' is given by the graph shown next. In Exercises 9–12, use the given value of f, an area approximation, and the Fundamental Theorem to approximate the requested value.

*Graph of f', the **derivative** of f*

9. Given that $f(3) = 2$, approximate $f(4)$.

10. Given that $f(0) = -1$, approximate $f(2)$.

11. Given that $f(2) = 3$, approximate $f(-2)$.

12. Given that $f(-1) = 2$, approximate $f(1)$.

Calculate each definite integral in Exercises 13–14, using (a) the definition of the definite integral as a limit of Riemann sums, (b) the definite integral formulas from Theorem 7.13, and (c) the Fundamental Theorem of Calculus. Then show that your three answers are the same.

13. $\displaystyle\int_1^2 3x^2 \, dx$

14. $\displaystyle\int_0^4 (3x + 2)^2 \, dx$

15. In the proof of the Fundamental Theorem of Calculus we encounter a telescoping sum. Find the values of the following sums, which are also telescoping.

   (a) $\displaystyle\sum_{k=1}^{100} \left(\frac{1}{k} - \frac{1}{k+1} \right)$

   (b) $\displaystyle\sum_{k=1}^{100} \frac{k^2 - (k-1)^2}{2}$

16. Determine whether or not each statement that follows is equivalent to the Fundamental Theorem of Calculus. Assume that all functions here are integrable.

   (a) If $f'(x) = F(x)$, then $\int_a^b F(x)\,dx = \left[f(x)\right]_a^b$.

   (b) $\int_a^b G(x)\,dx = [G'(x)]_a^b$.

   (c) If $h(x)$ is the derivative of $g(x)$, then $\int_a^b h(x)\,dx = g(b) - g(a)$.

   (d) $\left[\int w'(x)\,dx\right]_a^b = \int_a^b w(x)\,dx$.

   (e) $\left[\int s'(x)\,dx\right]_a^b = [s''(x)]_a^b$.

17. Determine whether or not each statement that follows is equivalent to the Fundamental Theorem of Calculus. Assume that all functions here are integrable.

   (a) $\int_a^b f(x)\,dx = g(b) - g(a)$, where $g'(x) = f(x)$.

   (b) If $f'(x) = F(x)$, then $\int_a^b f(x)\,dx = [F(x)]_a^b$.

   (c) If $g(x)$ is any antiderivative of $h(x)$, then $\int_a^b h(x)\,dx = g(a) - g(b)$.

   (d) $\int_a^b h''(x)\,dx = [h'(x)]_a^b$.

   (e) $\left[\int p(x)\,dx\right]_a^b = \int_a^b p(x)\,dx$.

18. In the proof of the Fundamental Theorem of Calculus, the Mean Value Theorem is used to choose values x_k^* in each subinterval $[x_{k-1}, x_k]$. Use the Mean Value Theorem in the same way to find the corresponding values x_k^* for a Riemann sum approximation of $\int_0^2 x^2\,dx$ with four rectangles.

Skills

Use the Fundamental Theorem of Calculus to find the exact values of each of the definite integrals in Exercises 19–64. Use a graph to check your answer. (*Hint: The integrands that involve absolute values will have to be considered piecewise.*)

19. $\int_{-1}^1 (x^4 + 3x + 1)\,dx$

20. $\int_{-3}^3 (x-1)(x+3)\,dx$

21. $\int_1^4 3(2^x)\,dx$

22. $\int_0^3 (x^2 - 4)^2\,dx$

23. $\int_2^5 \dfrac{1}{\sqrt{x^5}}\,dx$

24. $\int_{-\pi}^\pi \sin(3x)\,dx$

25. $\int_1^3 \dfrac{\sqrt{x}-1}{\sqrt{x}}\,dx$

26. $\int_1^3 2x\sqrt{x^2+1}\,dx$

27. $\int_0^{\pi/4} \sec x \tan x\,dx$

28. $\int_0^{\pi/4} \sec^2 x\,dx$

29. $\int_0^{\sqrt{2}/4} \dfrac{1}{\sqrt{1-4x^2}}\,dx$

30. $\int_0^{1/2} \dfrac{1}{1+4x^2}\,dx$

31. $\int_{-\pi}^\pi (1 + \sin x)\,dx$

32. $\int_2^4 3e^{2x-4}\,dx$

33. $\int_0^1 \dfrac{1}{2e^x}\,dx$

34. $\int_2^4 \dfrac{1}{4-3x}\,dx$

35. $\int_0^1 \dfrac{1}{1+x^2}\,dx$

36. $\int_0^1 \dfrac{x}{1+x^2}\,dx$

37. $\int_{-1}^1 \dfrac{2^x}{4^x}\,dx$

38. $\int_0^\pi (\sin^2 x + \cos^2 x)\,dx$

39. $\int_{-3}^2 \dfrac{1}{x+5}\,dx$

40. $\int_{-3}^2 \dfrac{1}{(x+5)^2}\,dx$

41. $\int_{-3}^2 \dfrac{2x}{x^2+5}\,dx$

42. $\int_{-3}^2 \dfrac{x^2+5}{2x}\,dx$

43. $\int_1^4 \dfrac{2x+3}{x^2+3x+4}\,dx$

44. $\int_{\pi/4}^{\pi/2} \csc x \cot x\,dx$

45. $\int_0^{3/2} \dfrac{1}{\sqrt{9-x^2}}\,dx$

46. $\int_{-3}^3 2\cos(\pi x)\,dx$

47. $\int_1^{e-3} \dfrac{3\pi}{2x+6}\,dx$

48. $\int_1^e 2(\ln x)\left(\dfrac{1}{x}\right)\,dx$

49. $\int_0^1 (x^{2/3} - x^{1/3})\,dx$

50. $\int_0^4 \dfrac{1}{(2x+1)^{5/2}}\,dx$

51. $\int_0^{\pi/2} \sin x(1 + \cos x)\,dx$

52. $\int_2^3 \dfrac{\ln x}{x}\,dx$

53. $\int_{-1}^1 \dfrac{e^x - xe^x}{e^{2x}}\,dx$

54. $\int_1^2 \dfrac{x^2 e^x - 2xe^x}{x^4}\,dx$

55. $\int_{\pi/4}^{\pi/2} x\csc^2(x^2)\,dx$

56. $\int_{-\pi/4}^{\pi/4} x^2 \sec^2(x^3)\,dx$

57. $\int_2^4 \dfrac{3x}{1-x^2}\,dx$

58. $\int_0^1 \dfrac{2^x - 3^x}{4^x}\,dx$

59. $\int_{-2}^5 |x-2|\,dx$

60. $\int_{-1}^3 |4-x^2|\,dx$

61. $\int_{-2}^4 |2x^2 - 5x - 3|\,dx$

62. $\int_0^6 |(x-1)(x-4)|\,dx$

63. $\int_0^{5\pi/4} |\sin x|\,dx$

64. $\int_{-1}^1 |e^{2x} - 1|\,dx$

Applications

65. Suppose a large oil tank develops a hole that causes oil to leak from the tank at a rate of $r(t) = 0.05t$ gallons per hour. Construct and then solve a definite integral to determine the amount of oil that has leaked from the tank after one day.

66. Suppose another oil tank with a hole in it takes three days to leak 11 gallons of oil. Assuming that the rate of leakage at time t is given by $r(t) = kt$ for some constant k, set up and solve an equation involving a definite integral to find k.

67. A figurine that is part of the display in a mechanical music box moves back and forth along a straight track as shown next at the left. The mechanics of the music box are programmed to move the figurine back and forth with velocity

$$v(t) = 0.00015t(t-5)(t-10)(t-15)(t-20)$$

inches per second over the 20 seconds that the music program plays; see the graph next at the right.

Velocity v(t) of the figurine

(a) Use the graph just shown to describe the direction of motion of the figurine during various parts of the music program.
(b) Determine the net distance travelled by the figurine in the first 5 seconds, the first 10 seconds, and the full 20-second music program.
(c) Determine the total amount of distance travelled by the figurine in either direction along the track.

68. Water is evaporating from an open vase. The rate of evaporation changes according to the amount of exposed surface area at the top of the container in such a way that the rate of change of the volume of water in the vase is given by

$$V'(t) = -0.05\pi(10 - 0.2t)^2$$

cubic centimeters per hour. The vase initially contains 200 cubic centimeters of water. Use the Net Change Theorem to determine the amount of time it will take for all of the water in the vase to evaporate.

69. Many email filters can be trained how to recognize spam by having a user identify spam messages from a lineup. Suppose you have looked at 12 messages with the word "opportunity" in the subject line, and identified 8 of them to the filter as spam. Now you get a new message with that word in the subject line, and your filter must compute the probability p that the message is spam. One way to do that is to identify p as the expected value of the Beta distribution

$$f(x) = 6435x^8(1-x)^4.$$

The expected value of this distribution is

$$\int_0^1 xf(x)\,dx.$$

Evaluate the probability that the next message you receive with the word "opportunity" in the subject line will be spam.

70. During World War II, a handful of German submarines were captured by the Allies to get access to German codes. On one of these submarines, the submarine's depth had been plotted carefully as $D(t)$, in meters, where t measures the number of hours after noon on that day. Unfortunately, during the capture, the information about $D(t)$ was lost. However, the following graph of the rate of change $D'(t)$ of the submarine's depth was recovered:

Rate of change D'(t)
of the depth of a submarine

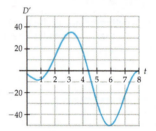

(a) At 4:00 P.M., was the submarine rising or falling? Was it speeding up or slowing down?
(b) Was the submarine closer to the surface at 2:00 P.M. or at 5:00 P.M.?
(c) At the end of the first two hours of the voyage, had the submarine risen or fallen from its position at the beginning? By how much?
(d) Approximately how much did the submarine rise or fall in the two minutes after 6:00 P.M.?
(e) At what time was the submarine diving at the fastest rate?
(f) When was the submarine at its highest point?
(g) Suppose the submarine was on the surface when it was at its highest point. What was the greatest depth of the submarine during its voyage that day?

71. A specialty bookshop is open from 8:00 A.M. to 6:00 P.M. Customers arrive at the shop at a rate $r(t)$ arrivals per hour. A graph of $r(t)$ for the duration of the business day is shown here, where t is the number of hours after the shop opens at 8:00 A.M. As a group, the salespeople in the bookshop can serve customers at a rate of 30 customers per hour. If there are no salespeople available, customers have to wait in line until they can be served, and the customers are willing to wait as long as is necessary in this line. Use the graph of $r(t)$ to estimate answers to the questions that follow.

Rate r(t) of arrivals at a bookshop

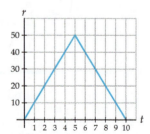

(a) At noon, approximately how many people had entered the bookshop so far? Is there a line yet, and if so, how long is it?

(b) At what rate is the line at the bookshop growing (or shrinking) at 1:00 P.M.?

(c) At approximately what time of day is the line the longest, and how many people are in the line at that time?

(d) At 6:00 P.M. when the bookshop closes, has every customer been served or is there still a line at closing time?

Proofs

Use the Fundamental Theorem of Calculus to give alternative proofs of the integration facts shown in Exercises 72–76. You may assume that all functions here are integrable.

72. $\displaystyle\int_a^b c\, dx = c(b-a)$

73. $\displaystyle\int_a^b x\, dx = \frac{1}{2}(b^2 - a^2)$

74. $\displaystyle\int_a^b x^2\, dx = \frac{1}{3}(b^3 - a^3)$

75. $\displaystyle\int_a^b kf(x)\, dx = k\int_a^b f(x)\, dx$

76. $\displaystyle\int_a^b (f(x) + g(x))\, dx = \int_a^b f(x)\, dx + \int_a^b g(x)\, dx$

77. Prove the Fundamental Theorem of Calculus in your own words. Use the proof in this section as a guide.

78. Prove that the Net Change Theorem (Theorem 7.25) is equivalent to the Fundamental Theorem of Calculus.

79. Prove that if two functions F and G differ by a constant, then $[F(x)]_a^b = [G(x)]_a^b$.

80. Use evaluation notation and the Fundamental Theorem of Calculus to prove Theorem 7.24: $\int_a^b f(x)\, dx = \left[\int f(x)\, dx\right]_a^b$.

Thinking Forward

Functions Defined by Integrals: Suppose $A(x)$ is the function that for each $x > 0$ is equal to the area under the graph of $f(t) = t^2$ from 0 to x.

▶ Write an equation that defines $A(x)$ in terms of a definite integral. (*Hint: Let the variable inside the integral be t, so as not to confuse it with the variable x.*)

▶ Calculate $A(0)$, $A(1)$, $A(2)$, $A(3)$, and $A(4)$.

▶ Sketch a graph of $f(x) = x^2$ and a graph of $A(x)$ on the same set of axes.

▶ Use the Fundamental Theorem of Calculus to find an equation for $A(x)$ that does not involve an integral.

▶ Use the equation for $A(x)$ that you just found to find $A'(x)$. What do you notice about $A'(x)$ and $f(x)$?

▶ Use the graphs of $A(x)$ and $f(x)$ that you drew earlier to illustrate the relationship between $A'(x)$ and $f(x)$ that you just discovered. Use the words *increasing, decreasing, positive,* and *negative* in your discussion.

7.6   AREAS AND AVERAGE VALUES

▶ Definite integrals of functions whose values are sometimes negative

▶ Using definite integrals to find the area between two curves

▶ Average value as a definite integral, and the Mean Value Theorem for Integrals

The Absolute Area Between a Graph and the *x*-axis

We have seen that Riemann sums, and thus definite integrals, automatically count area above the *x*-axis positively and area below the *x*-axis negatively. But what if we want to count the true area of a region, counting both area above and below the *x*-axis positively? The definite integral will not do this automatically, but we can use absolute values to express such an **absolute area** with a definite integral.

THEOREM 7.26   **Signed Area and Absolute Area**

For any integrable function f on an interval $[a, b]$,

(a) The signed area between the graph of f and the x-axis from $x = a$ to $x = b$ is given by the definite integral $\int_a^b f(x)\, dx$.

(b) The absolute area between the graph of f and the x-axis from $x = a$ to $x = b$, counting all areas positively, is given by the definite integral $\int_a^b |f(x)|\, dx$.

The first part of this theorem follows directly from the definition of signed area and the definite integral. The second part of the theorem follows from the fact that the graph of the absolute value of a function can be obtained by reflecting all the negative portions of the graph over the x-axis.

For example, consider the region between the graph of $f(x) = x - 1$ and the x-axis on $[0, 3]$. This region consists of a small triangle below the x-axis with area $1/2$ and a larger triangle above the x-axis with area 2; see the figure next at the left. The signed area of this region is $(-1/2) + 2 = 3/2$, and the absolute area of the region is $(1/2) + 2 = 5/2$. The figure next at the right shows the graph of $y = |x - 1|$, in which the triangle in the fourth quadrant has been reflected into the first quadrant.

$$\int_0^3 (x - 1)\, dx = (-1/2) + 2 = 3/2 \qquad\qquad \int_0^3 |x - 1|\, dx = (1/2) + 2 = 5/2$$

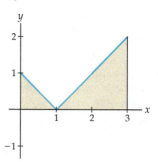

Although the second part of Theorem 7.26 gives us an expression for absolute area in terms of a definite integral, we cannot immediately calculate this type of definite integral with the Fundamental Theorem of Calculus. The reason is that we do not in general know how to antidifferentiate functions that involve absolute values. However, since we can write absolute values in terms of piecewise-defined functions, we write any definite integral of the form $\int_a^b |f(x)|\, dx$ as a sum of definite integrals that do not involve absolute values. Specifically, to calculate the integral of $|f(x)|$, we will split up the interval $[a, b]$ into smaller intervals on which f is either always above or always below the x-axis and then add negative signs as needed to turn negative values into positive values. For example, the definite integral described earlier can be split as

$$\int_0^3 |x - 1|\, dx = -\int_0^1 (x - 1)\, dx + \int_1^3 (x - 1)\, dx = -(-1/2) + 2 = 5/2.$$

Areas Between Curves

We can think of the area between the graph of a function f and the x-axis on an interval $[a, b]$ as the area between two graphs, namely, the graph of f and the graph of $y = 0$, on the interval $[a, b]$. What about the area between the graph of f and the graph of some other function g on $[a, b]$? For example, consider the region between the curves $f(x) = x + 2$ and $g(x) = x^2$ on the interval $[-1, 2]$, as shown here:

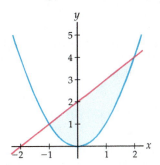

Region between $f(x) = x + 2$
and $g(x) = x^2$ on $[-1, 2]$

One way to find the area of this region is to calculate the area under the upper graph ($f(x) = x+2$) on $[-1, 2]$ and subtract the area under the lower graph ($f(x) = x^2$) on $[-1, 2]$. These two areas are shown in the first and second figures here:

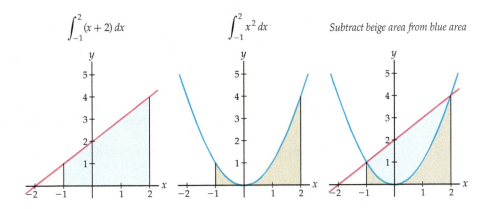

$$\int_{-1}^{2} (x+2)\, dx \qquad\qquad \int_{-1}^{2} x^2\, dx \qquad\qquad \textit{Subtract beige area from blue area}$$

From these pictures and the properties of definite integrals, we can express the area between f and g as a definite integral that we could easily calculate:

$$\text{Area between } f(x) = x+2 \atop \text{and } g(x) = x^2 \text{ on } [-1, 2] = \int_{-1}^{2} (x+2)\, dx - \int_{-1}^{2} x^2\, dx = \int_{-1}^{2} ((x+2) - x^2)\, dx.$$

The following definition generalizes this method of finding the area between two curves:

DEFINITION 7.27 **The Area Between Two Graphs**

If f and g are integrable functions, then the **area between the graphs of f and g from $x = a$ to $x = b$** is

$$\int_{a}^{b} |f(x) - g(x)|\, dx.$$

On any interval $[c, d]$ where $f(x) \geq g(x)$, we will have $|f(x) - g(x)| = f(x) - g(x)$. The area between the graphs of f and g on such an interval will be simply $\int_{c}^{d} (f(x) - g(x))\, dx$.

Similarly, on any interval $[r, s]$ where $g(x) \geq f(x)$, we will have $|f(x) - g(x)| = g(x) - f(x)$, and the area between the graphs of f and g on this interval will be $\int_r^s (g(x) - f(x))\, dx$. In general, sometimes $f(x)$ will be greater than $g(x)$, and sometimes it will be less than $g(x)$. In such cases we will have to split up the interval $[a, b]$ into smaller intervals on which $f(x)$ is always greater than $g(x)$–or vice versa–so we can express the area without using any absolute values, as will be shown in Example 2.

> **⊗ CAUTION**   The phrase "area between the graphs of f and g" always refers to an absolute area. The definite integral in Definition 7.27 does *not* consider the area between f and g to be positive or negative depending on whether it is, respectively, above or below the x-axis.

The Average Value of a Function on an Interval

So far we have used definite integrals to find areas; we will now investigate how definite integrals can be used to find the **_average value_** of a function on an interval. We will motivate our development of the definition of average value with a discrete approximation. Suppose we wish to approximate the average height of a plant during the first four days of its growth, given that the function $f(x) = 0.375x^2$ describes the height of the plant, in centimeters, x days after it breaks through the soil. We can approximate the average height of this plant by choosing a discrete number of x-values, for example $x = 1$, $x = 2$, $x = 3$, and $x = 4$, and averaging the heights $f(x)$ of the plant at these times:

$$\frac{f(1) + f(2) + f(3) + f(4)}{4} = \frac{0.375(1)^2 + 0.375(2)^2 + 0.375(3)^2 + 0.375(4)^2}{4}$$
$$= 2.8125 \text{ cm.}$$

This answer seems reasonable given the graph of the plant height function $f(x)$ shown next; the height $y = 2.8125$ does seem to be approximately the average height of the graph over $[0, 4]$.

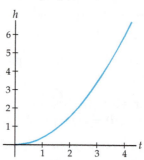

Height of the plant after
x days is $f(x) = 0.375x^2$

Of course, during the first four days the plant has more than just four heights. The height of the plant changes *continuously* as x changes, and we took into account only a *discrete* number of plant heights. To make a better approximation of the average height of the plant, we could average over a larger number of heights; for example, we could consider the height of the plant every quarter day (at $x = 0.25$, $x = 0.5$, $x = 0.75$, and so on until $x = 4$) and average those 16 heights:

$$\frac{f(0.25) + f(0.5) + f(0.75) + \cdots + f(3.75) + f(4)}{16} \approx 2.1914 \text{ cm.}$$

We can make as good an approximation as we like for the average value by choosing a large enough sample size n. To make n height measurements at equally spaced times, we would measure every $\frac{b-a}{n} = \frac{4}{n}$ days, at times $x_k = 0 + k\left(\frac{4}{n}\right)$ for $k = 1$ to $k = n$. The average value of the height function over those n measurements is

$$\frac{f(x_1) + f(x_2) + f(x_3) + \cdots + f(x_n)}{n} = \frac{\sum_{k=1}^{n} f(x_k)}{n} = \frac{1}{b-a} \sum_{k=1}^{n} f(x_k) \Delta x.$$

In the last step of the calculation, we used the fact that $\Delta x = \frac{b-a}{n}$. This equation should look awfully familiar to you. As we let n approach infinity we obtain the *exact* average value on $[a, b]$. But as $n \to \infty$ we are taking a limit of Riemann sums, which is a definite integral!

This discussion suggests the following general definition of average value:

DEFINITION 7.28

The Average Value of a Function on an Interval

The *average value* of an integrable function f on an interval $[a, b]$ is defined to be $\frac{1}{b-a}$ times the definite integral of f on $[a, b]$:

$$\text{Average value of } f \text{ from } x = a \text{ to } x = b = \frac{1}{b-a} \int_a^b f(x)\, dx.$$

Definition 7.28 gives us another way to utilize the power of the Fundamental Theorem of Calculus. Because we can express the exact average value of a function f on an interval in terms of a definite integral, we can find that average value as long as we can antidifferentiate the function f; see part (a) of Example 3.

The Mean Value Theorem for Integrals

The Mean Value Theorem (Theorem 3.5) guaranteed that for any function $f(x)$ that is continuous on a closed interval $[a, b]$ and differentiable on (a, b), there is always some point c in $[a, b]$ where the slope of f is equal to the average rate of change of f on $[a, b]$. We can now say a similar thing about finding a point c where the height of a function f is equal to its average value on an interval $[a, b]$.

Intuitively, the average value of a function f on $[a, b]$ should be between the highest and lowest values of f on that interval. By the Intermediate Value Theorem from Section 1.4, the function f has to achieve that average value at some point $c \in (a, b)$. This property suggests the following theorem, which you will prove in Exercise 82:

THEOREM 7.29

The Mean Value Theorem for Integrals

If f is continuous on a closed interval $[a, b]$, then there exists some $c \in (a, b)$ such that

$$f(c) = \frac{1}{b-a} \int_a^b f(x)\, dx.$$

Note that the Mean Value Theorem for Integrals tells us only that a certain value $x = c$ must exist, not what that value *is*. If we want to actually find the values $x = c$ for which the height of a function is equal to its absolute value, we have to use the Fundamental Theorem of Calculus and solve the equation in Theorem 7.29 for $x = c$.

Our construction in Definition 7.28 means that the average value of a function on an interval is related to the signed area under the same function on the same interval. Specifically, if $f(c) = K$ is the average value of f on an interval $[a, b]$, then

$$K = \frac{1}{b-a} \int_a^b f(x)\, dx \quad \Rightarrow \quad K(b-a) = \int_a^b f(x)\, dx.$$

This means that the signed area under the graph of f from $x = a$ to $x = b$ is equal to $K(b-a)$, which is the area of a rectangle of height K and width $b - a$. One way to think of this area is to imagine the region between the graph of f and the x-axis as the cross section of a wave of water, as if we are looking, from the side, at water sloshing in a glass tank. When the water settles, it will have a height of K, which means that its height will be the average value of the function f on the interval.

For example, consider the following graphs:

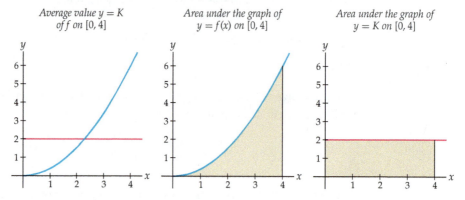

| Average value $y = K$ of f on $[0, 4]$ | Area under the graph of $y = f(x)$ on $[0, 4]$ | Area under the graph of $y = K$ on $[0, 4]$ |

Notice that the area of the shaded region in the middle figure is the same as the area of the shaded region in the last figure. In addition, the value $x = c$ guaranteed by the Mean Value Theorem for Integrals is approximately $x = 2.3$ in this example, since, according to the figure at the left this is the value at which the height of the function f is equal to the average value shown at height $y = 2$.

Examples and Explorations

EXAMPLE 1   Using definite integrals to find signed and absolute areas

Consider the region between the graph of a function f and the x-axis on $[-2, 4]$ as shown here. Express (a) the signed area and (b) the absolute area of this region in terms of definite integrals of the function f that do not involve absolute values.

Region between f and x-axis on $[-2, 4]$

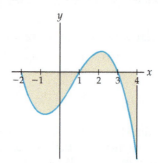

SOLUTION

(a) The signed area of this region is simply the definite integral of f from $x = -2$ to $x = 4$; in other words,

$$\text{signed area} = \int_{-2}^{4} f(x)\, dx.$$

(b) The absolute area of the region (counting all areas positively) is $\int_{-2}^{4} |f(x)|\, dx$. Unfortunately, we do not know how to calculate a definite integral of a function with an absolute value. However, as usual we can get rid of an absolute value if we are willing to work piecewise. We will express the absolute area without using absolute values by using three definite integrals. From $x = -2$ to $x = 1$ the function f is negative, and thus on the interval $[-2, 1]$ we have $|f(x)| = -f(x)$. From $x = 1$ to $x = 3$ the function is positive, so $|f(x)| = f(x)$ on this interval. Finally, from $x = 3$ to $x = 4$ the function is negative, so $|f(x)| = -f(x)$ on $[3, 4]$. Therefore the absolute area of the region is

$$\int_{-2}^{4} |f(x)|\, dx = -\int_{-2}^{1} f(x)\, dx + \int_{1}^{3} f(x)\, dx - \int_{3}^{4} f(x)\, dx.$$

Notice that in this expression, $\int_{-2}^{1} f(x)\, dx$ is a *negative* number, and thus $-\int_{-2}^{1} f(x)\, dx$ will be a *positive* number. We have essentially just flipped the signs of the definite integrals that measure area negatively. □

⊗ CAUTION   In general, absolute values do *not* commute with definite integrals. In other words, $\int_{a}^{b} |f(x)|\, dx$ is not in general the same as $\left|\int_{a}^{b} f(x)\, dx\right|$. This means that to find the absolute area it is not enough just to take the absolute value of the signed area.

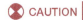

EXAMPLE 2   Using definite integrals to find the area between two graphs

Find the area of the region between the graphs of $f(x) = x^2 - x - 2$ and $g(x) = 4 - x^2$ on the interval $[-2, 3]$.

SOLUTION

According to Definition 7.27, we need to calculate

$$\int_{-2}^{3} |(x^2 - x - 2) - (4 - x^2)|\, dx.$$

However, we do not know how to calculate a definite integral that involves an absolute value. Once again if we are willing to work piecewise, we can get around this problem. Our first task is to find the points where f and g intersect. The region whose area we are interested in calculating is as follows (with f shown in blue and g in red):

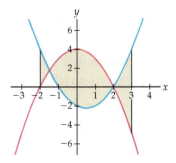

Region between $f(x) = x^2 - x - 2$
and $g(x) = 4 - x^2$ on $[-2, 3]$

The solutions of the equation $x^2 - x - 2 = 4 - x^2$ are $x = -\dfrac{3}{2}$ and $x = 2$, so these are the two points of intersection shown in the graph. Notice that f is above g on $\left[-2, -\dfrac{3}{2}\right]$, g is above f on $\left[-\dfrac{3}{2}, 2\right]$, and f is again above g on $[2, 3]$. On each of these smaller intervals we can use a definite integral to calculate the area between the graphs without requiring an absolute value. Each such definite integral is easy to calculate with the Fundamental Theorem of Calculus. Using the fact that $f(x) - g(x) = 2x^2 - x - 6$, we see that the total area of the region between the graphs of f and g on $[-2, 3]$ is

$$\int_{-2}^{3} |f(x) - g(x)| \, dx = \int_{-2}^{-3/2} (f(x) - g(x)) \, dx + \int_{-3/2}^{2} (g(x) - f(x)) \, dx + \int_{2}^{3} (f(x) - g(x)) \, dx$$

$$= \int_{-2}^{-3/2} (2x^2 - x - 6) \, dx + \int_{-3/2}^{2} -(2x^2 - x - 6) \, dx + \int_{2}^{3} (2x^2 - x - 6) \, dx$$

$$= \left[\frac{2}{3}x^3 - \frac{1}{2}x^2 - 6x\right]_{-2}^{-3/2} + \left[-\left(\frac{2}{3}x^3 - \frac{1}{2}x^2 - 6x\right)\right]_{-3/2}^{2} + \left[\frac{2}{3}x^3 - \frac{1}{2}x^2 - 6x\right]_{2}^{3}$$

$$= \frac{23}{24} + \frac{343}{24} + \frac{25}{6} = \frac{466}{24}.$$

□

EXAMPLE 3   The difference between average value and average rate of change

Suppose the function $f(x) = 0.375x^2$ describes the height, in centimeters, of a growing plant after x days.

(a) Find the average height of the plant during the first four days of its growth.

(b) Find the average rate of growth of the plant over those four days.

SOLUTION

(a) The average height of the plant during the first four days is the average value of the function $f(x) = 0.375x^2$ from $x = 0$ to $x = 4$, which by Definition 7.28 is

$$\frac{1}{4 - 0} \int_{0}^{4} 0.375x^2 \, dx = \frac{1}{4}\left[\frac{0.375}{3}x^3\right]_{0}^{4} = \frac{1}{4}\left(\frac{0.375}{3}(4)^3 - \frac{0.375}{3}(0)^3\right) = 2 \text{ cm}.$$

Notice that we reduced the problem of finding the average height of f to the problem of solving a definite integral. This is something that is simple to do with the Fundamental Theorem of Calculus, as long as we can find an antiderivative of f. Compare this result with our earlier approximations for the average height of the plant.

(b) In contrast, the average rate of growth of the plant over the first four days is the average rate of change of $f(x) = 0.375x^2$ on $[0, 4]$, which is equal to

$$\frac{f(4) - f(0)}{4 - 0} = \frac{0.375(4)^2 - 0.375(0)^2}{4} = 1.5\frac{\text{cm}}{\text{day}}.$$

Note that this answer makes sense, since the plant grew six centimeters in four days, and thus its average rate of growth was $\frac{6}{4} = 1.5$ centimeters per day.

□

EXAMPLE 4   Interpreting the Mean Value Theorem for Integrals

Answer the following questions about the function $f(x) = x^2$ on the interval $[1, 3]$:

(a) Why does the Mean Value Theorem for Integrals apply to this function on this interval? What can we conclude from it? Interpret this conclusion in terms of an area and in terms of an average value.

(b) Find a value c that satisfies the conclusion of the Mean Value Theorem for Integrals in this case.

SOLUTION

(a) The theorem applies because $f(x) = x^2$ is continuous on $[1, 3]$. The conclusion is that there is some point $c \in (1, 3)$ for which $f(c) = \frac{1}{3-1}\int_1^3 x^2\, dx$. This means that the average value of $f(x) = x^2$ on $[1, 3]$ is equal to the height c^2 of the function at the point c. Multiplying by $3 - 1$ we have $(3 - 1)f(c) = \int_1^3 x^2\, dx$, which says that the area under the graph of $f(x) = x^2$ on $[1, 3]$ is equal to the area of a rectangle with height c^2 and width $3 - 1$.

(b) Of course the Mean Value Theorem for Integrals does not tell us how to find a point $c \in (1, 3)$ at which f achieves its average value, only that such a point must exist. However, we can easily find the average value by applying the Fundamental Theorem of Calculus to the expression of average value as a definite integral:

$$\frac{1}{3-1}\int_1^3 x^2\, dx = \frac{1}{2}\left[\frac{1}{3}x^3\right]_1^3 = \frac{1}{2}\left(\frac{1}{3}(3)^3 - \frac{1}{3}(1)^3\right) = \frac{13}{3}.$$

Now it is easy to find a number $c \in (1, 3)$ so that $f(c) = c^2$ is equal to this average value:

$$c^2 = \frac{13}{3} \quad \Longrightarrow \quad c = \pm\sqrt{\frac{13}{3}}.$$

Note that $-\sqrt{\frac{13}{3}}$ is *not* in the interval $(1, 3)$, so the only value $c \in (1, 3)$ for which $f(c)$ is the average value $\frac{13}{3}$ is $c = \sqrt{\frac{13}{3}}$.   □

CHECKING THE ANSWER

In the previous example, we found that the average value of $f(x) = x^2$ on $[1, 3]$ should occur at $x = \sqrt{\frac{13}{3}} \approx 2.08$, and should be $f(x) = \frac{13}{3} \approx 4.33$. This means that the area under the graph of f on $[1, 3]$ should be equal to the area of the rectangle on $[1, 3]$ with height approximately 4.33, a result that looks reasonable considering the following two graphs:

Area under $f(x) = x^2$ on $[1, 3]$

Area under height $f(2.08) = 4.33$ on $[1, 3]$

TEST YOUR UNDERSTANDING

▶ What is the difference between signed area and absolute area? Which one does the definite integral automatically calculate?

▶ What can you say about f if the signed area between f and the x-axis is zero on an interval $[a, b]$? What can you say if the *absolute* area is zero on that interval?

▶ Why do we have to know where two graphs f and g intersect in order to calculate the area between f and g on an interval?

▶ How can it be that the plant described in the reading grows from 0 to 6 centimeters in the first 4 days and yet has an average height of only 2 centimeters, rather than 3? *(Hint: Use the shape of the graph.)*

▶ What is the difference between the average value of a function on an interval and the average rate of change of a function on an interval?

EXERCISES 7.6

Thinking Back

▶ *Finding sign intervals:* Consider the function $f(x) = x^3 - 4x^2 - 12x$. Find the intervals on which f is positive and the intervals on which f is negative.

▶ *Finding intersection points:* Find the x-coordinates of the points where the graphs of $f(x) = x^3 - 9x^2$ and $g(x) = x - 9$ intersect.

▶ *The Intermediate Value Theorem:* State the Intermediate Value Theorem, and illustrate the theorem with a graph.

▶ *The Mean Value Theorem:* State the Mean Value Theorem, and illustrate the theorem with a graph.

Absolute values as piecewise-defined functions: Write out each of the following functions as a piecewise-defined function that does not involve absolute values.

▶ $f(x) = |x|$

▶ $f(x) = |2 - 3x|$

▶ $f(x) = |x^2 - 3x - 4|$

▶ $f(x) = ||x - 1| - 1|$

Concepts

0. *Problem Zero:* Read the section and make your own summary of the material.

1. *True/False:* Determine whether each of the statements that follow is true or false. If a statement is true, explain why. If a statement is false, provide a counterexample.

(a) *True or False:* The absolute area between the graph of f and the x-axis on $[a, b]$ is equal to $|\int_a^b f(x)\,dx|$.

(b) *True or False:* The area of the region between $f(x) = x - 4$ and $g(x) = -x^2$ on the interval $[-3, 3]$ is negative.

(c) *True or False:* The signed area between the graph of f on $[a, b]$ is always less than or equal to the absolute area on the same interval.

(d) *True or False:* The area between any two graphs f and g on an interval $[a, b]$ is given by $\int_a^b (f(x) - g(x))\,dx$.

(e) *True or False:* The average value of the function $f(x) = x^2 - 3$ on $[2, 6]$ is $\frac{f(6) + f(2)}{2} = \frac{33 + 1}{2} = 17$.

(f) *True or False:* The average value of the function $f(x) = x^2 - 3$ on $[2, 6]$ is $\frac{f(6) - f(2)}{4} = \frac{33 - 1}{4} = 8$.

(g) *True or False:* The average value of f on $[1, 5]$ is equal to the average of the average value of f on $[1, 2]$ and the average value of f on $[2, 5]$.

(h) *True or False:* The average value of f on $[1, 5]$ is equal to the average of the average value of f on $[1, 3]$ and the average value of f on $[3, 5]$.

2. *Examples:* Construct examples of the thing(s) described in the following. Try to find examples that are different than any in the reading.

(a) A function f for which the signed area between f and the x-axis on $[0, 4]$ is zero, and a different function g for which the absolute area between g and the x-axis on $[0, 4]$ is zero.

(b) A function f whose signed area on $[0, 5]$ is less than its signed area on $[0, 3]$.

(c) A function f whose average value on $[-1, 6]$ is negative while its average rate of change on the same interval is positive.

3. Without using absolute values, how many definite integrals would we need in order to calculate the absolute area between $f(x) = \sin x$ and the x-axis on $\left[-\frac{\pi}{2}, 2\pi\right]$? Will the absolute area be positive or negative, and why? Will the signed area will be positive or negative, and why?

4. Without using absolute values, how many definite integrals would we need in order to calculate the area between the graphs of $f(x) = \sin x$ and $g(x) = \frac{1}{2}$ on $\left[-\frac{\pi}{2}, 2\pi\right]$?

5. Suppose f is positive on $(-\infty, -1]$ and $[2, \infty)$ and negative on the interval $[-1, 2]$. Write (a) the signed area and (b) the absolute area between the graph of f and the x-axis on $[-3, 4]$ in terms of definite integrals that do not involve absolute values.

6. Suppose $f(x) \geq g(x)$ on $[1, 3]$ and $f(x) \leq g(x)$ on $(-\infty, 1]$ and $[3, \infty)$. Write the area of the region between the graphs of f and g on $[-2, 5]$ in terms of definite integrals without using absolute values.

7. If f is negative on $[-3, 2]$, is the definite integral $\int_{-3}^2 f(x)\,dx$ positive or negative? What about the definite integral $-\int_{-3}^2 f(x)\,dx$?

8. Shade in the regions between the two functions shown here on the intervals (a) $[-2, 3]$; (b) $[-1, 2]$; and (c) $[1, 3]$. Which of these regions has the largest area? The smallest?

9. Describe an example that illustrates that $\int_a^b |f(x)|\,dx$ is not equal to $|\int_a^b f(x)\,dx|$.

10. Use the definition of absolute value to explain why the absolute area $\int_a^b |f(x)|\, dx$ counts area positively regardless of whether it is above or below the x-axis. Include graphs that support your explanation.

11. Consider the region between f and the x-axis on $[-2, 4]$ as in the graph next at the left. (a) Draw the rectangles of the left-sum approximation for the area of this region, with $n = 8$. Then express (b) the signed area and (c) the absolute area of the region with definite integrals that do not involve absolute values.

Graph for Exercise 11

Graph for Exercise 12

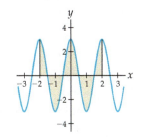

12. Repeat Exercise 11 for the function f shown above at the right, on the interval $[-2, 2]$.

13. Consider the region between f and g on $[0, 4]$ as in the graph next at the left. (a) Draw the rectangles of the left-sum approximation for the area of this region, with $n = 8$. Then (b) express the area of the region with definite integrals that do not involve absolute values.

Graph for Exercise 13

Graph for Exercise 14

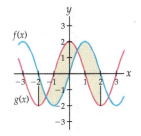

14. Repeat Exercise 13 for the function f shown above at the right, on the interval $[-2, 2]$.

15. Consider the function f shown in the graph next at the right. Use the graph to make a rough estimate of the average value of f on $[-4, 4]$, and illustrate this average value as a height on the graph.

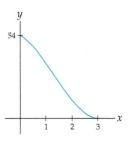

16. Without calculating any sums or definite integrals, determine the values of the described quantities. (*Hint: Sketch graphs first.*)

   (a) The signed area between the graph of $f(x) = \cos x$ and the x-axis on $[-\pi, \pi]$.

   (b) The average value of $f(x) = \cos x$ on $[0, 2\pi]$.

   (c) The area of the region between the graphs of $f(x) = \sqrt{4 - x^2}$ and $g(x) = -\sqrt{4 - x^2}$ on $[-2, 2]$.

17. Suppose f is a function whose average value on $[-2, 5]$ is 10 and whose average rate of change on the same interval is -3. Sketch a possible graph for f. Illustrate the average value and the average rate of change on your graph of f.

18. Suppose f is a function whose average value on $[-3, 1]$ is -2 and whose average rate of change on the same interval is 4. Sketch a possible graph for f. Illustrate the average value and the average rate of change on your graph of f.

19. State the Mean Value Theorem for Integrals, and explain what this theorem means. Include a picture with your explanation. What does the Mean Value Theorem for Integrals have to do with average values?

20. Explain what the Mean Value Theorem for Integrals has to do with the Intermediate Value Theorem.

21. Explain why the Mean Value Theorem for Integrals applies to the function $f(x) = x(x - 6)$ on the interval $[1, 5]$. Next state the conclusion of the Mean Value Theorem for Integrals in this particular case, and sketch a graph illustrating your conclusion. Then find all the values $c \in [1, 5]$ for which $f(c)$ is equal to the average value of f on $[1, 5]$, and indicate these values on your graph.

22. Explain why the Mean Value Theorem for Integrals applies to the function $f(x) = x^2$ on the interval $[-2, 5]$. Next state the conclusion of the Mean Value Theorem for Integrals in this particular case, and sketch a graph illustrating your conclusion. Then find all the values $c \in [-2, 5]$ for which $f(c)$ is equal to the average value of f on $[-2, 5]$, and indicate these values on your graph.

Skills

For each function f and interval $[a, b]$ in Exercises 23–25, use at least eight rectangles to approximate (a) the signed area and (b) the absolute area between the graph of f and the x-axis from $x = a$ to $x = b$. Your work should include a graph of f together with the rectangles that you used.

23. $f(x) = x^2 - 3x - 4$, $[-5, 5]$

24. $f(x) = \cos x$, $[-2\pi, 2\pi]$

25. $f(x) = 1 - e^x$, $[-1, 3]$

For each function f and interval $[a, b]$ in Exercises 26–37, use definite integrals and the Fundamental Theorem of Calculus to find the exact values of (a) the signed area and (b) the absolute area of the region between the graph of f and the x-axis from $x = a$ to $x = b$.

26. $f(x) = x^2$, $[-2, 2]$

27. $f(x) = 3 - x$, $[0, 5]$

28. $f(x) = x^2 - 1$, $[-1, 3]$

29. $f(x) = 2x^2 - 7x + 3$, $[0, 4]$

30. $f(x) = 1 - x^2$, $[0, 3]$

31. $f(x) = 3x^2 + 5x - 2$, $[-3, 2]$

32. $f(x) = \cos x$, $[a, b] = [-\pi, \pi]$

33. $f(x) = (2x - 1)^2 - 4$, $[a, b] = [-2, 4]$

34. $f(x) = 2 - e^{-x}$, $[a, b] = [-1, 0]$

35. $f(x) = \dfrac{1}{2x^3}$, $[a, b] = [1, 3]$

36. $f(x) = \dfrac{3 - 2x}{x^2}$, $[a, b] = [1, 4]$

37. $f(x) = \dfrac{1}{1 + x^2}$, $[a, b] = [-1, 1]$

For each pair of functions f and g and each interval $[a, b]$ in Exercises 38–40, use at least eight rectangles to approximate the area of the region between the graphs of f and g on the interval $[a, b]$. Your work should include graphs of f and g together with the rectangles that you used.

38. $f(x) = \sin x$, $g(x) = \cos x$, $[0, \pi]$

39. $f(x) = x$, $g(x) = 2$, $[-4, 4]$

40. $f(x) = e^x$, $g(x) = \ln x$, $[0.5, 2.5]$

For each pair of functions f and g and interval $[a, b]$ in Exercises 41–52, use definite integrals and the Fundamental Theorem of Calculus to find the exact area of the region between the graphs of f and g from $x = a$ to $x = b$.

41. $f(x) = 1 + x$, $g(x) = 2 - x$, $[0, 3]$

42. $f(x) = 1 + x$, $g(x) = x + 2$, $[0, 3]$

43. $f(x) = x^2$, $g(x) = x + 2$, $[-2, 2]$

44. $f(x) = x^2$, $g(x) = x - 2$, $[-2, 2]$

45. $f(x) = x^2$, $g(x) = x + 2$, $[-3, 3]$

46. $f(x) = x^2 - x - 1$, $g(x) = 5 - x^2$, $[-2, 3]$

47. $f(x) = x - 1$, $g(x) = x^2 - 2x - 1$, $[a, b] = [-1, 3]$

48. $f(x) = \sin x$, $g(x) = \cos x$, $[a, b] = \left[-\dfrac{\pi}{2}, \dfrac{\pi}{2}\right]$

49. $f(x) = x^3$, $g(x) = (x - 2)^2$, $[a, b] = [-1, 2]$

50. $f(x) = 2^x$, $g(x) = x^2$, $[a, b] = [0, 3]$

51. $f(x) = \dfrac{2}{1 + x^2}$, $g(x) = 1$, $[a, b] = [0, \sqrt{3}]$

52. $f(x) = \dfrac{3}{x - 2}$, $g(x) = 6 - x$, $[a, b] = [3, 7]$

For each function f and interval $[a, b]$ in Exercises 53–55, approximate the average value of f from $x = a$ to $x = b$, using a sample size of at least eight.

53. $f(x) = 2^x$, $[-1, 1]$

54. $f(x) = \ln x$, $[1, 5]$

55. $f(x) = \sin x$, $[0, 2\pi]$

For each function f and interval $[a, b]$ in Exercises 56–67, use definite integrals and the Fundamental Theorem of Calculus to find the exact average value of f from $x = a$ to $x = b$. Then use a graph of f to verify that your answer is reasonable.

56. $f(x) = x - 1$, $[-1, 3]$

57. $f(x) = 3x + 1$, $[0, 4]$

58. $f(x) = 4 - x^2$, $[-2, 2]$

59. $f(x) = 4$, $[-37.2, 103.75]$

60. $f(x) = (x + 2)^2 - 5$, $[-5, 0]$

61. $f(x) = x^2 - 2x - 1$, $[0, 3]$

62. $f(x) = 4x^{3/2}$, $[a, b] = [0, 2]$

63. $f(x) = (e^x)^2$, $[a, b] = [-1, 1]$

64. $f(x) = \dfrac{1}{3x + 1}$, $[a, b] = [2, 5]$

65. $f(x) = 2 - \sqrt[3]{x}$, $[a, b] = [1, 8]$

66. $f(x) = x^2 \sin(x^3 + 1)$, $[a, b] = [-1, 2]$

67. $f(x) = \sin x + x \cos x$, $[a, b] = [-\pi, \pi]$

For each function f and interval $[a, b]$ given in Exercises 68–73, find a real number $c \in (a, b)$ such that $f(c)$ is the average value of f on $[a, b]$. Then use a graph of f to verify that your answer is reasonable.

68. $f(x) = 3x + 1$, $[a, b] = [2, 6]$

69. $f(x) = 2x^2 + 1$, $[a, b] = [-1, 2]$

70. $f(x) = 9 - x^2$, $[a, b] = [0, 3]$

71. $f(x) = 1 + x + 2x^2$, $[a, b] = [-2, 2]$

72. $f(x) = 100(1 - x)$, $[a, b] = [0, 10]$

73. $f(x) = (x + 1)^2$, $[a, b] = [-1, 2]$

Applications

74. You slam on your brakes and come to a full stop exactly at a stop sign. Your distance from the sign after t seconds is $s(t) = 3t^3 - 12t^2 - 9t + 54$ feet, as shown in the following graph:

Distance from a stop sign
$s(t) = 3t^3 - 12t^2 - 9t + 54$

(a) What does the Mean Value Theorem from Section 3.1 say about your distance and/or velocity from the stop sign during the time that you are applying your brakes?

(b) What does the Mean Value Theorem for Integrals say about your distance from the stop sign during the time that you are applying the brakes?

(c) What does the Mean Value Theorem for Integrals say about your velocity during the time that you are applying the brakes?

75. Suppose the height, in centimeters, of a growing plant t days after it breaks through the soil is given by $f(t) = 0.36t^2$, as shown next at the left.

Height of a growing plant
$h(t) = .36t^2$

Height of a growing tree
$h(t) = .25t^2 + 1$

(a) What does the Mean Value Theorem from Section 3.1 say about the height of the plant during the first four days of its growth?

(b) What does the Mean Value Theorem for Integrals say about the height of the plant during the first four days of its growth?

76. Suppose the height of a growing tree t years after it is planted is $h(t) = 0.25t^2 + 1$ feet, as shown earlier at the right.

(a) Approximate the average height of the tree during the first five years of its growth, using a sample size of $n = 5$ times and then $n = 10$ times.

(b) Find the exact average height of the tree during the first five years of its growth.

(c) What was the average rate of growth of the tree over the first five years?

77. Wally's Burger Shack wants to put up a giant sign by Interstate 81. According to local sign ordinances, any sign visible from the interstate must have a frontal square footage of 529 feet or less. The entire sign will be a gigantic W cut out from billboard material, as shown in the graph that follows, where the top edges of the W are at a height of 55 feet and the boundaries of the W are given by the functions

$f(x) = 0.5(x - 12)^2$

$g(x) = 0.5(x - 24)^2$

$r(x) = (x - 12)^2 + 10$

$s(x) = (x - 24)^2 + 10$

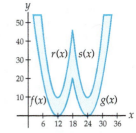

Shape of Wally's sign

(a) Write the total area of the front of the W sign in terms of definite integrals. You will need to find the solutions of $f(x) = 55$, $r(x) = 55$, $s(x) = 55$, and $g(x) = 55$ as part of your work. Be careful about how you divide up the region.

(b) Use your answer to part (a) to calculate the exact frontal square footage of the W sign. Will Wally's sign meet the local square footage requirements?

Proofs

78. Prove that for the region between the graph of a function f and the x-axis on an interval $[a, b]$, the absolute area is always greater than or equal to the signed area.

79. Prove that for all real numbers a and b with $a < b$, we have $|\int_a^b f(x)\, dx| \le \int_a^b |f(x)|\, dx$.

80. In this exercise you will use two different methods to prove that, for any real numbers a, b, c, and k,

$$\int_{-k}^{k} (ax^2 + c)\, dx = 2 \int_0^k (ax^2 + c)\, dx.$$

(a) Prove this equality by using a geometric argument that involves signed area.

(b) Now prove the equality a different way, by using an algebraic argument and the Fundamental Theorem of Calculus.

81. Suppose f is an integrable function on $[a, b]$ and $x_k = a + k\left(\dfrac{b-a}{n}\right)$.

(a) Use the definition of the definite integral as a limit of Riemann sums to show that

$$\lim_{n \to \infty} \frac{\sum_{k=1}^{n} f(x_k)}{n} = \frac{1}{b-a} \int_a^b f(x)\, dx.$$

(b) Why is it algebraically sensible that the left-hand side of the equation is a calculation of average value?

(c) Why is it graphically sensible that the right-hand side of the equation is a calculation of average value?

82. Prove the Mean Value Theorem for Integrals by following these steps:

    (a) Use the Extreme Value Theorem to argue that f has a maximum value M and a minimum value m on the interval $[a, b]$.

    (b) Use an upper sum and a lower sum with one rectangle to argue that $m(b-a) \le \int_a^b f(x)\,dx \le M(b-a)$ and thus that the average value of f on $[a, b]$ is between m and M.

    (c) Use the Intermediate Value Theorem to argue that there is some point $c \in (a, b)$ for which $f(c)$ is equal to the average value of f on $[a, b]$.

Thinking Forward

The Second Fundamental Theorem of Calculus: Just as you are driving past the big oak tree on Main Street, you notice a new stop sign 54 feet ahead of you. You slam on your brakes and end up coming to a full stop exactly at the stop sign.

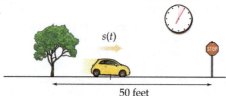

s(t)

50 feet

Suppose that your distance from the stop sign (in feet) t seconds after stepping on the brakes is given by the function $s(t) = 3t^3 - 12t^2 - 9t + 54$. By working through the following five problems you will see another argument that the distance travelled is related to the signed area under the velocity curve; this set of problems previews the Second Fundamental Theorem of Calculus, which we will see in Section 7.7.

▶ How long did it take you to come to a full stop?

▶ What was your average distance from the stop sign from the time that you first saw it to the time that you came to a stop?

▶ Use an average rate of change to find your average velocity during the time that you were trying to stop the car. (Assume that the average velocity is the average rate of change of position.)

▶ Find your average velocity from the oak tree to the stop sign another way, as follows: Differentiate the formula for position to get a formula $v(t)$ for your velocity, in feet per second, t seconds after hitting the brakes. Then use a definite integral to find the average velocity during the time that you were trying to stop the car. (Your final answer should of course be the same as the average velocity you found in the last part!)

▶ You have just calculated your average velocity in two ways, once using the formula $\dfrac{s(b) - s(a)}{b - a}$ for the average rate of change of position and once using the definition $\dfrac{1}{b-a} \int_a^b v(t)\,dt$ of the average value of $v(t)$ on $[a, b]$, where $a = 0$ and b is the amount of time it took you to stop the car. Use the fact that these two quantities are equal to discuss the relationship between the area under the graph of your velocity $v(t)$ on $[a, b]$ and the total distance that you travelled while trying to stop the car.

7.7   FUNCTIONS DEFINED BY INTEGRALS

▶ Functions that measure accumulated area

▶ The Second Fundamental Theorem of Calculus and constructing antiderivatives

▶ Defining the natural logarithm function with an integral

Area Accumulation Functions

We have seen many types of functions in this course, most of them being arithmetic combinations and/or compositions of powers of x, exponential functions, logarithmic functions, trigonometric functions, and inverse trigonometric functions. Of course there are many more functions than the ones we have named and notated throughout this course; any assignment of real numbers that associates exactly one real number to each number in the domain is a function. For example, we could define a function f so that for each real number x, the value of $f(x)$ is given by the greatest integer less than x. This function is called the *floor function*, as it rounds down to the next-lowest integer. As another example, we could define a function g so that for each real number x, the value of $g(x)$ is given by the first perfect square greater than x. We can't express these functions in terms of combinations of simple functions, but they are functions nonetheless.

In this section we examine a new way of defining functions, where for each input x the output is given by an accumulation of area. For example, consider the function $f(t)$ on $[a, b]$ shown next. We can associate with each x in $[a, b]$ a unique area, namely, the area under the curve on $[a, x]$. Since each input x corresponds to exactly one area, this relationship defines a function whose value at any real number x is given by $A(x) = \int_a^x f(t)\, dt$.

Each input x gives a unique area A(x) as output

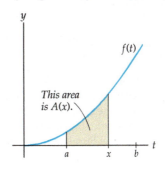

DEFINITION 7.30

Area Accumulation Functions

Suppose f is a continuous function on $[a, b]$. Then the ***area accumulation function*** for f on $[a, b]$ is the function that, for $x \in [a, b]$, is equal to the signed area between the graph of f and the x-axis on $[a, x]$:

$$A(x) = \int_a^x f(t)\, dt.$$

Notice that we have named the variable in the integrand t to distinguish it from the variable x that represents the input of the area accumulation function. The variable t is called a "dummy variable," because it will not appear in the output of our function $A(x)$. We are using the function f only as a means to define our new function $A(x)$.

For example, for $f(t) = \sqrt{t}$ on $[0, 4]$, the area accumulation function is

$$A(x) = \int_0^x \sqrt{t}\, dt = \left[\tfrac{2}{3} t^{3/2}\right]_0^x = \tfrac{2}{3} x^{3/2} - \tfrac{2}{3} 0^{3/2} = \tfrac{2}{3} x^{3/2}.$$

Using this equation for $A(x)$, we can easily calculate values such as $A(1)$, $A(2)$, and $A(3)$, illustrated, respectively, in the three graphs that follow. As x moves from left to right, $A(x)$ accumulates more and more area under the graph of $f(t) = \sqrt{t}$.

$$A(1) = \int_0^1 \sqrt{t}\, dt \approx 0.67 \qquad A(2) = \int_0^2 \sqrt{t}\, dt \approx 1.89 \qquad A(3) = \int_0^3 \sqrt{t}\, dt \approx 3.46$$

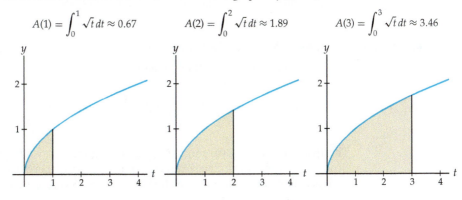

In this example, the area accumulation function turned out to be equal to one of the elementary functions we have studied in this course (namely, $\frac{2}{3}x^{3/2}$). This is not always the case. For example, the area accumulation function $A(x) = \int_0^x e^{-t^2}\,dt$ does not have such an elementary representation; note that in fact we do not even know how to antidifferentiate e^{-t^2}, so even if there were such an elementary representation of $A(x)$, we would be unable to identify it.

The Second Fundamental Theorem of Calculus

In the example we just discussed, you may have noticed something interesting about how A and f are related. Since $A(x) = \frac{2}{3}x^{3/2}$, we have $A'(x) = \left(\frac{2}{3}\right)\left(\frac{3}{2}\right)x^{1/2} = x^{1/2}$, which is equal to $f(x)$. In other words, the derivative of the area accumulation function is the function whose area we were considering! Said another way, the area accumulation function A is an antiderivative of the function f. This is true in general: Every area accumulation function for a continuous function f is an antiderivative of f. The relationship is so important that it is called the Second Fundamental Theorem of Calculus.

THEOREM 7.31   **The Second Fundamental Theorem of Calculus**

If f is continuous on $[a, b]$, and we define $F(x) = \int_a^x f(t)\,dt$ for all $x \in [a, b]$, then

(a) F is continuous on $[a, b]$ and differentiable on (a, b);

(b) F is an antiderivative of f, that is, $F'(x) = f(x)$.

We will postpone the proof of the Second Fundamental Theorem until the end of this section, so that we can focus on some of its consequences first.

The Second Fundamental Theorem guarantees that we can *always* construct an antiderivative for any continuous function f simply by defining an area accumulation function for f. For example, consider the function $f(x) = \sin(e^x)$. It is not immediately apparent how we would antidifferentiate this function. However, we can construct an antiderivative of f by using the Second Fundamental Theorem: If we define $F(x) = \int_0^x \sin(e^t)\,dt$, then we can see that this function F is an antiderivative of f, since $F'(x) = f(x)$ by part (b) of the theorem. We can construct other antiderivatives of f by choosing different values of a; for example, $G(x) = \int_3^x \sin(e^t)\,dt$ is also an antiderivative of $f(x) = \sin(e^x)$.

The second part of the Second Fundamental Theorem immediately gives us a formula for differentiating certain functions defined by integrals:

THEOREM 7.32   **Differentiating Area Accumulation Functions**

If f is continuous on $[a, b]$, then for all $x \in [a, b]$,

$$\frac{d}{dx}\int_a^x f(t)\,dt = f(x).$$

This theorem illustrates yet another way that integration and differentiation can be thought of as inverse operations: Given a function f and a real number a, if we integrate f from a to x and then differentiate the resulting function with respect to x, we arrive back at the function

f we started with. Said another way, the derivative of an area accumulation function for any function f is equal to the original function f.

The part of the Second Fundamental Theorem expressed in Theorem 7.32 follows immediately from the (first) Fundamental Theorem from Section 7.5, since if F is a function with $F' = f$, then

$$\frac{d}{dx} \int_a^x f(t)\, dt = \frac{d}{dx} \left[F(t) \right]_a^x = \frac{d}{dx} (F(x) - F(a)) = F'(x) = f(x).$$

The nontrivial part of the Second Fundamental Theorem is that such an antiderivative function F exists in the first place and that moreover it is continuous and differentiable.

We must be careful when using Theorem 7.32. In particular, in order for us to apply the theorem, the upper limit of the integral *must* be x and the lower limit of the integral must be a constant. If the upper limit is instead a function of x, then we will need to apply the chain rule. The following theorem illustrates how to do this:

THEOREM 7.33

Differentiating a Composition Involving an Area Accumulation Function

If f is continuous on $[a, b]$ and $u(x)$ is differentiable on $[a, b]$, then for all $x \in [a, b]$,

$$\frac{d}{dx} \int_a^{u(x)} f(t)\, dt = f(u(x)) u'(x).$$

The right-hand side of the equation in Theorem 7.33 looks like the chain rule, with the important exception that it begins with $f(u(x))$ rather than $f'(u(x))$. In fact, it *is* the chain rule, and $f(x)$ is the derivative of the area accumulation function $F(x) = \int_a^x f(t)\, dt$.

Proof. The proof involves recognizing $\int_a^{u(x)} f(t)\, dt$ as a composition and then applying the chain rule. If $F(x) = \int_a^x f(t)\, dt$, then $\int_a^{u(x)} f(t)\, dt$ is the composition $F(u(x))$. By the Second Fundamental Theorem we know that $F'(x) = f(x)$. Thus by the chain rule, we have

$$\frac{d}{dx} \int_a^{u(x)} f(t)\, dt = \frac{d}{dx} (F(u(x))) = F'(u(x)) u'(x) = f(u(x)) u'(x). \qquad \blacksquare$$

Defining the Natural Logarithm Function with an Integral

At the beginning of this section we saw that the function $A(x) = \int_0^x \sqrt{t}\, dt$ is equal to the elementary function $\frac{2}{3} x^{3/2}$. Many area accumulation functions can be written as elementary functions. One in particular can be used as the *definition* of a function we already know quite well:

DEFINITION 7.34

The Natural Logarithm Function

The *natural logarithm function* is the function that for $x > 0$ is defined by

$$\ln x = \int_1^x \frac{1}{t}\, dt.$$

For example, we define $\ln 2$ to be the area between the graph of $\frac{1}{t}$ and the x-axis from $t = 1$ to $t = 2$, or the number $\int_1^2 \frac{1}{t}\,dt$. Since $\frac{1}{t}$ is a continuous function, its definite integral on $[1, 2]$ is a well-defined real number that we can choose to call "$\ln 2$."

Of course we have already discussed logarithmic functions as the inverses of exponential functions in Chapter 5. However, our previous development of exponential functions was not entirely rigorous. In particular, our definition of irrational powers relied on a loose argument that we could consider them to be limits of rational powers. We can get around this problem by instead defining $\ln x$ with an integral, as just shown, and then defining exponential functions as the inverses of logarithmic functions. All of the properties of $\ln x$ that you have come to know and love follow directly from this new definition of $\ln x$ as an area accumulation function:

THEOREM 7.35

Properties of the Natural Logarithm Function

(a) $\ln x$ is continuous on $(0, \infty)$

(b) $\ln x$ is differentiable on $(0, \infty)$

(c) $\frac{d}{dx}(\ln x) = \frac{1}{x}$

(d) $\ln 1 = 0$

(e) $\ln x < 0$ on $(0, 1)$, $\ln x > 0$ on $(1, \infty)$

(f) $\ln x$ is increasing on $(0, \infty)$

(g) $\ln x$ is concave down on $(0, \infty)$

(h) $\ln x$ is one-to-one on $(0, \infty)$

Proof. Properties (a)–(c) follow from the Second Fundamental Theorem of Calculus. Properties (d) and (e) are easily proved by using properties of definite integrals. Properties (f) and (g) can be proved by considering the first and second derivatives of $\ln x$. We leave the proofs of these properties to Exercises 71–74 and prove only property (h) here.

To show that $F(x) = \ln x$ is one-to-one, we must show that for all $a, b \in (0, \infty)$, if $F(a) = F(b)$, then $a = b$. Now if $F(a) = F(b)$, then $\ln a = \ln b$, and by Definition 7.34 we have $\int_1^a \frac{1}{t}\,dt = \int_1^b \frac{1}{t}\,dt$. Subtracting all terms to one side and applying parts (b) and (c) of Theorem 7.12 gives

$$0 = \int_1^a \frac{1}{t}\,dt - \int_1^b \frac{1}{t}\,dt = -\int_a^1 \frac{1}{t}\,dt - \int_1^b \frac{1}{t}\,dt = -\int_a^b \frac{1}{t}\,dt.$$

Since $f(t) = \frac{1}{t}$ is positive for all $t > 0$, the integral $\int_a^b \frac{1}{t}\,dt$ can be zero only if a is equal to b, which is what we wanted to show.   ∎

Because $\ln x$ is one-to-one on its domain, it is invertible. As you might expect, the name that we will give to its inverse is e^x. This means that $e^x = y$ if and only if $\ln y = x$. Since $\ln x$ has domain $(0, \infty)$ and range $\mathbb{R}$ (as you will argue in Exercise 21), its inverse e^x has domain $\mathbb{R}$ and range $(0, \infty)$. We will define general logarithmic functions $\log_b x$ and general exponential functions b^x or e^{kx} in a similar fashion, in Exercises 24–26. Every single property of logarithmic and exponential functions can be derived from this new definition of logarithms. You will explore some of these properties in Exercises 75–78.

The Proof of the Second Fundamental Theorem

The proof of the Second Fundamental Theorem of calculus is lengthy, but involves little more than the definition of the derivative and the Squeeze Theorem for limits.

Proof. Suppose f is continuous on $[a, b]$, and define $F(x) = \int_a^x f(t)\,dt$ for $x \in [a, b]$. We'll leave the proof that F is continuous on $[a, b]$ to Exercise 66 and show that for all $x \in (a, b)$, F is differentiable with derivative $F'(x) = f(x)$. We calculate the right derivative of F (the calculation for the left derivative is similar):

$$F'(x) = \lim_{h \to 0^+} \frac{F(x+h) - F(x)}{h} = \lim_{h \to 0^+} \frac{\int_a^{x+h} f(t)\,dt - \int_a^x f(t)\,dt}{h} = \lim_{h \to 0^+} \frac{\int_x^{x+h} f(t)\,dt}{h}.$$

Note that since we are considering only the right derivative here, h is positive. The last step in the calculation follows from the fact that $a < x < x + h$, and thus

$$\int_a^{x+h} f(t)\,dt = \int_a^x f(t)\,dt + \int_x^{x+h} f(t)\,dt.$$

Let's examine the quantity $\int_x^{x+h} f(t)\,dt$ for a small value h. This definite integral represents the signed area between the graph of $f(t)$ and the x-axis from $t = x$ to $t = x + h$, for example, as shown in the first figure that follows. For each $h > 0$, define m_h to be a point in $[x, x + h]$ for which $f(t)$ has a minimum value and define M_h to be a point in $[x, x + h]$ for which $f(t)$ has a maximum value, as shown in the second and third figures. Note that the values m_h and M_h depend on h and may change as $h \to 0^+$.

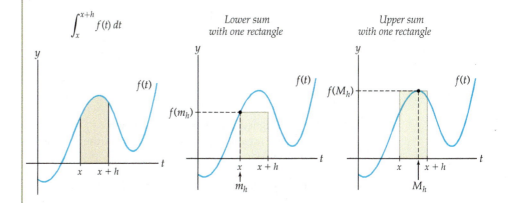

The area of the rectangle in the second of these figures is $f(m_h)h$, and the area of the rectangle in the third figure is $f(M_h)h$. Because $m_h \le f(x) \le M_h$ for all $x \in [x, x + h]$, we have

$$f(m_h)h \le \int_x^{x+h} f(t)\,dt \le f(M_h)h.$$

Dividing both sides by h (which in this case is positive), this inequality becomes

$$f(m_h) \le \frac{\int_x^{x+h} f(x)\,dt}{h} \le f(M_h).$$

Note that the quantity in the middle is exactly the quantity whose limit we need to find to calculate $F'(x)$. Since our results hold for all $h > 0$, we have

$$\lim_{h \to 0^+} f(m_h) \le \lim_{h \to 0^+} \frac{\int_x^{x+h} f(t)\,dt}{h} \le \lim_{h \to 0^+} f(M_h).$$

Because f is continuous, both $f(m_h)$ and $f(M_h)$ approach $f(x)$ as $h \to 0^+$ and the interval $[x, x + h]$ shrinks. Therefore both sides of the double inequality approach $f(x)$ as $h \to 0^+$, so by the Squeeze Theorem,

$$\lim_{h \to 0^+} \frac{\int_x^{x+h} f(t)\,dt}{h} = f(x).$$

Since we started this proof by showing that $F'(x)$ is equal to this limit, we have now shown that $F'(x) = f(x)$, as desired.

Examples and Explorations

EXAMPLE 1 **Graphically interpreting an area accumulation function**

Suppose f is the function shown here, and define the area accumulation function $A(x) = \int_1^x f(t)\, dt$. Use the graph of f to answer the following questions:

(a) Which is larger, $A(2)$ or $A(3)$? Which is larger, $A(3)$ or $A(6)$?

(b) List the intervals on which $A(x)$ is increasing and decreasing.

(c) Sketch a rough graph of the function $A(x)$ on $[1, 8]$.

(d) Use the graphs of f and A to verify that one of these functions is the derivative of the other.

The function f on $[1, 8]$

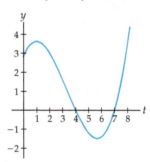

SOLUTION

(a) $A(2)$ is the signed area between the graph of f and the x-axis on $[1, 2]$, while $A(3)$ is the signed area from $t = 1$ to $t = 3$; see the first two graphs that follow. Clearly $A(3) > A(2)$.

$$A(2) = \int_1^2 f(t)\, dt \qquad\qquad A(3) = \int_1^3 f(t)\, dt \qquad\qquad A(6) = \int_1^6 f(t)\, dt$$

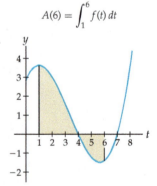

Similarly, $A(6)$ is the signed area between the graph of f and the x-axis on $[1, 6]$, as shown at the right, and differs from $A(3)$ by the area accumulated between $t = 3$ and $t = 6$. The area on $[3, 4]$ is positive, but the area on $[4, 6]$ is negative and larger. Therefore between $t = 3$ and $t = 6$ the accumulation function $A(x)$ actually *decreases*, and thus $A(6) < A(3)$.

(b) $A(x)$ is the accumulation of signed area as x moves from the left to the right. When f is positive, the area function A will increase; when f is negative, the area function A will decrease. From the graph of f we see that f is positive on $(1, 4) \cup (7, 8)$ and negative on $(4, 7)$. Therefore $A(x)$ is increasing on $(1, 4) \cup (7, 8)$ and decreasing on $(4, 7)$.

(c) To sketch the graph of $A(x)$, first notice that $A(1) = \int_1^1 f(t)\, dt = 0$. The height $A(x)$ is the signed area between the graph of f and the x-axis on $[1, x]$, as shown next at the left. Using the information about where A is increasing and where it is decreasing, we can sketch the rough graph of A shown at the right.

The function $f(t)$ on $[1, 8]$

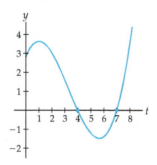

The area accumulation function $A(x)$ on $[1, 8]$

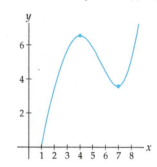

(d) By the Second Fundamental Theorem of Calculus, A is an antiderivative of f, that is, $A' = f$. In the graphs just shown, we can see that for each x, the slope of A at x is equal to the height of f at x. In particular, A has horizontal tangent lines at approximately $x = 4$ and $x = 7$, and its derivative f has roots at those same points. When A is increasing, its derivative f is above the x-axis, and when A is decreasing, f is below the x-axis.   □

EXAMPLE 2   Differentiating area accumulation functions

Find the derivatives of each of the following functions:

(a) $F(x) = \displaystyle\int_1^x \dfrac{\cos(\ln t)}{e^{\sin(t^2)}}\, dt$ $\qquad$ **(b)** $F(x) = \displaystyle\int_x^2 \ln t\, dt$

SOLUTION

(a) In this case we can apply Theorem 7.32 directly:

$$F'(x) = \frac{d}{dx}\int_1^x \frac{\cos(\ln t)}{e^{\sin(t^2)}}\, dt = \frac{\cos(\ln x)}{e^{\sin(x^2)}}.$$

(b) Here we cannot apply Theorem 7.32 immediately, because the lower limit of integration is the variable x, rather than a constant. However, we can easily fix this problem before differentiating:

$$F'(x) = \frac{d}{dx}\int_x^2 \ln t\, dt = \frac{d}{dx}\left(-\int_2^x \ln t\, dt\right) = -\frac{d}{dx}\int_2^x \ln t\, dt = -\ln x.$$ □

EXAMPLE 3   Differentiating compositions that involve area accumulation functions

Find the derivatives of each of the following functions:

(a) $F(x) = \displaystyle\int_0^{x^2} \sec t\, dt$ $\qquad$ **(b)** $F(x) = \displaystyle\int_0^{3x} \dfrac{x}{t^2+1}\, dt$

SOLUTION

(a) We cannot apply Theorem 7.32 directly to this function, because the upper limit of integration is x^2, not x. Instead we need to use the chain rule since the function is a composition, with outside function $A(x) = \int_0^x \sec t\, dt$ and inside function $u = x^2$. In other words, $F(x) = A(x^2)$. By the chain rule and the fact that $A'(x) = \sec x$,

Theorem 7.33 tells us that

$$F'(x) = \frac{d}{dx} \int_0^{x^2} \sec t \, dt = \frac{d}{dx}(A(x^2)) = A'(x^2)(2x) = \sec(x^2)(2x).$$

It may have occurred to you that we could have solved the integral $\int_0^x \sec t \, dt$ first and then found the derivative afterwards. However, in this example we cannot do that, because we do not know how to integrate the function $\sec t$. Our only option is to use Theorem 7.33.

(b) In this example the integrand $\frac{x}{t^2+1}$ involves *both* the variable x and the dummy variable t. Since the integral varies with t, x can be factored out of the integral:

$$F(x) = \int_0^{3x} \frac{x}{t^2+1} \, dt = x \int_0^{3x} \frac{1}{t^2+1} \, dt.$$

It is now clear that F is really the *product* of two functions, namely, x and $\int_0^{3x} \frac{1}{t^2+1} \, dt$. We must use the product rule to differentiate F. We will also need the chain rule, since $\int_0^{3x} \frac{1}{t^2+1} \, dt$ is the composition of $A(x) = \int_0^x \frac{1}{t^2+1}$ with $3x$. We have

$$F'(x) = \frac{d}{dx}\left(x \int_0^{3x} \frac{1}{t^2+1} \, dt \right) \qquad \leftarrow \text{factor out } x$$

$$= 1\left(\int_0^{3x} \frac{1}{t^2+1} \, dt \right) + x\left(\frac{d}{dx} \int_0^{3x} \frac{1}{t^2+1} \, dt \right) \qquad \leftarrow \text{product rule}$$

$$= \int_0^{3x} \frac{1}{t^2+1} \, dt + x\left(\frac{1}{(3x)^2+1} \right)(3). \qquad \leftarrow \text{chain rule} \qquad \square$$

EXAMPLE 4   **Getting comfortable with the integral definition of ln$x$**

Use pictures and properties of definite integrals to illustrate that $\ln(0.5)$ is negative, $\ln 3$ is positive, and $\ln 4$ is greater than $\ln 3$.

SOLUTION

By Definition 7.34, $\ln(0.5) = \int_1^{0.5} \frac{1}{t} \, dt = -\int_{0.5}^1 \frac{1}{t} \, dt$, and since the left-hand figure that follows shows that the signed area under $\frac{1}{t}$ on $[0.5, 1]$ is positive, it follows that $\ln(0.5)$ is negative. Similarly, the middle figure shows that $\ln 3 = \int_1^3 \frac{1}{t} \, dt$ is positive, and the right-hand figure shows that $\ln 4$ represents a larger area than $\ln 3$ does.

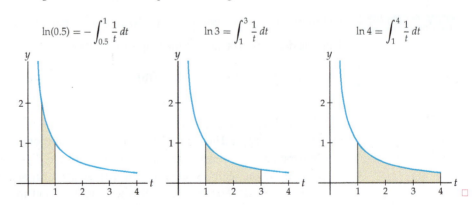

EXAMPLE 5	**Approximating logarithms with Riemann sums**

Use Riemann sums with four rectangles to find upper and lower bounds for $\ln 3$.

SOLUTION

Since $\frac{1}{t}$ decreases as t increases, a left sum for $\ln 3 = \int_1^3 \frac{1}{t}\, dt$ will be an over-approximation and a right sum an under-approximation, as shown here:

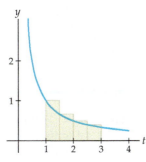

Left sum with four rectangles

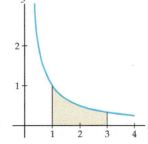

This area is $\ln 3$

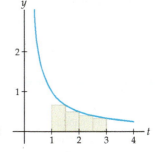

Right sum with four rectangles

The left sum just shown uses the values of $\frac{1}{t}$ at $t = 1, 1.5, 2,$ and 2.5 to define the heights of the rectangles and will be greater than the actual area:

$$\ln 3 = \int_1^3 \frac{1}{t}\, dt \approx \frac{1}{1}(0.5) + \frac{1}{1.5}(0.5) + \frac{1}{2}(0.5) + \frac{1}{2.5}(0.5) \approx 1.28333.$$

In contrast, the right sum just shown is less than the actual area:

$$\ln 3 = \int_1^3 \frac{1}{t}\, dt \approx \frac{1}{1.5}(0.5) + \frac{1}{2}(0.5) + \frac{1}{2.5}(0.5) + \frac{1}{3}(0.5) = 0.95.$$

Therefore $0.95 < \ln 3 < 1.28333$. We could, of course, get better upper and lower bounds for $\ln 3$ by using more rectangles. □

TEST YOUR UNDERSTANDING

▶ What is the difference between the variable x and the variable t in Definition 7.30?

▶ If $A(x) = \int_1^x t\, dt$, does it make sense to calculate $A(0) = \int_1^0 f(t)\, dt$?

▶ If $F(x) = \int_\pi^x \cos t\, dt$ and $u(x) = x^2$, what is the composition $F(u(x))$?

▶ In Example 1, why is $A(x)$ always positive, even though f is sometimes negative? Why do you think the graph of A is concave down until about $x = 5.8$?

▶ What is the difference between the development of $\ln x$ and e^x in Chapter 5 and the development presented in this section?

EXERCISES 7.7

Thinking Back

The relationship between a function and its derivative: Fill in the blanks if a relationship exists. If there is nothing of substance to say, write "not applicable."

▶ If f has a zero, then f' _____ .

▶ If f' has a zero, then f _____ .

▶ If f is increasing, then f' is _____ .

▶ If f is positive, then f' is _____ .

▶ If f' is negative, then f is _____ .

▶ If f'' is negative, then f is _____ .

▶ If f'' is negative, then f' is _____ .

Functions and their properties: Give formal mathematical definitions for each of the following functions.

▶ A *function f* from a set A to a set B

▶ A *positive* function on $[a, b]$

▶ An *increasing* function on $[a, b]$

Properties of logarithms and exponents: Determine whether each of the following statements is true or false.

▶ $\ln(a + b) = \ln(ab)$

▶ $\ln\left(\dfrac{a}{b}\right) = \ln a - \ln b$

▶ $e^{\ln x} = x$

▶ $\ln(a^x) = x \ln a$

▶ $e^{a/b} = e^a - e^b$

▶ $\ln(e^x) = 1$

Concepts

0. *Problem Zero:* Read the section and make your own summary of the material.

1. *True/False:* Determine whether each of the statements that follow is true or false. If a statement is true, explain why. If a statement is false, provide a counterexample.

(a) *True or False:* $\dfrac{d}{dv} \int_2^u f(v)\, du = f(u)$

(b) *True or False:* $\dfrac{d}{dx} \int_a^{x^2} \sin t\, dt = \sin x^2$

(c) *True or False:* If f is continuous on $[a, b]$, then there is exactly one $c \in [a, b]$ with $f(c) = \dfrac{1}{b-a} \int_a^b f(x)\, dx$.

(d) *True or False:* The function $A(x) = \int_0^x (3 - t)\, dt$ is positive and increasing on $[0, 3]$.

(e) *True or False:* The function $A(x) = \int_0^x (3 - t)\, dt$ is concave up.

(f) *True or False:* Every continuous function has an antiderivative.

(g) *True or False:* Every continuous function has more than one antiderivative.

(h) *True or False:* $\ln 5 = \int_0^5 \dfrac{1}{t}\, dt$.

2. *Examples:* Construct examples of the thing(s) described in the following. Try to find examples that are different than any in the reading.

(a) A function f whose area accumulation function is negative on $[0, 5]$.

(b) A function f whose area accumulation function is decreasing on $[0, 5]$.

(c) Three antiderivatives of $e^{\sin^2 x}$.

3. Use a definite integral to express the function whose output at any real number x is the signed area between the graph of $f(t) = t^2$ and the t-axis on $[2, x]$.

4. What, if anything, is wrong with the expression for a function F given by $F(x) = \int_2^x x^2\, dx$?

5. Consider the function $A(x) = \int_a^x f(t)\, dt$. What is the independent variable of this function? What does the dependent variable represent? Explain why we say that t is a "dummy" variable.

6. Suppose $A(x) = \int_0^x f(t)\, dt$, where f is positive and decreasing for $x > 0$. Is $A(x)$ increasing or decreasing, and why?

7. Suppose $A(x) = \int_0^x f(t)\, dt$, where f is positive and decreasing for $x > 0$.

(a) Explain graphically why the graph of $A(x)$ is concave down.

(b) Find $A''(x)$ and argue that if $f(x)$ is decreasing, then $A(x)$ must be concave down.

8. State the Second Fundamental Theorem of Calculus. Why is this theorem important?

9. The functions $A(x) = \int_0^x t^2\, dt$ and $B(x) = \int_3^x t^2\, dt$ differ by a constant. Explain why this is so in three ways, as follows:

(a) By comparing the derivatives of $A(x)$ and $B(x)$.

(b) By showing algebraically that $A(x) - B(x)$ is a constant.

(c) By interpreting $A(x) - B(x)$ graphically.

10. The functions $A(x) = \int_{-\pi}^x \cos x\, dx$ and $B(x) = \int_\pi^x \cos x\, dx$ differ by a constant. Explain why this is so in three ways, using Exercise 9 as a guide.

11. Write the function $F(x) = \int_3^{x^2} \sin t\, dt$ as a composition of two functions. What is the inside function? What is the outside function?

12. Let f be the function shown here, and define $A(x) = \int_0^x f(t)\, dt$. List the following quantities in order from smallest to largest:

(a) $A(1)$ (b) $A(3)$ (c) $A(6)$ (d) $A(7)$

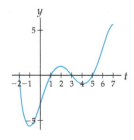

13. Let f be the function graphed in Exercise 12, and define $A(x) = \int_0^x f(t)\, dt$. List the following quantities in order from smallest to largest:

(a) $A(0)$ (b) $A(-1)$ (c) $A(-2)$ (d) $A(5)$

14. Let f be the function shown next at the left, and define $A(x) = \int_0^x f(t)\, dt$. On which interval(s) is A positive? Negative? Increasing? Decreasing? Sketch a rough graph of A.

Graph for Exercise 14   Graph for Exercise 15

     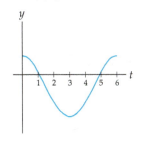

15. Let f be the function shown earlier at the right, and define $A(x) = \int_0^x f(t)\, dt$. On which interval(s) is A positive? Negative? Increasing? Decreasing? Sketch a rough graph of A.

16. Explain how we get the inequality

$$f(m_h)\, h \le \int_x^{x+h} f(t)\, dt \le f(M_h)\, h$$

in the proof of the Second Fundamental Theorem of Calculus. Make sure you define m_h and M_h clearly.

17. Explain precisely how the formula $\dfrac{d}{dx} \int_a^{u(x)} f(t)\, dt = f(u(x))u'(x)$ in Theorem 7.33 is an application of the chain rule.

18. Are indefinite integrals the "inverse" of differentiation? In other words, does one undo the other? Simplify each of the following to answer this question:

(a) $\displaystyle\int f'(x)\, dx$ (b) $\dfrac{d}{dx} \displaystyle\int f(x)\, dx$

19. Are definite integrals the "inverse" of differentiation? In other words, does one undo the other? Simplify each of the following to answer this question:

(a) $\displaystyle\int_a^b f'(x)\, dx$ (b) $\dfrac{d}{dx} \displaystyle\int_a^b f(x)\, dx$

20. Are area accumulation functions the "inverse" of differentiation? In other words, does one undo the other? Simplify each of the following to answer this question:

(a) $\displaystyle\int_a^x f'(t)\, dt$ (b) $\dfrac{d}{dx} \displaystyle\int_a^x f(t)\, dt$

21. Use the new definition of $\ln x$ from Definition 7.34 to argue that

(a) $\ln x$ has domain $(0, \infty)$ and range $\mathbb{R}$.
(b) e^x has domain $\mathbb{R}$ and range $(0, \infty)$.

22. Express the signed area between the graph of $y = \dfrac{1}{x}$ and the x-axis from $x = 0.25$ to $x = 1$ in terms of logarithms.

23. Express the signed area between the graph of $y = \dfrac{1}{x}$ and the x-axis from $x = e$ to $x = 10$ in terms of logarithms.

Now that we have defined $\ln x$ with an integral, we can define a general logarithm with base b as $\log_b x = \dfrac{1}{\ln b} \ln x$. In Exercises 24–26 you will investigate this definition of $\log_b x$.

24. For which b does the definition of $\log_b x$ just given make sense, and why? Is this the same allowable range of values for b that we saw in our old definition of $\log_b x$ from Chapter 5?

25. Use the definition of $\log_b x$ just given and the new definition of $\ln x$ from Definition 7.34 to express $\log_b x$ as an integral.

26. We can now define general exponential functions b^x as the inverses of the general logarithmic functions $\log_b x$. What can you say about $\log_b(b^y)$? Use this information to simplify $\dfrac{1}{\ln b} \int_1^{b^y} \dfrac{1}{t}\, dt$ so that it is written without an integral.

Skills

For each area accumulation function A in Exercises 27–30, (a) illustrate $A(2)$ graphically, (b) calculate $A(2)$ and $A(5)$, and (c) find an explicit elementary formula for $A(x)$.

27. $A(x) = \displaystyle\int_1^x (t^2 + 1)\, dt$ **28.** $A(x) = \displaystyle\int_{-\pi}^x \sin t\, dt$

29. $A(x) = \displaystyle\int_3^x \dfrac{1}{t}\, dt$ **30.** $A(x) = \displaystyle\int_{-x}^x t^3\, dt$

Use the Second Fundamental Theorem of Calculus to write down three antiderivatives of each function f in Exercises 31–34.

31. $f(x) = \sin^2(3^x)$ **32.** $f(x) = \dfrac{1}{\sqrt{1 + e^{4x}}}$

33. $f(x) = e^{(x^2)}$ **34.** $f(x) = \dfrac{\ln(\sin x)}{x^2 - 1}$

Use the Second Fundamental Theorem of Calculus, if needed, to calculate each the derivatives expressed in Exercises 35–48. (Recall that the operator $\dfrac{d^2}{dx^2}$ indicates that you should find the second derivative.)

35. $\dfrac{d}{dx} \displaystyle\int_x^4 e^{t^2+1}\, dt$ **36.** $\dfrac{d}{dx} \displaystyle\int_2^x \dfrac{\sin t}{t}\, dt$

37. $\dfrac{d}{dx} \displaystyle\int_0^{x^2} \cos t\, dt$ **38.** $\dfrac{d}{dx} \displaystyle\int_0^x e^{-t^2}\, dt$

39. $\dfrac{d}{dx} \displaystyle\int_1^{\sqrt{x}} x \ln t\, dt$ **40.** $\dfrac{d}{dx} \left(\left(\displaystyle\int_{\sin x}^{\pi} \dfrac{t}{\cos t}\, dt \right)^3 \right)$

41. $\dfrac{d}{dx} \displaystyle\int_1^8 \ln t\, dt$ **42.** $\dfrac{d}{dx} \displaystyle\int_1^x \ln t\, dt$

43. $\dfrac{d^2}{dx^2} \displaystyle\int_{x^2}^{1} \ln|t|\, dt$

44. $\dfrac{d^2}{dx^2} \displaystyle\int_{1}^{e^x} f(t)g(t)\, dt$

45. $\dfrac{d^2}{dx^2} \displaystyle\int_{2}^{3x} (t^2 + 1)\, dt$

46. $\dfrac{d}{dx}\left(x^2 \displaystyle\int_{0}^{\sin x} \sqrt{t}\, dt \right)$

47. $\dfrac{d}{dx} \displaystyle\int_{x}^{x+2} \sin(t^2)\, dt$

48. $\dfrac{d}{dx} \displaystyle\int_{x}^{\sqrt{x}} \dfrac{e^x}{x}\, dt$

In Exercises 49–54, find a function f that has the given derivative f' and value $f(c)$. Find an antiderivative of f' by hand, if possible; if it is not possible to antidifferentiate by hand, use the Second Fundamental Theorem of Calculus to write down an antiderivative.

49. $f'(x) = \dfrac{1}{2x - 1}, \quad f(1) = 3$

50. $f'(x) = \dfrac{1}{x^2 + 1}, \quad f(-1) = 0$

51. $f'(x) = \dfrac{1}{x^3 + 1}, \quad f(2) = 0$

52. $f'(x) = 2\sin(\pi x), \quad f(2) = 4$

53. $f'(x) = e^{-x^2}, \quad f(1) = 0$

54. $f'(x) = \sin(x^2), \quad f(0) = 0$

For each logarithmic value in Exercises 55–58, use Riemann sums with at least four rectangles to find an over-approximation and an under-approximation for the value. (*Note: In Exercise 58, you will need to use the fact that* $\log_2 x = \dfrac{1}{\ln 2}\ln x$.)

55. $\ln 10$

56. $\ln 0.4$

57. $\ln e$

58. $\log_2 4$

Applications

59. Suppose a very strange particle moves back and forth along a straight path in such a way that its velocity after t seconds is given by $v(t) = \sin(0.1\,t^2)$, measured in feet per second. Consider positions left of the starting position to be in the negative direction and positions right of the starting position in the positive direction.

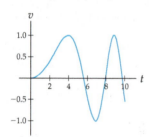

Velocity of very strange particle
$$v(t) = \sin(0.1\,t^2)$$

(a) Write down an expression for the position $s(t)$ of the very strange particle, measured in feet left or right of the starting position. Your expression for $s(t)$ will involve an integral.

(b) Use your answer to part (a) and a Riemann sum with 10 rectangles to approximate the position of the particle after 10 seconds of motion.

(c) Verify that the units in your calculation to part (b) make sense. Why is it clear that the Riemann sum will have units measured in feet?

(d) What happens to the velocity of the very strange particle after a long time? What does this mean about how the particle is moving after, say, about 100 seconds?

60. Jimmy is doing some arithmetic problems. As the evening wears on, he gets less and less effective, so that $t \geq 1$ minutes after the start of his study session he can solve $r(t) = 1 + \dfrac{1}{t}$ problems per minute. Assume that it takes 1 minute for Jimmy to set up his work area; thus he is not doing arithmetic problems in the first minute.

Problem-solving rate
$$r(t) = 1 + \dfrac{1}{t}$$

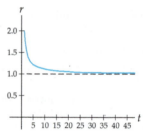

(a) How many arithmetic problems per minute can Jimmy do when he first begins his studies at time $t = 1$? What about at $t = 4$ and $t = 20$?

(b) Make a rough estimate of the number of arithmetic problems Jimmy will have completed after 4 minutes.

(c) Use an integral to express the number of arithmetic problems completed after t minutes, and interpret this definite integral as a logarithm. Calculate the number of problems Jimmy can complete in 10 minutes and in 20 minutes.

(d) Approximately how long will it take for Jimmy to finish his arithmetic homework if he must complete 40 problems?

61. The median rate of flow of the Lochsa River (in Idaho) at the Lowell gauge can be modeled surprisingly accurately as

$$f(t) = 800 + 3.1p(t),$$

where $p(t)$ is a function given by $(t - 90)(195 - t)$ for $t \in (90, 195)$ and is zero otherwise. The time t is measured in days after the beginning of the year, and the flow is measured in cubic feet of water per second.

(a) Given a starting day t_0, use the Fundamental Theorem of Calculus to find the function $F_{t_0}(t)$ that gives the total amount of water that has flowed past the Lowell gauge since that day. Note that there are 86400 seconds per day.

(b) In a non-leap year, March 31 is day 90 of the year and July 14 is day 195. Use the function you found in part (a) to compute how many cubic feet of water flowed past the gauge between those days.

(c) Use your function to compute how much water flows down the Lochsa River in a full year. What fraction of that amount flows by during the span of time you examined in part (b)?

62. Ian is climbing every day, using a camp at the base of a snowfield. His only supply of water is a trickle that comes out of the the the snowfield. The trickle dries at night, because the temperature drops and the snow stops melting.

Since he has run out of books, to entertain himself he uses measurements of the water in his cooking pot to model the flow as

$$w(t) = 15 \left(1 + \cos\left(\frac{2\pi(t-16)}{24}\right)\right),$$

where t is the time in hours after midnight and $w(t)$ is the rate at which snow is melting into his pot, in gallons per hour.

(a) About how long does it take Ian to fill his 2-quart pot at 4 in the afternoon?

(b) What is the total amount of water that flows out of the snowfield in a single day?

(c) Write an expression for the total amount of water that flows from the snowfield between any starting time t_0 and an ending time t.

(d) If Ian stashes his pot under the trickle at 5 in the morning, how long must he wait until he comes back to a full pot? *(Hint: You will need to do this numerically.)*

Proofs

63. Prove that if f has an antiderivative (say, G), then the function $A(x) = \int_0^x f(t)\,dt$ must also be an antiderivative of f. *(Hint: Use the Fundamental Theorem of Calculus.)*

64. Suppose f is continuous on all of $\mathbb{R}$. Prove that for all real numbers a and b, the functions $A(x) = \int_a^x f(t)\,dt$ and $B(x) = \int_b^x f(t)\,dt$ differ by a constant. Interpret this constant graphically.

65. The proof of the latter part of the Second Fundamental Theorem of Calculus in the reading covered only the case $h \to 0^+$. Rewrite this proof in your own words, and then write a proof of what happens as $h \to 0^-$.

66. Prove that if f is continuous on $[a, b]$ and we define $F(x) = \int_a^x f(t)\,dt$, then F is continuous on the closed interval $[a, b]$. *(Hint: Argue that F is continuous on (a, b), left-continuous at a, and right-continuous at b.)*

67. Prove Theorem 7.32: If f is continuous on $[a, b]$, then for all $x \in [a, b]$, $\frac{d}{dx}\int_a^x f(t)\,dt = f(x)$. The proof follows directly from the Second Fundamental Theorem of Calculus.

68. Prove Theorem 7.33 in your own words: If f is continuous on $[a, b]$ and $u(x)$ is a differentiable function, then for all $x \in [a, b]$, $\int_a^{u(x)} f(t)\,dt = f(u(x))u'(x)$. Be especially clear about how you use the chain rule.

69. Follow the given steps to give an alternative proof of the Mean Value Theorem for Integrals (Theorem 7.29): If f is continuous on a closed interval $[a, b]$, then there exists some $c \in (a, b)$ such that:

$$f(c) = \frac{1}{b-a}\int_a^b f(x)\,dx.$$

(a) Define $F(x) = \int_a^x f(t)\,dt$. What three things does the Second Fundamental Theorem of Calculus say about F?

(b) Why does the Mean Value Theorem apply to F on $[a, b]$, and what conclusion can we obtain from the Mean Value Theorem?

(c) Show that, for the value c that is guaranteed by the Mean Value Theorem, $f(c) = \frac{1}{b-a}\int_a^b f(t)\,dt$.

70. Prove in your own words the last part of Theorem 7.35: If we define $\ln x = \int_1^x \frac{1}{t}\,dt$ for $x > 0$, then $\ln x$ is one-to-one on $(0, \infty)$.

Prove each statement in Exercises 71–78, using the new definition of $\ln x$ as an integral and e^x as the inverse of $\ln x$.

71. Prove that $\ln x$ is continuous and differentiable on its entire domain $(0, \infty)$.

72. Prove that $\frac{d}{dx}(\ln x) = \frac{1}{x}$.

73. Prove that $\ln x$ is zero if $x = 1$, negative if $0 < x < 1$, and positive if $x > 1$.

74. Prove that $\ln x$ is increasing and concave down on its entire domain $(0, \infty)$.

75. Prove that $\ln(ab) = \ln a + \ln b$ for any $a, b > 0$ by following these steps:

(a) Use Theorem 7.33 to show that $\frac{d}{dx}(\ln ax) = \frac{1}{x}$ for any real number $a > 0$.

(b) Use your answer to part (a) to argue that $\ln ax = \ln x + C$. *(Hint: Think about Theorem 7.14.)*

(c) Solve for the constant C in the equation from the previous part, by evaluating the equation at $x = 1$. Use your answer to show that $\ln ax = \ln x + \ln a$. Why does this argument complete the proof?

76. Prove that $\ln x^a = a \ln x$ for any $x > 0$ and any a by following these steps:

(a) Use Theorem 7.33 to show that $\frac{d}{dx}(\ln x^a) = \frac{a}{x}$.

(b) Compare the derivatives of $\ln x^a$ and $a \ln x$ to argue that $\ln x^a = a \ln x + C$.

(c) Use part (b) with $x = 1$ and $a = 1$ to show that $C = 0$, and then complete the proof.

77. Use the previous two exercises to prove that for any a, $b > 0$, $\ln\left(\frac{a}{b}\right) = \ln a - \ln b$.

78. Prove that $e^{a+b} = e^a e^b$ for any real numbers a and b by following the steps in parts (a)–(c). You may use any properties of logarithms proved in the reading or the exercises.

(a) Show that if $y = e^{a+b}$, then $\ln y = a + b$.

(b) Show that if $z = e^a e^b$, then $\ln z = a + b$.

(c) Use parts (a) and (b) to prove that $y = z$, and therefore that $e^{a+b} = e^a e^b$.

Thinking Forward

Differential Equations: A **differential equation** is simply an equation that involves derivatives. The solution of a differential equation is the family of functions that make the differential equation true.

▶ Consider the differential equation $f'(x) = g(x)$. What has to be true about $g(x)$ for the Second Fundamental Theorem of Calculus to guarantee that a function f exists that satisfies $f'(x) = g(x)$?

▶ Solve the differential equation $f'(x) = \sin x$. Remember, the solution will be a *family* of functions.

▶ Solve the differential equation $f'(x) = \sin(e^x)$. The solution will be a family of nonelementary functions.

▶ What family of functions satisfies the differential equation $f'(x) = f(x)$? Notice that a function f satisfying this differential equation has the property that it is its own derivative.

Initial-Value Problems: An **initial-value problem** is a differential equation together with an initial condition that specifies one value of a function.

▶ If $f_1(x)$ and $f_2(x)$ are both solutions of the differential equation $f'(x) = g(x)$, where $g(x)$ is continuous, then what can you say about the relationship between f_1 and f_2? What could you say if, in addition, f_1 and f_2 agreed on any given value?

▶ Find the solution f to the initial-value problem $f'(x) = \sin x$, $f(\pi) = 0$.

▶ Find the solution f to the initial-value problem $f'(x) = \sin(e^x)$, $f(\pi) = 0$.

▶ Find the solution f to the initial-value problem $f'(x) = f(x)$, $f(0) = 3$.

CHAPTER REVIEW, SELF-TEST, AND CAPSTONES

Before you progress to the next chapter, be sure you are familiar with the definitions, concepts, and basic skills outlined here. The capstone exercises at the end bring together ideas from this chapter and look forward to future chapters.

Definitions

Give precise mathematical definitions or descriptions of each of the concepts that follow. Then illustrate the definition with a graph or algebraic example, if possible.

▶ a *Riemann sum* for a function f on an interval $[a, b]$, including the definitions of Δx, x_k, and x_k^*

▶ the geometric interpretations of the n-rectangle *left sum*, *right sum*, and *midpoint sum* for a function f on an interval $[a, b]$

▶ the geometric interpretations of the n-rectangle *upper sum* and *lower sum* for a function f on an interval $[a, b]$

▶ the geometric interpretations of the n-rectangle *trapezoid sum* for a function f on an interval $[a, b]$

▶ the *definite integral* of a function f on an interval $[a, b]$ as a limit of Riemann sums, including the definitions of Δx, x_k, and x_k^*

▶ the *indefinite integral* of a continuous function f

▶ what we mean when we refer to an *integrand*

▶ the *signed area* between the graph of a function f and the x-axis on $[a, b]$

▶ the *absolute area* between the graph of a function f and the x-axis on $[a, b]$

▶ the *area between two curves* f and g on $[a, b]$

▶ the *average value* of a function f on $[a, b]$

▶ the *area accumulation function* for a function f on an interval $[a, b]$

▶ the formal definition of the *natural logarithm function* in terms of an accumulation integral

Theorems

Fill in the blanks to complete each of the following theorem statements:

▶ If F and G are _____ functions, then $F'(x) = G'(x)$ if and only if $G(x) =$ _____ .

▶ *The Fundamental Theorem of Calculus*: If f is _____ on $[a, b]$, and if F is any _____ of f, then

$$\int_a^b f(x)\, dx = \underline{\hspace{1cm}}.$$

▶ *The Net Change Theorem*: If f is _____ on $[a, b]$ and its derivative f' is _____ on $[a, b]$, then

$$\int_a^b f'(x)\, dx = \underline{\hspace{1cm}}.$$

▶ *The Fundamental Theorem of Calculus in Evaluation Notation*: If f is _____ on $[a, b]$, then

$$\int_a^b f(x)\, dx = [\underline{\hspace{1cm}}]_a^b.$$

▶ *The Mean Value Theorem for Integrals*: If f is _____ on $[a, b]$, then there is some $c \in$ _____ such that $f(c) =$ _____ .

▶ *The Second Fundamental Theorem of Calculus*: Suppose f is _____ on $[a, b]$ and, for all $x \in [a, b]$, we define

$$F(x) = \int_a^x f(t)\, dt.$$

Then F is _____ on $[a, b]$ and _____ on (a, b), and F is an _____ of f, or in other words, _____ = _____ .

▶ If f is _____ on $[a, b]$, then for all $x \in [a, b]$,

$$\frac{d}{dx}\int_a^x f(t)\, dt = \underline{\hspace{1cm}}.$$

▶ If f is _____ on $[a, b]$ and $u(x)$ is _____ on $[a, b]$, then for all $x \in [a, b]$,

$$\frac{d}{dx}\int_a^{u(x)} f(t)\, dt = \underline{\hspace{1cm}}.$$

▶ The function

$$\ln x = \int_1^x \frac{1}{t}\, dt$$

is continuous and differentiable on the interval _____ . Moreover, the following properties follow from this definition (fill in the blanks or select the correct options):

▶ $\dfrac{d}{dx}(\ln x) =$ _____ and $\ln 1 =$ _____ .

▶ $\ln x < 0$ on the interval _____ , and $\ln x > 0$ on the interval _____ .

▶ $\ln x$ is (increasing)/(decreasing) on its entire domain and (concave up)/(concave down) on its entire domain

▶ $\ln x$ is (one-to-one)/(not one-to-one) on its domain

Notation, Formulas, and Algebraic Rules

Notation: Describe the meanings of each of the following mathematical expressions or how they are commonly used in this chapter:

▶ $\displaystyle\sum_{k=m}^{n} a_k$ ▶ $\displaystyle\sum_{i=1}^{m} b_i$ ▶ Δx

▶ x_k ▶ $[x_{k-1}, x_k]$ ▶ x_k^*

▶ $f(x_k^*)$ ▶ $f(x_k^*)\Delta x$ ▶ $\displaystyle\int_a^b f(x)\, dx$

▶ $\displaystyle\int f(x)\, dx$ ▶ $\displaystyle\int_a^b f(x)\, dx$ ▶ $[F(x)]_a^b$

The algebra of sums: Fill in the blanks to complete the sum rules that follow. You may assume that a_k and b_k are functions defined for nonnegative integers k and that c is any real number.

▶ $\displaystyle\sum_{k=1}^{n} ca_k = \underline{\hspace{1cm}}$ ▶ $\displaystyle\sum_{k=1}^{n}(a_k + b_k) = \underline{\hspace{1cm}}$

▶ $\displaystyle\sum_{k=m}^{p-1} a_k + \sum_{k=p}^{n} a_k = \underline{\hspace{1cm}}$

Sum formulas: Fill in the blanks to complete each sum formula:

▶ $\displaystyle\sum_{k=1}^{n} 1 = \underline{\hspace{1cm}}$ ▶ $\displaystyle\sum_{k=1}^{n} k = \underline{\hspace{1cm}}$

▶ $\displaystyle\sum_{k=1}^{n} k^2 = \underline{\hspace{1cm}}$ ▶ $\displaystyle\sum_{k=1}^{n} k^3 = \underline{\hspace{1cm}}$

Formulas for Riemann sums: Express each of the types of Riemann sums that follow in general sigma notation and also as an expanded sum. You may assume that f is a function defined on $[a, b]$, n is a positive integer, $\Delta x = \dfrac{b-a}{n}$, and $x_k = a + k\Delta x$.

▶ left sum   ▶ right sum

▶ midpoint sum   ▶ trapezoid sum

▶ upper sum   ▶ lower sum

The algebra of definite integrals: Fill in the blanks to complete the definite integral rules that follow. You may assume that f and g are integrable functions on $[a, b]$, that $c \in [a, b]$, and that k is any real number.

▶ $\displaystyle\int_a^b kf(x)\, dx = \underline{\hspace{1cm}}$ ▶ $\displaystyle\int_a^b (f(x) + g(x))\, dx = \underline{\hspace{1cm}}$

▶ $\displaystyle\int_a^a f(x)\,dx =$ ____

▶ $\displaystyle\int_b^a f(x)\,dx =$ ____

▶ $\displaystyle\int_a^c f(x)\,dx + \int_c^b f(x)\,dx =$ ____

Integral Formulas: Fill in the blanks to complete each of the following integration formulas.

▶ For $k \neq -1$, $\displaystyle\int x^k\,dx =$ ____

▶ $\displaystyle\int \frac{1}{x}\,dx =$ ____

▶ For $k \neq 0$, $\displaystyle\int e^{kx}\,dx =$ ____

▶ For $b > 0$ and $b \neq 1$, $\displaystyle\int b^x\,dx =$ ____

▶ $\displaystyle\int \sin x\,dx =$ ____

▶ $\displaystyle\int \cos x\,dx =$ ____

▶ $\displaystyle\int \sec^2 x\,dx =$ ____

▶ $\displaystyle\int \csc^2 x\,dx =$ ____

▶ $\displaystyle\int \sec x\,\tan x\,dx =$ ____

▶ $\displaystyle\int \csc x\,\cot x\,dx =$ ____

▶ $\displaystyle\int \frac{1}{1-x^2}\,dx =$ ____

▶ $\displaystyle\int \frac{1}{1+x^2}\,dx =$ ____

▶ $\displaystyle\int \frac{1}{|x|\sqrt{x^2-1}}\,dx =$ ____

Indefinite integrals of combinations: Fill in the blanks to complete the integration rules that follow. You may assume that f and g are continuous functions and that k is any real number.

▶ $\displaystyle\int kf(x)\,dx =$ ____

▶ $\displaystyle\int (f(x) + g(x))\,dx =$ ____

▶ $\displaystyle\int (f'(x)g(x) + f(x)g'(x))\,dx =$ ____

▶ $\displaystyle\int \frac{f'(x)g(x) - f(x)g'(x)}{(g(x))^2}\,dx =$ ____

▶ $\displaystyle\int f'(g(x))g'(x)\,dx =$ ____

Skill Certification: Sums, Integrals, and the Fundamental Theorem

Calculating sums: Determine the value of each of the sums that follow. Some can be computed directly, some require the use of sum formulas, and for some you will have to also compute a limit.

1. $\displaystyle\sum_{k=1}^{10} \frac{1}{k^2}$

2. $\displaystyle\sum_{k=1}^{4} \left(1 + \frac{k}{2}\right)^2 \left(\frac{1}{2}\right)$

3. $\displaystyle\sum_{k=1}^{200} (k^2 + 1)$

4. $\displaystyle\sum_{k=5}^{50} (k+1)^3$

5. $\displaystyle\lim_{n\to\infty} \sum_{k=1}^{n} \frac{k^2 + 1}{n^3}$

6. $\displaystyle\lim_{n\to\infty} \sum_{k=1}^{n} \frac{(k+1)^3}{n^4 + n + 1}$

Riemann sums: Calculate each of the following Riemann sum approximations for the definite integral of f on $[a, b]$, using the given value of n.

7. The left sum for $f(x) = \sqrt{x}$ on $[0, 4]$, $n = 4$.

8. The right sum for $f(x) = \sqrt{x}$ on $[0, 4]$, $n = 8$.

9. The midpoint sum for $f(x) = 9 - x^2$ on $[0, 3]$, $n = 3$.

10. The trapezoid sum for $f(x) = 9 - x^2$ on $[0, 3]$, $n = 6$.

11. The upper sum for $f(x) = 4x - x^2$ on $[0, 3]$, $n = 3$.

12. The lower sum for $f(x) = 4x - x^2$ on $[0, 3]$, $n = 6$.

Calculating definite integrals with limits of Riemann sums: Calculate the exact value of each the following definite integrals by setting up a general Riemann sum and then taking the limit as $n \to \infty$.

13. $\displaystyle\int_3^5 (2x + 4)\,dx$

14. $\displaystyle\int_0^4 x^2\,dx$

15. $\displaystyle\int_{-2}^3 (4 - x^2)\,dx$

16. $\displaystyle\int_1^4 (3x + 1)^2\,dx$

Indefinite integrals: Use integration formulas, algebra, and educated guess-and-check strategies to find the following integrals.

17. $\displaystyle\int (3x^4 - x^3 + 2)\,dx$

18. $\displaystyle\int \frac{1 - x + x^7}{x^2}\,dx$

19. $\displaystyle\int e^x(1 - e^{3x})\,dx$

20. $\displaystyle\int 2\cos 3x\,dx$

21. $\displaystyle\int \frac{3}{1+4x}\,dx$

22. $\displaystyle\int \frac{3}{1+4x^2}\,dx$

23. $\displaystyle\int (2x3^x + x^2(\ln 3)3^x)\,dx$

24. $\displaystyle\int x^3 \sin x^4\,dx$

25. $\displaystyle\int 3\operatorname{sech}^2 x\,dx$

26. $\displaystyle\int \frac{1}{1 - 2x^2}\,dx$

Definite integrals: Use the Fundamental Theorem of Calculus to find the exact values of each of the definite integrals that follow. Sketch the areas described by these definite integrals to determine whether your answers are reasonable.

27. $\int_{-2}^{2} (x^2 - 1)\, dx$ **28.** $\int_{1}^{3} \frac{1}{x}\, dx$

29. $\int_{-\pi}^{\pi} \sin 2x\, dx$ **30.** $\int_{0}^{2} e^{-3x}\, dx$

31. $\int_{0}^{4} (x - 1)(x - 2)\, dx$ **32.** $\int_{-\pi/4}^{\pi/4} \sec^2 x\, dx$

33. $\int_{-1}^{1} \frac{x}{1 + x^2}\, dx$ **34.** $\int_{-1}^{1} \frac{-1}{1 + x^2}\, dx$

Calculating areas and average values: Express each of the following areas or average values with a definite integral, and then use definite integral formulas to compute the exact value of the area or average value.

35. The signed area between the graph of $f(x) = 3x^2 - 7x + 2$ and the x-axis on $[0, 4]$.

36. The absolute area between the graph of $f(x) = 3x^2 - 7x + 2$ and the x-axis on $[0, 4]$.

37. The average value of the function $f(x) = \sin 2x$ on $[0, \pi]$.

38. The area between the graphs of $f(x) = 4 - x^2$ and $1 - 2x$ on $[-4, 4]$.

Combining derivatives and integrals: Simplify each of the following as much as possible.

39. $\dfrac{d}{dx} \int x^3\, dx$ **40.** $\dfrac{d}{dx} \int_{0}^{x} t^3\, dt$

41. $\int_{-2}^{2} \dfrac{d}{dx} (\ln(x^2 + 1))\, dx$ **42.** $\int_{-2}^{x} \dfrac{d}{dx} (\ln(x^2 + 1))\, dx$

43. $\dfrac{d}{dx} \int_{0}^{3} e^{-t^2}\, dt$ **44.** $\dfrac{d}{dx} \int_{0}^{x} e^{-t^2}\, dt$

45. $\dfrac{d}{dx} \int_{0}^{\ln x} \sin^3 t\, dt$ **46.** $\dfrac{d}{dx} \int_{x}^{3} \sin^3 t\, dt$

Capstone Problems

A. *The area under a velocity curve:* Return to the very start of this chapter, and review the discussion of driving down a straight road with stoplights. Describe in your own words the relationship velocity, distance, and accumulation functions illustrated in that discussion. Then use what you know from the material you learned in this chapter to calculate the exact distance travelled in that situation.

B. *Is integration the opposite of differentiation?* In what sense do derivatives "undo" integrals? In what sense do integrals "undo" derivatives? In what sense do they not? In your answer, be sure to consider indefinite integrals, definite integrals, and accumulation functions defined by integrals.

C. *The Fundamental Theorem of Calculus:* Why is the Fundamental Theorem of Calculus so fundamental? What does

it allow us to calculate, and what concepts does it relate? Give an overview outline of the proof of this important theorem.

D. *Defining logarithms with integrals:* In this chapter we defined the natural logarithm function as the accumulation integral

$$\ln x = \int_{0}^{x} \frac{1}{t}\, dt.$$

(a) Use the graph of $y = \dfrac{1}{x}$ and this definition to describe the graphical features of $y = \ln x$.

(b) Given this definition of $\ln x$, how would we define the natural exponential function e^x? Why is this a better definition for e^x than the one we introduced in Definition 1.25?

CHAPTER 8

Techniques of Integration

8.1   INTEGRATION BY SUBSTITUTION

▶ Reversing the chain rule to obtain a formula for integration by substitution

▶ Strategies for choosing a useful substitution

▶ Two ways of finding definite integrals with integration by substitution

The Art of Integration

In this chapter we will explore a series of techniques for solving integrals. At this point we know how to calculate simple integrals involving basic antidifferentiation, such as

$$\int x^3 \, dx = \tfrac{1}{4}x^4 + C, \quad \int \cos x \, dx = \sin x + C, \quad \text{and} \quad \int \frac{1}{1+x^2} \, dx = \tan^{-1} x + C.$$

We have also seen how to use algebra and intelligent guess-and-check methods to calculate certain types of integrals. However, at this point we do not know how to calculate the vast majority of integrals.

For example, consider the integrals

$$\int \frac{e^x}{1+e^x} \, dx, \quad \int x^2 \cos x \, dx, \quad \text{and} \quad \int \sin^3 x \cos^3 x \, dx.$$

Although differentiating any of these integrands would be quite simple, it is not clear how to *antidifferentiate* them. Throughout the chapter we will discuss techniques for calculating integrals such as the three we just listed, by rewriting the integrals in terms of simpler ones that we can solve by straightforward antidifferentiation. Even after learning these techniques, it will still take creativity and problem-solving skills to solve many integrals. Integration is not a rote calculational activity like differentiation; it is more of an *art*.

Undoing the Chain Rule

We will begin with a simple technique for rewriting integrals that is based on the chain rule. Since the derivative of a composition $f(u(x))$ is $f'(u(x))u'(x)$, we have the following theorem:

THEOREM 8.1   **A Formula for Integration by Substitution**

If f and u are functions such that $f'(u(x))u'(x)$ is integrable, then

$$\int f'(u(x)) \, u'(x) \, dx = f(u(x)) + C.$$

For example, with $f(x) = \sin x$ and $u(x) = x^2$, we have $\frac{d}{dx}(\sin x^2) = (\cos x^2)(2x)$, and therefore $\int (\cos x^2)(2x) \, dx = \sin x^2 + C$. The trouble is, the pattern $f'(u(x))u'(x)$ is not always easy to recognize, so it is not always clear when or how to apply Theorem 8.1 to an integral. In this section we will formalize the procedure of undoing the chain rule with a method known as **integration by substitution**, also known as **u-substitution**.

The key to integration by substitution is a **change of variables**, a technique by which we change an integral in terms of a variable x into an integral in terms of a different variable u. To do this, we must pay close attention to the **differential** dx that appears in integrals. Consider, for example, a function f whose independent variable is called u. Since the derivative of $f(u)$ with respect to u is $f'(u)$, we have

$$\int f'(u)\, du = f(u) + C.$$

Notice that this is exactly the familiar equation $\int f'(x)\, dx = f(x) + C$, but with the variable x renamed as u. Now if $u = u(x)$ is a function of x, comparing the equation we just displayed with the one in Theorem 8.1 tells us that we must have $du = u'(x)\, dx$. This is exactly what we would expect from the notation $\frac{du}{dx} = u'(x)$ if we imagine multiplying both sides by dx. In other words, the differentials dx and du act precisely the way the Leibniz notation $\frac{du}{dx}$ would suggest, even though Leibniz notation does not represent an actual numerical quotient. This property motivates the following definition:

DEFINITION 8.2

The Differential of a Function

If $u(x)$ is a differentiable function of x, then the **differential** du is equal to

$$du = u'(x)\, dx.$$

In particular, note that we *cannot* just replace dx with du when we change variables from x to $u = u(x)$. When changing variables in an integral, we will also have to change the differentials, and this will involve the derivative $u'(x)$. As a simple example, consider again the integral $\int (\cos x^2)(2x)\, dx = \int 2x \cos x^2\, dx$, and let $u(x) = x^2$. We must differentiate u to find du:

$$\boxed{u = x^2} \quad \implies \quad \frac{du}{dx} = 2x \quad \implies \quad \boxed{du = 2x\, dx}.$$

Using the boxed equations to substitute u for x^2 and du for $2x\, dx$, integrating with respect to u, and then writing again in terms of our original variable x, we have

$$\int 2x \cos x^2\, dx = \int \cos u\, du = \sin u + C = \sin x^2 + C.$$

Notice that in this example, the substitution $u = x^2$ changed our original integral into one that was easier to integrate. Integration by substitution is not a method for antidifferentiating; it is a method for changing a difficult integral into a simpler one.

Choosing a Useful Substitution

The first step in trying to solve an integral is to choose a method of attack. Very few integrals can be solved immediately from memory with antidifferentiation formulas. Other integrals can be simplified with an appropriate u-substitution, recognized as backwards quotient or product rule problems, or simplified with some clever algebra. If we choose to attack an integral with the method of integration by substitution, our next step will be to try to find a useful choice for the substitution variable u.

So what makes a good choice for u? The first thing to remember is that we are trying to match the integrand to the pattern $f'(u(x))u'(x)$. Therefore it is a good idea to try choices for $u(x)$ that are inside a composition of functions and whose derivatives $u'(x)$ appear elsewhere in the integrand as a multiplicative factor. For these reasons, we would expect $u = \sin x$ to be a good choice for each of the following integrals:

$$\int \cos x\, e^{\sin x}\, dx, \quad \int \sin^5 x \cos x\, dx, \quad \text{and} \quad \int \frac{\cos x}{\sin x}\, dx.$$

With this choice of u we have

$$\boxed{u = \sin x} \quad \implies \quad \frac{du}{dx} = \cos x \quad \implies \quad \boxed{du = \cos x\, dx},$$

and these integrals would be transformed into, respectively:

$$\int e^u \, du, \quad \int u^5 \, du, \quad \text{and} \quad \int \frac{1}{u} \, du,$$

each of which can be solved immediately with antidifferentiation formulas. In contrast, $u = \sin x$ would *not* be a good choice for the integral $\int \frac{\sin x}{\cos x} \, dx$, because in this case the derivative $u' = \cos x$ would not be a multiplicative part of the integrand. With $u = \sin x$ this integral has the form $\int \frac{u(x)}{u'(x)} \, dx$, which does not match the pattern in Theorem 8.1.

Another good rule of thumb is to choose $u(x)$ to be as much of the integrand as possible while still having things work out with the change in differential from dx to du. This helps change the integral into one that is as simple as possible. For example, with the integral

$$\int x^2 \sin(x^3 + 1) \, dx,$$

a choice of $u = x^3 + 1$ is better than a choice of $u = x^3$. Once we choose u, we must change the entire rest of the integral so that it is in terms of the new variable. After substituting $u = x^3 + 1$, we will still have $x^2 \, dx$ to replace, which we can do by considering the differentials as follows:

$$\boxed{u = x^3 + 1} \quad \Longrightarrow \quad \frac{du}{dx} = 3x^2 \quad \Longrightarrow \quad du = 3x^2 \, dx \quad \Longrightarrow \quad \boxed{\frac{1}{3} du = x^2 \, dx}.$$

Now by substituting with the boxed expressions, our integral is changed into something that is easy to integrate:

$$\int x^2 \sin(x^3 + 1) \, dx = \int \sin u \cdot \frac{1}{3} \, du = \frac{1}{3} \int \sin u \, du = \frac{1}{3}(-\cos u) + C = -\frac{1}{3} \cos(x^3 + 1) + C.$$

It is simple to check this calculation by differentiating:

$$\frac{d}{dx}\left(-\frac{1}{3}\cos(x^3 + 1)\right) = -\frac{1}{3}(-\sin(x^3 + 1))(3x^2) = x^2 \sin(x^3 + 1).$$

Note also that in this example, the original integrand was not *exactly* in the form $f'(u(x))u'(x)$, since we had $u'(x) = 3x^2$ but the constant multiple 3 did not appear in our original integral. However, the method of u-substitution helped us determine the constant $\frac{1}{3}$ that was needed to make everything work out.

Occasionally some preliminary algebra can help us find a useful u-substitution. This is the case for the following two basic integral formulas:

THEOREM 8.3

Integrals of the Tangent and Cotangent Functions

(a) $\int \tan x \, dx = \ln|\sec x| + C$ **(b)** $\int \cot x \, dx = -\ln|\csc x| + C$

Proof. The integral $\int \tan x \, dx$ looks too simple for applying integration by substitution, until we rewrite $\tan x$ as $\frac{\sin x}{\cos x}$. We can then choose

$$\boxed{u = \cos x} \quad \Longrightarrow \quad \frac{du}{dx} = -\sin x \quad \Longrightarrow \quad \boxed{-du = \sin x \, dx}.$$

This substitution changes the integral into

$$\int \tan x \, dx = \int \frac{\sin x}{\cos x} \, dx = -\int \frac{1}{u} \, du = -\ln|u| + C = -\ln|\cos x| + C.$$

Although we have found the integral of $\tan x$, it is not in the form we see in Theorem 8.3. Using properties of logarithms and trigonometric functions, we can rewrite our answer as desired:

$$-\ln|\cos x| + C = \ln\left|\frac{1}{\cos x}\right| + C = \ln|\sec x| + C.$$

The proof of part (b) is similar and is left to Exercise 90.

Finding Definite Integrals by Using Substitution

There are two methods for calcuating a definite integral $\int_a^b f(x)\,dx$ with the use of integration by substitution. Perhaps the most straightforward way is to use the Fundamental Theorem of Calculus as expressed in Theorem 7.24, which, when considered with the method of u-substitution, says the following:

THEOREM 8.4

Evaluating a Definite Integral After a Substitution

If $f(x) = g(u(x))$ is continuous on $[a, b]$, $u(x)$ is differentiable on (a, b), and $G(u)$ is an antiderivative of $g(u)$, then

$$\int_a^b f(x)\,dx = \int_{x=a}^{x=b} g(u)\,du = \left[G(u)\right]_{x=a}^{x=b} = \left[G(u(x))\right]_a^b = G(u(b)) - G(u(a)).$$

Theorem 8.4 tells us that we can simply use u-substitution (or any integration method, for that matter) to find $\int f(x)\,dx$ in terms of x and then evaluate this antiderivative from $x = a$ to $x = b$. For example, repeating our first example with a definite integral, we have

$$\int_0^\pi 2x\cos x^2\,dx = \int_{x=0}^{x=\pi}\cos u\,du = \left[\sin u\right]_{x=0}^{x=\pi} = \left[\sin x^2\right]_0^\pi = \sin(\pi^2) - \sin 0.$$

Notice that we are careful to write the limits of integration as $x = 0$ and $x = \pi$ when we change variables to u, so that we remember to substitute back $u = x^2$ before evaluating from 0 to π.

Alternatively, we could change the limits of integration so that they themselves are in terms of the new variable u and then not have to substitute back $u = u(x)$ at the end of the calculation:

THEOREM 8.5

Another Way to Evaluate a Definite Integral After a Substitution

If $f(x) = g(u(x))$ is continuous on $[a, b]$, $u(x)$ is differentiable on (a, b), and $G(u)$ is an antiderivative of $g(u)$, then

$$\int_a^b f(x)\,dx = \int_{u(a)}^{u(b)} g(u)\,du = \left[G(u)\right]_{u(a)}^{u(b)} = G(u(b)) - G(u(a)).$$

For example, if we repeat our earlier calculation with this second method, we have $u = x^2$ and thus $u(0) = 0^2 = 0$ and $u(\pi) = \pi^2$. Calculating the same definite integral gives

$$\int_0^\pi 2x\cos x^2\,dx = \int_0^{\pi^2}\cos u\,du = \left[\sin u\right]_0^{\pi^2} = \sin(\pi^2) - \sin 0,$$

which of course is the same as the answer we got earlier. In general, you may choose either of the two methods according to your preference. In some cases the first method will be easier, and in some cases the second. Both will always give the same answer.

Examples and Explorations

EXAMPLE 1   **Using integration by substitution when a constant multiple is missing**

Use integration by substitution to find $\int x^6 e^{5x^7-2}\, dx$.

SOLUTION

Since $5x^7 - 2$ is the inside function of a composition in the integrand and the x^6 part of its derivative $35x^6$ is a multiplicative part of the integrand, we'll try setting $u = 5x^7 - 2$ and see what happens. With this choice of u we have

$$\boxed{u = 5x^7 - 2} \quad\Longrightarrow\quad \frac{du}{dx} = 35x^6 \quad\Longrightarrow\quad du = 35x^6\, dx \quad\Longrightarrow\quad \boxed{\frac{1}{35}\, du = x^6\, dx}.$$

Using the boxed equations, we can now change variables and differentials to rewrite the integral in terms of u and du; moreover, this new integral is easy to solve:

$$\int x^6 e^{5x^7-2}\, dx = \int e^{5x^7-2}(x^6\, dx) = \int e^u \left(\frac{1}{35}\, du\right) = \frac{1}{35}\int e^u\, du = \frac{1}{35}e^u + C = \frac{1}{35}e^{5x^7-2} + C.$$

Although it was not necessary, we rewrote the integral in the first step of the calculation to make sure that the substitution of $x^6\, dx$ by $\frac{1}{35}\, du$ was clear. The antidifferentiation step in the calculation is the step $\int e^u\, du = e^u + C$, which is true because $\frac{d}{du}(e^u) = e^u$. Notice that u-substitution helped us rewrite the original integral into something that was very easy to antidifferentiate.   □

 CHECKING THE ANSWER   We can differentiate with the chain rule to check the answer we just found. Since we can see that

$$\frac{d}{dx}\left(\frac{1}{35}e^{5x^7-2}\right) = \frac{1}{35}e^{5x^7-2}(35x^6) = x^6 e^{5x^7-2},$$

we know that we have integrated correctly.

EXAMPLE 2   **Deciding when to use integration by substitution**

For each integral, determine whether integration by substitution is appropriate, and if so, determine a useful substitution $u(x)$.

(a) $\displaystyle\int x^4\sqrt{x^5+1}\, dx$ **(c)** $\displaystyle\int \cos x\, \sin^2 x\, dx$ **(e)** $\displaystyle\int \frac{e^x}{1+e^x}\, dx$

(b) $\displaystyle\int x^3\sqrt{x^5+1}\, dx$ **(d)** $\displaystyle\int \cos x\, \tan x\, dx$ **(f)** $\displaystyle\int \frac{1+e^x}{e^x}\, dx$

SOLUTION

(a) Integration by substitution will work for this integral. If we choose $u = x^5 + 1$, then $\frac{du}{dx} = 5x^4$, so $x^4\, dx = \frac{1}{5}du$. We can rewrite the integral as $\frac{1}{5}\int \sqrt{u}\, du$, which we know how to integrate.

(b) Integration by substitution won't work here. The most reasonable choice for u would be $u = x^5 + 1$, which makes $\frac{du}{dx} = 5x^4$. We cannot write the integral entirely in terms of u and du with this choice of u. For this integral, no choice of u will work.

(c) If we choose $u = \sin x$, then $\frac{du}{dx} = \cos x$, so $du = \cos x\, dx$. This substitution changes the integral into $\int u^2\, du$, which is easy to integrate. The choice $u = \cos x$ would *not* work in this example; why not?

(d) Neither $u = \cos x$ nor $u = \tan x$ work as substitutions here, because their derivatives are not present in the integrands. However, $\cos x \tan x = \cos x \left(\frac{\sin x}{\cos x} \right)$ is equal to $\sin x$, which we know how to integrate.

(e) At first glance, it does not appear that integration by substitution will work for this problem. However, remember that the derivative of e^x is itself—and in fact the derivative of $1 + e^x$ is also e^x. Thus we can let u be the denominator $u = 1 + e^x$ and get $\frac{du}{dx} = e^x$, that is, $du = e^x\, dx$. This substitution changes the integral into $\int \frac{1}{u}\, du$, which is easy to integrate.

(f) The method we used in part (e) won't work here: We cannot use $u = 1 + e^x$ and $du = e^x\, dx$ to change this integral into a new integral involving only u and du. The problem with this substitution is that the derivative e^x is in the denominator; it is not a *multiplicative* factor of the integrand. This integral can be solved by using algebra to write the integrand $\frac{1+e^x}{e^x}$ as $\frac{1}{e^x} + \frac{e^x}{e^x} = e^{-x} + 1$. Try it!   □

EXAMPLE 3	**Choosing a substitution**

Use integration by substitution to find (a) $\displaystyle\int \frac{\ln x}{x}\, dx$ and (b) $\displaystyle\int \frac{\sin \sqrt{x}}{\sqrt{x}}\, dx$.

SOLUTION

(a) Clearly $u = x$ will not be a good substitution, since it would not simplify the integral. Perhaps $u = \ln x$ will work. With this choice of u, we have

$$\boxed{u = \ln x} \quad \Longrightarrow \quad \frac{du}{dx} = \frac{1}{x} \quad \Longrightarrow \quad \boxed{du = \frac{1}{x}\, dx}.$$

At first you might think that there is no $\frac{1}{x}$ in the original integrand; we only have $\ln x$ and x. But notice that the quotient $\frac{\ln x}{x}$ is actually equal to the product $(\ln x)\left(\frac{1}{x}\right)$; we *do* have $\frac{1}{x}$ in the integrand. Changing variables gives us

$$\int \frac{\ln x}{x}\, dx = \int (\ln x)\left(\frac{1}{x}\, dx\right) = \int u\, du = \frac{1}{2} u^2 + C = \frac{1}{2}(\ln x)^2 + C.$$

As always, it is a good idea to check by differentiating. Since $\frac{d}{dx}\left(\frac{1}{2}(\ln x)^2\right) = \frac{1}{2}(2 \ln x)\left(\frac{1}{x}\right) = \frac{\ln x}{x}$, we know we did the integration correctly.

(b) A sensible first choice for u would be $u = \sqrt{x}$, since this is the inside function of a composition in the integrand. With this choice of u we have

$$\boxed{u = \sqrt{x}} \quad \Longrightarrow \quad \frac{du}{dx} = \frac{1}{2} x^{-1/2} = \frac{1}{2\sqrt{x}} \quad \Longrightarrow \quad \boxed{2\, du = \frac{1}{\sqrt{x}}\, dx}.$$

We solved for the expression $\frac{1}{\sqrt{x}}\, dx$ because that is the part of the integrand we need to rewrite in terms of du. With this change of variables we have

$$\int \frac{\sin \sqrt{x}}{\sqrt{x}}\, dx = \int (\sin \sqrt{x})\left(\frac{1}{\sqrt{x}}\, dx\right) = \int (\sin u)(2\, du)$$

$$= 2 \int \sin u\, du = 2(-\cos u) + C = -2 \cos \sqrt{x} + C.$$

Again, we can easily check this answer: $\frac{d}{dx}(-2\cos\sqrt{x}) = -2(-\sin\sqrt{x})\left(\frac{1}{2}x^{-1/2}\right) = \frac{\sin\sqrt{x}}{\sqrt{x}}$.

EXAMPLE 4   **Integration by substitution followed by back-substitution**

Use integration by substitution and algebra to find $\int x\sqrt{x-1}\,dx$.

SOLUTION

This is an example in which integration by substitution works even though the integrand is not at all in the form $f'(u(x))u'(x)$. A clever change of variables will allow us to rewrite the integral so that it can be algebraically simplified. We'll start by choosing $u = x - 1$, which is sensible because this choice of u is the inside function of a composition in the integrand:

$$\boxed{u = x - 1} \quad\Rightarrow\quad \frac{du}{dx} = 1 \quad\Rightarrow\quad \boxed{du = dx}.$$

We can now replace $\sqrt{x-1}$ with $\sqrt{u}$ and dx with du. However, if we do this, there will be an x left over in our integrand that we haven't written in terms of u or du. Luckily, we can use back-substitution to solve for x; since $u = x - 1$, we have

$$\boxed{x = u + 1}.$$

By using the three boxed equations, we have

$$\int x\sqrt{x-1}\,dx = \int (u+1)\sqrt{u}\,du = \int (u^{3/2} + u^{1/2})\,du \qquad \leftarrow \text{substitute and simplify}$$

$$= \frac{2}{5}u^{5/2} + \frac{2}{3}u^{3/2} + C \qquad\qquad \leftarrow \text{antidifferentiate}$$

$$= \frac{2}{5}(x-1)^{5/2} + \frac{2}{3}(x-1)^{3/2} + C.$$

The original integrand could not be algebraically expanded, but after our clever change of variables we had an integral that we could multiply out, since the sum was then outside, rather than inside, the square root.

 CHECKING THE ANSWER

As usual we can check our integration work by differentiating, and in particular, we know that the chain rule will be involved. In this case a bit of algebra is needed after differentiating, to show that $\frac{d}{dx}\left(\frac{2}{5}(x-1)^{5/2} + \frac{2}{3}(x-1)^{3/2}\right)$ is equal to $x\sqrt{x-1}$. Try it!

EXAMPLE 5   **Using integration by substitution to solve a definite integral**

Use integration by substitution to find $\displaystyle\int_1^4 \frac{x}{3x^2 + 1}\,dx$.

SOLUTION

A sensible substitution to try first is

$$\boxed{u = 3x^2 + 1} \quad\Longrightarrow\quad \frac{du}{dx} = 6x \quad\Longrightarrow\quad \boxed{\frac{1}{6}\,du = x\,dx}.$$

A definite integral consists of an integrand *and* limits of integration. When making our substitution we have to be mindful of both:

$$\int_1^4 \frac{x}{3x^2+1}\, dx = \int_1^4 \frac{1}{3x^2+1}(x\, dx) = \int_{x=1}^{x=4} \frac{1}{u}\left(\frac{1}{6}\, du\right) = \frac{1}{6}\int_{x=1}^{x=4} \frac{1}{u}\, du.$$

Note that this is not the same as writing 1 and 4 as the limits of integration on the new integral: Since the new integral is in terms of u and du, that would implicitly mean $u = 1$ and $u = 4$, which is not what we want. Our new definite integral is easy to evaluate:

$$\frac{1}{6}\int_{x=1}^{x=4} \frac{1}{u}\, du = \frac{1}{6}\Big[\ln|u|\Big]_{x=1}^{x=4} = \frac{1}{6}\Big[\ln|3x^2+1|\Big]_1^4 = \frac{1}{6}(\ln 49 - \ln 4).$$ □

EXAMPLE 6   Changing the limits of integration when using substitution

Repeat the calculation of $\displaystyle\int_1^4 \frac{x}{3x^2+1}\, dx$, this time replacing the limits of integration during the substitution.

SOLUTION

The choice of u will still be the same as in Example 5. Again we have

$$\boxed{u = 3x^2 + 1} \quad\Longrightarrow\quad \boxed{\frac{du}{dx} = 6x} \quad\Longrightarrow\quad \boxed{\frac{1}{6}\, du = x\, dx}.$$

We will now write the limits of integration ($x = 1$ and $x = 4$) in terms of the new variable u. When $x = 1$ we have $u = 3(1)^2 + 1 = 4$, and when $x = 4$ we have $u = 3(4)^2 + 1 = 49$; in other words,

$$\boxed{u(1) = 4} \quad\text{and}\quad \boxed{u(4) = 49}.$$

Using the information in all four boxed equations, we can change variables completely:

$$\int_1^4 \frac{x}{3x^2+1}\, dx = \int_1^4 \frac{1}{3x^2+1}(x\, dx) = \int_4^{49} \frac{1}{u}\left(\frac{1}{6}\, du\right) = \frac{1}{6}\int_4^{49} \frac{1}{u}\, du.$$

This time we do not have to write the antiderivative in terms of x before the evaluation step, because the limits of integration are already written in terms of u:

$$\frac{1}{6}\int_4^{49} \frac{1}{u}\, du = \frac{1}{6}\Big[\ln|u|\Big]_4^{49} = \frac{1}{6}(\ln 49 - \ln 4).$$ □

CHECKING THE ANSWER   As usual, we can check our antidifferentiation steps by differentiating. In addition, we can check the numerical answer of a definite integral by remembering that it represents a signed area under a curve. In both examples, our numerical answer was $\frac{1}{6}(\ln 49 - \ln 4) \approx 0.417588$, which we can see is reasonable by using a calculator to approximate the signed area between the graph of $f(x) = \frac{x}{3x^2+1}$ and the x-axis on $[1, 4]$.

? TEST YOUR UNDERSTANDING

▶ What would go wrong if we tried to use the substitution $u = x^2$ to solve the integral $\int x^2 \cos x^2\, dx$?

▶ When we change variables in an integral with a substitution $u = u(x)$, how is the differential dx related to the new differential du?

▶ Why is $u = \sin x$ not a good choice of substitution for solving the integral $\int \frac{\sin x}{\cos x} dx$?

▶ Show that $\frac{1}{6} \int_1^4 \frac{1}{u} du = \frac{\ln 4}{6} \approx 0.23104$. Why is this answer different from the value of the definite integral $\frac{1}{6} \int_{x=1}^{x=4} \frac{1}{u} du$ that we computed in Example 5?

▶ Compare the calculations in Examples 5 and 6. How are they the same? How are they different? What work did you have to do in Example 5 that you did not have to do in Example 6? Where was the equivalent work done in Example 6?

EXERCISES 8.1

Thinking Back

Algebra review: Simplify each algebraic expression as much as possible, until the expression is something that would be easy to antidifferentiate.

▶ $\sqrt{x}(x^{-2} - 5x^{2/3})$

▶ $\ln(5x^3) + 3\ln(2^x)$

▶ $\dfrac{x+1}{2\sqrt[3]{x}}$

▶ $\dfrac{e^{3x} - 5e^{-2x}}{3e^{2x}}$

▶ $\sec x \tan x \cos^2 x$

▶ $\sqrt{1 - \sin^2 x}$

▶ $\dfrac{2\sqrt{x} - 5}{3\sqrt{x}}$

▶ $2^x(1 - 3^x)$

Differentiation review: Differentiate each of the functions that follow. Simplify your answers as much as possible.

▶ $f(x) = -\dfrac{1}{2}\cot x^2$

▶ $f(x) = \dfrac{1}{3}\ln|3x + 1|$

▶ $f(x) = \dfrac{1}{2\pi}\sin^2 \pi x$

▶ $f(x) = \dfrac{2}{3}(\ln x)^{3/2}$

▶ $f(x) = (\sin x^2)^{1/2}$

▶ $f(x) = \dfrac{1}{2}(\sin^{-1} x)^2$

▶ $f(x) = \cos(\sin^2 x^3)$

▶ $f(x) = \ln(\ln(\ln x))$

Antidifferentiation review: Antidifferentiate each of the following functions.

▶ $f(x) = \sqrt{x} + \sqrt[3]{x}$

▶ $f(x) = \sec^2 5x$

▶ $f(x) = \dfrac{2}{3x}$

▶ $f(x) = \dfrac{\sin \pi x}{2}$

▶ $f(x) = \dfrac{3}{x^2 + 1}$

▶ $f(x) = \dfrac{-2}{\sqrt{1 - x^2}}$

Expressing geometric quantities with integrals: Express each of the given geometric quantities in terms of definite integrals. You do not have to solve the integrals.

▶ The signed area of the region between the graph of $f(x) = \sin x$ and the x-axis on $\left[0, \frac{3\pi}{2}\right]$.

▶ The absolute area of the region between the graph of $f(x) = \sin x$ and the x-axis on $\left[0, \frac{3\pi}{2}\right]$.

▶ The signed area of the region between the graphs of x^2 and 2^x on $[-5, 5]$.

▶ The average value of the function $f(x) = \frac{1}{x}$ from $x = 1$ to $x = 10$.

Concepts

0. *Problem Zero:* Read the section and make your own summary of the material.

1. *True/False:* Determine whether each of the statements that follow is true or false. If a statement is true, explain why. If a statement is false, provide a counterexample.

(a) *True or False:* $\int g'(h(x)) h'(x) dx = g(h(x)) + C$.

(b) *True or False:* If $v = u^2 + 1$, then $\int \sqrt{u^2 + 1} du = \int \sqrt{v} dv$.

(c) *True or False:* If $u = x^3$, then $\int x \sin(x^3) dx = \frac{1}{3x} \int \sin u \, du$.

(d) *True or False:* $\int_0^3 u^2 du = \int_{x=0}^{x=3} (u(x))^2 du$.

(e) *True or False:* $\int_0^1 x^2 dx = \int_0^1 u^2 du$.

(f) *True or False:* $\int_2^4 xe^{x^2-1} dx = \frac{1}{2} \int_2^4 e^u du$.

(g) *True or False:* $\int_2^3 f(u(x))u'(x)dx = \int_{u(2)}^{u(3)} f(u)du$.

(h) *True or False:* $\int_0^6 f(u(x))u'(x)dx = \left[\int f(u)du\right]_0^6$.

2. *Examples:* Construct examples of the thing(s) described in the following. Try to find examples that are different than any in the reading.

(a) Five integrals that can be solved with the method of integration by substitution.

(b) Five integrals that cannot be solved with the method of integration by substitution.

(c) Three relatively simple integrals that we cannot solve with any of the methods we now know.

3. Explain why $\int \frac{2x}{x^2+1}\,dx$ and $\int \frac{1}{x\ln x}\,dx$ are essentially the same integral after a change of variables.

4. List some things which would suggest that a certain substitution $u(x)$ could be a useful choice. What do you look for when choosing $u(x)$?

For each integral in Exercises 5–8, write down three integrals that will have that form after a substitution of variables.

5. $\displaystyle\int \sin u\,du$

6. $\displaystyle\int u^2\,du$

7. $\displaystyle\int \frac{1}{\sqrt{u}}\,du$

8. $\displaystyle\int e^u\,du$

For each function $u(x)$ in Exercises 9–12, write the differential du in terms of the differential dx.

9. $u(x) = x^2 + x + 1$

10. $u(x) = x^2 + 1$

11. $u(x) = \sin x$

12. $u(x) = \dfrac{1}{x}$

13. Suppose $u(x) = x^2$. Calculate and compare the values of the following definite integrals:

$$\int_{-1}^{5} u^2\,du,\quad \int_{x=-1}^{x=5} u^2\,du,\quad \text{and}\quad \int_{u(-1)}^{u(5)} u^2\,du.$$

14. Find three integrals in Exercises 21–68 in which the denominator of the integrand is a good choice for a substitution $u(x)$.

15. Find three integrals in Exercises 21–68 that we can anti-differentiate immediately after algebraic simplification.

16. Consider the integral $\int \sin x \cos x\,dx$.
   (a) Solve this integral by using u-substitution with $u = \sin x$ and $du = \cos x\,dx$.

(b) Solve the integral another way, using u-substitution with $u = \cos x$ and $du = -\sin x\,dx$.
   (c) How must your two answers be related? Use algebra to prove this relationship.

17. Consider the integral $\int x(x^2 - 1)^2\,dx$.
   (a) Solve this integral by using u-substitution.
   (b) Solve the integral another way, using algebra to multiply out the integrand first.
   (c) How must your two answers be related? Use algebra to prove this relationship.

18. Consider the integral $\int \frac{x^{-2}-4}{x^3}$.
   (a) Solve this integral by using u-substitution.
   (b) Solve the integral another way, using algebra to simplify the integrand first.
   (c) How must your two answers be related? Use algebra to prove this relationship.

19. Consider the integral $\int_{-2}^{2} e^{5-3x}\,dx$.
   (a) Solve this integral by using u-substitution while keeping the limits of integration in terms of x.
   (b) Solve the integral again with u-substitution, this time changing the limits of integration to be in terms of u.

20. Consider the integral $\int_{2}^{4} \frac{x}{x^2-1}\,dx$.
   (a) Solve this integral using u-substitution while keeping the limits of integration in terms of x.
   (b) Solve this integral again with u-substitution, this time changing the limits of integration to be in terms of u.

Skills

Solve each of the integrals in Exercises 21–68. Some integrals require substitution, and some do not.

21. $\displaystyle\int (3x + 1)^2\,dx$

22. $\displaystyle\int x(x^2 - 1)^2\,dx$

23. $\displaystyle\int \frac{8x}{x^2 + 1}\,dx$

24. $\displaystyle\int \left(\frac{2}{x-1} - \frac{3}{2x+1}\right)dx$

25. $\displaystyle\int \frac{\sqrt{x}+3}{2\sqrt{x}}\,dx$

26. $\displaystyle\int \frac{x^{2/3}+1}{\sqrt[3]{x}}\,dx$

27. $\displaystyle\int x\csc^2 x^2\,dx$

28. $\displaystyle\int 3\cos(\pi x)\,dx$

29. $\displaystyle\int \frac{1}{3x+1}\,dx$

30. $\displaystyle\int x^3 \cos x^4\,dx$

31. $\displaystyle\int \sin \pi x \cos \pi x\,dx$

32. $\displaystyle\int \frac{x}{\sqrt{3x^2+1}}\,dx$

33. $\displaystyle\int \sec 2x \tan 2x\,dx$

34. $\displaystyle\int \frac{\sqrt{x}+1}{x}\,dx$

35. $\displaystyle\int \cot^5 x \csc^2 x\,dx$

36. $\displaystyle\int \sin x\, e^{\cos x}\,dx$

37. $\displaystyle\int x^4(x^3 + 1)^2\,dx$

38. $\displaystyle\int 2xe^{3x^2}\,dx$

39. $\displaystyle\int x^{1/4} \sin x^{5/4}\,dx$

40. $\displaystyle\int \frac{\cos(\ln x)}{x}\,dx$

41. $\displaystyle\int x(2^{x^2+1})\,dx$

42. $\displaystyle\int x\ln(e^{x^2+1})\,dx$

43. $\displaystyle\int \frac{\sqrt{\ln x}}{x}\,dx$

44. $\displaystyle\int \frac{e^{\sqrt{x}}}{\sqrt{x}}\,dx$

45. $\displaystyle\int \frac{\ln \sqrt{x}}{x}\,dx$

46. $\displaystyle\int \frac{(\cos x + 1)^{3/2}}{\csc x}\,dx$

47. $\displaystyle\int \frac{3x+1}{x^2+1}\,dx$

48. $\displaystyle\int \frac{1-x}{1+3x^2}\,dx$

49. $\displaystyle\int \frac{e^x}{2-e^x}\,dx$

50. $\displaystyle\int (x^2+1)\sqrt{x}\,dx$

51. $\displaystyle\int \frac{2 - e^x}{e^x}\, dx$

52. $\displaystyle\int x^2 \sqrt{x+1}\, dx$

53. $\displaystyle\int \sin x \cos^7 x\, dx$

54. $\displaystyle\int \tan^3 x \sec^2 x\, dx$

55. $\displaystyle\int \frac{x \cos x^2}{\sqrt{\sin x^2}}\, dx$

56. $\displaystyle\int \frac{x^2}{\sqrt{x+1}}\, dx$

57. $\displaystyle\int \frac{\sin(1/x)}{x^2}\, dx$

58. $\displaystyle\int \frac{1}{x \ln x}\, dx$

59. $\displaystyle\int x\sqrt{x^2+1}\, dx$

60. $\displaystyle\int \frac{\sec^2 x}{\tan x + 1}\, dx$

61. $\displaystyle\int x\sqrt{x+1}\, dx$

62. $\displaystyle\int x(x+1)^{1/3}\, dx$

63. $\displaystyle\int \frac{x}{\sqrt{3x+1}}\, dx$

64. $\displaystyle\int \frac{1}{\sqrt{x}\sec\sqrt{x}}\, dx$

65. $\displaystyle\int \frac{e^x}{\csc e^x}\, dx$

66. $\displaystyle\int \cot x \ln(\sin x)\, dx$

67. $\displaystyle\int x^2(x-1)^{100}\, dx$

68. $\displaystyle\int \frac{\sin^{-1}x}{\sqrt{1-x^2}}\, dx$

Solve each of the definite integrals in Exercises 69–76. Some integrals require substitution, and some do not.

69. $\displaystyle\int_0^3 x(x^2+1)^{1/3}\, dx$

70. $\displaystyle\int_{-3}^{-1} \frac{1}{\sqrt{1-3x}}\, dx$

71. $\displaystyle\int_{-\pi}^{\pi} \sin x \cos x\, dx$

72. $\displaystyle\int_0^{\sqrt{4\pi}} x \sin x^2\, dx$

73. $\displaystyle\int_2^3 \frac{1}{x\sqrt{\ln x}}\, dx$

74. $\displaystyle\int_2^4 \frac{e^{2x}}{1+e^{2x}}\, dx$

75. $\displaystyle\int_0^1 \frac{x}{1+x^4}\, dx$

76. $\displaystyle\int_0^1 x\sqrt{1-x}\, dx$

77. Consider the function $f(x) = \dfrac{x}{x^2+1}$.

   (a) Find the signed area between the graph of $f(x)$ and the x-axis on $[-1, 3]$ shown next at the left.

   (b) Find the area between the graph of $f(x)$ and the graph of $g(x) = \dfrac{1}{4}x$ on $[-1, 3]$ shown next at the right.

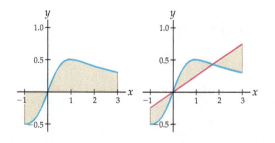

78. Consider the function $f(x) = 4xe^{-x^2}$.

   (a) Find the signed area of the region between the graph of $f(x)$ and the x-axis on $[-1, 2]$ shown here:

   (b) Find the absolute area of the same region.

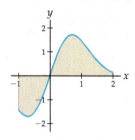

79. Consider the function $f(x) = \dfrac{\ln x}{x}$ shown here:

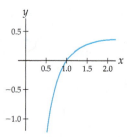

   (a) Find the average value of $f(x)$ on $\left[\frac{1}{2}, 2\right]$.

   (b) Find a value $c \in \left[\frac{1}{2}, 2\right]$ at which $f(x)$ achieves its average value.

80. Consider the function $f(x) = \sin x \cos x$.

   (a) Find the area between the graphs of $f(x)$ and $g(x) = \sin x$ on $[0, \pi]$ shown next at the left.

   (b) Find the area between the graphs of $f(x)$ and $h(x) = \sin 2x$ on $[0, \pi]$ shown next at the right.

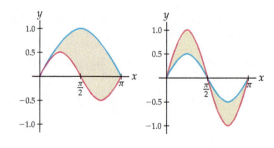

Applications

81. Your local weatherman predicts that there will be a rain-storm this afternoon for three hours and that the rate of rainfall will be $r(t) = 0.6t^2\sqrt{3-t}$ inches per hour from $t = 0$ to $t = 3$, as shown here. Find the total amount of rainfall predicted for the storm.

Rate of rainfall
$$r(t) = 0.6t^2\sqrt{3-t}$$

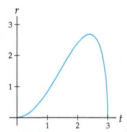

82. Rajini bends a metal rod into a curve that lines up with the graph of $y = 4x^{3/2}$ from $x = 0$ to $x = 2$, as shown here. Given that the length of a curve $y = f(x)$ from $x = a$ to $x = b$ can be calculated with the formula $\int_a^b \sqrt{1 + (f'(x))^2}\, dx$, find the length of the thin metal rod.

Shape of metal rod
$y = 4x^{3/2}$ on $[0, 2]$

83. One of Dr. Geek's favorite beakers is exactly like the shape obtained by revolving the graph of

$$y = 2\left(\frac{\ln x}{x}\right)^{1/2}$$

from $x = 1$ to $x = 10$ around the x-axis, as shown in the figure and measured in inches. Given that the volume of the shape obtained by revolving f around the x-axis on $[a, b]$ can be calculated with the formula $\pi \int_a^b (f(x))^2\, dx$, about how much liquid can the beaker hold?

Shape of beaker

84. A mass hanging at the end of a spring oscillates up and down from its equilibrium position with velocity

$$v(t) = 3\cos\left(\frac{3t}{\sqrt{2}}\right) - 3\sqrt{2}\sin\left(\frac{3t}{\sqrt{2}}\right)$$

centimeters per second, as shown in the figure. The mass is at its equilibrium at $t = 0$. Use definite integrals to determine whether the mass will be above or below its equilibrium position at times $t = 4$ and $t = 5$.

Velocity $v(t)$ of spring

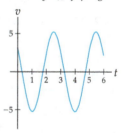

Proofs

85. Prove, in the following two ways, that for any integer k, the signed area under the graph of the function $f(x) = \sin(2(x - (\pi/4)))$ on the interval $[0, k\pi]$ is always zero:
    (a) by calculating a definite integral;
    (b) by considering the period and graph of the function $f(x) = \sin(2(x - (\pi/4)))$.
86. Prove, in the following two ways, that the signed area under the graph of the function $f(x) = \sin x \cos^2 x$ on an interval $[-a, a]$ centered about the origin is always zero:
    (a) by calculating a definite integral;
    (b) by considering the symmetry of the graph of the function $f(x) = \sin x \cos^2 x$.

87. Use the chain rule to prove the formula for integration by substitution:

$$\int f'(u(x))u'(x)\, dx = f(u(x)) + C.$$

88. Use the chain rule and the Fundamental Theorem of Calculus to prove the integration-by-substitution formula for definite integrals:

$$\int_a^b f'(u(x))u'(x)\, dx = f(u(b)) - f(u(a)).$$

89. Prove the integration formula

$$\int \tan x\, dx = \ln|\sec x| + C$$

(a) by using algebra and integration by substitution to find $\int \tan x\, dx$;

(b) by differentiating $\ln|\sec x|$.

90. Prove the integration formula

$$\int \cot x\, dx = -\ln|\csc x| + C$$

(a) by using algebra and integration by substitution to find $\int \cot x\, dx$;

(b) by differentiating $-\ln|\csc x|$.

Thinking Forward

Trigonometric integrals: The integrals that follow can be solved by using algebra to write the integrands in the form $f'(u(x))u'(x)$ so that u-substitution will apply.

▶ Solve $\int \sin^5 x\, dx$ by using the Pythagorean identity $\sin^2 x + \cos^2 x = 1$ to rewrite the integrand as $(1 - \cos^2 x)^2 \sin x$ and then applying substitution with $u = \cos x$.

▶ Solve $\int \sin^3 x \cos^4 x\, dx$ by using the Pythagorean identity $\sin^2 x + \cos^2 x = 1$ to rewrite the integrand as $(1 - \cos^2 x) \cos^4 x \sin x$ and then applying substitution with $u = \cos x$.

▶ Solve $\int \sec^4 x \tan^3 x\, dx$ by using the Pythagorean identity $\tan^2 x + 1 = \sec^2 x$ to rewrite the integrand as $(\tan^2 x + 1) \tan^3 x \sec^2 x$ and then applying substitution with $u = \tan x$.

8.2   INTEGRATION BY PARTS

▶ Reversing the product rule to arrive at the formula for integration by parts

▶ Strategies for deciding when and how to apply integration by parts

▶ Using integration by parts to calculate definite integrals

Undoing the Product Rule

As we saw in Theorem 7.21, since the derivative of a product $u(x)v(x)$ of differentiable functions is $u'(x)v(x) + u(x)v'(x)$, we have the following theorem:

THEOREM 8.6   **Reversing the Product Rule**

If u and v are functions such that $u'(x)v(x) + u(x)v'(x)$ is integrable, then

$$\int (u'(x)v(x) + u(x)v'(x))\, dx = u(x)v(x) + C.$$

For example, with $u(x) = x$ and $v(x) = \sin x$, we have $\frac{d}{dx}(x\sin x) = \sin x + x\cos x$, and thus $\int (\sin x + x\cos x)\, dx = x\sin x + C$.

The pattern $u'(x)v(x) + u(x)v'(x)$ does not often appear in the integrals we will be concerned with; however, products will routinely appear in them. With a little bit of rearranging we can rewrite the formula in Theorem 8.6 to obtain a formula for a technique called *integration by parts*, which will help us solve integrals that involve products. By splitting up the integral in the theorem and then solving for one of the smaller integrals, we obtain

$$\int u(x)v'(x)\, dx = u(x)v(x) - \int v(x)u'(x)\, dx.$$

You may have noticed that we just dropped the "$+C$" that had appeared in the formula from Theorem 8.6. This is allowable because there are indefinite integrals on both sides of the new equation, and thus both sides already represent families of antiderivatives.

Moreover, since we can write the differentials for u and v as $du = u'(x)\,dx$ and $dv = v'(x)\,dx$, we can write the new formula as follows:

THEOREM 8.7

Formula for Integration by Parts

If $u = u(x)$ and $v = v(x)$ are differentiable functions, then

$$\int u\,dv = uv - \int v\,du.$$

The parts referred to in Theorem 8.7 are u and dv. When applying integration by parts we will choose u and dv, use them to find du and v, and then apply the integration by parts formula. For example, to solve $\int x\cos x\,dx$ we could choose $u = x$ and $dv = \cos x\,dx$. We can find $du = u'(x)\,dx$ by differentiating u, and we can find v by antidifferentiating $dv = v'(x)\,dx$. Since differentiation intuitively seems like a forward operation, whereas antidifferentiation goes backwards, a sensible way to organize this work is as follows:

$$u = x \qquad \longrightarrow \qquad du = 1\,dx$$

$$v = \sin x \qquad \longleftarrow \qquad dv = \cos x\,dx$$

Now the formula for integration by parts tells us that $\int x\cos x\,dx = x\sin x - \int \sin x\,dx$. This is fortunate because it gives us a way to relate an integral that we cannot directly antidifferentiate to an easier integral that is simple to antidifferentiate. Notice that like integration by substitution, integration by parts does not solve integrals; rather, it is a method for rewriting a difficult integral into what is presumably a simpler integral.

Strategies for Applying Integration by Parts

Deciding whether or not an integral can be improved by applying the integration-by-parts formula for some choices of u and dv can sometimes be difficult. In general it is best to try algebraic simplification and u-substitution before attempting integration by parts. However, integrands that are certain simple products, such as $x\cos x$, $x^2 e^x$, or $e^x \sin x$, are often good candidates for integration by parts.

Note that integrals that look similar can sometimes require different methods of solution. For example, consider the integrals

$$\int x e^x\,dx, \quad \int x^2 e^x\,dx, \quad \int x e^{(x^2)}\,dx, \quad \text{and} \quad \int e^{(x^2)}\,dx.$$

The first integral can be integrated by parts, with $u = x$ and $v = e^x$, and the second can be integrated by applying parts twice. However, the third integral is better suited for u-substitution with $u = x^2$, and the fourth integrand does not even have an elementary antiderivative.

Once we do decide to use integration by parts, how do we decide what to choose for u and dv? A good rule of thumb is that u should be something that gets simpler when differentiated, and/or dv should be something that gets simpler (or at least not more complicated) when integrated. In other words, we want to choose u and dv so that the associated du and v are simpler than what we started with. This makes sense if you remember that the original integral in terms of u and dv will be replaced by an integral in terms of v and du that we hope will be simpler.

Let's return again to the example $\int x \cos x \, dx = \int u \, dv$. There are four possible ways to split the integrand into the parts u and dv. The possible choices for u are $u = x$, $u = \cos x$, $u = x \cos x$, and $u = 1$. The following table lists each of these choices, the corresponding dv, the resulting du and v, and the expressions for $uv - \int v \, du$:

Choice of u and dv and resulting du and v	New integral $uv - \int v \, du$
$u = x \quad \longrightarrow \quad du = 1 \, dx$ $v = \sin x \quad \longleftarrow \quad dv = \cos x \, dx$	$x \sin x - \int \sin x \, dx$
$u = \cos x \quad \longrightarrow \quad du = -\sin x \, dx$ $v = \dfrac{1}{2} x^2 \quad \longleftarrow \quad dv = x \, dx$	$\dfrac{1}{2} x^2 \cos x + \dfrac{1}{2} \int x^2 \sin x \, dx$
$u = x \cos x \quad \longrightarrow \quad du = (\cos x - x \sin x) \, dx$ $v = x \quad \longleftarrow \quad dv = dx$	$x^2 \cos x - \int (x \cos x - x^2 \sin x) \, dx$
$u = 1 \quad \longrightarrow \quad du = 0 \, dx$ $v = ??? \quad \longleftarrow \quad dv = x \cos x \, dx$	*not possible to apply formula*

At the beginning of this section we solved the integral $\int x \cos x \, dx$ by choosing $u = x$ and $dv = \cos x \, dx$, as shown in the first row of the table, and that worked out well. If we had chosen instead $u = \cos x$ and $dv = x \, dx$, as shown in the second row, we would not have been so lucky: In that case we would have rewritten our integral $\int x \cos x \, dx$ in terms of the more complicated integral $\int x^2 \sin x \, dx$. The third possible case is to choose u to be everything but the dx. In this example that does not work out so well. The fourth case is to choose dv to be the entire integrand. This choice is *never* going to work, because we would have to antidifferentiate dv to find v, and antidifferentiating the integrand is what we couldn't do in the first place.

 CAUTION However you choose u and dv, you must do it so that the integrand is the *product* of u and dv. For example, with $\int \dfrac{x}{e^x} \, dx$, we could *not* choose $u = x$ and $dv = e^x \, dx$, since, with this second choice, the integrand is not the product of u and dv. We could, however, choose $u = x$ and $dv = \dfrac{1}{e^x} \, dx$.

There are instances in which choosing u to be the entire integrand and $dv = dx$ is a useful choice. In fact, the method of choosing u to be the entire integrand yields a formula for integrating the natural logarithm function:

THEOREM 8.8 **Integral of the Natural Logarithm Function**

$$\int \ln x \, dx = x \ln x - x + C.$$

Proof. At first glance the integral of $\ln x$ might not look like it could be a good candidate for integration by parts, since $\ln x$ does not look like a product. If, however, we choose $u = \ln x$ to be the entire integrand and $dv = dx$, then we have

$$u = \ln x \quad \longrightarrow \quad du = \frac{1}{x} \, dx$$
$$v = x \quad \longleftarrow \quad dv = dx.$$

Now the formula for integration by parts gives us

$$\int \ln x \, dx = (\ln x)(x) - \int x \left(\frac{1}{x} \right) dx = x \ln x - \int dx = x \ln x - x + C.$$ ∎

As you will show in Exercises 88 and 89, a similar method yields formulas for integrating two inverse trigonometric functions:

THEOREM 8.9

Integrals of Inverse Sine and Inverse Tangent

(a) $\displaystyle\int \sin^{-1} x\, dx = x\sin^{-1} x + \sqrt{1-x^2} + C$

(b) $\displaystyle\int \tan^{-1} x\, dx = x\tan^{-1} x - \frac{1}{2}\ln(x^2 + 1) + C$

Finding Definite Integrals by Using Integration by Parts

When using integration by parts to solve a definite integral $\int_a^b f(x)\, dx$, we must evaluate the *entire* right-hand side of the integration-by-parts formula from $x = a$ to $x = b$. The integration-by-parts formula for definite integrals follows directly from the formula for indefinite integrals, the Fundamental Theorem of Calculus, and the definition of evaluation notation:

THEOREM 8.10

Integration by Parts for Definite Integrals

If $u = u(x)$ and $v = v(x)$ are differentiable functions on $[a, b]$, then

$$\int_a^b u\, dv = \left[\int u\, dv\right]_a^b = \left[uv\right]_a^b - \int_a^b v\, du.$$

For example, we have already seen that $\int x\cos x\, dx = x\sin x - \int \sin x\, dx$. This means that, for instance, on $[0, \pi]$ we have

$$\int_0^\pi x\cos x\, dx = \left[x\sin x\right]_0^\pi - \int_0^\pi \sin x\, dx = \left[x\sin x\right]_0^\pi - \left[-\cos x\right]_0^\pi.$$

We now have a choice: We can evaluate the expressions $x\sin x$ and $-\cos x$ on $[0, \pi]$ separately, or we can combine them first. Here we choose the latter approach for finishing the calculation:

$$\left[x\sin x + \cos x\right]_0^\pi = (\pi\sin\pi + \cos\pi) - (0\sin 0 + \cos 0) = -1 - 1 = -2.$$

Examples and Explorations

EXAMPLE 1

Choosing u and dv when applying integration by parts

Use integration by parts to find

(a) $\displaystyle\int x e^{2x}\, dx$

(b) $\displaystyle\int x^2 \ln x\, dx$

SOLUTION

(a) Since x gets simpler when differentiated and e^{2x} at least does not get worse when antidifferentiated, a sensible choice for parts is $u = x$ and $dv = e^{2x}\, dx$. We can use u

and dv to solve for du and v:

$$u = x \qquad \longrightarrow \qquad du = dx$$

$$v = \frac{1}{2}e^{2x} \qquad \longleftarrow \qquad dv = e^{2x}\,dx$$

We can now apply the formula for integration by parts to write our integral in terms of an integral that we hope will be easier to antidifferentiate:

$$\int x e^{2x}\,dx = x\left(\frac{1}{2}e^{2x}\right) - \int \frac{1}{2}e^{2x}\,dx \qquad \leftarrow \text{integration-by-parts formula}$$

$$= \frac{1}{2}x e^{2x} - \frac{1}{2}\int e^{2x}\,dx \qquad \leftarrow \text{simplify}$$

$$= \frac{1}{2}x e^{2x} - \frac{1}{4}e^{2x} + C. \qquad \leftarrow \text{antidifferentiate}$$

(b) This time there are six different ways to choose u and dv. For example we could choose $u = x^2$, which does indeed get simpler when differentiated. But with that choice of u we would have $dv = \ln x\,dx$, whose antiderivative is the relatively messy function $x\ln x - x$. A better choice is

$$u = \ln x \qquad \longrightarrow \qquad du = \frac{1}{x}\,dx$$

$$v = \frac{1}{3}x^3 \qquad \longleftarrow \qquad dv = x^2\,dx.$$

Notice that $u = \ln x$ gets *much* simpler when differentiated: It goes from being a transcendental function to being an algebraic function! Even though $v = \frac{1}{3}x^3$ is a little more complicated than $dv = x^2\,dx$, it is worth it to get rid of the logarithm. There is also some nice cancelling in the integral $\int v\,du$:

$$\int x^2 \ln x\,dx = (\ln x)\left(\frac{1}{3}x^3\right) - \int \left(\frac{1}{3}x^3\right)\left(\frac{1}{x}\right)\,dx \qquad \leftarrow \text{integration-by-parts formula}$$

$$= \frac{1}{3}x^3 \ln x - \frac{1}{3}\int x^2\,dx \qquad \leftarrow \text{simplify}$$

$$= \frac{1}{3}x^3 \ln x - \frac{1}{3}\left(\frac{1}{3}x^3\right) + C. \qquad \leftarrow \text{antidifferentiate} \qquad \square$$

EXAMPLE 2   **Deciding when to use integration by parts**

For each integral, determine whether integration by parts is appropriate, and if so, determine useful choices for u and dv.

(a) $\displaystyle\int x \sin x^2\,dx$ **(c)** $\displaystyle\int x \ln x\,dx$ **(e)** $\displaystyle\int \frac{x+1}{e^x}\,dx$

(b) $\displaystyle\int x^2 \sin x\,dx$ **(d)** $\displaystyle\int \frac{\ln x}{x}\,dx$ **(f)** $\displaystyle\int \frac{e^x}{e^x + 1}\,dx$

SOLUTION

(a) Although the integrand here is a product, integration by parts is not the best strategy for this integral. This is a classic u-substitution problem, since $u = x^2$ is the inside of a composition and its derivative $2x$ (without the constant 2) is a multiplicative factor in the integrand. This substitution gives us the integral $\frac{1}{2}\int \sin u\,du$, which is easy to solve.

(b) This time u-substitution cannot help us, because the x^2 is outside of the sine function while its derivative is inside. A good choice for integration by parts here is $u = x^2$ and $dv = \sin x\, dx$, since the x^2 will become simpler upon differentiating. In this case we will have to apply integration by parts twice, as we will do in part (a) of Example 3.

(c) This integrand is well suited for integration by parts because it is the product of two unrelated functions. Although we often like to choose $u = x$ in such integrals (e.g., with $\int x \sin x\, dx$ or $\int xe^x\, dx$), in this case that would force us to choose $dv = \ln x\, dx$, which becomes much more complicated after antidifferentiating. If we instead choose $u = \ln x$ and $dv = x\, dx$, then we can successfully apply integration by parts.

(d) Although we could certainly think of the integrand as a product $(\ln x)\left(\frac{1}{x}\right)$, and there is a choice of parts that will work here (see Exercise 24), this integral is best solved with u-substitution. With $u = \ln x$, we have $du = \frac{1}{x}\, dx$ and therefore can change variables to get the very simple integral $\int u\, du$.

(e) To apply integration by parts it is best to rewrite the integrand as the product $e^{-x}(x + 1)$. The most sensible choice of parts is $u = x + 1$ (since this expression gets simpler after differentiating) and $dv = e^{-x}\, dx$ (since this expression stays about the same after antidifferentiating).

(f) This integral is best solved with u-substitution, with $u = e^x + 1$. Then we have $du = e^x\, dx$ and can transform to the simpler integral $\int \frac{1}{u}\, du$.   □

| **EXAMPLE 3** | **Applying integration by parts multiple times** |

Use integration by parts to find

(a) $\displaystyle\int x^2 e^x\, dx$ **(b)** $\displaystyle\int e^x \sin x\, dx$

SOLUTION

(a) A sensible choice of parts here is $u = x^2$ and $dv = e^x\, dx$, since differentiation will improve x^2 and integration will not make e^x any worse. This choice of parts results in a simpler integral, but one that itself requires an application of integration by parts to solve:

$$\int x^2 e^x\, dx = x^2 e^x - 2\int xe^x\, dx \qquad \leftarrow \text{parts with } u = x^2 \text{ and } dv = e^x\, dx$$

$$= x^2 e^x - 2\left(xe^x - \int e^x\, dx\right) \qquad \leftarrow \text{parts with } u = x \text{ and } dv = e^x\, dx$$

$$= x^2 e^x - 2(xe^x - e^x) + C. \qquad \leftarrow \text{antidifferentiation}$$

The first application of integration by parts takes us from an integral involving $x^2 e^x$ to one involving the simpler xe^x. The second application of integration by parts further reduces the power of x until we are left with an integral involving simply e^x. If the original integral had been $\int x^5 e^x\, dx$, then we could use the same method, integrating by parts five times to eventually end up with $\int e^x\, dx$.

(b) In some cases, integrating by parts twice brings us back to the original integral. Although sometimes this is no help at all, in this case it will enable us to construct an equation involving the original integral, and it will be a simple matter to solve that equation. Applying parts to the resulting integral with $u = e^x$ and $dv = \sin x\, dx$, and then applying parts again with $u = e^x$ and $dv = \cos x\, dx$, we get

$$\int e^x \sin x\, dx = -e^x \cos x + \int e^x \cos x\, dx = -e^x \cos x + \left(e^x \sin x - \int e^x \sin x\, dx\right).$$

It might seem as if we have gone in a circle, but a sign change has made this into a very useful equation. For a moment let I represent the integral $\int e^x \sin x\, dx$ that we are trying to find. Then the calculation we just did implies that $I = -e^x \cos x + e^x \sin x - I$. It is now a simple matter to solve this equation for $I = \frac{1}{2}(-e^x \cos x + e^x \sin x) + C$. The "$+C$" must be there because I represents a family of antiderivatives. We have solved our integral, without even having to do a final antidifferentiation step!   □

CHECKING THE ANSWER

As usual, we can check our answers by differentiating. In both cases of the previous example we applied parts twice, so to differentiate we need to apply the product rule twice. We have

$$\frac{d}{dx}(x^2 e^x - 2(xe^x - e^x)) = 2xe^x + x^2 e^x - 2(e^x + xe^x - e^x) = x^2 e^x \text{ and}$$

$$\frac{d}{dx}\left(\frac{1}{2}(-e^x \cos x + e^x \sin x)\right) = \frac{1}{2}(-e^x \cos x + e^x \sin x + e^x \sin x + e^x \cos x) = e^x \sin x.$$

EXAMPLE 4

Solving a definite integral with integration by parts

Calculate $\displaystyle\int_0^{\pi/4} x \sec^2 x\, dx$.

SOLUTION

Notice that $\sec^2 x$ is easy to antidifferentiate, since it is the derivative of $\tan x$. Therefore we choose

$$u = x \qquad \longrightarrow \qquad du = dx$$
$$v = \tan x \qquad \longleftarrow \qquad dv = \sec^2 x\, dx$$

Integrating by parts, we have

$$\int_0^{\pi/4} x \sec^2 x\, dx = \left[x \tan x\right]_0^{\pi/4} - \int_0^{\pi/4} \tan x\, dx.$$

We can solve the new integral $\int_0^{\pi/4} \tan x\, dx = \int_0^{\pi/4} \frac{\sin x}{\cos x}\, dx$ with integration by substitution. To avoid confusion, since we've already used the letter u, let's use the letter w here:

$$\boxed{w = \cos x} \quad \Longrightarrow \quad \frac{dw}{dx} = -\sin x \quad \Longrightarrow \quad \boxed{-dw = \sin x\, dx}.$$

Continuing the calculation, we have

$$\int_0^{\pi/4} x \sec^2 x\, dx = \left[x \tan x\right]_0^{\pi/4} + \int_{x=0}^{x=\pi/4} \frac{1}{w}\, dw$$

$$= \left[x \tan x\right]_0^{\pi/4} + \left[\ln |w|\right]_{x=0}^{x=\pi/4} \qquad \leftarrow \text{antidifferentiate}$$

$$= \left[x \tan x\right]_0^{\pi/4} + \left[\ln |\cos x|\right]_0^{\pi/4} \qquad \leftarrow \text{since } w = \cos x$$

$$= \left(\frac{\pi}{4} \tan \frac{\pi}{4} - 0 \tan 0\right) + \left(\ln \left|\cos \frac{\pi}{4}\right| - \ln |\cos 0|\right) \qquad \leftarrow \text{evaluate}$$

$$= \left(\frac{\pi}{4} - 0\right) + \left(\ln \frac{\sqrt{2}}{2} - 0\right)$$

$$= \frac{\pi}{4} + \ln \left(\frac{\sqrt{2}}{2}\right).$$

□

CHECKING THE ANSWER

You can check the integration steps in the previous example by differentiating:

$$\frac{d}{dx}(x\tan x + \ln|\cos x|) = \tan x + x\sec^2 x + \frac{1}{\cos x}(-\sin x) = x\sec^2 x.$$

You can check the numerical answer by using a graphing calculator to verify that the signed area under the graph of $y = x\sec^2 x$ from $x = 0$ to $x = \frac{\pi}{4} \approx 0.785$ is approximately $\frac{\pi}{4} + \ln\left(\frac{\sqrt{2}}{2}\right) \approx 0.439$:

TEST YOUR UNDERSTANDING

▶ What types of functions get simpler when differentiated? What types of functions get simpler when antidifferentiated? What types of functions have derivatives and/or antiderivatives of about the same complexity?

▶ What types of integrals are usually amenable to integration by parts? How does integration by parts help you calculate an integral?

▶ After choosing $u = e^x$ and $dv = \sin x\, dx$ in Example 3(b), we could have chosen $u = \cos x$ and $dv = e^x\, dx$ in the second application of integration by parts. Would this work?

▶ In Example 3(b) we could have instead chosen $u = \sin x$ and $dv = e^x\, dx$ in the first integration by parts and then a similar u and dv in the second integration by parts. Would this work?

▶ Now that we know how to integrate $\ln x$, are there any other elementary functions that we do not yet know how to antidifferentiate?

EXERCISES 8.2

Thinking Back

Differentiation review: Differentiate each of the functions that follow. Simplify your answers as much as possible.

▶ $f(x) = xe^x - e^x$

▶ $f(x) = \frac{1}{2}x^2\ln x - \frac{1}{4}x^2$

▶ $f(x) = -e^{-x}(x^2 + 1) - 2xe^{-x} - 2e^{-x}$

▶ $f(x) = \frac{1}{3}x^3 - 2(xe^x - e^x) + \frac{1}{2}e^{2x}$

▶ $f(x) = -x\cot x + \ln|\sin x|$

▶ $f(x) = x^3\sin x + 3x^2\cos x - 6x\sin x - 6\cos x$

Review of integration by substitution: Use u-substitution to find each of the following integrals.

▶ $\displaystyle\int e^{3x+1}\, dx$

▶ $\displaystyle\int \frac{\ln x}{x}\, dx$

▶ $\displaystyle\int \tan x\sec^2 x\, dx$

▶ $\displaystyle\int \sin x\sqrt{\cos x}\, dx$

▶ $\displaystyle\int xe^{x^2+1}\, dx$

▶ $\displaystyle\int \frac{1}{x\ln x}\, dx$

▶ $\displaystyle\int \frac{1}{1+x^2}\, dx$

▶ $\displaystyle\int e^x\sin e^x\, dx$

Concepts

0. *Problem Zero:* Read the section and make your own summary of the material.

1. *True/False:* Determine whether each of the statements that follow is true or false. If a statement is true, explain why. If a statement is false, provide a counterexample.

   (a) *True or False:* If $u = x^2 + 1$, then $du = 2x$.

   (b) *True or False:* If $dv = x^2\,dx$, then $v = \frac{1}{3}x^3$.

   (c) *True or False:* We can apply integration by parts with $u = \ln x$ and $dv = x\,dx$ to the integral $\int \frac{\ln x}{x}\,dx$.

   (d) *True or False:* We can apply integration by parts with $u = x$ and $dv = \ln x\,dx$ to the integral $\int \frac{\ln x}{x}\,dx$.

   (e) *True or False:* Integration by parts has to do with reversing the product rule.

   (f) *True or False:* Integration by parts is a good method for any integral that involves a product.

   (g) *True or False:* In applying integration by parts, it is sometimes a good idea to choose u to be the entire integrand and let $dv = dx$.

   (h) *True or False:* $\int_0^3 xe^x\,dx = xe^x - \int_0^3 e^x\,dx$.

2. *Examples:* Construct examples of the thing(s) described in the following. Try to find examples that are different than any in the reading.

   (a) Five integrals for which u-substitution is a better strategy than integration by parts. List a good choice for u in each case.

   (b) Five integrals for which integration by parts is a better strategy than u-substitution. List good choices for u and dv in each case.

   (c) Three integrals that we cannot integrate with only the techniques we have learned so far.

3. State the integration-by-parts formula for indefinite integrals, (a) using the notation du and dv and (b) without using the notation du and dv.

4. State the integration-by-parts formula for definite integrals, (a) using the notation du and dv and (b) without using the notation du and dv.

5. Write down an integral that can be solved by using integration by parts with $u = x$ and another integral that can be solved by using integration by parts with $dv = x\,dx$.

6. Write down an integral that can be solved by using integration by parts with $u = \sin x$ and another integral that can be solved by using integration by parts with $dv = \sin x\,dx$.

7. Write down an integral that can be solved with integration by parts by choosing u to be the entire integrand and $dv = dx$.

8. Suppose $v(x)$ is a function of x. Explain why the integral of dv is equal to v (up to a constant).

9. Explain why choosing $u = 1$ (and thus choosing dv to be the entire integrand, including dx) is *never* a good choice for integration by parts.

10. Find three integrals in Exercises 27–68 for which either algebra or u-substitution is a better strategy than integration by parts.

11. Find three integrals in Exercises 27–68 for which a good strategy is to use integration by parts with $u = x$ (and dv the remaining part).

12. Find three integrals in Exercises 27–68 for which a good strategy is to apply integration by parts twice.

13. If $u(x) = \sin 3x$ and $v(x) = x$, what are du and dv? Write down $\int u\,dv$ and $\int v\,du$ in this situation. Which of these integrals would be easier to find? What does this exercise have to do with integration by parts?

14. Provide a justification for each equality in the statement of the integration-by-parts formula for definite integrals from Theorem 8.10.

15. Explain why $[g(x)]_a^b - [h(x)]_a^b = [g(x) - h(x)]_a^b$. What does this equation have to do with calculations of definite integrals with integration by parts?

For each pair of functions $u(x)$ and $v(x)$ in Exercises 16–18, fill in the blanks to complete each of the following:

   (a) $\dfrac{d}{dx}(u(x)v(x)) = $ _____

   (b) $\displaystyle\int$ _____ $dx = u(x)v(x) + C$

   (c) $\displaystyle\int u\,dv = $ _____

16. $u(x) = x,\ v(x) = \cos 2x$

17. $u(x) = \ln x,\ v(x) = x$

18. $u(x) = x^3,\ v(x) = e^{3x}$

Each expression in Exercises 19–23 is the result of a calculation that uses integration by parts. That is, each is an expression of the form $uv - \int v\,du$ for some functions u and v. Identify u, v, du, and dv, and determine the original integral $\int u\,dv$.

19. $\dfrac{x\,2^x}{\ln 2} - \dfrac{1}{\ln 2}\displaystyle\int 2^x\,dx$

20. $-x^2\cos x + 2\displaystyle\int x\cos x\,dx$

21. $-\dfrac{\ln x}{x^2} + \displaystyle\int \dfrac{1}{x^3}\,dx$

22. $\dfrac{1}{3}xe^{3x} - \dfrac{1}{3}\displaystyle\int e^{3x}\,dx$

23. $x\tan^{-1} x - \displaystyle\int \dfrac{x}{x^2 + 1}\,dx$

24. Consider the integral $\int \dfrac{\ln x}{x}\,dx$.

   (a) Solve this integral by using integration by parts with $u = \ln x$ and $dv = \dfrac{1}{x}\,dx$.

   (b) Now solve the integral another way, by using u-substitution with $u = \ln x$.

   (c) How must your answers to parts (a) and (b) be related? Use algebra to prove that this is so.

25. Consider the integral $\int x^2 \ln(x^3)\, dx$.
  (a) Solve this integral by using integration by parts with $u = \ln(x^3)$ and $dv = x^2\, dx$.
  (b) Now solve the integral another way, by using u-substitution with $u = x^3$.
  (c) How must your answers to parts (a) and (b) be related? Use algebra to prove this relationship.

26. Consider the integral $\int x(x+1)^{100}\, dx$.
  (a) Solve this integral by using integration by parts with $u = x$ and $dv = (x+1)^{100}\, dx$.
  (b) Now solve the integral another way, by using u-substitution with $u = x+1$ and back-substitution.
  (c) How must your answers to parts (a) and (b) be related? Use graphs to prove this relationship.

Skills

Solve each of the integrals in Exercises 27–68. Some integrals require integration by parts, and some do not.

27. $\displaystyle\int xe^x\, dx$

28. $\displaystyle\int x\sin x\, dx$

29. $\displaystyle\int x\ln x\, dx$

30. $\displaystyle\int x\cos x\, dx$

31. $\displaystyle\int x\sin x^2\, dx$

32. $\displaystyle\int x^2\cos x\, dx$

33. $\displaystyle\int x^2 e^{3x}\, dx$

34. $\displaystyle\int (3-x)e^{2x}\, dx$

35. $\displaystyle\int \frac{x}{e^x}\, dx$

36. $\displaystyle\int \frac{x^2}{e^x}\, dx$

37. $\displaystyle\int 3xe^{x^2}\, dx$

38. $\displaystyle\int \ln(3x)\, dx$

39. $\displaystyle\int \ln(x^3)\, dx$

40. $\displaystyle\int x\ln(x^2)\, dx$

41. $\displaystyle\int \frac{x^2+1}{e^x}\, dx$

42. $\displaystyle\int \frac{x}{e^{x^2+1}}\, dx$

43. $\displaystyle\int \frac{3x+1}{x}\, dx$

44. $\displaystyle\int \frac{3x+1}{\sec x}\, dx$

45. $\displaystyle\int (x-e^x)^2\, dx$

46. $\displaystyle\int \ln\sqrt{x}\, dx$

47. $\displaystyle\int \sqrt{x}\ln x\, dx$

48. $\displaystyle\int x\ln\sqrt{x}\, dx$

49. $\displaystyle\int x^3 e^{x^2}\, dx$

50. $\displaystyle\int x^3 e^{-x}\, dx$

51. $\displaystyle\int x\csc^2 x\, dx$

52. $\displaystyle\int x^5\cos x^3\, dx$

53. $\displaystyle\int x^3\cos x\, dx$

54. $\displaystyle\int \frac{x^2}{\sqrt{x-1}}\, dx$

55. $\displaystyle\int x^3\sin x^2\, dx$

56. $\displaystyle\int \arcsin 2x\, dx$

57. $\displaystyle\int \tan^{-1} 3x\, dx$

58. $\displaystyle\int \sin \pi x\cos \pi x\, dx$

59. $\displaystyle\int e^{2x}\sin x\, dx$

60. $\displaystyle\int e^x\ln(e^{2x})\, dx$

61. $\displaystyle\int \frac{2^x}{3^x}\, dx$

62. $\displaystyle\int \frac{\sin x}{3^x}\, dx$

63. $\displaystyle\int e^{-x}\cos x\, dx$

64. $\displaystyle\int x^3 e^{x^2}\, dx$

65. $\displaystyle\int \frac{x}{\cos^2 x}\, dx$

66. $\displaystyle\int \sec^4 x\, dx$

67. $\displaystyle\int x\tan^{-1} x\, dx$

68. $\displaystyle\int x^2\tan^{-1} x\, dx$

Solve each of the definite integrals in Exercises 69–76. Some integrals require integration by parts, and some do not.

69. $\displaystyle\int_1^2 \ln x\, dx$

70. $\displaystyle\int_0^\pi x\sin x\, dx$

71. $\displaystyle\int_{-1}^1 \frac{x}{e^x}\, dx$

72. $\displaystyle\int_{-\pi/4}^{-\pi/4} x\sec x\tan x\, dx$

73. $\displaystyle\int_0^\pi e^x\sin x\, dx$

74. $\displaystyle\int_{-1}^1 2^x\cos x\, dx$

75. $\displaystyle\int_{\pi/4}^{\pi/2} x\csc^2 x\, dx$

76. $\displaystyle\int_{-1/2}^{1/2} \sin^{-1} x\, dx$

77. Consider the function $f(x) = xe^{-x}$.
  (a) Find the signed area of the region between the graph of f and the x-axis on $[-1, 1]$ shown in the next figure.
  (b) Find the absolute area of the same region.

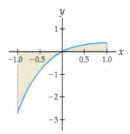

78. Consider the function $f(x) = 2 + \tan^{-1} x$ shown in the next figure.
  (a) Find the average value of $f(x)$ on $[-4, 4]$.
  (b) Find a value $c \in [-4, 4]$ at which $f(x)$ achieves its average value.

79. Consider the function $f(x) = x \cos 2x$.

    (a) Find the signed area between the graph of $f(x)$ and the x-axis on $[0, 3\pi/4]$ shown in the figure next at the left.

    (b) Find the area between the graph of $f(x)$ and the graph of $g(x) = -\cos 2x$ on $[0, 3\pi/4]$ shown in the figure next at the right.

80. Consider the function $f(x) = \ln x$.

    (a) Find the signed area between the graph of $f(x)$ and the x-axis on $[1/2, 4]$ shown in the figure next at the left.

    (b) Find the area between the graph of $f(x)$ and the graph of $g(x) = 1$ on $[1/2, 4]$ shown in the figure next at the right.

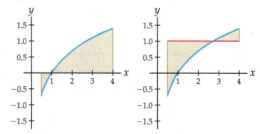

Applications

81. Your local weatherman predicts that there will be a rainstorm this afternoon for three hours and that the rate of rainfall will be given by $r(t) = 0.6t^2\sqrt{3-t}$ for those three hours from $t = 0$ to $t = 3$, as shown in the figure. Use integration by parts (twice) to solve the definite integral that represents the total amount of rainfall predicted for the storm. Compare this exercise with Exercise 81 in Section 8.1, where you used u-substitution to solve the same problem.

Rate of rainfall
$r(t) = 0.6t^2\sqrt{3-t}$

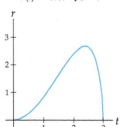

82. Suppose the rate at which a population of 100 weasels catches weasel-mumps is $r(t) = 8 - 2\ln(t+1)$ weasels per day, where $t > 0$ measures the number of days after the initial outbreak. Find a formula that represents the number $W(t)$ of weasels that have contracted weasel-mumps t days after the initial outbreak. Approximately how long will it be until all but one of the weasels are infected?

Rate of weasel-mump infection
$r(t) = 8 - 2\ln(t+1)$

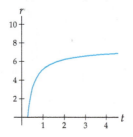

83. At Dr. Geek's house, the wine goblets are shaped exactly like the shape obtained by revolving the graph of $y = \ln x$ from $x = 0.5$ to $x = 5$ around the x-axis, as shown in the figure and measured in inches. Given that the volume of the shape obtained from revolving f around the x-axis on $[a, b]$ can be calculated with the formula $\pi \int_a^b (f(x))^2\, dx$, about how much wine can each glass hold?

Shape of wine goblet

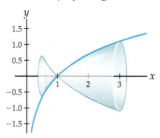

84. Dr. Geek also owns some champagne flutes, which are shaped exactly like the shape obtained by revolving the graph of $y = \dfrac{\ln x}{x}$ from $x = 0.6$ to $x = 5$ around the x-axis, as shown in the figure and measured in inches. Given that the volume of the shape obtained from revolving f around the x-axis on $[a, b]$ can be calculated with the formula $\pi \int_a^b (f(x))^2\, dx$, about how much champagne can each glass hold?

Shape of champagne flute

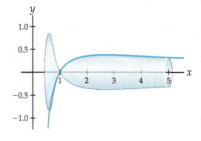

85. Suppose a very odd particle moves back and forth along a straight path in such a way that its velocity after t seconds is given by $v(t) = 5t \sin(\pi t)$, measured in feet per second, as shown in the figure. Consider positions left of the starting position to be in the negative direction and positions right of the starting position to be in the positive direction.

Velocity of odd particle
$v(t) = 5t \sin(\pi t)$

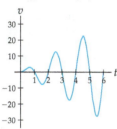

(a) Use integration by parts to find an equation for the position $s(t)$ of the very odd particle, measured in feet left or right of the starting position.

(b) Sketch a graph of $s(t)$, and describe the behavior of the particle over the first 3 seconds.

(c) Where is the particle at $t = 3$ seconds, and what it is doing at that time?

(d) What happens to the position and velocity of the very odd particle after a long time, say, 100 seconds?

Proofs

86. Use the product rule to derive the formula for integration by parts in Theorem 8.7.

87. Prove the integration formula

$$\int \ln x \, dx = x \ln x - x + C$$

(a) by applying integration by parts to $\int \ln x \, dx$;

(b) by differentiating $x \ln x - x$.

88. Prove the integration formula

$$\int \sin^{-1} x \, dx = x \sin^{-1} x + \sqrt{1 - x^2} + C$$

(a) by applying integration by parts to $\int \sin^{-1} x \, dx$;

(b) by differentiating $\sqrt{1 - x^2} + x \sin^{-1} x$.

89. Prove the integration formula

$$\int \tan^{-1} x \, dx = x \tan^{-1} x - \frac{1}{2} \ln(x^2 + 1) + C$$

(a) by applying integration by parts to $\int \tan^{-1} x \, dx$;

(b) by differentiating $x \tan^{-1} x - \frac{1}{2} \ln(x^2 + 1)$.

90. Show that choosing a different antiderivative $v + C$ of dv will yield an equivalent formula for integration by parts, as follows:

(a) Explain why what we need to show is that $uv - \int v \, du = u(v + C) - \int (v + C) \, du$.

(b) Rewrite the equation from part (a), substituting $u = u(x)$, $v = v(x)$, and $du = u'(x) \, dx$.

(c) Prove the equality you wrote down in part (b).

Thinking Forward

Volumes of surfaces of revolution: Consider the region between the graph of $f(x) = \ln x$ and the x-axis on $[1, 3]$ shown next at the left. If we revolve this region around the x-axis, then we get the three-dimensional solid shown at the right. In the problems that follow you will investigate two different methods of finding the volume of this solid.

▶ One way to think of this volume is as an accumulation of disks as x varies from 1 to 3, as shown next at the left. The disk at a given $x \in [1, 3]$ has radius $r = f(x)$ and thus area $\pi (f(x))^2$. As we will see in Section 9.1, the definite integral

$$\pi \int_1^3 (f(x))^2 \, dx$$

calculates the volume of the solid. Use integration by parts to calculate this volume.

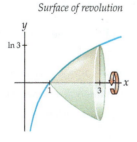

$f(x) = \ln x$ on $[1, 3]$ *Surface of revolution*

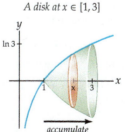

A disk at $x \in [1, 3]$ A shell at $y \in [0, \ln 3]$

▶ Another way of thinking of the volume in question is as an accumulation of nested open cylinders, or "shells," as y varies from 0 to $\ln 3$, as shown previously at the right. The shell at a given $y \in [0, \ln 3]$ has radius $r = y$ and length $h = 3 - f^{-1}(y)$, and thus surface area $2\pi y(3 - f^{-1}(y))$, where $f^{-1}(y) = e^y$ here is the inverse function of $f(x) = \ln x$. As we will see in Section 9.2, the definite integral

$$2\pi \int_0^{\ln 3} y\,(3 - f^{-1}(y))\,dy$$

calculates the volume of the same solid shown in the previous problem. Use integration by parts to calculate this volume.

▶ Since we just calculated the same volume two different ways, obviously we should have obtained the same answer both times. Verify that this is the case for the previous two volume calculations.

Centroid of a shape: Suppose you cut a shape out of a piece of cardboard and attempt to balance the shape on the end of a pencil. The point in the interior at which it balances perfectly is called the **centroid**. If f is positive and continuous on $[a, b]$, then the centroid of the region between the graph of $f(x)$ and the x-axis on $[a, b]$ is the point $(\bar{x}, \bar{y})$, where

$$\bar{x} = \frac{\int_a^b x f(x)\, dx}{\int_a^b f(x)\, dx} \quad \text{and} \quad \bar{y} = \frac{(1/2)\int_a^b (f(x))^2\, dx}{\int_a^b f(x)\, dx}.$$

▶ Find the coordinates of the centroid of the region between the graph of $f(x) = x^2$ and the x-axis on $[0, 3]$.

▶ Find the coordinates of the centroid of the region between $f(x) = \cos x$ and the x-axis on $[-\pi/2, \pi/2]$.

8.3 PARTIAL FRACTIONS AND OTHER ALGEBRAIC TECHNIQUES

▶ Writing improper rational functions by using polynomial long division

▶ The method and form of partial fractions

▶ Rewriting quadratics by the method of completing the square

Proper and Improper Rational Functions

The methods of integration by substitution and integration by parts are useful, but they can't help us solve every integration problem. Sometimes what we need is some good old-fashioned algebra. We have already used algebra extensively as part of our integration toolbox; for example the integral $\int \frac{x^3+1}{x}\, dx$ is easy to solve if we first apply algebra to rewrite it as $\int \left(x^2 + \frac{1}{x}\right) dx$. In this section we will examine some fancier algebraic tools for rewriting and simplifying integrals that involve rational functions.

Let's start by considering the integral

$$\int \frac{3x^3 + x^2 - 14x - 5}{x^2 - 5}\, dx.$$

The integrand is an example of an **improper** rational function, meaning that the degree of its numerator is greater than or equal to the degree of its denominator. It turns out that for many integration problems it is easier to work with **proper** rational functions, that is, rational functions for which the degree of the numerator is strictly less than the degree of the denominator.

Luckily, in much the same way that an improper fraction like $\frac{3}{2}$ can be written as the sum of an integer and a proper fraction, namely, $1 + \frac{1}{2}$, we can rewrite any improper rational function as the sum of a polynomial and a proper rational function. As we will see in Example 1, we have

$$\int \frac{3x^3 + x^2 - 14x - 5}{x^2 - 5}\, dx = \int \left((3x + 1) + \frac{x}{x^2 - 5}\right) dx.$$

This puts the integral into a form that we know how to deal with, since $3x + 1$ is easy to integrate and $\frac{x}{x^2-5}$ can be integrated with the simple substitution of $u = x^2 - 5$, $du = 2x\,dx$.

It is easy to combine fractions to show that $(3x + 1) + \frac{x}{x^2-5}$ is equal to $\frac{3x^3+x^2-14x-5}{x^2-5}$. But of course, what we need to be able to do is go the other way and turn an improper rational function into the sum of a polynomial and a proper rational function. As we saw in Section 4.1, the technique that will do this is polynomial long division. The polynomial long-division algorithm provides a method to divide any polynomial $p(x)$ by another polynomial $q(x)$. This algorithm always terminates with a remainder $r(x)$ whose degree is strictly less than that of $q(x)$, enabling us to rewrite any improper rational function as the sum of a polynomial and a proper rational function:

THEOREM 8.11

Rewriting Improper Rational Functions

Suppose $\frac{p(x)}{q(x)}$ is an improper rational function where $p(x)$ and $q(x)$ are polynomial functions with $\deg(p(x)) = n$ and $\deg(q(x)) = m < n$. Then

$$\frac{p(x)}{q(x)} = s(x) + \frac{r(x)}{q(x)},$$

for some polynomial $s(x)$ of degree $n - m$ and some proper rational function $\frac{r(x)}{q(x)}$ with $\deg(r(x)) < m$.

Partial Fractions

Let's think for a moment about the types of proper rational functions that are easy for us to integrate. Here are some examples of integrals that we can solve:

$$\int \frac{1}{x-3}\,dx, \quad \int \frac{1}{(x-3)^4}\,dx, \quad \int \frac{1}{x^2+1}\,dx, \quad \int \frac{x}{x^2+1}\,dx, \quad \int \frac{x}{(x^2+1)^3}\,dx.$$

The first is an integral involving a logarithm, the second can be done with a simple linear substitution and a power rule, the third concerns the inverse tangent, and the fourth and fifth can be solved after the substitution $u = x^2 + 1$ and $du = 2x\,dx$. There are many other examples, but in these five we see a pattern: Each integrand shown has a denominator that is some power of a linear or quadratic function.

What about more complicated rational functions? For example, we do not yet have a method that helps us solve the integral

$$\int \frac{8}{(x-1)^2(x^2+3)}\,dx.$$

What would be helpful at this point is if we could use algebra to rewrite the integrand $\frac{8}{(x-1)^2(x^2+3)}$ in terms of proper rational functions that we do know how to integrate, that is, in terms of proper rational functions whose denominators are powers of linear or quadratic functions. Obvious candidates for these denominators in this example are $x - 1$, $(x - 1)^2$, and $x^2 + 3$, since those are factors of the denominator of our integrand.

As you will see in Example 3, we can rewrite our integral as follows:

$$\int \frac{8}{(x-1)^2(x^2+3)}\,dx = \int \left(\frac{-1}{x-1} + \frac{2}{(x-1)^2} + \frac{x-1}{x^2+3} \right) dx.$$

Again note that it is easy to verify that this decomposition works, simply by combining the three terms on the right side to show that they are equal to our original proper rational function on the left. The question is how to find this rewritten form in the first place! We will examine two different ways to do that in Examples 3 and 4. Note that we know how to integrate each of the three parts of the decomposition; the first uses a logarithm, the second is a power rule calculation, and the third can be split into an integral we can integrate by substitution and an integral involving the inverse tangent.

It turns out that this type of rewriting is always possible if we use what are called **partial fractions**:

THEOREM 8.12

Rewriting Proper Rational Functions with Partial Fractions

Every proper rational function $\frac{p(x)}{q(x)}$ can be written as a sum of **partial fractions**, as follows:

(a) If $q(x)$ has a linear factor $x - c$ with multiplicity m, then for some constants $A_1, A_2, \ldots, A_m$, the sum will include terms of the form

$$\frac{A_1}{x - c} + \frac{A_2}{(x - c)^2} + \cdots + \frac{A_m}{(x - c)^m}.$$

(b) If $q(x)$ has an irreducible quadratic factor $x^2 + bx + c$ with multiplicity m, then for some constants $B_1, B_2, \ldots, B_m$ and $C_1, C_2, \ldots, C_m$, the sum will include terms of the form

$$\frac{B_1 x + C_1}{x^2 + bx + c} + \frac{B_2 x + C_2}{(x^2 + bx + c)^2} + \cdots + \frac{B_m x + C_m}{(x^2 + bx + c)^m}.$$

Proving that such a decomposition always exists requires some detailed algebra that is not within the scope of this course, so we will omit the proof. Such general theory is not our concern here; we are interested only in finding these partial-fraction decompositions in particular examples. Given a particular proper rational function $\frac{r(x)}{q(x)}$, we need only factor $q(x)$ and then set up systems of equations to solve for the constants A_i or B_i and C_i. Theorem 8.12 guarantees that in each case we will be able to succeed in finding such a decomposition.

Algebraic Techniques

Various algebraic techniques that we covered in Section 4.1 can be useful when integrating rational functions. In particular, you might find it worthwhile to review the Integer Root Theorem and the synthetic division algorithm, because those techniques are useful for factoring denominators before applying partial fractions. You should also review the method of polynomial long division, for writing improper rational functions in terms of polynomials and proper rational functions.

Another technique that can be useful when attempting to integrate rational functions is the method known as **completing the square**. The goal of this method is to write a quadratic of the form $f(x) = x^2 + bx + c$ in the form $(x - k)^2 + C$ for some numbers k and C; note that this rewritten form has a perfect square in it, from where the method gets its name. The completed-square form is sometimes more suitable for integration problems.

THEOREM 8.13

Completing the Square

Every quadratic function $x^2 + bx + c$ can be rewritten in the form

$$x^2 + bx + c = (x - k)^2 + C,$$

where $k = -\frac{b}{2}$ and $C = c - \frac{b^2}{4}$.

In practice, we do not actually memorize the forms of k and C in this theorem. Instead, when confronted with a quadratic $x^2 + bx + c$, we simply add and subtract the quantity $(b/2)^2$ and then rewrite in the form of a perfect square. For example, the quadratic $x^2 + 6x + 10$ has

$b = 6$ and thus $(b/2)^2 = 9$; to complete the square we calculate

$$x^2 + 6x + 10 = (x^2 + 6x + 10) + 9 - 9 = (x^2 + 6x + 9) + (10 - 9) = (x + 3)^2 + 1.$$

The proof of Theorem 8.13 is precisely like this calculation; we leave the details to Exercise 63. This particular example is used in an integration problem in Example 5.

You have most likely noticed by now that many integrals need some kind of rewriting before we can antidifferentiate or apply substitution or integration by parts. Deciding what kind of algebra to apply is a skill that can be developed only by doing many practice problems. One thing to keep in mind as you work through such problems is that when we use algebra it is always our goal to rewrite the integrand in a form that is easier to integrate than the original. This means that we don't just apply algebra because we *can*; we do it to satisfy a certain goal. Make sure you have such a goal in mind before you apply any algebra.

Examples and Explorations

EXAMPLE 1

Using polynomial long division to rewrite an improper rational function

Solve the integral

$$\int \frac{3x^3 + x^2 - 14x - 5}{x^2 - 5} \, dx$$

by writing the improper rational function in the integrand as the sum of a polynomial and a proper rational function.

SOLUTION

Following the algorithm for polynomial long division to divide $p(x) = 3x^3 + x^2 - 14x - 5$ by $q(x) = x^2 - 5$, we have

$$
\begin{array}{r}
3x + 1 \\
x^2 + 0x - 5 \,\overline{)\, 3x^3 + x^2 - 14x - 5} \\
-3x^3 + 0x^2 + 15x \\
\hline
x^2 + x - 5 \\
-x^2 + 0x + 5 \\
\hline
x
\end{array}
$$

We know to stop the algorithm at this point because the degree of the remainder polynomial x is less than the degree of the divisor polynomial $q(x) = x^2 + 5$. Using the result of this calculation, we can rewrite our integral and then perform the integration steps:

$$\int \frac{3x^3 + x^2 - 14x - 5}{x^2 - 5} \, dx = \int \left((3x + 1) + \frac{x}{x^2 - 5} \right) dx \quad \leftarrow \text{polynomial long division}$$

$$= \frac{3}{2}x^2 + x + \frac{1}{2} \ln |x^2 - 5| + C. \quad \leftarrow \text{substitution } u = x^2 - 5 \;\square$$

CHECKING THE ANSWER

As usual, we can check our work by differentiating. To verify that we arrive back at the original integrand, we will also have to do some algebra:

$$\frac{d}{dx}\left(\frac{3}{2}x^2 + x + \frac{1}{2} \ln |x^2 - 5| \right) = \left(\frac{3}{2} \right) 2x + 1 + \left(\frac{1}{2} \right) \frac{1}{x^2 - 5}(2x) \quad \leftarrow \text{differentiate}$$

$$= \frac{(3x + 1)(x^2 - 5)}{x^2 - 5} + \frac{x}{x^2 - 5} \quad \leftarrow \text{algebra}$$

$$= \frac{3x^3 - 15x + x^2 - 5 + x}{x^2 - 5} \quad \leftarrow \text{algebra}$$

$$= \frac{3x^3 + x^2 - 14x - 5}{x^2 - 5}. \quad \leftarrow \text{algebra}$$

EXAMPLE 2   **Determining the form of a partial-fraction decomposition**

Write each of the proper rational functions that follow as a sum of partial fractions with coefficients A_i, B_i, and/or C_i. Do not solve for these coefficients; just write out the form of the partial-fraction decomposition.

(a) $\dfrac{2x}{(x-1)(3x^2-2x-1)}$ (b) $\dfrac{1}{x^3-6x^2+11x-6}$ (c) $\dfrac{x^2-1}{x^7+2x^5+x^3}$

SOLUTION

(a) To find the partial-fraction decomposition we must first split the denominator $(x-1)(3x^2-2x-1)$ into linear and irreducible quadratic factors. Although the denominator is currently expressed as a product of a linear and a quadratic function, the quadratic $3x^2-2x-1$ is not irreducible and in fact can be factored as $(x-1)(3x+1)$. Therefore we have

$$\frac{2x}{(x-1)(3x^2-2x-1)} = \frac{2x}{(x-1)(x-1)(3x+1)} = \frac{2x}{(x-1)^2(3x+1)}.$$

By Theorem 8.12, the right-hand side has partial-fraction decomposition of the form

$$\frac{2x}{(x-1)^2(3x+1)} = \frac{A_1}{x-1} + \frac{A_2}{(x-1)^2} + \frac{A_3}{3x+1}.$$

(b) In this case the factorization of the denominator is not particularly easy, but by guessing and verifying that $x=1$ is a root (since $1^3-6(1)^2+11(1)-6=0$) and then using either synthetic division or polynomial long division, we see that

$$\frac{1}{x^3-6x^2+11x-6} = \frac{1}{(x-1)(x^2-5x+6)} = \frac{1}{(x-1)(x-2)(x-3)}.$$

From the right-hand fraction we obtain the partial-fraction decomposition

$$\frac{1}{(x-1)(x-2)(x-3)} = \frac{A_1}{x-1} + \frac{A_2}{x-2} + \frac{A_3}{x-3}.$$

(c) Again we must begin by factoring the denominator:

$$\frac{x^2-1}{x^7+2x^5+x^3} = \frac{x^2-1}{x^3(x^4+2x^2+1)} = \frac{x^2-1}{x^3(x^2+1)^2}.$$

The fraction to the right of the equals sign can be written as a partial-fraction decomposition of the form

$$\frac{x^2-1}{x^3(x^2+1)^2} = \frac{A_1}{x} + \frac{A_2}{x^2} + \frac{A_3}{x^3} + \frac{B_1x+C_1}{x^2+1} + \frac{B_2x+C_2}{(x^2+1)^2}.$$

□

EXAMPLE 3   **Solving for the coefficients in a partial-fractions decomposition**

Solve for the coefficients A_1, A_2, B, and C that satisfy the partial-fraction decomposition

$$\frac{8}{(x-1)^2(x^2+3)} = \frac{A_1}{x-1} + \frac{A_2}{(x-1)^2} + \frac{Bx+C}{x^2+3}.$$

SOLUTION

By using a common denominator to combine fractions on the right-hand side of the equation, we have

$$\frac{8}{(x-1)^2(x^2+3)} = \frac{A_1(x-1)(x^2+3) + A_2(x^2+3) + (Bx+C)(x-1)^2}{(x-1)^2(x^2+3)}.$$

After some tedious algebra to expand and then collect terms in the numerator according to the degrees of x, we can show that the fraction on the right is equal to

$$\frac{(A_1 + B)x^3 + (-A_1 + A_2 - 2B + C)x^2 + (3A_1 + B - 2C)x + (-3A_1 + 3A_2 + C)}{(x-1)^2(x^2+3)}.$$

Since this fraction must be equal to our original proper rational function $\frac{8}{(x-1)^2(x^2-3)}$, its numerator must be equal to 8. This means that the terms in the numerator have to be equal to the terms in $0x^3 + 0x^2 + 0x + 8$. Equating the corresponding coefficients of these terms gives us a system of equations that we can solve for A_1, A_2, B, and C:

$$\begin{cases} A_1 + B = 0 \\ -A_1 + A_2 - 2B + C = 0 \\ 3A_1 + B - 2C = 0 \\ -3A_1 + 3A_2 + C = 8 \end{cases} \implies \begin{cases} A_1 = -B \\ A_2 = 2B \\ C = -B \\ 8B = 8 \end{cases} \implies \begin{cases} A_1 = -1 \\ A_2 = 2 \\ C = -1 \\ B = 1. \end{cases}$$

Putting these coefficients back into the partial-fractions decomposition, we have

$$\frac{8}{(x-1)^2(x^2+3)} = \frac{-1}{x-1} + \frac{2}{(x-1)^2} + \frac{x-1}{x^2+3}.$$

EXAMPLE 4   **Another way to solve for the coefficients in a partial-fractions decomposition**

Solve for the coefficients A_1, A_2, and A_3 that satisfy the partial-fraction decomposition

$$\frac{1}{(x-1)(x-2)(x-3)} = \frac{A_1}{x-1} + \frac{A_2}{x-2} + \frac{A_3}{x-3}.$$

SOLUTION

Again using a common denominator to combine fractions, we have

$$\frac{1}{(x-1)(x-2)(x-3)} = \frac{A_1(x-2)(x-3) + A_2(x-1)(x-3) + A_3(x-1)(x-2)}{(x-1)(x-2)(x-3)}.$$

At this point we could solve for the coefficients A_1, A_2, and A_3 by the same method we used in the previous example: expanding the numerator and collecting terms according to powers of x. However, for partial fractions with many linear factors there is a faster way, which we outline here.

In order for this partial-fractions decomposition to be valid, we must have $A_1(x-2)(x-3) + A_2(x-1)(x-3) + A_3(x-1)(x-2) = 1$ for all values of x. By examining this equality for clever values of x, namely, those which make certain factors equal to zero, we can quickly solve for the coefficients A_i. For example, when $x = 1$, all of the $x - 1$ terms are zero and thus we must have

$$A_1(-1)(-2) + A_2(0)(-2) + A_3(0)(-1) = 1 \implies 2A_1 = 1 \implies A_1 = \frac{1}{2}.$$

Now, for $x = 2$, we have

$$A_1(0)(-1) + A_2(-1)(1) + A_3(1)(0) = 1 \implies -A_2 = 1 \implies A_2 = -1.$$

And finally, for $x = 3$, we have

$$A_1(1)(0) + A_2(2)(0) + A_3(2)(1) = 1 \implies 2A_3 = 1 \implies A_3 = \frac{1}{2}.$$

Therefore the partial-fraction decomposition is

$$\frac{1}{(x-1)(x-2)(x-3)} = \frac{1/2}{x-1} + \frac{-1}{x-2} + \frac{1/2}{x-3}.$$

Notice that each of the terms in this decomposition is easy to integrate. In fact we can very quickly compute the integral to be $\frac{1}{2}\ln|x-1| - \ln|x-2| + \frac{1}{2}\ln|x-3| + C$.

CHECKING THE ANSWER

It is easy to check the answers to the previous two examples by using common denominators to combine the terms in the partial-fraction decompositions. For example,

$$\frac{1/2}{x-1} + \frac{-1}{x-2} + \frac{1/2}{x-3} = \frac{(1/2)(x-2)(x-3) - (x-1)(x-3) + (1/2)(x-1)(x-2)}{(x-1)(x-2)(x-3)}$$

$$= \frac{(1/2)(x^2 - 5x + 6) - (x^2 - 4x + 3) + (1/2)(x^2 - 3x + 2)}{(x-1)(x-2)(x-3)}$$

$$= \frac{1}{(x-1)(x-2)(x-3)}.$$

EXAMPLE 5   Combining algebraic techniques to integrate a rational function

Solve the integral $\displaystyle\int \frac{x^3 + 4x^2 - 21}{x^2 + 6x + 10}\, dx$.

SOLUTION

Since neither substitution nor integration by parts seems to be a useful approach to solving this integral, and there is no obvious algebraic simplification trick, we will begin by rewriting the improper rational integrand with the use of polynomial long division. After using the polynomial long-division algorithm to divide $x^3 + 4x^2 - 21$ by $x^2 + 6x + 10$, we obtain

$$\int \frac{x^3 + 4x^2 - 21}{x^2 + 6x + 10}\, dx = \int \left((x-2) + \frac{2x-1}{x^2 + 6x + 10} \right) dx.$$

Obviously the $x - 2$ portion of the integrand is easy to deal with, so let's focus on the proper rational function $\frac{2x-1}{x^2+6x+10}$. Substitution does not seem to be immediately useful here, so we might try partial fractions. However, $x^2 + 6x + 10$ is an irreducible quadratic, since its discriminant $b^2 - 4ac = 36 - 4(1)(10) = -4$ is negative; thus, because the denominator does not factor any further, we cannot decompose this rational function into partial fractions.

Let's go back to substitution for a moment. The obvious choice to try is $u = x^2 + 6x + 1$, which gives $du = (2x + 6)\, dx$. This du is not exactly what we need for our numerator of $2x - 1$, but we can rewrite the numerator by adding and subtracting 6 so that the integral becomes

$$\int (x-2)\, dx + \int \frac{2x+6}{x^2 + 6x + 10}\, dx + \int \frac{-7}{x^2 + 6x + 10}\, dx.$$

Now the first two terms are easy to integrate, so we need only to deal with the third. We can use the method of completing the square to rewrite this third integral in a form whose antiderivative is related to the inverse tangent function. By the example following Theorem 8.13, we have $x^2 + 6x + 10 = (x+3)^2 + 1$. Therefore the integral expression we are working with is equal to

$$\int (x-2)\, dx + \int \frac{2x+6}{x^2 + 6x + 10}\, dx + \int \frac{-7}{(x+3)^2 + 1}\, dx.$$

Finally, with the substitution $u = x^2 + 6x + 10$ and $du = (2x + 6)\, dx$ in the second integral, and the substitution $w = x + 3$ and $dw = dx$ in the third, we can integrate to obtain

$$\frac{1}{2}x^2 - 2x + \ln|x^2 + 6x + 10| - 7\tan^{-1}(x+3) + C.$$

CHECKING THE ANSWER

We can check by differentiating the function $\frac{1}{2}x^2 - 2x + \ln|(x+3)^2 + 1| - 7\tan^{-1}(x+3)$ and algebraically simplifying to get

$$x - 2 + \frac{2(x+3)}{(x+3)^2 + 1} - \frac{7}{(x+3)^2 + 1} = \frac{(x-2)((x+3)^2 + 1) + 2(x+3) - 7}{(x+3)^2 + 1}$$

$$= \frac{x^3 + 4x^2 - 21}{x^2 + 6x + 10}.$$

TEST YOUR UNDERSTANDING

▶ Is $\frac{x^2+1}{x^2-1}$ a proper rational function or an improper rational function? Why?

▶ What is the form for the partial-fraction decomposition of the rational function $\frac{x^2-4}{(x-1)^6}$? What about $\frac{x^2-4}{(x^2+x+1)^6}$?

▶ What are two ways that we could solve for the coefficients in a partial-fractions decomposition?

▶ Why is the method of partial fractions a useful technique for solving some integrals? What do we hope to gain by rewriting an integrand into the form of a partial-fraction decomposition?

▶ What does it mean to complete the square of a quadratic expression?

EXERCISES 8.3

Thinking Back

Factoring polynomials: Factor each of the given polynomials into a product of linear and irreducible quadratic factors. For the third and fourth polynomials you will have to guess a root and use synthetic division or other factoring methods.

▶ $x^3 - 2x^2 - x + 2$

▶ $x^3 - 4x^2 + x + 6$

▶ $x^3 - 1$

▶ $x^3 - x^2 - 5x - 3$

Polynomial long division: Write each of the following rational functions as the sum of a polynomial function and a proper rational function.

▶ $\dfrac{2x^3 - 9x^2 + 7x + 6}{2x + 1}$

▶ $\dfrac{(x-1)(x^2 - 2x - 2)}{x - 3}$

▶ $\dfrac{2x^4 - 8x^3 + 9x - 4}{x^2 - 4x + 1}$

▶ $\dfrac{2x^4 - 3x^3 + x^2 - 4x - 5}{x^3 + 2x + 1}$

Integration: Calculate the following integrals.

▶ $\displaystyle\int \frac{1}{(x-1)^2}\,dx$

▶ $\displaystyle\int \frac{x-1}{x^2-1}\,dx$

▶ $\displaystyle\int \frac{1}{x^2+1}\,dx$

▶ $\displaystyle\int \frac{x}{x^2+1}\,dx$

▶ $\displaystyle\int \frac{x^2+1}{x^3}\,dx$

▶ $\displaystyle\int \frac{x+1}{x^2+1}\,dx$

Concepts

0. *Problem Zero:* Read the section and make your own summary of the material.

1. *True/False:* Determine whether each of the statements that follow is true or false. If a statement is true, explain why. If a statement is false, provide a counterexample.

(a) *True or False:* $f(x) = \frac{x+1}{x-1}$ is a proper rational function.

(b) *True or False:* Every improper rational function can be expressed as the sum of a polynomial and a proper rational function.

(c) *True or False:* After polynomial long division of $p(x)$ by $q(x)$, the remainder $r(x)$ has a degree strictly less than the degree of $q(x)$.

(d) *True or False:* Polynomial long division can be used to divide two polynomials of the same degree.

(e) *True or False:* If a rational function is improper, then polynomial long division must be applied before using the method of partial fractions.

(f) *True or False:* The partial-fraction decomposition of $\frac{x^2+1}{x^2(x-3)}$ is of the form $\frac{A}{x^2} + \frac{B}{x-3}$.

(g) *True or False:* The partial-fraction decomposition of $\frac{x^2+1}{x^2(x-3)}$ is of the form $\frac{Bx+C}{x^2} + \frac{A}{x-3}$.

(h) *True or False:* Every quadratic function can be written in the form $A(x-k)^2 + C$.

2. *Examples:* Construct examples of the thing(s) described in the following. Try to find examples that are different than any in the reading.

(a) A rational function that is its own partial-fraction decomposition.

(b) A rational function whose partial-fraction decomposition has five terms.

(c) A rational function $\frac{p(x)}{q(x)}$ that is of the form $(3x + 1) + \frac{1}{x^2-1}$ after polynomial long division.

3. What is a rational function? What does it mean for a rational function to be proper? Improper?

4. Explain how to use long division to write the improper fraction $\frac{172}{5}$ as the sum of an integer and a proper fraction.

5. Suppose you use polynomial long division to divide $p(x)$ by $q(x)$, and after doing your calculations you end up with the polynomial $x^2 - x + 3$ as the quotient above the top line, and the polynomial $3x - 1$ at the bottom as the remainder. Then $p(x) =$ _____ and $\frac{p(x)}{q(x)} =$ _____ .

6. Describe two ways in which the long-division algorithm for polynomials is similar to the long-division algorithm for integers and then two ways in which the two algorithms are different.

7. Show that the equation $p(x) = q(x)m(x) + R(x)$ is equivalent to the equation $\frac{p(x)}{q(x)} = m(x) + \frac{R(x)}{q(x)}$ for all x for which $q(x)$ is nonzero.

8. Verify that $(x^2 - 1) + \frac{x^2}{x^4 + 3x^2 + 4}$ is a valid result for the polynomial long-division calculation performed to divide the polynomial $x^6 + 2x^4 + 2x^2 - 4$ by the polynomial $x^4 + 3x^2 + 4$.

In Exercises 9–11, suppose that you apply polynomial long division to divide a polynomial $p(x)$ by a polynomial $q(x)$ with the goal of obtaining an expression of the form

$$\frac{p(x)}{q(x)} = m(x) + \frac{R(x)}{q(x)}.$$

9. If the degree of $p(x)$ is two greater than the degree of $q(x)$, what can you say about the degrees of $m(x)$ and $R(x)$?

10. If the degree of $p(x)$ is equal to the degree of $q(x)$, what can you say about the degrees of $m(x)$ and $R(x)$?

11. If the degree of $p(x)$ is less than the degree of $q(x)$, what happens? Does polynomial long division work? Why or why not?

In Exercises 12–14, suppose that you want to obtain a partial-fraction decomposition of a rational function $\frac{p(x)}{q(x)}$ according to Theorem 8.12.

12. What happens if the degree of $p(x)$ is greater than or equal to the degree of $q(x)$? Try to perform a partial-fraction decomposition of the improper rational function $\frac{x^3}{(x-1)(x-2)}$, and see what goes wrong.

13. If $q(x)$ is an irreducible quadratic, what can you say about its partial-fraction decomposition? What if $q(x)$ is a reducible quadratic? Consider the functions $q(x) = x^2 + 1$ and $q(x) = (x-1)(x-2)$ to find your answer.

14. Why do we need to put linear numerators of the form $B_i x + C_i$ in every term of a partial-fraction decomposition which involves irreducible quadratics? Think about the example $\frac{p(x)}{q(x)} = \frac{1}{(x^2+1)^2}$, and figure out what goes wrong if we attempt to make a decomposition of the form $\frac{C_1}{x^2+1} + \frac{C_2}{(x^2+1)^2}$.

15. Use the method of completing the square and what we know about transformations of graphs to show that the vertex of the parabola $f(x) = x^2 - 3x + 5$ is the point $(3/2, 11/4)$.

16. Theorem 8.13 tells us that if a quadratic $ax^2 + bx + c$ is *monic*, meaning that its leading coefficient is $a = 1$, then we can express that quadratic in the form $(x-k)^2 + C$ with $k = -\frac{b}{2}$ and $C = c - \frac{b^2}{4}$. What happens in the general case for non-monic quadratics? If $ax^2 + bx + c = (Ax - k)^2 + C$, then how can we express A, k and C in terms of a, b, and c?

Skills

Calculate each of the integrals in Exercises 17–46. For some integrals you may need to use polynomial long division, partial fractions, factoring or expanding, or the method of completing the square.

17. $\displaystyle\int \frac{3}{(x-2)(x+1)}\, dx$

18. $\displaystyle\int \frac{6(x+1)}{x(x-3)}\, dx$

19. $\displaystyle\int \frac{x+1}{(x-1)^2}\, dx$

20. $\displaystyle\int \frac{x+1}{(x-1)^3}\, dx$

21. $\displaystyle\int \frac{(x+3)(x-2)}{x^2}\, dx$

22. $\displaystyle\int \frac{x^2}{(x+3)(x-2)}\, dx$

23. $\displaystyle\int \frac{5x^3}{(x+3)(x-2)}\, dx$

24. $\displaystyle\int \frac{25x^2}{(x+3)(x-2)^2}\, dx$

25. $\displaystyle\int \frac{x}{(x^2+3)(x-2)}\, dx$

26. $\displaystyle\int \frac{7(x^2+1)}{(x^2+3)(x-2)}\, dx$

27. $\displaystyle\int \frac{3+3x}{x^3-1}\, dx$

28. $\displaystyle\int \frac{3x^3}{x^3-1}\, dx$

29. $\displaystyle\int \frac{2(x-1)}{x^2-4x+5}\, dx$

30. $\displaystyle\int \frac{2(x-1)}{x^2-2x+5}\, dx$

31. $\displaystyle\int \frac{3x^2-13x+13}{(x-2)^3}\, dx$

32. $\displaystyle\int \frac{x+7}{x^2-x-2}\, dx$

33. $\displaystyle\int \frac{-5x-5}{3x^2-8x-3}\, dx$

34. $\displaystyle\int \frac{x^2+4x+1}{x^3+x^2}\, dx$

35. $\displaystyle\int \frac{x^2+2x+3}{x^4+4x^2+3}\, dx$

36. $\displaystyle\int \frac{x^2+x+1}{(x^2+1)^2}\, dx$

37. $\displaystyle\int \frac{1-2x}{x^2(x^2+1)}\, dx$

38. $\displaystyle\int \frac{16x^2(x^2+1)}{1-2x}\, dx$

39. $\displaystyle\int \frac{2x^2+4x}{x^3+x^2+x+1}\, dx$

40. $\displaystyle\int \frac{x^3}{x^3+x^2+x+1}\, dx$

41. $\displaystyle\int \frac{2x(x^2-1)}{x^2-4}\, dx$

42. $\displaystyle\int \frac{(x-1)(7x-11)}{(x-2)(x-3)(x+1)}\, dx$

43. $\displaystyle\int \frac{2x^3 - 9x^2 + 7x + 6}{2x + 1}\, dx$

44. $\displaystyle\int \frac{(x - 1)(x^2 - 2x - 2)}{x - 3}\, dx$

45. $\displaystyle\int \frac{2x^4 - 8x^3 + 9x - 4}{x^2 - 4x + 1}\, dx$

46. $\displaystyle\int \frac{2x^4 - 3x^3 + x^2 - 4x - 5}{x^3 + 2x + 1}\, dx$

Calculate each of the definite integrals in Exercises 47–52. Some integrals require partial fractions or polynomial long division, and some do not.

47. $\displaystyle\int_3^4 \frac{4}{x^2 - 4}\, dx$

48. $\displaystyle\int_{-1}^1 \frac{4(x^2 - 1)}{x^2 - 4}\, dx$

49. $\displaystyle\int_1^3 \frac{1}{x(x + 1)}\, dx$

50. $\displaystyle\int_1^3 \frac{1}{x^2(x + 1)}\, dx$

51. $\displaystyle\int_{-1}^1 \frac{x^2 - 2x - 4}{x - 3}\, dx$

52. $\displaystyle\int_1^2 \frac{1}{x(x + 2)(x - 3)}\, dx$

Calculate each of the integrals in Exercises 53–56. Each integral requires substitution or integration by parts as well as the algebraic methods described in this section.

53. $\displaystyle\int \frac{\sin x}{\cos x + \cos^2 x}\, dx$

54. $\displaystyle\int \frac{e^x}{e^{3x} - 2e^{2x}}\, dx$

55. $\displaystyle\int \frac{\ln x}{x(1 + \ln x)^2}\, dx$

56. $\displaystyle\frac{\ln x + 1}{x((\ln x)^2 - 4)}$

57. Consider the function $f(x) = \dfrac{1}{x^2 - 4}$.

(a) Find the signed area between the graph of $f(x)$ and the x-axis on $[-1, 1]$ shown next at the left.

(b) Find the area between the graph of $f(x)$ and the graph of $g(x) = \dfrac{1}{4}(x - 1)$ on $[-1, 1]$ shown next at the right.

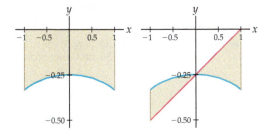

58. Consider the function $f(x) = \dfrac{2}{x^3 - 1} + 1$.

(a) Find the signed area of the region between the graph of $f(x)$ and the x-axis on $\left[-2, \dfrac{1}{2}\right]$ shown in the figure.

(b) Find the absolute area of the same region.

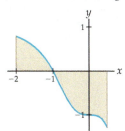

59. Consider the function $f(x) = \dfrac{2x^3 + x}{(x^2 + 1)^2}$.

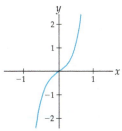

(a) Find the average value of $f(x)$ on $[0, 2]$.

(b) Graphically approximate a value of x for which $f(x)$ is equal to its average value on $[0, 2]$, and use this value to verify that your answer from part (a) is reasonable.

60. Consider the function $f(x) = \dfrac{16}{(x + 1)(x - 3)^2}$.

(a) Find the area between the graphs of $f(x)$ and $g(x) = 4 - 2x$ on $[0, 2]$ shown next at the left.

(b) Find the area between the graphs of $f(x)$ and $h(x) = \dfrac{4}{x + 1}$ on $[0, 2]$ shown next at the right.

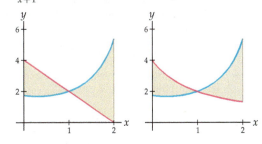

Applications

61. In her work as a population biologist for Idaho Fish and Game, Leila is assigned the task of modeling a population $p(t)$ of prairie dogs that ranchers in the eastern part of the state are complaining about. She uses the so-called **logistic growth model**, which quickly results in an expression of the form

$$r(t) = \int_{p_0}^{p(t)} \frac{1}{q\left(1 - \dfrac{q}{K}\right)}\, dq,$$

where K represents the carrying capacity of the land where the prairie dogs live, r is the rate of reproduction, and p_0 is the initial population of the prairie dogs. Use the Fundamental Theorem of Calculus to calculate $r(t)$ and then use your answer to solve for $p(t)$ in terms of $r(t)$, assuming that p_0 and $p(t)$ both lie between 0 and K.

62. Leila's model for the prairie dogs in the previous exercise was so successful that she was invited to assist a University of Idaho professor in modeling sage grouse populations. In contrast to prairie dogs, sage grouse show evidence of an inverse density dependence: If the number g of grouse falls too low, then the population is no longer sustainable and dies off. This relationship can be represented in a model that satisfies

$$-rt = \int_{g(0)}^{g(t)} \frac{1}{\gamma(1 - \gamma/K)(1 - \gamma/m)} \, d\gamma,$$

where $r > 0$, $0 < m < K$, and $g(t)$ represents the population of grouse at time t.

(a) Evaluate the integral in the case that t satisfies $0 < g(t) < g(0) < m$.

(b) It is difficult to solve for $g(t)$, but given that t satisfies $0 < g(t) < g(0) < m < K$, differentiate both sides of the equation with respect to t and use the Fundamental Theorem of Calculus in solving for $g'(t)$, to show that $\lim_{t \to \infty} g(t) = 0$. You may assume that all of the limits involved exist. This means that if the grouse are hunted to a population lower than m, then the grouse eventually all die off.

Proofs

63. Prove that every quadratic function of the form $f(x) = x^2 + bx + c$ can be written in the form $f(x) = (x - k)^2 + C$, with $k = -\frac{b}{2}$ and $C = c - \frac{b^2}{4}$.

64. Use the method of completing the square and what we know about transformations of graphs to prove that

for all real numbers b and c, the vertex of the parabola with equation $f(x) = x^2 + bx + c$ is located at the point $\left(-\frac{b}{2}, c - \frac{b^2}{4} \right)$.

Thinking Forward

Separable differential equations: Suppose a population $P = P(t)$ grows in such a way that its rate of growth $\frac{dP}{dt}$ obeys the equation

$$\frac{dP}{dt} = P(100 - P).$$

This is called a **differential equation** because it is an equation that involves a derivative. In the series of steps that follow, you will find a function $P(t)$ that behaves according to this differential equation.

▶ A differential equation of this form is **separable** to the following equivalent equation involving integrals:

$$\int \frac{1}{P(100 - P)} \, dP = \int 1 \, dt.$$

▶ Use partial fractions to show that the integral on the left side of this equation is equal to

$$\frac{1}{100} (\ln |P| - \ln |100 - P|) + C_1.$$

▶ Use properties of logarithms to simplify the answer you found in the previous step.

▶ Set the answer from the previous step equal to $\int 1 \, dt = t + C_2$, and solve for $P = P(t)$. Along the way you can combine unknown constants into new constants; for example, if you encounter $C_2 - C_1$, then you could just rename that constant C and proceed from there. At the end of your calculations you should have

$$P(t) = \frac{100Ae^{100t}}{1 + Ae^{100t}}$$

for some constant A.

▶ What happens to the function $P(t)$ as $t \to \infty$? What does that mean about this particular population?

8.4 TRIGONOMETRIC INTEGRALS

▶ Using Pythagorean identities to set up trigonometric integrals for simple u-substitutions

▶ Using trigonometric identities to simplify integrals involving powers of trigonometric functions

▶ The integrals of secant and cosecant

Using Pythagorean Identities to Set Up a Substitution

Integration by substitution is a powerful tool, but as we have seen, it can be difficult to recognize an integrand as being in the form $f'(u(x))u'(x)$. If we are lucky, we can use algebra to rewrite an integrand to set it up for a particular substitution. In the case of trigonometric integrands, this algebra often involves the Pythagorean identities, which we repeat here for convenience:

THEOREM 8.14   **Pythagorean Identities**

For all values of x at which the following equations are defined,

(a) $\sin^2 x + \cos^2 x = 1$ **(b)** $\tan^2 x + 1 = \sec^2 x$ **(c)** $1 + \cot^2 x = \csc^2 x$

It will help you remember these formulas if you notice that the second identity follows from the first by dividing both sides by $\cos^2 x$ and that the third identity follows from the first by dividing both sides by $\sin^2 x$.

With the Pythagorean identities we can freely write even powers of $\sin x$, $\tan x$, and $\cot x$ in terms of even powers of $\cos x$, $\sec x$, and $\csc x$, and vice versa. For example, using the first Pythagorean identity, we can write $\cos^4 x$ in terms of sine as

$$\cos^4 x = (\cos^2 x)^2 = (1 - \sin^2 x)^2.$$

With this equation in mind, consider the integral $\int \cos^5 x\, dx$. As it is written, we do not know how to solve the integral. However, we can convert four of the cosines to sines and have one cosine left over:

$$\int \cos^5 x\, dx = \int \cos^4 x \, \cos x\, dx = \int (\cos^2 x)^2 \, \cos x\, dx = \int (1 - \sin^2 x)^2 \, \cos x\, dx.$$

In this new form we have an integrand that can be easily solved with u-substitution with $u = \sin x$ and $du = \cos x\, dx$:

$$\int (1 - \sin^2 x)^2 \, \cos x\, dx = \int (1 - u^2)^2 \, du = \int (1 - 2u^2 + u^4) \, du$$

$$= u - \frac{2}{3}u^3 + \frac{1}{5}u^5 + C = \sin x - \frac{2}{3}\sin^3 x + \frac{1}{5}\sin^5 x + C.$$

In general, we would like to rewrite trigonometric integrands so that they are expressed in terms of a single trigonometric function multiplied by the derivative of that trigonometric function. Whether or not we can do this for a particular integral depends on the trigonometric functions and powers involved. For example, we can use the method for some integrals of the form $\int \sin^m x \cos^n x\, dx$ when either m or n is odd, since in that case we can convert all but one sine or cosine into the other function and apply a substitution. For instance, we could set up $\int \sin^8 x \cos^3 x\, dx$ for substitution with $u = \sin x$ by writing

$$\int \sin^8 x \cos^3 x\, dx = \int \underbrace{\sin^8 x(1 - \sin^2 x)} \cos x\, dx.$$

We could now perform the substitution $u = \sin x$, which would change the bracketed portion of the integral to simply $u^8(1 - u^2)$, which is easy to multiply out and integrate.

Similarly, we can use this method for some integrals of the form $\int \sec^m x \tan^n x \, dx$. When m is even, all but two of the secants can be converted into tangents. When n is odd, we can save a $\sec x \tan x$ and rewrite the rest of the integrand in terms of the secant. Both of these conditions hold for the integral $\int \sec^6 x \tan^5 x \, dx$, so we can rewrite this integral in two ways:

$$\int \sec^6 x \tan^5 x \, dx = \int \underbrace{(\tan^2 x + 1)^2 \tan^5 x}_{} \, \sec^2 x \, dx,$$

$$\int \sec^6 x \tan^5 x \, dx = \int \underbrace{\sec^5 x (\sec^2 x - 1)^2}_{} \, \sec x \tan x \, dx.$$

In both cases the integral is set up for a simple substitution, with $u = \tan x$ in the first case and $u = \sec x$ in the second case. The bracketed portions of the integrands become functions of u that are easy to antidifferentiate.

The table that follows summarizes some common situations in which we can use Pythagorean identities to set up a simple substitution in a trigonometric integral. In this table it is assumed that m and n are positive integers.

Form	Rewritten Form	Choose
$\int \sin^n x \, dx,\ \ n$ odd	$\int$ *(expression in* $\cos x$*)* $\sin x \, dx$	$u = \cos x,\ du = -\sin x \, dx$
$\int \sin^m x \cos^n x \, dx,\ \ n$ odd	$\int$ *(expression in* $\sin x$*)* $\cos x \, dx$	$u = \sin x,\ du = \cos x \, dx$
$\int \sin^m x \cos^n x \, dx,\ \ m$ odd	$\int$ *(expression in* $\cos x$*)* $\sin x \, dx$	$u = \cos x,\ du = -\sin x \, dx$
$\int \sec^m x \tan^n x \, dx,\ \ m$ even	$\int$ *(expression in* $\tan x$*)* $\sec^2 x \, dx$	$u = \tan x,\ du = \sec^2 x \, dx$
$\int \sec^m x \tan^n x \, dx,\ \ n$ odd	$\int$ *(expression in* $\sec x$*)* $\sec x \tan x \, dx$	$u = \sec x,\ du = \sec x \tan x \, dx$

Similar rows could be made for integrals that involve powers of cosine or products of powers of cosecant and cotangent.

You do not need to memorize the preceding table! Instead, notice that in each case we rewrite the original integral in such a way as to save part of the integrand for the *du* that corresponds to our desired choice of u-substitution. Whether or not we can do this depends whether the exponents involved in the original integral are even or odd. Integrals that are not in one of the forms in the table will require different solution techniques that are the focus of the remainder of this section.

The Pythagorean identities can also be applied in less obvious cases. For example, we can effectively reduce the integral of a power of a tangent by converting two copies of the tangent at a time. As an illustration of this procedure, consider the integral $\int \tan^5 x \, dx$. We can write

$$\tan^5 x = \tan^3 x \tan^2 x = \tan^3 x \, (\sec^2 x - 1) = \tan^3 x \sec^2 x - \tan^3 x.$$

Notice that we have written $\tan^5 x$ as the sum of a function we can integrate with the techniques discussed earlier and a function that is a lower power of the tangent. Repeating this kind of reduction will eventually solve this integral, as you will see in Example 4.

Using Double-Angle Formulas to Reduce Powers

Not all integrals involving trigonometric functions can be solved by using Pythagorean identities and u-substitution. For example, that technique would not work for the integrals $\int \sin^6 x \, dx$, $\int \sin^2 x \cos^2 x \, dx$, and $\int \tan^5 x \, dx$, because of the parity of the exponents involved, that is, whether certain exponents are even or odd. The first two of these integrals can be simplified with the use of the double-angle identities, which we repeat here:

THEOREM 8.15   **Double-Angle Identities**

For all real numbers x,

(a) $\sin^2 x = \dfrac{1}{2}(1 - \cos 2x)$ (b) $\cos^2 x = \dfrac{1}{2}(1 + \cos 2x)$

For example, we can write the integral $\int \sin^2 x\, dx$ as

$$\int \sin^2 x\, dx = \frac{1}{2}\int (1 - \cos 2x)\, dx = \frac{1}{2}\left(x - \frac{1}{2}\sin 2x\right) + C.$$

For higher even powers of sine or cosine we apply double-angle identities repeatedly, reducing the power by half each time. For example, since $\sin^4 x = (\sin^2 x)^2$, we have

$$\sin^4 x = \left(\frac{1}{2}(1 - \cos 2x)\right)^2 = \frac{1}{4}(1 - 2\cos 2x + \cos^2 2x) = \frac{1}{4}\left(1 - 2\cos 2x + \frac{1}{2}(1 + \cos 4x)\right).$$

Although it seems like we made the expression $\sin^4 x$ much more complicated, in fact the messy expression on the right-hand side of the equation is easy to integrate, piece by piece.

Integrating Secants and Cosecants

Believe it or not, at this point we still do not have formulas for antidifferentiating two of the six basic trigonometric functions. To finish our investigation of trigonometric integrals, we give formulas for these integrals:

THEOREM 8.16   **Integrals of Secant and Cosecant**

(a) $\displaystyle\int \sec x\, dx = \ln|\sec x + \tan x| + C$ (b) $\displaystyle\int \csc x\, dx = -\ln|\csc x + \cot x| + C$

Proof. We prove part (a) and leave the proof of part (b) to Exercise 79. Neither integration by substitution nor integration by parts can immediately improve the integral $\int \sec x\, dx$. However, there is a particular technique that happens to apply to this integral. If we multiply the integrand by a certain very special form of 1, we will have an integral that can be evaluated by substitution:

$$\int \sec x\, dx = \int \sec x \left(\frac{\sec x + \tan x}{\sec x + \tan x}\right) dx = \int \frac{\sec^2 x + \sec x \tan x}{\sec x + \tan x}\, dx.$$

It was not at all obvious why we multiplied by $\frac{\sec x + \tan x}{\sec x + \tan x}$ just now. However, now that we have done so, the following choice of substitution works out surprisingly well:

$$\boxed{u = \sec x + \tan x} \quad \Longrightarrow \quad \frac{du}{dx} = \sec x \tan x + \sec^2 x \quad \Longrightarrow \quad \boxed{du = (\sec x \tan x + \sec^2 x)\, dx}$$

With this change of variables, the integral becomes

$$\int \sec x\, dx = \int \frac{1}{u}\, du = \ln|u| + C = \ln|\sec x + \tan x| + C.$$

■

What about integrating powers of $\sec x$ and $\csc x$? Of course, it is easy to integrate $\sec^2 x$, since it is the derivative of $\tan x$. This fact can be exploited when integrating higher powers of $\sec x$, since we will be able to choose $dv = \sec^2 x \, dx$ and integrate it to get $v = \tan x$. For example, consider the integral $\int \sec^4 x \, dx$, with parts chosen as follows:

$$u = \sec^2 x \quad \longrightarrow \quad du = 2 \sec x (\sec x \tan x) \, dx$$
$$v = \tan x \quad \longleftarrow \quad dv = \sec^2 x \, dx.$$

Applying the formula for integration by parts and simplifying a bit, we have

$$\int \sec^4 x \, dx = \sec^2 x \tan x - 2 \int \tan^2 x \sec^2 x \, dx.$$

The new integral on the right can be solved by substitution, with $u = \tan x$ and $du = \sec^2 x \, dx$. This method of choosing parts with $dv = \sec^2 x \, dx$ is effective for simplifying any integrand that is a power of a secant, since it reduces the power each time the method is applied.

Examples and Explorations

EXAMPLE 1

Setting up substitutions with Pythagorean identities

Use algebra and u-substitution to find each of the following integrals:

(a) $\displaystyle\int \sin^3 x \cos^4 x \, dx$ **(b)** $\displaystyle\int \csc^4 x \cot^3 x \, dx$ **(c)** $\displaystyle\int \sec^3 x \tan^3 x \, dx$

SOLUTION

(a) Using the Pythagorean identity $\sin^2 x = 1 - \cos^2 x$, we can write the integrand so that it involves only cosines (the bracketed portion) except for a single copy of $\sin x$:

$$\int \sin^3 x \cos^4 x \, dx = \int \sin^2 x \sin x \cos^4 x \, dx = \int \underbrace{(1 - \cos^2 x) \cos^4 x}\ \sin x \, dx$$

Now with the substitution $u = \cos x$ we have $-du = \sin x \, dx$, and our integral becomes

$$-\int (1 - u^2) u^4 \, du = -\int (u^4 - u^6) \, du = -\frac{1}{5} \cos^5 x + \frac{1}{7} \cos^7 x + C.$$

Notice that this method worked only because the original power of $\sin x$ was odd. If both powers had been even, we would have needed to use another method, as we will in Example 2.

(b) We will set up the integral so that we can apply the substitution $u = \cot x$. This means that we must write the integral in terms of cotangents (the bracketed portion), with one copy of $\csc^2 x$ saved for the differential du:

$$\int \csc^4 x \cot^3 x \, dx = \int \csc^2 x \csc^2 x \cot^3 x \, dx = \int \underbrace{(1 + \cot^2 x) \cot^3 x}\ \csc^2 x \, dx$$

Now with $u = \cot x$ and thus $-du = \csc^2 x \, dx$, substitution gives us the new integral

$$-\int (u^2 + 1) u^3 \, du = -\int (u^5 + u^3) \, du = -\frac{1}{6} \cot^6 x - \frac{1}{4} \cot^4 x + C.$$

This method worked only because the original power of $\csc x$ was even, which allowed us to apply a Pythagorean identity to convert all but two of the cosecants into cotangents.

(c) This time the power of secant is odd, so the method used in part (b) will not work. Instead, we will convert everything to secants (the bracketed portion), with one copy of $\sec x \tan x$ saved for the differential:

$$\int \sec^3 x \tan^3 x \, dx = \int \sec^2 x \sec x \, \tan^2 x \tan x \, dx = \int \underbrace{\sec^2 x \, (\sec^2 x - 1)} \, \sec x \tan x \, dx.$$

Now with the substitution $u = \sec x$ and thus $du = \sec x \tan x \, dx$, our integral becomes

$$\int u^2 \, (u^2 - 1) \, du = \int (u^4 - u^2) \, du = \frac{1}{5} \sec^5 x - \frac{1}{3} \sec^3 x + C.$$

The method we just applied works only if the power of secant (respectively, cosecant) is odd. In part (b) we saw a method that works only if the power of tangent (respectively, cotangent) is even. If neither of these conditions holds, then we can apply a reduction method; see Example 4. □

EXAMPLE 2 Reducing powers with double-angle identities

Use double-angle identities to find each of the following integrals:

(a) $\displaystyle \int \cos^6 x \, dx$ **(b)** $\displaystyle \int \sin^2 x \, \cos^2 x \, dx$

SOLUTION

(a) Notice that we cannot use the method of Pythagorean identities as in the previous example, because the power of cosine is even. Instead, we apply the double-angle identity for cosine to rewrite $\cos^6 x$ as

$$\cos^6 x = (\cos^2 x)^3 = \left(\frac{1}{2}(1 + \cos 2x)\right)^3 = \frac{1}{8} + \frac{3}{8} \cos 2x + \frac{3}{8} \cos^2 2x + \frac{1}{8} \cos^3 2x.$$

We have now traded the problem of integrating $\cos^6 x$ for the problem of integrating four simpler functions: $\frac{1}{8}$ and $\frac{3}{8} \cos 2x$, which are easy to antidifferentiate; $\frac{3}{8} \cos^2 2x$, which requires another application of the double-angle formula; and $\frac{1}{8} \cos^3 2x$, which can be solved with a Pythagorean identity and the substitution $u = \sin 2x$, $du = 2 \cos 2x \, dx$. We have

$$\int \cos^6 x \, dx = \int \frac{1}{8} \, dx + \int \frac{3}{8} \cos 2x \, dx + \int \frac{3}{8} \cos^2 2x \, dx + \int \frac{1}{8} \cos^3 2x \, dx$$

$$= \frac{1}{8}x + \frac{3}{8}\left(\frac{1}{2}\sin 2x\right) + \frac{3}{8}\int \frac{1}{2}(1 + \cos 4x) \, dx + \frac{1}{8}\int (1 - \sin^2 2x) \cos 2x \, dx$$

$$= \frac{1}{8}x + \frac{3}{8}\left(\frac{1}{2}\sin 2x\right) + \frac{3}{16}\left(x + \frac{1}{4}\sin 4x\right) + \frac{1}{8}\left(\frac{1}{2}\right)\int (1 - u^2) \, du$$

$$= \frac{1}{8}x + \frac{3}{8}\left(\frac{1}{2}\sin 2x\right) + \frac{3}{16}\left(x + \frac{1}{4}\sin 4x\right) + \frac{1}{16}\left(u - \frac{1}{3}u^3\right) + C$$

$$= \frac{1}{8}x + \frac{3}{8}\left(\frac{1}{2}\sin 2x\right) + \frac{3}{16}\left(x + \frac{1}{4}\sin 4x\right) + \frac{1}{16}\left(\sin 2x - \frac{1}{3}\sin^3 2x\right) + C.$$

(b) One way to simplify this integral is to apply double-angle identities to write both $\sin^2 x$ and $\cos^2 x$ in terms of $\cos 2x$ and then apply a double-angle identity *again* to reduce powers even further:

$$\int \sin^2 x \cos^2 x \, dx = \int \left(\tfrac{1}{2}(1 - \cos 2x)\right)\left(\tfrac{1}{2}(1 + \cos 2x)\right) dx \quad \leftarrow \text{double-angle identities}$$

$$= \tfrac{1}{4}\int (1 - \cos^2 2x) \, dx \qquad\qquad \leftarrow \text{simplify}$$

$$= \tfrac{1}{4}\int \left(1 - \tfrac{1}{2}(1 + \cos 4x)\right) dx \qquad \leftarrow \text{double-angle identity}$$

$$= \tfrac{1}{4}\int \left(\tfrac{1}{2} - \tfrac{1}{2}\cos 4x\right) dx \qquad\qquad \leftarrow \text{simplify}$$

$$= \tfrac{1}{4}\left(\tfrac{1}{2}x - \tfrac{1}{8}\sin 4x\right) + C. \qquad\qquad \leftarrow \text{antidifferentiate} \qquad \square$$

EXAMPLE 3 **Using integration by parts to integrate a power of the secant**

Find $\displaystyle\int \sec^3 x \, dx$.

SOLUTION

In this case, Pythagorean identities would not be helpful. However, since we know how to integrate $\sec^2 x$, we can apply integration by parts with $dv = \sec^2 x \, dx$:

$$u = \sec x \quad \longrightarrow \quad du = \sec x \tan x \, dx$$
$$v = \tan x \quad \longleftarrow \quad dv = \sec^2 x \, dx.$$

Applying the integration-by-parts formula and then a Pythagorean identity, we have

$$\int \sec^3 x \, dx = \sec x \tan x - \int \sec x \tan^2 x \, dx. \qquad \leftarrow \text{integration by parts}$$

$$= \sec x \tan x - \int \sec x (\sec^2 x - 1) \, dx \qquad \leftarrow \text{Pythagorean identity}$$

$$= \sec x \tan x - \int \sec^3 x \, dx + \int \sec x \, dx \qquad \leftarrow \text{expand}$$

$$= \sec x \tan x - \int \sec^3 x \, dx + \ln|\sec x + \tan x| \qquad \leftarrow \text{Theorem 8.16}$$

We are now in a fortunate situation, since all we have to do is solve for $\int \sec^3 x \, dx$ in the equation $\int \sec^3 x \, dx = \sec x \tan x - \int \sec^3 x \, dx + \ln|\sec x + \tan x|$. Doing so gives us

$$\int \sec^3 x \, dx = \tfrac{1}{2}(\sec x \tan x + \ln|\sec x + \tan x|) + C. \qquad \square$$

EXAMPLE 4 **Using Pythagorean identities to reduce to simpler cases and smaller powers**

Find each of the following integrals:

(a) $\displaystyle\int \tan^5 x \, dx$

(b) $\displaystyle\int \sec x \tan^2 x \, dx$

SOLUTION

(a) The key here is to repeatedly apply a Pythagorean identity to decrease the powers of the tangent until we have a sum of expressions that we can integrate. We can write

$$\tan^5 x = \tan^3 x \,(\sec^2 x - 1) \qquad\qquad \leftarrow \text{Pythagorean identity}$$

$$= \tan^3 x \sec^2 x - \tan^3 x \qquad\qquad \leftarrow \text{expand}$$

$$= \tan^3 x \sec^2 x - \tan x \,(\sec^2 x - 1) \qquad \leftarrow \text{Pythagorean identity}$$

$$= \tan^3 x \sec^2 x - \tan x \sec^2 x + \tan x \qquad \leftarrow \text{expand}$$

The first two expressions can be integrated with the substitution $u = \tan x$ and $du = \sec^2 x \, dx$, and the third is an expression we already know how to antidifferentiate, as follows:

$$\int \tan^5 x \, dx = \int u^3 \, du - \int u^2 \, du + \int \tan x \, dx$$

$$= \frac{1}{4} u^4 - \frac{1}{3} u^3 + \ln|\cos x| + C = \frac{1}{4} \tan^4 x - \frac{1}{3} \tan^3 x + \ln|\cos x| + C.$$

(b) Again we will apply Pythagorean identities to reduce the powers of the tangents to expressions that we know how to integrate. We have

$$\sec x \tan^2 x = \sec x (\sec^2 x - 1) = \sec^3 x - \sec x.$$

We saw how to integrate $\sec^3 x$ in Example 3 and how to integrate $\sec x$ in Theorem 8.16. Putting these results together, we have

$$\int \sec x \tan^2 x \, dx = \frac{1}{2} \sec x \tan x - \frac{1}{2} \ln|\sec x + \tan x| + C. \qquad\qquad \square$$

? TEST YOUR UNDERSTANDING

► Why can't we use Pythagorean identities and u-substitution to solve the integral $\int \sin^4 x \cos^4 x \, dx$? What method could we use instead?

► In Example 1(b) we used a Pythagorean identity followed by substitution with $u = \cot x$ to solve the integral $\int \csc^4 x \cot^3 x \, dx$. Can you find another way to solve this integral?

► When we integrated $\int \sec^3 x \, dx$ in Example 3, we used integration by parts and identities to return to the original integral. Would the same thing happen if the power of the secant were even? Why or why not?

► The technique used in Example 4 also works if the power of the tangent is even. Use this technique to find $\int \tan^6 x \, dx$. What is different about the last steps of the computation when k is even?

► In solving trigonometric integrals by the techniques in this section, it is often difficult to check answers by differentiating. Why?

EXERCISES 8.4

Thinking Back

Pythagorean identities: Prove each of the following Pythagorean identities in the manner specified.

▶ Prove that $\sin^2 x + \cos^2 x = 1$ for acute angles x, using the right-triangle definitions of sine and cosine. Why is this called a *Pythagorean* identity?

▶ Prove that $\sin^2 x + \cos^2 x = 1$ for any angle x, using the unit-circle definitions of sine and cosine.

▶ Use the fact that $\sin^2 x + \cos^2 x = 1$ to prove that $\tan^2 x + 1 = \sec^2 x$.

▶ Use the fact that $\sin^2 x + \cos^2 x = 1$ to prove that $1 + \cot^2 x = \csc^2 x$.

Double-angle identities: Prove each of the following double-angle identities by applying the sum identity for the cosine followed by a Pythagorean identity.

▶ $\sin^2 x = \dfrac{1}{2}(1 - \cos 2x)$ ▶ $\cos^2 x = \dfrac{1}{2}(1 + \cos 2x)$

Concepts

0. *Problem Zero:* Read the section and make your own summary of the material.

1. *True/False:* Determine whether each of the statements that follow is true or false. If a statement is true, explain why. If a statement is false, provide a counterexample.

(a) *True or False:* $\int \cos^8 x\, dx = \dfrac{1}{9}\cos^9 x + C$.

(b) *True or False:* $\int \cos^8 x\, dx = \dfrac{1}{9}\sin^9 x + C$.

(c) *True or False:* $\int \cos^8 x\, dx = -\dfrac{1}{9}\sin^9 x + C$.

(d) *True or False:* If k is positive and odd, then $\int \sin^k x\, dx$ can be rewritten so that integration by substitution is possible with $u = \sin x$.

(e) *True or False:* The best method for integrating $\int \sin^3 x \cos^2 x\, dx$ is to apply double-angle identities.

(f) *True or False:* The best method for integrating $\int \sec^4 x \cos^5 x\, dx$ is to apply Pythagorean identities.

(g) *True or False:* The best method for integrating $\int \csc^2 x \cot^3 x\, dx$ is to apply substitution with $u = \cot x$.

(h) *True or False:* We now know how to integrate all six of the basic trigonometric functions.

2. *Examples:* Construct examples of the thing(s) described in the following. Try to find examples that are different than any in the reading.

(a) Three integrals that can be rewritten by using the Pythagorean identity $\sin^2 x + \cos^2 x = 1$ so that integration by substitution applies.

(b) Three integrals that can be rewritten by using the Pythagorean identity $\tan^2 x + 1 = \sec^2 x$ so that integration by substitution applies.

(c) Three integrals that can be improved by using one or both of the double-angle identities.

3. In Example 1(a) we showed that $\int \sin^3 x \cos^4 x\, dx = -\dfrac{1}{5}\cos^5 x + \dfrac{1}{7}\cos^7 x + C$. Check that this answer is correct by differentiating and applying trigonometric identities.

4. Compile lists of (a) the derivatives and (b) the integrals of the six basic trigonometric functions: $\sin x$, $\cos x$, $\tan x$, $\sec x$, $\csc x$, and $\cot x$.

5. Given the algebraic trick for integrating $\sec x$ used in the proof of Theorem 8.16, what do you think is the algebraic trick used for integrating $\csc x$?

6. Find three integrals in Exercises 21–62 that can be solved by applying Pythagorean identities to set up simple u-substitutions.

7. Find three integrals in Exercises 21–62 that can be solved by the application of double-angle formulas.

Describe strategies for solving the types of integrals given in Exercises 8–18.

8. $\displaystyle\int \cos^k x\, dx, \ k$ odd

9. $\displaystyle\int \cos^k x\, dx, \ k$ even

10. $\displaystyle\int \cot^k x\, dx, \ k$ odd

11. $\displaystyle\int \cot^k x\, dx, \ k$ even

12. $\displaystyle\int \csc^k x\, dx, \ k = 2$

13. $\displaystyle\int \csc^k x\, dx, \ k > 2$

14. $\displaystyle\int \sin^m x \cos^n x\, dx, \ $ one of m and n odd

15. $\displaystyle\int \sin^m x \cos^n x\, dx, \ $ both m and n even

16. $\displaystyle\int \sec^m x \tan^n x\, dx, \ m$ even

17. $\displaystyle\int \sec^m x \tan^n x\, dx, \ n$ odd

18. $\displaystyle\int \sec^m x \tan^n x\, dx, \ m$ odd and n even

19. Consider the integral $\int \sin^2 x \cos^2 x\, dx$. We solved this integral in Example 2(b) by applying double-angle identities at the very beginning.

(a) Solve this integral by applying the identity $\sin^2 x = 1 - \cos^2 x$.

(b) Solve this integral another way, by applying the identity $\sin x \cos x = \dfrac{1}{2}\sin 2x$.

20. Consider the integral $\int \sec^2 x \tan^3 x\, dx$.

(a) Solve this integral by using integration by substitution with $u = \tan x$.

(b) Solve this integral another way, by using integration by substitution with $u = \sec x$.

Skills

Solve each of the integrals in Exercises 21–62. Some of the integrals require the methods presented in this section, and some do not.

21. $\displaystyle\int \sin^2 3x\, dx$

22. $\displaystyle\int \cos^2 x\, dx$

23. $\displaystyle\int \cos^5 x\, dx$

24. $\displaystyle\int \sin^3 x\, dx$

25. $\displaystyle\int \tan 2x\, dx$

26. $\displaystyle\int \csc \pi x\, dx$

27. $\displaystyle\int \sec 4x\, dx$

28. $\displaystyle\int \tan^6 x\, dx$

29. $\displaystyle\int \cot^5 x\, dx$

30. $\displaystyle\int \csc^2 x\, dx$

31. $\displaystyle\int \sec^3 2x\, dx$

32. $\displaystyle\int \sec^6 x\, dx$

33. $\displaystyle\int \cos^4 x\, dx$

34. $\displaystyle\int \sin^6 x\, dx$

35. $\displaystyle\int \csc^4 x\, dx$

36. $\displaystyle\int \sin x \cos^4 x\, dx$

37. $\displaystyle\int \cos^3 x \sin^4 x\, dx$

38. $\displaystyle\int \sin^5 x \cos^2 x\, dx$

39. $\displaystyle\int \sec x \tan^3 x\, dx$

40. $\displaystyle\int \sin^2 x \cos^3 x\, dx$

41. $\displaystyle\int \sin^2 3x \cos^2 3x\, dx$

42. $\displaystyle\int \sec^2 x \tan^5 x\, dx$

43. $\displaystyle\int \sec^8 x \tan x\, dx$

44. $\displaystyle\int \tan^2 4x \sec 4x\, dx$

45. $\displaystyle\int \sec^3 x \tan^5 x\, dx$

46. $\displaystyle\int \csc^4 x \cot^2 x\, dx$

47. $\displaystyle\int \sin^{13} x \cos^5 x\, dx$

48. $\displaystyle\int \sec^3 x \tan^2 x\, dx$

49. $\displaystyle\int \csc x \cot^2 x\, dx$

50. $\displaystyle\int (\cot x \csc x)^3\, dx$

51. $\displaystyle\int \frac{\tan x}{\cos^3 x}\, dx$

52. $\displaystyle\int \frac{\cos^3 x}{\csc^8 x}\, dx$

53. $\displaystyle\int \sin x \sec x\, dx$

54. $\displaystyle\int \cos^6 x \sin^3 x\, dx$

55. $\displaystyle\int \cos^3 x \sec^2 x\, dx$

56. $\displaystyle\int \sin x \tan x\, dx$

57. $\displaystyle\int \tan^2 x \csc x\, dx$

58. $\displaystyle\int \tan x \cos^5 x\, dx$

59. $\displaystyle\int \frac{\cos x \ln(\cos x)}{\csc x}\, dx$

60. $\displaystyle\int \frac{3}{\sin^2(\pi x)}\, dx$

61. $\displaystyle\int (\sin x \sqrt{\cos x})^3\, dx$

62. $\displaystyle\int \frac{\sin^3 x}{\cos x}\, dx$

Solve each of the definite integrals in Exercises 63–72.

63. $\displaystyle\int_0^\pi \sin^5 x\, dx$

64. $\displaystyle\int_0^{3\pi} \sin^5 x\, dx$

65. $\displaystyle\int_{-\pi/2}^{\pi/2} \cos^3 x\, dx$

66. $\displaystyle\int_{-\pi}^{\pi} \sin^2 x\, dx$

67. $\displaystyle\int_0^\pi \sin^3 x \cos^2 x\, dx$

68. $\displaystyle\int_{-\pi}^{\pi} \sin^2 x \cos^2 x\, dx$

69. $\displaystyle\int_0^{\pi/4} \sec^4 x \tan^2 x\, dx$

70. $\displaystyle\int_{\pi/4}^{\pi/2} \csc x \cot^3 x\, dx$

71. $\displaystyle\int_0^{\pi/4} \frac{\sin^2 x}{\cos^4 x}\, dx$

72. $\displaystyle\int_0^{\pi/4} \frac{\sin x}{\cos^2 x}\, dx$

73. Consider the function $f(x) = 4\sin^3 x \cos^2 x$.
  (a) Find the signed area between the graph of $f(x)$ and the x-axis on $[-\pi, \pi]$, shown next at the left.
  (b) Find the absolute area between the graph of $f(x)$ and the x-axis on $[-\pi, \pi]$.
  (c) Find the average value of $f(x)$ on $\left[0, \frac{\pi}{2}\right]$, shown next at the right.

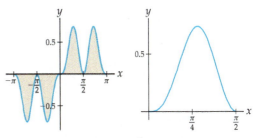

74. Consider the function $f(x) = \sin^5 x$.
  (a) Find the area between the graphs of $f(x)$ and $g(x) = \sin x$ on $[0, \pi]$, shown next at the left.
  (b) Find the area between the graphs of $f(x)$ and $h(x) = \sin^5 2x$ on $[0, \pi]$, shown next at the right. Note that the intersection point of the two graphs is at $x = \frac{\pi}{3}$.

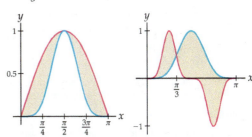

75. Consider the function $f(x) = \sin^2(\pi x)$.
  (a) Find the area between the graphs of $f(x)$ and $f(x) + \frac{1}{2}$ on $[0, 3]$, shown next at the left.
  (b) Find the area between the graphs of $f(x)$ and $-f(x) + \frac{1}{2}$ on $[0, 3]$, shown next at the right.

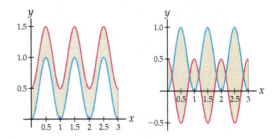

Applications

76. An office building has a thermostat that keeps the temperature between 20 and 26 degrees Celsius for 24 hours a day. As the heating and cooling units alternate, the temperature starting from midnight is given by

$$T(t) = 20\sin^3\left(\frac{\pi}{3}x\right)\cos^3\left(\frac{\pi}{3}x\right) + 22.5,$$

as shown in the figure. Find the average temperature in the office building between the hours of 9am and 5pm.

Temperature of office building

77. A wire filament is shaped to match the graph of the function of $y = \ln(2\cos x)$ from $x = -\frac{\pi}{3}$ to $x = \frac{\pi}{3}$, measured in inches, as shown in the figure. Given that the length of a curve $y = f(x)$ from $x = a$ to $x = b$ can be calculated with the formula

$$\int_a^b \sqrt{1 + (f'(x))^2}\, dx,$$

find the length of the filament.

Shape of wire filament

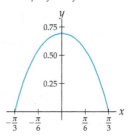

Proofs

78. Prove the integration formula

$$\int \sec x\, dx = \ln|\sec x + \tan x| + C$$

(a) by multiplying the integrand by a form of 1 and then applying a substitution;

(b) by differentiating $\ln|\sec x + \tan x|$.

79. Prove the integration formula

$$\int \csc x\, dx = -\ln|\csc x + \cot x| + C$$

(a) by multiplying the integrand by a form of 1 and then applying a substitution;

(b) by differentiating $-\ln|\csc x + \cot x|$.

80. In this problem we will see how to solve integrals of the form $\int \sin mx \cos nx\, dx$ by using the trigonometric identity

$$\sin\alpha\cos\beta = \frac{1}{2}(\sin(\alpha - \beta) + \sin(\alpha + \beta)).$$

(a) Explain why the techniques covered earlier in the reading do not apply to $\int \sin 2x \cos 3x\, dx$.

(b) Use the given trigonometric identity to solve $\int \sin 2x \cos 3x\, dx$.

(c) Use the sum and difference identities for the sine function to prove the given trigonometric identity.

81. It can be tedious to integrate powers of the sine, especially for powers that are large and even. The following *reduction formula* can help: For $k > 2$,

$$\int \sin^k x\, dx = -\frac{1}{k}\sin^{k-1} x \cos x + \frac{k-1}{k}\int \sin^{k-2} x\, dx.$$

(a) Use the given reduction formula to find $\int \sin^4 x\, dx$ and $\int \sin^8 x\, dx$. You may have to apply the formula more than once.

(b) Why is this formula called a *reduction* formula?

(c) Use integration by parts to prove the reduction formula. (*Hint: Choose $u = \sin^{n-1} x$.*)

82. It can be tedious to integrate powers of the secant, especially for powers that are large. The following reduction formula can help: For $k > 2$,

$$\int \sec^k x\, dx = \frac{1}{k-1}\sec^{k-2} x \tan x + \frac{k-2}{k-1}\int \sec^{k-2} x\, dx.$$

(a) Use the given reduction formula to find $\int \sec^3 x\, dx$ and $\int \sec^7 x\, dx$. You may have to apply the formula more than once.

(b) Use integration by parts to prove the reduction formula. (*Hint: Choose $dv = \sec^2 x\, dx$.*)

Thinking Forward

Trigonometric substitution: Some integrals that don't start out involving trigonometric functions can be cleverly converted into ones that do, and then we can bring the techniques of this section to bear. In the next few problems this is how we will solve the integral $\int \frac{1}{x\sqrt{1-x^2}}\, dx$.

▶ The non-intuitive substitution $u = \sin^{-1} x$ is equivalent to the backwards substitution $x = \sin u$. How are dx and du related?

▶ Rewrite the given integral by substituting $x = \sin u$ and $dx = \cos u$.

▶ Use algebra to show that your new integral is equivalent to $\int \csc u\, dx$.

▶ Solve the integral and write your answer in terms of x. Your answer will involve both trigonometric and inverse trigonometric functions.

8.5 TRIGONOMETRIC SUBSTITUTION

▶ Using inverse trigonometric functions as substitutions to solve integrals

▶ Domain considerations when applying trigonometric substitution

▶ Writing a trigonometric and inverse trigonometric composition as an algebraic function

Backwards Trigonometric Substitutions

As we have seen, integration is not always a simple, straightforward process. Sometimes, similar integrals require different solution strategies. For example, consider the four integrals

$$\int \frac{1}{\sqrt{1-x^2}}\,dx, \quad \int \frac{x}{\sqrt{1-x^2}}\,dx, \quad \int \frac{x^3}{\sqrt{1-x^2}}\,dx, \quad \text{and} \quad \int \frac{1}{x^2\sqrt{1-x^2}}\,dx.$$

The first three can be solved with methods we have used so far: the first by antidifferentiating, the second with the substitution $u = 1 - x^2$, and the third with the same substitution along with a back-substitution. The fourth integral, however, is more difficult. Fortunately, a nonintuitive substitution will save the day:

$$\boxed{u = \sin^{-1} x} \quad \Longrightarrow \quad \frac{du}{dx} = \frac{1}{\sqrt{1-x^2}} \quad \Longrightarrow \quad \boxed{du = \frac{1}{\sqrt{1-x^2}}\,dx}.$$

The differential du looks promising, since $\frac{1}{\sqrt{1-x^2}}$ does appear in our integrand. The choice of $u = \sin^{-1} x$ does not appear in the integrand, but we will be able to use back-substitution to fix that problem, with $x = \sin u$. With this back-substitution we can rewrite our integral as one that involves trigonometric functions but is *much* simpler:

$$\int \frac{1}{x^2\sqrt{1-x^2}}\,dx = \int \frac{1}{x^2}\left(\frac{1}{\sqrt{1-x^2}}\,dx\right) = \int \frac{1}{\sin^2 u}\,du = \int \csc^2 u\,du.$$

We can achieve the same result with a slightly different method of back-substitution called ***trigonometric substitution***. Instead of starting with the substitution $u = \sin^{-1} x$, we will start immediately with the substitution $x = \sin u$. Starting over with this choice, we have

$$\boxed{x = \sin u} \quad \Longrightarrow \quad \frac{dx}{du} = \cos u \quad \Longrightarrow \quad \boxed{dx = \cos u\,du}.$$

Compare this substitution with the one we did previously; the two substitutions are equivalent, but the one starting with $x = \sin u$ is simpler. Notice also that we are thinking of x as a function of u. With this new version of our substitution and a Pythagorean identity, we obtain the same simplification as before:

$$\int \frac{1}{x^2\sqrt{1-x^2}}\,dx = \int \frac{\cos u}{\sin^2 u \sqrt{1-\sin^2 u}}\,du = \int \frac{1}{\sin^2 u}\,du = \int \csc^2 u\,du.$$

Although the second calculation involves more algebra, at least it does not involve inverse trigonometric functions. The rewritten integral is easy to antidifferentiate, and since we have $x = \sin u$ and thus $u = \sin^{-1} x$, the integral is equal to $-\cot u + C = -\cot(\sin^{-1} x) + C$.

Domains and Simplifications with Trigonometric Substitutions

One key thing that made the substitution $x = \sin u$ work in our earlier calculation was that the Pythagorean identity $\sin^2 x + \cos^2 x = 1$ allowed us to simplify the expression

$\sqrt{1 - x^2} = \sqrt{1 - \sin^2 u}$. The other two Pythagorean identities allow similar simplifications. In each case we must be very careful concerning the domains of any variables that we use. Every substitution we make must be compatible with the domain of the integrand.

The three types of trigonometric substitutions and their allowable domains are described in the following theorem:

THEOREM 8.17

Three Types of Trigonometric Substitutions

Suppose a is a positive real number. The following substitutions can be useful in solving certain integrals:

(a) If $x = a \sin u$, then $\sqrt{a^2 - x^2} = a \cos u$.

This substitution is valid for $x \in [-a, a]$ and $u \in \left[-\dfrac{\pi}{2}, \dfrac{\pi}{2}\right]$.

(b) If $x = a \tan u$, then $x^2 + a^2 = a^2 \sec^2 u$.

This substitution is valid for $x \in (-\infty, \infty)$ and $u \in \left(-\dfrac{\pi}{2}, \dfrac{\pi}{2}\right)$.

(c) If $x = a \sec u$, then $\sqrt{x^2 - a^2} = \begin{cases} a \tan u, & \text{if } x > a \\ -a \tan u, & \text{if } x < -a. \end{cases}$

This substitution is valid for $x \in (-\infty, -a] \cup [a, \infty)$ and $u \in \left[0, \dfrac{\pi}{2}\right) \cup \left(\dfrac{\pi}{2}, \pi\right]$.

The domain considerations in Theorem 8.17 are essential. For example, if $x = a \sin u$, then x must be between $-a$ and a. Therefore the substitution $x = a \sin u$ makes sense only for integrands whose domain is contained in $[-a, a]$. Notice that the domain of $\sqrt{a^2 - x^2}$ is precisely $[-a, a]$, so integrals involving $\sqrt{a^2 - x^2}$ are well suited for trigonometric substitution with $x = a \sin u$. Furthermore, if $x = a \sin u$ and we want to be able to solve for $u = \sin^{-1}(x/a)$, then we must have u in the range of $\sin^{-1} x$. This is why we require that $u \in [-\pi/2, \pi/2]$ in part (a). The domain restrictions for u in the theorem come from the restricted domains of the sine, secant, and tangent, respectively.

Proof. To prove part (a), suppose $x \in [-a, a]$, $u \in [-\pi/2, \pi/2]$, and $x = a \sin u$. Using the Pythagorean identity $a^2 \sin^2 u + a^2 \cos^2 u = a^2$, we have

$$\sqrt{a^2 - x^2} = \sqrt{a^2 - a^2 \sin^2 u} = \sqrt{a^2 \cos^2 u} = |a \cos u| = a \cos u.$$

The last couple of steps in this calculation need some technical discussion. An absolute value appears in the second-to-last step because for any number A, we have $\sqrt{A^2} = |A|$. Luckily, in this case we can drop the absolute value sign because we are assuming $a > 0$ and because $\cos u$ is always positive for $u \in [-\pi/2, \pi/2]$.

The proof of part (b) is similar to the proof of part (a) and is left to Exercise 90. In this case there is no domain restriction on x, because $x = a \tan u$ can be any real number. This means that trigonometric substitution with $x = a \tan u$ can be used even for integrands that do not involve a square root.

When using secant for a trigonometric substitution as in part (c), we run into a technical point. Suppose $x \in (-\infty, a] \cup [a, \infty)$, $u \in [0, \pi/2) \cup (\pi/2, \pi]$, and $x = a \sec u$. Using the Pythagorean identity $a^2 \tan^2 u + a^2 = a^2 \sec^2 u$, we can write

$$\sqrt{x^2 - a^2} = \sqrt{a^2 \sec^2 u - a^2} = \sqrt{a^2 \tan^2 u} = |a \tan u| = \begin{cases} -a \tan u, & \text{if } x < -a \\ a \tan u, & \text{if } x > a. \end{cases}$$

Once again, absolute values appear because we are taking the square root of a square. However, this time the absolute values are significant, since $\tan u$ is positive for u in the first quadrant $[0, \pi/2)$ and negative for u in the second quadrant $(\pi/2, \pi]$. The former case happens when $x \in [a, \infty)$, and the latter case when $x \in (-\infty, -a]$. The piecewise-defined expression for $\sqrt{x^2 - a^2}$ means that we will have to consider the cases $x > a$ and $x < -a$ separately when using substitutions of the form $x = a \sec u$. ■

The point of all the work we just did boils down essentially to this: When making trigonometric substitutions with the sine or tangent, we need not worry about quadrants. But when making a trigonometric substitution with secant, there will be cases to consider.

Rewriting Trigonometric Compositions

Previously, we used trigonometric substitution to show that

$$\int \frac{1}{x^2 \sqrt{1 - x^2}}\, dx = -\cot(\sin^{-1} x) + C.$$

Although we have solved this integral, the form of its answer can be simplified. In fact, as we will see in a moment, we can always rewrite compositions of trigonometric functions and inverse trigonometric functions as algebraic functions.

To solve the integral listed above, we used the substitution $x = \sin u$. The domain of our integrand is $x \in (-1, 1)$, and considering this and the restricted domain of sine, we have $u \in (-\pi/2, \pi/2)$. Since $x = \sin u$, u is an angle in the first or fourth quadrant whose reference triangle has altitude x, as shown in the figure that follows. By the Pythagorean theorem, the length of the remaining side of the reference triangle is $\sqrt{1 - x^2}$. This expression should look familiar: It's part of our original integrand!

Reference triangles for $x = \sin u$ in the first and fourth quadrants

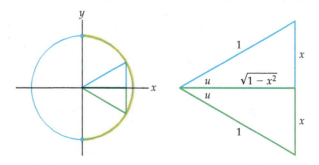

The quantity x is positive for the first-quadrant triangle and negative for the fourth-quadrant triangle, but in either case the cotangent of the angle $u = \sin^{-1} x$ is given by the ratio of the adjacent side to the opposite side.

$$\cot(\sin^{-1} x) = \cot u = \frac{\text{adj}}{\text{opp}} = \frac{\sqrt{1 - x^2}}{x}.$$

Combining this with our original work solving the integral, we have

$$\int \frac{1}{x^2 \sqrt{1 - x^2}}\, dx = -\cot(\sin^{-1} x) + C \doteq \frac{-\sqrt{1 - x^2}}{x} + C.$$

Luckily, it turns out that when using the substitutions $x = a \sin u$ and $x = a \tan u$, choice of quadrants will never be an issue and we will not have to consider the unit circle. From this point forward we will only need to look at one reference triangle in those cases.

Unfortunately, we will have to take quadrants into account when we use the substitution $x = a \sec u$; we will see this in Example 4, where the choice of quadrant will make a sign difference.

Examples and Explorations

EXAMPLE 1   Deciding when to use trigonometric substitution

Which of these integrals would be good candidates for trigonometric substitution?

(a) $\displaystyle \int \frac{x}{1+x^2}\, dx$ **(b)** $\displaystyle \int \frac{x^2}{1+x^2}\, dx$ **(c)** $\displaystyle \int \frac{1+x^2}{x^2}\, dx$

SOLUTION

(a) Trigonometric substitution is not necessary here because we can apply regular u-substitution with $u = 1 + x^2$. This technique simplifies the integral to $\frac{1}{2} \int \frac{1}{u}\, du$, which is easy to solve.

(b) Neither u-substitution nor integration by parts seems useful here. This integral is a good candidate for trigonometric substitution, with $x = \tan u$. This choice simplifies the integral to $\int \tan^2 u\, du$, which we can solve by the techniques of Section 8.4. Alternatively, we could use polynomial long division or other algebraic methods to write $\frac{x^2}{1+x^2}$ as $1 - \frac{1}{1+x^2}$, which is easy to antidifferentiate.

(c) Trigonometric substitution is not necessary here. Applying algebra first, we can write $\frac{1+x^2}{x^2}$ as $\frac{1}{x^2} + \frac{x^2}{x^2} = x^{-2} + 1$, which is easy to antidifferentiate. ☐

EXAMPLE 2   An integral involving $\sqrt{a^2 - x^2}$ and the substitution $x = a \sin u$

Use trigonometric substitution to solve the integral $\displaystyle \int (4 - x^2)^{-3/2}\, dx$.

SOLUTION

First, notice that the integrand can be expressed as $\dfrac{1}{\left(\sqrt{4-x^2}\right)^3}$ and therefore does involve the expression $\sqrt{a^2 - x^2}$, where $a = 2$. Using the substitution $x = 2\sin u$ and the Pythagorean identity $\sin^2 u + \cos^2 u = 1$, we have

$$(4 - x^2)^{-3/2} = (4 - 4\sin^2 u)^{-3/2} = \left(\sqrt{4\cos^2 u}\right)^{-3} = (2|\cos u|)^{-3} = \frac{1}{8\cos^3 u}.$$

Theorem 8.17 explains why the absolute values can be ignored in the preceding calculation. Now with the substitution $x = 2\sin u$, we have $\sqrt{4 - x^2} = 2\cos u$ and the differential $dx = 2\cos u\, du$, and thus we can write

$$\int (4 - x^2)^{-3/2}\, dx = \int \frac{1}{8\cos^3 u} (2\cos u)\, du$$

$$= \frac{1}{4} \int \sec^2 u\, du = \frac{1}{4}\tan u + C = \frac{1}{4}\tan\left(\sin^{-1}\left(\frac{x}{2}\right)\right) + C.$$

The last step in the calculation is a back-substitution obtained by solving $x = 2\sin u$ for $u = \sin^{-1}(x/2)$.

It now remains only to rewrite the answer as an algebraic function. Because we already know that quadrants are not relevant when using trigonometric substitution with the sine, we need to consider only one reference triangle. Since $x = 2\sin u$, we have $\sin u = \frac{x}{2}$. Thinking about "opposite over hypotenuse," we label the angle u and label two sides of the triangle with x and 2. The remaining leg of the triangle is determined by the Pythagorean theorem to be $\sqrt{4 - x^2}$, as follows:

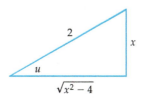

Now thinking about "opposite over adjacent," we can easily calculate that

$$\tan\left(\sin^{-1}\left(\tfrac{x}{2}\right)\right) = \tan u = \frac{x}{\sqrt{4 - x^2}}.$$

Therefore the integral in question is equal to

$$\int (4 - x^2)^{-3/2}\, dx = \frac{1}{4}\tan\left(\sin^{-1}\left(\tfrac{x}{2}\right)\right) + C = \frac{x}{4\sqrt{4 - x^2}} + C.$$

CHECKING THE ANSWER

Since we took the time to rewrite the trigonometric composition as an algebraic function, we can check the answer to the previous example by using the quotient rule and some algebra:

$$\frac{d}{dx}\left(\frac{x}{4\sqrt{4 - x^2}}\right) = \frac{(1)(4\sqrt{4 - x^2}) - (x)\left(4\left(\frac{1}{2}\right)(4 - x^2)^{-1/2}(-2x)\right)}{16(4 - x^2)}$$

$$= \frac{4\sqrt{4 - x^2} + \dfrac{4x^2}{\sqrt{4 - x^2}}}{16(4 - x^2)}\left(\frac{\sqrt{4 - x^2}}{\sqrt{4 - x^2}}\right)$$

$$= \frac{4(4 - x^2) + 4x^2}{16(4 - x^2)^{3/2}} = \frac{16}{16(4 - x^2)^{3/2}} = (4 - x^2)^{-3/2}.$$

EXAMPLE 3

An integral involving $x^2 + a^2$ and the substitution $x = a\tan u$

Use trigonometric substitution to solve the integral $\displaystyle\int (x^2 + 5)^{-3/2}\, dx$.

SOLUTION

This example is similar to the previous one, except that this time the substitution $x = \sqrt{5}\tan u$ is suggested. Using a Pythagorean identity and some basic algebra, we have

$$(x^2 + 5)^{-3/2} = (5\tan^2 u + 5)^{-3/2} = \sqrt{5\sec^2 u}^{\,-3} = (\sqrt{5}|\sec u|)^{-3} = \frac{1}{(\sqrt{5})^3 \sec^3 u}.$$

With the substitution $x = \sqrt{5}\tan u$ we also have $dx = \sqrt{5}\sec^2 u\, du$ and $u = \tan^{-1}(x/\sqrt{5})$, and thus can write

$$\int (x^2 + 5)^{-3/2}\, dx = \int \frac{1}{(\sqrt{5})^3 \sec^3 u}(\sqrt{5}\sec^2 u)\, du$$

$$= \frac{1}{5}\int \cos u\, du = \frac{1}{5}\sin u + C = \frac{1}{5}\sin\left(\tan^{-1}\left(\frac{x}{\sqrt{5}}\right)\right) + C.$$

Once again, we construct a reference triangle to rewrite our answer as an algebraic function. We have $\tan u = \frac{x}{\sqrt{5}}$, so relative to the angle u we label the opposite side x and the adjacent side $\sqrt{5}$. By the Pythagorean theorem, the corresponding hypotenuse has length $\sqrt{x^2 + 5}$:

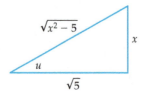

Using this triangle we have

$$\sin\left(\tan^{-1}\left(\frac{x}{\sqrt{5}}\right)\right) = \sin u = \frac{x}{\sqrt{x^2 + 5}}.$$

Therefore our integral is equal to

$$\int (x^2 + 5)^{-3/2}\, dx = \frac{1}{5}\sin\left(\tan^{-1}\left(\frac{x}{\sqrt{5}}\right)\right) + C = \frac{x}{5\sqrt{x^2 + 5}} + C.$$

The astute reader will notice that this example is quite similar to the previous one. To get a better understanding, compare and contrast the two examples.   □

EXAMPLE 4

An integral involving $\sqrt{x^2 - a^2}$ and the substitution $x = a\sec u$

Use trigonometric substitution to solve the integral $\displaystyle\int \frac{1}{\sqrt{x^2 - 4}}\, dx$.

SOLUTION

With the substitution $x = 2\sec u$ we have

$$\frac{1}{\sqrt{x^2 - 4}} = \frac{1}{\sqrt{4\sec^2 u - 4}} = \frac{1}{\sqrt{4\tan^2 u}} = \frac{1}{2|\tan u|}.$$

Unfortunately, at this point a problem arises that we didn't encounter in the previous two examples. The difficulty is that in this example the integrand $\frac{1}{\sqrt{x^2-4}}$ is defined for $x > 2$ and for $x < -2$, but not in between. Since we are using the substitution $x = 2\sec u$, we need u to be in the restricted domain of the secant, which is the first two quadrants. Let's examine the two cases.

If $x > 2$, then $x = 2\sec u$ means that $\sec u > 1$, which happens when u is in the first quadrant. The tangent function is positive in that quadrant, so if $x > 2$, then $\tan u > 0$ and thus $|\tan u| = \tan u$. However, if $x < -2$, then $\sec u < 1$ and thus u is in the second quadrant, where the tangent is negative. In that case we have $|\tan u| = -\tan u$. Unlike the solution in the previous two examples, the choice of quadrant will be significant. The problem here is that our original integrand was defined on a disconnected domain, and we have to deal with the two pieces of the domain separately.

If $x > 2$, then $|\tan u| = \tan u$, and thus with $x = 2\sec u$, $dx = 2\sec u \tan u\, du$, and a known antidifferentiation formula, we have

$$\int \frac{1}{\sqrt{x^2 - 4}}\, dx = \int \frac{1}{2\tan u}(2\sec u \tan u)\, du = \int \sec u\, du$$

$$= \ln|\sec u + \tan u| + C = \ln\left|\sec\left(\sec^{-1}\left(\frac{x}{2}\right)\right) + \tan\left(\sec^{-1}\left(\frac{x}{2}\right)\right)\right| + C.$$

Still considering the case where $x > 2$, we see that our reference triangle is in the first quadrant. Since $\sec u = \frac{x}{2}$, we have a hypotenuse of length x and an adjacent leg of length 2, relative to the angle u. By the Pythagorean theorem, the remaining side has length $\sqrt{x^2 - 4}$, as follows:

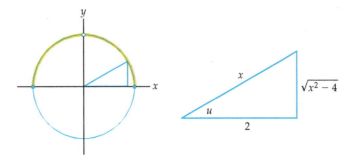

We clearly have $\sec(\sec^{-1}(x/2)) = x/2$. Using the triangle shown we see that

$$\tan\left(\sec^{-1}\left(\frac{x}{2}\right)\right) = \tan u = \frac{\sqrt{x^2 - 4}}{2}.$$

Therefore in the case where $x > 2$, we have

$$\int \frac{1}{\sqrt{x^2 - 4}}\, dx = \ln\left|\sec\left(\sec^{-1}\left(\frac{x}{2}\right)\right) + \tan\left(\sec^{-1}\left(\frac{x}{2}\right)\right)\right| + C = \ln\left|\frac{x}{2} + \frac{\sqrt{x^2 - 4}}{2}\right| + C.$$

In the case where $x < -2$, there are two places where our answer could differ from the answer for $x > 2$. First, if $x < -2$ and $x = 2\sec u$, then $\sec u$ is negative, which means that u is an angle terminating in the second quadrant. The tangent function is negative in that quadrant, so in this case $|\tan u| = -\tan u$. This means that before integrating, we have to insert a negative sign. By the end of the integration step we still have a negative sign, so for $x < -2$, we have

$$\int \frac{1}{\sqrt{x^2 - 4}}\, dx = -\ln\left|\sec\left(\sec^{-1}\left(\frac{x}{2}\right)\right) + \tan\left(\sec^{-1}\left(\frac{x}{2}\right)\right)\right| + C.$$

Second, when we use a reference triangle to write the answer as an algebraic function, we will have to consider quadrants. The diagrams that follow show the angle u in the unit circle and its corresponding reference triangle v.

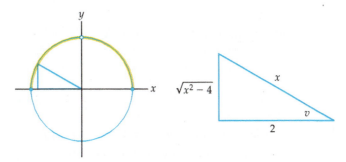

Since the tangent is negative in the second quadrant,

$$\tan\left(\sec^{-1}\left(\frac{x}{2}\right)\right) = \tan u = -\frac{\sqrt{x^2 - 4}}{2}.$$

Therefore for $x < -2$, the solution of our integral is

$$\int \frac{1}{\sqrt{x^2 - 4}}\, dx = -\ln\left|\sec\left(\sec^{-1}\left(\frac{x}{2}\right)\right) + \tan\left(\sec^{-1}\left(\frac{x}{2}\right)\right)\right| + C = -\ln\left|\frac{x}{2} - \frac{\sqrt{x^2 - 4}}{2}\right| + C.$$

CHECKING THE ANSWER

In the previous example we got different answers depending on whether we had $x > 2$ or $x < -2$. We can see graphically why this has to be the case, by plotting the integrand $f(x)$ and one of its antiderivatives $F(x)$ (we chose $C = 0$ in each case), as shown in the following graphs:

$$f(x) = \frac{1}{\sqrt{x^2 - 4}}$$

$$F(x) = \begin{cases} \ln\left|\dfrac{x}{2} + \dfrac{\sqrt{x^2 - 4}}{2}\right|, & \text{if } x > 2 \\[3mm] -\ln\left|\dfrac{x}{2} - \dfrac{\sqrt{x^2 - 4}}{2}\right|, & \text{if } x < -2 \end{cases}$$

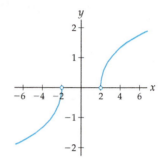

You should be able to convince yourself from looking at these graphs that it is at least reasonable that the derivative of $F(x)$ is equal to $f(x)$. Notice also that it makes sense that we had to look at two separate cases to integrate $f(x)$, since that function is defined on two different intervals (and not between).

EXAMPLE 5   **Choosing an appropriate trigonometric substitution**

Determine appropriate trigonometric substitutions for each of the integrals that follow. Apply the substitution and simplify until the integral is one that you know how to solve.

(a) $\displaystyle \int \sqrt{x^2 - 7}\, dx$ **(b)** $\displaystyle \int \sqrt{4x^2 + 9}\, dx$

SOLUTION

(a) This integrand has the form $\sqrt{x^2 - a^2}$, with $a = \sqrt{7}$. According to Theorem 8.17, this suggests that we apply trigonometric substitution with $x = \sqrt{7}\sec u$. With that substitution we have $dx = \sqrt{7}\sec u \tan u\, du$, so we can rewrite the integral as

$$\int \sqrt{x^2 - 7}\, dx = \int \sqrt{7\sec^2 u - 7}\ \sqrt{7}\sec u \tan u\, du.$$

Notice that the domain of the integrand $\sqrt{x^2 - 7}$ is $(-\infty, -\sqrt{7}\,] \cup [\sqrt{7}, \infty)$. According to Theorem 8.17, we must handle the cases $x < -\sqrt{7}$ and $x > \sqrt{7}$ separately. Keeping this in mind, we can apply a Pythagorean identity to write

$$\sqrt{7\sec^2 u - 7} = \sqrt{7\tan^2 u} = \sqrt{7}\,|\tan u| = \begin{cases} -\sqrt{7}\,\tan u, & \text{if } x < -\sqrt{7} \\[2mm] \sqrt{7}\,\tan u, & \text{if } x > \sqrt{7}. \end{cases}$$

This means that for $x < -\sqrt{7}$ our integral becomes $7\int \sec u \tan^2 u\, du$, and for $x > \sqrt{7}$ our integral becomes $-7\int \sec u \tan^2 u\, du$. Both of these integrals can be solved with the methods of Section 8.4.

(b) This integrand involves an expression that is the sum of two squares: $(2x)^2 + 3^2$. This suggests that we should apply trigonometric substitution with $2x = 3 \tan u$, or equivalently, $x = \frac{3}{2} \tan u$. With this choice we have $dx = \frac{3}{2} \sec^2 u\, du$, so our integral can be written as

$$\int \sqrt{4x^2 + 9}\, dx = \int \sqrt{4 \left(\frac{3}{2} \tan u \right)^2 + 9} \left(\frac{3}{2} \sec^2 u \right) du.$$

After a little bit of algebra the latter integral simplifies to $\frac{9}{2} \int \sec^3 u\, du$, which we know how to solve by the methods of the previous section. $\square$

EXAMPLE 6   **Completing the square before trigonometric substitution**

Determine an appropriate trigonometric substitution for the following integral:

$$\int \frac{1}{\sqrt{x^2 - 6x + 13}}\, dx$$

Then apply the substitution and simplify until the integral is one that you know how to solve.

SOLUTION

To use trigonometric substitution, we need to have an expression of the form $\sqrt{a^2 - x^2}$, $x^2 + a^2$, or $\sqrt{x^2 - a^2}$ in the integrand. To obtain a sum or difference of perfect squares we will apply the method of **completing the square** to the quadratic $x^2 - 6x + 13$:

$$x^2 - 6x + 13 = (x^2 - 6x + 9) - 9 + 13 = (x - 3)^2 + 4.$$

We added and subtracted 9 from the quadratic so that it would involve the perfect square $x^2 - 6x + 9 = (x - 3)^2$. In general, recall from Section 8.3 that to complete the square of a quadratic $x^2 + bx + c$, we add and subtract $(b/2)^2$. This enables us to write the quadratic in terms of the perfect square $(x - (b/2))^2$ and some other constant. Now that we have completed the square, we have

$$\int \frac{1}{\sqrt{x^2 - 6x + 13}}\, dx = \int \frac{1}{\sqrt{(x - 3)^2 + 4}}\, dx.$$

The part of the integrand inside the radical is *almost* of the form $x^2 + a^2$. Instead, it is of the form $(x - 3)^2 + a^2$ (with $a = 2$). Therefore we will choose the substitution $x - 3 = 2 \tan u$. This means that $dx = 2 \sec^2 u\, du$, and our integral becomes

$$\int \frac{1}{\sqrt{x^2 - 6x + 13}}\, dx = \int \frac{1}{\sqrt{(x - 3)^2 + 4}} = \int \frac{1}{\sqrt{4 \tan^2 u + 4}} (2 \sec^2 u)\, du,$$

which simplifies easily to $\int \sec u\, du$, an integral that we know how to solve. $\square$

TEST YOUR UNDERSTANDING

▸ In the first subsection we did one example two different ways. In the first way we had $du = \frac{1}{\sqrt{1 - x^2}}\, dx$, and in the second we had $dx = \cos u\, du$. Why are these the same if $x = \sin u$?

▸ What are the three trigonometric identities that come in handy for simplifying after performing a trigonometric substitution?

▸ Why would trigonometric substitution not be a good strategy for solving the integral $\int \frac{1}{x^2(1 - x^2)}\, dx$?

▶ Why do we have to consider cases for different domains and quadrants when doing a trigonometric substitution with secant, but not when we use trigonometric substitution with sine or tangent?

▶ We can use the unit circle and reference triangles to write certain trigonometric compositions as algebraic functions. What will go wrong if we use that technique to simplify $\sin(2\cos^{-1}x)$, and how can we fix it?

EXERCISES 8.5

Thinking Back

Domains and ranges of inverse trigonometric functions: For each function that follows, (a) list the domain and range, (b) sketch a labeled graph, and (c) discuss the domains and ranges in the context of the unit circle.

▶ $f(x) = \sin^{-1}x$ ▶ $f(x) = \tan^{-1}x$

▶ $f(x) = \sec^{-1}x$ ▶ $f(x) = \sin^{-1}\left(\dfrac{x}{3}\right)$

Working with inverse trigonometric functions: Solve each of the following problems by hand, using reference triangles in the unit circle.

▶ If $x = \tan u$ and $u \in \left(-\dfrac{\pi}{2}, \dfrac{\pi}{2}\right)$, find $\sin u$.

▶ If $x = 3\sin u$ and $u \in \left[-\dfrac{\pi}{2}, \dfrac{\pi}{2}\right]$, find $\cot u$.

▶ If $x = \sec u$ and $u \in \left[0, \dfrac{\pi}{2}\right)$, find $\tan u$.

▶ If $x = \sec u$ and $u \in \left(\dfrac{\pi}{2}, \pi\right]$, find $\tan u$.

Review of trigonometric integrals: Use the methods of the previous section to find each of the following integrals.

▶ $\displaystyle\int \sec x\, dx$ ▶ $\displaystyle\int \sec^2 x\, dx$

▶ $\displaystyle\int \sec^3 x\, dx$ ▶ $\displaystyle\int \sec x\tan^2 x\, dx$

Review of other integration methods: Use any method *except* trigonometric substitution to find each of the following integrals.

▶ $\displaystyle\int \dfrac{1}{\sqrt{1-x^2}}\, dx$ ▶ $\displaystyle\int \dfrac{x}{\sqrt{x^2-9}}\, dx$

▶ $\displaystyle\int \dfrac{9+x^2}{x}\, dx$ ▶ $\displaystyle\int \dfrac{1+x^4}{1+x^2}\, dx$

▶ $\displaystyle\int \dfrac{1}{1+4x^2}\, dx$ ▶ $\displaystyle\int \dfrac{x^3}{\sqrt{x^2+1}}\, dx$

Concepts

0. *Problem Zero:* Read the section and make your own summary of the material.

1. *True/False:* Determine whether each of the statements that follow is true or false. If a statement is true, explain why. If a statement is false, provide a counterexample.

(a) *True or False:* The substitution $x = 2\sec u$ is a suitable choice for solving $\int \dfrac{1}{x^2-4}\, dx$.

(b) *True or False:* The substitution $x = 2\sec u$ is a suitable choice for solving $\int \dfrac{1}{\sqrt{x^2-4}}\, dx$.

(c) *True or False:* The substitution $x = 2\tan u$ is a suitable choice for solving $\int \dfrac{1}{x^2+4}\, dx$

(d) *True or False:* The substitution $x = 2\sin u$ is a suitable choice for solving $\int (x^2+4)^{-5/2}\, dx$

(e) *True or False:* Trigonometric substitution is a useful strategy for solving any integral that involves an expression of the form $\sqrt{x^2-a^2}$.

(f) *True or False:* Trigonometric substitution doesn't solve an integral; rather, it helps you rewrite integrals as ones that are easier to solve by other methods.

(g) *True or False:* When using trigonometric substitution with $x = a\sin u$, we must consider the cases $x > a$ and $x < -a$ separately.

(h) *True or False:* When using trigonometric substitution with $x = a\sec u$, we must consider the cases $x > a$ and $x < -a$ separately.

2. *Examples:* Construct examples of the thing(s) described in the following. Try to find examples that are different than any in the reading.

(a) An integral to which we could reasonably apply trigonometric substitution with $x = \tan u$.

(b) An integral to which we could reasonably apply trigonometric substitution with $x = 4\sec u$.

(c) An integral to which we could reasonably apply trigonometric substitution with $x - 2 = 3\sin u$.

3. Show that if $x = \tan u$, then $dx = \sec^2 u\, du$, in the following two ways: (a) by using implicit differentiation, thinking of u as a function of x and (b) by thinking of x as a function of u.

4. Show by differentiating (and then using algebra) that $-\cot(\sin^{-1}x)$ and $-\dfrac{\sqrt{1-x^2}}{x}$ are both antiderivatives of $\dfrac{1}{x^2\sqrt{1-x^2}}$. How can these two very different-looking functions be antiderivatives of the same function?

5. Consider the integral $\int \frac{1}{x^2\sqrt{1-x^2}}\,dx$ from the reading at the beginning of the section.

   (a) Use the inverse trigonometric substitution $u = \sin^{-1} x$ to solve this integral.

   (b) Use the trigonometric substitution $x = \sin u$ to solve the integral.

   (c) Compare and contrast the two methods used in parts (a) and (b).

6. Give an example of an integral for which trigonometric substitution is possible but an easier method is available. Then give an example of an integral that we still don't know how to solve given the techniques we know at this point.

7. Why don't we ever have cause to use the trigonometric substitution $x = \cos u$?

8. Explain how to know when to use the trigonometric substitutions $x = a \sin u$, $x = a \tan u$, and $x = a \sec u$. Describe the trigonometric identity and the triangle that will be needed in each case. What are the possible values for x and u in each case?

9. Why don't we need to have a square root involved in order to apply trigonometric substitution with the tangent? In other words, why can we use the substitution $x = a \tan u$ when we see $x^2 + a^2$, even though we can't use the substitution $x = a \sin u$ unless the integrand involves the square root of $a^2 - x^2$? (Hint: Think about domains.)

10. Explain why it makes sense to try the trigonometric substitution $x = \sec u$ if an integrand involves the expression $\sqrt{x^2 - 1}$.

11. Which of the integrals that follow would be good candidates for trigonometric substitution? If trigonometric substitution is a good strategy, name the substitution. If another method is a better strategy, explain that method.

   (a) $\int \frac{4+x^2}{x}\,dx$　　　(b) $\int \frac{x}{4+x^2}\,dx$

   (c) $\int \frac{x^2}{4+x^2}\,dx$　　　(d) $\int \frac{16-x^4}{4+x^2}\,dx$

12. Explain why using trigonometric substitution with $x = a \tan u$ often involves a triangle with side lengths a and x and hypotenuse of length $\sqrt{x^2 + a^2}$.

13. Explain why, if $x = a \sin u$, then $\sqrt{a^2 - x^2} = a \cos u$. Your explanation should include a discussion of domains and absolute values.

14. Explain why, if $x = a \tan u$, then $\sqrt{x^2 + a^2} = a \sec u$. Your explanation should include a discussion of domains and absolute values.

15. Explain why, if $x = a \sec u$, then $\sqrt{a^2 \sec^2 u - a^2}$ is $-a \tan u$ if $x < -a$ and is $a \tan u$ if $x > a$. Your explanation should include a discussion of domains and absolute values.

16. Why is it okay to use a triangle without thinking about the unit circle when simplifying expressions that result from a trigonometric substitution with $x = a \sin u$ or $x = a \tan u$? Why do we need to think about

the unit circle after trigonometric substitution with $x = a \sec u$?

17. Why doesn't the definite integral $\int_2^3 \sqrt{1-x^2}\,dx$ make sense? (Hint: Think about domains.)

18. Find three integrals in Exercises 39–74 that can be solved by using a trigonometric substitution of the form $x = a \tan u$.

19. Find three integrals in Exercises 39–74 that can be solved without using trigonometric substitution.

In Exercises 20–27, use reference triangles and the unit circle to write the given trigonometric compositions as algebraic functions.

20. $\cot(\sec^{-1} x)$

21. $\cos(\sin^{-1} x)$

22. $\tan\left(\sin^{-1}\left(\frac{x}{3}\right)\right)$

23. $\sin\left(\tan^{-1}\left(\frac{x}{2}\right)\right)$

24. If $x + 3 = 4 \tan u$, then write $\sec u$ as an algebraic function of x.

25. If $x - 5 = 3 \sin u$, then write $\tan^2 u$ as an algebraic function of x.

26. Write $\tan(2 \sec^{-1} x)$ as an algebraic function.

27. Write $\sin(2 \cos^{-1} x)$ as an algebraic function.

Complete the square for each quadratic in Exercises 28–33. Then describe the trigonometric substitution that would be appropriate if you were solving an integral that involved that quadratic.

28. $x^2 - 4x - 8$

29. $x^2 + 6x - 2$

30. $2x^2 - 4x + 1$

31. $x^2 - 5x + 1$

32. $x - 3x^2$

33. $2(x + 2)^2$

34. Solve $\int \frac{x^2}{1+x^2}\,dx$ the following two ways:

   (a) with the substitution $u = \tan^{-1} x$;

   (b) with the trigonometric substitution $x = \tan u$.

35. Solve $\int \frac{1}{x^2+9}\,dx$ the following two ways:

   (a) with the trigonometric substitution $x = 3 \tan u$;

   (b) with algebra and the derivative of the arctangent.

36. Solve $\int \frac{x+2}{(x^2+4x)^{3/2}}\,dx$ the following two ways:

   (a) with the substitution $u = x^2 + 4x$;

   (b) by completing the square and then applying the trigonometric substitution $x + 2 = 2 \sec u$.

37. Solve $\int \frac{x}{4+x^2}\,dx$ the following two ways:

   (a) with the substitution $u = 4 + x^2$;

   (b) with the trigonometric substitution $x = 2 \tan u$.

38. Solve the integral $\int x^3\sqrt{x^2-1}\,dx$ three ways:

   (a) with the substitution $u = x^2 - 1$, followed by back-substitution;

   (b) with integration by parts, choosing $u = x^2$ and $dv = x\sqrt{x^2-1}\,dx$;

   (c) with the trigonometric subsitution $x = \sec u$.

Skills

Solve each of the integrals in Exercises 39–74. Some integrals require trigonometric substitution, and some do not. Write your answers as algebraic functions whenever possible.

39. $\displaystyle\int \frac{\sqrt{4-x^2}}{x^2}\,dx$

40. $\displaystyle\int x^3\sqrt{x^2+1}\,dx$

41. $\displaystyle\int \frac{x}{\sqrt{x^2-1}}\,dx$

42. $\displaystyle\int x\sqrt{x^2+1}\,dx$

43. $\displaystyle\int \frac{1}{\sqrt{3-x^2}}\,dx$

44. $\displaystyle\int \frac{1}{(x^2+4)^{3/2}}\,dx$

45. $\displaystyle\int \frac{\sqrt{9-x^2}}{x}\,dx$

46. $\displaystyle\int \frac{x^2+9}{x^{3/2}}\,dx$

47. $\displaystyle\int \sqrt{3-x}\sqrt{x-1}\,dx$

48. $\displaystyle\int \sqrt{4-x}\sqrt{x-2}\,dx$

49. $\displaystyle\int \frac{1}{x^2\sqrt{x^2-9}}\,dx$

50. $\displaystyle\int \frac{\sqrt{x^2-4}}{2x}\,dx$

51. $\displaystyle\int (1-x^2)^{-3/2}\,dx$

52. $\displaystyle\int \frac{1}{x^2\sqrt{x^2+3}}\,dx$

53. $\displaystyle\int \frac{1}{x^2-4x+13}\,dx$

54. $\displaystyle\int \frac{x}{\sqrt{1-4x^2}}\,dx$

55. $\displaystyle\int \frac{3+x}{\sqrt{9-4x^2}}\,dx$

56. $\displaystyle\int \frac{1}{\sqrt{x^2+9}}\,dx$

57. $\displaystyle\int \sqrt{x^2-8x+25}\,dx$

58. $\displaystyle\int \sqrt{x^2+6x+18}\,dx$

59. $\displaystyle\int (3-x^2)^{3/2}\,dx$

60. $\displaystyle\int \frac{x^3}{3x^2+5}\,dx$

61. $\displaystyle\int \sqrt{2-x^2}\,dx$

62. $\displaystyle\int \frac{1}{(x^2+1)^{5/2}}\,dx$

63. $\displaystyle\int \frac{\sqrt{x^2-1}}{x}\,dx$

64. $\displaystyle\int \frac{1}{x^2\sqrt{4-x^2}}\,dx$

65. $\displaystyle\int \frac{1}{\sqrt{x^2-2}}\,dx$

66. $\displaystyle\int \sqrt{x^2-2}\,dx$

67. $\displaystyle\int x^3\sqrt{x^2+1}\,dx$

68. $\displaystyle\int x^5\sqrt{x^2-1}\,dx$

69. $\displaystyle\int \ln(x^2+1)\,dx$

70. $\displaystyle\int \ln(4+x^2)\,dx$

71. $\displaystyle\int e^{2x}\sqrt{e^{2x}+1}\,dx$

72. $\displaystyle\int e^{2x}(1-e^{4x})^{3/2}\,dx$

73. $\displaystyle\int \frac{1}{e^x\sqrt{e^{2x}+3}}\,dx$

74. $\displaystyle\int x\sin^{-1}x\,dx$

Solve each of the integrals in Exercises 75–78 by using polynomial long division to rewrite the integrand. This is one way that you can sometimes avoid using trigonometric substitution; moreover, sometimes it works when trigonometric substitution does not apply.

75. $\displaystyle\int \frac{x^3}{x^2+4}\,dx$

76. $\displaystyle\int \frac{x^2-1}{x^2+1}\,dx$

77. $\displaystyle\int \frac{x^4-3}{2+3x^2}\,dx$

78. $\displaystyle\int \frac{x^3-3x^2+2x-3}{x^2+1}\,dx$

Solve each of the definite integrals in Exercises 79–86. Some integrals require trigonometric substitution, and some do not.

79. $\displaystyle\int_0^4 x\sqrt{x^2+4}\,dx$

80. $\displaystyle\int_3^5 \sqrt{x^2-9}\,dx$

81. $\displaystyle\int_4^5 \frac{1}{x\sqrt{x^2+9}}\,dx$

82. $\displaystyle\int_1^2 \frac{1}{x\sqrt{9-x^2}}\,dx$

83. $\displaystyle\int_{1/4}^{1/2} \frac{1}{x^2\sqrt{1-x^2}}\,dx$

84. $\displaystyle\int_{-1}^1 x^3\sqrt{9x^2-1}\,dx$

85. $\displaystyle\int_1^2 \frac{3}{(9-2x^2)^{3/2}}\,dx$

86. $\displaystyle\int_0^4 x^3\sqrt{x^2+4}\,dx$

Applications

87. A pharmaceutical company is designing a new drug whose shape in tablet form is obtained by rotating the

Shape of a tablet

graph of $f(x) = 2(42-x^2)^{1/4}$ on the interval $[-6, 6]$ around the x-axis, where units are measured in millimeters.

The volume of such a solid is given by $\pi \int_a^b (f(x))^2\,dx$, where a and b are the places where the graph of $f(x)$ meets the x-axis.

(a) Write down a specific definite integral that represents the amount of material, in cubic millimeters, required to make the tablet.

(b) Use trigonometric substitution to solve the definite integral and determine the amount of material needed.

88. The main cable on a certain suspension bridge follows a parabolic curve with equation $f(x) = (0.025x)^2$, measured in feet, as shown in the following figure:

Main cable of a suspension bridge

The length of a curve $f(x)$ from $x = a$ to $x = b$ can be calculated from the formula $\int_a^b \sqrt{1 + (f'(x))^2}\, dx$.

(a) Write down a specific definite integral that represents the length of the main cable of the suspension bridge.

(b) Use trigonometric substitution to solve the definite integral and determine the length of the cable.

Proofs

89. Prove that the area of a circle of radius r is πr^2, as follows:
(a) Write down a definite integral that represents the area of the circle of radius r centered at the origin. (*Hint: The equation of such a circle is $x^2 + y^2 = r^2$.*)
(b) Use trigonometric substitution to solve the definite integral you found in part (a). (*Hint: Change the limits of integration to match your substitution.*)

90. Prove part (b) of Theorem 8.17: For $x \in (-\infty, \infty)$ and $u \in (-\pi/2, \pi/2)$, the substitution $x = a\tan u$ gives $x^2 + a^2 = a\sec u$. Your proof should include a discussion of domains and a consideration of absolute values.

Thinking Forward

Arc length of a curve: Suppose we want to find the length of the curve traced out by the graph of a function $f(x)$ on an interval $[a, b]$. The obvious way to approximate this length is to use a series of straight line segments such as the one shown in the figure.

Approximating length with a line segment

By the distance formula, the length of the kth segment is $\sqrt{(\Delta x)^2 + (\Delta y_k)^2}$, where Δy_k will be different for each segment we use.

▶ Use algebra to show that the distance of the kth segment can be rewritten as

$$\sqrt{1 + \left(\frac{\Delta y_k}{\Delta x}\right)^2}\, \Delta x.$$

▶ Approximate the length of the graph of $f(x) = \sqrt{9 - x^2}$ on $[-3, 3]$, using four line segments and the distance formula. Then make a better approximation with eight line segments.

▶ The definite integral

$$L = \int_a^b \sqrt{1 + (f'(x))^2}\, dx$$

calculates the length of the curve of $f(x)$ from $x = a$ to $x = b$. Explain intuitively what this might have to do with the distance formula calculation.

▶ Use the formula just given for arc length to write down a specific definite integral that represents the length of $f(x) = \sqrt{9 - x^2}$ on $[-3, 3]$. Then use trigonometric substitution to solve that definite integral.

▶ Verify your previous answer by using the formula for the circumference of a circle.

Surface area of a solid of revolution: If $f(x)$ is a differentiable function with a continuous derivative, then the surface area of the solid of revolution obtained by rotating the region under the graph of $f(x)$ on the interval $[a, b]$ around the x-axis is given by the definite integral

$$S = 2\pi \int_a^b f(x)\sqrt{1 + (f'(x))^2}\, dx.$$

▶ Calculate the surface area of the solid of revolution obtained by revolving the region between $f(x) = 2x^3$ and the x-axis on $[0, 3]$ around the x-axis.

▶ Use the formula just given for surface area to write down and solve a specific definite integral to show that the surface area of a sphere of radius r is given by the formula $S = 4\pi r^2$.

8.6   IMPROPER INTEGRALS

▶ Defining improper integrals over unbounded intervals and for unbounded functions

▶ Determining the convergence or divergence of improper integrals of power functions

▶ The comparison test for improper integrals

Integrating over an Unbounded Interval

So far we have examined definite integrals $\int_a^b f(x)\,dx$ only for continuous functions $f(x)$ over finite intervals $[a, b]$, since those are the conditions under which the Fundamental Theorem of Calculus applies. If $f(x)$ fails to be continuous on the interval or if the interval itself is unbounded, then the Fundamental Theorem of Calculus does not apply and we say that the integral is ***improper***. In this section we develop definitions and strategies for examining such improper integrals.

For example, consider the functions $\dfrac{1}{\sqrt{x}}$, $\dfrac{1}{x}$, and $\dfrac{1}{x^2}$ on the interval $[1, \infty)$, as shown in the three figures that follow. Perhaps unsurprisingly, for the first two graphs there is an infinite amount of area between the graph of the function and the x-axis on this unbounded interval. However, you might be shocked to learn that for the third graph this area turns out to be finite!

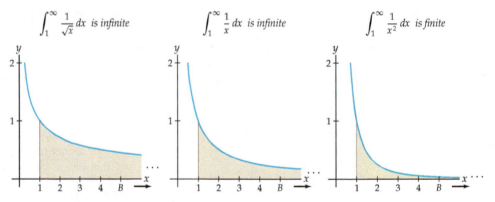

$\displaystyle\int_1^\infty \frac{1}{\sqrt{x}}\,dx$ *is infinite* $\displaystyle\int_1^\infty \frac{1}{x}\,dx$ *is infinite* $\displaystyle\int_1^\infty \frac{1}{x^2}\,dx$ *is finite*

How could the graph of $\dfrac{1}{x^2}$ be positive on all of $[1, \infty)$ and yet collect only a finite amount of area on this interval? Let's look at the area under each graph for intervals of the form $[1, B)$, with B getting larger and larger:

	$B = 5$	$B = 10$	$B = 50$	$B = 100$	$B = 500$	$B = 1000$	$B \to \infty$
$\displaystyle\int_1^B \frac{1}{\sqrt{x}}\,dx$	6.787	20.415	235.036	666.000	7,452.893	21,081.184	$\to \infty$?
$\displaystyle\int_1^B \frac{1}{x}\,dx$	1.609	2.302	3.912	4.605	6.215	6.908	$\to \infty$??
$\displaystyle\int_1^B \frac{1}{x^2}\,dx$	0.800	0.900	0.980	0.990	0.998	0.999	$\to 1$???

As B increases, the area under the graph of $\dfrac{1}{\sqrt{x}}$ on $[1, B)$ seems to grow without bound. The same seems true of $\dfrac{1}{x}$, but the growth is much slower. For $\dfrac{1}{x^2}$, however, the area under the graph on $[1, B)$ seems to stay below 1 no matter how large a value we take for B. The graph of $\dfrac{1}{x^2}$ decreases in some fundamentally faster way than the other two graphs, so much so that it collects only a finite amount of area.

We can investigate these areas algebraically over finite intervals $[1, B)$ by using the Fundamental Theorem of Calculus. For $B > 1$ we have

$$\int_1^B \frac{1}{\sqrt{x}}\, dx = \left[2\sqrt{x}\right]_1^B = 2\sqrt{B} - 2\sqrt{1} = 2\sqrt{B} - 2,$$

$$\int_1^B \frac{1}{x}\, dx = \left[\ln|x|\right]_1^B = \ln B - \ln 1 = \ln B,$$

$$\int_1^B \frac{1}{x^2}\, dx = \left[-\frac{1}{x}\right]_1^B = -\frac{1}{B} + \frac{1}{1} = 1 - \frac{1}{B}.$$

We can now see that the values of the first two definite integrals will grow without bound as $B \to \infty$, since $\displaystyle\lim_{B\to\infty} (2\sqrt{B} - 2) \to 2(\infty) - 2 = \infty$ and $\displaystyle\lim_{B\to\infty} \ln B = \infty$. In contrast, since $\displaystyle\lim_{B\to\infty} (1 - (1/B)) = 1 - \frac{1}{\infty} = 1$, the value of the third definite integral approaches 1.

In general, we define integrals over unbounded intervals as limits of integrals over ever-larger finite intervals:

DEFINITION 8.18   **Improper Integrals over Unbounded Intervals**

Suppose a and b are any real numbers and $f(x)$ is a function.

  (a) If f is continuous on $[a, \infty)$, then $\displaystyle\int_a^\infty f(x)\, dx = \lim_{B\to\infty} \int_a^B f(x)\, dx.$

  (b) If f is continuous on $(-\infty, b]$, then $\displaystyle\int_{-\infty}^b f(x)\, dx = \lim_{A\to-\infty} \int_A^b f(x)\, dx.$

  (c) If f is continuous on $(-\infty, \infty)$, then define the integral over $(-\infty, \infty)$ as a sum of the first two cases, splitting at any real number c:

$$\int_{-\infty}^\infty f(x)\, dx = \lim_{A\to-\infty} \int_A^c f(x)\, dx + \lim_{B\to\infty} \int_c^B f(x)\, dx$$

In all three cases, if the limits involved exist, then we say that the improper integral **converges** to the value determined by those limits, and if the limits are infinite or fail to exist, then we say that the improper integral **diverges**.

Remember that a limit exists if it is equal to a finite number, so an improper integral converges if it is equal to some finite number and diverges if it is infinite or for some other reason does not have a well-defined limit. In Exercise 81 you will use properties of definite integrals to prove that we can in fact split the integral in the third part of the theorem at any point c that we like. In the terminology of this theorem, the improper integrals $\int_1^\infty \frac{1}{\sqrt{x}}\, dx$ and $\int_1^\infty \frac{1}{x}\, dx$ diverge while the improper integral $\int_1^\infty \frac{1}{x^2}\, dx$ converges to 1.

Integrating Unbounded Functions

We now know how to extend our theory of definite integrals to unbounded intervals. What about over finite intervals, but where the function itself is unbounded? The Fundamental Theorem of Calculus does not apply to integrands that have discontinuities such as vertical asymptotes, but we can again use limits to determine whether integrals of unbounded functions converge or diverge.

For example, consider the same three functions we were working with earlier, but over a different graphing window to illustrate their behavior on $[0, 1]$, as shown in the next three figures. Each graph has a vertical asymptote at $x = 0$. In the second and third figures the area between the graph and the x-axis on $[0, 1]$ will turn out to be infinite, and in the first figure the graph of $\frac{1}{\sqrt{x}}$ will turn out to have a finite area over $[0, 1]$.

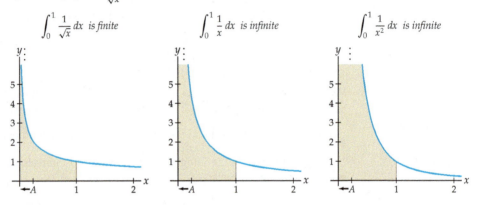

$$\int_0^1 \frac{1}{\sqrt{x}} \, dx \text{ is finite} \qquad \int_0^1 \frac{1}{x} \, dx \text{ is infinite} \qquad \int_0^1 \frac{1}{x^2} \, dx \text{ is infinite}$$

As suggested by the labels under the axes in each figure, what we should think about is the area under these curves on intervals of the form $(A, 1)$ as A gets closer and closer to 0 from the right. For $0 < A < 1$ the Fundamental Theorem of Calculus tells us that

$$\int_A^1 \frac{1}{\sqrt{x}} \, dx = \left[2\sqrt{x} \right]_A^1 = 2\sqrt{1} - 2\sqrt{A} = 2 - 2\sqrt{A},$$

$$\int_A^1 \frac{1}{x} \, dx = \left[\ln |x| \right]_A^1 = \ln 1 - \ln A = -\ln A,$$

$$\int_A^1 \frac{1}{x^2} \, dx = \left[-\frac{1}{x} \right]_A^1 = -\frac{1}{1} + v1A = \frac{1}{A} - 1.$$

As $A \to 0^+$, we have $\sqrt{A} \to 0$, $\ln A \to -\infty$, and $\frac{1}{A} \to \infty$, so the first integral will approach 2 and the second and third will both approach ∞. Therefore, of these three functions, only $\frac{1}{\sqrt{x}}$ has a finite area over $[0, 1]$.

The preceding example suggests how we should define improper integrals $\int_a^b f(x) \, dx$ for which $f(x)$ has a discontinuity at an endpoint of $[a, b]$. In cases where $f(x)$ is discontinuous somewhere in the interior of $[a, b]$, we have to split the integral at the discontinuity:

DEFINITION 8.19

Improper Integrals over Discontinuities

Suppose a and b are any real numbers and $f(x)$ is a function.

(a) If f is continuous on $(a, b]$ but not at $x = a$, then $\displaystyle\int_a^b f(x) \, dx = \lim_{A \to a^+} \int_A^b f(x) \, dx$.

(b) If f is continuous on $[a, b)$ but not at $x = b$, then $\displaystyle\int_a^b f(x) \, dx = \lim_{B \to b^-} \int_a^B f(x) \, dx$.

(c) If f is continuous on $[a, c) \cup (c, b]$ but not at $x = c$, then we can define the integral over $[a, b]$ as the sum of the first two cases:

$$\int_a^b f(x) \, dx = \lim_{B \to c^-} \int_a^B f(x) \, dx + \lim_{A \to c^+} \int_A^b f(x) \, dx$$

In all three cases, if the limits involved exist, then we say that the improper integral **converges** to the value determined by those limits, and if the limits are infinite or fail to exist, then we say that the improper integral **diverges**.

Improper Integrals of Power Functions

The function $\frac{1}{x}$ is a kind of boundary between power functions whose integrals converge and power functions whose integrals diverge. On both $[0, 1]$ and $[1, \infty)$ the improper integral of $\frac{1}{x}$ diverges. A power function of the form $\frac{1}{x^p}$ has an improper integrals that converges or diverges on an interval according to whether that power function is less than or greater than $\frac{1}{x}$ on that interval. We use the letter p in our notation here because the integrals we are studying now are closely related to certain infinite sums that we will later call *p-series* in later calculus courses.

More precisely, in Exercises 74 and 75 you will use limits of proper definite integrals to show that, for improper integrals over $[1, \infty)$, we have the following theorem:

THEOREM 8.20

Improper Integrals of Power Functions on $[1, \infty)$

(a) If $0 < p \leq 1$, then $\frac{1}{x^p} \geq \frac{1}{x}$ for $x \in [1, \infty)$ and $\int_1^\infty \frac{1}{x^p}\, dx$ diverges.

(b) If $p > 1$, then $\frac{1}{x^p} < \frac{1}{x}$ for $x \in [1, \infty)$ and $\int_1^\infty \frac{1}{x^p}\, dx$ converges to $\frac{1}{p-1}$.

Notice that we do not bother to consider power functions like x^2 or x^3 in this theorem, since those functions are increasing on $[1, \infty)$ and therefore have improper integrals that obviously diverge over that interval. The statements in the theorem about whether $\frac{1}{x^p}$ is greater or less than $\frac{1}{x}$ are there to help us remember whether we have convergence or divergence in each case. Theorem 8.20 will be a key tool in our investigation of improper integrals.

In addition, in Exercises 76 and 77 you will prove that improper integrals of power functions over $[0, 1]$ also either converge or diverge according to how closely they compare to $\frac{1}{x}$, as follows:

THEOREM 8.21

Improper Integrals of Power Functions on $[0, 1]$

(a) If $0 < p < 1$, then $\frac{1}{x^p} < \frac{1}{x}$ for $x \in [0, 1]$ and $\int_0^1 \frac{1}{x^p}\, dx$ converges to $\frac{1}{1-p}$.

(b) If $p \geq 1$, then $\frac{1}{x^p} \geq \frac{1}{x}$ for $x \in [0, 1]$ and $\int_0^1 \frac{1}{x^p}\, dx$ diverges.

Again we do not consider power functions such as x^2 or x^3 in the theorem because their integrals over $[0, 1]$ are not improper. (See Example 2 for an illustration of the four cases in the two previous theorems.)

Determining Convergence or Divergence with Comparisons

We now have a complete understanding of the convergence or divergence of improper integrals of power functions. These special examples of improper integrals will be the yardsticks against which we compare many other types of improper integrals. This comparison will enable us to determine the convergence or divergence of many improper integrals without actually doing any integration or antidifferentiation.

Loosely speaking, the key idea that will allow us to compare two improper integrals is this: Improper integrals that are smaller than convergent ones will also converge, and improper integrals that are larger than divergent ones will also diverge. This idea makes intuitive sense because a (positive) quantity that is smaller than a finite number will also be finite and a quantity that is larger than an infinite quantity will also be infinite. More precisely, we have the following theorem:

THEOREM 8.22

Comparison Test for Improper Integrals

Suppose f and g are functions that are continuous on an interval I.

(a) If the improper integral of f on I converges and $0 \leq g(x) \leq f(x)$ for all $x \in I$ where the functions are defined, then the improper integral of g on I also converges.

(b) If the improper integral of f on I diverges and $0 \leq f(x) \leq g(x)$ for all $x \in I$ where the functions are defined, then the improper integral of g on I also diverges.

Theorem 8.22 applies to all the types of improper integrals we have studied. For example, for improper integrals on an unbounded interval $[a, \infty)$, the theorem says that

$$\int_a^\infty f(x)\, dx \text{ converges and}$$
$$0 \leq g(x) \leq f(x) \text{ for all } x \in [a, \infty) \implies \int_a^\infty g(x)\, dx \text{ also converges;}$$

$$\int_a^\infty f(x)\, dx \text{ diverges and}$$
$$g(x) \geq f(x) \text{ for all } x \in [a, \infty) \implies \int_a^\infty g(x)\, dx \text{ also diverges.}$$

It is important to note that Theorem 8.22 says nothing about improper integrals that are larger than convergent ones or about improper integrals that are smaller than divergent ones.

For example, consider the improper integrals $\int_1^\infty \frac{1}{x^2}\, dx$ and $\int_1^\infty \frac{\sin^2 x}{x^2}\, dx$. We already know that the first of these converges. Considering the fact that $-1 \leq \sin x \leq 1$, we have $0 \leq \sin^2 x \leq 1$ and therefore $0 \leq \frac{\sin^2 x}{x^2} \leq \frac{1}{x^2}$. These facts taken together tell us that $\int_1^\infty \frac{\sin^2 x}{x^2}\, dx$ must also converge, even without our having to integrate the function $\frac{\sin^2 x}{x^2}$.

Proof. We will prove the theorem in the case for convergence of improper integrals over unbounded intervals of the form $[a, \infty)$ and leave similar proofs of other cases to Exercises 78–80. Suppose that $f(x)$ is a function for which the improper integral $\int_a^\infty f(x)\, dx$ converges. If $g(x)$ is a function satisfying $0 \leq g(x) \leq f(x)$ for all $x \in [a, \infty)$, then for all $B > a$, we have the following inequality of definite integrals:

$$0 \leq \int_a^B g(x)\, dx \leq \int_a^B f(x)\, dx.$$

Now taking the limit as $B \to \infty$ on each side, we get

$$0 \leq \lim_{B \to \infty} \int_a^B g(x)\, dx \leq \lim_{B \to \infty} \int_a^B f(x)\, dx.$$

By the definition of improper integrals on unbounded intervals, this double inequality is equivalent to

$$0 \leq \int_a^\infty g(x)\, dx \leq \int_a^\infty f(x)\, dx.$$

We are given that $\int_a^\infty f(x)\, dx$ converges, say, to some real number L. Then the inequality we found shows that $\int_a^\infty g(x)\, dx$ must converge to a real number between 0 and L. ∎

Examples and Explorations

EXAMPLE 1   Writing improper integrals as limits of definite integrals

Express each improper integral that follows as a sum of limits of proper definite integrals. Do not calculate any integrals or limits; just write them down.

(a) $\displaystyle\int_0^1 (3x-1)^{-2/3}\,dx$ **(b)** $\displaystyle\int_0^1 \frac{1}{x\ln x}\,dx$ **(c)** $\displaystyle\int_1^\infty \frac{1}{(x-2)^2}\,dx$

SOLUTION

The graphs of the three functions we are attempting to integrate are shown here:

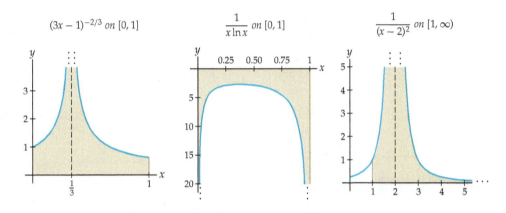

$(3x-1)^{-2/3}$ on $[0,1]$ $\dfrac{1}{x\ln x}$ on $[0,1]$ $\dfrac{1}{(x-2)^2}$ on $[1,\infty)$

We have to split each improper integral into a sum of integrals so that in each integral only one endpoint requires a limit. In parts (a) and (b) this will mean splitting into two integrals, and in part (c) we will require three integrals.

(a) The function $y = (3x-1)^{-2/3}$ is continuous everywhere except at $x = \frac{1}{3}$, where it has a vertical asymptote. Since $x = \frac{1}{3}$ is in the interior of the interval $[0,1]$, we must split the integral into two improper integrals:

$$\int_0^1 (3x-1)^{-2/3}\,dx = \int_0^{1/3} (3x-1)^{-2/3}\,dx + \int_{1/3}^1 (3x-1)^{-2/3}\,dx.$$

Each of these improper integrals can be written as a limit of definite integrals, which gives us

$$\int_0^1 (3x-1)^{-2/3}\,dx = \lim_{B\to 1/3^-} \int_0^B (3x-1)^{-2/3}\,dx + \lim_{A\to 1/3^+} \int_A^1 (3x-1)^{-2/3}\,dx.$$

If either of these limits of integrals diverges, then the entire improper integral diverges. If both limits of integrals converge, then the original improper integral is equal to the sum of those limits. (See Exercise 29.)

(b) The function $y = \dfrac{1}{x\ln x}$ has vertical asymptotes at both endpoints of the interval $[0,1]$, so we have to split this improper integral somewhere between $x = 0$ and $x = 1$. The point $x = \frac{1}{2}$ is as good a place as any to make the split:

$$\int_0^1 \frac{1}{x\ln x}\,dx = \int_0^{1/2} \frac{1}{x\ln x}\,dx + \int_{1/2}^1 \frac{1}{x\ln x}\,dx.$$

Now taking limits at the endpoints where there are vertical asymptotes, we have

$$\int_0^1 \frac{1}{x \ln x}\, dx = \lim_{A \to 0^+} \int_A^{1/2} \frac{1}{x \ln x}\, dx + \lim_{B \to 1^-} \int_{1/2}^B \frac{1}{x \ln x}\, dx.$$

You will determine whether this improper integral converges or diverges in Exercise 47.

(c) The graph of $y = \frac{1}{(x-2)^2}$ has a vertical asymptote at $x = 2$, and we need to express the improper integral over $[1, \infty)$. This means that we will need to split the integral at $x = 2$, as well as at some later point, say, $x = 3$, in order to separately consider the limit as $x \to \infty$. This approach forces a split into three integrals:

$$\int_1^\infty \frac{1}{(x-2)^2}\, dx = \int_1^2 \frac{1}{(x-2)^2}\, dx + \int_2^3 \frac{1}{(x-2)^2}\, dx + \int_3^\infty \frac{1}{(x-2)^2}\, dx.$$

Taking limits at the points where we have vertical asymptotes and as $x \to \infty$, we have

$$\int_1^\infty \frac{1}{(x-2)^2}\, dx = \lim_{B \to 2^-} \int_1^B \frac{1}{(x-2)^2}\, dx + \lim_{A \to 2^+} \int_A^3 \frac{1}{(x-2)^2}\, dx + \lim_{C \to \infty} \int_3^C \frac{1}{(x-2)^2}\, dx.$$

We used the letter C in the last limit only to distinguish it from the variable B that we used in the first limit. For calculations involving these improper integrals, see Exercises 30 and 31. ☐

EXAMPLE 2 **Calculating improper integrals of power functions**

For each power function $f(x)$ that follows, use limits of definite integrals to determine whether $\int_0^1 f(x)\, dx$ and $\int_1^\infty f(x)\, dx$ converge or diverge. Find the values of those improper integrals which converge.

(a) $f(x) = \dfrac{1}{\sqrt[3]{x}}$ 　　　　　(b) $f(x) = \dfrac{1}{x}$ 　　　　　(c) $f(x) = \dfrac{1}{x^3}$

SOLUTION

(a) Since $f(x) = \frac{1}{\sqrt[3]{x}}$ is continuous on $(0, 1]$ but has a vertical asymptote at $x = 0$, Definition 8.19 and the Fundamental Theorem of Calculus tell us that the improper integral on $[0, 1]$ converges:

$$\int_0^1 \frac{1}{\sqrt[3]{x}}\, dx = \lim_{A \to 0^+} \int_A^1 \frac{1}{\sqrt[3]{x}}\, dx = \lim_{A \to 0^+} \left[\frac{3}{2} x^{2/3}\right]_A^1 = \lim_{A \to 0^+} \left(\frac{3}{2} - \frac{3}{2} A^{2/3}\right) = \frac{3}{2}.$$

$f(x) = \frac{1}{\sqrt[3]{x}}\, dx$ is also continuous on $[1, \infty)$, and therefore Definition 8.18 and the Fundamental Theorem of Calculus tell us that the improper integral on $[1, \infty)$ diverges:

$$\int_1^\infty \frac{1}{\sqrt[3]{x}}\, dx = \lim_{B \to \infty} \int_1^B \frac{1}{\sqrt[3]{x}}\, dx = \lim_{B \to \infty} \left[\frac{3}{2} x^{2/3}\right]_1^B = \lim_{B \to \infty} \left(\frac{3}{2} B^{2/3} - \frac{3}{2}\right) = \infty.$$

(b) Earlier in the reading we used limits of definite integrals to determine that $\int_0^1 \frac{1}{x}\, dx$ and $\int_1^\infty \frac{1}{x}\, dx$ both diverge.

(c) Again we apply the definitions of improper integrals to get

$$\lim_{A\to 0^+}\int_A^1 \frac{1}{x^3}\,dx = \lim_{A\to 0^+}\left[-\frac{1}{2}x^{-2}\right]_A^1 = \lim_{A\to 0^+}\left(-\frac{1}{2}+\frac{1}{2}A^{-2}\right) = -\frac{1}{2}+\frac{1}{2(0^+)} = \infty,$$

$$\lim_{B\to\infty}\int_1^B \frac{1}{x^3}\,dx = \lim_{B\to\infty}\left[-\frac{1}{2}x^{-2}\right]_1^B = \lim_{B\to\infty}\left(-\frac{1}{2}B^{-2}+\frac{1}{2}\right) = -\frac{1}{2(\infty)}+\frac{1}{2} = \frac{1}{2}. \qquad \square$$

> ✔ **CHECKING THE ANSWER**

First, notice that the preceding two answers do match with the statements in Theorems 8.20 and 8.21. For $p = \frac{1}{3}$ only the integral on $[0, 1]$ converges, and it converges to $\frac{1}{1-p} = \frac{1}{1-(1/3)} = \frac{3}{2}$; for $p = 1$ neither improper integral converges; and for $p = 3$ only the integral on $[1, \infty)$ converges, to $\frac{1}{p-1} = \frac{1}{3-1} = \frac{1}{2}$.

In addition, we can examine the graphs of $\frac{1}{\sqrt[3]{x}}$, $\frac{1}{x}$, and $\frac{1}{x^3}$. Of course we cannot tell whether the areas of regions are infinite or finite just by visual inspection, but when we look at all three graphs together, the answers do seem plausible:

$\frac{1}{\sqrt[3]{x}}$ on $[0, 1]$ and $[1, \infty)$ $\quad$ $\frac{1}{x}$ on $[0, 1]$ and $[1, \infty)$ $\quad$ $\frac{1}{x^3}$ on $[0, 1]$ and $[1, \infty)$

  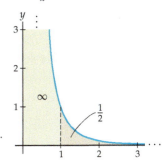

> **EXAMPLE 3**

Calculating improper integrals

Use limits of definite integrals to calculate each of the following improper integrals:

(a) $\displaystyle\int_0^\infty e^{-x}\,dx$ $\qquad$ **(b)** $\displaystyle\int_0^5 \frac{1}{(x-3)^{4/5}}\,dx$ $\qquad$ **(c)** $\displaystyle\int_{-\infty}^\infty \frac{1}{1+x^2}\,dx$

SOLUTION

(a) The function $y = e^{-x}$ has no vertical asymptotes and is in fact continuous on $[0, \infty)$. Therefore we have

$$\int_0^\infty e^{-x}\,dx = \lim_{B\to\infty}\int_0^B e^{-x}\,dx \qquad \leftarrow \text{definition of improper integral}$$

$$= \lim_{B\to\infty}\left[-e^{-x}\right]_0^B \qquad \leftarrow \text{antidifferentiate}$$

$$= \lim_{B\to\infty}\left(-e^{-B}+1\right) \qquad \leftarrow \text{evaluate}$$

$$= -\frac{1}{\infty}+1 = 1. \qquad \leftarrow \text{calculate limit}$$

(b) The function $y = \frac{1}{(x-3)^{4/5}}$ has a vertical asymptote at $x = 3$ but is continuous on $[0, 3) \cup (3, 5]$. We can solve this improper integral by splitting at $x = 3$:

$$\int_0^5 \frac{1}{(x-3)^{4/5}}\, dx = \int_0^3 \frac{1}{(x-3)^{4/5}}\, dx + \int_3^5 \frac{1}{(x-3)^{4/5}}\, dx \qquad \leftarrow \text{split at } x = 3$$

$$= \lim_{B \to 3^-} \int_0^B \frac{1}{(x-3)^{4/5}}\, dx + \lim_{A \to 3^+} \int_A^5 \frac{1}{(x-3)^{4/5}}\, dx \qquad \leftarrow \text{use limits}$$

$$= \lim_{B \to 3^-} \left[5(x-3)^{1/5}\right]_0^B + \lim_{A \to 3^+} \left[5(x-3)^{1/5}\right]_A^5 \qquad \leftarrow \text{antidifferentiate}$$

$$= \lim_{B \to 3^-} \left(5(B-3)^{1/5} - 5(-3)^{1/5}\right) + \lim_{A \to 3^+} \left(5(2)^{1/5} - 5(A-3)^{1/5}\right) \qquad \leftarrow \text{evaluate}$$

$$= 0 + 5(3^{1/5}) + 5(2^{1/5}) - 0 = 5(3^{1/5} + 2^{1/5}). \qquad \leftarrow \text{calculate limit}$$

(c) The function $y = \frac{1}{1+x^2}$ is continuous on all of $(-\infty, \infty)$, so we need only split this improper integral into two pieces so that we can handle the limits to ∞ and $-\infty$ separately. We might as well make the split at $x = 0$:

$$\int_{-\infty}^\infty \frac{1}{1+x^2}\, dx = \int_{-\infty}^0 \frac{1}{1+x^2}\, dx + \int_0^\infty \frac{1}{1+x^2}\, dx \qquad \leftarrow \text{split at } x = 0$$

$$= \lim_{A \to -\infty} \int_A^0 \frac{1}{1+x^2}\, dx + \lim_{B \to \infty} \int_0^B \frac{1}{1+x^2}\, dx \qquad \leftarrow \text{use limits}$$

$$= \lim_{A \to -\infty} \left[\tan^{-1} x\right]_A^0 + \lim_{B \to \infty} \left[\tan^{-1} x\right]_0^B \qquad \leftarrow \text{antidifferentiate}$$

$$= \lim_{A \to -\infty} (0 - \tan^{-1} A) + \lim_{B \to \infty} (\tan^{-1} B - 0) \qquad \leftarrow \text{evaluate}$$

$$= \frac{\pi}{2} + \frac{\pi}{2} = \pi. \qquad \leftarrow \text{calculate limit}$$

EXAMPLE 4   **Using the comparison test to determine convergence or divergence**

Determine whether each of the first two improper integrals listed here converges or diverges by applying the comparison test with simpler improper integrals. For the third improper integral, explain why the comparison test with $\frac{1}{x^2}$ does not give information; then determine convergence or divergence by using limits of definite integrals.

(a) $\displaystyle\int_1^\infty \frac{1}{xe^x}\, dx$ $\qquad$ **(b)** $\displaystyle\int_0^1 \frac{1+x}{x^3}\, dx$ $\qquad$ **(c)** $\displaystyle\int_2^\infty \frac{1}{x^2 - x}\, dx$

SOLUTION

(a) To use the comparison test we have two choices: If we think that this improper integral converges, then we should try to find a function larger than $\frac{1}{xe^x}$ on $[1, \infty)$, which we know converges on that interval. If instead we think that the improper integral diverges, then we should try to find a function smaller than $\frac{1}{xe^x}$ on $[1, \infty)$, which we know diverges.

A good first try might be to notice that $xe^x \geq x$ for all $x \in [1, \infty)$, from which we can deduce that $0 < \frac{1}{xe^x} \leq \frac{1}{x}$ on $[1, \infty)$. But we know from Theorem 8.20 that $\int_1^\infty \frac{1}{x}\, dx$ diverges, and knowing that our improper integral is less than something infinite does not tell us whether it is infinite or finite.

As another try, notice that since $e^x \geq x$ for all $x \in [1, \infty)$, we must have $xe^x \geq x^2$ and thus $0 < \frac{1}{xe^x} \leq \frac{1}{x^2}$. Now we are getting somewhere: Since $\int_1^\infty \frac{1}{x^2}\, dx$ converges by Theorem 8.20, we know from Theorem 8.22 that the smaller improper integral $\int_1^\infty \frac{1}{xe^x}\, dx$ must also converge. In symbols, we have

$$\int_1^\infty \frac{1}{xe^x}\, dx \leq \int_1^\infty \frac{1}{x^2}\, dx,$$

which is finite, and thus $\int_1^\infty \frac{1}{xe^x}\, dx$ must also be finite. Note that the comparison test has not told us the value of our improper integral, only that it must converge to some finite value.

(b) For all $x \in (0, 1]$ we have

$$1 + x \geq x \quad \Longrightarrow \quad \frac{1+x}{x^3} \geq \frac{x}{x^3} \quad \Longrightarrow \quad \frac{1+x}{x^3} \geq \frac{1}{x^2}.$$

(Alternatively, we could obtain this inequality by noticing that $\frac{1+x}{x^3} = \frac{1}{x^3} + \frac{1}{x^2} \geq \frac{1}{x^2}$ on $(0, 1]$.) We also know by Theorem 8.21 that the improper integral of $\frac{1}{x^2}$ diverges on $[0, 1]$. Combining these facts, we have

$$\int_0^1 \frac{1+x}{x^3}\, dx \geq \int_0^1 \frac{1}{x^2}\, dx = \infty,$$

and thus, by the comparison test, $\int_0^1 \frac{1+x}{x^3}\, dx$ is also infinite and thus diverges.

(c) First notice that since $0 \leq x^2 - x \leq x^2$ for all $x \in [2, \infty)$, it follows that $\frac{1}{x^2-x} \geq \frac{1}{x^2}$ on that interval. But a comparison with this function whose improper integral on $[2, \infty)$ converges can tell us only that our improper integral is greater than something that is finite, which does not tell us whether our improper integral is finite or infinite.

Resorting to calculation instead, we can use a limit of definite integrals that we can solve by the method of partial fractions:

$$\int_2^\infty \frac{1}{x^2 - x}\, dx = \lim_{B \to \infty} \int_2^B \frac{1}{x^2 - x}\, dx \qquad \leftarrow \text{definition of improper integral}$$

$$= \lim_{B \to \infty} \int_2^B \frac{1}{x(x-1)}\, dx \qquad \leftarrow \text{algebra}$$

$$= \lim_{B \to \infty} \int_2^B \left(-\frac{1}{x} + \frac{1}{x-1}\right) dx \qquad \leftarrow \text{partial fractions}$$

$$= \lim_{B \to \infty} \left[-\ln |x| + \ln |x - 1|\right]_2^B \qquad \leftarrow \text{antidifferentiate}$$

$$= \lim_{B \to \infty} (-\ln B + \ln(B - 1) + \ln 2 - \ln 1) \qquad \leftarrow \text{evaluate}$$

$$= \lim_{B \to \infty} \left(\ln\left(\frac{B-1}{B}\right) + \ln 2\right) \qquad \leftarrow \text{properties of logarithms}$$

$$= \ln 1 + \ln 2 = \ln 2. \qquad \leftarrow \text{calculate limit}$$

From this calculation we see that the improper integral converges. □

TEST YOUR
? UNDERSTANDING

▶ What types of power functions have improper integrals that converge on $[1, \infty)$? That diverge? What about on $[0, 1]$?

▶ Why do we have to split the improper integral $\int_{-\infty}^\infty \frac{1}{1+x^2}\, dx$ at some point $x = c$ in order to calculate it in terms of limits of definite integrals?

▶ Why do we have to split the improper integral $\int_0^2 \frac{1}{(x-1)^2}\, dx$ at $x = 1$ in order to calculate it in terms of limits of definite integrals?

▶ Why does the Fundamental Theorem of Calculus not apply directly to improper integrals? What hypotheses of the Fundamental Theorem of Calculus fail to be true for improper integrals?

▶ Why is it not helpful to know that an improper integral has a larger value than a convergent improper integral or a smaller value than a divergent improper integral?

EXERCISES 8.6

Thinking Back

Limits: Solve each of the following limits.

▶ $\lim\limits_{x\to\infty} x^{-2/3}$

▶ $\lim\limits_{x\to\infty} x^{-4/3}$

▶ $\lim\limits_{x\to 0^+} x^{-2/3}$

▶ $\lim\limits_{x\to 0^+} x^{-4/3}$

▶ $\lim\limits_{x\to 1^-} \dfrac{1}{x\ln x}$

▶ $\lim\limits_{x\to 0^+} \dfrac{1}{x\ln x}$

▶ $\lim\limits_{x\to\infty} e^{-x^2}$

▶ $\lim\limits_{x\to\infty} xe^{-x^2}$

▶ $\lim\limits_{x\to \pi/2^-} \sec x$

▶ $\displaystyle\int xe^{-x^2}\,dx$

▶ $\displaystyle\int \dfrac{\ln x}{x}\,dx$

▶ $\displaystyle\int \dfrac{1}{x(x-1)}\,dx$

▶ $\displaystyle\int \sec x\,dx$

The Fundamental Theorem of Calculus: Solve each of the following definite integrals.

Integration: Solve each of the following indefinite integrals.

▶ $\displaystyle\int (3x-1)^{-2/3}\,dx$

▶ $\displaystyle\int \dfrac{1}{x\ln x}\,dx$

▶ $\displaystyle\int_1^{100} \dfrac{1}{\sqrt[3]{x^4}}\,dx$

▶ $\displaystyle\int_1^{100} \dfrac{1}{\sqrt[4]{x^3}}\,dx$

▶ $\displaystyle\int_{0.01}^1 \dfrac{1}{\sqrt[3]{x^4}}\,dx$

▶ $\displaystyle\int_{0.01}^1 \dfrac{1}{\sqrt[4]{x^3}}\,dx$

Concepts

0. *Problem Zero:* Read the section and make your own summary of the material.

1. *True/False:* Determine whether each of the statements that follow is true or false. If a statement is true, explain why. If a statement is false, provide a counterexample.

(a) *True or False:* The Fundamental Theorem of Calculus applies to $\int_0^2 \dfrac{1}{1-x}\,dx$.

(b) *True or False:* The Fundamental Theorem of Calculus applies to $\int_2^\infty \dfrac{1}{1-x}\,dx$.

(c) *True or False:* If $f(x)$ is positive and decreasing, then the area between the graph of $f(x)$ and the x-axis on $[1,\infty)$ must be finite.

(d) *True or False:* If $f(x)$ is positive and decreasing, then the area between the graph of $f(x)$ and the x-axis on $[1,\infty)$ must be infinite.

(e) *True or False:* If $p > 1$, then $\int_1^\infty \dfrac{1}{x^p}\,dx = \dfrac{1}{p-1}$.

(f) *True or False:* If $p < 1$, then $\int_0^1 \dfrac{1}{x^p}\,dx = \dfrac{1}{1-p}$.

(g) *True or False:* If $\int_1^\infty f(x)\,dx$ converges and $0 \le f(x) \le g(x)$ for all $x \in [1,\infty)$, then $\int_1^\infty g(x)\,dx$ must diverge.

(h) *True or False:* If $\int_1^\infty f(x)\,dx$ diverges and $0 \le f(x) \le g(x)$ for all $x \in [1,\infty)$, then $\int_1^\infty g(x)\,dx$ must diverge.

2. *Examples:* Construct examples of the thing(s) described in the following. Try to find examples that are different than any in the reading.

(a) Three power functions whose improper integrals on $[1,\infty)$ converge.

(b) Three power functions whose improper integrals on $[0,1]$ converge.

(c) A function with the property that both its improper integral on $[1,\infty)$ and its improper integral on $[0,1]$ diverge.

3. What sorts of situations should you look for in order to determine whether an integral is improper?

4. What do we mean when we say that an improper integral converges? What do we mean when we say that an improper integral diverges?

5. Why is it very easy to conclude that $\int_0^1 \dfrac{1}{e^x}\,dx$ converges without making any integration calculations?

6. Why is it very easy to conclude that $\int_1^\infty x^2\,dx$ diverges without making any integration calculations?

7. The Fundamental Theorem of Calculus does not apply to the integral $\int_0^5 (x-3)^{-4/3}\,dx$; why not? What would the Fundamental Theorem of Calculus calculate for this integral if we applied it anyway? Would it give the correct answer?

8. Why does it make sense that $\int_1^\infty \dfrac{1}{x^p}\,dx$ diverges when $0 < p \le 1$? Consider how $\dfrac{1}{x^p}$ compares with $\dfrac{1}{x}$ in this case.

9. Why does it make sense that $\int_0^1 \dfrac{1}{x^p}\,dx$ diverges when $p \ge 1$? Consider how $\dfrac{1}{x^p}$ compares with $\dfrac{1}{x}$ in this case.

10. Draw pictures to illustrate why the comparison test for improper integrals makes intuitive sense for both convergence comparisons and for divergence comparisons.

11. What, if anything, does the divergence of $\int_1^\infty \frac{1}{x}\,dx$ and the comparison test tell you about the convergence or divergence of $\int_1^\infty \frac{1}{x+1}\,dx$, and why?

12. What, if anything, does the convergence of $\int_1^\infty \frac{1}{x^2}\,dx$ and the comparison test tell you about the convergence or divergence of $\int_1^\infty \frac{1}{x^2+1}\,dx$, and why?

13. Why can't the comparison test make conclusions about $\int_1^\infty \frac{1}{x+1}\,dx$ based on the divergence of $\int_1^\infty \frac{1}{x}\,dx$? What graphical way could you argue for divergence instead? (*Hint: Think about transformations of graphs.*)

14. Why can't the comparison test make conclusions about $\int_2^\infty \frac{1}{(x-1)^2}\,dx$ based on the convergence of $\int_2^\infty \frac{1}{x^2}\,dx$?

What graphical way could you argue for the convergence instead?

Express each improper integral in Exercises 15–20 as a sum of limits of proper definite integrals. Do not calculate any integrals or limits; just write them down.

15. $\int_1^\infty \frac{1}{x-1}\,dx$

16. $\int_1^\infty \frac{1}{x-2}\,dx$

17. $\int_1^5 \frac{1}{2x^2-10x+12}\,dx$

18. $\int_0^4 \frac{1}{3x^2-4x+1}\,dx$

19. $\int_{-\infty}^\infty \frac{e^{-x^2}}{x^2}\,dx$

20. $\int_2^\infty \frac{1}{x\ln(x-2)}\,dx$

Skills

Use limits of definite integrals to calculate each of the improper integrals in Exercises 21–56.

21. $\int_1^\infty \frac{1}{\sqrt[3]{x^4}}\,dx$

22. $\int_1^\infty \frac{1}{\sqrt[4]{x^3}}\,dx$

23. $\int_0^1 \frac{1}{\sqrt[3]{x^4}}\,dx$

24. $\int_0^1 \frac{1}{\sqrt[4]{x^3}}\,dx$

25. $\int_0^1 x^{-0.99}\,dx$

26. $\int_0^1 x^{-1.01}\,dx$

27. $\int_1^\infty x^{-0.99}\,dx$

28. $\int_1^\infty x^{-1.01}\,dx$

29. $\int_0^1 (3x-1)^{-2/3}\,dx$

30. $\int_3^\infty \frac{1}{(x-2)^2}\,dx$

31. $\int_1^\infty \frac{1}{(x-2)^2}\,dx$

32. $\int_1^\infty \frac{1+x}{x^2}\,dx$

33. $\int_0^4 \frac{2x-4}{x^2-4x+3}\,dx$

34. $\int_4^\infty \frac{2x-4}{x^2-4x+3}\,dx$

35. $\int_0^\infty xe^{-x^2}\,dx$

36. $\int_0^\infty xe^{-x}\,dx$

37. $\int_0^\infty x^2 e^{-x}\,dx$

38. $\int_0^1 \frac{1}{\sqrt{x}\,e^{\sqrt{x}}}\,dx$

39. $\int_0^\infty \frac{1}{\sqrt{x}\,e^{\sqrt{x}}}\,dx$

40. $\int_0^\infty \frac{x}{(x^2+1)^2}\,dx$

41. $\int_{-\infty}^\infty \frac{x}{(x^2+1)^2}\,dx$

42. $\int_0^\infty \frac{x}{(x^2+1)}\,dx$

43. $\int_0^{\pi/2} \sec x\,dx$

44. $\int_0^{\pi/2} \tan x\,dx$

45. $\int_0^1 \frac{\ln x}{x}\,dx$

46. $\int_1^\infty \frac{\ln x}{x}\,dx$

47. $\int_0^1 \frac{1}{x\ln x}\,dx$

48. $\int_1^\infty \frac{1}{x\ln x}\,dx$

49. $\int_e^\infty \frac{1}{x(\ln x)^2}\,dx$

50. $\int_1^e \frac{1}{x(\ln x)^2}\,dx$

51. $\int_{1/\pi}^\infty \frac{\sin(1/x)}{x^2}\,dx$

52. $\int_0^1 \frac{1}{x(x-1)}\,dx$

53. $\int_1^2 \frac{1}{x(x-1)}\,dx$

54. $\int_2^\infty \frac{1}{x(x-1)}\,dx$

55. $\int_0^2 \frac{x-2}{x(x-1)}\,dx$

56. $\int_2^\infty \frac{x-2}{x(x-1)}\,dx$

Determine the convergence or divergence of each improper integral in Exercises 57–64 by comparing it to simpler improper integrals whose convergence or divergence is known or can be found directly.

57. $\int_1^\infty \frac{x}{1+\sin^2 x}\,dx$

58. $\int_1^\infty \frac{1+x}{x^2}\,dx$

59. $\int_1^\infty \frac{1}{x^2+1}\,dx$

60. $\int_1^\infty \frac{x+2}{x^2}\,dx$

61. $\int_1^\infty \frac{1+\ln x}{x}\,dx$

62. $\int_1^\infty \frac{x+\ln x}{x^3}\,dx$

63. $\int_1^\infty \frac{1+\ln x}{x^3}\,dx$

64. $\int_1^\infty \frac{1+\ln x}{x^2}\,dx$

The Limit Comparison Test: If two functions $f(x)$ and $g(x)$ behave in a similar way on $[a, \infty)$, then their improper integrals on $[a, \infty)$ should converge or diverge together. More specifically, the **limit comparison test** says that if

$$\lim_{x\to\infty} \frac{f(x)}{g(x)} = L$$

for some positive real number L, then the improper integrals $\int_a^\infty f(x)\,dx$ and $\int_a^\infty g(x)\,dx$ either both converge or both diverge. Use this test to determine the convergence or divergence of the integrals in Exercises 65–70.

65. $\int_2^\infty \dfrac{1}{(x-1)^2}\,dx$

66. $\int_1^\infty \dfrac{1}{x+0.001}\,dx$

67. $\int_1^\infty \dfrac{1}{3x^2-1}\,dx$

68. $\int_1^\infty \dfrac{x^2-x+3}{x^4+1}\,dx$

69. $\int_1^\infty \dfrac{x}{\sqrt[4]{x^3}+100}\,dx$

70. $\int_1^\infty \dfrac{1+e^x}{xe^x}\,dx$

Applications

For Exercises 71 and 72, suppose that in the course of Emmy's job as a civil engineer at Hanford, she needs to find out how much radioactive cesium was present in the mixture in a certain tank when it was first filled exactly 12 years ago. Suppose $Q(0)$ is the number of pounds of radioactive cesium that was initially in the tank 12 years ago, at time $T=0$. The amount of cesium in the tank at any time $T \geq 0$ is the amount that remains to decay, which for some constant k is equal to

$$Q(t) = \int_T^\infty ke^{-0.023t}\,dt.$$

71. If Emmy measures 0.95 lb of cesium in the tank now, how much cesium was in the tank when it was first filled 12 years ago?

72. In another location in the tank farm, there is a tank that was filled with a different, known mixture that contained 1.5 lb of radioactive cesium when the tank was filled. However, there is no record of when this tank was filled.

Emmy measures 0.85 lb of cesium in the tank now. How long ago was the tank filled?

73. Frank is evaluating electric motors to drive automated mixing for some waste tanks that he must maintain. One pump is advertised to have a probability of failure that follows the exponential distribution

$$f(t) = 0.31e^{-0.31t},$$

where the time $t > 0$ is measured in years. Frank knows that the expected time of failure for something following this distribution is

$$\int_0^\infty tf(t)\,dt.$$

How long can he expect one of these pumps to last?

Proofs

Prove each statement in Exercises 74–77, using limits of definite integrals for general values of p.

74. If $0 < p \leq 1$, then $\int_1^\infty \dfrac{1}{x^p}\,dx$ diverges.

75. If $p > 1$, then $\int_1^\infty \dfrac{1}{x^p}\,dx$ converges to $\dfrac{1}{p-1}$.

76. If $0 < p < 1$, then $\int_0^1 \dfrac{1}{x^p}\,dx$ converges to $\dfrac{1}{1-p}$.

77. If $p \geq 1$, then $\int_0^1 \dfrac{1}{x^p}\,dx$ diverges.

Prove each statement in Exercises 78–80, using the definition of improper integrals as limits of proper definite integrals.

78. If $\int_a^\infty f(x)\,dx$ diverges and $0 \leq f(x) \leq g(x)$ for all $x \in [a, \infty)$, then $\int_a^\infty g(x)\,dx$ also diverges.

79. Suppose $f(x)$ and $g(x)$ are both continuous on $(a, b]$ but not at $x = a$. If $\int_a^b f(x)\,dx$ converges and $0 \leq g(x) \leq f(x)$ for all $x \in (a, b]$, then $\int_a^b g(x)\,dx$ also converges.

80. Suppose $f(x)$ and $g(x)$ are both continuous on $(a, b]$ but not at $x = a$. If $\int_a^b f(x)\,dx$ diverges and $0 \leq f(x) \leq g(x)$ for all $x \in (a, b]$, then $\int_a^b g(x)\,dx$ also diverges.

81. Suppose $f(x)$ is continuous on $\mathbb{R}$ and that for some real number c, both $\int_{-\infty}^c f(x)\,dx$ and $\int_c^\infty f(x)\,dx$ exist. Use properties of definite integrals to prove that for all real numbers d,

$$\int_{-\infty}^c f(x)\,dx + \int_c^\infty f(x)\,dx$$

is equal to

$$\int_{-\infty}^d f(x)\,dx + \int_d^\infty f(x)\,dx.$$

Thinking Forward

Gabriel's Horn: If we rotate the graph of $f(x) = \frac{1}{x}$ on $[1, \infty)$ around the x-axis, we get what is known as **Gabriel's Horn**, as shown in the figure.

Graph of $f(x) = \frac{1}{x}$ rotated around x-axis

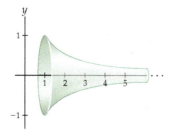

We already know that the area $\int_1^\infty \frac{1}{x}\,dx$ under the graph of $\frac{1}{x}$ on $[1, \infty)$ is infinite. Does this mean that the volume of Gabriel's Horn will also be infinite? The answer, surprisingly, is no. Let's find out why.

It turns out that the volume of the solid of revolution obtained by rotating the graph of a function $y = f(x)$ around the x-axis

on an interval $[a, b]$ is given by the formula

$$\text{Volume} = \pi \int_a^b (f(x))^2\,dx.$$

It also turns out that the surface area of such a solid of revolution is given by the formula

$$\text{Surface Area} = 2\pi \int_a^b f(x)\sqrt{1 + (f'(x))^2}\,dx.$$

▶ Set up an improper integral that describes the volume of Gabriel's Horn. Then solve the improper integral to show that the volume of Gabriel's Horn is not infinite, but rather is equal to π.

▶ Get ready to be surprised again: Although the volume of Gabriel's Horn is finite, the surface area of Gabriel's Horn is infinite! That means that it would take an infinite amount of paint to coat the outside of Gabriel's Horn, even though it takes only a finite amount of paint to fill up the inside. Set up an improper integral that describes the surface area of Gabriel's Horn. Then use the limit comparison test (see Exercises 65–70) with $\frac{1}{x}$ to show that the surface area of Gabriel's Horn is infinite.

8.7   NUMERICAL INTEGRATION*

▶ Error bounds for certain left, right, trapezoid, and midpoint sums

▶ Simpson's Rule for approximating areas under curves

▶ Using error bounds to calculate definite integrals to within specified degrees of accuracy

Approximations and Error

Even with a whole chapter of integration techniques under our belt, there are many integrals that we do not know how to solve. For example, the definite integrals

$$\int_0^{1.5} e^{-x^2}\,dx, \quad \int_1^2 \frac{x}{e^x - 1}\,dx, \quad \text{and} \quad \int_0^{\pi/2} \cos(\cos x)\,dx$$

have relatively simple integrands, although it turns out that their antiderivatives cannot be expressed in **closed form**, that is, as combinations of the types of functions we have studied in this course. However, each of the integrands in these integrals is continuous and thus is integrable, which means that each of the given definite integrals is equal to some real number that we can approximate by using Riemann sums.

Approximating is easy; the tough part is knowing when your approximation is "good." For example, suppose we want to get a rough estimate of the number of seconds in a day. The actual answer is of course that there are $(60 \text{ seconds})(60 \text{ minutes})(24 \text{ hours}) = 86{,}400$ seconds in each day, but if you wanted to calculate quickly in your head you might instead consider the simpler product $(60)(60)(20) = 72{,}000$. This approximation is good for knowing roughly what order of magnitude we are talking about, but not good enough for most practical applications. In this case we can see that our simple approximation is off by an error of 4,400 seconds. But how do we know how good or bad an approximation is without knowing the exact value we are approximating?

The **error** of an approximation is the difference between the approximated value and the actual value of what is being calculated. When we are interested only in the magnitude of such error and not in whether our approximation is more or less than the actual value, we consider the absolute value of the error:

$$|error| = |true\ value - approximation|.$$

When we use a Riemann sum, we are approximating an exact signed area under a curve. The resulting error will typically be large if we use only a few rectangles, but becomes smaller when we use a large number of rectangles. In general, the error $E(n)$ incurred by using a given n-rectangle Riemann sum is the difference between the actual area and the approximated area:

$$|E(n)| = \left| \int_a^b f(x)\,dx - \sum_{k=1}^n f(x_k^*)\,\Delta x \right|.$$

But how can we possibly find this error without knowing the exact value of $\int_a^b f(x)\,dx$ in the first place? It turns out that in specific situations we can get bounds on the error even without knowing the value of the definite integral. This means that in those situations we will be able to approximate definite integrals to whatever degree of accuracy we choose.

Error in Left and Right Sums

Since we will be calculating and comparing many types of Riemann sums in this section, it will be useful to have some descriptive notation for such sums. Suppose we wish to examine the signed area between the graph of a function f and the x-axis on $[a, b]$. We will denote the left-sum and right-sum n-rectangle approximations for this area, respectively, as

$$\text{LEFT}(n) = \sum_{k=1}^n f(x_{k-1})\Delta x = (f(x_0) + f(x_1) + f(x_2) + \cdots + f(x_{n-1}))\,\Delta x,$$

$$\text{RIGHT}(n) = \sum_{k=1}^n f(x_k)\Delta x = (f(x_1) + f(x_2) + f(x_3) + \cdots + f(x_n))\,\Delta x.$$

Sometimes these sums are over-approximations of the actual area, and sometimes they are under-approximations, depending on the behavior of the function f. If f happens to be **monotonic** on $[a, b]$, in other words either always increasing on $[a, b]$ or always decreasing on $[a, b]$, we can say more. For example, consider the monotonically decreasing function f shown following at the left. For this function, the four-rectangle left sum is an over-approximation and the four-rectangle right sum is an under-approximation.

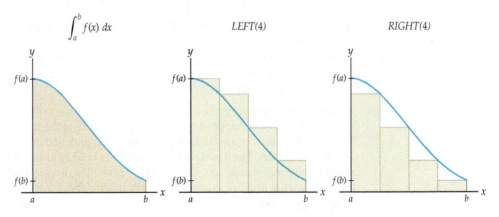

In fact, for any number n of rectangles, the area under a monotonically increasing or decreasing graph will be between the left-sum and right-sum approximations, as described in the following theorem:

THEOREM 8.23

Left and Right Sums of Monotonic Functions

Suppose f is integrable on $[a, b]$ and let n be a positive integer.

(a) If f is increasing on $[a, b]$, then $\text{LEFT}(n) \leq \int_a^b f(x)\, dx \leq \text{RIGHT}(n)$.

(b) If f is decreasing on $[a, b]$, then $\text{RIGHT}(n) \leq \int_a^b f(x)\, dx \leq \text{LEFT}(n)$.

Proof. We will prove part (b) and leave the entirely similar proof of part (a) to Exercise 58. Suppose we partition $[a, b]$ into n rectangles over subintervals of the form $[x_{k-1}, x_k]$. If f is monotonically decreasing, then for each k we have $f(x_{k-1}) \geq f(x) \geq f(x_k)$ for all $x \in [x_{k-1}, x_k]$, and thus the area under f on $[x_{k-1}, x_k]$ will be less than the area of the kth left-sum rectangle and greater than the area of the kth right-sum rectangle.   ■

The fact that the actual area under a monotonic function is always sandwiched between the left sum and the right sum can actually tell us something about the amount of error in those Riemann sum approximations. The shaded rectangles in the figures that follow represent the combined error from the left sum and the right sum. These errors join up to form a rectangle that can be stacked into a larger rectangle whose area $|f(b) - f(a)|\Delta x$ is the difference $|\text{LEFT}(n) - \text{RIGHT}(n)|$, regardless of the value of n:

Difference between LEFT(4) and RIGHT(4)   *Difference between LEFT(8) and RIGHT(8)*

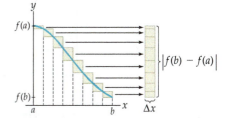

Since the stack of rectangles represents the combined error from the left and right sums, it follows that the error from either of these approximations must itself be less than $|f(b) - f(a)|\Delta x$. This argument holds for any monotonically decreasing function, and we can say similar things when f is monotonically increasing. Therefore we have proved the following theorem:

THEOREM 8.24

Error Bounds for Left and Right Sums

Suppose f is integrable and monotonic on $[a, b]$ and n is a positive integer. Then we can bound the error $E_{\text{LEFT}(n)}$ of the left sum as follows:

$$|E_{\text{LEFT}(n)}| \leq |\text{LEFT}(n) - \text{RIGHT}(n)| = |f(b) - f(a)|\,\Delta x.$$

The error $|E_{\text{RIGHT}(n)}|$ of the corresponding right sum has the same bound.

As we will see in Example 1, we can use these error bounds to ensure that our left- or right-sum approximations of area are to within any desired degree of accuracy.

Error in Trapezoid and Midpoint Sums

What about bounding error for other types of Riemann sums? Recall from Section 7.2 the trapezoid and midpoint sums for n subdivisions, which can be expressed, respectively, as

$$\text{TRAP}(n) = \sum_{k=1}^{n} \frac{f(x_{k-1}) + f(x_k)}{2} \Delta x = \left(f(x_0) + 2f(x_1) + 2f(x_2) + \cdots + f(x_n) \right) \left(\frac{\Delta x}{2} \right),$$

$$\text{MID}(n) = \sum_{k=1}^{n} f\left(\frac{x_{k-1} + x_k}{2} \right) \Delta x, = \left(f\left(\frac{x_0 + x_1}{2} \right) + f\left(\frac{x_1 + x_2}{2} \right) + \cdots + f\left(\frac{x_{n-1} + x_n}{2} \right) \right) \Delta x.$$

It turns out that we can guarantee that these sums are either over-approximations or under-approximations depending on whether the function f is concave up or concave down. For example, consider the concave-up function f shown following at the left. It is immediately obvious that the trapezoid sum approximation must be greater than the actual area under the curve, as shown in the middle figure.

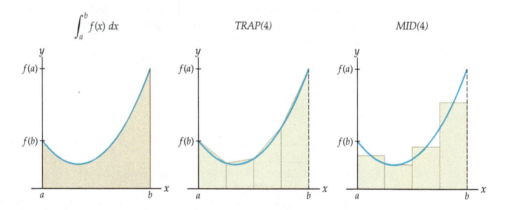

What is less obvious is that the midpoint sum of a concave-up function, as shown in the third figure, will always be an under-approximation of the actual area. For concave-down functions the situation is reversed, with the actual area sandwiched between the smaller trapezoid sum and the larger midpoint sum, as stated in the following theorem:

THEOREM 8.25

Trapezoid and Midpoint Sums of Functions with Consistent Concavity

Suppose f is integrable on $[a, b]$ and let n be a positive integer.

(a) If f is concave up on $[a, b]$, then $\text{MID}(n) \leq \int_a^b f(x)\, dx \leq \text{TRAP}(n)$.

(b) If f is concave down on $[a, b]$, then $\text{TRAP}(n) \leq \int_a^b f(x)\, dx \leq \text{MID}(n)$.

Proof. We will use geometric arguments to prove part (a) and leave the similar proof of part (b) to Exercise 59. That the trapezoid sum is an over-approximation for concave-up functions is clear from our earlier middle diagram. Specifically, upwards concavity guarantees that the secant line from $(x_{k-1}, f(x_{k-1}))$ to $(x_k, f(x_k))$ will always be completely above the graph of f on $[x_{k-1}, x_k]$.

That the midpoint sum is an under-approximation for concave-up functions becomes clear if we reinterpret each midpoint rectangle as a trapezoid. Given a subinterval $[x_{k-1}, x_k]$, consider the **midpoint trapezoid** on the interval to be the trapezoid whose top is formed by the line tangent to the graph of f at the midpoint of the subinterval. From the figure it is clear that the midpoint trapezoid will have the same area as the midpoint rectangle whose height is given by the height of $f(x)$ at the midpoint of the subinterval. Now look at the figure on the right; since f is concave up, the midpoint tangent line on each subinterval will be below the graph. Therefore the area of each midpoint trapezoid will be less than the area under the curve on that subinterval.

Midpoint trapezoid

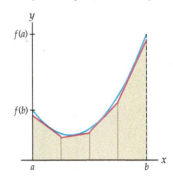

Reinterpretation of MID(4) with trapezoids

Because the area under the graph of a consistently concave-up or concave-down function will always be between the trapezoid sum and the midpoint sum, the error for either of those sums is less than the difference of those sums. This is entirely analogous to what we saw in Theorem 8.24. We can actually get an even better bound if we are able to restrict the behavior of the second derivative:

THEOREM 8.26

Error Bounds for Trapezoid and Midpoint Sums

Suppose f is integrable and either always concave up or always concave down on $[a, b]$, and let n be a positive integer. Then we can bound both $E_{\text{TRAP}(n)}$ and $E_{\text{MID}(n)}$ as follows:

(a) $|E_{\text{TRAP}(n)}| \leq |\text{TRAP}(n) - \text{MID}(n)|$ **(b)** $|E_{\text{MID}(n)}| \leq |\text{TRAP}(n) - \text{MID}(n)|$

If we suppose further that f'' is bounded on $[a, b]$, that is, that there is some positive real number M such that $|f''(x)| \leq M$ for all $x \in [a, b]$, then we can say that:

(c) $|E_{\text{TRAP}(n)}| \leq \dfrac{M(b-a)^3}{12n^2}$ **(d)** $|E_{\text{MID}(n)}| \leq \dfrac{M(b-a)^3}{24n^2}$

The proof of the second part of this theorem can be found in many numerical analysis texts. You will explore why these bounds are reasonable in Exercise 18. In Example 2 you will use the same bounds to calculate trapezoid and midpoint sums to specified degrees of accuracy.

Simpson's Rule

So far all of our approximations of a definite integral $\int_a^b f(x)\,dx$ have utilized piecewise-linear approximations of f. In other words, after subdividing $[a, b]$, we approximated each piece of f with a line. For example, the left sum just represents the area under the graph of a piecewise-linear step function that agrees with f on the left-hand side of each subinterval. The trapezoid sum uses a continuous piecewise-linear function whose pieces are secant lines of the graph of f, connecting the heights of the function at the subdivision points.

One way to get a more accurate approximation of a definite integral is to use a non-linear approximation instead. For example, we could approximate each piece of f with a parabola instead of a line. The first figure that follows shows some function f on $[a, b]$, and the second figure shows how we could approximate the area under the graph of f with three parabola-topped rectangles. We choose the parabola $p(x)$ on each subinterval to be the unique parabola that passes through the leftmost height, the midpoint height, and the rightmost height on that subinterval; see the third figure.

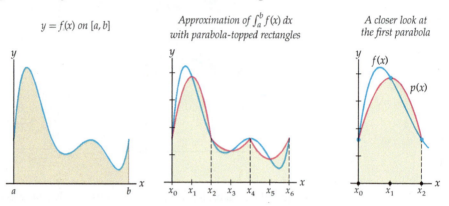

| $y = f(x)$ on $[a, b]$ | Approximation of $\int_a^b f(x)\, dx$ with parabola-topped rectangles | A closer look at the first parabola |

The example we just discussed was specifically chosen so that the parabolas would be easily distinguishable from the original function. In other words, the example does not show a very accurate approximation! However, in general, approximation by parabolas can be extremely accurate; imagine using 20 parabolas along the graph of f, and consider how closely such an approximation would match the function.

So how do we calculate a particular parabola-based area approximation? One way would be to actually find the equations of each little quadratic parabola piece and then use definite integral formulas for quadratics to calculate the area under each such piece. Doing this in practice, however, would be quite time-consuming. The next theorem can help make the process faster; it provides a specific formula for calculating the area under a quadratic on an interval that uses only the values of the quadratic at the endpoints and the midpoint of the interval. This nice theorem will allow us to bypass the step of actually finding the equation of each parabola.

THEOREM 8.27

The Definite Integral of a Quadratic in Terms of Endpoints and Midpoint

If $p(x)$ is a quadratic function on $[A, B]$, then

$$\int_A^B p(x)\, dx = \frac{B - A}{6}\left(p(A) + 4p\left(\frac{A+B}{2}\right) + p(B)\right).$$

The proof of this convenient formula is just an application of known definite integral formulas and a bunch of algebra and is left to Exercise 62.

It turns out to be convenient to construct our parabolas on *pairs* of intervals. For example, in the middle figure just shown we divided $[a, b]$ into $n = 6$ subintervals and then constructed three parabolas to approximate the curve $y = f(x)$. In Exercise 63 you will show that this process leads to the following summation formula:

THEOREM 8.28   ## Simpson's Rule

Suppose f is integrable on $[a, b]$ and n is a positive even integer. Let $\Delta x = \frac{b-a}{n}$ and $x_k = a + k\Delta x$. Then we can approximate $\int_a^b f(x)\, dx$ with $\frac{n}{2}$ parabola-topped rectangles by using the following sum, which is known as *Simpson's Rule*:

$$\text{SIMP}(n) = (f(x_0) + 4f(x_1) + 2f(x_2) + 4f(x_3) + \cdots + 2f(x_{n-2}) + 4f(x_{n-1}) + f(x_n))\left(\frac{\Delta x}{3}\right).$$

Intuitively, it makes sense that, for most functions, an approximation using pieces of parabolas could be more accurate than an approximation with pieces of lines. The following theorem backs up this intuition with a bound on the error from Simpson's Rule that is clearly better than the error bounds we found for left, right, trapezoid, and midpoint sums:

THEOREM 8.29   ## Error Bound for Simpson's Rule

Suppose f is an integrable function on $[a, b]$ whose fourth derivative is bounded. Then there is some positive real number M such that $|f^{(4)}(x)| \leq M$ for all $x \in [a, b]$. In this situation we can bound the error $E_{\text{SIMP}(n)}$ from Simpson's Rule as follows:

$$|E_{\text{SIMP}(n)}| \leq \frac{M(b-a)^5}{180n^4}.$$

The proof of Theorem 8.29 can be found in many numerical analysis texts, and we will not concern ourselves with it here.

Notice that all of the error bounds we have discussed have a requirement that either f or some derivative of f be bounded. For the left and right sums we required f to be monotonic, which implies that f is bounded. This restriction in turn puts some kind of limit on how much the height function f can change on $[a, b]$. For the trapezoid and midpoint sums we required f'' to be bounded, which in a sense puts a limit on the curviness of f on $[a, b]$. Notice that curvier functions will tend to have less accurate trapezoid sum approximations. For Simpson's Rule we required a bound on the fourth derivative, which also puts some limits on how wildly behaved f can be on $[a, b]$ and therefore limits how much error can arise by using parabolas to approximate the function on that interval.

Examples and Explorations

EXAMPLE 1   ### Guaranteeing a level of accuracy from a right sum

Suppose we wish to approximate the definite integral $\int_0^{1.5} e^{-x^2}\, dx$ with an n-rectangle right sum so that we can be sure that our estimate is within 0.005 of the actual answer. How large will we have to make n?

SOLUTION

As we mentioned at the start of this section, the integral $\int_0^{1.5} e^{-x^2}\, dx$ has no closed-form solution. This means that we cannot calculate its actual value with the Fundamental Theorem of Calculus, because we would be unable to perform the antidifferentiation step. Therefore in this example we have no choice but to approximate the value of the definite integral, and in addition we do not know the value of the actual answer we are attempting to approximate.

The graph of f on $[a, b]$ that appears first in this section happens to be the graph of $f(x) = e^{-x^2}$ on the interval $[a, b] = [0, 1.5]$ that we are concerned with in this example. The function $f(x) = e^{-x^2}$ is monotonically decreasing on $[0, 1.5]$ because its derivative $f'(x) = -2xe^{-x^2}$ is always negative on that interval. Therefore Theorem 8.24 applies. We have $a = 0$, $b = 1.5$, and $\Delta x = \frac{1.5-0}{n} = \frac{1.5}{n}$, so the magnitude of the error from using an n-rectangle right sum has the following bound:

$$|E_{\text{RHS}(n)}| \leq |f(b) - f(a)|\Delta x = \left|e^{-(1.5)^2} - e^{-0^2}\right|\left(\frac{1.5}{n}\right) \approx \frac{1.3419}{n}.$$

We want to find n such that $|E_{\text{RHS}(n)}|$ is less than 0.005. To do this we just have to find n such that the error bound is less than 0.005, so we need:

$$\frac{1.3419}{n} < 0.005 \implies \frac{1.3419}{0.005} < n \implies n > 268.38.$$

This means that $n = 269$ is the first positive integer for which the n-rectangle right sum will be within 0.005 of the actual area under the curve.

In case you're interested, the actual value of the definite integral is approximately 0.856188, and the right-sum approximation with 269 rectangles is approximately 0.853693; notice that our approximation is indeed within 0.005 of the actual answer. □

EXAMPLE 2   **Finding bounds for the error from trapezoid and midpoint sums**

Consider the signed area between the graph of $f(x) = x^2 - 2x + 2$ and the x-axis on the interval $[0, 3]$.

(a) Find trapezoid and midpoint sum approximations for this area with $n = 4$.

(b) Describe error bounds for the approximations from part (a) in two different ways (with Theorem 8.25 and with Theorem 8.26).

(c) Use the Fundamental Theorem of Calculus to find the actual area, and then verify that the error bounds computed in part (b) are accurate.

SOLUTION

(a) The function $f(x) = x^2 - 2x + 2$ on $[a, b] = [0, 3]$ just happens to be the function f on $[a, b]$ that we used in our discussion of the trapezoid and midpoint sums before Theorem 8.25. We begin by dividing the interval $[0, 3]$ into $n = 4$ subintervals of width $\Delta x = \frac{3}{4}$, with subdivision points $x_0 = 0$, $x_1 = \frac{3}{4}$, $x_2 = \frac{3}{2}$, $x_3 = \frac{9}{4}$, and $x_4 = 3$. Using the equation $f(x) = x^2 - 2x + 2$, we find that the $n = 4$ trapezoid sum approximation is

$$\text{TRAP}(4) = \left(f(0) + 2f\left(\tfrac{3}{4}\right) + 2f\left(\tfrac{3}{2}\right) + 2f\left(\tfrac{9}{4}\right) + f(3)\right)\left(\tfrac{3/4}{2}\right) = \tfrac{201}{32} \approx 6.28125.$$

To find MID(4), we must first find the midpoints of the four subintervals; these midpoints are $x_1^* = \frac{3}{8}$, $x_2^* = \frac{9}{8}$, $x_3^* = \frac{15}{8}$, and $x_4^* = \frac{21}{8}$. Therefore the $n = 4$ midpoint sum approximation is

$$\text{MID}(4) = \left(f\left(\tfrac{3}{8}\right) + f\left(\tfrac{9}{8}\right) + f\left(\tfrac{15}{8}\right) + f\left(\tfrac{21}{8}\right)\right)\left(\tfrac{3}{4}\right) = \tfrac{375}{64} \approx 5.85938.$$

(b) Theorem 8.25 applies because $f'(x) = 2x - 2$ is positive on all of $[0, 3]$ and thus f is concave up on that entire interval. Therefore the actual area under f on $[0, 3]$ is greater than $\text{MID}(4) = \frac{375}{64} \approx 5.85938$ and smaller than $\text{TRAP}(4) = \frac{201}{32} \approx 6.28125$. In particular, by the first part of Theorem 8.26 this means that

$$|E_{\text{TRAP}(4)}| \text{ and } |E_{\text{MID}(4)}| \leq \left|\tfrac{201}{32} - \tfrac{375}{64}\right| = \tfrac{27}{64} \approx 0.421875.$$

Accordingly, we can expect both the $n = 4$ trapezoid sum and the $n = 4$ midpoint sum to be within about 0.421875 of the actual area under f on $[0, 3]$.

The second part of Theorem 8.26 has the potential to give us a much better error bound in the sense that we should be confident in our $n = 4$ approximations to an even greater degree of accuracy. Let's see if that is the case in this example. We first have to get a bound M on the second derivative that is as good as possible. The second derivative of $f(x) = x^2 - 2x + 2$ is $f''(x) = 2$, and thus clearly $|f''(x)| \leq 2$ for all values $x \in [0, 3]$, so we can take $M = 2$. Note that we are lucky to have an *extremely* good bound for f'' in this example, since we know that it is constantly 2. We can now say that

$$|E_{\text{TRAP}(4)}| \leq \frac{M(b-a)^3}{12n^2} = \frac{2(3-0)^3}{12(4^2)} = \frac{9}{32} \approx 0.28125$$

and

$$|E_{\text{MID}(4)}| \leq \frac{M(b-a)^3}{24n^2} = \frac{2(3-0)^3}{24(4^2)} = \frac{9}{64} \approx 0.140625.$$

From the $n = 4$ trapezoid sum approximation, it follows that the actual value of $\int_0^3 f(x)\, dx$ is within $\frac{9}{32}$ of $\frac{201}{32}$ (i.e., between $\frac{192}{32} = 6$ and $\frac{210}{32} \approx 6.5625$). With the $n = 4$ midpoint sum we can guarantee even better accuracy, namely, that the actual value of $\int_0^3 f(x)\, dx$ is within $\frac{9}{64}$ of $\frac{375}{64}$ (i.e., between $\frac{366}{64} \approx 5.71875$ and $\frac{384}{64} = 6$).

(c) In general, we use approximations with error bounds to investigate definite integrals that we are unable to solve exactly. However, in this example the definite integral $\int_0^3 (x^2 - 2x + 2)\, dx$ is easy to solve exactly, so we can actually verify that the error bounds we just found are accurate. Using the Fundamental Theorem of Calculus, we obtain

$$\int_0^3 (x^2 - 2x + 2)\, dx = \left[\frac{1}{3}x^3 - x^2 + 2x\right]_0^3 = 6.$$

In part (b) we used Theorem 8.25 to argue that the value of this definite integral was within 0.421875 of both the trapezoid sum approximation of 6.28125 and the midpoint sum approximation of 5.85938. This meant that we could guarantee that the value of the definite integral was both in the interval $[5.859375, 6.703125]$ and in the interval $[5.437505, 6.281255]$, which is true, because we just showed that the exact value was 6. We then used Theorem 8.26 to obtain better error bounds, which allowed us to guarantee that the value of the definite integral was in both the interval $[6, 6.5625]$ and the interval $[5.71875, 6]$. In fact, notice that the only number in both of those intervals happens to be the exact answer of 6. $\square$

EXAMPLE 3

Arriving at the Simpson's Rule formula when $n = 6$

Show that the Simpson's Rule formula in Theorem 8.28 results from repeated application of Theorem 8.27 in the case where $n = 6$ with $\frac{n}{2} = 3$ parabolas.

SOLUTION

Suppose f is an integrable function on an interval $[a, b]$. In this case we have three quadratic functions $p_1(x)$, $p_2(x)$, and $p_3(x)$ that approximate $f(x)$ on the double subintervals $[x_0, x_2]$, $[x_2, x_4]$, and $[x_4, x_6]$, and agreeing with $f(x)$ at the endpoints and midpoints of these subintervals. Note that the midpoints of the parabola pieces are the odd subdivision points x_1, x_3, and x_5. We can approximate the area under f on $[a, b]$ by adding up the areas under these three parabola pieces $p_1(x)$, $p_2(x)$, and $p_3(x)$:

$$\int_a^b f(x)\, dx \approx \int_{x_0}^{x_2} p_1(x)\, dx + \int_{x_2}^{x_4} p_2(x)\, dx + \int_{x_4}^{x_6} p_3(x)\, dx.$$

By applying Theorem 8.27 to the first double subinterval $[x_0, x_2]$ we see that

$$\int_{x_0}^{x_2} p_1(x)\, dx = \frac{x_2 - x_0}{6}(p_1(x_0) + 4p_1(x_1) + p_1(x_2)).$$

Similarly, on the second and third double subintervals $[x_2, x_4]$ and $[x_4, x_6]$ we have

$$\int_{x_2}^{x_4} p_1(x)\, dx = \frac{x_4 - x_2}{6}(p_1(x_2) + 4p_1(x_3) + p_1(x_4)),$$

$$\int_{x_4}^{x_6} p_1(x)\, dx = \frac{x_6 - x_4}{6}(p_1(x_4) + 4p_1(x_5) + p_1(x_6)).$$

By construction, the values of $p_1(x)$ at x_0, x_1, and x_2 are the same as the corresponding values of $f(x)$. Similarly, the values of $p_2(x)$ and $p_3(x)$ agree with the values of $f(x)$ at the corresponding endpoints and midpoints. Moreover, the width of each double subinterval is $2\Delta x$, and thus $\frac{x_2 - x_0}{6}$, $\frac{x_4 - x_2}{6}$, and $\frac{x_6 - x_4}{6}$ are each equal to $\frac{\Delta x}{3}$. Therefore $\int_a^b f(x)\, dx$ is approximated with Simpson's Rule by the sum

$$(f(x_0) + 4f(x_1) + f(x_2))\frac{\Delta x}{3} + (f(x_2) + 4f(x_3) + f(x_4))\frac{\Delta x}{3} + (f(x_4) + 4f(x_5) + f(x_6))\frac{\Delta x}{3}.$$

A little algebra shows that Simpson's Rule indeed says what is predicted by Theorem 8.28:

$$\int_a^b f(x)\, dx \approx (f(x_0) + 4f(x_1) + 2f(x_2) + 4f(x_3) + 2f(x_4) + 4f(x_5) + f(x_6))\frac{\Delta x}{3}.$$

Note that if we knew the function f and the interval $[a, b]$, then we could easily calculate the preceding sum. ☐

? TEST YOUR UNDERSTANDING

▶ What does it mean for a function to be monotonic on an interval?

▶ Why do we require that f be monotonic on all of $[a, b]$ in Theorem 8.23?

▶ Why do we require that f be either always concave up or always concave down on $[a, b]$ in Theorem 8.25?

▶ As we saw in the reading, a trapezoid sum with n trapezoids can be written in the form $(f(x_0) + 2f(x_1) + 2f(x_2) + \cdots + f(x_n))\left(\frac{\Delta x}{2}\right)$. Why are $f(x_0)$ and $f(x_n)$ not multiplied by 2 in this expression?

▶ Simpson's Rule approximates the area under a curve by approximating a function with pieces of parabolas. How do we define the parabolas that are used on each subinterval?

EXERCISES 8.7

Thinking Back

Monotonicity and concavity: For each function f and interval $[a, b]$, use derivatives to determine whether or not f is monotonic on $[a, b]$ and whether or not f has consistent concavity on $[a, b]$.

▶ $f(x) = x^3 - 2x^2$, $[a, b] = [1, 3]$

▶ $f(x) = x^3 - 2x^2$, $[a, b] = [0, 1]$

▶ $f(x) = \sin\left(\frac{\pi}{2}x\right)$, $[a, b] = [-1, 1]$

Bounding the second derivative: Consider the function $f(x) = x^2(x - 3)$.

▶ To the nearest tenths place, find the smallest number M so that $|f''(x)| \le M$ for all $x \in [0, 2]$.

▶ To the nearest tenths place, find the smallest number M so that $|f''(x)| \le M$ for all $x \in [0, 4]$.

Determining quadratics: For each of the following problems, find the equation of the unique quadratic function $p(x)$ with the given properties.

▶ The graph of $p(x)$ is a parabola that agrees with $f(x) = x^3 + 1$ at $f(-1), f(0)$, and $f(1)$.

▶ The graph of $p(x)$ is a parabola that agrees with the function $f(x) = \frac{1}{x}$ at $x = 1, x = 2$, and $x = 3$.

▶ The graph of $p(x)$ is a parabola that agrees with the function $f(x) = \sin x$ at $x = 0, x = \frac{\pi}{2}$, and $x = \pi$.

Concepts

0. *Problem Zero:* Read the section and make your own summary of the material.

1. *True/False:* Determine whether each of the statements that follow is true or false. If a statement is true, explain why. If a statement is false, provide a counterexample.

(a) *True or False:* If we knew the value of the error from a given Riemann sum calculation, then we would know the actual value of the signed area under the curve.

(b) *True or False:* We can sometimes get a bound on the error from a given Riemann sum calculation even without knowing the actual value of the signed area under the curve.

(c) *True or False:* A left sum with $n = 10$ rectangles will always have a larger error than the corresponding left sum with $n = 100$ rectangles.

(d) *True or False:* A right sum with $n = 10$ rectangles for a monotonically decreasing function will always have a larger error than the corresponding right sum with $n = 100$ rectangles.

(e) *True or False:* If f is monotonically increasing on $[a, b]$, then the error incurred from using the left sum with n rectangles will always be less than the difference of the right and left sums with n rectangles.

(f) *True or False:* If f is concave down on $[a, b]$, then every trapezoid sum on that interval will be an under-approximation.

(g) *True or False:* If f is concave down on $[a, b]$, then every midpoint sum on that interval will be an over-approximation.

(h) *True or False:* Simpson's Rule with $n = 10$ subdivisions (and thus 5 parabolas) will always be more accurate than the trapezoid sum with $n = 10$ trapezoids.

2. *Examples:* Construct examples of the thing(s) described in the following. Try to find examples that are different than any in the reading.

(a) A function f on an interval $[a, b]$ for which the left sum and the trapezoid sum are over-approximations for every n.

(b) A function f on an interval $[a, b]$ for which the right sum and the trapezoid sum are over-approximations for every n.

(c) A function f on an interval $[a, b]$ for which Simpson's Rule has no error at all.

3. Suppose f is monotonicically increasing on $[a, b]$. Use pictures to explain why the error incurred from using the left sum with n rectangles to approximate $\int_a^b f(x)\, dx$ will always be less than or equal to the difference of the right sum with n rectangles and the left sum with n rectangles.

4. Suppose f is concave down on $[a, b]$. Use pictures to explain why the error incurred from using the trapezoid sum with n trapezoids to approximate $\int_a^b f(x)\, dx$ will always be less than or equal to the difference of the trapezoid sum with n trapezoids and the midpoint sum with n rectangles.

5. Sketch an example that shows that the inequality $\text{LEFT}(n) \le \int_a^b f(x)\, dx \le \text{RIGHT}(n)$ is not necessarily true if f is not monotonically increasing on $[a, b]$.

6. Sketch an example that shows that the inequality $\text{MID}(n) \le \int_a^b f(x)\, dx \le \text{TRAP}(n)$ is not necessarily true if f is not concave up on all of $[a, b]$.

7. Sketch an example that shows that the left-sum error bound $|E_{\text{LEFT}(n)}| \le |f(b) - f(a)|\, \Delta x$ does not necessarily hold for functions f that fail to be monotonic on $[a, b]$.

8. If f is monotonically increasing on $[a, b]$, which of the given approximations is *guaranteed* to be an over-approximation for $\int_a^b f(x)\, dx$? (Select *all* that apply.)

(a) left sum   (b) right sum

(c) trapezoid sum   (d) midpoint sum

9. If f is concave down on all of on $[a, b]$, which of the given approximations is *guaranteed* to be an over-approximation for $\int_a^b f(x)\, dx$? (Select *all* that apply.)

(a) left sum   (b) right sum

(c) trapezoid sum   (d) midpoint sum

10. The error bounds for right and left sums in Theorem 8.24 apply only to monotonic functions. Suppose f is a positive integrable function that is increasing on $[a, c]$ and decreasing on $[c, b]$, with $a < c < b$. How could you use a right or left sum to make an estimate of $\int_a^b f(x)\, dx$ and still get a bound on the error? Draw a picture to help illustrate your answer.

11. The error bounds for trapezoid and midpoint sums in Theorem 8.26 apply only to functions with consistent concavity. Suppose f is a positive integrable function that is concave down on $[a, c]$ and concave up on $[c, b]$, with $a < c < b$. How could you use a trapezoid or midpoint sum to make an estimate of $\int_a^b f(x)\, dx$ and still get a bound of the error? Draw a picture to help illustrate your answer.

12. Explain what we mean when we say that the right, left, trapezoid, and midpoint sum approximations for $\int_a^b f(x)\, dx$ involve piecewise-linear approximations of the function f.

13. What is a "midpoint trapezoid," and what does it have to do with the midpoint sum? How do midpoint trapezoids help us determine whether the midpoint sum will be an over-approximation or under-approximation?

14. Theorem 8.26 allows us to estimate $\text{MID}(n)$ with twice the accuracy (i.e., half the error) of $\text{TRAP}(n)$. Is this what you would have expected from looking at graphical examples? Why do you think the midpoint sum is so accurate? *(Hint: Think about "midpoint trapezoids" and how they are defined.)*

15. Suppose f is a function with $f(x) > 0$, $f'(x) \le 0$, and $f''(x) \le 0$.

(a) Put the right-sum, left-sum, trapezoid-sum, and midpoint-sum approximations of $\int_a^b f(x)\, dx$ in order from smallest to largest. (The order will not depend on the number n of rectangles.) Explain your answer with a picture.

(b) The true value of $\int_a^b f(x)\,dx$ must lie between two of the approximations that you listed in part (a). Which two, and why?

16. The following table describes 12 different approximations of the value of $\int_1^2 \frac{1}{x}\,dx$:

Method	$n = 4$	$n = 8$	$n = 16$
Right sum	0.634524	0.662872	0.677766
Left sum	0.759524	0.725372	0.709016
Trapezoid sum	0.697024	0.694122	0.693391
Midpoint sum	0.693147	0.693147	0.693147

(a) The actual value of $\int_1^2 \frac{1}{x}\,dx$ is $\ln 2 \approx 0.69314718$. Create a table showing the errors corresponding to each of the 12 approximations. What do you notice as n increases in each row?

(b) What do you notice about the sign of the errors of the right and left sums? What about the signs of the errors of the trapezoid and midpoint sums?

(c) How do the errors of the right and left sums compare with the errors of the trapezoid and midpoint sums for each n? How do the errors from the trapezoid sums compare with the errors from the midpoint sums?

17. The table that follows describes four different approximations of the value of $\int_2^5 f(x)\,dx$ for some differentiable function f. Use the weighted average formula given in Exercise 61 to find an estimate of $\int_2^5 f(x)\,dx$ that you would expect to be more accurate than any of the estimates in the table, and explain your answer.

Method	$n = 1000$
Right sum	1.09794591
Left sum	1.09927925
Trapezoid sum	1.09861258
Midpoint sum	1.09861214

18. Suppose f is a positive, increasing, concave-up function on an interval $[a, b]$, and let c be the midpoint of the interval. Let A be the point $(a, 0)$, B be the point $(b, 0)$, C be the point $(c, f(c))$, D be the point $(b, f(b))$, E be the point $(a, f(a))$, R be the point $(b, f(c))$, and S be the point $(a, f(c))$.

(a) Draw a picture illustrating the situation described. What does the area of trapezoid $ABDE$ represent?

(b) Draw a line through point C that is parallel to the line connecting D and E. Let P be the point on the line with x-coordinate b. Let Q be the point on the line with x-coordinate a. What does the area of the rectangle $ABRS$ represent? What can you say about the areas of the rectangle $ABRS$ and the trapezoid $ABPQ$?

(c) Let A_1 be the area between the line PQ and the graph of f. Let A_2 be the area between the line DE and the graph of f. How do these areas compare in size? What do the areas represent? How do your answers to these questions help motivate the statement of Theorem 8.26?

19. Verify that the formula in Theorem 8.27 is true in the case where $p(x) = 3x^2 - x + 4$ and $[A, B] = [1, 5]$.

20. Sketch the graph of some function f on $[2, 5]$ for which SIMP(6) (i.e., with $\frac{n}{2} = 3$ parabola-topped rectangles) is an under-approximation for $\int_a^b f(x)\,dx$.

21. Show that the summation expression for Simpson's Rule in Theorem 8.28 follows from repeated applications of the formula for the definite integral of a quadratic in Theorem 8.27, in the case where $n = 4$ (i.e., where there are $\frac{n}{2} = 2$ parabolas).

22. In this exercise we investigate two cases where an approximation with Simpson's Rule happens to find the exact value of a definite integral.

(a) Explain why Simpson's Rule finds the *exact* value of $\int_a^b f(x)\,dx$ if f is a quadratic function. (*Hint: Think about Theorem 8.27.*)

(b) Explain why Simpson's Rule finds the *exact* value of $\int_a^b f(x)\,dx$ if f is a cubic function. (*Hint: Think about M in the formula for the error bound in Theorem 8.29.*)

Skills

Calculate each definite integral approximation in Exercises 23–40, and then find an error bound for your approximation. If it is possible to calculate the definite integral exactly, then do so and verify that the error bounds you found are accurate.

23. $\int_0^4 (x^2 + x)\,dx$, right sum, $n = 4$

24. $\int_0^4 (x^2 + x)\,dx$, right sum, $n = 8$

25. $\int_1^3 e^{-x}\,dx$, left sum with $n = 6$

26. $\int_1^3 e^{-x}\,dx$, left sum, $n = 10$

27. $\int_1^3 e^{-x}\,dx$, trapezoid sum, $n = 6$

28. $\int_1^3 e^{-x}\,dx$, midpoint sum, $n = 6$

29. $\int_1^7 \ln x\,dx$, midpoint sum, $n = 12$

30. $\int_1^7 \ln x\,dx$, trapezoid sum, $n = 12$

31. $\int_0^2 \sqrt{1+x^3}\, dx$, trapezoid sum with $n = 8$

32. $\int_0^2 \sqrt{1+x^3}\, dx$, right sum, $n = 8$

33. $\int_0^3 \cos(x^2)\, dx$, left sum, $n = 9$

34. $\int_0^3 \cos(x^2)\, dx$, midpoint sum, $n = 9$

35. $\int_{-2}^4 x(x+1)(x-2)\, dx$, Simpson's Rule, $n = 12$

36. $\int_{-2}^4 x(x+1)(x-2)\, dx$, Simpson's Rule, $n = 24$

37. $\int_{-1}^2 (x^4 - 4x^3 + 4x^2)\, dx$, Simpson's Rule, $n = 4$

38. $\int_{-1}^2 (x^4 - 4x^3 + 4x^2)\, dx$, Simpson's Rule, $n = 8$

39. $\int_{-\pi}^{\pi} \cos x\, dx$, Simpson's Rule, $n = 8$

40. $\int_{-\pi}^{\pi} \cos x\, dx$, Simpson's Rule, $n = 12$

Approximate each definite integral in Exercises 41–52 with the indicated method and to the given degree of accuracy. Then calculate the definite integral exactly, and verify that the error bounds you found are accurate.

41. $\int_{-1}^1 xe^x\, dx$, left sum, within 0.5

42. $\int_{-1}^1 xe^x\, dx$, right sum, within 0.5

43. $\int_2^4 \frac{1}{x}\, dx$, right sum, within 0.05

44. $\int_2^4 \frac{1}{x}\, dx$, left sum, within 0.05

45. $\int_0^3 x^2\, dx$, midpoint sum, within 0.1

46. $\int_0^3 x^2\, dx$, trapezoid sum, within 0.1

47. $\int_1^4 x^{2/3}$, trapezoid sum, within 0.005

48. $\int_1^4 x^{2/3}$, midpoint sum, within 0.005

49. $\int_0^2 10e^{-x}\, dx$, Simpson's Rule, within 0.01

50. $\int_0^2 10e^{-x}\, dx$, Simpson's Rule, within 0.001

51. $\int_0^{\pi} \sin x\, dx$, Simpson's Rule, within 0.0005

52. $\int_0^{\pi} \sin x\, dx$, Simpson's Rule, within 0.00005

Applications

53. Dad's casserole surprise is hot out of the oven. It cools at a rate of $T'(t) = 15e^{-0.5t}$ degrees per minute. The change in temperature $T(t)$ over an interval is equal to the area under the graph of $T'(t)$ over that interval.

   (a) Estimate the change in temperature of the casserole during its first 4 minutes out of the oven, using LEFT(4), RIGHT(4), MID(4), TRAP(4), and SIMP(4).

   (b) Give a bound on the errors involved in each approximation.

54. The table that follows shows the velocity in meters per second of a parachutist at various times. The distance travelled by the parachutist over an interval of time is equal to the area under the velocity curve on the same interval.

t	1	2	3	4
$v(t)$	9	17	23	28

   (a) The parachutist's acceleration is decreasing because of air resistance. What does this fact imply about $v(t)$?

   (b) Assuming that the velocity curve is increasing and concave down as the data suggests, find the best possible upper and lower bounds on the distance the parachutist fell from time $t = 0$ to $t = 4$ seconds.

55. The rate r of increase of the gross domestic product (GDP), in percent per year, for the United States from 1993 to 2001 is recorded in the table shown here. For example, in 1993 the GDP rose 2.7% from January 1 to December 31. You may assume that there is no growth from December 31 of one year to January 1 of the next year.

Year	1993	1994	1995	1996	1997	1998	1999	2000	2001
r	2.7	4.0	2.7	3.6	4.4	4.3	4.1	3.8	0.3

   (a) Use Simpson's Rule to estimate the total percentage change in GDP from January 1, 1993, to December 31, 2001.

   (b) If the GDP was $6,880 billion on January 1, 1993, what was the GDP on December 31, 2001? Use the best approximating tool available to you.

   (c) Estimate a bound for the fourth derivative of r over the period from 1993 to 2001, and use that information to determine an error bound for your estimate of the total change in r.

56. Annie is building a kayak using wood and fabric. She needs to know the length of a sheet of polyurethane fabric to buy, which depends on the arc length of the longest surface of her kayak. The bottom of her kayak follows the curve

$$h(x) = (6 \times 10^{-4})\, x^4,$$

where h is the height of the bottom of the kayak above the ground when it is resting upright, and x is the distance from the center of the kayak. The kayak is 13 feet long; that is, it extends 6.5 feet from the center in either direction.

(a) Approximate the length of the curve along the bottom of Annie's kayak by using nine equally spaced points along the bottom of the kayak, and straight line segments between each adjacent pair of points.

(b) Make another approximation of the length of the curve along the bottom of Annie's kayak by Simpson's rule on the same nine equally spaced points, given

that the length of this curve is given by the definite integral

$$\int_{-6.5,6.5} \sqrt{1 + (h'(x))^2}\; dx.$$

57. The flow (in cubic feet per second) down the Lochsa River over the course of a year is given in the table that follows, where t measures days after January 1 and $r(t)$ measures the rate of flow at time t. Use Simpson's Rule to approximate the total amount of water that flows down the river in a year. What did you do to account for the fact that a year has 365 days in it?

t	0	60	120	180	240	300	360
$r(t)$	700	1000	6300	4000	500	650	700

Proofs

58. Prove part (a) of Theorem 8.23: If f is integrable and monotonically increasing on $[a, b]$, then, for any positive integer n, $\text{LEFT}(n) \leq \int_a^b f(x)\, dx \leq \text{RIGHT}(n)$.

59. Use geometric arguments to support part (b) of Theorem 8.25: If f is integrable and concave down on $[a, b]$ and n is any positive integer, then $\text{TRAP}(n) \leq \int_a^b f(x)\, dx \leq \text{MID}(n)$.

60. Prove that the trapezoid sum is equivalent to the average of the left and right sums:

$$\text{TRAP}(n) = \frac{\text{LEFT}(n) + \text{RIGHT}(n)}{2}.$$

61. Prove that Simpson's Rule is equivalent to the following weighted average of the trapezoid and midpoint sums:

$$\text{SIMP}(2n) = \frac{1}{3}\text{TRAP}(n) + \frac{2}{3}\text{MID}(n).$$

62. Use steps (a) and (b) that follow to prove Theorem 8.27: if $p(x)$ is a quadratic function on $[A, B]$, then

$$\int_A^B p(x)\, dx = \frac{B - A}{6}\left(p(A) + 4p\left(\frac{A+B}{2}\right) + p(B) \right).$$

(a) First, calculate the integral of an arbitrary quadratic $p(x) = a_0 + a_1 x + a_2 x^2$ over an interval $[A, B]$.

(b) Now use the fact that the parabola $p(x)$ passes through the points $(A, p(A))$, $\left(\frac{A+B}{2}, p\left(\frac{A+B}{2}\right)\right)$, and $(B, p(B))$ to show that what you found in part (a) is equivalent to the desired formula.

63. Prove that by repeated application of the formula for the definite integral of a quadratic in Theorem 8.27, we can arrive at the Simpson's Rule approximation given in Theorem 8.28. (*Hint: Apply Theorem 8.27 to the subintervals* $[x_k, x_{k+2}]$ *where k is even.*)

Thinking Forward

Approximating with power series: For values of x in the interval $(-1, 1)$, the rational function $f(x) = \frac{1}{1-x}$ turns out to be equal to

$$\lim_{n \to \infty} \sum_{k=1}^{n} x^k.$$

▶ Calculate the preceding sum when $x = \frac{1}{2}$ for $n = 5$, $n = 10$, and $n = 20$, and compare these numbers with the quantity $f\left(\frac{1}{2}\right)$. What is the error in each case?

▶ Repeat the previous problem for $x = 2$, and argue that the sum does *not* provide a good approximation of $f(2)$.

CHAPTER REVIEW, SELF-TEST, AND CAPSTONES

Before you progress to the next chapter, be sure you are familiar with the definitions, concepts, and basic skills outlined here. The capstone exercises at the end bring together ideas from this chapter and look forward to future chapters.

Definitions

Give precise mathematical definitions or descriptions of each of the concepts that follow. Then illustrate the definition with a graph or algebraic example, if possible.

▶ the *differential* of a differentiable function $u(x)$

▶ what it means for a rational function to be *improper*

▶ what it means for a definite integral to be *improper*

▶ what it means for an improper integral to *converge* or to *diverge*

▶ the *error* $|E(n)|$ of an approximation of a definite integral by a Riemann sum

▶ the geometric interpretations of the $\frac{n}{2}$-parabola *Simpson's Rule* approximation for the definite integral of a function f on an interval $[a, b]$

Theorems

Fill in the blanks to complete each of the following theorem statements:

▶ If f and u are functions such that $f'(u(x))u'(x)$ is _____ , then we can use integration by substitution to solve

$$\int f'(u(x))u'(x)\, dx = \underline{\quad} .$$

▶ If $f(x) = g(u(x))$ is _____ on $[a, b]$, $u(x)$ is _____ on (a, b), and $G(u)$ is an _____ of $g(u)$, then integration by substitution for definite integrals gives the following string of equalities (fill in the upper and lower limits that are marked by asterisks; there are two possible ways to do this):

$$\int_a^b f(x)\, dx = \int_*^* g(u)\, du = \left[G(u) \right]_*^*$$
$$= \left[G(u(x)) \right]_*^* = G(u(b)) - G(u(a)).$$

▶ If u and v are functions such that $u'(x)v(x) + u(x)v'(x)$ is _____ , then we can use the product rule in reverse to obtain

$$\int (u'(x)v(x) + u(x)v'(x))\, dx = \underline{\quad} .$$

▶ If $u = u(x)$ and $v = v(x)$ are differentiable functions, then the following integration-by-parts formula holds:

$$\int u\, dv = \underline{\quad} - \underline{\quad} .$$

▶ If $u = u(x)$ and $v = v(x)$ are differentiable functions on $[a, b]$, then the following integration-by-parts formula holds for definite integrals:

$$\int_a^b u\, dv = [\underline{\quad}]_a^b = \underline{\quad} - \underline{\quad} .$$

▶ Suppose $\dfrac{p(x)}{q(x)}$ is an _____ rational function with $\deg(p(x)) = n$ and $\deg(q(x)) = m < n$. Then we can write

$$\frac{p(x)}{q(x)} = \underline{\quad} + \underline{\quad} ,$$

for some polynomial $s(x)$ of degree $n-m$ and some proper rational function $\dfrac{r(x)}{q(x)}$ with $\deg(r(x)) < \underline{\quad} .$

▶ In a partial-fractions decomposition of a proper rational function $\dfrac{p(x)}{q(x)}$, if $q(x)$ has a linear factor $x - c$ with multiplicity m, then for some constants $A_1, A_2, \ldots, A_m$, the sum will include terms of the forms _____ .

▶ In a partial-fractions decomposition of a proper rational function $\dfrac{p(x)}{q(x)}$, if $q(x)$ has an irreducible quadratic factor $x^2 + bx + c$ with multiplicity m, then for some constants $B_1, B_2, \ldots, B_m$ and $C_1, C_2, \ldots, C_m$, the sum will include terms of the forms _____ .

▶ By applying polynomial long division to divide a polynomial $p(x)$ by a polynomial $q(x)$ of equal or lower degree, we can obtain polynomials $m(x)$ and $R(x)$ such that the following equation is true:

$$p(x) = \underline{\quad} + \underline{\quad} .$$

Equivalently, we have the following equation:

$$\frac{p(x)}{q(x)} = \underline{\quad} + \underline{\quad} .$$

▶ Every quadratic function $x^2 + bx + c$ can be rewritten in the form $x^2 + bx + c = (x - k)^2 + C$, where $k = \underline{\quad}$ and $C = \underline{\quad} .$

▶ If a is a positive real number, and if $x \in \underline{\quad}$ and $u \in \underline{\quad}$, then the substitution $x = a\sin u$ gives us $\sqrt{a^2 - x^2} = \underline{\quad} .$

▶ If a is a positive real number, and if $x \in \underline{\quad}$ and $u \in \underline{\quad}$, then the substitution $x = a\tan u$ gives us $x^2 + a^2 = \underline{\quad} .$

▶ If a is a positive real number, and if $x \in \underline{\quad}$ and $u \in \underline{\quad}$, then the substitution $x = a\sec u$ gives us $\sqrt{x^2 - a^2} = \underline{\quad}$ if $x < -a$ and $\sqrt{x^2 - a^2} = \underline{\quad}$ if $x > a$.

▶ If the improper integral of a function f on an interval I converges, and if _____ for all $x \in I$ where the functions f and g are defined, then the improper integral of $g(x)$ on I also converges.

▶ If the improper integral of f on I diverges, and $0 \le f(x) \le g(x)$ for all $x \in I$ where the functions are defined, then the improper integral of g on I must ____ .

▶ If f is integrable and increasing on $[a, b]$ and n is a positive integer, then

$$\text{LEFT}(n) \le \underline{\quad} \le \text{RIGHT}(n).$$

▶ If f is integrable and monotonic on $[a, b]$ and n is a positive integer, then

$$|E_{\text{LEFT}(n)}| \le |\underline{\quad} - \underline{\quad}| = |\underline{\quad}|\Delta x.$$

▶ If f is integrable and either always concave up or always concave down on $[a, b]$ and n is a positive integer, then

$$|E_{\text{TRAP}(n)}| \le \underline{\quad} \quad \text{and} \quad |E_{\text{MID}(n)}| \le \underline{\quad} .$$

▶ If f is integrable and either always concave up or always concave down on $[a, b]$, n is a positive integer, and f'' is bounded on $[a, b]$ so that there is some positive real number M such that ____ $\le M$ for all $x \in [a, b]$, then

$$|E_{\text{TRAP}(n)}| \le \underline{\quad} \quad \text{and} \quad |E_{\text{MID}(n)}| \le \underline{\quad} .$$

▶ If $p(x)$ is a quadratic function on $[A, B]$, then we can express $\int_A^B p(x)\, dx$ in terms of $p(A)$, $p\left(\frac{A+B}{2}\right)$, and $p(B)$ as follows: ____ .

▶ If f is integrable and $f^{(4)}$ is bounded on $[a, b]$ so that there is some positive real number M such that ____ $\le M$ for all $x \in [a, b]$, then

$$|E_{\text{SIMP}(n)}| \le \underline{\quad} .$$

Rules of Algebra and Integration

Pythagorean identities: Use Pythagorean identities to rewrite each of the following trigonometric expressions.

▶ $\sin^2 x = $ ____

▶ $\cos^4 x = $ ____

▶ $\tan^2 x = $ ____

▶ $\sec^6 x = $ ____

▶ $\cot^2 x = $ ____

▶ $\csc^8 x = $ ____

Double-angle identities: Use double-angle identities to rewrite each of the following trigonometric expressions until no exponents are involved.

▶ $\sin^2 x = $ ____

▶ $\cos^2 x = $ ____

▶ $\sin^4 x = $ ____

▶ $\cos^6 x = $ ____

Integration formulas: Fill in the blanks to complete each of the following integration formulas.

▶ $\int \tan x\, dx = $ ____

▶ $\int \cot x\, dx = $ ____

▶ $\int \ln x\, dx = $ ____

▶ $\int \sin^{-1} x\, dx = $ ____

▶ $\int \tan^{-1} x\, dx = $ ____

▶ $\int \sec x\, dx = $ ____

▶ $\int \csc x\, dx = $ ____

Defining improper integrals: Fill in the blanks, using limits and proper definite integrals to express each of the following types of improper integral.

▶ If f is continuous on $[a, \infty)$, then

$$\int_a^\infty f(x)\, dx = \underline{\quad} .$$

▶ If f is continuous on $(-\infty, b]$, then

$$\int_{-\infty}^b f(x)\, dx = \underline{\quad} .$$

▶ If f is continuous on $(-\infty, \infty)$, then for any real number c,

$$\int_{-\infty}^\infty f(x)\, dx = \underline{\quad} + \underline{\quad} .$$

▶ If f is continuous on $(a, b]$ but not at $x = a$, then

$$\int_a^b f(x)\, dx = \underline{\quad} .$$

▶ If f is continuous on $[a, b)$ but not at $x = b$, then

$$\int_a^b f(x)\, dx = \underline{\quad} .$$

▶ If f is continuous on $[a, c) \cup (c, b]$ but not at $x = c$, then

$$\int_a^b f(x)\, dx = \underline{\quad} + \underline{\quad} .$$

Convergence and divergence of basic improper integrals: Determine whether each of the given improper integrals converges or diverges. For those that converge, give the exact solution of the integral.

▶ $\int_0^1 \frac{1}{x^p}\, dx$, for $0 < p < 1$

▶ $\int_1^\infty \frac{1}{x^p}\, dx$, for $0 < p < 1$

▶ $\int_0^1 \frac{1}{x^p}\, dx$, for $p = 1$

▶ $\int_1^\infty \frac{1}{x^p}\, dx$, for $p = 1$

▶ $\int_0^1 \frac{1}{x^p}\, dx$, for $p > 1$

▶ $\int_1^\infty \frac{1}{x^p}\, dx$, for $p > 1$

Skill Certification: Integration Techniques

Integration techniques: Use whatever method you like to solve each of the given definite and indefinite integrals. These integrals are neither in order of difficulty nor in order of technique. Many of the integrals can be solved in more than one way.

1. $\displaystyle\int \cos^7 x \, dx$

2. $\displaystyle\int (e^{3\ln x})^2 \, dx$

3. $\displaystyle\int x \sec x \tan x \, dx$

4. $\displaystyle\int \frac{x^2 \sin x + 1}{x} \, dx$

5. $\displaystyle\int \frac{\ln 2x}{x^2} \, dx$

6. $\displaystyle\int \sin^{-1} x \, dx$

7. $\displaystyle\int \cot^6 x \, dx$

8. $\displaystyle\int x^3 \sin x \, dx$

9. $\displaystyle\int 2\pi x(8 - x^{3/2}) \, dx$

10. $\displaystyle\int \csc^4 x \cot^4 x \, dx$

11. $\displaystyle\int_0^1 \ln(1 + 3x) \, dx$

12. $\displaystyle\int_0^{\pi/2} \sin^2 x \cos^5 x \, dx$

13. $\displaystyle\int \frac{\sqrt{x}}{1 - x\sqrt{x}} \, dx$

14. $\displaystyle\int e^x \cos 2x \, dx$

15. $\displaystyle\int \sin(\ln x) \, dx$

16. $\displaystyle\int \sin^4 3x \, dx$

17. $\displaystyle\int \ln\left(\frac{x^3}{2x+1}\right) dx$

18. $\displaystyle\int \frac{\sin^3 x}{\sqrt{\cos x}} \, dx$

19. $\displaystyle\int (\ln x)^3 \, dx$

20. $\displaystyle\int \csc^3 x \, dx$

21. $\displaystyle\int \tan^3 x \sec^3 x \, dx$

22. $\displaystyle\int (x \ln x)^{-1} \, dx$

23. $\displaystyle\int_1^2 \frac{x^2 + 3x + 7}{\sqrt{x}} \, dx$

24. $\displaystyle\int \tan x \ln(\cos x) \, dx$

25. $\displaystyle\int \csc 3x \, dx$

26. $\displaystyle\int (\sin x + e^x)^2 \, dx$

27. $\displaystyle\int_1^3 \frac{e^{1/x}}{x^2} \, dx$

28. $\displaystyle\int x^3 \cos x^2 \, dx$

29. $\displaystyle\int x(x + 5)^{7/2} \, dx$

30. $\displaystyle\int e^x \tan e^x \, dx$

31. $\displaystyle\int \frac{x+1}{x^4 + 3x^2} \, dx$

32. $\displaystyle\int \sec^5 x \tan^3 x \, dx$

33. $\displaystyle\int \frac{x^3}{e^{x^2}} \, dx$

34. $\displaystyle\int x\sqrt{1-x}\sqrt{x-5} \, dx$

35. $\displaystyle\int 2^x e^x \, dx$

36. $\displaystyle\int \frac{x^4}{(x-1)^2(x+2)} \, dx$

37. $\displaystyle\int \frac{7x^2 + 6x + 5}{x^3 + x^2 + x - 3} \, dx$

38. $\displaystyle\int \frac{1}{e^x\sqrt{4 + e^{2x}}} \, dx$

39. $\displaystyle\int \frac{1}{(9 - 4x^2)^{3/2}} \, dx$

40. $\displaystyle\int \frac{2x^3 + x^2 + 2}{x^3 + 1} \, dx$

41. $\displaystyle\int \frac{\ln(\tan^{-1} x)}{x^2 + 1} \, dx$

42. $\displaystyle\int \frac{\sqrt{x}}{1 + x} \, dx$

43. $\displaystyle\int x \,\text{sech}^2 x^2 \, dx$

44. $\displaystyle\int \sinh^2 x \cosh^3 x \, dx$

45. $\displaystyle\int \frac{\sinh^{-1} x}{\sqrt{x^2 + 1}} \, dx$

46. $\displaystyle\int \frac{x^2}{(x^2 - 4)^{3/2}} \, dx$

Areas and average values: Set up and solve definite integrals to calculate the exact values of each of the given geometric quantities. Verify that your answers are reasonable with graphs.

47. The signed area between the graph of $f(x) = \frac{x^2 - 1}{x^2 + 1}$ and the x-axis on the interval $[-4, 4]$.

48. The absolute area between the graph of $f(x) = \sin^2 x \cos x$ and the x-axis on the interval $[0, \pi]$.

49. The area between the graphs of $f(x) = \sin^2 x$ and $g(x) = \frac{1}{4}$ on the interval $[0, \pi]$.

50. The area between the graphs of $f(x) = \ln x$ and $g(x) = (\ln x)^2$ on the interval $\left[\frac{1}{2}, 5\right]$.

51. The average value of the function $f(x) = xe^{-x}$ on the interval $[-1, 3]$.

52. The average value of the function $f(x) = \frac{2x-1}{(x+2)(x^2+1)}$ on the interval $[0, 10]$.

Improper integrals: Set up and solve limits of definite integrals to calculate each of the following improper integrals.

53. $\displaystyle\int_0^8 x^{-1/3} \, dx$

54. $\displaystyle\int_1^\infty x^{-1/4} \, dx$

55. $\displaystyle\int_0^\infty x \ln x \, dx$

56. $\displaystyle\int_0^1 \frac{1}{x(\ln x)^2} \, dx$

57. $\displaystyle\int_0^{\pi/2} \sec x \tan^3 x \, dx$

58. $\displaystyle\int_1^\infty \frac{1}{x^2\sqrt{x^2 - 1}} \, dx$

Numerical integration: Approximate each of the following definite integrals with the indicated method and to the given degree of accuracy. (*These problems assume that you have covered Section 8.7*)

59. $\displaystyle\int_{-1}^1 e^{-x^2} \, dx$, right sum, within 0.5

60. $\displaystyle\int_0^\pi \sin x^2 \, dx$, midpoint sum, within 0.1

61. $\displaystyle\int_0^3 \sqrt{1 + x^3} \, dx$, Simpson's Rule, within 0.01

Capstone Problems

A. *Integration strategies:* Make a flowchart or diagram that describes a strategy for choosing integration techniques. Given a particular integral, what techniques might you try first? What features of the integrand would lead you to choose one technique over another?

B. *Impossible integrals:* Even with all of the techniques of this chapter, there are still plenty of integrals that we cannot solve.

  (a) List three functions $f(x)$ that we cannot integrate with the techniques that we have learned so far. Describe why each of our integration techniques fails or does not apply to each of your three integrals.

  (b) For each of your functions $f(x)$, use the Second Fundamental Theorem of Calculus (Theorem 7.31) to write down an antiderivative of $f(x)$.

  (c) For each of your functions $f(x)$, choose an interval $[a, b]$ over which $f(x)$ is defined and continuous and consider the definite integral $\int_a^b f(x)\,dx$. Use Riemann sums to approximate the values of each of these definite integrals to within an accuracy of 0.05. (See Section 8.7.)

C. *Comparing polynomial long division and synthetic division:* There are two methods that we can use to divide a polynomial by a linear function. In this exercise you will compare those two methods.

  (a) Find the quotient that results by dividing the polynomial $x^3 - 4x^2 - 7x + 10$ by the linear function $x - 5$, using polynomial long division.

  (b) Repeat the same calculation, this time using synthetic division.

  (c) Use a comparison of the coefficients that appear in your two calculations to investigate how synthetic division is just a shorthand for polynomial long division.

D. *Improper integrals and infinite sums:* Consider the improper integral

$$\int_1^\infty \frac{x}{e^x}\,dx$$

  (a) Show, by calculating a limit of definite integrals, that this improper integral converges.

  (b) Sketch a right sum with $\Delta x = 1$ for the given improper integral.

  (c) Use the sketch you made in part (b) to argue that

$$\sum_{k=2}^\infty \frac{k}{e^k}$$

converges, that is, that the sum of infinitely many numbers $\frac{2}{e^2} + \frac{3}{e^3} + \frac{4}{e^4} + \cdots$ is a finite real number.

Applications of Integration

9.1   VOLUMES BY SLICING

▶ Approximating volumes of solids of revolution with disks and washers

▶ Finding exact volumes by accumulating cross-sectional area

▶ Using definite integrals to describe volumes of solids of revolution

Approximating Volume by Slicing

Some geometric quantities are easy to calculate, such as the area of a rectangle, length of a line segment, or volume of a cylinder. Harder to calculate are things like the area of a region with curved boundary, length of a curve, or volume of a curvy solid. In Chapter 7 we calculated the area under a curve by using a "subdivide, approximate, and add" strategy and then taking a limit to obtain a definite integral. We now apply the same strategy to volumes. Let's start by trying to find the volume of a sphere.

Of course, you probably already know that the formula for the volume V of a sphere with radius r is $V = \frac{4}{3}\pi r^3$. But why? Let's approximate the volume of a sphere of radius 2 to find out. We can construct this sphere by considering the region bounded by the graph of $\sqrt{4 - x^2}$ and the x-axis on $[-2, 2]$ and rotating this region around the x-axis, as shown in the first figure that follows. In the second figure we show an approximation of this region with rectangles. If we rotate these rectangles around the x-axis along with the region as shown in the third figure, then they become cylinders that approximate the sphere.

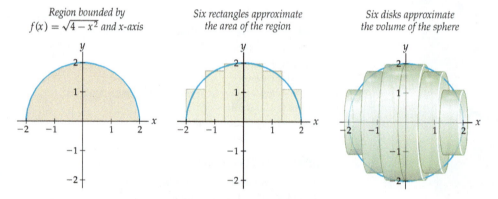

Region bounded by
$f(x) = \sqrt{4 - x^2}$ and x-axis

Six rectangles approximate
the area of the region

Six disks approximate
the volume of the sphere

We can approximate the volume of the sphere by adding up the volumes of these cylinders, which we will call **disks**. We have $n = 6$ disks, each of width $\Delta x = \frac{2}{3}$. These values define subdivision points $x_k = -2 + k\left(\frac{2}{3}\right)$, for $k = 0, 1, 2, 3, 4$, and 5. The heights of the rectangles in the middle figure are defined by the values of $f(x)$ at the midpoints x_k^* of each subinterval $[x_{k-1}, x_k]$. Each disk has radius $f(x_k^*)$ and thus volume $\pi f(x_k^*)^2 \left(\frac{2}{3}\right)$. The volume of the sphere is therefore approximately

$$\sum_{k=1}^{6} \pi f(x_k^*)^2 \left(\frac{2}{3}\right) = \sum_{k=1}^{6} \pi \left(\sqrt{4 - (x_k^*)^2}\right)^2 \left(\frac{2}{3}\right) = \pi \sum_{k=1}^{6} (4 - (x_k^*)^2)\left(\frac{2}{3}\right).$$

To make our approximation better, we could start with more rectangles and get more cylinders, each of which is a better approximation to its part of the sphere. For n disks and

$\Delta x = \frac{2-(-2)}{n} = \frac{4}{n}$, we have

$$\sum_{k=1}^{n} \pi f(x_k^*)^2 \left(\tfrac{4}{n}\right) = \sum_{k=1}^{n} \pi \left(\sqrt{4-(x_k^*)^2}\right)^2 \left(\tfrac{4}{n}\right) = \pi \sum_{k=1}^{n} (4-(x_k^*)^2) \left(\tfrac{4}{n}\right).$$

To find the exact volume we would try to take the limit as n approaches infinity. This should sound very familiar, because it is a limit of Riemann sums. That is fantastic news, because it means that we will be able to write down definite integrals that represent volumes, and definite integrals are much easier to calculate than limits of sums.

Volume as a Definite Integral of Cross-Sectional Area

As a warm-up, recall the definition of the area under a curve. For any continuous, nonnegatively-valued function $f(x)$ on $[a, b]$, we defined the area under the graph of f on $[a, b]$ to be

$$\text{Area} = \lim_{n \to \infty} \sum_{k=1}^{n} f(x_k^*) \Delta x = \int_a^b f(x)\, dx,$$

where $\Delta x = \frac{b-a}{n}$, $x_k = a + k\Delta x$, and x_k^* is a point in $[x_{k-1}, x_k]$. Here $f(x^*)$ represents a height, Δx is a width, and their product $f(x_k^*)\Delta x$ is the area of a rectangle that approximates a slice of the area under the graph of f, as shown here in the leftmost figure:

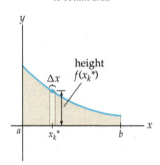

Accumulate heights $f(x_k^*)$ to obtain area

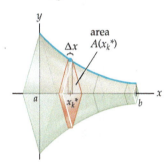

Accumulate cross-sectional areas $A(x_k^*)$ to obtain volume

So how should we define volume? This is a similar situation. Suppose we have a three-dimensional solid with cross sections that are defined in some homogeneous way, such as that shown in the rightmost figure. We subdivide the solid by slicing it, and then we approximate the volume of each slice by multiplying a cross-sectional area $A(x_k^*)$ at some point x_k^* by the width Δx. If the cross sections of the solid vary continuously, then we can get a formula for $A(x_k^*)$. In the preceding rightmost graph, the cross-sectional areas happen to be squares whose diagonals depend on $f(x_k^*)$.

DEFINITION 9.1

Volume as a Definite Integral of Cross-Sectional Areas

If S is a solid whose cross sections between $x = a$ and $x = b$ have areas given by a continuous function $A(x)$ on $[a, b]$, then the **volume** of S is

$$\text{Volume} = \lim_{n \to \infty} \sum_{k=1}^{n} A(x_k^*) \Delta x = \int_a^b A(x)\, dx,$$

where $\Delta x = \frac{b-a}{n}$, $x_k = a + k\Delta x$, and $x_k^* \in [x_{k-1}, x_k]$.

Of course if we sliced instead in the y-direction, then we could make a similar definition, but with area as a function of y instead of x and integrating from some point $y = A$ to some point $y = B$ instead of from $x = a$ to $x = b$.

Volumes by Disks and Washers

A ***solid of revolution*** is a three-dimensional object obtained by rotating a planar region around an axis, or line. For example, consider the planar region $\mathbb{R}$ bounded by $f(x) = x^2$, the x-axis, and the line $x = 2$, as shown in the first figure that follows. By rotating this region around different lines we can obtain many different solids of revolution. For example, the second figure shows the solid of revolution obtained by rotating $\mathbb{R}$ around the x-axis, and the third figure shows the solid of revolution obained by rotating $\mathbb{R}$ around the y-axis.

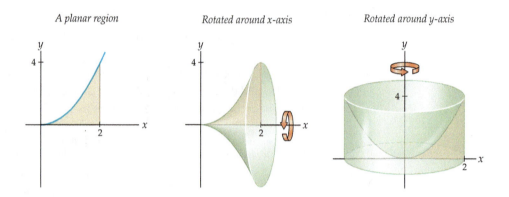

A planar region *Rotated around x-axis* *Rotated around y-axis*

Solids of revolution usually have cross sections that are either disks or ***washers***, which are large disks with a smaller disk removed from the center. For example, the solid shown in the second figure has disk cross sections if we slice along the x-axis, and the solid in the third figure has washer cross sections if we slice along the y-axis. If the cross sections are disks, then we can define volume as the accumulation of the areas of those disks:

DEFINITION 9.2   **Volume as a Definite Integral Using Disks**

Suppose S is a solid of revolution obtained by rotating a region on $[a, b]$ around a horizontal axis. If the cross sections of S are disks with radii given by a continuous function $r(x)$ on $[a, b]$, then the ***volume*** of S is

$$\begin{matrix} \text{Volume of solid with} \\ \text{disk cross sections} \end{matrix} = \lim_{n \to \infty} \sum_{k=1}^{n} \pi (r_k^*)^2 \Delta x = \pi \int_a^b (r(x))^2 \, dx,$$

where $\Delta x = \frac{b-a}{n}$, $x_k = a + k\Delta x$, $x_k^* \in [x_{k-1}, x_k]$, and $r_k^* = r(x_k^*)$.

In this definition, the cross section at each x_k^* is a disk with radius r_k^* as shown next at the left, so its cross-sectional area is $A(x_k^*) = \pi (r_k^*)^2$. If the cross section of a solid of revolution at each x_k^* is a washer with large radius R_k^* and small radius r_k^*, as shown next at the right, then its cross-sectional area is $A(x_k^*) = \pi ((R_k^*)^2 - (r_k^*)^2)$ and its volume can be defined as the accumulation of those areas.

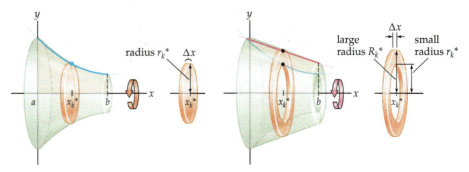

Representative disk in the x direction

radius r_k^* Δx

Representative washer in the x direction

Δx

large radius R_k^* small radius r_k^*

DEFINITION 9.3

Volume as a Definite Integral Using Washers

Suppose S is a solid of revolution obtained by rotating a region on $[a, b]$ around a horizontal axis. If the cross sections of S are washers with large radii given by a continuous funtion $R(x)$ and small radii given by a continuous function $r(x)$ on $[a, b]$, then the *volume* of S is

$$\text{Volume of solid with} \atop \text{washer cross sections} = \lim_{n\to\infty} \sum_{k=1}^{n} \pi((R_k^*)^2 - (r_k^*)^2)\Delta x = \pi \int_a^b ((R(x))^2 - (r(x))^2)\, dx,$$

where $\Delta x = \dfrac{b-a}{n}$, $x_k = a + k\Delta x$, $x_k^* \in [x_{k-1}, x_k]$, $R_k^* = R(x_k^*)$, and $r_k^* = r(x_k^*)$.

We have similar definitions for solids of revolution obtained by rotation around a vertical axis on an interval $[A, B]$ of y-values. The only differences in that case are that the radius functions will be functions of y rather than x and that we will integrate from A to B with respect to y instead of from a to b with respect to x, as shown in the following two figures:

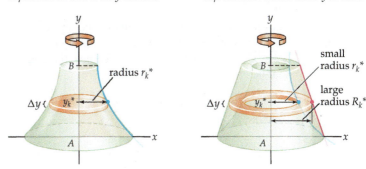

Representative disk in the y direction

radius r_k^*

Δy

Representative washer in the y direction

small radius r_k^*

large radius R_k^*

Δy

The analogs of Definitions 9.2 and 9.3 for integrating along the y-axis are then

$$\text{Volume of solid with} \atop \text{disk cross sections} = \lim_{n\to\infty} \sum_{k=1}^{n} \pi(r_k^*)^2 \Delta y = \pi \int_A^B (r(x))^2\, dy,$$

$$\text{Volume of solid with} \atop \text{washer cross sections} = \lim_{n\to\infty} \sum_{k=1}^{n} \pi((R_k^*)^2 - (r_k^*)^2)\Delta y = \pi \int_A^B ((R(x))^2 - (r(x))^2)\, dy.$$

Distances Defined by a Point on a Graph

In setting up definite integrals to represent volumes by disks or washers, the first step is to determine the radius functions. When we slice across the x direction, radius functions often depend on the function f that was rotated to obtain the solid or on some boundary x-value. When we slice across the y direction, radius functions often depend on f^{-1} or a boundary y-value. As a handy reference, the following two figures show the vertical and horizontal distances defined by a point on a graph:

Vertical and horizontal distances in terms of $y = f(x)$

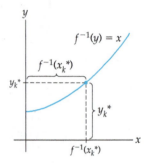

Vertical and horizontal distances in terms of $x = f^{-1}(y)$

Notice that to find the distance between the y-axis and a point on the graph in the second figure, we need to use $f^{-1}(y_k^*)$. This means that f has to be an invertible function (at least on a restricted domain) and that we have to be able to actually figure out what its inverse is.

Examples and Explorations

EXAMPLE 1   **Using disks to approximate volume**

Approximate the volume of a sphere of radius 2 with six disks.

SOLUTION

Let's use the six disks that were defined in the reading at the start of this section. Recall that a sphere of radius 2 can be obtained by rotating the region under the graph of $f(x) = \sqrt{4 - x^2}$ around the x-axis. For reference, the diagrams of these disks and the rectangles that define them are repeated here:

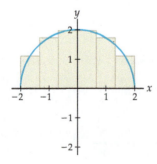

Six rectangles approximate the area under $f(x) = \sqrt{4 - x^2}$

Six disks approximate the volume of the sphere

We have $n = 6$ disks, and thus $\Delta x = \frac{2}{3}$. The six subintervals are

$$\left[-2, -\tfrac{4}{3}\right], \ \left[-\tfrac{4}{3}, -\tfrac{2}{3}\right], \ \left[-\tfrac{2}{3}, 0\right], \ \left[0, \tfrac{2}{3}\right], \ \left[\tfrac{2}{3}, \tfrac{4}{3}\right], \text{ and } \left[\tfrac{4}{3}, 2\right].$$

The radii of the six disks shown in the figure earlier at the right are the heights of the six rectangles in the graph shown earlier at the left, which are obtained by evaluating the function $f(x) = \sqrt{4 - x^2}$ at the midpoints of the six subintervals. Thus the radii of the six disks are

$$f\left(-\frac{5}{3}\right), f(-1), f\left(-\frac{1}{3}\right), f\left(\frac{1}{3}\right), f(1), \text{ and } f\left(\frac{5}{3}\right).$$

The volume of a cylinder with radius r and height h is $\pi r^2 h$, and our disks are cylinders with radii as given earlier and heights (for us they appear as widths) each $\Delta x = \frac{2}{3}$. The sum of their volumes is

$$\pi \left(f\left(-\frac{5}{3}\right)\right)^2 \left(\frac{2}{3}\right) + \pi (f(-1))^2 \left(\frac{2}{3}\right) + \pi \left(f\left(-\frac{1}{3}\right)\right)^2 \left(\frac{2}{3}\right)$$
$$+ \pi \left(f\left(\frac{1}{3}\right)\right)^2 \left(\frac{2}{3}\right) + \pi (f(1))^2 \left(\frac{2}{3}\right) + \pi \left(f\left(\frac{5}{3}\right)\right)^2 \left(\frac{2}{3}\right).$$

Evaluating with $f(x) = \sqrt{4 - x^2}$ and simplifying, we find that the combined volume of the six disks is

$$\frac{22}{27}\pi + 2\pi + \frac{70}{27}\pi + \frac{70}{27}\pi + 2\pi + \frac{22}{27}\pi \approx 33.9758 \text{ cubic units.}$$

This approximation is close to what our old formula for the volume of a sphere predicts: $\frac{4}{3}\pi r^3 = \frac{4}{3}\pi(2)^3 \approx 33.5103$ cubic units.   □

EXAMPLE 2   **Using disks to construct a definite integral for volume**

Find the exact volume of a sphere of radius 2 by using a definite integral to calculate the volume of the solid of revolution obtained by rotating the area under the graph of $f(x) = \sqrt{4 - x^2}$ on $[-2, 2]$ around the x-axis.

SOLUTION

We can rotate a representative rectangle at x_k^* to obtain a representative disk at x_k^*. The radius of this disk is given by the height $f(x_k^*)$ of the function, as shown in the following figure:

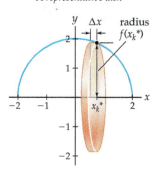

A representative disk

Therefore the volume of this representative disk is

$$\pi(f(x_k^*))^2 \Delta x = \pi \left(\sqrt{4 - (x_k^*)^2}\right)^2 \Delta x = \pi(4 - (x_k^*)^2)\Delta x.$$

We need to accumulate the volumes of these disks from $x = -2$ to $x = 2$, so, converting to definite integral form, we find that the volume of the sphere is

$$\pi \int_{-2}^{2} (4 - x^2)\, dx = \pi \left[4x - \frac{1}{3}x^3\right]_{-2}^{2} = \pi \left(\left(4(2) - \frac{1}{3}(2)^3\right) - \left(4(-2) - \frac{1}{3}(-2)^3\right)\right) = \frac{32\pi}{3}.$$

Note that of course our answer here exactly matches the answer we get from the formula for the volume of a sphere: $V = \frac{4}{3}\pi(2)^3 = \frac{32\pi}{3}$.   □

EXAMPLE 3   **Using washers to construct a definite integral for volume**

Find the volume of the solid of revolution obtained by rotating the region between the graphs of $f(x) = x + 1$ and $g(x) = x^2 - 4x + 5$ on $[1, 4]$ around the x-axis.

SOLUTION

The leftmost figure that follows shows the region in question, and the middle figure shows the solid of revolution whose volume we wish to find. Once again we can take a representative rectangle for the region at x_k^* and rotate it around the x-axis to obtain a representative slice of the solid at x_k^*, which in this case is washer shaped, as shown in the rightmost figure.

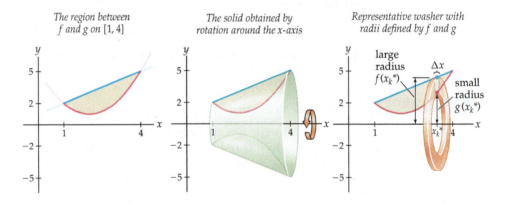

The representative washer has large radius given by $f(x_k^*)$ and small radius given by $g(x_k^*)$. Since the volume of a washer with large radius R, small radius r, and thickness Δx is $\pi R^2 \Delta x - \pi r^2 \Delta x = \pi(R^2 - r^2)\Delta x$, the volume of the representative washer at x_k^* is given by $\pi((f(x_k^*))^2 - (g(x_k^*))^2)\Delta x$.

We must accumulate these washers from $x = 1$ to $x = 4$, so, using the volume formula we just constructed and the equations for $f(x)$ and $g(x)$, we find that the volume of the solid is

$$\pi \int_1^4 ((x+1)^2 - (x^2 - 4x + 5)^2)\, dx = \pi \int_1^4 (-x^4 + 8x^3 - 25x^2 + 42x - 24)\, dx$$

$$= \pi \left[-\frac{1}{5}x^5 + \frac{8}{4}x^4 - \frac{25}{3}x^3 + \frac{42}{2}x^2 - 24x \right]_1^4$$

$$= \frac{117}{5}\pi \approx 73.513 \text{ cubic units.}   □$$

EXAMPLE 4   **Finding volume by integrating along the y-axis**

Find the volume of the cone obtained by rotating the region between $f(x) = \frac{3}{2}x$ and $g(x) = 3$ on $[0, 2]$ around the y-axis.

SOLUTION

The first figure that follows shows the region bounded by $f(x) = \frac{3}{2}x$ and $g(x) = 3$ from $x = 0$ to $x = 2$, and the middle figure shows this region rotated around the y-axis. Since we rotated around the y-axis rather than the x-axis, we will choose a representative rectangle in the horizontal rather than the vertical direction. Then this rectangle will become a representative disk when we rotate it around the y-axis, as shown in the right-most figure.

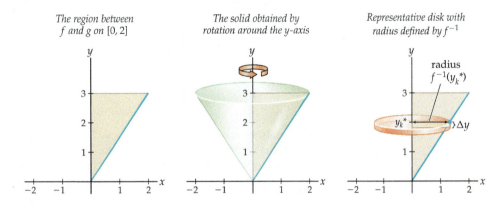

The region between   The solid obtained by   Representative disk with
f and g on [0, 2]   rotation around the y-axis   radius defined by f^{-1}

The representative disk at height y_k^* has thickness Δy and radius $f^{-1}(y_k^*)$. Since $f(x) = \frac{3}{2}x$, we have $y = \frac{3}{2}x$ and thus $x = \frac{2}{3}y$. Therefore $f^{-1}(y) = \frac{2}{3}y$, and the volume of the representative disk at y_k^* is

$$\pi\,(f^{-1}(y_k^*))^2\,\Delta y = \pi\left(\frac{2}{3}y_k^*\right)^2 \Delta y = \frac{4}{9}\pi\,(y_k^*)^2\,\Delta y.$$

To find the volume of the cone we must accumulate disks like the one just described, *vertically* from $y = 0$ to $y = 3$. This gives us a definite integral in terms of y and dy instead of x and dx, but all of our previous solving techniques still work, no matter what letter we happen to be using. The volume of the cone is therefore given by

$$\frac{4}{9}\pi \int_0^3 y^2\,dy = \frac{4}{9}\pi\left[\frac{1}{3}y^3\right]_0^3 = \frac{4\pi}{9}\left(\frac{1}{3}(3)^3 - \frac{1}{3}(0)^3\right) = 4\pi.$$

Notice that 4π is the volume predicted by the formula for the volume of a cone of radius $r = 2$ and height $h = 3$: $V = \frac{1}{3}\pi r^2 h = \frac{1}{3}\pi\,(2)^2(3) = 4\pi$. In Exercise 63 you will use a general version of the preceding calculation to prove the cone volume formula.   ☐

EXAMPLE 5

Setting up more complicated volume integrals

Use definite integrals to express the volumes of the following solids of revolution:

(a) The solid obtained from the region bounded by the graph of $f(x) = 3\ln x$, the line $y = 2$, and the x- and y-axes by rotating around the x-axis.

(b) The solid obtained from the region bounded by the graph of $f(x) = x^2 + 2$ and the line $y = 2$ on the interval $[0, 1]$ by rotating around the line $y = 1$.

SOLUTION

(a) The region and the solid it defines when rotated around the x-axis are shown in the first figure. The graph of $f(x) = 3\ln x$ intersects the line $y = 0$ when $x = 1$ and intersects the line $y = 2$ when $x = e^{2/3}$. The vertical cross sections of this solid are sometimes disks and sometimes washers, as shown in the second and third figures.

The solid obtained by rotating the region around the x-axis

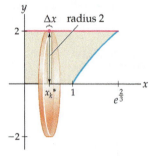

Disk cross sections on [0, 1]

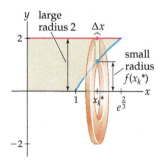

Washer cross sections on $[1, e^{2/3}]$

Since the cross sections change character at $x = 1$, we have to use two definite integrals to compute this volume. We must accumulate disks of radius 2 from $x = 0$ to $x = 1$ and then accumulate washers with large radius 2 and small radius $f(x_k^*) = 3 \ln x_k^*$ from $x = 1$ to $x = e^{2/3}$. The volume of the solid is therefore represented by

$$\pi \int_0^1 2^2 \, dx + \pi \int_1^{e^{2/3}} (2^2 - (3 \ln x)^2) \, dx = \pi \int_0^1 4 \, dx + \pi \int_1^{e^{2/3}} (4 - 9(\ln x)^2) \, dx.$$

(b) The region and the line that we plan to rotate about are shown in the first figure that follows. The second figure shows the solid of revolution that results. To get a good picture of a representative washer, draw the mirror image of the region over the line $y = 1$. Then draw a representative rectangle in the region, and imagine rotating this rectangle down to its mirror image. The resulting washer is shown in the third figure.

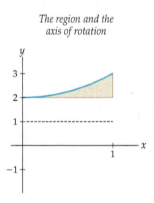

The region and the axis of rotation

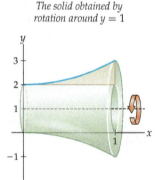

The solid obtained by rotation around $y = 1$

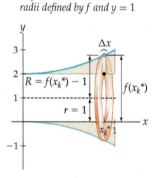

Representative washer with radii defined by f and $y = 1$

Because we rotated around a line other than one of the coordinate axes, determining the large and small radii of the representative washer takes some concentration. In the figure at the right we have labeled the associated distances very carefully. Notice that the large radius of the washer is $R = f(x_k^*) - 1$ and the small radius is constantly $r = 1$. Accumulating these washers from $x = 0$ to $x = 1$, we see that the volume of the solid is represented by the definite integral

$$\pi \int_0^1 ((f(x) - 1)^2 - 1^2) \, dx = \pi \int_0^1 ((x^2 + 1)^2 - 1) \, dx = \pi \int_0^1 (x^4 + 2x^2) \, dx. \quad \square$$

? TEST YOUR UNDERSTANDING

▶ For a solid S defined by revolving a region $\mathbb{R}$ around an axis or a line, how is an approximation of R by rectangles related to an approximation of S by disks and/or washers?

▶ If S is a solid whose cross-sectional area at $x \in [a, b]$ is $A(x)$, what has to be true about $A(x)$ for volume approximations to converge to the definite integral $\int_a^b A(x) \, dx$?

► What type of situation would require us to integrate along the y-axis instead of the x-axis to use disks or washers, and why?

► What is the vertical distance between the x-axis and a point on the graph of $y = f(x)$? What is the horizontal distance between the y-axis and a point on the graph?

► In Example 3, would we get the same answer if we subtracted the volume of the solid obtained by rotating $g(x) = x^2 - 4x + 5$ around the x-axis from the volume of the solid obtained by rotating $f(x) = x + 1$ around the x-axis? Why or why not?

EXERCISES 9.1

Thinking Back

Definite integrals: Calculate each of the following definite integrals, using integration techniques and the Fundamental Theorem of Calculus.

► $\displaystyle\int_{-3}^{3} (9 - x^2)\, dx$

► $\displaystyle\int_{0}^{1} (x^4 + 2x^2)\, dx$

► $\displaystyle\int_{0}^{1} 4\, dx$

► $\displaystyle\int_{1}^{e^{2/3}} (4 - 9(\ln x)^2)\, dx$

► $\displaystyle\int_{0}^{2} (4y^2 - y^4)\, dy$

► $\displaystyle\int_{0}^{1} (2 - \sqrt{y})^2\, dy$

Definite integrals for geometric quantities: Let $f(x) = \sqrt{x}$ and $g(x) = 2 - x$. Express each of the given quantities in terms of definite integrals. Do not solve the integrals; just set them up.

► The area under the graph of $f(x)$ on $[0, 4]$.

► The absolute area under the graph of $g(x)$ on $[0, 4]$.

► The area between $f(x)$ and $g(x)$ on $[0, 4]$.

► The average value of $f(x)$ on $[0, 4]$.

Concepts

0. *Problem Zero:* Read the section and make your own summary of the material.

1. *True/False:* Determine whether each of the statements that follow is true or false. If a statement is true, explain why. If a statement is false, provide a counterexample.

  (a) *True or False:* Every sum is a Riemann sum and can be turned into a definite integral.

  (b) *True or False:* Every sum involving only continuous functions is a Riemann sum and can be turned into a definite integral.

  (c) *True or False:* The volume of a disk can be obtained by multiplying its thickness by the circumference of a circle of the same radius.

  (d) *True or False:* The volume of a disk can be obtained by multiplying its thickness by the area of a circle of the same radius.

  (e) *True or False:* The volume of a cylinder can be obtained by multiplying the height of the cylinder by the area of a circle of the same radius.

  (f) *True or False:* The volume of a washer can be expressed as the difference of the volume of two disks.

  (g) *True or False:* The volume of a right cone is exactly one third of the volume of a cylinder with the same radius and height.

  (h) *True or False:* The volume of a sphere of radius r is $V = \frac{4}{3}\pi r^3$.

2. *Examples:* Construct examples of the thing(s) described in the following. Try to find examples that are different than any in the reading.

  (a) A region that, when revolved around the x-axis, has both disk and washer cross sections.

  (b) A region that, when revolved around the y-axis, has both disk and washer cross sections.

  (c) A solid of revolution for which it is not possible to use the disk or washer method.

3. Consider the rectangle bounded by $y = 3$ and $y = 0$ on the x-interval $[2, 2.25]$.

  (a) What is the volume of the disk obtained by rotating this rectangle around the x-axis?

  (b) What is the volume of the washer obtained by rotating this rectangle around the line $y = 5$?

4. Consider the rectangle bounded by $x = 1$ and $x = 4$ on the y-interval $[3, 3.5]$.

  (a) What is the volume of the disk obtained by rotating this rectangle around the line $x = 4$?

  (b) What is the volume of the washer obtained by rotating this rectangle around the y-axis?

5. Consider the region between $f(x) = 5 - x^2$ and the x-axis between $x = 0$ and $x = 4$. Draw a Riemann sum approximation of the area of this region, using a midpoint sum with four rectangles, and explain how it is related to a four-disk approximation of the solid obtained by rotating the region around the x-axis.

6. For a four-disk approximation of the volume of the solid obtained from the region between $f(x) = 5 - x^2$ and the x-axis between $x = 0$ and $x = 4$ by rotating around the x-axis, illustrate and calculate

  (a) Δx and each x_k;

  (b) some x_k^* in each subinterval $[x_{k-1}, x_k]$;

  (c) each $f(x_k^*)$;

  (d) the volume of the second disk.

7. For a four-washer approximation of the volume of the solid obtained from the region between $f(x) = x^2$ and the y-axis between $y = 0$ and $y = 4$ by rotating around the x-axis, illustrate and calculate

(a) Δx and each x_k;

(b) some x_k^* in each subinterval $[x_{k-1}, x_k]$;

(c) each $f(x_k^*)$;

(d) the volume of the second washer.

8. For a four-disk approximation of the volume of the solid obtained from the region between $f(x) = x^2$ and the y-axis between $y = 0$ and $y = 4$ by rotating around the y-axis, illustrate and calculate

(a) Δy and each y_k;

(b) some y_k^* in each subinterval $[y_{k-1}, y_k]$;

(c) each $f^{-1}(y_k^*)$;

(d) the volume of the second disk.

Write each of the limits in Exercises 9–11 in terms of definite integrals, and identify a solid of revolution whose volume is represented by that definite integral.

9. $\displaystyle\lim_{n\to\infty} \sum_{k=1}^{n} \pi(1 + x_k^*)^2 \left(\frac{1}{n}\right)$, with $x_k^* = x_k = 2 + k\left(\frac{1}{n}\right)$

10. $\displaystyle\lim_{n\to\infty} \sum_{k=1}^{n} \pi(1 + y_k^*)^2 \left(\frac{3}{n}\right)$, with $y_k^* = y_k = 1 + k\left(\frac{3}{n}\right)$

11. $\displaystyle\lim_{n\to\infty} \sum_{k=1}^{n} \pi(4 - (x_k^*)^2)\left(\frac{2}{n}\right)$, with $x_k^* = x_k = k\left(\frac{2}{n}\right)$

12. Suppose that for some $b > 0$, the region between $y = \sqrt{x}$ and $y = 0$ on $[0, b]$, rotated around the x-axis, has volume $V = 8\pi$. Without solving any integrals, find the volume of solid obtained by rotating the region bounded by $y = \sqrt{x}$, $y = \sqrt{b}$, and $x = 0$ around the x-axis.

For each pair of definite integrals in Exercises 13–18, decide which, if either, is larger, without computing any integrals.

13. $\displaystyle\pi \int_0^{\pi/2} \cos^2 x \, dx$ and $\pi \int_{\pi/2}^{\pi} \cos^2 x \, dx$

14. $\displaystyle\pi \int_0^1 e^{2x} \, dx$ and $\pi \int_0^1 (e^{2x} - 1) \, dx$

15. $\displaystyle\pi \int_0^2 x^4 \, dx$ and $\pi \int_0^2 (16 - x^4) \, dx$

16. $\displaystyle\pi \int_0^3 x^2 \, dx$ and $\pi \int_0^3 (9 - x^2) \, dx$

17. $\displaystyle\pi \int_0^2 x^4 \, dx$ and $\pi \int_0^4 y \, dy$

18. $\displaystyle\pi \int_0^4 y \, dy$ and $\pi \int_4^8 (8 - y) \, dy$

Each of the definite integrals in Exercises 19–24 represents the volume of a solid of revolution obtained by rotating a region around either the x- or y-axis. Find this region.

19. $\displaystyle\pi \int_1^3 (16 - (x+1)^2) \, dx$ **20.** $\displaystyle\pi \int_0^{\pi} \sin^2 x \, dx$

21. $\displaystyle\pi \int_1^3 (x^2 - 2x + 1) \, dx$ **22.** $\displaystyle\pi \int_0^2 y \, dy$

23. $\displaystyle\pi \int_1^5 \left(\frac{y-1}{2}\right)^2 dy$ **24.** $\displaystyle\pi \int_0^4 (2^2 - (\sqrt{y})^2) \, dy$

Write the volume of the two solids of revolution that follow in terms of definite integrals that represent accumulations of disks and/or washers. Do not compute the integrals.

25.

26.

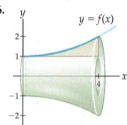

Skills

Consider the region between $f(x) = \sqrt{x}$ and the x-axis on $[0, 4]$. For each line of rotation given in Exercises 27–30, use four disks or washers based on the given rectangles to approximate the volume of the resulting solid.

27. Around the x-axis   **28.** Around the y-axis

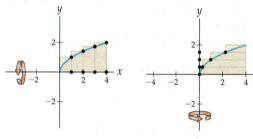

29. Around the line $y = -1$ **30.** Around the line $x = 5$

Consider the region between the graph of $f(x) = \dfrac{3}{x}$ and the x-axis on $[1, 3]$. For each line of rotation given in Exercises 31–34, use definite integrals to find the volume of the resulting solid.

31. *Around the x-axis*

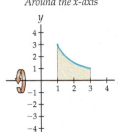

32. *Around the line y = −1*

33. *Around the y-axis*

34. *Around the line x = 1*

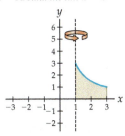

Consider the region between the graph of $f(x) = x - 2$ and the x-axis on $[2, 5]$. For each line of rotation given in Exercises 35–40, use definite integrals to find the volume of the resulting solid.

35. *Around the y-axis*

36. *Around the x-axis*

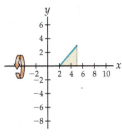

37. *Around the line y = 3*

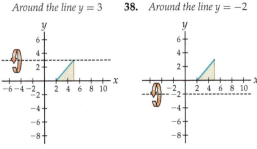

38. *Around the line y = −2*

39. *Around the line x = 2*

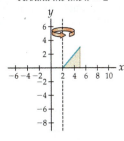

40. *Around the line x = 6*

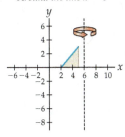

Consider the region between the graph of $f(x) = 4 - x^2$ and the line $y = 5$ on $[0, 2]$. For each line of rotation given in Exercises 41–44, use definite integrals to find the volume of the resulting solid.

41. *Around the y-axis*

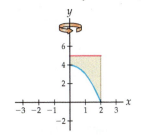

42. *Around the line x = 2*

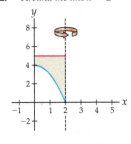

43. *Around the line x = 3*

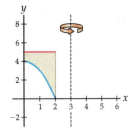

44. *Around the line y = 5*

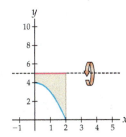

Consider the region between the graphs of $f(x) = 5 - x$ and $g(x) = 2x$ on $[1, 4]$. For each line of rotation given in Exercises 45 and 46, use definite integrals to find the volume of the resulting solid.

45. *Around the x-axis*

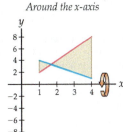

46. *Around the line x = 1*

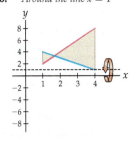

Consider the region between the graphs of $f(x) = x^2$ and $g(x) = 2x$ on $[0, 2]$. For each line of rotation given in Exercises 47–50, use definite integrals to find the volume of the resulting solid.

47. *Around the y-axis*   **48.** *Around the x-axis*

  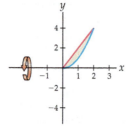

49. *Around the line $y = -1$*   **50.** *Around the line $x = 3$*

  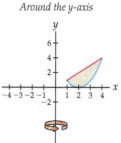

Consider the region between the graphs of $f(x) = (x-2)^2$ and $g(x) = x$ on $[1, 4]$. For each line of rotation given in Exercises 51–54, use definite integrals to find the volume of the resulting solid.

51. *Around the x-axis*   **52.** *Around the y-axis*

  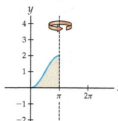

53. *Around the line $y = -1$*   **54.** *Around the line $x = 1$*

  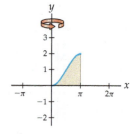

Consider the region between the graph of $f(x) = 1 - \cos x$ and the x-axis on $[0, \pi]$. For each line of rotation given in Exercises 55–58, write down definite integrals that represent the volume of the resulting solid and then use a calculator or computer to approximate the integrals.

55. *Around the x-axis*   **56.** *Around the y-axis*

57. *Around the line $x = \pi$*   **58.** *Around the line $y = 2$*

  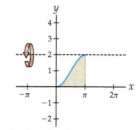

Applications

59. Rebecca plans to visit the Great Pyramid of Giza in Cairo, Egypt. The Great Pyramid has a square base with each side approximately 756 feet in length.

    (a) An interesting fact about the Great Pyramid of Giza is that the perimeter of its base is equal to the circumference of a circle whose radius is equal to the height of the pyramid. Use this fact to find the height of the Great Pyramid, rounded to the nearest integer.

    (b) The volume of a square-based pyramid with base area A and height h is given by the formula $V = \frac{1}{3}Ah$. Use this formula to find the volume of the Great Pyramid.

    (c) Find the volume of the Great Pyramid another, much more difficult way: Set up a definite integral that represents the volume, and solve it. Use the given diagram as a starting point, and justify your answer.

Great Pyramid of Giza

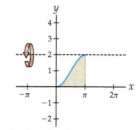

60. Two of Dr. Geek's friends got married this year, one in June and one in July. His traditional gift for such occasions is a fancy crystal bowl, whose volume corresponds to how much he likes the future spouse of the recipient.

Crystal hexagonal bowl

Crystal octagonal bowl

The area of a hexagon with radius r is $A = \dfrac{3\sqrt{3}}{2}r^2$.

What is the volume of this bowl, in cubic inches?

(b) For the July wedding, Dr. Geek gave a fancy octagonal crystal bowl with base radius 4, top radius 8, and height 6 inches. The area of an octagon with radius r is $A = 2\sqrt{2}r^2$. What is the volume of this bowl, in cubic inches?

(a) For the June wedding, Dr. Geek gave a fancy hexagonal crystal bowl with base hexagon of radius 4 inches, top hexagon of radius 8 inches, and height of 8 inches.

Proofs

61. Use a definite integral to prove that the volume formula $V = \dfrac{1}{3}\pi r^2 h$ holds for a cone of radius 3 and height 5.

62. Use a definite integral to prove that the volume formula $V = \dfrac{4}{3}\pi r^3$ holds for a sphere of radius 3.

63. Use a definite integral to prove that a cone of radius r and height h has volume given by the formula $V = \dfrac{1}{3}\pi r^2 h$.

64. Use a definite integral to prove that a sphere of radius r has volume given by the formula $V = \dfrac{4}{3}\pi r^3$.

Thinking Forward

Slicing to obtain a definite integral for work: The "subdivide, approximate, and add" strategy works for more than just geometric quantities like area, length, and volume. In this problem you will investigate the use of a slicing process to write work as a definite integral.

▶ The **work** involved in lifting an object is the product of the weight of the object and the distance it must be lifted. Suppose a flat, wide cylinder that weighs 50 pounds must be lifted from ground level to a height of 10 feet. How much work does this require, in foot-pounds?

▶ Now suppose we want to pump all of the water out of the top of a 4-foot-high cylindrical tank with radius 10 feet. It takes more work to pump out the water from

the bottom of the tank than it does to pump it out from the top of the tank. With the given figure as a guide, use horizontal slices to set up and solve a definite integral that represents the work required to pump all of the water out of the top of the tank.

9.2 VOLUMES BY SHELLS

▶ Approximating volumes of solids of revolution with shells

▶ Setting up Riemann sums for shells by using an alternative volume formula

▶ Using definite integrals to describe volumes of solids of revolution with shells

Approximating Volume by Shells

We now develop an alternative method for calculating the volume of a solid of revolution, using "shells" rather than slices. This new method will help us find some volumes that are inconvenient to find by using disks or washers.

For example, consider the region bounded by the graph of $f(x) = -x^2 - x + 8$ and the x-axis from $x = 0$ to $x = 2$, rotated about the y-axis. To use disks to approximate the volume of this solid we could slice the region into horizontal rectangles, for example as shown in the figure that follows at the left. When we revolve these rectangles about the y-axis, they produce four disks stacked on top of one another, as shown in the figure at the right.

Four horizontal rectangles

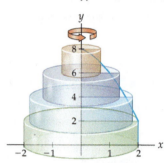

A stack of four disks

Unfortunately, it is difficult to find the radii of these disks, since it is not easy to solve the equation $y = -x^2 - x + 8$ for x. This makes it difficult to use the slicing/disk method here.

What happens if we instead try to use vertical rectangles, as shown next at the left? When we revolve these rectangles around the y-axis, each rectangle makes a *shell* rather than a disk or washer, as shown at the right. The green shell is a washer that is wrapped around the other shells. The blue shell is an even taller washer that wraps around an even taller purple washer, which in turn wraps around a very tall pink cylinder. Instead of slicing the solid horizontally or vertically, we have sliced it in layers from the inside out, like the layers of an onion.

Four vertical rectangles

Four nested shells

To get an approximation for the volume of the solid we just add up the volumes of these four shells. In this example, shells are easier to work with than disks because the heights of the shells are determined by $f(x)$ rather than the hard-to-compute $f^{-1}(x)$.

Finding a Riemann Sum for Volumes by Shells

Before we can construct a definite integral to represent the exact volume of a solid of revolution with the shell method, we need to know how to calculate the volume of a representative shell such as the one shown in the next figure. This shell has height $f(x_k^*)$ and thickness Δx.

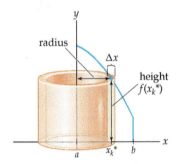

So what is the radius of this shell? Really, there are two radii, an inner radius r_1 and an outer radius r_2, where $r_2 - r_1$ is the thickness Δx of the shell. The shell can be thought of as a cylinder of radius r_2 and height $h = f(x_k^*)$ with a smaller cylinder of radius r_2 and height $h = f(x_k^*)$ removed. Therefore its volume is

$$V = \pi r_2^2 h - \pi r_1^2 h = \pi (r_2^2 - r_1^2) f(x_k^*).$$

This formula is not in the right form for constructing a definite integral, because it does not explicitly involve Δx, which we need in our expression if we want to construct a Riemann sum. We can get around this problem by calculating the volume of a shell in terms of its average radius.

THEOREM 9.4   **The Volume of a Shell**

A shell with inner radius r_1, outer radius r_2, and height h is said to have **average radius** $r = \frac{r_1 + r_2}{2}$ and **thickness** $\Delta x = r_2 - r_1$. Such a shell has volume $V = 2\pi rh\,\Delta x$.

The volume formula in this theorem makes intuitive sense, because a shell is a thickened cylinder and the formula is the product of the lateral surface area $2\pi rh$ and the thickness Δx.

Proof. Given a shell with inner radius r_1, outer radius r_2, and height h, define $r = \frac{r_1 + r_2}{2}$ and $\Delta x = r_2 - r_1$. With this notation we have

$$2\pi rh\,\Delta x = 2\pi \left(\frac{r_1 + r_2}{2}\right) h(r_2 - r_1) = \pi (r_1 + r_2)(r_1 - r_2)h = \pi (r_2^2 - r_1^2)h,$$

which is the volume of the shell.   ∎

Now suppose we are given a partition $x_0, x_1, \ldots, x_n$ of $[a, b]$. If the kth shell has width Δx, average radius $r_k = \frac{x_k + x_{k-1}}{2}$, and height $f(x_k^*)$, then the sum of the volumes of the n shells is given by

$$\sum_{k=1}^{n} 2\pi\, r_k f(x_k^*)\,\Delta x.$$

This is almost a Riemann sum, since the summand is the product of a function $2\pi r_k f(x_k^*)$ of the partition points and a factor of Δx. As $n \to \infty$ and thus $\Delta x \to 0$, the radius r_k that is the midpoint of each subinterval will get closer and closer to x_k^*, and the limit of this sum will converge to a definite integral.

A Definite Integral for Volume by Shells

In general, the heights of a representative shell may be given by something more complicated than just the height $f(x_k^*)$ of some function, depending on the region that is rotated. But in any case, if the shell height function is continuous, then we can use shells to express the volume of the solid of revolution as a definite integral.

DEFINITION 9.5

Volume as a Definite Integral Using Shells

Suppose S is a solid of revolution obtained by rotating a region on $[a, b]$ around a vertical axis. If S can be represented with nested shells whose average radii and heights are given by continuous functions $r(x)$ and $h(x)$ on $[a, b]$, then the **volume** of S is

$$\text{Volume of solid with nested shells} = \lim_{n \to \infty} \sum_{k=1}^{n} 2\pi \, r_k h_k^* \, \Delta x = 2\pi \int_a^b r(x) \, h(x) \, dx,$$

where $\Delta x = \dfrac{b-a}{n}$, $x_k = a + k\Delta x$, $x_k^* \in [x_{k-1}, x_k]$, $r_k = \dfrac{x_k + x_{k-1}}{2}$, and $h_k^* = h(x_k^*)$.

Of course, there is a similar definition for the volume of a solid of revolution obtained by rotating around a horizontal axis and integrating along an interval $[A, B]$ of y-values. In this case the radius and height functions for the shells will be functions of y rather than x. The following two figures illustrate examples of shells to be integrated in the x and y directions:

Shells in the x direction

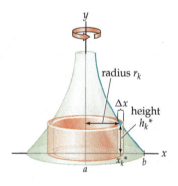

Shells in the y direction

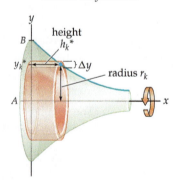

When integrating by shells in the x direction, we will have the radius function in terms of x and the height function in terms of some function of x. When integrating by shells in the y direction, we will have the radius function in terms of y and the height function in terms of some function of y, usually given by boundary values or some inverse function. The formula that is analogous to Definition 9.5 for integrating along the y-axis is

$$\text{Volume of solid with nested shells} = \lim_{n \to \infty} \sum_{k=1}^{n} 2\pi \, r_k h_k^* \, \Delta y = 2\pi \int_A^B r(x) \, h(x) \, dy.$$

Examples and Explorations

EXAMPLE 1	**Approximating a volume with shells**

Consider the solid obtained by rotating the region between $f(x) = -x^2 - x + 8$ and the x-axis on $[0, 2]$ around the y-axis. Approximate the volume of this region with four shells.

SOLUTION

Let's use the four shells that were defined at the start of the reading, whose heights were defined at the midpoints of each subinterval. For reference, we repeat the diagrams here:

Four rectangles approximate
the area under $f(x) = -x^2 - x + 8$

Four shells approximate
the volume of the solid

The centermost shell is the tall pink disk defined on $[0, 0.5]$. This disk has radius $r = 0.5$ and height defined by the value of the function at the midpoint $x = 0.25$ of the interval: $h = f(0.25) = -(0.25)^2 - 0.25 + 8 = 7.8125$. Therefore the first, innermost shell has volume

$$\pi r^2 h = \pi (0.5)^2 f(0.25) \approx 6.04 \text{ cubic units.}$$

The next shell is the purple shell defined on $[0.5, 1]$ with outer radius $r_2 = 1$, inner radius $r_1 = 0.5$, and height defined by the function at the midpoint $x = 0.75$ of the interval: $h = f(0.75) = -(0.75)^2 - (0.75) + 8 = 6.6875$. Therefore the volume of the second, purple shell is

$$\pi (r_2^2 - r_1^2) h = \pi (1^2 - 0.5^2) f(0.75) \approx 15.76 \text{ cubic units.}$$

We could have instead used the shell volume formula described just before Theorem 9.4 to compute the volume of the solid. The purple shell has average radius $r = 0.75$, thickness $\Delta x = \frac{1}{2}$, and height $h = f(0.75) = 6.6875$. Using the new formula, we see that once again the volume of the purple shell is

$$2\pi r h \, \Delta x = 2\pi (0.75) f(0.75) \left(\frac{1}{2}\right) \approx 15.76 \text{ cubic units.}$$

Using similar methods, we can calculate that the volumes of the blue and green shells are approximately 20.37 and 17.52 cubic units, respectively. The volume V of the solid of revolution is approximately the sum of the volumes of the four shells:

$$V \approx 6.04 + 15.76 + 20.37 + 17.52 = 59.69 \text{ cubic units.}$$

✔ CHECKING THE ANSWER

To verify that our answer is reasonable, we can compare the solid (shown next at the left) with a very rough approximation whose volume is easy to calculate (shown at the right).

Solid of revolution

Very rough approximation

The solid in the second figure is just a disk of radius 2 and height 2 with a cone of radius 2 and height 6 on top, so its volume is

$$\pi 2^2(2) + \frac{1}{3}\pi 2^2(6) = 16\pi \approx 50.2655.$$

This number does make our earlier approximation of 59.69 seem reasonably accurate, since the actual solid is somewhat larger than our cone-on-disk approximation.

EXAMPLE 2

Using shells to construct a definite integral for volume

Find the exact volume of the solid obtained by rotating the region between the graph of $f(x) = -x^2 - x + 8$ and the x-axis on $[0, 2]$ around the y-axis.

SOLUTION

The first figure that follows shows the region in question, and the second figure shows the resulting solid of revolution. If we take a vertical representative rectangle in the region and rotate it around the y-axis, then we obtain a representative shell for the solid at x_k^*, as shown in the third figure.

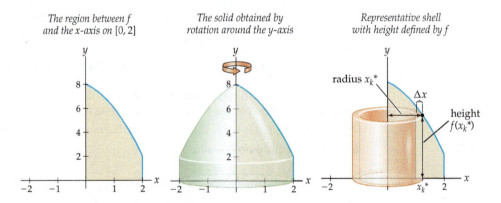

The region between f and the x-axis on $[0, 2]$

The solid obtained by rotation around the y-axis

Representative shell with height defined by f

If we imagine x_k^* to be the midpoint of the subinterval $[x_{k-1}, x_k]$, then this representative shell has average radius x_k^*, height $f(x_k^*)$ and thickness Δx. Therefore the volume of the representative shell at x_k^* is $2\pi x_k^* f(x_k^*)\Delta x$.

Accumulating these shells from the inside out, from $x = 0$ at the center to $x = 2$ on the outside, and applying the function $f(x) = -x^2 - x + 8$, we can write the volume of the solid as the definite integral

$$2\pi \int_0^2 x\,(-x^2 - x + 8)\,dx = 2\pi \int_0^2 (-x^3 - x^2 + 8x)\,dx$$

$$= 2\pi \left[-\frac{1}{4}x^4 - \frac{1}{3}x^3 + \frac{8}{2}x^2 \right]_0^2$$

$$= 2\pi \left(\left(-\frac{1}{4}(2)^4 - \frac{1}{3}(2)^3 + 4(2)^2\right) - \left(-\frac{1}{4}(0)^4 - \frac{1}{3}(0)^3 + 4(0)^2\right) \right)$$

$$= \frac{56}{3}\pi.$$

Notice that this volume of $\frac{56}{3}\pi \approx 58.6431$ cubic units is not far off from the four-shell approximation we did in the previous example.

EXAMPLE 3

Finding volume by integrating along the *y*-axis with shells

Find the volume of the solid obtained by rotating the region bounded by the graph of $f(x) = 3\ln x$, the line $y = 2$, and the *x*- and *y*-axes around the *x*-axis.

SOLUTION

The three figures that follow illustrate, respectively, the region in question, the resulting solid, and a representative shell obtained by rotating a horizontal rectangle around the *x*-axis. Note that this is the same solid whose volume we computed by using disks and washers in the first part of Example 5 of Section 9.1.

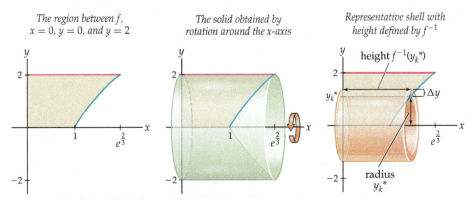

The region between f, $x = 0$, $y = 0$, and $y = 2$ | The solid obtained by rotation around the *x*-axis | Representative shell with height defined by f^{-1}

The representative shell at y_k^* has average radius y_k^*, (horizontal) height $f^{-1}(y_k^*)$, and thickness Δy. Therefore the volume of this representative shell is $2\pi y_k^* f^{-1}(y_k^*)\,\Delta y$. Since $f(x) = y = 3\ln x$, we have $f^{-1}(y) = e^{y/3}$. Accumulating the shells from the inside at $y = 0$ to the outside at $y = 2$, we get the volume integral

$$2\pi \int_0^2 y\,e^{y/3}\,dy.$$

This definite integral can be solved by using integration by parts with $u = y$ and $dv = e^{y/3}\,dy$:

$$2\pi \int_0^2 y\,e^{y/3}\,dy = 2\pi \left(\left[3y e^{y/3}\right]_0^2 - 3\int_0^2 e^{y/3}\,dy \right)$$

$$= 2\pi \left(\left[3y e^{y/3}\right]_0^2 - \left[9 e^{y/3}\right]_0^2 \right)$$

$$= 2\pi \left((6e^{2/3} - 0) - (9e^{2/3} - 9e^0) \right)$$

$$= 2\pi \left(-3e^{2/3} + 9 \right).$$

The volume of $2\pi(-3e^{2/3} + 9) \approx 19.83$ cubic units does seem appropriate given the dimensions of the solid in question and given that the volume of the cylinder that encloses our solid is $\pi(2)^2 e^{2/3} \approx 24.48$ cubic units.   ☐

EXAMPLE 4   Setting up a variety of volume integrals

Consider the region bounded by the graph of $f(x) = e^x$ and the lines $y = 0$, $y = 3$, $x = 0$, and $x = 2$. Use definite integrals to express the volumes of the following solids of revolution:

(a) The solid obtained by rotating the region around the x-axis. Do this problem in two ways: first by integrating in the x direction and then by integrating in the y direction.

(b) The solid obtained by rotating the region around the y-axis. Again, do this problem in two ways: first by integrating in the x direction and then by integrating in the y direction.

SOLUTION

The region in question is shown in the first figure that follows. This region changes character at $x = \ln 3$, since that is the solution of the equation $e^x = 3$. The second figure shows the solid resulting from rotating this region around the x-axis, and the third figure shows the result of rotating the region around the y-axis.

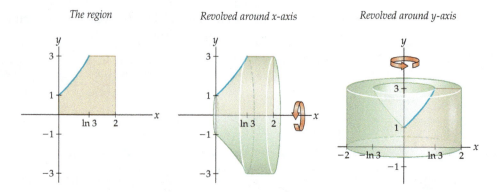

The region   Revolved around x-axis   Revolved around y-axis

(a) Integrating in the x direction, we can use disks to write the volume of the solid in the center graph as the sum of two definite integrals. From $x = 0$ to $x = \ln 3$ the radius of the disk at x_k^* will be given by $e^{x_k^*}$, and from $x = \ln 3$ to $x = 2$ the radius of the disk at x_k^* is 3. The volume can be written as

$$\pi \int_0^{\ln 3} (e^x)^2 \, dx + \pi \int_{\ln 3}^2 (3)^2 \, dx.$$

Integrating in the y direction, we can use shells. From $y = 0$ to $y = 1$, each shell will have a (horizontal) height of $2 - 0 = 2$. From $y = 1$ to $y = 3$, the (horizontal) height of the shell at y_k^* will be the difference $2 - \ln y_k^*$. Therefore the volume of the solid is

$$2\pi \int_0^1 y\,(2) \, dy + 2\pi \int_1^3 y\,(2 - \ln y) \, dy.$$

Of course, the two definite integral expressions we just constructed should find the same volume. Calculate them and see!

(b) We now turn to the solid shown earlier at the right. Integrating in the x direction produces shells in this case. From $x = 0$ to $x = \ln 3$ the shell at x_k^* will have a height of $e^{x_k^*}$, and from $x = \ln 3$ to $x = 2$ the shell will have a height of 3. Therefore the volume of the solid can be written as the following sum of definite integrals:

$$V_2 = 2\pi \int_0^{\ln 3} x e^x \, dx + 2\pi \int_{\ln 3}^2 x(3) \, dx.$$

To integrate in the y direction, we have to use disks and washers. From $y = 0$ to $y = 1$ the disk at y_k^* will have a radius of 2, and from $y = 1$ to $y = 3$ a rectangle at y_k^* will produce a washer with inner radius given by $\ln(y_k^*)$ (since $y = e^x$, we have $x = \ln y$) and an outer radius of 2. Therefore, we can write the volume of the solid as

$$V_2 = \pi \int_0^1 (2)^2 \, dy + \pi \int_1^3 (2^2 - (\ln y)^2) \, dy.$$

Again, the two definite integral expressions we found should give the same volume. Calculating a volume in two ways, when possible, can be a good method for checking your work. □

TEST YOUR UNDERSTANDING

▶ Why was the shell volume formula $2\pi r h \, \Delta x$ more useful than the shell volume formula $\pi(r_2^2 - r_1^2)h$ for constructing a Riemann sum?

▶ With shells, why do we integrate from the center of a solid to the outside, instead of from one side to the other?

▶ What is the analog of Definition 9.5 for using the shell method in the y direction?

▶ Why, geometrically, is the answer found in Example 3 reasonable?

▶ What do the disks, washers, and shells used in the various parts of Example 4 look like?

EXERCISES 9.2

Thinking Back

Definite integrals: Calculate each of the following definite integrals, using integration techniques and the Fundamental Theorem of Calculus.

▶ $\displaystyle\int_0^1 (3 - x)(x^2 + 1) \, dx$

▶ $\displaystyle\int_0^6 x((x - 3)^2 + 2) \, dx$

▶ $\displaystyle\int_0^1 (e^x)^2 \, dx$

▶ $\displaystyle\int_1^5 y e^y \, dy$

▶ $\displaystyle\int_0^\pi y \sin y \, dy$

▶ $\displaystyle\int_0^4 x(2 - \sqrt{x}) \, dx$

▶ $\displaystyle\int_1^2 x e^{x/3} \, dx$

▶ $\displaystyle\int_2^4 x \ln x \, dx$

▶ $\displaystyle\int_1^e (\ln y)^2 \, dy$

▶ $\displaystyle\int_2^4 y\sqrt{6 - y} \, dy$

Finding volumes geometrically: Use volume formulas to find the exact volumes of the solids pictured here.

Concepts

0. *Problem Zero:* Read the section and make your own summary of the material.

1. *True/False:* Determine whether each of the statements that follow is true or false. If a statement is true, explain why. If a statement is false, provide a counterexample.

(a) *True or False:* The thickness of a shell with inner radius r and outer radius R is $R - r$.

(b) *True or False:* The average radius of a shell with inner radius r and outer radius R is $\dfrac{r+R}{2}$.

(c) *True or False:* A shell with height h and average radius r has volume $\pi r^2 h \Delta x$.

(d) *True or False:* A shell with height h, average radius r, and thickness Δx has volume $\pi r^2 h \Delta x$.

(e) *True or False:* A shell with height h, outer radius r, and thickness Δx has volume $2\pi r h \Delta x$.

(f) *True or False:* Rotating a vertical rectangle around the y-axis results in a shell.

(g) *True or False:* Rotating a horizontal rectangle around the y-axis results in a shell.

(h) *True or False:* Rotating a horizontal rectangle around the x-axis results in a shell.

2. *Examples:* Construct examples of the thing(s) described in the following. Try to find examples that are different than any in the reading.

(a) A solid of revolution around the x-axis whose volume can be computed with disks/washers but not shells.

(b) A solid of revolution around the y-axis whose volume can be computed with shells but not disks/washers.

(c) A solid of revolution whose volume can be computed both with shells and with disks/washers.

3. Consider the rectangle bounded by $y = 1$ and $y = 4$ on the x-interval $[1, 1.5]$. Calculate the volume of the geometric object obtained by rotating this rectangle around

(a) the x-axis   (b) the y-axis

(c) the line $x = 1$ (d) the line $y = 1$

(e) the line $x = 2$ (f) the line $y = -1$

4. Consider the rectangle bounded by $x = 1$ and $x = 3$ on the y-interval $[2, 2.5]$. Calculate the volume of the geometric object obtained by rotating this rectangle around

(a) the x-axis   (b) the y-axis

(c) the line $x = 1$ (d) the line $y = 2$

(e) the line $x = -1$ (f) the line $y = 3$

5. In what way are shells the same as washers? How are they different?

6. Consider the region between $f(x) = x^2$ and the x-axis between $x = 0$ and $x = 2$. Draw a Riemann sum approximation of the area of this region, using a midpoint sum with four rectangles, and explain how this Riemann sum approximation is related to a four-shell approximation of the solid obtained by rotating the region around the y-axis.

7. For a four-shell approximation of the volume of the solid obtained from the region between $f(x) = x^2$ and the x-axis between $x = 0$ and $x = 2$ by rotating the region around the y-axis, illustrate and calculate

(a) Δx and each x_k;

(b) the midpoint x_k^* of each subinterval $[x_{k-1}, x_k]$;

(c) each $f(x_k^*)$;

(d) the volume of the second shell.

8. For a four-shell approximation of the volume of the solid obtained from the region between $f(x) = x^2$ and the x-axis between $x = 0$ and $x = 2$ by rotating the region around the x-axis, illustrate and calculate

(a) Δy and each y_k;

(b) the midpoint y_k^* of each subinterval $[y_{k-1}, y_k]$;

(c) each $f^{-1}(y_k^*)$;

(d) the volume of the second shell.

Write each of the limits in Exercises 9 and 10 as a definite integral, and identify the corresponding solid of revolution, given that x_k^* and y_k^* represent midpoints of subintervals.

9. $\displaystyle \lim_{n \to \infty} \sum_{k=1}^{n} 2\pi (y_k^*)^3 \left(\dfrac{2}{n}\right)$, with $y_k = \dfrac{2k}{n}$

10. $\displaystyle \lim_{n \to \infty} \sum_{k=1}^{n} 2\pi (x_k^*)^{3/2} \left(\dfrac{4}{n}\right)$, with $x_k = \dfrac{4k}{n}$

Each of the definite integrals in Exercises 11–16 represents the volume of a solid of revolution obtained by rotating a region around either the x- or y-axis, computed with the shell method. Find this region.

11. $2\pi \displaystyle\int_0^{\pi/4} x \cos x \, dx$

12. $2\pi \displaystyle\int_0^1 x(x^2) \, dx$

13. $2\pi \displaystyle\int_4^5 y\sqrt{y-4} \, dy$

14. $2\pi \displaystyle\int_0^1 (x^2 - x^3) \, dx$

15. $2\pi \displaystyle\int_1^e y \ln y \, dy$

16. $2\pi \displaystyle\int_0^1 (3y - 3y^2) \, dy$

Write the volumes of the solids of revolution shown in Exercises 17–20 in terms of definite integrals that represent accumulations of shells. Do not solve the integrals.

17.

18.

19.

20.
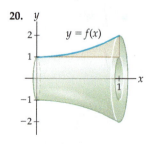

For each solid described in Exercises 21–24, set up volume integrals using both the shell and disk/washer methods. Which method produces an easier integral in each case, and why? Do not solve the integrals.

21. The region between the graph of $f(x) = |x|$ and the x-axis on $[-2, 2]$, revolved around the x-axis.

22. The region between the graph of $f(x) = \frac{1}{x}$ and the lines $y = 0, y = 1, x = 0$, and $x = 2$, revolved around the y-axis.

23. The region between the graph of $f(x) = x^2 + 1$ and the x-axis on $[0, 1]$, revolved around the line $x = 3$.

24. The region between the graph of $f(x) = x^2 + 1$ and the x-axis on $[0, 1]$, revolved around the line $y = -2$.

Skills

Consider the region between $f(x) = \sqrt{x}$ and the x-axis on $[0, 4]$. For each line of rotation given in Exercises 25–28, use four shells based on the given rectangles to approximate the volume of the resulting solid.

25. *Around the x-axis*   **26.** *Around the y-axis*

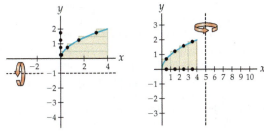

27. *Around the line $y = -1$*   **28.** *Around the line $x = 5$*

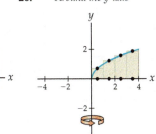

Consider the region between $f(x) = 4 - x^2$ and the x-axis on $[0, 2]$. For each line of rotation given in Exercises 29–32, use the shell method to construct definite integrals to find the volume of the resulting solid.

29. *Around the y-axis*   **30.** *Around the x-axis*

31. *Around the line $x = 2$*   **32.** *Around the line $y = 4$*

Consider the region between $f(x) = x - 2$ and the x-axis on $[2, 5]$. For each line of rotation given in Exercises 33–38, use the shell method to construct definite integrals to find the volume of the resulting solid.

33. *Around the y-axis*   **34.** *Around the x-axis*

 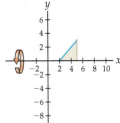

35. *Around the line $y = 3$*   **36.** *Around the line $y = -2$*

 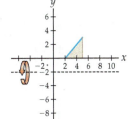

37. *Around the line $x = 2$*   **38.** *Around the line $x = 6$*

 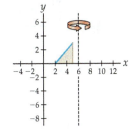

Consider the region between $f(x) = 4 - x^2$ and the line $y = 5$ on $[0, 2]$. For each line of rotation given in Exercises 39–42, use the shell method to construct definite integrals to find the volume of the resulting solid.

39. *Around the y-axis*   **40.** *Around the line $x = 2$*

 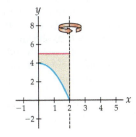

41. *Around the line $x = 3$*   **42.** *Around the line $y = 5$*

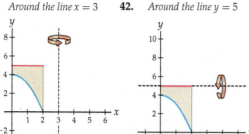

Consider the region between $f(x) = x + 1$ and the line $y = -2$ on $[1, 5]$. For each line of rotation given in Exercises 43 and 44, use the shell method to construct definite integrals to find the volume of the resulting solid.

43. *Around the line $x = 5$*   **44.** *Around the y-axis*

Consider the region between $f(x) = (x - 2)^2$ and $g(x) = x$ on $[1, 4]$. For each line of rotation given in Exercises 45–48, use the shell method to construct definite integrals to find the volume of the resulting solid.

45. *Around the x-axis*   **46.** *Around the y-axis*

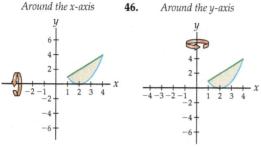

47. *Around the line $y = -1$*   **48.** *Around the line $x = 1$*

Use definite integrals to find the volume of each solid of revolution described in Exercises 49–61. (It is your choice whether to use disks/washers or shells in these exercises.)

49. The region between the graph of $f(x) = 4 - x^2$ and the line $y = 4$ on $[0, 2]$, revolved around the y-axis.

50. The region between the graph of $f(x) = e^{2x}$ and the x-axis on $[0, 4]$, revolved around the x-axis.

51. The region between the graph of $f(x) = (x - 3)^2 + 2$ and the x-axis on $[0, 6]$, revolved around the y-axis.

52. The region between the graph of $f(x) = x^2$ and the x-axis on $[0, 3]$, revolved around the line $y = 10$.

53. The region between the graphs of $f(x) = \sqrt{x}$ and $g(x) = x^2$ on $[0, 1]$, revolved around the x-axis.

54. The region between the graph of $f(x) = e^x$ and the line $y = e$ on $[0, 1]$, revolved around the x-axis.

55. The region between the graph of $f(x) = 9 - x^2$ and the x-axis on $[0, 3]$, revolved around the x-axis.

56. The region between the graph of $f(x) = (x - 2)^2$ and the x-axis on $[0, 4]$, revolved around the x-axis.

57. The region between the graph of $f(x) = x^2 - 4x + 4$ and the x-axis on $[0, 2]$, revolved around the y-axis.

58. The region bounded by the graphs of $f(x) = 2 \ln x$, $y = 0$, $y = 3$, and $x = 0$, revolved around the x-axis.

59. The region bounded by the graphs of $f(x) = \dfrac{2}{x}$, $y = 0$, $y = 3$, $x = 0$, and $x = 2$, revolved around the x-axis.

60. The region between the graph of $f(x) = x^2 + 2$ and the x-axis on $[-1, 3]$, revolved around the line $x = -1$.

61. The region bounded by the graphs of $f(x) = |x|$ and $y = 2$ on $[-2, 2]$, revolved around the line $x = 3$.

Applications

62. Mark carves napkin rings out of wooden spheres. His napkin rings have height $h = 8$ and radius $R = 5$, measured in centimeters, as shown next at the left. Alina also carves napkin rings out of spheres, but she likes bigger napkins, so she carves them with height $h = 8$ and radius $R = 6$, as shown at the right. Use the shell method to show that Mark's and Alina's napkin rings use exactly the same amount of wood. (This is an example of the famous *Napkin Ring Theorem*; see Exercise 70).

Napkin ring   *Bigger napkin ring*

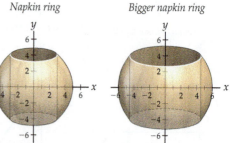

63. Alan makes beads that have the shape of flattened spheres with holes drilled through for stringing. The beads can be modeled by revolving the curve $f(x) = 3 - 0.15x^2$ on $[-4, 4]$ around the x-axis, as shown in the next figure, in which units are in centimeters. Use the shell method to determine the amount of material required to make 100 beads of this shape.

Shape of a bead

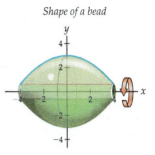

Proofs

64. The disk and shell methods for calculating volume of solids of revolution always give the same answer (when they are both computable). Prove that this is so for any solid of revolution defined by revolving the region between a linear function $y = mx + c$ and the x-axis on $[a, b]$, where m, c, a, and b are positive.

65. Prove that a shell with average radius r, height h, and thickness Δx has volume $V = 2\pi rh\,\Delta x$.

66. Use a definite integral with the shell method to prove that the volume formula $V = \frac{1}{3}\pi r^2 h$ holds for a cone of radius 3 and height 5.

67. Use a definite integral with the shell method to prove that the volume formula $V = \frac{4}{3}\pi r^3$ holds for a sphere of radius 3.

68. Use a definite integral with the shell method to prove that a cone of radius r and height h has volume given by the formula $V = \frac{1}{3}\pi r^2 h$.

69. Use a definite integral with the shell method to prove that a sphere of radius r has volume given by the formula $V = \frac{4}{3}\pi r^3$.

70. Prove the *Napkin Ring Theorem:* If a napkin ring is made by removing a cylinder of height h from a sphere, then the volume of the resulting shape does not depend on the radius of the sphere! Look at the figures in Exercise 62 to see why this result is surprising.

Thinking Forward

More "subdivide, approximate, and add": We have already seen how to use limits and a "subdivide, approximate, and add" strategy to calculate area by approximating with rectangles and to calculate volume by using disks, washers, or shells. What else can we calculate in this way?

▶ Consider the curve traced out by the graph of a function $y = f(x)$ on an interval $[a, b]$. Come up with a proposal for approximating the length of such a curve with a "subdivide, approximate, and add" strategy. What would play the role of rectangles, disks, washers, or shells in this case?

▶ Consider the surface of revolution obtained by revolving the graph of a function $y = f(x)$ around the x-axis on an interval $[a, b]$. Come up with a proposal for approximating the surface area of such an object with a "subdivide, approximate, and add" strategy. What do you think would play the role of rectangles, disks, washers, or shells in this case?

▶ Suppose you wish to find the mass of a metal rod that is made up of a material whose density increases linearly from the left end to the right end. Mass is equal to density times volume, but since density is varying in this example, we need to somehow use a "subdivide, approximate, and add" strategy. Come up with a proposal for how to use this strategy to calculate the mass of the rod. Again, consider what the best choice would be for the role of the pieces in this example.

9.3   ARC LENGTH AND SURFACE AREA

▶ Approximating arc length with a sum of line-segment lengths

▶ Approximating surface area with a sum of frustum surface areas

▶ Using definite integrals to calculate exact arc lengths and surface areas

Approximating Arc Length

We have now used the "subdivide, approximate, and add" strategy to define definite integrals for areas under curves and for volumes of solids of revolution. We will now use the same strategy to find lengths of curves and surface areas of solids of revolution. To illustrate the process in the context of lengths of curves, let's try to find the length of part of a circle.

You probably already know that the circumference C of a circle of radius r is given by $C = 2\pi r$. But where does that magical formula come from? If we didn't know this formula, how could we approximate the circumference of a circle, say, of radius 2? Since such a circle is given by the equation $x^2 + y^2 = 4$, the function $f(x) = \sqrt{4 - x^2}$ on $[-2, 2]$ traces out its top half, as shown here:

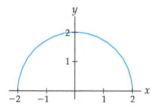

$$f(x) = \sqrt{4 - x^2} \ on \ [-2, 2]$$

The most obvious way to approximate the length of this curve is with line segments, much like the way that you would approximate the length of a curvy highway on a map by using straight segments. For example, we could divide the interval $[-2, 2]$ into four subintervals and approximate the length of the curve on each subinterval with a line segment, as shown in the first figure that follows. To get a better approximation of the length of the curve, we could use more line segments, as shown, for example, in the second figure.

Four-segment approximation

Eight-segment approximation

In order to use our "subdivide, approximate, and add" strategy to find the *exact* length of a curve, we will need to take a limit, letting the size of the subintervals get smaller and smaller. Before we can apply such a limit, we need to set up some general notation, much like the notation we used when we defined the area under a curve.

Suppose we wish to approximate the length of a curve $f(x)$ on an interval $[a, b]$. If we divide $[a, b]$ into n subintervals of equal width, then each subinterval will have width

$\Delta x = \frac{b-a}{n}$. If we define $x_k = a + k\Delta x$ for $k = 0, 1, \ldots, n$, then the kth subinterval is the interval $[x_{k-1}, x_k]$. Each subinterval will have width Δx, and the line segment on the kth subinterval will change vertically by $\Delta y_k = f(x_k) - f(x_{k-1})$ units on that subinterval, as illustrated in the next figure. Notice that Δx does not depend on the choice of k, but Δy_k does, since the graph of f increases and decreases at different rates as we move from $x = a$ to $x = b$.

Notation for the kth line segment

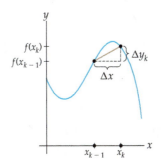

We can use the distance formula to calculate the length l_k of each line segment:

$$l_k = \text{length of } k\text{th line segment} = \sqrt{(\Delta x)^2 + (\Delta y_k)^2}.$$

With just a little bit of algebra we can factor out a Δx, as follows:

$$l_k = \sqrt{(\Delta x)^2 + (\Delta y_k)^2} = \sqrt{(\Delta x)^2 \left(1 + \frac{(\Delta y_k)^2}{(\Delta x)^2}\right)} = \sqrt{1 + \left(\frac{\Delta y_k}{\Delta x}\right)^2}\, \Delta x.$$

As we will see in Theorem 9.7, this second form for the distance will be much more useful. Now given a function $f(x)$, an interval $[a, b]$, and a number of subdivisions n, we have defined n subintervals, each with a line segment that approximates a small piece of the curve. Adding up the lengths of these line segments gives us an approximation for the length of the curve:

$$\text{length of } f(x) \text{ on } [a, b] \approx \sum_{k=1}^{n} l_k = \sum_{k=1}^{n} \sqrt{1 + \left(\frac{\Delta y_k}{\Delta x}\right)^2}\, \Delta x.$$

You will calculate such an approximation by hand in Example 1. For any individual approximation all this notation is not particularly necessary, but as we will see in a moment, such notation will enable us to obtain a Riemann sum, and thus a definite integral, for arc length.

A Definite Integral for Arc Length

So what is the *exact* length of the curve? For sufficiently well behaved functions the approximation we just developed gets better as we let the number n of line segments get larger and larger, since this makes the width Δx of the subintervals get smaller and smaller. Therefore to get the exact length we need to take the limit of our n-subinterval approximation as n approaches infinity. For nice-enough functions this limit will converge to a real number, and we define that number to be the ***arc length*** of $f(x)$ on $[a, b]$.

DEFINITION 9.6

The Arc Length of a Function on an Interval

Suppose $f(x)$ is a differentiable function with a continuous derivative. Then the **arc length** of $f(x)$ from $x = a$ to $x = b$ is

$$\text{Arc length of } f(x) \atop \text{from } x = a \text{ to } x = b} = \lim_{n \to \infty} \sum_{k=1}^{n} \sqrt{1 + \left(\frac{\Delta y_k}{\Delta x}\right)^2} \, \Delta x,$$

where $\Delta x = \frac{b-a}{n}$, $x_k = a + k\Delta x$, and $\Delta y_k = f(x_k) - f(x_{k-1})$ for all $k = 0, 1, 2, \ldots, n$.

The limit of sums in this definition should look familiar to you. It is a kind of Riemann sum, and therefore it can be written as a definite integral.

THEOREM 9.7

Arc Length as a Definite Integral

Suppose $f(x)$ is a differentiable function with a continuous derivative on $[a, b]$. Then the arc length of $f(x)$ from $x = a$ to $x = b$ can be represented by the definite integral

$$\text{Arc length of } f(x) \atop \text{from } x = a \text{ to } x = b} = \int_a^b \sqrt{1 + (f'(x))^2} \, dx.$$

If you think about it for a moment, you will see that Theorem 9.7 says something very interesting: The length of the curve traced out by the graph of a function $f(x)$ on an interval $[a, b]$ is the same as the *area* under the graph of $\sqrt{1 + (f'(x))^2}$ on that same interval. There is a nice consequence to this: We already know how to calculate definite integrals, and therefore Theorem 9.7 makes it simple for us to calculate certain exact arc lengths.

Considering the definition of the definite integral, we can see how the limit of sums in Definition 9.6 becomes the definite integral in Theorem 9.7, at least if we believe that $\frac{\Delta y_k}{\Delta x}$ becomes $f'(x)$ as $n \to \infty$. The key to showing the latter, and thus the key to the proof of Theorem 9.7, is the Mean Value Theorem.

Proof. If $f(x)$ is a differentiable function, then the Mean Value Theorem applies to the function $f(x)$ on each subinterval $[x_{k-1}, x_k]$ and guarantees that there exists some point $x_k^* \in (x_{k-1}, x_k)$ at which

$$f'(x_k^*) = \frac{f(x_k) - f(x_{k-1})}{x_k - x_{k-1}} = \frac{\Delta y_k}{\Delta x},$$

where $\Delta y_k = f(x_k) - f(x_{k-1})$. Therefore, there is a point x_k^* in each subinterval such that the definition of arc length can be written as

$$\lim_{n \to \infty} \sum_{k=1}^{n} \sqrt{1 + \left(\frac{\Delta y_k}{\Delta x}\right)^2} \, \Delta x = \lim_{n \to \infty} \sum_{k=1}^{n} \sqrt{1 + (f'(x_k^*))^2} \, \Delta x.$$

Since the derivative $f'(x)$ is assumed to be continuous, so is the function $\sqrt{1 + (f'(x))^2}$. In addition, $\Delta x = \frac{b-a}{n}$ and $x_k = a + k\,\Delta x$. Therefore this limit of sums represents the definite integral of $\sqrt{1 + (f'(x))^2}$ on the interval $[a, b]$. In other words,

$$\lim_{n \to \infty} \sum_{k=1}^{n} \sqrt{1 + (f'(x_k^*))^2} \, \Delta x = \int_a^b \sqrt{1 + (f'(x))^2} \, dx.$$

Approximating Surface Area

The tools we have developed for calculating arc length can be adapted to define the area of a **surface of revolution**. Given the graph of a function $f(x)$ on an interval $[a, b]$, we can revolve just the graph–not the area under the graph–around some axis to obtain a surface. As we did for arc length, we can use a piecewise-linear approximation of the graph of $y = f(x)$ and a "subdivide, approximate, and add" strategy to find the area of such a surface.

For example, consider the graph shown next at the left. Revolving this graph around the x-axis produces the surface in the middle figure. One way to approximate the area of this surface is by approximating the graph with two straight line segments and then revolving these line segments around the x-axis to obtain truncated cones, as illustrated in the rightmost figure.

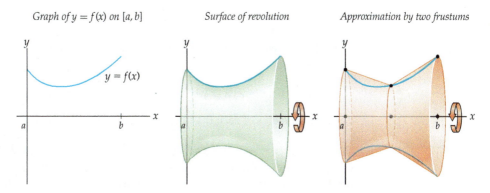

| *Graph of $y = f(x)$ on $[a, b]$* | *Surface of revolution* | *Approximation by two frustums* |

The truncated cones in this example are called **frustums**. In Exercise 81 you will use the fact that a frustum is obtained by removing the top of a cone to prove that the surface area of a frustum is given by the following formula:

THEOREM 9.8

The Surface Area of a Frustum

A frustum with radii p and q and **slant length** s, as shown in the figure, has surface area

$$S = 2\pi rs,$$

where $r = \frac{p+q}{2}$ is the **average radius** of the frustum.

To remember the formula $S = 2\pi rs$, it may help to notice that it is a lot like the formula $S = 2\pi rl$ for the surface area of a cylinder, except that in the case of a frustum the length is measured at a slant and the radius is an average radius.

Now suppose we wish to use frustums to approximate the surface area of the surface obtained by revolving the graph of a function $y = f(x)$ around the x-axis on the interval

$[a, b]$. Just as we did with arc length, we subdivide $[a, b]$ into n subintervals $[x_{k-1}, x_k]$, each of width $\Delta x = \frac{b-a}{n}$, with corresponding vertical changes $\Delta y_k = f(x_k) - f(x_{k-1})$ on each subinterval as shown here:

Notation for the kth frustum

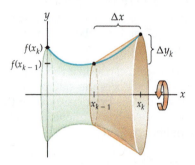

We must find the surface area of each frustum and then add up all those surface areas. Following the same algebra as we used for arc length, we can calculate the slant length of each frustum:

$$s_k = \text{slant length of } k\text{th frustum} = \sqrt{(\Delta x)^2 + (\Delta y_k)^2} = \sqrt{1 + \left(\frac{\Delta y_k}{\Delta x}\right)^2}\, \Delta x.$$

The average radius of each frustum depends on the function $y = f(x)$:

$$r_k = \text{average radius of } k\text{th frustum} = \frac{f(x_{k-1}) + f(x_k)}{2}.$$

Using the formula for the surface area of a frustum and adding up these surface areas, we obtain an approximation for the surface area of the surface of revolution:

$$\text{surface area of } S \approx \sum_{k=1}^{n} 2\pi r_k s_k = \sum_{k=1}^{n} 2\pi \left(\frac{f(x_{k-1}) + f(x_k)}{2}\right) \sqrt{1 + \left(\frac{\Delta y_k}{\Delta x}\right)^2}\, \Delta x.$$

You will calculate such an approximation by hand in Example 5, and just as happened for arc length, we can use the notation we have developed to obtain a definite integral for surface area.

A Definite Integral for Surface Area

We have already done the "legwork" needed to define surface area as a limit of sums of frustums:

DEFINITION 9.9

The Area of a Surface of Revolution

Suppose $f(x)$ is a positive, differentiable function with a continuous derivative. The **surface area** of the solid of revolution obtained by revolving $f(x)$ around the x-axis from $x = a$ to $x = b$ is

$$\begin{array}{l}\text{Surface area under } f(x) \\ \text{from } x = a \text{ to } x = b\end{array} = \lim_{n \to \infty} \sum_{k=1}^{n} 2\pi \left(\frac{f(x_{k-1}) + f(x_k)}{2}\right) \sqrt{1 + \left(\frac{\Delta y_k}{\Delta x}\right)^2}\, \Delta x,$$

where $\Delta x = \frac{b-a}{n}$, $x_k = a + k\Delta x$, and $\Delta y_k = f(x_k) - f(x_{k-1})$ for all $k = 0, 1, 2, \ldots, n$.

Intuitively, as $n \to \infty$, the average radius $\frac{f(x_{k-1}) + f(x_k)}{2}$ will be given by the height $f(x)$ of the function and the quotient $\frac{\Delta y_k}{\Delta x}$ will be given by the derivative $f'(x)$. This gives us the following definite integral formula for surface area:

THEOREM 9.10   **Surface Area as a Definite Integral**

Suppose $f(x)$ is a positive, differentiable function with a continuous derivative. The **surface area** of the solid of revolution obtained by revolving $f(x)$ around the x-axis from $x = a$ to $x = b$ is

$$\text{Surface area under } f(x) \atop \text{from } x = a \text{ to } x = b} = 2\pi \int_a^b f(x)\sqrt{1 + (f'(x))^2}\, dx.$$

Proof. To obtain a definite integral we will attempt to write the expression in Definition 9.9 as a Riemann sum. That means that we need the terms of the sum to be in the form $g(x_k^*)\Delta x$ for some function g and some choice of point x_k^* in each subinterval. This will be a two-step process.

As we saw in the proof of Theorem 9.7, the Mean Value Theorem guarantees that on each subinterval $[x_{k-1}, x_k]$ we can find some point x_k^* at which $f'(x_k^*) = \frac{\Delta y_k}{\Delta x}$.

Now consider the expression $\frac{f(x_{k-1}) + f(x_k)}{2}$ for the average radius. Since f is continuous and this average lies between $f(x_{k-1})$ and $f(x_k)$, the Intermediate Value Theorem guarantees that there is some point x_k^{**} at which $f(x_k^{**})$ is equal to $\frac{f(x_{k-1}) + f(x_k)}{2}$.

Although in general x_k^* and x_k^{**} will not be the same number in $[x_{k-1}, x_k]$, as $n \to \infty$ and thus $\Delta x \to 0$ these two numbers will get closer and closer to each other. Thus we *almost* have a Riemann sum. There are some technical details here that we are omitting, but given that $x_k^* \approx x_k^{**}$ and using the fact that $f(x)$ and $f'(x)$ are assumed to be continuous, we obtain

$$\lim_{n \to \infty} \sum_{k=1}^{n} 2\pi f(x_k^{**})\sqrt{1 + (f'(x_k^*))^2}\, \Delta x = 2\pi \int_a^b f(x)\sqrt{1 + (f'(x))^2}\, dx. \qquad \blacksquare$$

We now have a way to use a definite integral to calculate the exact surface area of a surface of revolution. However, this method is practical only if the resulting definite integral is one that we know how to solve. In the examples and exercises we will choose functions very carefully so that the definite integrals are solvable with the techniques we have covered in this course.

Examples and Explorations

EXAMPLE 1   **Using line segments to approximate the length of a curve**

Approximate the length of the curve $f(x) = \sqrt{4 - x^2}$ from $x = -2$ to $x = 2$ with four line segments. Then find a better approximation that uses eight line segments.

SOLUTION

We begin by dividing the interval $[-2, 2]$ into four subintervals of equal size: $[-2, -1]$, $[-1, 0]$, $[0, 1]$, and $[1, 2]$. The four line segments in question are shown next at the left.

Four-segment approximation   *Eight-segment approximation*

It is easy to calculate the lengths of these line segments with the distance formula; for example, the first line segment starts at the point $(-2, f(-2)) = (-2, 0)$ and ends at the point $(-1, f(-1)) = (-1, \sqrt{3})$ and therefore has length

$$\sqrt{(-1 - (-2))^2 + (\sqrt{3} - 0)^2} = \sqrt{1 + 3} = 2.$$

Similarly, by using the coordinates of $(-1, f(-1))$, $(0, f(0))$, $(1, f(1))$, and $(2, f(2))$ we can calculate that the lengths of the remaining three line segments are 1.035, 1.035, and 2. The sum of these four lengths approximates the length of the curve:

$$2 + 1.035 + 1.035 + 2 = 6.070.$$

Clearly this approximation should be an under-approximation of the actual length, since the lengths of the lines are shorter than the lengths of the curves they approximate. Because the curve in question is half of the circumference of a circle of radius 2, the length of the curve should be $\frac{1}{2}(2\pi(2)) = 2\pi \approx 6.28319$. Our four-segment approximation is an under-approximation by about 0.21263 unit.

To get a better approximation with the eight line segments shown earlier at the right, we need to make eight distance formula calculations. The first line segment extends from $(-2, 0)$ to $(-1.5, \sqrt{1.75})$ and thus has length

$$\sqrt{(-1.5 - (-2))^2 + (\sqrt{1.75} - 0)^2} = \sqrt{0.25 + 1.75} = \sqrt{2} \approx 1.41421.$$

The remaining seven line segments can be calculated similarly, and the sum of the eight lengths gives us an eight-segment approximation for the arc length of the curve:

$$1.414 + 0.646 + 0.540 + 0.504 + 0.504 + 0.540 + 0.646 + 1.414 \approx 6.208.$$

Note that the eight-segment approximation is indeed closer to the length we would expect from the circumference formula and in fact under-approximates the exact value of $2\pi \approx 6.28319$ by only about 0.07519 unit.   □

EXAMPLE 2   Approximating arc length in general notation

Approximate the length of one period of the function $f(x) = \sin x$ by using the formula and notation in Definition 9.6, with $n = 4$. For each k, describe x_k, y_k, and Δy_k.

SOLUTION

We will approximate the arc length of $f(x) = \sin x$ from $x = 0$ to $x = 2\pi$. Since we are using $n = 4$ line segments, as shown in the following figure, we have $\Delta x = \frac{\pi}{2}$:

Arc length of $f(x) = \sin x$ on $[0, 2\pi]$ approximated by four line segments

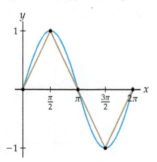

We could use the distance formula to find the length of each of these line segments without bothering with notation at all, as in the previous example. However, the purpose of this example is to practice using the notation from Definition 9.6. We have

$$x_0 = 0, \quad x_1 = \frac{\pi}{2}, \quad x_2 = \pi, \quad x_3 = \frac{3\pi}{2}, \quad \text{and} \quad x_4 = 2\pi.$$

Evaluating the function $f(x) = \sin x$ at these points gives

$$f(x_0) = f(0) = \sin 0 = 0,$$

$$f(x_1) = f\left(\frac{\pi}{2}\right) = \sin \frac{\pi}{2} = 1,$$

$$f(x_2) = f(\pi) = \sin \pi = 0,$$

$$f(x_3) = f\left(\frac{3\pi}{2}\right) = \sin \frac{3\pi}{2} = -1, \text{ and}$$

$$f(x_4) = f(2\pi) = \sin 2\pi = 0.$$

Although $\Delta x = \frac{\pi}{2}$ is the same on each subinterval, the changes in the y-values vary:

$$\Delta y_1 = f(x_1) - f(x_0) = 1 - 0 = 1,$$

$$\Delta y_2 = f(x_2) - f(x_1) = 0 - 1 = -1,$$

$$\Delta y_3 = f(x_3) - f(x_2) = -1 - 0 = -1, \text{ and}$$

$$\Delta y_4 = f(x_4) - f(x_3) = 0 - (-1) = 1.$$

In the notation of Definition 9.6, the lengths of the four line segments are thus

$$\sqrt{1 + \left(\frac{\Delta y_1}{\Delta x}\right)^2}\, \Delta x = \sqrt{1 + \left(\frac{1}{\pi/2}\right)^2}\left(\frac{\pi}{2}\right) \approx 1.8621,$$

$$\sqrt{1 + \left(\frac{\Delta y_2}{\Delta x}\right)^2}\, \Delta x = \sqrt{1 + \left(\frac{-1}{\pi/2}\right)^2}\left(\frac{\pi}{2}\right) \approx 1.8621,$$

$$\sqrt{1 + \left(\frac{\Delta y_3}{\Delta x}\right)^2}\, \Delta x = \sqrt{1 + \left(\frac{-1}{\pi/2}\right)^2}\left(\frac{\pi}{2}\right) \approx 1.8621, \text{ and}$$

$$\sqrt{1 + \left(\frac{\Delta y_4}{\Delta x}\right)^2}\, \Delta x = \sqrt{1 + \left(\frac{1}{\pi/2}\right)^2}\left(\frac{\pi}{2}\right) \approx 1.8621.$$

In this example, all four of the line segments happen to be the same length; of course that is not usually the case! The reason it is here is that $\sin x$ just happens to have the same shape of arc on each of the four pieces we are examining. Adding up all of these lengths, we get the following approximation for the length of one period of the sine function:

$$\sum_{k=1}^{4} \sqrt{1 + \left(\frac{\Delta y_k}{\Delta x}\right)^2}\, \Delta x \approx 1.8621 + 1.8621 + 1.8621 + 1.8621 \approx 7.4484.$$

□

CHECKING THE ANSWER

The answer of 7.4484 we just found does seem roughly reasonable for the length of one period of the sine function, considering its graph. To get a better check on our approximation, we could compute the definite integral

$$\int_0^{2\pi} \sqrt{1 + (f'(x))^2}\, dx = \int_0^{2\pi} \sqrt{1 + \left(\frac{d}{dx}(\sin x)\right)^2}\, dx = \int_0^{2\pi} \sqrt{1 + \cos^2 x}\, dx.$$

Unfortunately, we do not know how to calculate this integral. However, it represents an area, and we can use a graphing calculator to approximate area. The area in question is shown in the following figure:

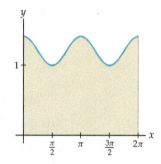

The area under the graph of
$\sqrt{1 + (f'(x))^2} = \sqrt{1 + \cos^2 x}$ *on* $[0, 2\pi]$

This area is *not* the area under the graph of $f(x) = \sin x$. Rather, it is the area under the graph of $\sqrt{1 + (f'(x))^2} = \sqrt{1 + \cos^2 x}$ from $x = 0$ to $x = 2\pi$. Using a graphing calculator, we can approximate this area as 7.6404; therefore the arc length of $f(x) = \sin x$ from $x = 0$ to $x = 2\pi$ is approximately 7.6404. This verifies that 7.4484 was indeed a reasonable approximation of the arc length of one period of $f(x) = \sin x$.

EXAMPLE 3

Using a definite integral to calculate exact arc length

Construct and solve a definite integral to find the exact length of the curve traced out by $f(x) = x^{3/2}$ on the interval $[0, 3]$.

SOLUTION

By the arc length formula in Theorem 9.7, the arc length of $f(x) = x^{3/2}$ on $[0, 3]$ is given by

$$\int_0^3 \sqrt{1 + \left(\frac{d}{dx}(x^{3/2})\right)^2}\, dx = \int_0^3 \sqrt{1 + \left(\frac{3}{2}x^{1/2}\right)^2}\, dx = \int_0^3 \sqrt{1 + \frac{9}{4}x}\, dx.$$

Applying the simple substitution $u = 1 + \frac{9}{4}x$, $du = \frac{9}{4}\, dx$, we can write this definite integral as

$$\frac{4}{9}\int_{x=0}^{x=3} \sqrt{u}\, du = \frac{4}{9}\left[\frac{2}{3}u^{3/2}\right]_{x=0}^{x=3} = \frac{4}{9}\left[\frac{2}{3}\left(1 + \frac{9}{4}x\right)^{3/2}\right]_{x=0}^{x=3}$$

$$= \frac{8}{27}\left(\left(1 + \frac{9}{4}(3)\right)^{3/2} - \left(1 + \frac{9}{4}(0)\right)^{3/2}\right)$$

$$= \frac{8}{27}\left(\left(\frac{31}{4}\right)^{3/2} - 1\right) \approx 6.096.$$

CHECKING THE ANSWER

The answer we just found appears reasonable if we consider a rough estimate of the length of the curve traced out by the graph of $f(x) = x^{3/2}$ from $x = 0$ to $x = 3$, as shown in the next figure. By the distance formula, the diagonal line from $(0, 0)$ to $(3, 5)$ has length $\sqrt{3^2 + 5^2} \approx 5.83$, just a little less than our approximation of 6.096 for the length of the curve.

$$f(x) = x^{3/2} \text{ on } [0, 3]$$

EXAMPLE 4

Recovering the formula for the circumference of a circle

Construct and solve a definite integral to find the circumference of a circle of radius 2.

SOLUTION

As we have already discussed in this section, to calculate the circumference of a circle of radius 2 we can instead calculate the arc length of $f(x) = \sqrt{4 - x^2}$ on the interval $[-2, 2]$ and then multiply that answer by 2. Using the definite integral formula for arc length, we have

$$\int_{-2}^{2} \sqrt{1 + (f'(x))^2}\, dx = \int_{-2}^{2} \sqrt{1 + \left(\frac{d}{dx}\left(\sqrt{4 - x^2}\right)\right)^2}\, dx \qquad \leftarrow \text{definition of } f(x)$$

$$= \int_{-2}^{2} \sqrt{1 + \left(\frac{1}{2}(4 - x^2)^{-1/2}(-2x)\right)^2}\, dx \qquad \leftarrow \text{differentiate } f(x)$$

$$= \int_{-2}^{2} \sqrt{1 + \frac{x^2}{4 - x^2}}\, dx = \int_{-2}^{2} \sqrt{\frac{(4 - x^2) + x^2}{4 - x^2}}\, dx \quad \leftarrow \text{algebra}$$

$$= \int_{-2}^{2} \sqrt{\frac{4}{4 - x^2}}\, dx = 2\int_{-2}^{2} \frac{1}{\sqrt{4 - x^2}}\, dx. \qquad \leftarrow \text{algebra}$$

Fortunately, $\dfrac{1}{\sqrt{4 - x^2}}$ is a function whose integral we can find by using either algebra and the antiderivative of the arcsine (Section 6.4) or trigonometric substitution (Section 8.5). With those techniques we find that

$$\int \frac{1}{\sqrt{4 - x^2}}\, dx = \sin^{-1}\frac{x}{2} + C.$$

Less fortunately, since $\dfrac{1}{\sqrt{4 - x^2}}$ is not defined at $x = \pm 2$, the definite integral in question is actually an improper integral. To solve it we will have to split the integral and consider limits at the endpoints $x = \pm 2$ (see Section 8.6):

$$2\int_{-2}^{2} \frac{1}{\sqrt{4 - x^2}}\, dx = 2\left(\lim_{A \to -2^+} \int_{A}^{0} \frac{1}{\sqrt{4 - x^2}}\, dx + \lim_{B \to 2^-} \int_{0}^{B} \frac{1}{\sqrt{4 - x^2}}\, dx\right)$$

$$= 2\left(\lim_{A \to -2^+} \left[\sin^{-1}\frac{x}{2}\right]_{A}^{0} + \lim_{B \to 2^-} \left[\sin^{-1}\frac{x}{2}\right]_{0}^{B}\right)$$

$$= 2\left(\lim_{A \to -2^+} \left(\sin^{-1} 0 - \sin^{-1}\frac{A}{2}\right) + \lim_{B \to 2^-} \left(\sin^{-1}\frac{B}{2} - \sin^{-1} 0\right)\right).$$

Since $\sin^{-1} 0 = 0$, $\sin^{-1} 1 = \frac{\pi}{2}$, and $\sin^{-1} -1 = -\frac{\pi}{2}$, we have

$$2\int_{-2}^{2} \frac{1}{\sqrt{4 - x^2}}\, dx = 2\left(\left(0 - \left(-\frac{\pi}{2}\right)\right) + \left(\frac{\pi}{2} - 0\right)\right) = 2\pi.$$

Therefore, the length of the graph of $f(x) = \sqrt{4 - x^2}$ from $x = -2$ to $x = 2$ is 2π units. This arc length is *half* of the arc length of the circumference of a circle with radius 2, so the circumference of a circle of radius 2 is $2(2\pi) = 4\pi$ units. This is exactly the answer we expected, since, by the formula for the circumference of a circle, we should have $C = 2\pi r = 2\pi(2) = 4\pi$.

□

TECHNICAL POINT Although it may seem as if the preceding example could be used to re-derive the formula for the circumference of a circle, note that solving the definite integral involved the use of inverse trigonometric functions, which are defined from trigonometric functions, which are in turn defined with the use of the unit circle and the fact that the unit circle has circumference 2π.

EXAMPLE 5

Approximating the surface area of a sphere with frustums

Use four frustums to approximate the surface area of the sphere of radius 2 centered at the origin. This sphere can be realized by rotating the curve $f(x) = \sqrt{4 - x^2}$ from $x = -2$ to $x = 2$ around the x-axis.

SOLUTION

Subdividing the interval $[-2, 2]$ into four subintervals and then rotating a corresponding piecewise-linear approximation of $y = \sqrt{4 - x^2}$ around the x-axis yields four frustums (two are actually cones):

Approximating a sphere with four frustums

In our subdivision notation we have $x_0 = -2$, $x_1 = -1$, $x_2 = 0$, $x_3 = 1$, and $x_4 = 2$, with $\Delta x = 1$. We can treat the cones at the end as cones or as frustums; either surface area formula will work. For simplicity we will treat them as frustums.

Since $f(-2) = \sqrt{4 - (-2)^2} = 0$ and $f(-1) = \sqrt{4 - (-1)^2} = \sqrt{3}$, the first frustum (a cone) has average radius and slant length given by

$$r_1 = \frac{f(-2) + f(-1)}{2} = \frac{\sqrt{3}}{2},$$

$$s_1 = \sqrt{(-2 - (-1))^2 + (f(-2) - f(-1))^2} = \sqrt{1 + 3} = 2.$$

Therefore the surface area of the first frustum is $2\pi r_1 s_1 = 2\pi\left(\frac{\sqrt{3}}{2}\right)(2)$.

In a similar fashion we can find the volumes of the remaining three frustums to obtain the following approximation for the surface area S of a sphere:

$$S = 2\pi r_1 \, s_1 + 2\pi r_2 \, s_2 + 2\pi r_3 \, s_3 + 2\pi r_4 \, s_4$$

$$= 2\pi \left(\frac{\sqrt{3}}{2}\right)(2) + 2\pi \left(\frac{\sqrt{3}+2}{2}\right)\left(\sqrt{8 - 4\sqrt{3}}\right)$$

$$+ 2\pi \left(\frac{\sqrt{3}+2}{2}\right)\left(\sqrt{8 - 4\sqrt{3}}\right) + 2\pi \left(\frac{\sqrt{3}}{2}\right)(2)$$

$$\approx 46.04.$$

CHECKING THE ANSWER

In this case we happen to know the exact value of the surface area of a sphere of radius 2, since the surface area of a sphere of radius r is given by the formula $S = 4\pi r^2$. Thus the exact answer to the preceding problem is $4\pi (2)^2 = 16\pi \approx 50.265$.

EXAMPLE 6   Using a definite integral to calculate exact surface area

Construct and solve a definite integral to find the exact surface area of the surface of revolution obtained by revolving the graph of $f(x) = x^3$ around the x-axis on $[0, 1]$.

SOLUTION

Since the derivative of $f(x) = x^3$ is $f'(x) = 3x^2$, the surface area formula in Theorem 9.10 gives us the following definite integral to solve:

$$2\pi \int_0^1 x^3 \sqrt{1 + 9x^4} \, dx.$$

This integral is simple to solve by using substitution, with $u = 1 + 9x^4$ and $du = 36x^3$. With this substitution the integral becomes

$$2\pi \left(\frac{1}{36}\right) \int_{x=0}^{x=1} \sqrt{u} \, du = \frac{\pi}{18} \left[\frac{2}{3} u^{3/2}\right]_{x=0}^{x=1}$$

$$= \frac{\pi}{18} \left(\frac{2}{3}(1 + 9(1)^4)^{3/2} - \frac{2}{3}(1 + 9(0)^4)^{3/2}\right)$$

$$= \frac{\pi}{18} \left(\frac{2}{3}\sqrt{1000} - \frac{2}{3}\right).$$

This number is approximately equal to 3.56, which is a reasonable size for the surface area obtained by revolving $f(x) = x^3$ around the x-axis on $[0, 1]$.

TEST YOUR UNDERSTANDING

▶ In the four-segment approximation of the arc length of $f(x) = \sqrt{4 - x^2}$ on $[-2, 2]$, what is it about the graph of the function that causes some line segments to be longer than others, even though we are dividing the interval $[-2, 2]$ into equal pieces?

▶ Why would you expect an arc length approximation with six segments to be a better approximation than one with three segments?

▶ What is the difference between Definition 9.6 and Theorem 9.7? What needs to be proved to show Theorem 9.7?

▶ What parts of the construction of the definite integral for surface area are generalized from the construction of the definite integral for arc length?

▶ How are the Mean Value Theorem and the Intermediate Value Theorem involved in the proof of Theorem 9.10?

EXERCISES 9.3

Thinking Back

Approximating area with rectangles: Approximate the area represented by each of the following definite integrals by using a right sum with $n = 4$ rectangles.

▶ $\displaystyle\int_0^2 (\sqrt{x} + 1)\, dx$

▶ $\displaystyle\int_3^4 \frac{1}{x-2}\, dx$

Riemann sums: For each given definite integral, write the integral as a Riemann sum of the form $\lim_{n\to\infty} \sum_{k=1}^n f(x_k^*)\, \Delta x$, using a right sum (i.e., using the rightmost point of each subinterval as x_k^*).

▶ $\displaystyle\int_1^e \ln x\, dx$

▶ $\displaystyle\int_{-\pi}^\pi \sin x\, dx$

Review of integration techniques: Solve each of the following definite integrals.

▶ $\displaystyle\int_0^4 \sqrt{3 + 16x}\, dx$

▶ $\displaystyle\int_{-2}^2 \sqrt{26}\, dx$

▶ $\displaystyle\int_0^3 \frac{3}{\sqrt{9 - x^2}}\, dx$

▶ $\displaystyle\int_{\pi/4}^{\pi/2} \csc x\, dx$

Approximating definite integrals: Use a calculator or computer algebra system to approximate the values of the following definite integrals.

▶ $\displaystyle\int_{-1}^3 \sqrt{1 + e^{2x}}\, dx$

▶ $\displaystyle\int_0^{2\pi} \sqrt{1 + \sin^2 x}\, dx$

Concepts

0. *Problem Zero:* Read the section and make your own summary of the material.

1. *True/False:* Determine whether each of the statements that follow is true or false. If a statement is true, explain why. If a statement is false, provide a counterexample.

  (a) *True or False:* If we subdivide an interval $[a, b]$ into n subintervals, then each subinterval has width $\Delta x = \frac{b-a}{n}$.

  (b) *True or False:* If we subdivide an interval $[a, b]$ into n subintervals at points $x_0 = a, x_1, \ldots, x_{n-1}, x_n = b$, then the kth subdivision point x_k is given by $a + k$.

  (c) *True or False:* The distance formula says that if $\Delta x = x_k - x_{k-1}$ and $\Delta y_k = f(x_k) - f(x_{k-1})$, then the distance from $(x_{k-1}, f(x_{k-1}))$ to $(x_k, f(x_k))$ is given by $\sqrt{(\Delta x)^2 + (\Delta y_k)^2}$.

  (d) *True or False:* For any real numbers a and b, $\sqrt{a^2 + b^2} = a^2 \sqrt{1 + \left(\frac{b}{a}\right)^2}$.

  (e) *True or False:* We cannot use the definite integral formula for arc length for a function $f(x)$ on $[a, b]$ if $f(x)$ has a discontinuity on an interval $[a, b]$.

  (f) *True or False:* An approximation of arc length by line segments can sometimes be an over-approximation.

  (g) *True or False:* An approximation of surface areas by frustums can sometimes be an over-approximation.

  (h) *True or False:* The area of a frustum can be expressed in terms of average radius and slant length.

2. *Examples:* Construct examples of the thing(s) described in the following. Try to find examples that are different than any in the reading.

  (a) A graph of a function $f(x)$ on $[a, b]$ whose arc length is poorly approximated by four line segments.

  (b) An equation of a function $f(x)$ on $[a, b]$ that gives rise to an arc length integral that we do not know how to solve.

  (c) An equation of a surface of revolution defined by revolving a function $f(x)$ on $[a, b]$ around the x-axis that gives rise to a surface area integral that we do not know how to solve.

3. What is the definition of the arc length of a continuous function $f(x)$ on an interval $[a, b]$? Explain this definition in words in terms of a limit of sums, and explain the meaning of the notation used in the definition.

4. Suppose $f(x)$ is a differentiable function with a continuous derivative. Describe the arc length of $f(x)$ on the interval $[a, b]$ in terms of a definite integral. Why is this definite integral equal to the definition of arc length for $f(x)$ on $[a, b]$?

5. How is the Mean Value Theorem involved in proving that the arc length of a function on an interval can be represented by a definite integral?

6. Fill in the blank: The length of the graph of $y = x^2$ on $[2, 4]$ is equal to the area under the graph of the function _____ on the same interval.

7. Fill in the blank: The area under the graph of $y = \sqrt{1 + e^{2x}}$ on $[0, 2]$ is equal to the length of the graph of the function _____ on the same interval.

8. In this exercise you will investigate whether approximating arc length with line segments results in an under-approximation or an over-approximation.

  (a) Sketch graphs of the functions $f(x) = 4 - x^2$, $g(x) = e^x$, and $h(x) = 3x + 1$ on $[-2, 2]$.

  (b) Suppose you wanted to approximate the arc length of each of these functions on $[-2, 2]$ by using eight line segments. Draw these line segments on each graph.

  (c) Determine graphically whether each approximation is greater than, less than, or equal to the actual arc length for each function listed in part (a).

  (d) Explain why approximating the length of a curve with line segments will *always* yield either an under-approximation or an exact answer.

9. Suppose you want to approximate the arc length of $f(x) = \tan x$ on the interval $[0, \pi]$. Would you get a good approximation if you subdivided the interval $[0, \pi]$ into five subintervals and added up the lengths of the line segments over each subinterval? Why or why not? (*Hint: Draw the line segments.*)

10. Use definite integrals to show that the graphs of $f(x) = \dfrac{1}{x}$ and $g(x) = -\dfrac{1}{x}$ have the same arc length on $[1, 3]$. (*Hint: You do* not *have to solve the integrals!*) Why does this equality make sense in terms of graphical transformations?

11. Use definite integrals to show that $f(x) = x^2 - 3$ and $g(x) = 5 - x^2$ have the same arc length on $[0, 2]$. (*Hint: You do* not *have to solve the integrals!*) Why does this equality make sense in terms of transformations?

12. Do you think that the function $f(x) = 2x^2$ has twice the arc length of $g(x) = x^2$ on the interval $[0, 3]$? Why or why not?

13. Write down a definite integral that describes the circumference of a circle of radius 5. Don't try to solve the integral; just write it down.

14. Write down a definite integral that describes the arc length around an ellipse with horizontal radius of 3 units and vertical radius of 2 units. The ellipse in this exercise extends 6 units across and 4 units vertically and has equation $\dfrac{x^2}{9} + \dfrac{y^2}{4} = 1$. Don't try to solve the integral; just write it down.

15. In this exercise you will approximate the arc length of $f(x) = \cos x$ on $[0, \pi]$ in two ways and compare your answers.

(a) By using four line segments and the distance formula.

(b) By using a right sum with four rectangles to approximate the area under the graph of the function $y = \sqrt{1 + \sin^2 x}$ on $[0, 2]$.

(c) Why do the approximations you found in parts (a) and (b) both approximate the arc length of $f(x) = \cos x$ on $[0, \pi]$? Which, if either, do you think might be a better approximation?

16. In this exercise you will approximate the arc length of $f(x) = x^2$ on $[0, 2]$ in two ways and compare your answers.

(a) By using four line segments and the distance formula.

(b) By using a right sum with four rectangles to approximate the area under the graph of the function $y = \sqrt{1 + 4x^2}$ on $[0, 2]$.

(c) Why do the approximations you found in parts (a) and (b) both approximate the arc length of $f(x) = x^2$ on $[0, 2]$? Which, if either, do you think might be a better approximation?

17. Suppose you wish to use n frustums to approximate the area of the surface obtained by revolving the graph of a function $y = f(x)$ around the x-axis on $[a, b]$. Use a labeled graph to explain why the slant length s_k of the kth frustum is given by
$$s_k = \sqrt{(\Delta x)^2 + (\Delta y_k)^2}.$$

18. Suppose you wish to use n frustums to approximate the area of the surface obtained by revolving the graph of a function $y = f(x)$ around the x-axis on $[a, b]$. Use a labeled graph to explain why the average radius r_k of the kth frustum is given by
$$r_k = \frac{f(x_{k-1}) + f(x)}{2}.$$

Skills

In Exercises 19–24, approximate the arc length of $f(x)$ on $[a, b]$, using n line segments and the distance formula. Include a sketch of $f(x)$ and the line segments.

19. $f(x) = x^3$, $[a, b] = [-2, 2]$, $n = 4$

20. $f(x) = \cos x$, $[a, b] = [0, \pi]$, $n = 3$

21. $f(x) = \ln x$, $[a, b] = [1, 3]$, $n = 4$

22. $f(x) = e^x$, $[a, b] = [-1, 3]$, $n = 4$

23. $f(x) = \sqrt{9 - x^2}$, $[a, b] = [-3, 3]$, $n = 3$

24. $f(x) = \sqrt{9 - x^2}$, $[a, b] = [-3, 3]$, $n = 6$

In Exercises 25–30, approximate the arc length of $f(x)$ on $[a, b]$, using the approximation $\sum_{k=1}^{n} \sqrt{1 + \left(\frac{\Delta y_k}{\Delta x}\right)^2}\, \Delta x$ with the given value of n. In each problem list the values of Δy_k for $k = 1, 2, \ldots, n$.

25. $f(x) = x^3$, $[a, b] = [-2, 2]$, $n = 4$

26. $f(x) = \cos x$, $[a, b] = [0, \pi]$, $n = 3$

27. $f(x) = \ln x$, $[a, b] = [1, 3]$, $n = 4$

28. $f(x) = e^x$, $[a, b] = [-1, 3]$, $n = 4$

29. $f(x) = \sqrt{9 - x^2}$, $[a, b] = [-3, 3]$, $n = 3$

30. $f(x) = \sqrt{9 - x^2}$, $[a, b] = [-3, 3]$, $n = 6$

Find the *exact* value of the arc length of each function $f(x)$ in Exercises 31–46 on $[a, b]$ by writing the arc length as a definite integral and then solving that integral.

31. $f(x) = 3x + 1$, $[a, b] = [-1, 4]$

32. $f(x) = 4 - x$, $[a, b] = [2, 5]$

33. $f(x) = 2x^{3/2} + 1$, $[a, b] = [1, 3]$

34. $f(x) = x^{3/2}$, $[a, b] = [0, 2]$

35. $f(x) = (2x + 3)^{3/2}$, $[a, b] = [-1, 1]$

36. $f(x) = 2(1 - x)^{3/2} + 3$, $[a, b] = [-2, 0]$

37. $f(x) = \sqrt{9 - x^2}$, $[a, b] = [-3, 3]$

38. $f(x) = \sqrt{1 - x^2}$, $[a, b] = [-1, 1]$

39. $f(x) = \dfrac{1}{3}x^{3/2} - x^{1/2}$, $[a, b] = [0, 1]$

40. $f(x) = (1 - x^{2/3})^{3/2}$, $[a, b] = [0, 1]$

41. $f(x) = (4 - x^{2/3})^{3/2}$, $[a, b] = [0, 2]$

42. $f(x) = x^2$, $[a, b] = [-1, 1]$

43. $f(x) = x^2 - \dfrac{1}{8}\ln x$, $[a, b] = [1, 2]$

44. $f(x) = \ln(\cos x)$, $[a, b] = \left[0, \dfrac{\pi}{4}\right]$

45. $f(x) = \ln(\sin x)$, $[a, b] = \left[\dfrac{\pi}{4}, \dfrac{3\pi}{4}\right]$

46. $f(x) = \dfrac{x^4 + 3}{6x}$, $[a, b] = [1, 3]$

Each definite integral in Exercises 47–52 represents the arc length of a function $f(x)$ on an interval $[a, b]$. Determine the function and interval.

47. $\displaystyle\int_{-2}^{5} \sqrt{1 + 9e^{6x}}\, dx$

48. $\displaystyle\int_{0}^{\pi} \sqrt{1 + \sin^2 x}\, dx$

49. $\displaystyle\int_{2}^{3} \sqrt{\dfrac{x^4 + 1}{x^4}}\, dx$

50. $\displaystyle\int_{0}^{2} \sqrt{1 + 36x}\, dx$

51. $\displaystyle\int_0^{\pi/4} \sec x\, dx$ **52.** $\displaystyle\int_0^1 \sqrt{1 + 9x^4}\, dx$

You may have noticed that even very simple functions give rise to arc length integrals that we have no idea how to compute. In Exercises 53–56, use a graphing calculator to approximate a definite integral that represents the arc length of the given function $f(x)$ on the interval $[a, b]$.

53. $f(x) = x^3$, $[a, b] = [-1, 1]$

54. $f(x) = x^2 + 1$, $[a, b] = [0, 4]$

55. $f(x) = 3x^2 - 1$, $[a, b] = [1, 2]$

56. $f(x) = \ln x$, $[a, b] = [1, 3]$

In Exercises 57–62, use n frustums to approximate the area of the surface of revolution obtained by revolving the curve $y = f(x)$ around the x-axis on the interval $[a, b]$.

57. $f(x) = x^2$, $[a, b] = [0, 4]$, $n = 2$

58. $f(x) = x^2$, $[a, b] = [0, 4]$, $n = 4$

59. $f(x) = x^3$, $[a, b] = [1, 3]$, $n = 4$

60. $f(x) = \ln x$, $[a, b] = [1, 4]$ $n = 3$

61. $f(x) = \sin x$, $[a, b] = [0, \pi]$, $n = 2$

62. $f(x) = \sin x$, $[a, b] = [0, \pi]$ $n = 4$

In Exercises 63–72, set up and solve a definite integral to find the exact area of each surface of revolution obtained by revolving the curve $y = f(x)$ around the x-axis on the interval $[a, b]$.

63. $f(x) = x$, $[a, b] = [0, 2]$

64. $f(x) = \dfrac{1}{x}$, $[a, b] = [1, 10]$

65. $f(x) = \sqrt{x}$, $[a, b] = [0, 4]$

66. $f(x) = \sqrt{25 - x^2}$, $[a, b] = [-5, 5]$

67. $f(x) = \sqrt{3x + 1}$, $[a, b] = [0, 3]$

68. $f(x) = e^{-x}$, $[a, b] = [-1, 1]$

69. $f(x) = e^{4x}$, $[a, b] = [-1, 1]$

70. $f(x) = x^2$, $[a, b] = [-1, 1]$

71. $f(x) = \sin x$, $[a, b] = [0, \pi]$

72. $f(x) = \cos x$, $[a, b] = \left[-\dfrac{\pi}{2}, \dfrac{\pi}{2}\right]$

Applications

73. Suppose a drinking fountain expels water so that it falls in the shape of the parabola $f(x) = 3 - \frac{1}{2}x^2$, measured in inches, as shown in the given figure. Set up and solve a definite integral that measures the distance that water travels starting from when it leaves the fountain (at $x = 0$) and ending when it hits the surface of the fountain (when $f(x) = 0$).

Path of water from a drinking fountain

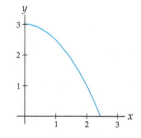

74. Dr. Geek built a skate ramp in his backyard modeled after the function $f(x) = 11 \sin\left(\frac{1}{42}x^2\right)$, measured in feet, as shown in the next figure. Set up a definite integral that measures the distance required to skate from one side of the ramp to the other. The starting and ending x-values should both satisfy $f(x) = 11$ feet. Use a graphing calculator to approximate the value of this definite integral.

Curve of skate ramp

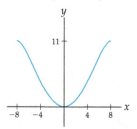

Proofs

75. Suppose $f(x)$ is a continuous function on an interval $[a, b]$, and let n be a positive integer. With the notation Δx, x_k, and Δy_k given in Definition 9.6, prove that the sum $\sum_{k=1}^{n} \sqrt{(x_k - x_{k-1})^2 + (f(x_k) - f(x_{k-1}))^2}$ is equal to the sum $\sum_{k=1}^{n} \sqrt{1 + \left(\frac{\Delta y_k}{\Delta x}\right)^2}\, \Delta x$.

76. Use Definition 9.6, the Mean Value Theorem, and the definition of the definite integral to prove Theorem 9.7: The arc length of a sufficiently well behaved function $f(x)$ on an interval $[a, b]$ can be represented by the definite integral $\int_a^b \sqrt{1 + (f'(x))^2}\, dx$.

77. Prove, in two ways, that the arc length of a linear function $f(x) = mx + c$ on an interval $[a, b]$ is equal to $(b - a)\sqrt{1 + m^2}$:

(a) by using the distance formula;
(b) by using Theorem 9.7.

78. Use Theorem 9.7 to prove that a circle of radius 5 has circumference 10π.

79. Use Theorem 9.7 to prove that a circle of radius r has circumference $2\pi r$.

80. Prove that if f is continuous on $[a, b]$ and C is any real number, then $f(x)$ and $f(x) + C$ have the same arc length on $[a, b]$. Then explain why this makes sense graphically.

81. In this exercise you will prove that the surface area of a frustum with radii p and q and slant length s is equal to $SA = 2\pi rs$, where $r = \frac{p+q}{2}$ is the average radius of the frustum. The argument will hinge on the fact that a frustum is a cone with its top removed, as shown here:

(a) Use similar triangles to prove that, in the notation of the given diagram, $pt + ps = qt$.

(b) We know from Theorem 3.12 that the surface area of a cone with radius A and height B is given by $\pi A\sqrt{A^2 + B^2}$. Use the Pythagorean theorem to prove that this expression implies that the surface area of a cone with radius A and slant height C is given by πAC.

(c) Express the surface area of the frustum in the figure as the difference of the area of the larger cone and the area of the smaller cone, and use the previous two parts of this problem to prove that the surface area of the frustum is $SA = \pi(p+q)s$.

(d) Finally, use the relationship $r = \frac{p+q}{2}$ to prove that the surface area of the frustum is $S = 2\pi rs$.

82. Use Theorem 9.10 to prove that a sphere of radius 5 has surface area 100π.

83. Use Theorem 9.10 to prove that a sphere of radius r has surface area $4\pi r^2$.

Thinking Forward

Arc length of parametric curves: A curve in the plane is defined by *parametric equations* if we can express the points on the curve in the form $(x(t), y(t))$ for some pair of continuous functions $x(t)$ and $y(t)$. You should think of t as a *parameter* and $x(t)$ and $y(t)$ as tracing out the x- and y-coordinates of the curve as t changes. A classic example of a parametric curve is $(\cos t, \sin t)$, which traces out the unit circle as t increases from 0 to 2π.

▶ Calculate the coordinates $(\cos t, \sin t)$ for the values $t = 0, \frac{\pi}{4}, \frac{5\pi}{6}, \frac{3\pi}{2}$, and 2π, and verify that these co-ordinates define the points pictured in the following diagram:

$(\cos t, \sin t)$ traces out a circle

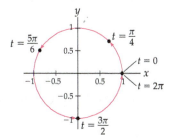

▶ To approximate the arc length of a parametric curve on some interval from $t = a$ to $t = b$, we subdivide $[a, b]$ into n pieces to define $t_1 = a, t_2, t_3, \ldots, t_n$, and consider the line segments connecting $(x(t_{k-1}), y(t_{k-1}))$ and $(x(t_k), y(t_k))$ for each $k = 1, 2, \ldots, n$. Define Δx_k and Δy_k for the kth line segment in terms of the coordinates shown in the following figure:

Notation for the kth line segment

▶ Mimic the argument in the reading for this section to argue that a reasonable definition for the arc length of a parametric curve $(x(t), y(t))$ from $t = a$ to $t = b$ is

$$\lim_{n \to \infty} \sum_{k=1}^{n} \sqrt{(\Delta x_k)^2 + (\Delta y_k)^2},$$

where Δx_k and Δy_k are defined as in the previous problem.

▶ Use algebra and the concept of Riemann sums to argue that the arc length of a parametric curve $(x(t), y(t))$ from $t = a$ to $t = b$ should be represented by the definite integral

$$\int_a^b \sqrt{(x'(t))^2 + (y'(t))^2}\, dt.$$

▶ Use this new description of arc length to show that the circumference of the unit circle is 2π, by thinking of the unit circle as the parametric curve $(\cos t, \sin t)$ from $t = 0$ to $t = 2\pi$ and applying the definite integral formula from the previous problem.

9.4   REAL-WORLD APPLICATIONS OF INTEGRATION

▶ Definitions of mass, weight, density, work, force, and hydrostatic force

▶ Using definite integrals to represent accumulations of varying quantities

▶ Examples of real-world physics problems

Mass and Density

We now examine applications from physics that can be solved by our "subdivide, approximate, and add" strategy, and thus with definite integrals. The first concept that will concern us is **mass**. Mass is different from, but related to, **weight**. Mass is an inherent quantity that does not depend on which gravitational force is acting on the object. Mass measures the amount of matter in an object, while weight measures the force of gravity on the object.

A small object that is very dense (like a small rock) can have the same mass as a larger object that is less dense (for example, a very large serving of cotton candy). **Density** is mass per unit volume, and is traditionally labeled by the letter ρ, the lowercase Greek letter "rho." The mass of an object is then the product of its density and its volume:

DEFINITION 9.11

Mass Given a Constant Density
The **mass** of an object with density ρ and volume V is $m = \rho V$.

For example, gold has a density of approximately 19.3 grams per cubic centimeter, so the mass of a gold coin of radius 1.5 centimeters and thickness 0.2 centimeter is

$$m = \rho V = \left(19.3 \, \tfrac{\text{grams}}{\text{cm}^3}\right)(\pi (1.5)^2 (0.2) \text{cm}^3) \approx 27.28 \text{ grams.}$$

A silver coin of the same dimensions has a smaller mass, since silver has a lower density of approximately 10.5 grams per cubic centimeter:

$$m = \rho V = \left(10.5 \, \tfrac{\text{grams}}{\text{cm}^3}\right)(\pi (1.5)^2 (0.2) \text{cm}^3) \approx 14.84 \text{ grams.}$$

This simple mass formula works only for objects that are homogeneous, with same density everywhere. When density varies, finding mass is more complicated. The key will be to slice the object in such a way that for each piece, we can assume that the density is approximately constant, as we will soon see.

Work and Force

The second concept from physics that we will be interested in is **work**. Clearly it takes more work to lift a heavy object (say, a car) than it takes to lift a lighter object (say, a book). It also takes more work to lift an object a long distance (such as lifting a car 30 feet) than a short distance (such as lifting a car 1 inch off the ground). The work required to lift an object is thus the product of its weight and the distance that it must be moved.

DEFINITION 9.12   **Work Given a Constant Distance**

The **work** required to lift an object with a weight of F through a distance d is $W = Fd$.

We use the letter F for weight because the weight of an object is the force exerted on it by gravity. This force is the product $F = \omega V$ of the object's weight per unit volume, or its **weight-density**, and its volume.

For example, water has a weight-density of $\omega = 62.4$ pounds per cubic foot. How much work does it take to lift a full bucket of water with radius 1 foot and height 2 feet to a height of 3 feet off the ground? The water in the bucket weighs

$$F = \omega V = 62.4(\pi (1)^2(2)) = 124.8\,\pi \text{ pounds.}$$

Assuming that the weight of the bucket is negligible, the work required to lift such a full bucket of water a distance of 3 feet off the ground is

$$W = Fd = (124.8\,\pi \text{ pounds})(3 \text{ feet}) \approx 1176 \text{ foot-pounds.}$$

Note that work involves units called *foot-pounds*; this is because work is the product of force (measured in pounds) and distance (measured in feet). In physics, one might see force measured in *newtons* and distance measured in meters, in which case work would be measured in *newton-meters* which are also called *joules*.

The simple work formula just presented applies only when force and displacement are constant. If either the weight of an object or the distance that it is to be moved is variable, then finding work will require integrals, as we will soon see.

Hydrostatic Force

Our final physics concept is **hydrostatic force**, which is the force exerted on a surface by a volume of water or some other liquid. The force exerted on the bottom, horizontal surface of any tank or container is equal to the weight of the water above it. Force due to gravity is equal to weight, which is equal to weight-density times volume, which, for water in a container, is equal to weight-density, times area, times depth and thus:

DEFINITION 9.13   **Hydrostatic Force Given a Constant Depth**

The **hydrostatic force** exerted by a liquid of weight-density ω and depth d on a horizontal wall of area A is given by $F = \omega Ad$.

For example, in a rectangular swimming pool measuring 10 feet by 30 feet that has a constant depth of 5 feet, the hydrostatic force on the bottom surface of the pool is

$$F = \omega Ad = \left(62.4\,\frac{\text{lbs}}{\text{ft}^3}\right)(300\,\text{ft}^2)(5\,\text{ft}) = 93{,}600 \text{ pounds.}$$

In fact, the same principle applies to the force exerted against *any* wall (horizontal, vertical, or otherwise) by a mass of water. The reason is that the pressure, or force per square foot of area, exerted by the water is the same in all directions. However, a non-horizontal wall will be at different depths at various points on the wall, and the force from the water will be different at these various points. The water in a pool will exert a great deal of force near the bottom of a side wall, but less force near the top of that side wall. We will have to use a slicing technique–and thus, eventually, definite integrals–to find the hydrostatic force exerted on nonhorizontal walls.

Representing Varying Quantities with Definite Integrals

Mass, work, and hydrostatic force have simple formulas when all the associated quantities are constant. But what happens when things vary? How do we calculate the mass of an object of varying density, or the work required to pump water of varying depths out of a tank, or the hydrostatic force on the side wall of a pool?

The key, of course, is to consider slices on which any varying quantities can be approximated with *constant* values. Then the simple formulas we developed earlier will apply on each slice, and we can add up all the slice-approximations to find an approximation for the total mass, work, or hydrostatic force. Subdivide, approximate, and add ... and of course if we set things up right, then we will be able to express our approximation as a Riemann sum, take a limit, and obtain a definite integral for the exact answer!

Let's walk through this process with a particular example. Consider the task of pumping all the water out of the top of a cylindrical hot tub that is 4 feet deep and has a radius of 3.5 feet. The water at the bottom of the tub has to be moved farther than the water at the top of the tub, so the distance involved is variable. If we consider a thin horizontal slice of the hot tub, we can assume without too much error that all the water in that slice has to be moved the same upward distance; see the following figure:

Horizontal slice of hot tub

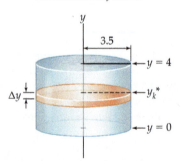

This figure introduces some needed notation. We are imagining that the bottom of the hot tub is at $y = 0$ and and the top is at $y = 4$. We divide the interval from $y = 0$ to $y = 4$ into n subintervals $[y_{k-1}, y_k]$ of thickness Δy. A representative slice of the hot tub at some height $y_k^* \in [y_{k-1}, y_k]$ is shown in the figure.

The slice at y_k^* needs to be moved upwards $d_k = 4 - y_k^*$ units in order to be pumped out of the tub. This slice is a disk with volume $V_k = \pi r^2 h = \pi (3.5)^2 \Delta y$ cubic feet. Water weighs $\omega = 62.4$ pounds per cubic foot. Since for constant quantities we have $W = Fd = \omega V d$, the work W_k required to pump the kth slice of water out of the hot tub is

$$W_k = F_k d_k = \omega V_k d_k = 62.4(\pi (3.5)^2 \Delta y)(4 - y_k^*) \text{ foot-pounds.}$$

By repeating this approximation for each horizontal slice and then adding all these approximations, we can get an approximation for the total work W required to pump all of the water out of the hot tub:

$$W \approx \sum_{k=1}^{n} (62.4)(\pi)(3.5)^2(4 - y_k^*)\Delta y.$$

You should recognize this as a Riemann sum. As n gets larger, we make thinner and thinner slices and better approximations. In the limit, as $n \to \infty$, this approximation approaches the actual work required. Accumulating slices of work from $y = 0$ to $y = 4$, the actual amount of work required is given by the definite integral

$$W = \int_0^4 (62.4)(\pi)(3.5)^2(4 - y)\, dy.$$

With basic integration techniques it is easy to solve this integral to find that the total amount of work required to pump all of the water out of the top of the hot tub is $6115.2\,\pi \approx 19,211.5$ foot-pounds.

Let's review what just happened from a more general perspective. The sum we found earlier was a Riemann sum that converged to a definite integral because it was of the form $\sum_{k=1}^{n} f(y_k^*)\Delta y$ for some continuous function $f(y)$. In this case the function was

$$f(y) = 62.4\pi(3.5)^2(4 - y) = (\text{weight-density})(\text{cross-sectional area})(\text{distance}) = \omega A d,$$

where $d = d(y)$ is a function of y. Compare this with the formula $W = Fd = \omega V d$ for work. The difference is that $V = A\Delta y$; we extracted this Δy so that we could form a Riemann sum. In the definite integral, Δy becomes dy.

In other examples we will follow the same basic strategy: Subdivide, or slice, so that quantities are constant; use physics formulas to approximate quantities on each slice; extract a Δx or Δy to write the sum as a Riemann sum; and then take the limit to get a definite integral that we can solve.

Centroids and Centers of Mass

As a geometric application of the techniques of this section, we will investigate the **centroid**, otherwise known as the **center of mass**, of a thin plate of material. Let's consider the simple case where the plate is defined by the region between a positive function $f(x)$ and the x-axis, as shown next at the left. The centroid of a thin plate of material in the shape of this region is the point $(\tilde{x}, \tilde{y})$ where the plate would balance horizontally, as shown at the right.

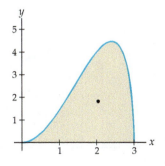

A region in the plane and its centroid

Balancing horizontally on the centroid

As one might expect, the centroid of a rectangular region is easy to find; it is found at the center of the rectangle. It is also easy to find the centroid of a collection of rectangles, because the centroid of a composite object is a weighted average of the centroids of its pieces. Specifically, suppose a region with area A is divided into a collection of regions with areas $A_1, A_2, \ldots A_n$ and centroids $(\bar{x}_1, \bar{y}_1), (\bar{x}_2, \bar{y}_2), \ldots, (\bar{x}_n, \bar{y}_n)$. Then the centroid $(\bar{x}, \bar{y})$ of the combined region can be obtained by taking the average of the centroids of the rectangles, weighted by their areas, with coordinates as follows:

$$\bar{x} = \frac{\bar{x}_1 A_1 + \bar{x}_2 A_2 + \cdots + \bar{x}_n A_n}{A},$$

$$\bar{y} = \frac{\bar{y}_1 A_1 + \bar{y}_2 A_2 + \cdots + \bar{y}_n A_n}{A}.$$

An example of this situation with three rectangular regions is shown next at the left, with the centroid of each rectangle marked with an "x" and the centroid of the overall shape marked with a dot. Note that in our weighted average, the third "x" will count the most, since it is the centroid of the largest rectangle. This particular set of rectangles was not arbitrary; the figure at the right shows the same collection of rectangles set against the region we illustrated at the start of this subsection. It is reasonable to imagine that the centroid of the collection of rectangles is an approximation for the centroid of this region. (See Example 6.)

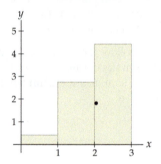

The centroid of a collection of rectangles

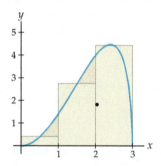

Approximating the centroid of a region

We are now on very familiar ground. Suppose we want to find the centroid of a region between the graph of some function $f(x)$ and the x-axis on an interval $[a, b]$. Following our usual "subdivide, approximate, and add" strategy, we begin by subdividing the interval $[a, b]$ into n subintervals of the form $[x_{k-1}, x_k]$, where $x_k = a + k\,\Delta x$ and $\Delta x = \frac{b-a}{n}$. For each k, we then choose x_k^* to be the midpoint of the subinterval $[x_{k-1}, x_k]$ and consider the rectangle on that subinterval whose height is $f(x_k^*)$ as shown here:

The centroid of a representative rectangle

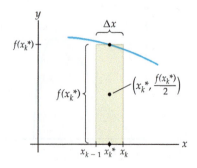

The centroid of this rectangle is at its center, which is the point

$$(\bar{x}_k, \bar{y}_k) = \left(x_k^*, \frac{1}{2}f(x_k^*)\right).$$

The area of this kth rectangle is

$$A_k = f(x_k^*)\Delta x.$$

Using our weighted average formula from earlier, we find that the centroid of the collection of all n rectangles has coordinates

$$\bar{x} = \frac{\sum_{k=1}^{n} \bar{x}_k A_k}{\sum_{k=1}^{n} A_k} = \frac{\sum_{k=1}^{n} x_k^* f(x_k^*)\Delta x}{\sum_{k=1}^{n} f(x_k^*)\Delta x},$$

$$\bar{y} = \frac{\sum_{k=1}^{n} \bar{y}_k A_k}{\sum_{k=1}^{n} A_k} = \frac{\sum_{k=1}^{n} \frac{1}{2} f(x_k^*) f(x_k^*)\Delta x}{\sum_{k=1}^{n} f(x_k^*)\Delta x}.$$

Finally, we notice that the numerators and denominators of $\bar{x}$ and $\bar{y}$ are Riemann sums! Taking the limit as $n \to \infty$ allows us to describe the exact centroid of the region in terms of definite integrals:

THEOREM 9.14

The Centroid of a Region Under a Curve

Let f be an positive, integrable function on $[a, b]$. Then the centroid $(\bar{x}, \bar{y})$ of the region between the graph of $f(x)$ and the x-axis on the interval $[a, b]$ is the point

$$(\bar{x}, \bar{y}) = \left(\frac{\int_a^b x f(x)\, dx}{\int_a^b f(x)\, dx}, \frac{\frac{1}{2}\int_a^b f(x)^2\, dx}{\int_a^b f(x)\, dx}\right).$$

By construction, this formula works only for finding centroids of regions between the graph of some function $f(x)$ and the x-axis. A similar construction allows us to describe the centroid between two graphs $f(x)$ and $g(x)$ on an interval $[a, b]$ (see Example 7 for this construction):

THEOREM 9.15

The Centroid of a Region Between Two Curves

Let f and g be integrable functions on $[a, b]$. The centroid $(\bar{x}, \bar{y})$ of the region between the graphs of $f(x)$ and $g(x)$ on the interval $[a, b]$ is the point

$$(\bar{x}, \bar{y}) = \left(\frac{\int_a^b x|f(x) - g(x)|\, dx}{\int_a^b |f(x) - g(x)|\, dx}, \frac{\frac{1}{2}\int_a^b |f(x)^2 - g(x)^2|\, dx}{\int_a^b |f(x) - g(x)|\, dx}\right).$$

The absolute values in these expressions play the same role as those we saw in the calculations of the area between two curves as those we saw in Section 7.6. Specifically, they allow us to write the centroid as one formula regardless of whether $f(x) \geq g(x)$ or $f(x) \leq g(x)$. Just as in Section 7.6, when calculating centroids you must divide the interval $[a, b]$ into subintervals according to these inequalities.

Examples and Explorations

EXAMPLE 1

Finding work when distance and volume vary

Find the work involved in pumping all of the water out of the top of a cone-shaped tank that is 4 feet tall and has a radius of 3.5 feet at the top.

The first thing we have to do is establish some notation. Suppose that the bottom of the tank is at height $y = 0$ and the top of the tank is at height $y = 4$. A representative horizontal slice at some point y_k^* is shown next at the left.

Horizontal slice of conical tank   *Use the law of similar triangles*

We can use the law of similar triangles (see the diagram at the right) to find the radius r_k of this slice:

$$\frac{r_k}{y_k^*} = \frac{3.5}{4} \quad \Longrightarrow \quad r_k = \frac{3.5}{4}y_k^*.$$

The volume of the slice is $V_k = \pi r^2 \Delta y$, the weight-density of water is $\omega = 62.4$ pounds per cubic foot, and the vertical distance that the water in the kth slice must move is approximately $d_k = 4 - y_k^*$. Since for constant quantities we have $W = Fd = \omega Vd$, the work involved in pumping out the representative slice of water is

$$W_k = F_k d_k = \omega V_k d_k = (62.4)(\pi)\left(\frac{3.5}{4}y_k^*\right)^2 \Delta y \cdot (4 - y_k^*) \text{ foot-pounds.}$$

Thus the work W required to pump all of the water out of the top of the conical tank is approximately

$$W \approx \sum_{k=1}^{n} (62.4)(\pi)\left(\frac{3.5}{4}\right)^2 (y_k^*)^2 (4 - y_k^*) \Delta y.$$

This is a Riemann sum, so as $n \to \infty$ we obtain a definite integral. Accumulating slices from $y = 0$ to $y = 4$, we find that the amount of work required to pump all of the water out of the top of the tank is

$$W = \int_0^4 (62.4)(\pi)\left(\frac{3.5}{4}\right)^2 y^2 (4 - y) \, dy$$

$$= 47.775\,\pi \int_0^4 (4y^2 - y^3) \, dy = 47.775\,\pi \left[\frac{4}{3}y^3 - \frac{1}{4}y^4\right]_0^4$$

$$= 3201.91\,\pi \text{ foot-pounds.} \qquad \square$$

EXAMPLE 2   **Finding mass when density varies**

Suppose a cylindrical rod with a radius of 2 centimeters and a length of 24 centimeters is made of a combination of silver and gold in such a way that the density of the rod x centimeters from the left end is given by the function $\rho(x) = 10.5 + 0.01527x^2$ grams per cubic centimeter. Find the total mass of the rod.

SOLUTION

The following diagram shows a thin representative slice of the rod at some point x_k^* centimeters from the left end:

Representative slice of rod with varying mass

Each representative slice is a disk of radius 2 and thickness Δx, and therefore volume $\pi(2)^2\Delta x$. We can assume without too much error that the density throughout the entire thin slice is $\rho_k = \rho(x_k^*) = 10.5 + 0.01527(x_k^*)^2$ grams per cubic centimeter. Since for constant values we have $m = \rho V$, the mass of the slice is

$$m_k = \rho_k V_k = (10.5 + 0.01527(x_k^*)^2)(\pi(2)^2\Delta x).$$

The approximate mass of the entire rod is thus given by the sum

$$m \approx \sum_{k=1}^{n} 4\pi(10.5 + 0.01527(x_k^*)^2)\,\Delta x.$$

As $n \to \infty$, this Riemann sum approaches the actual mass of the rod. Accumulating mass from $x = 0$ to $x = 24$, we see that the total mass of the rod is

$$m = \int_0^{24} 4\pi(10.5 + 0.01527x^2)\,dx$$

$$= 4\pi\left[10.5x + \frac{0.01527}{3}x^3\right]_0^{24} = 1289.46\,\pi \approx 4050.95 \text{ grams.} \qquad \square$$

EXAMPLE 3

Finding mass when density and volume vary

Alina has made a gelatin mold that is 4.5 inches tall, 6 inches across, and in the shape of the top of a downwards-pointing parabola that has been revolved around the y-axis. She put strawberry pieces into the gelatin mold, but they have tended to settle down to the bottom. This causes the density of the gelatin mold to vary linearly with height in such a way that the density at the very top of the mold is 0.25 ounce per cubic inch, while the density at the very bottom of the mold is 1.3 ounces per cubic inch. Find the mass of the entire gelatin mold.

SOLUTION

We first need to find functions that describe the shape and density of the gelatin mold. This is not difficult to do, using the information given in the problem. It turns out that the shape of the gelatin mold is the solid of revolution obtained by rotating the graph of $y = 4.5 - 0.5x^2$ on $[0, 3]$ around the x-axis and that the density of the gelatin mold at height y is approximately given by the function $\rho(y) = 1.3 - 0.233y$. (See Exercises 10 and 11.)

In this example, both the density and the volume vary with the slice of the object. The next figure shows a representative slice at some height y_k^*. We can assume without introducing too much error that the entire slice has a constant density of $\rho_k = \rho(y_k^*) = 1.3 - 0.233y_k^*$ ounces per cubic inch.

Strawberries settle to the bottom

According to this graph, if r_k is the radius corresponding to the height y_k^*, then $y_k^* = 4.5 - 0.5r_k^2$, and thus $r_k = \sqrt{9 - 2y_k^*}$. The volume of the representative slice is therefore

$$V_k = \pi r_k^2 \,\Delta y = \pi \left(\sqrt{9 - 2y_k^*} \right)^2 \Delta y = \pi (9 - 2y_k^*)\Delta y.$$

Since for constant values we have $m = \rho V$, the mass of our representative slice is

$$m_k = \rho_k V_k = (1.3 - 0.233y_k^*)(\pi)(9 - 2y_k^*)\Delta y.$$

The total mass m of the gelatin mold can be approximated by the sum of the approximate masses of each of the slices:

$$m \approx \sum_{k=1}^{n} \pi (1.3 - 0.233y_k^*)(9 - 2y_k^*)\Delta y.$$

This is a Riemann sum, and as $n \to \infty$ the sum converges to a definite integral. Accumulating slices from $y = 0$ to $y = 4.5$, we get the total mass of the gelatin mold:

$$\int_0^{4.5} \pi (1.3 - 0.233y)(9 - 2y)\,dy = \pi \int_0^{4.5} (11.7 - 4.697y + 0.466y^2)\,dy$$

$$= \pi \left[11.7y - \frac{4.697}{2}y^2 + \frac{0.466}{3}y^3 \right]_0^{4.5} \approx 60.4682 \text{ ounces.} \quad \square$$

EXAMPLE 4   **Finding hydrostatic force when depth varies**

Consider a swimming pool that measures 10 feet by 30 feet, with a constant depth of 5 feet. Find the hydrostatic force exerted on one of the shorter side walls of the pool.

SOLUTION

The force exerted by the water on the side walls varies with depth. However, if we consider a very thin horizontal slice of the wall, then we can assume without too much error that the entire slice is at the same depth. The following diagram shows a representative slice of the side wall at some height y_k^* from the bottom of the pool:

Representative slice of side wall

The area of the representative wall slice is $A_k = 10\Delta y$ square feet. Water has a weight-density of $\omega = 62.4$ pounds per cubic foot. When values are constant, hydrostatic force is given by $F = \omega A d$. Therefore, assuming that the entire thin slice of wall is at a depth of $d_k = 5 - y_k^*$ feet, we find that the hydrostatic force on that slice of wall is given by

$$F_k = \omega A_k d_k = 62.4(10\Delta y)(5 - y_k^*).$$

The hydrostatic force on the entire side wall is thus approximately

$$F \approx \sum_{k=1}^{n} 62.4(10)(5 - y_k^*)\,\Delta y.$$

As n approaches ∞ and we accumulate slices from $y = 0$ to $y = 5$, this Riemann sum approaches a definite integral. The hydrostatic force on a shorter side wall of the pool is

$$\int_0^5 62.4(10)(5 - y)\,dy = (62.4)(10)\left[5y - \frac{1}{2}y^2\right]_0^5 = 7800 \text{ pounds.} \qquad \square$$

EXAMPLE 5

Finding hydrostatic force when depth and area vary

Consider a rectangular swimming pool that is 30 feet long and 10 feet wide, with a depth of 4 feet at the shallow end and 8 feet at the deep end. Suppose the shallow end of the pool makes up the first 10 feet of the pool and is followed by 10-foot linear ramp along which the depth of the pool increases from 4 feet to 8 feet and then a final length of 10 feet for the deep end. Find the hydrostatic force exerted on one of the long sides of the pool.

SOLUTION

The next diagram shows a side view of the pool, as well as two types of horizontal slices. If y_k^* is between $y = 0$ and $y = 4$, then the length of the slice varies with the depth of the pool. If y_k^* is between $y = 4$ and $y = 8$, then the length of the slice is always 30 feet.

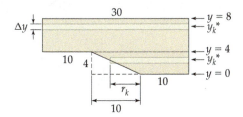

For y_k^* between $y = 0$ and $y = 4$, the kth slice has width Δy and length $10 + r_k$, where r_k is the distance that the slice extends into the ramp part of the pool. By the law of similar triangles and the triangles whose legs are shown as dotted lines in the figure, we have

$$\frac{r_k}{y_k^*} = \frac{10}{4} \quad \Longrightarrow \quad r_k = \frac{5}{2}y_k^*.$$

The area of the kth slice, between $y = 0$ and $y = 4$, is $A_k = \left(10 + \frac{5}{2}y_k^*\right)\Delta y$. Again, the weight-density of water is $\omega = 62.4$ pounds per cubic foot. When values are constant, hydrostatic force is given by $F = \omega A d$. Since we are assuming that the entire slice is at depth $d_k = 8 - y_k^*$, the hydrostatic force on each slice between $y = 0$ and $y = 4$ is

$$F_k = \omega A_k d_k = 62.4\left(\left(10 + \frac{5}{2}y_k^*\right)\Delta y\right)(8 - y_k^*).$$

The total hydrostatic force on the part of the side wall from $y = 0$ to $y = 4$ (the lower half of the wall) is thus approximately

$$F_{\text{lower}} \approx \sum_{k=1}^{n} 62.4 \left(10 + \frac{5}{2}y_k^*\right)(8 - y_k^*)\Delta y.$$

As $n \to \infty$ and we accumulate slices from $y = 0$ to $y = 4$, this sum converges to a definite integral that represents the hydrostatic force F_{lower} on the bottom half of the wall:

$$F_{\text{lower}} = \int_0^4 62.4 \left(10 + \frac{5}{2}y\right)(8 - y)\,dy = 21,632 \text{ pounds}.$$

By an argument similar to the one used in the previous example, we can show that the hydrostatic force from $y = 4$ to $y = 8$ (the upper half of the side wall) is given by the definite integral

$$F_{\text{upper}} = \int_4^8 62.4(30)(8 - y)\,dy = 59,904 \text{ pounds}.$$

(We leave it to you to solve these integrals.) The total hydrostatic force exerted on a long side wall is the sum of these two forces:

$$F = F_{\text{lower}} + F_{\text{upper}} = 21,632 \text{ pounds} + 59,904 \text{ pounds} = 81,536 \text{ pounds}.$$

Notice that in this problem we never needed to use the fact that the pool was 10 feet wide! Surprisingly, the answer is the same whether the pool is 10 feet wide or 100 feet wide; the hydrostatic force exerted on the side wall does not depend on how far the pool extends from that wall.   □

EXAMPLE 6   Approximating the centroid of a region

Use three rectangles to approximate the centroid of the region between the graph of $f(x) = x^2\sqrt{3 - x}$ and the x-axis on $[0, 3]$.

SOLUTION

Subdividing $[0, 3]$ into three intervals and then defining rectangles whose heights match the function at the midpoints of those subintervals gives the following picture:

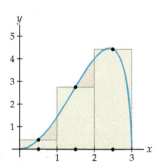

The centroids of the rectangles are at their centers, and, given the graph and equation for $f(x)$, these centroids are

$$(\tilde{x}_1, \tilde{y}_1) = \left(0.5, \frac{1}{2}f(0.5)\right) = \left(0.5, \frac{1}{2}((0.5)^2\sqrt{3 - 0.5})\right) \approx (0.5, 0.1976),$$

$$(\tilde{x}_2, \tilde{y}_2) = \left(1.5, \frac{1}{2}f(1.5)\right) = \left(1.5, \frac{1}{2}((1.5)^2\sqrt{3 - 1.5})\right) \approx (1.5, 1.3778),$$

$$(\tilde{x}_3, \tilde{y}_3) = \left(2.5, \frac{1}{2}f(2.5)\right) = \left(2.5, \frac{1}{2}((2.5)^2\sqrt{3 - 2.5})\right) \approx (2.5, 2.2097).$$

Note that these coordinates do seem reasonable compared with the points marked "x" in the preceding figure.

The three rectangles each have $\Delta x = 1$ and height defined by $f(x)$ at their midpoints $x = 0.5, 1.5,$ and 2.5, and therefore have the following areas:

$$A_1 = f(0.5)\,\Delta x = ((0.5)^2\sqrt{3-0.5})(1) \approx 0.3953,$$

$$A_2 = f(1.5)\,\Delta x = ((1.5)^2\sqrt{3-1.5})(1) \approx 2.7557,$$

$$A_3 = f(2.5)\,\Delta x = ((2.5)^2\sqrt{3-2.5})(1) \approx 4.4194.$$

Note again that these values seem reasonable compared with the three rectangles we graphed earlier. The area of the three rectangles together is naturally $A = A_1 + A_2 + A_3 \approx 7.5704$.

Finally, to find the coordinates of the centroid of the collection of three rectangles we must take the average of the three centroids we found, weighted by the areas of the corresponding rectangles. This gives us the following approximation for the centroid $(\bar{x}, \bar{y})$ of the region:

$$\bar{x} \approx \frac{\bar{x}_1 A_1 + \bar{x}_2 A_2 + \bar{x}_3 A_3}{A} \approx \frac{0.5(0.3953) + 1.5(2.7557) + 2.5(4.4194)}{7.5704} \approx 2.0316,$$

$$\bar{y} \approx \frac{\bar{y}_1 A_1 + \bar{y}_2 A_2 + \bar{y}_3 A_3}{A} \approx \frac{0.1976(0.3953) + 1.3778(2.7557) + 2.2097(4.4194)}{7.5704} \approx 1.8018.$$

Therefore the centroid of the region between $f(x) = x^2\sqrt{3-x}$ and the x-axis on $[0, 3]$ is approximately $(\bar{x}, \bar{y}) \approx (2.0316, 1.8018)$. You will find a finer approximation with $n = 6$ rectangles in Exercise 32 and find this centroid exactly in Exercise 36.   □

EXAMPLE 7   **Expressing a centroid in terms of Riemann sums and definite integrals**

Consider the region between the graphs of $f(x) = e^{1-x}$ and $g(x) = x^2$ on the interval $[0, 1]$. Express the centroid of this region in terms of (a) Riemann sums and (b) definite integrals.

SOLUTION

(a)   The region whose centroid we wish to calculate is shown next at the left. We can already visually judge that this centroid should be close to, say, $(0.3, 1)$. The figure at the right shows a representative rectangle at some point x_k^* in $[0, 1]$.

 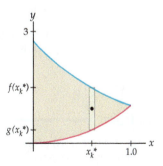

Note that the top of this rectangle is at height $f(x_k^*) = e^{1-x_k^*}$ and the bottom is at height $g(x_k^*) = (x_k^*)^2$. Thus the height of the marked point in the center of the rectangle is the average $\frac{f(x_k^*) + g(x_k^*)}{2}$, and therefore the centroid of the rectangle is

$$(\bar{x}_k, \bar{y}_k) = \left(x_k^*, \frac{f(x_k^*) + g(x_k^*)}{2}\right).$$

The height of the rectangle is the difference $f(x_k^*) - g(x_k^*)$, and therefore the area of the rectangle is

$$A_k = (f(x_k^*) - g(x_k^*))\Delta x.$$

We can now approximate the coordinates of the centroid of the region in terms of Riemann sums by taking a weighted average of the centroids of the rectangles:

$$\bar{x} \approx \frac{\sum_{k=1}^{n} x_k^* \left(f(x_k^*) - g(x_k^*)\right)\Delta x}{\sum_{k=1}^{n} \left(f(x_k^*) - g(x_k^*)\right)\Delta x},$$

$$\bar{y} \approx \frac{\sum_{k=1}^{n} \frac{f(x_k^*) + g(x_k^*)}{2} \left(f(x_k^*) - g(x_k^*)\right)\Delta x}{\sum_{k=1}^{n} \left(f(x_k^*) - g(x_k^*)\right)\Delta x} = \frac{\sum_{k=1}^{n} \frac{1}{2} \left(f(x_k^*)^2 - g(x_k^*)^2\right)\Delta x}{\sum_{k=1}^{n} \left(f(x_k^*) - g(x_k^*)\right)\Delta x}.$$

(b) Taking the limit as $n \to \infty$ and using the equations $f(x) = e^{1-x}$ and $g(x) = x^2$, we obtain an expression for the coordinates of the centroid of the region in terms of definite integrals:

$$\bar{x} = \frac{\int_0^1 x(f(x) - g(x))\, dx}{\int_0^1 (f(x) - g(x))\, dx} = \frac{\int_0^1 x(e^{1-x} - x^2)\, dx}{\int_0^1 (e^{1-x} - x^2)\, dx},$$

$$\bar{y} = \frac{\frac{1}{2}\int_0^1 (f(x)^2 - g(x)^2)\, dx}{\int_0^1 (f(x) - g(x))\, dx} = \frac{\frac{1}{2}\int_0^1 ((e^{1-x})^2 - (x^2)^2)\, dx}{\int_0^1 (e^{1-x} - x^2)\, dx}.$$

Calculating these definite integrals (some require preliminary algebra or integration by parts) gives us a centroid of $(\bar{x}, \bar{y}) \approx (0.338, 1.081)$, which does seem compatible with our initial visual approximation. □

? TEST YOUR UNDERSTANDING

▶ Would the work required to pump the water out of the top of a hot tub be the same as the work required to pump the water out of the bottom? Why or why not?

▶ Would the work required to pump the water out of a buried conical tank be the same as the work required to lift the entire conical tank up to ground level? Why or why not?

▶ Why did we have to consider two types of horizontal slices in Example 5?

▶ Where does the expression $|f(x)^2 - g(x)^2|$ come from in the formula for the centroid of a region between two curves in Theorem 9.15?

▶ Each example in this section followed the same general procedure. Can you explain this procedure in your own words?

EXERCISES 9.4

Thinking Back

Definite integrals: Calculate each of the following definite integrals, using integration techniques and the Fundamental Theorem of Calculus.

▶ $\int_0^{10} 4\pi \left(11.2 + 0.072x^2\right) dx$

▶ $\int_0^4 (62.4\pi)(3.5)^2(4 - y)\, dy$

▶ $\int_0^4 62.4 \left(10 + \frac{5}{2}y\right)(8 - y)\, dy$

▶ $\pi \int_0^{100} (0.75x + 250)(200 - x)\, dx$

▶ $\pi \int_0^6 (-2y + 1.12)\left(100 - \frac{25}{2}y\right) dy$

▶ *Setting up and solving a word problem involving arc length:* Calvin wants to hang some Christmas lights along the edge of the front side of his garage roof. The edge of the front side of the roof of his garage is a curve in the shape of a downwards-pointing parabola extending 3 feet above the ceiling of the garage and 12 feet across. How long a string of Christmas lights does Calvin need?

▶ *Setting up and solving a word problem involving volume:* A specialty glass flask has a bulb-shaped bottom half and a cylindrical top half. The bulb consists of a sphere with a radius of 10 centimeters whose top and bottom are truncated by 2 centimeters. The cylindrical tube has a radius just large enough to connect to the bulb and is 7 centimeters tall. Find the volume of the flask.

Concepts

0. *Problem Zero:* Read the section and make your own summary of the material.

1. *True/False:* Determine whether each of the statements that follow is true or false. If a statement is true, explain why. If a statement is false, provide a counterexample.

   (a) *True or False:* Mass and weight are the same.

   (b) *True or False:* A large object can have less mass than a smaller object.

   (c) *True or False:* An object can have one mass on Earth and another mass on the moon.

   (d) *True or False:* An object can have one weight on Earth and another weight on the moon.

   (e) *True or False:* Water has a weight-density of 62.4 foot-pounds.

   (f) *True or False:* It always takes more work to lift a heavy object than a light object for the same distance.

   (g) *True or False:* Hydrostatic force can be measured only on horizontal walls.

   (h) *True or False:* Hydrostatic force is stronger at greater depths than it is at lesser depths.

2. *Examples:* Construct examples of the thing(s) described in the following. Try to find examples that are different than any in the reading.

   (a) An object whose density is constant throughout, and another whose density varies.

   (b) A situation where work can be calculated without an integral, and another situation where a definite integral is needed to calculate work.

   (c) An example of something that is measured in foot-pounds.

3. What is the difference between mass and weight?

4. What is the difference between density and weight-density?

5. Why is a definite integral needed to calculate the hydrostatic force exerted on a vertical wall of a swimming pool, but no integral is needed to calculate the hydrostatic force on the bottom of the pool?

6. Describe the "subdivide, approximate, and add" strategy that is used in each of the examples in this section. Why is it necessary? How does it work? What do Riemann sums and definite integrals have to do with it?

7. Explain how the values of Δx and x_k (or Δy and y_k) in a Riemann sum determine the limits of integration on the definite integral obtained by taking the limit of the Riemann sum as $n \to \infty$.

8. What do we mean when we say that a sum is a Riemann sum? What has to be true about a Riemann sum for us to be able to guarantee that it converges to a definite integral when we take the limit as $n \to \infty$?

9. Why is the task of pumping all of the water out of the top of a hot tub different from the task of lifting the entire hot tub 4 feet in the air (even assuming that the hot tub itself doesn't weigh anything)? Which task requires more work?

10. Use the information in the description of the problem in Example 3 to show that the height of the gelatin mold is given by the function $y = 4.5 - 0.5x^2$ and that the density of the gelatin mold at height y is given by the linear function $\rho(y) = 1.3 - 0.233y$.

11. If the approximating sum in Example 3 runs from $k = 1$ to $k = n$, why does the definite integral have 0 and 4.5 as its limits of integration?

12. Make a list of all of the physics formulas introduced in this section. Give an example of an application of each formula, together with appropriate units.

13. Strangely, the hydrostatic force exerted on the side wall of a pool does not depend on how much water is behind that wall, but only on the depth of the water. Explain why this makes sense.

14. Mass is the product of density and volume. When we integrate to find accumulated mass, we integrate the product of density and cross-sectional area. Express the previous two sentences in mathematical notation. Then explain why volume is used in one, but area in the other.

15. Work is the product of weight-density, volume, and distance. When we integrate to find accumulated work, we integrate the product of weight-density, cross-sectional area, and distance. Express the previous two sentences in mathematical notation. Then explain why volume is used in one, but area in the other.

16. Hydrostatic force is the product of weight-density, area, and depth. When we integrate to find accumulated hydrostatic force, we integrate the product of weight-density, length, and depth. Express the previous two sentences in mathematical notation. Then explain why area is used in one, but length in the other.

17. Consider a region in the plane divided into disjoint rectangles with areas $A_1, A_2, \ldots, A_n$ and individual centroids $(\bar{x}_1, \bar{y}_1), (\bar{x}_2, \bar{y}_2), \ldots, (\bar{x}_n, \bar{y}_n)$. The centroid of this region is

$$(\bar{x}, \bar{y}) = \left(\frac{\sum_{k=1}^{n} x_k A_k}{\sum_{k=1}^{n} A_k}, \frac{\sum_{k=1}^{n} y_k A_k}{\sum_{k=1}^{n} A_k} \right).$$

Explain what it means to say that this is a "weighted" average of the coordinates of the centroids of the rectangles.

18. In Theorem 9.14 there are four definite integrals in the expression for the centroid of a region under a curve. Explain where each of these definite integrals comes from in the construction of this centroid by Riemann sums. Then give a similar explanation for the four definite integrals in the expression for the centroid of a region between two curves given in Theorem 9.15. In this second theorem, why are there absolute values?

Skills

Use physics formulas to calculate each quantity in Exercises 19–24.

19. The mass of an object with density 12 grams per cubic centimeter in the shape of a cylinder with radius 4 centimeters and height 10 centimeters.

20. The mass of an object with density 600 grams per cubic centimeter in the shape of a cone with radius 0.5 meters and height 1 meter.

21. The work required to lift an object that weighs 20 pounds to a height of 100 feet.

22. The work required to lift an object 2 feet into the air, given that the object has a weight-density of 50 pounds per cubic foot and that the object is a cylinder with radius 2 inches and height 6 inches.

23. The hydrostatic force exerted by water on the bottom of a full, 2-foot deep rectangular tank with a 3×4 foot base.

24. The hydrostatic force exerted by a liquid of weight-density 70 pounds per cubic foot on the bottom of a cylindrical tank with height 8 feet and radius 2 feet.

Use four slices to construct an approximation of each of the quantities described in Exercises 25–30. In each case include a labeled diagram and an explicit list of the values of Δx and each x_k and x_k^* (or Δy and each y_k and y_k^*).

25. The mass of a cylindrical rod with radius of 2 centimeters and length of 24 centimeters whose density at a point x centimeters from the left end is $\rho(x) = 10.5 + 0.01527x^2$ grams per cubic centimeter.

26. The mass of the solid of revolution obtained by rotating the graph of $y = 4.5 - 0.5x^2$ on $[0, 3]$ around the y-axis and whose density at height y is $\rho(y) = 1.3 - 0.233y$ ounces per cubic inch.

27. The work required to pump all of the water out of a cone with top radius 3.5 feet and height 4 feet.

28. The work required to pump all of the water out of a cylinder with radius 3.5 feet and depth 4 feet.

29. The hydrostatic force exerted by water on one of the sides of a small cubical tank that is 4 feet to a side.

30. The hydrostatic force exerted by water on one of the shorter side walls of a rectangular container that is 3 feet long, 4 feet wide, and 2 feet deep.

In Exercises 31–34, use a weighted average over n rectangles to approximate the centroid of the region described. (*Hint: It may help to draw a picture.*)

31. The region between $f(x) = \sqrt{x}$ and the x-axis on $[a, b] = [1, 9]$, with $n = 2$.

32. The region between $f(x) = x^2\sqrt{3 - x}$ and the x-axis on $[a, b] = [0, 3]$, with $n = 6$.

33. The region between $f(x) = x^3$ and the line $y = 8$ on $[a, b] = [0, 2]$, with $n = 4$.

34. The region between $f(x) = x^2$ and $g(x) = 64 - x^2$ on $[a, b] = [0, 8]$, with $n = 4$.

In Exercises 35–40, use definite integrals to calculate the centroid of the region described. Use graphs to verify that your answers are reasonable.

35. The region between $f(x) = \sqrt{x}$ and the x-axis on $[a, b] = [1, 9]$. (Compare with Exercise 31.)

36. The region between $f(x) = x^2\sqrt{3 - x}$ and the x-axis on $[a, b] = [0, 3]$. (Compare with Exercise 32.)

37. The region between $f(x) = x^3$ and the line $y = 8$ on $[a, b] = [0, 2]$. (Compare with Exercise 33.)

38. The region between $f(x) = \ln x$ and $g(x) = 2 - \ln x$ on $[a, b] = [1, e^2]$.

39. The region between $f(x) = \sin x$ and the line $y = \frac{1}{2}$ on $[a, b] = [0, \pi]$.

40. The region between $f(x) = x^2$ and $g(x) = 64 - x^2$ on $[a, b] = [0, 8]$. (Compare with Exercise 34.)

Applications

Find each mass described in Exercises 41–48.

41. The mass of a block of copper that has a square base with sides measuring 20 centimeters and a height of 13 centimeters. Copper has a density of 8.93 grams per cubic centimeter.

42. The mass of a gold walking stick in the shape of a tall, thin cylinder that is 4 feet high and 3 inches wide. Gold has a density of 19.3 grams per cubic centimeter. (*Caution: Watch your units! You will need to use the conversion 2.54 cm = 1 inch.*)

43. The mass of a 12-inch rod with a square cross section of side length 1.5 inches, with density x inches from the left end given by $\rho(x) = 4.2 + 0.4x - 0.03x^2$ grams per cubic inch.

44. The mass of a cylindrical rod with a radius of 3 centimeters and a length of 20 centimeters, made of a combination of copper and aluminum in such a way that the density of the rod x centimeters from the left end is $\rho(x) = 8.93 - 0.015x$ grams per cubic centimeter.

45. The mass of a 10-centimeter rod with square cross section of side length 1 centimeter, made of a combination of metals in such a way that the density of the rod x centimeters from the left end is $\rho(x) = 15 - 0.5x$ grams per cubic centimeter.

46. The mass of a rectangular block of metal that is 30 centimeters wide, 21 centimeters long, and 14 centimeters tall, given that its density y centimeters from its base is $\rho(y) = 6.8 + 0.14y$ grams per cubic centimeter.

47. Last week Jack made an upside-down pudding cake that was 8 inches tall and 10 inches across, in the shape of the top of a downwards-pointing parabola that has been revolved around the y-axis. The raisins he put in the cake have tended to drift towards the bottom of the cake over the past week (nobody seems to want to eat this cake!) in such a way that the density of the cake at the very top of the mold is 0.15 ounce per cubic inch, while the density at the very bottom of the cake is 1.12 ounces per cubic inch.

(a) Describe Jack's upside-down pudding cake as a solid of revolution, by identifying the region to be revolved and the axis of revolution.

(b) Describe the density of Jack's upside-down pudding cake as a linear function of the height y of a point in the cake.

(c) Find the weight of Jack's upside-down pudding cake.

48. A solid cone of metal with a radius of 10 centimeters and a height of 15 centimeters is made of a combination of metals in such a way that, if the cone is pictured with its point up, then the density of the metal in the cone increases by 0.35 gram per cubic centimeter for each centimeter of height.

(a) Describe the metal cone as a solid of revolution, by identifying the region to be revolved and the axis of revolution.

(b) Given that the density at the bottom of the cone is 7.25 grams per cubic centimeter, find a function $\rho(y)$ that describes the density of the cone y centimeters above its base.

(c) Using the density function you found in part (b), find the total mass of the cone.

Find the work described in each of Exercises 49–60.

49. The work required to lift a full cylindrical soda fountain cup 1 foot into the air, given that the cup is full and has a radius of 3 inches and a height of 6 inches. You may assume that soda has the same weight-density as water. *(Caution: Watch your units!)*

50. The work required to lift a bathtub full of water 30 feet into the air, given that the bathtub itself weighs 150 pounds and it holds a rectangular volume of water measuring 5 feet by 2 feet by 2 feet.

51. The work required to pump all of the water out of the top of a giant rectangular hot tub that is 5 feet wide, 8 feet long, and 3 feet deep.

52. The work required to pump all of the water out of the top of a spherical tank with a radius of 27 feet.

53. The work required to pump all of the water out of the top of an upright conical tank that is 10 feet high and has a radius of 4 feet at the top.

54. The work required to pump all of the water out of the top of an upright conical tank that is 12 feet high and has a radius of 6 feet at the top.

55. The work required to pump all the water out of the top of a tank and up to ground level, given that the tank is a large upright cylinder with a radius of 4 feet and a height of 13 feet, buried so that its top is 3 feet below the surface.

56. The work required to pump all the liquid out of the top of a tank and up to 4 feet above ground level, given that the tank is a giant upright cylinder with a radius of 8 feet and a height of 20 feet, buried so that its top is 2 feet below the surface. The tank is only two-thirds full and contains a strange liquid that weighs 71.8 pounds per cubic foot.

57. The work required to pump all of the liquid out of the top of a cylindrical tank on a tanker truck (so that the cylinder is on its side) with a length of 100 feet and a radius of 8 feet. The tank is full of a liquid that weighs 42.3 pounds per cubic foot. Be careful; in this problem the slices are not cylinders!

58. Buffy has tied a rope to a 50-pound sword so that she can hoist it up to her upstairs bedroom window, 23 feet off the ground. How much work will it take for Buffy to lift the sword up to her window, assuming that the weight of the rope is negligible?

59. Willow has tied a rope to a 20-pound mallet so that she can hoist it up to her upstairs bedroom window, 18 feet off the ground. How much work will it take for Willow to lift the mallet up to her window, assuming that the rope weighs 0.25 pound per foot?

60. Xander has tied a rope to a 100-pound block of ice so that he can hoist it up to his upstairs bedroom window, 22 feet off the ground. How much work will it take for Xander to lift the block of ice up to his window, assuming that the weight of the rope is negligible and the block of ice melts off 2 pounds of water for each foot it is lifted?

Find the hydrostatic force described in each of Exercises 61–69.

61. The hydrostatic force on the bottom of a full cylindrical water glass with a radius of 2 inches and a height of 7 inches.

62. The hydrostatic force on the bottom of a full cylindrical beer stein with a radius of 6 centimeters and a height of 20 centimeters. You may assume that the weight-density of beer is the same as for water.

63. The hydrostatic force exerted on one of the long sides of a giant rectangular hot tub that is 5 feet wide, 8 feet long, and 3 feet deep.

64. The hydrostatic force exerted on one of the long sides of a rectangular swimming pool that is 8 feet long, 6 feet wide, and 4.5 feet deep.

65. The hydrostatic force exerted by the water in a birdbath on its cylindrical side, given that the birdbath is in the shape of a cylinder that is 1 foot deep with a radius of 0.4 foot.

66. The hydrostatic force exerted on a dam in the shape of an isosceles triangle whose top is 200 feet wide and whose total height is 100 feet, given that the dam is holding back a body of water that reaches to the top of the dam.

67. The hydrostatic force exerted on a dam in the shape of a trapezoid whose top is 400 feet long and whose base is 250 feet long, given that the dam is holding back a body of water that reaches all 200 feet from the bottom to the top of the dam.

68. The hydrostatic force exerted on one of the long sides of a swimming pool that is 24 feet long and 10 feet wide, with a shallow end that has a depth of 3 feet (for the first 4 feet of the pool) and a deep end that has a depth of 15 feet (for the last 8 feet of the pool). The middle portion of the bottom of the pool is a straight ramp leading down from the shallow depth to the depth at the deep end.

69. The hydrostatic force exerted on one of the long sides of a pool that is 20 feet long and 8 feet wide, with a shallow end that has a depth of 4 feet (for the first 6 feet of the pool) and a deep end that has a depth of 12 feet (for the last 6 feet of the pool), and with a middle portion that is a straight ramp.

Use centroids to solve the real-world problems in Exercises 70–71.

70. Ian is making some snow flukes, which are metal sheets that can be used to prevent a fall on steep snow. The flukes are in the shape of a polygon with vertices $(0, 8)$, $(6, 8)$, $(6, 2)$, $(3, 0.5)$, and $(0, 8)$ with all distances given in inches. Ian needs to clip a cable to the centroid of the fluke. Where should he attach the cable?

71. Leila needs to find a location for a radio repeater that will monitor the signals from a pack of radio-collared wolves. In order to get the strongest signal, she wants to locate the repeater at the centroid of the range of the wolf pack. The valley in which the wolves live encompasses roughly the union of the rectangle $0 \leq x \leq 8, 0 \leq y \leq 4$ and the part of the ellipse

$$\frac{(x - 6)^2}{4} + \frac{(y - 2)^2}{100} = 1$$

that lies above the x-axis, where all distances are in miles. Where should Leila put the repeater?

Proofs

72. Use the physics formulas given in this text to prove that the work required to lift a cube of liquid with weight-density ω and side length s through a distance of d feet is $W = \omega s^2 d$.

73. Suppose an object has horizontal cross sections for $x \in [a, b]$, so that the functions for density $\rho(x)$ and area $A(x)$ are continuous. Use physics formulas and the definition of the definite integral to prove that the accumulated mass of such an object is

$$m = \int_a^b \rho(x) A(x) \, dx.$$

74. Suppose an object whose particles are to be displaced has vertical cross sections for $y \in [a, b]$, so that the functions for weight-density $\omega(y)$, area $A(y)$, and displacement $d(y)$ are continuous. Use physics formulas and the definition of the definite integral to prove that the accumulated work required is

$$W = \int_a^b \omega(y) A(y) d(y) \, dy.$$

75. Suppose the wall of a container has vertical cross sections for $y \in [a, b]$, so that the functions for weight-density $\rho(y)$, length $l(y)$, and depth $d(x)$ are continuous.

Use physics formulas and the definition of the definite integral to prove that the accumulated hydrostatic force on such a wall is

$$F = \int_a^b \omega(y) \, l(y) \, d(y) \, dy.$$

76. Use a Riemann sum argument to prove that the centroid $(\bar{x}, \bar{y})$ of the region between two curves $f(x)$ and $g(x)$ on an interval $[a, b]$ has coordinates

$$\bar{x} = \frac{\int_a^b x |f(x) - g(x)| \, dx}{\int_a^b |f(x) - g(x)| \, dx},$$

$$\bar{y} = \frac{\frac{1}{2} \int_a^b |f(x)^2 - g(x)^2| \, dx}{\int_a^b |f(x) - g(x)| \, dx}.$$

As part of your argument you will need to explain the reason for the four definite integrals in these expressions, as well as the absolute value signs that appear in the expression. Your argument should also contain the words "rectangle," "weighted average," and "limit."

Thinking Forward

▶ *The volume of a torus:* A **torus** is a donut-shaped object that is obtained by rotating a circle around an axis. For example, we could revolve the circle with equation $y^2 + (x - 3)^2 = 1$ (i.e., the circle with radius 1 and center at $(0, 3)$) around the y-axis to obtain a torus. Use a "subdivide, approximate, and add" strategy to find the volume of this torus. (*Hint: The functions $\sqrt{1 - y^2} + 3$ and $-\sqrt{1 - y^2} + 3$ will be involved; why? You can use disks/washers or shells to find this volume.*)

▶ *Pappus's Centroid Theorem:* The volume of a solid obtained by revolving a region in the plane around an axis that does not intersect that region is equal to $V = 2\pi r A$, where r is the distance from the centroid of the region to the axis and A is the area of the region. Use this theorem to find the volume of the torus described in the previous problem without having to set up any definite integrals. Then discuss why the theorem makes intuitive sense in terms of an alternative "subdivide, approximate, and add" strategy for finding volume in this case.

9.5   DIFFERENTIAL EQUATIONS*

▶ Using antidifferentiation and separation of variables to solve differential equations

▶ Using slope fields and Euler's method to approximate solutions of differential equations

▶ Common forms of initial-value problems and their applications

Differential Equations and Initial-Value Problems

Recall that an **equation** is nothing more than two mathematical expressions connected by an equals sign. For example, $x^2 - 4 = 0$ is a single-variable equation, and its set of **solutions** is the set the values of x that make the equation true, namely, $x = 2$ and $x = -2$. Similarly, $y = x^2 - 4$ is a two-variable equation, and its set of solutions is the set of ordered pairs (x, y) that make the equation true, namely, the ordered pairs of the form $(x, y) = (x, x^2 - 4)$ that, when graphed together, make up the parabola that is the graph of $y = x^2 - 4$.

If an equation involves the derivative $\frac{dy}{dx}$ of some function $y(x)$, then we say that the equation is a **differential equation**. For example, two examples of differential equations are

$$\frac{dy}{dx} = x^2 \quad \text{and} \quad \frac{dy}{dx} = y.$$

Once again, the solutions of these equations are the objects that make the equations true. For example, the solutions of $\frac{dy}{dx} = x^2$ are the functions $y(x)$ whose derivatives equal x^2, namely, the functions of the form $y(x) = \frac{1}{3}x^3 + C$. The solutions of $\frac{dy}{dx} = y$ are the functions $y(x)$ that are equal to their own derivatives, namely, the functions $y(x) = Ae^x$. In general, a **solution** of a differential equation is a function that makes the equation true.

A **first-order** differential equation is an equation that involves the first derivative $\frac{dy}{dx}$, a function y, and independent variable x. Higher order differential equations involve higher order derivatives. Many of the first-order differential equations we will study are of the form $\frac{dy}{dx} = g(x, y)$, where $g(x, y)$ is some function of the variables x and y. The simplest case occurs when g involves only the variable x. Such a differential equation can be solved by antidifferentiation:

THEOREM 9.16   **Differential Equations Whose Solutions Are Antiderivatives**

The solution of a differential equation of the form $\frac{dy}{dx} = g(x)$ is the family $y(x) = G(x) + C$ of antiderivatives of $g(x)$.

Proof. If $G(x)$ is an antiderivative of $g(x)$, then $G'(x) = g(x)$, which means that $y(x) = G(x)$ is a solution of the differential equation $\frac{dy}{dx} = g(x)$. Moreover, we know that two functions of x have the same derivative precisely when they differ by a constant. This means that *every* solution of the differential equation $\frac{dy}{dx} = f(x)$ must be of the form $y(x) = G(x) + C$. ∎

A differential equation of the form $\frac{dy}{dx} = g(x, y)$ with initial condition $y(x_0) = y_0$ is called an **initial-value problem**. Provided that various continuity, differentiability, and boundedness conditions are satisfied, an initial-value problem will have one and only one solution $y(x)$. The details can be found in any standard differential equations textbook. For our purposes it is sufficient to say that when we consider an initial-value problem in this book, it will be one for which these conditions are met and it will have one and only one solution.

For example, the differential equation $\frac{dy}{dx} = x^2$ has many solutions, all of the form $y(x) = \frac{1}{3}x^3 + C$. Only one of these solutions also satisfies the initial condition $y(0) = 2$, namely, the function $y(x) = \frac{1}{3}x^3 + 2$. Therefore $y(x) = \frac{1}{3}x^3 + 2$ is the unique solution of the initial-value problem $\frac{dy}{dx} = x^2$ and $y(0) = 2$. This solution is shown next at the left, highlighted amongst other solutions of the differential equation $\frac{dy}{dx} = x^2$. Similarly, the function $y(x) = 2e^x$ is the unique solution of the initial-value problem $\frac{dy}{dx} = y$ and $y(0) = 2$, as illustrated at the right.

$y(x) = \frac{1}{3}x^3 + 2$ *is the only solution*

of $\frac{dy}{dx} = x^2$ *that satisfies* $y(0) = 2$

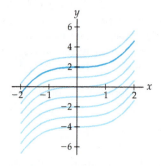

$y(x) = 2e^x$ *is the only solution*

of $\frac{dy}{dx} = y$ *that satisfies* $y(0) = 2$

Separable Differential Equations

We have seen that differential equations of the form $\frac{dy}{dx} = g(x)$ can be solved if we can antidifferentiate the function $g(x)$. How do we solve differential equations that involve the independent variable y? The answer is that sometimes we can and sometimes we cannot. One particular technique is useful for differential equations for which the y-related information can be separated from the x-related information. Note that in this section we will be assuming that the functions $y(x)$ we work with are continuous on their domains.

Let's consider the simple example $\frac{dy}{dx} = y$. Our strategy will be to move all the y-related parts of the differential equation to the side with the derivative $\frac{dy}{dx}$ and then integrate both sides:

$$\frac{dy}{dx} = y \quad \Longrightarrow \quad \frac{1}{y}\frac{dy}{dx} = 1 \quad \Longrightarrow \quad \int \frac{1}{y}\frac{dy}{dx}\,dx = \int 1\,dx \quad \Longrightarrow \quad \int \frac{1}{y}\,dy = \int 1\,dx.$$

The last step followed from integration by substitution, with $y(x)$ playing the role of what we would normally call $u(x)$ when substituting. Notice that Leibniz notation makes it easy for us to remember this substitution, since it has the effect of "cancelling the dx" in the notation.

After considering the work we just did, we now have a way to move towards finding the solutions $y(x)$ to the differential equation: We simply perform the integration and solve for $y(x)$. This gives us

$$\ln|y| + C_1 = x + C_2 \quad \Longrightarrow \quad \ln|y| = x + C \quad \Longrightarrow \quad |y| = e^{x+C} = e^C e^x \quad \Longrightarrow \quad y = Ae^x.$$

Along the way we just defined C_1 and C_2 as arbitrary constants, with $C = C_2 - C_1$ as their difference and $A = \pm e^C$. All that matters by the end of this calculation is that A is some constant. If we knew a particular initial value for $y(x)$, then we could solve uniquely for this constant A.

For a more complicated differential equation like $\frac{dy}{dx} = y + xy$, we can use a similar technique, rearranging the equation so that all the y-related expressions are on the left side with a multiplier of $\frac{dy}{dx}$ and all the x-related expressions are on the right, as follows:

$$\frac{dy}{dx} = y + xy \quad \Longrightarrow \quad \frac{dy}{dx} = y(1+x) \quad \Longrightarrow \quad \frac{1}{y}\frac{dy}{dx} = 1 + x.$$

We could then integrate both sides and solve for y to find the family of solutions of the differential equation; see Example 1.

This technique is called *separation of variables*, because the initial step involves separating the y-related and x-related parts of the differential equation. We say that a differential equation $\frac{dy}{dx} = g(x, y)$ is *separable* if we can write it in the form $\frac{dy}{dx} = p(x)q(y)$ for some functions $p(x)$ and $q(x)$. With such a separable differentiable equation we have

$$\frac{1}{q(y)}\frac{dy}{dx} = p(x).$$

Integrating both sides with respect to x and applying u-substitution with $u = y$ and $du = dy = y'(x)\,dx = \frac{dy}{dx}\,dx$, we have the following theorem:

THEOREM 9.17

Solving a Differential Equation by Separation of Variables

Given a separable differential equation $\frac{dy}{dx} = p(x)q(y)$, it follows that

$$\int \frac{1}{q(y)}\,dy = \int p(x)\,dx.$$

Solving both of these integrals and then solving for $y(x)$, if possible, yields a solution of the differential equation.

Of course, in some cases one or both integration steps will be impossible, or perhaps after integrating we might be unable to solve for the function $y(x)$. But if everything works out, then this method gives us a nice way to solve more complicated differential equations with our antidifferentiation skills.

Slope Fields

Every differential equation $\frac{dy}{dx} = g(x, y)$ is at its heart a statement about the slopes of some function $y(x)$ at various points (x, y) in the plane. For example, a solution of the differential equation $\frac{dy}{dx} = x^2$ is a function $y(x)$ whose slope at each of its graphed points (a, b) is equal to a^2. If we draw small line segments at each point (a, b) in the plane, each with slope a^2, we get the *slope field* shown next at the left. The solutions of $\frac{dy}{dx} = x^2$ are all of the form $y(x) = \frac{1}{3}x^3 + C$, and each of these functions flows along with the small line segments in the slope field. Similarly, a slope field for the differential equation $\frac{dy}{dx} = y$ consists of line segments whose slope at (a, b) is equal to b, as shown at the right. The curves that flow through this slope field are the solutions $y(x) = Ae^{kx}$.

Slope at (a, b) is equal to a^2

Slope at (a, b) is equal to b

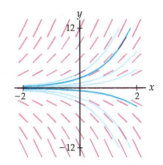

In the first slope field, the slopes are the same across each column because the differential equation $\frac{dy}{dx} = x^2$ does not involve the variable y. Similarly, in the second slope field the slopes are the same in each row because $\frac{dy}{dx} = y$ does not involve the variable x. In Example 2 you will construct a slope field that depends on both x and y. In general, we can sketch a slope field for any differential equation:

DEFINITION 9.18

The Slope Field of a Differential Equation

Given a differential equation $\frac{dy}{dx} = g(x, y)$ and a lattice of points in the plane, the associated **slope field** is a collection of line segments drawn at each point (x, y) in the lattice, with the property that the slope of the segment drawn at (x, y) is equal to $g(x, y)$.

Euler's Method

The method of tracing through a slope field to approximate a solution of a differential equation can be put into algebraic form with the technique known as **Euler's method**. Consider for example the initial-value problem consisting of the differential equation $\frac{dy}{dx} = y$ and the initial value $y(0) = 2$. We will use slope field information to construct a sequence of points going forward from this initial condition that approximate values of the solution of the given differential equation.

By construction, the solution $y(x)$ we would like to approximate must pass through the initial point $(x_0, y_0) = (0, 2)$. We also know from the differential equation $\frac{dy}{dx} = y$ that the slope of the function $y(x)$ at this point is given by $\frac{dy}{dx}\big|_{(0,2)} = 2$. Near the point $(0, 2)$ we would expect the function $y(x)$ to behave a lot like its tangent line, so we will follow this slope for a step of $\Delta x = 0.5$ to the right, as illustrated here:

Getting from (x_0, y_0) to (x_1, y_1)

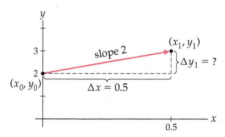

So what are the coordinates of our new points (x_1, y_1)? Clearly we have $x_1 = x_0 + \Delta x = 0 + 0.5 = 0.5$. To find the new y-coordinate y_1 we must find the change Δy_1 between y_0 and y_1. Using the slope information, we have

$$2 = \frac{\Delta y_1}{0.5} \quad \Longrightarrow \quad \Delta y_1 = 1.$$

Since $y_0 = 2$, it follows that $y_1 = y_0 + \Delta y_1 = 2 + 1 = 3$.

We have just followed the slope $\frac{dy}{dx}\big|_{(x_0, y_0)}$ from our initial point (x_0, y_0) to a new point (x_1, y_1). This new point will in general not be on the actual graph of the solution $y(x)$ to the differential equation, but it will be a reasonable approximation to a point on the graph of the solution. We can repeat the process, following the slope $\frac{dy}{dx}\big|_{(x_1, y_1)}$ to get from the point (x_1, y_1) to another new point (x_2, y_2), and so on. This produces a sequence of points that we can connect to form an approximation of the solution function $y(x)$. In the next figure at the left is the result of continuing the process with $\Delta x = 0.5$ for four steps. At the right is the result of following the same process with $\Delta x = 0.25$ for eight steps; see Example 3 for the details.

Euler's method with $\Delta x = 0.5$

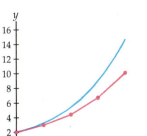

Euler's method with $\Delta x = 0.25$

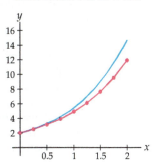

In each of these figures the Euler's method approximation is shown, compared with the actual solution $y(x) = 2e^x$ of the initial-value problem $\frac{dy}{dx} = y$, $y(0) = 2$. Notice that the approximation is better when we use a smaller value of Δx and therefore update our slope information more frequently. Notice also that Euler's method approximations start out close to the solution graph but then tend to drift away as errors compound in each step of the process. In general, Euler's method provides us with a useful way of approximating solutions of initial-value problems that are too complicated to solve algebraically:

DEFINITION 9.19

Euler's Method of Approximating the Solution of an Initial-Value Problem

Given an initial-value problem with differential equation $\frac{dy}{dx} = g(x, y)$ and initial condition $y(x_0) = y_0$, and given a value $\Delta x > 0$, define a sequence of points (x_k, y_k) by the iterative formula

$$(x_{k+1}, y_{k+1}) = (x_k + \Delta x, y_k + \Delta y_k), \quad \text{where } \Delta y_k = g(x_k, y_k) \, \Delta x.$$

This sequence of points forms an **Euler's method** approximation to the solution $y(x)$ of the initial-value problem.

Applications of Initial-Value Problems

Many real-world situations can be modeled by initial-value problems. When we have information about the rate of change and initial value of a quantity, we can often set up a model of an initial-value problem whose solution will be the function that describes that quantity. We will describe three such models next, each of which results in a differential equation that can be solved exactly by the technique of separation of variables.

One of the simplest situations is when a quantity $Q(t)$ grows at a rate proportional to itself. This happens often with simple growth and decay models, as you will see in Exercises 65–67.

THEOREM 9.20

Exponential Growth and Decay Models

If a quantity $Q(t)$ changes over time at a rate proportional to its value, then that quantity is modeled by a differential equation of the form

$$\frac{dQ}{dt} = kQ$$

for some proportionality constant k, where $k > 0$ represents growth and $k < 0$ represents decay. The solution of this differential equation that satisfies the initial value $Q(0) = Q_0$ is

$$Q(t) = Q_0 e^{kt}.$$

The proof of this theorem requires that we solve the differential equation $\frac{dQ}{dt} = kQ$ with initial condition $Q(0) = Q_0$. The work is similar to what we did in our motivating example for the technique of separation of variables, so the details of the proof are left to Exercise 75.

Of course, real-life situations are often more complicated. For example, it makes sense to imagine that a population of animals on an island might grow at a rate proportional to the number of animals on the island; when there are more animals on the island, there are also more animals around to make offspring. However, at some point there may be too many animals on the island to be supported by the available food and shelter. In this case we say that the island has a **carrying capacity** of some maximum number L of animals. The following **logistic model** describes such a situation, where for small population values the growth rate is approximately exponential with some **natural growth rate** k, but for population values near the carrying capacity L the growth rate levels off to zero:

THEOREM 9.21

Logistic Growth Model

If a population $P(t)$ has natural growth rate k and is restricted by a carrying capacity L, then it can be modeled by the differential equation

$$\frac{dP}{dt} = kP\left(1 - \frac{P}{L}\right).$$

The solution of this differential equation that satisfies the initial value $P(0) = P_0$ is

$$P(t) = \frac{LP_0}{P_0 + (L - P_0)e^{-kt}}.$$

To prove this theorem we must solve the initial-value problem that represents logistic growth and show that we obtain the given solution. This can be done by separation of variables followed by partial fractions, either in general (see Exercise 76) or on a case-by-case basis (see Exercises 68 and 69).

We can also use differential equations to model the changing temperature of an object. If a hot or cold object is in an environment with a constant ambient temperature, then the object will cool down or heat up until its temperature matches that of its environment. In an ideal situation this happens in such a way that the rate of change of the temperature of the object is proportional to the difference of the ambient temperature and the current temperature of the object. In this situation we have the following differential equation, attributed to Newton:

THEOREM 9.22

Newton's Law of Cooling and Heating

If an object with temperature $T(t)$ is placed in a location with a constant ambient temperature A, then the temperature of the object can be modeled by a differential equation of the form

$$\frac{dT}{dt} = k(A - T)$$

for some proportionality constant k. The solution of this differential equation that satisfies the initial value $T(0) = T_0$ is

$$T(t) = A - (A - T_0)e^{-kt}.$$

It is not difficult to solve this initial-value problem by the technique of separation of variables; you will work out the details of the proof in general in Exercise 78 and in specific cases in Exercises 70 and 71. Note that in practice we might encounter a situation where we know the temperature of the object not at time $t = 0$, but at some other point in time. In those situations the equation for $T(t)$ in Theorem 9.22 will involve different constants, which are not difficult to solve for after performing separation of variables to solve the differential equation; see Example 4.

Examples and Explorations

EXAMPLE 1 Using separation of variables to solve initial-value problems

Solve each of the following initial-value problems:

(a) $\frac{dy}{dx} = \sin x, \ \ y(0) = 2$ **(b)** $\frac{dy}{dx} = y^2, \ \ y(0) = 3$ **(c)** $\frac{dy}{dx} = y + xy, \ \ y(0) = 2$

SOLUTION

(a) The differential equation $\frac{dy}{dx} = \sin x$ does not involve the dependent variable at all, so technically it is already separated. By antidifferentiating, the solutions of this differential equation are of the form

$$y(x) = -\cos x + C.$$

Using the initial condition $y(0) = 2$, we get

$$2 = -\cos 0 + C \quad \Longrightarrow \quad 2 = -1 + C \quad \Longrightarrow \quad C = 3.$$

Therefore the solution of the initial-value problem is the function $y(x) = -\cos x + 3$.

(b) We can write this differential equation in separable form as $\frac{dy}{dx} = (1)(y^2)$; in the notation of Theorem 9.17 we have $p(x) = 1$ and $q(x) = y^2$. Dividing both sides by the y^2 expression and integrating both sides, we obtain

$$\frac{dy}{dx} = y^2 \quad \Longrightarrow \quad \frac{1}{y^2}\frac{dy}{dx} = 1 \quad \Longrightarrow \quad \int y^{-2}\,dy = \int 1\,dx \quad \Longrightarrow \quad -y^{-1} = x + C.$$

Notice that we did not add a constant after integrating the left side of the equation; this is because any constant on that side can be absorbed into the constant on the right side. All that remains now is to solve for y:

$$-\frac{1}{y} = x + C \quad \Longrightarrow \quad y = -\frac{1}{x + C}.$$

We can use the initial condition $y(0) = 3$ to solve for C:

$$y(0) = 3 \quad \Longrightarrow \quad 3 = -\frac{1}{0 + C} \quad \Longrightarrow \quad C = -\frac{1}{3}.$$

Therefore the solution of the initial-value problem is $y(x) = -\dfrac{1}{x - (1/3)}$.

(c) The differential equation $\frac{dy}{dx} = y + xy$ involves both x and y. As we saw earlier in the reading, we can separate the variables and then integrate:

$$\frac{dy}{dx} = y + xy \quad \Longrightarrow \quad \frac{dy}{dx} = y(1 + x) \quad \Longrightarrow \quad \int \frac{1}{y}\,dy = \int 1 + x\,dx$$

$$\Longrightarrow \quad \ln|y| = x + \frac{1}{2}x^2 + C.$$

Now solving for y, we have

$$|y| = e^{x + (1/2)x^2 + C} \quad \Longrightarrow \quad y = Ae^{x + (1/2)x^2}.$$

Notice that C is an arbitrary constant from the integration step, and we have defined $A = \pm e^C$ to simplify the expression and take care of the absolute value. We are given the initial value $y(0) = 2$, which yields:

$$y(0) = 2 \implies 2 = Ae^0 \implies A = 2.$$

Therefore the solution we seek is $y(x) = 2e^{x+(1/2)x^2}$.

CHECKING THE ANSWER To check our answers to the previous example we need only differentiate and plug in values to compare against the original initial-value problems. Let's do this for part (c). If $y(x) = 2e^{x+(1/2)x^2}$, then we have $\frac{dy}{dx} = 2e^{x+(1/2)x^2}(1+x) = y(1+x) = y + xy$, as we started with in that problem. Moreover, $y(0) = 2e^{0+(1/2)(0)} = 2(1) = 2$, as desired.

EXAMPLE 2 Sketching a slope field and tracing solution curves

Sketch a slope field for the differential equation $\frac{dy}{dx} = 2xy$ on a lattice of points contained within $-2 \leq x \leq 2$ and $-10 \leq y \leq 10$. Then use the slope field to sketch graphs of three approximate solutions of the differential equation.

SOLUTION

Let's start by looking at the slopes at a few chosen points in the lattice. We have $\frac{dy}{dx} = g(x, y)$ with $g(x, y) = 2xy$. At each point (a, b) in the lattice we want to sketch a line segment with slope $g(a, b) = 2ab$. For example, at the point $(0, 0)$ we draw a segment with slope $2(0)(0)$, and at the point $(1, 2)$ we draw a segment with slope $2(1)(2) = 4$. Clearly, for positive a and b, the slope $2ab$ will be positive. Likewise, if a and b are both negative, then the slope $2ab$ will be positive, and if a and b have opposite signs, then the slope $2ab$ will be negative. In all cases, the magnitude of the slope will grow larger as a and/or b increase, and grow smaller as a and/or b decrease. After plotting a few of these line segments to get an initial idea, we can sketch the slope field shown here at the left:

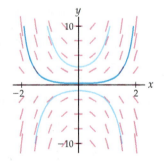

Slope field for $\frac{dy}{dx} = 2xy$ Three solution curves in the slope field

Three curves through the slope field are shown in the figure at the right. Each of these is an approximate solution of the differential equation $\frac{dy}{dx} = 2xy$.

CHECKING THE ANSWER Sometimes we sketch a slope field because we cannot solve a differential equation by hand. That is not the case in Example 2, and so we can check whether our slope field is reasonable by actually solving the differential equation $\frac{dy}{dx} = 2xy$. A straightforward separation-of-variables calculation (see Exercise 21) shows that the solutions are each of the form $y(x) = Ae^{x^2}$ for some constant A. This means that the graphs of, for example, $y(x) = 3e^{x^2}$, $y(x) = \frac{1}{4}e^{x^2}$, and $y(x) = -e^{x^2}$ should flow through the slope field. And in fact, these are precisely the three functions that we traced through the slope field earlier at the right.

EXAMPLE 3

Applying Euler's method

Consider the differential equation $\frac{dy}{dx} = y$ with initial condition $y(0) = 2$. Use Euler's method with $\Delta x = 0.25$ to approximate the solution $y(x)$ to this initial-value problem on the interval $[0, 2]$.

SOLUTION

We want to approximate the solution of $\frac{dy}{dx} = y$ that passes through the point $(x_0, y_0) = (0, 2)$. Euler's method with $\Delta x = 0.25$ allows us to construct a sequence of eight more co-ordinate points $(x_1, y_1), \ldots, (x_8, y_8)$ over the interval $[0, 2]$. Each (x_{k+1}, y_{k+1}) is obtained from the previous point (x_k, y_k) by stepping over Δx and up or down $g(x_k, y_k)\Delta x$ according to the value of $\frac{dy}{dx}\big|_{(x_k, y_k)} = g(x_k, y_k) = y_k$:

$(x_0, y_0) = (0, 2);$

$(x_1, y_1) = (x_0 + \Delta x, y_0 + g(x_0, y_0)\Delta x) = (0 + 0.25, 2 + 2(0.25)) = (0.25, 2.5);$

$(x_2, y_2) = (x_1 + \Delta x, y_1 + g(x_1, y_1)\Delta x) = (0.25 + 0.25, 2.5 + 2.5(0.25)) = (0.5, 3.125);$

$(x_3, y_3) = (x_2 + \Delta x, y_2 + g(x_2, y_2)\Delta x) = (0.5 + 0.25, 3.125 + 3.125(0.25)) = (0.75, 3.91);$

$(x_4, y_4) = (x_3 + \Delta x, y_3 + g(x_3, y_3)\Delta x) = (0.75 + 0.25, 3.91 + 3.91(0.25)) = (1, 4.89);$

$(x_5, y_5) = (x_4 + \Delta x, y_4 + g(x_4, y_4)\Delta x) = (1 + 0.25, 4.89 + 4.89(0.25)) = (1.25, 6.11);$

$(x_6, y_6) = (x_5 + \Delta x, y_5 + g(x_5, y_5)\Delta x) = (1.25 + 0.25, 6.11 + 6.11(0.25)) = (1.5, 7.64);$

$(x_7, y_7) = (x_6 + \Delta x, y_6 + g(x_6, y_6)\Delta x) = (1.5 + 0.25, 7.64 + 7.64(0.25)) = (1.75, 9.55);$

$(x_8, y_8) = (x_7 + \Delta x, y_7 + g(x_7, y_7)\Delta x) = (1.75 + 0.25, 9.55 + 9.55(0.25)) = (2, 11.94).$

When plotted and connected by line segments, this sequence of points produces a piecewise-linear approximation to the solution of the initial-value problem $\frac{dy}{dx} = y$ with initial condition $y(0) = 2$. This is the same piecewise-linear approximation that we sketched in the rightmost figure preceding Definition 9.19.   □

EXAMPLE 4

Using Newton's Law of Cooling

Suppose your teacher sets down a fresh cup of coffee onto a table in a classroom that is kept at a constant temperature of $68°$ Fahrenheit. After giving a fascinating 45-minute lecture on differential equations she remembers her cup of coffee, which has now cooled to a lukewarm $75°$ Fahrenheit. Assuming that the proportionality constant for Newton's Law of Cooling in this case is $k = 0.065$, how hot was the coffee before the lecture started?

SOLUTION

Newton's Law of Cooling tells us that the coffee should get cooler until it matches the temperature of the $68°$ room. We are told that for this particular cup of coffee and room environment, the rate of this cooling will be proportional to the difference in temperatures of the coffee and the room, with proportionality constant $k = 0.065$. In other words, if $T(t)$ is the temperature of the coffee after t minutes, then we have the differential equation

$$\frac{dT}{dt} = 0.065(68 - T).$$

Notice that T starts out hotter than the $68°$ room, so the rate of change $\frac{dT}{dt} = 0.065(68 - T)$ is negative, as we would expect.

Let us start by solving this differential equation. After applying separation of variables, we have

$$\int \frac{1}{68 - T}\, dT = \int 0.065\, dt.$$

Integrating both sides gives us the equation

$$-\ln |68 - T| = 0.065t + C,$$

and after solving this equation for T, we have

$$T = 68 - Be^{-0.065t},$$

where $B = e^{-C}$ is some constant.

We do not know the initial temperature $T(0)$ of the coffee at the start of the teacher's lecture, but we do know that the temperature of the coffee after 45 minutes is $T(45) = 75°$. Substituting this number into our equation for $T(t)$ results in

$$75 = 68 - Be^{-0.065(45)} \quad \Rightarrow \quad B = \frac{-7}{e^{-0.065(45)}} \approx -130.44.$$

Therefore the temperature of the coffee is given by the equation $T(t) = 68 + 130.44e^{-0.065t}$.

Now that we have an equation for the temperature $T(t)$ of the coffee, we can easily calculate $T(0)$ to obtain the initial temperature of the teacher's coffee. This initial temperature was approximately $T(0) = 68 + 130.44e^{-0.065(0)} = 198.44$ degrees Fahrenheit.   □

✔ **CHECKING THE ANSWER**

As a reality check, let's graph the function $T(t) = 68 + 130.44e^{-0.065t}$ and verify that it is a good model for the temperature of the coffee as it cools:

Temperature $T(t)$ of coffee

Notice that the coffee starts out at nearly $200°$, and then its temperature declines, first quickly and then more slowly, approaching the ambient temperature of the room. According to this model the coffee's temperature after $t = 45$ minutes is $75°$, as we wanted. Notice also that as $t \to \infty$ the temperature of the coffee becomes asymptotically close to the temperature of the $68°$ classroom, as we would expect. This model function $T(t)$ does indeed seem to match with what we would expect from the teacher's cooling cup of coffee.

? **TEST YOUR UNDERSTANDING**

▶ What does it mean for a function $y(x)$ to be a solution of a differential equation? What about a solution of an initial-value problem?

▶ What does it mean for a differential equation to be separable? Can you give an example of a differential equation that is not separable?

▶ What is a slope field associated with a differential equation, and how are the slopes of the line segments in a slope field related to the differential equation?

▶ When using Euler's method, how do we use slope information to determine how to get each point (x_{k+1}, y_{k+1}) from the previous point (x_k, y_k)?

▶ Why does the differential equation for Newton's Law of cooling make real-world sense?

EXERCISES 9.5

Thinking Back

Integration: Solve each of the following indefinite integrals.

▶ $\int \dfrac{1}{y^2} \, dy$

▶ $\int \dfrac{1}{1-y} \, dy$

▶ $\int \dfrac{1}{y^2+1} \, dy$

▶ $\int \dfrac{y}{y^2+1} \, dy$

▶ $\int \dfrac{1}{y(1-y)} \, dy$

▶ $\int \dfrac{1}{y(1-3y)} \, dy$

Local linearity: In each problem that follows we are given values for $f(c)$ and $f'(c)$ for some c. Use this information to approximate $f(c+h)$ for the given value of h.

▶ $f(0) = 2,\ f'(0) = 1,\ h = 1$

▶ $f(1) = -1,\ f'(1) = -3,\ h = 0.5$

▶ $f(-2) = -3,\ f'(-2) = 2,\ h = 0.25$

▶ $f(1.5) = 6.2,\ f'(1.5) = 2.2,\ h = 0.1$

Concepts

0. *Problem Zero:* Read the section and make your own summary of the material.

1. *True/False:* Determine whether each of the statements that follow is true or false. If a statement is true, explain why. If a statement is false, provide a counterexample.

   (a) *True or False:* A solution of an initial-value problem is a function that satisfies both the differential equation and the initial value.

   (b) *True or False:* If $y_1(x)$ and $y_2(x)$ are both solutions of some differential equation $\frac{dy}{dx} = g(x, y)$, then $y_1(x) - y_2(x) = C$ for some constant C.

   (c) *True or False:* Solving a differential equation $\frac{dy}{dx} = g(x)$ by antidifferentiating is just a special case of solving by separation of variables.

   (d) *True or False:* The differential equation $\frac{dy}{dx} = x + y$ is separable.

   (e) *True or False:* In a slope field for a differential equation $\frac{dy}{dx} = g(x, y)$, the slope of the line segment at (a, b) is given by $\frac{dy}{dx}\big|_{(a,b)}$.

   (f) *True or False:* In a slope field for a differential equation $\frac{dy}{dx} = g(y)$, the slope at $(2, b)$ will be the same as the slope at $(3, b)$.

   (g) *True or False:* Euler's method is a way to construct a piecewise-linear approximation of the solution of an initial-value problem.

   (h) *True or False:* For populations well below carrying capacity, the logistic growth model is similar to an exponential model.

2. *Examples:* Construct examples of the thing(s) described in the following. Try to find examples that are different than any in the reading.

   (a) Two differential equations that are separable and two that are not separable.

   (b) A slope field that describes solution functions of the form $f(x) = x^2 + C$.

   (c) Three real-world situations that could be modeled with initial-value problems.

3. What is the difference between a solution of a differential equation and a solution of an initial-value problem?

4. Suppose you solve an initial-value problem of the form $\frac{dy}{dx} = g(x, y)$ with $y(0) = y_0$ and obtain an explict solution function $y(x)$. How could you check that your function $y(x)$ is indeed a solution of the initial-value problem?

5. Verify that $y(x) = \sqrt{x + C}$ is a solution of the differential equation $\frac{dy}{dx} = \frac{1}{2y}$ for every constant C.

6. Verify that $y(x) = \sqrt{t + 9}$ is a solution of the initial-value problem $\frac{dy}{dx} = \frac{1}{2y}$ and $y(0) = 3$.

7. A falling object accelerates downwards due to gravity at a rate of 9.8 meters per second squared. Suppose an object is dropped and falls to the ground from a height of 100 meters.

   (a) Set up and solve a first-order initial-value problem whose solution is the velocity $v(t)$ of the object at time t.

   (b) Use your answer to part (a) to set up and solve a first-order initial-value problem whose solution is the position $s(t)$ of the object at time t.

8. Explain, using the chain rule and/or u-substitution, why

$$\int \frac{1}{q(y)} \frac{dy}{dx} \, dx = \int \frac{1}{q(y)} \, dy.$$

9. Why do we use the terminology "separable" to describe a differential equation that can be written in the form $\frac{dy}{dx} = p(x)q(y)$? Once we have a differential equation in this form, how does the technique of separation of variables work?

10. In the process of solving $\frac{dy}{dx} = y$ by separation of variables, we obtain the equation $|y| = e^{x+C}$. After solving for y, this equation becomes $y = Ae^x$. How is A related to C? What happened to the absolute value?

11. In the process of solving the differential equation $\frac{dy}{dx} = 1 - y$ by separation of variables, we obtain the equation $-\ln|1 - y| = x + C$. After solving for y, this equation becomes $y = 1 - Ae^{-x}$. Given that $y > 1$, how is A related to C?

12. Explain in your own words how the slopes of the line segments in a slope field for a differential equation are related to the differential equation.

13. How does a slope field help us to understand the solutions of a differential equation? How can a slope field help us sketch an approximate solution of an initial-value problem?

14. Given an initial-value problem, we can apply Euler's method to generate a sequence of points (x_0, y_0), (x_1, y_1), (x_2, y_2), and so on. How are these coordinate points related to the solution of the initial-value problem?

15. Explain how the equality $\frac{dy}{dx}\big|_{(x_k, y_k)} = \frac{\Delta y_k}{\Delta x}$ is relevant to Euler's method.

16. Suppose your bank account grows at 3 percent interest yearly, so that your bank balance after t years is $B(t) = B_0(1.03)^t$.
    (a) Show that your bank balance grows at a rate proportional to the amount of the balance.
    (b) What is the proportionality constant for the growth rate, and what is the corresponding differential equation for the exponential growth model of $B(t)$?

17. Suppose a population $P(t)$ of animals on a small island grows according to a logistic model of the form $\frac{dP}{dt} = kP\left(1 - \frac{P}{500}\right)$ for some constant k.
    (a) What is the carrying capacity of the island under this model?
    (b) Given that the population $P(t)$ is growing and that $0 < P(t) < 500$, is the constant k positive or negative, and why?
    (c) Explain why $\frac{dP}{dt} \approx kP$ for small values of P.
    (d) Explain why $\frac{dP}{dt} \approx 0$ for values of P that are close to the carrying capacity.

18. Suppose an object is heating up according to a model for Newton's Law of Cooling with temperature satisfying $\frac{dT}{dt} = k(350 - T)$ for some constant k.
    (a) What is the ambient temperature of the environment under this model?
    (b) Given that the temperature $T(t)$ is increasing and that $0 < T < 350$, is the constant k positive or negative, and why?
    (c) Use the differential equation to argue that the object's temperature changes faster when it is much cooler than the ambient temperature than when it is close to the ambient temperature.
    (d) Part (c) is the basis for the oft-misunderstood saying "Cold water boils faster." Why?

Skills

Use antidifferentiation and/or separation of variables to solve each of the differential equations in Exercises 19–28. Your answers will involve unsolved constants.

19. $\frac{dy}{dx} = (x^3 + 4)^2$

20. $\frac{dy}{dx} = \frac{x+1}{\sqrt{x}}$

21. $\frac{dy}{dx} = 2xy$

22. $\frac{dy}{dx} = -3xy$

23. $\frac{dy}{dx} = x^2 y$

24. $\frac{dy}{dx} = xy^2$

25. $\frac{dy}{dx} = \frac{3x+1}{xy}$

26. $\frac{dy}{dx} = y \sin x$

27. $\frac{dy}{dx} = xe^{-y}$

28. $\frac{dy}{dx} = x^2 e^{-y}$

Use antidifferentiation and/or separation of variables to solve each of the initial-value problems in Exercises 29–52.

29. $\frac{dy}{dx} = 3y$, $y(0) = 4$

30. $\frac{dy}{dx} = 3x$, $y(0) = 4$

31. $\frac{dy}{dx} = -6x^2$, $y(1) = 3$

32. $\frac{dy}{dx} = -5y$, $y(0) = 2$

33. $\frac{dy}{dx} = 1 - y$, $y(0) = 4$

34. $\frac{dy}{dx} = \frac{1}{y}$, $y(0) = 3$

35. $\frac{dy}{dx} = \frac{3}{y}$, $y(0) = 2$

36. $\frac{dy}{dx} = \frac{6}{x}$, $y(1) = 7$

37. $\frac{dy}{dx} = xy$, $y(0) = -1$

38. $\frac{dy}{dx} = \sqrt{4y}$, $y(1) = 1$

39. $\frac{dy}{dx} = e^{x+y}$, $y(0) = 2$

40. $\frac{dy}{dx} = \frac{2x}{y^2}$, $y(0) = 4$

41. $\frac{dy}{dx} = \frac{x}{1+x^2}$, $y(0) = 4$

42. $\frac{dy}{dx} = 10 - 7y$, $y(0) = \frac{1}{11}$

43. $\frac{dy}{dx} = 9.8 - \frac{0.3}{150}y$, $y(0) = 1000$

44. $\frac{dy}{dx} = 0.4y - 100$, $y(0) = 0$

45. $\frac{dy}{dx} = -2(1 - 3y)$, $y(0) = 2$

46. $\frac{dy}{dx} = -2y(1 - 3y)$, $y(0) = 2$

47. $\frac{dy}{dx} = 3y(1 - y)$, $y(0) = 1$

48. $\frac{dy}{dx} = xy + x + y + 1$, $y(0) = c$

49. $\frac{dy}{dx} = y^2 \cos x$, $y(\pi) = 1$

50. $\frac{dy}{dx} = y^3 \cos x$, $y(\pi) = 1$

51. $\dfrac{dy}{dx} = \dfrac{x^2 - 2}{y^2}$, $y(0) = 8$

52. $\dfrac{dy}{dx} = \dfrac{y+1}{x^2+1}$, $y(0) = 1$

For each initial-value problem $\dfrac{dy}{dx} = g(x, y)$ with $y(x_0) = y_0$ in Exercises 53–58, use Euler's method with the given Δx to approximate four additional points on the graph of the solution $y(x)$. Use these points to sketch a piecewise-linear approximation of the solution.

53. $\dfrac{dy}{dx} = -2y$, $y(0) = 3$; $\Delta x = 0.25$

54. $\dfrac{dy}{dx} = \dfrac{1}{x}$, $y(1) = 1$; $\Delta x = 0.25$

55. $\dfrac{dy}{dx} = x + y$, $y(0) = 0$; $\Delta x = 0.5$

56. $\dfrac{dy}{dx} = \dfrac{x}{1-y}$, $y(0) = 2$; $\Delta x = 0.1$

57. $\dfrac{dy}{dx} = \dfrac{x}{y}$, $y(0) = 2$; $\Delta x = 0.5$.

58. $\dfrac{dy}{dx} = x^2 - y$, $y(1) = 0$; $\Delta x = 0.25$

Sketch slope fields for each of the differential equations in Exercises 59–64, and within each slope field sketch four different approximate solutions of the differential equation.

59. $\dfrac{dy}{dx} = -y$

60. $\dfrac{dy}{dx} = 1 - x$

61. $\dfrac{dy}{dx} = x + y$

62. $\dfrac{dy}{dx} = x - y$

63. $\dfrac{dy}{dx} = y - 2x$

64. $\dfrac{dy}{dx} = e^x - y$

Applications

The situations in Exercises 65–67 concern exponential growth models of the form $\dfrac{dQ}{dt} = kQ$. When the constant of proportionality k (also called the **continuous growth rate**, when expressed as a percentage) happens to be negative rather than positive, this differential equation models exponential decay.

65. Suppose that the current population of Freedonia is 1.08 million people and that the continuous growth rate of the population is 1.39%.

(a) Set up a differential equation describing $\dfrac{dP}{dt}$, and solve it to get a formula for the population $P(t)$ of Freedonia t years from now.

(b) According to your model, how long ago was it that the population of Freedonia was just half a million people?

(c) How long will it take for the population of Freedonia to double from its current size?

66. Suppose $X(t)$ is the number of milligrams of the drug Xenaphoril that is present in the body t hours after it is ingested. As the drug is absorbed, the quantity of the drug decreases at a rate proportional to the amount of the drug in the body.

(a) Set up a differential equation describing $\dfrac{dX}{dt}$, and solve it to get a formula for $X(t)$. Your answer will involve two constants.

(b) The **half-life** of a drug is the number of hours that it takes for the quantity of the drug to decrease by half. In an exponential decay model, the half-life will be the same no matter when we start measuring the amount of the drug. If Xenaphoril has a half-life of 4 hours, what is the constant of proportionality for this model?

(c) Given the constant of proportionality you found in part (b), how much of a 20-mg dose of Xenaphoril remains after 10 hours?

67. The amount of the radioactive isotope carbon-14 present in small quantities can be measured with a Geiger counter. Carbon-14 is replenished in live organisms, and after an organism dies the carbon-14 in it decays at a rate proportional to the amount of carbon-14 present in the body. Suppose $C(t)$ is the amount of carbon-14 in a dead organism t years after it dies.

(a) Set up a differential equation describing $\dfrac{dC}{dt}$, and solve it to get a formula for $C(t)$. Your answer will involve two constants.

(b) The half-life of carbon-14 is 5730 years. (See part (b) of the previous problem for the definition of half-life.) Use this half-life to find the value of the proportionality constant for the model you found in part (a).

(c) Suppose you find a bone fossil that has 10% of its carbon-14 left. How old would you estimate the fossil to be?

The situations in Exercises 68 and 69 concern logistic growth models of the form $\dfrac{dP}{dt} = kP\left(1 - \dfrac{P}{L}\right)$ for some constant of proportionality k (also called the **natural growth rate**) and some carrying capacity L.

68. Suppose that the country of Freedonia has a carrying capacity of 5 million people, with natural growth rate and initial population as given in Exercise 65.

(a) Set up a differential equation describing $\dfrac{dP}{dt}$, and solve it to get a formula for the population $P(t)$ of Freedonia t years from now.

(b) How long will it be before the population of Freedonia is half of the carrying capacity?

(c) How fast is Freedonia's population changing when the population is at half of carrying capacity? What about when the population has reached 90% of carrying capacity?

69. Suppose 100 rabbits are shipwrecked on a deserted island and their population $P(t)$ after t years is determined by a logistic growth model, where the natural growth rate of the rabbits is $k = 0.1$ and the carrying capacity of the island is 1000 rabbits.

(a) Set up a differential equation describing $\dfrac{dP}{dt}$, and solve it to get a formula for the population $P(t)$ of rabbits on the island in t years.

(b) Sketch a graph of the population $P(t)$ of rabbits on the island over the next 100 years.

(c) It turns out that a population governed by a logistic model will be growing fastest when the population is equal to exactly half of the carrying capacity. In how many years will the population of rabbits be growing the fastest?

The situations in Exercises 70 and 71 concern Newton's Law of Cooling and Heating and models of the form $\frac{dT}{dt} = k(A - T)$ for some proportionality constant k and constant ambient temperature A.

70. A cold drink is heating up from an initial temperature $T(0) = 2°C$ to room temperature of $22°C$ according to Newton's Law of Heating with constant of proportionality $0.05°C$.

   (a) Set up a differential equation describing $\frac{dT}{dt}$, and solve it to get a formula for the temperature of the drink after t minutes.

   (b) Use the differential equation and/or its solution to determine the units of the constant of proportionality.

   (c) How long will it take for the drink to warm up to within 1 degree of room temperature?

71. A crime scene investigator finds a body at 8 P.M., in a room that is kept at a constant temperature of $70°F$. The temperature of the body is $88.8°F$ at that time. Thirty minutes later the temperature of the body is $87.5°F$.

   (a) Set up a differential equation describing $\frac{dT}{dt}$, and solve it to get a formula for the temperature of the body t minutes after the time of death. Your answer will involve a proportionality constant k.

   (b) Use the information in the problem to determine k.

   (c) Assuming that the body had a normal temperature of $98.6°F$ at the time of death, estimate the time of death of the victim.

In Exercises 72 and 73 you will set up and solve differential equations to model different population growth scenarios.

72. A colony of bacteria is growing in a large petri dish in such a way that the shape of the colony is a disk whose area $A(t)$ after t days grows at rate proportional to the diameter d of the disk. Suppose the colony has an area of 2 cm² on the first day and 5 cm² on the third day.

   (a) Set up a differential equation that describes this situation, and solve it to get an equation for the area $A(t)$ of the colony after t days. Your answer will involve a proportionality constant k.

   (b) Use the information in the problem to determine k.

   (c) Given that the petri dish has a diameter of 6 inches and that the colony started in the exact center of the dish, how long will it take for the colony to fill the entire petri dish?

73. Another model for population growth is what is called *supergrowth*. It assumes that the rate of change in a population is proportional to a higher power of P than P^1. For example, suppose the rate of change of the world's human population $P(t)$ is proportional to $P^{1.1}$.

   (a) Set up a differential equation that describes $\frac{dP}{dt}$, and solve it to get an equation for the world population $P(t)$. Your answer will involve two unsolved constants.

   (b) Given that the world population was estimated to be 6.451 billion in 2005 and 6.775 billion in 2010, what is the value of the proportionality constant for this supergrowth model?

   (c) According to your model, when will the world population be 7 billion?

Proofs

74. Show that if $y_1(x)$ and $y_2(x)$ are both solutions of the differential equation $\frac{dy}{dx} = ky$, then the sum $y_1(x) + y_2(x)$ is also a solution of the differential equation.

75. Prove Theorem 9.20 by solving the initial-value problem $\frac{dQ}{dt} = kQ$ with $Q(0) = Q_0$, where k is a constant.

76. Prove Theorem 9.21 by solving the initial-value problem $\frac{dP}{dt} = rP\left(1 - \frac{P}{K}\right)$ with $P(0) = P_0$, where r and K are constants. (*Hint: After separation of variables you will need to use partial fractions to solve the integral that concerns P. Before solving for P, use properties of logarithms to simplify the expression.*)

77. Use the solution of the logistic model $\frac{dP}{dt} = kP\left(1 - \frac{P}{L}\right)$ to prove that as $t \to \infty$, the population $P(t)$ approaches the carrying capacity L. Assume that the constant k is positive.

78. Prove Theorem 9.22 by solving the initial-value problem $\frac{dT}{dt} = k(A - T)$ with $T(0) = T_0$, where k and A are constants.

79. Use the solution of the differential equation $\frac{dT}{dt} = k(A-T)$ for the Newton's Law of Cooling and Heating model to prove that as $t \to \infty$, the temperature $T(t)$ of an object approaches the ambient temperature A of its environment. The proof requires that we assume that k is positive. Why does this make sense regardless of whether the model represents heating or cooling?

Thinking Forward

Implicit solution curves: Sometimes the solution of a differential equation will be an implicit solution, meaning that y will be given as an implicit function of x, not an explicit function of x. For each differential equation $\frac{dy}{dx} = g(x, y)$ that follows, find an equation that gives y as an implicit function of x. Then draw the continuous curve that satisfies this differential equation and passes through the point $(2, 0)$.

   ▶ $\dfrac{dy}{dx} = \dfrac{-x}{y}$

   ▶ $\dfrac{dy}{dx} = \dfrac{x}{y}$

CHAPTER REVIEW, SELF-TEST, AND CAPSTONES

Be sure you are familiar with the definitions, concepts, and basic skills outlined here. The capstone exercises at the end bring together ideas from this chapter.

Definitions

Give precise mathematical definitions or descriptions of each of the concepts that follow. Then illustrate the definition with a graph or algebraic example, if possible.

▶ the *volume of a solid with cross-sectional area function* given by $A(x)$ on $[a, b]$, both as a limit of Riemann sums and as a definite integral

▶ the *volume of a solid with disk cross sections* whose radii are given by $r(x)$ on $[a, b]$, both as a limit of Riemann sums and as a definite integral

▶ the *volume of a solid with washer cross sections* whose outer and inner radii are given, respectively, by $R(x)$ and $r(x)$ on $[a, b]$, as a limit of Riemann sums

▶ the *volume of a solid with nested shell sections* whose heights and average radii are given, respectively, by $h(x)$ and $r(x)$ on $[a, b]$, as a limit of Riemann sums

▶ the *arc length* of the graph of a continuous, differentiable function $y = f(x)$ whose derivative is also continuous, from $x = a$ to $x = b$, as a limit of Riemann sums

▶ if f is a continuous, differentiable function with a continuous derivative, the *area of the surface of revolution* obtained by revolving $y = f(x)$ around the x-axis on $[a, b]$, as a limit of Riemann sums

▶ the definition of a *differential equation* and what it means for a function to be a *solution of a differential equation*

▶ the definition of an *initial-value problem* and what it means for a function to be a *solution of an initial-value problem*

▶ what it means for a differential equation to be *separable*

▶ the definition of a *slope field* for a differential equation, including the specific property that holds at each lattice point (x, y)

▶ the method of approximation known as *Euler's method*, including the iterative formula that describes a new coordinate point (x_{k+1}, y_{k+1}) in terms of the previous point (x_k, y_k)

Theorems

Fill in the blanks to complete each of the following theorem statements.

▶ The volume of a shell with height h, average radius r, and thickness Δx is $V =$ _____ .

▶ The surface area of a frustum with average radius r and slant length s is $S =$ _____ .

▶ If a quantity $Q(t)$ changes over time at a rate proportional to its value, then that quantity is modeled by a differential equation of the form $\dfrac{dQ}{dt} =$ _____ , with solution $Q(t) =$ _____ .

▶ If a population $P(t)$ changes over time with natural growth rate k and carrying capacity L, then it can be modeled by the differential equation $\dfrac{dP}{dt} =$ _____ , with solution $P(t) =$ _____ .

▶ If an object with temperature $T(t)$ is in a location with constant ambient temperature A, then the temperature of the object can be modeled by a differential equation of the form $\dfrac{dT}{dt} =$ _____ , with solution $T(t) =$ _____ .

Formulas and Geometric Quantities

Definite integral formulas for geometric quantities: Write down definite integrals to express each of the given geometric quantities. You may assume that $f(x)$ is continuous and differentiable, with a continuous derivative.

▶ The volume of the solid obtained by revolving $f(x)$ on $[a, b]$ around the x-axis, by the disk method.

▶ The volume of the solid obtained by revolving $f(x)$ on $[a, b]$ around the y-axis, by the disk method.

▶ The volume of the solid obtained by revolving $f(x)$ on $[a, b]$ around the x-axis, by the shell method.

▶ The volume of the solid obtained by revolving $f(x)$ on $[a, b]$ around the y-axis, by the shell method.

▶ The arc length of the curve formed by $y = f(x)$ on $[a, b]$.

▶ The area of the surface obtained by revolving $f(x)$ on $[a, b]$ around the x-axis.

▶ The centroid $(\bar{x}, \bar{y})$ of the region between the graph of an integrable function f and the x-axis on an interval $[a, b]$.

▶ The centroid $(\bar{x}, \bar{y})$ of the region between the graphs of two integrable functions f and g on an interval $[a, b]$.

Finding distances related to graphs: In setting up definite integrals for volume problems, it is often necessary to describe various distances in terms of f, f^{-1}, x_k^*, and/or y_k^*. Describe these distances for each figure that follows.

▶ For the following figure, write the distances A and B in terms of f or f^{-1} and the point x_k^*.

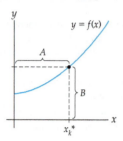

▶ For the following figure, write the distances A and B in terms of f or f^{-1} and the point y_k^*.

▶ For the following figure, write the distances A, B, and C in terms of f or f^{-1}, the point x_k^*, and K.

▶ For the following figure, write the distances A, B, and C in terms of f or f^{-1}, the point y_k^*, and K.

Concepts from physics: Fill in the blanks to complete the descriptions of each of the following physical quantities and units.

▶ The *mass* of an object with density ρ and volume V is given by $m =$ _____ .

▶ If density is measured in grams per cubic centimeter and volume is measured in cubic centimeters, then mass is measured in units of _____ .

▶ The *work* required to lift an object with a weight F through a distance d is given by $W =$ _____ .

▶ If weight is measured in pounds and distance is measured in feet, then work is measured in units of _____ .

▶ The *hydrostatic force* exerted by a liquid of weight-density ω and depth d on a horizontal wall of area A is given by $F =$ _____ .

▶ If weight-density is measured in pounds per cubic foot and distance is measured in feet, then hydrostatic force is measured in units of _____ .

Skill Certification: Definite Integrals for Geometry and Applications

Sketching disks, washers, and shells: Sketch the three disks, washers, or shells that result from revolving the rectangles shown in the given figures around the given lines.

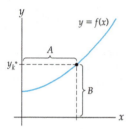

1. the x-axis
2. the y-axis
3. the line $x = -1$
4. the line $y = -1$

Sketching a representative disk, washer, or shell: Sketch a representative disk, washer, or shell for the solid obtained by revolving the regions shown in the given figures around the given lines.

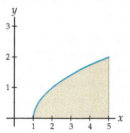

5. the x-axis
6. the y-axis
7. the line $x = 7$
8. the line $y = 3$

Finding geometric quantities with definite integrals: Set up and solve definite integrals to find each volume, surface area, or arc length that follows. Solve each volume problem both with disks/washers and with shells, if possible.

9. The volume of the solid obtained by revolving the region between the graph of $f(x) = x^2$ and the y-axis for $0 \le x \le 2$ around (a) the x-axis and (b) the y-axis.

10. The volume of the solid obtained by revolving the region between the graph of $f(x) = 9 - x^2$ and the x-axis on $[-3, 3]$ around (a) the x-axis and (b) the line $y = -3$.

11. The volume of the solid obtained by revolving the region between the graphs of $f(x) = \sqrt{x}$ and $g(x) = x^3$ on $[0, 1]$ around (a) the y-axis and (b) the line $x = 2$.

12. The arc length of the curve traced out by the graph of $f(x) = \ln(\csc x)$ on the interval $\left[\dfrac{\pi}{4}, \dfrac{\pi}{2}\right]$.

13. The area of the surface obtained by revolving the curve $f(x) = \sin(\pi x)$ around the x-axis on $[-1, 1]$.

14. The centroid of the region between the graph of $f(x) = x^2$ and the x-axis on $[0, 2]$.

15. The centroid of the region between the graphs of $f(x) = \sqrt{x}$ and $g(x) = 4 - x$ on $[0, 4]$.

Real-world applications of definite integrals: Set up and solve definite integrals to answer each of the following questions.

16. Find the mass of a 30-centimeter rod with square cross sections of side length 2 centimeters, given that the density of the rod x centimeters from the left end is $\rho(x) = 10.5 + 0.01527x^2$ grams per cubic centimeter.

17. Find the work required to pump all of the water out of the top of a cylindrical hot tub that is 8 feet in diameter and 3 feet deep.

18. Find the hydrostatic force exerted on one of the long sides of a rectangular swimming pool that is 20 feet long, 12 feet wide, and 6 feet deep.

Differential equations: Solve each of the following initial-value problems. *(These problems assume that you covered Section 9.5.)*

19. $\dfrac{dy}{dx} = 5y, \quad y(0) = 1$ **20.** $\dfrac{dy}{dx} = 3 - 4y, \quad y(0) = 2$

21. $\dfrac{dy}{dx} = 2y(1 - 5y), \quad y(0) = 1$

22. $\dfrac{dy}{dx} = y^2 \sin x, \quad y(\pi) = 1$

Capstone Problems

A. *Proving geometric properties of spheres:* Set up and solve definite integrals to prove each of the following geometric formulas concerning the sphere with radius r.

(a) Volume: $V = \dfrac{4}{3}\pi r^3$ (with disks)

(b) Volume: $V = \dfrac{4}{3}\pi r^3$ (with shells)

(c) Equator: $C = 2\pi r$

(d) Surface area: $S = 4\pi r^2$

B. *Approximating volumes with frustums:* In Section 9.3 we used frustums to approximate surface areas. We could also use frustums, instead of disks, to approximate volumes.

(a) Consider the solid of revolution obtained by revolving the graph of $f(x) = x^2$ around the x-axis on $[0, 2]$. Use four frustums to approximate the volume of this solid.

(b) Suppose we used a limit of volumes of frustums to develop a definite integral formula for volume. Would we obtain a different definite integral formula? Why or why not?

C. *Center of mass when density varies:* In Section 9.4 we saw both how to find the mass of a rod whose density varies and how to find the center of mass of a planar region with constant density. Combine these two ideas in a "subdivide, approximate, and add" strategy to construct a formula for determining the center of mass of a planar region whose density varies. You may assume that the density at any point in the region is given by some function $\rho(x)$ that depends only on the x-coordinate.

D. *A differential equation for escape velocity:* To get a satellite up into space, it needs to escape the gravitational pull of the earth. The velocity it must attain to break free from the earth is called its **escape velocity**. Let $s = s(t)$ be the position of the satellite above the surface of the earth. As an object moves away from the earth, the acceleration due to gravity is reduced. In fact, the acceleration due to gravity on an object s kilometers from the surface of the earth is

$$a = a(t) = \frac{gR^2}{(R + s)^2},$$

where R is the radius of the earth and g is the acceleration due to gravity on the surface of the earth. The sign on g is negative, pointing in the direction toward the earth. Knowing that $a = \dfrac{dv}{dt}$, we can rewrite our equation for acceleration due to gravity as a differential equation:

$$\frac{dv}{dt} = \frac{gR^2}{(R + s)^2}.$$

(a) Use the chain rule to show that

$$v\frac{dv}{ds} = \frac{gR^2}{(R + s)^2}.$$

(b) Use separation of variables to solve the differential equation.

(c) Find the escape velocity of the satellite, that is, the smallest value of $v(0)$ for which the velocity $v(t)$ will never be zero.

Chapter 0

Section 0.1

1. T, F, T, F, T, T, F, F.　　3. Yes; yes; yes; yes.

5. No; it does not terminate or repeat a sequence of digits.

7. Yes; $\dfrac{1.7}{2.3} = \dfrac{17}{23}$.

9. (a) "is an element of"; (b) intersection; (c) union; (d) the empty set.

11. The interval $(3, 5)$ and the point $(3, 5)$ in the coordinate plane.

13. No; this set is not a connected subset of the real line. It is a union of two intervals: $(-\infty, -2] \cup [2, \infty)$.

15. (a) $|5 - 3| = |2| = 2$ and $|3 - 5| = |-2| = 2$;
    (b) $|5 - (-3)| = |8| = 8$ and $|-3 - 5| = |-8| = 8$;
    (c) $|-5 - 3| = |-8|$ and $|3 - (-5)| = |8| = 8$;
    (d) $|-5 - (-3)| = |-2| = 2$ and $|-3 - (-5)| = |2| = 2$.

17. Geometrically, it is the length of the line segment connecting the points P and Q in the plane. Algebraically, it is given by $\sqrt{(x - u)^2 + (y - v)^2}$.

19. Yes; yes; no.

21. No; for example $(0, 1) \cup (2, 3)$ is not connected, and thus not an interval. Yes; the intersection of two connected subsets of the real-number line is always connected.

23. $\dfrac{17}{100}$

25. $\dfrac{1}{100,000}$

27. $\dfrac{-117}{10}$

29. $\dfrac{0}{1}\left(\text{or } \dfrac{0}{8}, \text{ etc.}\right)$

31. $\dfrac{-2}{3}$

33. $-2b$

35. $-ab^3$

37. $-(ab - 1)$

39. (a) $\{x \mid -7 < x < 1\}$; (b) $(-7, 1)$; (c) below.

41. (a) $\{x \mid x \in \mathbb{R}\}$; (b) $(-\infty, \infty)$; (c) below.

43. (a) $\{x \mid x > 5 \text{ or } x < 1\}$; (b) $(-\infty, 1) \cup (5, \infty)$; (c) below.

45. (a) $\{-1, -3, -5, -7, \dots\}$; (b) inconvenient to write in interval notation; (c) inconvenient to express on the real-number line.

47. (a) $\{2, 4\}$; (b) we don't usually write such sets in interval notation, but we could write $\{2\} \cup \{4\}$; (c) below.

49. (a) $\{x \mid 0 < |x - 3| < 5\} = \{x \mid -2 < x < 8 \text{ but } x \neq 3\}$;
    (b) $(-2, 3) \cup (3, 8)$; (c) below.

51. (a) $\{(x, y) \mid y > x + 1\}$; (b) not possible; (c) below.

53. $|x - 0.5| \le 3.5$　　　　55. $0 < |x + 1| < 4$

57. The length is $\sqrt{13} \approx 3.60555$ and the midpoint is $\left(-\dfrac{1}{2}, -2\right)$.

59. The length is $\sqrt{58} \approx 7.61577$ and the midpoint is $(-1.5, 3.5)$.

61. (a) $\{x \mid -1 < x \le 3\}$; (b) $(-1, 3]$; (c) below.

63. (a) $\{x \mid -1 \le x\}$; (b) $[-1, \infty)$; (c) below.

65. (a) $\{0\}$; (b) $\{0\}$; (c) below.

67. (a) $\{x \mid x < -2 \text{ or } -1 < x \le 3\}$;
(b) $(-\infty, -2) \cup (-1, 3]$; (c) below.

69. (a) $\{x \mid x < -2 \text{ or } -1 < x \le 0\}$;
(b) $(-\infty, -2) \cup (-1, 0]$; (c) below.

71. (a) $\{x \mid x < -2 \text{ or } -1 \le x < 2\}$;
(b) $(-\infty, -2) \cup [-1, 2)$; (c) below.

73.

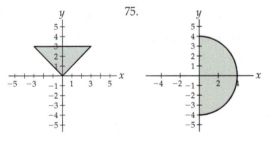

75.

77. The taxi traveled

$$|5-9| + |9-3| + |3-12| + |12-8| + |8-1| + |1-5| = 34$$

blocks.

79. No lunch price is acceptable to all. Lunches that cost between $5 and $6 will make Bekah and Calvin happy, and lunches that cost between $9 and $12 will make Anne and Calvin happy.

81. Every integer n can be expressed as $\frac{n}{1}$.

83. Suppose $\frac{p}{q}$ and $\frac{r}{s}$ are rational. Their product is

$$\left(\frac{p}{q}\right)\left(\frac{r}{s}\right) = \frac{pr}{qs}.$$ Since $p, q, r,$ and s are integers, we know that pr and qs are integers; thus $\frac{pr}{qs}$ is rational.

Section 0.2

1. F, F, F, F, T, F, T, T.

3. Substitute each proposed solution into the equation and verify that it makes the equation true. If two or fewer variables are involved, you can also graph the equation and verify that your solution points lie on the graph.

5. Three of the many solutions are $(-2, 0)$, $(0, -1.5)$, and $(1, 0)$. Three points that are not solutions are $(0, 1)$, $(2, 2)$, and $(3, 0)$.

7. (a) $x = 1$ does not make the equation true, so it is not a solution; $x = 2$ does satisfy the equation and thus is a solution. (b) By finding the x-coordinates of the points on the graph of $y = x^3 - 4x^2 + x + 6$ with y-coordinate equal to zero, we estimate that the solutions are $x = -1, x = 2,$ and $x = 3$.

9. (a) $(1, 4)$ makes the equation true and thus is a solution, but $(2, 1)$ does not. (b) Three of the many possible solutions are $(0, 2)$, $(1, 4)$, $(-1, 0)$. (c) and (d) See figure below, with our choices from part (c) indicated on the graph.

11. (a) $(1, -1)$ is not a solution to the system because it is not a solution to the first or the second equation. $(-2, -1)$ is a solution to the system because $-2 - (-1) = -1$ and $-1 + (-2)^2 = 3$. (b) From the graph below we approximate that $(-2, -1)$ and $(1, 2)$ are the only solutions to the system.

13. No, since the left side of the equation has not been written as a product; only part of the expression has been factored.

15. $\pm a(x - r_1)(x - r_2)$, $\pm a(x - r_1)^2$

17. When $x = -2$ the value of the original does not exist. After cancellation the value is $-\frac{1}{4}$. This is not the same; the cancellation only preserves the value of the expression when $x + 2 \neq 0$.

19. $2x(3x - 5)(x + 2)$ 21. $(2x - 9)(2x + 9)$

23. $(9x^2 + 1)(3x + 1)(3x - 1)$ 25. $(x + 1)(x + 1)(x + 1)$

27. $(3 - 4x)(16x^2 + 12x + 9)$; the quadratic in the second factor has no real roots by the quadratic formula.

29. $(x - 1)(x^4 + x^3 + x^2 + x + 1)$; the second factor may factor further, but we do not yet know how to do that.

31. $(t - x)(t^6 + t^5x + t^4x^2 + t^3x^3 + t^2x^4 + tx^5 + x^6)$; the second factor may factor further, but we do not yet know how to do that.

33. $-x(x - 1)(x + 1)^2$, if $x \neq 0$ and $x \neq -1$.

35. $\frac{(x-2)(x-1)}{x+1}$, if $x \neq -2$ 37. $\frac{x(5 - 4x)}{x^2 - 1}$

39. $x = 1, x = 6$ 41. $x = 0, x = -\frac{1}{2}, x = 2$

43. $x = -1, x = \frac{3}{2}$ 45. $x = \frac{3}{2}$

47. $x = 0, x = \frac{1}{3}, x = 5$

49. $x = -1 - \sqrt{6}, x = -1 + \sqrt{6}$

51. $x = \frac{1 - \sqrt{7}}{2}, x = \frac{1 + \sqrt{7}}{2}$ 53. $x = 2 - \sqrt{2}, x = 2 + \sqrt{2}$

55. $x = -3, x = 0, x = 3$ 57. $x = \frac{3}{2}, x = -\frac{3}{2}$

59. $x = 0, x = \frac{3}{2}, x = -\frac{3}{2}$ 61. $x = -2$

63. $x = 0$ and $x = -2$ 65. $x = -2$

67. $\emptyset$ 69. $(x, y) = (8, -12)$

71. $(a, b) = \left(-\frac{4}{7}, \frac{5}{7}\right)$

73. $(t, u) = (-1, 0)$ and $(t, u) = (2, 3)$

75. $(a, b, c) = (-3, 1, -7)$ 77. $\emptyset$

79. By the quadratic formula, the solutions of the equation are $r = \frac{-6\pi \pm \sqrt{750\pi + 36\pi^2}}{5\pi}$. These solutions are approximately equal to -4.51501 and 2.11501. Only the positive solution makes sense for the problem.

81. 7, 12, and 23

83. $(a - b)(a^2 + ab + b^2) = a^3 + a^2b + ab^2 - a^2b - ab^2 - b^3 = a^3 - b^3$.

85. We can write $c = \frac{c}{1}$. Then using the second rule from Theorem 0.14 we have $c\left(\frac{a}{b}\right) = \left(\frac{c}{1}\right)\left(\frac{a}{b}\right) = \frac{ca}{1b} = \frac{ac}{b}$.

Section 0.3

1. F, F, F, F, T, F, T, F.

3. $x = 2$ is a solution, because it makes the inequality true; $x = 3$ is not a solution.

5. The first set is $(-\infty, 2) \cup (0, \infty) = (-\infty, \infty)$ and the second set is $(-\infty, 2) \cap (0, \infty) = (0, 2)$.

7. $(A > 0$ and $B > 0)$ or $(A < 0$ and $B < 0)$ or $A = 0$ or $B = 0$

9. $(A < 0$ and $B > 0$ and $C > 0)$ or $(A > 0$ and $B < 0$ and $C > 0)$ and $(A > 0$ and $B > 0$ and $C < 0)$

11. $A \geq \delta$ or $A \leq -\delta$

13. In the first case $A < D$ because $<$ is the weakest link in the chain; in the second case we cannot say anything about how A is related to D.

15. (a) $|x - 1.5| < 2.5$; (b) given; (c) $[-1, 4]$; (d) the distance between x and 1.5 is less than or equal to 2.5; (e) below.

17. (a) $|x + 3| > 2$; (b) $x < -5$ or $x > -1$; (c) $(-\infty, 5) \cup (-1, \infty)$; (d) given; (e) below.

19. $\left[\frac{3}{2}, \infty\right)$ 21. $\left[\frac{1}{3}, \infty\right)$

23. $\left[\frac{1}{2}, 3\right]$ 25. $(-\infty, 2] \cup [0, 2]$

27. $(-\infty, -2) \cup (2, \infty)$ 29. $(-\infty, -2) \cup [1, \infty)$

31. $(-\infty, 2) \cup [7, \infty)$ 33. $(-\infty, -1) \cup (1, 3)$

35. $\left[-\frac{1}{3}, 1\right]$ 37. $(-\infty, -3] \cup (-2, -1]$

39. $(-2, 1) \cup (4, \infty)$ 41. $(-\infty, -6) \cup (2, \infty)$

43. $[-9, -7]$ 45. $(-\infty, -1) \cup (5, \infty)$

47. $(-\infty, \infty)$ 49. $(-\infty, -2] \cup [2, \infty)$

51. $[0, 2]$

53. System is $(x - 4 > 0$ and $x + 4 > 0)$ or $(x - 4 < 0$ and $x + 4 < 0)$ or $(x = \pm 4)$; solution is $(-\infty, -4] \cup [4, \infty)$.

55. System is $(3x + 1 > 0$ and $x - 2 < 0)$ or $(3x + 1 < 0$ and $x - 2 > 0)$; solution is $\left(-\frac{1}{3}, 2\right)$.

57. System is $(x < 0$ and $x + 2 > 0$ and $x - 3 > 0)$ or $(x > 0$ and $x + 2 < 0$ and $x - 3 > 0)$ or $(x > 0$ and $x + 2 > 0$ and $x - 3 < 0)$ or $x = 0$ or $x = -2$ or $x = 3$; solution is $(-\infty, -2] \cup [0, 3]$.

59. $(-\infty, 1.7247]$, using a graphing utility to graph $y = x^5 - 2x^3 - 5$ as below.

61. $(-\infty, -0.72005] \cup [0.8233, 2.987]$, using a graphing utility to graph $y = 3x^4 - 1 - x^5$ as below.

63. (a) $0 < C < 100$; (b) $68 < F < 77$.

65. $|W - 20| \le 0.5$

67. $(x + c)(x - c) < 0$ exactly when $(x + c > 0$ and $x - c < 0)$ or $(x + c < 0$ and $x - c > 0)$, i.e. when $(x > -c$ and $x < c)$ or $(x < -c$ and $x > c)$. Since $c > 0$, the solution set is $((-c, \infty) \cap (-\infty, c)) \cup ((-\infty, c) \cap (c, \infty)) = (-c, c) \cup \emptyset = (-c, c)$.

Section 0.4

1. F, F, F, T, F, F, T, T.

3. See the reading at the beginning of this section.

5. Domain$(f) = \{x \in \mathbb{R} \mid f(x)$ is defined$\}$; for $f(x) = \sqrt{x}$ the domain is Domain$(f) = \{x \in \mathbb{R} \mid x \ge 0\}$.

7. (a) Yes: $\sqrt{3 + 1} = 2$; (b) No: $\sqrt{1 + 1} \ne 1$; (c) No: $\sqrt{-5 + 1}$ is undefined.

9. (a) $f(2) = 5$; (b) There is no x with $f(x) = 0$; (c) You can solve $y = x^2 + 1$ for x if and only if $y \in [1, \infty)$.

11. One example is $f(2) = 3, f(4) = 1, f(6) = 3, f(8) = 2, f(10) = 1$.

13. Yes, it could be a function; however, it would not be one-to-one.

15. (a) This is well-defined as the function $f(x) = x$, for $x \in \mathbb{R}$. (b) Not well-defined, since we would need $f(0) = 0$, and $f(3) = 3$, but 0 and 3 are not in the target set $(0, 3)$. (c) This is well-defined as the multivariable function $f(x, y, z) = (x, y, z)$, for $(x, y, z) \in \mathbb{R}^3$.

17. See the reading near Theorem 0.27.

19. $f(2) = f(-2)$

21. (a) $f(x) < 0$ for all $x \in I$; (b) $f(b) < f(a)$ for all $b > a$ in I

23. Average rate of change of f on $[a, b]$ is given by $\frac{f(b) - f(a)}{b - a}$. Here $f(b) - f(a)$ is the difference Δy of y-values, or the "rise." Similarly, $b - a$ is the difference Δx of x-values, or the "run."

25. The largest possible δ for the local maximum at $x = 2$ is approximately $\delta = 1.5$.

27. Domain $[1, \infty)$, range $[0, \infty)$

29. Domain $x \ne -2$, range $y \ne 0$

31. Domain $\mathbb{R}$, range $(0, 1]$

33. $(-\infty, 0] \cup [2, \infty)$

35. $(-\infty, -3] \cup [1, \infty)$

37. $(-\infty, -3) \cup (1, \infty)$

39. $(-\infty, -3) \cup (-3, -1] \cup [1, 3) \cup (3, \infty)$

41. $(1, 2) \cup (2, \infty)$

43. (a) 17; (b) $a^6 + 1$; (c) $(x^2 + 1)^2 + 1 = x^4 + 2x^2 + 2$.

45. (a) $\sqrt{25 + 9 + 4} = \sqrt{38}$; (b) $\sqrt{9 + 0 + 16} = 5$; (c) $\sqrt{x^2 + y^2 + z^2}$

47. (a) $(9, -3, 13)$; (b) $(17, 2, 8)$; (c) $(3a + b, a, b)$.

49. $f(-1) = -2, f(0) = 1, f(1) = 4, f(2) = 8$

51. One good window for this graph is $x \in [-1, 1]$, $y \in [-0.3, 1]$.

53.

55.

57. (a) Domain $\mathbb{R}$, range $[-1, \infty)$; (b) roots at $x = 0$ and $x = 2$, y-intercept at $y = 0$, local and global minimum at $x = 1$ (with a value of $f(1) = -1$), no local or global maxima, no inflection points; (c) positive on $(-\infty, 0) \cup (2, \infty)$, negative on $(0, 2)$, decreasing on $(-\infty, 1)$, increasing on $(1, \infty)$, concave up on $\mathbb{R}$, concave down nowhere; (d) no asymptotes.

59. (a) Domain $\mathbb{R}$, range $[-4.75, \infty)$; (b) roots at $x \approx -1.6$ and $x = 0$, y-intercept at $y = 0$, local and global minimum at $x = -1$ (with a value of $f(-1) = -4.75$), local minimum at $x = 2$ (with value $f(2) = 2$), local maximum at $x = 1$ (with value $f(1) = 3.25$), inflection points at $x \approx 0$ and $x \approx 1.5$; (c) positive on $(-\infty, 1.6) \cup (0, \infty)$, negative on $(-1.6, 0)$, decreasing on $(-\infty, -1) \cup (1, 2)$, increasing on $(-1, 1) \cup (2, \infty)$, concave up on $(-\infty, 0) \cup (1.5, \infty)$, concave down on $(0, 1.5)$; (d) no asymptotes.

61. (a) Domain $\mathbb{R}$, range $(1, 2]$; (b) no roots, y-intercept at $y = 2$, local and global maximum at $x = 0$ (with a value of $f(0) = 2$), no local or global minima, inflection points at $x = -1$ and $x = 1$; (c) positive on $(-\infty, \infty)$, negative nowhere, increasing on $(-\infty, 0)$, decreasing on $(0, \infty)$, concave up on $(-\infty, -1) \cup (1, \infty)$, concave down on $(-1, 1)$; (d) horizontal asymptote at $y = 1$.

63. One example is the graph of $f(x) = 2^{-x}$.

65. Not possible.

67.

69. One example is the graph of $f(x) = \dfrac{(x^2 - 1)(x^2 - 4)}{x}$.

71.

73. 4.2

75. $\dfrac{\sqrt{10} - \sqrt{2}}{8}$

77. $-\dfrac{100}{99}$

79.

81.

83. (a) Independent variables a and b, dependent variable H, function equation $H(a, b) = \sqrt{a^2 + b^2}$; (b) independent variables x, y, and z, dependent variable V, function equation $V(x, y, z) = xyz$.

85. 356 groundhogs

87. (a) $I(t) = \begin{cases} 36{,}000, & \text{if } 0 \le t < 4 \\ 38{,}500, & \text{if } 4 \le t < 6 \\ 49{,}000, & \text{if } t \ge 6. \end{cases}$

(b) $M(t) = \begin{cases} 36{,}000t, & \text{if } 0 \le t < 4 \\ 144{,}000 + 38{,}500(t - 4), & \text{if } 4 < t \le 6 \\ 221{,}000 + 49{,}000(t - 6), & \text{if } t \ge 6. \end{cases}$

(c) You will have earned a total of one million dollars after approximately 21.9 years.

89. The range of f is the set of real numbers y such that there is some x with $3x - 1 = y$. This range is $\mathbb{R}$, since for any real number y, the number $x = \dfrac{y+1}{3}$ has the property that $f(x) = 3\left(\dfrac{y+1}{3}\right) - 1 = y$.

91. It suffices to show that f is not one-to-one if and only if it fails the Horizontal Line Test. Not one-to-one means there exists some $a \ne b$ in the domain of f such that $f(a) = f(b)$. This is precisely what it means for f to intersect the horizontal line $y = f(a)$ in two places, namely, $(a, f(a))$ and $(b, f(b)) = (b, f(a))$.

93. For all a and b in $(-\infty, \infty)$, we have $a < b \Rightarrow -3a > -3b \Rightarrow 1 - 3a > 1 - 3b \Rightarrow f(a) > f(b)$.

95. For all real numbers a and b,
$$\dfrac{f(b) - f(a)}{b - a} = \dfrac{(-2b + 4) - (-2a + 4)}{b - a} = \dfrac{-2(b - a)}{b - a} = -2.$$

Section 0.5

1. T, F, T, F, T, T, T, F.

3. 3, 18.

5. $\Delta y = -12$

7. A horizontal line has slope 0, because it has a vertical change of 0 units for any change in horizontal units. A vertical line has undefined slope, because its "run" of zero units forces a zero in the denominator of the average rate of change formula.

9. Graphs of f, g, and h are shown below. Three points on the graph of $y = f(x)$ are $(-1, -4)$, $(0, -1)$, and $(2, 5)$. Three points on $y = g(x)$ are $\left(-2, \dfrac{9}{5}\right)$, $(0, 2)$, and $\left(2, \dfrac{11}{5}\right)$. Three points on $y = h(x)$ are $(-1, -10)$, $(0, 0)$, and $(1, 10)$.

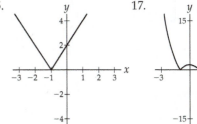

11. The missing entries in the first row are 4 and 9. The missing entries in the second row are -17 and -26.

13. f could not be proportional because it does not have a y-intercept of 0. It could be linear, since the average rate of change between any two of the given points is equal to 1. If it were linear its equation would thus be $f(x) = x + 1$.

15.

17.

19. If $x \geq 0$ then $\sqrt{x^2} = x$, and if $x < 0$ then $\sqrt{x^2} = -x$.

21. $f(x) = \frac{1}{5}x^2$ is a power function with $A = \frac{1}{5}$ and $k = 2$, and a polynomial function with $n = 2$, $a_2 = \frac{1}{5}$, $a_1 = 0$, and $a_0 = 0$, and a rational function with $p(x) = x^2$ and $q(x) = 5$ (a constant polynomial).

23. f is not linear, a power function, or a polynomial function, but it is rational since $x^{-2} + x^{-3} = \frac{x+1}{x^3}$.

25. f is rational with $p(x) = x^2 - 1$ and $q(x) = x^3 + x + 2$

27. Linear with slope $-\frac{1}{8}$, y-intercept $\frac{3}{8}$

29. Not linear

31. Linear with slope 2, y-intercept 13

33. $y + 2 = -(x - 3)$ 35. $y = 6x - 3$

37. $y = 2x - 2$ 39. $y - 4 = 2(x + 1)$

41. One possible answer is $f(x) = 2x^2$.

43. One possible answer is $f(x) = x(x - 2)(x - 4)$.

45. Three points on the graph are approximately $(-1, 1.82)$, $(0, 3)$, and $(2, 8.15)$.

47. Three points on the graph are $(1, 0)$, $(2, 1)$, and approximately $(3, 1.58)$.

49. Three points on the graph are $(0, 0)$, $\left(\frac{\pi}{2}, 1\right)$, and $(\pi, 0)$.

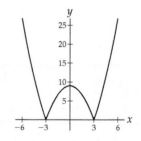

51. Three points on the graph are approximately $(-1, -0.785)$, $(0, 0)$, and $(2, 1.107)$.

53. $|5 - 3x| = \begin{cases} 5 - 3x, & \text{if } x \leq \frac{5}{3} \\ 3x - 5, & \text{if } x > \frac{5}{3} \end{cases}$

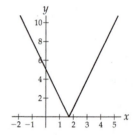

55. $|x^2 + 1| = x^2 + 1$ for all real numbers x

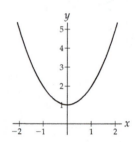

57. $|9 - x^2| = \begin{cases} 9 - x^2, & \text{if } -3 \leq x \leq 3 \\ -(9 - x^2), & \text{if } x < -3 \text{ or } x > 3. \end{cases}$

59. $|x^2 - 3x - 4| = \begin{cases} x^2 - 3x - 4, & \text{if } x \leq -1 \\ -(x^2 - 3x - 4), & \text{if } -1 < x < 4 \\ x^2 - 3x - 4, & \text{if } x \geq 4. \end{cases}$

61. Let h be the number of hours each week spent watching television, and let d represent the amount, in dollars, of credit card debt. $h(d) = kd$ for some constant k.

63. Let t be the number of years after 1985, and let P represent the number of people on the island. $P(t) = \dfrac{56}{15}t + 35$. Thus $P(t) = 200$ at $t \approx 44.2$ years after 1985, or sometime in 2029.

65. $\dfrac{1}{50}$

67. The average rate of change between any two data points is 3 CDs per month. Let N be the number of CDs, and let t be time in months, with $t = 0$ representing January. Then $N(t) = 3t + 51$. We have $N = 200$ when $t \approx 49.667$; since December is $t = 11$, it will be 39 months after December before Alina has 200 CDs.

69. Paul is losing at a constant rate of one pound every four days across all of the recorded data, so this could be a linear function. His weight after t days is $W(t) = -\dfrac{1}{4}t + 195$. We have $W(t) = 155$ after $t = 160$ days.

71. $f(x+1) = m(x+1) + b = mx + m + b = (mx + b) + m = f(x) + m$.

73. See the proof in the reading.

75. (a) If a linear function has slope m, then the average rate of change between the point (x_0, y_0) and any other point (x, y) on the line is given by $m = \dfrac{\Delta y}{\Delta x} = \dfrac{y - y_0}{x - x_0}$.
(b) Simply solve for y to obtain the formula.

77. Suppose lines l_1, l_2, and l_3 have slopes m_1, m_2, and m_3, respectively. l_2 perpendicular to l_1 means that $m_2 = -\dfrac{1}{m_1}$. l_3 perpendicular to l_2 means that $m_3 = -\dfrac{1}{m_2}$. Therefore $m_3 = -\dfrac{1}{m_2} = -\dfrac{1}{-1/m_1} = m_1$, so l_1 is parallel to l_3.

Section 0.6

1. F, T, T, F, F, F, F, F.

3. $(f - g)(x) = (f + (-g))(x) = f(x) + (-g)(x) = f(x) - g(x)$.

5. (a) One answer is $(1, 2)$; (b) one answer is $f\left(\dfrac{1}{2}x\right)$;
(c) $(-1, 4)$; (d) $(2, -5)$.

7. (a) The domain is the set of x such that $x \in [2, \infty)$ and $f(x) \in [-10, \infty)$. Since the range of f, $[-3, 3]$, always satisfies the second condition, the domain of $g \circ f$ is $[2, \infty)$. (b) We would need to know which x-values guarantee that $g(x) \in [2, \infty)$. (c) Not enough information for $f \circ f$; the domain of $g \circ g$ is $[-10, \infty)$.

9. $f(2x)$

11. If f is odd then $f(0)$ is either 0 or undefined.

13. (a) $f(-2) = 1$, $f(-1) = -2$, $f(3) = 4$ ($f(0)$ could be anything); (b) $f(-2) = -1$, $f(-1) = 2$, $f(0) = 0$, and $f(3) = -4$.

15. $(f^{-1})^{-1} = f$.

17. f^{-1} has domain $(-\infty, 3]$ and range $[-1, 1)$.

19. The missing entries in the first column are $-3, -2, 0, 0$. The missing entries in the second column are $-2, 3, 3, -6$. The missing entries in the third column are $-1, 2, 2, -7$.

21. Domain $x \neq 2$; $(f + g)(x) = x^2 + 1 + \dfrac{1}{x-2}$; $(f + g)(1) = 1$.

23. Domain $(0, 2) \cup (2, \infty)$; $\left(\dfrac{g}{h}\right)(x) = \left(\dfrac{1}{x-2}\right)/\sqrt{x}$; $\left(\dfrac{g}{h}\right)(1) = -1$.

25. Domain $x \neq 2, x \neq \dfrac{5}{2}$; $(g \circ g)(x) = \dfrac{1}{\dfrac{1}{x-2} - 2}$; $(g \circ g)(1) = -\dfrac{1}{3}$.

27. Domain $x \neq 7$; $g(x - 5) = \dfrac{1}{(x-5)-2}$; $g(1 - 5) = -\dfrac{1}{6}$.

29. Domain $x \geq -\dfrac{1}{3}$; $h(3x + 1) = \sqrt{3x + 1}$; $h(3(1) + 1) = 2$

31. The new row is $3, 5, 9, 7, 9, 3, 7$.

33. The new row is $2, 3, 2, 2, 3, 2, 3$.

35. The new row is $0, 1, 3, 2, 3, 0, 2$.

37. For $x = 1, 2, 3, 4, 5, 6, 7$ we have $g(x - 1) = 1, 0, 1, 1, 0, 1, 0$, respectively.

39.

41.

43.

45.

47.

49.

51. $(f+g)(-1) = 0$, $(f+g)(0) = 1$, $(f+g)(1) = 0$, $(f+g)(2) = 9$, $(f+g)(3) = 14$;
$$(f+g)(x) = \begin{cases} x+1, & \text{if } x \le 0 \\ x^2 - x, & \text{if } 0 < x < 2 \\ x^2 + 5, & \text{if } x \ge 2. \end{cases}$$

53. $(g \circ f)(-1) = 1$, $(g \circ f)(0) = -1$, $(g \circ f)(1) = -1$, $(g \circ f)(2) = 5$, $(g \circ f)(3) = 5$;
$$(g \circ f)(x) = \begin{cases} -(2x+1), & \text{if } x \le 0 \\ -x^2, & \text{if } 0 < x < \sqrt{2} \\ 5, & \text{if } x \ge \sqrt{2}. \end{cases}$$

55. $f(-1-1) = -3$, $f(0-1) = -1$, $f(1-1) = 1$, $f(2-1) = 1$, $f(3-1) = 4$;
$$f(x-1) = \begin{cases} 2x - 1, & \text{if } x \le 1 \\ (x-1)^2, & \text{if } x > 1. \end{cases}$$

57. One way: $g(x) = 3x + 5$, $h(x) = x^2$; another way: $g(x) = x + 5$, $h(x) = 3x^2$.

59. One way: $g(x) = \dfrac{6}{x}$, $h(x) = x + 1$; another way: $g(x) = 6x$, $h(x) = \dfrac{1}{x+1}$.

61. Even, since $f(-x) = (-x)^4 + 1 = f(x)$.

63. Neither, since $f(-x) \ne f(x)$ and $f(-x) \ne -f(x)$.

65. Odd, since $f(-x) = -f(x)$.

67. $f(g(x)) = 2 - 3\left(-\dfrac{1}{3}x + \dfrac{2}{3}\right) = x$ and
$g(f(x)) = -\dfrac{1}{3}(2 - 3x) + \dfrac{2}{3} = x$.

69. $f(g(x)) = f\left(\dfrac{x}{1+x}\right) = \dfrac{x/(1+x)}{1 - (x/(1+x))} = x$ (for $x \ne -1$);
similarly, $g(f(x)) = x$ for $x \ne 1$.

71.

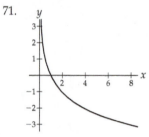

73. f is invertible on the restricted domain $[0, 2]$, and its inverse has the following graph:

75. $f^{-1}(x) = \dfrac{1-2x}{5}$

77. $f^{-1}(x) = \dfrac{1}{x-1}$

79. $f^{-1}(x) = \dfrac{-x-1}{x-1}$

81. (a) $S(x) = x^2$; (b) $C(S) = 4.25S + 200$;
(c) $C(x) = 4.25x^2 + 200$; $C(x)$ is the composition $C(S(x))$.

83. (a) $C(n) = 20 + 12n$. (b) Suppose there were a and b in the domain $(-\infty, \infty)$ of $C(n)$ with $C(a) = C(b)$. In this case $20 + 12a = 20 + 12b$, so $12a = 12b$, and therefore we must have $a = b$. (c) $n(C) = \dfrac{C-20}{12}$. This function describes the number n of shirts that you can have made for C dollars. (d) $n(150) \approx 10.83$, so you can make 10 full shirts.

85. $f(2x) = (2x)^k = 2^k x^k = 2^k f(x)$.

87. (a) If (x, y) is on the graph of f, then $y = f(x)$, so $kf(x) = ky$; thus (x, ky) is a point on the graph of $kf(x)$.
(b) If (x, y) is on the graph of f, then $y = f(x)$, so $f\left(k\left(\dfrac{1}{k}x\right)\right) = f(x) = y$, so $\left(\dfrac{1}{k}x, y\right)$ is on the graph of $f(kx)$.

89. If f is odd then for all x in the domain of f we have $f(-x) = -f(x)$. In particular, $f(-0) = -f(0)$, so $f(0) = -f(0)$. This means we must have $f(0) = 0$.

91. Since $f^{-1}(f(x)) = x$ for all x in the domain of f, we know that f^{-1} is defined for all values $f(x)$ in the range of f. Therefore the domain of f^{-1} is the range of f. Similarly, since $f(f^{-1}(x)) = x$, the range of f^{-1} is in the domain of f.

93. If $f(a) = b$ then $f^{-1}(f(a)) = f^{-1}(b)$ and therefore $a = f^{-1}(b)$.

Section 0.7

1. F, T, T, F, T, T, F, F.

3. If C is true, then D must be true. If C is false, then D may or may not be true.

5. "For all $x > 0$, we have $x > -2$." and "If $x > 0$, then $x > -2$."

7. The original statement is true. The converse is "Every rectangle is a square," which is false. The contrapositive is "Everything that is not a rectangle is not a square," which is true.

9. The contrapositive is "Not(Q)$\Rightarrow$Not(P)," which is logically equivalent to $P \Rightarrow Q$.

11. False.   13. True.

15. True.

17. True. The negation is "For all real numbers x, $x \le 2$ and $x \ge 3$."

19. True. The negation is "There exists a real number that is both rational and irrational."

21. True. The negation is "There exists x such that, for all y, $y \ne x^2$." (In other words, "There exists x for which there is no y with $y = x^2$.)"

23. True. The negation is "There is some integer x greater than 1 for which $x < 2$."

25. False. One counterexample is $x = 1.35$.

27. False.

29. False. One counterexample is $x = -1$.

31. True. One example is $x = 3$.

33. True. The negation is "There exist real numbers a and b such that $a < b$ but $3a + 1 \geq 3b + 1$."

35. True.   37. False.

39. True.

41. True. One example is $x = -1$, since for all y we have $|y| > -1$.

43. True.

45. False. The only counterexample where the two sides of the double implication are not equivalent is $x = 0$, $y = 0$.

47. (a) $B \Rightarrow$ (Not A); (b) (Not B) $\Rightarrow A$

49. (a) (Not A) $\Rightarrow$ (Not B); (b) $A \Rightarrow B$

51. (a) $C \Rightarrow (A$ and B); (b) Not(C) $\Rightarrow$ (Not A) or (Not B)

53. (a) (B and C) $\Rightarrow A$; (b) ((Not B) or (Not C)) $\Rightarrow$ (Not A)

55. (a) The converse is "If x is rational, then x is a real number." (b) The contrapositive is "If x is irrational, then x is not a real number." (c) $x = \pi$ is a counterexample to the original and the contrapositive.

57. (a) The converse is "If $x \geq 3$, then $x > 2$." (b) The contrapositive is "If $x < 3$, then $x \leq 2$." which is false. (c) $x = 2.5$ is a counterexample to both the original and the contrapositive.

59. (a) The converse is "If $\sqrt{x}$ is not a real number, then x is negative." (b) The contrapositive is "If $\sqrt{x}$ is a real number, then x is nonnegative. (c) No possible counterexamples for any of the statements.

61. (a) The converse is "If $|x| = -x$, then $x \leq 0$." (b) The contrapositive is "If $|x| \neq -x$, then $x > 0$." (c) No possible counterexamples for any of the statements.

63. (a) The converse is "If $x^2 > x$, then x is not zero." (b) The contrapositive is "If $x^2 \leq x$, then $x = 0$." (c) $x = 0.5$ is a counterexample to the original statement and its contrapositive; the converse happens to be true.

65. (a) The converse is "If there is some integer n such that $x = 2n + 1$, then x is odd." (b) The contrapositive is "If there is no integer n such that $x = 2n + 1$, then x is not odd." (c) No possible counterexamples for any of the statements.

67. Linda wears a red hat, Alina wears a blue hat, Phil wears a green hat, Stuart wears a yellow hat.

69. Liz and Rein are liars, and Zubin tells the truth. Here is an argument that proves this solution is correct: If Liz told the truth then everyone would, which would contradict Rein's and Zubin's statements. Therefore Liz must be a liar. Now if Rein is telling the truth, then by Rein's statement, Zubin must tell the truth also, but Zubin's statement would be false; another contradiction. Therefore Rein must be lying. This makes Zubin's statement true, so Zubin is telling the truth.

71. Kate and Hyun are liars and Jaan tells the truth. Here's why: Seeking a contradiction, suppose Kate tells the truth. Then by her statement both Hyun and Jaan lie. If Hyun lies then her statement means that Kate also lies, which is a contradiction. Therefore we can assume that Kate lies. Thus Hyun's statement is false, so Hyun is a liar. But by Kate's false statement, at least one of Hyun or Jaan tells the truth. Therefore Jaan tells the truth.

73. Suppose x is irrational and r is rational. Seeking a contradiction, suppose $x - r$ is rational. Since the sum of two rational numbers is rational, we know that $(x - r) + r = x$ must be rational. But this contradicts the hypothesis that x is irrational; therefore $x - r$ must be an irrational number.

75. If n is even and m is odd then there are integers k and l such that $n = 2k$ and $m = 2l + 1$. Then $n + m = (2k) + (2l + 1) = 2(k + l) + 1$. Since $k + l$ is an integer, this means that $n + m$ is odd.

77. Let $n = 2k$ and $m = 2l + 1$ and compute n times m.

79. If $M = \left(\frac{x_1 + x_2}{2}, \frac{y_1 + y_2}{2} \right)$, the distance formula easily shows that dist$(M, P) = $ dist(M, Q).

81. If a, b, and c are any real numbers with $b \neq 0$ and $c \neq 0$, then $\frac{a/b}{c} = \frac{a/b}{c/1} = \frac{a(1)}{bc} = \frac{a}{bc}$.

83. Follow the outline given in the problem.

85. $|a - b| + |b| \geq |(a - b) + b| = |a|$ so $|a - b| \geq |a| - |b|$.

87. $|x - c| > \delta$ if and only if $x - c > \delta$ or $x - c < -\delta$, which is equivalent to having $x > c + \delta$ or $x < c - \delta$, meaning that $x \in (c + \delta, \infty) \cup x \in (-\infty, c - \delta)$.

For answers to odd-numbered Skill Certification exercises in the Chapter Review, please visit the Book Companion Web Site at www.whfreeman.com/integcalculus.

Chapter 1

Section 1.1

1. T, F, T, F, F, T, F, T.

3. $\lim\limits_{x \to 1} f(x) = 5$; cannot say anything about $f(1)$

5. $\lim\limits_{x \to 2^-} f(x)$ is not equal to 8.

7. f has a horizontal asymptote at $y = 3$ and a vertical asymptote at $x = 1$.

9. (a) The terms get smaller; $\lim\limits_{k \to \infty} \frac{1}{3^k} = 0$. (b) For every $k > 8$ the terms will be less than 0.0001.

11. (a) The terms get larger and larger as more numbers are added. (b) For every $k > 44$ the terms will be greater than 1000.

13. (a) By solving $s(t) = 0$ we see that the orange hits the surface of the moon at about $t = 2.75$ seconds. The average rate of change on $[1.75, 2.75]$ is -11.925 feet per second; on $[2.25, 2.75]$ it is -13.25, on $[2.5, 2.75]$ it is -13.9125, and on $[2.625, 2.75]$ it is -14.2438. (b) We might estimate -14.5 feet per second on impact.

15. (a) 6.25 for four rectangles, and 5.8125 for eight. (b) If we did this for more and more rectangles, the area approximation would decrease to become closer and closer to the actual area under the curve. We might estimate 5.5 square units from our earlier work. (The actual answer turns out to be about 5.33.)

17. Let the graph of the function escape out to $\pm\infty$ at $x = 2$ with a vertical asymptote at the last minute.

19. (a) $g(1)$ is of the form $\frac{0}{0}$ and therefore is not defined. A calculator graph is shown here. It is not immediately clear that $g(x)$ is not defined at $x = 1$. Tracing along the graph near $x = 1$ might reveal a hole in the graph, depending on your calculator. (b) Even though $g(1)$ is undefined, the function values do approach 0 as $x \to 1$, so the limit exists and is equal to 0.

21.

23.

25.

27.

29.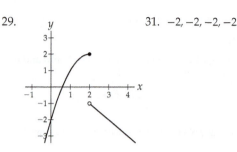

31. $-2, -2, -2, -2$

33. $-2, 2$, DNE, 2

35. $\infty, -\infty$, DNE, undefined

37. $-1, -1, -1, -2$

39. 7

41. DNE: $-\infty$ from the left and ∞ from the right.

43. $\frac{1}{7}$

45. -3

47. 0

49. DNE: $\sin x$ oscillates between -1 and 1 as $x \to \infty$.

51. 7

53. 0

55. DNE: $-\infty$ from the left and ∞ from the right.

57. 2

59. DNE: 4 from the left and -5 from the right.

61. 2

63. -93

65. DNE: $-\infty$ from the left and ∞ from the right.

67. 0

69. 1

71. (a) $S(0) = \frac{12}{3} = 4$; (b) 17, 19, 21; (c) 22; (d) 22.

73. (a) Your y-range will have to be quite large; (b) ∞; (c) population explosion past what any finite-sized planet can handle; (d) in this model no world would be left after 2027.

75. For $k = 101$ we have $\frac{101(101+1)}{2} = 5151$, and for larger k the quantity $\frac{k(k+1)}{2}$ will be still larger.

77. By the previous problem, if x is within 0.01 of $x = 1$ then $(x - 1)^2$ is in the interval $(0, 0.0001)$, and thus $0 < (x - 1)^2 < 0.0001$. Therefore $\frac{1}{(x-1)^2} > \frac{1}{0.0001} = 10,000$.

Section 1.2

1. F, F, F, F, F, F, T, T.

3. See the reading at the beginning of the section.

5. (a) $(1, 1.5) \cup (1.5, 2)$; (b) $(0, 0.25) \cup (0.25, 0.5)$; (c) $(0, 1) \cup (1, 2)$

7. If $x \in (2 - \delta, 2) \cup (2, 2 + \delta)$, then $f(x) \in (5 - \varepsilon, 5 + \varepsilon)$.

9. If $x \in (N, \infty)$, then $f(x) \in (2 - \varepsilon, 2 + \varepsilon)$.

11. If $x \in (1, 1 + \delta)$, then $f(x) \in (M, \infty)$.

13. Combine the two figures above Theorem 1.8. The two blue one-sided δ-bars combine to make a punctured δ-interval around $x = c$.

15. $(1.75, 6.25)$

17. There is some $M>0$ for which there is no $N>0$ that would guarantee that if $x>N$, then $\frac{1000}{x}>M$.

19. For all $\epsilon > 0$, there exists a $\delta > 0$ such that if $x \in (-3 - \delta, -3) \cup (-3, -3 + \delta)$, then $\sqrt{x+7} \in (2 - \epsilon, 2 + \epsilon)$.

21. For all $\epsilon > 0$, there exists a $\delta > 0$ such that if $x \in (-1 - \delta, -1) \cup (-1, -1 + \delta)$, then $x^3 - 2 \in (-3 - \epsilon, -3 + \epsilon)$.

23. For all $\epsilon > 0$, there exists a $\delta > 0$ such that if $x \in (2 - \delta, 2) \cup (2, 2 + \delta)$, then $\frac{x^2-4}{x-2} \in (4 - \epsilon, 4 + \epsilon)$.

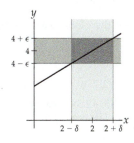

25. For all $\epsilon > 0$, there exists a $\delta > 0$ such that if $x \in (0, \delta)$, then $\sqrt{x} \in (-\epsilon, \epsilon)$.

27. For all $M > 0$, there exists a $\delta > 0$ such that if $x \in (-2, -2 + \delta)$, then $\frac{1}{x+2} \in (M, \infty)$.

29. For all $M > 0$, there exists a $\delta > 0$ such that if $x \in (3 - \delta, 3)$, then $\frac{1}{3-x} \in (M, \infty)$.

31. For all $\epsilon > 0$, there is some $N > 0$ such that if $x \in (N, \infty)$, then $\frac{x}{1-2x} \in (-0.5 - \epsilon, -0.5 + \epsilon)$.

33. For all $M > 0$, there is some $N > 0$ such that if $x \in (N, \infty)$, then $x^3 + x + 1 \in (M, \infty)$.

35. For all $\epsilon > 0$, there exists a $\delta > 0$ such that if $h \in (-\delta, 0) \cup (0, \delta)$, then $\frac{(2+h)^2 - 4}{h} \in (4 - \epsilon, 4 + \epsilon)$.

37. For all $M < 0$, there exists some $\delta > 0$ such that if $x \in (c, c + \delta)$, then $f(x) \in (-\infty, M)$.

39. For all $\epsilon > 0$ there is some $N < 0$ such that if $x \in (-\infty, N)$, then $f(x) \in (L - \epsilon, L + \epsilon)$.

41. For all $M > 0$, there is some $N < 0$ such that if $x \in (-\infty, N)$, then $f(x) \in (M, \infty)$.

43. $\delta = (8.5)^{1/3} - 2 \approx 0.0408$ 45. $\delta = 3$

47. $\delta = -2 + \sqrt{4.5} \approx 0.12132$ 49. $\delta = 0.25$

51. $\delta \approx 0.016662$ 53. $\delta = 1$

55. $\delta \approx 0.0005$ 57. $N = 5$

59. $N = 250{,}002$ 61. $N = -2$

63. $N = \sqrt{104} \approx 10.198$

65. (a) 149.3 months; (b) yes, since $\lim_{t \to \infty} 50t = \infty$.

67. (a) \$18.68; (b) between \$16.20 and \$21.50; (c) $L = 6000$, $c = 18.68$, $\epsilon = 1000$, $\delta = 2.48$.

69. (a) For all $\epsilon > 0$, there exists $\delta > 0$ such that if $x \in (2 - \delta, 2) \cup (2, 2 + \delta)$, then $7 - x \in (5 - \epsilon, 5 + \epsilon)$.
(b) $2 - \delta < x < 2 + \delta$ and $x \neq 2$ means that $-\delta < x - 2 < \delta$ and $x \neq 2$. This means that $|x - 2| < \delta$ and $x - 2 \neq 0$, and thus $0 < |x - 2| < \delta$.
(c) $5 - \epsilon < 7 - x < 5 + \epsilon$ means that $-\epsilon < 2 - x < \epsilon$, and thus that $|2 - x| < \epsilon$, which is equivalent to $|x - 2| < \epsilon$. (d) For a given $\epsilon > 0$, choose $\delta = \epsilon$.
(e) Given any $\epsilon > 0$, let $\delta = \epsilon$. If $x \in (2 - \delta, 2) \cup (2, 2 + \delta)$, then $0 < |x - 2| < \delta$ by part (b). Thus by part (d), $|x - 2| < \epsilon$, and therefore $7 - x \in (5 - \epsilon, 5 + \epsilon)$ by part (c).

71. (a) For all $\epsilon > 0$, there exists $N > 0$ such that if $x \in (N, \infty)$, then $\frac{1}{x} \in (0 - \epsilon, 0 + \epsilon)$. (b) $x \in (N, \infty)$ is equivalent to $x > N$. (c) $\frac{1}{x} \in (-\epsilon, \epsilon)$ means that $-\epsilon < \frac{1}{x} < \epsilon$. For the limit we are considering, $\frac{1}{x}$ approaches $y = 0$ from above, so we actually wish to show that we can guarantee $0 < \frac{1}{x} < \epsilon$. (d) For a given $\epsilon > 0$, choose $N = \frac{1}{\epsilon}$. (e) Given any $\epsilon > 0$, let $N = \frac{1}{\epsilon}$. If $x \in (N, \infty)$ then $x > N > 0$ by part (b). Thus by our choice from part (d), $0 < \frac{1}{\epsilon} < x$, and thus $0 < \frac{1}{x} < \epsilon$. By part (c) this means $-\epsilon < \frac{1}{x} < \epsilon$ and thus $\frac{1}{x} \in (0 - \epsilon, 0 + \epsilon)$.

Section 1.3

1. T, T, F, T, T, F, F, F.

3. For all $\epsilon > 0$, there exists a $\delta > 0$ such that if $0 < |x + 2| < \delta$, then $|\frac{3}{x+1} + 3| < \epsilon$.

5. $0 < |x - 5| < 0.01$; $|f(x) + 2| < 0.5$

7. $x \in (1.9, 2) \cup (2, 2.1)$

9. $x^2 - 1 \in (-3.5, -2.5)$ (note that there are in fact no x-values with this property)

11. $f(x) \in (L - \varepsilon, L + \varepsilon)$

13. The statement is false. One counterexample is $x = 1.5$, since $x = 1.5$ satisfies $0 < |x - 2| < 1$, but not $|x^2 - 4| < 0.5$, since $(1.5)^2 = 2.25$.

15. The statement is false. One counterexample is $x = 0.74$; this value satisfies $0 < |x - 0| < 0.75$ but not $|x^2 - 0| < 0.5$, since $(0.74)^2 = 0.5476$.

17. The statement is true. Sketch the graph of $y = x^2$ and show an ϵ interval around $y = 4$ with $\epsilon = 0.4$ (i.e., show the horizontal "bar" between $y = 3.6$ and $y = 4.4$). Then draw the δ interval around $x = -2$ with $\delta = 0.075$, excluding $x = -2$ (i.e., show the vertical "bar" between $x = -2.075$ and $x = -1.925$, excluding $x = -2$). Indicate on your graph that every x-value in $(-2.075, -2) \cup (-2, -1.925)$ has an $f(x)$-value in $(3.6, 4.4)$.

19. (a) $\delta = \frac{5}{325}$; (b) $\delta = \frac{.01}{325}$; (c) $\delta = 1$

21. See the proof of Theorem 1.11.

23. $\delta = \frac{0.25}{3}$

25. $\delta = 0.25$

27. $N = \sqrt{1000} \approx 31.62$

29. $\delta = \epsilon$

31. $\delta = \frac{\epsilon}{4}$

33. $\delta = \sqrt{\frac{\epsilon}{5}}$

35. $\delta = \sqrt{\epsilon}$

37. $\delta = \min(1, 2\epsilon)$

39. $\delta = \min(1, \epsilon/5)$

41. $\delta = \varepsilon^2$

43. $\delta = \frac{1}{M}$

45. With $C(r) = 0.25(2\pi r(5)) + 0.5(2\pi r^2) + 0.1(2(2\pi r) + 5)$, we have (a) $r \approx 1.94$ inches; (b) $|C(r) - 30| < 10$; (c) $r \in (1.43, 2.38)$, so $|r - 1.94| < 0.44$.

47. Given $\epsilon > 0$, choose $\delta = \epsilon/2$. Then if $0 < |x - 1| < \delta$, we have $|(2x + 4) - 6| = 2|x - 1| < 2\delta = \epsilon$.

49. Given $\epsilon > 0$, choose $\delta = \epsilon$. Then if $0 < |x + 6| < \delta$, we have $|(x + 2) - (-4)| = |x + 6| < \delta = \epsilon$.

51. Given $\epsilon > 0$, choose $\delta = \frac{\epsilon}{6}$. Then if $0 < |x - 4| < \delta$, we have
$|(6x - 1) - (23)| = |6x - 24| = 6|x - 4| < 6\delta = \epsilon$.

53. Given $\epsilon > 0$, choose $\delta = \sqrt{\epsilon/3}$. Then if $0 < |x| < \delta$, we have $|(3x^2 + 1) - 1| = |3x^2| = 3|x|^2 < 3\delta^2$ $= 3(\sqrt{\epsilon/3})^2 = \epsilon$.

55. Given $\epsilon > 0$, choose $\delta = \sqrt{\frac{\epsilon}{2}}$. Then if $0 < |x - 1| < \delta$, we have $|(2x^2 - 4x + 3) - 1| = |2x^2 - 4x + 2| = 2|x^2 - 2x + 1| = 2|x - 1|^2 < 2\delta^2 = \epsilon$.

57. Given $\epsilon > 0$, choose $\delta = \epsilon$. Then if $0 < |x - 1| < \delta$, we have $\left|\frac{x^2 - 1}{x - 1} - 2\right| = \left|\frac{x^2 - 2x + 1}{x - 1}\right| = |x - 1| < \delta = \epsilon$.

59. Given $\epsilon > 0$, choose $\delta = \epsilon^2$. Then if $5 < x < 5 + \delta$, we have $0 < x - 5 < \delta$ and thus
$$|\sqrt{x - 5} - 0| = \sqrt{x - 5} < \sqrt{\delta} = \epsilon.$$

61. Given $M > 0$, choose $\delta = 1/M$. Then if $x \in (-2, -2 + \delta)$, we have $0 < x + 2 < \delta$, and thus $\frac{1}{x + 2} > \frac{1}{\delta} = \frac{1}{1/M} = M$.

63. Given $\epsilon > 0$, choose $N = \frac{1}{\epsilon}$. Then if $x > N$, we have $\left|\frac{2x - 1}{x} - 2\right| = \left|\frac{2x - 1 - 2x}{x}\right| = \frac{1}{|x|} < \frac{1}{N} = \epsilon$.

65. Given $M > 0$, choose $N = \frac{M + 5}{3}$. Then if $x > N$, we have $3x - 5 > 3N - 5 = M$.

67. Given $\epsilon > 0$, choose $\delta = \min(1, \epsilon/5)$. Then if $0 < |x - 3| < \delta$, we have $|(x^2 - 2x - 3) - 0| = |x^2 - 2x - 3| = |x - 3||x + 1| < \delta|x + 1| \leq \delta(5) \leq \epsilon$.

69. Given $\epsilon > 0$, choose $\delta = \min(1, \epsilon/5)$. Then if $0 < |x - 5| < \delta$, we have $|(x^2 - 6x + 7) - 2| = |x^2 - 6x + 5| = |x - 5||x - 1| < \delta|x - 1| \leq \delta(5) = \epsilon$.

71. Given $\epsilon > 0$, choose $\delta = \min(1, \epsilon/3)$. Then if $0 < |x - 2| < \delta$, we have
$$\left|\frac{4}{x^2} - 1\right| = \frac{|2 - x||2 + x|}{x^2} < \delta\frac{|2 + x|}{x^2} \leq \delta(3) = \epsilon.$$

Section 1.4

1. T, T, F, F, F, F, T, F.

3. $\lim_{x \to 1} f(x) = f(1)$

5. See the reading.

7. Define $f(1)$ to be 0.

9. Removable discontinuity, neither left nor right continuous. One possible graph is:

11. Jump discontinuity, neither left nor right continuous. One possible graph is:

13. For all $\epsilon > 0$, there exists $\delta > 0$ such that whenever $x \in (c - \delta, c + \delta)$, we also have $f(x) \in (f(c) - \epsilon, f(c) + \epsilon)$.

15. For all $\epsilon > 0$, there exists $\delta > 0$ such that whenever $x \in (c, c + \delta)$, we also have $f(x) \in (f(c) - \epsilon, f(c) + \epsilon)$.

17. See the reading.

19. The graph below is not continuous on $[1, 3]$, and there is no minimum value on that interval.

21.

23. Continuous on $(-\infty, 2) \cup (2, \infty)$, removable discontinuity at $x = 2$, neither left- nor right continuous at $x = 2$.

25. Continuous on $(-\infty, -1) \cup (-1, 1] \cup (1, \infty)$, removable discontinuity at $x = -1$, infinite discontinuity at $x = 1$, left continuous at $x = 1$.

27. Not possible.

29.

31. Not possible.   33. 6

35. -17

37. Can't do this yet, since 0 is not in the domain of x^{-3}.

39. Continuous at $x = 3$.

41. Jump discontinuity and left continuous at $x = 2$.

43. Continuous at $x = 1$, jump discontinuity and right continuous at $x = 2$.

45. Not continuous at $x = 2$.

47. Continuous at $x = 1$.

49. (a) Since f is continuous on $[-2, 2]$, the Extreme Value Theorem guarantees that f will attain global maximum and minimum values on $[-2, 2]$. (b) The maximum value on that interval occurs twice, at $x = -2$ and at $x = 2$, and the minimum value occurs both at $x \approx -1.22$ and at $x \approx 1.22$.

51. (a) Since f is continuous on $[-1, 1]$, the Extreme Value Theorem guarantees that f will attain global maximum and minimum values on $[-1, 1]$. (b) The maximum value on that interval occurs at $x = 0$, and the minimum value occurs twice, at $x = -1$ and $x = 1$.

53. (a) Since f is continuous on $[0, 2]$, the Extreme Value Theorem guarantees that f will attain global maximum and minimum values on $[0, 2]$. (b) The minimum value on that interval occurs at $x = \dfrac{4}{3}$, and the maximum value occurs twice, at $x = 0$ and $x = 2$.

55. (a) Since f is continuous on $[0, 2]$ and $K = 0$ is between $f(0) = 5$ and $f(2) = -11$, the Intermediate Value Theorem guarantees that there is some $c \in (0, 2)$ such that $f(c) = 0$. (b) Using a graph we can approximate that in $(0, 2)$ there is one such value of c, namely, $c \approx 1.5$.

57. (a) Since f is continuous on $[-2, 4]$ and $K = -4$ is between $f(-2) = -22$ and $f(4) = 14$, the Intermediate Value Theorem guarantees that there is some $c \in (-2, 4)$ such that $f(c) = -4$. (b) Using a graph we can approximate that in $(-2, 4)$ there are three such values of c, namely, $c = 1$, $c \approx -0.73$, and $c \approx 2.73$.

59. (a) Since f is continuous on $[2, 4]$ and $K = -4$ is between $f(2) = -6$ and $f(4) = 14$, the Intermediate Value Theorem guarantees that there is some $c \in (2, 4)$ such that $f(c) = -4$. (b) Using a graph we can approximate that in $(2, 4)$ there is one such value of c, namely, $c \approx 2.73$.

61. (a) Since f is continuous everywhere, the Intermediate Value Theorem applies on any interval $[a, b]$. For example, since $f(-3) = -25 < -15$ and $f(0) = 2 > -15$, the Intermediate Value Theorem guarantees that there is some $c \in (-3, 0)$ for which $f(c) = -15$. (b) Using a graph we can approximate that $c \approx -2.57$ is such a value.

63. (a) Since f is continuous everywhere, the Intermediate Value Theorem applies on any interval $[a, b]$. For example, since $f(1) = 14 > -1$ and $f(3) = -2 < -1$, the Intermediate Value Theorem guarantees that there is some $c \in (1, 3)$ for which $f(c) = -1$. (b) Since $f(2) = -1$, $c = 2$ is such a value in $(1, 3)$.

65. (a) Since f is continuous everywhere, the Intermediate Value Theorem applies on any interval $[a, b]$. For example, since $f\left(-\dfrac{1}{3}\right) = 0 < 1$ and $f(2) = 7 > 1$, the Intermediate Value Theorem guarantees that there is some $c \in (0, 7)$ for which $f(c) = 1$. (b) $c = 0$ is such a value.

67. Positive on $(-\infty, -2) \cup \left(-\dfrac{1}{2}, \infty\right)$, negative on $\left(-2, -\dfrac{1}{2}\right)$

69. Positive on $(-\infty, -2) \cup (-1, 1) \cup (2, \infty)$, negative on $(-2, -1) \cup (1, 2)$

71. Positive on $(2, \infty)$, negative on $(-\infty, 2)$

73. Positive on $(0, 2]$, negative on $(-\infty, 0) \cup (2, \infty)$

75. The length $H(t)$ of Alina's hair is a continuous function because her hair can't suddenly get longer or shorter without going through a continuous change, so the theorems apply. We are given that $H(0) = 2$ and $H(6) = 42$. The Extreme Value Theorem says that Alina's hair had a maximum and a minimum length sometime in the last 6 years. The Intermediate Value Theorem says that Alina's hair has been every length between 2 and 42 inches at some point during the last 6 years.

77. The Extreme Value Theorem says that at some point in the past year, there was a time when Phil's wagon contained the most gas and a time when it contained the least. The Intermediate Value Theorem says that for every amount between 0 and 19 gallons of gas, there was at least one time in the past year that Phil's wagon contained that amount of gas.

79. (a) $M(v) = \begin{cases} 8,500, & \text{if } 0 \le v < 30 \\ 10,000, & \text{if } 30 \le v < 60 \\ 11,500, & \text{if } 60 \le v < 90 \\ 13,000, & \text{if } v = 90. \end{cases}$
(b) $M(0) = 8,500$, $M(30) = 10,000$, $M(59) = 10,000$, $M(61) = 11,500$, and $M(90) = 13,000$; (c) $M(v)$ fails to be continuous at $v = 30$, $v = 60$, and $v = 90$.

81. Given $\epsilon > 0$, let $\delta = \epsilon/3$. If $0 < |x - 2| < \delta$, then $|(3x - 5) - (3(2) - 5)| = |3x - 6| = 3|x - 2| < 3\delta = \epsilon$.

83. Given $\epsilon > 0$, choose $\delta = \epsilon$. If $0 < |x - c| < \delta$, then $||x| - |c|| < |x - c| < \delta = \epsilon$.

85. f is not continuous at $c = 2$. For example, consider $\epsilon = 1$. No matter how small we take $\delta > 0$, there are irrational values of x with $0 < |x - 2| < \delta$ for which $|f(x) - f(c)| = |x^2 - 0| = x^2$ is outside of the interval $(0 - \epsilon, 0 + \epsilon) = (-1, 1)$.

87. If $A > 0$ then for large magnitude N we will have $f(-N) < 0 < f(N)$, and if $A < 0$ then for large magnitude N we will have $f(-N) > 0 > f(N)$. In either case the Intermediate Value Theorem applies to the continuous cubic function $f(x)$ on $[-N, N]$ and therefore $f(x)$ has at least one real root on $[-N, N]$.

89. Suppose $c \ge 0$; the proof for $c \le 0$ is similar. Given $\epsilon > 0$, let $\delta = \min\left(1, \dfrac{\epsilon}{2c+1}\right)$. If $0 < |x - c| < \delta$, then since $\delta \le 1$ we have $c - 1 < x < c + 1$. Then $|x^2 - c^2| = |x - c| \cdot |x + c| < \delta \cdot |x + c| \le \delta|(c + 1) + c| = \delta(2c + 1) = \dfrac{\epsilon}{2c+1}(2c + 1) = \epsilon$.

Section 1.5

1. T, T, F, F, T, F, F, T.   3. See Theorem 1.20.

5. Consider the limit $\lim\limits_{x \to 1}\dfrac{x-1}{x^2-1}$.

7. $f(x) = x$, $g(x) = \dfrac{1}{x}$, $c = 0$ is one example. This does not contradict the product rule for limits because one of the limits does not exist as $x \to 0$.

9. For all $\epsilon > 0$, there exists $\delta > 0$ such that if $0 < |x - c| < \delta$, then $|(f(x) - g(x)) - (L - M)| < \epsilon$, where $L = \lim\limits_{x \to c} f(x)$ and $M = \lim\limits_{x \to c} g(x)$.

11. To get this function we add the power function $\sqrt{x}$ and the constant function 1, and then compose the result with the power function x^3. Thus by the sum and composition rules for limits, and the fact that power and constant functions are continuous, we know that this function f is continuous and thus we can calculate its limits at domain points by evaluation.

13. -2

15. Not enough information.

17. $\dfrac{1}{2}$

19. The graph of g looks just like the graph of f but with a hole at $x = 1$. $f(1) = 2$ but $g(1)$ is undefined, while both functions approach 2 as x approaches 1.

21. Using the constant multiple rule followed by the continuity of linear functions, we have $\lim\limits_{x \to 1} 15(3 - 2x) = 15 \lim\limits_{x \to 1}(3 - 2x) = 15(3 - 2(1)) = 15$.

23. Applying the sum rule, the product rule, the constant multiple rule, and then the continuity of power and linear functions, in that order, we have $\lim\limits_{x \to 3}(3x + x^2(2x + 1)) = \lim\limits_{x \to 3}(3x) + \lim\limits_{x \to 3}(x^2(2x + 1)) = \lim\limits_{x \to 3}(3x) + \lim\limits_{x \to 3} x^2 \cdot \lim\limits_{x \to 3}(2x + 1) = 3(3) + (3^2)(2(3) + 1) = 72$.

25. -1 27. 4

29. 2   31. 2

33. $-\dfrac{1}{10}$ 35. $-\dfrac{1}{3}$

37. $-\dfrac{1}{4}$ 39. 0

41. $\dfrac{4}{7}$ 43. 0

45. 3   47. -1

49. 6   51. 3

53. -1 55. -1

57. Continuous on $\mathbb{R}$.

59. Continuous on $(-\infty, -2)$ and on $(-2, \infty)$; neither left nor right continuous at $x = -2$.

61. Continuous on $(-\infty, 1)$ and on $(1, \infty)$; neither left nor right continuous at $x = 1$.

63. Continuous on $(-\infty, 2)$ and $(2, \infty)$; left but not right continuous at $x = 2$.

65. (a) $T(63,550) = 14381.50$; (b) this value for $T(63, 550)$ matches up with the value of the next piece $14,381.5 + 0.31(m - 63, 550)$ when $m = 63, 550$. This means that when you change tax brackets, the higher tax rate only applies to the amount over $63, 550$.

67. If $f(x) = a_n x^n + a_{n-1}x^{n-1} + \cdots + a_1 x + a_0$ is a polynomial function and c is a real number, then by the sum rule, constant multiple rule, and limit of a constant, we have $\lim\limits_{x \to c} f(x) = a_n \lim\limits_{x \to c} x^n + a_{n-1} \lim\limits_{x \to c} x^{n-1} + \cdots + a_1 \lim\limits_{x \to c} x + a_0$. Since power functions with positive integer powers are continuous everywhere, we can solve each of the component limits by evaluation, which gives us $\lim\limits_{x \to c} f(x) = a_n c^n + a_{n-1} c^{n-1} + \cdots + a_1 x + a_0 = f(c)$. Therefore the polynomial function f is continuous at $x = c$.

69. Given $\epsilon > 0$, we can choose $\delta > 0$ so that if $x \in (c - \delta, c) \cup (c, c + \delta)$, then $f(x)$ is within $\dfrac{\epsilon}{|k|}$ of L. Then $|kf(x) - kL| = |k||f(x) - L| < |k|(\epsilon/|k|) = \epsilon$.

71. If $\lim_{x \to c} f(x)$ exists and $\lim_{x \to c} g(x)$ exists and is nonzero, then by the product and reciprocal rules for limits we have

$$\lim_{x \to c} \frac{f(x)}{g(x)} = \lim_{x \to c} f(x) \frac{1}{g(x)} = \lim_{x \to c} f(x) \lim_{x \to c} \frac{1}{g(x)} =$$

$$\lim_{x \to c} f(x) \frac{1}{\lim_{x \to c} g(x)} = \lim_{x \to c} f(x) \lim_{x \to c} g(x).$$

Section 1.6

1. T, T, F, T, T, T, F, F.

3. $\lim_{x \to c} \frac{f(x)}{g(x)} = \frac{L}{M}$

5. $\lim_{x \to c} \frac{f(x)}{g(x)}$ is infinite

7. $\infty + \infty \to \infty$, $\infty - \infty$ is indeterminate, $\infty + 1 \to \infty$, and $0 + \infty \to \infty$.

9. See Theorems 1.27 and 1.28.

11. (a) The numerator approaches 1 while the denominator gets larger and larger, which makes the quotient get smaller and smaller; (b) see the reading; (c) the denominator approaches 1 while the numerator gets smaller and smaller, which makes the quotient get smaller and smaller.

13. (a) The numerator approaches 1 while the denominator gets smaller and smaller, which makes the quotient get larger and larger; (b) the numerator gets larger and larger, making the quotient get larger and larger, while at the same time the denominator gets smaller and smaller, making the quotient get larger and larger still; (c) the denominator approaches 1 while the numerator gets larger and larger, which makes the quotient get larger and larger.

15. (a) $\lim_{x \to 2} \frac{(x-2)^2}{x-2}$; (b) $\lim_{x \to 2} \frac{2x-2}{x-2}$; (c) $\lim_{x \to 2} \frac{x-2}{(x-2)^3}$

17. (a) $\lim_{x \to \infty} x \left(\frac{1}{x^2} \right)$; (b) $\lim_{x \to \infty} x \left(\frac{1}{x} \right)$; (c) $\lim_{x \to \infty} x^2 \left(\frac{1}{x} \right)$

19. Root at $x = 0$, no holes, vertical asymptote at $x = -3$, horizontal asymptote at $y = -3$.

21. Root at $x = 0$, no holes, vertical asymptote at $x = 2$, no horizontal asymptotes.

23. Root at $x = -1$, hole at $x = 3$, no asymptotes.

25. $\lim_{x \to 0^+} -4x^{-3} = -\infty$, $\lim_{x \to 0^-} -4x^{-3} = \infty$

27. 0

29. $-\infty$

31. $-\infty$

33. -2

35. ∞

37. DNE; $\lim_{x \to 1^-} \frac{1}{1-x} = \infty$, $\lim_{x \to 1^+} \frac{1}{1-x} = -\infty$

39. $-\frac{1}{3}$

41. DNE; $\lim_{x \to 0^-} \frac{x^2+1}{x(x-1)} = \infty$, $\lim_{x \to 0^+} \frac{x^2+1}{x(x-1)} = -\infty$

43. -1

45. -9

47. $-\infty$

49. 0

51. 0

53. 0

55. 0

57. 0

59. ∞; population explosion impossible for any finite-sized planet to contain.

61. (a) $\lim_{t \to \infty} s(t)$ does not exist, since both sine and cosine oscillate between -1 and 1 as $x \to \infty$. (b) There is no friction to damp the oscillations of the spring.

(c) $s(t) = \sqrt{2} \sin \left(\sqrt{\frac{9}{2}} t \right) + 2\cos \left(\sqrt{\frac{9}{2}} t \right)$

63. Mimic the proof of the first part of Theorem 1.24, but using $-M < 0$.

65. Given $\epsilon > 0$, choose $N_1 > 0$ to get $f(x)$ within $\frac{\epsilon}{2}$ of L and choose $N_2 > 0$ to get $g(x)$ within $\frac{\epsilon}{2}$ of M. Then for $N = \max(N_1, N_2)$ and $x > N$ we have $(L + M) - \epsilon < f(x) + g(x) < (L + M) + \epsilon$.

67. By Theorem 1.25 and the previous exercise,

$$\lim_{x \to \infty} x^{-k} = \lim_{x \to \infty} \frac{1}{x^k} \to \frac{1}{\infty} \to 0.$$

For answers to odd-numbered Skill Certification exercises in the Chapter Review, please visit the Book Companion Web Site at www.whfreeman.com/integcalculus.

Chapter 2

Section 2.1

1. T, T, F, F, T, T, T, T.

3. We only know one point on the line.

5. d is the difference $s(b) - s(a)$ in position, t is the difference $b - a$ in time, and $r = \frac{d}{t}$.

7. They are all the same.

9. The secant lines have slopes 0.995, 0.9995, and 0.99995, so again 1 is a good estimate for the slope of the tangent line.

11. (a) In order, the x-coordinate of the first dot, the x-coordinate of the second dot, the y-coordinate of the first dot, and the y-coordinate of the second dot. (b) The horizontal and vertical distances, respectively, between the first and second dot. (c) The slope of the line connecting the two dots, and the slope of the line tangent to the graph at the first dot.

13. The answers are the same as those in Exercise 11.

15. (c) < (d) < (a) < (b)

17. $x = 0.5; x = -1$ (or $x = 2$); $x = -0.4$ and $x = 1.3$

19.
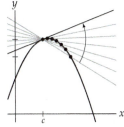

21. At $x = 1$ your graph should have a height of 2 and a horizontal tangent line. At $x = 3$ your graph should be drawn so that its tangent line has a slope of 2.

23. Your graph should go through the points $(-1, 2)$ and $(1, -2)$, and the tangent line at each of those points should have slope 3.

25.
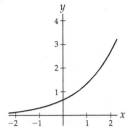

27. Your graph of $y = f'(x)$ should have zeroes at $x = -1.5$, $x = 0$, and $x = 1.5$, should be positive on $(-\infty, -1.5)$ and $(0, 1.5)$, and should be negative on $(-1.5, 0)$ and $(1.5, \infty)$.

29. Clearly the graph of f' should have a root at $x = 0$ and should be positive on $(-\infty, 0)$ and negative on $(0, \infty)$. There is one more thing to notice here, however; the graph of f gets very "flat" at its ends, so its tangent lines have very shallow slopes as $x \to \infty$ and as $x \to -\infty$. Therefore, the graph of the derivative should be approaching zero as $x \to \infty$ and as $x \to -\infty$. Thus your graph of f' should also have a two-sided horizontal asymptote at $y = 0$

31. Your graph should have horizontal tangent lines at $x = -2$ and at $x = 0$. Moreover, the tangent lines to the graph of your function should have a negative slope everywhere on $(-\infty, -2)$, a positive slope everywhere on $(-2, 0)$, and a negative slope everywhere on $(0, \infty)$.

33.

35. For $h = 1$, $h = 0.5$, $h = 0.25$, and $h = 0.1$, respectively, we have secant lines with slopes -3, -2.5, -2.25, and -2.1. One estimate of $f'(1)$ is -2.

37. For $h = 1$, $h = 0.5$, $h = 0.25$, and $h = 0.1$, respectively, we have secant lines with slopes 2, 1.25, 1.0625, and 1.01. One estimate of $f'(0)$ is 1.

39. For $h = 1$, $h = 0.5$, $h = 0.25$, and $h = 0.1$, respectively, we have secant lines with slopes 0, 0, 0, and 0. One estimate of $f'(5)$ is 0.

41. For $h = 1$, $h = 0.5$, $h = 0.25$, and $h = 0.1$, respectively, we have secant lines with slopes 7, 6.5, 6.25, and 6.1. One estimate of $f'(3)$ is 6.

43. For $h = 1$, $h = 0.5$, $h = 0.25$, and $h = 0.1$, respectively, we have secant lines with slopes 1, 1, 1, and 1. One estimate of $f'(3)$ is 1.

45. Average velocities on $[0, 0.5]$, $[0, 0.25]$, $[0, 0.1]$ are -8, -4, and -1.6 feet per second, respectively. We would expect the initial velocity to be 0.

47. Average velocities on $[2, 2.1]$ and $[2, 2.01]$ are -65.6 and -64.16 feet per second. Average velocities on $[1.9, 2]$ and $[1.99, 2]$ are -62.4 and -63.84 feet per second, respectively. We might expect the velocity at time $t = 2$ to be about 64 feet per second.

49. Although this book does know what you did today, we will not repeat it here.

51. (a) Her average rate of change is zero on $[0, 30]$; this means that after 30 minutes she is exactly as far away from the oak tree as she was at time $t = 0$. (b) The second 10 minutes. (c) One approximation might be $\frac{140 - 280}{10 - 5} = -28$ feet per minute. The fact that the sign is negative indicates that she is moving *toward* the oak tree. (d) $t \approx 4$ minutes, $t \approx 12$ minutes, $t \approx 25$ minutes. These are the times she changes direction. (e) $(4, 12)$ and $(25, 30)$. These are the times she is moving toward the oak tree.

53. (a) $h(12)$ is the average height of a 12-year-old, in feet. $h'(12)$ is measured in feet per year and represents the instantaneous rate of change of the height of a 12-year-old person; that is, $h'(12)$ is the rate at which an average 12-year-old is growing (in feet per year). (b) $h(12)$ is positive; $h'(12)$ is positive. (c) $h(t)$ might have a maximum at $t \approx 60$ years (people may tend to slouch or get shorter as they get older); $h'(t)$ might have a maximum at $t \approx 14$ years (growth spurt).

55. 5 miles per hour

57. (a) About 139.163 billion dollars; (b) about 3.445 billion dollars per year; (c) about 3.479 billion dollars per year; (d) something a little bit less than the previous answer (the actual answer turns out to be about 3.436 billion dollars a year).

59. If $f(x) = mx + c$ with $m > 0$, then on any interval $[a, b]$ the average rate of change of f is
$$\frac{f(b) - f(a)}{b - a} = \frac{(mb + c) - (ma + c)}{b - a} = m,$$ which is positive by assumption. This means that for all $b > a$ we have $f(b) = f(a) + m(b - a)$, and thus $f(b) > f(a)$; this the definition of what it means for f to be increasing.

Section 2.2

1. F, F, F, F, F, F, F, T.

3. (a) $f'(5) = \lim\limits_{h \to 0} \dfrac{f(5+h) - f(5)}{h} = \lim\limits_{z \to 5} \dfrac{f(z) - f(5)}{z - 5}$;

   (b) $f'(x) = \lim\limits_{h \to 0} \dfrac{f(x+h) - f(x)}{h} = \lim\limits_{z \to x} \dfrac{f(z) - f(x)}{z - x}$;

   (c) $f'_+(-2) = \lim\limits_{h \to 0^+} \dfrac{f(-2+h) - f(-2)}{h} = \lim\limits_{z \to -2^+} \dfrac{f(z) - f(-2)}{z + 2}$.

5. Let $z = x + h$. Then $h \to 0$ is equivalent to $z \to x$, and $z - x = h$.

7. $\lim\limits_{x \to 2} (4x^3 - 5x + 1) = 23$ and $\lim\limits_{h \to 0} \dfrac{(4(2+h)^3 - 5(2+h) + 1) - 23}{h}$ exists.

9. f is differentiable, and thus continuous, at $x = c$.

11. Continuous yes, differentiable no, left and right differentiable yes.

13. See the graphs in the reading in the discussion about differentiability.

15. (a) If $f(x) = 3x^2 - 1$ then $f'(-4) = -24$.

   (b) If $f(x) = 3x^2 - 1$ then $\dfrac{df}{dx}\Big|_{-4} = -24$.

   (c) $\dfrac{d}{dx}\Big|_{-4}(3x^2 - 1) = -24$.

17. $3x^2 - 2$, $3x^2 - 2$, $3x^2 - 2$, and $(3x^2 - 2)(x^3 - 2x + 1)$; the last notation represents a product of the derivative and the function.

19. For all $\epsilon > 0$, there exists $\delta > 0$ such that if $h \in (-\delta, 0) \cup (0, \delta)$, then $\dfrac{(3+h)^2 - 9}{h} \in (f'(3) - \epsilon, f'(3) + \epsilon)$.

21. Since $f'(4) = \dfrac{1}{2\sqrt{4}} = \dfrac{1}{4}$, the tangent line to f at $x = 4$ has equation $y = f(4) + f'(4)(x - 4) = 2 + \dfrac{1}{4}(x - 4)$. The value of $f(x)$ at $x = 4.1$ should be close to the value of this linear function at $x = 4.1$, which is $2 + \dfrac{1}{4}(4.1 - 4) = 2.025$. This is rather close to the calculator approximation of $\sqrt{4.1} = 2.02485$.

23. $\lim\limits_{h \to 0} \dfrac{(1 - 5(2+h)) - (-9)}{h} = -5$; $\lim\limits_{z \to 2} \dfrac{(1 - 5z) + 9}{z - 2} = -5$

25. $\lim\limits_{h \to 0} \dfrac{(4(-2+h) + 3) - (4(-2) + 3)}{h} = 4$; $\lim\limits_{z \to -2} \dfrac{(4z + 3) + 5}{z + 2} = 4$

27. $\lim\limits_{h \to 0} \dfrac{(-3+h)^2 - (-3)^2}{h} = -6$; $\lim\limits_{z \to -3} \dfrac{z^2 - 9}{z - 3} = -6$

29. $\lim\limits_{h \to 0} \dfrac{\frac{1}{-1+h} + 1}{h} = -1$; $\lim\limits_{z \to -1} \dfrac{\frac{1}{z} + 1}{z + 1} = -1$

31. $\lim\limits_{h \to 0} \dfrac{(1 - (-1+h)^3) - (1 - (-1)^3)}{h} = -3$; $\lim\limits_{z \to -1} \dfrac{(1 - z^3) - 2}{z + 1} = -3$

33. $\lim\limits_{h \to 0} \dfrac{(5(x+h) - 2) - (5x - 2)}{h} = 5$

35. $\lim\limits_{h \to 0} \dfrac{-1 - (-1)}{h} = \lim\limits_{h \to 0} 0 = 0$

37. $\lim\limits_{h \to 0} \dfrac{(-2(x+h)^2) - (-2x^2)}{h} = -4x$

39. $\lim\limits_{h \to 0} \dfrac{((x+h)^3 + 2) - (x^3 + 2)}{h} = 3x^2$

41. $\lim\limits_{z \to x} \dfrac{\frac{2}{z+1} - \frac{2}{x+1}}{z - x} = -2\dfrac{1}{(x+1)^2}$

43. $\lim\limits_{z \to x} \dfrac{\frac{1}{z^2} - \frac{1}{x^2}}{z - x} = \dfrac{-2}{x^3}$

45. $6x^2$, $12x$, and 12.

47. $6x$ and 12

49. $y = -6x - 9$

51. $y + 1 = -3(x - 1)$

53. $y - 2 = -(x - 3)$

55. Continuous everywhere except at $x = 2$, and neither left nor right continuous at that point. Not differentiable, but left and right differentiable, at $x = -1$. Not differentiable, and not left or right differentiable, at $x = 2$.

57. Continuous everywhere except at $x = -1$, where it is only left continuous. Not differentiable at $x = -1$. Left but not right differentiable at $x = -1$.

59. $\lim\limits_{x \to 0} \dfrac{1}{x}$ does not exist, and neither does $\lim\limits_{h \to 0} \dfrac{\frac{1}{0+h} - \frac{1}{0}}{h}$ since $\dfrac{1}{0}$ is undefined. Therefore f is neither continuous nor differentiable at $x = 0$, from either side.

61. $\lim\limits_{h \to 0^-} \dfrac{|(2+h)^2 - 4| - 0}{h} = -4$, $\lim\limits_{h \to 0^+} \dfrac{|(2+h)^2 - 4| - 0}{h} = 4$, so f is both left and right differentiable at $x = 2$, but not differentiable at $x = 2$.

63. $f'_-(2) = \lim\limits_{h \to 0^-} \dfrac{((2+h) + 4) - 3(2)}{h} = 1$; $f'_+(2) = \lim\limits_{h \to 0^+} \dfrac{3(2+h) - 3(2)}{h} = 3$; $f'(2)$ does not exist.

65. $f'_-(1) = \lim\limits_{h \to 0^-} \dfrac{(1+h)^2 - (1)^2}{h} = 2$; $f'_+(1) = \lim\limits_{h \to 0^+} \dfrac{(2(1+h) - 1) - (1)^2}{h} = 2$; $f'(1) = 2$.

67. Continuous but not differentiable at $x = 0$.

69. Not continuous and not differentiable at $x = 1$.

71. $f(1) = -4$ is less than zero and $f(3) = 4$ is greater than zero. Since f is continuous on $[1, 3]$, the Intermediate Value Theorem tells us that f must have a root between $x = 1$ and $x = 3$.

73. $f(0) = 1$ is greater than zero and $f(1) = -1$ is less than zero. Since f is continuous on $[0, 1]$, the Intermediate Value Theorem tells us that f must have a root between $x = 0$ and $x = 1$.

75. $f(-2) = -7$ is less than zero and $f(1) = 2$ is greater than zero. Since f is continuous on $[-2, 1]$, the Intermediate Value Theorem tells us that f must have a root between $x = -2$ and $x = 1$.

77. $\lim\limits_{h\to 0}\dfrac{s(1+h)-s(1)}{h}=-32.$

79. (a) The graph has a bump/hill shape with roots at $x=0$ and at $x=2$. The physical interpretation is that your velocity is always positive and that you start out with a zero velocity, speed up, and then eventually slow back down to a zero velocity. (b) The graph has a root at about $x=1$ and is positive on $[0,1)$ and negative on $(1,2]$ (for example, a line with negative slope passing through $(1,0)$). This represents the fact that you were speeding up, and then slowing down, during your trip.

81. (a) $S(3)=200+8(10)(3)=\$440,$
$S(6)=200+8(10)(6)=\$680,$ and
$S(8)=200+8(10)(6)+11.5(10)(2)=\$910.$ $S'(3)=$
80 dollars per week, $S'(8)=11.5(10)=\$115$ dollars per week. Cannot compute $S'(6)$ because the rate at which you are paid changes at that point.
(b) $S(t)=200+8(10)t$, if $t\le 6$ and
$S(t)=680+11.5(10)(t-6)$, if $t>6$. (c) Continuous but not differentiable. (d) $\lim\limits_{t\to 6}s(t)=s(6)=680$, but

$\lim\limits_{h\to 0^+}\dfrac{s(6+h)-s(6)}{h}=115$ while $\lim\limits_{h\to 0^-}\dfrac{s(6+h)-s(6)}{h}=80.$

83. $\lim\limits_{h\to 0}\dfrac{(m(x+h)+b)-(mx+b)}{h}=\lim\limits_{h\to 0}\dfrac{mh}{h}=\lim\limits_{h\to 0}m=m.$

85. The tangent line to f at $x=c$ has slope $f'(c)$ and passes through the point $(c,f(c))$. Putting this slope and point into the point-slope form of a line gives the desired answer.

87. $f'(c)$ exists when $\lim\limits_{h\to 0}\dfrac{f(c+h)-f(c)}{h}$ exists, which is true precisely when $f'_+(c)=\lim\limits_{h\to 0^+}\dfrac{f(c+h)-f(c)}{h}$ and $f'_-(c)=\lim\limits_{h\to 0^-}\dfrac{f(c+h)-f(c)}{h}$ exist and are equal.

Section 2.3

1. T, F, T, F, F, T, T, T.

3. $\dfrac{d}{dx}(kf(x))=k\dfrac{df}{dx}$; $\dfrac{d}{dx}(f(x)+g(x))=\dfrac{df}{dx}+\dfrac{dg}{dx}$; $\dfrac{d}{dx}(f(x)-g(x))=\dfrac{df}{dx}-\dfrac{dg}{dx}$

5. See the first paragraph in the reading.

7. The power rule only applies to functions of the form x^k, and 3^x cannot be written in this form because there is a variable in the exponent.

9. The rule is $f'(x)g(x)h(x)+f(x)g'(x)h(x)+f(x)g(x)h'(x)$. Then $y'=(2)(x^2+x+1)(1-3x^4)+(2x-1)(2x+1)(1-3x^4)+(2x-1)(x^2+x+1)(-12x^3).$

11. 5   13. 8

15. f has negative slope for $x<0$ and positive slope for $x>0$, so $f'(x)$ is negative on $x<0$ and positive on $x>0$. Then f' has positive slope everywhere, so $f''(x)$ is positive everywhere. Continue this type of argument for the remaining graphs.

17. As you can see, this algebra is no fun. If you remember to simplify functions before differentiating then you will not have this problem.

19. No; we can say f is continuous at $x=2$, but it might approach that breakpoint $x=2$ with different slopes from the left and the right.

21. $a=-2, b=6$ 23. 4

25. 1   27. $\dfrac{27}{2}$

29. (a) $f'(x)=(2x)(x+1)+(x^2)(1).$ (b) Multiply out first to get $f(x)=x^3+x^2$; then $f'(x)=3x^2+2x.$ (c) Multiply out the answer in part (a) to get the answer in part (b).

31. (a) $f'(x)=\dfrac{7}{2}x^{5/2}(2-5x^3)+x^{7/2}(-15x^2).$ (b) Multiply out first to get $f(x)=2x^{7/2}-5x^{13/2}$; then $f'(x)=7x^{5/2}-\dfrac{65}{2}x^{11/2}.$ (c) Multiply out the answer in part (a) to get the answer in part (b).

33. (a) $f'(x)=\dfrac{(2x-3x^2)\sqrt{x}-(x^2-x^3)(1/2)x^{-1/2}}{x}.$ (b) Expand first to get $f(x)=x^{1.5}-x^{2.5}$; then $f'(x)=1.5x^{0.5}-2.5x^{1.5}.$ (c) Multiply out the answer in part (a) to get the answer in part (b).

35. $f'(x)=-3(7x^6)$ 37. $f'(x)=12x$

39. $f'(x)=2(3)-4(5x^4)$ 41. $f'(x)=2x+1$

43. $f'(x)=81x^2+108x+36$ 45. $f'(x)=-6x^2$

47. $f'(x)=\dfrac{d}{dx}(x-1)=1$, for $x\ne -1$

49. $f'(x)=0$ 51. $f'(x)=\dfrac{8}{5}x^{3/5}$

53. $f'(x)=-\dfrac{8}{5}x^{-13/5}-2$

55. $f'(x)=\dfrac{(7x^6-15x^4)(1-3x^4)-(x^7-3x^5+4)(-12x^3)}{(1-3x^4)^2}$

57. $f'(x)=-3x^{-5/2}$ 59. $f'(x)=\dfrac{23}{(4+5x)^2}$

61. It is easier to differentiate after simplifying to $f(x)=\dfrac{1}{x^2-5x+6}.$ We have $f'(x)=\dfrac{0(x^2-5x+6)-(1)(2x-5)}{(x^2-5x+6)^2}=\dfrac{-2x+5}{(x^2-5x+6)^2}.$

63. $f'(x)=\dfrac{2x(x^3+5x^2-3x)-x^2(3x^2+10x-3)}{(x^3+5x^2-3x)^2}$

65. $f'(x)=\begin{cases}-1, & \text{if }x<0\\ \text{DNE}, & \text{if }x=0\\ 1, & \text{if }x>0.\end{cases}$

67. $f'(x)=\begin{cases}-2, & \text{if }x<\dfrac{1}{2}\\ \text{DNE}, & \text{if }x=\dfrac{1}{2}\\ 2, & \text{if }x>\dfrac{1}{2}.\end{cases}$

69. $f'(x)=\begin{cases}3x^2, & \text{if }x<1\\ \text{DNE}, & \text{if }x=1\\ 1, & \text{if }x>1.\end{cases}$

71. $f'(x)=\begin{cases}-2x, & \text{if }x\le 0\\ 2x, & \text{if }x>0.\end{cases}$

73. $f(x)=\dfrac{1}{2}x^6-\dfrac{2}{3}x^3+4x+1$

75. $f(x) = -\frac{4}{7}x^7 + x + \frac{18}{7}$

77. $f(x) = -\frac{3}{10}x^{10} + 4x^6 + \frac{1}{5}x^5 - 8x + 2$

79. (a) $a(t) = 0.024t$, $s(t) = 0.004t^3 + 400t$ (assuming initial position is $s_0 = 0$ at Venus). $a_0 = 0$ thousands of miles per hour per hour, $v_0 = 400$ miles per hour, $s_0 = 0$ miles. (b) Yes, since the velocity is always positive. (c) No, the acceleration is increasing (since $a(t) = 0.024t$ grows larger as t increases). (d) Approximately 140 hours, at which time it will be going about 635.2 thousand miles per hour.

81. If $a(t) = -32$ then (by antidifferentiating) $v(t) = -32t + C$, and $v(0) = v_0$ implies that $C = v_0$. Now (by antidifferentiating) $s(t) = -16t^2 + v_0 t + K$, and $s(0) = s_0$ implies that $K = s_0$.

83. (a) $s(t) = -\frac{2}{5}t^3 - 17t^2 + 575$; (b) $s(5) = 100$, $s'(5) = -200$, $s''(5) = -46$, $s'(0) = 0$; (c) 575 feet;

(d) $a(t) = -\frac{12}{5}t - 34$ is *not* constant.

85. Suppose k is a positive integer. $\frac{d}{dx}(x^k) = \lim_{h \to 0} \frac{(x+h)^k - x^k}{h}$
$= \lim_{h \to 0} \frac{(x^k + khx^{k-1} + h^2 \cdot (\text{lower-order terms})) - x^k}{h}$
$= \lim_{h \to 0} \frac{khx^{k-1} + h^2 \cdot (\text{lower-order terms})}{h}$
$= \lim_{h \to 0} (kx^{k-1} + h \cdot (\text{lower-order terms})) = kx^{k-1}$.

87. $(f - g)'(x) = \frac{d}{dx}(f + (-g))$
$= \frac{d}{dx}(f) + \frac{d}{dx}(-g) = \frac{d}{dx}(f) + (-1)\frac{d}{dx}(g)f'(x) - g'(x)$.

89. $\left(\frac{f}{g}\right)'(x) = \lim_{h \to 0} \frac{\frac{f(x+h)}{g(x+h)} - \frac{f(x)}{g(x)}}{h}$
$= \lim_{h \to 0} \frac{\left(\frac{f(x+h)g(x) - f(x)g(x+h)}{g(x+h)g(x)}\right)}{h}$
$= \lim_{h \to 0} \frac{f(x+h)g(x) - f(x)g(x+h)}{hg(x+h)g(x)}$
$= \lim_{h \to 0} \frac{f(x+h)g(x) - f(x)g(x) + f(x)g(x) - f(x)g(x+h)}{hg(x+h)g(x)}$
$= \lim_{h \to 0} \frac{g(x)(f(x+h) - f(x)) - f(x)(g(x+h) - g(x))}{hg(x+h)g(x)}$
$= \lim_{h \to 0} \frac{\left(g(x)\frac{f(x+h) - f(x)}{h} - f(x)\frac{g(x+h) - g(x)}{h}\right)}{g(x+h)g(x)}$
$= \frac{g(x)\left(\lim_{h \to 0} \frac{f(x+h) - f(x)}{h}\right) - f(x)\left(\lim_{h \to 0} \frac{g(x+h) - g(x)}{h}\right)}{\lim_{h \to 0}(g(x+h)g(x))}$
$= \frac{g(x)f'(x) - f(x)g'(x)}{\lim_{h \to 0}(g(x+h)g(x))} = \frac{f'(x)g(x) - f(x)g'(x)}{(g(x))^2}$.

91. (a) Since any two functions with the same derivative differ by a constant, it suffices to prove that the derivative of $\frac{1}{k+1}x^{k+1}$ is x^k. By the power rule and the

constant multiple rule we have $\frac{d}{dx}\left(\frac{1}{k+1}x^{k+1}\right) =$
$\frac{1}{k+1}\frac{d}{dx}(x^{k+1}) = \frac{1}{k+1}(k+1)x^{(k+1)-1} = x^k$. (b) If $k = -1$ then there will be division by zero in the calculation from part (a). Therefore, this proof does not work in the case where $f(x) = x^{-1}$.

Section 2.4

1. T, F, F, F, T, T, T, F.

3. (a) $(g(h(x)))' = g'(h(x))h'(x)$ and (b) $\frac{dg}{dx} = \frac{dg}{dh}\frac{dh}{dx}$.

5. (a) $(f(u(v(w(x)))))' = f'(u(v(w(x))))u'(v(w(x)))v'(w(x))w'(x)$;

(b) $\frac{df}{dx} = \frac{df}{du}\frac{du}{dv}\frac{dv}{dw}\frac{dw}{dx}$

7. (a) $\frac{d}{dx}((3x + \sqrt{x})^2) = 2(3x + \sqrt{x})\left(3 + \frac{1}{2}x^{-1/2}\right)$.

(b) $\frac{d}{dx}((3x + \sqrt{x})(3x + \sqrt{x})) =$
$\left(3 + \frac{1}{2}x^{-1/2}\right)(3x + \sqrt{x}) + (3x + \sqrt{x})\left(3 + \frac{1}{2}x^{-1/2}\right)$.

(c) $\frac{d}{dx}(9x^2 + 6x^{3/2} + x) = 18x + 9x^{1/2} + 1$.

9. -2 11. 6

13. 4 15. -12

17. If $y = y(x)$ then $\frac{d}{dx}(y^3) = \frac{d}{dx}((y(x))^3) = 3(y(x))^2 y'(x)$.

19. $(x + 1)(y^2 + y - 1) = 1$ is shown in the right graph, and $xy^2 + y = 1$ is shown in the left graph.

21. $f'(x) = -(x^3 + 1)^{-2}(3x^2)$

23. $f'(x) = (3x^2 + 1)^9 + 54x^2(3x^2 + 1)^8$

25. $f'(x) = -\frac{1}{2}(x^2 + 1)^{-3/2}(2x)$

27. $f'(x) = -2(x\sqrt{x+1})^{-3}\left(\sqrt{x+1} + x\left(\frac{1}{2}(x+1)^{-1/2}\right)\right)$

29. $f'(x) = \frac{(-x^{-2} - 6x)(x^5 - x^{-1/2}) - (x^{-1} - 3x^2)\left(5x^4 + \frac{1}{2}x^{-3/2}\right)}{(x^5 - x^{-1/2})^2}$

31. $-1(x^{1/3} - 2x)^{-2}\left(\frac{1}{3}x^{-2/3} - 2\right)$

33. $-\frac{1}{2}x^{-3/2}(x^2 - 1)^3 + x^{-1/2}3(x^2 - 1)^2(2x)$

35. $f'(x) = \frac{1}{2}(3x - 4(2x + 1)^6)^{-1/2}(3 - 24(2x + 1)^5(2))$

37. $f'(x) = 100(5(3x^4 - 1)^3 + 3x - 1)^{99}(15(3x^4 - 1)^2(12x^3) + 3)$

39. $f'(x) = 3\left(-\frac{2}{3}\right)((x^2 + 1)^8 - 7x)^{-5/3}(8(x^2 + 1)^7(2x) - 7)$

41. $f'(x) = 7(5x^4 - 3x^2)^6(20x^3 - 6x)(2x^3 + 1) + (5x^4 - 3x^2)^7(6x^2)$

43. $-9((2x + 1)^{-10} - 1)^{-10}(-10(2x + 1)^{-6})$

45. $f'(x) = 8(x^4 - \sqrt{3 - 4x})^7\left(4x^3 - \frac{1}{2}(3 - 4x)^{-1/2}(-4)\right) + 5$

47. $\dfrac{(-3)(x^5+2x^4+x^3)-(-2-3x)(5x^4+8x^3+3x^2)}{(x^5+2x^4+x^3)^2}$

49. 0

51. 162

53. $3s^2s'$

55. 0

57. s^2+2rss'

59. Combine the graphs of $y=\sqrt{9-\dfrac{1}{4}x^2}$ and

$y=-\sqrt{9-\dfrac{1}{4}x^2}$.

61. Combine the graphs of $y=\sqrt{\dfrac{1}{3}x^2-\dfrac{16}{3}}$ and

$y=-\sqrt{\dfrac{1}{3}x^2-\dfrac{16}{3}}$.

63. $f(x)=(x^2+1)^5$

65. $f(x)=\dfrac{1}{10}(x^2+1)^5-\dfrac{1}{10}$

67. $f(x)=\dfrac{2}{9}(3x+1)^{3/2}+\dfrac{7}{9}$

69. $\dfrac{dy}{dx}=\dfrac{4x}{y}$

71. $\dfrac{dy}{dx}=\dfrac{-6x-y^2}{2xy}$

73. $\dfrac{dy}{dx}=\dfrac{-3(y^2-y+6)}{(3x+1)(2y-1)}$

75. $\dfrac{dy}{dx}=\dfrac{5y}{(3/2)(3y-1)^{-1/2}-5x}$

77. $\dfrac{dy}{dx}=\dfrac{(3y-1)^2}{2y(3y-1)-3(y^2+1)}$

79. $\dfrac{dy}{dx}=\dfrac{2x^3y^2(y+1)-y^2(y+1)^2}{x^4y^2-x^2(y+1)^2}$

81. (a) Slope is $-\dfrac{1}{\sqrt{3}}$ at the point $\left(\dfrac{1}{2},\dfrac{\sqrt{3}}{2}\right)$ and $\dfrac{1}{\sqrt{3}}$ at the

point $\left(\dfrac{1}{2},-\dfrac{\sqrt{3}}{2}\right)$. (b) Slope is -1 at the point

$\left(\dfrac{\sqrt{2}}{2},\dfrac{\sqrt{2}}{2}\right)$, and 1 at the point $\left(-\dfrac{\sqrt{2}}{2},\dfrac{\sqrt{2}}{2}\right)$.

(c) $(1,0)$ and $(-1,0)$.

(d) $\left(\dfrac{\sqrt{2}}{2},\dfrac{\sqrt{2}}{2}\right)$ and $\left(-\dfrac{\sqrt{2}}{2},-\dfrac{\sqrt{2}}{2}\right)$

83. (a) $(1,-1)$; (b) $(-3,1)$; (c) none; (d) $(-3,1)$

85. If Linda sells magazines at a rate of $\dfrac{dm}{dt}=12$ magazine

subscriptions per week, and she makes $\dfrac{dD}{dm}=4$ dollars
per magazine, then by the chain rule, the amount
of money she makes each week is
$\dfrac{dD}{dt}=\dfrac{dD}{dm}\dfrac{dm}{dt}=4(12)=48$ dollars per week.

87. (a) $\dfrac{dA}{dr}=2\pi r$. (b) No; Yes.

(c) $\dfrac{dA}{dt}=\dfrac{d}{dt}(\pi(r(t))^2)=2\pi r(t)r'(t)=2\pi r\dfrac{dr}{dt}$.

(d) Yes; Yes. (e) $\dfrac{dA}{dt}\big|_{r=24}=2\pi(24)(2)=96\pi$.

89. $\dfrac{d}{dx}\left(\dfrac{f(x)}{g(x)}\right)=\dfrac{d}{dx}(f(x)(g(x))^{-1})$

$=\dfrac{d}{dx}(f(x))\cdot(g(x))^{-1}+f(x)\cdot\dfrac{d}{dx}((g(x))^{-1})$

$=f'(x)\cdot(g(x))^{-1}+f(x)\cdot(-g(x))^{-2}g'(x)$

$=\dfrac{f'(x)}{g(x)}-\dfrac{f(x)g'(x)}{(g(x))^2}=\dfrac{f'(x)g(x)-f(x)g'(x)}{(g(x))^2}$.

91. Mimic the proof in Example 5(a).

93. Let $y=x^{-k}$, where $-k$ is a negative integer. Then
$x^ky=1$, so by implicit differentiation and the product
rule, we have $kx^{k-1}y+x^ky'=0$, and therefore

$y'=\dfrac{-kx^{k-1}x^{-k}}{x^k}=-kx^{-k-1}$.

For answers to odd-numbered Skill Certification exercises
in the Chapter Review, please visit the Book Companion
Web Site at www.whfreeman.com/integcalculus.

Chapter 3

Section 3.1

1. T, T, T, F, F, T, F, F.

3. $f'(1)$ is either 0 or does not exist. If f is differentiable at
$x=1$ then $f'(1)$ must equal 0.

5. $x=-2,x=0,x=4,x=5$.

7. f' has at least two zeroes in the interval $[-4,2]$.

9. There is some $c\in(-2,4)$ with $f'(c)=-\dfrac{1}{3}$.

11. If f is continuous on $[a,b]$, differentiable on (a,b), and if
$f(a)=f(b)=0$, then there exists at least one value
$c\in(a,b)$ for which the tangent line to f at $x=c$ is
horizontal.

13. The graph of $f(x)=(x-2)(x-6)$ is one example.

15. Your graph should have roots at $x=-2$ and $x=2$ and
horizontal tangent lines at three places between these
roots.

17. One example is an "upside-down V" with roots at
$x=-2$ and $x=2$ where the top point of the V occurs at
$x=-1$.

19. One example is the function f that is equal to $x+3$ for
$-3\le x<-1$ and equal to 0 for $x=-1$.

21. Draw a graph that happens to have a slight "cusp" just
at the place where its tangent line would have been
equal to the average rate of change.

23. $f'(x)=0$ at $x=\dfrac{3}{2}$ and $f'(x)$ does not exist at $x=3$. f has

a local maximum at $x=\dfrac{3}{2}$ and a local minimum at
$x=3$.

25. $f'(x)=0$ at $x\approx0.5$, $x\approx2$, and $x\approx3.5$. f has a local
minimum at $x\approx0.5$ and a local minimum at $x=3.5$.
There is neither a maximum nor a minimum at $x\approx2$.

27. One critical point at $x=\dfrac{3}{2}$, a local minimum.

29. One critical point at $x=1$, neither a local minimum nor
a local maximum.

31. One critical point at $x=-0.65$, a local minimum.

33. Critical points $x=-3,x=0,x=1$, a local minimum,
maximum, and minimum, respectively.

35. From the graph, f appears to be continuous on $[-3,1]$
and differentiable on $(-3,1)$, and moreover
$f(-3)=f(1)=0$, so Rolle's Theorem applies. Therefore
there is some $c\in(-3,1)$ such that $f'(c)=0$. In this
example there are three such values of c, namely
$c\approx-2.3$, $c=-1$, and $c=0.3$.

37. From the graph, f appears to be continuous on $[0, 4]$ and differentiable on $(0, 4)$, and moreover $f(0) = f(4) = 0$, so Rolle's Theorem applies. Therefore there is some $c \in (0, 4)$ such that $f'(c) = 0$. In this example there are three such values of c, namely $c \approx 0.5$, $c \approx 2$, and $c \approx 3.5$.

39. Because the function $f(x) = x^2 - 3x - 4$ is continuous and differentiable everywhere, and because $f(-1) = f(4) = 0$, it satisfies the hypotheses of Rolle's Theorem, so we can conclude that there is some $c \in (-1, 4)$ such that $f'(c) = 0$. By solving $f'(x) = 0$ we can see that there is exactly one such value, $c = \dfrac{3}{2}$.

41. f is not differentiable at $x = 1$, and therefore not differentiable on $(0, 3)$. Therefore Rolle's Theorem does not apply.

43. f is continuous and differentiable everywhere, and $f(0) = f(3) = 0$, so Rolle's Theorem applies; $c = \dfrac{1}{3}(4 - \sqrt{7})$ and $c = \dfrac{1}{3}(4 + \sqrt{7})$.

45. f is continuous and differentiable everywhere, and $f(-2) = f(2) = 0$, so Rolle's Theorem applies; $c \approx -1.27279$, $c \approx 0$, and $c \approx 1.27279$.

47. f appears continuous on $[0, 2]$ and differentiable on $(0, 2)$; there is one value $x = c$ that satisfies the conclusion of the Mean Value Theorem, roughly at $x \approx 1.2$.

49. f appears continuous on $[-3, 0]$ and differentiable on $(-3, 0)$; there are two values $x = c$ that satisfy the conclusion of the Mean Value Theorem, at $c \approx -2.8$ and $c \approx -0.9$.

51. f is continuous and differentiable everywhere, so the Mean Value Theorem applies; $c = 1$.

53. f is continuous and differentiable everywhere; thus by MVT there is some $c \in (-3, 2)$ such that $f'(c) = 1$ (the average rate of change of f on $[-3, 2]$). Solving $f'(x) = 1$ we see that there is exactly one such value, namely $x = -\dfrac{1}{2}$.

55. f is not continuous or differentiable on $[-3, 2]$, so the Mean Value Theorem does not apply.

57. f is continuous and differentiable on $[-2, 3]$; $c \approx -0.5275$ and $c \approx 2.5275$.

59. $C(h)$ is a differentiable function, and $C'(4) = 0.6 \neq 0$, so $C(h)$ cannot have a local minimum at $h = 4$.

61. (a) Let $s(t)$ be Alina's distance from the grocery store (in miles) at time t. Then $s(0) = 20$ and $s(0.5) = 0$; since $s(t)$ is continuous and differentiable, the MVT applies, and tells you that there is some time $c \in (0, 0.5)$ where $s'(c) = \dfrac{0 - 20}{0.5 - 0} = -40$ miles per hour (the negative sign means she is travelling towards the store).
(b) Questions to think about: Could Alina possibly have been travelling under 40 miles per hour for her whole trip? Or over 40 miles an hour for her whole trip? If her velocity was less than 40 miles per hour at some point, and more than 40 miles per hour at some other point, can she have avoided going *exactly* 40 miles per hour at some time during her trip? Why or why not?

63. See the proof in the reading.

65. Suppose r_1, r_2, and r_3 are the roots of f. Since f is continuous and differentiable everywhere, Rolle's Theorem guarantees that f' will have at least one root on $[r_1, r_2]$, and at least one root on $[r_2, r_3]$; therefore f' has at least two roots.

67. $g(x)$ is continuous on $[a, b]$ and differentiable on (a, b) because f is. Moreover, $g(a) = f(a) - f(a) = 0$ and $g(b) = f(b) - f(a) = f(a) - f(a) = 0$, so Rolle's Theorem applies to $g(x)$. Therefore we can conclude that there is some $c \in (a, b)$ such that $g'(c) = 0$. Since $g'(x) = f'(x) - 0 = f'(x)$, this also means that $f'(c) = 0$, as desired.

Section 3.2

1. T, T, T, F, F, F, F, F.

3. See the first paragraph in the reading.

5. Yes; yes; no.

7. A function can only change sign at roots and discontinuities or non-domain points, and the critical points of f are exactly these types of points for f'.

9.

11. If $a < b \leq 0$ then $a^4 > b^4$, and if $0 < a < b$ then $a^4 < b^4$. With derivatives, we have $f'(x) = 4x^3$, which is negative for $x < 0$ and positive for $x > 0$.

13. $x = \pm\sqrt{15}$;

15. There is some constant C such that $g(x) - h(x) = C$ for all x in their domains.

17. f is shown in Graph II and f' is shown in Graph I.

19.

21. Your graph of f' should have roots at $x = 0$ and $x = 2$, and should be positive on $(0, \infty)$ and negative on $(-\infty, 0)$. In particular this means that your graph of f' should "bounce" off the x-axis at $x = 2$.

23.

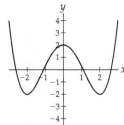

25. Your graph of f should have horizontal tangent lines at $x = 1$ at $x = 3$, and should be decreasing on $(-\infty, 1)$ and $(1, 3)$ and increasing on $(3, \infty)$. The graph of f has an inflection point, but not a local extremum, at $x = 1$, and a local minimum at $x = 3$.

27. Always increasing.

29. Increasing on $\left(-\infty, \frac{-2-\sqrt{13}}{3}\right)$ and $\left(\frac{-2+\sqrt{13}}{3}, \infty\right)$; decreasing on $\left(\frac{-2-\sqrt{13}}{3}, \frac{-2+\sqrt{13}}{3}\right)$.

31. Increasing on $(-\infty, 0)$ and $(3, \infty)$; decreasing on $(0, 3)$.

33. Decreasing on $(-\infty, -2)$ and $(2, \infty)$; increasing on $(-2, 2)$.

35. No local extrema

37. Local minimum at $x = \frac{5}{2}$

39. Local maximum at $x = 0$, local minimum at $x = 2$

41. Local maximum at $x = -\frac{1}{2}$

43. Possible pictures of f and f', respectively:

45. Possible pictures of f and f', respectively:

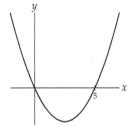

47. Your graph of f should be increasing on $(-\infty, 1)$, $(3, 8)$, and $(8, \infty)$ and decreasing on $(1, 3)$. It should also have horizontal tangent lines at $x = 1$ (a local maximum) and $x = 8$ (not an extremum), and be nondifferentiable at

at $x = 3$ (a local minimum). Your graph of f' should be positive on $(-\infty, 1)$, $(3, 8)$, and $(8, \infty)$ and negative on $(1, 3)$. It should also have roots at $x = 1$ and $x = 8$ and be undefined at $x = 3$.

49. Defined everywhere, roots at $x = -\frac{1}{3}$ and $x = 2$. Positive on $\left(-\infty, -\frac{1}{3}\right) \cup (2, \infty)$ and negative elsewhere. Local minimum at $x = \frac{5}{6}$. Increasing on $\left(\frac{5}{6}, \infty\right)$ and decreasing elsewhere. $\lim\limits_{x \to -\infty} f(x) = \infty$ and $\lim\limits_{x \to \infty} f(x) = \infty$.

51. Defined everywhere, roots at $x = -1$ and $x = 1$. Positive on $(-1, 1)$ and negative elsewhere. Local minimum at $x = 1$. Increasing on $\left(-\infty, -\frac{1}{3}\right)$ and $(1, \infty)$ and decreasing elsewhere. $\lim\limits_{x \to \infty} f(x) = \infty$ and $\lim\limits_{x \to -\infty} f(x) = -\infty$.

53. Defined everywhere, root at $x = 0$. Positive on $(0, \infty)$ and negative elsewhere. Critical point at $x = 2$ that is *not* a local extremum. Increasing everywhere (yes, even including $x = 2$ – although answering $(-\infty, 2) \cup (2, \infty)$ is sufficient for our purposes here). $\lim\limits_{x \to -\infty} f(x) = -\infty$ and $\lim\limits_{x \to \infty} f(x) = \infty$.

55. Defined everywhere, root at $x = -\frac{11}{2}$. Positive on $\left(-\frac{11}{2}, \infty\right)$ and negative elsewhere. Local maximum at $x = -2$, local minimum at $x = -\frac{5}{3}$. Increasing on $(-\infty, -2) \cup \left(-2, -\frac{5}{3}\right)$ and decreasing elsewhere. $\lim\limits_{x \to -\infty} f(x) = -\infty$ and $\lim\limits_{x \to \infty} f(x) = \infty$.

57. Defined everywhere, roots at $x = \pm 1$. Positive on $(-1, 1)$ and negative elsewhere. Local maximum at $x = 0$. Increasing on $(-\infty, 0)$ and decreasing elsewhere. $\lim\limits_{x \to -\infty} f(x) = -\infty$ and $\lim\limits_{x \to \infty} f(x) = -\infty$.

59. Defined everywhere except at $x = 1$, where there is a vertical asymptote; root at $x = -1$. Positive on $(-\infty, -1) \cup (1, \infty)$ and negative elsewhere it is defined. No local extrema. Always decreasing. $\lim\limits_{x \to -\infty} f(x) = 0$ and $\lim\limits_{x \to \infty} f(x) = 0$, so there is a horizontal asymptote at $y = 0$.

61. Defined everywhere; no roots. Positive everywhere. Local minimum at $x = 0$. Increasing on $(0, \infty)$ and decreasing elsewhere. $\lim\limits_{x \to \infty} f(x) = \infty$ and $\lim\limits_{x \to -\infty} f(x) = \infty$.

63. Defined except at $x = -3$ and $x = 2$ where there are vertical asymptotes; root at $x = 1$. Positive on $(-\infty, -3) \cup (2, \infty)$ and negative elsewhere. Local maximum at $x = 1$, local minimum at $x = \frac{11}{3}$. Horizontal asymptote at $y = 1$.

65. Defined except at $x = 2$ where there is a vertical asymptote; roots at $x = 0$ and $x = 1$. Positive on $(1, 2) \cup (2, \infty)$ and negative elsewhere. Local maximum at $x = 0$, local minima at $x = 3 - \sqrt{5}$ and $x = 3 + \sqrt{5}$. Increasing on $(-\infty, 0) \cup (3 - \sqrt{5}, 2) \cup (3 + \sqrt{5}, \infty)$ and decreasing elsewhere. No horizontal asymptote.

67. (a) $(0, 2.5)$; (b) $t = 2.5$; (c) $(0, 1)$ and $(3.5, 4)$; (d) $(3.5, 3.8)$ is such an interval; Bubbles is moving towards the left side of the tunnel and is slowing down while doing so.

69. Given that $f(x) = mx + b$ where $m \neq 0$ (so f is nonconstant), show that f' is either always positive or always negative. Since $f'(x) = m$ is a (nonzero) constant, this is clearly the case.

71. Mimic the proof from the reading of part (a) of the theorem.

73. Mimic the proof from the reading of part (a) of the theorem.

Section 3.3

1. F, F, F, T, T, F, F, F.

3. See the discussion following Definition 3.9.

5. The converse is "If f is concave up on I, then f'' is positive on I," which is *false* ($f''(x)$ might be zero at some point).

7. The graph of $f(x) = -x^3$ has this property at $x = 0$.

9. $f''(x) = 30x^4$, so $f''(0) = 0$. However, $f''(-1) > 0$ and $f''(1) > 0$, so f'' does not change sign at $x = 0$; therefore $x = 0$ is not an inflection point of f.

11. Linear

13. 5, 4.

15. Draw an arc connecting $(0, 5)$ and $(3, 1)$ that is concave up.

17. Draw an arc connecting $(0, -5)$ and $(3, -1)$ that is concave up.

19. The function $f(x) = -e^x$ has these properties on all of $\mathbb{R}$.

21. Possible graph of f' followed by possible graph of f'':

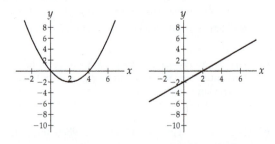

23. Your graph of f' should have roots at $x = -3$ and at $x = 0$ and should be positive on $(-\infty, -3)$ and negative on $(-3, 0)$ and $(0, \infty)$. Your graph of f'' should have roots at $x \approx -2$ and $x = 0$ and should be negative on $(-\infty, -2) \cup (0, \infty)$ and positive on $(-2, 0)$.

25. Possible graph of f followed by possible graph of f''

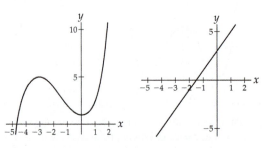

27. Your graph of f' should be decreasing on $(-\infty, 2)$ and increasing on $(2, \infty)$, with a local minimum at $x = 2$. Your graph of f should be concave down on $(-\infty, 2)$ and concave up on $(2, \infty)$, with an inflection point at $x = 2$.

29. No local extrema

31. Local minimum at $x = \dfrac{5}{2}$.

33. Local maximum at $x = 0$, local minimum at $x = 2$

35. Local maximum at $x = -\dfrac{1}{2}$

37. Concave up on $\left(-\infty, \dfrac{2}{3}\right)$, concave down on $\left(\dfrac{2}{3}, \infty\right)$; inflection point at $x = \dfrac{2}{3}$.

39. Concave up on $\left(-\infty, \dfrac{3-\sqrt{3}}{3}\right) \cup \left(\dfrac{3+\sqrt{3}}{3}, \infty\right)$, concave down on $\left(\dfrac{3-\sqrt{3}}{3}, \dfrac{3+\sqrt{3}}{3}\right)$; inflection points at $x = \dfrac{3 \pm \sqrt{3}}{3}$.

41. Always concave up; no inflection points.

43. Concave up on $(-\infty, 0) \cup (1, \infty)$, concave down on $(0, 1)$; inflection points at $x = 0$ and $x = 1$.

45. Concave up on $(-\infty, -1) \cup (0, \infty)$, concave down on $(-1, 0)$; inflection points at $x = -1$ and $x = 0$.

47.

49.

51.

53. Among other things, your graph should have roots at $x = -3$ and $x = -1$, a local minimum at $x = -2$, and inflection points at $x = -1$ and $x = 0$.

55. Among other things, your graph should have roots at $x = -3$, $x = 1$, and $x = 4$, a local maximum at $x = -2$, a local minimum at $x = 3$, and inflection points at $x = -1$, $x = 1$, and $x = 2$.

57. Defined everywhere, roots at $x = -3$ and $x = 0$. Positive on $(-\infty, -3) \cup (0, \infty)$ and negative elsewhere. Local minimum at $x = -\frac{3}{2}$. Increasing on $\left(-\frac{3}{2}, \infty\right)$ and decreasing elsewhere. No inflection points. Concave up everywhere. $\lim\limits_{x \to \infty} f(x) = \infty$ and $\lim\limits_{x \to -\infty} f(x) = \infty$.

59. Defined everywhere, roots at $x = -3$ and $x = 0$. Positive on $(-3, 0) \cup (0, \infty)$ and negative elsewhere. Local maximum at $x = -2$, local minimum at $x = 0$. Increasing on $(-\infty, -3) \cup (0, \infty)$ and decreasing elsewhere. Inflection point at $x = -1$. Concave up on $(-1, \infty)$ and concave down elsewhere. $\lim\limits_{x \to \infty} f(x) = \infty$ and $\lim\limits_{x \to -\infty} f(x) = -\infty$.

61. Defined everywhere, roots at $x = -1$ and $x = 0$. Positive on $(0, \infty)$ and negative elsewhere. Local maximum at $x = \frac{1}{3}$, local minimum at $x = 1$. Increasing on $(-\infty, -1) \cup \left(-\frac{1}{3}, \infty\right)$ and decreasing elsewhere. Inflection point at $x = -\frac{2}{3}$. Concave up on $\left(-\frac{2}{3}, \infty\right)$ and concave down elsewhere. $\lim\limits_{x \to \infty} f(x) = \infty$ and $\lim\limits_{x \to -\infty} f(x) = -\infty$.

63. Defined everywhere, roots at $x = -2$ and $x = 0$. Positive on $(-\infty, -2) \cup (0, \infty)$ and negative elsewhere. Local minimum at $x = -\frac{3}{2}$. Increasing on $\left(-\frac{3}{2}, \infty\right)$ and decreasing elsewhere. Inflection points at $x = -1$ and $x = 0$. Concave up on $(-\infty, -1) \cup (0, \infty)$ and concave down elsewhere.

65. Defined for $x \geq 0$, roots at $x = 0$ and $x = 4$. Positive on $(0, 4)$ and negative elsewhere it is defined. Local maximum at $x = \frac{4}{3}$. Increasing on $\left(0, \frac{4}{3}\right)$ and decreasing elsewhere it is defined. No inflection points. Always concave down. $\lim\limits_{x \to \infty} f(x) = -\infty$.

67. Defined everywhere except at $x = 1$ and $x = 4$; root at $x = -1$. Positive on $(4, \infty)$ and negative elsewhere it is defined. No local extrema. Decreasing everywhere. No inflection points. Concave up on $(4, \infty)$ and concave down elsewhere it is defined. $\lim\limits_{x \to 1} f(x) = -\frac{2}{3}$ but $f(1)$ is not defined, so f has a hole at $x = 1$. $\lim\limits_{x \to 4} f(x)$ is infinite (∞ from the right and $-\infty$ from the left), so f has a vertical asymptote at $x = 4$. $\lim\limits_{x \to \pm\infty} f(x) = 1$, so f has a horizontal asymptote at $x = 1$.

69. f has a local maximum at $x = -2$, a local minimum at $x = 1$, and an inflection point at $x = -\frac{1}{2}$.

71. f has a local minimum at $x = 0$, no local maxima, and no inflection points.

73. (a) $[0, 15]$; he is north of the corner, walking north and slowing down. (b) $[60, 80]$; he is south of the corner, walking south and speeding up. (c) $[15, 40]$; he is walking south and speeding up (first north and then south of the corner). (d) At $t = 40$; he is south of the corner, walking south; he just finished speeding up (in the southward direction) and is now just starting to slow down (but still walking in the southward direction).

75. Mimic the proof of part (a) of the same theorem.

77. If $f(x) = ax^2 + bx + c$, then $f''(x) = 2a$. If $a > 0$ then $f''(x) > 0$ for all x, so f is always concave up. If $a < 0$ then $f''(x) < 0$ for all x, so f is always concave down.

79. If $f''(x) = 0$ on an interval I, then $f'(x) = k$ must be constant, and therefore $f(x) = kx + c$ must be linear.

Section 3.4

1. F, T, T, F, F, F, F, F.

3. The derivative may or may not be zero (or fail to exist) at the endpoints of the interval.

5. There will not be any endpoint extrema; there may or may not be a global maximum (or minimum) on the interval.

7. For example, f might have a vertical asymptote on I.

9. (a) Global minimum at $x = 0$, no global maximum. (b) Global minimum at $x = 5$, global maximum at $x = 2$. (c) Global minimum at $x = 0$, no global maximum. (d) Global minimum at $x = 0$, no global maximum. (e) Global minimum at $x = 0$, no global maximum. (f) Global minimum at $x = 0$, no global maximum.

11. (a) Global maximum at $x = -2$ and at $x = 3$ with value $f(-2) = f(3) = 0$, global minimum at $x = \frac{1}{2}$ with value $f\left(\frac{1}{2}\right) = -\frac{25}{4}$. (b) Global maximum at $x = 4$ with value $f(4) = 6$, global minimum at $x = \frac{1}{2}$ with value $f\left(\frac{1}{2}\right) = -\frac{25}{4}$. (c) No global maximum, global minimum at $x = 1$ with value -6. (d) No global maximum, global minimum at $x = \frac{1}{2}$ with value $f\left(\frac{1}{2}\right) = -\frac{25}{4}$.

13. (a) Global maximum at $x = -1$ with value $f(-1) = 7$, global minimum at $x = -3$ with value $f(-3) = -45$. (b) Global maximum at $x = 0$ with value $f(0) = 0$, global minimum at $x = 2$ with value $f(2) = -20$. (c) No global maximum, global minimum at $x = 2$ with value $f(2) = -20$. (d) Global maximum at $x = -1$ with value $f(-1) = 7$, no global minimum.

15. (a) Global maximum at $x = -1$ with value $f(-1) = 11$, global minimum at $x = 1$ with value $f(1) = -5$. (b) No global maximum or global minimum. (c) Global maximum at $x = -1$ with value $f(-1) = 11$, no global minimum. (d) Global maximum at $x = 0$ with value $f(0) = 0$, global minimum at $x = 3$ with value $f(3) = -117$.

17. (a) Global maximum at $x = 0$ with value $f(0) = 1$, global minimum at $x = 3$ with value $f(3) = \frac{1}{1+\sqrt{3}}$. (b) No global maximum or global minimum. (c) Global maximum at $x = 1$ with value $f(1) = \frac{1}{2}$, global minimum at $x = 2$ with value $f(2) = \frac{1}{1+\sqrt{2}}$. (d) Global maximum at $x = 0$ with value $f(0) = 1$, no global minimum.

19. The most optimal solution is to let the horizontal straight edges have length zero, resulting in a circle of radius $\frac{60}{\pi}$.

21. The horizontal top and bottom edges should have length ≈ 19.6903 units, and the vertical edges should have length 13.1268 units.

23. $x = 18$ and $y = 18$

25. $a = 50$, $b = 50$

27. 200

29. $\left(-\frac{1}{5}, \frac{2}{5}\right)$

31. $\left(\frac{2-\sqrt{6}}{2}, 1.75\right)$ or $\left(\frac{2+\sqrt{6}}{2}, 1.75\right)$

33. A square pen with sides of length $\frac{175}{2}$ feet will produce the maximum area of 7656.25 square feet.

35. Four parallel north-south fences of length 125 feet, and two east-west fences of length 250 feet. The resulting pen will have an area of $31,250$ square feet.

37. 180 feet of border fencing

39. The base of the box should have sides of length 6.26 inches, and the height of the box should be approximately 10.65 inches. The resulting largest possible volume is approximately 417.48 cubic inches.

41. The largest possible volume is $11,664$ cubic inches; the largest possible surface area is $3,332.6$ square inches.

43. The largest possible volume is $14,851$ cubic inches; the largest possible surface area is 462.6 square inches.

45. At $t = 0$ minutes; at the right endpoint of the model (one such endpoint could be $t = 5$).

47. The minimum area occurs when you cut the wire so that 4.399 inches of wire are used to make the circle. The maximum area occurs when you don't cut the wire at all, and use all 10 inches to make the circle.

49. The minimum area occurs when you cut the wire so that 0.577 inches of wire are used to make the circle. The maximum area occurs when you don't cut the wire at all, and use all 10 inches to make the circle.

51. To get the minimum surface area, the oil drums should be constructed so they are $\frac{40}{\pi 20^{2/3}} \approx 3.0767$ feet high with a radius of $\frac{20^{1/3}}{\pi} \approx 1.8533$ feet. There is no global maximum.

53. To minimize the cost, the cans should be made with a radius of $2\left(\frac{5}{\pi}\right)^{1/3} \approx 2.335$ inches, and a height of ≈ 11.6754 inches. It is not possible to maximize the cost.

55. The steam pipe should be buried along the 800-foot side of the parking lot for 425 feet, and then diagonally under the parking lot to the opposite corner.

57. (a) You would sell the most (975 paintings) at $5.00 apiece. You would sell the least (15 paintings) at $45.00 apiece. (b): $R(c) = c(0.6c^2 - 54c + 1230)$. (c) You would earn the most money ($8327.10) by selling the paintings for $15.28 apiece. You would earn the least money ($672.90) by selling the paintings for $44.72 apiece. (d) Although you sell the most paintings when charging $5.00 apiece, you only earn $5.00 for each of those paintings.

59. If the sides of the rectangle have length x and y, then $P = 2x + 2y$, so $y = \frac{1}{2}(P - 2x)$. Therefore the area is $A = xy = x\left(\frac{1}{2}(P - 2x)\right)$, which has derivative $A'(x) = \frac{1}{2}P - 2x$. Thus the only critical point of $A(x)$ is $x = \frac{P}{4}$. Since $A'\left(\frac{P}{4} - 1\right) > 0$ and $A'\left(\frac{P}{4} + 1\right) < 0$, the first-derivative test tells us that $A(x)$ has a local maximum at $x = \frac{P}{4}$; in fact it is the global maximum of $A(x)$ on $\left[0, \frac{P}{2}\right]$. Therefore A is maximized when $x = \frac{P}{4}$ $\left(\text{and thus } y = \frac{P}{4}\right)$, i.e., when the rectangle is a square.

Section 3.5

1. F, T, T, F, T, T, T, F.

3. $V = \pi y^2 s$, $SA = 2\pi ys + 2\pi y^2$

5. $V = \frac{\pi}{4}h^3$, $SA = \frac{3}{2}\pi h^2$

7. See Theorem 3.13(a) for the statement of the theorem. A triangle with legs of lengths 3 and 4 will have a hypotenuse of length $\sqrt{3^2 + 4^2} = \sqrt{25} = 5$.

9. $\frac{dV}{dt} = 4\pi r^2 \frac{dr}{dt}$

11. $\frac{dV}{dt} = \frac{E^2}{2\pi^2}\frac{dE}{dt}$, or equivalently, $\frac{dV}{dt} = 2r^2\frac{dE}{dt}$.

13. (a) $\frac{dV}{dt} = 4\pi r^2\frac{dr}{dt}$; (b) $\frac{dS}{dt} = 8\pi r\frac{dr}{dt}$.

15. $\frac{dV}{dt} = 2\pi rh\frac{dr}{dt}$

17. $\frac{dV}{dt} = \frac{3\pi}{4}h^2\frac{dh}{dt}$

19. $A = \frac{P^2}{16}$

21. $SA = 4\pi h + 8\pi$

23. $SA = \pi r\sqrt{r^2 + 25} + \pi r^2$

25. $A = \frac{c^2}{4}$

27. $f' = 2uu' + kv'$

29. $f' = v + tv' + kv'$

31. $f' = 2v'\sqrt{u + w} + \frac{v(u' + w')}{\sqrt{u+w}}$

33. $f' = w'(u + t)^2 + 2w(u + t)(u' + 1)$

35. $f' = \frac{1}{k}(u't + u + w')$

37. $\left.\frac{dA}{dt}\right|_{r=12} = 96\pi \,\frac{in^2}{sec}$; $\left.\frac{dA}{dt}\right|_{r=24} = 192\pi \,\frac{in^2}{sec}$; $\left.\frac{dA}{dt}\right|_{r=100} = 800\pi \,\frac{in^2}{sec}$.

39. $\left.\dfrac{dV}{dt}\right|_{s=8} = 6(8)^2 = 384$ cubic inches per minute.

41. $\left.\dfrac{dV}{dt}\right|_{x=\sqrt[3]{55}} = 6(\sqrt[3]{55})^2 \approx 86.77$ cubic inches per minute.

43. $\left.\dfrac{dr}{dt}\right|_{r=12} = \dfrac{5}{24\pi}\dfrac{\text{in}}{\text{sec}}$ 45. $16\dfrac{\text{in}^2}{\text{sec}}$

47. Suppose x is Stuart's distance from the streetlight, l is the length of his shadow, and $y = x + l$ is the distance from the tip of his shadow to the streetlight. Then $\left.\dfrac{dy}{dt}\right|_{x=10} = -\dfrac{40}{7}\dfrac{\text{ft}}{\text{sec}}$; this does not depend on x.

49. $\dfrac{2}{\sqrt{128}} \approx 0.177$ feet per second downwards.

51. The area is growing by 6 square feet per second at that particular moment.

53. $\left.\dfrac{dr}{dt}\right|_{h=4} = \dfrac{1}{8\pi}\dfrac{\text{in}}{\text{sec}}$ 55. $\left.\dfrac{dh}{dt}\right|_{h=4} = \dfrac{1}{12\pi}\dfrac{\text{in}}{\text{sec}}$

57. $\left.\dfrac{dh}{dt}\right|_{h=3} = -\dfrac{25}{72\pi}\dfrac{\text{in}}{\text{min}}$

59. (a) $\dfrac{dA}{dt} = 2b - a$ for all times t. (b) $1 \le a < 21$ and $0 < b \le 10$, with $0 \le t < 5$. (c) The area is increasing when $2b > a$ and decreasing when $2b < a$. In terms of time, the area is increasing from time $t = 0$ to time $t = 2.375$ and then decreasing from time $t = 2.375$ to time $t = 5$.

61. (a) We see that $x(t)/(13200 - 11710) = \tan(\pi/2 - \theta(t))$. It follows that $x' = -1490\sec^2(\pi/2 - \theta)\theta'$. (b) Now $\theta' = -(11\pi/180)/60$ radians per minute (negative because the sun is going down), and when it hits Ian the shadow makes $\tan\theta = 1490/3100$. It follows that $x' = 25.4$ feet per minute.

63. $\dfrac{d}{dr}\left(\dfrac{4}{3}\pi r^3\right) = 4\pi r^2$

65. Mimic the work in Example 1(c).

For answers to odd-numbered Skill Certification exercises in the Chapter Review, please visit the Book Companion Web Site at www.whfreeman.com/integcalculus.

Chapter 4

Section 4.1

1. T, T, T, T, F, F, F, T.

3. A function that can be expressed as an arithmetic combination of expressions of the form x^k where k is rational. x^3, $\sqrt{x}$, and $\dfrac{1}{1+\sqrt{x}}$ are algebraic, but $\ln x$ and $\sin x$ are not.

5. $|x| = \sqrt{x^2}$

7. (a) It is not automatically clear what would it mean to multiply a number by itself "zero times," so we choose to define $x^0 = 1$ if $x \ne 0$. (b) If x^0 were zero, then $\dfrac{x^2}{x^2} = x^{2-2} = x^0$ would equal zero (instead of what we want it to equal, namely, 1). (c) If x^0 were zero, then

we'd have $x^2 = x^{2+0} = x^2 x^0 = x^2(0) = 0$, and clearly it isn't true that $x^2 = 0$ for all x. (d) Because the rules generalize what happens when we work with positive integer exponents.

9. (a) $\lim\limits_{x\to 0} x^0 = \lim\limits_{x\to 0} 1 = 1$ (b) $\lim\limits_{k\to 0} 0^k = \lim\limits_{k\to 0} 0 = 0$

11. (a) $f(x) = x(2x^2 - 3x + 8)$. The quadratic term is irreducible because $b^2 - 4ac = (-3)^2 - 4(2)(8) < 0$. (b) $f(x) = 2x\left(x^2 - \dfrac{3}{2}x + 4\right)$

13. $f(-1) = 0$, $f(1) = 0$, and $f\left(\dfrac{1}{2}\right) = 0$, but $f(2) = 9 \ne 0$ and $f(-3) = 224 \ne 0$.

15. $\pm 1, \pm 2, \pm 3, \pm 4, \pm 6$, and ± 12

17. $(x - c)g(x)$

19. Polynomial long division is an algorithm for dividing a polynomial by a another polynomial of equal or less degree. Look at the polynomial long-division calculation in the reading and compare it to the algorithm of long division of numbers.

21. $p(x) = q(x)(x^2 - x + 3) + (3x + 1)$; $\dfrac{p(x)}{q(x)} = (x^2 - x + 3) + \dfrac{3x-1}{q(x)}$

23. (a) $\dfrac{27}{8}$; (b) undefined; (c) $\dfrac{4}{9}$; (d) 0.

25. $[-1, 0] \cup [1, \infty)$ 27. $[2, \infty)$

29. $[0, \infty)$ 31. Domain $\mathbb{R}$, zero at $x = 0$

33. Domain $(0, \infty)$, never zero

35. Domain $(-\infty, -1) \cup (1, \infty)$, never zero

37. Domain $(-\infty, -1) \cup (-1, 3]$, zero at $x = 3$

39. Domain $(-3, -1] \cup (2, \infty)$, zero at $x = -1$

41. $(x + 1)(x - 1)(x - 3)$

43. $(x - 1)(x + 1)(x + 2)(x + 5)$

45. $(x + 1)(x - 3)^2$ 47. $(1 + x^2)(x - 2)$

49. $-(4x - 1)(2x - 3)$ 51. $(x + 4)(x - 4)(1 + 2x)$

53. $(x - 3)(x - 1)(x + 2)$ 55. $(x - 1)^2(x + 6)$

57. $2(x - 1)(x + 1)(4 + x^2)$ 59. $2x(x - 2)^2(x + 1)^3$

61. Possible rational roots: $\pm 4, \pm 2, \pm 1, \pm\dfrac{1}{2}$; $f(x) = (x^2 + 4)(2x + 1)$

63. Possible rational roots: $\pm 1, \pm\dfrac{1}{2}, \pm\dfrac{1}{4}, \pm\dfrac{1}{8}$, and $\pm\dfrac{1}{16}$; $f(x) = (2x - 1)^2(4x + 1)$

65. $x^2 + 5x + 14 + \dfrac{37}{x-3}$

67. $(-x^3 + 3x^2 - 9x + 27) - \dfrac{80}{x+3}$

69. $(5x^3 - 5x^2 + 5x - 8) + \dfrac{10}{x+1}$

71. $f(x) = x^2 + \dfrac{5}{x-2}$

73. $f(x) = x^3 - 2x - 4$ (when $x \ne \pm\sqrt{3}$)

75. $f(x) = (-x^2 + 2x - 5) + \dfrac{12x-4}{x^2+2x-1}$

77. $f(x) = \left(\frac{1}{2}x^2 - \frac{7}{4}\right) + \dfrac{\frac{15}{4}}{2x^2+1}$

79. $f(x) = (x^5 + 2x^4 + 3x^3 + 4x^2 + 5x + 6) + \dfrac{7x-7}{x^2-2x+1}$

81. No extrema, no inflection points.

83. Local minimum at $x = 3$; no inflection points.

85. Local maximum at $x = 0$, local minimum at $x = 6$, no inflection points.

87. Using integer root-guessing and synthetic division we have $s(t) = (t - 1)^2(t - 3)(t - 6)$, so the object is also in the center point at time $t = 3$.

89. The idea of the proof is that $(x^a)^b$ is b copies of a copies of x, i.e., ab copies of x.

91. $\left(\dfrac{x}{y}\right)^a = \left(x \cdot \dfrac{1}{y}\right)^a = x^a \left(\dfrac{1}{y}\right)^a = x^a(y^{-1})^a = x^a y^{-a} = x^a \dfrac{1}{y^a} = x^a y^a$.

93. If $x^a x^b = x^{a+b}$, then $x^2 x^0 = x^{2+0} = x^2$ for any nonzero real number x. Since x is nonzero, so is x^2, and thus we can divide both sides of the equation by x^2 to obtain $x^0 = 1$.

Section 4.2

1. T, F, F, T, F, T, F, F.

3. None of them can be written in the form Ax^k for some real number A and rational number k.

5. The graphs look like:

7.

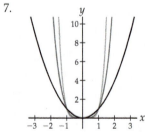

9. yes; $\dfrac{1}{f(x)} = \dfrac{1}{Ax^k} = \dfrac{1}{A}x^{-k}$, so $C = \dfrac{1}{A}$ and $r = -k$.

11. $f(x) = x^{1/2}$ is continuous on the interval $[0, \infty)$; in other words, it is continuous on $(0, \infty)$ and right but not left continuous at $x = 0$.

13. (a) $\lim\limits_{x\to 0} 3x^{1/3} = 0$; (b) For all $\varepsilon > 0$, there exists $\delta > 0$ such that if $0 < |x| < \delta$, then $|3x^{1/3}| < \varepsilon$.

15. The derivative of a power function is always a power function, but the antiderivative of a power function x^k is only a power function when $k \neq -1$.

17. Any function of the form $f(x) = \frac{8}{3}x^{3/2} + C$, where C is a constant, will work.

19. Domain is $(-\infty, \infty)$, range is $(-\infty, \infty)$. f is positive, decreasing, and concave down on $(-\infty, 0)$ and is negative, decreasing, and concave up on $(0, \infty)$. f has a vertical tangent line at $x = 0$. $\lim\limits_{x\to\infty} f(x) = -\infty$ and $\lim\limits_{x\to-\infty} f(x) = \infty$.

21. (a) $\sqrt[k]{A}$; (b) $\dfrac{1}{\sqrt[k]{A}}$

23. $f(x) = x^{-3/10}$

25. $f(x) = x^{-2}$

27. Not a power function

29. $f(x) = \frac{1}{16}x^2$ (for $x \neq 0$)

31.

33.

35.

37.

39.

41.

43. $y = x^4$ is the graph with the "flattest" bottom part; $y = 2x^2$ is the graph that is closest to the y-axis; $y = x^2$ is the remaining graph.

45. $y = x^{2/3}$ is the graph that is closest to the x-axis; $y = 2x^{2/5}$ is the graph that bends in closest to the y-axis; $y = 2x^{2/3}$ is the remaining graph.

47. $f(x) = 2x^4$

49. $f(x) = 2x^{1/4}$

51. $f(x) = 3x^{1/3}$

53. $-\infty$

55. 0

57. Does not exist, because $x^{3/4}$ is not defined for negative values of x.

59. $-\infty$

61. 0

63. $\lim\limits_{h \to 0} \dfrac{\sqrt{9+h}-3}{h} = \dfrac{1}{6}; \lim\limits_{z \to 9} \dfrac{\sqrt{z}-3}{z-9} = \dfrac{1}{6}$

65. $f'(0) = \lim\limits_{h \to 0} \dfrac{(0+h)^{6/5} - 0^{6/5}}{h} = 0$

67. $\lim\limits_{h \to 0} \dfrac{(0+h)^{1/3} - 0^{1/3}}{h} = \infty; f$ has a vertical tangent line at $x = 0$.

69. $\lim\limits_{h \to 0} \dfrac{3\sqrt{x+h}-3\sqrt{x}}{h} = \dfrac{3}{2\sqrt{x}}$

71. $\lim\limits_{z \to x} \dfrac{\sqrt{2z+1}-\sqrt{2x+1}}{z-x} = \dfrac{1}{\sqrt{1+2x}}$

73. (a) Global maximum at $x = 4$ with value $f[4] = 56$, global minimum at $x = 1$ with value $f(1) = -2$. (b) No global maximum, global minimum at $x = 1$ with value $f(1) = -2$. (c) Global maximum at $x = 0$ with value $f(0) = 0$, no global minimum. (d) No global maximum, no global minimum.

75. (a) Global minimum at $x = 0$ and $x = 4$ with value $f(0) = f(4) = \dfrac{1}{\sqrt{3}}$; no global maximum. (b) Global minimum at $x = 10$ with value $f(10) = \dfrac{1}{3\sqrt{7}}$, no global maximum. (c) Global minimum at $x = 0$ with value $f(0) = \dfrac{1}{\sqrt{3}}$, no global maximum. (d) No global minimum, no global maximum.

77. $f(x) = -2x^{-1/2} + \dfrac{5}{3}$

79. $f(x) = \dfrac{15}{2}x^{1/5} - \dfrac{11}{2}$

81. (a) $s(t) = 6t^{1/2} + 100$; (b) $a(t) = -\dfrac{3}{2}t^{-3/2}$.

83. If $f(x) = Ax^k$ and $g(x) = Bx^l$ are power functions, then $(f \circ g)(x) = f(Bx^l) = A(Bx^l)^k = (AB^k)x^{lk}$ is also a power function.

85. We need only exhibit a counterexample: if $f(x) = x^2$ and $g(x) = x^3$ then $f(x) + g(x) = x^2 + x^3$ cannot be written in the form Ax^k, and is therefore not a power function.

87. Mimic the proof in the reading.

89. Mimic the proof in the reading.

91. Mimic the proof in the reading.

93. Suppose $c > 0$; the proof for $c < 0$ is similar. Given $\epsilon > 0$, let $\delta = \min\left(1, \epsilon \cdot \dfrac{(c-1)^2 c^2}{2c+1}\right)$. If $0 < |x - c| < \delta$, then since $\delta \le 1$ we have $c - 1 < x < c + 1$. Then
$$\left|\dfrac{1}{x^2} - \dfrac{1}{c^2}\right| = \dfrac{|c^2 - x^2|}{|x^2 c^2|} = \dfrac{|c-x| \cdot |c+x|}{x^2 c^2} < \dfrac{\delta|c+x|}{x^2 c^2} \le$$
$$\dfrac{\delta|c + (c+1)|}{(c-1)^2 c^2} = \epsilon.$$

95. Suppose $c > 0$. Given $\epsilon > 0$, let
$$\delta = \min\left(1, \epsilon \cdot \dfrac{\sqrt{c-1}+\sqrt{c}}{2c+1}\right). \text{ If } 0 < |x - c| < \delta, \text{ then}$$
since $\delta \le 1$ we have $c - 1 < x < c + 1$. Then $|\sqrt{x} - \sqrt{c}| =$
$$\dfrac{|x^2 - c^2|}{\sqrt{x}+\sqrt{c}} = \dfrac{|x-c| \cdot |x+c|}{\sqrt{x}+\sqrt{c}} < \dfrac{\delta|x+c|}{\sqrt{x}+\sqrt{c}} \le \dfrac{\delta|(c+1)+c|}{\sqrt{c-1}+\sqrt{c}} = \epsilon.$$

Section 4.3

1. T, F, T, T, F, F, T, F.

3. Leading coefficient 3, leading term $3x^4$, degree 4, constant term 0, coefficients $a_1 = 4$, $a_3 = -11$.

5. If $b^2 - 4ac < 0$, then $\sqrt{b^2 - 4ac}$ is not a real number, so by the quadratic formula the quadratic has no real roots. If $f(x) = 3x^2 + 2x + 6$, then $b^2 - 4ac = 2^2 - 4(3)(6) = -68$ is negative.

7. $(x - c)^2 g(x)$

9. Some possible answers are (a) $f(x) + 2$; (b) $f(x) + 1.5$; (c) $f(x)$; (d) $f(x) - 0.4$ or $f(x) - 1.8$; (e) $f(x) - 1$.

11.   13.

15. Such a polynomial cannot exist.

17. $f(x) = (x^2 + 1)(x - 2)(x - 3)(x - 4)(x - 5)^2$

19. $f(x) = (x - 2)(x - 3)(x^2 + 1)^2$

21. $f(x) = (x + 1)(x - 1)(x - 3) + 10$

23. $-\infty$ 25. $-\infty$

27.   29.

31. Your graph should have a quadruple root at $x = -2$, a root at $x = 1$, and a double root at $x = 3$, and it should point "up" at the left end and "down" at the right end. There is a y-intercept at $(0, 288)$.

33.

35. $f(x) = \frac{5}{4}(x-1)^2(x-2)^2$ 37. $f(x) = (x-2)^3$

39. $f(x) = x^3 - x^2 - 3x - 5$ 41. $f(x) = x^2 + 3x - 5$

43. Local minimum at $x = -2$; inflection points at $x = \pm 1$

45. Local minimum at $x = -2$; no inflection points

47. Local minima at $x = -1$ and $x = 3$, local maximum at $x = 1$; inflection points at $x = \frac{3 \pm 2\sqrt{3}}{3}$

49. Root at $x = 2$, local maximum at $x = \frac{1}{3}$, local minimum at $x = 1$, inflection point at $x = \frac{2}{3}$:

51. Root at $x = 0$, no local extrema, inflection points at $x = 0$, $x = \frac{1}{2}$, and $x = 1$:

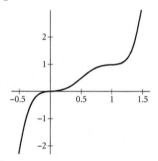

53. No roots, local minimum at $x = 0$, no inflection points.

55. $f(x) = 2(x+1)(x-2)(x-3)$

57. $f(x) = -2(x+2)^2 x \left(x - \frac{5}{2}\right)$

59. $f(x) = -(x+2)(x-1)^3$

61. $f(x) = (x-2)^2 + 2$

63. (a) $x = -2$, $x = 0$, $x = 2$, and $x = 4$; 4; negative.
(b) $f'(x) = -x(x+2)(x-2)(x-4)$;
(c) $f(x) = -\frac{1}{5}x^5 + x^4 + \frac{4}{3}x^3 - 8x^2 + C$; (d) $C = 25$

65. $s(t) = 10t^4 - \frac{100}{3}t^3 + 25t^2 + 45t + 550$,
$a(t) = 120t^2 - 200t + 50$.

67. The minimum possible degree is 3.

69. If $f(x) = mx + b$ and $g(x) = nx + c$ then
$f(g(x)) = f(nx + c) = m(nx + c) + b = (mn)x + (mc + b)$.

71. If $f(x) = Ax^3 +$ lower-degree terms and
$g(x) = Bx^3 +$ lower-degree terms, then
$f(x)g(x) = ABx^6 +$ lower-degree terms. Since f and g are
cubic we know that A and B are nonzero. Thus AB must
also be nonzero, and therefore fg is of degree 6.

73. $\lim_{x \to \infty} (a_n x^n + a_{n-1}x^{n-1} + \cdots + a_1 x + a_0)$
$= \lim_{x \to \infty} \left(a_n x^n \left(1 + \frac{a_{n-1}}{a_n x} + \cdots + \frac{a_1}{a_n x^{n-1}} + \frac{a_0}{a_n x^n} \right) \right)$
$= \left(\lim_{x \to \infty} a_n x^n \right)(1 + 0 + \cdots + 0 + 0) = \lim_{x \to \infty} a_n x^n$.

75. See the proof in the reading of part (c) of Theorem 4.10.

77. See the discussion in the reading above Theorem 4.11.

79. See what happens when you differentiate
$f(x) = a_n x^n + a_{n-1}x^{n-1} + \cdots + a_2 x^2 + a_1 x + a_0$
repeatedly for a total of n times.

Section 4.4

1. T, F, F, T, F, T, F, T.

3. A function that can be written as a quotient of
polynomials, where the polynomial in the denominator
is not constantly zero.

5. After reducing, $f(x)$ becomes $\frac{x-1}{x+2}$, which has a vertical
asymptote at $x = -2$.

7. One answer is $f(x) = \frac{3x^2+1}{(x+2)(x-2)}$.

9. (a) Graph the functions given in part (b) to see three
possible answers; (b) $f(x) = \frac{-(x-1)(x-3)(x+1)}{(x-2)^2(x+1)}$;
$f(x) = \frac{-(x-1)(x-3)(x+1)^2}{(x-2)^2(x+1)^2}$; $f(x) = \frac{-(x-1)^2(x-3)(x+1)}{(x-2)(x+1)(x^2+1)}$.

11. (a) $n \leq m$; (b) $n < m$; (c) $n = m + 1$; (d) $n \geq m + 2$;
(e) $n = m + 4$.

13.

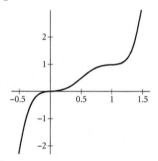

15. A common irreducible quadratic factor in the numerator
and the denominator will not affect a rational function,
since an irreducible quadratic factor is never zero.

17. (a) If there is no cancellation in the expression
$p'(x)q(x) - p(x)q'(x)$, then the numerator of f' will have
degree $n + m - 1$. If there is cancellation, then the
numerator of f' may have a lower degree. In any case,
the denominator of f' will have degree $2m$. (b) Yes,
since if $n < m$, then
$n + m - 1 < m + m - 1 = 2m - 1 < 2m$. (c) Yes, at $y = 0$.

19. Root at $x = -1$, hole at $x = 2$, vertical asymptote at
$x = -2$, horizontal asymptote at $y = 1$.

21. Root at $x = -1$, no holes, vertical asymptotes at $x = \pm 2$, horizontal asymptote at $y = 0$.

23. No roots; holes at $x = 2$ and $x = -2$; vertical asymptote at $x = 0$; horizontal asymptote at $y = -1$.

25. Roots at $x = 1$ and $x = -\dfrac{3}{2}$; no holes; vertical asymptote at $x = -1$; slant asymptote with equation $y = 2x - 1$.

27. Root at $x = -\dfrac{3}{2}$; hole at $x = \dfrac{3}{2}$; curve asymptote with equation $y = -8x^3 - 12x^2 - 18x - 27$

29. Roots at $x = \pm 2$; no holes; vertical asymptote at $x = 1$; horizontal asymptote at $y = \dfrac{1}{2}$.

31. Roots at $x = \pm 1$ and $x = -\dfrac{2}{3}$; no holes; slant asymptote with equation $y = 3x - 1$

33. Root at $x = 0$; hole at $x = 1$; vertical asymptote at $x = -1$; curve asymptote with equation $y = x^3 - 2x^2 + 2x - 2$.

35. The graph of f looks like the graph of $y = x + 3$ with a hole at $x = 2$.

37. The graph of f has roots at $x = 0$ and $x = -1$, vertical asymptotes at $x = 5$ and $x = -2$, and a horizontal asymptote at $y = 0$.

39. The graph of f has roots at $x = \pm 1$, vertical asymptotes at $x = -\dfrac{1}{2}$ and $x = 2$, and a horizontal asymptote at $y = \dfrac{1}{2}$.

41. The graph of f has roots at $x = 0$, $x = -3$, and $x = 1$, vertical asymptotes at $x = \pm 2$, and a slant asymptote with equation $y = 2x + 4$.

43.

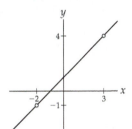

45. The graph of f has a hole at $x = -1$, roots at $x = -\dfrac{3}{2}$ and $x = 1$, a vertical asymptote at $x = 3$, and a slant asymptote with equation $y = 2x + 7$.

47.

49. The graph of f has no roots or holes, a vertical asymptote at $x = 2$, and a horizontal asymptote at $y = 3$; moreover, the graph of f is always positive. f has a local minimum at $x = -\dfrac{2}{5}$ and an inflection point at $x = -\dfrac{8}{5}$.

51. The graph of f has a hole at $x = -1$, roots at $x = 0$ and $x = 1$, a vertical asymptote at $x = 2$, and a slant asymptote with equation $y = x + 1$. f has a local maximum at $x = 2 - \sqrt{2}$, a local minimum at $x = 2 + \sqrt{2}$, and no inflection points. $\lim\limits_{x \to 2^-} f(x) = -\infty$, and $\lim\limits_{x \to 2^+} f(x) = \infty$.

53. The graph of f has a hole at $x = 1$ with a height of $y = 0$, a double root at $x = -1$, a vertical asymptote at $x = -\dfrac{3}{2}$, and no horizontal asymptotes. The graph has a curve asymptote of $y = \dfrac{1}{3}x^2 - \dfrac{1}{9}x - \dfrac{11}{27}$. This graph has a local minimum at $x = -1$ and an inflection point at $x = \dfrac{\sqrt[3]{5} - 2}{3}$.

55. $f(x) = \dfrac{-4(x^2 + 1)(x - 4)}{(x + 2)(x - 2)(x - 4)}$

57. $f(x) = \dfrac{-4(x - 1)(x + 1)}{(x^2 + 1)^2}$

59. $f(x) = \dfrac{(x^2 + 3x - 2)(x + 4)(x - 2)}{(x + 4)(x - 2)}$

61. $f(x) = (x + 1) - \dfrac{1}{x + 1}$

63. $f(x) = \dfrac{-6x(x - 3)(x + 3)}{(x + 2)(x - 2)}$

65. $f(x) = |2(x + 1)(x - 3)|$

67. (a) Over 10 years the dryer costs you $800 + 10(55) = 1350$ dollars, which is a yearly cost of \$135. Over 20 years the dryer costs $800 + 20(55) = 1900$ dollars, which is a yearly cost of \$95. (b) $Y(n) = \dfrac{800 + n(55)}{n}$. (c) The limit is 55. This makes sense because if you owned the dryer for a very long period of time, the initial cost of the dryer would be spread over many years, and your yearly costs would be close to the yearly electricity costs. (d) $Y(n) = \dfrac{800}{n} + 55$, which makes sense because the yearly cost of the dryer is \$55 per year plus one nth of the initial cost of the dryer.

69. No, since polynomials increase or decrease without bound and do not have asymptotes. A rational function might be a better choice.

71. If $f(x) = \dfrac{p(x)}{q(x)}$ and $g(x) = \dfrac{r(x)}{s(x)}$ where $p(x)$, $q(x)$, $r(x)$, and $s(x)$ are polynomials, then $f(x) + g(x) = \dfrac{p(x)}{q(x)} + \dfrac{r(x)}{s(x)} = \dfrac{p(x)s(x) + q(x)r(x)}{q(x)s(x)}$. Since sums and products of polynomials are also polynomials, $\dfrac{f(x)}{g(x)}$ is a quotient of two polynomials, and thus a rational function.

73. The domain of a quotient $f(x) = \dfrac{p(x)}{q(x)}$ of functions is $\{x \mid x \in \text{Domain}(p(x)) \cap \text{Domain}(q(x))\text{ and }q(x) \neq 0\}$. Since $p(x)$ and $q(x)$ are polynomials, they are defined on $(-\infty, \infty)$; thus the domain of f is $\{x \mid q(x) \neq 0\}$.

75. See the proofs in the reading.

77. If f is a rational function with numerator of degree n and denominator of degree m, then its derivative f' is a rational function with numerator of degree at most $n + m - 1$ and denominator of degree $2m$. If f has a horizontal asymptote, then $n \leq m$, which implies that the degree of the numerator of f' is less than the degree of the denominator of f', since $n + m - 1 \leq m + m - 1 = 2m - 1 < 2m$. Therefore, f' must have a horizontal asymptote.

For answers to odd-numbered Skill Certification exercises in the Chapter Review, please visit the Book Companion Web Site at www.whfreeman.com/integcalculus.

Chapter 5

Section 5.1

1. F, T, T, T, T, F, T, T.

3. A function is exponential if it can be written in the form $f(x) = Ab^x$; the variable is in the exponent and a constant is in the base. For a power function, the situation is reversed. x^x is neither a power nor an exponential function because a variable appears in both the base and the exponent.

5. $\sqrt{3} \approx 1.73205$. We have $2^{1.7} \approx 3.2490$, $2^{1.73} \approx 3.3173$, $2^{1.732} \approx 3.3219$, $2^{1.7320} \approx 3.3219$, $2^{1.73205} \approx 3.3220$, and so on. Each of these approximations gets closer to the value of $2^{\sqrt{3}}$.

7. $f(x) = 3(e^{-2})^x \approx 3(0.135)^x$, $g(x) = -2e^{(\ln 3)x} \approx -2e^{1.0986x}$.

9. $f(3) = 2^9 = 512$ and $g(3) = 8^2 = 64$. The solutions of $2^{(x^2)} = (2^x)^2$ are $x = 0$ and $x = 2$.

11. We must have $b > 0$ and $b \neq 1$, since those conditions are necessary for b^x to be an exponential function.

13. The graph that passes through $(2, 1)$ is $\log_2 x$; the graph that passes through $\left(2, \frac{1}{2}\right)$ is $\log_4 x$.

15. $f(x) = 2\ln x$ is the graph of $\ln x$ stretched vertically by a factor of 2; $g(x) = -\ln x$ is the graph of $\ln x$ reflected over the x-axis.

17. $\log_2 x = \log_3 x \iff \dfrac{\ln x}{\ln 2} = \dfrac{\ln x}{\ln 3} \iff (\ln 3)(\ln x)$
$= (\ln 2)(\ln x) \iff (\ln x)(\ln 3 - \ln 2)$
$= 0 \iff \ln x = 0 \iff x = 1$.

19. (a) $3^{x-1} = 2 \Rightarrow \ln(3^{x-1}) = \ln 2 \Rightarrow (x - 1)(\ln 3)$
$= (\ln 2) \Rightarrow x = \dfrac{\ln 2}{\ln 3} + 1$. (b) $3^{x-1} = 2 \Rightarrow \log_3(3^{x-1})$
$= \log_3 2 \Rightarrow x - 1 = \log_3 2 \Rightarrow x = (\log_3 2) + 1$. The answers are the same by the base conversion formula.

21. $k = \ln b$, $b = e^k$

23. $f(x) = 3\left(\dfrac{1}{2}\right)^x$ and $f(x) = 3e^{(\ln(1/2))x} \approx 3e^{-0.693x}$

25. $f(x) = \dfrac{2}{e^4}(e^3)^x \approx 0.03663(20.0855)^x$

27. $f(x) = 8\left(\dfrac{5}{2}\right)^x$ and $f(x) = 8e^{(\ln(5/2))x} \approx 8e^{0.916x}$

29. 2

31. 6

33. -2

35. 4

37. $(-\infty, \infty)$

39. $(-\infty, 0) \cup (0, \infty)$

41. $(2, 3) \cup (3, \infty)$

43. $(2, \infty)$

45. $(-\infty, -1) \cup (2, \infty)$

47. $[-1, 0) \cup (0, \infty)$

49. $x = 2$, $x = -3$

51. $x = 0$ and $x = 3$

53. $x = \pm\sqrt{\dfrac{e^2 + 1}{3}} \approx \pm 1.67223$

55. $x = \dfrac{\ln 3}{\ln(3/2)} \approx 2.70951$

57. $x = -\dfrac{17}{15}$

59. $x = 0$ and $x = -2$

61. $x = \dfrac{\ln(5/3)}{\ln 3}$

63. Exponential growth graph with no root, y-intercept at $(0, 2)$, and horizontal asymptote at $y = 0$.

65. Start with the graph of $y = \left(\dfrac{1}{2}\right)^x$, then reflect over the x-axis and shift up by 10 units.

67. 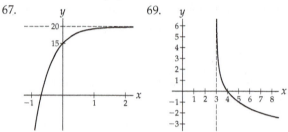 69.

71. $f(x) = 2e^{-x} - 3$

73. $f(x) = -5e^{-x} + 10$

75. $f(x) = 9e^{\ln(2/3)/2x} \approx 9e^{-0.2027x}$

77. $f(x) = -3e^{\ln(7/3)/5x} \approx -3e^{0.16946x}$

79. $f(x) = 2e^{-(\ln 32/5)x} \approx 2e^{-0.693147x}$

81. 1999, that is, after $t \approx 9.006$ years.

83. (a) $I(t) \approx 10,000e^{0.08t}$; (b) $I(t) = 10,000(1.085)^t$.

85. (a) $S(t) = 250e^{-\ln 2/29t} \approx 250e^{-0.0239t}$, or equivalently, $S(t) = 250(0.97638)^t$; (b) 2.39%; (c) 156 years

87. Seeking a contradiction, suppose that A and b are nonzero real numbers with $Ab^x = 0$ for some real number x. Since $A \neq 0$ we know $b^x = 0$, and therefore that $x(\ln b) = \ln(b^x) = \ln 0$. But this is a contradiction because $\ln 0$ is undefined, so the product $x(\ln b)$ of real numbers cannot equal $\ln 0$.

89. $\log_2 x = \log_3 x \Leftrightarrow \dfrac{\ln x}{\ln 2} = \dfrac{\ln x}{\ln 3} \Leftrightarrow (\ln 3)(\ln x)$
$= (\ln 2)(\ln x) \Leftrightarrow (\ln x)(\ln 3 - \ln 2) = 0 \Leftrightarrow \ln x = 0$
$\Leftrightarrow x = 1$.

91. $y = \log_b x$ if and only if $b^y = x$. Since the only solution of $b^y = 1$ is $y = 0$ (if $b \neq 0$), we know that $\log_b 1 = 0$.

93. Since $x = b^{\log_b x}$, we have $\log_b(x^a) = \log_b((b^{\log_b x})^a) = \log_b(b^{(\log_b x)a}) = (\log_b x)a = a\log_b x$.

95. (a) $\log_b\left(\frac{1}{x}\right) = \log_b(x^{-1}) = -\log_b x$; (b)
$\log_b\left(\frac{x}{y}\right) = \log_b(xy^{-1}) = \log_b x - \log_b y$.

Section 5.2

1. T, T, T, T, F, T, T, T.

3. (a) $\lim\limits_{x\to 1} 2e^{3x} = 2e^3$; (b) $\lim\limits_{x\to\infty} 2e^{3x} = \infty$;
   (c) $\lim\limits_{x\to -\infty} 2e^{3x} = 0$.

5. (a) 0; (b) 0; (c) $(0, \infty)$; (d) $(0, 1)$

7. For all $\epsilon > 0$, there exists $\delta > 0$ such that whenever $x \in (2 - \delta, 2) \cup (2, 2 + \delta)$ we have $e^x \in (e^2 - \epsilon, e^2 + \epsilon)$.

9. For all $\epsilon > 0$, there exists $\delta > 0$ such that whenever $x \in (1 - \delta, 1) \cup (1, 1 + \delta)$ we have $\ln x \in (-\epsilon, \epsilon)$.

11. Hint: Think about reflecting over the line $y = x$.

13. See the proof of Theorem 5.9.

15. (a) $\lim\limits_{x\to -\infty} \frac{(1/2)^x - 1}{(3/4)^x} \to \frac{\infty}{\infty}$, which is indeterminate; (b)
$\lim\limits_{x\to -\infty} 3^x\left(\left(\frac{2}{3}\right)^x - \left(\frac{4}{3}\right)^x\right)3^x = \infty$.

17. -1

19. ∞

21. DNE: ∞ from the left and $-\infty$ from the right.

23. $-\dfrac{1}{4}$

25. $\dfrac{7}{27}$

27. 1

29. $\log_3 21$

31. $-\infty$

33. $\dfrac{1}{2}$

35. $\dfrac{2}{5}$

37. $-\infty$

39. 0

41. $3e^{1.7(4)} + 1$

43. $\dfrac{1}{4}$

45. 2

47. -3

49. 0

51. ∞

53. 0

55. $-\infty$

57. ∞

59. 4

61. 0

63. e^2

65. e^3

67. 0

69. Root at $x = \dfrac{\ln 25}{\ln 0.5}$, horizontal asymptote at $y = 25$.

71. No roots, no holes, no vertical asymptotes, horizontal asymptotes at $y = 0$ on the left and $y = \dfrac{1}{2}$ on the right.

73. Root at $x = 0$, no holes, no asymptotes.

75. Root at $x = e^{3/4} - 1$, vertical asymptote at $x = -1$.

77. $N = e^{100}$

79. $N = \dfrac{\ln(1/2)}{\ln 3} \approx -0.63093$

81. (a) 70 degrees; (b) 350 degrees; (c) 21 minutes.

83. Mimic the proof of Theorem 5.10.

85. Given $c \in \mathbb{R}$, $\lim\limits_{x\to c} f(x) = \lim\limits_{x\to c} Ab^x = A\lim\limits_{x\to c}(e^{\ln b})^x =$
$A\lim\limits_{x\to c} e^x \ln b = A(e^c)^{\ln b} = A(e^{\ln b})^c = Ab^c = f(c)$.

87. $\lim\limits_{x\to\infty} \log_b x = \lim\limits_{x\to\infty} \frac{\ln x}{\ln b} = \left(\frac{1}{\ln b}\right)\lim\limits_{x\to\infty} \ln x \to \frac{1}{\ln b}\cdot\infty$. If
$b > 1$ then $\ln b$ is positive and thus $\frac{1}{\ln b})\cdot\infty = \infty$. If
$0 < b < 1$ then $\ln b$ is negative and thus $\frac{1}{\ln b}\cdot\infty = -\infty$.

89. By Theorem 1.25 and the previous exercise,
$\lim\limits_{x\to\infty} e^{-x} = \lim\limits_{x\to\infty} \frac{1}{e^x} \to \frac{1}{\infty} \to 0$.

Section 5.3

1. T, T, F, F, T, T, T, F.

3. No, since the base is a variable. No, since the exponent is a variable.

5. When $k = 1$ we have $\dfrac{d}{dx}(e^x) = \dfrac{d}{dx}(e^{1x}) = 1e^{1x} = e^x$.
   When $b = e$ we have
   $\dfrac{d}{dx}(b^x) = \dfrac{d}{dx}(e^x) = (\ln e)e^x = (1)e^x = e^x$.

7. $\ln x$ is of the form $\log_b x$ with $b = e$.

9. The graph passing through $(0, 2)$ is $2(2^x)$; the graph passing through $(0, 1)$ and $(1, 4)$ is 4^x; the graph passing through $(2, 4)$ is 2^x.

11. Logarithmic differentiation is the process of applying $\ln|x|$ to both sides of an equation $y = f(x)$ and then differentiating both sides in order to solve for $f'(x)$. It is useful for finding derivatives of functions that involve multiple products or quotients, and functions that have variables in both a base and an exponent.

13. For $h = 1$, $h = 0.5$, $h = 0.25$, and $h = 0.1$, respectively, we have secant lines with slopes 0.6931, 0.4463, 0.2425, and 0.0995. One estimate of $f'(0)$ is 0.

15. For $h = 0.1$, $h = 0.01$, $h = 0.001$, and $h = 0.0001$ respectively we have secant lines with slopes 5.7419, 5.5644, 5.5471, and 5.5454. An estimate for $f'(3)$ is 5.54.

17. $f'(x) = -(2 - e^{5x})^{-2}(-5e^{5x})$

19. $f'(x) = 6xe^{-4x} - 12x^2 e^{-4x}$

21. $f'(x) = \dfrac{(-1)e^x - (1-x)e^x}{e^{2x}}$

23. $e^x(x^2 + 3x - 1) + e^x(2x + 3)$

25. $f'(x) = 3x^2$, with $x > 0$

27. $f'(x) = e^{(e^x)}e^x$

29. $f'(x) = e^{e x + 1}$

31. $f'(x) = 0$

33. $f'(x) = -3x^{-4}e^{2x} + x^{-3}(2e^{2x})$.

35. $f'(x) = 2x\log_2 x + \dfrac{x}{\ln 2} + 3x^2$

37. $f'(x) = \dfrac{1}{x^2 + e^{\sqrt{x}}}\left(2x + e^{\sqrt{x}}\left(\frac{1}{2}x^{-1/2}\right)\right)$

39. $f'(x) = \frac{1}{2}(\log_2(3x-5))^{-(1/2)}\left(\frac{1}{\ln 2}\frac{1.}{3^x-5}((\ln 3)3^x\right)$

41. $f'(x) = 2x\ln(\ln x) + x^2\frac{1}{\ln x}\cdot\frac{1}{x}$

43. $f'(x) = (\ln 2)2^{x^2}(2x)$

45. One critical point at $x = \ln\left(\frac{3}{2}\right)$, a local maximum.

47. Undefined at $x = 0$. One critical point at $x = \frac{e}{2}$, a local maximum.

49. Local minimum at $x = 2$.

51. Local minimum at $x = 1$.

53. Decreasing on $(-\infty, 1)$ and increasing on $(1, \infty)$.

55. Decreasing on $(-\infty, 0)$ and increasing on $(0, \infty)$.

57. Concave up on $(-\infty, \ln\left(\frac{9}{16}\right)$, concave down on $(\ln\left(\frac{9}{16}\right), \infty)$; inflection point at $\ln\left(\frac{9}{16}\right)$.

59. Concave up on $\left(-\infty, 1 - \frac{\sqrt{2}}{2}\right) \cup \left(1 + \frac{\sqrt{2}}{2}, \infty\right)$, concave down on $\left(1 - \frac{\sqrt{2}}{2}, 1 + \frac{\sqrt{2}}{2}\right)$; inflection points at $x = 1 \pm \frac{\sqrt{2}}{2}$.

61. Domain is $(-\infty, \infty)$, always increasing and concave down, $\lim_{x\to\infty} f(x) = 400$, $\lim_{x\to-\infty} f(x) = -\infty$.

63. Domain is $(-\infty, \infty)$, always increasing, concave up on $(-\infty, -0.3054)$, concave down on $(-0.3054, \infty)$, $\lim_{x\to\infty} f(x) = 2$, $\lim_{x\to-\infty} f(x) = 0$.

65. Domain $(0, \infty)$, local minimum at $x = \frac{1}{e}$, always concave up, $\lim_{x\to 0^+} f(x) = 0$, $\lim_{x\to\infty} f(x) = \infty$.

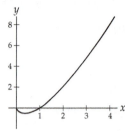

67. Domain is $(-\infty, 0]$, always decreasing and concave down, $\lim_{x\to-\infty} f(x) = \infty$.

69. $f'(x) = \frac{1}{2}(x\ln|2^x + 1|)^{-(1/2)}\left(\ln|2^x + 1| + \frac{(\ln 2)x2^x}{2^x+1}\right)$

71. $f'(x) = \frac{2^x\sqrt{x^3-1}}{\sqrt{x}(2x-1)}\left(\ln 2 + \frac{3x^2}{2(x^3-1)} - \frac{1}{2x} - \frac{2}{2x-1}\right)$

73. $f'(x) = x^{\ln x}(2)(\ln x)\left(\frac{1}{x}\right)$

75. $f'(x) = \left(\frac{x}{x-1}\right)^x\left(\ln x - \ln(x-1) + 1 - \frac{x}{x-1}\right)$

77. $f'(x) = (\ln x)^{\ln x}\left(\frac{\ln(\ln x)}{x} + \frac{1}{x}\right)$

79. $f(x) = \frac{2}{3}e^{3x} + \frac{1}{3}$

81. $f(x) = \ln x + 4$

83. $f(x) = \ln(x^2 + 1) - 2$

85. The number of fruit flies in the population.

87. (a) $A(t) = 1000e^{0.077t}$; (b) $A(30) \approx \$10,074.42$; (c) $t \approx 18$ years.

89. (a) Calculate the derivative $P'(t)$ and show that P' is always positive (hence P is always increasing). Then calculate the second derivative $P''(t)$ and show that P'' is positive for $x < \frac{25\ln 35}{3}$ and negative for $x > \frac{25\ln 35}{3}$. (b) $\frac{25\ln 35}{3} \approx 29.6$ days

91. $\frac{d}{dx}(e^{kx}) = e^{kx}\cdot\frac{d}{dx}(kx) = e^{kx}\cdot k = ke^{kx}$.

93. If $f(x) = Ae^{kx}$ is exponential, then $f'(x) = Ake^{kx} = k(Ae^{kx}) = kf(x)$, so $f'(x)$ is proportional to $f(x)$.

95. For $x > 0$ we have $|x| = x$ and therefore $\frac{d}{dx}(\ln|x|) = \frac{d}{dx}(\ln x) = \frac{1}{x}$. For $x < 0$ we have $|x| = -x$ and therefore by the chain rule we have $\frac{d}{dx}(\ln|x|) = \frac{d}{dx}(\ln(-x)) = \frac{1}{-x}(-1) = \frac{1}{x}$.

97. Follow the proof outline after Theorem 5.16. Specifically, calculate $f(0)$, $\lim\limits_{x\to\infty} e^{kx}$, $\lim\limits_{x\to-\infty} e^{kx}$, and the signs of $f'(x)$ and $f''(x)$. Throughout your argument you will need to use the fact that k is assumed to be positive.

Section 5.4

1. F, F, F, T, F, T, F, T.   3. $Q(t)$

5. Constant; exponential   7. 1.16583%

9. 15% compounded daily is better.

11. $Q(t) = Q_0\left(1+\dfrac{r}{n}\right)^{nt}$ 13. $\lim\limits_{n\to\infty}\left(1+\dfrac{r}{n}\right)^n = e^r$.

15. $Q(17) = \dfrac{1}{2}Q_0$, $Q(51) = \dfrac{1}{8}Q_0$, $Q(-34) = 4Q_0$

17. Yes, since $\dfrac{dW}{dt}$ is proportional to $W(t)$, and thus $W(t)$ is an exponential function.

19. e^6 21. $\dfrac{1}{e}$

23. $e^{-\frac{3}{2}}$ 25. $\ln(1.24) \approx 0.215$

27. 31.37%   29. $\dfrac{1}{32}\ln\left(\dfrac{1}{2}\right) \approx -0.02166$

31. $\dfrac{\ln 2}{\ln 1.13} \approx 5.67$ years   33. $\dfrac{6.5\ln 3}{\ln 2} \approx 10.3$ years

35. $\dfrac{\ln 0.5}{\ln 0.96} \approx 16.98$ years   37. $\ln(0.96) \approx -0.0408$

39. $\left(1+\dfrac{0.18}{12}\right)^{12} \approx 1.1956$, so your account would increase at about 19.56 percent a year; if daily, then $\left(1+\dfrac{0.18}{365}\right)^{365} \approx 1.1972$, so your account would increase at about 19.72 percent a year.

41. If yearly, $98,711.43; if monthly, $131,571.79.

43. $135,563.19

45. If daily, 24.5994% a year; if continuously, 24.6077% a year.

47. If daily, 24.09 years; if continuously, 24.08 years.

49. $\dfrac{\ln(840/600)}{3} \approx 0.11216$ 51. 66.14 days

53. 2.5237 hours   55. 72,800%

57. $\dfrac{\ln(1/2)}{29} \approx -0.0239$ 59. 53.6 years ago

61. 125.34 years

63. If $Q(12) = 2Q_0$ then $Q_0 e^{k(12)} = 2Q_0$, and thus $e^{12k} = 2$. Then $Q(t+12) = Q_0 e^{k(t+12)} = Q_0 e^{kt+12k} = Q_0 e^{kt} e^{12k} = Q_0 e^{kt}(2) = 2(Q_0 e^{kt}) = 2Q(t)$.

65. $k = \ln(1+r)$, because if $Q_0 e^{kt} = Q_0(1+r)^t$, then $kt = t\ln(1+r)$ and therefore $k = \ln(1+r)$.

67. We will show that $\alpha = \dfrac{\ln 2}{k}$. If α is the doubling time, and $Q(t) = Q_0 e^{kt}$, then $2Q_0 = Q_0 e^{k\alpha}$. Solving for α, we have $2 = e^{k\alpha} \Rightarrow \ln 2 = k\alpha \Rightarrow \alpha = \dfrac{\ln 2}{k}$.

Section 5.5

1. F, F, F, F, T, F, T, F.

3. See the discussion at the beginning of the section.

5. $y = 2(x-1)$ is the line tangent to f at $x = 1$, and $y = x - 1$ is the line tangent to $g(x)$ at $x = 1$. Very close to $x = 1$ the graphs of f and g are very close to the graphs of these tangent lines.

7. $\lim\limits_{x\to\infty}\dfrac{x}{2^x} \to \dfrac{\infty}{\infty}$ and $\lim\limits_{x\to\infty}\dfrac{1/2^x}{1/x} \to \dfrac{0}{0}$; the first form works better.

9. $\lim\limits_{x\to\infty}\dfrac{\ln x}{x^2} \to \dfrac{\infty}{\infty}$ and $\lim\limits_{x\to\infty}\dfrac{x^{-2}}{(1/\ln x)} \to \dfrac{0}{0}$; the first form works better.

11. $\lim\limits_{x\to\infty} e^{-x}\dfrac{1}{\ln x} \to \dfrac{0}{0}$ and $\lim\limits_{x\to\infty}\dfrac{\ln x}{e^x} \to \dfrac{\infty}{\infty}$; the second form works better.

13. The only error is in the second application of L'Hôpital's Rule (it does not apply).

15. (a) $\lim\limits_{x\to 1}\dfrac{2x+1}{1} = 3$; (b) $\lim\limits_{x\to 1}(x+2) = 3$

17. (a) $\lim\limits_{x\to\infty}\dfrac{1}{-6x} = 0$; (b) $\lim\limits_{x\to\infty}\dfrac{\frac{1}{x}-\frac{1}{x^2}}{\frac{2}{x^2}-3} = \dfrac{0-0}{0-3} = 0$

19. (a) $\lim\limits_{x\to\infty}\dfrac{3e^{3x}}{-2e^{2x}} = \lim\limits_{x\to\infty}\dfrac{3e^x}{-2} = -\infty$; (b) $\lim\limits_{x\to\infty}\dfrac{1}{\frac{1}{e^{3x}}-\frac{1}{e^x}} = -\infty$

21. $-8\ln 2$ 23. $\dfrac{1}{2}$

25. 0   27. 0

29. ∞ 31. 0

33. 0   35. 0

37. 1   39. 1

41. ∞ 43. 1

45. e 47. 1

49. 0   51. 1

53. $\lim\limits_{x\to\infty}\dfrac{x+100}{x} = 1$, so neither dominates.

55. $\lim\limits_{x\to\infty}\dfrac{2x^{100}}{100x^2} = \infty$, so $v(x)$ dominates $u(x)$.

57. $\lim\limits_{x\to\infty}\dfrac{e^x}{2^x} = \infty$, so $v(x)$ dominates $u(x)$.

59. Neither function dominates.

61. $\lim\limits_{x\to\infty}\dfrac{0.001e^{0.001x}}{100x^{100}} = \infty$, so $u(x)$ dominates $v(x)$.

63. 0   65. ∞

67. ∞

69. On I, f has a global maximum at $x = 1$ and a global minimum at $x = \dfrac{1}{e}$, since $\lim\limits_{x\to 0^+} f(x) = 0$. On J, f has no global maximum and a global minimum at $x = \dfrac{1}{3}$, since $\lim\limits_{x\to\infty} f(x) = \infty$.

71. On I, f has a global maximum at $x = 3$ and a global minimum at $x = 0$. On J, f has a global maximum at $x = 3$ but no global minimum.

73. $\dfrac{\Delta w}{\Delta t} = \dfrac{w(t) - w(0)}{t - 0} = \dfrac{w(t)}{t}$.

75. If $f(x) = Ae^{kx}$ is exponential and $g(x) = x^2$, then by applying L'Hôpital's rule twice, we have
$$\lim_{x \to \infty} \frac{Ae^{kx}}{x^2} = \lim_{x \to \infty} \frac{Ake^{kx}}{2x} = \lim_{x \to \infty} \frac{Ak^2 e^{kx}}{2}, \text{ which in the limit}$$
is equal to ∞.

77. If $f(x) = Ae^k x$ is exponential and $g(x) = Bx^r$ is a power function, then by applying L'Hôpital's rule r times, we have
$$\lim_{x \to \infty} \frac{Ae^{kx}}{Bx^r} = \lim_{x \to \infty} \frac{Ake^{kx}}{Brx^{r-1}} = \cdots = \lim_{x \to \infty} \frac{Ak^r e^{kx}}{Br(r-1)(r-2)\cdots(1)},$$
which is of the form $\dfrac{\infty}{\text{constant}} \to \infty$.

79. $\displaystyle\lim_{x \to \infty} \ln\left(\left(1 + \frac{r}{x}\right)^x\right) = \lim_{x \to \infty} x \ln\left(1 + \frac{r}{x}\right) =$
$$\lim_{x \to \infty} \frac{\ln\left(1 + \frac{r}{x}\right)}{\frac{1}{x}} = \lim_{x \to \infty} \frac{\frac{1}{1+(r/x)}\left(\frac{-r}{x^2}\right)}{\frac{-1}{x^2}} = \lim_{x \to \infty} \frac{r}{1 + \frac{r}{x}} = r,$$
and thus $\displaystyle\lim_{x \to \infty} \left(1 + \frac{r}{x}\right)^x = e^r$.

For answers to odd-numbered Skill Certification exercises in the Chapter Review, please visit the Book Companion Web Site at www.whfreeman.com/integcalculus.

Chapter 6

Section 6.1

1. T, F, F, F, T, F, T, T.

3. The length of the hypotenuse of a triangle is always greater than the length of either leg of the triangle; thus $\sin \theta = \dfrac{\text{opp}}{\text{hyp}} < 1$ for all triangles.

5. $\left(\dfrac{\sqrt{2}}{2}\right)^2 + \left(\dfrac{\sqrt{2}}{2}\right)^2 = 1^2$ and $\left(\dfrac{\sqrt{3}}{2}\right)^2 + \left(\dfrac{1}{2}\right)^2 = 1^2$.

7. $\dfrac{5}{13}; \dfrac{12}{13}; \dfrac{5}{12}; \dfrac{13}{12}; \dfrac{13}{5}; \dfrac{12}{5}$

9. (a) False; (b) true; (c) false.

11. 0.981627, 3.23607, 0.017455

13. If θ is an angle in standard position, then $\sin \theta$ the vertical coordinate y of the point (x, y) where the terminal edge of θ intersects the unit circle.

15. The terminal edges of the angles $\dfrac{\pi}{4}$, $\dfrac{9\pi}{4}$, and $-\dfrac{7\pi}{4}$ all meet the unit circle at the same point (and in particular, at the same y-coordinate).

17. $\cos \theta = x$ is $-\dfrac{\sqrt{8}}{3}$ if θ is in the third quadrant;

$\cos \theta = \dfrac{\sqrt{8}}{3}$ if θ is in the fourth quadrant; θ cannot be in the first or second quadrant.

19. They are *all* equal to $\cot \theta$.

21. Use the equation $x^2 + y^2 = 1$ to verify that the point is on the unit circle. $\sin \theta = \dfrac{\sqrt{15}}{4}$; $\cos \theta = -\dfrac{1}{4}$; $\tan \theta = -\sqrt{15}$.

23. For any angle x we can obtain a unique terminal edge, which meets the unit circle in exactly one point, which has exactly one horizontal coordinate; thus $\cos x$ is defined for all angles x. Each of these horizontal coordinates is between -1 and 1 inclusive, and thus $\cos x$ has range $[-1, 1]$.

25. $1/2$

27. 2

29. 1

31. $\sqrt{2}$

33. $\dfrac{1}{\sqrt{3}}$

35. $\dfrac{5}{\sqrt{3}}$ (leg); $\dfrac{10}{\sqrt{3}}$ (hypotenuse)

37. $x \approx 3.38095$ (short leg); $x \approx 7.2505$ (long leg)

39. $x \approx 6.085$ (leg); $x \approx 9.275$ (hypotenuse)

41. $150°$

43. The reference angle is $30°$:

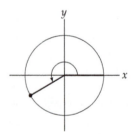

45. The reference angle is $45°$:

47. $\left(-\dfrac{\sqrt{2}}{2}, -\dfrac{\sqrt{2}}{2}\right)$

49. 0

51. 0

53. -1

55. $\sqrt{2}$

57. $-\dfrac{\sqrt{3}}{2}$

59. Does not exist

61. Does not exist

63. $\dfrac{\sqrt{2}}{2}$

65. $-\dfrac{1}{\sqrt{3}}$

67. $-\dfrac{\sqrt{24}}{5}$

69. $-\sqrt{8}$

71. $x = \pi k$, where k is any integer

73. $\theta = \frac{\pi}{2} + 2\pi k$ or $\theta = \frac{7\pi}{6} + 2\pi k$, where k is any integer

75. No solutions

77. $x \neq \frac{\pi}{4} + \frac{\pi}{2}k$, for all integers k.

79. $x \neq \frac{\pi}{2}k$, for any integer k

81. $\ldots \cup \left(-\frac{5\pi}{2}, -\frac{3\pi}{2}\right) \cup \left(-\frac{\pi}{2}, \frac{\pi}{2}\right) \cup \left(\frac{5\pi}{2}, \frac{7\pi}{2}\right) \cup \ldots$

83. 4

85. $\frac{\pi}{2}$

87. 1

89. $A = \frac{9\sqrt{3}}{4}$ square units

91. 211.97 feet

93. about 143 feet of border fencing

95. See the proof in the reading.

97. See the argument in the reading.

99. We have $\sin \theta = 0$ if and only if θ is an angle in standard position whose terminal edge meets the unit circle at a point whose y-coordinate is 0. This happens exactly when θ is an integer multiple of π.

Section 6.2

1. F, T, T, T, T, F, F, T.

3. Cosine and secant are even functions; the remaining four are odd.

5. $\theta = 0$ is one of many counterexamples.

7. $\theta = 0$ is one of many counterexamples.

9. $\left(-\frac{\sqrt{3}}{2}\right)^2 + \left(-\frac{1}{2}\right)^2 = 1$

11. $\left(-\frac{1}{2}\right) = 1 - 2\left(-\frac{\sqrt{3}}{2}\right)^2$

13. $-\frac{\sqrt{2}}{2} = (-1)\left(\frac{\sqrt{2}}{2}\right) + \left(-\frac{\sqrt{2}}{2}\right)(0)$

15. $-\frac{\sqrt{2}}{2} = (0)\left(\frac{\sqrt{2}}{2}\right) - (-1)\left(-\frac{\sqrt{2}}{2}\right)$

17. $\sin(\theta + 2\pi) = \sin \theta$, $\sin(\theta + \pi) = -\sin \theta$, and $\sin\left(\theta + \frac{\pi}{2}\right) = \cos \theta$.

19. The way the proof is written, it says that *if* $\sin 2\theta = 2\sin \theta \cos \theta$ is true (which is what we meant to *prove*, not *assume*), then something obvious is true (namely that $2 \sin \theta \cos \theta = 2 \sin \theta \cos \theta$). This is bad logic; rewrite the proof so that it is a chain of equalities starting with $\sin 2\theta$ and ending with $2 \sin \theta \cos \theta$.

21. See the discussion in the reading.

23. $\frac{\sqrt{35}}{6}$

25. $-\frac{17}{18}$

27. Negative

29. $\frac{\sqrt{2} - \sqrt{6}}{4}$ $\left(\text{use} - \frac{\pi}{12} = \frac{\pi}{4} - \frac{\pi}{3}\right)$

31. $\sqrt{3} - 2$

33. $\frac{2\sqrt{2}}{1 - \sqrt{3}}$

35. $\frac{2\sqrt{3}}{3}$

37. $2 - \sqrt{3}$

39. Neither: $f(-\theta) \neq f(\theta), f(-\theta) \neq -f(\theta)$

41. Odd: $f(-\theta) = -f(\theta)$

43. Even: $f(-\theta) = f(\theta)$

45. This is an identity, since $\sin \theta \sec \theta = \frac{\sin \theta}{\cos \theta} = \tan \theta$ for every angle θ.

47. This is true for every θ, since
$$\csc(-\theta) = \frac{1}{\sin(-\theta)} = \frac{1}{-\sin \theta} = -\csc \theta.$$

49. This is true for every θ, since it is equivalent to the double-angle identity $\cos^2 \theta = \frac{1 + \cos 2\theta}{2}$.

51. $\frac{100 \sin(64.55°)}{\sin(93.99° - 64.55°)} \approx 183.66$ feet

53. Mimic the proof of part (a) of Theorem 6.6.

55. Mimic the proof of part (c) of Theorem 6.7.

57. Given that $\cos 2\theta = \cos^2 \theta - \sin^2 \theta$ we know by the Pythagorean identity that $\cos 2\theta = \cos^2 \theta - (1 - \cos^2 \theta)$. Rewriting this gives us $\cos 2\theta = 2 \cos^2 \theta - 1$.

59. $(\sin \theta + \cos \theta)^2 = \sin^2 \theta + 2 \sin \theta \cos \theta + \cos^2 \theta = 2 \sin \theta \cos \theta + 1 = \sin 2\theta + 1$.

61. $\sec \theta \cot \theta = \frac{1}{\cos \theta} \frac{\cos \theta}{\sin \theta} = \frac{1}{\sin \theta} = \frac{\sin \theta}{\sin^2 \theta} = \frac{\sin \theta}{1 - \cos^2 \theta}$

63. $\tan(\alpha - \beta) = \frac{\sin(\alpha - \beta)}{\cos(\alpha - \beta)} = \frac{\sin \theta \cos \beta - \sin \beta \cos \alpha}{\cos \alpha \cos \beta + \sin \alpha \sin \beta}$. Multiply numerator and denominator by $\frac{1}{\cos \alpha \cos \beta}$ to obtain the desired answer.

Section 6.3

1. F, F, T, T, T, F, T, F.

3. Since $\sec x$ is continuous on its domain, the limit will equal $\sec c$.

5. $\lim_{x \to 0}(\cos x - 1) = 0$ implies that $\lim_{x \to 0} \cos x = 1$; therefore $\lim_{x \to 0} \cos x = \cos 0$, so $\cos x$ is continuous at $x = 0$.

7. 2

9. Limits as $x \to c$ are not determined by function values at the point $x = c$, only with the behavior of the functions as x approaches c.

11. Consider a picture like the one on the right below the Squeeze Theorem for limits in the reading, but upside-down.

13. It is incorrect to use L'Hôpital's rule at all, since the original limit is of the form $\frac{0}{1-0}$ and so the limit is 0. Bad differentiation in the denominator when applying L'H; also $\frac{-\sin(\pi/2)}{-\cos \pi} = -1$. The correct calculation shows that the limit is equal to 0.

15. For $h = 1$, $h = 0.5$, $h = 0.25$, and $h = 0.1$, respectively, we have secant lines with slopes -0.4597, -0.2448, -0.1244, and -0.04996. One estimate of $f'\left(\frac{\pi}{2}\right)$ is 0.

17. See the proof in the reading.

19. (a) If x is in degrees, then the slope of the graph of $\sin x$ at $x = 0$ is very small, and in particular not equal to $\cos 0 = 1$. To convince yourself of this, graph $\sin x$ (in degrees) together with the line $y = x$ (which has slope 1 at $x = 0$).

21. (a) $\cos(3x^2)$ is a composition, not a product, but the product rule was applied; (b) the chain rule was applied incorrectly, with the derivative of $3x^2$ written on the inside instead of the outside.

23. DNE: ∞ from the left and $-\infty$ from the right.

25. 0   27. 1

29. 0

31. $-\infty$ from the left, ∞ from the right

33. 0   35. ∞

37. 1   39. 0

41. Does not exist (oscillates)   43. $f'(x) = 3\sin(3x)$

45. $f'(x) = 3x^2 \cos x - x^3 \sin x$

47. $f'(x) = -3\csc\cot(3x + 1)$

49. $f'(x) = \dfrac{2x\cos x + (x^2 + 1)\sin x}{\cos^2 x}$

51. $f'(x) = -\csc^2 x + \csc x \cot x$

53. $f'(x) = 0$

55. $f'(x) = 3\sec x \sec^2 x + 3\sec x \tan^2 x$

57. $f'(x) = \cos(\cos(\sec x))(-\sin(\sec x))(\sec x \tan x)$

59. $f'(x) = e^{\csc^2 x}(2\csc x)(-\csc x \cot x)$

61. $f'(x) = \dfrac{-(\ln 2)2^x(5x\sin x) + 2^x(5\sin x + 5x\cos x)}{25x^2\sin^2 x}$

63. $f'(x) = \sqrt{\sin x \cos x} + \dfrac{1}{2}x(\sin x \cos x)^{-(1/2)}(\cos^2 x - \sin^2 x)$

65. $f'(x) = \dfrac{(6x\ln x + 3x)\tan x - 3x^2\ln x\sec^2 x}{\tan^2 x}$

67. $f'(x) = \cos(\ln x)\left(\dfrac{1}{x}\right)$

69. Critical points at points of the form $x = \pi k$ where k is an integer; local minima at the odd multiples $x = \pi(2k + 1)$, local maxima at the even multiples $x = \pi(2k)$.

71. Local maxima at all odd integers, local minima at all even integers.

73. Increasing on intervals of the form $[2k + 1, 2k + 3]$ where k is an integer, and decreasing elsewhere.

75. Increasing on intervals of the form $\left[-\dfrac{\pi}{4} + \pi k, \dfrac{\pi}{4} + \pi k\right]$, decreasing elsewhere.

77. Concave down on intervals of the form $\left(\dfrac{\pi}{4} + 2\pi k, \dfrac{5\pi}{4} + 2\pi k\right)$ for integers k, concave up elsewhere; inflection points at points of the form $\dfrac{\pi}{4} + \pi k$ for integers k.

79. Concave down on intervals of the form $\left(\pi k, \pi k + \dfrac{\pi}{2}\right)$ for integers k, and concave up elsewhere; inflection points at $x = \dfrac{k\pi}{2}$ for integers k.

81. $\delta = \dfrac{\pi}{4}$

83. $\delta \approx 1.11$, using a calculator approximation to solve $\dfrac{1 - \cos x}{x} = \dfrac{1}{2}$.

85. $\lim\limits_{h \to 0} \dfrac{\sin h - 0}{h} = 1$; $\lim\limits_{z \to 0} \dfrac{\sin z - 0}{z - 0} = 1$

87. $\lim\limits_{h \to 0} \dfrac{\tan h - 0}{h} = 1$; $\lim\limits_{z \to 0} \dfrac{\tan z - 0}{z - 0} = 1$

89. $f'(x) = (\sin x)^x \left(\ln(\sin x) + \dfrac{x\cos x}{\sin x}\right)$

91. $f'(x) = (\sin x)^{\cos x}\left(-\sin x \ln(\sin x) + \dfrac{\cos^2 x}{\sin x}\right)$

93. (a) To simplify things, let $C = \dfrac{\sqrt{4km - f^2}}{2m}$; since k, m, and f are all constants, so is C. Then follow the given hint. (b) Set $s(0) = s_0$ and $s'(0) = v_0$ and solve for A and B.

95. Mimic the proof in the reading that $\sin x$ is continuous everywhere.

97. If $x = c$ is in the domain of $\sec x$, then c is not a multiple of $\dfrac{\pi}{2}$, and $\cos c \neq 0$. Therefore $\lim\limits_{x \to c} \sec x = \lim\limits_{x \to c} \dfrac{1}{\cos x}$ is the quotient of $\lim\limits_{x \to c} 1$ by $\lim\limits_{x \to c} \cos x = \cos c$, or $\dfrac{1}{\cos c} = \sec c$.

99. For any integer k,
$$\lim_{x \to (\pi/2) + \pi k} \tan x = \lim_{x \to (\pi/2) + \pi k} \dfrac{\sin x}{\cos x} \to \dfrac{\pm 1}{0}, \text{ which is}$$
infinite.

101. Mimic the proof in the reading for $\dfrac{d}{dx}(\sin x)$, except using a sum identity for cosine in the second step.

103. $\dfrac{d}{dx}(\csc x) = \dfrac{d}{dx}\left(\dfrac{1}{\sin x}\right) = \dfrac{(0)(\sin x) - (1)(\cos x)}{(\sin x)^2} = \dfrac{-\cos x}{\sin^2 x} = -\left(\dfrac{1}{\sin x}\right)\left(\dfrac{\cos x}{\sin x}\right) = -\csc x \cot x.$

105. $\dfrac{d}{dx}(\cos x) = \dfrac{d}{dx}\left(\sin\left(x + \dfrac{\pi}{2}\right)\right) = \cos\left(x + \dfrac{\pi}{2}\right)(1) = \cos x \cos\dfrac{\pi}{2} - \sin x \sin\dfrac{\pi}{2} = (\cos x)(0) - (\sin x)(1) = -\sin x.$

Section 6.4

1. T, F, F, F, F, T, F, F.

3. See Definition 6.14 for the restricted domains of the trigonometric functions, and thus the ranges of the inverse trigonometric functions. The domain of $\arcsin x$ is $[-1, 1]$, the domain of $\arctan x$ is all of $\mathbb{R}$, and the domain of $\text{arcsec}\, x$ is $(-\infty, -1] \cup [1, \infty)$. Their ranges are the restricted domains of $\sin x$, $\tan x$, and $\sec x$, respectively.

5. Only (a) and (c) are defined.

7. (a) negative; (b) negative; (c) positive; (d) positive

9. $y = \sin^{-1} x$ is an angle. $x = \sin y$ is the vertical coordinate of the point on the unit circle corresponding to the angle y.

11. $\frac{5\pi}{6}$ is not in the range of $\sin^{-1} x$.

13. To illustrate the restricted the domain of sine, color in the boundary of the unit circle from $-\frac{\pi}{2}$ to $\frac{\pi}{2}$ in the fourth and first quadrants. The possible sines of these angles are y-coordinates on this shaded part of the unit circle, and thus are all in the interval $[-1, 1]$.

15. To illustrate the restricted domain of secant, color in the boundary of the unit circle from 0 to π in the first and second quadrants, putting a hole at $\frac{\pi}{2}$. The possible secants of these angles are reciprocals of y-coordinates on this shaded part of the unit circle, and thus are all reciprocals of numbers in the interval $[-1, 1]$. This means that the possible secants are all in the set $-\infty, -1 \cup 1, \infty$.

17. No, because to differentiate $\sin^{-1}(\sin x) = x$ we would first need to know how to differentiate $\sin^{-1} x$, which is exactly what we would be trying to prove.

19. The expressions are equal; the algebraic expression is clearly easier to evaluate at a given number.

21. $y = \sin x$ has horizontal tangent lines at $\left(-\frac{\pi}{2}, -1\right)$ and $\left(\frac{\pi}{2}, 1\right)$, so $y = \sin^{-1} x$ has vertical tangent lines at $\left(-1, -\frac{\pi}{2}\right)$ and $\left(1, \frac{\pi}{2}\right)$.

23. 0

25. $\frac{\pi}{2}$

27. $-\frac{\pi}{3}$

29. π

31. $\frac{3\pi}{4}$

33. $\frac{2\pi}{3}$

35. $\frac{\pi}{3}$

37. $\frac{\pi}{5}$

39. 0.201358

41. Not defined

43. 1.77215

45. $\frac{\pi}{6}$

47. $\frac{2}{\pi}$

49. ∞

51. 1

53. $f'(x) = \dfrac{6x}{\sqrt{1-9x^4}}$

55. $f'(x) = 2x \arctan x^2 + x^2 \left(\dfrac{2x}{1+x^4}\right)$

57. $f'(x) = \dfrac{2x}{|x^2|\sqrt{x^4-1}}$

59. $f'(x) = \dfrac{1}{\sqrt{1-\sec^4 x}}(2\sec x)(\sec x \tan x)$

61. $\dfrac{\frac{\tan^{-1}x}{\sqrt{1-x^2}} - \frac{\sin^{-1}x}{1+x^2}}{(\tan^{-1}x)^2}$

63. $f'(x) = \dfrac{1}{\mathrm{arcsec}(\sin^2 x)} \dfrac{1}{\sin^2 x \sqrt{\sin^4 x - 1}} (2\sin x \cos x)$

65. $f'(x) = \sec(1 + \tan^{-1} x)\tan(1 + \tan^{-1} x) \times \left(\dfrac{1}{1+x^2}\right)$

67. $\sqrt{1-x^2}$

69. $|x|\sqrt{1 - \dfrac{1}{x^2}}$

71. $\sqrt{1-x^2}$

73. $x^2 + 1$

75. $f(x) = \sin^{-1} 2x$

77. $f(x) = \frac{1}{6}\ln(1 + 9x^2)$

79. $f(x) = \sin^{-1}\left(\frac{3x}{2}\right)$

81. $f(x) = \ln(1 + x^2) + C$

83. No local extrema.

85. Local maximum at $x = 0$

87. 0

89. If $s(t)$ is the height of the penny at time t, then $\alpha(t) = \arctan\left(\frac{s(t)}{250}\right)\left(\frac{180}{\pi}\right) = \arctan\left(\frac{-16t^2 + 1250}{250}\right)\left(\frac{180}{\pi}\right)$ degrees.

91. If $\theta = \sec^{-1} x$, then $\sec\theta = x$, and thus $\frac{1}{\cos\theta} = x$. Therefore $\cos\theta = \frac{1}{x}$, so $\theta = \cos^{-1}\left(\frac{1}{x}\right)$.

93. Mimic the proof of part (a) of Theorem 6.6.

95. Differentiating both sides of $\tan(\tan^{-1} x) = x$ gives us $\sec^2(\tan^{-1} x)\frac{d}{dx}(\tan^{-1} x) = 1$, and therefore $\frac{d}{dx}(\tan^{-1} x) = \dfrac{1}{\sec^2(\tan^{-1} x)}$. By one of the Thinking Back problems in this section, we have $\sec^2(\tan^{-1} x) = 1 + x^2$, and therefore $\frac{d}{dx}(\tan^{-1} x) = \dfrac{1}{1+x^2}$.

For answers to odd-numbered Skill Certification exercises in the Chapter Review, please visit the Book Companion Web Site at www.whfreeman.com/integcalculus.

Chapter 7

Section 7.1

1. T, T, T, F, F, F, F, F.

3. The terms in the sum do not appear to have a consistent pattern.

5. The sum of the squares of the integers greater than or equal to 3 and less than or equal to 87.

7. (a) $b_p + b_{p+1} + \cdots + b_{q-1} + b_q$. (b) Index is i, starting value is p, ending value is q, b_i describes each term in the sum. (c) p and q must be nonnegative integers; b_i can be any real number; also, we must have $p \le q$.

9. One possible answer is $a_k = k^2$, $m = 3$, $n = 7$.

11. The first sum is $1 + \frac{1}{2} + \frac{1}{5} + \frac{1}{10} + \cdots + \frac{1}{65}$; the second sum is $2\left(\frac{1}{2} + \frac{1}{4} + \frac{1}{10} + \frac{1}{20} + \cdots + \frac{1}{130}\right)$.

13. (a) $\sum_{k=1}^{5} \frac{2}{k^2}$; (b) $\sum_{k=2}^{6} \frac{2}{(k-1)^2}$; (c) $\sum_{k=0}^{4} \frac{2}{(k+1)^2}$.

15. (a) $\sum_{k=1}^{2} k = 1 + 2 = 3$, and $2(2+1)2 = \frac{6}{2} = 3$; (b) $\sum_{k=1}^{8} k = 1+2+3+4+5+6+7+8 = 36$, and $8(8+1)2 = \frac{72}{2} = 36$; (c) $\sum_{k=1}^{9} k = 1+2+3+4+5+6+7+8+9 = 45$, and $\frac{9(9+1)}{2} = \frac{90}{2} = 45$.

17. See Theorem 7.4.

19. The area under f' should be related to the function f, since in the example the area under $v(t) = s'(t)$ is related to the distance travelled $s(t)$.

21. $\sum_{k=1}^{8} 3$; $m = 1$, $n = 8$, $a_k = 3$

23. $\sum_{k=1}^{5} \frac{k+2}{k^3}$ and $\sum_{k=3}^{7} \frac{k}{(k-2)^3}$ are the most obvious ways.

25. $\sum_{k=2}^{10}(k^2 + 1)$ 27. $\sum_{k=1}^{n} \frac{k}{n}$

29. $16 + 25 + 36 + 49 + 64 + 81 = 271$

31. $0 + \frac{1}{8} + \frac{1}{2} + \frac{9}{8} + 2 + \frac{25}{8} = \frac{55}{8}$

33. Approximately
$0.17321 + 0.17607 + 0.17889 + 0.18166 + 0.18439 + 0.18708 + 0.18973 + 0.19235 + 0.19494 + 0.19748 \approx 1.8558$

35. $3n - \frac{n(n+1)}{2}$; -4750; $-123,750$; $-497,500$.

37. $\frac{n(n+1)(2n+1)}{6} + 2\frac{n(n+1)}{2} + n - 13$; $348,537$; $42,042,737$; $334,835,487$.

39. $\frac{1}{n^4}\left(\frac{n^2(n+1)^2}{4} - n\right)$; 0.255024; 0.251001; 0.2505

41. $2a_0 + 2a_1 + 2a_2 - a_{101} + \sum_{k=3}^{100} a_k$

43. $\sum_{k=2}^{25}(3k^2 + 2k - 1) - 52$ 45. 48

47. $\lim_{n \to \infty} \frac{2n^3 + 6n^2 + 10n}{6n^3} = \frac{1}{3}$

49. $\lim_{n \to \infty} \frac{n + n(n+1) + \frac{1}{6}n(n+1)(2n+1)}{n^3 - 1} = \frac{1}{3}$

51. $\lim_{n \to \infty}\left(1 + \frac{n^2 + n}{n^2} + \frac{2n^3 + 3n^2 + n}{6n^3}\right) = 1 + 1 + \frac{2}{6} = \frac{7}{3}$

53. Answers can vary depending on what you choose to use as your velocity in each subinterval. Using the initial velocity of each 5-second chunk we approximate that the distance travelled is $d \approx 5v(0) + 5v(5) + 5v(10) + 5v(15) + 5v(20) + 5v(25) + 5v(30) + 5v(35) = 2310$ feet.

55. (a) In any year t, $B(t)$ is the sum of the increases $I(1), I(2), \ldots I(t)$. (b) The graph of $I(t)$ is a series of rectangles with heights $I(1)$, $I(2)$, and so on. For any year t, $B(t)$ is the sum of the areas of the first through tth rectangles. (c) In any given year, if I is positive then B will increase that year; if I is negative then B will decrease that year. It seems like I could be related to the derivative of B.

57. (a) $A(n) = \sum_{k=1}^{n} 10(0.7)^k$; (b) $A(3) = 15.33$ billion, $A(4) = 17.73$ billion, and $A(5) = 19.41$ billion; (c) $A(10) = 22.67$ billon, so the total might be about 23 billion dollars.

59. $\sum_{k=0}^{n} 3a_k = 3a_0 + 3a_1 + 3a_2 + \cdots + 3a_n = 3(a_0 + a_1 + a_2 + \cdots + a_n) = 3\sum_{k=0}^{n} a_k$.

61. Pairing the first and last terms in the sum $1 + 2 + 3 + \ldots + n$ gives a sum of $n + 1$. Pairing the second and second-to-last terms also gives a sum of $n + 1$. Continuing in this fashion we obtain $\frac{n}{2}$ pairs whose sum is $n + 1$.

Section 7.2

1. F, T, F, F, F, T, T, F.

3. (a) Counting partial squares as half-squares, the blob has an approximate area of $46\left(\frac{1}{4}\right) + 36\left(\frac{1}{8}\right) = 16$ square units. (b) Two possible ways to improve the approximation are to count partial squares more accurately depending on their sizes, or to use a finer grid.

5. (a) The right sum is an over-approximation, and the left sum an under-approximation, in this example. Both the midpoint sum and the trapezoid sum look like very close approximations. (b) The midpoint sum was off by the smallest amount, with an error of only about 0.0417.

7. (a) 7.85; (b) No; (c) upper sum is ≥ 8.2, lower sum is ≤ 7.5; (d) No; (e) No.

9. The heights of the rectangles used with the upper sum are by definition always greater than or equal to the heights of the rectangles used for any other Riemann sum.

11. (a) $M_1 = 0$, $M_2 = 2$; (b) $M_1 = 0$, $M_2 = \frac{2}{3}$ (or $\frac{4}{3}$), $M_3 = 2$; (c) $M_1 = 0$, $M_2 = \frac{1}{2}$, $M_3 = \frac{3}{2}$, $M_4 = 2$.

13. (a) Meters; (b) the area of a rectangle; (c) looks like distance should be related to the area under the velocity curve.

15. Part of the sum indicates that we have $\Delta x = \frac{1}{2}$, whereas another part tells us that $\Delta x = 0.25$.

17. Right sum with $f(x) = x^2$, $N = 4$, $\Delta x = \frac{1}{2}$, and $x_k = 1 + \frac{k}{2}$; thus $a = 1$ and $b = 3$.

19. Trapezoid sum with $f(x) = \sin x$, $N = 100$, $\Delta x = 0.05$, and $x_k = 0 + 0.05k$ (so $x_{k-1} = 0 + 0.05(k - 1)$); thus $a = 0$ and $b = 0 + 0.05(100) = 5$.

21. 1.219   23. 51.88

25. $\frac{8}{3}$ 27. (a) 5; (b) 6.875

29. (a) 1.16641; (b) 1.26997

31. (a) 51.3434; (b) 52.9562

33. (a) 2; (b) 2.148; (c) 2.25

35. Exact $= 26$; LS $= 24.5$; RS $= 27.5$; midpoint $= 26$; upper $= 27.5$; lower $= 24.5$; trapezoid $= 26$

37. Exact $= \frac{\pi}{2}$; LHS ≈ 1.366; RHS ≈ 1.366; midpoint ≈ 1.630; upper ≈ 1.866; lower ≈ 0.866; trapezoid ≈ 1.366.

39. $\sum_{k=1}^{4} \frac{1}{4}\sqrt{\frac{k}{4} + 1}$ 41. $\sum_{k=1}^{6} e^{(2k+3)/4}\left(\frac{1}{2}\right)$

43. $\sum_{k=1}^{4} \frac{\sin\left(\frac{\pi}{4}(k-1)\right) + \sin\left(\frac{\pi}{4}k\right)}{2}\left(\frac{\pi}{4}\right)$

45. (a) 1320 feet; (b) 3080 feet

47. With a right sum we approximate that the change in temperature is -24.2385 degrees.

49. (a) $c_1 \approx 6114.15$, $c_2 \approx 8722.35$; (b) plotting $g(t) = 6114.15 + 8722.35(\sin((t-90)\pi/105) - 2/\pi)$ with the four data points between 90 and 195 days gives a relatively good fit.

51. Since f is increasing, we have $x_{k-1} \le x_k$; now apply Definition 7.6.

53. Draw a picture of a concave up function with a trapezoid sum approximation. Compare the trapezoid to the area under the curve. What do you notice?

Section 7.3

1. F, T, F, F, F, F, T, F.

3. $\int_a^b f(x)dx$; the definite integral of f on $[a, b]$

5. $\int_a^b f(x)dx = \lim_{n \to \infty} \sum_{k=1}^n f(x_k^*)\Delta x$, where $\Delta x = \dfrac{b-a}{n}$, $x_k = a + k\Delta x$, and x_k^* is a point in the kth subinterval $[x_{k-1}, x_k]$.

7. The region between the graph of f and the x-axis from $x = a$ to $x = a$ has a width of zero.

9. A larger portion of the region between the graph of $f(x) = x^2 - 4$ and the x-axis on $[-3, 3]$ is negative, so the definite integral is negative.

11. 0

13. (a) Negative; (b) negative

15. (a) $\int_a^b (f(x) + g(x))dx = \int_a^b f(x)dx + \int_a^b g(x)dx$, but $\int_a^b f(x) \cdot g(x)dx \ne \int_a^b f(x)dx \cdot \int_a^b g(x)dx$. (b) With $f(x) = x$ and $g(x) = 1$, $\int_0^1 f(x)dx = \dfrac{1}{2}$, $\int_0^1 g(x)dx = 1$, and $\int_0^1 (f(x) + g(x))dx = \dfrac{3}{2}$. (c) With $f(x) = x$ and $g(x) = x$, $\int_0^1 f(x)dx = \dfrac{1}{2}$ and $\int_0^1 g(x)dx = \dfrac{1}{2}$, but $\int_0^1 f(x)g(x) = \dfrac{1}{3}$.

17. (a) $\sum_{k=1}^n \left(\dfrac{3k}{n}\right)^2 \left(\dfrac{3}{n}\right) = \dfrac{27}{n^3}\left(\dfrac{n(n+1)(2n+1)}{6}\right)$
(b) $\sum_{k=1}^n \left(\dfrac{3}{n}(k-1)\right)^2 \left(\dfrac{3}{n}\right) = $
$\dfrac{27}{n^3}\left(\dfrac{n(n+1)(2n+1)}{6} - 2\dfrac{n(n+1)}{2} + n\right)$ (c) For $n = 100$, RHS ≈ 9.13545 and LHS ≈ 8.86545; for $n = 1000$, RHS ≈ 9.0135 and LHS ≈ 8.9865. (d) Both expressions approach 9 as $n \to \infty$.

19. (a) Infinite discontinuity (vertical asymptote) at $x = 0$; (b) the area accumulated would be infinite; (c) a sequence of Riemann sum approximations would not converge to a real number as $n \to \infty$; (d) compare the two graphs and speculate.

21. 2

23. 264

25. $\dfrac{\pi}{2}$

27. $\dfrac{1}{2}$

29. 5

31. 14

33. Not enough information.

35. Not enough information.

37. -8

39. $-\dfrac{52}{3}$

41. (a) $\sum_{k=1}^n \left(5 - \left(2 + \dfrac{3k}{n}\right)\right)\left(\dfrac{3}{n}\right) = \dfrac{3}{n}(3n) - \dfrac{9}{n^2}\left(\dfrac{n(n+1)}{2}\right)$.
(b) 4.455, 4.4955; (c) $\dfrac{9}{2}$.

43. (a) $\sum_{k=1}^n 2\left(\dfrac{k}{n}\right)^2 \left(\dfrac{1}{n}\right) = \dfrac{2}{n^3}\left(\dfrac{n(n+1)(2n+1)}{6}\right)$.
(b) 0.6767, 0.667667; (d) $\dfrac{2}{3}$.

45. (a) $\sum_{k=1}^n \left(\left(2 + \dfrac{k}{n}\right) + 1\right)^2 \left(\dfrac{1}{n}\right) = $
$\dfrac{1}{n}(9n) + \dfrac{6}{n^2}\left(\dfrac{n(n+1)}{2}\right) + \dfrac{1}{n^3}\left(\dfrac{n(n+1)(2n+1)}{6}\right)$.
(b) 12.3684, 12.3368; (c) $\dfrac{37}{3}$.

47. $\dfrac{62}{3}$

49. -93

51. $\dfrac{136}{3}$

53. (a) $\int_0^{40}(-0.22t^2 + 8.8t)dt$; (b) 2346.67 feet.

55. $\int_a^b cf(x)dx = \lim_{n \to \infty} \sum_{k=1}^n cf(x_k^*)\Delta x = c \lim_{n \to \infty} \sum_{k=1}^n f(x_k^*)\Delta x = c\int_a^b f(x)dx.$

57. If $\int_a^b f(x)dx = \lim_{n \to \infty} \sum_{k=1}^n f(x_k)\Delta x$, then $\int_b^a f(x)dx = \lim_{n \to \infty} \sum_{k=1}^n f(x_{n-k})(-\Delta x) = -\lim_{n \to \infty} \sum_{k=1}^n f(x_k)\Delta x = -\int_a^b f(x)dx.$

59. The area in question is a trapezoid with heights a and b and width $b - a$, therefore area $\dfrac{a+b}{2}(b-a) = \dfrac{1}{2}(b^2 - a^2)$.

61. Simplify $\sum_{k=1}^n \left(a + \dfrac{k(b-a)}{n}\right)^2 \left(\dfrac{b-a}{n}\right)$, apply sum formulas, and then take the limit as $n \to \infty$. Remember that a and b are constants.

Section 7.4

1. T, F, F, T, F, F, F, F.

3. An antiderivative of f is just one function whose derivative is f, while the indefinite integral $\int f(x)dx$ is the family of all antiderivatives of f.

5. See Definitions 7.9 and 7.15. The definite integral is a number while the indefinite integral is a family of functions; the definite integral concerns area while the indefinite integral concerns antiderivatives; we calculate definite integrals using Riemann sums, but we calculate indefinite integrals by antidifferentiating.

7. All blanks should contain $\dfrac{1}{7}x^7$.

9. At this point we don't know of a function whose derivative is $\sec x$, but we do know of a function whose derivative is $\sin x$ (it is $-\cos x$).

11. See the theorems in this section.

13. $\dfrac{d}{dx}\left(\dfrac{1}{2}e^{(x^2)}(x^2 - 1)\right) = x^3 e^{(x^2)}$

15. $\dfrac{d}{dx}(x(\ln x - 1)) = 1(\ln x - 1) + x\left(\dfrac{1}{x} - 0\right) = \ln x$

17. One easy example is $f(x) = x$, $g(x) = x^2$.

19. $F'(x) = \csc^2 x$ for both $x < 0$ and $x > 0$. The piecewise antiderivative $F(x)$ here differs from the one in Theorem 7.18 by being shifted up 100 units on the right of the y-axis. Notice from the graphs that this does transformation does not affect the slope of the graph of $F(x)$.

21. $\frac{1}{3}x^3 - \frac{1}{2}x^6 - 7x + C$

23. $\frac{3}{4}x^4 + \frac{5}{3}x^3 - \frac{3}{2}x^2 - 5x + C$

25. $\frac{1}{3}x^3 + \frac{1}{\ln 2}(2^x) + 4x + C$

27. $\frac{2}{3}x^{3/2} + 2x^{1/2} + C$

29. $\frac{3.2}{\ln(1.43)}(1.43)^x - 50x + C$

31. $2e^{2x-6} + C$

33. $\frac{1}{3}\sec(3x) + C$

35. $3(\tan x - x) + C$

37. $\pi x + C$

39. $\frac{7}{2}\sin^{-1}(2x) + C$

41. $\frac{1}{3}\sec^{-1}(3x) + C$

43. $\ln|1 + x^2| + C$

45. $x^2 e^x + C$

47. $\ln|3x^2 + 1| + C$

49. $e^x - e^{2x} + C$

51. $\frac{1}{18}(x^3 + 1)^6 + C$

53. $\frac{x^2}{\ln x} + C$

55. $e^x \sqrt{x} + C$

57. $\ln x \cos x + C$

59. $\frac{a}{b}e^{bx+c} + C$

61. $\frac{1}{ab}\sec(ax) + C$

63. $\frac{a}{b}\tan^{-1}(bx) + C$

65. (a) $v(t) = -32t + 42$ feet per second; (b) $s(t) = -16t^2 + 42t + 4$ feet.

67. See the proof in the reading.

69. $\frac{d}{dx}\left(\frac{1}{k}e^{kx}\right) = \left(\frac{1}{k}\right)(k)e^{kx} = e^{kx}$, and $\frac{d}{dx}\left(\frac{1}{\ln b}b^x\right) = \left(\frac{1}{\ln b}\right)(\ln b)b^x = b^x$.

71. Each of the integrands is the derivative of the corresponding inverse trigonometric function in its solution.

73. See the proof in the reading.

75. By the sum rule for differentiation, and the fact that the derivative of a constant is zero, $\frac{d}{dx}(G(x)) = \frac{d}{dx}(F(x) + C)$ implies that $G'(x) = F'(x) + 0$.

Section 7.5

1. F, F, T, F, F, T, T, F.

3. The Fundamental Theorem of Calculus relates the concept of signed area under a curve to antiderivatives, which enables us to quickly find the exact areas under curves, and connects the important concepts of definite integrals and indefinite integrals.

5. The Mean Value Theorem

7. $\frac{1}{b-a}\int_a^b v(t)dt$ is equal to $\frac{s(b)-s(a)}{b-a}$, so $\int_a^b v(t)dx = s(b) - s(a)$. Since $s(t)$ is an antiderivative of $v(t)$, this is an example of the equation in the Fundamental Theorem.

9. $f(4) \approx 4.125$

11. $f(-2) \approx 2.5$

13. (a) $\lim_{n\to\infty}\left(\frac{3}{n}(n) + \frac{6}{n^2}\frac{n(n+1)}{2} + \frac{3}{n^3}\frac{n(n+1)(2n+1)}{6}\right) = 7$; (b) $3\left(\frac{1}{3}\right)(2^3 - 1^3) = 7$; (c) $\left[3\left(\frac{1}{3}\right)x^3\right]_1^2 = 7$

15. (a) $\frac{100}{101}$; (b) 5000

17. Equivalent to FTC: (a), (d), and (e).

19. $\frac{12}{5}$

21. $\frac{42}{\ln 2}$

23. $-\frac{2}{3}(5)^{-3/2} + \frac{2}{3}(2)^{-3/2}$

25. $4 - 2\sqrt{3}$

27. $\sqrt{2} - 1$

29. $\frac{\pi}{8}$

31. 2π

33. $-\frac{1}{2e} + \frac{1}{2}$

35. $\frac{\pi}{4}$

37. $\frac{3}{\ln 4}$

39. $\ln 7 - \ln 2$

41. $\ln 9 - \ln 14$

43. $\ln(32) - \ln(8)$

45. $\frac{\pi}{6}$

47. $\frac{3\pi}{2} - \pi \ln 8$

49. $-\frac{3}{20}$

51. $\frac{3}{2}$

53. $\frac{1}{e} + e$

55. $\frac{1}{2}\cot\left(\frac{\pi^2}{16}\right) - \frac{1}{2}\cot\left(\frac{\pi^2}{4}\right)$

57. $-\frac{3}{2}\ln 5$

59. $\frac{25}{2}$

61. $\frac{343}{12}$

63. $3 - \frac{1}{\sqrt{2}}$

65. $\int_0^{24} 0.05t\,dt = \left[\frac{0.05}{2}t^2\right]_0^{24} = 14.4$ gallons.

67. (a) Supposing that the "positive" direction is to the right, the figurine moves right along the track for the first five seconds, then left for the next five seconds, then right again, then left again. (b) approximately 5.27 inches, 3.125 inches, and 0 inches. (c) approximately 14.84 inches.

69. $\frac{9}{14}$

71. (a) 80, and there is a line of about 5 people at noon; (b) 50 customers per hour; (c) at 3:00 P.M. there are about 40 people in the line; (d) every customer has been served.

73. $\int_a^b x\,dx = \left[\frac{1}{2}x^2\right]_a^b = \frac{1}{2}b^2 - \frac{1}{2}a^2 = \frac{1}{2}(b^2 - a^2)$.

75. If F is an antiderivative of f, then kF is an antiderivative of kf. Therefore $\int_a^b kf(x)dx = kF(b) - kF(a) = k(F(b) - F(a)) = k[F(x)]_a^b = k\int_a^b f(x)dx$.

77. See the proof in the reading.

79. If $G(x) = F(x) + C$ then $[G(x)]_a^b = [F(x) + C]_a^b = (F(b) + C) - (F(a) + C) = F(b) - F(a) = [F(x)]_a^b$.

Section 7.6

1. F, F, T, F, F, F, F, T.

3. Three (although two of them will be equal); positive; negative

5. (a) $\int_{-3}^{4} f(x)dx$; (b) $\int_{-3}^{-1} f(x)dx - \int_{-1}^{2} f(x)dx + \int_{2}^{4} f(x)dx$

7. negative; positive

9. One possible example is $f(x) = x$ on $[-2, 2]$.

11. (a) See graph that follows; (b) $\int_{-2}^{4} f(x)dx$; (c) $\int_{-2}^{-1} f(x)dx - \int_{-1}^{3} f(x)dx + \int_{3}^{4} f(x)dx$.

13. (a) See graph that follows; (b) $\int_{0}^{1}(g(x) - f(x))dx + \int_{1}^{3}(f(x) - g(x))dx + \int_{3}^{4}(g(x) - f(x))dx$.

15. Approximately $-\dfrac{10}{3}$

17. Draw any graph where the average height of the graph on $[-2, 5]$ is 10, and the slope from $(-2, f(-2))$ to $(5, f(5))$ on your graph is -3.

19. See the reading.

21. f is continuous on $[1, 5]$, so the Mean Value Theorem for Integrals applies and says that there is some $c \in (1, 5)$ for which $f(c) = c(c - 6)$ is equal to the average value $\dfrac{1}{5-1} \int_{1}^{5} x(x - 6)dx$ of the function f. You can compute this average value; it is equal to $-\dfrac{23}{3}$. Therefore, there is some $c \in (1, 5)$ such that $f(c) = -\dfrac{23}{3}$. We can solve the equation $f(c) = -\dfrac{23}{3}$ to find such a value of c. Actually there are two such values: $c \approx 1.8453$ and $c \approx 4.1547$.

23. Using a left sum with $n = 10$, we have (a) 43.333; (b) 80.

25. Using a left sum with $n = 8$, we have (a) -11.1973; (b) 12.2229.

27. (a) $\dfrac{5}{2}$; (b) $\dfrac{13}{2}$

29. (a) $-\dfrac{4}{3}$; (b) $\dfrac{109}{12}$

31. (a) $\dfrac{25}{2}$; (b) 25.2

33. (a) 54, (b) $\dfrac{194}{3}$

35. (a) $\dfrac{2}{9}$, (b) $\dfrac{2}{9}$

37. (a) $\dfrac{\pi}{2}$, (b) $\dfrac{\pi}{2}$

39. Using a left sum with $n = 8$, we have 22.

41. $\dfrac{13}{2}$

43. $\dfrac{38}{6}$

45. 15

47. $\dfrac{19}{3}$

49. $\dfrac{145}{12}$

51. $-2 + \sqrt{3} + \dfrac{\pi}{3}$

53. $(f(-1) + f(-0.75) + f(-0.5) + f(-0.25) + f(0) + f(0.25) + f(0.5) + f(0.75)) \left(\dfrac{1}{4}\right) \approx 1.981$

55. One possible approximation is $\left(f\left(\dfrac{\pi}{4}\right) + f\left(\dfrac{\pi}{2}\right) + \cdots + f(2\pi) \right) \left(\dfrac{\pi}{4}\right) = 0$

57. 7

59. 4

61. -1

63. $\dfrac{1}{2}(e^2 - e^{-2})$

65. $\dfrac{11}{28}$

67. 0

69. $c = 1$

71. $c = \dfrac{-3 + \sqrt{201}}{12}$

73. $-1 + \sqrt{3}$

75. (a) There is some time $c \in (0, 4)$ when $f'(c) = \dfrac{f(4) - f(0)}{4 - 0} = 1.44$; therefore there is some time c in the first four days when the rate of growth of the plant is 1.44 centimeters per day. (b) There is some time c during the first four days when the height of the plant is equal to the average height of the plant over those four days. Since this average height is 1.92 feet, we know that there is some $c \in (0, 4)$ such that $f(c) = 1.92$. We can solve for $c = 2.3094$.

77. (a) $\int_{1.51}^{5.29}(55 - f(x))dx + \int_{5.29}^{18}(r(x) - f(x))dx + \int_{18}^{30.71}(s(x) - g(x))dx + \int_{30.71}^{34.49}(55 - g(x))dx$; (b) 561.9 square feet (too big!)

79. Write out the Riemann sum for each side; the absolute values on the right-hand side factor out of both the sum and the limit.

81. See the reading concerning the development of average value in the plant height example.

Section 7.7

1. F, F, F, T, F, T, T, F.

3. $\int_{2}^{x} t^2 dt$

5. x is the independent variable; $A(x)$ is the dependent variable, which represents the signed area under the graph of $f(t)$ from $t = 0$ to $t = x$; t does not affect the dependent or independent variable.

7. (a) As x increases, the signed area $A(x)$ accumulated also increases, but at a rate that decreases. Thus the rate of change of the rate of change of $A(x)$ is negative, so $A(x)$ is concave down).
   (b) $A''(x) = \dfrac{d}{dx}\left(\dfrac{d}{dx}\int_{0}^{x} f(t)dt\right) = \dfrac{d}{dx}(f(x)) = f'(x)$; therefore, f decreasing $\Rightarrow f'$ negative $\Rightarrow A''$ negative $\Rightarrow A$ concave down.

9. (a) $A'(x)$ and $B'(x)$ are both equal to x^2, so $A(x)$ and $B(x)$ differ by a constant.
   (b) $A(x) - B(x) = \int_{0}^{x} t^2 dt - \int_{3}^{x} t^2 dt = \int_{0}^{3} t^2 dt$, which is a constant. (c) $\int_{0}^{3} t^2 dt$ is the signed area under the graph of $y = x^2$ from $x = 0$ to $x = 3$.

11. $F(x) = A(g(x))$, where $A(x) = \int_3^x \sin t\, dt$ is outside and $g(x) = x^2$ is inside.

13. $A(5) < A(0) < A(-1) < A(-2)$

15. $A(x)$ is positive on approximately $[0, 2]$ and negative on $[2, 6]$. $A(x)$ increases on $[0, 1]$, then decreases on $[1, 5]$, then increases again on $[5, 6]$. The graph of A has the following shape:

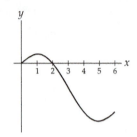

17. See the proof of Theorem 7.35.

19. (a) $f(b) - f(a)$; (b) 0.

21. (a) $\ln x = \int_1^x \frac{1}{t}\, dt$ is defined on $(0, \infty)$ because $\frac{1}{x}$ is continuous on $(0, \infty)$. Moreover, $\ln x = \int_1^x \frac{1}{t}\, dt$ is zero for $x = 1$, increases without bound as $x \to \infty$, and decreases without bound as $x \to 0^+$; therefore $\ln x$ has range $(-\infty, \infty)$. (b) Follows from the fact that we define e^x to be the inverse of $\ln x$.

23. $\ln 10 - 1$

25. $\log_b x = \frac{1}{\ln b} \int_1^x \frac{1}{t}\, dt$, or $\log_b x = \int_1^x \frac{1}{(\ln b)t}\, dt$

27. (a) $A(2)$ is the signed area under the graph of $t^2 + 1$ from $t = 1$ to $t = 2$; (b) $A(2) = \frac{10}{3}$, $A(5) = \frac{136}{3}$; (c) $A(x) = \frac{1}{3}x^3 + x - \frac{4}{3}$

29. (a) $A(x)$ is the signed area under the graph of $\frac{1}{t}$ from $t = 3$ to $t = x$; (b) $A(2) = \ln 2 - \ln 3$, $A(5) = \ln 5 - \ln 3$; (c) $A(x) = \ln x - \ln 3$

31. $\int_0^x \sin^2(3^t)\, dt$, $\int_2^x \sin^2(3^t)\, dt$, $\int_{-1}^x \sin^2(3^t)\, dt$

33. $\int_0^x e^{(t^2)}\, dt$, $\int_{-2}^x e^{(t^2)}\, dt$, $\int_{10}^x e^{(t^2)}\, dt$

35. $-e^{x^2+1}$

37. $\cos(x^2)(2x)$

39. $\int_1^{\sqrt{x}} \ln t\, dt + \frac{1}{2}\sqrt{x} \ln \sqrt{x}$

41. 0

43. $-2 \ln |x^2| - 4$

45. $54x$

47. $\sin((x+2)^2) - \sin(x^2)$

49. $f(x) = \frac{1}{2} \ln |2x - 1| + 3$

51. $f(x) = \int_2^x \frac{1}{t^3+1}\, dt$

53. $f(x) = \int_1^x e^{-t^2}\, dt$

55. Using left and right sums with 9 rectangles, $1.92897 < \ln 10 < 2.82897$.

57. Using left and right sums with four rectangles, $0.877269 < \ln e < 1.14881$.

59. (a) $s(t) = \int_0^t \sin(0.1w^2)\, dw$; (b) about 2.41143 feet to the right of the starting position; (c) each rectangle has a height measured in feet per second and a width measured in seconds, thus an area that is measured in feet; (d) by looking at the graph of $v(t) = \sin(0.1t^2)$, the velocity oscillates faster and faster between 1 and -1 feet per second.

61. (a) $F_{t_0}(t) = \int_{t_0}^t 86400 f(x)\, dx = 86400[-1.033x^3 + 441.75x^2 - 53605x]_{t_0}^t$; (b) $F_{90}(195) \approx 58.9$ billion cubic feet; (c) about 17.971 billion cubic feet, about 76% of which flows through between days 90 and 195.

63. If G is an antiderivative of f, then by the Fundamental Theorem of Calculus we have $\int_a^b f(x)\, dx = G(b) - G(a)$. Therefore, $A(x) = \int_0^x f(t)\, dt = G(x) - G(0)$. Since A and G differ by a constant, they have the same derivative, namely, f; thus A is an antiderivative of f.

65. See the proof given in the reading.

67. This is a just a rewording of the Second Fundamental Theorem, since $F(x) = \int_a^x f(t)\, dt$ and the Second Fundamental Theorem tells us that the derivative of F is f.

69. (a) $F' = f$, F is continuous on $[a, b]$, and F is differentiable on (a, b). (b) There exists a number $c \in (a, b)$ such that $F'(c) = \frac{F(b)-F(a)}{b-a}$. (c) $f(c) = F'(c) = \frac{F(b)-F(a)}{b-a} = \int_a^b f(t)\, dt - \int_a^a f(t)\, dtb - a = \int_a^b f(t)\, dtb - a$.

71. Since $\frac{1}{t}$ is continuous for $t > 0$, the first part of the Second Fundamental Theorem guarantees that if $\ln x = \int_1^x \frac{1}{t}\, dt$, then $\ln x$ is continuous and differentiable on $(0, \infty)$.

73. $\ln 1 = \int_1^1 \frac{1}{t}\, dt = 0$. Since the signed area under $\frac{1}{t}$ is positive over any interval in $(0, \infty)$, if $0 < x < 1$, then $\ln x = \int_1^x \frac{1}{t}\, dt = -\int_x^1 \frac{1}{t}\, dt$ is negative. Similarly, if $x > 1$, then $\ln x = \int_1^x \frac{1}{t}\, dt$ is positive.

75. (a) $\frac{d}{dx}(\ln ax) = \frac{d}{dx} \int_1^{ax} \frac{1}{t}\, dt = \frac{1}{ax}(a) = \frac{1}{x}$. (b) $\ln ax$ and $\ln x$ have the same derivative, so they must differ by a constant, that is, $\ln ax = \ln x + C$. (c) $\ln a = \ln 1 + C$, so $\ln a = 0 + C$, so $C = \ln a$. Then just rename x to be b to complete the proof.

77. $\ln \frac{a}{b} = \ln(ab^{-1}) = \ln a + \ln(b^{-1}) = \ln a - \ln b$

For answers to odd-numbered Skill Certification exercises in the Chapter Review, please visit the Book Companion Web Site at www.whfreeman.com/integcalculus.

Chapter 8

Section 8.1

1. T, F, F, F, T, F, T, F.

3. Both integrals turn into $\int \frac{1}{u} du$ after a change of variables; $u = x^2 + 1$ in the first case, $u = \ln x$ in the second.

5. Three possible answers are $\int 3x^2 \sin(x^3)dx$, $\int \frac{\sin(\ln x)}{x} dx$, and $\int \frac{\sin(\sqrt{x}+2)}{2\sqrt{x}} dx$.

7. Three possible answers are $\int \frac{3}{\sqrt{3x+1}} dx$, $\int \frac{\cos x}{\sqrt{\sin x}} dx$, and $\int \frac{1}{x\sqrt{\ln x}} dx$.

9. $du = (2x + 1)dx$ 11. $du = \cos x dx$

13. $\int_{-1}^{5} u^2 du = \frac{126}{3} = 42$;

$\int_{x=-1}^{x=5} u^2 du = \frac{1}{3}(5)^6 - \frac{1}{3}(-1)^6 = 5208$;

$\int_{u(-1)}^{u(5)} u^2 du = \int_{1}^{25} u^2 du = \frac{1}{3}(25)^3 - \frac{1}{3}(1)^3 = 5208$

15. Three of the many such integrals are in Exercises 34, 37, and 51.

17. (a) $\frac{1}{2}\int u^2 du = \frac{1}{6}(x^2 - 1)^3 + C$;

(b) $\int (x^5 - 2x^3 + x) dx = \frac{1}{6}x^6 - \frac{1}{2}x^4 + \frac{1}{2}x^2 + C$;

(c) the answers differ by a constant:

$\frac{1}{6}(x^2 - 1)^3 = \frac{1}{6}x^6 - \frac{1}{2}x^4 + \frac{1}{2}x^2 - \frac{1}{6}$.

19. (a) $-\frac{1}{3}\int_{x=-2}^{x=2} e^u du = -\frac{1}{3}[e^{5-3x}]_{-2}^{2} = -\frac{1}{3}(e^{-1} - e^{11})$;

(b) $-\frac{1}{3}\int_{11}^{-1} e^u du = -\frac{1}{3}[e^u]_{11}^{-1} = -\frac{1}{3}(e^{-1} - e^{11})$

21. $\frac{1}{9}(3x + 1)^3 + C$ 23. $4\ln(x^2 + 1) + C$

25. $\frac{1}{2}(\sqrt{x} + 3)^2 + C$ 27. $-\frac{1}{2}\cot x^2 + C$

29. $\frac{1}{3}\ln|3x + 1| + C$ 31. $\frac{1}{2\pi}\sin^2 \pi x + C$

33. $\frac{1}{2}\sec 2x + C$ 35. $-\frac{1}{6}\cot^6 x + C$

37. $\frac{1}{11}x^{11} + \frac{1}{4}x^8 + \frac{1}{5}x^5 + C$ 39. $-\frac{4}{5}\cos(x^{5/4}) + C$

41. $\frac{1}{2\ln 2}2^{x^2+1} + C$ 43. $\frac{2}{3}(\ln x)^{3/2} + C$

45. $\frac{1}{4}(\ln x)^2 + C$

47. $\frac{3}{2}\ln(x^2 + 1) + \tan^{-1}x + C$

49. $-\ln|2 - e^x| + C$ 51. $-2e^{-x} - x + C$

53. $-\frac{1}{8}\cos^8 x + C$ 55. $(\sin x^2)^{1/2} + C$

57. $\cos\left(\frac{1}{x}\right) + C$ 59. $\frac{1}{3}(x^2 + 1)^{3/2} + C$

61. $\frac{2}{5}(x + 1)^{5/2} - \frac{2}{3}(x + 1)^{3/2} + C$

63. $\frac{2}{27}(3x + 1)^{3/2} - \frac{2}{9}(3x + 1)^{1/2} + C$

65. $-\cos(e^x) + C$

67. $\frac{1}{103}(x - 1)^{103} + \frac{2}{102}(x - 1)^{102} + \frac{1}{101}(x - 1)^{101} + C$

69. $\frac{3}{8}(10^{4/3} - 1)$ 71. 0

73. $2\sqrt{\ln 3} - 2\sqrt{\ln 2}$ 75. $\frac{\pi}{8}$

77. $1/4 + (5/2)\ln 2 - (1/2)\ln 10$

79. (a) 0; (b) $c = 1$.

81. 4.27569 inches

83. 33.2128 cubic inches

85. (a) $\int_0^{k\pi} \sin\left(2\left(x - \frac{\pi}{4}\right)\right) dx = -\frac{1}{2}\sin(2k\pi) = 0$ for any integer k; (b) f is a general sine function with period π whose graph is half above and half below the x-axis.

87. Since $\frac{d}{dx}(f(u(x))) = f'(u(x))u'(x)$, we have $\int f'(u(x))u'(x)dx = f(u(x)) + C$.

89. See the proof of Theorem 8.3.

Section 8.2

1. F, T, F, F, T, F, T, F.

3. $\int u dv = uv - \int v du$; (b) $\int u(x)v'(x)dx = u(x)v(x) - \int v(x)u'(x)dx$

5. $\int x \sin x dx$; $\int x \ln x dx$

7. $\int \ln x dx$

9. If $u = 1$, then integrating dv is equivalent to solving the original integral.

11. Exercises 28, 27, and 30 are three such examples.

13. $du = 3\cos 3x dx$ and $dv = dx$; $\int u dv = \int \sin 3x dx$, and $\int v du = \int 3x \cos 3x dx$. The first integral is clearly easier, and the second integral can be rewritten in terms of the first using integration by parts.

15. $[g(x)]_a^b - [h(x)]_a^b = (g(b) - g(a)) - (h(b) - h(a)) = (g(b) - h(b)) - (g(a) - h(a)) = [g(x) - h(x)]_a^b$.

17. (a) $\frac{d}{dx}(x \ln x) = \ln x + 1$; (b) $\int(\ln x + 1)dx = x \ln x + C$; (c) $\int \ln x dx = x \ln x - \int 1 dx$

19. $u = x, v = \frac{1}{\ln 2}2^x$ 21. $u = \ln x, v = -x^{-2}$

23. $u = \tan^{-1}x, dv = dx$

25. (a) $x^3 \ln x - \frac{1}{3}x^3 + C$; (b) $\frac{1}{3}(x^3 \ln x^3 - x^3) + C$;

(c) Your answers should differ by a constant, which in this case is $C = 0$.

27. $xe^x - e^x + C$ 29. $\frac{1}{2}x^2 \ln x - \frac{1}{4}x^2 + C$

31. $-\frac{1}{2}\cos x^2 + C$

33. $\frac{1}{3}x^2 e^{3x} - \frac{2}{9}xe^{3x} + \frac{2}{27}e^{3x} + C$

35. $-xe^x - e^x + C$

37. $\frac{3}{2}e^{x^2} + C$

39. $3(x\ln x - x)$

41. $-e^{-x}(x^2 + 1) - 2xe^{-x} - 2e^{-x} + C$

43. $3x + \ln x + C$

45. $\frac{1}{3}x^3 - 2(xe^x - e^x) + \frac{1}{2}e^{2x} + C$

47. $\frac{2}{3}x^{3/2}\ln x - \frac{4}{9}x^{3/2} + C$

49. $\frac{1}{2}x^2 e^{x^2} - \frac{1}{2}e^{x^2} + C$

51. $-x\cot x + \ln|\sin x| + C$

53. $x^3 \sin x + 3x^2 \cos x - 6x\sin x - 6\cos x + C$

55. $-\frac{1}{2}x^2 \cos x^2 + \frac{1}{2}\sin x^2 + C$

57. $x\tan^{-1} 3x - \frac{1}{6}\ln|1 + 9x^2| + C$

59. $-\frac{1}{5}e^{2x}\cos x + \frac{2}{5}e^{2x}\sin x + C$

61. $\frac{1}{\ln(2/3)}(2/3)^x + C$

63. $\frac{1}{2}e^{-x}\sin x - \frac{1}{2}e^{-x}\cos x + C$

65. $x\tan x - \ln|\sec x| + C$

67. $\frac{1}{2}x^2 \tan^{-1} x + \frac{1}{2}\tan^{-1} x - \frac{1}{2}x + C$

69. $2\ln 2 - 1$

71. $-\frac{2}{e}$

73. $\frac{1}{2}e^\pi + \frac{1}{2}$

75. $\frac{1}{4}(\pi + 2\ln 2)$

77. (a) $-\frac{2}{e}$; (b) $1 + \frac{-2+e}{e}$

79. (a) $-\frac{1}{4} - \frac{3\pi}{8}$; (b) $\frac{5\pi}{8}(2 + \pi)$

81. 4.27569 inches

83. About 15.47 cubic inches

85. (a) $-\frac{5}{\pi}\cos(\pi t) + \frac{5}{\pi^2}\sin(\pi t)$; (b) the particle moves right, then left, then right again, a greater distance each time – see the picture of $s(t)$ that follows; (c) at $t = 3$ seconds it is more than 4 feet to the right, and it is just turning around to come back to the starting position; (d) after a long time the particle oscillates back and forth by greater and greater distances, moving faster and faster to cover more distance in the same time period.

87. (a) See the proof of Theorem 8.8;

(b) $\frac{d}{dx}(x\ln x - x) = \ln x + x\left(\frac{1}{x}\right) - 1 = \ln x.$

89. (a) Let $u = \tan^{-1} x$ and $dv = dx$ to get

$x\tan^{-1} x - \int \frac{x}{x^2 + 1}\,dx$, and then use substitution with

$w = x^2 + 1$ to turn this into $x\tan^{-1} x - \frac{1}{2}\int \frac{1}{w}\,dw =$

$x\tan^{-1} x - \frac{1}{2}\ln|w| + C = x\tan^{-1} x - \frac{1}{2}\ln(x^2 + 1) + C.$

(b) $\frac{d}{dx}(x\tan^{-1} x - \frac{1}{2}\ln(x^2 + 1)) =$

$\tan^{-1} x + x\left(\frac{1}{x^2 + 1}\right) - \frac{1}{2}\left(\frac{1}{x^2 + 1}\right)(2x) = \tan^{-1} x.$

Section 8.3

1. F, T, T, T, T, F, F, T.

3. A rational function is a function that can be expressed as a quotient of polynomials. A rational function is proper if the degree of its numerator is strictly less than the degree of its denominator, and improper otherwise.

5. $p(x) = q(x)(x^2 - x + 3) + (3x + 1);$
   $\frac{p(x)}{q(x)} = (x^2 - x + 3) + \frac{3x - 1}{q(x)}$

7. Divide both sides by $q(x)$ and simplify.

9. $m(x)$ will have degree 2 and the degree of $R(x)$ will be strictly less than the degree of $q(x)$.

11. Hint: Try an example, and then think about it in general.

13. If $q(x)$ is an irreducible quadratic then $\frac{p(x)}{q(x)}$ is its own partial fractions decomposition; there is nothing further we can decompose. If $q(x)$ is a reducible quadratic then $q(x)$ is a product $l_1(x)l_2(x)$ of two linear functions, and we can obtain a partial fractions decomposition of the form $\frac{A_1}{l_1(x)} + \frac{A_2}{l_2(x)}$; in the example given, we have $l_1(x) = x - 1$ and $l_2(x) = x - 2$.

15. $y = x^2 - 3x + 5 = \left(x - \frac{3}{2}\right)^2 + \frac{11}{4}$ can be obtained from the graph of $y = x^2$ by translating $\frac{3}{2}$ units to the right and $\frac{11}{4}$ units up.

17. $\ln|x - 2| - \ln|x + 1| + C$

19. $\ln|x - 1| - 2(x - 1)^{-1} + C$

21. $x + \ln|x| + \frac{6}{x} + C$

23. $\frac{5}{2}x^2 - 5x + 8\ln|x - 2| + 27\ln|x + 3| + C$

25. $\frac{1}{7}\left(2\ln|x - 2| - \ln|x^2 + 3| + \sqrt{3}\tan^{-1}\left(\frac{1}{\sqrt{3}}x\right)\right) + C$

27. $2\ln|x - 1| - \ln|x^2 + x + 1| + C$

29. $\ln|x^2 - 4x + 5| + 2\tan^{-1}(x - 2) + C$

31. $3\ln|x - 2| + (x - 2)^{-1} + \frac{1}{2}(x - 2)^{-2} + C.$

33. $\frac{1}{3}\ln|3x + 1| - 2\ln|x - 3| + C$

35. $\frac{1}{2}\ln|x^2 + 1| + \tan^{-1} x - \frac{1}{2}\ln|x^2 + 3| + C$

37. $-x^{-1} - 2\ln|x| + \ln|x^2 + 1| - \tan^{-1} x + C$

39. $-\ln|x+1| + \frac{3}{2}\ln(x^2+1) + \tan^{-1}x + C$

41. $x^2 + 3\ln|x^2-4| + C$

43. $\frac{1}{3}x^3 - \frac{5}{2}x^2 + 6x + C$

45. $\frac{2}{3}x^3 - 2x + \frac{1}{2}\ln|x^2-4x+1| + C$

47. $\ln 2 - \ln 6 + \ln 5$ 　　　49. $\ln 3 - \ln 4 + \ln 2$

51. $2 + \ln 4 - \ln 2$

53. $-\ln|\cos x| + \frac{1}{1+\ln x} + C$

55. $\ln|1 + \ln x| + \frac{1}{1+\ln x} + C$

57. (a) $-\frac{\ln 3}{2}$; (b) $\frac{1}{4}$

59. (a) $\frac{1}{2}\left(\ln 5 - \frac{2}{5}\right) \approx 0.605$; (b) $x \approx 0.669$; note that $f(0.668) \approx 0.605$.

61. $r(t) = \ln\left(\frac{p(t)(K-p_0)}{p_0(K-p(t))}\right)$, and thus $p(t) = \frac{p_0 K e^{r(t)}}{K - p_0 + p_0 e^{r(t)}}$.

63. Either mimic the example after Theorem 8.13 by adding and subtracting $\left(\frac{b}{2}\right)^2$, or else just simply multiply out $f(x) = \left(x + \frac{b}{2}\right)^2 + \left(c - \frac{b^2}{4}\right)$ and show that it is equal to $x^2 + bx + c$.

Section 8.4

1. F, F, F, F, F, F, T, T.

3. $\frac{d}{dx}\left(-\frac{1}{5}\cos^5 x + \frac{1}{7}\cos^7 x\right) = \cos^4 x \sin x - \cos^6 x \sin x = \cos^4 x \sin x(1-\cos^2 x) = \cos^4 x \sin x \sin^2 x = \sin^3 x \cos^4 x$.

5. Multiply the integrand by $\frac{\csc x + \cot x}{\csc x + \cot x}$.

7. Three possible examples are the integrals in Exercises 21, 22, and 41.

9. Reduce powers with double-angle identities

11. Apply the identity $\cot^2 x = \csc^2 x - 1$ repeatedly until the integrand is converted into a sum of expressions that we can integrate. The power of cotangent will eventually decrease to zero.

13. Use parts with $u = \csc^{k-2} x$ and $dv = \csc^2 x$.

15. Reduce powers with double-angle identities.

17. Use $\tan^2 x = \sec^2 x - 1$ to rewrite the integrand entirely in terms of secant, saving one copy of $\sec x \tan x$ for the differential of the substitution $u = \sec x$.

19. (a) Rewrite as $\int(1-\cos^2 x)\cos^2 x\, dx = \int\cos^2 x\, dx - \int\cos^4 x\, dx$ and then apply double-angle identities to reduce powers;
(b) $\int\left(\frac{1}{2}\sin 2x\right)^2 dx = \frac{1}{8}\left(x - \frac{1}{4}\sin 4x\right) + C$.

21. $\frac{1}{2}x - \frac{1}{12}\sin(6x) + C$

23. $\sin x - \frac{2}{3}\sin^3 x + \frac{1}{5}\sin^5 x + C$

25. $-\frac{1}{2}\ln|\cos 2x| + C$

27. $\frac{1}{4}\ln|\sec 4x + \tan 4x| + C$

29. $-\frac{1}{4}\cot^4 x + \frac{1}{2}\cot^2 x + \ln|\sin x| + C$

31. $\frac{1}{4}\sec 2x \tan 2x + \frac{1}{4}\ln|\sec 2x + \tan 2x| + C$

33. $\frac{3}{8}x + \frac{1}{4}\sin 2x + \frac{1}{32}\sin 4x + C$

35. One way is to use parts with $u = \csc^2 x$ and $dv = \csc^2 x\, dx$ to get $-\csc^2 x \cot x - 2\int \csc^2 x \cot^2 x\, dx$. Then use substitution with $u = \cot x$ to get $-\csc^2 x \cot x + \frac{2}{3}\cot^3 x + C$.

37. $\frac{1}{5}\sin^5 x - \frac{1}{7}\sin^7 x + C$

39. $\frac{1}{3}\sec^3 x - \sec x + C$

41. $-\frac{1}{8}x - \frac{1}{96}\sin(12x) + C$ 　　43. $\frac{1}{8}\sec^8 x + C$

45. $\frac{1}{7}\sec^7 x - \frac{2}{5}\sec^5 x + \frac{1}{3}\sec^3 x + C$

47. $\frac{1}{14}\sin^{14} x - \frac{1}{8}\sin^{16} x + \frac{1}{18}\sin^{18} x + C$

49. $-\frac{1}{2}\csc x \cot x + \frac{1}{2}\ln|\csc x + \cot x| + C$

51. $\frac{1}{3}\sec^3 x + C$ 　　　　　53. $-\ln|\cos x| + C$

55. $\sin x + C$ 　　　　　　57. $\frac{1}{\cos x} + C$

59. $-\frac{1}{2}\cos^2 x \ln(\cos x) + \frac{1}{4}\cos^2 x + C$

61. $\frac{2}{9}\cos^{9/2} x - \frac{2}{5}\cos^{5/2} x + C$

63. $\frac{16}{15}$ 　　　　　　65. $\frac{4}{3}$

67. $\frac{4}{15}$ 　　　　　　69. $\frac{8}{15}$

71. $\frac{1}{3}$ 　　　73. (a) 0; (b) $\frac{32}{15}$; (c) $\frac{16}{15\pi}$

75. (a) 1.5; (b) $\frac{1}{2} + \frac{3\sqrt{3}}{\pi}$ 　　77. $\ln(7 + 4\sqrt{3})$ inches

79. See the proof of Theorem 8.17 in the reading.

81. (a) $\int \sin^4 x\, dx = -\frac{1}{4}\sin^3 x \cos x + \frac{3}{8}x - \frac{3}{16}\sin 2x + C$, and $\int \sin^8 x\, dx = -\frac{1}{8}\sin^7 x \cos x - \frac{7}{48}\sin^5 x \cos x - \frac{35}{192}\sin^3 x \cos x + \frac{35}{128}x - \frac{35}{256}\sin 2x$. (b) The formula reduces the power of $\sin x$ in the integrand by two every time we apply it. (c) Apply integration by parts with $u = \sin^{k-1} x$ and $dv = \sin x\, dx$, so that $v = -\cos x$ and $du = (k-1)\sin^{k-2} x \cos x\, dx$. After applying the integration by parts formula, multiply out the integrand to obtain a term of the form $(k-1)\int \sin^k x\, dx$. Then solve for $\int \sin^k x\, dx$.

Section 8.5

1. T, T, T, F, F, T, F, T.

3. (a) $1 = \sec^2 u \frac{du}{dx}$, so $dx = \sec^2 u\,du$; (b) $\frac{dx}{du} = \sec^2 u$, so $dx = \sec^2 u\,du$

5. See the reading.

7. In terms of the Pythagorean identity, sine and cosine behave the same way; we can always use $x = \sin u$ instead of $x = \cos u$.

9. $x = a \sin u$ makes sense only for $x \in [-1, 1]$, but the domain of $a^2 - x^2$ is $(-\infty, \infty)$. On the other hand, $x = a \tan u$ makes sense for $x \in (-\infty, \infty)$, i.e., everywhere that $x^2 + a^2$ is defined.

11. (a) Use algebra; (b) use conventional substitution with $u = 4 + x^2$; (c) use trigonometric substitution with $x = 2 \tan u$; (d) use algebra after writing $16 - x^4 = (4 - x^2)(4 + x^2)$.

13. See the proof of Theorem 8.18.

15. See the proof of Theorem 8.18.

17. $\sqrt{1 - x^2}$ is not defined on $[2, 3]$; its domain is $[-1, 1]$.

19. Three of the many such integrals are in Exercises 41, 42, and 46.

21. $\sqrt{1 - x^2}$

23. $\frac{x}{\sqrt{x^2 + 4}}$

25. $\tan^2 u = \frac{(x-5)^2}{9 - (x-5)^2}$

27. $\sin(2 \cos^{-1} x) = 2x\sqrt{1 - x^2}$

29. $(x + 3)^2 - 11$; $x + 3 = \sqrt{11} \sec u$

31. $\left(x - \frac{5}{2}\right)^2 - \frac{21}{4}$; $x - \frac{5}{2} = \frac{\sqrt{21}}{2} \sec u$

33. $12 - 2(x + 2)^2$; $x + 2 = \sqrt{6} \sin u$

35. (a) $\displaystyle\int \frac{1}{9\tan^2 u + 9}(3 \sec^2 u)\,du = \frac{1}{3} \tan^{-1} \frac{x}{3} + C$;

    (b) $\displaystyle\frac{1}{9}\int \frac{1}{\left(\frac{x}{3}\right)^2 + 1}\,dx = \frac{1}{3} \tan^{-1} \frac{x}{3} + C.$

37. (a) $\displaystyle\frac{1}{2}\int \frac{1}{u}\,du = \frac{1}{2} \ln|4 + x^2| + C$

    (b) $\displaystyle\int \frac{2\tan u}{4 + 4\tan^2 u}(2 \sec^2 u)\,du = -\ln\left|\frac{2}{\sqrt{x^2 + 4}}\right| + C$

39. $-\frac{\sqrt{4 - x^2}}{x} - \sin^{-1} \frac{x}{2} + C$

41. $(x^2 - 1)^{1/2} + C$

43. $\sin^{-1} \frac{x}{\sqrt{3}} + C$

45. $\sqrt{9 - x^2} - 3\ln\left(\frac{\sqrt{9 - x^2} + 3}{x}\right) + C$

47. $\frac{1}{2}(x - 2)\sqrt{1 - (x - 2)^2} + \frac{1}{2} \sin^{-1}(x - 2) + C$

49. $\frac{\sqrt{x^2 - 9}}{9x} + C$

51. $\frac{x}{\sqrt{1 - x^2}} + C$

53. $\frac{1}{3} \tan^{-1}\left(\frac{x - 2}{3}\right) + C$

55. $-\frac{1}{4}\sqrt{9 - 4x^2} + \frac{3}{2} \sin^{-1} \frac{2x}{3} + C$

57. $\frac{1}{2}(x - 4)\sqrt{(x - 4)^2 + 9} + \frac{9}{2} \ln\left|\frac{1}{3}\sqrt{(x - 4)^2 + 9} + \frac{1}{3}(x - 4)\right| + C$

59. $\sqrt{3 - x^2}\left(\frac{15}{8}x - \frac{1}{4}x^3\right) + \frac{27}{8} \sin^{-1}\left(\frac{x}{\sqrt{3}}\right) + C$

61. $\frac{1}{2}x\sqrt{2 - x^2} + \sin^{-1}\left(\frac{x}{\sqrt{2}}\right) + C$

63. If $x > 1$, we have $\sqrt{x^2 - 1} - \sec^{-1} x + C$; if $x < -1$, we have $\sqrt{x^2 - 1} + \sec^{-1} x + C$.

65. $\ln(2(\sqrt{x^2 - 2} + x)) + C$

67. $\frac{1}{5}(x^2 + 1)^{5/2} - \frac{1}{3}(x^2 + 1)^{3/2}$

69. $x\ln(1 + x^2) - 2x + 2\tan^{-1} x + C$

71. $\frac{1}{3}(e^{2x} + 1)^{3/2} + C$

73. $-\frac{1}{3} \frac{\sqrt{e^{2x} + 3}}{e^x} + C$

75. $\frac{1}{2}x^2 - 2\ln(x^2 + 4) + C$

77. $-\frac{2}{9}x + \frac{1}{9}x^3 - \frac{23}{9\sqrt{6}} \tan^{-1}\left(\sqrt{\frac{3}{2}}x\right) + C$

79. $\frac{1}{3}(20)^{3/2} - \frac{1}{3}(4)^{3/2}$

81. $\frac{1}{3} \ln 10 - \frac{1}{3} \ln(3 + \sqrt{34})$

83. $-\sqrt{3} + \sqrt{15}$

85. $\frac{2}{3} - \frac{1}{3\sqrt{7}}$

87. (a) $4\pi \displaystyle\int_{-\sqrt{42}}^{\sqrt{42}} \sqrt{42 - x^2}\,dx$; (b) using trigonometric substitution with $x = \sqrt{42} \sin u$, we obtain a volume of approximately 829 cubic millimeters.

89. (a) $2 \displaystyle\int_{-r}^{r} \sqrt{r^2 - x^2}\,dx$

    (b) $2 \displaystyle\int_{-(\pi/2)}^{\pi/2} \sqrt{r^2 - r^2 \sin^2 u}(r \cos u)\,du =$
    $2r^2 \displaystyle\int_{-\pi/2}^{\pi/2} \cos^2 u\,du = 2r^2\left(\frac{1}{2}\right)\int_{-\pi/2}^{\pi/2}(1 + \cos 2u)\,du =$
    $r^2\left[u + \frac{1}{2} \sin 2u\right]_{-\pi/2}^{\pi/2} = \pi r^2$

Section 8.6

1. F, F, F, F, T, T, F, T.

3. The interval over which we are integrating could be infinite, or the function to be integrated could have a discontinuity or a vertical asymptote on the interval.

5. The integrand $\frac{1}{e^x}$ is continuous on $[0, 1]$, in particular with no vertical asymptotes; this integral is not improper.

7. The Fundamental Theorem of Calculus does not apply here because $(x - 3)^{-4/3}$ is not continuous at $x = 3$, inside the interval $[0, 5]$. Applying it anyway would give the incorrect answer $\frac{-3}{2^{1/3}} + \frac{3}{(-3)^{1/3}}$, when the correct answer is that the improper integral diverges.

9. If $p \geq 1$ and $x \in [0, 1]$ then $\frac{1}{x^p}$ is greater than $\frac{1}{x}$, whose improper integral on $[0, 1]$ is known to diverge.

11. Since $\frac{1}{x+1} < \frac{1}{x}$ for $x \geq 1$ and $\int_1^\infty \frac{1}{x} dx$ diverges, this comparison does not tell us anything.

13. $\frac{1}{x+1} \leq \frac{1}{x}$, but knowing that an improper integral is less than a divergent one does not give us any information. Instead, note that the graph of $y = \frac{1}{x+1}$ is just a horizontal translation of the graph of $y = \frac{1}{x}$. This means their improper integrals are related; how?

15. We can split at any point in $[1, \infty)$, for example at $x = 2$ to get: $\int_1^2 \frac{1}{x-1} dx + \int_2^\infty \frac{1}{x-1} dx$.

17. $\int_0^2 \frac{1}{2x^2-10x+12} dx + \int_2^{2.5} \frac{1}{2x^2-10x+12} dx$
$+ \int_{2.5}^3 \frac{1}{2x^2-10x+12} dx + \int_3^5 \frac{1}{2x^2-10x+12} dx$

19. Choosing $x = \pm 1$ for the splits, we have
$\int_{-\infty}^{-1} \frac{e^{-x^2}}{x^2} dx + \int_{-1}^0 \frac{e^{-x^2}}{x^2} dx + \int_0^1 \frac{e^{-x^2}}{x^2} dx + \int_1^\infty \frac{e^{-x^2}}{x^2} dx.$

21. 3

23. Diverges

25. 100

27. Diverges

29. $2^{1/3} + 1$

31. Diverges

33. Diverges

35. $\frac{1}{2}$

37. 2

39. 2

41. 0

43. Diverges

45. Diverges

47. Diverges

49. 1

51. 2

53. Diverges

55. Diverges

57. Diverges by comparison with $\frac{1}{x}$

59. Converges by comparison with $\frac{1}{x^2}$

61. Diverges by comparison with $\frac{1}{x}$

63. Converges by comparison with $\frac{1}{x^2}$

65. Converges by limit comparison with $\frac{1}{x^2}$

67. Converges by limit comparison with $\frac{1}{x^2}$

69. Diverges by comparison with $\frac{1}{x^{1/3}}$

71. About 1.25 pounds.

73. About 3.23 years.

75. If $p > 1$, then
$$\int_1^\infty \frac{1}{x^p} dx = \lim_{B \to \infty} \int_1^B x^{-p} dx = \lim_{B \to \infty} \left[\frac{1}{-p+1} x^{-p+1} \right]_1^B =$$
$$\lim_{B \to \infty} \left(\frac{1}{1-p} B^{1-p} - \frac{1}{1-p} \right) = \frac{1}{p-1}.$$ Note that we used the fact that $p > 1$ to take the limit, since in that case we have $1 - p < 0$ and $B^{1-p} \to 0$ as $B \to \infty$.

77. If $p \geq 1$, then $\int_0^1 \frac{1}{x^p} dx = \lim_{A \to 0^+} \int_A^1 x^{-p} dx =$
$$\lim_{A \to 0^+} \left[\frac{1}{-p+1} x^{-p+1} \right]_A^1 = \lim_{A \to 0^+} \left(\frac{1}{1-p} - \frac{1}{1-p} A^{1-p} \right)$$
diverges.

79. Mimic the proof of the case of Theorem 8.23 in the reading.

81. We will suppose that $d > c$; the case where $d < c$ is similar. We have $\int_{-\infty}^d f(x) dx + \int_d^\infty f(x) dx =$
$\left(\int_{-\infty}^c f(x) dx + \int_c^d f(x) dx \right) + \int_d^\infty f(x) dx = \int_{-\infty}^c f(x) dx +$
$\left(\int_c^d f(x) dx + \int_d^\infty f(x) dx \right) = \int_{-\infty}^c f(x) dx + \int_c^\infty f(x) dx.$

Section 8.7

1. T, T, F, T, T, T, T, F.

3. Use pictures similar to the three immediately before Theorem 8.24, but with an increasing function.

5. One example is $f(x) = -x^2 + 2x + 3$ on $[0, 3]$, with $n = 3$ left and right sums.

7. The same example from Problem 5 works here as well.

9. Midpoint sum

11. Do the approximation separately on $[a, c]$ and on $[c, b]$ and calculate bounds for the error on each piece using Theorem 8.27. The total error will be less than or equal to the sum of the absolute values of the two errors you calculated.

13. See the discussion in the reading prior to Theorem 8.27.

15. (a) LEFT$(n) <$ MID$(n) <$ TRAP$(n) <$ RIGHT(n);
(b) between TRAP(n) and MID(n).

17. Using the weighted average from Exeric se 61, the $n = 1000$ Simpson's Rule approximation is 1.09861229, which we would expect to be an even better approximation.

19. By the old formula we have
$\int_1^5 (3x^2 - x + 4) dx = 3 \int_1^5 x^2 dx - \int_1^5 x dx + \int_1^5 4 dx =$
$3 \left(\frac{1}{3} \right) (5^3 - 1^3) - \left(\frac{1}{2} \right) (5^2 - 1^2) + 4(5 - 1) = 128$
and by the new formula we also have
$\int_1^5 (3x^2 - x + 4) dx = \frac{5-1}{6} ((3(1)^2 - 1 + 4) + 4(3(3)^2 - 3 + 4) + (3(5^2) - 5 + 4)) = 128.$

21. Mimic Example 3 but with $n = 4$ instead of $n = 6$.

23. RIGHT(4) = 40. Since $f'(x) = 2x + 1$ is positive on $[0, 4]$ we know that $f(x) = x^2 + x$ is monotonically increasing on $[0, 4]$. Therefore we can guarantee that $|E_{\text{RIGHT}(4)}| \leq |f(4) - f(0)| \left(\frac{4-0}{4}\right) = 20$. The actual area is about 29.33, which is within the error bounds of our approximation.

25. LEFT(6) = 0.37404. Since $f'(x) = -e^{-x}$ is negative on $[1, 3]$ we know that $f(x) = e^{-x}$ is monotonically decreasing on $[1, 3]$. Therefore we can guarantee that $|E_{\text{LEFT}(6)}| \leq |f(3) - f(1)| \left(\frac{3-1}{6}\right) = |e^{-3} - e^{-1}| \left(\frac{1}{3}\right) = 0.106031$. The actual area is approximately 0.318092, which is within the error bounds for our approximation.

27. TRAP(6) = 0.321033. Since $f''(x) = e^{-x}$ is positive on $[1, 3]$ we know that $f(x) = e^{-x}$ is concave up on $[1, 3]$. Moreover, $|f''(x)| = |e^{-x}| \leq e^{-1}$ for all $x \in [1, 3]$. Therefore we can guarantee that $|E_{\text{TRAP}(6)}| \leq e^{-1}(3 - 1)^3 12(6)^2 = 0.00681258$.

29. MID(12) = 7.63016. Since $f''(x) = -\frac{1}{x^2}$ is negative on $[1, 7]$ we know that $f(x) = \ln x$ is concave down on $[1, 7]$. Moreover $|f''(x)| = |-\frac{1}{x^2}| \leq \frac{1}{1^2} = 1$ for all $x \in [1, 7]$. Therefore $|E_{\text{MID}(12)}| \leq 1(7 - 1)^3 24(12)^2 = 0.0625$.

31. TRAP(8) = 3.25174; using the fact that $|f''(x)| \leq 2$ for all $x \in [0, 2]$ (this is not the best possible bound), we can guarantee that $|E_{\text{TRAP}(8)}| \leq 0.0208333$.

33. LEFT(9) $\approx$ 0.996; $|E_{\text{LEFT}(9)}| \leq 0.484$.

35. SIMP(12) = 24. Since $f^{[4]} = 0$ for all $x \in [-2, 4]$ we can guarantee that $E_{\text{SIMP}(12)} \leq 0$, i.e., that the estimate is in fact exact.

37. SIMP(4) = 3.7266. Since $f'[4] = 24$ for all $x \in [-1, 2]$ we can guarantee that $E_{\text{SIMP}(4)} \leq 0.1266$.

39. SIMP(8) = 0. Since $f'[4] \leq 1$ for all $x \in [-\pi, \pi]$ we can guarantee that $E_{\text{SIMP}(8)} \leq 0.013$.

41. Since $f(x) = xe^x$ is monotonically increasing on $[-1, 1]$, Theorem 8.25 guarantees we can have $|E_{\text{LEFT}(n)}| \leq 0.5$ if we choose $n = 13$ or higher. We have LEFT(13) = 0.509077.

43. Since $f(x) = \frac{1}{x}$ is monotonically decreasing on $[2, 4]$, Theorem 8.25 guarantees we can have $|E_{\text{LEFT}(n)}| \leq 0.05$ if we choose $n = 10$ or higher. We have RIGHT(10) = 0.668771.

45. Since $f(x) = x^2$ is concave up on $[0, 3]$ with $f''(x) = 2$ for all x, Theorem 8.27 guarantees we can have $|E_{\text{MID}(n)}| \leq 0.1$ if we choose $n = 5$ or higher. We have MID(5) = 8.91.

47. Since $f(x) = x^{2/3}$ is concave down on $[1, 4]$ with $|f''(x)| \leq \frac{2}{9}$ for all $x \in [1, 4]$, Theorem 8.27 guarantees we can have $|E_{\text{MID}(n)}| \leq 0.005$ if we choose $n = 10$ or higher. We have TRAP(10) = 5.44578.

49. Since $f(x) = 10e^{-x} dx$ has $|f^{[4]}(x)| \leq 10$ for all $x \in [0, 2]$, Theorem 8.30 guarantees we can have $|E_{\text{SIMP}(n)}| \leq 0.01$ if we choose $n = 4$ or higher, meaning that we should use 2 or more parabola-topped rectangles. We have SIMP(4) = 8.64956.

51. Since $f(x) = \sin x$ has $|f^{[4]}(x)| \leq 1$ for all x, Theorem 8.30 guarantees we can have $|E_{\text{SIMP}(n)}| \leq 0.0005$ if we choose $n = 8$ or higher, meaning that we should use 4 or more parabola-topped rectangles. We have SIMP(8) = 2.00027.

53. (a) LEFT(4) = 32.96 degrees, RIGHT(4) = 19.99 degrees, MID(4) = 25.6717 degrees, TRAP(4) = 26.48 degrees, and SIMP(4) = 25.94 degrees. (b) $|E_{\text{LEFT}(4)}| < 12.97$, $|E_{\text{RIGHT}(4)}| < 12.97$, $|E_{\text{MID}(4)}| < 0.625$, $|E_{\text{TRAP}(4)}| < 1.25$, and $|E_{\text{SIMP}(4)}| < 0.208$.

55. Hints: For (a) you need to use four parabolas, over the intervals [1993, 1995], [1995, 1997], [1997, 1999], and [1999, 2001]. In part (b) note that the area under the graph of r on an interval is the change in the GDP over that interval. For part (c) you can make a table of difference quotients to estimate r', and then repeat that process to estimate higher derivatives.

57. Remember that the period of time must be in seconds. A discrete integral for the total flow is $\phi \approx (1/3)(60 \times 60 \times 60 \times 24(700 + 4 \times 1000 + 2 \times 6300 + 4 \times 4000 + 2 \times 500 + 4 \times 650 + 700)) + 60(24(700 \times 5) \times 60) = 65{,}275{,}200{,}000$. The answer is in cubic feet of water. Inasmuch as the last flow is 700 cfs, as is the first, and the flow is periodic, it seems obvious that we should take the flow to be a constant 700 cfs for the last five days of the year, which is where that last term comes from.

59. Mimic the proof of part (a) of the theorem given in the reading.

61. Use the expanded sum expressions for the trapezoid sum with n subdivision points called $x_0, x_2, x_4, \ldots, x_{2n}$, the midpoint sum with the same n subdivision points and thus midpoints $x_1, x_3, x_5, \ldots, x_{2n-1}$, and compare to Simpson's Rule with all $2n$ subdivision points $x_0, x_1, x_2, \ldots, x_{2n}$.

63. Generalize the method shown for $n = 6$ in Example 3 to arbitrary n.

For answers to odd-numbered Skill Certification exercises in the Chapter Review, please visit the Book Companion Web Site at www.whfreeman.com/integcalculus.

Chapter 9

Section 9.1

1. F, F, F, T, T, T, T, T.

3. (a) 2.25π; (b) 5.25π

5. Mimic the example and illustrations in the reading at the beginning of the section.

7. (a) $\Delta x = \frac{1}{2}$, subdivision points $x_0 = 0$, $x_1 = 0.5$, $x_2 = 1$, $x_3 = 1.5$, and $x_4 = 2$; (b) using midpoints, $x_1^* = 0.25$, $x_2^* = 0.75$, $x_3^* = 1.25$, and $x_4^* = 1.75$; (c) corresponding values $f(x_1^*) = 0.0625$, $f(x_2^*) = 0.5625$, $f(x_3^*) = 1.5625$, and $f(x_4^*) = 3.0625$;
(d) $(16\pi - \pi((0.75)^2)^2) \left(\frac{1}{2}\right) \approx 24.6357$.

9. $\pi \int_2^3 (1+x)^2 dx$; the solid obtained by rotating the region between the graph of $f(x) = 1+x$ and the x-axis on $[2, 3]$ around the x-axis.

11. $\pi \int_0^2 (4-x^2) dx$; the solid obtained by rotating the region between the graph of $f(x) = \sqrt{4-x^2}$ and the x-axis on $[0, 2]$ around the x-axis.

13. They are equal.   15. The second one is larger.

17. The second one is larger.

19. The region between $f(x) = x+1$ and the line $y = 4$ from $x = 1$ to $x = 3$, rotated around the x-axis.

21. The region between $f(x) = x-1$ and the x-axis on $[1, 3]$, rotated around the x-axis.

23. The region between the graph of $f(x) = 2x+1$ and the line $y = 5$ on the x-interval $[0, 2]$, rotated about the y-axis.

25. $\pi \int_{-3}^0 2^2 dy + \pi \int_0^3 (2^2 - (f^{-1}(y))^2) dy$

27. $\pi + 2\pi + 3\pi + 4\pi = 10\pi$

29. $3\pi + (2+2\sqrt{2})\pi + (3+2\sqrt{3})\pi + 8\pi \approx 70.03$

31. $\pi \int_1^3 \left(\frac{3}{x}\right)^2 dx = 6\pi$

33. $\pi \int_0^1 (3^2 - 1^2) dy + \pi \int_1^3 \left(\left(\frac{3}{y}\right)^2 - 1^2\right) dy = 8\pi + 4\pi = 12\pi$

35. $\pi \int_0^3 (5^2 - (y+2)^2) dy = 36\pi$

37. $\pi \int_2^5 (3^2 - (3-(x-2))^2) dx = 18\pi$

39. $\pi \int_0^3 (3^2 - ((y+2)-2)^2) dy = 18\pi$

41. $\pi \int_0^4 (2^2 - (\sqrt{4-y})^2) dy + \pi \int_4^5 (2^2) dy = 8\pi + 4\pi = 12\pi$

43. $\pi \int_0^4 ((3-(\sqrt{4-y}))^2 - 1^2) dy + \pi \int_4^5 (3^2 - 1^2) dy = 8\pi + 8\pi = 16\pi$

45. $\pi \int_1^{5/3} ((5-x)^2 - (2x)^2) dx + \pi \int_{5/3}^4 ((2x)^2 - (5-x)^2) dx = \frac{112\pi}{27} = \frac{1813\pi}{27} = \frac{1925\pi}{27}$

47. $\pi \int_0^4 \left((\sqrt{y})^2 - \left(\frac{y}{2}\right)^2\right) dy = \frac{8\pi}{3}$

49. $\pi \int_0^2 ((2x+1)^2 - (x^2+1)^2) dx = \frac{104\pi}{15}$

51. $\pi \int_1^4 (x^2 - ((x-2)^2)^2) dx = \frac{72\pi}{5}$

53. $\pi \int_1^4 ((x+1)^2 - ((x-2)^2+1)^2) dx = \frac{117\pi}{5}$

55. $\pi \int_0^\pi (1-\cos x)^2 dx \approx 14.80$

57. $\pi \int_0^2 (\pi - \cos^{-1}(1-y))^2 dy \approx 18.4399$

59. (a) 481 feet; (b) 91,636,272 cubic feet; (c) $\int_0^{481} \left(756 - \frac{756}{481}x\right)^2 dx = 91,636,272.$

61. The cone can be obtained by rotating the region under $y = 3/5x$ around the x-axis; this solid has volume $\pi \int_0^5 \left(\frac{3}{5}x\right)^2 dx = 15\pi.$

63. A cone of radius r can be obtained by rotating the region under the graph of $y = \frac{r}{h}x$ around the x-axis; this solid has volume $\pi \int_0^h \left(\frac{r}{h}x\right)^2 dx = \frac{1}{3}\pi r^2 h.$

Section 9.2

1. T, T, F, F, F, T, F, T.

3. (a) Washer of volume $\frac{15\pi}{2}$; (b) shell of volume $\frac{15\pi}{4}$; (c) cylinder of volume $\frac{3\pi}{4}$; (d) disk of volume $\frac{9\pi}{2}$; (e) shell of volume $\frac{9\pi}{4}$; (f) washer of volume $\frac{21\pi}{2}$.

5. Shells have the same types of dimensions as washers, but what we think of as a height on a shell is considered a thickness on a washer, and vice-versa.

7. (a) $\Delta x = 0.5$, subdivision points $x_0 = 0$, $x_1 = 0.5$, $x_2 = 1$, $x_3 = 1.5$, and $x_4 = 2$; (b) midpoints $x_1^* = 0.25$, $x_2^* = 0.75$, $x_3^* = 1.25$, and $x_4^* = 1.75$; (c) corresponding values $f(x_1^*) = 0.0625$, $f(x_2^*) = 0.5625$, $f(x_3^*) = 1.5625$, and $f(x_4^*) = 3.0625$; (d) $2\pi rh\Delta x = 2\pi \cdot \frac{3}{4} \cdot \frac{9}{16} \cdot \frac{1}{2} = \frac{27}{64}\pi$

9. $\int_0^2 2\pi y^3 dy = 2\pi \int_0^2 y(y^2) dy$; the solid obtained by rotating the region between the graph of $f(x) = \sqrt{x}$, the y-axis, and the line $y = 2$ around the x-axis

11. The region between the graph of $f(x) = \cos x$ and the x-axis on the interval $\left[0, \frac{\pi}{4}\right]$, rotated around the y-axis

13. The region between $y = x^2 + 4$ and $y = 5$ from $x = 0$ to $x = 1$, rotated around the x-axis

15. The region between $y = e^x$ and $y = e$ from $x = 0$ to $x = 1$, rotated around the x-axis

17. $2\pi \int_0^2 y(3 - f^{-1}(y)) dy$ 19. $2\pi \int_0^2 x(f(x)+3) dx$

21. With shells: $2(2\pi \int_0^2 y(2-y) dy)$; with disks: $2(\pi \int_0^2 x^2 dx)$; disks are easier in this case.

23. With shells: $2\pi \int_0^1 (3-x)(x^2+1) dx$; with washers: $\pi \int_0^1 (3^2 - 2^2) dy + \pi \int_1^2 ((3-\sqrt{y-1})^2 - 2^2) dy$; shells are easier.

25. $3.093 + 8.099 + 9.572 + 5.154 = 25.92$

27. $10.953 + 14.082 + 16.671 + 18.426 = 60.132$

29. $2\pi \int_0^2 x(4-x^2) dx = 8\pi$

31. $2\pi \int_0^2 (2-x)(4-x^2) dx = \frac{40\pi}{3}$

33. $2\pi \int_2^5 x(x-2) dx = 36\pi$

35. $2\pi \int_0^3 (3-y)(5-(y+2)) dy = 18\pi$

37. $2\pi \int_2^5 (x-2)(x-2) dx = 18\pi$

39. $2\pi \int_0^2 x(5-(4-x^2)) dx = 12\pi$

41. $2\pi \int_0^2 (3-x)(5-(4-x^2)) dx = 16\pi$

43. $2\pi \int_1^5 (5-x)((x+1)+2) dx = \frac{256\pi}{3}$

45. $2\pi \int_0^1 y((2+\sqrt{y})-(2-\sqrt{y}))dy+2\pi \int_1^4 y((2+\sqrt{y})-y)dy$

$= \frac{8\pi}{5} + \frac{64\pi}{5} = \frac{72\pi}{5}$

47. $2\pi \int_0^1 (y+1)((2+\sqrt{y})-(2-\sqrt{y}))dy$

$+2\pi \int_1^4 (y+1)((2+\sqrt{y})-y)dy = \frac{64\pi}{15} + \frac{287\pi}{15} = \frac{117\pi}{5}$

49. $2\pi \int_0^2 x(4-(4-x^2))dx = 2\pi \int_0^2 x^3 dx = 8\pi$

51. $2\pi \int_0^6 x((x-3)^2+2)dx = 180\pi$

53. $2\pi \int_0^1 y(\sqrt{y}-y^2)dy = \frac{3\pi}{10}$

55. With disks: $\pi \int_0^3 (9-x^2)^2 dx = \frac{648\pi}{5}$

57. With shells: $2\pi \int_0^2 x(x^2-4x+4)dx = \frac{8\pi}{3}$

59. With disks: $\pi \int_0^{2/3} 9dx + \pi \int_{2/3}^2 (\frac{2}{x})^2 dx = 10\pi$; with

shells: $2\pi \int_0^1 2ydy + 2\pi \int_1^3 (\frac{2}{y})ydy = 10\pi$

61. With shells:

$2\pi \int_{-2}^0 (3-x)(2+x)dx+2\pi \int_0^2 (3-x)(2-x)dx = \frac{44\pi}{3} + \frac{28\pi}{3}$

63. $2\pi \int_{0.6}^3 y\left(2\sqrt{\frac{1}{0.15}(3-y)}\right)dy \approx 125.463$ cubic

centimeters.

65. See the proof of Theorem 9.4.

67. The sphere can be obtained by rotating the region under $y = \sqrt{9-x^2}$ around the x-axis; with the shell method, this solid has volume $2\pi \int_0^3 x(2\sqrt{9-x^2})dx$. We can solve this integral with the substitution $u = 9 - x^2$, $du = -2xdx$ to obtain a volume of 36π, just as predicted by the volume formula.

69. A sphere of radius r can be obtained by rotating the region between the graph of $y = \sqrt{r^2-x^2}$ and the x-axis on $[-r, r]$ around the x-axis. Using integration by substitution with $u = r^2 - x^2$, $du = -2x\,dx$, the volume of this solid is $2\pi \int_0^r x(\sqrt{r^2-x^2} - (-\sqrt{r^2-x^2}))dx =$ $2\pi \int_0^r x(2\sqrt{r^2-x^2}) = \frac{4}{3}\pi r^3$. Technical note: Usually we can't do the shell method along the x-axis when we've rotated a region around the x-axis; it works here only because the solid is a perfect sphere, so it is equivalent to rotating the region between the graph of $x = \sqrt{r^2-y^2}$ and the y-axis from $y = -r$ to $y = r$ around the y-axis.

Section 9.3

1. T, F, T, F, T, F, F, T.   3.   See Definition 9.6.

5. See the proof of Theorem 9.7.

7. $y = e^x$

9. No; $f(x)$ has a vertical asymptote at $x = \frac{\pi}{2}$.

11. Both arc lengths are given by the definite integral $\int_0^2 \sqrt{1+4x^2}dx$, so they are equal. This makes sense because $f(x) = x^2 - 3$ and $g(x) = 5 - x^2$ differ by a vertical flip and shift only; these transformations should not affect the arc length.

13. $2\int_{-5}^5 \sqrt{1+\left(\frac{-x}{\sqrt{25-x^2}}\right)^2}dx$

15. (a) 3.79009; (b) 3.81994; (c) it is unclear which approximation might be the best.

17. See the reading before Definition 9.9.

19. 16.97   21. 2.3

23. $2+4\sqrt{3} \approx 8.9282$

25. $\Delta y_1 = 7$, $\Delta y_2 = 1$, $\Delta y_3 = 1$, $\Delta y_4 = 7$; 16.97

27. $\Delta y_1 \approx 0.405$, $\Delta y_2 \approx 0.288$, $\Delta y_3 \approx 0.223$, $\Delta y_4 \approx 0.182$; approximately 2.3

29. $\Delta y_1 = 2\sqrt{2}$, $\Delta y_2 = 0$, $\Delta y_3 = -2\sqrt{2}$; $2+4\sqrt{3} \approx 8.9282$

31. $\int_{-1}^4 \sqrt{10}dx = 5\sqrt{10}$

33. $\int_1^3 \sqrt{1+9x}dx = \frac{2}{27}(28)^{3/2} - \frac{2}{27}(10)^{3/2}$

35. $\int_{-1}^1 \sqrt{28+18x}dx = \frac{1}{27}(46)^{3/2} - \frac{1}{27}(10)^{3/2}$

37. $\int_{-3}^3 \frac{3}{\sqrt{9-x^2}}dx = 3\pi$ 39. $\frac{1}{2}\int_0^1 \frac{1+x}{\sqrt{x}}dx = \frac{4}{3}$

41. $2\int_0^2 x^{-1/3}dx = 3(2^{2/3})$

43. $\int_1^2 \frac{1+16x^2}{8x}dx = 3 + \frac{\ln 2}{8} \approx 3.087$

45. $\int_{\pi/4}^{3\pi/4} \csc x dx = \ln(2\sqrt{2}+3)$

47. $f(x) = e^{3x}$, $[a, b] = [-2, 5]$

49. $f(x) = \frac{1}{x}$, $[a, b] = [2, 3]$

51. $f(x) = \ln|\cos x|$, $[a, b] = [0, \frac{\pi}{4}]$

53. One approximation is 3.09573

55. One approximation is 9.05755

57. $8\pi \sqrt{5} + 40\pi \sqrt{37}$

59. $\frac{\pi}{64}(35\sqrt{377} + 91\sqrt{1385} + 189\sqrt{3737} + 341\sqrt{8297})$

61. $\pi \sqrt{\pi^2+4}$ 63. $4\sqrt{2\pi}$

65. $\frac{\pi}{6}(17\sqrt{17}-1)$ 67. $\frac{\pi}{18}(343 - 13^{3/2})$

69. $\frac{\pi}{16}(4e^4\sqrt{16e^8+1} + \ln(4e^4 + \sqrt{16e^8+1}) - 4e^{-4}\sqrt{16e^{-8}+1} - \ln(4e^{-4} + \sqrt{16e^{-8}+1})$

71. $2\pi\sqrt{2} + \pi\ln(\sqrt{2}+1) - \pi\ln(\sqrt{2}-1)$

73. $\int_0^{\sqrt{6}} \sqrt{1+x^2}dx = \sqrt{\frac{21}{2}} + \frac{1}{2}\ln(\sqrt{6}+\sqrt{7}) \approx 4.0545$ inches

75. See the discussion in the reading preceding Definition 9.6.

77. (a) The distance from $(a, f(a)) = (a, ma+c)$ to $(b, f(b)) = (b, mb+c)$ is given by

$\sqrt{(b-a)^2 + ((mb+c)-(ma+c))^2} =$ $\sqrt{(b-a)^2 + m^2(b-a)^2} = (b-a)\sqrt{1+m^2}$. Since the graph of $f(x) = mx + c$ is a line, this is the exact arc length. (b) If $f(x) = mx + c$, then $f'(x) = m$, which is constant, so the arc length is given by $\int_a^b \sqrt{1+m^2}dx = [\sqrt{1+m^2}]_a^b = \sqrt{1+m^2}(b-a)$.

79. If $f(x) = \sqrt{r^2 - x^2}$ (the top half of the circle), then $f'(x) = -\dfrac{x}{\sqrt{r^2-x^2}}$. Thus the circumference of the circle is twice the arc length of the top half:

$$2\int_{-r}^{r} \sqrt{1 + \left(-\frac{x}{\sqrt{r^2-x^2}}\right)^2}\, dx = 2\int_{-r}^{r} \frac{r}{\sqrt{r^2-x^2}}\, dx.$$ Using trigonometric substitution with $x = r\sin u$ (and changing the limits of integration accordingly), this is equal to

$$2r\int_{-\pi/2}^{\pi/2} 1\, du = 2r[u]_{-\pi/2}^{\pi/2} = 2r\left(\frac{\pi}{2} - \left(-\frac{\pi}{2}\right)\right) = 2\pi r.$$

81. (a) Use the fact that $\dfrac{s+t}{q} = \dfrac{t}{p}$. (b) Use the fact that $C = \sqrt{A^2 + B^2}$. (c) Start with the fact that the surface area of the frustum is $S = \pi q(s + t) - \pi pt$. (d) Combine the given relationship with the equation $S = \pi(p + q)s$.

83. Apply the definite integral formula in Theorem 9.10 to the function $f(x) = \sqrt{r^2 - x^2}$ on the interval $[-r, r]$, and then solve the integral. Simplify the integrand as much as possible before attempting to integrate!

Section 9.4

1. F, T, F, T, F, T, F, T.

3. See the reading at the beginning of the section.

5. The depth at the bottom of the pool is constant, but the depth along the side walls varies at different parts of the wall.

7. If you know x_k, Δx, and N, then you can solve for a and b, since $x_k = a + k\Delta x$ and $\Delta x = \dfrac{b-a}{N}$.

9. It would require 38,422.9 foot-pounds of work to lift the entire full hot tub; it would take significantly less work to pump the water out of the tub, since only the water at the bottom of the tub has to move 4 feet (the rest of the water has a shorter distance to travel).

11. Hint: Think about the definitions of Δy and y_k.

13. Hydrostatic force is calculated using the *area* of a slice of the pool wall, not the volume of a slice of the entire pool.

15. See the paragraphs in the reading just before the Examples and Explorations subsection. The extra dimension Δx becomes the differential dx when we pass from a Riemann sum to a definite integral.

17. In the numerators each individual centroid coordinate is multiplied by the area of the corresponding rectangle, which makes centroid coordinates from larger rectangles count more than those from smaller rectangles.

19. 1920π grams   21. 2000 foot-pounds

23. 1497.6 pounds

25. $\Delta x = 6, x_0 = 0, x_1 = 6, x_2 = 12, x_3 = 18, x_4 = 24$. If we choose $x_1^* = 0, x_2^* = 6, x_3^* = 12$, and $x_4^* = 18$, then the mass is approximately 3746.95 grams.

27. $\Delta y = 1, y_0 = 0, y_1 = 1, y_2 = 2, y_3 = 3, y_4 = 4$. If we choose $y_1^* = 0.5, y_2^* = 1.5, y_3^* = 2.5$, and $y_4^* = 3.5$, then the work is approximately 3301.97 foot-pounds.

29. $\Delta y = 1, y_0 = 0, y_1 = 1, y_2 = 2, y_3 = 3, y_4 = 4$. If we choose $y_1^* = 0.5, y_2^* = 1.5, y_3^* = 2.5$, and $y_4^* = 3.5$, then the force is 7987.2 pounds.

31. $(\bar{x}, \bar{y}) \approx (5.417, 1.142)$ 33. $(\bar{x}, \bar{y}) \approx (0.819, 4.604)$

35. $(\bar{x}, \bar{y}) = \left(\dfrac{363}{65}, \dfrac{15}{13}\right)$ 37. $(\bar{x}, \bar{y}) = \left(\dfrac{4}{5}, \dfrac{32}{7}\right)$

39. $(\bar{x}, \bar{y}) \approx \left(\dfrac{\pi}{2}, 0.6\right)$ 41. 46,436 grams

43. $m = (1.5)^2 \int_0^{12} (4.2 + 0.4x - 0.03x^2)dx = 139.32$ grams

45. 125 grams

47. (a) The solid of revolution obtained by revolving the region between the graph of $y = 8 - \dfrac{2}{25}x^2$ and the x-axis on $[0, 10]$ around the y-axis; (b) $\rho(y) = -\dfrac{0.97}{8}y + 1.12$; (c)

$$m = \pi \int_0^8 \left(-\frac{0.97}{8}y + 1.12\right)\left(100 - \frac{25}{2}y\right) dy = 1001.12$$ ounces

49. 6.126 foot-pounds

51. $W = (62.4)(40)\int_0^3 (3 - y)dy = 11{,}232$ foot-pounds

53. $W = (62.4)\left(\dfrac{4\pi}{25}\right)\int_0^{10} y^2(10 - y)dy = 26{,}138.1$ foot-pounds

55. $W = (62.4)(16\pi)\int_0^{13}(16 - y)dy = 387{,}366$ foot-pounds

57. $W = (42.3)(100)(2)\int_0^8 \sqrt{64 - (8 - y)^2}(16 - y)dy + (42.3)(100)(2)\int_0^{16} \sqrt{64 - (16 - y)^2}(16 - y)dy \approx 6{,}289{,}650$ foot-pounds.

59. $W = (20)(18) + (.25)\int_0^{18}(18 - y)dy = 400.5$ foot-pounds.

61. 3.1765 pounds

63. $F = (62.4)(8)\int_0^3 (3 - y)dy = 2246.4$ pounds

65. $F = (62.4)(0.8\pi)\int_0^1 (1 - y)dy = 78{,}4142$ pounds

67. $F = 62.4\displaystyle\int_0^{200}\left(\dfrac{150}{200}y + 250\right)(200 - y)dy = 3.744 \times 10^8$ pounds

69. $F = 62.4\int_0^8 (6 + y)(12 - y)dy + 62.4\int_8^{12}(20)(12 - y)dy \approx 47{,}257$ pounds.

71. Leila should put the repeater approximately at coordinates $(4.85, 4.26)$.

73. $m = \rho V = \rho A\Delta x$. Adding up the mass on each cross section gives a Riemann sum of the form $\sum_{k=1}^{n} \rho(x_k^*)A(x_k^*)\Delta x$. Since ρ and A are continuous functions, this Riemann sum converges to a definite integral.

75. $F = \omega Ad = \omega l d\Delta y$. Adding up the hydrostatic force on each cross section gives a Riemann sum of the form $\sum_{k=1}^{n} \omega(x_k^*)l(x_k^*)d(x_k^*)\Delta y$. Since ω, l, and A are continuous functions, this Riemann sum converges to a definite integral.

Section 9.5

1. T, F, T, F, T, T, T, T.

3. A solution of a differential equation is one of possibly many functions that makes the differential equation true, while a solution of an initial-value problem is the unique function that makes both the differential equation and the initial value true.

5. If $y(x) = \sqrt{x + C}$ then
$$\frac{dy}{dx} = \frac{1}{2}(x + C)^{-1/2} = \frac{1}{2\sqrt{x+C}} = \frac{1}{2y}.$$

7. (a) The initial-value problem is $\frac{dv}{dt} = -9.8$ with $v(0) = 0$. The solution is $v(t) = -9.8t$. (b) The initial-value problem is $\frac{ds}{dt} = -9.8t$ with $s(0) = 100$. The solution is $s(t) = 100 - 4.9t^2$.

9. The x and y parts of the differential equation have been "separated" into factors. A differential equation in separable form can be solved by dividing both sides by $q(y)$, integrating both sides, and then solving for y, if each of these steps are possible.

11. $A = e^{-C}$

13. The solutions of the differential equation flow through the slope field; this means that we can trace out approximate solution curves in the slope field. For an initial-value problem we can begin at the initial value and then trace out a curve by flowing along with the slope field to obtain a sketch of the unique solution.

15. When using Euler's method to approximate a solution of $\frac{dy}{dx} = g(x, y)$ we have at each step a vertical change of $\Delta y_k = g(x_k, y_k)\Delta x$. The value $g(x_k, y_k)$ is the same as $\frac{dy}{dx}\Big|_{(x_k, y_k)}$.

17. (a) 500. (b) Since $0 < P < 500$ we have $0 < 1 - \frac{P}{500} < 1$; since P is growing we have $\frac{dP}{dt} > 0$. Therefore k must be positive. (c) When P is small, the quantity $\frac{P}{500}$ is also small, so $\frac{dP}{dt} \approx kP(1 - 0) = kP$. (d) When $P \approx 500$ we have $\frac{P}{500} \approx 1$, and therefore $\frac{dP}{dt} \approx kP(1 - 1) = 0$.

19. $y(x) = \frac{1}{7}x^7 + 2x^4 + 16x + C$

21. $y(x) = Ae^{x^2}$

23. $y(x) = Ae^{1/3x^3}$

25. $y = \pm\sqrt{6x + 2\ln|x| + C}$

27. $y(x) = \ln\left(\frac{1}{2}x^2 + C\right)$

29. $y(x) = 4e^{3x}$

31. $y(x) = -2x^3 + 5$

33. $y(x) = 3e^{-x} + 1$

35. $y(x) = \sqrt{6x + 4}$

37. $y(x) = -e^{1/2x^2}$

39. $y(x) = -\ln(-e^x + 1 + e^{-2})$

41. $y(x) = \frac{1}{2}\ln|1 + x^2| + 4$

43. $y = 4900 - 3900e^{-x/500}$

45. $y = \frac{1}{3}(1 + 5e^{6x})$

47. $y = \frac{2e^{3x}}{1 + 2e^{3x}}$

49. $y(x) = \frac{1}{1 - \sin x}$

51. $y(x) = \sqrt[3]{x^3 - 6x + 512}$

53. $(x_1, y_1) = (0.25, 1.5)$, $(x_2, y_2) = (0.5, 0.75)$, $(x_3, y_3) = (0.75, 0.375)$, $(x_4, y_4) = (1, 0.1875)$. A plot of these points gives:

55. $(x_1, y_1) = (0.5, 0)$, $(x_2, y_2) = (1, 0.25)$, $(x_3, y_3) = (1.5, 0.875)$, $(x_4, y_4) = (2, 2.0625)$. A plot of these points gives:

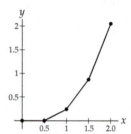

57. $(x_1, y_1) = (0.5, 2)$, $(x_2, y_2) = (1, 2.125)$, $(x_3, y_3) = (1.5, 2.36)$, $(x_4, y_4) = (2, 2.68)$. A plot of these points gives:

59.

61.

63.

65. (a) $\frac{dP}{dt} = 0.0139P$; with initial value $P(0) = 1.08$ million the solution is $P(t) = 1.08e^{0.0139t}$. (b) Between 55 and 56 years ago. (c) Just under 50 years.

67. (a) $\frac{dC}{dt} = -kC$ for some proportionality constant k. The solution is $C(t) = Ae^{-kt}$ for some constant A that represents the initial amount $C(0)$ of carbon when the organism died. (b) $k = \frac{\ln 2}{5730} \approx 0.00012$. (c) About 19,188 years old.

69. (a) $\frac{dP}{dt} = 0.1P\left(1 - \frac{P}{1000}\right)$, and by separation of variables followed by partial fractions (or the formula in Theorem 9.21) we obtain $P(t) = \frac{1000(100)e^{0.1t}}{1000 + 100(e^{0.1t} - 1)}$. (b) The graph is shown next. (c) According to the logistic model, the population will be growing the fastest in about 21.97 years.

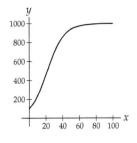

71. (a) We'll set $t = 0$ at 8 pm and measure time in hours. The differential equation is $\frac{dT}{dt} = k(70 - T)$, and its solution is $T(t) = 70 + 18.8e^{kt}$. (b) $k \approx 0.1433$. (c) Just after 5 pm.

73. (a) The differential equation is $\frac{dP}{dt} = kP^{1.1}$, and its solutions are of the form $P(t) = (-0.1kt + C)^{-10}$. (b) With $t = 0$ set at 2005 we have $P(t) = (-0.1(0.008)t + 0.83)^{-10}$. (c) In mid-2013.

75. By separation of variables we have $\int \frac{1}{Q} dQ = \int k\, dt$, which gives $\ln |Q| = kt + C$. Thus $|Q| = e^{kt+C}$ so $Q = Ae^{kt}$. Since $Q(0) = Q_0$ we have $Q_0 = Ae^{k(0)}$ and thus $A = Q_0$.

77. $\lim_{t \to \infty} P(t) = \lim_{t \to \infty} \frac{LP_0 e^{kt}}{L + P_0(e^{kt} - 1)} = \lim_{t \to \infty} \frac{LP_0}{Le^{-kt} + P_0 - P_0 e^{-kt}} = \frac{LP_0}{P_0} = L.$

79. $\lim_{t \to \infty} (A - (A - T_0)e^{-kt} = A - 0 = A$. If the model represents cooling, then $T > A$ and thus $A - T < 0$; in this case we wish the rate of change $\frac{dT}{dt} = k(A - T)$ to be negative, which it will be if $k > 0$. For a model of heating we have $T < A$ and thus $A - T > 0$ and we want a positive rate of change, which again requires $k > 0$.

For answers to odd-numbered Skill Certification exercises in the Chapter Review, please visit the Book Companion Web Site at www.whfreeman.com/integcalculus.